Choosing Blockbuster or Netflix

More and more people are getting their movies via the mail or over the Internet, rather than going to the video store. This new phenomenon can be analyzed by using linear equations. See the Group Activity: Working with Real Data on page 263.

Crowding on an Island

It might be obvious that larger Pacific islands have a greater variety of tropical birds. However, if one island is 10 times larger than another, does the larger island have 10 times as many species of birds? It turns out that in order to understand nature, we need logarithms. See Example 8 on page 711.

Waiting, Waiting, Waiting

Nobody likes to wait in long lines. However, waiting in lines or in congested traffic has become a way of life. Some people spend as much as 3 years of their life waiting. To understand how long lines quickly form, we need to understand the concept of a rational function. See the Chapter 6 Opener on page 397 and Example 6 on page 439.

Giving Up on Online Videos

Today's viewers won't hesitate to click on a different online video if they aren't entertained right away. In fact, if the average online video has 100,000 viewers, then 20,000 of these viewers abandon the video within the first 10 seconds. With rational exponents, we can make predictions about people's behaviors regarding online videos. See the Section 7.2 A Look Into Math on page 503 and Example 3 on page 505.

Setting Yourself Up for SUCCESS in MATH

Make the Most of Class

☐ Attend class regularly.

☐ Bring your textbook and a notebook to every class.

☐ Take notes in class, and write down everything that is on the board.

☐ Ask questions. (You won't be the only one who wants to know the answer!)

☐ Study with others in your class.

Get Help When You Need it

☐ Find your instructor's office, and write down the number and office hours.

☐ Find the location and hours of the math tutor resources on campus.

☐ Try to solve the problems and write down your questions before you ask for help.

☐ Use the online resources, such as videos, that accompany this book.

Master the Exam

☐ Schedule a time and place to study every day instead of "cramming" before the exam.

☐ Watch the Chapter Test Prep Videos on YouTube to learn how an instructor solves the practice test problems.

☐ Read all your notes and go over your homework assignments to review important concepts.

☐ Pick a few problems from your assignments at random and try to solve them without help.

☐ Carefully write out all your work on the exam whenever possible.

Relax and Visualize doing well.
You CAN Learn Math.

EDITION 4

Intermediate Algebra

with Applications and Visualization

Gary K. Rockswold

Minnesota State University, Mankato

Terry A. Krieger

Rochester Community and Technical College

PEARSON

Boston   Columbus   Indianapolis   New York   San Francisco   Upper Saddle River
Amsterdam   Cape Town   Dubai   London   Madrid   Milan   Munich   Paris   Montreal   Toronto
Delhi   Mexico City   Sao Paulo   Sydney   Hong Kong   Seoul   Singapore   Taipei   Tokyo

Editorial Director Chris Hoag
Executive Editor Cathy Cantin
Executive Content Editor Kari Heen
Senior Content Editor Lauren Morse
Editorial Assistant Kerianne Okie
Executive Director of Development Carol Trueheart
Senior Managing Editor Karen Wernholm
Associate Managing Editor Tamela Ambush
Senior Production Project Manager Sheila Spinney
Digital Assets Manager Marianne Groth
Supplements Production Coordinator Kerri Consalvo
Associate Media Producer Jonathan Wooding
Content Development Manager Rebecca E. Williams
Senior Content Developer Mary Durnwald
Marketing Manager Rachel Ross
Marketing Assistant Ashley Bryan
Senior Author Support/Technology Specialist Joe Vetere
Rights and Permissions Advisor Michael Joyce
Image Manager Rachel Youdelman
Procurement Manager Evelyn Beaton
Procurement Specialist Debbie Rossi
Media Procurement Specialist Ginny Michaud
Associate Director of Design, USHE North and West Andrea Nix
Senior Designer Barbara T. Atkinson
Text Design, Production Coordination, Composition, and Illustrations Cenveo Publisher
  Services/Nesbitt Graphics, Inc.
Cover Image Japanese garden waterfalls, slow shutter@szefi / Shutterstock
Cover Design Studio Montage
Art Director Barbara T. Atkinson

For permission to use copyrighted material, grateful acknowledgment is made to the copyright
holders on page A-56, which is hereby made part of this copyright page.

Many of the designations used by manufacturers and sellers to distinguish their products are claimed
as trademarks. Where those designations appear in this book, and Pearson was aware of a trademark
claim, the designations have been printed in initial caps or all caps.

Library of Congress Cataloging-in-Publication Data
Rockswold, Gary K.
  Intermediate algebra with applications and visualization / Gary K.
  Rockswold, Terry A. Krieger. –4th ed.
    p. cm.
  ISBN-13: 978-0-321-77331-9 (hardcover : alk. paper)
  ISBN-10: 0-321-77331-4 (hardcover : alk. paper)
    1. Algebra–Textbooks. I. Krieger, Terry A. II. Title.
QA152.3.R65 2013

512.9–dc22   2011007443

Copyright © 2013 Pearson Education, Inc.

3 4 5 6 7 8 9 10—RRDJC—14 13

ISBN 13: 978-0-321-77331-9
ISBN 10: 0-321-77331-4

Contents

5   Polynomial Expressions and Functions   314

6   Rational Expressions and Functions   397

7   Radical Expressions and Functions   491

8   Quadratic Functions and Equations   573

9   Exponential and Logarithmic Functions   654

Preface

Intermediate Algebra with Applications and Visualization, Fourth Edition, connects the real world to mathematics in ways that are both meaningful and motivational. Students using this textbook will have no shortage of realistic and convincing answers to the question, "Why is math important in my life?" Mathematical concepts are introduced by using relevant applications from students' lives that support their conceptual understanding. This textbook does not present math as a sequence of disconnected procedures to be memorized; rather, it shows that math does make sense.

In addition to presenting symbolic techniques, we present mathematical concepts in both visual and numerical ways whenever possible. Graphs, diagrams, and tables of values are often included to provide students with an opportunity to understand math in more than one way. Multiple representations (symbolic, graphical, and numerical) allow students to solve problems using three different methods. This textbook prepares students for future mathematics courses by teaching them both reasoning and problem-solving skills.

As mathematicians, we understand the profound impact that mathematics has on the world around us; as teachers with a combined experience of over 50 years, we know that our students often have difficulty both learning mathematics and recognizing its relevance. We believe that applications and visualization are essential pathways to empowering students with mathematics. Our goal in *Intermediate Algebra with Applications and Visualization* is to help students succeed.

This textbook is one of four textbooks in our series:

- *Prealgebra*
- *Beginning Algebra with Applications and Visualization*, Third Edition
- *Beginning & Intermediate Algebra with Applications and Visualization*, Third Edition
- *Intermediate Algebra with Applications and Visualization*, Fourth Edition.

New to this Edition

- **Updated Data** The data in hundreds of exercises and examples have been updated to keep the applications relevant and fresh.
- **More Visualization** Whenever appropriate, we have made mathematics more visual by using graphs, tables, charts, color, and diagrams to explain important concepts with fewer words.
  - We have added titles to the graphs and also comments on/near the graphs/tables to make them more (immediately) understandable. Now students don't always have to read the text to understand the table or graph.
  - We have added green equation labels, written in standard mathematical notation, to calculator graphs.
- **Teaching Examples** Every example in the text is now paired with a teaching example in the Annotated Instructor's Edition that instructors can use in class to promote further understanding.
- **New Vocabulary Checklist** At the start of every section that contains new mathematical definitions, a checklist of new vocabulary has been added to highlight the math concepts that are defined in the section.

- **Updated Glossary** The glossary has been updated to include all New Vocabulary words.
- **Reading Checks** To be sure that students are grasping important concepts, short questions about the material have been added throughout each section.
- **Study Tips** Throughout the text, study tips are now included to give not only general study advice but also pointers that relate to learning specific mathematical concepts.
- **New Application Topics** Students are more engaged when mathematics is tied to current and relevant topics. Several new examples have been added that discuss the mathematics of the Internet, social networks, tablet computers, and other contemporary topics.
- **Online Explorations** New exercises that invite students to use the Internet to explore real-world mathematics are now included throughout the text.
- **Refinement of Content**

CHAPTER 2: In the discussion of slope in Section 2.3, we put more emphasis on visualization to make this important topic more accessible for students.

CHAPTER 3: We put more emphasis on the visualization of the solutions to linear equations and inequalities in Sections 3.1 and 3.3.

CHAPTER 4: A new visual justification of the Fundamental Theorem of Linear Programming in Section 4.4 was added at the request of reviewers. A new subsection in Section 4.6 related to social networks and matrices was added to keep topics current for students.

CHAPTER 6: We added a new 4-step procedure for solving variation problems, which students can apply to all variation problems.

CHAPTER 7: At the request of reviewers, Section 7.1 was split into two sections so that radical expressions and rational exponents can be taught in two different sections. Section 7.1 was reorganized so that the square root and cube root functions can now be taught in Section 7.1 rather than later in the chapter. Greater emphasis has been given to converting between radical forms and forms with rational exponents.

CHAPTER 8: At the request of reviewers, Sections 8.1 and 8.2 have been reorganized to put topics in better order for student learning. The topic of increasing and decreasing functions is now optional in Section 8.2. In Section 8.5, more visuals for quadratic inequalities have been included to make this topic more accessible for students.

CHAPTER 9: Section 9.2 has a new subsection on percent change and exponential functions. Greater emphasis was given to percentages and how they relate to exponential functions. At the request of reviewers, the topic of compounding interest more than once a year was added in Section 9.2.

Hallmark Features

Math in the Real World

The Rockswold/Krieger series places an emphasis on teaching algebra in context. Students typically understand best when concepts are tied to the real world or presented visually. We believe that meaningful applications and visualization of important concepts are pathways to success in mathematics.

- **Chapter Openers** Each chapter opens with an application that motivates students by offering insight into the relevance of that chapter's mathematical concepts.
- **A Look into Math** Each section is introduced with a practical application of the math topic students are about to learn.
- **Real World Connection** Where appropriate, we expand on specific math topics and their connections to everyday life.
- **Applications** We integrate applications into both the discussions and the exercises to help students become more effective problem solvers.
- **Modeling Data** We provide opportunities for students to model real and relevant data with their own functions.
- **Online Exploration NEW** These exercises invite students to find data on the Internet and then use mathematics to analyze the data.

Understanding the Concepts

Conceptual understanding is critical to success in mathematics.

- **Math from Multiple Perspectives** Throughout the text, we present concepts by means of verbal, graphical, numerical, and symbolic representations to support multiple learning styles and problem-solving methods.
- **New Vocabulary** At the beginning of each section, we direct the students' attention to important terms before they are discussed in context.
- **Reading Check** Reading Check questions appear along with important concepts, ensuring that students understand the material they have just read.
- **Study Tips** Study Tips offer just-in-time suggestions to help students stay organized and focused on the material at hand.
- **Making Connections** Throughout the text, we help students understand the relationship between previously learned concepts and new concepts.
- **Critical Thinking** One or more Critical Thinking exercises are included in most sections. They pose questions that can be used for classroom discussion or homework assignments.
- **Putting It All Together** At the end of each section, we summarize the techniques and reinforce the mathematical concepts presented in the section.

Practice

Multiple types of exercises in this series support the application-based and conceptual nature of the content. These exercise types are designed to reinforce the skills students need to move on to the next concept.

- **Extensive Exercise Sets** The comprehensive exercise sets cover basic concepts, skill-building, writing, applications, and conceptual mastery. These exercise sets are further enhanced by several special types of exercises that appear throughout the text:
  - **Now Try Exercises** Suggested exercises follow every example for immediate reinforcement of skills and concepts.
  - **Checking Basic Concepts** These mixed review exercises appear after every other section and can be used for individual or group review. These exercises require 10–20 minutes to complete and are also appropriate for in-class work.
  - **Thinking Generally** These exercises appear in most exercise sets and offer open-ended conceptual questions that encourage students to synthesize what they have just learned.
  - **Writing About Mathematics** This exercise type is at the end of most sections. Students are asked to explain the concepts behind the mathematics procedures they have just learned.
- **Group Activities: Working with Real Data** This feature occurs once or twice per chapter, and provides an opportunity for students to work collaboratively on a problem that involves real-world data. Most activities can be completed with limited use of class time.
- **Graphing Calculator Exercises** The icon denotes an exercise that requires students to have access to a graphing calculator.
- **Technology Notes** Occurring throughout the book, optional Technology Notes offer students guidance, suggestions, and cautions on the use of a graphing calculator.

Mastery

By reviewing the material and putting their abilities to the test, students will be able to assess their level of mastery as they complete each chapter.

- **Chapter Summary** In a quick and thorough review, we combine key terms, topics, and procedures with illuminating examples to assist in test preparation.
- **Chapter Review Exercises** Students can work these exercises to gain confidence that they have mastered the material.
- **Extended And Discovery Exercises** These capstone projects at the end of every chapter challenge students to synthesize what they have learned.
- **Chapter Test** Students can reduce math anxiety by using these tests as a rehearsal for the real thing.
- **Chapter Test Prep Videos** Students can prepare for the chapter test by viewing videos of an instructor working through the complete solutions for all Chapter Test exercises for each chapter.
- **Cumulative Review Exercises** Starting with Chapter 2 and appearing in all subsequent chapters, Cumulative Review exercises help students see the big picture of math by reviewing topics and skills they have already learned.

Supplements

STUDENT RESOURCES

Student's Solutions Manual

- Contains solutions for the odd-numbered section-level exercises (excluding Writing About Mathematics and Group Activity exercises), and solutions to all Concepts and Vocabulary exercises, Checking Basic Concepts exercises, Chapter Review Exercises, Chapter Test exercises, and Cumulative Review Exercises.

  ISBNs: 0-321-74784-4, 978-0-321-74784-6

Worksheets

These lab- and classroom-friendly worksheets provide

- Learning objectives and key vocabulary terms for every text section, along with vocabulary practice problems.
- Extra practice exercises for every section of the text with ample space for students to show their work.
- Additional opportunities to explore multiple representation solutions.

  ISBNs: 0-321-74788-7, 978-0-321-74788-4

Video Resources

- A series of lectures correlated directly to the content of each section of the text.
- Material presented in a format that stresses student interaction, often using examples from the text.
- Ideal for distance learning and supplemental instruction.
- Video lectures include English captions.
- The Chapter Test Prep Videos let students watch instructors work through step-by-step solutions to all the Chapter Test exercises from the textbook. Chapter Test Prep videos are available on YouTube™ (search using author name and book title) as well as in MyMathLab.
- Available in MyMathLab.

INSTRUCTOR RESOURCES

Annotated Instructor's Edition

- Contains Teaching Tips, Teaching Examples, and answers to every exercise in the textbook excluding the Writing About Mathematics exercises.
- Answers that do not fit on the same page as the exercises themselves are supplied in the Instructor Answer Appendix at the back of the textbook.

  ISBNs: 0-321-74781-X, 978-0-321-74781-5

Instructor's Resource Manual with Tests and Mini Lectures (Download Only)

Includes resources designed to help both new and adjunct faculty with course preparation and classroom management:

- Teaching tips and additional exercises for selected content.
- Four sets of Cumulative Review Exercises that cover Chapters 1–3, 1–6, 1–9, and 1-11.
- Notes for presenting graphing calculator topics as well as supplemental activities.
- Three free-response alternate test forms and one multiple-choice test form per chapter; one free-response and one multiple-choice final exam.
- Available in MyMathLab® and on the Instructor's Resource Center.

Instructor's Solutions Manual (Download Only)

- Provides solutions to all section-level exercises (excluding Writing About Mathematics and Group Activity exercises), and solutions to all Checking Basic Concepts exercises, Chapter Review Exercises, Chapter Test exercises, and Cumulative Review Exercises.
- Available in MyMathLab and on the Instructor's Resource Center.

PowerPoint Slides

- Key concepts and definitions from the text.
- Available in MyMathLab or can be downloaded from the Instructor's Resource Center.

TestGen®

TestGen (www.pearsoned.com/testgen) enables instructors to build, edit, and print, and administer tests using a computerized bank of questions developed to cover all the objectives of the text. TestGen is algorithmically based, allowing instructors to create multiple but equivalent versions of the same question or test with the click of a button. Instructors can also modify test bank questions or add new questions. The software and testbank are available for download from Pearson Education's online catalog.

MathXL® Online Course (access code required)

MathXL is the homework and assessment engine that runs MyMathLab. (MyMathLab is MathXL plus a learning management system.) With MathXL, instructors can:

- Create, edit, and assign online homework and tests using algorithmically generated exercises correlated at the objective level to the textbook.
- Create and assign their own online exercises and import TestGen tests for added flexibility.
- Maintain records of all student work tracked in MathXL's online gradebook.

With MathXL, students can:

- Take chapter tests in MathXL and receive personalized study plans and/or personalized homework assignments based on their test results.
- Use the study plan and/or the homework to link directly to tutorial exercises for the objectives they need to study.
- Access supplemental animations and video clips directly from selected exercises.

MathXL is available to qualified adopters. For more information, visit www.mathxl.com, or contact your Pearson representative.

MyMathLab Online Course (access code required)

MyMathLab delivers **proven results** in helping individual students succeed.

- MyMathLab has a consistently positive impact on the quality of learning in higher education math instruction. MyMathLab can be successfully implemented in any environment–lab-based, hybrid, fully online, traditional–and demonstrates the quantifiable difference that integrated usage has on student retention, subsequent success, and overall achievement.
- MyMathLab's comprehensive online gradebook automatically tracks students' results on tests, quizzes, homework, and in the study plan. Instructors can use the gradebook to quickly intervene if their students have trouble or to provide positive feedback on a job well done. The data within MyMathLab is easily exported to a variety of spreadsheet programs, such as Microsoft Excel. Instructors can determine which points of data they want to export, and then analyze the results to determine success.

MyMathLab provides **engaging experiences** that personalize, stimulate, and measure learning for each student.

- **Tutorial Exercises:** The homework and practice exercises in MyMathLab and MyStatLab are correlated to the exercises in the textbook, and they regenerate algorithmically to give students unlimited opportunity for practice and mastery. The software offers immediate, helpful feedback when students enter incorrect answers.
- **Multimedia Learning Aids:** Exercises include guided solutions, sample problems, animations, videos, and eText clips for extra help at point-of-use.
- **Expert Tutoring:** Although many students describe the whole of MyMathLab as 'like having your own personal tutor," students using MyMathLab and MyStatLab do have access to live tutoring from Pearson, from qualified math and statistics instructors.

And, MyMathLab comes from a **trusted partner** with educational expertise and an eye on the future.

Knowing that you are using a Pearson product means knowing that you are using quality content. That means that our eTexts are accurate, that our assessment tools work, and that our questions are error-free. And whether you are just getting started with MyMathLab, or have a question along the way, we're here to help you learn about our technologies and how to incorporate them into your course.

To learn more about how MyMathLab combines proven learning applications with powerful assessment, visit **www.mymathlab.com** or contact your Pearson representative.

Acknowledgments

Many individuals contributed to the development of this textbook. We thank the following reviewers, whose comments and suggestions were invaluable in preparing *Intermediate Algebra with Applications and Visualization,* Fourth Edition.

Mary Kay Abbey, *Montgomery College—Takoma Park*
Raul Aparicio, *Blinn College—Brenham Campus*
David Atwood, *Rochester Community and Technical College*
Alina Coronel, *Miami-Dade College—Kendall Campus*
Emmett Dennis, *Southern Connecticut State University*
James Deveney, *Virginia Commonwealth University*
Christopher Donnelly, *Macomb Community College*
Debbie Garrison, *Valencia Community College—East Campus*
Joan Van Glabek, *Edison Community College—Collier Campus*
Cynthia Gubitose, *Southern Connecticut State University*
Shawna Haider, *Salt Lake Community College*
Kristy Hill, *Hinds Community College—Raymond Campus*
Amy Hlavacek, *Saginaw Valley State University*
Dr. Philip Kaatz, *Mesalands Community College*
Paul Kinion, *Rochester Community and Technical College*
Mike Kirby, *Tidewater Community College—Virginia Beach Campus*
Kim Luna, *Eastern New Mexico University*
Rhonda MacLeod, *Tallahassee Community College*
Toni McCall, *Angelina College*
Lauren McNamara, *Hillsborough Community College*
Benjamin Moulton, *Utah Valley University*
Glenn Noé, *Brookdale Community College*
Mario Scribner, *Tidewater Community College—Virginia Beach Campus*
Robert Shankin, *Santa Fe Community College*
Yvette Stepanian, *Virginia Commonwealth University*
Carol Walker, *Hinds Community College—Raymond Campus*
Elizabeth White, *Trident Technical College*

We would like to welcome Jessica Rockswold to our team. She has been instrumental in developing, writing, and proofing new applications, graphs, and visualizations for this text.

Paul Lorczak, Mitchell Johnson, Lynn Baker, and Mark Rockswold deserve special credit for their help with accuracy checking. Without the excellent cooperation from the professional staff at Pearson Education, this project would have been impossible. Thanks go to Greg Tobin and Maureen O'Connor for giving their support. Particular recognition is due Cathy Cantin who gave essential advice and assistance. The outstanding contributions of Lauren Morse, Sheila Spinney, Joe Vetere, Michelle Renda, Tracy Rabinowitz, Rachel Ross, Ashley Bryan, and Jonathan Wooding are greatly appreciated. Special thanks go to Kathy Diamond, who was instrumental in the success of this project.

Thanks go to Wendy Rockswold and Carrie Krieger, whose unwavering encouragement and support made this project possible. We also thank the many students and instructors who used the previous editions of this textbook. Their suggestions were insightful and helpful.

Please feel free to send us your comments and questions at either of the following e-mail addresses: gary.rockswold@mnsu.edu or terry.krieger@roch.edu. Your opinion is important to us.

Gary K. Rockswold
Terry A. Krieger

1

Real Numbers and Algebra

The human mind has never invented a labor-saving device greater than algebra.

—J. W. GIBBS

"**W**hy *do* I need to learn math?" This is a question commonly asked by students across the country. Despite the invention of computers and calculators, the need for people who understand numbers and mathematics has never been greater. With so many details to learn in mathematics classes, students sometimes lose sight of the increasing presence of mathematics in society. Mathematics has been used in science and engineering for centuries, but today mathematics is also being used to describe human behavior in areas such as economics, medicine, advertising, social networking, and the Internet. (See Section 1.3, Exercises 125 and 126, or Section 1.4, Exercises 89 and 90.)

The world is moving into a new age of numbers. Mathematics is the key to analyzing all types of data, especially the data that people openly share online. This information helps maximize the impact of advertisements for social networks, such as Facebook. Max Levchin is the CEO of Slide, a successful media company that writes applications for Facebook. When asked what set his company apart from others, he stated that, "Our competitive advantage is our *math skills*."

The importance of mathematics will only increase in the future. Whether or not you decide to be a math major, mathematics is essential for reaching your full potential in vocation, income level, and lifestyle.

1.1 Describing Data With Sets of Numbers

**Natural and Whole Numbers • Integers and Rational Numbers •
Real Numbers • Properties of Real Numbers**

A LOOK INTO MATH ▶

The need for numbers has existed in nearly every society. Numbers first occurred in the measurement of time, currency, goods, and land. As the complexity of a society increased, so did its numbers. One tribe that lived near New Guinea counted only to 6. Any number higher than 6 was referred to as "many." This number system met the needs of that society.

Have you ever thought about how we live by the numbers? In today's highly diverse culture, numbers impact almost every aspect of our lives, including money, webpage usage, speed limits, grade point averages, gas mileages, download speed, and medication dosages.

In this section we discuss numbers that are vital to our complex society. We also show how different types of data can be described with sets of numbers. (*Source: Historical Topics for the Mathematics Classroom, Thirty-first Yearbook, NCTM.*)

NEW VOCABULARY

☐ Natural numbers
☐ Set braces
☐ Whole numbers
☐ Integers
☐ Rational numbers
☐ Real numbers
☐ Irrational numbers
☐ Approximately equal
☐ Average
☐ Identity properties
☐ Commutative properties
☐ Associative properties
☐ Distributive properties

Natural and Whole Numbers

One important set of numbers found in most societies is the set of **natural numbers**. These numbers comprise the counting numbers and may be expressed as

$$N = \{1, 2, 3, 4, 5, 6, \ldots\}.$$

Set braces, { }, are used to enclose the elements of a set. Because there are infinitely many natural numbers, three dots show that the list continues in the same pattern without end. A second set of numbers, called the **whole numbers**, is given by

$$W = \{0, 1, 2, 3, 4, 5, \ldots\}.$$

▶ **REAL-WORLD CONNECTION** Natural numbers and whole numbers can be used when data are not broken into fractional parts. For example, Table 1.1 lists the number of bachelor's degrees awarded during selected academic years. Note that either natural numbers or whole numbers are appropriate to describe the data because a fraction of a degree cannot be awarded.

TABLE 1.1 Bachelor's Degrees Awarded

Year	1979–1980	1989–1990	1999–2000	2009–2010
Degrees	929,000	1,051,000	1,238,000	1,634,000

Source: Department of Education.

Integers and Rational Numbers

The set of **integers** is given by

$$I = \{\ldots, -3, -2, -1, 0, 1, 2, 3, \ldots\}.$$

The integers include both the natural numbers and the whole numbers. During the eighteenth century, negative numbers were not readily accepted by all mathematicians. Such numbers did not seem to have any physical meaning. However, today when a person opens a personal checking account for the first time, negative numbers quickly take on meaning. There is a significant difference between a positive and a negative balance.

A **rational number** is any number that can be expressed as the ratio of two integers: $\frac{p}{q}$, where q is not equal to 0 because we cannot divide by 0. Rational numbers can be written as fractions and include all integers. Some examples of rational numbers are

$$\frac{8}{1}, \quad \frac{2}{3}, \quad -\frac{3}{5}, \quad -\frac{7}{2}, \quad \frac{22}{7}, \quad 1.2, \quad \text{and} \quad 0.$$

Note that 1.2 and 0 are both rational numbers because they can be written as $\frac{12}{10}$ and $\frac{0}{1}$.

Rational numbers may be expressed in **decimal form** that either *repeats* or *terminates*. The fraction $\frac{1}{3}$ may be expressed as $0.\overline{3}$, a repeating decimal, and the fraction $\frac{1}{4}$ may be expressed as 0.25, a terminating decimal. The overbar indicates that $0.\overline{3} = 0.3333333\ldots$.

▶ **REAL-WORLD CONNECTION** Integers and rational numbers are used to describe things such as temperature. Table 1.2 lists equivalent temperatures in both degrees Fahrenheit and degrees Celsius. Note that both positive and negative numbers are used to describe temperature.

TABLE 1.2 Fahrenheit and Celsius Temperature

°F	°C	Observation
−89	$-67.\overline{2}$	Alcohol freezes
−40	−40	Mercury freezes
0	$-17.\overline{7}$	Snow and salt mixture freezes
32	0	Water freezes
100	$37.\overline{7}$	A very warm day
212	100	Water boils

EXAMPLE 1 **Classifying numbers**

Classify each number as one or more of the following: natural number, whole number, integer, or rational number.

(a) $\frac{6}{3}$ (b) −1 (c) 0 (d) $-\frac{11}{3}$

Solution
(a) Because $\frac{6}{3} = 2$, the number $\frac{6}{3}$ is a natural number, a whole number, an integer, and a rational number.
(b) The number −1 is an integer and a rational number but not a natural or a whole number.
(c) The number 0 is a whole number, an integer, and a rational number but not a natural number.
(d) The fraction $-\frac{11}{3}$ is a rational number as it is the ratio of two integers. However, it is not a natural number, a whole number, or an integer.

Now Try Exercises 1, 3, 29

Real Numbers

Real numbers can be represented by decimal numbers. Every fraction has a decimal form, so real numbers include rational numbers. However, some real numbers cannot be expressed by fractions. They are called **irrational numbers**. If a square root of a positive integer is not an integer, then it is an irrational number. The numbers $\sqrt{2}$, $\sqrt{15}$, and π are examples of irrational numbers. They can be expressed by decimals but *not* by decimals that either repeat or terminate. Examples of real numbers include

$$2, \quad -10, \quad 151\frac{1}{4}, \quad -131.37, \quad \frac{1}{3}, \quad -\sqrt{5}, \quad \text{and} \quad \sqrt{11}.$$

CALCULATOR HELP

To evaluate π and square roots, see Appendix A (page AP-1).

Any real number may be approximated by a terminating decimal. We use the symbol $\approx$, which means **approximately equal**, to denote an approximation. Each of the following real numbers has been approximated to three *decimal places*.

$$\pi \approx 3.142, \quad \frac{2}{3} \approx 0.667, \quad \sqrt{200} \approx 14.142$$

READING CHECK

Identify what numbers are in each set of numbers.
• Natural numbers
• Whole numbers
• Rational numbers
• Real numbers

Figure 1.1 shows the relationships among the different sets of numbers. *Each real number must be either a rational number or an irrational number but not both.* The natural numbers, whole numbers, and integers are contained in the set of rational numbers.

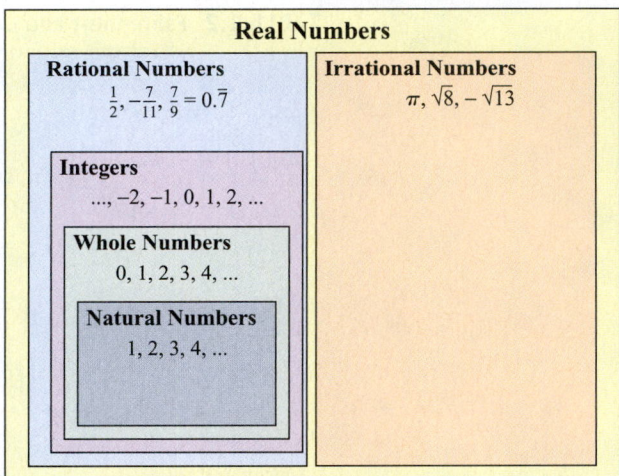

Figure 1.1 The Set of Real Numbers

EXAMPLE 2 **Classifying numbers**

Classify each real number as one or more of the following: a natural number, an integer, a rational number, or an irrational number.

$$5, \quad -1.2, \quad \frac{13}{7}, \quad -\sqrt{7}, \quad -12, \quad \sqrt{16}$$

Solution
Natural numbers: 5 and $\sqrt{16} = 4$
Integers: 5, -12, and $\sqrt{16} = 4$
Rational numbers: 5, -1.2, $\frac{13}{7}$, -12, and $\sqrt{16} = 4$
Irrational number: $-\sqrt{7}$

Now Try Exercise 23

Even though a data set may contain only integers, decimals are often needed to describe it. One common way to do so is to find the **average**. To calculate the average of a set of numbers, we find their sum and divide by the number of numbers in the set.

EXAMPLE 3   **Analyzing test scores**

A student obtains the following test scores: 81, 96, 79, and 82.
(a) Find the student's average test score.
(b) Is this average a natural, a rational, or a real number?

Solution
(a) To find the average, we find the sum of the four test scores and then divide by 4:

$$\frac{81 + 96 + 79 + 82}{4} = \frac{338}{4} = 84.5.$$

(b) The average of these four test scores is both a rational number *and* a real number but is not a natural number.

Now Try Exercise 93

Properties of Real Numbers

Several properties of real numbers are used in algebra. They are the identity properties, the commutative properties, the associative properties, and the distributive properties. These properties are essential to understanding algebra. If a person does not know "the rules of the game," then it is very difficult to use algebra correctly to solve problems.

CRITICAL THINKING

Is the sum of two irrational numbers ever a rational number? Explain.

IDENTITY PROPERTIES The **identity property of 0** states that if 0 is added to any real number a, the result is a. The number 0 is called the **additive identity**. For example,

$$-3 + 0 = -3 \quad \text{and} \quad 0 + 18 = 18.$$

The **identity property of 1** states that if any number a is multiplied by 1, the result is a. The number 1 is called the **multiplicative identity**. Examples include

$$-7 \cdot 1 = -7 \quad \text{and} \quad 1 \cdot 9 = 9.$$

We can summarize these results as follows.

STUDY TIP

Important concepts and definitions are written in a box like the one to the right. Be sure to pay special attention to these types of boxes.

IDENTITY PROPERTIES

For any real number a,

$$a + 0 = 0 + a = a$$

and

$$a \cdot 1 = 1 \cdot a = a.$$

COMMUTATIVE PROPERTIES The **commutative property for addition** states that two numbers, a and b, can be added together in any order and the result is the same. That is, $a + b = b + a$. For example, if a person is paid \$5 and then \$7 or paid \$7 and then \$5, the result is the same. Either way the person is paid a total of

$$5 + 7 = 7 + 5 = \$12.$$

There is also a **commutative property for multiplication**. It states that two numbers, a and b, can be multiplied in any order and the result is the same. That is, $a \cdot b = b \cdot a$. For example, 3 groups of 5 people or 5 groups of 3 people both contain

$$3 \cdot 5 = 5 \cdot 3 = 15 \text{ people.}$$

We can summarize these results as follows.

COMMUTATIVE PROPERTIES

For any real numbers a and b,

$$a + b = b + a$$

and

$$a \cdot b = b \cdot a.$$

ASSOCIATIVE PROPERTIES The commutative properties allow us to interchange the order of two numbers when we add or multiply. The associative properties allow us to change how numbers are grouped. For example, we may add the numbers 3, 4, and 2 as follows.

$$(\mathbf{3 + 4}) + 2 = \mathbf{7} + 2 = 9$$
$$3 + (\mathbf{4 + 2}) = 3 + \mathbf{6} = 9$$

In either case we obtain the same answer, which is the result of the **associative property for addition**. Note that we did not change the order of the numbers; rather, we changed how the numbers were grouped. There is also an **associative property for multiplication**, which is illustrated as follows.

$$(\mathbf{3 \cdot 4}) \cdot 2 = \mathbf{12} \cdot 2 = 24$$
$$3 \cdot (\mathbf{4 \cdot 2}) = 3 \cdot \mathbf{8} = 24$$

We can stack cubes as shown in Figure 1.2 to illustrate the associative property for multiplication. The stacks are shown with either 2 layers of 12 cubes or 3 layers of 8 cubes. In both cases there is a total of 24 cubes in each stack.

Associative Property

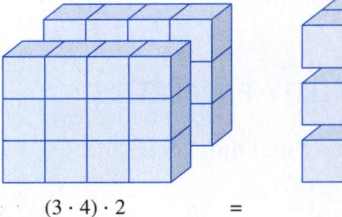

$(3 \cdot 4) \cdot 2$ = $3 \cdot (4 \cdot 2)$

Figure 1.2

We can summarize these results as follows.

ASSOCIATIVE PROPERTIES

For any real numbers a, b, and c,

$$(a + b) + c = a + (b + c)$$

and

$$(a \cdot b) \cdot c = a \cdot (b \cdot c).$$

NOTE: Sometimes we omit the multiplication dot. Thus $a \cdot b = ab$ and $5 \cdot x = 5x$.

EXAMPLE 4 | **Identifying properties of real numbers**

State the property of real numbers that justifies each statement.
(a) $4 \cdot (3x) = (4 \cdot 3)x$ (b) $(1 \cdot 5) \cdot 4 = 5 \cdot 4$ (c) $5 + ab = ab + 5$

Solution
(a) This equation illustrates the associative property for multiplication, with the grouping of the numbers changed. That is, $4 \cdot (3x) = (4 \cdot 3)x = 12x$.
(b) This equation illustrates the identity property of 1 because $1 \cdot 5 = 5$.
(c) This equation illustrates the commutative property for addition; the order of the terms 5 and ab changed.

▮ Now Try Exercises 35, 37, 45

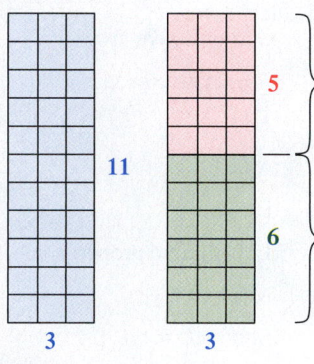

$3(11) = 3(6 + 5) = 3 \cdot 6 + 3 \cdot 5$

Figure 1.3

DISTRIBUTIVE PROPERTIES The **distributive properties** (see Figure 1.3) are used frequently in algebra to simplify expressions. An example of a distributive property is

$$3(6 + 5) = 3 \cdot 6 + 3 \cdot 5.$$

It is important to multiply the 3 by *both* the 6 and 5, not just the 6. This distributive property is valid when addition is replaced by subtraction. For example,

$$3(6 - 5) = 3 \cdot 6 - 3 \cdot 5.$$

We can summarize these results as follows.

> **DISTRIBUTIVE PROPERTIES**
>
> For any real numbers a, b, and c,
> $$a(b + c) = ab + ac$$
> and
> $$a(b - c) = ab - ac.$$

NOTE: Because multiplication is commutative, the distributive properties may be written as

$$(b + c)a = ba + ca \quad \text{and} \quad (b - c)a = ba - ca.$$

EXAMPLE 5 | **Applying the distributive properties**

Apply a distributive property to each expression.
(a) $5(4 + x)$ (b) $10 - (1 + a)$ (c) $9x - 5x$ (d) $5x + 2x - 3x$

Solution
(a) $5(4 + x) = 5 \cdot 4 + 5 \cdot x = 20 + 5x$
(b) $10 - (1 + a) = 10 - 1(1 + a) = 10 - (1 \cdot 1) - (1 \cdot a) = 9 - a$
(c) $9x - 5x = (9 - 5)x = 4x$
(d) Because each term contains an x, apply the distributive properties to all *three* terms:
$$5x + 2x - 3x = (5 + 2 - 3)x = 4x.$$

▮ Now Try Exercises 65, 69, 77, 83

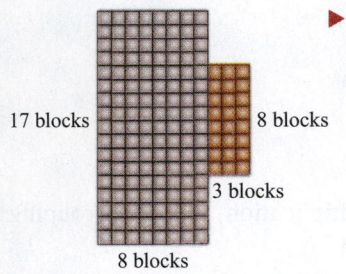

17 blocks   8 blocks

3 blocks

8 blocks

▶ **REAL-WORLD CONNECTION** Properties of real numbers can be used to simplify calculations, particularly when we are working *without* a calculator. For example, suppose we are making a patio having two rectangular sections. One rectangular section has 8 rows of 17 blocks each, and the other has 8 rows of only 3 blocks each. See the figure in the margin. One way to determine the total number of blocks is to apply a distributive property.

$$8 \cdot 17 + 8 \cdot 3 = 8(17 + 3) = 8 \cdot 20 = 160 \text{ blocks}$$

EXAMPLE 6   **Applying properties of real numbers**

Use properties of real numbers to simplify each expression.
(a) $52 + 37 + 48 + 63$ **(b)** $30 \cdot 97$

Solution

(a) $52 + 37 + 48 + 63$	$= 52 + 48 + 37 + 63$	Commutative property
	$= (52 + 48) + (37 + 63)$	Associative property
	$= 100 + 100$	Add.
	$= 200$	Add.
(b) $30 \cdot 97$	$= 30(100 - 3)$	Rewrite 97.
	$= 30 \cdot 100 - 30 \cdot 3$	Distributive property
	$= 3000 - 90$	Multiply.
	$= 2910$	Subtract.

STUDY TIP

Putting It All Together gives a summary of important concepts in each section.

Now Try Exercises 99, 101

1.1   Putting It All Together

Data and numbers play a central role in a diverse, complex society. Because of the variety of data, it has been necessary to develop different sets of numbers. Without numbers, data could be described qualitatively but not quantitatively. For example, we might say that the day seems hot, but we would not be able to give an actual number for the temperature. The following table summarizes some of the sets of numbers.

CONCEPT	COMMENTS	EXAMPLES
Natural Numbers	Are referred to as the *counting numbers*	$1, 2, 3, 4, 5, \ldots$
Whole Numbers	Include the natural numbers	$0, 1, 2, 3, 4, \ldots$
Integers	Include the natural numbers and the whole numbers	$\ldots, -2, -1, 0, 1, 2, \ldots$
Rational Numbers	Include integers and all fractions $\frac{p}{q}$, where p and q are integers (q is not equal to 0), and all repeating and terminating decimals	$\frac{1}{2}, -3, \frac{128}{6}, -0.335, 0,$ $0.25 = \frac{1}{4}, \text{ and } 0.\overline{3} = \frac{1}{3}$

CONCEPT	COMMENTS	EXAMPLES
Irrational Numbers	Include real numbers that cannot be expressed *exactly* by a fraction	$-\pi,\ \sqrt{2},$ and $\sqrt{7}$
Real Numbers	Any number that can be expressed in decimal form, including the rational numbers and the irrational numbers	$\pi,\ \sqrt{3},\ -\dfrac{4}{7},\ 0,\ -10,$ $0.\overline{6}=\dfrac{2}{3},\ 3,$ and $\sqrt{15}$

Real numbers have several important properties, which are summarized as follows.

PROPERTY	DEFINITION	EXAMPLES
Identity (0 and 1)	The identity for addition is 0 and the identity for multiplication is 1. For any real number a, $a + 0 = a$ and $a \cdot 1 = a$.	$5 + 0 = 5$ and $5 \cdot 1 = 5$
Commutative	For any real numbers a and b, $a + b = b + a$ and $a \cdot b = b \cdot a$.	$4 + 6 = 6 + 4$ and $4 \cdot 6 = 6 \cdot 4$
Associative	For any real numbers a, b, and c, $(a + b) + c = a + (b + c)$ and $(a \cdot b) \cdot c = a \cdot (b \cdot c)$.	$(3 + 4) + 5 = 3 + (4 + 5)$ and $(3 \cdot 4) \cdot 5 = 3 \cdot (4 \cdot 5)$
Distributive	For any real numbers a, b, and c, $a(b + c) = ab + ac$ and $a(b - c) = ab - ac$.	$5(x + 2) = 5x + 10,$ $5(x - 2) = 5x - 10,$ and $5x + 4x = (5 + 4)x = 9x$

1.1   Exercises

PRACTICE   WATCH   DOWNLOAD   READ   REVIEW

CONCEPTS AND VOCABULARY

1. Which numbers are in the set of natural numbers and which are in the set of whole numbers?

2. Which numbers are in the set of integers? Give an example of an integer that is not a natural number.

3. Which numbers are in the set of rational numbers? Give an example.

4. Which numbers are in the set of real numbers? If a number is a real number but not a rational number, what type of number must it be? Give an example of a real number that is not a rational number.

5. Give an example of the commutative property for addition.

6. Give an example of the commutative property for multiplication.

7. Why is the number 1 called the multiplicative identity?

8. Why is the number 0 called the additive identity?

9. Give an example of the associative property for multiplication.

10. Give an example of a distributive property.

11. The average of the numbers 6, 9, 11, and 22 is _____.

12. The average of the numbers 11, 5, and 8 is _____.

13. Earning $100 and then $50 is equivalent to earning $50 and then $100. What property of real numbers does this illustrate?

14. A marching band with 10 rows and 8 people in each row has the same number of members as a different marching band with 8 rows and 10 people in each row. What property of real numbers does this illustrate?

CLASSIFYING NUMBERS

Exercises 15–22: Classify the number as one or more of the following: natural number, integer, rational number, or real number.

15. 33,000 (Number of Subway franchises in 2010)

16. 3.8 (Percent of mathematics majors who go on to obtain a Ph.D.)

17. $\frac{89}{3687}$ (Fraction of 18- to 19-year-old males who are married in the United States)

18. 4 (Pounds of garbage the average person produces each day)

19. 7.5 (Average number of gallons of water a person uses each minute while taking a shower)

20. 2.57 (Average number of people per household in 2009)

21. $90\sqrt{2}$ (Distance in feet from home plate to second base in baseball)

22. −71 (Wind chill when the temperature is −30°F and the wind speed is 40 miles per hour)

Exercises 23–28: Classify each real number as one or more of the following: natural number, whole number, integer, rational number, or irrational number.

23. $-5, 6, \frac{1}{7}, \sqrt{7}, 0.2$　　**24.** $-3, \frac{2}{9}, \sqrt{9}, -1.37$

25. $\frac{3}{1}, -\frac{5}{8}, \sqrt{5}, 0.\overline{45}, \pi$　　**26.** $0, \frac{50}{10}, -\frac{23}{27}, 0.\overline{6}, -\sqrt{3}$

27. $-2, \frac{1}{2}, \sqrt{9}, 0.\overline{26}$　　**28.** $\sqrt{4}, \sqrt{6}, \frac{4}{2}, 0.26$

Exercises 29–34: For the measured quantity, state the set of numbers that is most appropriate to describe it. Choose from the natural numbers, integers, or rational numbers. Explain your answer.

29. Shoe sizes　　　　**30.** Populations of states

31. Speed limits　　　　**32.** Gallons of gasoline

33. Temperatures given in a winter weather forecast in Montana

34. Number of songs sold on iTunes

PROPERTIES OF REAL NUMBERS

Exercises 35–48: State whether the equation is the result of an identity, a commutative, an associative, or a distributive property.

35. $b + 0 = b$　　　　**36.** $1 \cdot 5 = 5$

37. $4 + a = a + 4$

38. $(5 + 1) + 8 = 5 + (1 + 8)$

39. $8(9x - 3) = 8 \cdot 9x - 8 \cdot 3$

40. $4(3 + 5a) = 4 \cdot 3 + 4 \cdot 5a$

41. $4 \cdot (10 \cdot 6) = (4 \cdot 10) \cdot 6$

42. $x \cdot 5 = 5x$

43. $3 \cdot (6 \cdot 2) = (6 \cdot 2) \cdot 3$

44. $bac = abc$

45. $4 \cdot (5x) = 20x$

46. $4(x - 3) = (x - 3)4$

47. $5(x - 3) - 2(x - 3) = 3(x - 3)$

48. $4 - (a - b) = 4 - a + b$

Exercises 49–56: Use a commutative property to write an equivalent expression.

49. $4 + a$　　　　**50.** ba

51. $a \cdot \frac{1}{3}$　　　　**52.** $100 + x$

53. $1 + x$　　　　**54.** $b \cdot 5$

55. yx　　　　**56.** $b + a$

Exercises 57–64: Use an associative property to write an equivalent expression.

57. $4 + (5 + b)$　　　　**58.** $(x + 2) + 3$

59. $5(10x)$　　　　**60.** $(a \cdot 5) \cdot 4$

61. $(x + y) + z$　　　　**62.** $-3(4x)$

63. $(x \cdot 3)4$　　　　**64.** $(a + 6) + 5$

Exercises 65–86: Use a distributive property to write an equivalent expression.

65. $4(x + y)$

66. $-3(a + 5)$

67. $(x - 7)5$

68. $(11 + b)a$

69. $-(x + 1)$

70. $-(a - 2)$

71. $ax - ay$

72. $4a + 4b$

73. $12 + 3x$

74. $2 - 4x$

75. $3x + 7x$

76. $\frac{1}{2}z + \frac{3}{2}z$

77. $8t - 2t$

78. $y - \frac{2}{3}y$

79. $x - \frac{3}{4}x$

80. $13r - 6r$

81. $3 - (1 - 2z)$

82. $5 - (1 - 4y)$

83. $3z + 4z + z$

84. $5z + 4z + 2z$

85. $10t - t - 4t$

86. $20t - 2t - t$

Exercises 87–92: Use a distributive property to evaluate the expression two different ways.

87. $7(11 - 3)$

88. $(5 + 9)12$

89. $13(16 + 23)$

90. $4(12 - 8)$

91. $5(19 - 7)$

92. $(8 + 12)9$

Exercises 93–98: Calculate the average of the list of numbers. Classify the result as a natural number, an integer, or a rational number.

93. 3, 4, 5, 8

94. 5, 8, 10, 23, 9

95. 45, 33, 52

96. 3.2, 7.5, 8.1, 12.8, 13.4

97. 121.5, 45.7, 99.3, 45.9

98. 99.88, 39.11, 85.67, 23.86, 19.11

Exercises 99–108: Use properties of real numbers to evaluate the expression mentally.

99. $8 + 3 + 2 + 7$

100. $52 + 103 + 48 + 97$

101. $50 \cdot 198$

102. $27 \cdot 30$

103. $\frac{2}{9} \cdot 8 \cdot 9 \cdot \frac{3}{8}$

104. $\frac{1}{2} \cdot \frac{1}{3} \cdot \frac{1}{4} \cdot 2 \cdot 3 \cdot 4$

105. $4 \cdot 16 - 4 \cdot 6$

106. $7 \cdot 12 + 7 \cdot 3 + 7 \cdot 5$

107. $\frac{4}{5}\left(\frac{5}{2} + 5 + \frac{15}{4}\right)$

108. $37 - 19 + 63 - 31$

109. Thinking Generally The fraction $\frac{ab}{ac}$ can be simplified as follows.

$$\frac{ab}{ac} = \frac{a}{a} \cdot \frac{b}{c} = \frac{b}{c}$$

Give a property of real numbers used in this simplification.

110. Thinking Generally Is the product of two irrational numbers always an irrational number? Explain.

APPLICATIONS

111. *Higher Education* The following table lists the total higher education enrollment for selected years.

Year	2012	2014	2016	2018
*Students (millions)	19.5	19.9	20.3	20.6

Source: Department of Education.
*projected

(a) What might the enrollment be in 2016?
(b) Mentally estimate the average of these 4 enrollments shown in the table.
(c) Calculate the average. Is your estimate from part (b) in reasonable agreement with your calculated result?

112. *Federal Budget* The following table lists the spending by the federal government for selected years.

Year	2011	2012	2013	2014	2015
*Budget ($ trillions)	3.8	3.8	3.9	4.1	3.6

Source: Office of Management and Budget.
*projected

(a) Give the budget for 2013.
(b) Mentally estimate the average budget for this 5-year period.
(c) Calculate the average. Is your estimate from part (b) in reasonable agreement with your calculated result?

113. *iPods* Would eight 4 GB iPods store more songs than four 8 GB iPods? Explain the property of real numbers that your calculation illustrates.

114. *Digital Images* If an image downloaded from the Internet is 240 pixels across by 360 pixels high and a different image is 360 pixels across by 240 pixels high, how do the total numbers of pixels in the images compare? What property of real numbers does this result illustrate?

115. *Circumference* The circumference C of a circle is given by $C = 2\pi r$, where r is the radius of the circle. State if the formula is valid for calculating C.
(a) $C = 2r\pi$
(b) $C = 2(\pi r)$
(c) $C = 2\pi + 2r$

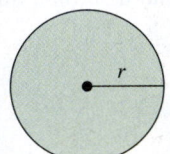

116. *Geometry* The area A of a rectangle equals the product of its length L and width W. What property states that either formula, $A = LW$ or $A = WL$, is correct?

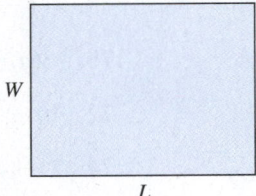

WRITING ABOUT MATHEMATICS

117. Is subtraction either commutative or associative? Explain, using examples.

118. Is division either commutative or associative? Explain, using examples.

Group Activity   Working with Real Data

Directions: Form a group of 2 to 4 people. Select someone to record the group's responses for this activity. All members of the group should work cooperatively to answer the questions. If your instructor asks for your results, each member of the group should be prepared to respond.

1. *Median Age of Marriage for U.S. Males* The following table lists the median age of first marriages for males during various years.

Median Age of First Marriage for Males

Year	1940	1950	1960	1970
Median Age	24.3	22.8	22.8	23.2

Year	1980	1990	2000	2010
Median Age	24.7	26.1	26.8	28.2

Source: U.S. Census Bureau.

(a) Find the median age of marriage for males in 1980.
(b) For each ten-year increase does the median age change by a constant amount? Explain.
(c) Discuss the general trend in the data.
(d) Estimate the median age of marriage for males in 1985.
(e) **Online Exploration** Look up the true value for part (d) on the Internet. How does this answer compare with your estimate?

2. *Median Age of Marriage for U.S. Females* The following table lists the median age of first marriages for females during various years.

Median Age of First Marriage for Females

Year	1940	1950	1960	1970
Median Age	21.5	20.3	20.3	20.8

Year	1980	1990	2000	2010
Median Age	22.0	23.9	25.1	26.1

Source: U.S. Census Bureau.

(a) Find the median age of marriage for females in 1980.
(b) For each ten-year increase does the median age change by a constant amount? Explain.
(c) Discuss the general trend in the data.
(d) Estimate the median age of marriage for females in 1985.
(e) **Online Exploration** Look up the true value for part (d) on the Internet. How does this answer compare with your estimate?

1.2 Operations on Real Numbers

The Real Number Line • Arithmetic Operations • Data and Number Sense

A LOOK INTO MATH ▶ Real numbers are used to describe data. To obtain information from data we frequently perform operations on real numbers. For example, exams are often assigned a score between 0 and 100. This step reduces each exam to a number or *data point*. We might obtain more information about the exams by calculating the average score. To do so we perform the arithmetic operations of addition and division.

In this section we discuss operations on real numbers and provide examples of where these computations occur in real life.

NEW VOCABULARY

☐ Origin
☐ Less/greater than
☐ Positive/negative number
☐ Absolute value
☐ Addends/sum
☐ Additive inverse (opposite)
☐ Difference
☐ Factors/product
☐ Multiplicative inverse
☐ Reciprocal
☐ Dividend/divisor
☐ Quotient
☐ Not equal to

The Real Number Line

We can visualize the real number system by using a number line, as shown in Figure 1.4. Each real number corresponds to a point on the number line. The point associated with the real number 0 is called the **origin**.

The Number Line

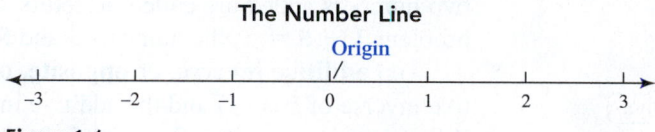

Figure 1.4

If a real number a is located to the left of a real number b on the number line, we say that a is **less than** b and write $a < b$. Similarly, if a real number a is located to the right of a real number b, we say that a is **greater than** b and write $a > b$. Thus $-3 < 2$ because -3 is located to the left of 2, and $2 > -3$ because 2 is located to the right of -3. If $a > 0$, then a is a **positive number**, and if $a < 0$, then a is a **negative number**. See Figure 1.5.

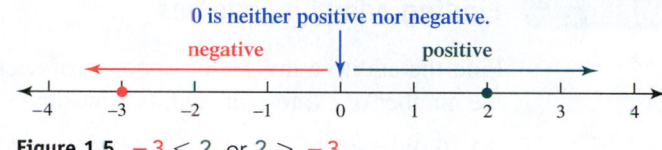

Figure 1.5 $-3 < 2$, or $2 > -3$

The **absolute value** of a real number a, written $|a|$, is equal to its distance from the origin on the number line. Distance may be either a positive number or zero, but it cannot be a negative number. Because the points corresponding to 2 and -2 are both 2 units from the origin, $|2| = 2$ and $|-2| = 2$, as shown in Figure 1.6. The absolute value of a real number is *never* negative.

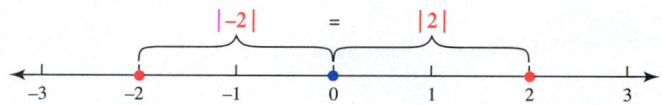

Figure 1.6 Absolute Value

EXAMPLE 1 **Finding the absolute values of real numbers**

Evaluate each expression.
(a) $|9.12|$ **(b)** $\left|-\frac{3}{4}\right|$ **(c)** $|-\pi|$
(d) $|-a|$, if a is a positive number
(e) $|-10| - |-8|$

Solution
(a) $|9.12| = \mathbf{9.12}$
(b) $\left|-\frac{3}{4}\right| = \frac{\mathbf{3}}{\mathbf{4}}$
(c) $|-\pi| = \boldsymbol{\pi}$ because $\pi \approx 3.14$.
(d) If a is positive, $-a$ is negative and $|-a| = \boldsymbol{a}$.
(e) $|-10| - |-8| = 10 - 8 = \mathbf{2}$

Now Try Exercises 19, 21, 23, 27, 33

The absolute value of a number can be defined as follows.

$$|a| = a \qquad \text{if } a > 0 \text{ or } a = 0, \text{ and}$$
$$|a| = -a \qquad \text{if } a < 0.$$

Arithmetic Operations

The four arithmetic operations are addition, subtraction, multiplication, and division.

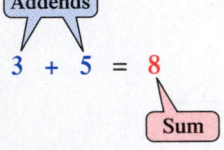

Figure 1.7

ADDITION AND SUBTRACTION OF REAL NUMBERS In an addition problem, the two numbers added are called **addends**, and the answer is called the **sum**. In the addition problem $3 + 5 = 8$, the numbers 3 and 5 are the addends and 8 is the sum. See Figure 1.7.

The **additive inverse**, or **opposite**, of a real number a is $-a$. For example, the additive inverse of 5 is -5 and the additive inverse of -1.5 is $-(-1.5)$, or 1.5. See Figure 1.8. When we add opposites, the result is 0. That is, $a + (-a) = \mathbf{0}$ for every real number a.

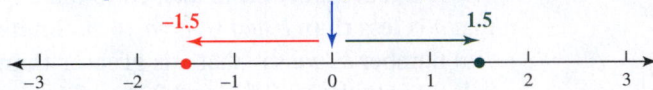

Figure 1.8 Additive Inverses, or Opposites

EXAMPLE 2   **Finding additive inverses**

Find the additive inverse or opposite of each number or expression. Then find the sum of the number or expression and its opposite.

(a) 10,961   (b) π (c) $-\dfrac{3}{4}$ (d) $6x - 2$

Solution
(a) The opposite of 10,961 is $-10{,}961$. Their sum is $10{,}961 + (-10{,}961) = 0$.
(b) The opposite of π is $-\pi$. Their sum is $\pi + (-\pi) = 0$.
(c) The opposite of $-\frac{3}{4}$ is $\frac{3}{4}$. Their sum is $-\frac{3}{4} + \frac{3}{4} = 0$.
(d) The opposite of $6x - 2$ is $-(6x - 2) = -6x + 2$. Their sum is

$$6x - 2 + (-6x) + 2 = 6x + (-6x) - 2 + 2 = 0.$$

Now Try Exercises 41, 45, 47

▶ **REAL-WORLD CONNECTION** When you add real numbers, it may be helpful to think of money. A positive number represents being paid an amount of money, whereas a negative number indicates a debt owed. The sum

$$8 + (-6) = 2$$

would represent being paid \$8 and owing \$6, resulting in \$2 being left over. Similarly,

$$-7 + (-5) = -12$$

would represent owing \$7 and owing \$5, resulting in a debt of \$12.

STUDY TIP

Do you know your instructor's name, office number, and office hours? Find out so you can get help when you need it.

To add two real numbers we may use the following rules.

ADDITION OF REAL NUMBERS

To add two numbers that are either *both positive* or *both negative*, add their absolute values. Their sum has the same sign as the two numbers.

To add two numbers with *opposite signs*, find the absolute value of each number. Subtract the smaller absolute value from the larger. Give the result the same sign as the number with the larger absolute value. The sum of two opposites is 0.

The next example illustrates addition of real numbers.

EXAMPLE 3   **Adding real numbers**

Evaluate each expression.
(a) $-3 + (-5)$ **(b)** $-4 + 7$ **(c)** $8.4 + (-9.5)$

Solution
(a) The addends are both negative, so we add the absolute values $|-3|$ and $|-5|$ to obtain 8. As the signs of the addends are both negative, the sum has the same sign and equals -8. That is, $-3 + (-5) = -8$. If we owe \$3 and then owe an additional \$5, the total amount owed is \$8.
(b) The addends have opposite signs, so we subtract their absolute values to obtain 3. The result is positive because $|7|$ is greater than $|-4|$. That is, $-4 + 7 = 3$. If we owe \$4 and are paid \$7, the result is that we have \$3 to keep.
(c) $8.4 + (-9.5) = -1.1$ because $|-9.5|$ is 1.1 more than $|8.4|$. If we are paid \$8.40 and owe \$9.50, we still owe \$1.10.

Now Try Exercises 65, 67

Addition of positive and negative numbers occurs at banks if deposits are represented by positive numbers and withdrawals are represented by negative numbers.

EXAMPLE 4   **Balancing a checking account**

The initial balance in a checking account is \$157. Find the final balance if the following represents a list of withdrawals and deposits: $-55, -19, 123, -98$.

Solution
We find the sum.

$$157 + (-55) + (-19) + 123 + (-98) = 102 + (-19) + 123 + (-98)$$
$$= 83 + 123 + (-98)$$
$$= 206 + (-98)$$
$$= 108$$

The final balance is \$108. This result may be supported by evaluating the expression with a calculator. See Figure 1.9, where the sum has been calculated two different ways. Instead of adding the opposite of a number, we can also subtract.

```
157+(-55)+(-19)+
123+(-98)
              108
157-55-19+123-98
              108
```

Figure 1.9

Now Try Exercise 75

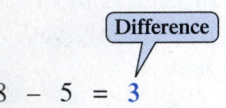

$157+(-55)+(-19)+$
$123+(-98)$
$\qquad\qquad 108$
$157-55-19+123-98$
$\qquad\qquad 108$

Figure 1.9 (Repeated)

Difference

$8 - 5 = 3$

Figure 1.10

TECHNOLOGY NOTE

Subtraction and Negation
On graphing calculators, two different keys typically represent subtraction and negation. In the first calculation in Figure 1.9, we used the negation key, and in the second calculation, we used the subtraction key.

The answer to a subtraction problem is called the **difference**. See Figure 1.10. When you're subtracting two real numbers, it sometimes helps to change the subtraction problem to an addition problem.

SUBTRACTION OF REAL NUMBERS

For any real numbers a and b,

$$a - b = a + (-b).$$

To subtract b from a, add a and the opposite of b.

EXAMPLE 5 **Subtracting real numbers**

Simplify each expression.

(a) $-12 - 7$ **(b)** $-5.1 - (-10.6)$ **(c)** $\frac{1}{2} - \left(-\frac{2}{3}\right)$ **(d)** $3t - 7t$

Solution
(a) $-12 - 7 = -12 + (-7) = -19$
(b) $-5.1 - (-10.6) = -5.1 + 10.6 = 5.5$

(c) $\frac{1}{2} - \left(-\frac{2}{3}\right) = \frac{1}{2} + \frac{2}{3} = \frac{3}{6} + \frac{4}{6} = \frac{7}{6}$

(d) $3t - 7t = (3 - 7)t = -4t$

Now Try Exercises 69, 71, 73, 95

MULTIPLICATION AND DIVISION OF REAL NUMBERS In a multiplication problem, the two numbers multiplied are called the **factors**, and the answer is called the **product**. In the problem $3 \cdot 5 = 15$, the numbers 3 and 5 are factors and 15 is the product. See Figure 1.11. The **multiplicative inverse**, or **reciprocal**, of a *nonzero* number a is $\frac{1}{a}$. The product of a nonzero number a and its reciprocal is $a \cdot \frac{1}{a} = 1$. For example, the reciprocal of -5 is $-\frac{1}{5}$ because $-5 \cdot -\frac{1}{5} = 1$, and the reciprocal of $\frac{2}{3}$ is $\frac{3}{2}$ because $\frac{2}{3} \cdot \frac{3}{2} = 1$. To multiply positive or negative numbers we may use the following rules.

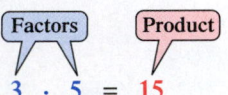

Factors   Product

$3 \cdot 5 = 15$

Figure 1.11

MULTIPLICATION OF REAL NUMBERS

The product of two numbers with *like* signs is positive. The product of two numbers with *unlike* signs is negative.

EXAMPLE 6 **Multiplying real numbers**

Evaluate each expression.

(a) $-11 \cdot 8$ **(b)** $\frac{3}{5} \cdot \frac{4}{7}$ **(c)** $-1.2(-10)$ **(d)** $(1.2)(5)(-7)$

Solution

(a) The product is negative because the factors -11 and 8 have unlike signs. Thus the product is $-11 \cdot 8 = -88$.

(b) The product is positive because both factors are positive. Thus $\frac{3}{5} \cdot \frac{4}{7} = \frac{3 \cdot 4}{5 \cdot 7} = \frac{12}{35}$.

(c) As both factors are negative, the product is positive. Thus $-1.2(-10) = 12$.

(d) $(1.2)(5)(-7) = (6)(-7) = -42$

Now Try Exercises 77, 79, 91

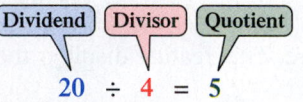

$$20 \div 4 = 5$$

Figure 1.12

In the division problem $20 \div 4 = 5$, the number **20** is the **dividend**, **4** is the **divisor**, and **5** is the **quotient**. See Figure 1.12. This division problem can also be written as $\frac{20}{4} = 5$. Division of real numbers can be defined in terms of multiplication and reciprocals.

DIVISION OF REAL NUMBERS

For real numbers a and b, with $b \neq 0$,

$$\frac{a}{b} = a \cdot \frac{1}{b}.$$

That is, to divide a by b, multiply a by the reciprocal of b.

NOTE: The quotient of two numbers with *like* signs is positive, and the quotient of two numbers with *unlike* signs is negative.

READING CHECK

Explain why division by 0 is undefined.

WHY WE NEVER DIVIDE BY 0. The expression $b \neq 0$ is read "b **not equal to** 0." Note that division by 0 is always *undefined*. For example, suppose that we try to define $12 \div 0$ to be equal to some number k. Then $\frac{12}{0} = k$ and k must satisfy $0 \cdot k = 12$ because a division problem can be checked by using multiplication. (For example, $\frac{12}{3} = 4$, so $3 \cdot 4 = 12$.) But the product of 0 and any number k is 0, not 12. Thus there is no reasonable value for k, so division by 0 is undefined.

EXAMPLE 7   **Dividing real numbers**

Evaluate each expression.

(a) $-12 \div \frac{1}{2}$ (b) $\dfrac{\frac{2}{3}}{-7}$ (c) $\dfrac{-4}{-24}$ (d) $6 \div 0$

Solution

(a) $-12 \div \dfrac{1}{2} = -12 \cdot \dfrac{2}{1} = -24$ Multiply by reciprocal: $\frac{2}{1}$.

(b) $\dfrac{\frac{2}{3}}{-7} = \dfrac{2}{3} \div (-7) = \dfrac{2}{3} \cdot \left(-\dfrac{1}{7}\right) = -\dfrac{2}{21}$ Multiply by reciprocal: $-\frac{1}{7}$.

(c) $\dfrac{-4}{-24} = -4 \cdot \left(-\dfrac{1}{24}\right) = \dfrac{4}{24} = \dfrac{1}{6}$ Multiply by reciprocal: $-\frac{1}{24}$.

(d) $6 \div 0$ is undefined because division by 0 is not possible.

Now Try Exercises 81, 87, 89

Many calculators have the capability to perform arithmetic on fractions and express the answer as either a decimal or a fraction. The next example illustrates this capability.

EXAMPLE 8 | **Performing arithmetic operations with technology**

Use a calculator to evaluate each expression as a decimal and as a fraction.

(a) $\dfrac{1}{3} + \dfrac{2}{5} - \dfrac{4}{9}$ **(b)** $\left(\dfrac{4}{9} \cdot \dfrac{3}{8}\right) \div \dfrac{2}{3}$

Solution

(a) From Figure 1.13,

$$\frac{1}{3} + \frac{2}{5} - \frac{4}{9} = 0.2\overline{8}, \quad \text{or} \quad \frac{13}{45}.$$

In Figure 1.13, the second calculation uses the "Frac" feature. This feature displays the answer as a fraction, rather than a decimal.

NOTE: Generally it is a good idea to put parentheses around fractions when you are using a calculator.

(b) From Figure 1.14, $\left(\frac{4}{9} \cdot \frac{3}{8}\right) \div \frac{2}{3} = 0.25$, or $\frac{1}{4}$.

CALCULATOR HELP

To express answers as fractions, see Appendix A (pages AP-1 and AP-2).

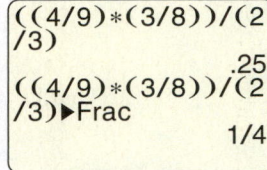

Figure 1.13   Figure 1.14

▌ **Now Try Exercises 101, 103**

Data and Number Sense

▶ **REAL-WORLD CONNECTION** In everyday life we commonly make approximations involving a variety of data. To make estimations we often use arithmetic operations on real numbers, as demonstrated in the next three examples.

EXAMPLE 9 | **Developing number sense**

A rectangular birthday cake serves exactly 24 pieces of cake. Suppose that a similar cake is twice as wide and twice as long. How many pieces does the second cake serve?

Solution

Let L be the length and W be the width of the first cake. Then the area of this rectangular cake is $A = LW$. The length of the second cake is $2L$ and the width is $2W$. The area of this cake is $A = 2L \cdot 2W = 4LW$, or four times the area of the first cake. Thus, the second cake can serve $4 \cdot 24 = 96$ pieces of cake.

▌ **Now Try Exercise 125**

EXAMPLE 10 | **Determining a reasonable answer**

It is 2823 miles from New York to Los Angeles. Determine mentally which of the following would best estimate the number of hours of *driving time* required to travel this distance in a car: 50, 100, or 120 hours.

Solution

Average speed equals distance divided by time. Dividing by 100 is easy, so start by dividing 100 hours into 2800 miles. The average speed would be $\frac{2800}{100} = 28$ miles per hour, which is too slow for most drivers. A more reasonable choice is 50 hours because then the average speed would be double, or about 56 miles per hour.

Now Try Exercise 129

EXAMPLE 11   **Estimating a numeric value**

Table 1.3 lists the percentage of the U.S. population having wireless phones for selected years. Estimate this percentage in 2010.

TABLE 1.3 Wireless Phone Ownership in the United States

Year	2006	2007	2008	2009	2010
Percentage	78%	84%	88%	91%	?

Source: Cellular Telecommunications Industry Association.

Solution
Overall the data are increasing from 2006 to 2009. However, the rate of growth is slowing. From 2006 to 2007 the increase was 6%, from 2007 to 2008 the increase was 4%, and from 2008 to 2009 the increase was 3%. It might be reasonable that the increase from 2009 to 2010 was 1% or 2%. Thus one estimate might be $91\% + 2\% = 93\%$. Other estimates are possible. Note that the true value is 93%.

Now Try Exercise 133

1.2   Putting It All Together

OPERATION	DEFINITION	EXAMPLES
Absolute Value of a Real Number	For real number a, $|a| = a$ if $a > 0$ or $a = 0$, and $|a| = -a$ if $a < 0$.	$|-5| = 5$, $|3.7| = 3.7$, and $|-4 + 4| = 0$
Additive Inverse (Opposite)	The opposite of a is $-a$.	4 and -4 are opposites. -5 and $-(-5) = 5$ are opposites.
Addition of Real Numbers	See the highlighted box: Addition of Real Numbers on page 15.	$3 + (-6) = -3$, $-2 + (-10) = -12$, $-1 + 3 = 2$, and $18 + 11 = 29$
Subtraction of Real Numbers	We can transform a subtraction problem into an addition problem: $$a - b = a + (-b).$$	$4 - 6 = 4 + (-6) = -2$, $-7 - (-8) = -7 + 8 = 1$, $-8 - 5 = -8 + (-5) = -13$, and $9 - (-1) = 9 + 1 = 10$
Multiplication of Real Numbers	The product of two numbers with *like* signs is positive. The product of two numbers with *unlike* signs is negative.	$3 \cdot 5 = 15$, $-4(-7) = 28$, $-8 \cdot 7 = -56$, and $5(-11) = -55$
Division of Real Numbers	For real numbers a and b with $b \neq 0$, $$\frac{a}{b} = a \cdot \frac{1}{b}.$$	$-3 \div \frac{3}{4} = -3 \cdot \frac{4}{3} = -4$ and $\frac{5}{2} \div \left(-\frac{7}{4}\right) = \frac{5}{2}\left(-\frac{4}{7}\right) = -\frac{20}{14} = -\frac{10}{7}$

1.2 Exercises

MyMathLab PRACTICE   WATCH    DOWNLOAD   READ   REVIEW

CONCEPTS AND VOCABULARY

1. If $a > b$, then a is located to the _____ of b on the number line.

2. If $a < 0$, then a is located to the _____ of the origin on the number line.

3. The product of two negative numbers is a(n) _____ number.

4. The sum of two negative numbers is a(n) _____ number.

5. The quotient of a positive number and a negative number is a(n) _____ number.

6. If $a < 0$ and $b > 0$, then the difference $a - b$ is a(n) _____ number.

7. The additive inverse of a is _____.

8. The multiplicative inverse, or reciprocal, of $\frac{a}{b}$ is _____ for $a \neq 0$ and $b \neq 0$.

9. If $a < 0$ and $b < 0$, then $\frac{a}{b}$ is a(n) _____ number.

10. The opposite of $a + b$ is _____.

11. The absolute value of -8 is _____.

12. If a is positive, the absolute value of a is _____.

THE REAL NUMBER LINE AND ABSOLUTE VALUE

Exercises 13–18: Plot each number on a number line. Be sure to include an appropriate scale.

13. $0, 4, -3, 1.5$

14. $10, 20, -5, -15$

15. $100, 300, -200, 50$

16. $-0.4, 0.2, 0.5, -0.1, 0$

17. $-\frac{3}{2}, \pi, \frac{1}{3}, 2$

18. $-2, -\frac{1}{2}, 1, \frac{7}{3}$

Exercises 19–34: Evaluate the expression.

19. $|-6.1|$

20. $|17|$

21. $\left|-\frac{2}{3}\right|$

22. $\left|-\frac{5}{4}\right|$

23. $|\pi - 1|$

24. $|4 - \pi|$

25. $-|-8|$

26. $-|3 + 4|$

27. $|-8| - |2|$

28. $|3| - |-3|$

29. $|8 - 11|$

30. $|2 \cdot 8 - 23|$

31. $|x|$, where $x > 0$

32. $|x|$, where $x < 0$

33. $|x - y|$, where $x > 0$ and $y < 0$

34. $|x + y|$, where $x < 0$ and $y < 0$

Exercises 35–40: State whether only positive numbers are typically used to measure the given quantity or whether both positive and negative numbers are used. Explain your reasoning.

35. Area

36. Distance

37. Temperature

38. A person's net worth

39. Gas mileage of a car

40. Elevation relative to sea level

ARITHMETIC OPERATIONS

Exercises 41–52: Find the additive inverse, or opposite, of the number or expression.

41. 56

42. $-\frac{5}{7}$

43. -6.9

44. 12.8

45. $-\pi + 2$

46. $a + b$

47. $a - b$

48. $-x + 3$

49. $-(x - y)$

50. $-1 - x$

51. $z - (2 - z)$

52. $2z - (1 + 2z)$

Exercises 53–64: Find the multiplicative inverse, or reciprocal, of the number or expression.

53. 3

54. $\frac{3}{8}$

55. $-\frac{2}{3}$

56. 1.5

57. π

58. $-\frac{x}{y}$

59. $a + 3$

60. $x - 1$

61. $-\frac{2a}{b}$

62. $-\frac{2b}{3a}$

63. $\frac{1}{x - 7}$

64. $\frac{3}{x + y}$

Exercises 65–98: Perform the following arithmetic operations and simplify, if possible.

65. $4 + (-6)$

66. $-10 + 14$

67. $-7.4 + (-9.2)$

68. $-8.4 - 10.3$

69. $-8 - 22$

70. $5 - (-4)$

71. $-1.1 - (-3.4)$

72. $-7 - (-3)$

73. $-\frac{3}{4} - \left(-\frac{1}{4}\right)$

74. $\frac{1}{5} - \left(-\frac{3}{10}\right)$

75. $-9 + 1 + (-2) + 5$

76. $-5 + 7 - (-2) + 3$

77. $-6 \cdot -12$

78. $-(8 \cdot -4)$

79. $-\frac{1}{2} \cdot \frac{5}{7}$

80. $-\frac{6}{7} \cdot -\frac{5}{3}$

81. $-\frac{1}{2} \div -\frac{3}{4}$

82. $-5 \div \frac{4}{5}$

83. $\frac{3}{4} \div (-2)$

84. $-\frac{4}{5} \div \frac{7}{10}$

85. $-\frac{8}{2}$

86. $-\frac{45}{9}$

87. $-25 \div -5$

88. $8 \div 0$

89. $\frac{1}{0}$

90. $\frac{0}{7}$

91. $-4 \cdot 7 \cdot (-5)$

92. $-2 \cdot -6 \cdot 6 \cdot -1$

93. $6 \cdot \frac{2}{3} \cdot 3 \cdot \left(-\frac{1}{6}\right)$

94. $\left(\frac{4}{5} \cdot \frac{5}{8}\right) \div \frac{1}{2}$

95. $4x - 9x$

96. $\frac{1}{2}x - \frac{3}{4}x$

97. $z - 4z + 2z$

98. $5z - 3z - 6z$

Exercises 99–106: Evaluate the expression.

99. $-23.1 + 45.7 - (-34.6)$

100. $102 - (-341) + (-112)$

101. $\frac{1}{2} + \frac{2}{3} - \frac{5}{7}$

102. $-\frac{8}{13} + \frac{1}{2} - \frac{2}{5}$

103. $-\frac{3}{4} \cdot \frac{4}{5} \div \frac{5}{3}$

104. $\left(\frac{3}{4} \div (-11)\right) - \frac{2}{5}$

105. $\frac{1}{2}\left(\frac{4}{11} + \frac{2}{5}\right)$

106. $-\frac{5}{13} + \left(\frac{3}{5} \div \frac{2}{17}\right)$

DATA AND NUMBER SENSE

Exercises 107–114: Mentally estimate the average of the list of numbers. Check your estimate.

107. $9, 5, 15, -9$

108. $12, 8, 27, -7$

109. $-2, 12, 7, -17$

110. $45, 55, 65, 35$

111. $101, 99, -42, 82$

112. $-4, 2, 5, -8, 5, 24$

113. Thinking Generally a, b, a, b, a, b

114. Thinking Generally $a, 2a, 3a, 4a, 5a$

Exercises 115–124: Mentally evaluate the expression.

115. $-2 + 8 + 3 + 2 - 11$

116. $-22 + 43 - 78 + 7$

117. $103 - 44 + 97 - 56$

118. $10 + 11 + 12 - 11 - 11 - 11$

119. $\frac{1}{5} \cdot \frac{2}{3} \cdot \frac{1}{7} \cdot \frac{1}{9} \cdot 5 \cdot \frac{3}{2} \cdot 7 \cdot 9$

120. $\frac{1}{2} \cdot \frac{1}{3} \cdot \frac{1}{4} \cdot (-4) \cdot (-3) \cdot (-2)$

121. $\left(\frac{1}{2} - \frac{1}{3}\right) + \left(\frac{1}{3} - \frac{1}{4}\right) + \left(\frac{1}{4} - \frac{1}{5}\right) + \left(\frac{1}{5} - \frac{1}{6}\right)$

122. $\frac{1}{2} \div \frac{1}{3} \cdot \frac{1}{3} \div \frac{1}{4} \cdot \frac{1}{4} \div \frac{1}{5} \cdot \frac{1}{5}$

123. Thinking Generally $a \cdot b \cdot c \cdot \frac{1}{a} \cdot \frac{1}{b} \cdot \frac{1}{c}$

124. Thinking Generally $b - a(b + c) + ab + ac$

APPLICATIONS

125. *Storage in a Box* A rectangular box contains 200 cubic inches. If a similar box has the same height but double the length and triple the width, how many cubic inches does this box contain? (*Hint:* Volume = Length × Width × Height.)

126. *Water in a Lake* A lake covers 400 acres, has an average depth of 20 feet, and contains 350 million cubic feet of water. Estimate how many cubic feet of water there are in a lake that covers 200 acres and has an average depth of 10 feet.

127. *Filling a Pool* A hose fills a rectangular swimming pool in 2 days. Suppose that a similar swimming pool has the same average depth but is four times longer and two times wider. If the same hose is used, how long will it take to fill this swimming pool?

128. *Fudge in a Pan* A single batch of fudge fills a circular pan as illustrated in the figure. How many batches of fudge are needed to fill a similar pan that has twice the diameter? (*Hint:* $A = \pi r^2$.)

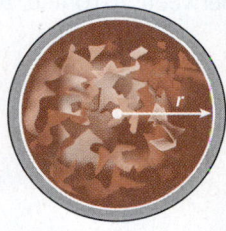

129. *Buying a Computer* An advertisement for a computer states that it costs $202 down and $98.99 per month for 24 months. Mentally estimate which of the following represents the cost of the computer if it is purchased under these terms: $2600, $3200, or $3800. Explain your reasoning. Find the actual cost.

130. *Leasing a Car* To lease a car costs $1497 down and $249 per month for 36 months. Mentally estimate the cost of the lease. Find the actual cost.

131. *Blogs and Bloggers* The attention given to blogs and bloggers in reputable news sources has risen dramatically. The following table lists the number of references to blogs or bloggers in the top 20 national magazines and newspapers for selected years.

Year	2004	2005	2006	2007	2008
References	795	2179	5591	9903	13,000

Source: Cision Media Research.

(a) Have the data increased by a fixed number each year? Is the number of references increasing?

(b) Find the average number of references over this 5-year period.

(c) Estimate the number of references in 2009. Discuss how you arrived at your answer.

132. *Cigarette Consumption* Although smoking in the United States has declined since 1980, it has continued to grow globally. The table lists the global cigarette consumption for selected years.

Year	1960	1970	1980	1990	2000	2010
Cigarettes (trillions)	2.2	3.1	4.4	5.4	5.5	6.3

Source: Department of Agriculture.

(a) In 1960 the world population was 3 billion people. How many cigarettes were consumed, on average, per person?

(b) In 2010 the world population was 6.8 billion people. How many cigarettes were consumed, on average, per person?

(c) Estimate cigarette consumption in 1985. Explain your reasoning.

(d) Is cigarette consumption likely to increase or decrease between 2010 and 2020? Explain your reasoning.

133. *Netflix Subscribers* Netflix is the world's largest subscription service for streaming movies over the Internet and sending DVDs by mail. The following table lists the number of Netflix subscribers in millions for selected years.

Year	2006	2007	2008	2009
Subscribers (millions)	6.3	7.5	9.4	12.3

Source: Netflix.

(a) Discuss any trends in the data.

(b) Estimate the number of subscribers in 2010. Explain your reasoning.

134. *Blockbuster Retail Stores* Blockbuster competes directly with Netflix. The following table lists the number of Blockbuster stores for selected years.

Year	2006	2007	2008	2009
Stores	5194	4855	4585	4356

Source: Blockbuster.

(a) Discuss any trends in the data.

(b) Estimate the number of stores in 2010. Explain your reasoning.

(c) Discuss how these data and the data found in Exercise 133 are related.

WRITING ABOUT MATHEMATICS

135. Suppose that $a \neq b$. Explain how a number line can be used to determine whether a is greater than b or whether a is less than b. Give examples.

136. Explain why a positive number times a negative number is a negative number.

SECTIONS
1.1 and 1.2

Checking Basic Concepts

1. Identify each number as one or more of the following: a natural number, an integer, a rational number, or a real number.
   (a) -9
   (b) $\frac{8}{4}$
   (c) $\sqrt{5}$
   (d) 0.5

2. Identify the property of real numbers that each equation illustrates. Write the property, using variables.
   (a) $3 + 4 = 4 + 3$
   (b) $-5 \cdot (4 \cdot 8) = (-5 \cdot 4) \cdot 8$

(c) $4(5 + 2) = 4 \cdot 5 + 4 \cdot 2$
(d) $4x - (5 - x) = 4x - 5 + x$

3. Evaluate each expression.
   (a) $-3 + 4 + (-5)$
   (b) $5.1 \cdot (-4) \cdot 2$
   (c) $-\frac{2}{3} \cdot \left(\frac{1}{4} \div \frac{2}{5} \right)$

4. A small lake covers 400 acres, has an average depth of 15 feet, and contains about 2 billion gallons of water. Estimate the number of gallons in a lake that covers 1200 acres and has an average depth of 30 feet.

1.3   Integer Exponents

Bases and Positive Exponents • Zero and Negative Exponents • Product, Quotient, and Power Rules • Order of Operations • Scientific Notation

A LOOK INTO MATH ▶

Modern society has brought with it the need for both small and large numbers. The size of an average virus is 5 millionths (0.000005) of a centimeter, whereas the distance to the nearest star, our sun, is 93 million (93,000,000) miles. To represent such numbers we often use exponents. In this section we discuss properties of integer exponents and some of their applications. (*Source:* C. Ronan, *The Natural History of the Universe.*)

NEW VOCABULARY

☐ Exponential expression
☐ Base
☐ Exponent
☐ Order of operations
☐ Scientific notation
☐ Standard (decimal) form

Bases and Positive Exponents

The area of a square that is 8 inches on a side is given by the expression

$$\underbrace{8 \cdot 8}_{\text{2 factors of 8}} = 8^2 = 64 \text{ square inches.}$$

The expression 8^2 is an **exponential expression** with **base** 8 and **exponent** 2. See Figures 1.15 and 1.16.

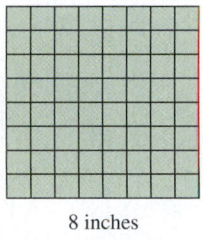

8 inches

8 inches
$8^2 = 64$ square inches

Figure 1.15

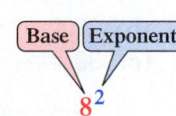

Base   Exponent

8^2

Figure 1.16

Exponential expressions occur frequently in a variety of applications. For example, suppose that an investment doubles its initial value 3 times. Then its final value is

$$\underbrace{2 \cdot 2 \cdot 2}_{\textbf{3 factors of 2}} = 2^3 = 8 \qquad \text{Doubles 3 times}$$

times as large as its original value. In general, if n is a positive integer and a is a real number, then

$$a^n = \underbrace{a \cdot a \cdot a \cdots \cdot a.}_{n \text{ factors of } a}$$

Table 1.4 contains examples of exponential expressions.

TABLE 1.4

Expression	Base	Exponent
2^3	2	3
6^4	6	4
7^1	7	1
0.5^2	0.5	2
x^3	x	3

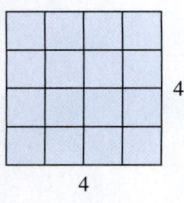

4

4 Squared

Figure 1.17

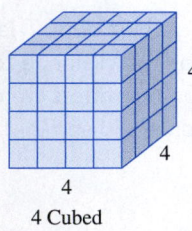

4

4

4

4 Cubed

Figure 1.18

Read 0.5^2 as "0.5 squared," 2^3 as "2 cubed," and 6^4 as "6 to the fourth power." The terms *squared* and *cubed* come from geometry. If the length of a side of a **square** is 4, then its area is

$$4 \cdot 4 = 4^2 = 16$$

square units, as illustrated in Figure 1.17. Similarly, if the length of an edge (side) of a **cube** is 4, then its volume is

$$4 \cdot 4 \cdot 4 = 4^3 = 64$$

cubic units, as shown in Figure 1.18.

EXAMPLE 1   **Writing numbers in exponential notation**

Using the given base, write each number as an exponential expression. Check your results with a calculator.
(a) 10,000 (base 10)
(b) 27 (base 3)
(c) 32 (base 2)

Solution
(a) $10,000 = 10 \cdot 10 \cdot 10 \cdot 10 = 10^4$
(b) $27 = 3 \cdot 3 \cdot 3 = 3^3$
(c) $32 = 2 \cdot 2 \cdot 2 \cdot 2 \cdot 2 = 2^5$

These values are supported in Figure 1.19 on the next page, where we have evaluated exponential expressions with a calculator, using the $\boxed{\wedge}$ key.

CALCULATOR HELP

To calculate exponential expressions, see Appendix A (page AP-1).

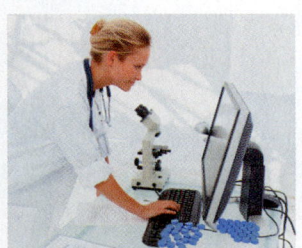

Figure 1.19

■ **Now Try Exercises 17, 19**

▶ **REAL-WORLD CONNECTION** Computer memory is often measured in bytes. A *byte* is capable of storing one letter of the alphabet. For example, the word "math" requires four bytes to store in a computer. Bytes of computer memory are often manufactured in amounts equal to powers of 2, as illustrated in the next example.

EXAMPLE 2 | **Using exponents to analyze computer memory**

In computer technology, 1 K (kilobyte) of memory is equal to 2^{10} bytes, and 1 MB (megabyte) of memory is equal to 2^{20} bytes. Determine whether 1 K of memory is equal to 1000 bytes and whether 1 MB is equal to 1,000,000 bytes. (*Source:* D. Horn, *Basic Electronics Theory.*)

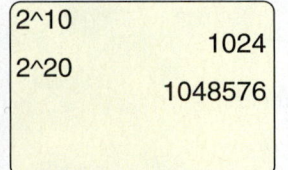

Solution

Figure 1.20 shows that $2^{10} = 1024$ and that $2^{20} = 1,048,576$. Thus 1 K represents slightly more than 1000 bytes and 1 MB is more than 1,000,000 bytes.

■ **Now Try Exercise 129**

Figure 1.20

Zero and Negative Exponents

Exponents can be defined for any integer. To illustrate these definitions for exponents, consider Table 1.5. Each time the exponent on the base 2 decreases by 1, the result in the next row down can be found by dividing the result in the previous row by 2. For example, $8 \div 2 = 4$, $4 \div 2 = 2$, $2 \div 2 = 1$, and so on. The results for bases 10 and x are found similarly in Tables 1.6 and 1.7, where it is assumed that $x \neq 0$.

TABLE 1.5	TABLE 1.6	TABLE 1.7
$2^3 = 8$	$10^3 = 1000$	$x^3 = x \cdot x \cdot x$
$2^2 = 4$	$10^2 = 100$	$x^2 = x \cdot x$
$2^1 = 2 \cdot$	$10^1 = 10$	$x^1 = x$
$2^0 = 1$	$10^0 = 1$	$x^0 = 1$
$2^{-1} = \dfrac{1}{2}$	$10^{-1} = \dfrac{1}{10}$	$x^{-1} = \dfrac{1}{x}$
$2^{-2} = \dfrac{1}{4}$	$10^{-2} = \dfrac{1}{100}$	$x^{-2} = \dfrac{1}{x \cdot x}$
$2^{-3} = \dfrac{1}{8}$	$10^{-3} = \dfrac{1}{1000}$	$x^{-3} = \dfrac{1}{x \cdot x \cdot x}$
At each step divide the result by 2.	At each step divide the result by 10.	At each step divide the result by x.

These tables help justify the following definition for a *nonzero* base a with a zero exponent.

$$a^0 = 1, a \neq 0 \qquad 0^0 \text{ is } undefined.$$

For example, $\left(\frac{1}{7}\right)^0 = 1$. In Table 1.5 on the previous page, note that

$$2^{-2} = \frac{1}{4} = \frac{1}{2^2} \quad \text{and} \quad 2^{-3} = \frac{1}{8} = \frac{1}{2^3}.$$

The pattern also continues in Tables 1.6 and 1.7 on the previous page for bases 10 and x, and helps motivate the following definition for negative exponents, where $a \neq 0$ and n is a positive integer.

$$a^{-n} = \frac{1}{a^n}.$$

Thus $5^{-4} = \frac{1}{5^4}$ and $y^{-2} = \frac{1}{y^2}$. Using the previous definition, we obtain

$$\frac{1}{a^{-n}} = \frac{1}{\frac{1}{a^n}} = \frac{a^n}{1} = a^n.$$

Thus $\frac{1}{2^{-5}} = 2^5$ and $\frac{1}{x^{-2}} = x^2$. If a and b are nonzero numbers, then

$$\frac{a^{-n}}{b^{-m}} = \frac{\frac{1}{a^n}}{\frac{1}{b^m}} = \frac{1}{a^n} \cdot \frac{b^m}{1} = \frac{b^m}{a^n}.$$

Thus $\frac{4^{-3}}{z^{-2}} = \frac{z^2}{4^3}$. This discussion leads to the following properties for integer exponents.

READING CHECK

Give an example of each of the five properties in the box to the right.

INTEGER EXPONENTS

Let a and b be nonzero real numbers and m and n be positive integers. Then

1. $a^n = a \cdot a \cdot a \cdot \cdots \cdot a$ (n factors of a)
2. $a^0 = 1$ (*Note:* 0^0 is undefined.)
3. $a^{-n} = \dfrac{1}{a^n}$ and $\dfrac{1}{a^{-n}} = a^n$
4. $\dfrac{a^{-n}}{b^{-m}} = \dfrac{b^m}{a^n}$
5. $\left(\dfrac{a}{b}\right)^{-n} = \left(\dfrac{b}{a}\right)^n$

EXAMPLE 3 **Simplifying expressions**

Simplify each expression. Use positive exponents.

(a) 3^{-4} **(b)** $\dfrac{1}{2^{-3}}$ **(c)** $\left(\dfrac{5}{7}\right)^{-2}$ **(d)** $\dfrac{1}{(xy)^{-1}}$ **(e)** $\dfrac{2^{-2}}{3t^{-3}}$

Solution

(a) $3^{-4} = \dfrac{1}{3^4} = \dfrac{1}{3 \cdot 3 \cdot 3 \cdot 3} = \dfrac{1}{81}$ **(b)** $\dfrac{1}{2^{-3}} = 2^3 = 2 \cdot 2 \cdot 2 = 8$

(c) $\left(\dfrac{5}{7}\right)^{-2} = \left(\dfrac{7}{5}\right)^2 = \dfrac{7}{5} \cdot \dfrac{7}{5} = \dfrac{49}{25}$ **(d)** $\dfrac{1}{(xy)^{-1}} = (xy)^1 = xy$

⎯⎯⎯⎯⎯⎯ Base is xy.

(e) Note that only t and *not* $3t$ is raised to the power of -3.

$$\frac{2^{-2}}{3t^{-3}} = \frac{t^3}{3(2^2)} = \frac{t^3}{3 \cdot 4} = \frac{t^3}{12}$$

⎯⎯ Base is t.

Now Try Exercises 25, 27, 31

Product, Quotient, and Power Rules

We can calculate products and quotients of exponential expressions *provided their bases are the same*. For example,

$$4^2 \cdot 4^3 = \underbrace{(4 \cdot 4)}_{\text{2 factors of 4}} \cdot \underbrace{(4 \cdot 4 \cdot 4)}_{\text{3 factors of 4}} = 4^5. \qquad 2 + 3 = 5$$

This expression has a total of $2 + 3 = 5$ factors of 4, so the result is 4^5. To multiply exponential expressions with like bases, add exponents. Thus

$$x^3 \cdot x^4 = \underbrace{(x \cdot x \cdot x)}_{\text{3 factors of } x} \cdot \underbrace{(x \cdot x \cdot x \cdot x)}_{\text{4 factors of } x} = x^7. \qquad 3 + 4 = 7$$

THE PRODUCT RULE

For any real number a and integers m and n,

$$a^m \cdot a^n = a^{m+n}.$$

Note that the product rule holds for negative exponents. For example,

$$10^5 \cdot 10^{-2} = 10^{5+(-2)} = 10^3.$$

EXAMPLE 4   **Using the product rule**

Multiply and simplify. Use positive exponents.
(a) $10^2 \cdot 10^4$ **(b)** $7^3 \, 7^{-4}$ **(c)** $x^3 x^{-2} x^4$ **(d)** $3y^2 \cdot 2y^{-4}$

Solution
(a) $10^2 \cdot 10^4 = 10^{2+4} = 10^6 = 1{,}000{,}000$ Add exponents.
(b) $7^3 \, 7^{-4} = 7^{3+(-4)} = 7^{-1} = \dfrac{1}{7}$
(c) $x^3 \, x^{-2} \, x^4 = x^{3+(-2)+4} = x^5$
(d) $3y^2 \cdot 2y^{-4} = 3 \cdot 2 \cdot y^2 \cdot y^{-4} = 6y^{2+(-4)} = 6y^{-2} = \dfrac{6}{y^2}$

Note that 6 is *not* raised to the power of -2 in the expression $6y^{-2}$.

Now Try Exercises 33, 35(a)

Consider division of exponential expressions with like bases. Here

$$\frac{6^5}{6^3} = \frac{6 \cdot 6 \cdot 6 \cdot 6 \cdot 6}{6 \cdot 6 \cdot 6} = \frac{6}{6} \cdot \frac{6}{6} \cdot \frac{6}{6} \cdot 6 \cdot 6 = 1 \cdot 1 \cdot 1 \cdot 6 \cdot 6 = 6^2. \qquad 5 - 3 = 2$$

Because there are "two more" 6s in the numerator, the result is $6^{5-3} = 6^2$. Thus, to divide exponential expressions with like bases, subtract exponents.

THE QUOTIENT RULE

For any nonzero real number a and integers m and n,

$$\frac{a^m}{a^n} = a^{m-n}.$$

Note that the quotient rule holds for negative exponents. For example,

$$\frac{2^{-6}}{2^{-4}} = 2^{-6-(-4)} = 2^{-2} = \frac{1}{2^2}.$$

However, you may want to evaluate this quotient as

$$\frac{2^{-6}}{2^{-4}} = \frac{2^4}{2^6} = \frac{1}{2^2}.$$

NOTE: The quotient rule can also be used to justify the conclusion that $a^0 = 1$. For example, $\frac{3^2}{3^2} = \frac{9}{9} = 1$, and by the quotient rule $\frac{3^2}{3^2} = 3^{2-2} = 3^0$. Thus $3^0 = 1$.

EXAMPLE 5 **Using the quotient rule**

Simplify each expression. Use positive exponents.

(a) $\dfrac{10^4}{10^6}$ **(b)** $\dfrac{x^5}{x^2}$ **(c)** $\dfrac{15x^2y^3}{5x^4y}$ **(d)** $\dfrac{3a^{-2}b^5}{9a^4b^{-3}}$

Solution

(a) $\dfrac{10^4}{10^6} = 10^{4-6} = 10^{-2} = \dfrac{1}{10^2} = \dfrac{1}{100}$ Subtract exponents.

(b) $\dfrac{x^5}{x^2} = x^{5-2} = x^3$

(c) $\dfrac{15x^2y^3}{5x^4y} = \dfrac{15}{5} \cdot \dfrac{x^2}{x^4} \cdot \dfrac{y^3}{y^1} = 3 \cdot x^{2-4}y^{3-1} = 3x^{-2}y^2 = \dfrac{3y^2}{x^2}$

(d) $\dfrac{3a^{-2}b^5}{9a^4b^{-3}} = \dfrac{3b^5b^3}{9a^4a^2} = \dfrac{b^8}{3a^6}$

Now Try Exercises 39, 41, 43

How should we evaluate $(4^3)^2$? To answer this question consider

$$(4^3)^2 = 4^3 \cdot 4^3 = 4^{3+3} = 4^6. \qquad \textcolor{red}{3 \cdot 2 = 6}$$

Similarly,

$$(x^4)^3 = x^4 \cdot x^4 \cdot x^4 = x^{4+4+4} = x^{12}. \qquad \textcolor{red}{4 \cdot 3 = 12}$$

These results suggest that to raise a power to a power, multiply the exponents.

RAISING POWERS TO POWERS

For any real number a and integers m and n,

$$(a^m)^n = a^{mn}.$$

EXAMPLE 6 **Raising powers to powers**

Simplify each expression. Use positive exponents.

(a) $(5^2)^3$ **(b)** $(2^4)^{-2}$ **(c)** $(b^{-7})^5$ **(d)** $\dfrac{(x^3)^{-2}}{(x^{-5})^2}$

Solution
(a) $(5^2)^3 = 5^{2 \cdot 3} = 5^6 = 15{,}625$ Multiply exponents.

(b) $(2^4)^{-2} = 2^{4(-2)} = 2^{-8} = \dfrac{1}{2^8} = \dfrac{1}{256}$

(c) $(b^{-7})^5 = b^{-7 \cdot 5} = b^{-35} = \dfrac{1}{b^{35}}$

(d) $\dfrac{(x^3)^{-2}}{(x^{-5})^2} = \dfrac{x^{-6}}{x^{-10}} = \dfrac{x^{10}}{x^6} = x^4$

■ **Now Try Exercises 45, 55**

How can we simplify the expression $(2x)^3$? Consider

$$(\textcolor{blue}{2x})^3 = 2x \cdot 2x \cdot 2x = \underbrace{(2 \cdot 2 \cdot 2)}_{\textcolor{red}{3 \text{ factors of } 2}} \cdot \underbrace{(x \cdot x \cdot x)}_{\textcolor{red}{3 \text{ factors of } x}} = \textcolor{blue}{2^3 x^3}.$$

This result suggests that to cube a product, cube each factor. The following rule generalizes this concept.

RAISING PRODUCTS TO POWERS

For any real numbers a and b and integer n,

$$(ab)^n = a^n b^n.$$

EXAMPLE 7   **Raising products to powers**

Simplify each expression. Use positive exponents.

(a) $(6y)^2$ **(b)** $(x^2 y)^{-2}$ **(c)** $(2xy^3)^4$ **(d)** $\dfrac{(2a^2 b^{-3})^2}{4(ab^3)^3}$

Solution

(a) $(\textcolor{blue}{6y})^2 = \textcolor{blue}{6^2 y^2} = 36y^2$ Base is $6y$.

(b) $(\textcolor{blue}{x^2 y})^{-2} = \dfrac{1}{(x^2 y)^2} = \dfrac{1}{(x^2)^2 y^2} = \dfrac{1}{x^4 y^2}$

(c) $(\textcolor{blue}{2xy^3})^4 = 2^4 x^4 (y^3)^4 = 16x^4 y^{12}$

(d) $\dfrac{(\textcolor{blue}{2a^2 b^{-3}})^2}{4(\textcolor{blue}{ab^3})^3} = \dfrac{2^2 a^4 b^{-6}}{4a^3 b^9} = \dfrac{4a^4}{4a^3 b^9 b^6} = \dfrac{a}{b^{15}}$

■ **Now Try Exercises 47, 57**

The expression $\left(\dfrac{a}{b}\right)^3$ can be simplified as

$$\left(\frac{a}{b}\right)^3 = \frac{a}{b} \cdot \frac{a}{b} \cdot \frac{a}{b} = \frac{a^3}{b^3}.$$

This result suggests the following rule.

RAISING QUOTIENTS TO POWERS

For nonzero real numbers a and b and any integer n,

$$\left(\frac{a}{b}\right)^n = \frac{a^n}{b^n}.$$

EXAMPLE 8   **Raising quotients to powers**

Simplify each expression. Use positive exponents.

(a) $\left(\dfrac{3}{x}\right)^3$ **(b)** $\left(\dfrac{1}{2^3}\right)^{-2}$ **(c)** $\left(\dfrac{3x^{-3}}{y^2}\right)^4$ **(d)** $\left(\dfrac{3x^2}{4y^2z}\right)^3$

Solution

(a) $\left(\dfrac{3}{x}\right)^3 = \dfrac{3^3}{x^3} = \dfrac{27}{x^3}$

(b) $\left(\dfrac{1}{2^3}\right)^{-2} = \left(\dfrac{2^3}{1}\right)^2 = \dfrac{(2^3)^2}{1^2} = 2^6 = 64$

(c) $\left(\dfrac{3x^{-3}}{y^2}\right)^4 = \dfrac{3^4(x^{-3})^4}{(y^2)^4} = \dfrac{81x^{-12}}{y^8} = \dfrac{81}{x^{12}y^8}$

(d) $\left(\dfrac{3x^2}{4y^2z}\right)^3 = \dfrac{3^3x^6}{4^3y^6z^3} = \dfrac{27x^6}{64y^6z^3}$

Now Try Exercises 49, 61, 63

EXAMPLE 9   **Simplifying expressions**

Write each expression using positive exponents. Simplify the result completely.

(a) $\left(\dfrac{y^{-3}}{3z^{-4}}\right)^{-2}$ **(b)** $\dfrac{(rt^3)^{-3}}{(r^2t^3)^{-2}}$

Solution

(a) $\left(\dfrac{y^{-3}}{3z^{-4}}\right)^{-2} = \left(\dfrac{3z^{-4}}{y^{-3}}\right)^2 = \left(\dfrac{3y^3}{z^4}\right)^2 = \dfrac{9y^6}{z^8}$

(b) $\dfrac{(rt^3)^{-3}}{(r^2t^3)^{-2}} = \dfrac{(r^2t^3)^2}{(rt^3)^3} = \dfrac{r^4t^6}{r^3t^9} = \dfrac{r}{t^3}$

Now Try Exercises 67, 69

Order of Operations

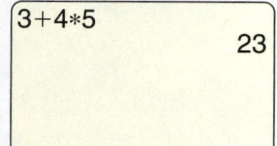

Figure 1.21

When we evaluate the expression $3 + 4 \cdot 5$, is the result 35 or 23? Figure 1.21 shows that a calculator gives a result of 23. This is because multiplication is performed before addition.

Because it is important that we evaluate arithmetic expressions consistently, the following rules are used. (Two people should evaluate the same expression the same way.)

ORDER OF OPERATIONS

Using the following **order of operations**, first perform all calculations within parentheses and absolute values, or above and below the fraction bar. Then use the same order of operations to perform the remaining calculations.

1. Evaluate all exponential expressions. Do any negations *after* evaluating exponents.
2. Do all multiplication and division from *left to right*.
3. Do all addition and subtraction from *left to right*.

Be sure to evaluate exponents before performing negation. For example,

$$-2^4 = -(2 \cdot 2 \cdot 2 \cdot 2) = -16, \quad \text{but} \quad (-2)^4 = (-2)(-2)(-2)(-2) = 16.$$

These results are supported by Figure 1.22.

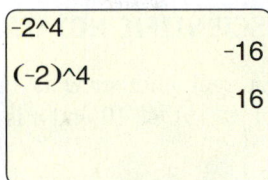

```
-2^4
                  -16
(-2)^4
                   16
```

Figure 1.22

EXAMPLE 10   **Evaluating arithmetic expressions**

Evaluate each expression. Use a calculator to support your results.

(a) $5 - 3 \cdot 2 - (4 + 5)$ **(b)** $-3^2 + \dfrac{5 + 7}{2 + 1}$ **(c)** $4^3 - 5(2 - 6 \cdot 2)$

Solution
(a) $\begin{aligned} 5 - 3 \cdot 2 - (4 + 5) &= 5 - 3 \cdot 2 - \mathbf{9} \\ &= 5 - \mathbf{6} - 9 \\ &= \mathbf{-1} - 9 \\ &= \mathbf{-10} \end{aligned}$

(b) Assume that both the numerator and the denominator of a fraction have parentheses around them.

$$\begin{aligned} -3^2 + \frac{5 + 7}{2 + 1} &= -3^2 + \frac{(5 + 7)}{(2 + 1)} \\ &= -3^2 + \frac{\mathbf{12}}{\mathbf{3}} \\ &= \mathbf{-9} + \frac{12}{3} \qquad \text{Do exponents before division.} \\ &= -9 + \mathbf{4} \\ &= \mathbf{-5} \end{aligned}$$

(c) $\begin{aligned} 4^3 - 5(2 - 6 \cdot 2) &= 4^3 - 5(2 - \mathbf{12}) \\ &= 4^3 - 5(\mathbf{-10}) \\ &= \mathbf{64} - 5(-10) \\ &= 64 + \mathbf{50} \\ &= \mathbf{114} \end{aligned}$

```
5-3*2-(4+5)
                  -10
-3²+(5+7)/(2+1)
                   -5
4^3-5(2-6*2)
                  114
```

Figure 1.23

These results are supported by Figure 1.23.

Now Try Exercises 85, 87, 91

Scientific Notation

▶ **REAL-WORLD CONNECTION** Numbers that are large or small in absolute value occur frequently in applications. For simplicity, these numbers are often expressed in scientific notation. As mentioned at the beginning of this section, the distance to the nearest star, our sun, is 93,000,000 miles. This number can be expressed in scientific notation as

$$9.3 \times 10^7.$$

In contrast, a typical virus is about 0.000005 of a centimeter in diameter. In scientific notation this number can be written as

$$5 \times 10^{-6}.$$

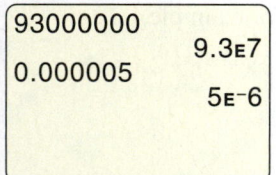

Figure 1.24

CALCULATOR HELP

To display numbers in scientific notation, see Appendix A (page AP-2).

A calculator set in *scientific mode* expresses these numbers in scientific notation, as illustrated in Figure 1.24. The letter E denotes a power of 10. That is, 9.3E7 = 9.3×10^7 and 5E−6 = 5×10^{-6}.

SCIENTIFIC NOTATION

A real number a is in **scientific notation** when a is written as $b \times 10^n$, where $1 \le |b| < 10$ and n is an integer.

Use the following steps to express a positive (rational) number a in scientific notation.

WRITING A POSITIVE NUMBER IN SCIENTIFIC NOTATION

1. Move the decimal point in a number a until it represents a number b such that $1 \le b < 10$.
2. Count the number of decimal places that the decimal point was moved. Let this positive integer be n. (If the decimal point is *not* moved, then $a = a \times 10^0$.)
3. If the decimal point was moved to the left, then $a = b \times 10^n$. If the decimal point was moved to the right, then $a = b \times 10^{-n}$.

NOTE: The scientific notation for a negative number a is the additive inverse of the scientific notation of $|a|$. For example, $5200 = 5.2 \times 10^3$, so $-5200 = -5.2 \times 10^3$.

Table 1.8 shows the values of some important powers of 10.

TABLE 1.8 Important Powers of 10

Number	10^{-3}	10^{-2}	10^{-1}	10^3	10^6	10^9	10^{12}
Value	Thousandth	Hundredth	Tenth	Thousand	Million	Billion	Trillion

EXAMPLE 11   Writing a number in scientific notation

Express each number in scientific notation.
(a) 117,000,000   (Active Facebook users in the United States during 2010)
(b) 0.00000538   (Time in seconds for light to travel 1 mile)
(c) 10,000,000,000   (Estimated world population in 2050)

Solution
(a) Move the assumed decimal point in 117,000,000 *eight places* to the *left* to obtain 1.17.

$$1.1\,7\,0\,0\,0\,0\,0\,0\,.$$

The scientific notation for 117,000,000 is 1.17×10^8.
(b) Move the decimal point in 0.00000538 *six places* to the *right* to obtain 5.38.

$$0.0\,0\,0\,0\,0\,5.38$$

The scientific notation for 0.00000538 is 5.38×10^{-6}.
(c) Move the decimal point in 10,000,000,000 *ten places* to the *left* to obtain 1.

$$1.0\,0\,0\,0\,0\,0\,0\,0\,0\,0\,.$$

The scientific notation for 10,000,000,000 is 1×10^{10}. Note that positive powers of 10 indicate a large number, whereas negative powers of 10 indicate a small number.

Now Try Exercises 97, 101

In the next example we convert numbers from scientific notation to **standard form**, or **decimal form.**

EXAMPLE 12 **Writing a number in standard form**

Write the number in standard (decimal) form.
(a) 1.2×10^8 (Number of people who suffer from depression worldwide)
(b) 2.04×10^{-3} (Fraction of population who died from heart disease)

Solution
(a) Because the exponent of 10 is *positive*, move the decimal point in 1.2 to the *right* 8 places to obtain 120,000,000.
(b) Because the exponent of 10 is *negative*, move the decimal point in 2.04 to the *left* 3 places to obtain 0.00204.

Now Try Exercises 105, 111

CALCULATOR HELP

To enter numbers in scientific notation, see Appendix A (page AP-2).

Arithmetic can be performed on expressions in scientific notation. For example, multiplication of the expressions 8×10^4 and 4×10^2 may be done by hand.

$$(8 \times 10^4) \cdot (4 \times 10^2) = (8 \cdot 4) \times (10^4 \cdot 10^2) \quad \text{Properties of real numbers}$$
$$= 32 \times 10^6 \quad \text{Add exponents and simplify.}$$
$$= (3.2 \times 10^1)(10^6) \quad \text{Write in scientific notation.}$$
$$= 3.2 \times 10^7 \quad \text{Add exponents.}$$

Division may be performed as follows.

$$\frac{8 \times 10^4}{4 \times 10^2} = \frac{8}{4} \times \frac{10^4}{10^2} \quad \text{Property of fractions}$$
$$= 2 \times 10^2 \quad \text{Subtract exponents.}$$

▶ **REAL-WORLD CONNECTION**   The next example illustrates how scientific notation can be used to analyze the federal debt.

EXAMPLE 13 **Analyzing the federal debt**

In 2010, the federal debt held by the public was 9.02 trillion dollars, and the population of the United States was 308 million. Approximate the national debt per person.

Solution
In scientific notation, 9.02 trillion equals 9.02×10^{12} and 308 million equals 3.08×10^8. The per person debt held by the public is given by

$$\frac{9.02 \times 10^{12}}{3.08 \times 10^8} \approx \$29,286.$$

Figure 1.25 supports this result.

Now Try Exercise 127

```
(9.02*10^12)/(3.
08*10^8)
        29285.71429
```

Figure 1.25

CRITICAL THINKING

Estimate the number of seconds that you have been alive.

1.3 Putting It All Together

The following table summarizes important properties of exponents, where a and b are non-zero real numbers and m and n are integers.

PROPERTY	DEFINITION	EXAMPLES
Bases and Exponents	In the expression a^n, a is the base and n is the exponent. *Note:* 0^0 is undefined.	$3^4 = 3 \cdot 3 \cdot 3 \cdot 3 = 81,\ 7^0 = 1,$ $2^{-3} = \dfrac{1}{2^3} = \dfrac{1}{2 \cdot 2 \cdot 2} = \dfrac{1}{8},$ and $-3^2 = -9$
The Product Rule	$a^m \cdot a^n = a^{m+n}$	$8^4 \cdot 8^2 = 8^{4+2} = 8^6$ and $5^6 \cdot 5^{-3} = 5^{6+(-3)} = 5^3$
The Quotient Rule	$\dfrac{a^m}{a^n} = a^{m-n}$	$\dfrac{6^7}{6^4} = 6^{7-4} = 6^3$ and $\dfrac{7^{-4}}{7^{-2}} = 7^{-4-(-2)} = 7^{-2} = \dfrac{1}{7^2}$
The Power Rules	**1.** $(a^m)^n = a^{mn}$ **2.** $(ab)^n = a^n b^n$ **3.** $\left(\dfrac{a}{b}\right)^n = \dfrac{a^n}{b^n}$	**1.** $(2^2)^3 = 2^6$ **2.** $(3y)^4 = 3^4 y^4$ **3.** $\left(\dfrac{x^2}{y}\right)^4 = \dfrac{x^8}{y^4}$
Quotients and Negative Exponents	**1.** $a^{-n} = \dfrac{1}{a^n}$ and $\dfrac{1}{a^{-n}} = a^n$ **2.** $\dfrac{a^{-n}}{b^{-m}} = \dfrac{b^m}{a^n}$ **3.** $\left(\dfrac{a}{b}\right)^{-n} = \left(\dfrac{b}{a}\right)^n$	**1.** $\dfrac{1}{x^{-2}} = x^2$ **2.** $\dfrac{z^{-3}}{y^{-5}} = \dfrac{y^5}{z^3}$ **3.** $\left(\dfrac{4}{t}\right)^{-2} = \left(\dfrac{t}{4}\right)^2 = \dfrac{t^2}{4^2}$
Scientific Notation	A positive number a is in scientific notation when a is written as $b \times 10^n$, where $1 \le b < 10$ and n is an integer.	$52{,}600 = 5.26 \times 10^4$ and $0.0068 = 6.8 \times 10^{-3}$

1.3 Exercises

CONCEPTS AND VOCABULARY

1. Identify the base and the exponent in 8^3.

2. Evaluate 97^0 and 2^{-1}.

3. Write 7 cubed, using symbols.

4. Write 5 squared, using symbols.

5. Are the expressions -4^2 and $(-4)^2$ equal? Explain.

6. Are the expressions 2^3 and 3^2 equal? Explain.

7. $7^{-n} = $ _____

8. $6^m \cdot 6^n = $ _____

9. $\dfrac{5^m}{5^n} = $ _____

10. $(3x)^k = $ _____

11. $(2^m)^k = $ _____

12. $\left(\dfrac{x}{y}\right)^m = $ _____

13. $\dfrac{1}{x^{-n}} = $ _____

14. $\dfrac{a^{-n}}{b^{-m}} = $ _____

15. $\left(\dfrac{y}{z}\right)^{-n} = $ _____

16. $5 \times 10^2 = $ _____

PROPERTIES OF EXPONENTS

Exercises 17–22: (Refer to Example 1.) Write the number as an exponential expression, using the base shown. Check your result with a calculator.

17. 8 (base 2)

18. 1000 (base 10)

19. 256 (base 4)

20. $\frac{1}{16}$ (base 2)

21. 1 (base 6)

22. $\frac{1}{125}$ (base 5)

Exercises 23–32: Evaluate each expression.

23. (a) 4^3

(b) 3^4

24. (a) 2^{-3}

(b) 7^{-2}

25. (a) $\dfrac{1}{4^{-2}}$

(b) $\dfrac{1}{3^{-3}}$

26. (a) 5^0

(b) $\dfrac{1}{6^0}$

27. (a) $\left(\dfrac{2}{3}\right)^3$

(b) $\left(-\dfrac{2}{3}\right)^{-3}$

28. (a) $\left(-\dfrac{1}{2}\right)^4$

(b) $\left(-\dfrac{3}{4}\right)^3$

29. (a) $\dfrac{3^{-2}}{2^{-4}}$

(b) $\dfrac{4^{-3}}{5^{-2}}$

30. (a) $\dfrac{4^{-2}}{2^3}$

(b) $\dfrac{3^3}{5^{-2}}$

31. (a) $\dfrac{1}{2x^{-3}}$

(b) $\dfrac{1}{(ab)^{-1}}$

32. (a) $\dfrac{1}{3b^{-2}}$

(b) $\dfrac{2^{-3}}{b^{-1}}$

Exercises 33–38: Use the product rule and positive exponents to simplify the expression.

33. (a) $3^5 \cdot 3^{-3}$

(b) $x^2 x^5$

34. (a) $10^{-5} \cdot 10^2$

(b) $y^4 y^{-3}$

35. (a) $(-3x^{-2})(5x^5)$

(b) $(ab)(a^2b^{-3})$

36. (a) $6z^2(-z^3)$

(b) $(a^2b^{-3})(2ab^3)$

37. (a) $5^{-2}5^3 2^{-4}2^3$

(b) $2a^3 \cdot b^2 \cdot 4a^{-4} \cdot b^{-5}$

38. (a) $2^{-3}3^4 3^{-2}2^5$

(b) $3x^{-4} \cdot 2x^2 \cdot 5y^4 \cdot y^{-3}$

Exercises 39–44: Use the quotient rule and other properties of exponents to simplify the expression. Use positive exponents to write your answer.

39. (a) $\dfrac{4^3}{4^2}$

(b) $\dfrac{10^{-3}}{10^{-5}}$

40. (a) $\dfrac{5^4}{5^{-7}}$

(b) $\dfrac{6^{-5}}{6}$

41. (a) $\dfrac{b^{-3}}{b^2}$

(b) $\dfrac{24x^3}{6x}$

42. (a) $\dfrac{x^0}{x^{-5}}$

(b) $\dfrac{10x^5}{5x^{-3}}$

43. (a) $\dfrac{12a^2b^3}{18a^4b^2}$

(b) $\dfrac{21x^{-3}y^4}{7x^4y^{-2}}$

44. (a) $\dfrac{-6x^7y^3}{3x^2y^{-5}}$

(b) $\dfrac{32x^3y}{-24x^5y^{-3}}$

Exercises 45–50: Use the power rules to simplify the expression. Use positive exponents to write your answer.

45. (a) $(3^2)^4$

(b) $(x^3)^{-2}$

46. (a) $(-2^2)^3$

(b) $(xy)^3$

47. (a) $(4y^2)^3$

(b) $(-2xy^3)^{-4}$

48. (a) $(-2a^2)^3$

(b) $(a^{-1}b^5)^{-2}$

49. (a) $\left(\dfrac{4}{x}\right)^3$

(b) $\left(\dfrac{2x}{z^4}\right)^{-5}$

50. (a) $\left(\dfrac{-3}{x^3}\right)^2$

(b) $\left(\dfrac{2xy}{3z^5}\right)^{-1}$

Exercises 51–74: Use rules of exponents to simplify the expression. Use positive exponents to write your answer.

51. $\dfrac{12m^2n^{-5}}{8mn^{-2}}$

52. $\dfrac{15m^{-1}n}{5m^{-2}n^3}$

53. $(2x^3y^{-2})^{-2}$

54. $(4x^{-4}y)^2$

55. $\dfrac{(b^2)^3}{(b^{-1})^2}$

56. $\dfrac{(a^3)^{-1}}{(a^{-3})^{-2}}$

57. $\dfrac{(-3ab^2)^3}{(a^2b)^2}$

58. $\dfrac{(-2ab)^3}{(ab)^2}$

59. $\dfrac{(-m^2n^{-1})^{-2}}{(mn)^{-1}}$ **60.** $\dfrac{(-mn^4)^{-1}}{(m^2n)^{-3}}$

61. $\left(\dfrac{2a^3}{6b}\right)^4$ **62.** $\left(\dfrac{-3a^2}{9b^3}\right)^4$

63. $\left(\dfrac{t^{-3}}{t^{-4}}\right)^2$ **64.** $\left(\dfrac{r^{-2}}{2r^{-1}}\right)^{-4}$

65. $\dfrac{8x^{-3}y^{-2}}{4x^{-2}y^{-4}}$ **66.** $\dfrac{6x^{-1}y^{-1}}{9x^{-2}y^3}$

67. $\left(\dfrac{2t}{-r^2}\right)^{-3}$ **68.** $\left(\dfrac{t^2}{3r}\right)^{-1}$

69. $\dfrac{(r^2t^2)^{-2}}{(r^3t)^{-1}}$ **70.** $\dfrac{(2rt)^2}{(rt^4)^{-2}}$

71. $\dfrac{4x^{-2}y^3}{(2x^{-1}y)^2}$ **72.** $\dfrac{(ab)^3}{a^4b^{-4}}$

73. $\left(\dfrac{-15r^2t}{3r^{-3}t^4}\right)^3$ **74.** $\left(\dfrac{4(xy)^2}{(2xy^{-2})^3}\right)^{-2}$

ORDER OF OPERATIONS

Exercises 75–96: (Refer to Example 10.) Evaluate.

75. $4 + 5 \cdot 6$ **76.** $4 - 5 - 9$

77. $2(4 + (-8))$ **78.** $500 - 10^3$

79. $5 \cdot 2^3$ **80.** $\dfrac{4 + 8}{2} - \dfrac{6 + 1}{3}$

81. $\dfrac{-2^4 - 3^2}{4} + \dfrac{1 + 2}{4}$ **82.** $\dfrac{(-4^2 + 1)}{\frac{2}{3}}$

83. $\dfrac{1 - 2 \cdot 4^2}{5^{-1}}$ **84.** $6 \div 4 \div 2$

85. $7^2 - 3(4 + 2 \cdot 5)$ **86.** $4 - 3^2(5 - 2 \cdot 4)$

87. $4 + 6 - 3 \cdot 5 \div 3$ **88.** $-3(25 - 2 \cdot 5^2) \div 5$

89. $\dfrac{-3^2 + 3}{3}$ **90.** $\dfrac{2 \cdot 4 - 7}{7}$

91. $-4^2 + \dfrac{15 + 2}{8 - 7}$ **92.** $1 - 3 \cdot 4^3$

93. $|7 - 2^2 \cdot 3|$ **94.** $\dfrac{4^2 - |5 - (-6)|}{-2^2}$

95. $\sqrt{4^2 + 3^2}$ **96.** $\sqrt{13^2 - 12^2}$

SCIENTIFIC NOTATION

Exercises 97–104: Write in scientific notation.

97. 5,900,000 (Annual number of pregnancies in U.S.)

98. 250,000 (Childhood polio deaths prevented by Rotary International)

99. 50 million (Tweets per day on Twitter in 2010)

100. 700 billion (Minutes spent on Facebook per month in 2010)

101. 0.032 (Fraction of United States population that is vegetarian)

102. 0.189 (Fraction of nonelderly people without health insurance)

103. 0.000001 (Approximate wavelength of light in meters)

104. 0.038 (Fraction of deaths worldwide caused by alcohol)

Exercises 105–112: Write in standard (decimal) form.

105. 5×10^5 **106.** -7.85×10^3

107. 9.3×10^6 **108.** 2.961×10^2

109. -6×10^{-3} **110.** 4.1×10^{-2}

111. 5.876×10^{-5} **112.** 9.9×10^{-1}

Exercises 113–118: Evaluate the expression. Write your answer in both scientific notation and standard form.

113. $(2 \times 10^4)(3 \times 10^2)$

114. $(5 \times 10^{-4})(4 \times 10^6)$

115. $(4 \times 10^{-4})(2 \times 10^{-2})$

116. $\dfrac{6 \times 10^4}{2 \times 10^2}$

117. $\dfrac{6.2 \times 10^3}{3.1 \times 10^{-2}}$

118. $\dfrac{2 \times 10^{-2}}{8 \times 10^{-5}}$

APPLICATIONS

119. *GPS Clocks* The Global Positioning System (GPS) is made up of 24 satellites that allow individuals with a GPS receiver, such as a smart phone, to pinpoint their positions on Earth. Every 1024 weeks the clocks in the GPS satellites reset to zero. The first time this resetting occurred was on August 21, 1999. (*Source:* Associated Press.)

(a) Find an exponent k so that $2^k = 1024$.

(b) Estimate the number of years in 1024 weeks.

120. *Astronomy* Light travels at 186,000 miles per second. The distance that light travels in 1 year is called a *light-year*.

(a) Calculate the number of miles in 1 light-year. Write your answer in scientific notation.

(b) Express your answer from part (a) in standard (decimal) notation.

(c) Except for the sun, the nearest star is Alpha Centauri. Its distance from Earth is about 4.27 light-years. How many miles is this?

(d) If a rocket flew at 50,000 miles per hour, how many years would it take to reach the star Alpha Centauri?

121. *Volume* If the sides of a cube have length a, then its volume is $V = a^3$. Find the volume of a cube with sides of length $2a$.

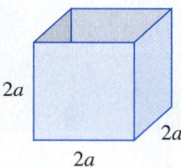

122. *Area* If the sides of a square have length a, then its area is $A = a^2$. Find the area of a square with sides of length $3a$.

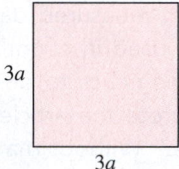

123. *Area* Find the area of the square.

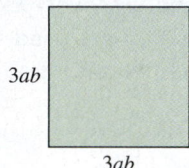

124. *Volume* Find the volume of the cube.

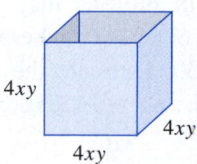

125. *YouTube Video Half-Life* After a typical YouTube video reaches peak viewing, the average time it takes for the number of people watching it per day to drop to half is 6 days. Determine the fraction of the number of viewers watching after
(a) 12 days.   **(b)** 30 days.

126. *Number of Facebook Users* In recent years, the number of Facebook users has doubled every 6 months. Determine the factor by which the number of Facebook users increased after
(a) 1 year.   **(b)** 18 months.

127. *Federal Debt* (Refer to Example 13.) In 2005, the federal debt held by the public was $4.72 trillion, and the population of the United States was 296 million. Approximate the national debt per person.

128. *Federal Debt* In 1980, the federal debt held by the public was $710 billion, and the debt per person was $3127. Approximate the population of the United States in 1980.

129. *Flash Drive Memory* (Refer to Example 2.) If a flash drive has 256 MB of memory, how many bytes is this? Express your answer in standard form.

130. *Computer Memory* One gigabyte of computer memory equals 2^{30} bytes. Write the number of bytes in 1 gigabyte, using standard notation.

131. *World Population* The following table lists populations of selected countries in 2000 and their projected populations in 2025. Rewrite the table below, expressing each population in scientific notation.

Country	2000	2025
China	1,268,900,000	1,480,000,000
Germany	82,200,000	80,900,000
India	1,004,100,000	1,330,200,000
Mexico	99,900,000	130,200,000
United States	282,100,000	332,500,000

Source: United Nations Population Fund.

132. *World Population* If current trends continue, world population P in billions may be modeled by the equation $P = 6(1.014)^x$, where x is the number of years after 2000. Estimate the world population in 2010 and 2025. (*Source:* United Nations Population Fund.)

WRITING ABOUT MATHEMATICS

133. Give the product and quotient rules for exponents and an example of each.

134. A student evaluates three expressions:
-4^2 as 16, $6 + 4 \cdot 2$ as 20,   and
$$20 \div 4 \div 2 \text{ as } 10.$$
Correct the errors and explain the mistakes.

Group Activity   Working with Real Data

Directions: Form a group of 2 to 4 people. Select someone to record the group's responses for this activity. All members of the group should work cooperatively to answer the questions. If your instructor asks for your results, each member of the group should be prepared to respond.

1. *Generating Trash* Every year the United States generates about 230 million tons of trash. There are about 308 million people in the United States.
  (a) How many tons of trash are generated per person each year?
  (b) How many pounds of trash are generated per person each day?

2. *Salary* Suppose that for full-time work a person earns 1¢ for the first week, 2¢ for the second week, 4¢ for the third week, 8¢ for the fourth week, and so on for 1 year.
  (a) Discuss whether you think this pay scale would be a good deal.
  (b) Estimate how much this person would make the last week of a 52-week year.

1.4   Variables, Equations, and Formulas

Basic Concepts • Modeling Data • Square Roots and Cube Roots • Tables and Calculators (Optional)

A LOOK INTO MATH ▶

Most of the mathematics that people have discovered throughout the ages can be derived by using only pencil and paper, not from science and measured data. However, one of the amazing aspects of mathematics is that it can be used in countless applications that improve our quality of life. In fact, this ability of mathematics to describe the real world is so amazing that Nobel Laureate Eugene Wigner wrote the article "The Unreasonable Effectiveness of Mathematics in the Natural Sciences." Without mathematics, we would not have DVD players, cars, warm buildings, social networks, or accurate weather forecasts. Mathematics can even be used to predict the increase in sea level if the polar ice caps were to melt. (See Exercise 1 in the Chapter 1 Extended and Discovery Exercises.) In this section we introduce some of the mathematical concepts that are used to model our world.
(*Source:* Eugene Wigner, *Communications of Pure and Applied Mathematics,* February 1960.)

NEW VOCABULARY

☐ Variable
☐ Algebraic expression
☐ Equation
☐ Formula
☐ Model
☐ Square root
☐ Principal square root
☐ Cube root

Basic Concepts

▶ **REAL-WORLD CONNECTION** Suppose that we want to calculate the distance traveled by a car moving at a constant speed of 30 miles per hour. One method would be to make a table of values, as shown in Table 1.9.

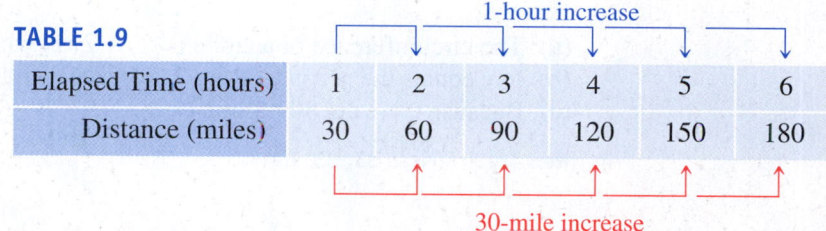

TABLE 1.9

Elapsed Time (hours)	1	2	3	4	5	6
Distance (miles)	30	60	90	120	150	180

Note that for each 1-hour increase the distance increases by 30 miles. Many times it is not possible to list all relevant values in a table. Instead, we use *variables* to describe data. For example, we might let elapsed time be represented by the variable t and let distance be represented by the variable d. If $t = 2$, then $d = 60$; if $t = 5$, then $d = 150$. In this example, the value of d is always equal to the value of t multiplied by 30. We can *model* this situation by using the *equation* or *formula* $d = 30t$. The values for distance in Table 1.9 can be calculated by letting $t = 1, 2, 3, 4, 5, 6$ in this formula.

A **variable** is a symbol, such as x, y, d, or t, used to represent any unknown number or quantity. An **algebraic expression** can consist of numbers, variables, arithmetic symbols $(+, -, \times, \div)$, exponents, and grouping symbols, such as parentheses, brackets, and square roots. Examples of algebraic expressions include

$$6, \quad 30t, \quad 4(t - 5) + 1, \quad \sqrt{x + 1}, \quad \text{and} \quad LW. \qquad \text{Expressions}$$

An **equation** is a statement that two algebraic expressions are equal. An equation *always* contains an equals sign. Examples of equations include

$$3 + 6 = 9, \quad x + 1 = 4, \quad d = 30t, \quad \text{and} \quad x + y = 20. \qquad \text{Equations}$$

The first equation contains only constants, the second equation contains one variable, and both the third and fourth equations contain two variables. A **formula** is an equation that can calculate one quantity by using a known value of another quantity. (Formulas can also contain known values of more than one quantity.) Formulas show relationships between variables. The formula $y = \frac{x}{3}$ computes the number of yards in x feet. If $x = $ **15**, then $y = \frac{\mathbf{15}}{3} = 5$. That is, in 15 feet there are 5 yards.

EXAMPLE 1 ▸ **Writing and using a formula**

If a car travels at a constant speed of 70 miles per hour, write a formula that calculates the distance d that the car travels in t hours. Evaluate your formula when $t = 1.5$ and interpret the result.

Solution

Traveling at 70 miles per hour, the car will travel a distance of $d = 70t$ miles in t hours. Evaluating this formula at $t = $ **1.5** results in

$$d = 70(\mathbf{1.5}) = 105.$$

After 1.5 hours the car has traveled 105 miles.

Now Try Exercise 83

EXAMPLE 2 ▸ **Writing formulas**

Write a formula that does the following.
(a) Finds the circumference C of a circle with radius r
(b) Calculates the pay P for working H hours at \$9 per hour
(c) Calculates the number of tweets T in millions on Twitter during a period of D days, if there are 50 million tweets each day.

Solution
(a) The circumference of a circle is $C = 2\pi r$, where $\pi \approx 3.14$.
(b) Pay equals the product of the hourly wage and the hours worked. Thus $P = 9H$.
(c) Because there are 50 million tweets per day, $T = 50D$.

▍ **Now Try Exercises 33, 35, 89**

EXAMPLE 3 **Evaluating formulas from geometry**

Evaluate each formula for the given value(s) of the variable(s). See Figure 1.26.
(a) $A = \pi r^2, r = 4$ Area of a circle
(b) $P = 2L + 2W, L = 8$ and $W = 4$ Perimeter of a rectangle
(c) $V = s^3, s = 3$ Volume of a cube

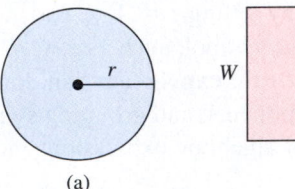 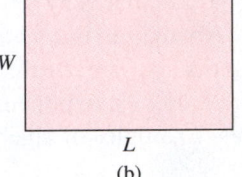 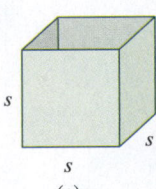

     (a)   (b)   (c)

Figure 1.26

Solution
(a) $A = \pi(4)^2 = \pi(16) = 16\pi$ Let $r = 4$.
(b) $P = 2(8) + 2(4) = 16 + 8 = 24$ Let $L = 8$ and $W = 4$.
(c) $V = 3^3 = 3 \cdot 3 \cdot 3 = 27$ Let $s = 3$.

▍ **Now Try Exercises 47, 53, 57**

Modeling Data

▶ **REAL-WORLD CONNECTION** Faster moving automobiles require more distance to stop. For example, at 60 miles per hour it takes about four times the distance to stop as it does at 30 miles per hour. Highway engineers have developed formulas to estimate the braking distance of a car.

EXAMPLE 4 **Calculating braking distance**

The braking distances in feet for a car traveling on wet, level pavement are shown in Table 1.10. Distances have been rounded to the nearest foot.

TABLE 1.10 Braking Distances

Speed (mph)	10	20	30	40	50	60	70
Distance (feet)	11	44	100	178	278	400	544

Source: L. Haefner, Introduction to Transportation Systems.

(a) If a car doubles its speed, what happens to the braking distance?
(b) If the speed is represented by the variable x and the braking distance by the variable d, then the braking distance may be calculated by the formula $d = \frac{x^2}{9}$. Verify the distance values in Table 1.10 for $x = 10, 30, 60$.
(c) Calculate the braking distance for a car traveling at 90 miles per hour. If a football field is 300 feet long, how many football field lengths does this braking distance represent?

Solution

(a) When the speed increases from 10 to 20 miles per hour, the braking distance increases by a factor of $\frac{44}{11} = 4$. Similarly, if the speed doubles from 30 to 60 miles per hour, the distance increases by a factor of $\frac{400}{100} = 4$. Thus it appears that if the speed of a car doubles, the braking distance quadruples.

(b) Let $x = \mathbf{10}, \mathbf{30}, \mathbf{60}$ in the formula $d = \frac{x^2}{9}$. Then

$$d = \frac{\mathbf{10}^2}{9} = \frac{100}{9} \approx 11 \text{ feet},$$

$$d = \frac{\mathbf{30}^2}{9} = \frac{900}{9} = 100 \text{ feet, and}$$

$$d = \frac{\mathbf{60}^2}{9} = \frac{3600}{9} = 400 \text{ feet.}$$

These values agree with the values in Table 1.10.

(c) If $x = 90$, then $d = \frac{90^2}{9} = 900$ feet. At 90 miles per hour the braking distance equals three football fields placed end to end.

Now Try Exercise 87

CRITICAL THINKING

Write a formula that calculates the time T for a bike rider to travel 100 miles moving at x miles per hour. Test your formula for different values of x.

In the next example we find a formula that models data in a table.

EXAMPLE 5 **Finding a formula**

The data in Table 1.11 can be modeled by the formula $y = ax$. Find a.

TABLE 1.11

x	y
1	3
2	6
3	9
4	12
5	15

Solution

Each value of y is 3 times the corresponding value of x, so $a = 3$. We can also find a symbolically. If $x = \mathbf{1}$, then $y = \mathbf{3}$. We can substitute these values into the equation.

$y = ax$	Given equation
$\mathbf{3} = a \cdot \mathbf{1}$	Let $x = 1$ and $y = 3$.
$3 = a$	Identity

Thus $a = 3$. The formula $y = 3x$ models the data in Table 1.11.

Now Try Exercise 65

Square Roots and Cube Roots

The number b is a **square root** of a (nonnegative) number a if $b^2 = a$. For example, one square root of 9 is 3 because $3^2 = 9$. The other square root of 9 is -3 because $(-3)^2 = 9$. We use the symbol $\sqrt{9}$ to denote the *positive* or **principal square root** of 9. That is, $\sqrt{9} = 3$. The following are examples of how to evaluate the square root symbol. A calculator is sometimes needed to approximate square roots.

$$\sqrt{16} = 4, \quad -\sqrt{100} = -10, \quad \sqrt{3} \approx 1.732, \quad \pm\sqrt{4} = \pm 2$$

The symbol "$\pm$" is read "plus or minus." Note that ± 2 represents the numbers 2 or -2.

EXAMPLE 6 **Finding square roots**

Evaluate each expression, if possible.

(a) $\pm\sqrt{36}$ **(b)** $\sqrt{3^2 + 4^2}$ **(c)** $\sqrt{5 - 9}$

Solution

(a) $\pm\sqrt{36} = \pm 6$ because $6^2 = 36$ and $(-6)^2 = 36$

(b) $\sqrt{3^2 + 4^2} = \sqrt{9 + 16} = \sqrt{25} = 5$

 NOTE: $\sqrt{3^2 + 4^2} \neq \sqrt{3^2} + \sqrt{4^2} = 3 + 4 = 7$

(c) $\sqrt{5 - 9} = \sqrt{-4}$, which is *not* a real number

 NOTE: There is no *real* number b such that $b^2 = -4$.

▌ **Now Try Exercises 19, 21, 25**

CALCULATOR HELP

To calculate square roots and cube roots, see Appendix A (page AP-1).

The number b is a **cube root** of a number a if $b^3 = a$. The cube root of 8 is 2 because $2^3 = 8$, which may be written as $\sqrt[3]{8} = 2$. Similarly, $\sqrt[3]{-8} = -2$ because $(-2)^3 = -8$. Each real number has *exactly one* real cube root.

EXAMPLE 7 | **Finding cube roots**

Evaluate each expression.

(a) $\sqrt[3]{64}$ (b) $\sqrt[3]{-2^3 - 19}$

Solution

(a) $\sqrt[3]{64} = 4$ because $4 \cdot 4 \cdot 4 = 4^3 = 64$.

(b) $\sqrt[3]{-2^3 - 19} = \sqrt[3]{-8 - 19} = \sqrt[3]{-27} = -3$ because $(-3)(-3)(-3) = -27$.

▌ **Now Try Exercises 27, 29**

EXAMPLE 8 | **Finding lengths of sides**

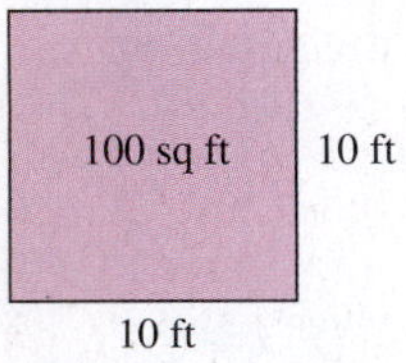

Find the length of a side s for each geometric shape.

(a) A square with area 100 square feet

(b) A cube with volume 125 cubic inches

100 sq ft   10 ft

10 ft

$s = \sqrt{100} = 10$ ft

Figure 1.27

Solution

(a) The area of a square is $A = s^2$. Thus $s = \sqrt{100} = 10$ feet because $10^2 = 100$. See Figure 1.27.

(b) The volume of a cube is $V = s^3$. Thus $s = \sqrt[3]{125} = 5$ inches because $5^3 = 125$. See Figure 1.26(c) on page 40.

▌ **Now Try Exercises 105, 107**

MAKING CONNECTIONS

Square Roots and Cube Roots

The square root of a negative number is *not* a real number. However, the cube root of a negative (or positive) number is a real number. For example, $\sqrt{-8}$ is not a real number, whereas $\sqrt[3]{-8} = -2$ is a real number.

▶ **REAL-WORLD CONNECTION** Roots of numbers often occur in biology, as illustrated in the next example.

EXAMPLE 9 | **Analyzing the walking speed of animals**

When smaller animals walk, they tend to take faster, shorter steps, whereas larger animals tend to take slower, longer steps. For example, a hyena is about 0.8 meter high at the shoulder and takes roughly 1 step per second when walking, whereas a giraffe 3 meters high at the shoulder takes 1 step every 2 seconds. If an animal is h meters high at the shoulder, then the frequency F in steps per second while it is walking can be estimated with the formula $F = \frac{0.87}{\sqrt{h}}$. The value of F is referred to as the animal's *stepping frequency*. (*Source:* C. Pennycuick, *Newton Rules Biology.*)

(a) A Thomson's gazelle is 0.6 meter high at the shoulder. Estimate its stepping frequency.
(b) An elephant is 2.7 meters high at the shoulder. Estimate its stepping frequency.
(c) What happens to the stepping frequency as h increases?

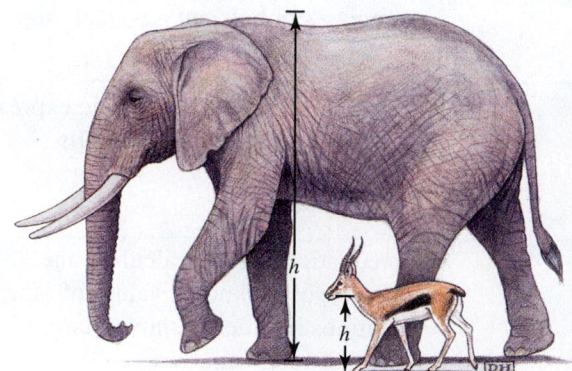

Solution
(a) $F = \frac{0.87}{\sqrt{0.6}} \approx 1.12$. A Thomson's gazelle takes about 1.12 steps per second when walking.
(b) $F = \frac{0.87}{\sqrt{2.7}} \approx 0.53$. An elephant takes roughly half a step per second when walking, or 1 step every 2 seconds.
(c) As h increases, the denominator of $\frac{0.87}{\sqrt{h}}$ also increases, so the ratio becomes smaller. Thus, as h increases, the stepping frequency decreases.

Now Try Exercise 97

Tables and Calculators (Optional)

Many calculators are able to generate tables. To generate a table, we specify the formula, the starting x-value (TblStart), and the increment (Δ Tbl) between x-values. The calculator generates the required table automatically, as demonstrated in the next example.

EXAMPLE 10 **Using the table feature**

Make a table for $y = \frac{x^2}{9}$, starting at $x = 10$ and incrementing by 10. Compare this table to Table 1.10 in Example 4.

Solution
In Figure 1.28 the desired table is generated. Note that if values are rounded to the nearest foot, the values in Figure 1.28(c) agree with those in Table 1.10.

CALCULATOR HELP

To display a table of values, see Appendix A (pages AP-2 and AP-3).

Plot1 Plot2 Plot3
\Y₁◼X^2/9
\Y₂=
\Y₃=
\Y₄=
\Y₅=
\Y₆=
\Y₇=

(a)

TABLE SETUP
TblStart=10
ΔTbl=10
Indpnt: **Auto** Ask
Depend: **Auto** Ask

(b)

X	Y₁
10	11.111
20	44.444
30	100
40	177.78
50	277.78
60	400
70	544.44

Y₁◼X^2/9

(c)

Figure 1.28

Now Try Exercise 79

1.4   Putting It All Together

CONCEPT	COMMENTS	EXAMPLES	
Variable	Represents an unknown quantity	x, y, z, A, V	
Algebraic Expression	Can consist of numbers, variables, operation symbols, exponents, and grouping symbols but *no* equals sign	$2x - 8$ $3 - (5y + 6)$ s^3 $2L + 2W$	
Equation	A statement that two algebraic expressions are equal; *always contains an equals sign*	$5x = 10$ $y = 2x + 1$ $n + 5 = 3 - n$ $z^2 + 1 = 17$	
Formula	An equation used to calculate one quantity, using known values of other quantities in order to show relationships between variables	$A = \pi r^2$ $C = 2\pi r$ $V = s^3$ $P = 2L + 2W$ $A = \frac{1}{2}bh$	Area of a circle Circumference of a circle Volume of a cube Perimeter of a rectangle Area of a triangle
Square Root	The *positive* or *principal square root* of a is written $\sqrt{a}$. The square root of a negative number is *not* a real number.	$\sqrt{25} = 5$ $\pm\sqrt{100} = \pm 10$ $\sqrt{-16}$ is not a real number.	
Cube Root	The cube root of a is written $\sqrt[3]{a}$.	$\sqrt[3]{-8} = -2$ because $(-2)^3 = -8$. $\sqrt[3]{64} = 4$ because $4^3 = 64$.	

1.4   Exercises

MyMathLab    PRACTICE   WATCH   DOWNLOAD   READ   REVIEW

CONCEPTS AND VOCABULARY

1. A(n) _____ is a symbol used to represent an unknown number or quantity.

2. A(n) _____ is a statement that says two algebraic expressions are equal.

3. An equation always contains a(n) _____ sign.

4. A(n) _____ is an equation that can be used to calculate one quantity by using known values of another quantity.

5. Identify the variables in the equation $x^2 + y^2 = 9$.

6. Give an example of an equation with no variables.

7. Give an example of an equation with one variable.

8. Give an example of an equation with two variables.

9. If $y = 2x$ and $x = 3$, then $y = $ _____.

10. If $A = s^2$ and $s = 2$, then $A = $ _____.

SQUARE ROOTS AND CUBE ROOTS

11. $\sqrt{9} = $ _____   **12.** $\sqrt[3]{-8} = $ _____

13. If $b > 0$, then $\sqrt{b^2} = $ _____.

14. $\sqrt[3]{b^3} = $ _____

15. If b is the square root of a, then $a = $ _____.

16. If b is the cube root of a, then $a = $ _____.

17. (True or False?) The square root of a negative number is a negative number.

18. (True or False?) The cube root of a negative number is a negative number.

Exercises 19–32: Evaluate each expression, if possible.

19. $-\sqrt{49}$ **20.** $-\sqrt{64}$

21. $\pm\sqrt{5^2 + 12^2}$ **22.** $\pm\sqrt{61^2 - 60^2}$

23. $\sqrt{10 - 2 \cdot 3}$ **24.** $\sqrt{10 - 3^2}$

25. $\sqrt{-4^2}$ **26.** $\sqrt{5 - 4 \cdot 2}$

27. $\sqrt[3]{27}$ **28.** $\sqrt[3]{125}$

29. $\sqrt[3]{1 - 3^2}$ **30.** $\sqrt[3]{-1000}$

31. $\sqrt[3]{2 \cdot 8 - 4^2}$ **32.** $\sqrt[3]{-64}$

WRITING FORMULAS

Exercises 33–42: Write a formula that does the following.

33. Converts x miles to y feet

34. Converts x quarts to y gallons

35. Finds the area A of a square with a side of length s

36. Finds the surface area A of a cube with a side of length s

37. Determines the number of seconds y in x hours

38. Determines the gas mileage G of a car that travels m miles on g gallons of gas

39. Determines the area A of a triangle with base b and height h

40. Determines the perimeter P of a square with side s

41. Finds the area A of a circle with diameter d

42. Finds the circumference C of a circle with diameter d

USING DATA, VARIABLES, AND FORMULAS

Exercises 43–60: Evaluate the formula for the given value(s) of the variable(s).

43. $y = 5x$ $x = 6$

44. $y = \frac{x}{10}$ $x = 30$

45. $y = x + 5$ $x = -3.1$

46. $d = 5 - 4t$ $t = -1.5$

47. $d = t^2 + 1$ $t = -3$

48. $z = 3k^2 - \frac{3}{4}$ $k = \frac{1}{4}$

49. $z = \sqrt{2k}$ $k = 18$

50. $y = \sqrt{5 - x}$ $x = 1$

51. $y = -\frac{1}{2}\sqrt[3]{x}$ $x = \frac{1}{8}$

52. $M = \sqrt[3]{1 - p}$ $p = 65$

53. $N = 3h^3 - 1$ $h = \frac{1}{3}$

54. $S = 1 - \frac{1}{2}w^3$ $w = -2$

55. $P = |5 - w|$ $w = 4.7$

56. $D = |2t - 5|$ $t = 2.5$

57. $A = \frac{1}{2}bh$ $b = 3, \quad h = 6$

58. $P = 2L + 2W$ $L = 9, \quad W = 7$

59. $V = \pi r^2 h$ $r = \frac{1}{2}, \quad h = 5$

60. $S = 2\pi r(r + h)$ $r = 2, \quad h = 8$

Exercises 61–64: Select the formula (i)–(iii) that best models the data in the table.

61.

x	1	2	3	4	5
y	2	4	6	8	10

(i) $y = x + 2$,   (ii) $y = 2x$,   (iii) $y = 4x - 2$

62.

x	-2	-1	0	1	2
y	4	1	0	1	4

(i) $y = x^2$,   (ii) $y = x + 6$,   (iii) $y = 2x$

63.

x	-4	-2	0	2	4
y	4	2	0	2	4

(i) $y = x^2$,   (ii) $y = x$,   (iii) $y = |x|$

64.

x	-27	-1	0	8	64
y	-3	-1	0	2	4

(i) $y = \sqrt[3]{x}$,   (ii) $y = \frac{x}{4}$,   (iii) $y = \sqrt{x}$

Exercises 65–68: (Refer to Example 5.) Find a value of the variable a so that the equation models the data.

65. $y = ax$

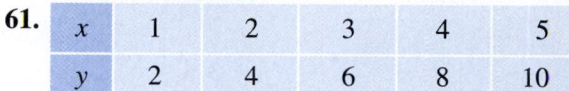

x	-2	-1	0	1	2
y	6	3	0	-3	-6

66. $y = ax$

x	-10	-5	5	10	15
y	-15	-7.5	7.5	15	22.5

67. $d = t - a$

t	0	1	2	3	4
d	-2	-1	0	1	2

68. $N = aw^2$

w	0	2	4	6	8
N	0	-4	-16	-36	-64

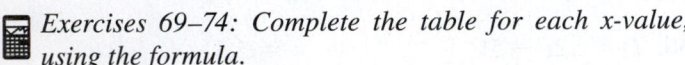 *Exercises 69–74: Complete the table for each x-value, using the formula.*

69. $y = 2.5x - 0.5$

x	0	2	4	6	8
y					

70. $y = \frac{1}{2}x^2$

x	0	2	4	6	8
y					

71. $y = |5x|$

x	-3	-1	1	3
y				

72. $y = \sqrt{x + 2}$

x	-2	-1	2	7
y				

73. $y = \sqrt[3]{x} - 2$

x	-1	0	1	8
y				

74. $y = x^3 - 4x$

x	-2	0	1	2
y				

Exercises 75 and 76: **Thinking Generally** *Use the formula to complete the table for each x-value. Assume that the variable a represents a positive number.*

75. $y = \sqrt{x} + 1$

x	a	$a - 1$	$a^2 - 1$
y			

76. $y = 2x + 1$

x	a	$\frac{3}{2}a$	$\frac{1}{2}a - \frac{1}{2}$
y			

Exercises 77–80: Use a calculator to make a table for the given formula. Let $x = 1, 2, 3, \ldots, 7$.

77. $y = x + \sqrt[3]{x}$

78. $y = \dfrac{1}{\sqrt{x}}$

79. $y = \sqrt{7 - x}$

80. $y = \dfrac{x^3 - 2x}{3}$

81. Online Exploration Find a formula for the volume of a pyramid with a square base. Estimate the volume of the Great Pyramid of Giza in cubic feet.

82. Online Exploration Find a formula for the volume of a sphere. Estimate the volume of an NBA basketball in cubic inches.

APPLICATIONS

83. *Modeling Motion* The following table lists the distance y traveled by a car in t hours. Find an equation that models these data. Evaluate y when $t = 6$.

t (hours)	1	2	3	4
y (miles)	60	120	180	240

84. *Modeling Motion* The following table lists the distance y traveled by a bicycle in t minutes. Find an equation that models these data. Evaluate y when $t = 3$.

t (minutes)	5	10	15	20
y (feet)	110	220	330	440

85. *Songs on an iPod* A 4-GB iPod can hold about 1000 songs, whereas a 60-GB iPod can store 15,000 songs. (*Source:* Apple Corporation.)
  (a) Find a formula that calculates the number of songs S that can be stored on x gigabytes.
  (b) How many songs can be stored on a 30-GB iPod?

86. *Diving and Water Pressure* The world's record for descending below the surface of the ocean on a single breath is in excess of 400 feet by Francisco Ferreras. During his descent his heart rate slows from 60 beats per minute on the surface to 4 beats per minute at 400 feet. He is able to hold his breath for 7 minutes. These dives are dangerous because of the extreme water pressure. The water pressure P in pounds per square inch at a depth of x feet can be calculated with the formula $P = 0.445x$.

(a) Calculate the water pressure at a depth of 400 feet.

(b) The world's deepest diving mammals are sperm whales, which can dive to a depth of 7000 feet. Calculate the water pressure at this depth. (*Source:* G. Carr, *Mechanics of Sport.*)

87. *Braking Distance* (Refer to Example 4.) The braking distance d for a car traveling at x miles per hour on *dry*, level pavement is $d = \frac{x^2}{12}$. (*Source:* L. Haefner.)

(a) Make a table of braking distances for speeds of 10 to 70 miles per hour in increments of 10 miles per hour.

(b) What is the braking distance for a car traveling at 40 miles per hour?

(c) What happens to the braking distance when the speed doubles?

(d) At 60 miles per hour how much farther does it take the car to stop on wet pavement than on dry pavement? (*Hint:* For wet pavement, $d = \frac{x^2}{9}$.)

88. *Bicycle Speed Records* The fastest speed attained on a bicycle is 223.3 feet per second. This speed was attained by the bicyclist following a pace vehicle. The bicyclist thus experienced less wind resistance and was carried along by the draft or "suction" created by the pace vehicle. The formula $M = \frac{15}{22}x$ converts feet per second to miles per hour. Find the speed of this record in miles per hour. (*Source:* G. Carr.)

89. *Facebook Usage* The average Facebook visitor views about 662 pages each month on this social network. (*Source:* The Nielsen Company.)

(a) Write a formula that calculates how many pages P have been viewed by the average visitor after x months.

(b) Calculate the average number of pages viewed in 2 years.

90. *MySpace Usage* The average MySpace visitor views about 262 pages each month on this social network. (*Source:* The Nielsen Company.)

(a) Write a formula that calculates how many pages P have been viewed by the average visitor after x months.

(b) Calculate the average number of pages viewed in a year and a half.

91. *Wing Span and Weight* A bird's weight W is frequently related to the length L of its wing span. For one species of bird the formula $W = 1.1L^3$ could be used to predict a bird's weight W in kilograms for a wing span of L meters. (*Source:* C. Pennycuick.)

(a) If a bird has a wing span of 0.75 meter, estimate its weight.

(b) If a bird has a wing span of 1.5 meters, estimate its weight.

(c) If the wing span of a bird doubles, what happens to its weight?

92. *Wing Span and Weight* (Refer to Exercise 91.) The formula $L = \sqrt[3]{16W}$ relates the weight W of one species of bird in pounds to its wing span L in feet.
(a) Estimate the wing span L when $W = 5$ pounds.
(b) Does a doubling of the weight W double the wing span? Explain.

93. *Inline Skating* During a strenuous skating workout, an athlete can burn 336 calories in 40 minutes. (*Source:* Runner's World.)
(a) Write a formula that calculates the calories C burned from skating 40 minutes a day for x days.
(b) How many calories could be burned in 30 days?

94. *Inline Skating* (Refer to the preceding exercise.) If a person loses 1 pound for every 3500 calories burned, write a formula that gives the number of pounds P lost in x days from skating 40 minutes per day. How many pounds could be lost in 100 days?

95. *Birds* The surface area of a bird's wings S is frequently related to its weight W. For one species of bird, the formula $S = 0.11\sqrt[3]{W^2}$ could be used to predict the surface area S in square meters of a bird's wings for a weight W in kilograms. (*Source:* C. Pennycuick.)
(a) If a bird weighs 0.5 kilogram, estimate the area of its wings.
(b) If a bird's weight doubles, does the area of its wings also double?

96. *Pollution and Lawn Mowers* A person using a gas-powered lawn mower for 1 hour pollutes the air as much as a person driving a car for 340 miles. (*Source:* Minnesota Pollution Control Agency.)
(a) Find a formula that calculates the hours H of lawn mowing that create the same amount of pollution as x miles of driving.
(b) How many hours of lawn mowing pollute the same as driving 850 miles?

97. *Stepping Frequency* (Refer to Example 9.) The formula $F = \frac{0.87}{\sqrt{h}}$ gives the stepping frequency F of an animal with a shoulder height h. Suppose that an animal has four times the shoulder height of another animal. How do their stepping frequencies compare?

98. *Animals and Trotting Speeds* (Refer to Example 9.) The relationship between the shoulder height h in meters and an animal's stepping frequency F in steps per second while *trotting* is given by $F = \frac{1.84}{\sqrt{h}}$. (*Source:* C. Pennycuick.)
(a) Estimate the stepping frequency for a trotting buffalo that is 1.5 meters high at the shoulders.
(b) What happens to an animal's stepping frequency while trotting as its shoulder height increases?

99. *Pulse Rate in Animals* According to one model, the rate at which an animal's heart beats varies with its weight. Smaller animals tend to have faster pulses, whereas larger animals tend to have slower pulses. The pulse rate of an animal can be modeled by the equation $N = \frac{885}{\sqrt{W}}$, where N is the number of beats per minute and W is the animal's weight in pounds. (*Source:* C. Pennycuick.)
(a) Estimate the pulse for a 25-pound dog.
(b) Estimate the pulse for a 1600-pound elephant.

100. *Pulse Rate in Animals* (Refer to the preceding exercise.) Suppose that an animal has half the pulse rate of another animal. How do the weights of these animals compare?

FORMULAS FROM GEOMETRY

101. *Area of an Equilateral Triangle* An equilateral triangle has three sides equal in length. Its area A is given by $A = \frac{\sqrt{3}}{4}s^2$, where s is the length of a side, as shown in the accompanying figure. Calculate the areas for the given values of s.

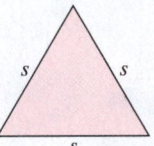

(a) $s = 2$ feet   (b) $s = 4$ meters

102. *Equilateral Triangle* (Refer to the preceding exercise.) What happens to the area of an equilateral triangle if the length of a side triples?

103. *Circumference of a Circle* The circumference C of a circle is given by $C = 2\pi r$, where r is the radius. Calculate the circumference of each circle with the given radius.
(a) $r = 14$ inches   (b) $r = 1.3$ miles

104. *Area of a Circle* The area A of a circle is given by $A = \pi r^2$, where r is the radius. Calculate the area of each circle with the given radius.
(a) $r = 12$ inches   (b) $r = 6$ feet

Exercises 105–108: Determine the length of a side for each geometric shape.

105. A square with an area of 81 square inches

106. A square with an area of 121 square meters

107. A cube with a volume of 27 cubic meters

108. A cube with a volume of 64 cubic feet

WRITING ABOUT MATHEMATICS

109. Explain the difference between an *algebraic expression* and an *equation*.

110. Give an example of a formula that models an application and identify each variable. Explain how to use the formula.

| SECTIONS 1.3 and 1.4 | **Checking Basic Concepts** |

1. Evaluate each expression.
   (a) 2^4 **(b)** $3^{-2} \cdot 2^0$

   (c) $\dfrac{2^4}{2^2 \cdot 2^{-3}}$ **(d)** $x^3 x^{-4} x^2$

   (e) $\left(\dfrac{2x^3}{y^{-4}} \right)^2$

2. Evaluate each expression.
   (a) $4 + 5 \cdot (-2)$ **(b)** $\dfrac{1 + 3}{-4 + 3}$
   (c) $2^3 - 5(2 - 3 \cdot 4)$

3. Express each number in scientific notation.
   (a) 103,000   **(b)** 0.000523   **(c)** 6.7

4. Express each number in standard (decimal) form.
   (a) 5.43×10^6 **(b)** 9.8×10^{-3}

5. *Indoor Air Pollution* Ventilation is an effective method for removing indoor air pollution. The formula $y = 900x$ calculates the cubic feet per hour of air that should be circulated in a classroom containing x people. (This formula assumes no one is smoking in the room!) Make a table showing the ventilation necessary for classes containing 10, 20, 30, and 40 people. How much ventilation is necessary per person? (*Source:* ASHRAE.)

1.5 Introduction to Graphing

Relations • The Cartesian Coordinate System • Scatterplots and Line Graphs • The Viewing Rectangle (Optional) • Graphing with Calculators (Optional)

A LOOK INTO MATH ▶ Computers, the Internet, and other types of electronic communication are creating large amounts of data. Before conclusions can be drawn, data must be analyzed. A powerful tool in this step is visualization. The map in Figure 1.29 shows the average date of the first 32°F temperature in autumn. Imagine trying to describe this map by using *only* words. In this section we discuss how graphs are used to visualize data. (*Source:* J. Williams, *The Weather Almanac.*)

NEW VOCABULARY

☐ Ordered pair
☐ Relation
☐ Domain/Range
☐ *xy*–plane
☐ *x*–axis
☐ *y*–axis
☐ Origin
☐ Quadrants
☐ Scatterplot
☐ Line graph

First Frost

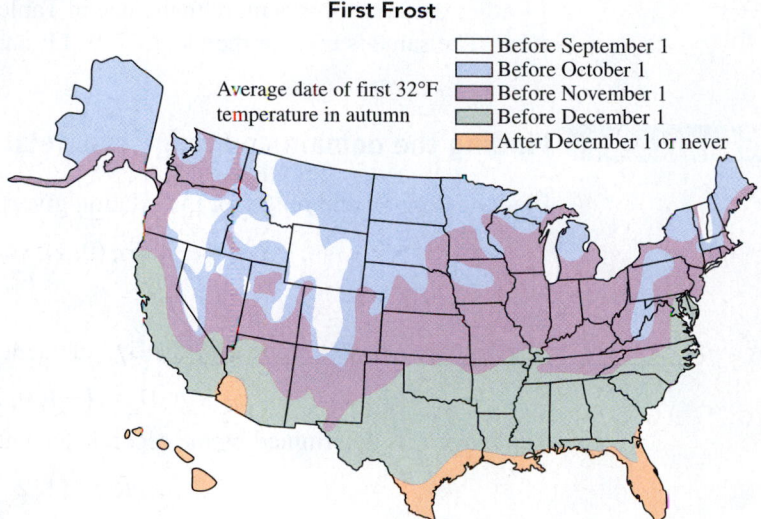

Average date of first 32°F temperature in autumn

☐ Before September 1
☐ Before October 1
☐ Before November 1
☐ Before December 1
☐ After December 1 or never

Figure 1.29

Relations

▶ **REAL-WORLD CONNECTION** Table 1.12 lists monthly average wind speeds in miles per hour for San Francisco. In this table January corresponds to 1, February to 2, and so on, until December is represented by 12. For example, in April the average wind speed is 12 miles per hour.

TABLE 1.12 Average Wind Speeds in San Francisco

Month	1	2	3	4	5	6	7	8	9	10	11	12
Wind Speed (mph)	7	9	11	12	13	14	14	13	11	9	8	7

Source: J. Williams, *The Weather Almanac.*

If we let x be the month and y be the wind speed, then the **ordered pair** (x, y) represents the average wind speed y during month x. For example, the ordered pair $(2, 9)$ indicates that in February the average wind speed is 9 miles per hour, whereas the ordered pair $(9, 11)$ indicates that the average wind speed in September is 11 miles per hour. *Order is important in an ordered pair.*

The data in Table 1.12 establish a *relation*; that is, each month is associated with a wind speed in an ordered pair (**month**, **wind speed**). This relation can be represented by a set S, which contains 12 ordered pairs:

$$S = \{(1, 7), (2, 9), (3, 11), (4, 12), (5, 13), (6, 14),$$
$$(7, 14), (8, 13), (9, 11), (10, 9), (11, 8), (12, 7)\}.$$

RELATION

A **relation** is a set of ordered pairs.

READING CHECK

Give an example of a relation. State its domain and range.

If we denote the ordered pairs in a relation (x, y), then the set of all x-values is called the **domain** of the relation and the set of all y-values is called the **range**. In Table 1.12 the domain is

$$D = \{1, 2, 3, 4, 5, 6, 7, 8, 9, 10, 11, 12\},$$

which corresponds to the 12 months. The range is

$$R = \{7, 8, 9, 11, 12, 13, 14\},$$

which corresponds to the monthly average wind speeds. Note that an average wind speed of 14 miles per hour occurs more than once in Table 1.12, but it is listed *only once* in the range set R. The same is true for the values 7, 9, 11, and 13.

EXAMPLE 1 Finding the domain and range of a relation

Find the domain and range for the relation given by

$$S = \{(-1, 5), (0, 1), (2, 4), (4, 2), (5, 1)\}.$$

Solution
The domain D is determined by the first element in each ordered pair, or

$$D = \{-1, 0, 2, 4, 5\}.$$

The range R is determined by the second element in each ordered pair, or

$$R = \{1, 2, 4, 5\}.$$

▌ **Now Try Exercise 9**

▶ **REAL-WORLD CONNECTION** We are all well aware that the cost of tuition and fees frequently goes up each year. In the next example, we describe this interaction between the year and this cost as a relation.

EXAMPLE 2 **Analyzing tuition and fees**

Table 1.13 lists the average cost of tuition and fees at public colleges from 2006 through 2009. Express this table as a relation S. Identify the domain and range of S.

TABLE 1.13 Tuition and Fees

Year	2006	2007	2008	2009
Cost	$5804	$6185	$6591	$7020

Source: The College Board.

Solution
Let the year be the first element in the ordered pair and the cost of tuition and fees be the second element. Then relation S is given by the following set of ordered pairs.

$$S = \{(2006, 5804), (2007, 6185), (2008, 6591), (2009, 7020)\}$$

The domain of S is

$$D = \{2006, 2007, 2008, 2009\}$$

and the range of S is

$$R = \{5804, 6185, 6591, 7020\}.$$

Now Try Exercise 17

NOTE: It is possible for a relation to contain *infinitely many* ordered pairs. For example, let the equation $y = 2x$ define a relation S, where x is *any* real number. Then S contains infinitely many ordered pairs of the form $(x, 2x)$, such as $(-2, -4)$, $(3, 6)$, and $(0.1, 0.2)$.

EXAMPLE 3 **Analyzing iPod memory**

The formula $S = 250m$ identifies a relationship between the memory m of an iPod in gigabytes and the number of songs S that can be stored on it. Give four ordered pairs in this relation.

Solution
One way to determine four ordered pairs is to assign the variable m four different values. For example, when $m = 4$, $S = 250(4) = 1000$. The ordered pair $(4, 1000)$ means that a 4-GB iPod can hold 1000 songs. Table 1.14 lists this ordered pair along with three additional ordered pairs: $(8, 2000)$, $(20, 5000)$, and $(30, 7500)$.

TABLE 1.14

m	4	8	20	30
S	1000	2000	5000	7500

Now Try Exercise 23

The Cartesian Coordinate System

We can use the **Cartesian coordinate system**, or *xy-plane*, to visualize or *graph* a relation. The horizontal axis is the *x*-**axis** and the vertical axis is the *y*-**axis**. The axes intersect at the **origin** and determine four regions called **quadrants**. They are numbered I, II, III, and IV counterclockwise, as illustrated in Figure 1.30. We can plot the ordered pair (*x*, *y*) by using the *x*-axis and the *y*-axis. For example, the point (**1**, **2**) is located in quadrant I, **1** unit to the right of the origin and **2** units above the *x*-axis, as shown in Figure 1.31.

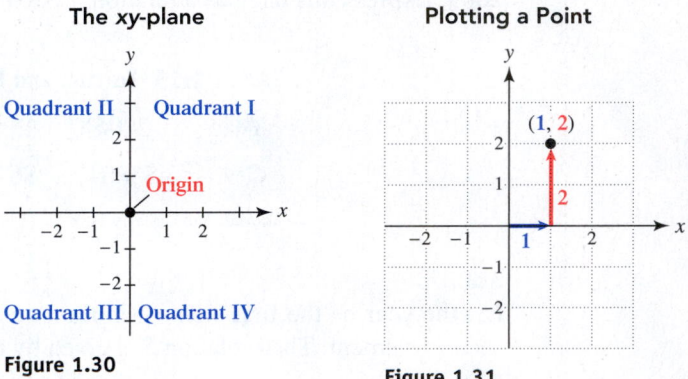

The *xy*-plane

Figure 1.30

Plotting a Point

Figure 1.31

READING CHECK

Explain how to plot points in the *xy*-plane. What are the coordinates of the origin?

Similarly, the ordered pair (−2, 3) is located in quadrant II, (−3, −3) is in quadrant III, and (3, −2) is in quadrant IV. See Figure 1.32. A point lying on a coordinate axis does not belong to any quadrant. The point (−2, 0) is located on the *x*-axis, whereas the point (0, −2) lies on the *y*-axis.

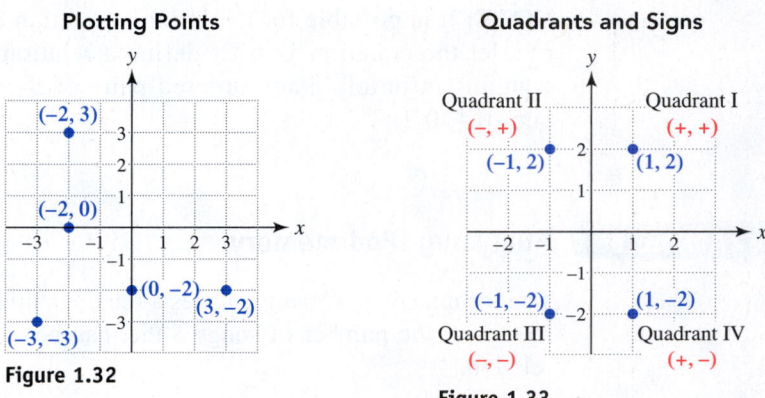

Plotting Points

Figure 1.32

Quadrants and Signs

Figure 1.33

For any point (*x*, *y*) with *x* and *y* nonzero, we can determine the quadrant in which it is located. For example, the point (1, 2) lies in quadrant I because both *x* and *y* are positive, whereas (−1, −2) lies in quadrant III because both *x* and *y* are negative. The point (−1, 2) lies in quadrant II where *x* is negative and *y* is positive, and (1, −2) lies in quadrant IV where *x* is positive and *y* is negative. These concepts are illustrated in Figure 1.33, where (+, +) indicates that $x > 0$ and $y > 0$ for any point (*x*, *y*) in quadrant I. Other quadrants and ordered pairs can be interpreted similarly.

CRITICAL THINKING

If *a* is negative, then what quadrant does the point (*a*, −*a*²) lie in?

EXAMPLE 4 | **Plotting points**

Plot the data listed in Table 1.15. For each point, state the quadrant containing the point or the axis on which the point lies.

TABLE 1.15

x	−3	0	1	4
y	1	4	−2	3

Solution

We plot the points $(-3, 1)$, $(0, 4)$, $(1, -2)$, and $(4, 3)$ in the xy-plane, as shown in Figure 1.34. The point $(-3, 1)$ is in quadrant II, $(1, -2)$ is in quadrant IV, and $(4, 3)$ is in quadrant I. The point $(0, 4)$ lies on the y-axis and does not belong to any quadrant.

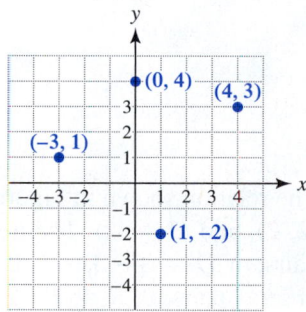

Figure 1.34

Now Try Exercise 19

EXAMPLE 5 | **Graphing points given an equation**

Evaluate $y = x^2 + 1$ for $x = -2, -1, 0, 1$, and 2. Plot the resulting ordered pairs.

Solution

Start by evaluating the formula $y = x^2 + 1$ for each x-value.

$$x = -2: \quad y = (-2)^2 + 1 = 5$$
$$x = -1: \quad y = (-1)^2 + 1 = 2$$
$$x = 0: \quad y = 0^2 + 1 = 1$$
$$x = 1: \quad y = 1^2 + 1 = 2$$
$$x = 2: \quad y = 2^2 + 1 = 5$$

The points $(-2, 5)$, $(-1, 2)$, $(0, 1)$, $(1, 2)$, and $(2, 5)$ are plotted in Figure 1.35.

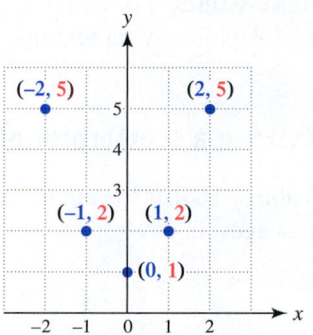

Figure 1.35

Now Try Exercise 49

EXAMPLE 6   **Determining the domain and range**

Use the graph in Figure 1.36 to determine the domain and range of the relation.

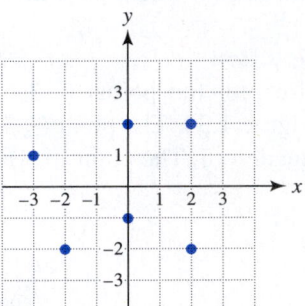

Figure 1.36

Solution

The relation shown in Figure 1.36 includes the points $(-3, 1)$, $(-2, -2)$, $(0, 2)$, $(0, -1)$, $(2, 2)$, and $(2, -2)$. The domain of this relation consists of the x-values of these ordered pairs, or $D = \{-3, -2, 0, 2\}$. The range of this relation consists of the y-values of these ordered pairs, or $R = \{-2, -1, 1, 2\}$.

Now Try Exercise 29

Scatterplots and Line Graphs

STUDY TIP

Reading graphs is important. Ask for help if you aren't sure how to read graphs.

If distinct points are plotted in the xy-plane, the resulting graph is called a **scatterplot**. Figure 1.36 above is an example. A scatterplot of a different relation is shown in Figure 1.37, where the points $(1, 2)$, $(2, 4)$, $(3, 5)$, $(4, 6)$, $(5, 4)$, and $(6, 3)$ have been plotted. Its domain is $D = \{1, 2, 3, 4, 5, 6\}$, and its range is $R = \{2, 3, 4, 5, 6\}$.

A Scatterplot

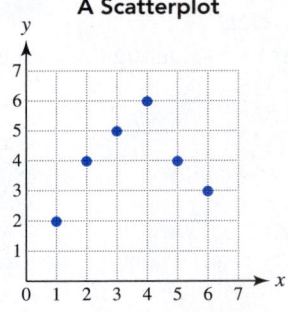

Figure 1.37

▶ **REAL-WORLD CONNECTION** The next example illustrates how to make a scatterplot from real data involving texting.

EXAMPLE 7   **Making a scatterplot of texting**

Table 1.16 lists the approximate number of text messages sent per month, by people of various ages, in 2010.

TABLE 1.16 Texts Sent per Month

Age	15	20	30	40	50
Texts	3500	1800	1000	700	300

Source: The Nielsen Company.

(a) Discuss the general trend in the data.

(b) Make a scatterplot of the data.

Solution

(a) The number of texts sent appears to decrease with age.

(b) Plot the five points (15, 3500), (20, 1800), (30, 1000), (40, 700), and (50, 300). The *x*-values vary between 15 and 50, so we label the *x*-axis from 0 to 60. The *y*-values vary from 300 to 3500, so we label the *y*-axis from 0 to 4000. (Note that other *x*- and *y*-scales are possible.) The scatterplot is shown in Figure 1.38, with the *x*- and *y*-axes labeled with the proper units.

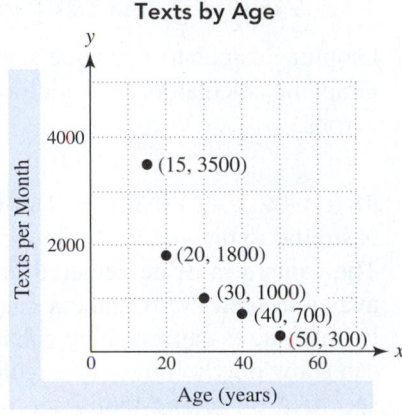

Figure 1.38

■ **Now Try Exercise 37**

▶ **REAL-WORLD CONNECTION** Sometimes it is helpful to connect the data points in a scatterplot with straight line segments. This type of graph emphasizes changes in the data and is called a **line graph**.

EXAMPLE 8 ▏ **Interpreting a line graph**

The line graph shown in Figure 1.39 depicts the total number of all types of college degrees awarded in millions for selected years. Note that the double hash marks // on the *x*-axis indicate that there is a break in the scale, which starts at 0 and then jumps to 1950. (*Source:* Department of Education.)

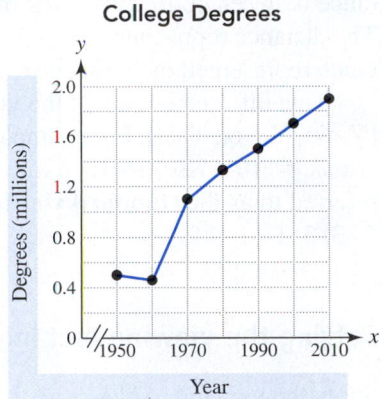

Figure 1.39

(a) Did the number of degrees ever decrease during this time period? Explain.

(b) Approximate the number of college degrees in the year 1970.

(c) Determine the 10-year period when the increase in the number of college degrees was greatest. What was this increase?

Solution

(a) Yes, the number of degrees decreased slightly between 1950 and 1960. For this time period, the line segment slopes slightly downward from left to right.

(b) In 1970, about 1.1 million degrees were awarded.

(c) The greatest increase corresponds to the line segment that slopes upward most from left to right. This increase occurred between 1960 and 1970 and was about $1.1 - 0.5 = 0.6$ million degrees.

■ **Now Try Exercise 77**

The Viewing Rectangle (Optional)

Graphing calculators provide several features beyond those found on scientific calculators. Graphing calculators have additional keys that can be used to create tables, scatterplots, and graphs.

▶ **REAL-WORLD CONNECTION** The **viewing rectangle**, or **window**, on a graphing calculator is similar to the viewfinder in a camera. A camera cannot take a picture of an entire scene. The camera must be centered on some object and can photograph only a portion of the available scenery. A camera can capture different views of the same scene by zooming in and out, as can graphing calculators. The xy-plane is infinite, but the calculator screen can show only a finite, rectangular region of the xy-plane. The viewing rectangle must be specified by setting minimum and maximum values for both the x- and y-axes before a graph can be drawn.

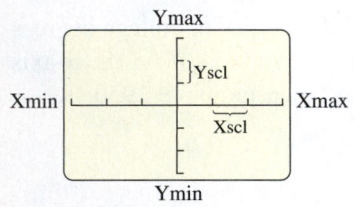

Figure 1.40

We use the following terminology regarding the size of a viewing rectangle. **Xmin** is the minimum x-value along the x-axis, and **Xmax** is the maximum x-value. Similarly, **Ymin** is the minimum y-value along the y-axis, and **Ymax** is the maximum y-value. Most graphs show an x-scale and a y-scale with tick marks on the respective axes. Sometimes the distance between consecutive tick marks is 1 unit, but at other times it might be 5 or 10 units. The distance represented by consecutive tick marks on the x-axis is called **Xscl**, and the distance represented by consecutive tick marks on the y-axis is called **Yscl** (see Figure 1.40).

This information about the viewing rectangle can be written as [Xmin, Xmax, Xscl] by [Ymin, Ymax, Yscl]. For example, [−10, 10, 1] by [−10, 10, 1] means that Xmin = −10, Xmax = 10, Xscl = 1, Ymin = −10, Ymax = 10, and Yscl = 1. This setting is referred to as the **standard viewing rectangle**. The window in Figure 1.40 is [−3, 3, 1] by [−3, 3, 1].

EXAMPLE 9 **Setting the viewing rectangle**

Show the viewing rectangle [−2, 3, 0.5] by [−100, 200, 50] on your calculator.

Solution

The window setting and viewing rectangle are displayed in Figure 1.41. Note that in Figure 1.41(b) there are 6 tick marks on the positive x-axis because its length is 3 units and the distance between consecutive tick marks is 0.5 unit.

CALCULATOR HELP

To set a viewing rectangle, see Appendix A (page AP-3).

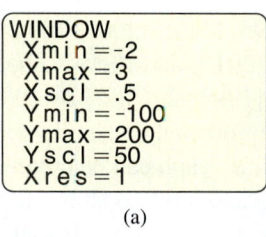

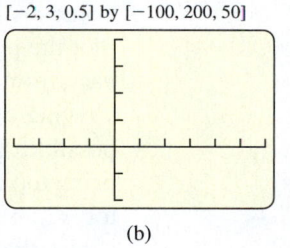

[−2, 3, 0.5] by [−100, 200, 50]

(a)   (b)

Figure 1.41

■ **Now Try Exercise 63**

Graphing with Calculators (Optional)

Many graphing calculators have the capability to create scatterplots and line graphs. The next example illustrates how to make a scatterplot with a graphing calculator.

EXAMPLE 10 ▌ **Making a scatterplot with a graphing calculator**

Plot the points $(-2, -2)$, $(-1, 3)$, $(1, 2)$, and $(2, -3)$ in $[-4, 4, 1]$ by $[-4, 4, 1]$.

Solution

We entered the points $(-2, -2)$, $(-1, 3)$, $(1, 2)$, and $(2, -3)$ shown in Figure 1.42(a), using the STAT EDIT feature. The variable L1 represents the list of x-values, and the variable L2 represents the list of y-values. In Figure 1.42(b) we set the graphing calculator to make a scatterplot with the STATPLOT feature, and in Figure 1.42(c) the points have been plotted. If you have a different model of calculator you may need to consult your owner's manual.

CALCULATOR HELP

To make a scatterplot, see Appendix A (pages AP-3 and AP-4).

[−4, 4, 1] by [−4, 4, 1]

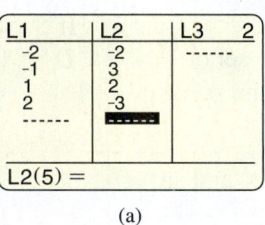

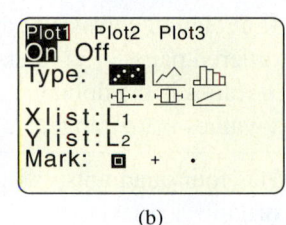

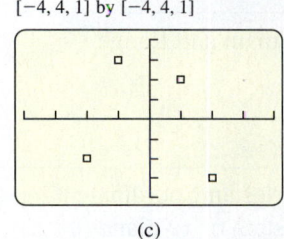

(a)   (b)   (c)

Figure 1.42

■ **Now Try Exercise 71**

▶ **REAL-WORLD CONNECTION** In the next example, a graphing calculator is used to create a line graph of wireless subscriber connections.

EXAMPLE 11 ▌ **Making a line graph with a graphing calculator**

Table 1.17 lists the number of wireless subscriber connections for selected years from 1995 through 2010. Make a line graph of these numbers in an appropriate viewing rectangle and then interpret the line graph.

TABLE 1.17 Wireless Subscriber Connections

Year	1995	2000	2005	2010
Connections (millions)	33.8	109.5	207.9	285.6

Source: Cellular Telecommunications Industry Association.

CALCULATOR HELP

To make a line graph, see
Appendix A (pages AP-3
and AP-4).

Solution

Plot the points (1995, 33.8), (2000, 109.5), (2005, 207.9), and (2010, 285.6). The *x*-values
vary from 1995 to 2010, and the *y*-values vary between 33.8 and 285.6. We selected the
viewing rectangle [1993, 2012, 5] by [0, 300, 50], although other viewing rectangles are
possible. The viewing rectangle should be large enough to show all four data points without
being too large. A line graph can be created by selecting this option on the graphing calculator. Figures 1.43(a) and 1.43(b) show the data entries and plotting scheme. Figure 1.43(c)
shows the resulting graph. It reveals that wireless subscribers have increased dramatically
during this time period.

[1993, 2012, 5] by [0, 300, 50]

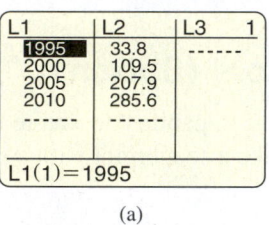

(a)

(b)

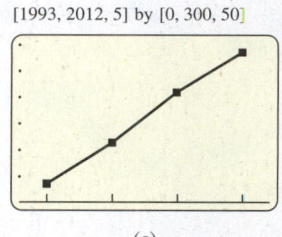

(c)

Figure 1.43

■ **Now Try Exercise 79**

1.5 Putting It All Together

CONCEPT	EXPLANATION	EXAMPLE
Relation	A set of ordered pairs	$S = \{(1, 2), (-2, 3), (4, 2)\}$
Domain and Range	If a relation consists of a set of ordered pairs (x, y), then the set of x-values is the domain and the set of y-values is the range.	If $S = \{(1, 2), (-2, 3), (4, 2)\}$, then $D = \{-2, 1, 4\}$ and $R = \{2, 3\}$.
Cartesian Coordinate System or *xy*-plane	Has four quadrants, two axes, and an origin Used to plot points and graphs	
Scatterplot	A scatterplot results when individual points are plotted in the *xy*-plane.	
Line Graph	A line graph is similar to a scatterplot except that line segments are drawn between consecutive points.	

1.5 Exercises

MyMathLab

CONCEPTS AND VOCABULARY

1. What is a relation?

2. What are the domain and range of a relation?

3. Sketch the *xy*-plane and identify each of the following: the *x*-axis, the *y*-axis, the origin, and the four quadrants.

4. Sketch an example of a scatterplot and a line graph.

5. Do the ordered pairs (1, 2) and (2, 1) represent the same point in the *xy*-plane?

6. In what quadrant does the point $(-1, -1)$ lie?

7. If the ordered pair (3, 4) belongs to a relation S, then _____ is an element of the domain of S.

8. If the ordered pair $(0, -2)$ belongs to a relation S, then _____ is an element of the range of S.

RELATIONS AND THE RECTANGULAR COORDINATE SYSTEM

Exercises 9–14: Identify the domain and range of S.

9. $S = \{(1, 2), (3, -4), (5, 6)\}$

10. $S = \{(0, 4), (0, 6), (3, -1), (4, 0)\}$

11. $S = \{(-2, 3), (-1, 2), (0, 1), (1, 0), (2, 1)\}$

12. $S = \left\{ \left(\frac{1}{2}, -\frac{3}{4}\right), \left(-\frac{5}{8}, \frac{4}{7}\right), \left(\frac{1}{2}, \frac{3}{4}\right), \left(\frac{8}{7}, \frac{3}{4}\right) \right\}$

13. $S = \{(41, 67), (87, 53), (41, 88), (96, 24)\}$

14. $S = \{(-1.2, -1.1), (0.8, 2.5), (1.5, -0.6)\}$

Exercises 15–18: Express the relation S in the table as a set of ordered pairs. Then identify the domain and range.

15.

x	1	3	5	7	9
y	3	7	11	15	19

16.

x	−2.1	−1.5	0.7	1.3	2.9
y	9.6	7.4	3.3	−2.0	−8.8

17. Registered users of Skype in millions

x	2006	2007	2008	2009	2010
y	139	217	325	474	560

18. U.S. population in millions

x	1800	1840	1880	1920	1960	2000
y	5	17	50	106	179	281

Source: U.S. Census Bureau.

Exercises 19–22: Plot the points in the table in the xy-plane. For each point, state the quadrant containing the point or the axis on which the point lies.

19.

x	1	−3	0	−1
y	2	0	−2	3

20.

x	2	−4	−2	0
y	6	−4	0	−5

21.

x	10	−30	50	−20
y	50	20	−25	−25

22.

x	0.2	0.4	0.6	0.8
y	3	1	−1	−3

Exercises 23–28: (Refer to Example 3.) Find four ordered pairs that belong to the relation determined by the given equation. Answers may vary.

23. $y = 4x$

24. $y = 3x + 5$

25. $A = 4 - t^2$

26. $R = 3t^2$

27. $z = \dfrac{1}{r^2 + 1}$

28. $z = \dfrac{2}{5 - r}$

Exercises 29–34: Express the relation shown in the graph or bar graph as a set of ordered pairs. Identify the domain and range.

29.

 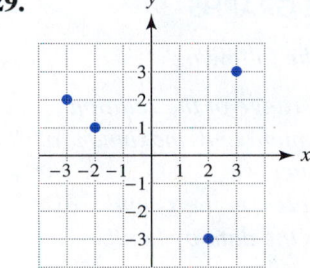

30.

 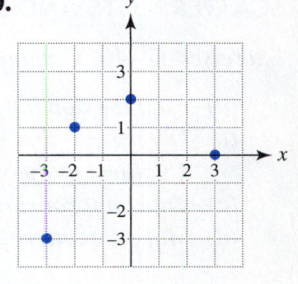

31.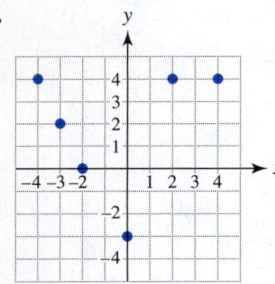

32.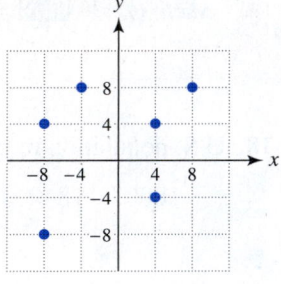

33. Monthly page views per visitor for social network sites (*Source:* The Nielsen Company.)

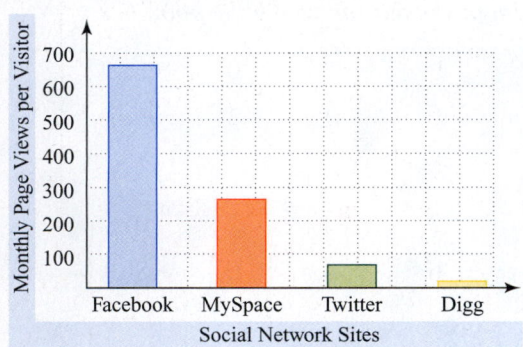

34. Capital expenditures in millions of dollars by mobile service provider (*Source:* The Nielsen Company.)

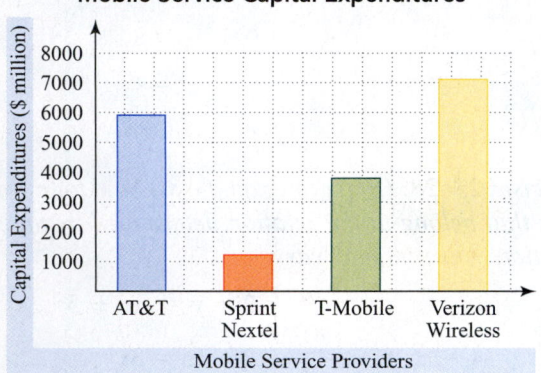

35. Thinking Generally If the ordered pairs (a, b) and (b, a) represent the same point in the xy-plane, what can be said about a and b?

36. Thinking Generally In what quadrant does the point $(-b, b)$ lie, if b is negative?

SCATTERPLOTS AND LINE GRAPHS

Exercises 37–40: Complete the following.

(a) *Find the domain and range of the relation.*
(b) *Determine the minimum and maximum of the x-values; of the y-values.*
(c) *Label appropriate scales on the x- and y-axes.*
(d) *Make a scatterplot of the data by hand.*

37. $\{(0, 2), (-3, 4), (-2, -4), (1, -3), (0, 0)\}$

38. $\{(1, 1), (3, 0), (-4, -4), (5, -2), (0, 3)\}$

39. $\{(10, 50), (-30, 40), (20, -50), (30, 20)\}$

40. $\{(5, 15), (25, 20), (10, 10), (-10, 30), (-20, -10)\}$

Exercises 41–44: Make a line graph of the data.

41. $(0, 2), (1, 4), (2, 5), (4, 4), (5, 2)$

42. $(-2, 4), (-1, 1), (0, 0), (1, 1), (2, 4)$

43. $(4, 4), (8, -4), (12, 8), (16, 0), (20, -8)$

44. $(10, 20), (20, 30), (30, 40), (40, 60), (50, 30)$

Exercises 45–48: Make a line graph of the data.

45.

x	0	1	2	3
y	-2	-1	0	3

46.

x	-2	-1	0	1	2
y	0	3	4	3	0

47.

x	0	1	2	3
y	-3	0	3	0

48.

x	0	1	4	9
y	0	1	2	3

Exercises 49–60: (Refer to Example 5.) Evaluate the formula for $x = -2, -1, 0, 1$, and 2. Plot the resulting ordered pairs.

49. $y = 3x$

50. $y = -2x$

51. $y = -x + 2$

52. $y = 2x + 4$

53. $y = -\frac{1}{2}x$

54. $y = \frac{1}{2}x + 1$

55. $y = x^2 - 1$

56. $y = \frac{1}{2}x^2$

57. $y = -x^2$

58. $y = 2x^2$

59. $y = |x|$

60. $y = \sqrt{x + 2}$

GRAPHING CALCULATORS

Exercises 61–66: Show the given viewing rectangle on your graphing calculator. Predict the number of tick marks on the positive x-axis and the positive y-axis

61. Standard viewing rectangle

62. $[-12, 12, 2]$ by $[-8, 8, 2]$

63. [0, 100, 10] by [−50, 50, 10]

64. [−30, 30, 5] by [−20, 20, 5]

65. [1980, 1995, 1] by [12000, 16000, 1000]

66. [1900, 1990, 10] by [1700, 2800, 100]

Exercises 67–70: Express the relation S shown in the graph as a set of ordered pairs.

67. [−3, 3, 1] by [−2, 2, 1]

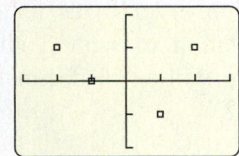

68. [−6, 6, 1] by [−4, 4, 1]

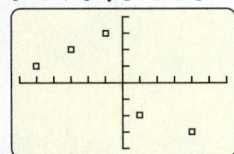

69. [−5, 5, 1] by [−3, 3, 1]

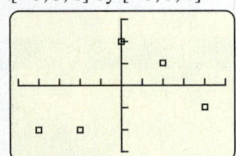

70. [1900, 2000, 20] by [0, 500, 100]

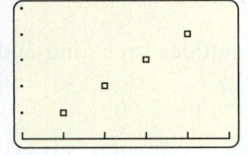

Exercises 71–76: Use your calculator to make a scatterplot of the relation after determining an appropriate viewing rectangle.

71. {(4, 3), (−2, 1), (−3, −3), (5, −2)}

72. {(5, 5), (2, 0), (−2, 7), (2, −8), (−1, −5)}

73. {(20, 40), (−25, −15), (−20, 25), (15, −25)}

74. {(−13, 12), (3, 10), (−15, −4), (12, −9)}

75. {(100, −100), (50, 200), (−150, −140), (−30, 80)}

76. {(−125, 75), (45, 65), (−53, −67), (150, −80)}

APPLICATIONS

*Exercises 77–82: **Graphing Real Data** Each table contains real data.*

(a) *Make a line graph of the data.*
(b) *Comment on any trends in the data.*

77. Text messages *y* sent in billions during year *x*

x	2005	2006	2007	2008	2009
y	57	113	241	601	1560

Source: The Nielsen Company.

78. Sales of CDs *y* in millions during year *x*

x	1990	1994	2000	2005	2009
y	290	660	1273	705	374

Source: Recording Industry Association of America.

79. Welfare beneficiaries *y* in millions during year *x*

x	1995	1997	1999	2003	2010
y	13.7	10.9	7.0	5.0	4.4

Source: Administration for Children and Families.

80. Google ad revenues *y* in billions of dollars during year *x*

x	2001	2003	2005	2008	2009
y	0	1	4	16	18

Source: Company report.

81. Asian-American population *y* in millions during year *x*

x	2000	2002	2004	2006	2008
y	11.2	12.0	12.8	14.0	14.9

Source: U.S. Census Bureau.

82. Internet usage in millions of users *y* during year *x*

x	1991	1997	2001	2004	2009
y	7.5	70	513	817	1802

Source: The Nielsen Company.

WRITING ABOUT MATHEMATICS

83. Explain how the domain and range of a relation can be used to determine an appropriate viewing rectangle for a scatterplot.

84. Explain the difference between a scatterplot and a line graph. Give an example of each.

Checking Basic Concepts

1. State the domain and range of the relation given by $S = \{(-5, 3), (1, 4), (2, 3), (1, -1)\}$.

2. Plot the following points in the xy-plane. State the quadrant containing each point or the axis on which each point lies.
   (a) $(1, 4)$ (b) $(0, -3)$
   (c) $(2, -2)$ (d) $(-2, 3)$

3. Evaluate $y = 2 - x^2$ for $x = -2, -1, 0, 1$, and 2. Plot the resulting ordered pairs.

4. *Mobile Data Traffic* The following table lists mobile data traffic (actual and projected) in petabytes for selected years. Make a line graph of these data and comment on any trends.

Year	2010	2011	2012	2013	2014
Petabytes	10	22	55	135	215

Source: The Nielsen Company.

5. **Online Exploration** (Refer to the preceding exercise.) Determine the number of bytes in a petabyte. Use scientific notation to write your answer. If this number were written in decimal form, how would you read it?

CHAPTER 1   Summary

SECTION 1.1 ■ DESCRIBING DATA WITH SETS OF NUMBERS

Sets of Numbers

Natural Numbers $N = \{1, 2, 3, 4, \ldots\}$

Whole Numbers $W = \{0, 1, 2, 3, \ldots\}$

Integers $I = \{\ldots, -2, -1, 0, 1, 2, \ldots\}$

Rational Numbers   Can be written as $\frac{p}{q}$, where p and $q \neq 0$ are integers; includes repeating and terminating decimals

     Examples: $\frac{1}{2}, -3, 6, \sqrt{9}, 0.\overline{7}, 0.123$

Irrational Numbers   A real number that is not rational

     Examples: $\pi, \sqrt{11}, -\sqrt{3}$

Real Numbers   Any number that can be written in decimal form

     Examples: $-\frac{2}{3}, 0, 12.6, \sqrt{11}, \pi$

Properties of Real Numbers

Identity Properties

     $a + 0 = a$ $a \cdot 1 = a$

Commutative Properties

     $a + b = b + a$ $a \cdot b = b \cdot a$

Associative Properties

     $(a + b) + c = a + (b + c)$ $(a \cdot b) \cdot c = a \cdot (b \cdot c)$

Distributive Properties

     $a(b + c) = ab + ac$ $a(b - c) = ab - ac$

SECTION 1.2 ■ OPERATIONS ON REAL NUMBERS

Real Number Line The number line is used to graph real numbers.

Origin

$-3 \quad -2 \quad -1 \quad 0 \quad 1 \quad 2 \quad 3$

Absolute Value $|a|$ equals a if $a > 0$ or $a = 0$, and $-a$ if $a < 0$.

Examples: $|-4| = 4, |7| = 7$, and $|\pi - 7| = 7 - \pi$ because $7 > \pi$

Arithmetic Operations

Addition **Examples:** $-3 + 4 = 1, \quad 3 + (-4) = -1, \quad$ and $\quad -3 + (-4) = -7$

Subtraction Use $a - b = a + (-b)$.

Examples: $-5 - 6 = -5 + (-6) = -11 \quad$ and $\quad 4 - (-3) = 4 + 3 = 7$

Multiplication The product of two numbers with like signs is positive. The product of two numbers with unlike signs is negative.

Examples: $3 \cdot (-4) = -12 \quad$ and $\quad -5 \cdot (-6) = 30$

Division Use $\dfrac{a}{b} = a \cdot \dfrac{1}{b}$.

Examples: $\frac{1}{2} \div -\frac{4}{5} = \frac{1}{2} \cdot -\frac{5}{4} = -\frac{5}{8} \quad$ and

$$-\frac{3}{2} \div 6 = -\frac{3}{2} \cdot \frac{1}{6} = -\frac{3}{12} = -\frac{1}{4}$$

SECTION 1.3 ■ INTEGER EXPONENTS

Exponential Expression

Base$\rightarrow$ 6^2 $\leftarrow$ Exponent

Integer Exponents Let n be a positive integer and a be a nonzero number.

$a^n = a \cdot a \cdot a \cdot \cdots \cdot a \quad$ (n factors of a)

$a^0 = 1$ (Note: 0^0 is undefined.)

$a^{-n} = \dfrac{1}{a^n}$

Examples: $4^3 = 4 \cdot 4 \cdot 4 = 64, \quad 5^0 = 1, \quad$ and $\quad 2^{-3} = \dfrac{1}{2^3}$

Properties of Exponents

Product Rule $a^m \cdot a^n = a^{m+n}$

Example: $z^3 \cdot z^5 = z^8$

Quotient Rule $\dfrac{a^m}{a^n} = a^{m-n}$

Example: $\dfrac{x^5}{x^7} = x^{-2} = \dfrac{1}{x^2}$

Power Rules $(a^m)^n = a^{mn}, \quad (ab)^n = a^n b^n, \quad$ and $\quad \left(\dfrac{a}{b}\right)^n = \dfrac{a^n}{b^n}$

Examples: $(5^2)^3 = 5^6, \quad (2x)^3 = 8x^3, \quad$ and $\quad \left(\dfrac{2x}{y}\right)^3 = \dfrac{8x^3}{y^3}$

Negative Exponents $\dfrac{1}{a^{-n}} = a^n, \quad \dfrac{a^{-n}}{b^{-m}} = \dfrac{b^m}{a^n}, \quad$ and $\quad \left(\dfrac{a}{b}\right)^{-n} = \left(\dfrac{b}{a}\right)^n$

Examples: $\dfrac{1}{2^{-3}} = 2^3, \quad \dfrac{x^{-4}}{y^{-3}} = \dfrac{y^3}{x^4}, \quad$ and $\quad \left(\dfrac{2}{5}\right)^{-4} = \left(\dfrac{5}{2}\right)^4$

Order of Operations

Using the following order of operations, first perform all calculations within parentheses and absolute values, or above and below the fraction bar. Then use the same order of operations to perform the remaining calculations.

1. Evaluate all exponential expressions. Do any negation *after* evaluating exponents.
2. Do all multiplication and division from *left to right*.
3. Do all addition and subtraction from *left to right*.

Example: $-2^4 - 2 \cdot 3 = \mathbf{-16} - 2 \cdot 3 = -16 - \mathbf{6} = \mathbf{-22}$

Scientific Notation   A number a written as $b \times 10^n$, where $1 \leq |b| < 10$ and n is an integer.

Examples: $23{,}400 = 2.34 \times 10^4$ and $0.0034 = 3.4 \times 10^{-3}$

SECTION 1.4 ■ VARIABLES, EQUATIONS, AND FORMULAS

Terminology

Variable	Symbol that represents an unknown quantity
	Examples: x, y, z, A, and T
Algebraic Expression	Can consist of numbers, variables, operation symbols, exponents, and grouping symbols but *no* equals sign
	Examples: $3z$, $(x - y)^3$, $4a + 3b$, 5, and $\|x - 2\|$
Equation	A statement that two algebraic expressions are *equal*; always contains an equals sign
	Examples: $2 + 4 = 6$, $2t = 8$, and $x^2 + 2 = 10$
Formula	An equation used to calculate one quantity by using known values of other quantities
	Examples: $P = 2W + 2L$ and $A = \pi r^2$

Square Root   The number b is a square root of a number a if $b^2 = a$.

Example: The square roots of 36 are 6 and -6.

Principal Square Root   The positive square root of a number, denoted $\sqrt{a}$

Examples: $\sqrt{4} = 2$, $\sqrt{100} = 10$,   and $\pm\sqrt{81} = \pm 9 = 9$ or -9

Cube Root   The number b is a cube root of a number a if $b^3 = a$.

Examples: $\sqrt[3]{8} = 2$, $\sqrt[3]{-27} = -3$,   and $\sqrt[3]{64} = 4$

SECTION 1.5 ■ INTRODUCTION TO GRAPHING

Relation   A set of ordered pairs

Example: $S = \{(-2, 3), (0, 3), (1, 2)\}$

Domain and Range   In a relation consisting of ordered pairs (x, y), the set of x-values is the domain and the set of y-values is the range.

Example: For $S = \{(-2, 3), (0, 3), (1, 2)\}$, $D = \{-2, 0, 1\}$ and $R = \{2, 3\}$.

The Cartesian Coordinate System (xy-plane)

Points	Plotted as (x, y) ordered pairs
Four Quadrants	I, II, III, and IV; the axes do *not* lie in a quadrant.
	NOTE: The point $(1, 0)$ does not lie in a quadrant.

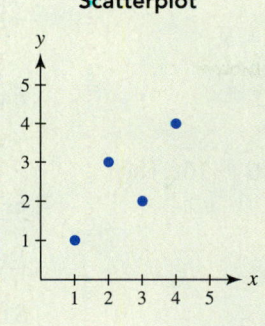

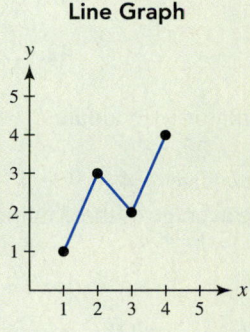

CHAPTER 1   Review Exercises

SECTION 1.1

Exercises 1 and 2: Classify each real number as one or more of the following: natural number, whole number, integer, rational number, or irrational number.

1. $-2, 9, \frac{2}{5}, \sqrt{11}, \pi, 2.68$

2. $\frac{6}{2}, -\frac{2}{7}, \sqrt{6}, 0.\overline{3}, \frac{0}{4}$

3. (True or False?) The number $\sqrt{3}$ is a rational number.

4. (True or False?) The number π is a real number.

5. $\sqrt{81} = $ _____

6. $\pm\sqrt{9} = $ _____

7. $\sqrt[3]{-64} = $ _____

8. $-\sqrt[3]{27} = $ _____

Exercises 9–14: State whether the equation illustrates an identity, commutative, associative, or distributive property.

9. $a \cdot 1 = a$

10. $4 \cdot x = x \cdot 4$

11. $(a + 1) + 4 = a + (1 + 4)$

12. $a(b + 2) = a \cdot b + a \cdot 2$

13. $3(t + 2) - r(t + 2) = (3 - r)(t + 2)$

14. $4x + 8x = 12x$

15. Use identity properties to simplify $1 \cdot (a + 0)$.

16. Use a commutative property to write $x \cdot \frac{1}{4}$ as an equivalent expression.

17. Use an associative property to write $8(10x)$ as an equivalent expression.

18. Use a distributive property to write $5z - 3z$ as an equivalent expression.

Exercises 19 and 20: Evaluate the expression two different ways by applying a distributive property.

19. $5(8 + 11)$

20. $3(9 - 5)$

Exercises 21 and 22: Calculate the average of the list of numbers.

21. 6, 9, 3, 11, 5, 20

22. 3.2, 6.8, 6.1, 10.8, 1.7

Exercises 23–26: Use properties of real numbers to evaluate the expression mentally.

23. $12 + 23 + (-2) + 7$

24. $\frac{3}{2} \cdot \frac{2}{5} \cdot \frac{5}{7} \cdot \frac{7}{3}$

25. $5 \cdot 23 + 5 \cdot 7$

26. $45 - 34 + 55 - 66$

SECTION 1.2

27. Plot the numbers $-3, 0, 2,$ and $\frac{7}{2}$ on a number line.

28. Evaluate the expression $|-7.2 + 4|$.

29. Find the additive inverse of $-\frac{2}{3}$.

30. Find the multiplicative inverse of $\frac{4}{5}$.

31. Find the opposite of $-2x + 3$.

32. Find the reciprocal of $-\frac{1}{a + b}$.

Exercises 33–42: Simplify the expression.

33. $-5 + (-7) + 8$

34. $-9 + 11$

35. $-12 - (-8)$

36. $\frac{1}{2} + (-2) + \frac{3}{4}$

37. $\frac{2}{3} \div (-4) - \frac{1}{3}$

38. $-\frac{7}{11} + \dfrac{\frac{1}{5}}{\frac{2}{9}}$

39. $-7 \cdot 4 \cdot -\frac{1}{7}$

40. $\frac{3}{4} \cdot \frac{4}{7} + 1$

41. $4x + 5x$

42. $2z - 3z + 8z$

43. Use a calculator to evaluate $\frac{1}{5} - \frac{3}{7} + \frac{\frac{2}{13}}{\frac{4}{5}}$.

44. Mentally evaluate $-4 + 9 + 4 + 11 - 6 + 16$. Then estimate the average of these numbers.

SECTION 1.3

45. Identify the base and the exponent for 4^{-2}.

46. Use a calculator to decide if 3^π and π^3 are equal.

Exercises 47–54: Evaluate the expression.

47. -2^4

48. $(-2)^4$

49. 9^0

50. $\left(\frac{2}{3}\right)^{-3}$

51. 4^{-3}

52. $\frac{1}{5^{-2}}$

53. $\frac{5^{-3}}{3^{-2}}$

54. $\frac{1}{2 \cdot 4^{-2}}$

Exercises 55–70: Simplify the expression. Write the result using positive exponents.

55. $4^3 \cdot 4^{-5}$

56. $10^4 \cdot 10^{-2}$

57. $x^7 x^{-2}$

58. $\frac{3^4}{3^{-7}}$

59. $\frac{5a^{-4}}{10a^2}$

60. $\frac{15a^4 b^3}{3a^2 b^6}$

61. $(2^2)^4$

62. $(x^{-3})^5$

63. $(4x^{-2}y^3)^2$

64. $(4a)^5$

65. $\left(\frac{5x^3}{3z^4}\right)^3$

66. $\left(\frac{-3x^4 y^3}{z}\right)^{-2}$

67. $\left(\frac{3a^{-4}}{4b^{-7}}\right)^2$

68. $\left(\frac{3m^2 n^{-4}}{9m^3 n}\right)^{-1}$

69. $\left(\frac{rt}{2r^3 t^{-1}}\right)^{-3}$

70. $\left(\frac{3r^2}{4t^{-3}}\right)^2$

Exercises 71–76: Evaluate each expression by hand.

71. $2 + 3 \cdot 9$

72. $4 - 1 - 6$

73. $5 \cdot 2^3$

74. $\frac{2+4}{2} + \frac{3-1}{3}$

75. $20 \div 4 \div 2$

76. $\frac{3^3 - 2^4}{4 - 3}$

Exercises 77 and 78: Write in scientific notation.

77. 186,000

78. 0.00034

Exercises 79 and 80: Write in standard (decimal) form.

79. 4.5×10^4

80. 9.23×10^{-3}

SECTION 1.4

Exercises 81 and 82: Write a formula for the following.

81. Converting x feet to y inches

82. Finding the total area A of 6 circles all with radius r

Exercises 83–88: Evaluate the formula for the given value(s) of the variable(s).

83. $y = 12x$ $x = 3$

84. $d = \sqrt{t - 3}$ $t = 67$

85. $N = h^2 - \frac{3}{4}$ $h = \frac{3}{2}$

86. $P = w^3 - 2$ $w = -2$

87. $A = \frac{1}{2}bh$ $b = 4, h = 5$

88. $V = a^2 b$ $a = 3, b = 3$

89. Select the formula (i)–(iii) that models the data in the table.

x	1	2	3	4	5
y	-1	1	3	5	7

(i) $y = x - 2$,   (ii) $y = 3x - 4$,
(iii) $y = 2x - 3$

90. Find a value for a so that $y = ax$ models the data.

x	-2	0	2	4	6
y	-3	0	3	6	9

91. Use $y = x^3 + 1$ to complete the table.

x	-2	-1	0	1	2
y					

92. Use a calculator to make a table of $y = \sqrt{\frac{x-1}{x+1}}$. Let $x = 1, 2, 3, \dots, 7$.

SECTION 1.5

93. Identify the domain and range of the relation given by $S = \{(-1, 1), (2, 3), (3, -6), (3, 7)\}$.

94. Express the relation shown in the graph as a set of ordered pairs. Identify the domain and range.

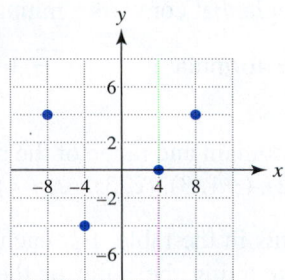

Exercises 95–98: Evaluate the formula for $x = -2,$ $-1, 0, 1,$ *and* 2. *Plot the resulting ordered pairs.*

95. $y = -3x$

96. $y = \frac{1}{2}x - 1$

97. $y = x^2$

98. $y = \dfrac{5}{x^2 + 1}$

Exercises 99 and 100: Plot the points in the table. For each point, state the quadrant containing the point or the axis on which the point lies.

99.

x	-2	-1	0	2
y	2	-3	1	-1

100.

x	-15	-5	10	20
y	-5	0	20	-10

Exercises 101 and 102: Show the viewing rectangle on your graphing calculator. Predict the number of tick marks on the positive x-axis and the positive y-axis.

101. $[-9, 9, 1]$ by $[-6, 6, 3]$

102. $[-20, 20, 5]$ by $[-12, 12, 4]$

Exercises 103 and 104: Make a scatterplot of the relation. Identify the domain and range.

103. $\{(0, 4), (-1, 2), (3, 3), (4, -1), (2, 0)\}$

104. $\{(-10, 10), (50, 20), (-20, -30), (45, -25)\}$

APPLICATIONS

105. *Richest People* The following table lists the net worth in billions of dollars of the four richest people in 2010.

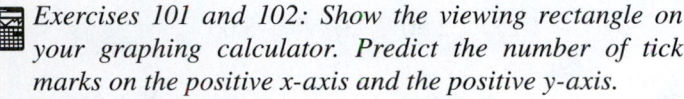

Net Worth ($ Billions)	29.0	47.0	53.0	53.5

(a) Is their average more or less than $50 billion?

(b) Find their average net worth.

106. *Calculating Interest* If P dollars are deposited in a savings account paying 1% annual interest, the amount A in the account after t years is $A = P(1.01)^t$. Find A for the given values of P and t.
(a) $P = \$2500, t = 1$ year
(b) $P = \$800, t = 2$ years

107. *Speed of Earth* Earth orbits the sun in a nearly circular orbit with a radius of 93,000,000 miles.
(a) Calculate the distance in miles traveled by Earth in 1 year. Write your answer in scientific notation. (*Hint:* $C = 2\pi r$.)
(b) Estimate the speed of Earth around the sun in miles per hour.

108. *Modeling Motion* The following table lists the distances d in miles traveled by a car for various elapsed times t. Find an equation that models these data.

t (hr)	2	4	6	8
d (mi)	80	160	240	320

109. *Heart Beat in Animals* The rate N at which an animal's heart beats varies with its weight. This relation can be modeled by $N = \frac{885}{\sqrt{W}}$, where N is in beats per minute and W is the animal's weight in pounds. Estimate the pulse for a 16-pound cat and a 144-pound person. (*Source:* C. Pennycuick, *Newton Rules Biology*.)

110. *Graphing Real Data* The following data show the poverty threshold y for a single person for selected years. Make a line graph and comment on any trends in the data.

x	1970	1980	1990	2000	2010
y	$1954	$4190	$6652	$8794	$10,830

Source: U.S. Census Bureau.

111. *Area* Find the area of a square whose sides have length $4ab$.

112. *Volume* Find the volume of a cube whose sides have length $5z$.

CHAPTER 1   Test

Step-by-step test solutions are found on the Chapter Test Prep Videos available in **MyMathLab** and on **YouTube** (search "RockswoldInterAlg" and click on "Channels").

1. Classify each of the real numbers as one or more of the following: natural number, whole number, integer, rational number, or irrational number.

$$-5, \frac{2}{3}, -\frac{1}{\sqrt{5}}, \sqrt{9}, \pi, -1.83$$

2. State whether each equation illustrates an identity, commutative, associative, or distributive property.
   (a) $a + 0 = a$ (b) $4(12x) = 48x$
   (c) $5(2 + 3x) = 5(3x + 2)$
   (d) $a(x - y) = ax - ay$

3. Use identity, commutative, associative, and/or distributive properties to simplify the expression.
   (a) $0 + (a \cdot 1) \cdot 1$ (b) $2x - 5x + 4x$
   (c) $2(x \cdot 3) + 2x$ (d) $(a + b)a - ab$

4. Calculate the average for 34, 15, 96, 11, and 0.

5. Plot the numbers $-1.5, 0, 3,$ and $\frac{3}{2}$ on a number line.

6. Evaluate each expression.
   (a) $\left| \frac{1}{2} + \frac{2}{3} - \frac{8}{3} \right|$
   (b) $12 + (-3) + 18 + (-7)$
   (c) $\frac{1}{2} \cdot \frac{2}{3} \cdot \frac{3}{4} \cdot \frac{3}{2} \cdot 2$
   (d) $\frac{1}{2} + \frac{2}{3} - \frac{3}{2} + \frac{1}{3}$

7. Evaluate each expression.
   (a) $\sqrt{36}$ (b) $\sqrt[3]{-8}$

8. Find the multiplicative inverse of each expression.
   (a) $-\frac{5}{4}$ (b) $\frac{2}{b + 1}$

9. Evaluate each expression.
   (a) $-\frac{1}{2} + \frac{2}{3} \div 3$ (b) $-4 + \frac{1 - 3}{2 + 8}$
   (c) $5 - 2 \cdot 5^2 \div 5$ (d) $(6 - 4 \cdot 5) \div (-7)$

10. Evaluate each expression.
    (a) 5^{-2} (b) π^0 (c) $\left(-\frac{2}{5} \right)^4$
    (d) $\frac{6^{-2}}{2^{-4}}$ (e) $\frac{1}{5^{-3}}$ (f) $2^2 \cdot 2^{-4}$

11. Simplify each expression. Use positive exponents.
    (a) $x^6 \cdot x^{-4} \cdot y^3$ (b) $\frac{16x^{-2}y^8}{6xy^{-7}}$
    (c) $(2yz^{-2})^3$ (d) $\left(\frac{15x^4}{10xy^{-2}} \right)^{-2}$

12. Write 5.2×10^{-4} in standard (decimal) form.

13. Write 3,400,000 in scientific notation.

14. Write a formula that converts x minutes to H hours.

15. Evaluate the formula $y = \sqrt{x + 1} + x$ for $x = 3$ and $x = -1$.

16. Identify the domain and range of the relation given by $S = \{(-3, 2), (-1, 3), (2, 3), (2, -4)\}$.

17. Plot the points in the table. For each point, state the quadrant containing the point or the axis on which the point lies.

x	-2	-1	0	1	2
y	3	-2	2	-1	1

18. Express the relation shown in the graph as a set of ordered pairs. Identify the domain and range.

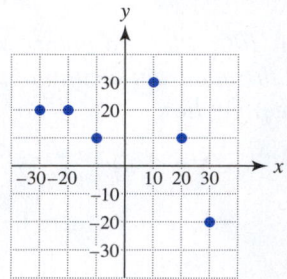

19. Make a scatterplot of the data. Write an equation that models the data.

x	1	2	3	4
y	1.25	2.50	3.75	5.00

20. Use your calculator to make a scatterplot of S after determining an appropriate viewing rectangle.
    $S = \{(-15, 5), (10, 20), (-10, 15), (0, 0), (5, 35)\}$

21. *Temperature* The formula $C = \frac{5}{9}(F - 32)$ can be used to convert degrees Fahrenheit F to degrees Celsius C. If the outside temperature is 5°F find the equivalent temperature in Celsius.

22. *Alcohol Consumption* In 2007, about 241 million people in the United States were age 14 or over. They consumed, on average, 2.14 gallons of alcohol per person. Use scientific notation to estimate the total gallons of alcohol consumed by this age group.
    (*Source:* Department of Health and Human Services.)

23. *Radius of a Circle* If a circle has area A, then its radius r is given by $r = \sqrt{\frac{A}{\pi}}$. Find the radius of a circle with an area of 25 square feet. Approximate this radius to the nearest hundredth of a foot.

24. *Modeling Motion* The following table lists the distances in miles M traveled by a car for various elapsed times t. Find an equation that models these data.

t (hr)	4	6	10
M (mi)	244	366	610

CHAPTER 1   Extended and Discovery Exercises

1. *Global Warming* If the global climate were to warm significantly as a result of the greenhouse effect or other climatic change, the Arctic ice cap would start to melt. This ice cap contains an estimated 680,000 cubic miles of water. More than 200 million people currently live on land that is less than 3 feet above sea level. In the United States several large cities have low average elevations. Three examples are Boston (14 feet), New Orleans (4 feet), and San Diego (13 feet). In this exercise you are to estimate the rise in sea level if this cap were to melt and determine whether this event would have a significant impact on people.

(a) The surface area of a sphere is given by the formula $4\pi r^2$, where r is its radius. Although the shape of the earth is not exactly spherical, it has an average radius of 3960 miles. Estimate the surface area of the earth.

(b) Oceans cover approximately 71% of the total surface area of the earth. How many square miles of the earth's surface are covered by oceans?

(c) Approximate the potential rise in sea level by dividing the total volume of the water from the ice cap by the surface area of the oceans. Convert your answer from miles to feet.

(d) Discuss the implications of your calculation. How would cities such as Boston, New Orleans, and San Diego be affected?

(e) The Antarctic ice cap contains 6,300,000 cubic miles of water. Estimate how much sea level would rise if this ice cap melted. (*Source:* Department of the Interior, Geological Survey.)

2. *Fatal Injuries at Work* The following table lists the fatal injuries per 100,000 full-time workers in private industry.

Year	2006	2007	2008	2009
Injuries	4.2	4.0	3.7	3.3

Source: Bureau of Labor Statistics.

Explain your reasoning for each of the following.

(a) Make a scatterplot of the data. Discuss how the injury rate has changed over this period of time.

(b) Estimate the injury rate in 2005.

(c) Assuming that trends continued, estimate the injury rate in 2010.

(d) Would it be valid to try to estimate when the accident rate will reach 0? Explain.

3. *Body Mass Index* Many studies have tried to find a recommended relationship between a person's height and weight. The following steps may be used to compute the body mass index (BMI). Federal guidelines suggest that $19 \leq \text{BMI} \leq 25$ is desirable. (*Source:* Associated Press.)

Step 1: Multiply the weight W in pounds by 0.455.
Step 2: Multiply the height H in inches by 0.0254.
Step 3: Square the result in Step 2.
Step 4: Divide the answer in Step 1 by the answer in Step 3.

The result is the person's BMI. Write a formula to calculate the BMI given the height H and weight W of a person. Let y represent the BMI.

Exercises 4–6: Body Mass Index (Refer to the previous exercise.) Compute the BMI for each individual.

4. 123 pounds, 5 feet 8 inches (Anna Kournikova, tennis player) (*Source:* Netglimse).

5. 160 pounds, 6 feet 1 inch (Venus Williams, tennis player) (*Source:* CBS SportsLine.)

6. 300 pounds, 7 feet 1 inch (Shaquille O'Neal, professional basketball player) (*Source:* The Topps Company, Inc.)

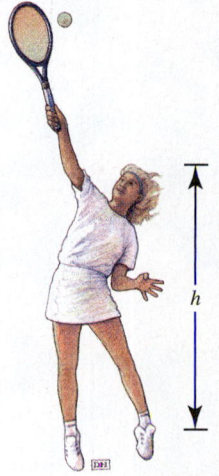

Exercises 7–10: **Modeling Data with Formulas** *Find values for a and b so that the formula models the data.*

7. $y = ax + b$

x	-3	-1	1	3
y	5	2	-1	-4

8. $y = a\sqrt{x} + b$

x	1	4	9	16
y	1	-1	-3	-5

9. $y = ax^2 + bx$

x	1	2	3	4
y	1	3	6	10

10. $y = b(a^x)$

x	0	1	2	3
y	3	6	12	24

2 Linear Functions and Models

> *"Our competitive advantage is our math skills, which is probably not something you would expect of a media company."*
>
> —MAX LEVCHIN,
> CEO OF SLIDE
>
> (Slide is the number one company for writing Facebook applications.)

Every day millions of people create trillions of bytes of information. The only way we can make sense out of these data and determine what is occurring within society is to use mathematics. One of the most important mathematical concepts used to discover trends and patterns is that of a *function*. A function typically receives an input (or question), performs a computation, and gives the output (or answer).

Functions have been used in science and engineering for centuries to answer questions related to things like eclipses, communication, and transportation. However, today functions are also being used to describe human behavior and to design social networks. (See Section 2.1, Exercise 79.) In fact, you may have noticed that new features available on Twitter and Facebook are sometimes referred to as applications or *functions*. People are creating thousands of new functions every day. *Math skills are essential* for writing successful applications and functions.

2.1 Functions and Their Representations

Basic Concepts • Representations of a Function • Definition of a Function • Identifying a Function • Tables, Graphs, and Calculators (Optional)

A LOOK INTO MATH ▶

In Chapter 1 we showed how to use numbers to describe data. For example, instead of simply saying that there are *a lot* of people on Twitter, we might say that there are about 50 million tweets per day. A number helps explain what "a lot" means. We also showed that data can be summarized with formulas and graphs. Formulas and graphs are sometimes used to represent *functions*, which are essential in mathematics. In this section we introduce functions and their representations.

Basic Concepts

NEW VOCABULARY

☐ Function
☐ Function notation
☐ Input/Output
☐ Name of the function
☐ Dependent variable
☐ Independent variable
☐ Verbal representation
☐ Numerical representation
☐ Symbolic representation
☐ Graphical representation
☐ Diagrams/Diagrammatic representation
☐ Domain/Range
☐ Vertical line test

▶ **REAL-WORLD CONNECTION** Functions are used to calculate many important quantities. For example, suppose that a person works for $7 per hour. Then we could use a function *named* f to calculate the amount of money the person earned after working x hours simply by multiplying the *input* x by 7. The result y is called the *output*. This concept is shown visually in the following diagram.

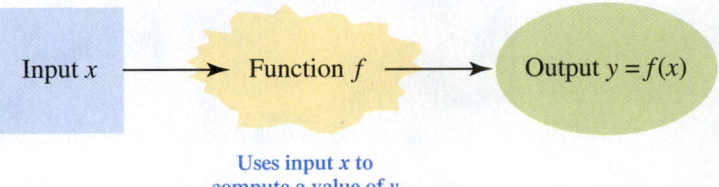

Input x → Function f → Output $y = f(x)$

Uses input x to
compute a value of y

For each valid input x, a function computes *exactly one* output y, which may be represented by the ordered pair (x, y). If the input is 5 hours, f outputs $7 \cdot 5 = \$35$; if the input is 8 hours, f outputs $7 \cdot 8 = \$56$. These results can be represented by the ordered pairs $(5, 35)$ and $(8, 56)$. Sometimes an input may not be valid. For example, if $x = -3$, there is no reasonable output because a person cannot work -3 hours.

We say that *y is a function of x* because the output y is determined by and *depends* on the input x. As a result, y is called the *dependent variable* and x is the *independent variable*. To emphasize that y is a function of x, we use the notation $y = f(x)$. The symbol $f(x)$ does not represent multiplication of a variable f and a variable x. The notation $y = f(x)$ is called *function notation*, is read "y equals f of x," and means that function f with input x produces output y. For example, if $x = 3$ hours, $y = f(3) = \$21$.

FUNCTION NOTATION

The notation $y = f(x)$ is called **function notation**. The **input** is x, the **output** is y, and the **name of the function** is f.

$$\text{Name}$$
$$\downarrow$$
$$y = f(x)$$
$$\text{Output} \quad \text{Input}$$

The variable y is called the **dependent variable** and the variable x is called the **independent variable**. The expression $f(4) = 28$ is read "f of 4 equals 28" and indicates that f outputs 28 when the input is 4. A function computes *exactly one* output for each valid input. The letters f, g, and h are often used to denote names of functions.

NOTE: Functions can be given *meaningful* names and variables. For example, function f could have been defined by $P(h) = 7h$, where function P calculates the pay after working h hours for \$7 per hour.

▶ **REAL-WORLD CONNECTION** Functions can be used to compute a variety of quantities. For example, suppose that a boy has a sister who is exactly 5 years older than he is. If the age of the boy is x, then a function g can calculate the age of his sister by adding 5 to x. Thus $g(4) = 4 + 5 = 9$, $g(10) = 10 + 5 = 15$, and in general $g(x) = x + 5$. That is, function g adds 5 to input x to obtain the output $y = g(x)$.

Functions can be represented by an input–output machine, as illustrated in Figure 2.1. This machine represents function g and receives input $x = 4$, adds 5 to this value, and then outputs $g(4) = 4 + 5 = 9$.

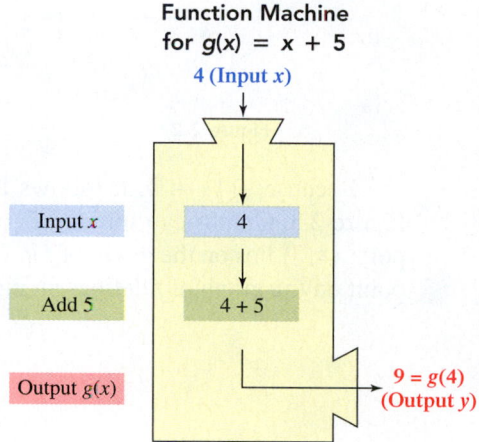

Figure 2.1

Representations of a Function

▶ **REAL-WORLD CONNECTION** A function f forms a relation between inputs x and outputs y that can be represented verbally, numerically, symbolically, and graphically. Functions can also be represented with diagrams. We begin by considering a function f that converts yards to feet.

VERBAL REPRESENTATION (WORDS) To convert x yards to y feet we multiply x by 3. Therefore, if function f computes the number of feet in x yards, a **verbal representation** of f is "Multiply the input x in yards by 3 to obtain the output y in feet."

TABLE 2.1

x (yards)	y (feet)
1	3
2	6
3	9
4	12
5	15
6	18
7	21

NUMERICAL REPRESENTATION (TABLE OF VALUES) A function f that converts yards to feet is shown in Table 2.1, where $y = f(x)$.

A *table of values* is called a **numerical representation** of a function. Many times it is impossible to list all valid inputs x in a table. On the one hand, if a table does not contain every x-input, it is a *partial* numerical representation. On the other hand, a *complete* numerical representation includes *all* valid inputs. Table 2.1 is a partial numerical representation of f because many valid inputs, such as $x = 10$ or $x = 5.3$, are not shown in it. Note that for each valid input x there is exactly one output y. *For a function, inputs are not listed more than once in a table.*

SYMBOLIC REPRESENTATION (FORMULA) A *formula* provides a **symbolic representation** of a function. The computation performed by f to convert x yards to y feet is expressed by $y = 3x$. A formula for f is $f(x) = 3x$, where $y = f(x)$. We say that function f is *defined by* or *given by* $f(x) = 3x$. Thus $f(2) = 3 \cdot 2 = 6$.

GRAPHICAL REPRESENTATION (GRAPH) A **graphical representation**, or **graph**, visually associates an x-input with a y-output. The ordered pairs

$$(1, 3), (2, 6), (3, 9), (4, 12), (5, 15), (6, 18), \text{ and } (7, 21)$$

from Table 2.1 are plotted in Figure 2.2(a). This scatterplot suggests a line for the graph f. For each real number x there is exactly one real number y determined by $y = 3x$. If we restrict inputs to $x \geq 0$ and plot all ordered pairs $(x, 3x)$, then a line with no breaks will appear, as shown in Figure 2.2(b).

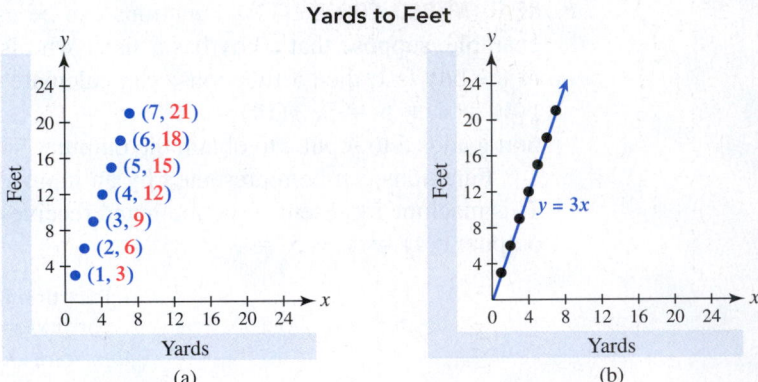

Figure 2.2

Because $f(1) = 3$, it follows that the point $(1, 3)$ lies on the graph of f, as shown in Figure 2.3. Graphs can sometimes be used to define a function f. For example, because the point $(1, 3)$ lies on the graph of f in Figure 2.3, we can conclude that $f(1) = 3$. That is, each point on the graph of f defines an input–output pair for f.

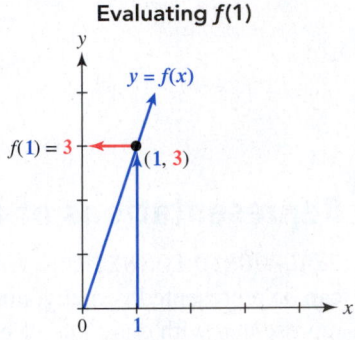

Figure 2.3

MAKING CONNECTIONS

Functions, Points, and Graphs

If $f(a) = b$, then the point (a, b) lies on the graph of f. Conversely, if the point (a, b) lies on the graph of f, then $f(a) = b$. See Figure 2.4(a). Thus each point on the graph of f can be written in the form $(a, f(a))$. See Figure 2.4(b).

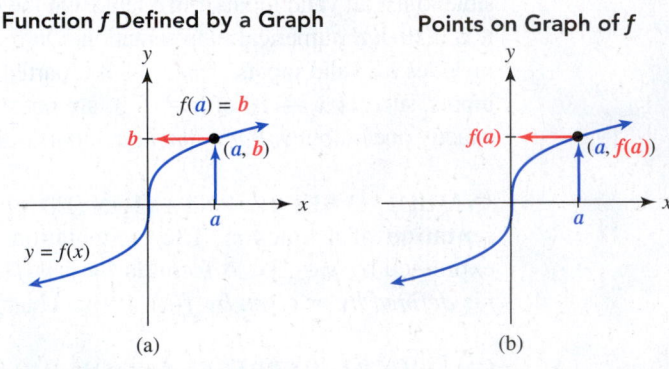

Figure 2.4

Yards to Feet

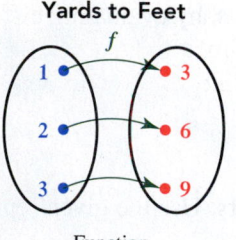

Function

Figure 2.5

DIAGRAMMATIC REPRESENTATION (DIAGRAM) Functions may be represented by **diagrams**. Figure 2.5 is a diagram of a function where an arrow is used to identify the output y associated with input x. For example, an arrow is drawn from input **2** to output **6**, which is written in function notation as $f(2) = 6$. That is, **2** yards are equivalent to **6** feet.

Figure 2.6(a) shows a (different) function f even though $f(1) = 4$ and $f(2) = 4$. Although two inputs for f have the same output, each valid input has exactly one output. In contrast, Figure 2.6(b) shows a relation that is *not* a function because input 2 results in two different outputs, 5 and 6.

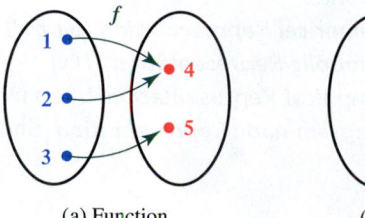   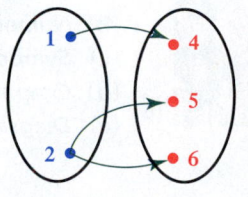

(a) Function   (b) Not a Function

Figure 2.6

MAKING CONNECTIONS

Four Representations of a Function

Symbolic Representation $f(x) = x + 1$

Numerical Representation

x	y
-2	-1
-1	0
0	1
1	2
2	3

Graphical Representation

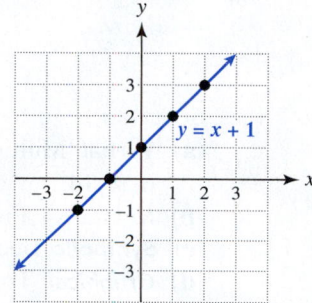

$y = x + 1$

Verbal Representation f adds 1 to an input x to produce an output y.

EXAMPLE 1   **Evaluating symbolic representations (formulas)**

Evaluate each function f at the given value of x.
(a) $f(x) = 3x - 7$ $x = -2$

(b) $f(x) = \dfrac{x}{x + 2}$ $x = 0.5$

(c) $f(x) = \sqrt{x - 1}$ $x = 10$

Solution
(a) $f(-2) = 3(-2) - 7 = -6 - 7 = -13$

(b) $f(0.5) = \dfrac{0.5}{0.5 + 2} = \dfrac{0.5}{2.5} = 0.2$

(c) $f(10) = \sqrt{10 - 1} = \sqrt{9} = 3$

Now Try Exercises **21, 23, 31**

▶ **REAL-WORLD CONNECTION** In the next example we calculate sales tax by evaluating different representations of a function.

EXAMPLE 2   **Calculating sales tax**

Let a function f compute a sales tax of 7% on a purchase of x dollars. Use the given representation to evaluate $f(2)$.

(a) *Verbal Representation* Multiply a purchase of x dollars by 0.07 to obtain a sales tax of y dollars.

(b) *Numerical Representation (partial)* Shown in Table 2.2

(c) *Symbolic Representation* $f(x) = 0.07x$

(d) *Graphical Representation* Shown in Figure 2.7

(e) *Diagrammatic Representation* Shown in Figure 2.8

TABLE 2.2

x	$f(x)$
$1.00	$0.07
$2.00	$0.14
$3.00	$0.21
$4.00	$0.28

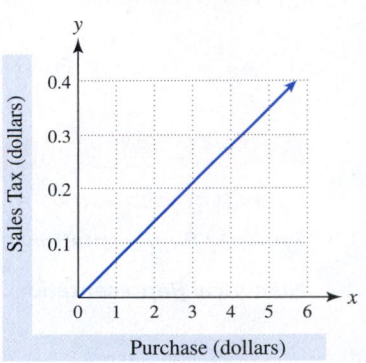

Sales Tax of 7%

Figure 2.7

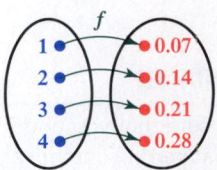

Figure 2.8

Solution

(a) *Verbal* Multiply the input 2 by 0.07 to obtain 0.14. The sales tax on a $2.00 purchase is $0.14.

(b) *Numerical* From Table 2.2, $f(2) = \$0.14$.

(c) *Symbolic* Because $f(x) = 0.07x$, $f(2) = 0.07(2) = 0.14$, or $0.14.

(d) *Graphical* To evaluate $f(2)$ with a graph, first find **2** on the x-axis in Figure 2.7. Then move vertically upward until you reach the graph of f. The point on the graph may be estimated as $(2, 0.14)$, meaning that $f(2) = 0.14$ (see Figure 2.9). Note that it may not be possible to find the exact answer from a graph. For example, one might estimate $f(2)$ to be 0.13 or 0.15 instead of 0.14.

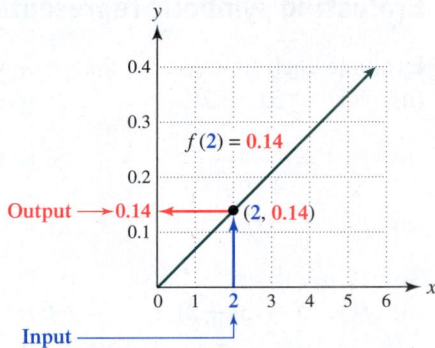

Evaluating a Function

Figure 2.9

(e) *Diagrammatic* In Figure 2.8, follow the arrow from **2** to **0.14**. Thus $f(2) = 0.14$.

▌ **Now Try Exercises 25, 33, 53, 59, 61**

▶ **REAL-WORLD CONNECTION** There are many examples of functions. To give more meaning to a function, sometimes we change both its name and its input variable. For instance, if we know the radius r of a circle, we can calculate its circumference by using $C(r) = 2\pi r$. The next example illustrates how functions are used in physical therapy.

EXAMPLE 3   **Computing crutch length**

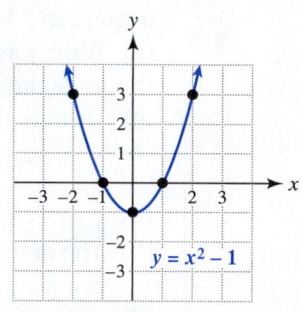

People who sustain leg injuries often require crutches. A proper crutch length can be estimated without using trial and error. The function L, given by $L(t) = 0.72t + 2$, outputs an appropriate crutch length in inches for a person t inches tall. (*Source: Journal of the American Physical Therapy Association.*)

(a) Find $L(60)$ and interpret the result.
(b) If one person is 70 inches tall and another person is 71 inches tall, what should be the difference in their crutch lengths?

Solution
(a) $L(\mathbf{60}) = 0.72(\mathbf{60}) + 2 = \mathbf{45.2}$. Thus a person 60 inches tall needs crutches that are about 45.2 inches long.
(b) From the formula $L(t) = 0.72t + 2$, we can see that each 1-inch increase in t results in a 0.72-inch increase in $L(t)$. For example,

$$L(71) - L(70) = 53.12 - 52.4 = 0.72.$$

Now Try Exercise 79

In the next example we find a formula and then sketch a graph of a function.

EXAMPLE 4   **Finding representations of a function**

Let function f square the input x and then subtract 1 to obtain the output y.
(a) Write a formula, or symbolic representation, for f.
(b) Make a table of values, or numerical representation, for f. Use $x = -2, -1, 0, 1, 2$.
(c) Sketch a graph, or graphical representation, of f.

Solution
(a) **Symbolic Representation** If we square x and then subtract 1, we obtain $x^2 - 1$. Thus a formula for f is $f(x) = x^2 - 1$.
(b) **Numerical Representation** Make a table of values for $f(x)$, as shown in Table 2.3. For example,

$$f(\mathbf{-2}) = (\mathbf{-2})^2 - 1 = 4 - 1 = \mathbf{3}.$$

(c) **Graphical Representation** To obtain a graph of $f(x) = x^2 - 1$, plot the points from Table 2.3 and then connect them with a smooth curve, as shown in Figure 2.10. Note that we need to plot enough points so that we can determine the overall shape of the graph.

TABLE 2.3

x	$f(x)$
-2	3
-1	0
0	-1
1	0
2	3

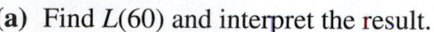

Figure 2.10

Now Try Exercise 63

READING CHECK

Give a verbal, numerical, symbolic, and graphical representation of a function that calculates the number of days in x weeks. Choose meaningful variables.

Definition of a Function

A function is a fundamental concept in mathematics. Its definition should allow for all representations of a function. *A function receives an input x and produces exactly one output y,* which can be expressed as an ordered pair:

$$(x, y).$$

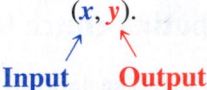

Input   Output

A relation is a set of ordered pairs, and a function is a special type of relation.

FUNCTION

A **function** f is a set of ordered pairs (x, y) where each x-value corresponds to exactly one y-value.

The **domain** of f is the set of all x-values, and the **range** of f is the set of all y-values. For example, a function f that converts 1, 2, 3, and 4 yards to feet could be expressed as

$$f = \{(1, 3), (2, 6), (3, 9), (4, 12)\}.$$

The domain of f is $D = \{1, 2, 3, 4\}$, and the range of f is $R = \{3, 6, 9, 12\}$.

MAKING CONNECTIONS

Relations and Functions

A relation can be thought of as a set of input–output pairs. A function is a special type of relation whereby each input results in exactly one output.

▶ **REAL-WORLD CONNECTION** In the next example, we see how education can improve a person's chances for earning a higher income.

EXAMPLE 5   ### Computing average income

A function f computes the average individual income in dollars in relation to educational attainment. This function is defined by $f(N) = 21,484$, $f(H) = 31,286$, $f(B) = 57,181$, and $f(M) = 70,181$, where N denotes no diploma, H a high school diploma, B a bachelor's degree, and M a master's degree. (*Source: 2010 Statistical Abstract.*)

(a) Write f as a set of ordered pairs.
(b) Give the domain and range of f.
(c) Discuss the relationship between education and income.

Solution

(a) $f = \{(N, 21484), (H, 31286), (B, 57181), (M, 70181)\}$
(b) The domain of function f is given by $D = \{N, H, B, M\}$, and the range of function f is given by $R = \{21484, 31286, 57181, 70181\}$.
(c) Education pays—the greater the educational attainment, the greater are annual earnings.

Now Try Exercise 105

| EXAMPLE 6 | **Finding the domain and range graphically** |

Use the graphs of f shown in Figures 2.11 and 2.12 to find each function's domain and range.

(a)

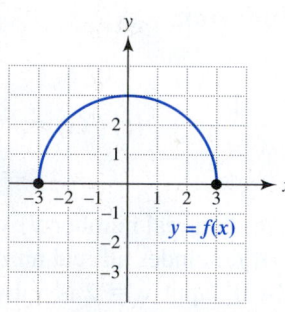

Figure 2.11

(b)

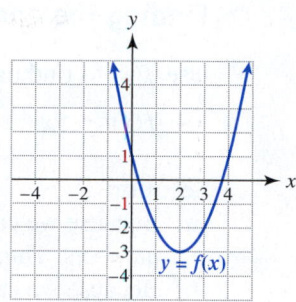

Figure 2.12

READING CHECK

Use the graph in
Figure 2.12 to evaluate $f(3)$.

Solution

(a) The domain is the set of all x-values that correspond to points on the graph of f. Figure 2.13 shows that the domain D includes all x-values satisfying $-3 \leq x \leq 3$. (Recall that the symbol $\leq$ is read "*less than or equal to*.") Because the graph is a semi-circle with no breaks, the domain includes all real numbers between and including -3 and 3. The range R is the set of y-values that correspond to points on the graph of f. Thus R includes all y-values satisfying $0 \leq y \leq 3$.

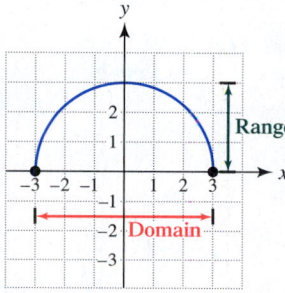

Figure 2.13

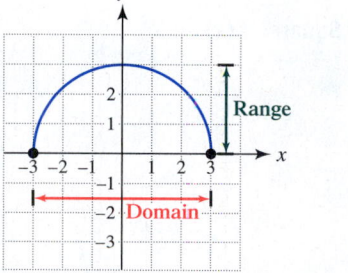

Figure 2.14

(b) The arrows on the ends of the graph in Figure 2.12 indicate that the graph extends indefinitely left and right, as well as upward. Thus D includes all real numbers. See Figure 2.14. The smallest y-value on the graph is $y = -3$, which occurs when $x = 2$. Thus the range R is $y \geq -3$. (Recall that the symbol $\geq$ is read "*greater than or equal to*.")

■ **Now Try Exercises 81, 85**

CRITICAL THINKING

Suppose that a car travels at 50 miles per hour to a city that is 250 miles away. Sketch a graph of a function f that gives the distance y traveled after x hours. Identify the domain and range of f.

The domain of a function is the set of all valid inputs. To determine the domain of a function from a formula, we must find x-values for which the formula is defined. To do this, we must determine if we can substitute any real number in the formula for $f(x)$. If we can, then the domain of f is *all real numbers*. However, there are situations in which we must limit the domain of f. For example, the domain must often be limited when there is either division or a square root in the formula for f. When division occurs, we must be careful to avoid values of the variable that result in division by 0, which is undefined. When a square root occurs, we must be careful to avoid values of the variable that result in the square root of a negative number, which is not a real number. This concept is demonstrated in the next example.

EXAMPLE 7   **Finding the domain of a function**

Use $f(x)$ to find the domain of f.

(a) $f(x) = 5x$ **(b)** $f(x) = \dfrac{1}{x - 2}$ **(c)** $f(x) = \sqrt{x}$

Solution

(a) Because we can always multiply a real number x by 5, $f(x) = 5x$ is defined for all real numbers. Thus the domain of f includes all real numbers.

(b) Because we cannot divide by 0, input $x = 2$ is not valid for $f(x) = \frac{1}{x-2}$. The expression for $f(x)$ is defined for all other values of x. Thus the domain of f includes all real numbers except 2, or $x \neq 2$.

(c) Because square roots of negative numbers are not real numbers, the inputs for $f(x) = \sqrt{x}$ cannot be negative. Thus the domain of f includes all nonnegative numbers, or $x \geq 0$.

■ **Now Try Exercises 91, 95, 99**

Symbolic, numerical, and graphical representations of three common functions are shown in Figure 2.15. Note that their graphs are not lines. For this reason they are called **nonlinear functions**. Use the graphs to find the domain and range of each function.

Absolute value: $f(x) = |x|$

x	-2	-1	0	1	2
$\|x\|$	2	1	0	1	2

Square: $f(x) = x^2$

x	-2	-1	0	1	2
x^2	4	1	0	1	4

Square root: $f(x) = \sqrt{x}$

x	0	1	4	9
$\sqrt{x}$	0	1	2	3

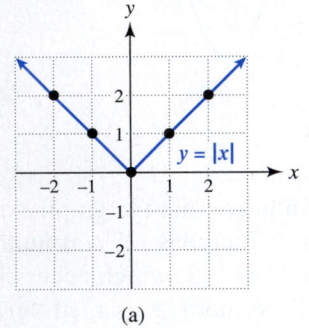

(a)

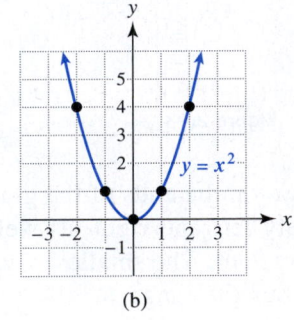

(b)

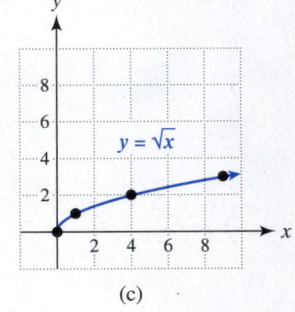

(c)

Figure 2.15

Identifying a Function

Recall that for a function each valid input x produces exactly one output y. In the next three examples we demonstrate techniques for identifying a function.

EXAMPLE 8   **Determining whether a set of ordered pairs is a function**

The set S of ordered pairs (x, y) represents the number of mergers and acquisitions y in 2010 for selected technology companies x.

$$S = \{(\text{IBM}, 12), (\text{HP}, 7), (\text{Oracle}, 5), (\text{Apple}, 5), (\text{Microsoft}, 0)\}$$

Determine if S is a function. (*Source:* cbinsights.)

Solution

The input x is the name of the technology company, and the output y is the number of mergers and acquisitions associated with that company. The set S *is* a function because each company x is associated with exactly one number y. Note that even though there were 5 mergers and acquisitions corresonding to both Oracle and Apple, S is nonetheless a function.

■ **Now Try Exercise 127**

| EXAMPLE 9 | **Determining whether a table of values represents a function** |

Determine whether Table 2.4 represents a function.

TABLE 2.4

x	y
1	−4
2	8
3	2
1	5
4	−6

Solution

The table does not represent a function because input $x = 1$ produces two outputs: -4 and 5. That is, the following two ordered pairs both belong to this relation.

Same input x

$(1, -4) \qquad (1, 5) \quad \leftarrow$ **Not a function**

Different outputs y

■ **Now Try Exercise 129**

VERTICAL LINE TEST To determine whether a graph represents a function, we must be convinced that it is impossible for an input x to have two or more outputs y. If two distinct points have the *same x*-coordinate on a graph, then the graph cannot represent a function. For example, the ordered pairs $(-1, 1)$ and $(-1, -1)$ could not lie on the graph of a function because input -1 results in *two* outputs: 1 and -1. When the points $(-1, 1)$ and $(-1, -1)$ are plotted, they lie on the same vertical line, as shown in Figure 2.16(a). A graph passing through these points intersects the vertical line twice, as illustrated in Figure 2.16(b).

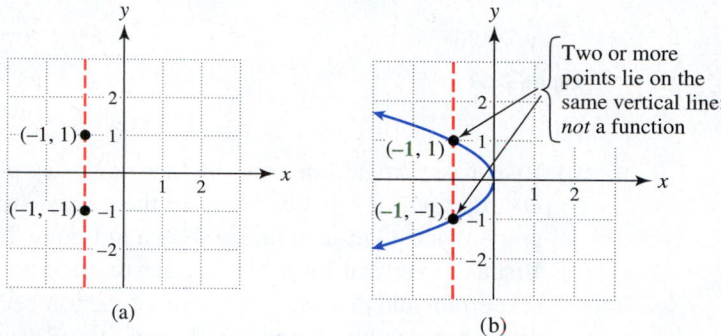

(a)   (b)

Figure 2.16

To determine whether a graph represents a function, visualize vertical lines moving across the *xy*-plane. If each vertical line intersects the graph *at most once*, then it is a graph of a function. This test is called the **vertical line test**. Note that the graph in Figure 2.16(b) fails the vertical line test and therefore does not represent a function.

READING CHECK

What is the vertical line test used for?

VERTICAL LINE TEST

If every vertical line intersects a graph at no more than one point, then the graph represents a function.

EXAMPLE 10   **Determining whether a graph represents a function**

Determine whether the graphs shown in Figure 2.17 represent functions.

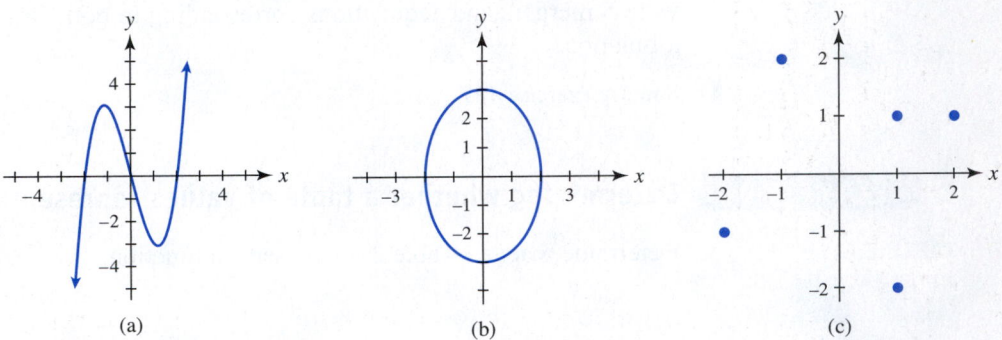

(a)   (b)   (c)

Figure 2.17

Solution

(a) Visualize vertical lines moving across the xy-plane from left to right. Any (red) vertical line will intersect the graph at most once, as depicted in Figure 2.18(a). Therefore the graph *does* represent a function.

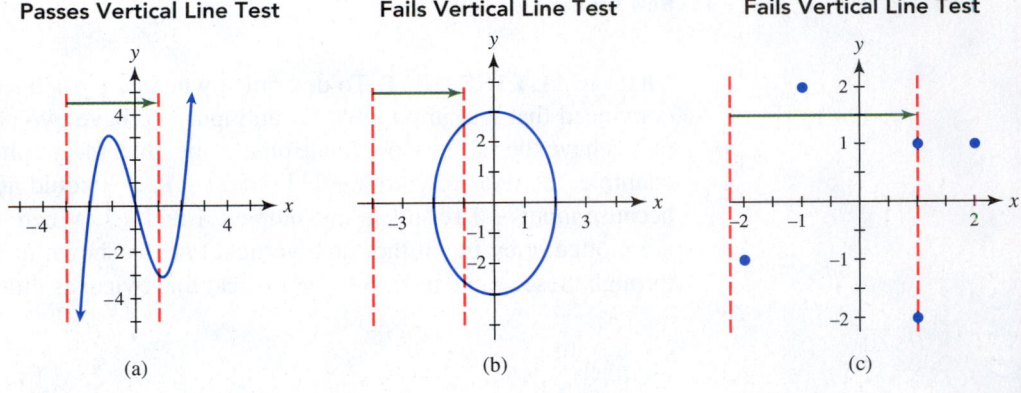

Passes Vertical Line Test   **Fails Vertical Line Test**   **Fails Vertical Line Test**

(a)   (b)   (c)

Figure 2.18

(b) Visualize vertical lines moving across the xy-plane from left to right. The graph *does not* represent a function because there exist (red) vertical lines that can intersect the graph twice. One such line is shown in Figure 2.18(b).

(c) Visualize vertical lines moving across the xy-plane from left to right. The graph is a scatterplot and *does not* represent a function because there exists one (red) vertical line that intersects two points: $(1, 1)$ and $(1, -2)$ with the same x-coordinate, as shown in Figure 2.18(c).

Now Try Exercises 115, 117, 123

Tables, Graphs, and Calculators (Optional)

We can use graphing calculators to create graphs and tables, usually more efficiently and reliably than with pencil-and-paper techniques. However, a graphing calculator uses the same techniques that we might use to sketch a graph. For example, one way to sketch a graph of $y = 2x - 1$ is first to make a table of values, as shown in Table 2.5.

We can plot these points in the xy-plane, as shown in Figure 2.19. Next we might connect the points, as shown in Figure 2.20.

TABLE 2.5

x	y
-1	-3
0	-1
1	1
2	3

Plotting Points

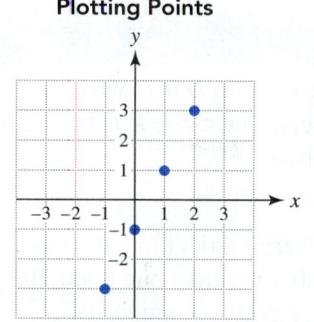

Figure 2.19

Graphing a Line

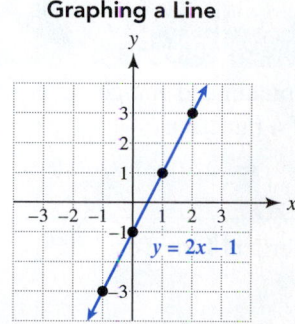

Figure 2.20

In a similar manner, a graphing calculator plots numerous points and connects them to make a graph. To create a similar graph with a graphing calculator, we enter the formula $Y_1 = 2X - 1$, set an appropriate viewing rectangle, and graph as shown in Figures 2.21 and 2.22. A table of values can also be generated as illustrated in Figure 2.23.

CALCULATOR HELP

To make a graph, see Appendix A (page AP-5). To make a table, see Appendix A (pages AP-2 and AP-3).

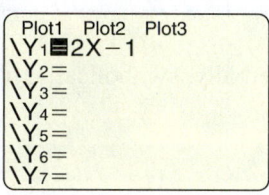

Figure 2.21

$[-10, 10, 1]$ by $[-10, 10, 1]$

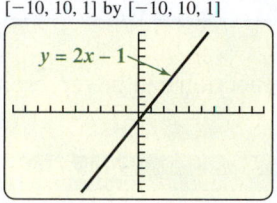

Figure 2.22

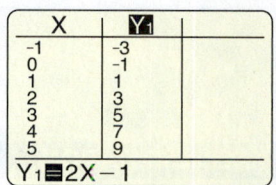

Figure 2.23

2.1 Putting It All Together

A function calculates exactly one output for each valid input and produces input–output ordered pairs in the form (x, y). A function typically computes something such as area, speed, or sales tax.

CONCEPT	EXPLANATION	EXAMPLES
Function	A set of ordered pairs (x, y), where each x-value corresponds to exactly one y-value	$f = \{(1, 3), (2, 3), (3, 1)\}$ $f(x) = 2x$ A graph of $y = x + 2$ A table of values for $y = 4x$
Independent Variable	The *input* variable for a function	*Function* *Independent Variable* $f(x) = 2x$ x $A(r) = \pi r^2$ r $V(s) = s^3$ s
Dependent Variable	The *output* variable of a function There is exactly one output for each valid input.	*Function* *Dependent Variable* $y = f(x)$ y $T = F(r)$ T $V = g(r)$ V

continued on next page

continued from previous page

CONCEPT	EXPLANATION	EXAMPLES
Domain and Range of a Function	The domain D is the set of all valid inputs. The range R is the set of all outputs.	For $S = \{(-1, 0), (3, 4), (5, 0)\}$, $D = \{-1, 3, 5\}$ and $R = \{0, 4\}$. For $f(x) = \frac{1}{x}$ the domain includes all real numbers except 0, or $x \neq 0$.
Vertical Line Test	If every vertical line intersects a graph at no more than one point, the graph represents a function.	This graph does *not* pass this test and thus does not represent a function. Two points lie on the same vertical line: *not* a function

A function can be represented verbally, symbolically, numerically, and graphically.

REPRESENTATION	EXPLANATION	COMMENTS
Verbal	Precise word description of what is computed	May be oral or written. Must be stated *precisely*
Symbolic	Mathematical formula	Efficient and concise way of representing a function (e.g., $f(x) = 2x - 3$)
Numerical	List of specific inputs and their outputs	May be in the form of a table or an explicit set of ordered pairs
Graphical, diagrammatic	Shows inputs and outputs visually	No words, formulas, or tables. Many types of graphs and diagrams are possible.

2.1 Exercises

MyMathLab   PRACTICE   WATCH   DOWNLOAD   READ   REVIEW

CONCEPTS AND VOCABULARY

1. The notation $y = f(x)$ is called _____ notation.

2. The notation $y = f(x)$ is read _____.

3. The notation $f(x) = x^2 + 1$ is a(n) _____ representation of a function.

4. A table of values is a(n) _____ representation of a function.

5. The set of valid inputs for a function is the _____.

6. The set of outputs for a function is the _____.

7. A function computes _____ output for each valid input.

8. (True or False?) The vertical line test is used to identify graphs of relations.

9. (True or False?) Four ways to represent functions are verbal, numerical, symbolic, and graphical.

10. If $f(3) = 4$, the point _____ is on the graph of f. If $(3, 6)$ is on the graph of f, then $f(\_\_\_\_) = \_\_\_\_$.

11. Thinking Generally If $f(a) = b$, the point _____ is on the graph of f.

12. Thinking Generally If (c, d) is on the graph of g, then $g(c) = \_\_\_\_$.

13. Thinking Generally If a is in the domain of f, then $f(a)$ represents how many outputs?

14. Thinking Generally If $f(x) = x$ for every x in the domain of f, then the domain and range of f are _____.

Exercises 15–20: Determine whether the phrase describes a function.

15. Calculating the square of a number

16. Calculating the low temperature for a day

17. Listing the students who passed a given math exam

18. Listing the children of parent x

19. Finding sales tax on a purchase

20. Naming the people in your class

REPRESENTING AND EVALUATING FUNCTIONS

Exercises 21–32: Evaluate $f(x)$ at the given values of x.

21. $f(x) = 4x - 2$ $x = -1, 0$

22. $f(x) = 5 - 3x$ $x = -4, 2$

23. $f(x) = \sqrt{x}$ $x = 0, \frac{9}{4}$

24. $f(x) = \sqrt[3]{x}$ $x = -1, 27$

25. $f(x) = x^2$ $x = -5, \frac{3}{2}$

26. $f(x) = x^3$ $x = -2, 0.1$

27. $f(x) = 3$ $x = -8, \frac{7}{3}$

28. $f(x) = 100$ $x = -\pi, \frac{1}{3}$

29. $f(x) = 5 - x^3$ $x = -2, 3$

30. $f(x) = x^2 + 5$ $x = -\frac{1}{2}, 6$

31. $f(x) = \dfrac{2}{x + 1}$ $x = -5, 4$

32. $f(x) = \dfrac{x}{x - 4}$ $x = -3, 1$

Exercises 33–38: Do the following.

   (a) *Write a formula for the function described.*
   (b) *Evaluate the function for input* 10 *and interpret the result.*

33. Function I computes the number of inches in x yards.

34. Function A computes the area of a circle with radius r.

35. Function M computes the number of miles in x feet.

36. Function C computes the circumference of a circle with radius r.

37. Function A computes the square feet in x acres. (*Hint:* There are 43,560 square feet in one acre.)

38. Function K computes the number of kilograms in x pounds. (*Hint:* There are about 2.2 pounds in one kilogram.)

Exercises 39–42: Write each function f as a set of ordered pairs. Give the domain and range of f.

39. $f(1) = 3, f(2) = -4, f(3) = 0$

40. $f(-1) = 4, f(0) = 6, f(1) = 4$

41. $f(a) = b, f(c) = d, f(e) = a, f(d) = b$

42. $f(a) = 7, f(b) = 7, f(c) = 7, f(d) = 7$

Exercises 43–52: Sketch a graph of f.

43. $f(x) = -x + 3$ **44.** $f(x) = -2x + 1$

45. $f(x) = 2x$ **46.** $f(x) = \frac{1}{2}x - 2$

47. $f(x) = 4 - x$ **48.** $f(x) = 6 - 3x$

49. $f(x) = x^2$ **50.** $f(x) = \sqrt{x}$

51. $f(x) = \sqrt{x + 1}$ **52.** $f(x) = \frac{1}{2}x^2 - 1$

Exercises 53–58: Use the graph of f to evaluate the given expressions.

53. $f(0)$ and $f(2)$ **54.** $f(-2)$ and $f(2)$

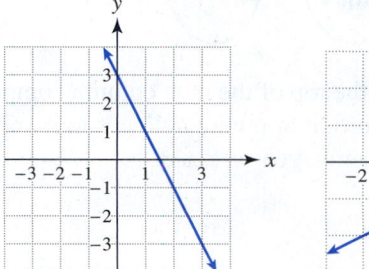
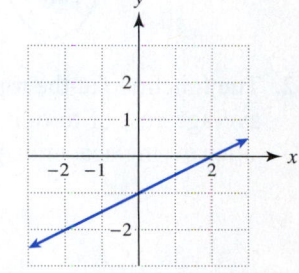

55. $f(-2)$ and $f(1)$　　　**56.** $f(-1)$ and $f(0)$

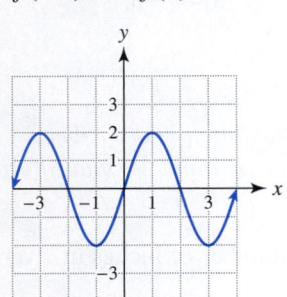　　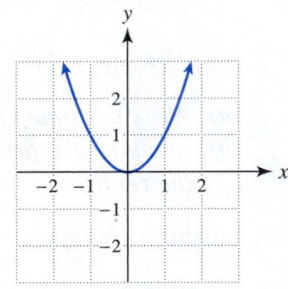

57. $f(1)$ and $f(2)$　　　**58.** $f(-1)$ and $f(4)$

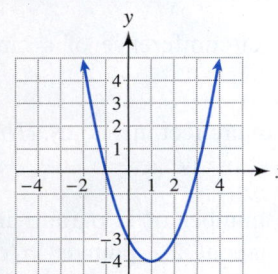　　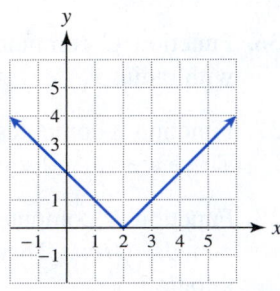

Exercises 59 and 60: Use the table to evaluate the given expressions.

59. $f(0)$ and $f(2)$

x	0	1	2	3	4
$f(x)$	5.5	4.3	3.7	2.5	1.9

60. $f(-10)$ and $f(5)$

x	-10	-5	0	5	10
$f(x)$	23	96	-45	-33	23

Exercises 61 and 62: Use the diagram to evaluate $f(1990)$. Interpret your answer.

61. The function f computes average fuel efficiency of new U.S. passenger cars in miles per gallon during year x. (*Source:* Department of Transportation.)

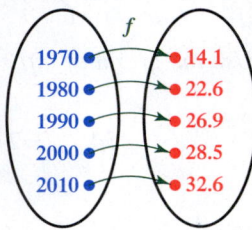

62. The function f at the top of the next column computes average cost of tuition at public colleges and universities during academic year x. (*Source:* The College Board.)

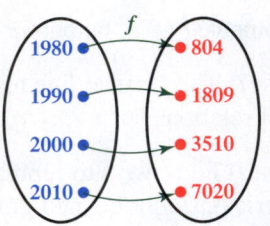

Exercises 63–66: Express the verbal representation for the function f numerically, symbolically, and graphically. Let $x = -3, -2, -1, \ldots, 3$ for the numerical representation (table), and let $-3 \le x \le 3$ for the graph.

63. Add 5 to the input x to obtain the output y.

64. Square the input x to obtain the output y.

65. Multiply the input x by 5 and then subtract 2 to obtain the output y.

66. Divide the input x by 2 and then add 3 to obtain the output y.

Exercises 67–70: Represent the function f numerically and graphically. Let $x = -3, -2, -1, \ldots, 3$ for your numerical representation and use the standard window for your graph.

67. $f(x) = \sqrt{x + 3}$　　　**68.** $f(x) = x^3 - \frac{1}{2}x^2$

69. $f(x) = \dfrac{5 - x}{5 + x}$　　　**70.** $f(x) = |2 - x| + \sqrt[3]{x}$

Exercises 71–76: Give a verbal representation for $f(x)$.

71. $f(x) = x - \frac{1}{2}$　　　**72.** $f(x) = \frac{3}{4}x$

73. $f(x) = \dfrac{x}{3}$　　　**74.** $f(x) = x^2 + 1$

75. $f(x) = \sqrt{x - 1}$　　　**76.** $f(x) = 1 - 3x$

77. *Cost of Driving* In 2010, the average cost of driving a new car in the United States was about 50 cents per mile. Symbolically, graphically, and numerically represent a function f that computes the cost in dollars of driving x miles. For the numerical representation (table) let $x = 10, 20, 30, \ldots, 70$. (*Source:* Associated Press.)

78. *Federal Income Taxes* In 2010, the lowest U.S. income tax rate was 10 percent. Symbolically, graphically, and numerically represent a function f that computes the tax on a taxable income of x dollars. For the numerical representation (table) let $x = 1000, 2000, 3000, \ldots, 7000$, and for the graphical representation let $0 \le x \le 10,000$. (*Source:* Internal Revenue Service.)

79. *Global Web Searches* The number of World Wide Web searches S in billions during year x can be approximated by $S(x) = 225x - 450{,}650$ from 2009 to 2012. Evaluate $S(2011)$ and interpret the result. (*Source:* RBC Capital Markets Corp.)

80. *Cost of Smartphones* The average cost difference D in dollars between smartphones and all other types of phones during year x can be approximated by $D(x) = -23.5x + 47{,}275$ from 2005 to 2009. Evaluate $D(2009)$ and interpret the result. (*Source:* Business Insider.)

IDENTIFYING DOMAINS AND RANGES

Exercises 81–88: Use the graph of f to estimate its domain and range.

81.
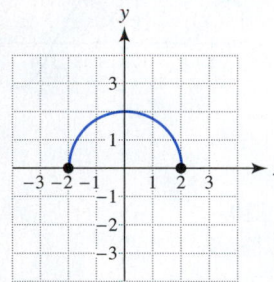

82.
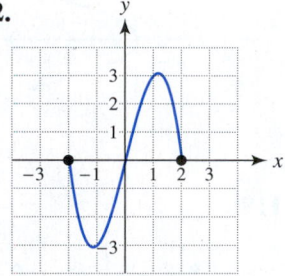

83.
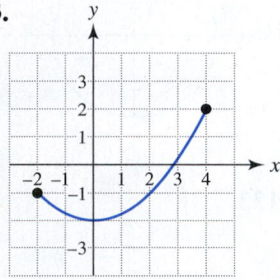

84.
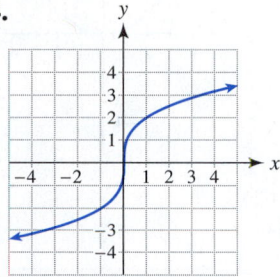

85.
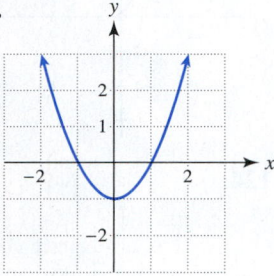

86.
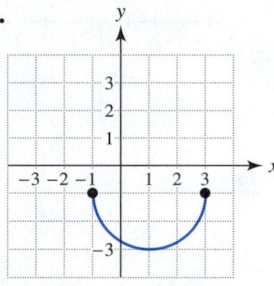

87.
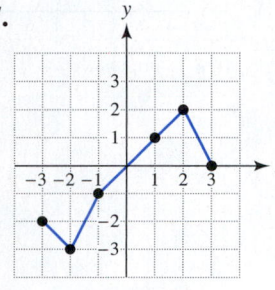

88.
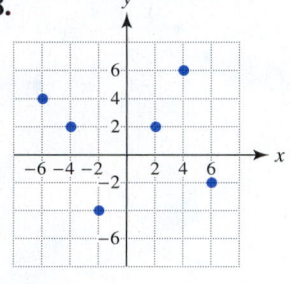

Exercises 89 and 90: Use the diagram to find the domain and range of f.

89.
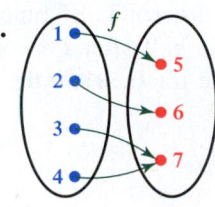

90.

Exercises 91–104: Find the domain.

91. $f(x) = 10x$

92. $f(x) = 5 - x$

93. $f(x) = x^2 - 3$

94. $f(x) = \frac{1}{2}x^2$

95. $f(x) = \dfrac{3}{x - 5}$

96. $f(x) = \dfrac{x}{x + 1}$

97. $f(x) = \dfrac{2x}{x^2 + 1}$

98. $f(x) = \dfrac{6}{1 - x}$

99. $f(x) = \sqrt{x - 1}$

100. $f(x) = |x|$

101. $f(x) = |x - 5|$

102. $f(x) = \sqrt{2 - x}$

103. $f(x) = \dfrac{1}{x}$

104. $f(x) = 1 - 3x^2$

105. *Humpback Whales* The number of humpback whales W sighted in Maui's annual whale census for year x is given by $W(2005) = 649$, $W(2006) = 1265$, $W(2007) = 959$, $W(2008) = 1726$, and $W(2009) = 1010$. (*Source:* Pacific Whale Foundation.)
(a) Evaluate $W(2008)$ and interpret the result.
(b) Identify the domain and range of W.
(c) Describe the pattern in the data.

106. *Digital Music Downloads* The percentage of digital music D that was purchased through downloads during year x is given by $D(2004) = 0.9$, $D(2005) = 5.7$, $D(2006) = 6.7$, $D(2007) = 11.2$, and $D(2008) = 12.8$. (*Source:* The Recording Industry Association of America.)
(a) Evaluate $D(2006)$ and interpret the result.
(b) Identify the domain and range of D.
(c) Describe the pattern in the data.

107. *Cost of Tuition* Suppose that a student can take from 1 to 20 credits at a college and that each credit costs \$200. If function C calculates the cost of taking x credits, determine the domain and range of C.

108. *Falling Ball* Suppose that a ball is dropped from a window that is 64 feet above the ground and that the ball strikes the ground after 2 seconds. If function H calculates the height of the ball after t seconds, determine a domain and range for H, while the ball is falling.

IDENTIFYING A FUNCTION

Exercises 109–112: Determine whether the diagram could represent a function.

109. 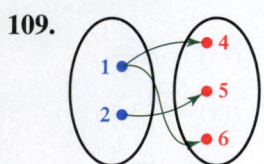   **110.**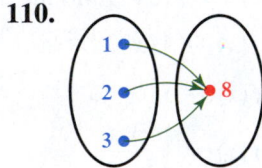

111. 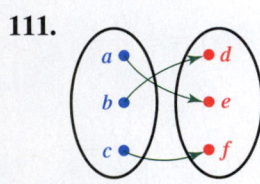   **112.**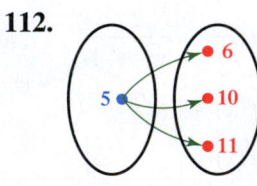

113. *Average Precipitation* The table lists the monthly average precipitation P in Las Vegas, Nevada, where $x = 1$ corresponds to January and $x = 12$ corresponds to December.

x (month)	1	2	3	4	5	6
P (inches)	0.5	0.4	0.4	0.2	0.2	0.1

x (month)	7	8	9	10	11	12
P (inches)	0.4	0.5	0.3	0.2	0.4	0.3

Source: J. Williams.

(a) Determine the value of P during May.
(b) Is P a function of x? Explain.
(c) If $P = 0.4$, find x.

114. *Wind Speeds* The table lists the monthly average wind speed W in Louisville, Kentucky, where $x = 1$ corresponds to January and $x = 12$ corresponds to December.

x (month)	1	2	3	4	5	6
W (mph)	10.4	12.7	10.4	10.4	8.1	8.1

x (month)	7	8	9	10	11	12
W (mph)	6.9	6.9	6.9	8.1	9.2	9.2

Source: J. Williams.

(a) Determine the month with the highest average wind speed.
(b) Is W a function of x? Explain.
(c) If $W = 6.9$, find x.

Exercises 115–126: Determine whether the graph represents a function. If it does, identify the domain and range.

115.   **116.**

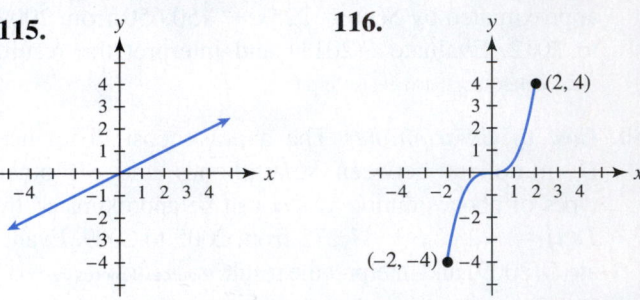

117.   **118.**

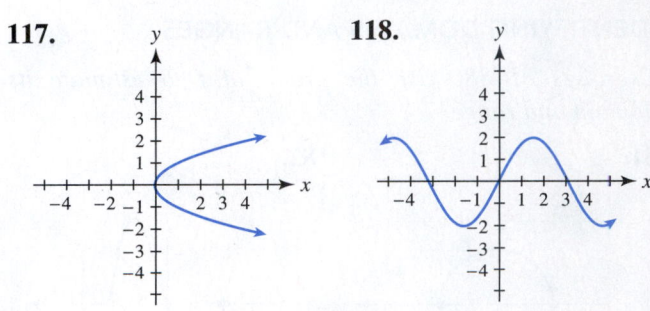

119.   **120.**

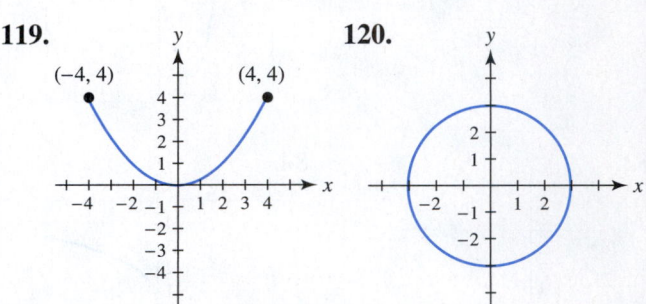

121.   **122.**

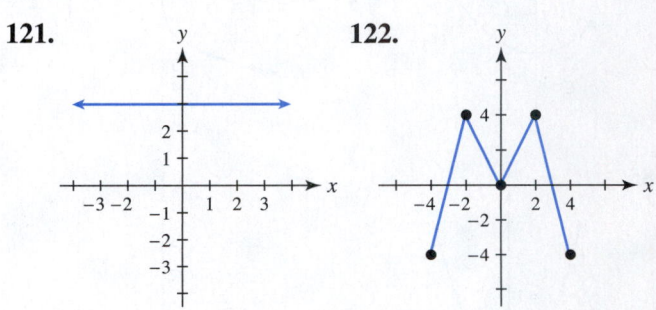

123.   **124.**

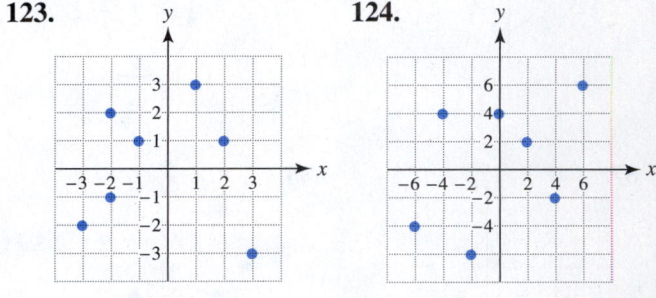

125.

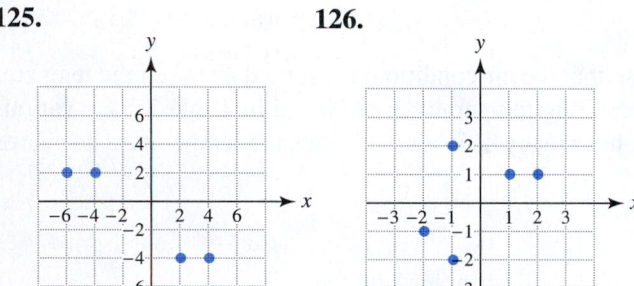

126.

131.

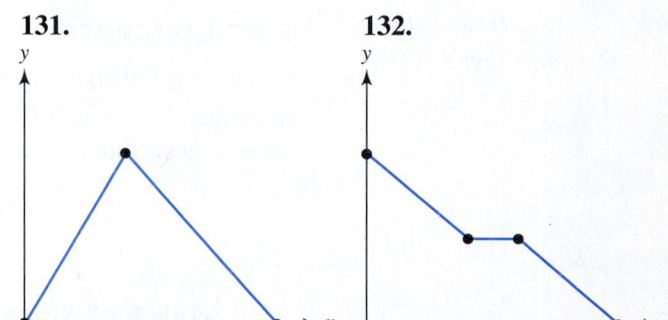

132.

Exercises 127–130: Determine whether S is a function.

127. $S = \{(1, 2), (4, 5), (7, 8), (5, 4), (2, 2)\}$

128. $S = \{(4, 7), (-2, 1), (3, 8), (4, 9)\}$

129. *S* is given by the table.

x	5	10	5
y	2	1	0

130. *S* is given by the table.

x	−3	−2	−1
y	10	10	10

GRAPHICAL INTERPRETATION

Exercises 131 and 132: The graph represents the distance that a person is from home while walking on a straight path. The x-axis represents time and the y-axis represents distance. Interpret the graph.

133. *Texting* The average 18- to 24-year-old person texts about 1500 messages per month. Sketch a graph that shows the total number of text messages sent over a period of 4 months. Assume that the same number of texts is sent each day. (*Source:* The Nielsen Company.)

134. *Computer Viruses* In 2000 there were about 50 thousand computer viruses. In 2010 there were about 1.6 million computer viruses. Sketch a graph of this increase from 2000 to 2010. Answers may vary. (*Source:* Symantec.)

WRITING ABOUT MATHEMATICS

135. Give an example of a function. Identify the domain and range of your function.

136. Explain in your own words what a function is. How is a function different from other relations?

137. Explain how to evaluate a function by using a graph. Give an example.

138. Give one difficulty that may occur when you use a table of values to evaluate a function.

2.2   Linear Functions

Basic Concepts • Representations of Linear Functions • Modeling Data with Linear Functions

A LOOK INTO MATH ▶

Functions are frequently used to model, or describe, the real world. For example, people are becoming more energy conscious. As a result, there is an increase in the number of *green* buildings that are being constructed. Table 2.6 lists estimated U.S. sales of green building material. Because sales increase by $5 billion each year, a *linear function* can be used to model these data. (See Example 7.) In this section we discuss this important type of function.

NEW VOCABULARY

☐ Linear function
☐ Rate of change
☐ Constant function

TABLE 2.6 Green Material Sales ($ billions)

Year	2010	2011	2012	2013
Sales	65	70	75	80

Source: Freedonia Group, Green Building Material.

Basic Concepts

▶ **REAL-WORLD CONNECTION** Suppose that the air conditioner is turned on when the temperature inside a house is 80°F. The resulting temperatures are listed in Table 2.7 for various elapsed times. Note that for each 1-hour increase in elapsed time, the temperature decreases by 2°F.

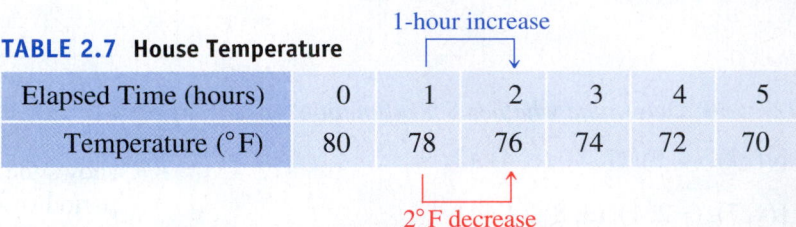

TABLE 2.7 House Temperature

Elapsed Time (hours)	0	1	2	3	4	5
Temperature (°F)	80	78	76	74	72	70

We want to determine a function f that models, or calculates, the house temperature after x hours. To do this, we will find numerical, graphical, verbal, and symbolic representations of f.

STUDY TIP

Be sure you understand what representations of a function are.

NUMERICAL REPRESENTATION (TABLE OF VALUES) We can think of Table 2.7 as a numerical representation (table of values) for the function f. A similar numerical representation that uses x and $f(x)$ is shown in Table 2.8.

TABLE 2.8 Numerical Representation of $f(x)$

x	0	1	2	3	4	5
$f(x)$	80	78	76	74	72	70

GRAPHICAL REPRESENTATION (GRAPH) To graph $y = f(x)$, we begin by plotting the points in Table 2.8, as shown in Figure 2.24. This scatterplot suggests that a line models these data, as shown in Figure 2.25. We call f a *linear function* because its graph is a *line*.

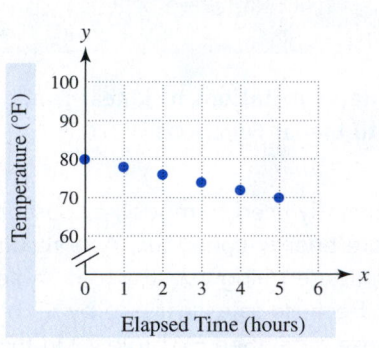

Figure 2.24 A Scatterplot

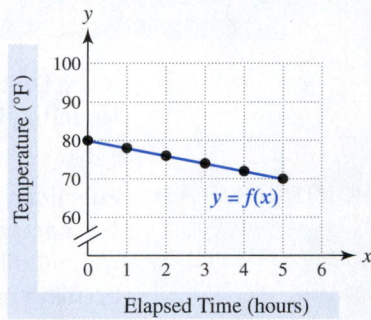

Figure 2.25 A Linear Function

Another graph of $y = f(x)$ with a different y-scale is shown in Figure 2.26. Because the y-values always decrease by the same amount for each 1-hour increase on the x-axis, we say that function f has a *constant rate of change*. In this example, the constant rate of change is -2°F per hour.

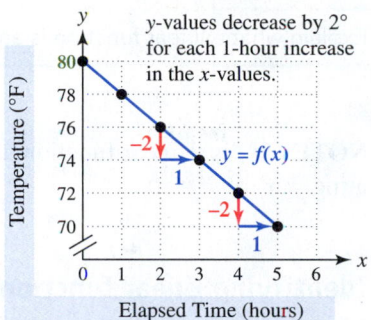

Figure 2.26

VERBAL REPRESENTATION (WORDS) Over a 5-hour period, the air conditioner lowers the initial temperature of 80°F by 2°F for each elapsed hour x. Thus a description of how to calculate the temperature is:

"Multiply x by -2°F and then add 80°F."   Verbal representation of $f(x)$

SYMBOLIC REPRESENTATION (FORMULA) Our verbal representation of $f(x)$ makes it straightforward for us to write a formula.

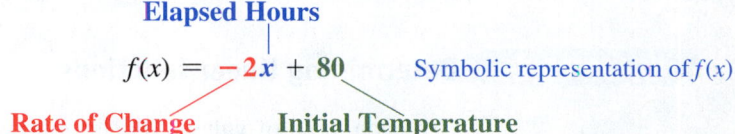

For example,

$$f(2.5) = -2(2.5) + 80 = 75$$

means that the temperature is **75**°F after the air conditioner has run for **2.5** hours. In this instance, it might be appropriate to *limit the domain* of f to x-values between 0 and 5, inclusive.

LINEAR FUNCTION

A function f defined by $f(x) = mx + b$, where m and b are constants, is a **linear function**.

For $f(x) = -2x + 80$, we have $m = -2$ and $b = 80$. The constant m represents the rate at which the air conditioner cools the building, and the constant b represents the initial temperature.

NOTE: In previous mathematics courses you may have studied slope m. In Section 2.3 we discuss further how m corresponds to the slope of a line and to a linear function's rate of change.

▶ **REAL-WORLD CONNECTION** In general, a linear function defined by $f(x) = mx + b$ changes by m units for each unit increase in x. This **rate of change** is an increase if $m > 0$ and a decrease if $m < 0$. For example, if new carpet costs \$20 per square yard, then the linear function defined by $C(x) = 20x$ gives the cost of buying x square yards of carpet. The value of $m = 20$ gives the cost (rate of change) for each additional square yard of carpet. For function C, the value of b is 0 because it costs \$0 to buy 0 square yards of carpet.

NOTE: If f is a linear function, then $f(0) = m(0) + b = b$. Thus b can be found by evaluating $f(x)$ at $x = 0$.

EXAMPLE 1 ▌ **Identifying linear functions**

Determine whether f is a linear function. If f is a linear function, find values for m and b so that $f(x) = mx + b$.
(a) $f(x) = 4 - 3x$ **(b)** $f(x) = 8$ **(c)** $f(x) = 2x^2 + 8$

Solution
(a) Let $m = -3$ and $b = 4$. Then $f(x) = -3x + 4$, and f is a linear function.
(b) Let $m = 0$ and $b = 8$. Then $f(x) = 0x + 8$, and f is a linear function.
(c) Function f is not linear because its formula contains x^2. The formula for a linear function cannot contain an x with an exponent other than 1.

▌ **Now Try Exercises 11, 13, 15**

EXAMPLE 2 ▌ **Determining linear functions**

Use each table of values to determine whether $f(x)$ could represent a linear function. If f could be linear, write a formula for f in the form $f(x) = mx + b$.

(a)

x	0	1	2	3
$f(x)$	10	15	20	25

(b)

x	-2	0	2	4
$f(x)$	4	2	0	-2

(c)

x	0	1	2	3
$f(x)$	1	2	4	7

(d)

x	-2	0	3	5
$f(x)$	7	7	7	7

Solution
(a) For each unit increase in x, $f(x)$ increases by 5 units, so $f(x)$ could be linear with $m = 5$. Because $f(0) = 10$, $b = 10$. Thus $f(x) = 5x + 10$.
(b) For each 2-unit increase in x, $f(x)$ decreases by 2 units. Equivalently, each unit increase in x results in a 1-unit decrease in $f(x)$, so $f(x)$ could be linear with $m = -1$. Because $f(0) = 2$, $b = 2$. Thus $f(x) = -x + 2$.
(c) Each unit increase in x does not result in a constant change in $f(x)$. Thus $f(x)$ does *not* represent a linear function.
(d) For any change in x, $f(x)$ does *not* change, so $f(x)$ could be linear with $m = 0$. Because $f(0) = 7$, let $b = 7$. Thus $f(x) = 0x + 7$, or $f(x) = 7$. (When $m = 0$, we say that f is a *constant function*. See Example 8.)

▌ **Now Try Exercises 23, 25, 27, 31**

Representations of Linear Functions

The graph of a linear function is a line. To graph a linear function f we can start by making a table of values and then plotting three or more points. We can then sketch the graph of f by drawing a line through these points, as demonstrated in the next example.

EXAMPLE 3   **Graphing a linear function by hand**

Sketch a graph of $f(x) = x - 1$. Use the graph to evaluate $f(-2)$.

Solution
Begin by making a table of values containing at least three points. Pick convenient values of x, such as $x = -1, 0, 1$.

$$f(-1) = -1 - 1 = -2$$
$$f(0) = 0 - 1 = -1$$
$$f(1) = 1 - 1 = 0$$

Display the results, as shown in Table 2.9.

Plot the points $(-1, -2)$, $(0, -1)$, and $(1, 0)$. Sketch a line through these points to obtain the graph of f. A graph of a line results when *infinitely* many points are plotted, as shown in Figure 2.27.

To evaluate $f(-2)$, first find $x = -2$ on the x-axis. See Figure 2.28. Then move downward to the graph of f. By moving across to the y-axis, we see that the corresponding y-value is -3. Thus $f(-2) = -3$.

TABLE 2.9

x	y
-1	-2
0	-1
1	0

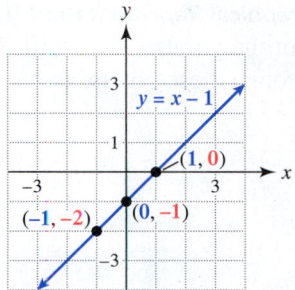

Figure 2.27

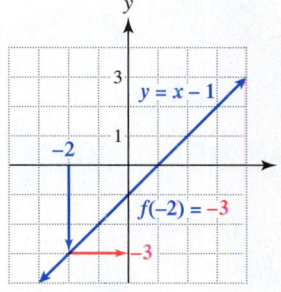

Figure 2.28

■ **Now Try Exercises 39, 57**

In the next example a graphing calculator is used to create a graph and table.

EXAMPLE 4   **Using a graphing calculator**

Give numerical and graphical representations of $f(x) = \frac{1}{2}x - 2$.

Solution
Numerical Representation To make a numerical representation, construct the table for $Y_1 = .5X - 2$, starting at $x = -3$ and incrementing by 1, as shown in Figure 2.29(a). (Other tables are possible.)
Graphical Representation Graph Y_1 in the standard viewing rectangle, as shown in Figure 2.29(b). (Other viewing rectangles may be used.)

CALCULATOR HELP

To make a table, see Appendix A (pages AP-2 and AP-3). To make a graph, see Appendix A (page AP-5).

$[-10, 10, 1]$ by $[-10, 10, 1]$

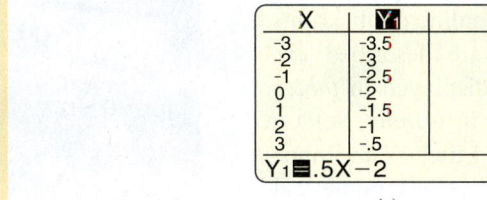

(a)

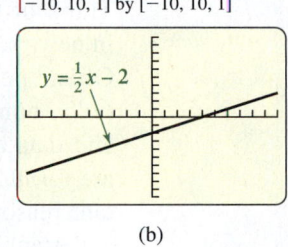

(b)

Figure 2.29

■ **Now Try Exercise 75**

CRITICAL THINKING

Two points determine a line. Why is it a good idea to plot at least three points when graphing a linear function by hand?

EXAMPLE 5   **Representing a linear function**

A linear function is given by $f(x) = -3x + 2$.
(a) Give a verbal representation of f.
(b) Make a numerical representation (table) of f by letting $x = -1, 0, 1$.
(c) Plot the points listed in the table from part (b). Then sketch a graph of $y = f(x)$.

Solution
(a) *Verbal Representation* Multiply the input x by -3 and add 2 to obtain the output.

(b) *Numerical Representation* Evaluate the formula $f(x) = -3x + 2$ at $x = -1, 0, 1$, which results in Table 2.10. Note that $f(-1) = 5$, $f(0) = 2$, and $f(1) = -1$.

(c) *Graphical Representation* To make a graph of f by hand without a graphing calculator, plot the points $(-1, 5)$, $(0, 2)$, and $(1, -1)$ from Table 2.10. Then draw a line passing through these points, as shown in Figure 2.30.

TABLE 2.10

x	$f(x)$
-1	5
0	2
1	-1

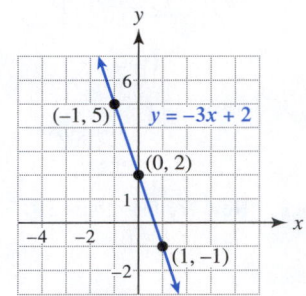

Figure 2.30

Now Try Exercise 71

NOTE: In the next section we use one point and slope m to graph lines rather than plotting three points.

MAKING CONNECTIONS

Mathematics in Newspapers

Think of the mathematics that you see in newspapers or in online publications. Often, percentages are described *verbally*, numbers are displayed in *tables*, and data are shown in *graphs*. Seldom are *formulas* given, which is an important reason to study verbal, numerical, and graphical representations.

Modeling Data with Linear Functions

▶ **REAL-WORLD CONNECTION** A distinguishing feature of a linear function is that when the input x increases by 1 unit, the output $f(x) = mx + b$ always changes by an amount equal to m. For example, the percentage of wireless households during year x from 2005 to 2010 can be modeled by the linear function

$$f(x) = 4x - 8013,$$

where x is the year. The value of $m = 4$ indicates that the percentage of wireless households has increased, on average, by 4% per year. (*Source: National Center for Health Statistics.*)

The following are other examples of quantities that are modeled by linear functions. Try to determine the value of the constant m.

• The wages earned by an individual working x hours at \$8 per hour
• The distance traveled by a jet airliner in x hours if its speed is 500 miles per hour
• The cost of tuition and fees when registering for x credits if each credit costs \$200 and the fees are fixed at \$300

When we are modeling data with a linear function defined by $f(x) = mx + b$, the following concepts are helpful to determine m and b.

MODELING DATA WITH A LINEAR FUNCTION

The formula $f(x) = mx + b$ may be interpreted as follows.

$$f(x) \quad = \quad mx \quad + \quad b$$

(New amount) = (Change) + (Fixed amount)

When x represents time, *change* equals (rate of change) × (time).

$$f(x) \quad = \quad m \quad \times \quad x \quad + \quad b$$

(Future amount) = (Rate of change) × (Time) + (Initial amount)

▶ **REAL-WORLD CONNECTION** These concepts are applied in the next three examples.

EXAMPLE 6   **Modeling growth of bamboo**

Bamboo is gaining popularity as a *green* building material because of its fast-growing, regenerative characteristics. Under ideal conditions, some species of bamboo grow at an astonishing 2 inches per hour. Suppose a bamboo plant is initially 6 inches tall. (*Source: Cali Bamboo.*)
(a) Find a function H that models the plant's height in inches under ideal conditions after t hours.
(b) Find $H(3)$ and interpret the result.

Solution
(a) The initial height is **6** inches and the rate of change is **2** inches per hour.

$$H(t) \quad = \quad 2 \quad \times \quad t \quad + \quad 6,$$

(Future height) = (Rate of change) × (Time) + (Initial height)

or $H(t) = 2t + 6$.
(b) $H(3) = 2(3) + 6 = 12$. After **3** hours the bamboo plant is **12** inches tall.

Now Try Exercise 89

EXAMPLE 7   **Modeling demand for building green**

Table 2.11 lists estimated sales of green building material in billions of dollars. (Refer to A Look Into Math at the beginning of this section.)

TABLE 2.11 Green Material Sales ($ billions)

Year	2010	2011	2012	2013
Sales	65	70	75	80

Source: Freedonia Group, Green Building Material.

(a) Make a scatterplot of the data and sketch the graph of a function f that models these data. Let x represent years after 2010. That is, let $x = 0$ correspond to 2010, $x = 1$ to 2011, and so on.
(b) What were the sales in 2010? What was the annual increase in sales each year?
(c) Find a formula for $f(x)$.
(d) Use your formula to estimate sales in 2014.

READING CHECK

How can you determine whether data in a table can be modeled by a linear function?

Solution
(a) In Figure 2.31 the scatterplot suggests that a linear function models the data. A line has been sketched with the data.

Green Building Material Sales

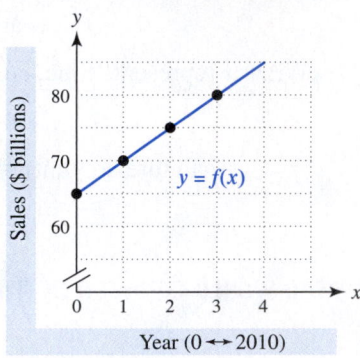

Figure 2.31 A Linear Model

(b) From Table 2.11, sales for green material were $65 billion in 2010, with sales increasing at a *constant rate* of $5 billion per year.
(c) From part (b) initial sales ($x = 0$) were $**65** billion, and sales increased by $**5** billion per year. Thus

$$f(x) \quad = \quad 5 \quad \times \quad x \quad + \quad 65,$$

(Future sales) = (Rate of change in sales) × (Time) + (Initial sales)

or $f(x) = 5x + 65$.
(d) Because $x = 4$ corresponds to 2014, evaluate $f(4)$.

$$f(4) = 5(4) + 65 = 85$$

This model estimates sales of green building material to be $85 billion in 2014.

Now Try Exercise 91

In the next example, we consider a simple function that models the speed of a car.

EXAMPLE 8 **Modeling with a constant function**

A car travels on a freeway with its speed recorded at regular intervals, as listed in Table 2.12.

TABLE 2.12 Speed of a Car

Elapsed Time (hours)	0	1	2	3	4
Speed (miles per hour)	70	70	70	70	70

(a) Discuss the speed of the car during this time interval.
(b) Find a formula for a function f that models these data.
(c) Sketch a graph of f together with the data.

Solution
(a) The speed of the car appears to be constant at 70 miles per hour.
(b) Because the speed is constant, the rate of change is 0. Thus

$$f(x) \quad = \quad 0x \quad + \quad 70$$

(Future speed) = (Change in speed) + (Initial speed)

and $f(x) = 70$. We call f a *constant function*.
(c) Because $y = f(x)$, graph $y = 70$ with the data points

$$(0, 70), (1, 70), (2, 70), (3, 70), \text{ and } (4, 70)$$

to obtain Figure 2.32.

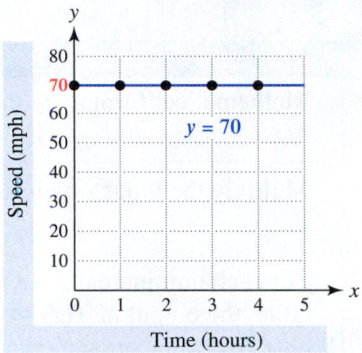

Figure 2.32 Speed of a Car

Now Try Exercise 85

The function defined by $f(x) = 70$ is an example of a *constant function*. A **constant function** *is a linear function* with $m = 0$ and can be written as $f(x) = b$. Regardless of the input, a constant function always outputs the same value, b. Its graph is a horizontal line. Its domain is all real numbers and its range is $R = \{b\}$.

CRITICAL THINKING

Find a formula for a function D that calculates the *distance* traveled by the car in Example 8 after x hours. What is the rate of change for $D(x)$?

▶ **REAL-WORLD CONNECTION** The following are three applications of constant functions.

- A thermostat calculates a constant function regardless of the weather outside by maintaining a set temperature.
- A cruise control in a car calculates a constant function by maintaining a fixed speed, regardless of the type of road or terrain.
- A constant function calculates the 1250-foot height of the Empire State Building.

2.2 Putting It All Together

CONCEPT	EXPLANATION	EXAMPLES
Linear Function	Can be represented by $f(x) = mx + b$. Graph is a line.	$f(x) = 2x - 6$, $m = 2$ and $b = -6$ $f(x) = 10$, $m = 0$ and $b = 10$
Constant Function	Can be represented by $f(x) = b$. Graph is a horizontal line.	$f(x) = -7$, $b = -7$ $f(x) = 22$, $b = 22$
Rate of Change for a Linear Function	The output of a linear function changes by a constant amount for each unit increase in the input.	$f(x) = -3x + 8$ decreases 3 units for each unit increase in x. $f(x) = 5$ neither increases nor decreases. The rate of change is 0.

REPRESENTATION	COMMENTS	EXAMPLE
Symbolic	Mathematical formula in the form $f(x) = mx + b$	$f(x) = 2x + 1$, where $m = 2$ and $b = 1$
Verbal	Multiply the input x by m and add b.	Multiply the input x by 2 and then add 1 to obtain the output.
Numerical (table of values)	For each unit increase in x in the table, the output of $f(x) = mx + b$ changes by an amount equal to m.	1-unit increase 	x	0	1	2	 	$f(x)$	1	3	5	 2-unit increase
Graphical	The graph of a linear function is a line. Plot at least 3 points and then sketch the line.	$y = 2x + 1$

2.2 Exercises

MyMathLab   Math XL PRACTICE   WATCH   DOWNLOAD   READ   REVIEW

CONCEPTS AND VOCABULARY

1. The formula for a linear function is $f(x) =$ _____.

2. The formula for a constant function is $f(x) =$ _____.

3. The graph of a linear function is a(n) _____.

4. The graph of a constant function is a(n) _____ line.

5. If $f(x) = 7x + 5$, each time x increases by 1 unit, $f(x)$ increases by _____ units.

6. If $f(x) = 5$, each time x increases by 1 unit, $f(x)$ increases by _____ units.

7. (True or False?) Every constant function is a linear function.

8. (True or False?) Every linear function is a constant function.

9. If $C(x) = 2x$ calculates the cost in dollars of buying x square feet of carpet, what does 2 represent in the formula? Interpret the fact that the point (10, 20) lies on the graph of C.

10. If $G(x) = 100 - 4x$ calculates the number of gallons of water in a tank after x minutes, what does -4 represent in the formula? Interpret the fact that the point (5, 80) lies on the graph of G.

IDENTIFYING LINEAR FUNCTIONS

Exercises 11–18: Determine whether f is a linear function. If f is linear, give values for m and b so that f may be expressed as $f(x) = mx + b$.

11. $f(x) = \dfrac{1}{2}x - 6$

12. $f(x) = x$

13. $f(x) = \dfrac{5}{2} - x^2$

14. $f(x) = \sqrt{x} + 3$

15. $f(x) = -9$

16. $f(x) = 1.5 - 7.3x$

17. $f(x) = -9x$

18. $f(x) = \dfrac{1}{x}$

Exercises 19–22: Determine whether the graph represents a linear function.

19.
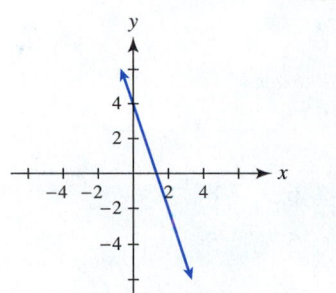

20.
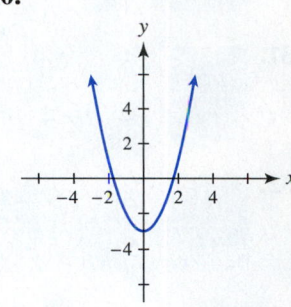

21.
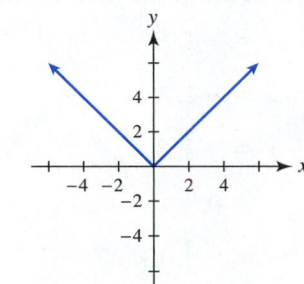

22.
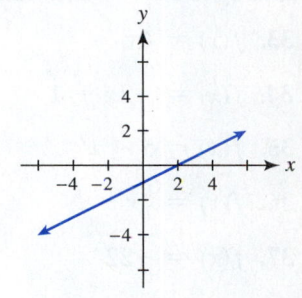

Exercises 23–32: (Refer to Example 2.) Use the table to determine whether $f(x)$ could represent a linear function. If it could, write $f(x)$ in the form $f(x) = mx + b$.

23.
x	0	1	2	3
$f(x)$	-6	-3	0	3

24.
x	0	2	4	6
$f(x)$	-2	2	6	10

25.
x	-2	0	2	4
$f(x)$	6	3	0	-3

26.
x	0	3	6	9
$f(x)$	8	4	2	1

27.
x	-2	-1	0	1
$f(x)$	-5	0	20	40

28.
x	-2	-1	0	1
$f(x)$	6	3	0	-3

29.

x	0	2	3	4
$f(x)$	0	4	6	8

30.

x	1	2	3	4
$f(x)$	0	1	3	7

31.

x	−1	0	1	2
$f(x)$	−4	−4	−4	−4

32.

x	2	5	6	8
$f(x)$	5	5	5	5

EVALUATING LINEAR FUNCTIONS

Exercises 33–38: Evaluate $f(x)$ at the given values of x.

33. $f(x) = 4x$ $\qquad$ $x = -4, 5$

34. $f(x) = -2x + 1$ $\qquad$ $x = -2, 3$

35. $f(x) = 5 - x$ $\qquad$ $x = -\frac{2}{3}, 3$

36. $f(x) = \frac{1}{2}x - \frac{1}{4}$ $\qquad$ $x = 0, \frac{1}{2}$

37. $f(x) = -22$ $\qquad$ $x = -\frac{3}{4}, 13$

38. $f(x) = 9x - 7$ $\qquad$ $x = -1.2, 2.8$

Exercises 39–44: Use the graph of f to evaluate the given expressions.

39. $f(-1)$ and $f(0)$ $\qquad$ **40.** $f(-2)$ and $f(2)$

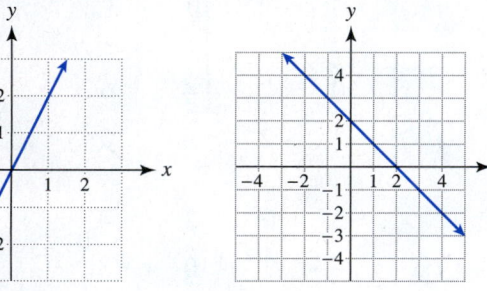

41. $f(-2)$ and $f(4)$ $\qquad$ **42.** $f(0)$ and $f(3)$

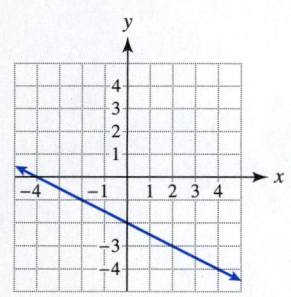 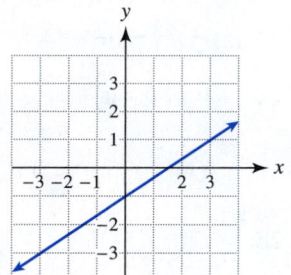

43. $f(-3)$ and $f(1)$ $\qquad$ **44.** $f(1.5)$ and $f(0.5\pi)$

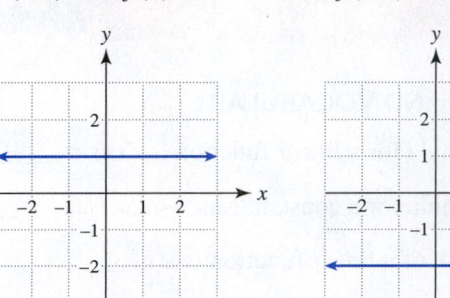

Exercises 45–48: Use the verbal description to write a formula for $f(x)$. Then evaluate $f(3)$.

45. Multiply the input by 6.

46. Multiply the input by −3 and add 7.

47. Divide the input by 6 and subtract $\frac{1}{2}$.

48. Output 8.7 for every input.

REPRESENTING LINEAR FUNCTIONS

Exercises 49–52: Match $f(x)$ with its graph (a.–d.).

49. $f(x) = 3x$ $\qquad$ **50.** $f(x) = -2x$

51. $f(x) = x - 2$ $\qquad$ **52.** $f(x) = 2x + 1$

a. $\qquad\qquad\qquad\qquad$ **b.**

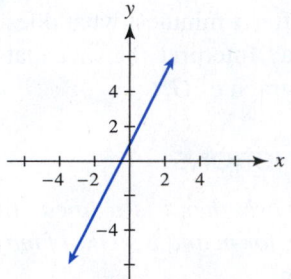 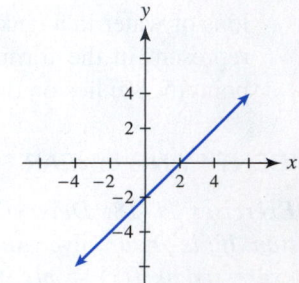

c. $\qquad\qquad\qquad\qquad$ **d.**

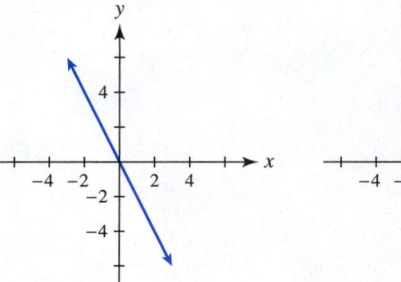 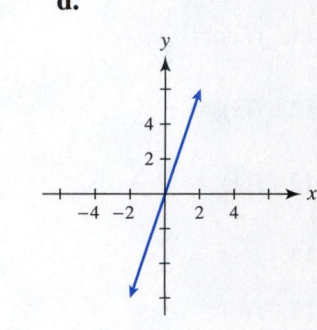

Exercises 53–62: Sketch a graph of $y = f(x)$.

53. $f(x) = 2$ $\qquad$ **54.** $f(x) = -1$

55. $f(x) = -\frac{1}{2}x$ $\qquad$ **56.** $f(x) = 2x$

57. $f(x) = x + 1$ $\qquad$ **58.** $f(x) = x - 2$

59. $f(x) = 3x - 3$

60. $f(x) = -2x + 1$

61. $f(x) = 3 - x$

62. $f(x) = \frac{1}{4}x + 2$

Exercises 63–68: Write a symbolic representation (formula) for a linear function f that calculates the following.

63. The number of pounds in x ounces

64. The number of dimes in x dollars

65. The distance traveled by a car moving at 65 miles per hour for t hours

66. The long-distance phone bill *in dollars* for calling t minutes at 10 cents per minute and a fixed fee of $4.95

67. The total number of hours in a day during day x

68. The total cost of downhill skiing x times with a $500 season pass

69. Thinking Generally For each 1-unit increase in x with $y = ax + b$ and $a > 0$, y increases by _____ units.

70. Thinking Generally For each 1-unit decrease in x with $y = cx + d$ and $c < 0$, y increases by _____ units.

Exercises 71–74: Do the following.

(a) *Give a verbal representation of f.*

(b) *Make a numerical representation (table) of f for $x = -2, 0, 2$.*

(c) *Plot the points listed in the table from part (b), then sketch a graph of f.*

71. $f(x) = -2x + 1$

72. $f(x) = 1 - x$

73. $f(x) = \frac{1}{2}x - 1$

74. $f(x) = \frac{3}{4}x$

 Exercises 75–78: Do the following.

(a) *Make a numerical representation (table) of f for $x = -3, -2, -1, \ldots, 3$.*

(b) *Graph f in the window $[-6, 6, 1]$ by $[-4, 4, 1]$.*

75. $f(x) = \frac{1}{3}x + \sqrt{2}$

76. $f(x) = -\frac{2}{3}x - \sqrt{3}$

77. $f(x) = \frac{x + 2}{5}$

78. $f(x) = \frac{2 - 3x}{7}$

MODELING

Exercises 79–82: Match the situation with the graph (a.–d.) at the top of the next column that models it best, where x-values represent time from 2000 to 2010.

79. The cost of college tuition

80. The cost of 1 gigabyte of computer memory

81. The distance between Chicago and Denver

82. The total distance traveled by a satellite that is orbiting Earth if the satellite was launched in 2000.

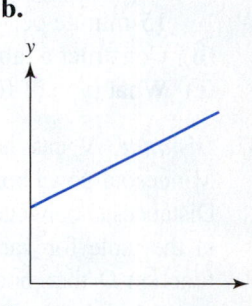

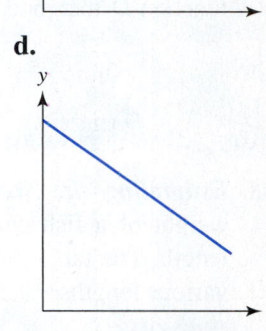

83. Online Exploration Look up the fuel efficiency E in miles per gallon for one of your favorite cars. (Answers will vary.)

(a) Find a function G that calculates the number of gallons required to travel x miles.

(b) If the cost of gasoline is $3 per gallon, find function C that calculates the cost of fuel to travel x miles.

84. Online Exploration Suppose that you would like to drive to Miami for spring break (if it is possible) in the car that you chose in Exercise 83. Calculate the gallons of gasoline needed for the trip.

APPLICATIONS

85. *Thermostat* Let $y = f(x)$ describe the temperature y of a room that is kept at $70° F$ for x hours.

(a) Represent f symbolically and graphically over a 24-hour period for $0 \leq x \leq 24$.

(b) Construct a table of f for $x = 0, 4, 8, 12, \ldots, 24$.

(c) What type of function is f?

86. *Cruise Control* Let $y = f(x)$ describe the speed y of an automobile after x minutes if the cruise control is set at 60 miles per hour.
  (a) Represent f symbolically and graphically over a 15-minute period for $0 \le x \le 15$.
  (b) Construct a table of f for $x = 0, 1, 2, \ldots, 6$.
  (c) What type of function is f?

87. *Distance* A car is initially 50 miles south of the Minnesota–Iowa border traveling south on Interstate 35. Distances D between the car and the border are recorded in the table for various elapsed times t. Find a linear function D that models these data.

t (hours)	0	2	3	5
D (miles)	50	170	230	350

88. *Estimating the Weight of a Bass* Sometimes the weight of a fish can be estimated by measuring its length. The table lists typical weights of bass having various lengths.

Length (inches)	12	14	16	18	20	22
Weight (pounds)	1.0	1.7	2.5	3.6	5.0	6.6

Source: Minnesota Department of Natural Resources.

  (a) Let x be the length and y be the weight. Make a line graph of the data.
  (b) Could the data be modeled accurately with a linear function? Explain your answer.

89. *Texting* In 2010, the average American under age 18 sent approximately 93 texts per day, whereas the average adult over age 65 sent approximately 1 text per day.
  (a) Find a formula for a function K that calculates the number of texts sent in x days by the average person under age 18.
  (b) Find a formula for a function A that calculates the number of texts sent in x days by the average person over age 65.
  (c) Evaluate $K(365)$ and $A(365)$. Interpret your results.

90. *Rain Forests* Rain forests are forests that grow in regions receiving more than 70 inches of rain per year. The world is losing about 49 million acres of rain forest each year. (*Source:* New York Times Almanac.)
  (a) Find a linear function f that calculates the change in the acres of rain forest in millions in x years.
  (b) Evaluate $f(7)$ and interpret the result.

91. *Car Sales* The table shows the number of U.S. Toyota vehicles sold in millions for past years.

Year	2000	2001	2002	2003	2004
Vehicles	1.6	1.7	1.8	1.9	2.0

Source: Autodata.

  (a) What were the sales in 2000?
  (b) What was the annual increase in sales?
  (c) Find a linear function f that models these data. Let $x = 0$ correspond to 2000, $x = 1$ to 2001, and so on.
  (d) Use f to estimate sales in 2006.

92. *Tuition and Fees* Suppose tuition costs $300 per credit and that student fees are fixed at $100.
  (a) Find a formula for a linear function T that models the cost of tuition and fees for x credits.
  (b) Evaluate $T(16)$ and interpret the result.

93. *Skype Users* The number of Skype users S in millions x years after 2006 can be modeled by the formula $S(x) = 110x + 123$.
  (a) How many users were there in 2010?
  (b) What does the number 123 indicate in the formula?
  (c) What does the number 110 indicate in the formula?

94. *Temperature and Volume* If a sample of a gas such as helium is heated, it will expand. The formula $V(T) = 0.147T + 40$ calculates the volume V in cubic inches of a sample of gas at temperature T in degrees Celsius.
  (a) Evaluate $V(0)$ and interpret the result.
  (b) If the temperature increases by $10°C$, by how much does the volume increase?
  (c) What is the volume of the gas when the temperature is $100°C$?

95. *Temperature and Volume* (Refer to the preceding exercise.) A sample of gas at $0°C$ has a volume V of 137 cubic centimeters, which increases in volume by 0.5 cubic centimeter for every $1°C$ increase in temperature T.
  (a) Write a formula $V(T) = aT + b$ that gives the volume of the gas at temperature T.
  (b) Find the volume of the gas when $T = 50°C$.

96. *Cost* To make a music video it costs $750 to rent a studio plus $5 for each copy produced.
  (a) Write a formula $C(x) = ax + b$ that calculates the cost of producing x videos.
  (b) Find the cost of producing 2500 videos.

97. *Weight Lifting* Lifting weights can increase a person's muscle mass. Each additional pound of muscle burns an extra 40 calories per day. Write a linear function that models the number of calories burned each day by x pounds of muscle. By burning an extra 3500 calories a person can lose 1 pound of fat. How many pounds of muscle are needed to burn 1 pound of fat in 30 days? (*Source: Runner's World.*)

98. *Wireless Households* The percentage P of wireless households x years after 2005 can be modeled by the formula $P(x) = 4x + 7$, where $0 \le x \le 7$.
(a) Evaluate $P(0)$ and $P(3)$. Interpret your results.
(b) Explain the meaning of 4 and 7 in the formula.

99. *Mobile Data Penetration* The table lists the percentage P of people with cell phones who also subscribed to a data package during year x. (For example, with a data package one can surf the Web and check email.)

Year	2007	2008	2009
Percentage	55%	59%	63%

(a) What was this percentage in 2007?
(b) By how much did this percentage change each year?
(c) Write a function P that models these data. Let x be years after 2007.
(d) Estimate this percentage in 2010.

100. *Wal-Mart Sales* The table shows Wal-Mart's share as a percentage of overall U.S. retail sales for past years. (This percentage excludes restaurants and motor vehicles.)

Year	1998	1999	2000	2001	2002
Share (%)	6	6.5	7	7.5	8

Source: Commerce Department, Wal-Mart.

(a) What was Wal-Mart's share in 1998?
(b) By how much (percent) did Wal-Mart's share increase each year?
(c) Find a linear function f that models these data. Let $x = 0$ correspond to 1998.
(d) Use f to estimate Wal-Mart's share in 2005.

WRITING ABOUT MATHEMATICS

101. Explain how you can determine whether a function is linear by using its
(a) symbolic representation,
(b) graphical representation, and
(c) numerical representation.

102. Describe one way to determine whether data can be modeled by a linear function.

Checking Basic Concepts

SECTIONS 2.1 and 2.2

1. Find a formula and sketch a graph for a function that squares the input x and then subtracts 1.

2. Use the graph of f to do the following.
(a) Find the domain and range of f.
(b) Evaluate $f(0)$ and $f(2)$.
(c) Is f a linear function? Explain.

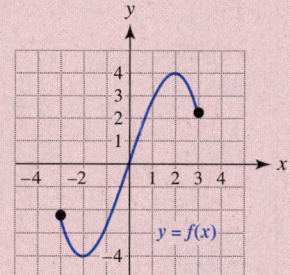

3. Determine whether f is a linear function.
(a) $f(x) = 4x - 2$
(b) $f(x) = 2\sqrt{x} - 5$
(c) $f(x) = -7$
(d) $f(x) = 9 - 2x + 5x$

4. Graph $f(x) = 4 - 3x$. Evaluate $f(-2)$.

5. Find a formula for a linear function that models the data.

x	0	1	2	3	4
$f(x)$	-1	$-\frac{1}{2}$	0	$\frac{1}{2}$	1

6. The median age in the United States from 1970 to 2010 can be approximated by
$$f(x) = 0.225x + 27.7,$$
where $x = 0$ corresponds to 1970, $x = 1$ to 1971, and so on.
(a) Evaluate $f(20)$ and interpret the result.
(b) Interpret the numbers 0.225 and 27.7.

2.3   The Slope of a Line

Slope • Slope–Intercept Form of a Line • Interpreting Slope in Applications

A LOOK INTO MATH ▶

Figure 2.33 shows some graphs of lines, where the x-axis represents time.

Which graph might represent the amount of gas in your car's tank while you are driving?
Which graph might represent the temperature inside a refrigerator?
Which graph might represent the amount of water in a hot tub that is being filled?

To answer these questions, you probably used the concept of slope. In mathematics, slope is a *number* that measures the "tilt" of a line in the xy-plane from left to right. We assume throughout the text that lines are always straight. In this section we discuss how slope relates to the graph of a linear function and how to interpret slope.

NEW VOCABULARY

☐ Rise/Change in *y*
☐ Run/Change in *x*
☐ Subscript
☐ Slope
☐ Positive slope
☐ Negative slope
☐ Undefined slope
☐ Zero slope
☐ *y*-intercept
☐ Slope–intercept form
☐ Rate of change

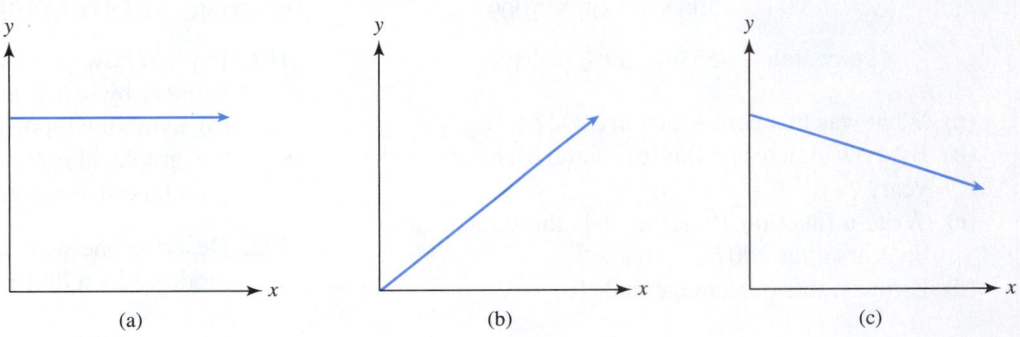

(a)   (b)   (c)

Figure 2.33

Slope

▶ **REAL-WORLD CONNECTION** Suppose we want to construct a stairway where each tread is 12 inches wide and the rise between steps is 6 inches high, as illustrated in Figure 2.34(a). We say that this stairway has a *rise* of 6 inches for each 12 inches of *run*. Next, if we draw a blue line along this stairway and include an x- and y-axis, as shown in Figure 2.34(b), then this blue line has a *slope* of $\frac{6}{12}$, or $\frac{1}{2}$. That is, the slope of a line equals *rise over run*. For a line, the ratio $\frac{\text{rise}}{\text{run}}$ equals the slope and is always the same regardless of where one is on the line.

Stairway   **Slope**

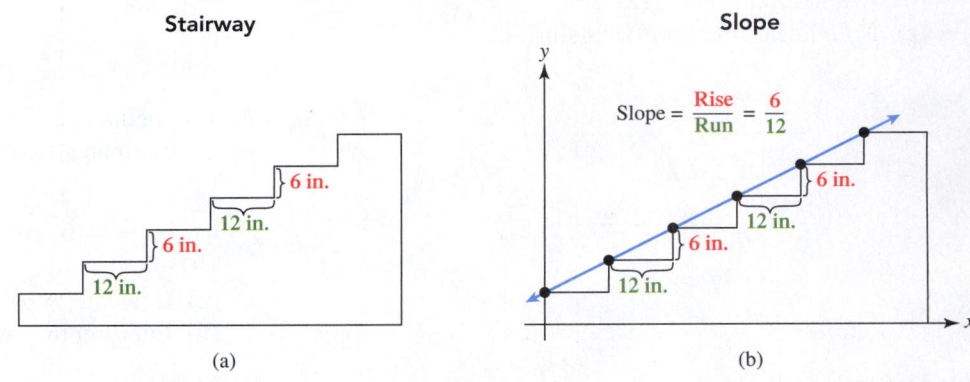

(a)   (b)

Figure 2.34

Slope can also be used to describe other things in real life. The graph shown in Figure 2.35 illustrates the cost of buying x pounds of candy. The graph tilts upward from left to right, which indicates that the cost increases as the number of pounds purchased increases. Note that for every 2 additional pounds purchased, the total cost increases by $3. The graph *rises* **3** units for every **2** units of *run*. The slope of this line is $\frac{3}{2}$, or 1.5. That is, for every unit of run along the *x*-axis, the graph rises 1.5 units. A slope of 1.5 indicates that candy costs $1.50 per pound.

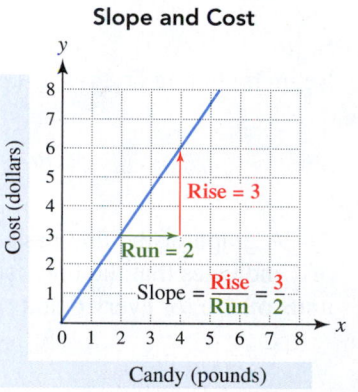

Figure 2.35 Cost of Candy

CALCULATING SLOPE If we know two points on a line, we can find its slope. A general case is shown in Figure 2.36, where a line passes through the points (x_1, y_1) and (x_2, y_2). The **rise** or **change in** y is $y_2 - y_1$, and the **run** or **change in** x is $x_2 - x_1$. The slope m of a line can be found using any of the following three formulas.

$$m = \frac{\textbf{rise}}{\textbf{run}}, \quad m = \frac{\textbf{Change in } y}{\textbf{Change in } x}, \quad \text{or} \quad m = \frac{y_2 - y_1}{x_2 - x_1}$$

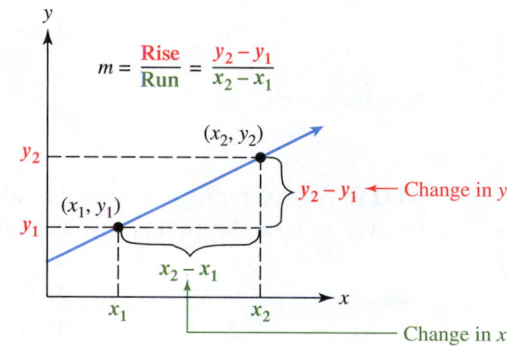

Figure 2.36

NOTE: The expression x_1 has a **subscript** of 1 and is read "*x* sub one" or "*x* one." Thus x_1 and x_2 denote two different *x*-values. Similar comments can be made about y_1 and y_2.

SLOPE

The **slope** m of the line passing through the points (x_1, y_1) and (x_2, y_2) is

$$m = \frac{y_2 - y_1}{x_2 - x_1},$$

where $x_1 \neq x_2$. That is, slope equals *rise over run*.

NOTE: *Change in x* is sometimes denoted Δx and equals $x_2 - x_1$. The expression Δx is read "delta x." Similarly, *change in y* is sometimes denoted Δy and equals $y_2 - y_1$. Using this notation, we can express slope as $m = \frac{\Delta y}{\Delta x} = \frac{y_2 - y_1}{x_2 - x_1}$.

EXAMPLE 1   Calculating the slope of a line

Find the slope of the line passing through the points $(-4, 1)$ and $(2, 4)$. Plot these points and graph the line. Interpret the slope.

Solution
Begin by letting $(x_1, y_1) = (-4, 1)$ and $(x_2, y_2) = (2, 4)$. The slope is

$$m = \frac{y_2 - y_1}{x_2 - x_1} = \frac{4 - 1}{2 - (-4)} = \frac{3}{6} = \frac{1}{2}.$$

A graph of the line passing through these two points is shown in Figure 2.37. A slope of $\frac{1}{2}$ indicates that the line rises 1 unit for every 2 units of run. This slope is equivalent to 3 units of rise for every 6 units of run.

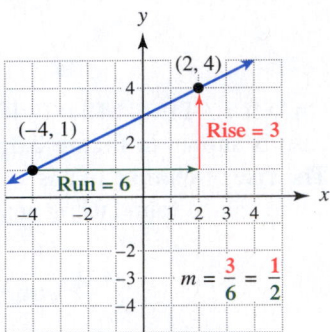

Figure 2.37

Now Try Exercise 35

NOTE: In Example 1 we would calculate the same slope if we let $(x_1, y_1) = (2, 4)$ and $(x_2, y_2) = (-4, 1)$. In this case the calculation would be

$$m = \frac{y_2 - y_1}{x_2 - x_1} = \frac{1 - 4}{-4 - 2} = \frac{-3}{-6} = \frac{1}{2}.$$

READING CHECK

Explain positive slope, negative slope, zero slope, and undefined slope.

If a line has **positive slope**, the line *rises* from *left to right*. In Figure 2.38 the rise is 2 units for each unit of run, so the slope is 2. If a line has **negative slope**, the line *falls* from left to right. In Figure 2.39 the line *falls* 1 unit for every 2 units of run, so the slope is $-\frac{1}{2}$. **Zero slope** indicates that a line is horizontal, as shown in Figure 2.40. If (x_1, y_1) and (x_2, y_2) are two points on a vertical line, $x_1 = x_2$. In Figure 2.41 the run is $x_2 - x_1 = 0$, so a vertical line has **undefined slope** because division by 0 is undefined.

Positive Slope	Negative Slope	Zero Slope	Undefined Slope
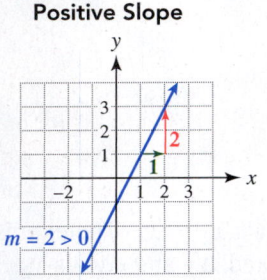	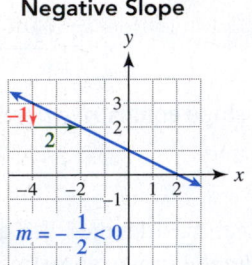	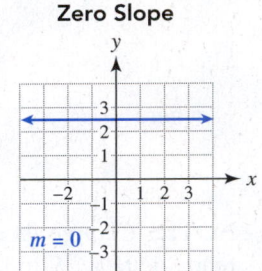	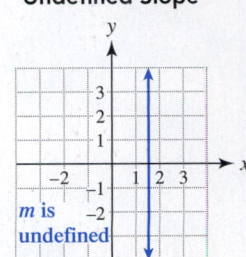
Figure 2.38	**Figure 2.39**	**Figure 2.40**	**Figure 2.41**

MAKING CONNECTIONS

Slope, Hills, and Driving

Suppose that one vehicle is going uphill while the other vehicle is going downhill. To avoid any confusion about "uphill" and "downhill," slope is *always calculated by moving left to right* along the *x-axis*. In Figure 2.42, the driver of the orange car is going uphill and experiencing a *positive* slope, while the driver of the purple car is going downhill and experiencing a *negative* slope. Slopes that have large absolute values correspond to steeper hills.

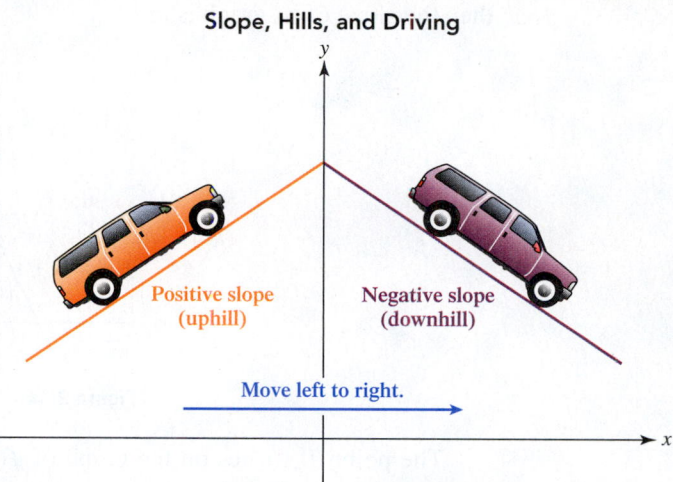

Figure 2.42

EXAMPLE 2   **Calculating slope**

Find the slope of the line passing through each pair of points, if possible.

(a) $(-1, 4), (8, 4)$ **(b)** $\left(-\frac{3}{4}, \frac{1}{4}\right), \left(-\frac{3}{4}, \frac{5}{4}\right)$ **(c)** $(a, 3b), (3a, 5b)$

Solution

(a) For $(-1, 4)$ and $(8, 4)$, $m = \dfrac{y_2 - y_1}{x_2 - x_1} = \dfrac{4 - 4}{8 - (-1)} = \dfrac{0}{9} = 0$.

(b) For $\left(-\frac{3}{4}, \frac{1}{4}\right)$ and $\left(-\frac{3}{4}, \frac{5}{4}\right)$, $m = \dfrac{y_2 - y_1}{x_2 - x_1} = \dfrac{\frac{5}{4} - \frac{1}{4}}{-\frac{3}{4} - \left(-\frac{3}{4}\right)} = \dfrac{1}{0}$, which is undefined.

(c) For $(a, 3b)$ and $(3a, 5b)$, $m = \dfrac{y_2 - y_1}{x_2 - x_1} = \dfrac{5b - 3b}{3a - a} = \dfrac{2b}{2a} = \dfrac{b}{a}$.

Now Try Exercises 21, 23, 31

EXAMPLE 3   **Sketching a line with a given slope**

Sketch a line passing through the point $(0, 4)$ and having slope $-\frac{2}{3}$.

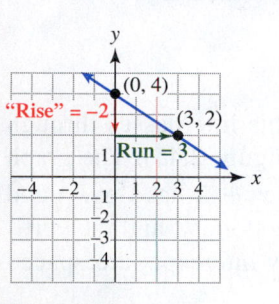

Figure 2.43

Solution

Start by plotting the point $(0, 4)$. Because $m = \dfrac{\text{change in } y}{\text{change in } x}$, a slope of $-\frac{2}{3}$ indicates that the *y*-values *decrease* **2** units each time the *x*-values increase by **3** units. That is, the line *falls* 2 units for every 3-unit increase in the run. The line passes through $(0, 4)$, so a 2-unit decrease in *y* and a 3-unit increase in *x* results in the line passing through the point $(0 + 3, 4 - 2)$ or $(3, 2)$, as shown in Figure 2.43.

Now Try Exercise 39

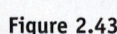

Slope–Intercept Form of a Line

Suppose we want to graph $f(x) = 2x + 3$. Because $f(0) = 3$ and $f(1) = 5$, the graph is a line that passes through the points $(0, 3)$ and $(1, 5)$, as shown in Figure 2.44. Therefore the slope of this line is

$$m = \frac{5 - 3}{1 - 0} = 2.$$

Note that slope **2** equals the coefficient of x in $f(x) = 2x + 3$. In general, if $f(x) = mx + b$, then the slope of its graph is m.

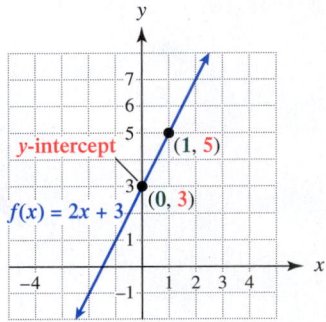

Figure 2.44

The point $(0, 3)$ lies on the graph of $f(x) = 2x + 3$ and is located on the y-axis. The y-value of **3** is called the *y-intercept*. A **y-intercept** is the y-coordinate of a point where a graph intersects the y-axis. Because every point on the y-axis has 0 for its x-coordinate, we can locate the y-intercept by letting $x = 0$ in the formula for $f(x)$. In general, if $f(x) = mx + b$, then the y-intercept is

$$f(0) = m \cdot 0 + b = b.$$

Thus if $f(x) = -4x + 7$, then the y-intercept is **7** and the slope is -4.

Because $y = f(x)$, any linear function is given by $y = mx + b$, where m is the slope and b is the y-intercept. The form $y = mx + b$ is called the *slope–intercept form* of a line.

SLOPE–INTERCEPT FORM

The line with slope m and y-intercept b is given by

$$y = mx + b,$$

the **slope–intercept form** of a line.

EXAMPLE 4 ▏ **Graphing lines**

Identify the slope and y-intercept for the three lines $y = \frac{1}{2}x - 2$, $y = \frac{1}{2}x$, and $y = \frac{1}{2}x + 2$. Graph and compare the lines.

Solution

The graph of $y = \frac{1}{2}x - 2$ has slope $\frac{1}{2}$ and y-intercept -2. This line passes through the point $(0, -2)$ and rises 1 unit for each 2 units of run (see Figure 2.45). The graph of $y = \frac{1}{2}x$ has slope $\frac{1}{2}$ and y-intercept 0, and the graph of $y = \frac{1}{2}x + 2$ has slope $\frac{1}{2}$ and y-intercept 2. These lines are parallel, because the lines all have slope $\frac{1}{2}$, and the vertical distance between adjacent lines is always 2 because their y-intercepts are spaced 2 units apart.

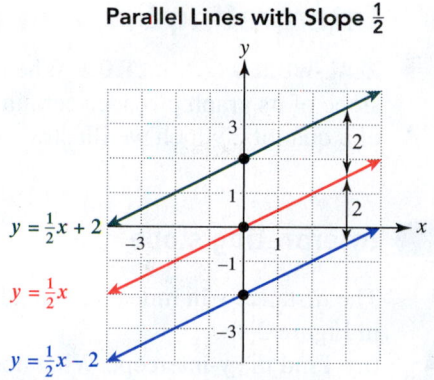

Figure 2.45

■ Now Try Exercise 49

EXAMPLE 5 | **Using a graph to write the slope–intercept form**

For the graphs shown in Figures 2.46 and 2.47, write the slope–intercept form of the line.

(a)

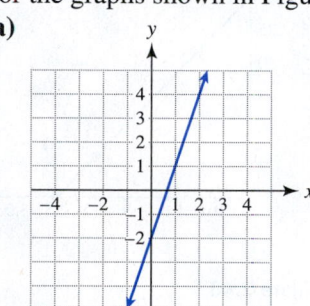

Figure 2.46

(b)

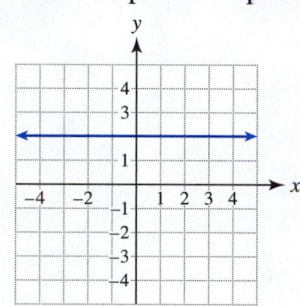

Figure 2.47

Solution

(a) The graph passes through $(0, -2)$, so the y-intercept is -2. Because the graph rises **3** units for each unit increase in x, the slope is 3. The slope–intercept form is $y = 3x - 2$.

(b) The graph passes through $(0, 2)$, so the y-intercept is **2**. Because the graph is a horizontal line, its slope is **0**. The slope–intercept form is $y = 0x + 2$, or more simply, $y = 2$.

■ Now Try Exercises 57, 59

EXAMPLE 6 | **Finding the slope–intercept form**

TABLE 2.13

x	y
-1	1
0	?
2	10

The points listed in Table 2.13 all lie on a line.

(a) Find the missing value in the table.

(b) Write the slope–intercept form of the line.

Solution

(a) The line passes through $(-1, 1)$ and $(2, 10)$, so its slope is

$$m = \frac{10 - 1}{2 - (-1)} = \frac{9}{3} = 3.$$

For each unit increase in x, y increases by 3. When x increases from -1 to 0, y increases from 1 to $1 + 3 = 4$. Therefore the missing value is 4.

(b) Because the line passes through the point $(0, 4)$, its y-intercept is **4**. The slope–intercept form is $y = 3x + 4$.

■ Now Try Exercise 73

Interpreting Slope in Applications

▶ **REAL-WORLD CONNECTION** When a linear function is used to model physical quantities, the slope of its graph provides certain information. Slope can be interpreted as a **rate of change** of a quantity, which we illustrate in the next four examples.

EXAMPLE 7 **Interpreting slope**

The distance y in miles that an athlete riding a bicycle is from home after x hours is shown in Figure 2.48.
(a) Find the y-intercept. What does the y-intercept represent?
(b) The graph passes through the point $(2, 6)$. Discuss the meaning of this point.
(c) Find the slope–intercept form of this line. Interpret the slope as a rate of change.

Distance from Home

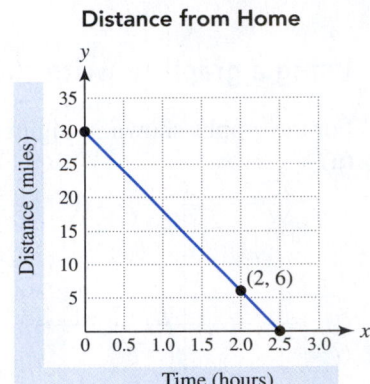

Figure 2.48

Solution
(a) The y-intercept is **30**, which indicates that the athlete is initially 30 miles from home.
(b) The point $(2, 6)$ means that after 2 hours the athlete is 6 miles from home.
(c) The line passes through the points $(0, 30)$ and $(2, 6)$. Thus its slope is

$$m = \frac{6 - 30}{2 - 0} = -12,$$

and the slope–intercept form is $y = -12x + 30$. A slope of -12 indicates that the athlete is traveling at 12 miles per hour *toward* home. The negative sign indicates that the distance between the athlete and home is *decreasing*.

▌ **Now Try Exercise 83**

READING CHECK

Explain how slope can be interpreted as a rate of change.

EXAMPLE 8 **Interpreting slope**

Water is being pumped into a 500-gallon tank during a 10-minute interval. The amount of water W in the tank after t minutes is given by $W(t) = 40t + 100$, where W is in gallons.
(a) Evaluate $W(0)$ and $W(10)$. Interpret each result.
(b) Graph W for $0 \le t \le 10$.
(c) What is the slope of the graph of W? Interpret this slope.

Solution
(a) $W(0) = 40(0) + 100 = 100$; $W(10) = 40(10) + 100 = 500$. Initially, there are 100 gallons of water in the tank. After 10 minutes, there are 500 gallons and the tank is full.
(b) Because $W(0) = 100$ and $W(10) = 500$, plot the points $(0, 100)$ and $(10, 500)$. Function W is linear, so connect these points with a line segment. See Figure 2.49.

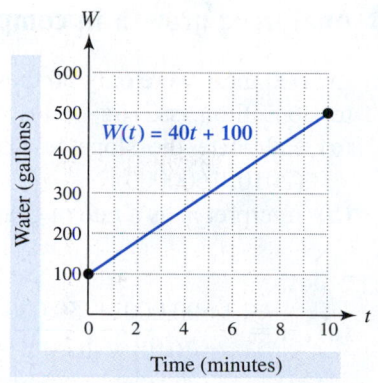

Figure 2.49

(c) The slope is 40, which indicates that water is being pumped into the tank at a rate of 40 gallons per minute.

■ **Now Try Exercise 89**

EXAMPLE 9 | **Analyzing growth of Wal-Mart**

Table 2.14 lists numbers of Wal-Mart employees.
(a) Make a line graph of the data.
(b) Find the slope of each line segment in the graph.
(c) Interpret these slopes as rates of change.

TABLE 2.14 Wal-Mart Employees (millions)

Year	Employees
1999	1.1
2002	1.4
2007	2.2
2010	2.1

Source: Wal-Mart.

Solution

(a) To make a line graph, start by plotting the points (1999, 1.1), (2002, 1.4), (2007, 2.2), and (2010, 2.1). Connecting these points with line segments results in Figure 2.50.

Wal-Mart Employees

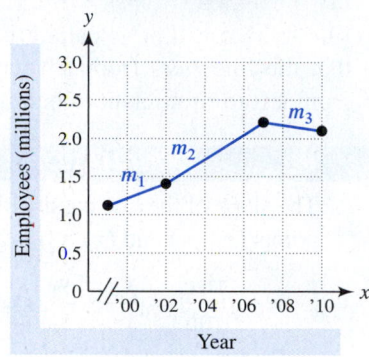

Figure 2.50

(b) The slopes of the line segments in Figure 2.50 are

$$m_1 = \frac{1.4 - 1.1}{2002 - 1999} = 0.1, \quad m_2 = \frac{2.2 - 1.4}{2007 - 2002} = 0.16, \text{ and}$$

$$m_3 = \frac{2.1 - 2.2}{2010 - 2007} = -\frac{1}{30} \approx -0.033.$$

(c) Slope $m_1 = 0.1$ means that, on average, the number of Wal-Mart employees increased by 0.1 million (or 100,000) per year between 1999 and 2002. Slopes m_2 and m_3 can be interpreted similarly. Note that m_3 is *negative* and indicates that the number of employees *decreased*.

■ **Now Try Exercise 79**

EXAMPLE 10   **Analyzing growth in computer viruses**

In 2000 there were only 50 thousand computer viruses, but by 2010 this number had grown to 1.6 million. (*Source*: Symantec.)
(a) Calculate the slope *m* of a line that passes through the points (2000, 50000) and (2010, 1600000).
(b) Interpret *m* as a rate of change.

Solution

(a) $m = \dfrac{1{,}600{,}000 - 50{,}000}{2010 - 2000} = \dfrac{1{,}550{,}000}{10} = 155{,}000$

(b) *On average*, the number of computer viruses has increased by 155,000 per year from 2000 to 2010.

Now Try Exercise 97

CRITICAL THINKING

An athlete runs away from home at 10 miles per hour for 30 minutes and then jogs back home at 5 miles per hour. Sketch a graph that shows the distance between the athlete and home.

2.3   Putting It All Together

The "tilt" of a line is called the slope and equals rise over run. A positive slope indicates that the line *rises* from left to right, whereas a negative slope indicates that the line *falls* from left to right. A horizontal line has zero slope and a vertical line has undefined slope.

CONCEPT	EXPLANATION	EXAMPLE
Slope	The slope of the line passing through the points (x_1, y_1) and (x_2, y_2) is given by $$m = \frac{\text{rise}}{\text{run}} = \frac{y_2 - y_1}{x_2 - x_1}, (x_1 \neq x_2).$$ rise = change in y run = change in x	The slope of the line passing through $(-2, 3)$ and $(1, 5)$ is $$m = \frac{5 - 3}{1 - (-2)} = \frac{2}{3}.$$ If the x-values increase by 3 units, the y-values increase by 2 units.
Slope–Intercept Form for a Line	The slope equals m, and the y-intercept equals b. $$y = mx + b$$	If $y = \frac{1}{2}x + 1$, the slope of the graph is $\frac{1}{2}$ and the y-intercept is 1.

CONCEPT	EXPLANATION	EXAMPLE
Slope as a Rate of Change	The slope of the graph of a linear function indicates the rate at which a quantity is either increasing or decreasing.	From 1981 to 2010, average public college tuition and fees can be modeled by $$f(x) = 175x + 683,$$ where $x = 1$ corresponds to 1981. The slope of the graph of f is $m = 175$ and indicates that, on average, tuition and fees increased by \$175 per year between 1981 and 2010.

2.3 Exercises

MyMathLab

PRACTICE   WATCH   DOWNLOAD   READ   REVIEW

CONCEPTS AND VOCABULARY

1. Slope is equal to *change in* _____ divided by *change in* _____.

2. If two points have the same y-coordinates but different x-coordinates, then the slope of the line passing through these points is _____.

3. A line with positive slope _____ from left to right.

4. A line with slope 0 is a(n) _____ line.

5. A line with undefined slope is a(n) _____ line.

6. The form $y = mx + b$ is called the _____ form for a line.

INTERPRETATION OF *y*-INTERCEPTS

Exercises 7–10: Explain the meaning of the y-intercept in each situation.

7. $C(x) = \frac{1}{3}x + 9$ models the cost in billions of dollars of pet health care x years after 1999.

8. $A(x) = 1000 - 5x$ models the gallons of water in a large tank.

9. $S(x) = 110x + 123$ models the number of Skype users in millions x years after 2006.

10. $K(x) = 93x$ models the number of texts sent in x days by the average American kid.

SLOPE

Exercises 11–16: Find the slope of the line.

11.

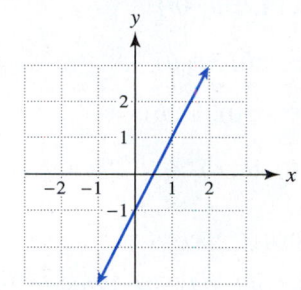

12.

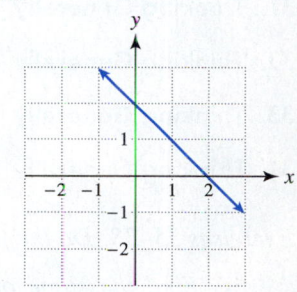

13.

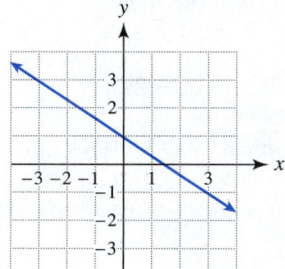

14.

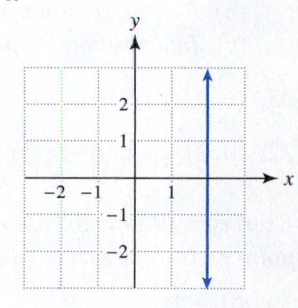

15.

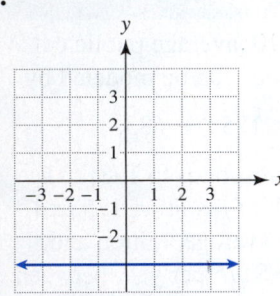

16.

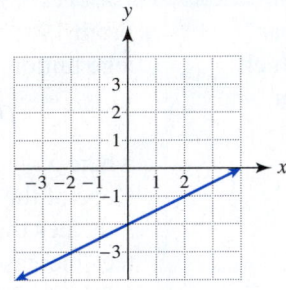

Exercises 17–34: Find the slope of the line passing through the two points.

17. $(1, 2), (2, 4)$

18. $(-3, 2), (2, -3)$

19. $(2, 1), (-1, 3)$

20. $\left(-\frac{3}{5}, -\frac{4}{5}\right), \left(\frac{4}{5}, \frac{3}{5}\right)$

21. $(-3, 6), (4, 6)$

22. $(-3, 5), (-5, 5)$

23. $\left(\frac{1}{2}, -\frac{2}{7}\right), \left(\frac{1}{2}, \frac{13}{17}\right)$

24. $(-1, 6), (-1, -4)$

25. $\left(\frac{1}{3}, -\frac{4}{3}\right), \left(\frac{1}{6}, \frac{1}{3}\right)$

26. $\left(-\frac{1}{2}, \frac{3}{2}\right), \left(2, \frac{1}{2}\right)$

27. $(1989, 10), (1999, 16)$

28. $(1950, 6.1), (2000, 10.6)$

29. $(2.1, 3.6), (-1.2, 4.3)$

30. $(12, -34), (14, 64)$

31. **Thinking Generally** $(2a, b), (3a, 4b)$

32. **Thinking Generally** $(-a, -b), (b, a)$

33. **Thinking Generally** $(a + b, 0), (a, b)$

34. **Thinking Generally** $(a, b), (b, a)$

Exercises 35–38: Do the following.

   (a) *Find the slope of the line passing through the given points.*
   (b) *Plot these points and graph the line.*
   (c) *Interpret the slope.*

35. $(-1, 2), (3, -1)$

36. $(-2, -2), (1, 3)$

37. $(0, 2), (-3, 0)$

38. $(-1, 2), (1, 2)$

Exercises 39–46: Sketch a line passing through the given point with slope m.

39. $(0, -2), m = 3$

40. $(0, 1), m = -1$

41. $(0, 4), m = -\frac{1}{2}$

42. $(0, -3), m = \frac{2}{3}$

43. $(-1, 1), m = -2$

44. $(2, 1), m = \frac{1}{2}$

45. $(2, -3), m = \frac{1}{2}$

46. $(-2, 3), m = -\frac{3}{5}$

SLOPE–INTERCEPT FORM

Exercises 47–56: Do the following.

   (a) *Find the slope and y-intercept of the line.*
   (b) *Graph the equation.*

47. $y = x + 2$

48. $y = x - 2$

49. $y = -3x + 2$

50. $y = \frac{1}{2}x - 1$

51. $y = \frac{1}{3}x$

52. $y = -2x$

53. $y = 2$

54. $y = -3$

55. $y = -x + 3$

56. $y = \frac{2}{3}x - 2$

Exercises 57–62: Use the graph to express the line in slope–intercept form.

57.

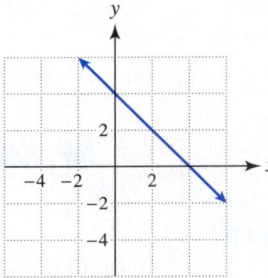

58.

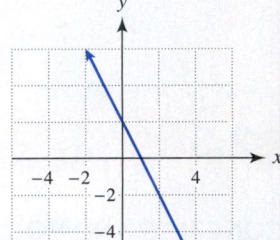

59.

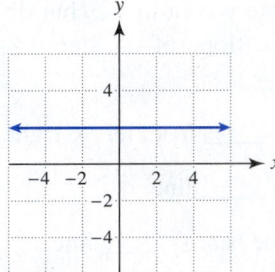

60.

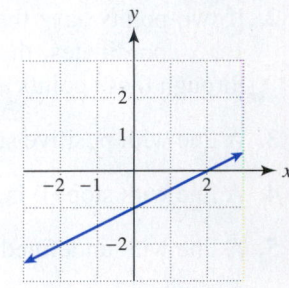

61.

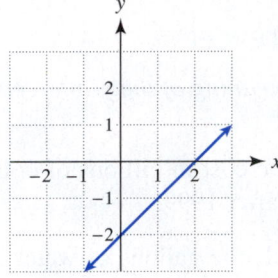

62.

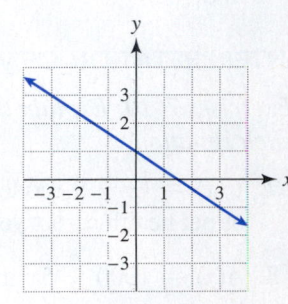

Exercises 63–66: Write the slope–intercept form for a line satisfying the following conditions.

63. Slope 3, *y*-intercept -5

64. Slope $-\frac{2}{3}$, *y*-intercept 7

65. Passing through $(1, 0)$ and $\left(0, -\frac{3}{2}\right)$

66. Passing through $(0, 4)$ and $(-2, 0)$

Exercises 67–70: Let f be a linear function. Use the table to find the slope and y-intercept of the graph of f.

67.

x	0	1	2	3
$f(x)$	-2	0	2	4

68.

x	-1	0	1	2
$f(x)$	5	10	15	20

69.

x	-2	-1	1	2
$f(x)$	18	11	-3	-10

70.

x	-4	-2	2	4
$f(x)$	6	3	-3	-6

Exercises 71–74: Let f(x) represent a linear function.

   (a) Find the missing value in the table.
   (b) Write the slope–intercept form for f.

71.

x	0	1	2
$f(x)$	-1	1	?

72.

x	1	2	3
$f(x)$	12	8	?

73.

x	-2	0	4
$f(x)$	2	?	11

74.

x	0	5	20
$f(x)$	?	10	40

INTERPRETING SLOPE

*Exercises 75–78: **Modeling** Choose the graph (a.–d.) at the top of the next column that models the situation best.*

75. Money that a person earns after x hours, working for $10 per hour

76. Total acres of rain forests in the world during the past 20 years

77. World population from 1980 to 2010

78. Square miles of land in Nevada from 1950 to 2010

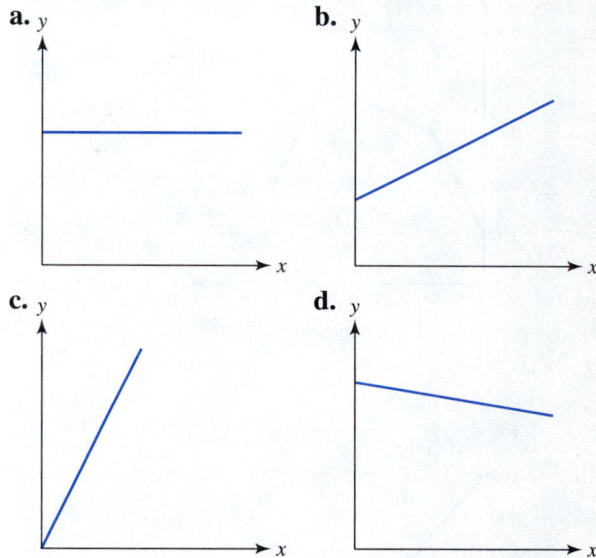

*Exercises 79–82: **Modeling** Each line graph represents the gallons of water in a small swimming pool after x hours. Assume that a pump can either add water to or remove water from the pool.*

   (a) Estimate the slope of each line segment.
   (b) Interpret each slope as a rate of change.
   (c) Describe what happened to the amount of water in the pool.

79.

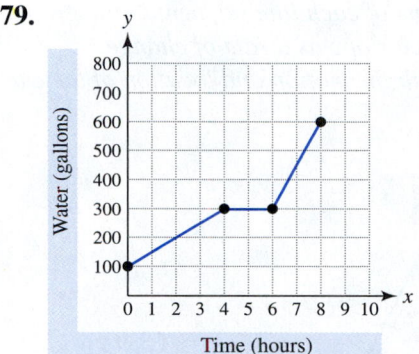

80.

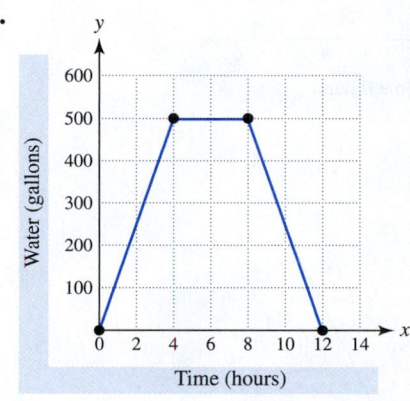

81.

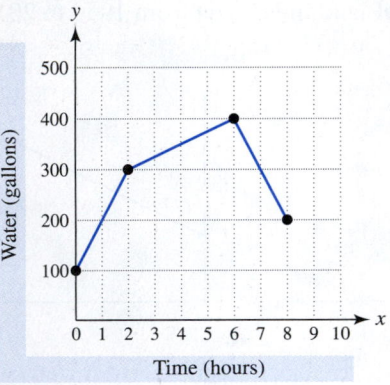

82.

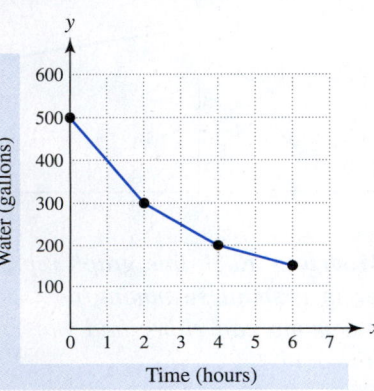

*Exercises 83 and 84: **Modeling Distance** An individual is driving a car along a straight road. The graph shows the distance that the driver is from home after x hours.*

(a) *Find the slope of each line segment in the graph.*
(b) *Interpret each slope as a rate of change.*
(c) *Describe both the motion and location of the car.*

83.

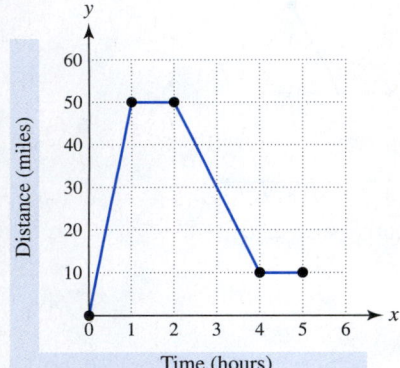

84.

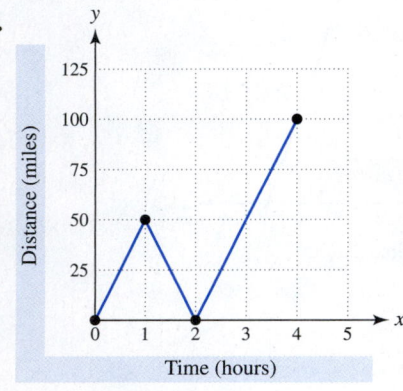

*Exercises 85–88: **Sketching a Model** Sketch a graph that models the given situation.*

85. The distance that a bicycle rider is from home if the rider is initially 20 miles away from home and arrives home after riding at a constant speed for 2 hours

86. The distance that an athlete is from home if the athlete runs away from home at 8 miles per hour for 30 minutes and then walks back home at 4 miles per hour

87. The distance that a person is from home if this individual drives (at a constant speed) to a mall, stays 2 hours, and then drives home, assuming that the distance to the mall is 20 miles and that the trip takes 30 minutes

88. The amount of water in a 10,000-gallon swimming pool that is filled at the rate of 1000 gallons per hour, left full for 10 hours, and then drained at the rate of 2000 gallons per hour

APPLICATIONS

89. *Online Ads* Revenues R in billions of dollars for online advertisements from 2005 to 2010 can be estimated by the formula

$$R(x) = \frac{10}{3}x - 6670,$$

where x is the year. (*Source:* Interactive Advertising Bureau.)
(a) Estimate the revenue in 2005 and in 2008.
(b) Graph $y = R(x)$.
(c) What is the slope of the graph?
(d) Interpret the slope as a rate of change.

90. *Newspaper Ads* Revenues R in billions of dollars for newspaper advertisements from 2005 to 2010 can be estimated by

$$R(x) = -\frac{13}{3}x + 8736,$$

where x is the year. (*Source:* Newspaper Association of America.)
(a) Estimate the revenue in 2005 and in 2008.
(b) Graph $y = R(x)$.
(c) What is the slope of the graph?
(d) Interpret the slope as a rate of change.

91. *Marathon Finishers* From 2000 to 2010, the number of marathon finishers M in thousands t years after 2000 can be estimated by

$$M(t) = 18.8t + 299.$$

(*Source:* Marathon Guide.)
(a) How many marathon finishers were there during 2009?
(b) What is the slope and y-intercept for the graph of the linear function M?

(c) Interpret the slope as a rate of change.
(d) Interpret the *y*-intercept.

92. *Number of Radio Stations* The number of radio stations on the air from 1950 to 2008 may be modeled by

$$f(x) = 201.7x - 390,477,$$

where *x* is the year. (*Source:* National Association of Broadcasters.)
(a) How many radio stations were on the air in 1980?
(b) What is the slope of the graph of *f*?
(c) Interpret the slope as a rate of change.

93. *Filling a Swimming Pool* A 20,000-gallon swimming pool is being filled at a constant rate. Over a 5-hour period the water in the pool increases from $\frac{1}{4}$ full to $\frac{5}{8}$ full. At what rate is water entering the pool?

94. *Speed of an Airplane* An airplane on a 1200-mile trip is flying at a constant rate. Over a 2-hour period the location of the plane changes from covering $\frac{1}{3}$ of the distance to covering $\frac{7}{8}$ of the distance. What is the speed of the airplane?

95. *Cost of Carpet* The graph shows the cost of buying *x* square yards of carpet. Find the slope of this graph and interpret the result.

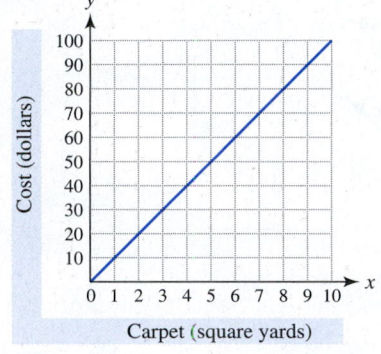

96. *Cost of Removing Snow* The graph shows the cost of removing *x* tons of snow. Find the slope of this graph and interpret the result.

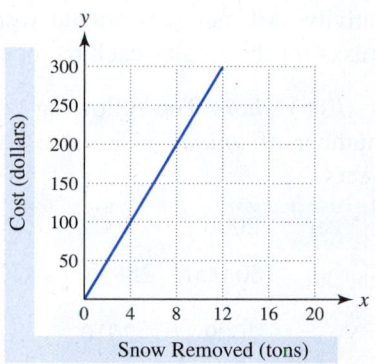

97. *Pet Health Care* In 1984 spending on health care for pets was $4 billion, and by 2008 it had increased to $12 billion. (*Source:* American Express Institute.)
(a) Calculate the slope *m* of the line that passes through the points (1984, 4) and (2008, 12).
(b) Interpret *m* as a rate of change.

98. *Manufacturing Employment* In 1956, employees of manufacturing firms accounted for 30% of the total employment in the United States. In 2010, this percentage was 9%. (*Source:* U.S. Census Bureau.)
(a) Calculate the slope *m* of the line that passes through (1956, 30) and (2010, 9).
(b) Interpret *m* as a rate of change.

99. *U.S. Median Family Income* In 1990, the median family income was about $35,000, and in 2007 it was about $60,000. (*Source:* Department of the Treasury.)
(a) Let *x* be the number of years after 1990. Find values for *m* and *b* so that $f(x) = mx + b$ models the data.
(b) Estimate the median family income in 2000.

100. *Minimum Wage* In 1980, the minimum wage was about $3.10 per hour, and in 2009 it increased to $7.25. (*Source:* Department of Labor.)
(a) Let *x* be the number of years after 1980. Find values for *m* and *b* so that $f(x) = ax + b$ models the data.
(b) If this trend continues, what should be the minimum wage in 2020?

WRITING ABOUT MATHEMATICS

101. Describe the information that the slope *m* of a line gives. Be as complete as possible.

102. Could one line have two different slope–intercept forms? Explain your answer.

Group Activity   Working with Real Data

Directions: Form a group of 2 to 4 people. Select someone to record the group's responses for this activity. All members should work cooperatively to answer the questions. If your instructor asks for the results, each member of the group should be prepared to respond.

U.S. Craigslist Visitors The following table lists the average number of *unique* visitors to Craigslist for selected years.

Year	2006	2007	2008
Visitors	180,000	288,000	420,000

Year	2009	2010
Visitors	516,000	624,000

Soruce: Citi Investment Research and Analysis.

(a) Make a scatterplot of the data. Let *x* represent the number of years after 2006. Discuss any trend in numbers of visitors to Craigslist.

(b) Estimate the slope of a line that could be used to model the data.

(c) Find an equation of a line $y = mx + b$ that models the data.

(d) Interpret the slope as a rate of change.

(e) Use your results to estimate numbers of *unique* visitors to Craigslist in 2012.

2.4   Equations of Lines and Linear Models

Point–Slope Form • Horizontal and Vertical Lines • Parallel and Perpendicular Lines

A LOOK INTO MATH ▶

In 2000, there were approximately 124 million Internet users in the United States, and this number grew to about 240 million in 2010. This growth is illustrated in Figure 2.51, where the line passes through the points (2000, 124) and (2010, 240). In this section we discuss how to find the equation of the line that models these data. To do so we start by discussing the *point–slope form* of a line.

NEW VOCABULARY

☐ Equation of a line
☐ Point–slope form
☐ Equivalent equations
☐ *x*-intercept
☐ Horizontal line
☐ Vertical line
☐ Parallel line
☐ Perpendicular line
☐ Negative reciprocals

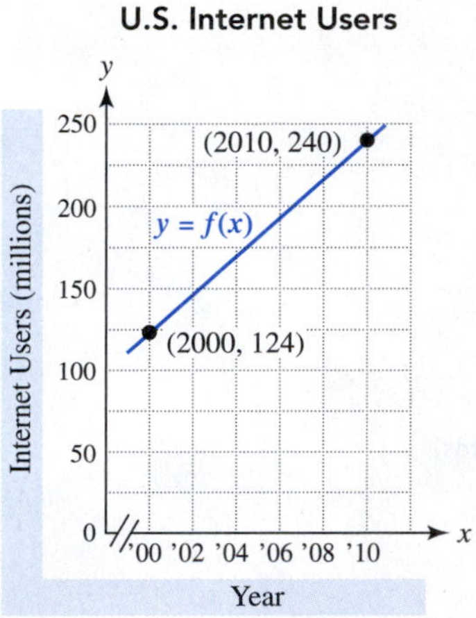

U.S. Internet Users

Figure 2.51

Point–Slope Form

If we know the slope m and y-intercept b of a line, we can write its slope–intercept form, $y = mx + b$. The slope–intercept form is an example of an **equation of a line**. The point–slope form is a different form of the equation of a line.

Suppose that a (nonvertical) line with slope m passes through the point (x_1, y_1). If (x, y) is a different point on this line, then $m = \dfrac{y - y_1}{x - x_1}$ (see Figure 2.52).

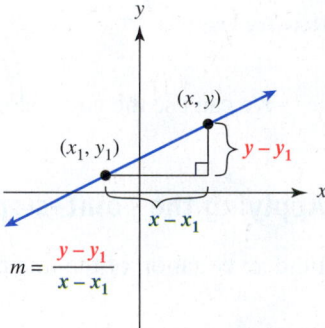

Figure 2.52

We can use this slope formula to find the point–slope form.

$$m = \frac{y - y_1}{x - x_1} \qquad \text{Slope formula}$$

$$m(x - x_1) = y - y_1 \qquad \text{Multiply each side by } (x - x_1).$$

$$y - y_1 = m(x - x_1) \qquad \text{Rewrite the equation.}$$

$$y = m(x - x_1) + y_1 \qquad \text{Add } y_1 \text{ to each side.}$$

The equation $y - y_1 = m(x - x_1)$ is traditionally called the *point–slope form*. We frequently think of y as being a function of x, written $y = f(x)$, so the equivalent form $y = m(x - x_1) + y_1$ is also referred to as the point–slope form.

> **POINT–SLOPE FORM**
>
> The line with slope m passing through the point (x_1, y_1) is given by
> $$y = m(x - x_1) + y_1,$$
> or, equivalently,
> $$y - y_1 = m(x - x_1),$$
> the **point–slope form** of a line.

EXAMPLE 1 **Using the point–slope form**

Sketch a line passing through the point $(1, 2)$ with slope -3. Find the point–slope form of this line. Does the point $(2, -1)$ lie on this line?

Solution

First, plot the point $(1, 2)$. Because the slope is -3, this line decreases 3 units along the y-axis for each 1-unit increase along the x-axis. See Figure 2.53.

To find the equation of this line, let $m = -3$ and $(x_1, y_1) = (1, 2)$ in the point–slope form.

$$y = m(x - x_1) + y_1 \qquad \text{Point–slope form}$$

$$y = -3(x - 1) + 2 \qquad \text{Substitute.}$$

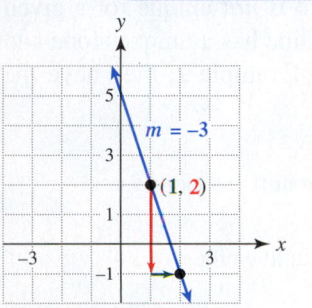

Figure 2.53 Slope $= -3$

To determine whether the point $(2, -1)$ lies on the line, substitute 2 for x and -1 for y in the equation $y = -3(x - 1) + 2$.

$$-1 \stackrel{?}{=} -3(2 - 1) + 2 \qquad \text{Let } x = 2 \text{ and } y = -1.$$

$$-1 \stackrel{?}{=} -3 + 2 \qquad \text{Simplify.}$$

$$-1 = -1 \checkmark \qquad \text{The point satisfies the equation.}$$

The point $(2, -1)$ lies on the line because it satisfies the point–slope form.

Now Try Exercise 39

We can use the point–slope form to find the equation of a line passing through two points.

| **EXAMPLE 2** | **Applying the point–slope form** |

Find an equation of the line passing through $(-2, 3)$ and $(6, -1)$.

Solution
Before we can apply the point–slope form, we must find the slope.

$$m = \frac{y_2 - y_1}{x_2 - x_1} \qquad \text{Slope formula}$$

$$= \frac{-1 - 3}{6 - (-2)} \qquad \text{Substitute.}$$

$$= -\frac{1}{2} \qquad \text{Simplify.}$$

We can use either $(-2, 3)$ or $(6, -1)$ for (x_1, y_1) in the point–slope form. If we choose $(-2, 3)$, the point–slope form becomes the following.

$$y = m(x - x_1) + y_1 \qquad \text{Point–slope form}$$

$$y = -\frac{1}{2}(x - (-2)) + 3 \qquad \text{Let } x_1 = -2 \text{ and } y_1 = 3.$$

$$y = -\frac{1}{2}(x + 2) + 3 \qquad \text{Simplify.}$$

READING CHECK

Give the slope–intercept and the point–slope forms of a line.

If we choose $(6, -1)$, the point–slope form with $x_1 = 6$ and $y_1 = -1$ becomes

$$y = -\frac{1}{2}(x - 6) - 1.$$

Note that, although the two point–slope forms appear to be different, they are **equivalent equations** because their graphs are identical.

Now Try Exercise 33

NOTE: Example 2 illustrates the fact that the point–slope form *is not* unique for a given line. However, the slope–intercept form *is* unique because each line has a unique slope and a unique y-intercept. If we simplify both point–slope forms in Example 2, they have the same slope–intercept form.

$$y = -\frac{1}{2}(x + 2) + 3 \qquad y = -\frac{1}{2}(x - 6) - 1 \qquad \text{Different point–slope forms}$$

$$y = -\frac{1}{2}x - 1 + 3 \qquad y = -\frac{1}{2}x + 3 - 1 \qquad \text{Distributive property}$$

$$y = -\frac{1}{2}x + 2 \qquad y = -\frac{1}{2}x + 2 \qquad \text{Identical slope–intercept forms}$$

We can use the *slope–intercept form*, $y = mx + b$, instead of the point–slope form to find the equation of a line, as illustrated in the next example.

EXAMPLE 3   **Applying the slope–intercept form**

Find the equation of the line that passes through the points $(-3, 9)$ and $(1, 1)$.

Solution
First, find the slope of the line.

$$m = \frac{y_2 - y_1}{x_2 - x_1} = \frac{1 - 9}{1 - (-3)} = -\frac{8}{4} = -2$$

Now substitute -2 for m and $(-3, 9)$ for x and y in the slope–intercept form. (The point $(1, 1)$ could be used instead.)

$y = mx + b$	Slope–intercept form
$9 = -2(-3) + b$	Let $y = 9$, $m = -2$, and $x = -3$.
$9 = 6 + b$	Simplify.
$3 = b$	Solve for b.

Thus the slope–intercept form is $y = -2x + 3$.

■ **Now Try Exercise 47**

▶ **REAL-WORLD CONNECTION** In the next example we model the data presented in A look Into Math for this section.

EXAMPLE 4   **Modeling growth in Internet users**

In 2000, there were approximately 124 million Internet users in the United States, and this number grew to about 240 million in 2010 (see Figure 2.51).
(a) Find values for m, x_1, and y_1 so that $f(x) = m(x - x_1) + y_1$ models these data.
(b) Interpret this equation.
(c) Use f to estimate Internet users in 2012.

Solution
(a) The slope of the line passing through $(2000, 124)$ and $(2010, 240)$ is

$$m = \frac{240 - 124}{2010 - 2000} = \frac{116}{10} = 11.6.$$

Thus by choosing the point $(2000, 124)$ for the point–slope form, we can write

$$f(x) = 11.6(x - 2000) + 124.$$

(b) From the point–slope form it is apparent that $m = 11.6$ and that $f(2000) = 124$. This means that in 2000 there were 124 million Internet users and that this number grew, on average, by 11.6 million per year.
(c) $f(2012) = 11.6(2012 - 2000) + 124 \approx 263$ million.

■ **Now Try Exercise 117**

Intercepts

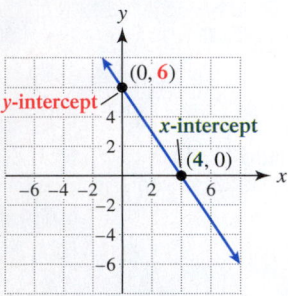

Figure 2.54 x-intercept 4; y-intercept 6

MAKING CONNECTIONS

Modeling and the Dependent Variable x

From Example 4, $f(x) = 11.6(x - 2000) + 124$ models the number of Internet users in the United States. In this formula x represents the actual year. We could also model the number of Internet users with $g(x) = 11.6x + 124$, where $x = 0$ corresponds to 2000, $x = 1$ to 2001, and so on. Then to estimate the number of Internet users in 2012, we let $x = 12$.

FINDING x- AND y-INTERCEPTS If a line intersects the y-axis at the point $(0, 6)$, the *y-intercept* is **6**. Similarly, if the line intersects the x-axis at the point $(4, 0)$, the *x-intercept* is **4**. This line and its intercepts are illustrated in Figure 2.54. The x-coordinate of a point where a graph intersects the x-axis is called the **x-intercept**. The next example interprets intercepts in a physical situation.

| EXAMPLE 5 | **Modeling water in a pool** |

A small swimming pool containing 6000 gallons of water is emptied by a pump removing 1200 gallons per hour.
(a) How long does it take to empty the pool?
(b) Sketch a function f that models the amount of water in the pool after x hours.
(c) Identify the x-intercept and the y-intercept. Interpret each intercept.
(d) Find the slope–intercept form of the line. Interpret the slope as a rate of change.
(e) What are the domain and range of f?

Solution

(a) The time needed to empty the pool is $\frac{6000}{1200} = 5$ hours.
(b) Initially the pool contains 6000 gallons, and after 5 hours the pool is empty. Because water is being removed at a constant rate the graph of f is a *line* (segment) passing through $(0, 6000)$ and $(5, 0)$, as shown in Figure 2.55.

Water in a Pool

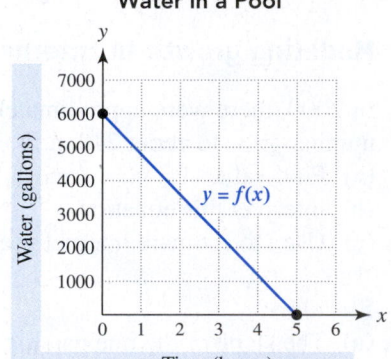

Figure 2.55

CRITICAL THINKING

Can the graph of a function have more than one y-intercept? Explain.

Can the graph of a function have more than one x-intercept? Explain.

(c) The x-intercept is 5, which means that after 5 hours the pool is empty. The y-intercept of 6000 means that initially (when $x = 0$) the pool contained 6000 gallons of water.
(d) To find the equation of the line shown in Figure 2.55, we first find the slope of the line passing through the points $(0, 6000)$ and $(5, 0)$.

$$m = \frac{y_2 - y_1}{x_2 - x_1} \qquad \text{Slope formula}$$

$$= \frac{0 - 6000}{5 - 0} \qquad \text{Substitute.}$$

$$= -1200 \qquad \text{Simplify.}$$

The slope is -1200 and the y-intercept is 6000, so the slope–intercept form is

$$y = -1200x + 6000.$$

The pump *removed* water at the rate of 1200 gallons per hour.

(e) The domain is $D: 0 \leq x \leq 5$, and the range is $R: 0 \leq y \leq 6000$.

❚ **Now Try Exercise 91**

In the next example we introduce a method for modeling linear data by hand.

EXAMPLE 6   **Modeling linear data by hand**

Find a line $y = mx + b$ that models the data in Table 2.15.

TABLE 2.15

x	10	20	30	40	50
y	15	24	30	39	45

Solution

STEP 1: *Carefully make a scatterplot of the data.* Be sure to properly label the x- and y-axes. For the data in Table 2.15 we can label each axis from 0 to 60 and have each hash mark represent 10 units. A scatterplot of the data is shown in Figure 2.56.

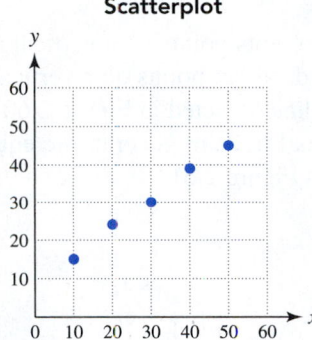

Scatterplot

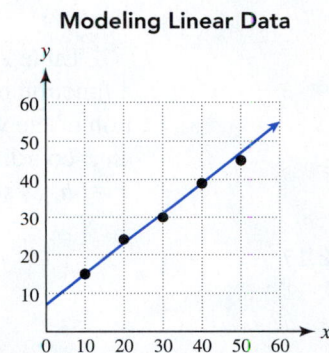

Modeling Linear Data

Figure 2.56   **Figure 2.57**

STEP 2: *Sketch a line that models the data.* You may want to use a ruler for this step. In Figure 2.57 a line is drawn that passes through the first and fourth data points. Your line may be slightly different. Note that the line does not have to pass through *any* of the data points.

STEP 3: *Choose two points on the line and find the equation of the line.* The line in Figure 2.57 passes through $(10, 15)$ and $(40, 39)$. Therefore its slope is

$$m = \frac{39 - 15}{40 - 10} = \frac{24}{30} = \frac{4}{5}.$$

The equation of the line is

$$y = \frac{4}{5}(x - 10) + 15 \quad \text{or} \quad y = \frac{4}{5}x + 7.$$

Note that answers may vary because the data are not exactly linear.

❚ **Now Try Exercise 103**

STUDY TIP

Finding equations of lines is an important skill in this and future math courses.

Modeling, Lines, and Linear Functions

If a set of data is modeled by $y = mx + b$, then the data are also modeled by the linear function defined by $f(x) = mx + b$ because $y = f(x)$. In Example 6 the data set is modeled by $y = \frac{4}{5}x + 7$, so it is also modeled by $f(x) = \frac{4}{5}x + 7$.

Horizontal and Vertical Lines

Table 2.16 represents the constant function $f(x) = 3$. Every output y is 3, regardless of the input x. The graph of a constant function is a horizontal line, and the graph of $f(x) = 3$ is a horizontal line with y-intercept **3**, as shown in Figure 2.58. Its equation may be expressed as $y = 3$, so every point on the line has a y-coordinate of **3**. In general, the equation $y = b$ represents a horizontal line with y-intercept b, as shown in Figure 2.59.

TABLE 2.16

x	y
-4	3
-2	3
0	3
2	3
4	3

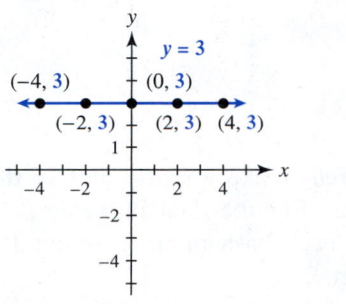

Figure 2.58

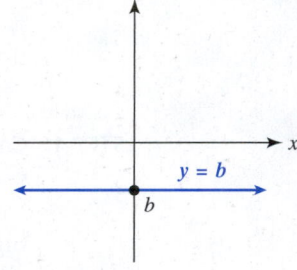

Figure 2.59

Table 2.17 represents points on a vertical line. A vertical line cannot be represented by a function because different points on a vertical line have the same x-coordinate. The equation of the vertical line depicted in Figure 2.60 is $x = 3$. Every point on the line $x = 3$ has an x-coordinate equal to **3**. In general, the equation of a vertical line with x-intercept h is $x = h$, as shown in Figure 2.61.

TABLE 2.17

x	y
3	-6
3	-4
3	-2
3	2
3	4
3	6

y is *not* a function of x.

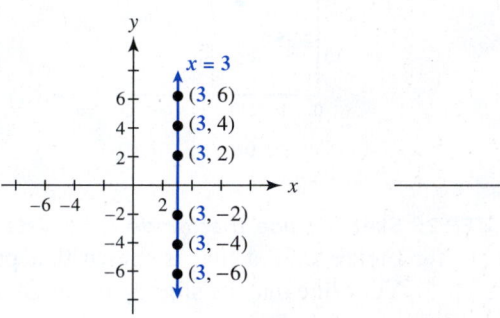

Figure 2.60

Figure 2.61

EQUATIONS OF HORIZONTAL AND VERTICAL LINES

The equation of a horizontal line with y-intercept b is $y = b$.

The equation of a vertical line with x-intercept h is $x = h$.

▶ **REAL-WORLD CONNECTION** Vertical lines are sometimes used in economics. For example, because there is no substitute for insulin, the demand for insulin remains *constant* even though the price may fluctuate. The demand for insulin is said to be *inelastic*. Demand

and price are shown graphically by placing the demand D on the horizontal axis and the price P on the vertical axis. See Figure 2.62, where the vertical line segment shows how the demand remains constant on the horizontal axis even though the price varies along the vertical axis. (See Exercise 95.)

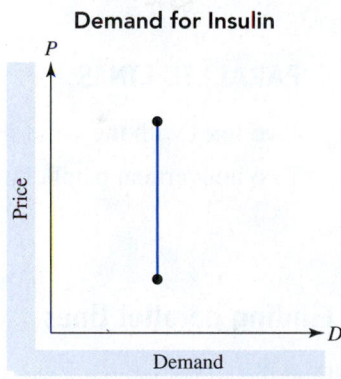

Figure 2.62 Inelastic Demand

EXAMPLE 7   **Finding equations of horizontal and vertical lines**

Find equations of the vertical and horizontal lines that pass through the point $(-3, 4)$. Graph these two lines.

Solution
The x-coordinate of the point $(-3, 4)$ is -3. The vertical line $x = -3$ passes through *every* point in the xy-plane with an x-coordinate of -3, including the point $(-3, 4)$. Similarly, the horizontal line $y = 4$ passes through *every* point with a y-coordinate of 4, including the point $(-3, 4)$. The lines $x = -3$ and $y = 4$ are graphed in Figure 2.63.

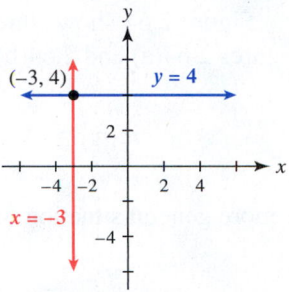

Figure 2.63

Now Try Exercises 79, 81

TECHNOLOGY NOTE:

CALCULATOR HELP

To graph a vertical line, see Appendix A (page AP-5).

Graphing Vertical Lines
The equation of a vertical line is $x = h$ and cannot be expressed on a graphing calculator in the form "$Y_1 =$". Some graphing calculators graph a vertical line by accessing the DRAW menu. The accompanying figure shows a calculator graph of Figure 2.63.

$[-6, 6, 1]$ by $[-6, 6, 1]$

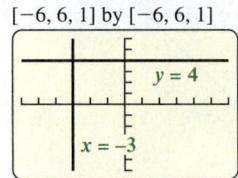

Parallel and Perpendicular Lines

Slope is important when we are determining whether two lines are parallel. For example, the lines $y = 2x$ and $y = 2x + 1$ are parallel because they both have slope 2.

PARALLEL LINES

Two lines with the same slope are parallel.

Two nonvertical parallel lines have the same slope.

EXAMPLE 8 **Finding parallel lines**

Find the slope–intercept form of a line parallel to $y = -2x + 5$, passing through $(-4, 3)$.

Solution
Because the line $y = -2x + 5$ has slope -2, any parallel line also has slope -2. The line passing through $(-4, 3)$ with slope -2 is determined as follows.

$$y = -2(x - (-4)) + 3 \quad \text{Point–slope form}$$
$$y = -2(x + 4) + 3 \qquad \text{Simplify.}$$
$$y = -2x - 8 + 3 \qquad \text{Distributive property}$$
$$y = -2x - 5 \qquad\quad \text{Slope–intercept form}$$

Now Try Exercise 59

Figure 2.64 shows three pairs of perpendicular lines with their slopes labeled. Note in Figures 2.64(a) and 2.64(b) that the product $m_1 m_2$ equals -1. That is,

$$m_1 m_2 = 1 \cdot (-1) = -1 \quad \text{and} \quad m_1 m_2 = 2 \cdot \left(-\frac{1}{2}\right) = -1.$$

A more general situation for two perpendicular lines is shown in Figure 2.64(c), where

$$m_1 m_2 = m_1 \cdot \left(-\frac{1}{m_1}\right) = -1.$$

That is, if two nonvertical lines are perpendicular, then the product of their slopes is -1.

Perpendicular Lines

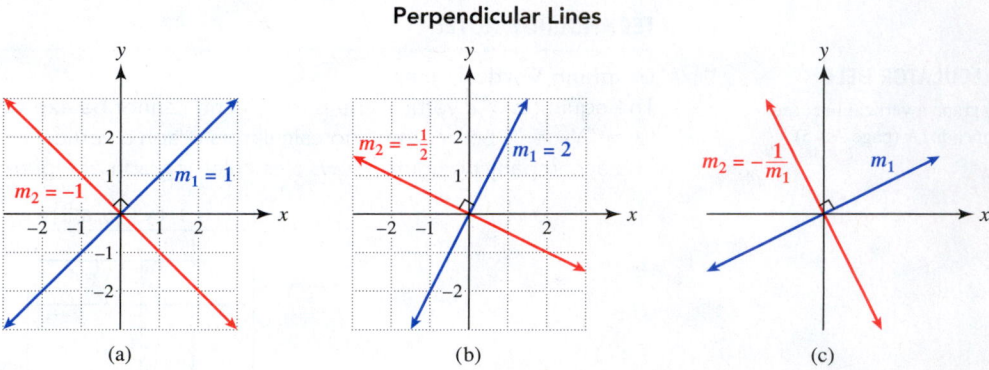

(a)   (b)   (c)

Figure 2.64

These results are summarized in the following box.

PERPENDICULAR LINES

Two lines with nonzero slopes m_1 and m_2 are perpendicular if $m_1 m_2 = -1$.

If two lines have slopes m_1 and m_2 such that $m_1 \cdot m_2 = -1$, then they are perpendicular.

A vertical line and a horizontal line are perpendicular.

READING CHECK

How are the slopes of two parallel lines related? How are the slopes of two perpendicular lines related?

Table 2.18 shows examples of slopes m_1 and m_2 that result in perpendicular lines because $m_1 m_2 = -1$. Note that m_1 and m_2 are **negative reciprocals** of each other; that is, $m_2 = -\frac{1}{m_1}$ and $m_1 = -\frac{1}{m_2}$.

TABLE 2.18 Slopes of Perpendicular Lines m_1 and m_2

m_1	1	$-\frac{1}{2}$	-4	$\frac{2}{3}$	$\frac{3}{4}$	$-\frac{5}{4}$
m_2	-1	2	$\frac{1}{4}$	$-\frac{3}{2}$	$-\frac{4}{3}$	$\frac{4}{5}$
$m_1 m_2$	-1	-1	-1	-1	-1	-1

EXAMPLE 9   **Finding perpendicular lines**

Find the slope–intercept form of the line perpendicular to $y = -\frac{1}{2}x + 1$, passing through the point $(3, 2)$. Graph the two lines.

Solution
The line $y = -\frac{1}{2}x + 1$ has slope $m_1 = -\frac{1}{2}$. From Table 2.18 the slope of a line that is perpendicular to this line is $m_2 = 2$. The slope–intercept form of a line having slope **2** and passing through $(\mathbf{3, 2})$ can be found as follows.

$$y = \mathbf{2}(x - \mathbf{3}) + \mathbf{2} \quad \text{Point–slope form}$$
$$y = 2x - 6 + 2 \quad \text{Distributive property}$$
$$y = 2x - 4 \quad \text{Slope–intercept form}$$

A graph of these perpendicular lines is shown in Figure 2.65.

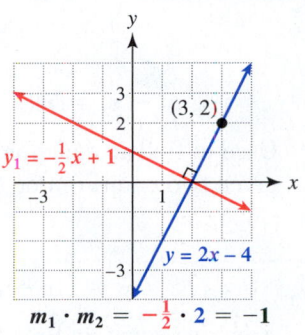

$$m_1 \cdot m_2 = -\tfrac{1}{2} \cdot 2 = -1$$

Figure 2.65

Now Try Exercise 71

CALCULATOR HELP

To set a square viewing rectangle, see Appendix A (pages AP-5 and AP-6).

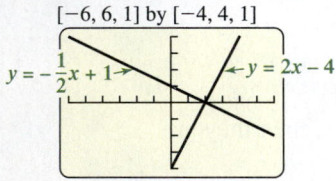

$[-6, 6, 1]$ by $[-4, 4, 1]$

$y = -\frac{1}{2}x + 1$ $y = 2x - 4$

TECHNOLOGY NOTE:

Square Viewing Rectangles

The figure to the left shows a square viewing rectangle in which the perpendicular lines from Example 9 intersect at 90°. Try graphing the two perpendicular lines in Example 9 by using the viewing rectangle $[-6, 6, 1]$ by $[-10, 10, 1]$. Do the lines appear perpendicular? For many graphing calculators, a square viewing rectangle results when the distance along the y-axis is about $\frac{2}{3}$ the distance along the x-axis. On some graphing calculators you can create a square viewing rectangle automatically by using the ZOOM menu.

EXAMPLE 10 **Equations of perpendicular lines**

Find the slope–intercept form of each line shown in Figure 2.66. Verify that the two lines are perpendicular.

Solution

In Figure 2.67, the graph of y_1 is shown to have slope $m_1 = \frac{2}{3}$ because the line rises 2 units for every 3 units of run. Its y-intercept is -2, and its slope–intercept form is $y_1 = \frac{2}{3}x - 2$. Similarly, the graph of y_2 is shown to have slope $m_2 = -\frac{3}{2}$ because the line falls 3 units for every 2 units of run. Its y-intercept is 1, and its slope–intercept form is $y = -\frac{3}{2}x + 1$. To be perpendicular, the product of their slopes must equal -1; that is,

$$m_1 \cdot m_2 = \frac{2}{3} \cdot \left(-\frac{3}{2}\right) = -1. \checkmark \quad \text{It checks.}$$

Figure 2.66 Perpendicular Lines

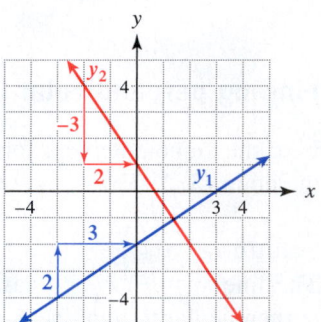

Figure 2.67

Now Try Exercise 75

2.4 Putting It All Together

CONCEPT	EXPLANATION	EXAMPLE
Point–Slope Form	$y = m(x - x_1) + y_1$ or $y - y_1 = m(x - x_1)$ Used to find an equation of a line, given two points or one point and the slope	Given two points $(1, 2)$ and $(3, 5)$, first compute $m = \dfrac{5 - 2}{3 - 1} = \dfrac{3}{2}.$ An equation of this line is $y = \dfrac{3}{2}(x - 1) + 2.$
Slope–Intercept Form	$y = mx + b$ A unique equation for a line, determined by the slope m and the y-intercept b	An equation of the line with slope $m = 3$ and y-intercept $b = -5$ is $y = 3x - 5.$

The graph of a linear function f is a line. Therefore linear functions can be represented by

$$f(x) = mx + b \quad \text{or} \quad f(x) = m(x - x_1) + y_1.$$

CONCEPT	EQUATION(S)	EXAMPLE
Horizontal Line	$y = b$, where b is a constant	A horizontal line passing through $(-3, 5)$ has the equation $y = 5$.
Vertical Line	$x = h$, where h is a constant	A vertical line passing through $(-3, 5)$ has the equation $x = -3$.
Parallel Lines	$y = m_1 x + b_1$ and $y = m_2 x + b_2$, where $m_1 = m_2$	The lines $y = 2x - 1$ and $y = 2x + 5$ are parallel because both have slope 2.
Perpendicular Lines	$y = m_1 x + b_1$ and $y = m_2 x + b_2$, where $m_1 m_2 = -1$	The lines $y = 3x - 5$ and $y = -\frac{1}{3}x + 2$ are perpendicular because $m_1 m_2 = 3\left(-\frac{1}{3}\right) = -1$.

2.4 Exercises

MyMathLab 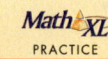 PRACTICE   WATCH   DOWNLOAD   READ   REVIEW

CONCEPTS AND VOCABULARY

1. How many lines are determined by two points?

2. How many lines are determined by a point and a slope?

3. Give the slope–intercept form of a line.

4. Give the point–slope form of a line.

5. Write an equation of a vertical line with x-intercept 5.

6. Write an equation of a horizontal line with y-intercept 2.

7. Give an equation of a horizontal line that has y-intercept b.

8. Give an equation of a vertical line that has x-intercept h.

9. If two parallel lines have slopes m_1 and m_2, what can be said about m_1 and m_2?

10. If two perpendicular lines have slopes m_1 and m_2, then $m_1 \cdot m_2 = $ _____.

11. Give an equation of a line that is parallel to the line $y = -\frac{3}{4}x + 1$.

12. Give an equation of a line that is perpendicular to the line $y = -\frac{3}{4}x$.

13. Give an equation of a line that is perpendicular to the x-axis.

14. Give an equation of a line that is perpendicular to the y-axis.

EQUATIONS OF LINES

Exercises 15–18: Use the labeled point and the slope to find the slope–intercept form of the line.

15.

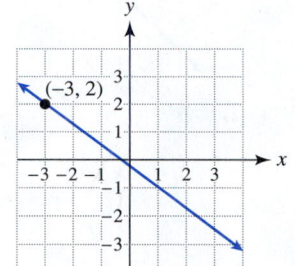

16.

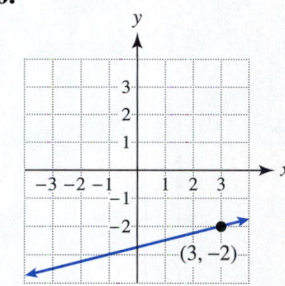

17.

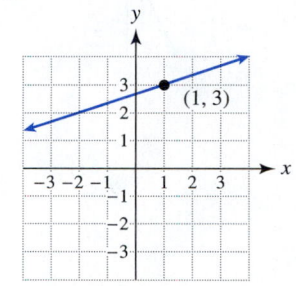

18.

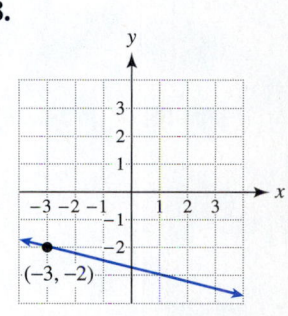

Exercises 19–22: Determine whether the given point lies on the line.

19. $(-3, 4)$ $y = -\frac{2}{3}x + 2$

20. $(4, 2)$ $y = \frac{1}{4}x - 1$

21. $(-4, 3)$ $y = \frac{1}{2}(x + 4) + 2$

22. $(1, -13)$ $y = 3(x - 5) - 1$

Exercises 23–28: Match the equation with its graph (a.–f.), where m and b are constants.

23. $y = mx + b, m > 0$ and $b \neq 0$

24. $y = mx + b, m < 0$ and $b \neq 0$

25. $y = mx, m < 0$

26. $y = b$

27. $y = mx, m > 0$

28. $x = h$

a.

b.

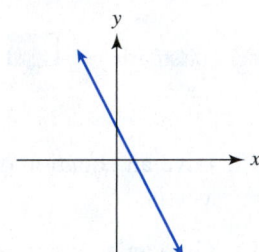

c.

d.

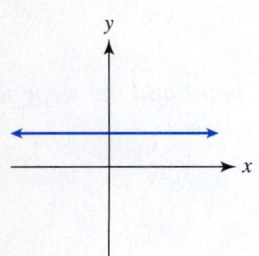

e.

f.

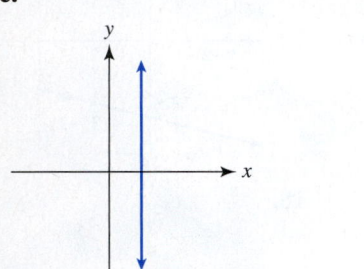

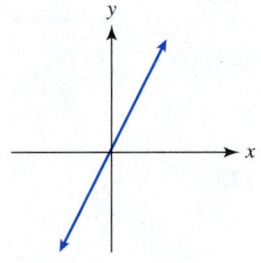

Exercises 29–38: Find a point–slope form of the line satisfying the given conditions.

29. Slope -2, passing through $(2, -3)$

30. Slope $\frac{1}{2}$, passing through $(-4, 1)$

31. Slope 1.3, passing through $(1990, 25)$

32. Slope 45, passing through $(1999, 103)$

33. Passing through $(1, 3)$ and $(-5, -1)$

34. Passing through $(-2, 4)$ and $(3, -1)$

35. Passing through $(1980, 5)$ and $(2000, 45)$

36. Passing through $(1990, 50)$ and $(2000, -10)$

37. Passing through $(6, 0)$ and $(0, 4)$

38. Passing through $(1980, 16)$ and $(2000, 66)$

39. Sketch a line passing through the point $(0, 2)$ with slope $-\frac{1}{2}$. Find the point–slope form of the line. Does the point $(-2, 3)$ lie on this line?

40. Sketch a line passing through the point $(2, 1)$ with slope 2. Find the point–slope form of the line. Does the point $(1, 2)$ lie on this line?

Exercises 41–46: Write in slope–intercept form.

41. $y = 2(x - 1) - 2$

42. $y = -3(x + 2) + 5$

43. $y = \frac{1}{2}(x + 4) + 1$

44. $y = -\frac{2}{3}(x - 3) - 6$

45. $y = 22(x - 1.5) - 10$

46. $y = -30(x + 3) + 106$

Exercises 47–50: (Refer to Example 3.) Use the slope–intercept form to find the equation of the line passing through the given points.

47. $(-3, 7), (2, -8)$

48. $(-6, -3), (3, 3)$

49. $(-4, 1), (4, -5)$

50. $(-1, 14), (4, -6)$

Exercises 51–68: Find the slope–intercept form of the line satisfying the given conditions.

51. Slope $-\frac{1}{3}$, passing through $(0, -5)$

52. Slope 5, passing through $(-1, 4)$

53. Passing through $(3, -2)$ and $(2, -1)$

54. Passing through $(8, 3)$ and $(-7, 3)$

55. x-intercept 2, y-intercept $-\frac{2}{3}$

56. x-intercept -3, y-intercept 4

57. x-intercept 1, y-intercept 3

58. x-intercept -4, y-intercept -5

59. Parallel to $y = 4x - 2$, passing through $(1, 3)$

60. Parallel to $y = -\frac{2}{3}x$, passing through $(0, -10)$

61. Passing through $(-3, 2)$ and parallel to the line passing through $(-2, 3)$ and $(1, -2)$

62. Passing through $\left(\frac{1}{2}, -1\right)$ and parallel to the line passing through $(1, -4)$ and $(-3, 1)$

63. Perpendicular to $y = -\frac{1}{3}x + 4$, passing through the point $(-3, 5)$

64. Perpendicular to $y = \frac{3}{4}(x - 2) + 1$, passing through the point $(-2, -3)$

65. Passing through $\left(-\frac{1}{2}, -2\right)$ and perpendicular to the line passing through $(-1, 6)$ and $(8, -4)$

66. Passing through $(4, -3)$ and perpendicular to the line passing through $(-4, -1)$ and $(2, 7)$

67. Thinking Generally Passing through $(0, b)$ and perpendicular to a line having slope c with $c \neq 0$

68. Thinking Generally Passing through $(a, 0)$ and perpendicular to $y = -mx$ with $m \neq 0$

Exercises 69–74: (Refer to Example 9.) Do the following.

(a) *Find the slope–intercept form of the line perpendicular to the given line, passing through the given point.*

(b) *Graph the two lines.*

69. $y = \frac{1}{2}x$, $(0, 2)$

70. $y = -3x$, $(0, -3)$

71. $y = -2x + 1$, $(-1, 2)$

72. $y = \frac{2}{3}x + 2$, $(-1, 0)$

73. $y = -\frac{1}{3}x + 2$, $(1, 1)$

74. $y = -\frac{4}{3}x + 2$, $(1, -1)$

Exercises 75–78: (Refer to Example 10.) Do the following.

(a) *Find the slope–intercept form of each line.*

(b) *Verify that the two lines are perpendicular.*

75.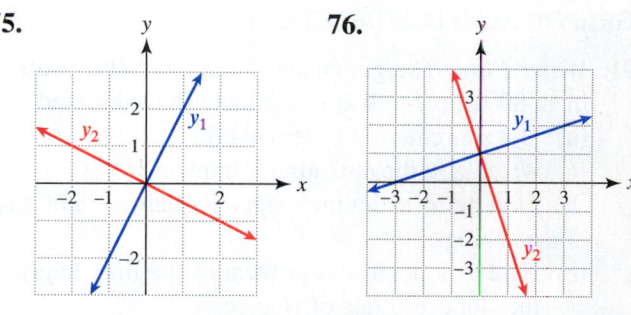

76.

77.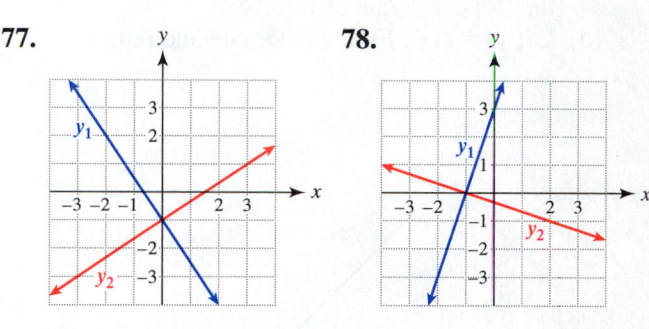

78.

Exercises 79–86: Find an equation of a line satisfying the given conditions.

79. Vertical, passing through $(-1, 6)$

80. Vertical, passing through $(2, -7)$

81. Horizontal, passing through $\left(\frac{3}{4}, -\frac{5}{6}\right)$

82. Horizontal, passing through $(5.1, 6.2)$

83. Perpendicular to $y = \frac{1}{2}$, passing through $(4, -9)$

84. Perpendicular to $x = 2$, passing through $(3, 4)$

85. Parallel to $x = 4$, passing through $\left(-\frac{2}{3}, \frac{1}{2}\right)$

86. Parallel to $y = -2.1$, passing through $(7.6, 3.5)$

Exercises 87–90: Decide whether the points in the table lie on a line. If they do, find the slope–intercept form of the line.

87.

x	1	2	3	4
y	-4	0	4	8

88.

x	-1	0	1	2
y	8	5	4	1

89.

x	-3	0	3	6
y	4	8	14	18

90.

x	-2	2	4	6
y	-6	0	3	6

GRAPHICAL INTERPRETATION

91. *Water Flow* The graph shows the amount of water y in an 80-gallon tank after x minutes have elapsed.
   (a) Is water entering or leaving the tank? How much water is in the tank after 3 minutes?
   (b) Find the x- and y-intercepts. Explain each of their meanings.
   (c) Find a slope–intercept form of the line. Interpret the slope as a rate of change.
   (d) Let $y = f(x)$. Find the domain and range.

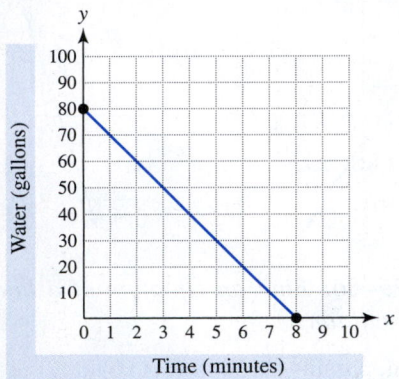

Time (minutes)

92. *Distance and Speed* A person is driving a car along a straight road. The graph shows the distance y in miles that the driver is from home after x hours.
   (a) Is the person traveling toward or away from home?
   (b) The graph passes through $(1, 35)$ and $(3, 95)$. Discuss the meaning of these points.
   (c) Find a point-slope form of the line. Interpret the slope and y-intercept.

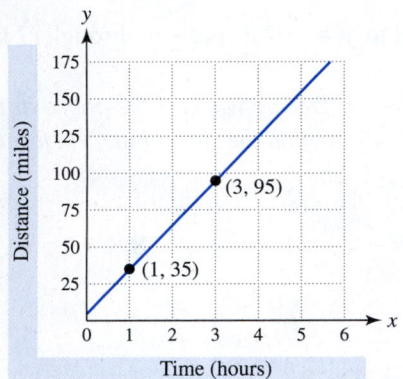

Time (hours)

93. *U.S. Population Density* The graph at the top of the next column shows the average number of people living on parcels of land of various sizes in 1960.
   (a) The graph passes through the points $(2, 100)$ and $(4, 200)$. Discuss the meaning of these points.
   (b) Explain why it is reasonable for the graph to pass through the point $(0, 0)$.
   (c) Find the slope–intercept form of the line.
   (d) Interpret the slope as a rate of change.

(e) Write the equation of this line as a linear function P that outputs the average number of people living on x acres of land.

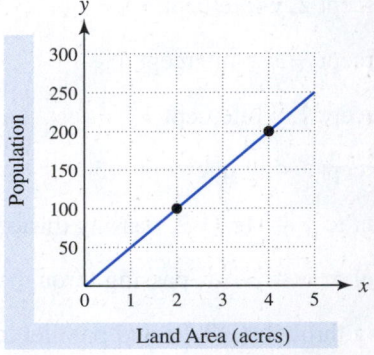

Land Area (acres)

94. *Cost of Rock* The graph shows the cost of purchasing landscape rock.
   (a) The graph passes through the points $(2, 48)$ and $(5, 120)$. Discuss the meaning of these points.
   (b) Explain why it is reasonable for the graph to pass through the point $(0, 0)$.
   (c) Find the slope–intercept form of the line.
   (d) Interpret the slope as a rate of change.
   (e) Write the equation of this line as a linear function C that outputs the cost of x tons of landscape rock.

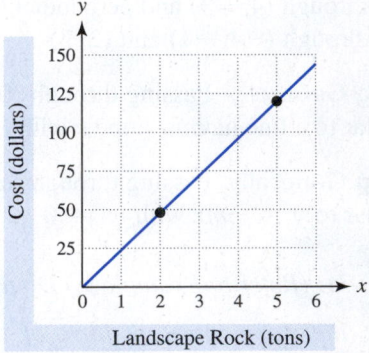

Landscape Rock (tons)

95. *Elastic Demand* (Refer to Figure 2.62 and the corresponding Real-World Connection.) If there are a number of gas stations located near each other, then prices for a gallon of gasoline are usually (nearly) equal. If one station were to raise its price by even a small amount, then sales (demand) would likely fall dramatically. In a small area, gasoline exhibits *elastic demand*.
   (a) Sketch a graph of elastic demand, where even though the demand may change, the price remains constant.
   (b) What type of line segment models elastic demand?

96. **Online Exploration** Find the population in 2000 of the city where you are currently residing. Next, find its current population. Use a linear function P to model these data and then use P to estimate the population in 2007.

APPLICATIONS

97. *Projected Cost of College* The annual cost of the average private college or university is shown in the table. This cost includes tuition, fees, room, and board.

Year	2006	2011
Cost	$28,000	$38,000

Source: Cerulli Associates.

(a) Find the slope–intercept form of a line that passes through these two data points.
(b) Interpret the slope as a rate of change.
(c) Estimate the cost of private college in 2015.

98. *Western Population* In 1950 the western region of the United States had a population of 20 million, and in 2000 it was 63 million. (*Source:* U.S. Census Bureau.)
(a) Find a function f defined by $f(x) = mx + b$ that models this population during year x.
(b) Use f to estimate this population in 1960 and 2010.
(c) Estimate this population in 1900 and discuss your result.

Exercises 99–102: ***Modeling Data*** *Find $f(x) = mx + b$ so that f models the data, where x is the year. Then use $f(x)$ to make the requested estimate.*

99. *Chicken Consumption* In 2005 the average American ate 57 pounds of chicken, and in 2010 this amount increased to 60 pounds. Estimate chicken consumption in 2008. (*Source:* Department of Agriculture.)

100. *Internet* In 2005 the average American spent 9 hours per week on the Internet, and in 2009 this time increased to 13 hours per week. Estimate the time spent on the Internet in 2012. (*Source:* UCLA Center for Communication Policy.)

101. *Life Expectancy* The life expectancy for a baby born in 1910 was 50 years, and in 2010 it was 78 years. Estimate the life expectancy in 1980. (*Source:* U.S. Census Bureau.)

102. *CPI* In 2000 the Consumer Price Index was 178, and in 2010 it was 218. Estimate what this index might be in 2018. (*Source:* Department of Labor.)

Exercises 103–106: ***Modeling Data*** *(Refer to Example 6.) Find the slope–intercept form of a line that models the data. Because the data are not exactly linear, answers may vary slightly.*

103.

x	1	2	3	4
y	1	4	10	13

104.

x	1	3	5	7
y	3	0	-5	-10

105.

x	1	2	3	4	5
y	8	3	0	-1	-6

106.

x	10	20	30	40	50
y	5	20	30	50	60

Exercises 107–110: ***Modeling Data*** *(Refer to Example 6.) Find the slope–intercept form of a line that models the data. Answers may vary.*

107. Population over 85 (millions)

x	1970	1980	1990	2000	2010
y	1.4	2.3	3.0	4.2	5.8

Source: U.S. Census Bureau.

108. Women in Marine Corps

x	1980	1990	2000	2010
y	6000	8600	9500	11,700

Source: Department of Defense.

109. Worldwide Cigarette Consumption (trillions)

x	1950	1970	1990	2010
y	1.7	3.1	5.4	7.1

Source: Department of Agriculture.

110. Global Online Betting Losses ($ billions)

x	2006	2007	2008	2009	2010
y	15	18	20	23	25

Source: Christiansen Capital Advisors.

111. *Cost of Driving* The cost of driving a car includes both fixed costs and mileage costs. Assume that it costs $189.20 per month for insurance and car payments and that it costs $0.30 per mile for gasoline, oil, and routine maintenance.
(a) Find values for a and b so that $f(x) = ax + b$ models the monthly cost of driving x miles.
(b) Interpret the y-intercept on the graph of f.

112. *HIV Infection Rates* In 1999, there were an estimated 24 million HIV infections worldwide, with an annual infection rate of 5.4 million. (*Source:* Centers for Disease Control and Prevention.)
(a) Determine values for m, x_1, and y_1 so that $f(x) = m(x - x_1) + y_1$ models the number of HIV infections in year x.
(b) Estimate the number of HIV infections in 2007.

113. *Ring Sizes* Ring sizes can be modeled by a linear function. If the circumference of a person's finger is 4.9 centimeters, then the ring size is 5; if the circumference is 5.4 centimeters, then the ring size is 7. (*Source*: Overstock.com.)
 (a) Find $R(c) = mc + b$ so that R calculates the ring size for a finger with circumference c.
 (b) A person's finger has a circumference of 6.16 centimeters. What is the ring size?

114. *Hat Sizes* Hat sizes can be modeled by a linear function. If the circumference of a person's head is $21\frac{7}{8}$ inches, the hat size is 7; if the circumference is 25 inches, the hat size is 8. (*Source*: Brentblack.com.)
 (a) Find $H(c) = mc + b$ so that H calculates the hat size for a head with circumference c.
 (b) A person's head has a circumference of 23 inches. What is the hat size to the nearest eighth?

115. *U.S. Population* The population in millions for selected years is shown in the table.

Year	1980	1990	2000	2010
Population	227	249	281	308

Source: U.S. Census Bureau.

 (a) Make a scatterplot of the data.
 (b) Find values for m, x_1, and y_1 so that
$$f(x) = m(x - x_1) + y_1$$
  models the data.
 (c) Use f to estimate the population in 2020.

116. *Sexual Partners* The table in the next column shows the percentages of people who had four or more sexual partners in the last 12 months for various years.

Year	2002	2004	2006	2008
Percentage	1.7	1.6	1.5	1.4

Source: National Opinion Research Center.

 (a) Find values for m, x_1, and y_1 so that
$$f(x) = m(x - x_1) + y_1$$
  models these data.
 (b) Assuming that trends continued, estimate this percentage in 2014.

117. *Pet Health Care* In 1984 spending on health care for pets was \$4 billion, and by 2008 it had increased to \$12 billion. (*Source*: American Express Institute.)
 (a) Find $f(x) = m(x - x_1) + y_1$ so that f models these data in billions of dollars during year x.
 (b) Estimate health care spending for pets in 2014.

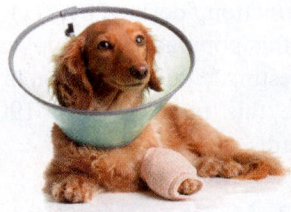

118. *Manufacturing Employment* In 1956, employees of manufacturing firms accounted for 30% of the total employment in the United States. In 2010 this percentage was 9%. (*Source*: U.S. Census Bureau.)
 (a) Find $f(x) = m(x - x_1) + y_1$ so that f models these data during year x.
 (b) Estimate the percentage of people employed by manufacturing in 2014.

WRITING ABOUT MATHEMATICS

119. Explain how you can recognize equations of parallel lines. How can you recognize equations of perpendicular lines?

120. Suppose that some real data can be modeled by a linear function f. Explain what the slope of the graph of f indicates about the data.

SECTIONS 2.3 and 2.4   **Checking Basic Concepts**

1. Find the slope and a point on each line.
 (a) $y = -3(x - 5) + 7$
 (b) $y = 10$
 (c) $x = -5$
 (d) $y = 5x + 3$

2. (a) Calculate the slope of the line passing through the points $(2, -4)$ and $(5, 2)$.
 (b) Find a point–slope form and the slope–intercept form of the line.
 (c) Find the x-intercept and the y-intercept of the line.

3. Find the equation of a vertical line and the equation of a horizontal line passing through $(-2, 5)$.

4. Find equations of lines that are perpendicular or parallel to $y = -\frac{1}{2}x + 3$, passing through the point $(2, -4)$.

5. The graph to the right shows the distance that a car is from home after x hours.
   (a) Is the car moving toward or away from home?
   (b) Find the slope of the line. Interpret the slope as a rate of change.
   (c) Find the x-intercept and the y-intercept. Explain the meaning of each.
   (d) Determine a and b so that $f(x) = ax + b$ models this situation.
   (e) Give the domain and range of f.

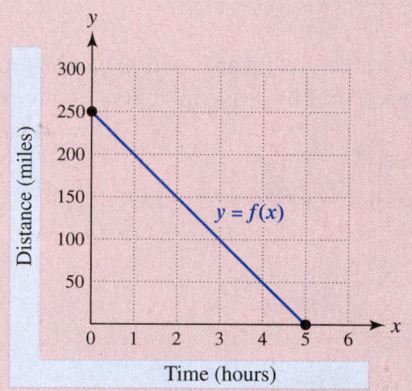

CHAPTER 2 Summary

SECTION 2.1 ■ FUNCTIONS AND THEIR REPRESENTATIONS

Function A function is a set of ordered pairs (x, y) where each x-value corresponds to exactly one y-value. A function takes a valid input x and computes exactly one output y, forming the ordered pair (x, y).

Domain and Range of a Function The domain D is the set of all valid inputs, or x-values, and the range R is the set of all outputs, or y-values.

Examples: $f = \{(1, 2), (2, 3), (3, 3)\}$ has $D = \{1, 2, 3\}$ and $R = \{2, 3\}$.

$f(x) = x^2$ has domain all real numbers and range $y \geq 0$. (See the graph below.)

Function Notation $y = f(x)$ and is read "y equals f of x."

Example: $f(x) = \frac{2x}{x-1}$ implies that $f(3) = \frac{2 \cdot 3}{3 - 1} = \frac{6}{2} = 3$. Thus the point $(3, 3)$ is on the graph of f.

Function Representations A function can be represented symbolically, numerically, graphically, or verbally.

Symbolic Representation (Formula) $f(x) = x^2$

Numerical Representation (Table)

x	y
-2	4
-1	1
0	0
1	1
2	4

Graphical Representation (Graph)

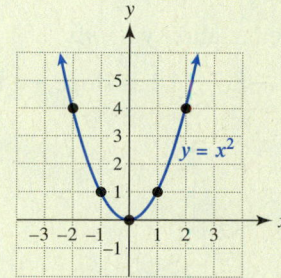

Verbal Representation (Words) f computes the square of the input x.

Vertical Line Test If every vertical line intersects a graph at most once, then the graph represents a function.

SECTION 2.2 ■ LINEAR FUNCTIONS

Linear Function A linear function can be represented by $f(x) = mx + b$. Its graph is a (straight) line. For each unit increase in x, $f(x)$ changes by an amount equal to m.

Example: $f(x) = 2x - 1$ represents a linear function with $m = 2$ and $b = -1$.

Numerical Representation   **Graphical Representation**

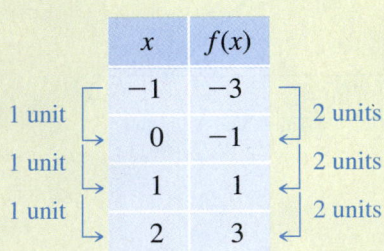

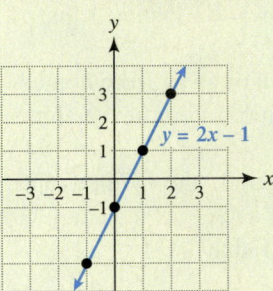

Each 1-unit increase in x results in a 2-unit increase in $f(x)$; thus $m = 2$.

NOTE: A numerical representation is a table of values of $f(x)$.

Modeling Data with Linear Functions When data have a constant rate of change, they can be modeled by $f(x) = mx + b$. The constant m represents the *rate of change*, and the constant b represents the *initial amount* or the value when $x = 0$. That is,

$$f(x) = (\text{Rate of change})x + (\text{Initial amount}).$$

Example: In the following table, the y-values decrease by 3 units for each 1-unit increase in x. When $x = 0$, $y = 4$. Thus the data are modeled by $f(x) = -3x + 4$.

x	-2	-1	0	1	2
y	10	7	4	1	-2

SECTION 2.3 ■ THE SLOPE OF A LINE

Slope The slope m of the line passing through the points (x_1, y_1) and (x_2, y_2) is

$$m = \frac{\text{rise}}{\text{run}} = \frac{y_2 - y_1}{x_2 - x_1}, \quad \text{where } x_1 \neq x_2.$$

Example: The slope of the line passing through $(-2, 3)$ and $(4, 0)$ is

$$m = \frac{0 - 3}{4 - (-2)} = \frac{-3}{6} = -\frac{1}{2}.$$

Slope–Intercept Form The equation $y = mx + b$ gives the slope m and y-intercept b of a line.

Example: The graph of $y = -\frac{1}{2}x + 1$ has slope $-\frac{1}{2}$ and y-intercept 1.

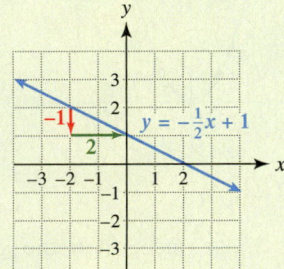

x	-2	0	2	4
y	2	1	0	-1

y decreases by 1 unit for each 2-unit increase in x.

Slope as a Rate of Change The slope of the graph of a linear function indicates the rate at which a function is increasing or decreasing.

Example: If $V(t) = 10t$ models the volume of water in a tank in gallons after t minutes, water is *entering* the tank at 10 gallons per minute.

SECTION 2.4 ■ EQUATIONS OF LINES AND LINEAR MODELS

Point–Slope Form

$$y = m(x - x_1) + y_1 \quad \text{or} \quad y - y_1 = m(x - x_1),$$

where m is the slope and (x_1, y_1) is a point on the line.

Example: The point–slope form of the line with slope 4 passing through $(2, -3)$ is

$$y = 4(x - 2) - 3.$$

Equations of Horizontal and Vertical Lines

$$y = b \text{ (horizontal)}, \quad x = h \text{ (vertical)}$$

Example: The equation of the horizontal line passing through $(2, 3)$ is $y = 3$. The equation of the vertical line passing through $(2, 3)$ is $x = 2$.

Parallel Lines
Two lines with the same slope are parallel.

Two nonvertical parallel lines have the same slope.

Example: The lines $y = 2x - 1$ and $y = 2x + 3$ are parallel, with slope 2.

Perpendicular Lines Two lines with nonzero slopes m_1 and m_2 are perpendicular if $m_1 m_2 = -1$.
If two lines have slopes m_1 and m_2 such that $m_1 \cdot m_2 = -1$, then they are perpendicular.

A vertical line and a horizontal line are perpendicular.

Example: The lines $y = 2x - 1$ and $y = -\frac{1}{2}x + 3$ are perpendicular because the product of their slopes equals -1. That is, $2\left(-\frac{1}{2}\right) = -1$.

CHAPTER 2   Review Exercises

SECTION 2.1

Exercises 1–4: Evaluate $f(x)$ for the given values of x.

1. $f(x) = 3x - 1$ $\qquad$ $x = -2, \frac{1}{3}$

2. $f(x) = 5 - 3x^2$ $\qquad$ $x = -3, 1$

3. $f(x) = \sqrt{x} - 2$ $\qquad$ $x = 0, 9$

4. $f(x) = 5$ $\qquad$ $x = -5, \frac{7}{5}$

Exercises 5 and 6: Do the following.

  (a) *Write a symbolic representation (formula) for the function described.*

  (b) *Evaluate the function for input 5 and interpret the result.*

5. Function P computes the number of pints in q quarts.

6. Function f computes 3 less than 4 times a number x.

7. If $f(3) = -2$, then the point _____ lies on the graph of f.

8. If $(4, -6)$ lies on the graph of f, then $f(\_\_\_\_) = \_\_\_\_$.

Exercises 9–12: Sketch a graph of f.

9. $f(x) = -2x$ $\qquad$ **10.** $f(x) = \frac{1}{2}x - \frac{3}{2}$

11. $f(x) = x^2 - 1$ $\qquad$ **12.** $f(x) = \sqrt{x + 1}$

Exercises 13 and 14: Use the graph of f to evaluate the given expressions.

13. $f(0)$ and $f(-3)$ **14.** $f(-2)$ and $f(1)$

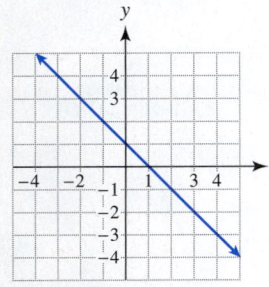   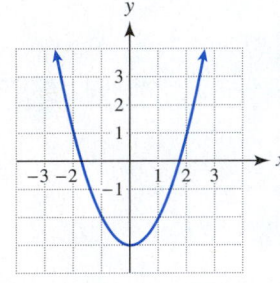

15. Use the table to evaluate $f(-1)$ and $f(3)$.

x	-1	1	3	5
$f(x)$	7	3	-1	-5

16. A function f is represented verbally by "Multiply the input x by 3 and then subtract 2." Give numerical, symbolic, and graphical representations for f. Let $x = -3, -2, -1, \ldots, 3$ in the table of values, and let $-3 \le x \le 3$ for the graph.

Exercises 17 and 18: Use the graph of f to estimate its domain and range.

17.   **18.**

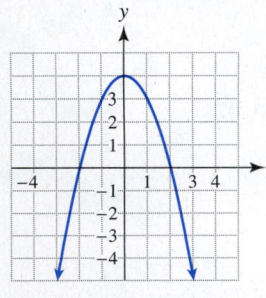   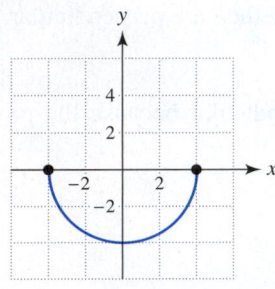

Exercises 19 and 20: Does the graph represent a function?

19.   **20.**

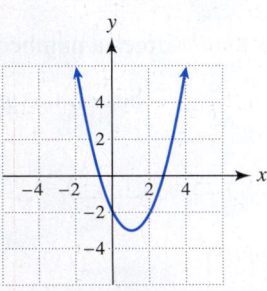   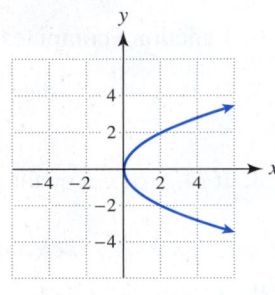

Exercises 21 and 22: Find the domain and range of S. Then state whether S defines a function.

21. $S = \{(-3, 4), (-1, 4), (2, 3), (4, -1)\}$

22. $S = \{(-1, 5), (0, 3), (1, -2), (-1, 2), (2, 4)\}$

Exercises 23–30: Find the domain.

23. $f(x) = -3x + 7$ **24.** $f(x) = \sqrt{x}$

25. $f(x) = \frac{3}{x}$ **26.** $f(x) = x^2 + 2$

27. $f(x) = \sqrt{5 - x}$ **28.** $f(x) = \dfrac{x}{x + 2}$

29. $f(x) = |2x + 1|$ **30.** $f(x) = x^3$

SECTION 2.2

Exercises 31 and 32: Does the graph represent a linear function?

31.   **32.**

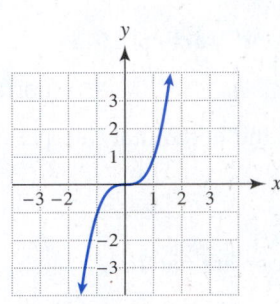   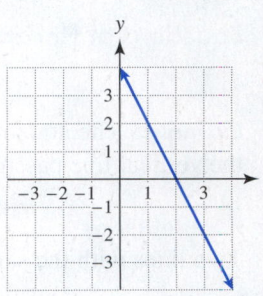

Exercises 33–36: Determine whether f is a linear function. If f is linear, give values for m and b so that f may be expressed as $f(x) = mx + b$.

33. $f(x) = -4x + 5$ **34.** $f(x) = 7 - x$

35. $f(x) = \sqrt{x}$ **36.** $f(x) = 6$

Exercises 37 and 38: Use the table to determine whether $f(x)$ could represent a linear function. If it could, write the formula for f in the form $f(x) = mx + b$.

37.

x	0	2	4	6
$f(x)$	-3	0	3	6

38.

x	-1	0	1	2
$f(x)$	-5	0	10	15

39. Evaluate $f(x) = \frac{1}{2}x + 3$ at $x = -4$.

40. Use the graph to evaluate $f(-2)$ and $f(1)$.

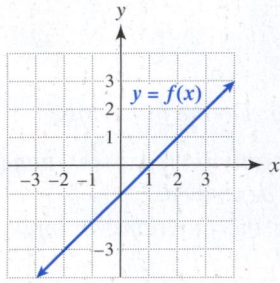

Exercises 41–44: Sketch a graph of y = f(x).

41. $f(x) = x + 1$ **42.** $f(x) = 1 - 2x$

43. $f(x) = -\frac{1}{3}x$ **44.** $f(x) = -1$

45. Write a symbolic representation (formula) for a linear function H that calculates the number of hours in x days. Evaluate $H(2)$ and interpret the result.

46. Let $f(x) = \sqrt{x + 2} - x^2$.

   (a) Make a numerical representation (table) for the function f with $x = 1, 2, 3, \ldots, 7$.
   (b) Graph f in the standard window. What is the domain of f?

SECTION 2.3

Exercises 47–50: Find the slope of the line.

47.

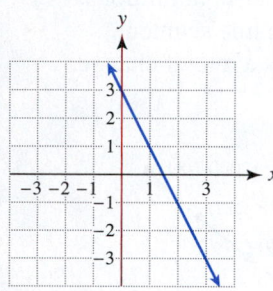

48.

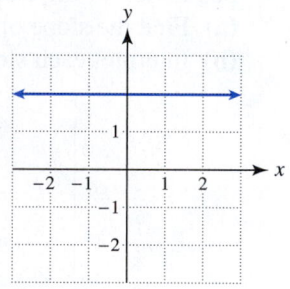

49.

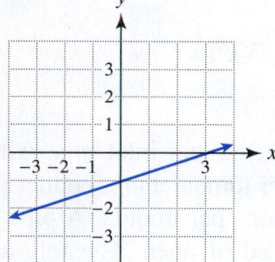

50.

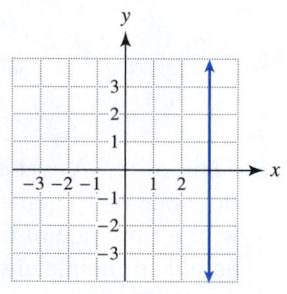

Exercises 51–54: Calculate the slope of the line passing through the given points, if possible.

51. $(-1, 2), (3, 8)$ **52.** $\left(-3, \frac{5}{2}\right), \left(1, -\frac{1}{2}\right)$

53. $(3, -4), (5, -4)$ **54.** $(-2, 6), (-2, 8)$

55. Sketch a line passing through $(1, 2)$ with slope $-\frac{1}{2}$.

56. Find the slope and y-intercept of the line $y = -\frac{2}{3}x$. Graph the line.

57. Use the graph to express the line in slope–intercept form.

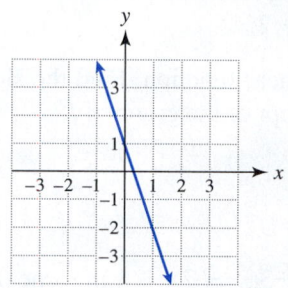

58. Write the slope–intercept form of a line passing through $(0, 2)$ and $(-1, 0)$.

59. Let f be a linear function. Use the table to find the slope and y-intercept of the graph of f.

x	-2	0	3
$f(x)$	3	1	-2

60. The line graph represents the gallons of water in a small swimming pool after x hours.
   (a) Estimate the slope of each line segment.
   (b) Interpret each slope as a rate of change.
   (c) Describe what happens to the amount of water in the pool.

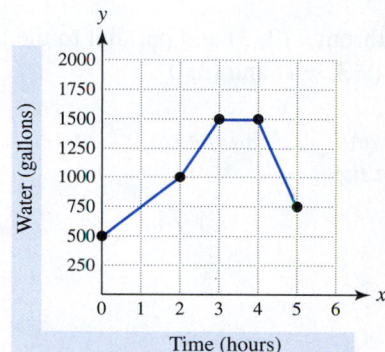

61. *Sketching a Model* A distance runner starts at home and runs at a constant speed. After 1.5 hours she is 12 miles from home. She then turns around and runs home at the same speed. Sketch a graph that shows the distance that the runner is from home after x hours.

62. Let $f(x) = \frac{1}{2}x - 2$.
   (a) Find the slope and y-intercept of the graph of f.
   (b) Sketch a graph of f.

SECTION 2.4

63. Determine whether the point (2, 1) lies on the graph of $y = \frac{3}{2}x - 2$.

64. Let f be a linear function. Find the slope, x-intercept, and y-intercept of the graph of f.

x	-1	0	1	2
$f(x)$	6	4	2	0

65. Write $y = -3(x - 2) + 1$ in slope–intercept form.

66. Write the point–slope form and the slope–intercept form of the line with slope -3, passing through the point $(-2, 3)$.

Exercises 67–72: Write the slope–intercept form of a line satisfying the given conditions.

67. x-intercept 2, y-intercept -3

68. Passing through $(-1, 4)$ and $(2, -2)$

69. Parallel to $y = 4x - 3$, passing through $\left(-\frac{3}{5}, \frac{1}{5}\right)$

70. Perpendicular to $y = \frac{1}{2}x$, passing through $(-1, 1)$

71. Passing through $(3, 1)$ and perpendicular to the line passing through $(-2, 3)$ and $(2, 1)$

72. Passing through $(3, 3)$ and parallel to the line passing through $(-3, -4)$ and $(3, 0)$

Exercises 73 and 74: Find the slope–intercept form of the line shown in the graph.

73.   **74.**

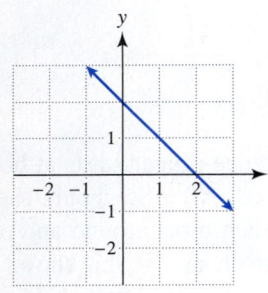   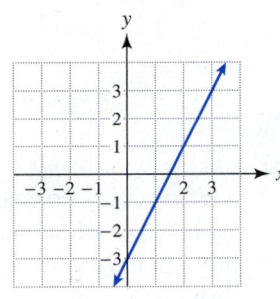

Exercises 75 and 76: Determine whether the given point lies on the line.

75. $(2, -1)$ $y = -2(x - 1) + 3$

76. $\left(4, -\frac{5}{2}\right)$ $y = \frac{1}{4}(x + 2) - 4$

Exercises 77–80: Find an equation of a line satisfying the given conditions.

77. Vertical, passing through $(-4, 14)$

78. Horizontal, passing through $\left(\frac{11}{13}, -\frac{7}{13}\right)$

79. Perpendicular to $x = -3$, passing through $(-2, 1)$

80. Parallel to $y = 5$, passing through $(4, -8)$

Exercises 81 and 82: Decide whether the points in the table lie on a line. If they do, find its slope–intercept form.

81.

x	1	2	3	4
y	6	7	8	7

82.

x	-2	0	2	4
y	1	5	9	13

APPLICATIONS

83. *U.S. Population* The following line graph shows the population of the United States in millions.
(a) Find the slope of each line segment.
(b) Interpret each slope as a rate of change.

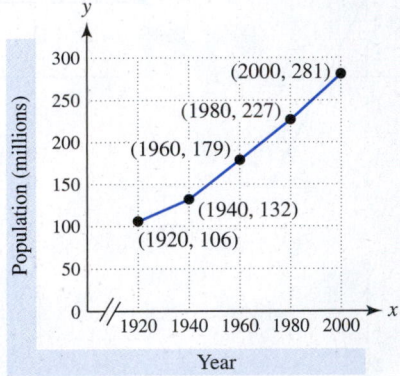

84. *Flow Rates* A water tank has an inlet pipe with a flow rate of 10 gallons per minute and an outlet pipe with a flow rate of 6 gallons per minute. A pipe can be either completely closed or open. The following graph shows the number of gallons of water in the tank after x minutes. Interpret the graph.

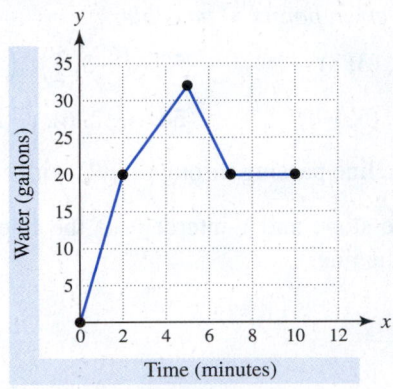

85. *Age at First Marriage* The median age at the first marriage for men from 1890 to 1960 can be modeled by $f(x) = -0.0492x + 119.1$, where x is the year.
(*Source:* National Center of Health Statistics.)

(a) Find the median age in 1910.

(b) Graph *f* in [1885, 1965, 10] by [22, 26, 1]. What happened to the median age?

(c) What is the slope of the graph of *f*? Interpret the slope as a rate of change.

86. *Marriages* From 2002 to 2008 the number of U.S. marriages in millions could be modeled by the formula $f(x) = 2.2$, where *x* is the year.
(a) Estimate the number of marriages in 2006.
(b) What information does *f* give about the number of marriages from 2002 to 2008?

87. *Fat Grams* A cup of milk contains 8 grams of fat.
(a) Give a formula for $f(x)$ that calculates the number of fat grams in *x* cups of milk.
(b) What is the slope of the graph of *f*?
(c) Interpret the slope as a rate of change.

88. *Birth Rate* The U.S. birth rate per 1000 people for selected years is shown in the table.

Year	1950	1970	1990	2010
Birth Rate	24.1	18.4	16.7	13.5

Source: U.S. Census Bureau.

(a) Make a scatterplot of the data.
(b) Model the data with $f(x) = mx + b$, where *x* is the year. Answers may vary.
(c) Use *f* to estimate the birth rate in 2000.

89. *Unhealthy Air Quality* The Environmental Protection Agency (EPA) monitors air quality in U.S. cities. The function *f* gives the annual number of days with unhealthy air quality in Los Angeles, California, for selected years.

x	1995	1999	2000	2003	2007
$f(x)$	113	56	87	88	100

Source: Environmental Protection Agency.

(a) Find $f(1995)$ and interpret your result.
(b) Identify the domain and range of *f*.
(c) Discuss the trend of air pollution in Los Angeles.

90. *Temperature Scales* The following table shows equivalent temperatures in degrees Celsius and degrees Fahrenheit.

°C	−40	0	15	35	100
°F	−40	32	59	95	212

(a) Plot the data. Let the *x*-axis correspond to the Celsius temperature and the *y*-axis correspond to the Fahrenheit temperature. What type of relation exists between the data?

(b) Find $f(x) = mx + b$ so that *f* receives the Celsius temperature *x* as input and outputs the corresponding Fahrenheit temperature. Interpret the slope of the graph of *f*.
(c) If the temperature is 20°C, what is the equivalent temperature in degrees Fahrenheit?

Exercises 91–96: Modeling Match the situation to the graph (a.–f.) that models it best.

91. The total federal debt *y* from 2001 to 2010

92. The distance *y* from New York City to Seattle, Washington, during year *x*

93. The amount of money *y* earned working for *x* hours at a fixed hourly rate

94. The sales of 8-millimeter movie projectors from 1970 to 1990

95. The average Celsius temperature *y* in a freezer

96. The height above sea level of a rocket launched from a submarine during the first minute of flight

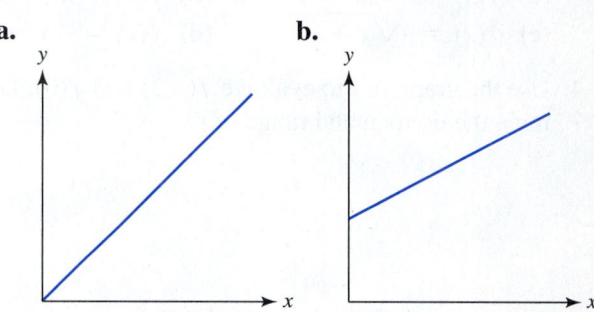
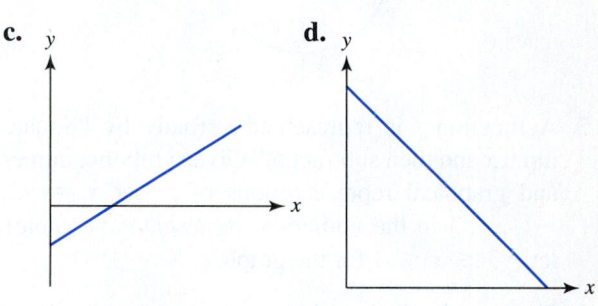
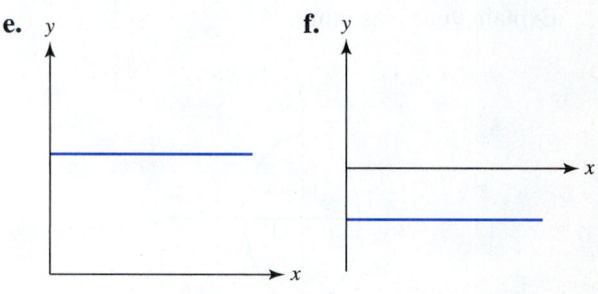

97. *Graphical Model* A 500-gallon water tank is initially full and then emptied at a constant rate of 50 gallons per minute. Ten minutes after the tank is empty, it is filled by a pump that supplies 25 gallons per minute. Sketch a graph that depicts the amount of water in the tank after x minutes.

98. *HIV/AIDS* In 2006 there were a total of 1.1 million people living with HIV/AIDS in the United States, with an annual infection rate of 56,000. (*Source:* Centers for Disease Control and Prevention.)

(a) Assuming that this trend continues, determine $f(x) = m(x - x_1) + y_1$ so that f models the number of AIDS cases to date during year x.

(b) Find $f(2015)$ and interpret the result.

CHAPTER 2   Test    Step-by-step test solutions are found on the Chapter Test Prep Videos available in *MyMathLab* and on You[Tube] (search "RockswoldInterAlg" and click on "Channels").

1. Evaluate $f(4)$ if $f(x) = 3x^2 - \sqrt{x}$. Give a point on the graph of f.

2. Write a symbolic representation (formula) for a function C that calculates the cost of buying x pounds of candy at \$4 per pound. Evaluate $C(5)$ and interpret your result.

3. Sketch a graph of f.
  (a) $f(x) = -2x + 1$ **(b)** $f(x) = x^2 + 1$
  (c) $f(x) = \sqrt{x + 3}$ **(d)** $f(x) = |x + 1|$

4. Use the graph of f to evaluate $f(-3)$ and $f(0)$. Determine the domain and range of f.

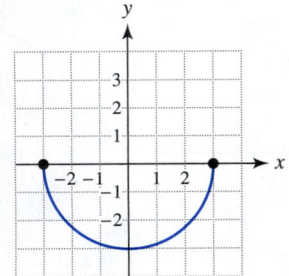

5. A function f is represented verbally by "Square the input x and then subtract 5." Give symbolic, numerical, and graphical representations of f. Let $x = -3, -2, -1, \ldots, 3$ in the numerical representation (table) and let $-3 \le x \le 3$ for the graph.

6. Determine whether the graph represents a function. Explain your reasoning.

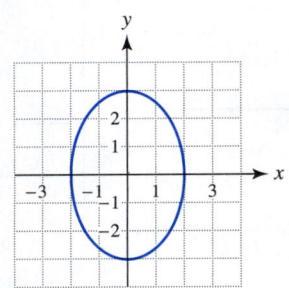

7. Find the domain of function f.
  (a) $f = \{(-2, 3), (-1, 5), (0, 3), (5, 7)\}$
  (b) $f(x) = \frac{3}{4}x - 5$
  (c) $f(x) = \sqrt{x + 4}$
  (d) $f(x) = 2x^2 - 1$
  (e) $f(x) = \frac{3x}{5 - x}$

8. Determine if $f(x) = 6 - 8x$ is a linear function. If it is, write it in the form $f(x) = mx + b$.

9. Find the slope of the line passing through $(-3, 7)$ and $(6, -2)$.

10. Determine the slope of the line shown in the graph.

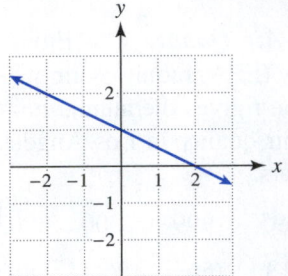

11. Write the slope–intercept form of a line satisfying the given conditions.
  (a) x-intercept -2, y-intercept -4
  (b) Passing through $(-5, 2)$ and perpendicular to the line passing through $(-2, 5)$ and $(-1, 3)$
  (c) Passing through $(1, -2)$ and $\left(-5, \frac{3}{2}\right)$

12. Let f be a linear function. Find the slope, x-intercept, and y-intercept of the graph of f.

x	-2	-1	0	2
$f(x)$	8	6	4	0

13. Give the slope–intercept form of a line parallel to $y = 1 - 3x$, passing through $\left(\frac{1}{3}, 2\right)$.

14. Find the slope–intercept form for the line shown in the graph. Then find the equation of a line that passes through the origin and is perpendicular to the given line.

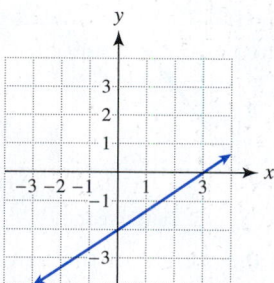

15. Find equations of the vertical line and the horizontal line passing through the point $\left(\frac{2}{3}, -\frac{1}{7}\right)$.

16. *Modeling* The line graph shows the number of welfare beneficiaries in millions for selected years. (*Source: Administration for Children and Families.*)
  (a) Find the slope of each line segment.
  (b) Interpret each slope as a rate of change.

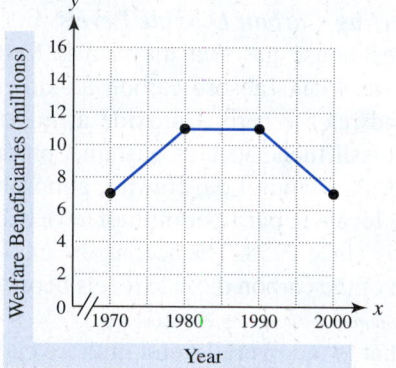

17. *Distance from Home* Starting at home, a driver travels away from home on a straight highway for 2 hours at 60 miles per hour, stops for 1 hour, and then drives home at 40 miles per hour. Sketch a graph that shows the distance between the driver and home.

18. *iPod Sales* The table shows the number of worldwide iPod sales in millions for selected years.

Year	2004	2005	2006	2007
Sales	5	22	39	52

Source: Apple Corporation.

  (a) Make a scatterplot of the data.
  (b) Determine values for m, x_1, and y_1 so that $f(x) = m(x - x_1) + y_1$ models these data.
  (c) Use f to estimate iPod sales in 2008.

CHAPTER 2   Extended and Discovery Exercises

1. *Developing a Model* Two identical cylindrical tanks, A and B, each contain 100 gallons of water. Tank A has a pump that begins removing water at a constant rate of 8 gallons per minute. Tank B has a plug removed from its bottom and water begins to flow out—faster at first and then more slowly.
  (a) Assuming that the tanks become empty at the same time, sketch a graph that models the amount of water in each tank. Explain your graphs.
  (b) Which tank is half empty first? Explain.

 2. *Modeling Real Data* Per capita personal incomes in the United States are listed in the following table.

Year	2002	2003	2004
Income	$31,461	$32,271	$33,881

Year	2005	2006	2007
Income	$35,424	$37,698	$39,458

Source: Department of Commerce.

  (a) Make a scatterplot of the data.
  (b) Find a function f that models the data. Explain your reasoning.
  (c) Use f to estimate per capita income in 2000.

3. *Graphing a Rectangle* One side of a rectangle has vertices (0, 0) and (5, 3). If the point (0, 5) lies on a different side of this rectangle, as shown in the figure, write the equation of each line.

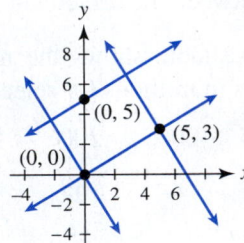

Exercises 4–6: (Refer to Exercise 3.) Graph the rectangle that satisfies the given conditions. Write the equation of each line.

4. Vertices (0, 0), (2, 2), and (1, 3)

5. Vertices (1, 1), (5, 1), and (1, 5)

6. Vertices (4, 0), (0, 4), (0, −4), and (−4, 0)

7. *Remaining Life Expectancy* The table lists the average *remaining* life expectancy E in years for females at age x.

x (year)	0	10	20	30	40
E (year)	72.3	69.9	60.1	50.4	40.9

x (year)	50	60	70	80
E (year)	31.6	23.1	15.5	9.2

Source: Department of Health and Human Services.

(a) Make a line graph of the data.

(b) Assume that the graph represents a function f. Calculate the slope of each line segment and interpret each slope as a rate of change.

(c) Determine the life expectancy (not the remaining life expectancy) of a 20-year-old woman. What is the life expectancy of a 70-year-old woman? Discuss reasons why these two expectancies are not equal.

8. *Weight of a Small Fish* The graph at the top of the next column shows a function f that models the weight in milligrams of a small fish, *Lebistes reticulatus*, during the first 14 weeks of its life. (*Source:* D. Brown and P. Rothery, *Models in Biology*.)

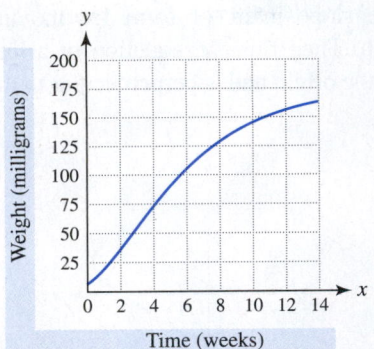

(a) Estimate the weight of the fish when it hatches, at 6 weeks, and at 12 weeks.

(b) If (x_1, y_1) and (x_2, y_2) are points on the graph of a function, the *average rate of change of f from x_1 to x_2* is given by $\frac{y_2 - y_1}{x_2 - x_1}$. Approximate the average rates of change of f from hatching to 6 weeks and from 6 weeks to 12 weeks.

(c) Interpret these rates of change.

(d) During which time period does the fish gain weight the fastest?

9. *Interpreting Carbon Dioxide Levels* Carbon dioxide is a greenhouse gas that may cause Earth's climate to change. Plants absorb carbon dioxide during daylight and release carbon dioxide at night. The burning of fossil fuels, such as gasoline, produces carbon dioxide. At Mauna Loa, Hawaii, atmospheric carbon dioxide levels in parts per million have been measured regularly since 1958. The accompanying figure shows a graph of the carbon dioxide levels between 1960 and 2000. (*Source:* A. Nilsson, *Greenhouse Earth*.)

(a) What is the overall trend in these carbon dioxide levels?

(b) Discuss what happens to these carbon dioxide levels each year.

(c) Give an explanation for the shape of this graph.

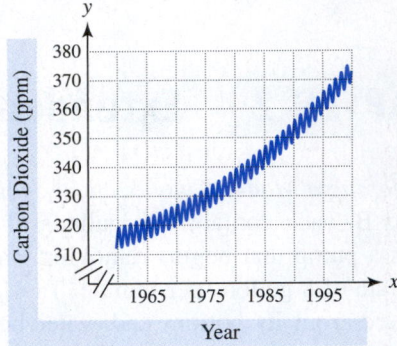

10. *Carbon Dioxide in Alaska* (Refer to Exercise 9.) The atmospheric carbon dioxide levels at Barrow, Alaska, in parts per million from 1970 to 2000 are shown in the figure to the right. (*Source:* M. Zeilik, S. Gregory, and D. Smith, *Introductory Astronomy and Astrophysics.*)

(a) Compare this graph with the graph in Exercise 9.

(b) Discuss possible reasons for their similarities and differences.

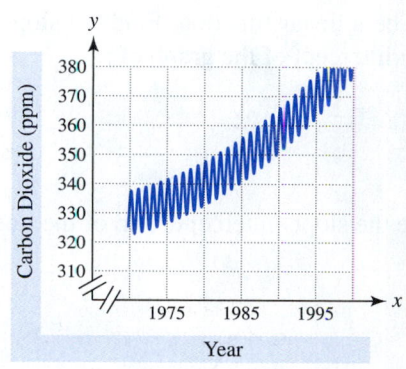

CHAPTERS 1–2   Cumulative Review Exercises

Exercises 1 and 2: Classify each real number as one or more of the following: natural number, whole number, integer, rational number, or irrational number.

1. $-3, \frac{3}{4}, \sqrt{7}, -5.8, \sqrt[3]{8}$

2. $0, -5, \frac{1}{2}, -\sqrt{9}, \frac{50}{10}, \sqrt{8}$

Exercises 3 and 4: State whether the equation illustrates an identity, commutative, associative, or distributive property.

3. $6x + 8x = 14x$

4. $a + (b + c) = (b + c) + a$

5. Evaluate $-5^2 + 3 + 4 \cdot 5$ by hand.

6. Write $(x + 1)x + (x + 1)5$ as a product of two factors by using the distributive property.

7. Find the opposite of $a - b$.

8. Find the reciprocal of $\frac{a + b}{c}$.

9. Evaluate $\frac{1}{3} \div \left(\frac{3}{4} \cdot \frac{7}{8}\right) + \frac{5}{6}$ with a calculator.

10. Simplify each expression. Write the result using positive exponents.

(a) $\left(\dfrac{2^{-3}}{3^{-2}}\right)^2$

(b) $\dfrac{(3x^2 y^{-3})^4}{x^3 (y^4)^{-2}}$

(c) $\left(\dfrac{ab^{-2}}{a^{-3} b^4}\right)^{-2}$

11. Write 9540 in scientific notation.

12. Find $f(4)$ if $f(x) = 2x^2 + \sqrt{x}$.

13. Evaluate $B = \frac{1}{2}bh^2$ for $b = 4$ and $h = 3$.

14. Identify the domain of each function f.

(a) $f = \{(-3, 4), (0, 3), (2, -1)\}$

(b) $f(x) = \dfrac{3}{x + 6}$

(c) $f(x) = \sqrt{x + 4}$

Exercises 15–20: Graph f by hand.

15. $f(x) = -2x + 1$

16. $f(x) = -\frac{1}{2}x^2$

17. $f(x) = |x - 2|$

18. $f(x) = \sqrt{-x}$

19. $f(x) = -1$

20. $f(x) = \frac{1}{3}x$

21. Use the graph of f to do the following.

(a) Determine if the graph represents a function.

(b) Identify the domain and range.

(c) Evaluate $f(-1)$ and $f(0)$.

(d) Identify the x- and y-intercepts.

(e) Find the slope of the line.

(f) Find a formula for $f(x)$.

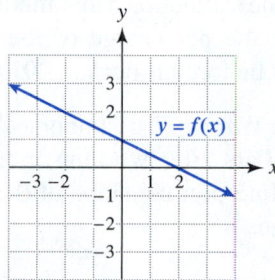

22. Find the slope–intercept form of the line that passes through $(-4, 5)$ and $(2, -4)$.

23. Find the slope–intercept form of the line that passes through $(-1, 2)$ and is perpendicular to the line that passes through $(1, -3)$ and $(2, 0)$.

24. Write an equation of a line that is perpendicular to the x-axis and passes through $(-2, 4)$.

25. Let f be a linear function. Find the slope, x-intercept, and y-intercept of the graph of f.

x	-2	-1	1	2
$f(x)$	9	6	0	-3

26. Write the slope–intercept form of the line.

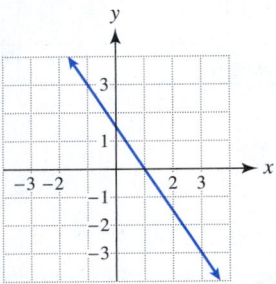

APPLICATIONS

27. *Earnings* Earning $120 one day and $80 the next day is equivalent to earning $80 the first day and $120 the second day. What property of real numbers is illustrated?

28. *Storage in a Box* A rectangular box contains 400 cubic inches. If a similar box has the same width but double the length and double the height, how many cubic inches does this box contain?

29. *Foreign-Born Population* The table shows the percentage of the U.S. population that is foreign-born for selected years.

Year	1970	1980	1990	2000	2010
Percentage	4.7	6.2	7.9	10.4	12.7

Source: U.S. Census Bureau.

(a) Find a linear function f that models the data.
(b) Estimate the percentage of the U.S. population that may be foreign-born in 2017.

30. *Calculating Wages* The amount A that a person makes in dollars after working x hours is given by $A(x) = 9x$. Interpret the slope of the graph of A as a rate of change.

31. *Distance* A person walks away from home at 4 miles per hour for 30 minutes, rests in a park for 1 hour, and then takes 1 hour to walk back home at a constant speed. Sketch a graph that models the distance y that the person is from home after x hours.

32. *Age in the United States* The median age of the population for each year x between 1970 and 2010 can be approximated by

$$f(x) = 0.243x - 450.8.$$

(*Source:* U.S. Census Bureau.)

(a) What type of function is f?
(b) Evaluate $f(2005)$ and interpret your result.
(c) What is the slope of the graph of f?
(d) Interpret this slope as a rate of change.

33. *Fat Grams* Some slices of pizza contain 10 grams of fat.
(a) Find a formula $f(x)$ that calculates the number of fat grams in x slices of pizza.
(b) Graph f for $0 \le x \le 6$.
(c) What is the slope of the graph of f?
(d) Interpret the slope as a rate of change.

34. *Two-Cycle Engines* Two-cycle engines used in many snowmobiles, jet skis, chain saws, and outboard motors require a mixture of gas and oil to run properly. For certain engines, the amount of oil in pints that should be added to x gallons of gasoline is computed by $f(x) = \frac{4}{25}x$. (*Source:* Johnson Outboard Motor Company.)
(a) How many pints of oil should be added to 6 gallons of gasoline?
(b) Graph f for $0 \le x \le 25$.
(c) What is the slope of the graph of f?
(d) Interpret the slope as a rate of change.

35. *Municipal Waste* From 1960 to 1995, municipal solid waste in millions of tons can be modeled by

$$f(x) = 3.4(x - 1960) + 87.8,$$

where x is the year. (*Source:* EPA.)
(a) Find the tons of waste in 1960.
(b) Graph f in the viewing rectangle [1960, 1995, 5] by [60, 220, 20].
(c) What is the slope of the graph of f? Interpret the slope as a rate of change.

36. *Modeling Distance* An individual is driving a car along a straight road. The graph shows the distance that the driver is from home after x hours.
(a) Find the slope of each line segment in the graph.
(b) Interpret each slope as a rate of change.
(c) Describe both the motion and location of the car.

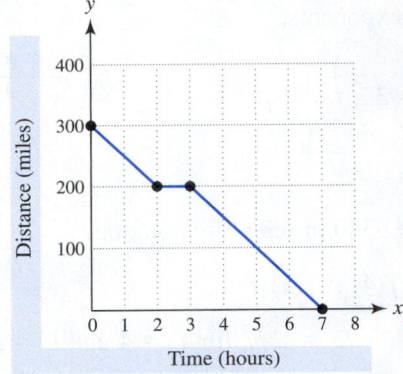

3

Linear Equations and Inequalities

Say it, I'll forget. Demonstrate it, I may recall. But if I do it, I'll understand.

—OLD CHINESE PROVERB

Global consumption of seafood has increased from 42 million tons in 1960 to 157 million tons in 2008. This trend is due to a dramatic increase in world population living near the ocean, to an increase in per capita consumption, and to unsustainable fishing practices. These data are modeled by using a linear function f in the figure.

Global Seafood Consumption

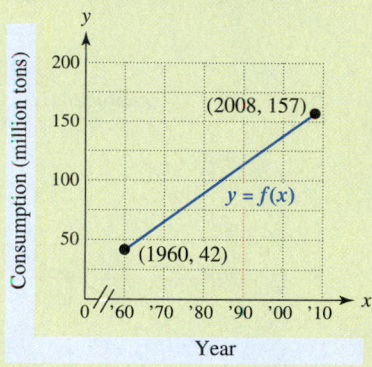

By 2027, worldwide consumption may increase by at least 45 million tons. How are predictions like this one made? Mathematics plays an essential role in predictions and in analyzing environmental issues. (See Section 3.1, Exercise 117.) Before we can make predictions of our own, we need to discuss linear equations. Linear equations are fundamental to understanding many concepts in mathematics.

Source: Lischewski, C., "The Global Seafood Industry: A Perspective on Consumption and Supply," 2008.

3.1   Linear Equations

Equations ● Symbolic Solutions ● Numerical and Graphical Solutions ● Identities and Contradictions ● Intercepts of a Line

A LOOK INTO MATH ▶   In Chapter 2, functions were introduced as a way to model a variety of data. Society uses functions to answer important questions and make predictions. To do this, equations need to be solved. As a result, a primary objective of mathematics is to solve equations. In this section we focus on methods for solving one type of equation called a *linear equation*.

Equations

NEW VOCABULARY

☐ Linear equation
☐ Solution/Solution set
☐ Addition property
☐ Multiplication property
☐ Symbolic solution
☐ Expression/Equation
☐ Graphical solution
☐ Numerical solution
☐ Identity/Contradiction
☐ Conditional equation
☐ Standard form

▶ **REAL-WORLD CONNECTION**   In 2010, sales of green building materials were $65 billion and growing at a constant rate of $5 billion per year. We modeled these sales in billions of dollars with the *linear function* $f(x) = 5x + 65$, where x is the number of years after 2010. (See Example 7, Section 2.2.)

Now suppose we want to predict when sales of green building materials might reach $80 billion. To answer this question, set the expression $5x + 65$ equal to 80 and form the following *linear equation*. (You may be able to solve this particular equation in your head, but many times equations are more complicated.)

Forming a Linear Equation

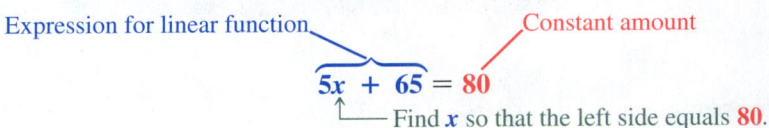

Expression for linear function   Constant amount

$$5x + 65 = 80$$

Find x so that the left side equals 80.

Finding the value of x that makes the left side of the equation equal to 80 is called *solving the equation*, and this value of x is called the *solution*. For this example, the solution tells us how many years after 2010 green building sales will reach $80 billion.

Equations like this one can be solved *symbolically*, *graphically*, and *numerically*. The basic idea behind these three methods is illustrated by the following. (These three methods will be explained more fully throughout this section.)

Symbolically

$5x + 65 = 80$
$5x = 15$
$x = 3$

Start by subtracting 65 from each side, and then divide each side by 5 to get the solution of 3.

Graphically

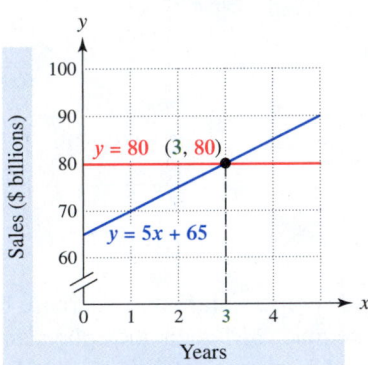

Graph each side of the equation. Find the point of intersection.

Numerically (table)

x	$5x + 65 \stackrel{?}{=} 80$	T or F
1	$5(1) + 65 \stackrel{?}{=} 80$	F
2	$5(2) + 65 \stackrel{?}{=} 80$	F
3	$5(3) + 65 \stackrel{?}{=} 80$	T ✓
4	$5(4) + 65 \stackrel{?}{=} 80$	F

Find an x-value that makes the equation true.

LINEAR EQUATION IN ONE VARIABLE

A **linear equation** in one variable is an equation that can be written in the form

$$ax + b = 0,$$

where a and b are constants with $a \neq 0$.

Examples of linear equations include

$$2x - 1 = 0, \quad -5x = 10 + x, \quad \text{and} \quad 3x + 8 = 2.$$

Although the second and third equations do not appear to be in the form $ax + b = 0$, they can be transformed by using properties of algebra, which we discuss later in this section.

To *solve* an equation means to find all values for a variable that make the equation a true statement. Such values are called **solutions**, and the set of all solutions is called the **solution set**. For example, substituting **2** for x in the equation $3x - 1 = 5$ results in $3(\mathbf{2}) - 1 = 5$, which is a true statement. The value 2 *satisfies* the equation $3x - 1 = 5$ and is the only solution. The solution set is $\{2\}$. Two equations are *equivalent* if they have the same solution set.

LINEAR EQUATIONS AND SOLUTIONS Any linear equation can be written as $ax + b = 0$ with $a \neq 0$, so every linear equation has exactly *one solution*. To understand why, consider the linear equation $\frac{1}{2}x - 1 = 0$. The graph of $y = \frac{1}{2}x - 1$ (Figure 3.1) is a line with slope $\frac{1}{2}$ that intersects the x-axis *once* at the point $(\mathbf{2}, \mathbf{0})$. Thus 2 is the solution because when $x = 2$, the y-value is 0 and the linear equation

$$\frac{1}{2}x - 1 = 0 \quad \text{simplifies to} \quad \frac{1}{2}(2) - 1 = \mathbf{0}. \quad \longleftarrow \text{True statement}$$

In fact, any line with *nonzero* slope must intersect the x-axis at exactly one point $(k, 0)$. This value of k is the only solution to the linear equation $ax + b = 0$.

Linear Equation: $\frac{1}{2}x - 1 = 0$

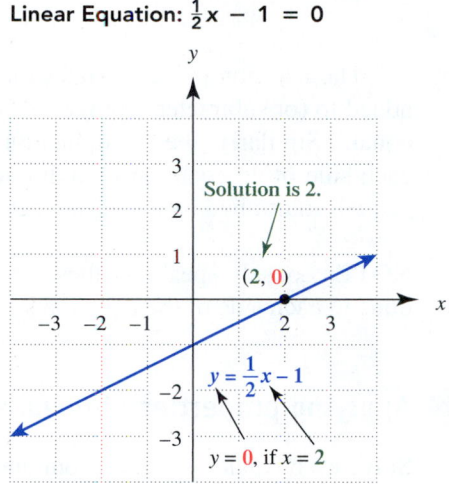

Figure 3.1 One Solution: **2**

READING CHECK

Give an example of a linear function. Use the formula for this function to write a linear equation.

MAKING CONNECTIONS

Linear Functions and Equations

A linear function can be written as $f(x) = \mathbf{a}x + \mathbf{b}$ ($a = $ slope m).

A linear equation can be written as $\mathbf{a}x + \mathbf{b} = 0$ ($a \neq 0$).

Symbolic Solutions

Linear equations can be solved symbolically, graphically, and numerically. One advantage of a symbolic method is that *the solution is always exact*. To solve an equation symbolically, we write a sequence of equivalent equations, using algebraic properties. For example, to solve $3x - 5 = 0$, we might add 5 to *each side* of the equation and then divide *each side* by 3 to obtain $x = \frac{5}{3}$.

$$3x - 5 = 0 \qquad \text{Given equation}$$

$$3x - 5 + \mathbf{5} = 0 + \mathbf{5} \qquad \text{Add 5 to each side.}$$

$$3x = 5 \qquad \text{Simplify.}$$

$$\frac{3x}{\mathbf{3}} = \frac{5}{\mathbf{3}} \qquad \text{Divide each side by 3.}$$

$$x = \frac{5}{3} \qquad \text{Simplify.}$$

The solution is $\frac{5}{3}$. That is, $3\left(\frac{5}{3}\right) - 5 = 0$ is a true statement.

Adding 5 to each side is an example of the *addition property of equality,* and dividing each side by 3 is an example of the *multiplication property of equality*. Note that dividing each side by 3 is equivalent to multiplying each side by $\frac{1}{3}$.

PROPERTIES OF EQUALITY

Addition Property of Equality
If a, b, and c are real numbers, then

$$a = b \quad \text{is equivalent to} \quad a + c = b + c.$$

Multiplication Property of Equality
If a, b, and c are real numbers with $c \neq 0$, then

$$a = b \quad \text{is equivalent to} \quad ac = bc.$$

The addition property states that an equivalent equation results if the same number is added to (or subtracted from) each side of an equation. That is, "equals added to equals are equal." Similarly, the multiplication property states that an equivalent equation results if each side of an equation is multiplied (or divided) by the same *nonzero* number. That is, "equals multiplied by equals are equal."

NOTE: Loosely speaking, these properties say that the same "thing" or "action" must be done to each side of the equation so that the equation "remains balanced."

EXAMPLE 1 **Applying properties of equality**

Solve each equation. Check your answer.
(a) $x - 4 = 5$ **(b)** $5x = 13$

Solution
(a) Isolate x in the equation $x - 4 = 5$ by applying the addition property of equality. To do this, we add **4** to *each side* of the equation.

$$x - 4 = 5 \qquad \text{Given equation}$$

$$x - 4 + \mathbf{4} = 5 + \mathbf{4} \qquad \text{Add 4 to each side.}$$

$$x = 9 \qquad \text{Simplify.}$$

Check: To check that 9 is the solution, we substitute **9** in the *given* equation.

$$x - 4 = 5 \qquad \text{Given equation}$$
$$9 - 4 \stackrel{?}{=} 5 \qquad \text{Let } x = 9.$$
$$5 = 5 \ \checkmark \qquad \text{It checks.}$$

(b) Isolate x in the equation $5x = 13$ by applying the multiplication property of equality. To do this, we multiply *each side* of the equation by $\frac{1}{5}$, the reciprocal of 5. Note that this step is equivalent to dividing *each side* of the equation by **5**.

$$5x = 13 \qquad \text{Given equation}$$
$$\frac{5x}{5} = \frac{13}{5} \qquad \text{Divide each side by 5.}$$
$$x = \frac{13}{5} \qquad \text{Simplify: } \tfrac{5}{5} = 1.$$

Check: To check that $\frac{13}{5}$ is the solution, we substitute $\frac{13}{5}$ in the *given* equation.

$$5x = 13 \qquad \text{Given equation}$$
$$5 \cdot \frac{13}{5} \stackrel{?}{=} 13 \qquad \text{Let } x = \tfrac{13}{5}.$$
$$13 = 13 \ \checkmark \qquad \text{It checks.}$$

▌ **Now Try Exercises 23, 29**

STUDY TIP

When you solve an equation, be sure that you can explain, or justify, each step.

Sometimes both properties are needed, as in the next example.

EXAMPLE 2 **Solving a linear equation symbolically**

Solve $2 - \frac{1}{2}x = 1$.

Solution

To solve, we write a sequence of equivalent equations by using the properties of equality.

$$2 - \frac{1}{2}x = 1 \qquad \text{Given equation}$$

$$-2 + 2 - \frac{1}{2}x = 1 + (-2) \qquad \text{Add } -2 \text{ to each side.}$$

$$-\frac{1}{2}x = -1 \qquad \text{Simplify.}$$

$$-2\left(-\frac{1}{2}x\right) = -1(-2) \qquad \text{Multiply each side by } -2, \text{ the reciprocal of } -\tfrac{1}{2}.$$

$$x = 2 \qquad \text{Simplify.}$$

The solution is 2.

▌ **Now Try Exercise 35**

MAKING CONNECTIONS

Expressions and Equations

Expressions and equations are *different* mathematical concepts. An expression does *not* contain an equals sign, whereas an equation *always* contains an equals sign. *An equation is a statement that two expressions are equal.* For example,

$$2x - 1 \quad \text{and} \quad x + 1 \qquad \text{Expressions}$$

are two different expressions, and

$$2x - 1 = x + 1 \qquad \text{Equation}$$

is an equation. We often *solve an equation* for a value of x that makes the equation a true statement, but we *do not solve an expression.* We sometimes *simplify* an expression.

READING CHECK

Explain the difference between an expression and an equation.

In the next example we use the distributive property to solve a linear equation.

EXAMPLE 3 **Solving a linear equation symbolically**

Solve $2(x - 1) = 4 - \frac{1}{2}(4 + x)$. Check your answer.

Solution
We begin by applying the distributive property.

$$2(x - 1) = 4 - \frac{1}{2}(4 + x) \qquad \text{Given equation}$$

$$2x - 2 = 4 - 2 - \frac{1}{2}x \qquad \text{Distributive property}$$

$$2x - 2 = 2 - \frac{1}{2}x \qquad \text{Simplify.}$$

Next, we move the constant terms to the right and the x-terms to the left.

$$2x - 2 + \mathbf{2} = 2 - \frac{1}{2}x + \mathbf{2} \qquad \text{Add 2 to each side.}$$

$$2x = 4 - \frac{1}{2}x \qquad \text{Simplify.}$$

$$2x + \frac{\mathbf{1}}{\mathbf{2}}x = 4 - \frac{1}{2}x + \frac{\mathbf{1}}{\mathbf{2}}x \qquad \text{Add } \tfrac{1}{2}x \text{ to each side.}$$

$$\frac{5}{2}x = 4 \qquad \text{Simplify.}$$

Finally, we multiply by $\frac{2}{5}$, which is the reciprocal of $\frac{5}{2}$.

$$\frac{\mathbf{2}}{\mathbf{5}} \cdot \frac{5}{2}x = 4 \cdot \frac{\mathbf{2}}{\mathbf{5}} \qquad \text{Multiply each side by } \tfrac{2}{5}.$$

$$x = \frac{8}{5} = 1.6 \qquad \text{Simplify.}$$

The solution is $\frac{8}{5}$.

To check our answer, we substitute $x = \frac{8}{5}$ in the *given* equation.

$$2(x - 1) = 4 - \frac{1}{2}(4 + x) \qquad \text{Given equation}$$

$$2\left(\frac{8}{5} - 1\right) \overset{?}{=} 4 - \frac{1}{2}\left(4 + \frac{8}{5}\right) \qquad \text{Let } x = \frac{8}{5}.$$

$$\frac{16}{5} - 2 \overset{?}{=} 4 - 2 - \frac{4}{5} \qquad \text{Distributive property}$$

$$\frac{6}{5} = \frac{6}{5} \checkmark \qquad \text{It checks.}$$

Now Try Exercise 53

The equation in Example 3 contained a fraction. Sometimes it is easier to avoid working with fractions. To clear fractions we can multiply each side by a common denominator.

EXAMPLE 4 **Solving equations involving fractions or decimals**

Solve each equation.

(a) $\frac{1}{3}(2z - 3) - \frac{1}{2}z = -2$ (b) $0.4t + 0.3 = 0.75 - 0.05t$

Solution

(a) The least common denominator of $\frac{1}{3}$ and $\frac{1}{2}$ is 6, so multiply each side (or equivalently, every term) of the equation by 6. This step will clear, or eliminate, fractions.

$$\frac{1}{3}(2z - 3) - \frac{1}{2}z = -2 \qquad \text{Given equation}$$

$$6\left(\frac{1}{3}(2z - 3) - \frac{1}{2}z\right) = 6(-2) \qquad \text{Multiply each side by 6.}$$

$$2(2z - 3) - 3z = -12 \qquad \text{Distributive property}$$

$$4z - 6 - 3z = -12 \qquad \text{Distributive property}$$

$$z - 6 = -12 \qquad \text{Combine like terms.}$$

$$z = -6 \qquad \text{Add 6 to each side.}$$

The solution is -6.

(b) The decimals 0.4, 0.3, 0.75, and 0.05 can be written as $\frac{4}{10}$, $\frac{3}{10}$, $\frac{75}{100}$, and $\frac{5}{100}$. A common denominator is 100, so multiply each side by 100.

$$0.4t + 0.3 = 0.75 - 0.05t \qquad \text{Given equation}$$

$$100(0.4t + 0.3) = 100(0.75 - 0.05t) \qquad \text{Multiply each side by 100.}$$

$$40t + 30 = 75 - 5t \qquad \text{Distributive property}$$

$$45t + 30 = 75 \qquad \text{Add } 5t \text{ to each side.}$$

$$45t = 45 \qquad \text{Subtract 30 from each side.}$$

$$t = 1 \qquad \text{Divide each side by 45.}$$

The solution is 1.

Now Try Exercises 57, 61

▶ **REAL-WORLD CONNECTION** In the next example, we model some real data with a linear function and then use this function to make an estimate.

EXAMPLE 5 **Modeling numbers of LCD screens**

Flat screens or LCD (liquid crystal display) screens are becoming increasingly popular. In 2002 about 30 million LCD screens were manufactured, and this number increased to 102 million in 2008.

(a) Find a linear function that models these data.

(b) Estimate the year when 150 million LCD screens were manufactured.

Solution

(a) The graph of this linear function must pass through (**2002**, **30**) and (**2008**, **102**). Its slope is

$$m = \frac{102 - 30}{2008 - 2002} = \frac{72}{6} = 12.$$

Using the point (**2002**, **30**) in the point–slope form gives

Expression for $f(x)$

$$f(x) = \mathbf{12}(x - \mathbf{2002}) + \mathbf{30}.$$

(b) To determine when 150 million LCD screens were manufactured, we can set the expression for $f(x)$ equal to **150**.

$12(x - 2002) + 30 = \mathbf{150}$	Equation to be solved
$12(x - 2002) = 120$	Subtract 30 from each side.
$x - 2002 = 10$	Divide each side by 12.
$x = 2012$	Add 2002 to each side.

In 2012, 150 million LCD screens may have been manufactured.

Now Try Exercise 117

Numerical and Graphical Solutions

Linear equations can also be solved numerically (with a table) and graphically. The disadvantage of using a table or graph is that the solution is often an estimate, rather than exact. In the next example these two methods are applied to the equation in Example 2.

EXAMPLE 6 **Solving an equation numerically and graphically**

Solve $2 - \frac{1}{2}x = 1$ numerically and graphically.

Solution

Numerical Solution Begin by constructing a table for the expression $2 - \frac{1}{2}x$, as shown in Table 3.1. This expression equals **1** when $x = \mathbf{2}$. That is, the solution to $2 - \frac{1}{2}x = \mathbf{1}$ is **2**.

TABLE 3.1 A Numerical Solution

x	0	1	**2**	3	4	5	6
$2 - \frac{1}{2}x$	2	1.5	**1**	0.5	0	−0.5	−1

Exact solution is 2.

A Numerical Solution

X	Y₁
0	2
1	1.5
2	1
3	.5
4	0
5	-.5
6	-1

Y₁◼2−(1/2)X

Figure 3.2

In Figure 3.2 a calculator has been used to create the same table.

A Graphical Solution

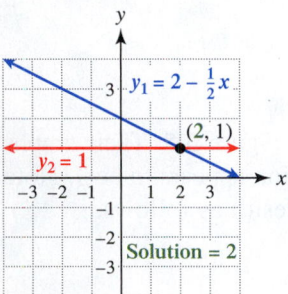

Figure 3.3

Graphical Solution One way to find a graphical solution is to let y_1 equal the left side of the equation, to let y_2 equal the right side of the equation, and then to graph $y_1 = 2 - \frac{1}{2}x$ and $y_2 = 1$, as shown in Figure 3.3. The graphs intersect at the point $(2, 1)$. We are seeking an x-value that satisfies $2 - \frac{1}{2}x = 1$, so **2** is the solution.

Now Try Exercise 91

TECHNOLOGY NOTE

Finding a Numerical Solution

The solution to the equation $2(x - 1) = 4 - \frac{1}{2}(4 + x)$ in Example 3 is the decimal number 1.6. It is possible to find it numerically. Let

$$Y_1 = 2(X - 1) \quad \text{and} \quad Y_2 = 4 - (1/2)(4 + X)$$

and increment the x-values by 1. There is no value where $y_1 = y_2$, as shown in the left-hand figure below. However, note that when $x = 1$, $y_1 < y_2$ and when $x = 2$, $y_1 > y_2$. This change indicates that there is a solution *between* $x = 1$ and $x = 2$. When x is incremented by 0.1, $y_1 = y_2$ when $x = 1.6$, as shown in the right-hand figure below.

$y_1 < y_2 \longrightarrow$
$y_1 > y_2 \longrightarrow$

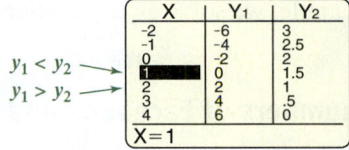

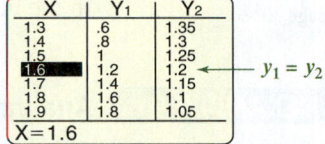

$\longleftarrow y_1 = y_2$

Sometimes a graphical solution gives an *approximate* solution because it depends on how accurately a graph can be read. To verify that a graphical solution is exact, check it by substituting it in the given equation, as is done in the next two examples.

EXAMPLE 7   **Solving a linear equation graphically**

Figure 3.4 shows graphs of $y_1 = 2x + 1$ and $y_2 = -x + 4$. Use the graphs to solve the equation $2x + 1 = -x + 4$. Check your answer.

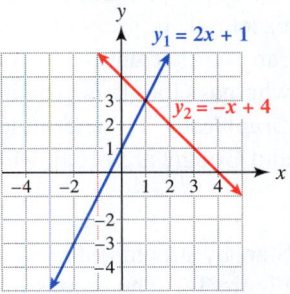

Figure 3.4

Solution

The graphs of y_1 and y_2 intersect at the point $(1, 3)$. Therefore **1** is the solution. We can check this solution by substituting $x = 1$ in the equation.

$$2x + 1 = -x + 4 \qquad \text{Given equation}$$
$$2(1) + 1 \stackrel{?}{=} -1 + 4 \qquad \text{Let } x = 1.$$
$$3 = 3 \checkmark \qquad \text{The answer checks.}$$

Now Try Exercise 77

EXAMPLE 8 **Solving a linear equation graphically**

Solve $3(1 - x) = 2$ graphically.

Solution

We begin by graphing $y_1 = 3(1 - x)$ and $y_2 = 2$, as shown in Figure 3.5. Their graphs intersect near the point $(0.3333, 2)$. Because $\frac{1}{3} = 0.\overline{3}$, the solution appears to be $\frac{1}{3}$. Note that this graphical solution is *approximate*. We can verify our result as follows.

$[-6, 6, 1]$ by $[-4, 4, 1]$

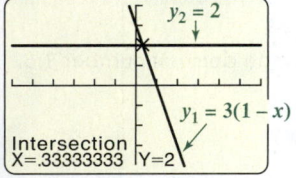

Figure 3.5

$$3(1 - x) = 2 \qquad \text{\color{blue}Given equation}$$

$$3\left(1 - \frac{1}{3}\right) \stackrel{?}{=} 2 \qquad \text{\color{blue}Substitute } x = \tfrac{1}{3}.$$

$$2 = 2 \checkmark \qquad \text{\color{blue}It checks.}$$

Now Try Exercise 87

CALCULATOR HELP

To find a point of intersection, see Appendix A (page AP-6).

Technology can be helpful when we are solving an equation that is more complicated. In the next example, we solve an application graphically and symbolically.

EXAMPLE 9 **Analyzing numbers of Facebook and Google visitors**

From 2009 to 2014, the number of unique visitors F to Facebook in millions x years after 2009 is expected to be $F(x) = 202x + 300$. Similarly, the number of unique visitors G to Google is expected to be $G(x) = 83x + 800$.

(a) Evaluate $F(3)$ and interpret the result.
(b) Does Facebook or Google have a greater rate of growth? Explain.
(c) Estimate graphically the year when Facebook and Google could have the same number of unique visitors.
(d) Solve part (c) symbolically. Do your answers agree?

Solution

(a) $F(3) = 202(3) + 300 = 906$. Facebook is expected to have 906 million unique visitors in the year $2009 + 3 = 2012$.

(b) Facebook is growing faster. From $F(x) = 202x + 300$ and $G(x) = 83x + 800$, we can see that unique Facebook visitors are growing at a rate of 202 million per year, whereas Google is growing at a rate of 83 million per year.

(c) *Graphically* To determine the year when the numbers of unique visitors to Facebook and Google are equal, we must find an x such that $\color{red}F(x) = G(x)$. That is, we must solve

$$\color{red}202x + 300 = 83x + 800.$$

Start by graphing the equations $y_1 = 202x + 300$ and $y_2 = 83x + 800$, as shown in Figure 3.6. The graphs intersect near $(4.2, 1149)$. This model predicts that during 2013 $(2009 + 4.2 = 2013.2)$, there will be about $1,149,000,000$ unique visitors to both Facebook and Google.

$[0, 8, 1]$ by $[0, 2000, 500]$

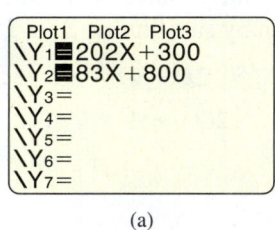

(a)

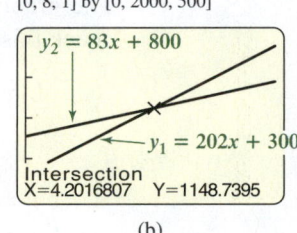

(b)

Figure 3.6

(d) *Symbolically* As in part (c), we must find an x such that $F(x) = G(x)$.

$$202x + 300 = 83x + 800 \qquad \text{Solve } F(x) = G(x).$$
$$202x = 83x + 500 \qquad \text{Subtract 300 from each side.}$$
$$119x = 500 \qquad \text{Subtract } 83x \text{ from each side.}$$
$$x = \frac{500}{119} \approx 4.2 \qquad \text{Divide each side by 119.}$$

Our graphical and symbolic answers agree. Note that the symbolic answer of $\frac{500}{119}$ is exact, while the graphical answer of 4.2 is approximate.

Now Try Exercise 123

Identities and Contradictions

A linear equation ($ax + b = 0, a \neq 0$) has exactly one solution. However, there are other types of equations, called *identities* and *contradictions*. An **identity** is an equation that is always *true*, regardless of the values of any variables. For example, $x + x = 2x$ is an identity, because it is true for all real numbers x. A **contradiction** is an equation that is always *false*, regardless of the values of any variables. For example, $x = x + 1$ is a contradiction because no real number x can be equal to itself plus 1. If an equation is true for some but not all values of any variable, then it is called a **conditional equation**. The equation $x + 1 = 4$ is conditional because 3 is the only solution. Thus far, we have only discussed conditional equations.

EXAMPLE 10   **Determining identities and contradictions**

Determine whether each equation is an identity, contradiction, or conditional equation.
(a) $0 = 1$ **(b)** $5 = 5$ **(c)** $2x = 6$ **(d)** $2(z - 1) + z = 3z + 2$

Solution
(a) The equation $0 = 1$ is always false. It is a contradiction.
(b) The equation $5 = 5$ is always true. It is an identity.
(c) The equation $2x = 6$ is true only when $x = 3$. It is a conditional equation.
(d) Simplify the equation to obtain the following.

$$2(z - 1) + z = 3z + 2 \qquad \text{Given equation}$$
$$2z - 2 + z = 3z + 2 \qquad \text{Distributive property}$$
$$3z - 2 = 3z + 2 \qquad \text{Combine like terms.}$$
$$-2 = 2 \ \text{✗} \qquad \text{Subtract } 3z \text{ from each side.}$$

Because $-2 = 2$ is always false, the given equation is a contradiction.

Now Try Exercises 95, 97

Intercepts of a Line

Equations of lines can be written in slope–intercept form or point–slope form. A third form for the equation of a line is called *standard form*, which is defined as follows.

STANDARD FORM OF A LINE

An equation for a line is in **standard form** when it is written as

$$ax + by = c,$$

where a, b, and c are constants with a and b not both 0.

Standard Form

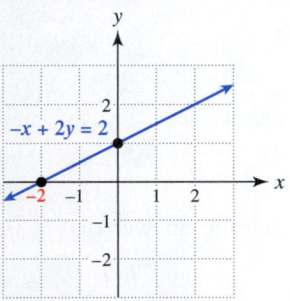

Figure 3.7

Examples of lines in standard form include

$$3x - 7y = -2, \quad -5x - 6y = 0, \quad 2y = 5, \quad \text{and} \quad \tfrac{1}{2}x + y = 4.$$

To graph a line in standard form, we often start by locating the x- and y-intercepts. For example, the graph of $-x + 2y = 2$ is shown in Figure 3.7. All points on the x-axis have a y-coordinate of 0. To find the x-intercept, we let $y = 0$ in the equation $-x + 2y = 2$ and then solve for x.

$$-x + 2(\mathbf{0}) = 2 \quad \text{or} \quad x = -\mathbf{2} \qquad \text{Let } y = 0.$$

The x-intercept is -2. Note that the graph intersects the x-axis at $(-2, 0)$. Similarly, all points on the y-axis have an x-coordinate of 0. To find the y-intercept, we let $x = 0$ in the equation $-x + 2y = 2$ and then solve for y to obtain 1. Note that the graph intersects the y-axis at the point $(0, 1)$. This discussion is summarized by the following.

FINDING INTERCEPTS

To find the x-intercept of a line, let $y = 0$ in its equation and solve for x.

To find the y-intercept of a line, let $x = 0$ in its equation and solve for y.

NOTE: In some texts, intercepts are defined to be points rather than real numbers. In this case the x-intercept in Figure 3.7 is $(-2, 0)$ and the y-intercept is $(0, 1)$.

EXAMPLE 11 **Finding intercepts of a line in standard form**

Let the equation of a line be $3x - 2y = 6$.
(a) Find the x- and y-intercepts.
(b) Graph the line.
(c) Solve the equation for y to obtain the slope–intercept form.

Solution
(a) *x-intercept:* Let $y = \mathbf{0}$ in $3x - 2y = 6$ to obtain $3x - 2(\mathbf{0}) = 6$, or $x = 2$. The x-intercept is 2.
    y-intercept: Let $x = \mathbf{0}$ in $3x - 2y = 6$ to obtain $3(\mathbf{0}) - 2y = 6$, or $y = -3$. The y-intercept is -3.
(b) Sketch a line passing through $(2, 0)$ and $(0, -3)$, as shown in Figure 3.8.
(c) To solve for y, start by subtracting $3x$ from each side.

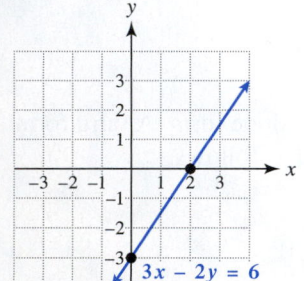

Figure 3.8

$$
\begin{array}{ll}
3x - 2y = 6 & \text{Given equation} \\
-2y = -3x + 6 & \text{Subtract } 3x \text{ from each side.} \\
-\dfrac{1}{2}(-2y) = -\dfrac{1}{2}(-3x + 6) & \text{Multiply each side by } -\tfrac{1}{2}. \\
y = \dfrac{3}{2}x - 3 & \text{Distributive property}
\end{array}
$$

The slope–intercept form is $y = \tfrac{3}{2}x - 3$.

■ **Now Try Exercise 107**

3.1 Putting It All Together

CONCEPT	EXPLANATION	EXAMPLES
Linear Equations	Can be written as $ax + b = 0$, where $a \neq 0$; has *one solution*	$3x - 4 = 0$, $2(x + 3) = -2$, and $2x = \dfrac{2}{3} - 5x$
Solution Set	The set of all solutions	The solution set for $x - 4 = 0$ is $\{4\}$ because 4 is the only solution to the equation.
Addition Property of Equality	$a = b$ is equivalent to $a + c = b + c$. "Equals added to equals are equal."	If 2 is added to each side of $x - 2 = 1$, the equation becomes $x = 3$.
Multiplication Property of Equality	$a = b$ is equivalent to $ac = bc$ for $c \neq 0$. "Equals multiplied by equals are equal."	If each side of $\frac{1}{2}x = 3$ is multiplied by 2, the resulting equation is $x = 6$.
Standard Form for a Line	$ax + by = c$, where a, b, and c are constants with a and b not both zero	$3x + 5y = 10$, $-2x + y = 0$, $3y = 18$, and $x = 4$
Finding Intercepts	To find the x-intercept, let $y = 0$ and solve for x. To find the y-intercept, let $x = 0$ and solve for y.	Let $2x + 4y = 8$. x-intercept: $2x + 4(0) = 8$, or $x = 4$ y-intercept: $2(0) + 4y = 8$, or $y = 2$

Linear equations can be solved symbolically, graphically, and numerically. The following illustrates how to solve the equation $2x - 1 = 3$ with each method.

Symbolic Solution

$$2x - 1 = 3$$
$$2x = 4$$
$$x = 2$$

The solution is **2**.

Graphical Solution

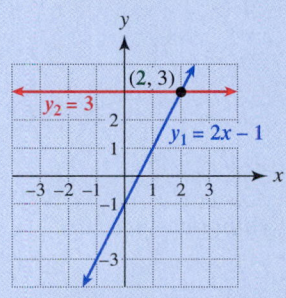

The solution is **2**.

Numerical Solution

x	$2x - 1$
0	-1
1	1
2	**3**
3	5

Because $2x - 1$ equals 3 when $x = 2$, the solution to $2x - 1 = 3$ is **2**.

3.1 Exercises

 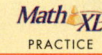

CONCEPTS AND VOCABULARY

1. Give the general form of a linear equation.

2. How many solutions does a linear equation have?

3. Is 1 the solution to $4x - 1 = 3x$?

4. Are $3x = 6$ and $x = 2$ equivalent equations?

5. The solution to $x + 6 = 10$ is _____.

6. The solution to $7x = -14$ is _____.

7. What symbol must occur in every equation?

8. Name three methods for solving a linear equation.

9. If a graphical solution to a linear equation results in the point of intersection $(3, 4)$, then the solution to the equation is _____.

10. The standard form for the equation of a line is _____.

11. The addition property of equality states that $a = b$ is equivalent to _____.

12. The multiplication property of equality states that $a = b$ is equivalent to _____.

Exercises 13–16: Classify each statement as true or false.

13. The equation $2x - 1 = 2x$ is a contradiction.

14. The equation $3(x - 5) = 3x - 15$ is an identity.

15. The equation $x + 7 = 9$ is an identity.

16. The equation $4x - 6 = x$ is a contradiction.

Exercises 17–22: Decide whether the given value for the variable is a solution to the equation.

17. $x - 6 = -2$ $x = 5$

18. $-\frac{1}{2}x + 1 = \frac{1}{3}x - \frac{2}{3}$ $x = 2$

19. $3(2t + 3) = \frac{13}{3} - t$ $t = -\frac{2}{3}$

20. $t - 1 = 2 - (t + 1)$ $t = 2$

21. $-(2z - 3) + 2z = 1 - z$ $z = -2$

22. $-\frac{1}{3}(4 - z) + \frac{2}{5}(z + 1) = -\frac{1}{5}$ $z = 1$

SYMBOLIC SOLUTIONS

Exercises 23–64: Solve the equation symbolically. Check your answer.

23. $x - 4 = 10$

24. $x - \frac{1}{2} = \frac{3}{2}$

25. $5 + x = 3$

26. $8 + x = -5$

27. $\frac{1}{2} - x = 2$

28. $10 - x = -5$

29. $-6x = -18$

30. $-5x = 15$

31. $-\frac{1}{3}x = 4$

32. $\frac{3}{4}x = 9$

33. $2x + 3 = 13$

34. $6x - 4 = -10$

35. $7 - 3x = -8$

36. $5 - 2x = -2$

37. $2x = 8 - \frac{1}{2}x$

38. $-7 = 3.5x$

39. $3x - 1 = 11(1 - x)$

40. $4 - 3x = -5(1 + 2x)$

41. $x + 4 = 2 - \frac{1}{3}x$

42. $2x - 5 = 6 - \frac{5}{2}x$ 43. $2(x - 1) = 5 - 2x$

44. $-(x - 4) = 4(x + 1) + 3(x - 2)$

45. $\dfrac{2x + 1}{3} = \dfrac{2x - 1}{2}$ 46. $\dfrac{3 - 4x}{5} = \dfrac{3x - 1}{2}$

47. $4.2x - 6.2 = 1 - 1.1x$

48. $8.4 - 2.1x = 1.4x$ 49. $\frac{1}{2}x - \frac{3}{2} = 4$

50. $5 - \frac{1}{3}x = x - 3$

51. $4(x - 1980) + 6 = 18$

52. $-5(x - 1900) - 55 = 145$

53. $2(y - 3) + 5(1 - 2y) = 4y + 1$

54. $4(y + 1) - (3 - 5y) = -2(y - 2)$

55. $-\left(5 - (z + 1)\right) = 9 - \left(5 - (2z - 3)\right)$

56. $-3(2 - 3z) + 2z = 1 - 3z$

57. $\frac{2}{3}(t - 3) + \frac{1}{2}t = 5$ 58. $\frac{1}{5}t + \frac{3}{10} = 2 - \frac{1}{2}t$

59. $\dfrac{3k}{4} - \dfrac{2k}{3} = \dfrac{1}{6}$ 60. $\dfrac{k + 1}{3} - \dfrac{1}{2} = \dfrac{3k - 3}{6}$

61. $0.2(n - 2) + 0.4n = 0.05$

62. $0.15(n + 1) = 0.1n + 1$

63. $0.7y - 0.8(y - 1) = 2$

64. $0.2y + 0.3(5 - 2y) = -0.2y$

65. **Thinking Generally** Solve the general linear equation $ax + b = 0$ for x, if $a \neq 0$.

66. **Thinking Generally** Solve the general linear equation $ax - b = cx$ for x, if $a \neq c$.

NUMERICAL SOLUTIONS

Exercises 67–70: Complete the table. Then use the table to solve the equation.

67. $-4x + 8 = 0$

x	1	2	3	4	5
$-4x + 8$	4				

68. $3x + 2 = 5$

x	-2	-1	0	1	2
$3x + 2$	-4				

69. $4 - 2x = x + 7$

x	-2	-1	0	1	2
$4 - 2x$	8				0
$x + 7$	5				9

70. $3(x - 1) = -2(1 - x)$

x	-2	-1	0	1	2
$3(x - 1)$	-9				
$-2(1 - x)$	-6				

Exercises 71–76: Solve the linear equation numerically.

71. $x - 3 = 7$ **72.** $2x + 1 = 11$

73. $2t - \frac{1}{2} = \frac{3}{2}$ **74.** $7 - 4t = -5$

 75. $3(z - 1) + 1.5 = 2z$

 76. $-3z - 6 = z + 2$

GRAPHICAL SOLUTIONS

Exercises 77–80: A linear equation is solved graphically by letting y_1 equal the left side of the equation and y_2 equal the right side of the equation. Find the solution. Check your answer.

77.   **78.**

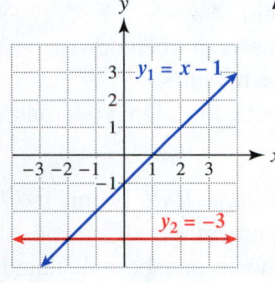 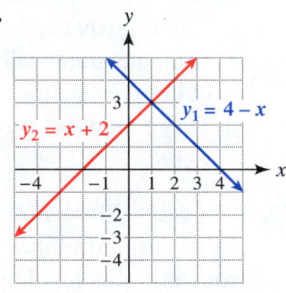

79.   **80.**

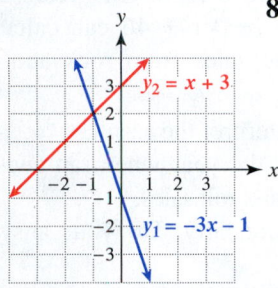 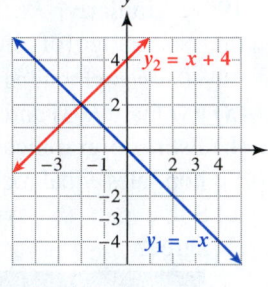

Exercises 81–90: Solve the equation graphically. Check your answer.

81. $5 - 2x = 7$ **82.** $-2x + 3 = -7$

83. $2x - (x - 2) = 2$ **84.** $x + 1 = 3 - x$

85. $3(x + 2) + 1 = x + 1$

86. $x + 5 = 2x + 2$

 87. $5(x - 1990) + 15 = 100$

88. $10(x - 1985) - 20 = 55$

89. $\sqrt{2} + 4(x - \pi) = \frac{1}{7}x$

 90. $\pi(x - \sqrt{5}) = \dfrac{7x + 5}{3}$

SOLVING LINEAR EQUATIONS BY MORE THAN ONE METHOD

*Exercises 91–94: Solve the equation (**a**) numerically, (**b**) graphically, and (**c**) symbolically.*

91. $2x - 1 = 13$ **92.** $9 - x = 3x + 1$

93. $3x - 5 - (x + 1) = 0$

94. $\dfrac{1}{2}x - 1 = \dfrac{5 - x}{2}$

IDENTITIES AND CONTRADICTIONS

Exercises 95–102: Determine whether the equation is an identity, contradiction, or conditional equation.

95. $2x + 3 = 2x$ **96.** $x + 5 = x - 4$

97. $2x + 3x = 5x$ **98.** $-3x = 4x - 7x$

99. $2(x - 1) = 2x - 2$

100. $3x - 10 = 2(x - 5) + x$

101. $5 - 7(4 - 3x) = x - 4(2 - 4x)$

102. $7 - 2(x + 3) - x = -4(x - 5) - 22$

103. Thinking Generally Let f and g be linear functions satisfying $f(-4) = 9$, $f(1) = -1$, $g(-3) = 1$, and $g(0) = 13$. Determine the point where their graphs intersect.

104. Thinking Generally Let f and g be linear functions where $f(-2) = -2$, $f(8) = -7$, $g(-4) = -11$, and $g(12) = 1$. Determine the point where their graphs intersect.

FINDING INTERCEPTS

Exercises 105–116: Do the following.

   (**a**) Find the x- and y-intercepts.
   (**b**) Graph the line.
   (**c**) Solve the equation for y to obtain the slope–intercept form.

105. $x + y = 4$

106. $x - y = -2$

107. $2x + 5y = 10$

108. $-3x + 4y = -6$

109. $2x - 3y = 9$

110. $\frac{1}{2}x - \frac{1}{4}y = 1$

111. $-\frac{2}{3}x - y = 2$

112. $-x - \frac{2}{5}y = 2$

113. $2x + 5y = 7$

114. $3x - 4y = 2$

115. $5x - 6y = -3$

116. $-8x + 9y = -10$

APPLICATIONS

117. *Global Seafood Consumption* (Refer to the chapter introduction.) In 1960 global seafood consumption was 42 million tons, and it increased to 157 million tons in 2008.
   (a) Determine $S(x) = m(x - x_1) + y_1$, where x is the year, so that S models these data.
   (b) Use $S(x)$ to estimate the amount of seafood that was consumed in 2000.
   (c) Assuming that trends continue, when might the 2008 amount increase by 45 million tons?

118. *Centenarians* In 1990 there were 37,306 centenarians, people age 100 or older, in the United States. The number of centenarians is estimated to increase by 8260 per year. (*Source:* U.S. Census Bureau.)
   (a) Find values for m, x_1, and y_1 so that the formula $f(x) = m(x - x_1) + y_1$ models the number of centenarians during year x.
   (b) Estimate the year when there will be 450,000 centenarians.

119. *State and Federal Inmates* From 1995 to 2009 the number of state and federal prison inmates in millions can be modeled by $f(x) = 0.0338x - 66.5$ during year x. (*Source:* Department of Justice.)
   (a) Determine symbolically the year when there were 1.3 million inmates.
   (b) Solve part (a) numerically.
   (c) Solve part (a) graphically.

120. *Asian-American Population* In 1998 there were 10.5 million Asian-Americans, and by 2009 this number grew to 14.9 million. (*Source:* U.S. Census Bureau.)
   (a) Find a linear function P that approximates this population in millions x years after 1998.
   (b) Interpret the y-intercept on the graph of P.
   (c) Estimate symbolically when this population was 13.7 million.
   (d) Solve part (c) graphically.

121. *College Degrees* In 1970 there were 1.1 million college degrees awarded. This number increased to 1.9 million in 2010. (*Source:* The College Board.)
   (a) Find a linear function D that approximates degree numbers in millions x years after 1970.
   (b) Interpret the y-intercept on the graph of D.
   (c) Estimate symbolically when this number was 1.7 million.
   (d) Solve part (c) graphically.

122. *Injured at Work* In 2006 there were 4.2 fatal work injuries per 100,000 full-time workers. By 2009 this number decreased to 3.3 per 100,000 full-time workers. (*Source:* Department of Labor.)
   (a) Find a linear function D that models these data x years after 2006.
   (b) If trends continue, estimate symbolically when this rate could be 2.1.
   (c) Solve part (b) graphically.

123. *Online Versus Newspaper Ads* From 2005 to 2012 the revenue in billions of dollars from newspaper ads N is given by $N(x) = -3.285x + 47.5$, where x is the number of years after 2005. Similarly, the revenue from online ads O can be calculated by $O(x) = 2.244x + 13$. (*Source:* Newspaper Association of America and Interactive Advertising Bureau.)
   (a) Evaluate $N(3)$ and interpret the result.
   (b) Do newspaper ads or online ads have a greater rate of growth? Explain.
   (c) Estimate graphically the year when newspaper and online ads had the same revenue.
   (d) Solve part (c) symbolically. Do your answers agree?

124. *Hip Hop Versus Rock and Roll* From 1987 to 2005 the (combined) percentage of music sales from rap and hip hop music can be calculated by $H(x) = \frac{83}{170}x + 3.8$, where x represents years after 1987. Similarly, $R(x) = -\frac{22}{17}x + 46$ can calculate the percentage of music sales from rock and roll music. (*Source:* Recording Industry of America.)
   (a) Evaluate $R(3)$ and interpret the result.
   (b) Estimate graphically the year when rap and hip hop sales equal rock and roll sales.
   (c) Solve part (b) symbolically. Do your answers agree?

125. *Smartphones* From 2008 to 2012 the percentage of mobile phone owners S with a smartphone can be modeled by $S(x) = 9.76x + 10$, where x is the number of years after 2008. Similarly, $F(x) = -9.76x + 90$ models the percentage of mobile phone owners with a feature phone. (*Source:* Nielsen Company.)

(a) Evaluate $S(2)$ and interpret the result.

(b) Interpret the y-intercept on the graph of S.

(c) Estimate the year when smartphone ownership will exceed feature phone ownership.

126. *Craigslist Visitors* From 2006 to 2011 the number of *monthly* unique visitors to Craigslist C can be modeled by $C(x) = 7660x + 13{,}000$, where x is the number of years after 2006. Similarly, $E(x) = -5176x + 75{,}000$ models the number of monthly unique visitors to eBay. (*Source:* Citi Investment Research Analysis.)

(a) Describe the rate of change in monthly unique visitors for Craigslist and eBay.

(b) Interpret the y-intercept on the graph of C.

(c) Estimate the year when Craigslist visitors could exceed eBay visitors.

127. *CD and LP Record Sales* From 1985 to 1990, sales of CDs (compact discs) in the United States can be modeled by $y_1 = 51.6(x - 1985) + 9.1$, and sales of LPs (vinyl records) can be modeled by $y_2 = -31.9(x - 1985) + 167.7$. All sales are in millions. Determine the year when sales of CDs and LPs were equal.

128. *U.S. Population Density* In 1900 there were, on average, 22 people per square mile in the United States. By 2008 the population density had increased to 86 people per square mile. (*Source:* U.S. Census Bureau.)

(a) Find a linear function f that models the population density during year x.

(b) Determine the year when the population density was 58 people per square mile.

129. **Online Exploration** Look up the U.S. divorce rate for two different years, such as 1990 and 2005. Find a linear function D that models these rates and then use D to estimate the year when the divorce rate was 5.0 per 1000 population.

130. **Online Exploration** Look up the federal debt for 2001 and the current year. Find a linear function that models these debts and use it to estimate when the debt was $8 trillion. Look up the true value.

WRITING ABOUT MATHEMATICS

131. Explain each of the following terms: linear equation, solution, solution set, and equivalent equations. Give an example of each.

132. Explain how to solve the equation $ax + b = 0$ symbolically and graphically. Use both methods to solve the equation $5x + 10 = 0$.

Group Activity   Working with Real Data

Directions: Form a group of 2 to 4 people. Select someone to record the group's responses for this activity. All members of the group should work cooperatively to answer the questions. If your instructor asks for your results, each member of the group should be prepared to respond.

Older Mothers The table lists the number of births per 1000 for women ages 40–44.

Year	1980	1990	1995	2000	2007
Births	3.9	5.5	6.6	7.9	9.5

Source: U.S. Census Bureau.

(a) Make a scatterplot of the data.

(b) Comment on any trends in the data.

(c) Find a linear function B in the form
$$B(x) = m(x - x_1) + y_1$$
that models the data. Answers may vary.

(d) Interpret the slope as a rate of change.

(e) If this trend continues, estimate when the birth rate might reach 12.

3.2 Introduction to Problem Solving

Solving a Formula for a Variable • Steps for Solving a Problem • Percentages

A LOOK INTO MATH ▶ Problem-solving skills are essential because every day we must solve problems—both small and large—for survival. A logical approach to a problem is often helpful, which is especially true in mathematics. In this section we discuss steps that can be used to solve mathematical problems. We begin the section by discussing how to solve a formula for a variable.

NEW VOCABULARY

☐ Percentage
☐ Percent form
☐ Decimal form

Solving a Formula for a Variable

The perimeter P of a rectangle with width W and length L is given by $P = 2W + 2L$, as illustrated in Figure 3.9. Suppose that we know the perimeter P and length L of this rectangle. Even though the formula is written to calculate P, we can still use it to find the width W. This technique is demonstrated in the next example.

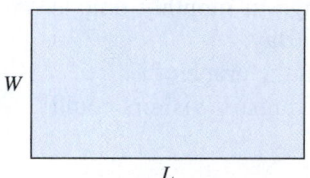

Figure 3.9 $P = 2W + 2L$

EXAMPLE 1 **Finding the width of a rectangle**

A rectangle has a perimeter of 80 inches and a length of 23 inches.
(a) Write a formula to find the width W of a rectangle with known perimeter P and length L.
(b) Use the formula to find W.

Solution
(a) We must solve $P = 2W + 2L$ for W.

$$P = 2W + 2L \qquad \text{Given formula}$$

$$P - 2L = 2W \qquad \text{Subtract } 2L \text{ from each side.}$$

$$\frac{1}{2}(P - 2L) = \frac{1}{2}(2W) \qquad \text{Multiply each side by } \tfrac{1}{2}.$$

$$\frac{P}{2} - L = W \qquad \text{Distributive property}$$

The required formula is $W = \frac{P}{2} - L$.
(b) Substitute $P = 80$ and $L = 23$ into this formula. The width is

$$W = \frac{80}{2} - 23 = 17 \text{ inches.}$$

Now Try Exercise 45

▶ **REAL-WORLD CONNECTION** In the next example we use a formula that converts degrees Fahrenheit to degrees Celsius to find a formula that converts degrees Celsius to degrees Fahrenheit.

EXAMPLE 2 **Converting temperature**

The formula $C = \frac{5}{9}(F - 32)$ is used to convert degrees Fahrenheit to degrees Celsius.
(a) Use this formula to convert $212°\,F$ to an equivalent Celsius temperature.
(b) Solve the formula for F to obtain a formula that can convert degrees Celsius to degrees Fahrenheit. Interpret the slope of the graph of F.
(c) Use this new formula to convert $25°\,C$ to an equivalent Fahrenheit temperature.

Solution
(a) $C = \frac{5}{9}(212 - 32) = \frac{5}{9}(180) = 100°C$

(b)

$$C = \frac{5}{9}(F - 32) \qquad \text{Given formula}$$

$$\frac{9}{5}C = \frac{9}{5} \cdot \frac{5}{9}(F - 32) \qquad \text{Multiply each side by } \frac{9}{5}.$$

$$\frac{9}{5}C = F - 32 \qquad \text{Simplify.}$$

$$\frac{9}{5}C + 32 = F \qquad \text{Add 32 to each side.}$$

Because $F = \frac{9}{5}C + 32$, the slope of the graph of F is $\frac{9}{5}$, which indicates that a $1°$ increase in Celsius temperature is equivalent to a $\frac{9°}{5}$ increase in Fahrenheit temperature.

(c) Substitute **25** for C in this formula to obtain $F = \frac{9}{5}(25) + 32 = 77°\,F.$

Now Try Exercise 49

Sometimes an equation in two variables can be used to determine the formula for a function. To do so, we must let one variable be the dependent variable and the other variable be the independent variable. For example, $2x - 5y = 10$ is the equation of a line written in standard form. Solving for y gives the following result.

$$-5y = -2x + 10 \qquad \text{Subtract } 2x \text{ from each side.}$$

$$-\frac{1}{5}(-5y) = -\frac{1}{5}(-2x + 10) \qquad \text{Multiply each side by } -\frac{1}{5}.$$

$$y = \frac{2}{5}x - 2 \qquad \text{Distributive property}$$

If y is the dependent variable, x is the independent variable, and $y = f(x)$, then

$$f(x) = \frac{2}{5}x - 2$$

defines a function f whose graph is the line determined by $2x - 5y = 10$.

EXAMPLE 3 **Writing a function defined by an equation**

Solve each equation for y and write a formula for a function f defined by $y = f(x)$.

(a) $4(x - 2y) = -3x$ **(b)** $\dfrac{x + y}{2} - 5 = 20$

Solution
(a)

$$4(x - 2y) = -3x \qquad \text{Given equation}$$

$$4x - 8y = -3x \qquad \text{Distributive property}$$

$$-8y = -7x \qquad \text{Subtract } 4x \text{ from each side.}$$

$$y = \frac{7}{8}x \qquad \text{Divide each side by } -8.$$

Thus $f(x) = \frac{7}{8}x.$

(b)

$$\frac{x + y}{2} - 5 = 20 \qquad \text{Given equation}$$

$$\frac{x + y}{2} = 25 \qquad \text{Add 5 to each side.}$$

$$\frac{\mathbf{2}(x + y)}{2} = \mathbf{2}(25) \qquad \text{Multiply each side by 2.}$$

$$x + y = 50 \qquad \text{Simplify.}$$

$$y = 50 - x \qquad \text{Subtract } x \text{ from each side.}$$

Thus $f(x) = 50 - x$.

Now Try Exercises 21, 23

Steps for Solving a Problem

Solving problems in mathematics can be challenging, especially when formulas and equations are not given to us. In these situations we have to write them, but to do so, we need a strategy. The following steps are often a helpful strategy. They are based on George Polya's (1888–1985) four-step process for problem solving.

STEPS FOR SOLVING A PROBLEM

STEP 1: Read the problem carefully to be sure that you understand it. (You may need to read the problem more than once.) Assign a variable to what you are being asked to find. If necessary, write other quantities in terms of this variable.

STEP 2: Write an equation that relates the quantities described in the problem. You may need to sketch a diagram, make a table, or refer to known formulas.

STEP 3: Solve the equation and determine the solution.

STEP 4: Look back and check your answer. Does it seem reasonable? Did you find the required information?

STUDY TIP

These four steps can make it easier to solve word problems.

In the next example we apply these steps to a word problem involving unknown numbers.

EXAMPLE 4 **Solving a number problem**

The sum of three consecutive *even* integers is 108. Find the three numbers.

Solution

STEP 1: *Start by assigning a variable to an unknown quantity.*

$$n: \text{smallest of the three integers}$$

Next, write the other two numbers in terms of n.

$$n + 2: \text{next consecutive } even \text{ integer}$$

$$n + 4: \text{largest of the three even integers}$$

STEP 2: *Write an equation that relates these unknown quantities.* As the sum of these three even integers is 108, the needed equation is

$$n + (n + 2) + (n + 4) = 108.$$

STEP 3: *Solve the equation in Step 2.*

$n + (n + 2) + (n + 4) = 108$	Equation to be solved
$(n + n + n) + (2 + 4) = 108$	Commutative and associative properties
$3n + 6 = 108$	Combine like terms.
$3n = 102$	Subtract 6 from each side.
$n = 34$	Divide each side by 3.

The smallest of the three numbers is 34, so the three numbers are 34, 36, and 38.

STEP 4: *Check your answer.* The sum of these three consecutive even integers is

$$34 + 36 + 38 = 108.$$

The answer checks.

▮ **Now Try Exercise 53**

In the next example we apply this procedure to find the dimensions of a room.

EXAMPLE 5 ▶ **Solving a geometry problem**

The length of a rectangular room is 2 feet more than its width. If the perimeter of the room is 80 feet, find the width and length of the room.

Solution
STEP 1: *Start by assigning a variable to an unknown quantity.* See Figure 3.10.

$$x: \text{width of the room}$$
$$x + 2: \text{length of the room}$$

STEP 2: *Write an equation that relates these unknown quantities.* The perimeter is the distance around the room and equals 80 feet. Thus

$$2x + 2(x + 2) = 80,$$

which simplifies to $4x + 4 = 80$.

STEP 3: *Solve the equation in Step 2.*

$4x + 4 = 80$	Equation to be solved
$4x = 76$	Subtract 4 from each side.
$\dfrac{4x}{4} = \dfrac{76}{4}$	Divide each side by 4.
$x = 19$	Simplify.

The width is 19 feet and the length is $19 + 2 = 21$ feet.

STEP 4: *Check your answer.* The perimeter is $2(19) + 2(21) = 80$ feet. The answer checks.

▮ **Now Try Exercise 57**

Figure 3.10 Perimeter = 80

x

$x + 2$

EXAMPLE 6 | **Solving a geometry problem**

A wire, 100 inches long, needs to be cut into two pieces so that it can be bent into a circle and a square. The length of a side of the square must equal the diameter of the circle. Approximate where the 100-inch wire should be cut.

Solution
STEP 1: *Assign a variable.* Let x be both the diameter of the circle and the length of a side of the square, as shown in Figure 3.11.

STEP 2: *Write an equation.* Because $C = \pi d$, the circumference of the circle is $C = \pi x$, and the perimeter of the square is $P = 4x$. The wire is 100 inches long, so

$$\pi x + 4x = 100.$$

STEP 3: *Solve the equation in Step 2.*

$$\pi x + 4x = 100 \qquad \text{Equation to be solved}$$
$$(\pi + 4)x = 100 \qquad \text{Distributive property}$$
$$x = \frac{100}{\pi + 4} \qquad \text{Divide by } \pi + 4.$$
$$x \approx 14 \qquad \text{Approximate.}$$

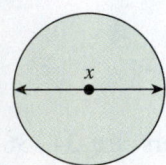
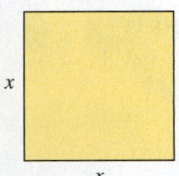

Figure 3.11

The wire should be cut so that the piece for the square is about $4 \times 14 = 56$ inches and the piece for the circle is about $100 - 56 = 44$ inches.

STEP 4: *Check the answer.* If the square's perimeter is 56 inches, then each of its sides is $\frac{56}{4} = 14$ inches. If the diameter of the circle is 14 inches, then its circumference is $C = \pi(14) \approx 43.98$, or about 44 inches. Because $56 + 44 = 100$, the answer checks.

▌ **Now Try Exercise 59**

EXAMPLE 7 | **Solving a motion problem**

Two cars are traveling in opposite directions on a freeway. Three hours after meeting, they are 420 miles apart. If one car is traveling 10 miles per hour faster than the other car, find the speeds of the two cars.

Solution
STEP 1: *Assign a variable.*

$$x: \text{speed of the slower car}$$
$$x + 10: \text{speed of the faster car}$$

STEP 2: *Write an equation.* The information is summarized in Table 3.2.

NOTE: Distance = Rate × Time.

TABLE 3.2

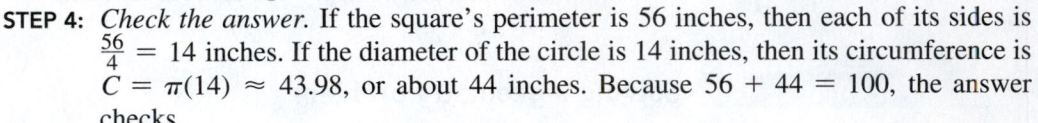

	Rate	Time	Distance
Slower Car	x	3	$3x$
Faster Car	$x + 10$	3	$3(x + 10)$
Total			420

Sum equals 420 miles.

The sum of the distances traveled by the cars is 420 miles. Thus

$$3x + 3(x + 10) = 420.$$

STEP 3: *Solve the equation in Step 2.*

$$3x + 3(x + 10) = 420 \quad \text{Equation to be solved}$$
$$3x + 3x + 30 = 420 \quad \text{Distributive property}$$
$$6x + 30 = 420 \quad \text{Combine like terms.}$$
$$6x = 390 \quad \text{Subtract 30 from each side.}$$
$$x = 65 \quad \text{Divide each side by 6.}$$

The speed of the slower car is 65 miles per hour and the speed of the faster car is 10 miles per hour faster, or 75 miles per hour.

STEP 4: *Check the answer.* The difference in the cars' speeds is $75 - 65 = 10$ miles per hour.

Distance traveled by the slower car: $65 \times 3 = 195$ miles

Distance traveled by the faster car: $75 \times 3 = 225$ miles

The total distance is $195 + 225 = 420$ miles, so it checks.

■ **Now Try Exercise 61**

▶ **REAL-WORLD CONNECTION** Many times we are called upon to solve a mixture problem. Such problems may involve a mixture of acid solutions, a mixture of loan amounts, or some other mixture. In the next example an athlete jogs at two different speeds, and mathematics is used to determine how much time is spent jogging at each speed.

EXAMPLE 8   **Solving a "mixture" problem**

An athlete begins jogging at 8 miles per hour and then jogs at 7 miles per hour, traveling 10.9 miles in 1.5 hours. How long did the athlete jog at each speed?

Solution

STEP 1: *Assign a variable.* Let t represent the time spent jogging at 8 miles per hour. Because total time spent jogging is 1.5 hours, the time spent jogging at 7 miles per hour must be $1.5 - t$.

STEP 2: *Write an equation.* A table is often helpful in solving a mixture problem. Because $d = rt$, Table 3.3 shows that the distance the athlete jogs at 8 miles per hour is $8t$. The distance traveled at 7 miles per hour is $7(1.5 - t)$, and the total distance traveled is 10.9 miles. From the "Distance" column, we can write the equation as

$$8t + 7(1.5 - t) = 10.9.$$

TABLE 3.3

	Rate	Time	Distance	
First Part	8	t	$8t$	← Sum equals
Second Part	7	$1.5 - t$	$7(1.5 - t)$	← 10.9 miles.
Total		1.5	10.9	

STEP 3: *Solve the equation in Step 2.*

$$8t + 7(1.5 - t) = 10.9 \quad \text{Equation to be solved}$$
$$8t + 10.5 - 7t = 10.9 \quad \text{Distributive property}$$
$$t + 10.5 = 10.9 \quad \text{Combine like terms.}$$
$$t = 0.4 \quad \text{Subtract 10.5 from each side.}$$

The athlete jogged 0.4 hour at 8 miles per hour and $1.5 - 0.4 = 1.1$ hours at 7 miles per hour.

STEP 4: *Check your answer.* Start by calculating the distance traveled at each speed.

$$8 \cdot 0.4 = 3.2 \text{ miles} \quad \text{Distance at 8 mph}$$
$$7 \cdot 1.1 = 7.7 \text{ miles} \quad \text{Distance at 7 mph}$$

The total distance is $3.2 + 7.7 = 10.9$ miles and the total time is $0.4 + 1.1 = 1.5$ hours. The answer checks.

Now Try Exercise 65

Percentages

A **percentage** can be written either in **percent form** or in **decimal form**. For example, the percent form of 15% can be also written in decimal form as 0.15. To change a percent form $R\%$ to a decimal form r we divide R by 100. That is, $r = \frac{R}{100}$.

EXAMPLE 9 **Writing percentages as decimals**

Write each percentage as a decimal.
(a) 37%   **(b)** 0.02%   **(c)** 320%   **(d)** -1.45% **(e)** $\frac{2}{5}\%$

Solution
(a) Let $R = 37$. Then $r = \frac{37}{100} = 0.37$.

(b) Let $R = 0.02$. Then $r = \frac{0.02}{100} = 0.0002$.

> **NOTE:** Dividing a number by 100 is equivalent to moving the decimal point two places to the *left*.

(c) Let $R = 320$. Then $r = \frac{320}{100} = 3.2$

(d) Let $R = -1.45$. Then $r = \frac{-1.45}{100} = -0.0145$. A negative percentage generally corresponds to a quantity decreasing rather than increasing.

(e) Let $R = \frac{2}{5}$. Then $r = \frac{2/5}{100} = \frac{2}{500} = 0.004$.

Now Try Exercises 35, 37, 39, 41, 43

▶ **REAL-WORLD CONNECTION** Applications involving percentages often make use of linear equations. Taking r percent of x is given by rx, where r is written in decimal form. For example, to calculate 35% of x, we compute $0.35x$. As a result, 35% of $150 is $0.35(150) = 52.5$, or $52.50.

> **NOTE:** The word *of* often indicates multiplication when working with percentages.

EXAMPLE 10 **Analyzing smoking data**

In 2010 an estimated 19.8% of Americans age 18 and older, or 46.2 million people, were smokers. Use these data to estimate the number of Americans age 18 and older in 2010.
(*Source:* Department of Health and Human Services.)

Solution
STEP 1: *Assign a variable.* Let x be the number of Americans age 18 and older.
STEP 2: *Write an equation.* Note that 19.8% of x equals 46.2 million. Thus

$$0.198x = 46.2.$$

STEP 3: *Solve the equation in Step 2.*

$$0.198x = 46.2 \qquad \text{Equation to be solved}$$

$$\frac{0.198x}{\mathbf{0.198}} = \frac{46.2}{\mathbf{0.198}} \qquad \text{Divide each side by 0.198.}$$

$$x \approx 233.3 \qquad \text{Approximate with a calculator.}$$

In 2010 there were about 233.3 million Americans age 18 and older.

STEP 4: *Check your answer.* If there were 233.3 million Americans age 18 and older, then 19.8% of 233.3 million would equal $0.198 \times 233.3 \approx 46.2$ million. The answer checks.

▌ **Now Try Exercise 77**

EXAMPLE 11 **Solving a percent problem**

In 2010 an estimated $280 billion was spent on advertising in the United States. This was an increase of 409% over the amount spent in 1980. Determine how much was spent on advertising in 1980. (*Source: Advertising Age.*)

Solution

STEP 1: *Assign a variable.* Let x be the amount spent on advertising in 1980.

STEP 2: *Write an equation.* Note that the increase was 409% of x, or $4.09x$.

$$x \quad + \quad 4.09x \quad = \quad 280$$

Amount in 1980 + Increase = Amount in 2010

STEP 3: *Solve the equation in Step 2.*

$$x + 4.09x = 280 \qquad \text{Equation to be solved}$$

$$5.09x = 280 \qquad \text{Add like terms.}$$

$$x = \tfrac{280}{5.09} \qquad \text{Divide each side by 5.09.}$$

$$x \approx 55 \qquad \text{Simplify.}$$

In 1980, about $55 billion was spent on advertising.

STEP 4: *Check your answer.* An increase of 409% of $55 billion is $4.09 \times 55 \approx \$225$ billion. Thus the amount spent in 2010 would be $55 + 225 = \$280$ billion. The answer checks.

▌ **Now Try Exercise 79**

EXAMPLE 12 **Solving an interest problem**

A college student takes out two loans to pay tuition. The first loan is for 3% annual interest and the second loan is for 7% annual interest. If the total of the two loans is $4000 and the student must pay $170 in interest at the end of the first year, determine the amount borrowed at each interest rate.

Solution

STEP 1: *Assign a variable.*

$$x: \text{the loan amount at 3%}$$

$$4000 - x: \text{the loan amount at 7%}$$

STEP 2: *Write an equation.* Note that the total interest is $170.

$$0.03x \quad + \quad 0.07(4000 - x) \quad = \quad 170$$

Interest at 3% + Interest at 7% = Total Interest

STEP 3: *Solve the equation in Step 2.*

$$0.03x + 0.07(4000 - x) = 170 \qquad \text{Equation to be solved}$$

$$3x + 7(4000 - x) = 17{,}000 \qquad \text{Multiply by 100 to clear decimals.}$$

$$\mathbf{3x} + 28{,}000 - \mathbf{7x} = 17{,}000 \qquad \text{Distributive property}$$

$$\mathbf{-4x} + 28{,}000 = 17{,}000 \qquad \text{Combine like terms.}$$

$$-4x = -11{,}000 \qquad \text{Subtract 28,000.}$$

$$x = 2750 \qquad \text{Divide by } -4.$$

The amount at 3% is $2750, and the amount at 7% is $4000 - 2750 = \$1250$.

STEP 4: *Check the answer.* The sum is $2750 + 1250 = \$4000$. Also, 3% of $2750 is $82.50 and 7% of $1250 is $87.50. The total interest is $82.50 + \$87.50 = \$170$. The answer checks.

■ **Now Try Exercise 69**

▶ **REAL-WORLD CONNECTION** In chemistry acids are frequently mixed. In the next example, percentages are used to determine how to mix an acid solution with a specified concentration.

EXAMPLE 13 | **Mixing acid**

A chemist mixed 2 liters of 20% sulfuric acid with another sample of 60% sulfuric acid to obtain a sample of 50% sulfuric acid. How much of the 60% sulfuric acid was used? See Figure 3.12.

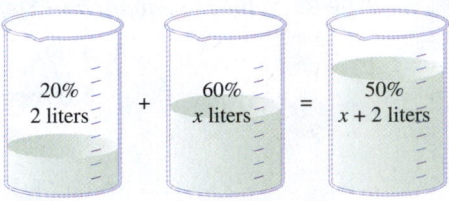

Figure 3.12 Mixing acid

Solution

STEP 1: *Assign a variable.* Let x be as follows.

$$x:\text{ liters of 60\% sulfuric acid}$$

$$x + 2:\text{ liters of 50\% sulfuric acid}$$

STEP 2: *Write an equation.* Table 3.4 can be used to organize our calculations. The total amount of pure sulfuric acid in the 20% and 60% samples must equal the amount of pure sulfuric acid in the final 50% acid solution.

TABLE 3.4 Mixing Acid

Concentration (as a decimal)	Solution Amount (liters)	Pure Acid (liters)
20% = 0.20	2	0.20(2)
60% = 0.60	x	0.60x
50% = 0.50	$x + 2$	0.50($x + 2$)

From the third column, we can write the equation

$$0.20(2) \qquad + \qquad 0.60x \qquad = \qquad 0.50(x + 2).$$

Pure acid in 20% solution + Pure acid in 60% solution = Pure acid in 50% solution

STEP 3: *Solve the equation in Step 2.*

$$0.20(2) + 0.60x = 0.50(x + 2) \quad \text{Equation to be solved}$$
$$2(2) + 6x = 5(x + 2) \quad \text{Multiply by 10.}$$
$$4 + 6x = 5x + 10 \quad \text{Distributive property}$$
$$x = 6 \quad \text{Subtract } 5x \text{ and 4 from each side.}$$

Six liters of the 60% acid solution were added to the 2 liters of 20% acid solution.

STEP 4: *Check your answer.* If 6 liters of 60% acid solution are added to 2 liters of 20% solution, then there will be **8** liters of acid solution containing

$$0.60(6) + 0.20(2) = \mathbf{4} \text{ liters}$$

of pure acid. The concentration is $\frac{4}{8} = 0.50$, or 50%. The answer checks.

■ **Now Try Exercise 63**

3.2   Putting It All Together

When solving an application problem, follow the *Steps for Solving a Problem* discussed on page 166. Be sure to read the problem carefully and check your answer. The following table summarizes some other important concepts from this section.

CONCEPT	EXPLANATION	EXAMPLES
Writing a Function	If possible, solve an equation for a variable. Express the result in function notation.	$5x + y = 10$ $y = -5x + 10$ If $y = f(x)$, then $f(x) = -5x + 10$.
Changing a Percentage to a Decimal	Move the decimal point two places to the left and remove the percent symbol.	$73\% = 0.73,\ 5.3\% = 0.053,$ and $125\% = 1.25$
Percent Problems	To find $P\%$ of a quantity Q, change $P\%$ to a decimal and multiply by Q.	To find 45% of $200, calculate $0.45 \times 200 = \$90$.

3.2   Exercises

MyMathLab
PRACTICE   WATCH   DOWNLOAD   READ   REVIEW

CONCEPTS AND VOCABULARY

1. If we solve $2y = x$ for y, we obtain $y = $ _____.

2. If $y = 4x - 7$ and $y = f(x)$, then $f(x) = $ _____.

3. If a rectangle has width W and length L, then its perimeter is $P = $ _____.

4. The first step in solving an application is to _____.

5. Write 43.1% as a decimal.

6. Find 20% of 50.

SOLVING FOR A VARIABLE

Exercises 7–14: Solve the equation for the given variable.

7. $4x + 3y = 12$ for y **8.** $6x - 3y = 6$ for y

9. $5(2x - 3y) = 2x$ for x

10. $-2(2x + y) = 3y$ for x

11. $S = 6ab$ for b 12. $S = 2\pi rh$ for r

13. $\dfrac{r + t}{2} = 7$ for t 14. $\dfrac{2 + a + b}{3} = 8$ for a

Exercises 15–24: Solve the equation for y. Let $y = f(x)$ and write a formula for $f(x)$.

15. $-3x + y = 8$ 16. $2x - 5y = 10$

17. $4x = 2\pi y$ 18. $\dfrac{x}{2} = \dfrac{y}{3}$

19. $\dfrac{3y}{8} = x$ 20. $\dfrac{3x + 2y}{4} = 3$

21. $3(2x + 3y) = -x$ 22. $-(x - 2y) = 5x + 1$

23. $\dfrac{2x - 3y}{5} = 4$ 24. $\dfrac{2y + 5x + 1}{5} = 2x$

WRITING AND SOLVING EQUATIONS

Exercises 25–34: Do the following.

(a) Translate the sentence into an equation by using the variable x.
(b) Solve the resulting equation.

25. The sum of a number and 2 is 12.

26. Twice a number plus 7 equals 9.

27. A number divided by 5 equals the number plus 1.

28. 25 times a number is 125.

29. If a number is increased by 5 and then divided by 2, the result equals 7.

30. A number subtracted from 8 is 5.

31. The quotient of a number and 2 is 17.

32. The product of 5 and a number equals 95.

33. The sum of three consecutive integers is 30.

34. A rectangle that is 5 inches longer than it is wide has a perimeter of 60 inches.

PERCENTAGES

Exercises 35–44: Write each percentages as a decimal.

35. 55%   36. 67%

37. 0.04%   38. 0.056%

39. 341%   40. 243%

41. −1.67%   42. −0.5%

43. $\frac{1}{4}$%   44. $\frac{3}{10}$%

APPLICATIONS

45. *Perimeter of a Rectangle* A rectangle has a perimeter of 86 feet and a width of 19 feet.
    (a) Write a formula to find the length L of a rectangle with perimeter P and width W.
    (b) Use the formula to find L.

46. *Perimeter of a Triangle* The perimeter P of a triangle with sides a, b, and c is $P = a + b + c$.

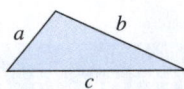

    (a) Solve this formula for a.
    (b) Find a if $b = 5$, $c = 7$, and $P = 15$.

47. *Area of a Cylinder* The area A of the *side* of a cylinder with height h and radius r is $A = 2\pi rh$. Solve this formula for h.

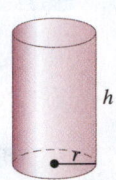

48. *Volume of a Cylinder* The volume V of a cylinder with height h and radius r is $V = \pi r^2 h$. Solve this formula for h.

49. *Temperature* (Refer to Example 2.) Solve the formula $F = \frac{9}{5}C + 32$ for C.

50. *Area of a Parallelogram* The area A of a parallelogram with base b and height h is $A = bh$. Solve this formula for b.

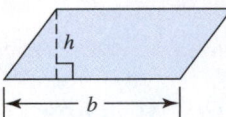

51. *Area of a Triangle* The area A of a triangle with base b and height h is $A = \frac{1}{2}bh$. Solve this formula for h.

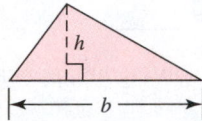

52. *Volume of a Box* The volume V of a box with length L, width W, and height H is $V = WLH$. Solve this formula for L.

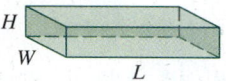

53. *Number Problem* The sum of three consecutive integers is 135. Find the three integers.

54. *Number Problem* Are there three consecutive *even integers* whose sum is 82? If so, find them.

55. *Number Problem* Twice the sum of three consecutive *odd integers* is 150. Find the three integers.

56. *Number Problem* Ten plus twice the sum of two consecutive integers is 196. Find the two integers.

57. *Geometry* Find the value of x in the figure if the perimeter of the room is 48 feet.

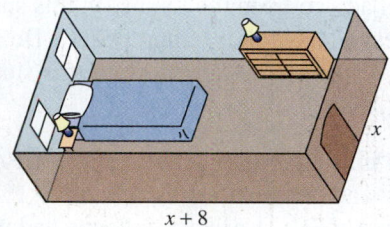

58. *Distance and Time* A train 100 miles west of St. Louis, Missouri, is traveling east at 60 miles per hour. How long will it take for the train to be 410 miles east of St. Louis?

59. *Geometry* A wire 41 inches long needs to be cut into three pieces so that it can be formed into two identical circles and a square. The length of the side of the square must equal the diameter of the circles. Approximate where the wire should be cut.

60. *Fencing* Three identical pens for dogs are to be enclosed with 120 feet of fence, as illustrated in the figure. If the length of each pen is twice its width plus 2 feet, find the dimensions of each pen.

61. *Distance and Time* Two cars are traveling in opposite lanes on a freeway. Two and a half hours after they meet they are 355 miles apart. If one car is traveling 6 miles per hour faster than the other car, find the speed of each car.

62. *Distance and Time* At first an athlete jogs at 6 miles per hour and then jogs at 5 miles per hour, traveling 7 miles in 1.3 hours. How long did the athlete jog at each speed?

63. *Chemical Mixture* A chemist mixed 3 liters of 30% sulfuric acid with a sample of 80% sulfuric acid to obtain a sample of 60% sulfuric acid. How much of the 80% sulfuric acid was used?

64. *Antifreeze Mixture* A radiator holds 4 gallons of fluid. If it is full with a 20% solution, how much fluid should be drained and replaced with a 70% antifreeze mixture to result in a 50% mixture of antifreeze?

65. *Distance and Time* A car travels at 74 miles per hour and then 63 miles per hour, traveling 248 miles in 3.5 hours. How long did the driver travel at each speed?

66. *Distance and Time* Two vehicles are beside each other on a freeway. After 1 hour and 30 minutes, the faster car is 12 miles ahead of the slower car. During this time the two cars travel a total of 198 miles. Find the speed of each car.

67. *Chemical Mixture* A chemist needs to mix 6 liters of a 50% acid solution with some 20% solution to obtain a 38% solution. How much of the 20% solution should the chemist use?

68. *Herbicide Solution* How much water should be added to 4 gallons of a 5% herbicide solution to dilute it to a 3.2% herbicide solution?

69. *Loan Interest* A student takes out two student loans, one at 6% annual interest and the other at 4% annual interest. The total amount of the two loans is $5000, and the total interest after 1 year is $244. Find the amount of each loan.

70. *Investment Interest* A person invests two amounts of money at 5% and 6% interest. The total amount of money invested is $2400 and the total interest after 1 year is $130.50. Find the amount invested at each rate.

71. *Angles in a Triangle* The measure of the largest angle in a triangle is twice the measure of the smallest angle. The third angle is 10° less than the largest angle. Find the measure of each angle.

72. *Angles in a Triangle* The sum of the degree measures of the angles in a triangle equals 180°. Find the measure of x in the triangle shown.

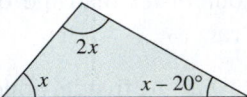

73. *Classroom Ventilation* Ventilation is effective at removing air pollutants. A classroom should have a ventilation rate of 900 cubic feet of air per hour for each person in the classroom. (*Source:* ASHRAE.)
(a) What ventilation rate should a classroom containing 40 people have?
(b) If a ventilation system moves 60,000 cubic feet of air per hour, find the maximum number of people that should be in the classroom.

74. *Lead Poisoning* According to the EPA, the maximum amount of lead that can be ingested by a person without becoming ill is 36.5 milligrams per year.
(a) Write an expression that gives the maximum amount of lead in milligrams that can be "safely" ingested over x years.
(b) Under these guidelines what is the minimum number of years that the consumption of 500 milligrams should be spread over?

75. *Hybrid Cars* The 2011 model of the Honda Civic Hybrid averages about 40 miles per gallon and has an annual fuel cost of $1050. If it is assumed that a Hybrid owner drives 15,000 miles per year, what is the average cost of a gallon of gasoline? (*Source:* Honda Corporation.)

76. *Hybrid Cars* The 2011 Toyota Camry Hybrid has an annual fuel cost of $1320. If it is assumed that gasoline costs $3.00 per gallon and that a Hybrid owner drives 15,000 miles per year, estimate the average mileage for this vehicle. (*Source:* Toyota Corporation.)

77. *Solar Energy in Nevada* The U.S. Department of Energy has estimated that solar power plants covering 9900 square miles, or 9% of Nevada, could generate enough electricity to meet the needs of the entire United States. Find the area of Nevada.

78. *Energy Efficiency Rating* The Energy Efficiency Rating (EER) of an air conditioner must be at least 8.0. Changing to an air conditioner with an EER of 10.0 can cut electrical bills by 20%. Suppose that a person's monthly electrical bill is $125 and that the more efficient air conditioner costs an extra $275. How many months will it take to recover this added cost? (*Source:* Energy Service Office, Springfield, Illinois.)

79. *Wireless Expenditures* In 2009, Verizon Wireless spent $7.1 billion on capital, such as new cell towers, a 492% increase over the amount that Sprint spent on capital in 2009. How much did Sprint spend on capital in 2009? (*Source:* Business Insider.)

80. *Watching Online Advertisements* In 2010 on theWB.com, each viewer watched about 20 ads per month, on average. This number is a 138% increase over the average number of ads each viewer watched on NBC.com in 2010. How many ads per month would someone view on NBC.com? (*Source:* comScore.)

81. *Meeting a Future Mate* In a survey, about 32% of adult Americans believe that they will meet their mates on the Internet. In this survey 480 respondents held this belief. Determine the number of people participating in the survey. (*Source:* Men's Health.)

82. *Financial Advice* In a survey, about 34% of adult Americans believe that they have gotten the best financial advice from the media. In this survey, about 5608 respondents held this belief. Determine the number of people participating in this survey. (*Source:* CNNMoney.com.)

83. *Grade Point Averages* The average GPA at public colleges and universities is 3.0, or 9.1% less than the average GPA at private colleges and universities. What is the average GPA at private colleges and universities? (*Source:* New York Times.)

84. *Grades* To receive an A in a college course a student must average 90 percent correct on four exams of 100 points each and on a final exam of 200 points. If a student scores 83, 87, 94, and 91 on the 100-point exams, what minimum score on the final exam is necessary for the student to receive an A?

WRITING ABOUT MATHEMATICS

85. A student solves the following equation for y. What is the student's mistake?

$$\frac{x + 2y}{2} = 5$$
$$x + y \stackrel{?}{=} 5$$
$$y \stackrel{?}{=} 5 - x$$

86. Explain how to find $P\%$ of a quantity Q. Give an example.

SECTIONS
3.1 and 3.2
Checking Basic Concepts

1. How many solutions does a linear equation have?

2. Solve $\frac{1}{2}z - (1 - 2z) = \frac{2}{3}z$.

3. Solve $2(3x + 4) + 3 = -1$
   (a) symbolically,
   (b) graphically, and
   (c) numerically.
   Do your answers agree?

4. At first an athlete jogs at 10 miles per hour and then jogs at 8 miles per hour, traveling 10.2 miles in 1.2 hours. How long did the athlete jog at each speed?

5. Find the x- and y-intercepts for the graph of the equation $5x - 4y = 20$.

6. Classify each equation as either an identity or a contradiction.
   (a) $3(x - 2) = 3x - 6$
   (b) $2x + 3x = 5x - 2$

7. The length of a rectangular room is 5 feet more than its width. If the perimeter of a room is 82 feet, find the length and width of the room.

3.3   Linear Inequalities

Basic Concepts • Symbolic Solutions • Numerical and Graphical Solutions • An Application

A LOOK INTO MATH ▶

On a freeway, the speed limit might be 70 miles per hour. A driver traveling x miles per hour is obeying the speed limit if $x \le 70$ and breaking the speed limit if $x > 70$. A speed of $x = 70$ represents the boundary between obeying the speed limit and breaking it. A posted speed limit, or *boundary*, allows drivers to easily determine whether they are speeding.

Solving linear inequalities is closely related to solving linear equations because equality is the boundary between *greater than* and *less than*. In this section we discuss techniques needed to solve linear inequalities.

NEW VOCABULARY

☐ Inequality
☐ Solution
☐ Solution set
☐ Set-builder notation
☐ Linear inequality
☐ Break-even point
☐ Properties of inequalities

Basic Concepts

An **inequality** results whenever the equals sign in an equation is replaced with any one of the symbols $<$, $\le$, $>$, or $\ge$. Examples of *linear equations* include

$$2x + 1 = 0, \quad 1 - x = 6, \quad \text{and} \quad 5x + 1 = 3 - 2x, \qquad \text{Equations}$$

and, therefore, examples of *linear inequalities* include

$$2x + 1 < 0, \quad 1 - x \ge 6, \quad \text{and} \quad 5x + 1 \le 3 - 2x. \qquad \text{Inequalities}$$

A **solution** to an inequality is a value of the variable that makes the statement true. The set of all solutions is called the **solution set**. Two inequalities are *equivalent* if they have the same solution set. Inequalities frequently have *infinitely many solutions*. For example, the solution set to the inequality $x - 5 > 0$ includes all real numbers greater than 5, which can be written as $x > 5$. Using **set-builder notation**, we can write the solution set as $\{x \mid x > 5\}$. This expression is read as "the set of all real numbers x such that x is greater than 5." The vertical line | in set-builder notation is read "such that." To summarize, we have the following.

Inequality	Set-Builder Notation	Meaning
$t \le 3$	$\{t \mid t \le 3\}$	The set of all real numbers t such that t is less than or equal to 3
$z > 8$	$\{z \mid z > 8\}$	The set of all real numbers z such that z is greater than 8

Next we more formally define a linear inequality in one variable.

LINEAR INEQUALITY IN ONE VARIABLE

A **linear inequality** in one variable is an inequality that can be written in the form

$$ax + b > 0,$$

where $a \ne 0$. (The symbol $>$ may be replaced with $\ge$, $<$, or $\le$.)

MAKING CONNECTIONS

Linear Functions, Equations, and Inequalities

Concept	*Can be written as:*
Linear function	$f(x) = ax + b$ (a = slope m)
Linear equation	$ax + b = 0$ ($a \ne 0$)
Linear inequality	$ax + b > 0$ ($a \ne 0$)

A linear equation has *one* solution, while a linear inequality has *infinitely many* solutions

▶ **REAL-WORLD CONNECTION** To better understand linear inequalities, suppose that it costs a student $\$100$ to make a large batch of candy. If the student sells bags of this candy for $\$5$ each, then the profit y is $y = 5x - 100$, where x represents the number of bags sold. The graph of y is a line with slope 5 and y-intercept -100, as shown in Figure 3.13.

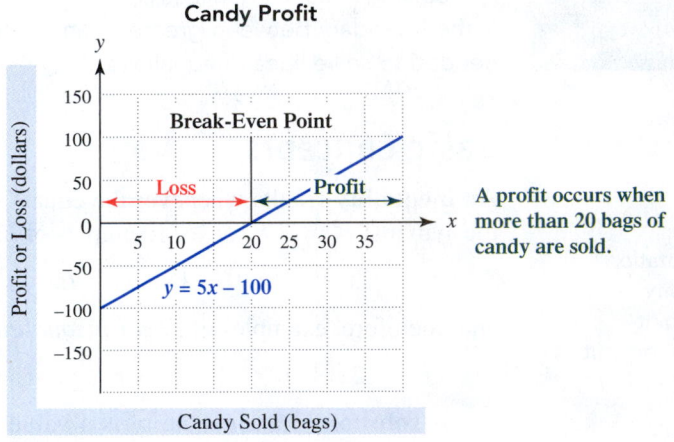

Candy Profit

Figure 3.13

Because the x-intercept is 20, the student will *break even* when $5x - 100 = 0$, or when 20 bags of candy are sold. (The **break-even point** occurs when the *revenue* from selling the candy equals the *cost* of making the candy.) Note that the student incurs a **loss** when the line is below the x-axis, or when $5x - 100 < 0$. This situation corresponds to selling less than 20 bags, or $x < 20$. A **profit** occurs when the graph is above the x-axis, or when $5x - 100 > 0$. This situation corresponds to selling more than 20 bags, or $x > 20$. Selling $x = 20$ bags of candy represents the *boundary* between **loss** and **profit**.

To better understand linear inequalities more generally, let the graph of $y = ax + b$ with slope $a \neq 0$ $(m = a)$ be a line that intersects the x-axis at $x = k$. Figure 3.14(a) shows this line with positive slope $(a > 0)$, and Figure 3.14(b) shows this line with negative slope $(a < 0)$. The solution to the *equation* $ax + b = 0$ is k, and the solution set to the *inequality* $ax + b > 0$ is either all x-values greater than k, as shown in Figure 3.14(a), or all x-values less than k, as shown in Figure 3.14(b). (See also Example 4.)

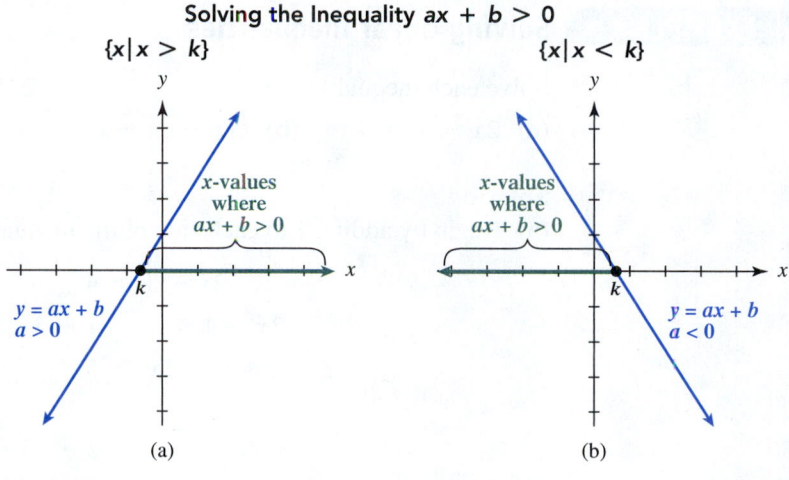

Figure 3.14

Symbolic Solutions

The inequality $3 < 5$ is equivalent to $3 + 1 < 5 + 1$. That is, we can add the same number to each side of an inequality. This is an example of one property of inequalities.

PROPERTIES OF INEQUALITIES

Let a, b, and c be real numbers.

1. $a < b$ and $a + c < b + c$ are equivalent.
   (The same number may be added to or subtracted from each side of an inequality.)
2. If $c > 0$, then $a < b$ and $ac < bc$ are equivalent.
   (Each side of an inequality may be multiplied or divided by the same positive number.)
3. If $c < 0$, then $a < b$ and $ac > bc$ are equivalent.
   (Each side of an inequality may be multiplied or divided by the same negative number provided the inequality symbol is reversed.)

Similar properties exist for the $\leq$ and $\geq$ symbols.

STUDY TIP

Be sure you understand Property 2 and Property 3.

When applying Property 2 or 3, we need to determine whether the inequality symbol should be reversed. For example, if each side of the inequality $3 < 5$ is multiplied by the *positive* number 2, we obtain

$$3 < 5 \quad \text{or} \quad 2 \cdot 3 < 2 \cdot 5 \quad \text{or} \quad 6 < 10,$$

which is a true statement. However, if each side of the inequality $3 < 5$ is multiplied by the *negative* number -2, we obtain

$$3 < 5 \quad \text{or} \quad -2 \cdot 3 > -2 \cdot 5 \quad \text{or} \quad -6 > -10,$$

$$\uparrow$$

Reverse the inequality symbol.

which is a true statement because the inequality symbol is reversed from $<$ to $>$.

To solve an inequality we apply properties of inequalities to find a simpler, equivalent inequality, as illustrated in the next example.

EXAMPLE 1 **Solving linear inequalities**

Solve each inequality.

(a) $2x - 1 > 4$ **(b)** $\frac{1}{2}(z - 3) - (2 - z) \leq 1$

Solution

(a) Begin by adding 1 to each side of the inequality.

$2x - 1 > 4$	Given inequality
$2x - 1 + 1 > 4 + 1$	Add 1 to each side.
$2x > 5$	Simplify.
$\dfrac{2x}{2} > \dfrac{5}{2}$	Divide by 2; do *not* reverse the inequality symbol because $2 > 0$.
$x > \dfrac{5}{2}$	Simplify.

The solution set is $\left\{ x \mid x > \frac{5}{2} \right\}$.

(b) To clear (eliminate) fractions multiply each term by 2.

$\dfrac{1}{2}(z - 3) - (2 - z) \leq 1$	Given inequality
$(z - 3) - 2(2 - z) \leq 2$	Multiply by 2; do *not* reverse the inequality symbol because $2 > 0$.
$z - 3 - 4 + 2z \leq 2$	Distributive property
$3z - 7 \leq 2$	Combine like terms.
$3z - 7 + 7 \leq 2 + 7$	Add 7 to each side.
$3z \leq 9$	Simplify.
$z \leq 3$	Divide by 3; do *not* reverse the inequality symbol because $3 > 0$.

The solution set is $\{z \mid z \leq 3\}$.

Now Try Exercises 19, 27

EXAMPLE 2 **Solving linear inequalities**

Solve each inequality.

(a) $5 - 3x \leq x - 3$

(b) $\dfrac{2t - 3}{5} \geq \dfrac{t + 1}{3} + 3t$

Solution

(a) Begin by subtracting 5 from each side of the inequality.

$5 - 3x \leq x - 3$	Given inequality
$5 - 3x - \mathbf{5} \leq x - 3 - \mathbf{5}$	Subtract 5 from each side.
$-3x \leq x - 8$	Simplify.
$-3x - \mathbf{x} \leq x - 8 - \mathbf{x}$	Subtract x from each side.
$-4x \leq -8$	Simplify.

Next divide each side by -4. As we are dividing by a *negative* number, Property 3 requires reversing the inequality by changing $\leq$ to $\geq$.

$\dfrac{-4x}{-4} \geq \dfrac{-8}{-4}$	Divide by -4; reverse the inequality because $-4 < 0$.
$x \geq 2$	Simplify.

The solution set is $\{x \mid x \geq 2\}$.

(b) To clear (eliminate) fractions multiply each term by 15.

$\dfrac{2t - 3}{5} \geq \dfrac{t + 1}{3} + 3t$	Given inequality
$\mathbf{15} \cdot \left(\dfrac{2t - 3}{5}\right) \geq \mathbf{15} \cdot \left(\dfrac{t + 1}{3}\right) + \mathbf{15} \cdot 3t$	Multiply each term by 15; do not reverse the inequality symbol because $15 > 0$.
$3(2t - 3) \geq 5(t + 1) + 45t$	Simplify: $\frac{15}{5} = 3$ and $\frac{15}{3} = 5$
$6t - 9 \geq \mathbf{5}t + 5 + \mathbf{45}t$	Distributive property
$6t - 9 + \mathbf{9} \geq 50t + 5 + \mathbf{9}$	Add 9; combine like terms.
$6t \geq 50t + 14$	Simplify.
$6t - \mathbf{50}t \geq 50t - \mathbf{50}t + 14$	Subtract $50t$ from each side.
$-44t \geq 14$	Simplify.
$t \leq \dfrac{14}{-44}$	Divide by -44; reverse the inequality symbol because $-44 < 0$.
$t \leq -\dfrac{7}{22}$	Simplify.

The solution set is $\left\{t \mid t \leq -\frac{7}{22}\right\}$.

▮ **Now Try Exercises 23, 29**

READING CHECK

When do you reverse the inequality symbol?

Numerical and Graphical Solutions

In Section 3.1 we solved linear equations with symbolic, numerical, and graphical methods. We can also use these methods to solve linear inequalities.

NOTE: Numerical and graphical solutions can sometimes be difficult to find if the *x*-value that determines equality is not an integer. Symbolic methods work well in such situations.

EXAMPLE 3 **Solving linear inequalities numerically**

Use Table 3.5 to find the solution set to each equation or inequality.

(a) $-\frac{1}{2}x + 1 = 0$ **(b)** $-\frac{1}{2}x + 1 > 0$ **(c)** $-\frac{1}{2}x + 1 < 0$

TABLE 3.5

x	-1	0	1	2	3	4	5
$-\frac{1}{2}x + 1$	$\frac{3}{2}$	1	$\frac{1}{2}$	0	$-\frac{1}{2}$	-1	$-\frac{3}{2}$

Solution
(a) The expression $-\frac{1}{2}x + 1$ equals 0 when $x = 2$. Thus the solution set is $\{x \mid x = 2\}$.
(b) The expression $-\frac{1}{2}x + 1$ is positive when $x < 2$. Thus the solution set is $\{x \mid x < 2\}$.
(c) The expression $-\frac{1}{2}x + 1$ is negative when $x > 2$. Thus the solution set is $\{x \mid x > 2\}$.

Now Try Exercise 53

EXAMPLE 4 **Solving linear inequalities graphically**

Graph $y = -\frac{1}{2}x + 1$ and find the solution set to each equation or inequality.
(a) $-\frac{1}{2}x + 1 = 0$ **(b)** $-\frac{1}{2}x + 1 > 0$ **(c)** $-\frac{1}{2}x + 1 < 0$

Solution
(a) The graph of $y = -\frac{1}{2}x + 1$ in Figure 3.15 crosses the x-axis at $x = 2$. Thus the solution set to $-\frac{1}{2}x + 1 = 0$ is $\{x \mid x = 2\}$.
(b) The graph is *above* the x-axis when $x < 2$. Thus the solution set is $\{x \mid x < 2\}$.
(c) The graph is *below* the x-axis when $x > 2$. Thus the solution set is $\{x \mid x > 2\}$.

Graphical Solution

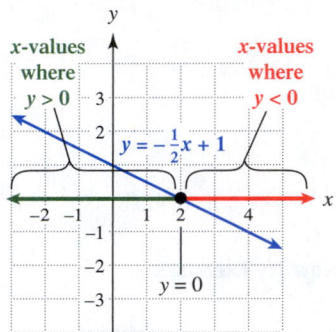

Figure 3.15

Now Try Exercise 59

CRITICAL THINKING

A linear equation has one solution that can be easily checked. A linear inequality has infinitely many solutions. Discuss ways that the solution set to a linear inequality could be checked.

Sometimes linear inequalities are written in the form $y_1 < y_2$, $y_1 \leq y_2$, $y_1 > y_2$, or $y_1 \geq y_2$, where both y_1 and y_2 represent expressions containing variables. These types of inequalities can also be solved graphically or numerically.

▶ **REAL-WORLD CONNECTION** Figure 3.16 shows the distances that two cars are from Chicago, Illinois, after x hours while traveling in the same direction on a freeway. The distance for Car 1 is denoted y_1, and the distance for Car 2 is denoted y_2.

Distance from Chicago

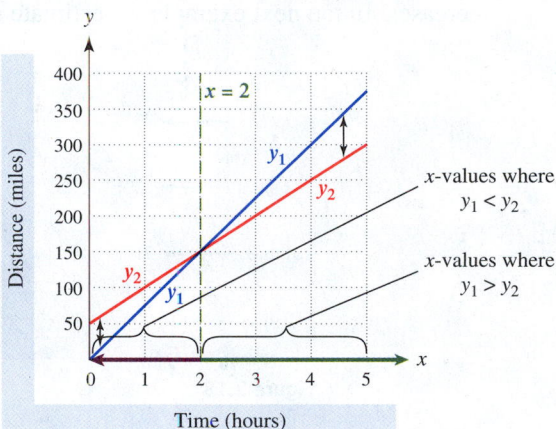

Figure 3.16

When $x = 2$ hours, $y_1 = y_2$ and both cars are 150 miles from Chicago. To the left of the dashed vertical line $x = 2$, the graph of y_1 is *below* the graph of y_2, so Car 1 is *closer* to Chicago than Car 2. Thus

$$y_1 < y_2 \quad \text{when} \quad x < 2.$$

To the right of the dashed vertical line $x = 2$, the graph of y_1 is *above* the graph of y_2, so Car 1 is *farther* from Chicago than Car 2. Thus

$$y_1 > y_2 \quad \text{when} \quad x > 2.$$

EXAMPLE 5   **Solving an inequality graphically**

Use a graph to solve $5 - 3x \geq x - 3$.

Solution
The graphs of $y_1 = 5 - 3x$ and $y_2 = x - 3$ intersect at the point $(2, -1)$, as shown in Figure 3.17(a). Equality, or $y_1 = y_2$, occurs when $x = 2$. The graph of y_1 is above the graph of y_2 to the left of the vertical line, or when $x < 2$. Thus $5 - 3x \geq x - 3$ is satisfied when $x \leq 2$. The solution set is $\{x \mid x \leq 2\}$. Figure 3.17(b) shows the same graph generated with a graphing calculator.

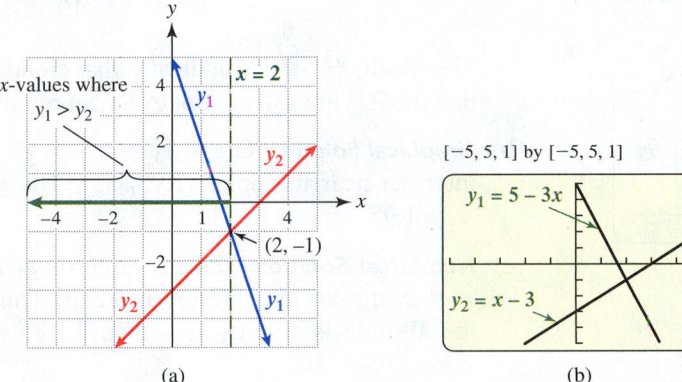

(a)   (b)

CRITICAL THINKING

Use the results from Example 5 to write the solution set for $5 - 3x \leq x - 3$.

Figure 3.17

▪ **Now Try Exercise 63**

An Application

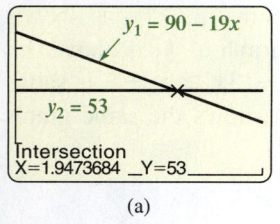

▶ **REAL-WORLD CONNECTION** Air generally becomes colder as the altitude increases. See Figure 3.18. One mile above Earth's surface the temperature is about 19°F colder than the ground-level temperature. As the air temperature cools, the chance of clouds forming increases. In the next example we estimate the altitudes at which clouds will *not* form.

3 miles ⟶ 33°F

2 miles ⟶ 52°F

1 mile ⟶ 71°F

0 miles ⟶ 90°F

Figure 3.18

EXAMPLE 6 **Finding the altitude of clouds**

If ground-level temperature is 90°F, the temperature T above Earth's surface is modeled by $T(x) = 90 - 19x$, where x is the altitude in miles. Suppose that clouds will form only if the temperature is 53°F or colder. (*Source:* A. Miller and R. Anthes, *Meteorology.*)

(a) Determine symbolically the altitudes at which there are no clouds.

(b) Give graphical support for your answer.

(c) Give numerical support for your answer.

Solution

(a) *Symbolic Solution* Clouds will not form at altitudes at which the temperature is greater than 53°F. Thus we must solve the inequality $T(x) > 53$.

$$90 - 19x > 53 \qquad \text{Inequality to be solved}$$

$$90 - 19x - 90 > 53 - 90 \qquad \text{Subtract 90 from each side.}$$

$$-19x > -37 \qquad \text{Simplify.}$$

$$\frac{-19x}{-19} < \frac{-37}{-19} \qquad \text{Divide by } -19 \text{; reverse inequality.}$$

$$x < \frac{37}{19} \qquad \text{Simplify.}$$

The result, $\frac{37}{19} \approx 1.95$, indicates that clouds will not form below about 1.95 miles. Note that models are usually not exact, so rounding values is appropriate.

(b) *Graphical Solution* Graph $y_1 = 90 - 19x$ and $y_2 = 53$. In Figure 3.19(a) their graphs intersect near the point (1.95, 53). The graph of y_1 is above the graph of y_2 when $x < 1.95$.

(c) *Numerical Solution* Make a table of y_1 and y_2, as shown in Figure 3.19(b). When $x = 2$, the air temperature is 52°F. Thus the air temperature would be 53°F when the altitude is slightly less than 2 miles, which supports our symbolic solution of $x \approx 1.95$.

[0, 3, 1] by [0, 100, 20]

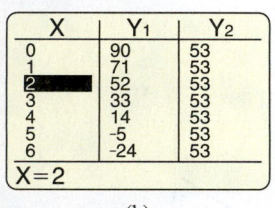

$y_1 = 90 - 19x$

$y_2 = 53$

Intersection
X=1.9473684 Y=53

(a)

X	Y₁	Y₂
0	90	53
1	71	53
2	52	53
3	33	53
4	14	53
5	-5	53
6	-24	53

X=2

(b)

Figure 3.19

CALCULATOR HELP

To find a point of intersection, see Appendix A (page AP-6).

Now Try Exercise 105

3.3 Putting It All Together

CONCEPT	EXPLANATION	EXAMPLES
Linear Inequalities in One Variable	Can be written as $$ax + b > 0,$$ where $a \neq 0$ and $>$ may be replaced by $<$, $\leq$, or $\geq$; has infinitely many solutions	$3x - 4 > 0,$ $-(x - 5) < 2,$ $2x \geq 5 - x,$ and $4 - 2(x + 1) \leq 7x - 1$
Set-Builder Notation	Used to express sets of real numbers	$\{x \mid x > 4\}$ represents the set of real numbers x such that x is greater than 4.
Properties of Inequalities	See the box on page 179. Be sure to note Property 3: When multiplying or dividing by a negative number, *reverse* the inequality symbol.	*Property 1:* (Add 3.) $x - 3 \geq 2$ is equivalent to $x \geq 5.$ *Property 2:* (Divide by 2.) $2x \leq 6$ is equivalent to $x \leq 3.$ *Property 3:* (Divide by -3) $-3x < 6$ is equivalent to $x > -2.$

Symbolic, graphical, and numerical methods are used to solve $2x - 4 \geq 0$.

Symbolic Solution

$2x - 4 \geq 0$

$2x \geq 4$

$x \geq 2$

Graphical Solution

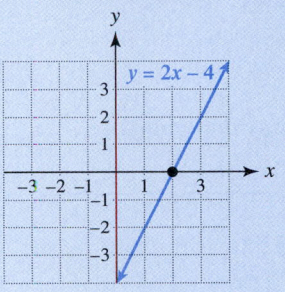

The graph of $y = 2x - 4$ is above or intersects the x-axis when $x \geq 2$.

Numerical Solution

x	$2x - 4$
0	-4
1	-2
2	0
3	2
4	4

The values of $2x - 4$ are greater than or equal to 0 when $x \geq 2$.

3.3 Exercises

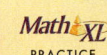

CONCEPTS AND VOCABULARY

1. Give an example of a linear inequality.

2. Can a linear inequality have infinitely many solutions? Explain.

3. Are $x + 5 < y + 5$ and $x < y$ equivalent?

4. Are $x - 7 \geq y - 7$ and $x \geq y$ equivalent?

5. Are $2x > 6$ and $x > 3$ equivalent? Explain.

6. Are $-4x < 8$ and $x < -2$ equivalent? Explain.

7. Distinguish between an equation and an inequality.

8. Name three methods for solving a linear inequality.

9. Write the set of real numbers x less than 6 in set-builder notation.

10. Write the set of real numbers x greater than or equal to -0.7 in set-builder notation.

Exercises 11–14: Decide whether the given value for the variable is a solution to the inequality.

11. $x - 5 \leq 3$ $\quad\quad\quad\quad x = 2$

12. $\frac{3}{2}x - \frac{1}{2} \geq 1 - x$ $\quad\quad x = -2$

13. $2t - 3 > 5t - (2t + 1)$ $\quad t = 5$

14. $2(z - 1) < 3(z + 1)$ $\quad z = \pi$

SYMBOLIC SOLUTIONS

Exercises 15–18: Solve each equation or inequality.

  (a) $y_1 = 0$ **(b)** $y_1 < 0$ **(c)** $y_1 > 0$

15. $y_1 = x - 2$

16. $y_1 = x + 5$

17. $y_1 = -2x + 6$

18. $y_1 = -6 - 6x$

Exercises 19–44: Solve the inequality symbolically.

19. $x + 3 \leq 5$ $\quad\quad$ **20.** $x - 5 \geq -3$

21. $\frac{1}{4}x > 9$ $\quad\quad\quad$ **22.** $14 < -3.5x$

23. $4 - 3x \leq -\frac{2}{3}$ $\quad\quad$ **24.** $4x - 2 \geq \frac{5}{2}$

25. $7 - \frac{1}{2}x < x - \frac{3}{2}$ $\quad$ **26.** $4x - 6 > 12 - 10x$

27. $\frac{5}{2}(2x - 3) < 6 - 2x$

28. $1 - \left(\frac{3}{2}x - 4\right) > \frac{1}{2}(x + 1)$

29. $\frac{3x - 2}{-2} \leq \frac{x - 4}{-5}$ $\quad$ **30.** $\frac{5 - 2x}{2} \geq \frac{2x + 1}{4}$

31. $3(x - 2000) + 15 < 45$

32. $-2(x - 1990) + 75 > 25$

33. $0.4x - 0.7 < 1.3$ $\quad$ **34.** $-0.3(x - 5) + 0.2 > 2$

35. $\frac{4}{5}x - \frac{1}{5} \geq -5$ $\quad$ **36.** $\frac{3x}{4} - \frac{1}{2} \leq \frac{1}{4}(2x + 1)$

37. $-\frac{1}{3}(z - 3) - \frac{1}{4} \geq \frac{1}{4}(5 - z)$

38. $2(3 - 2z) - (1 - z) \leq 4z - (1 + z)$

39. $\frac{3}{4}(2t - 5) \leq \frac{1}{2}(4t - 6) + 1$

40. $\frac{5}{6}x + (3 - x) \geq \frac{x - 3}{3}$

41. $\frac{1}{2}\left(4 - (x + 3)\right) + 4 > -\frac{1}{3}\left(2x - (1 - x)\right)$

42. $\frac{x - (2 - 3x)}{2} < \frac{1 - (4 - 5x)}{6}$

43. $0.05 + 0.08x < 0.01x - 0.04(3 - 4x)$

44. $-0.2(5x + 2) \geq 0.4 + 1.5x$

DOMAINS OF FUNCTIONS

Exercises 45–52: Write the domain of f in set-builder notation.

45. $f(x) = \sqrt{2x - 1}$ **46.** $f(x) = \sqrt{1 - x}$

47. $f(x) = \frac{x}{5 - 2x}$ **48.** $f(x) = \frac{x - 1}{4x + 1}$

49. $f(x) = \sqrt{5(x - 2) + 1}$

50. $f(x) = \sqrt{-(2x + 5) - x}$

51. $f(x) = \frac{1}{\sqrt{1 + 2x}}$

52. $f(x) = \frac{\sqrt{x}}{x + 1}$

NUMERICAL SOLUTIONS

Exercises 53 and 54: Complete the table. Then use the table to solve the inequality.

53. $-2x + 6 \leq 0$

x	1	2	3	4	5
$-2x + 6$	4				-4

54. $3x - 1 < 8$

x	0	1	2	3	4
$3x - 1$					

Exercises 55–58: Solve the inequality numerically.

55. $x - 3 > 0$ $\quad\quad\quad$ **56.** $2x < 0$

57. $2x - 1 \geq 3$ $\quad\quad$ **58.** $2 - 3x \leq -4$

GRAPHICAL SOLUTIONS

Exercises 59–62: Use the graph of y_1 to solve each equation or inequality.

(a) $y_1 = 0$ (b) $y_1 < 0$ (c) $y_1 > 0$

59.
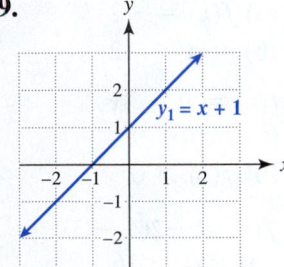

60.
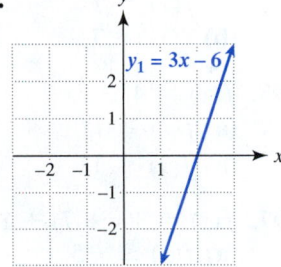

61.
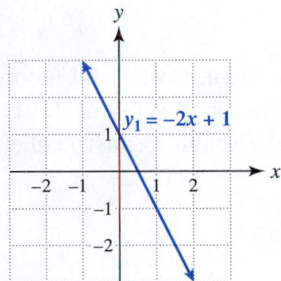

62.
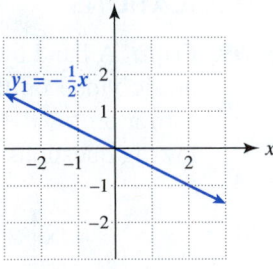

Exercises 63–66: Use the graph to solve the inequality.

63. $y_1 \le y_2$
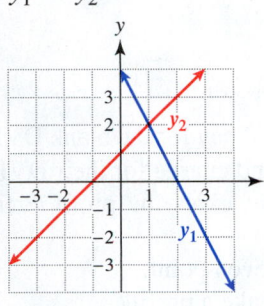

64. $y_2 > y_1$
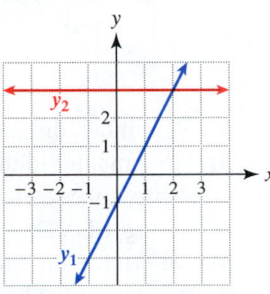

65. $y_1 > y_2$

66. $y_1 \le y_2$
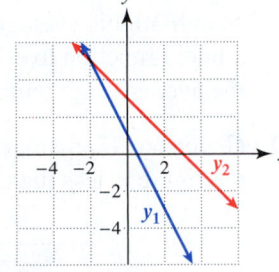

67. *Distance Between Cars* Car 1 and Car 2 are traveling in the same direction. Their distances in miles north of St. Louis, Missouri, after x hours are shown in the graph in the next column, where $x \le 8$.

(a) Which car is traveling faster? Explain.

(b) How many hours elapse before the two cars are the same distance from St. Louis? How far are they from St. Louis when this equality occurs?

(c) During what time interval is Car 2 farther from St. Louis than Car 1?

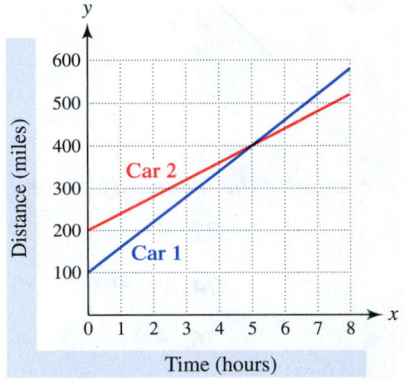

68. *Distance Between Cars* Two cars are traveling on a freeway. The distance in miles between Chicago and each car after x hours is shown in the graph.

(a) Are the cars moving toward or away from Chicago?

(b) Interpret the y-intercept of Car 2.

(c) When is Car 1 farther from Chicago than Car 2?

(d) When are Car 1 and Car 2 the same distance from Chicago?

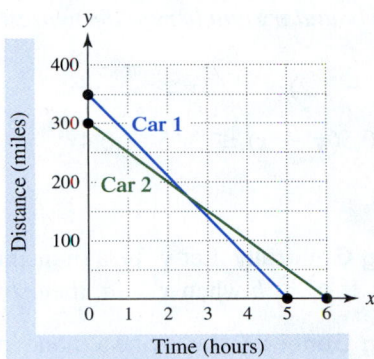

69. Use the following graph to solve each equation or inequality.

(a) $f(x) = g(x)$ (b) $f(x) < g(x)$

(c) $f(x) \ge g(x)$ (d) $f(x) \le g(x)$

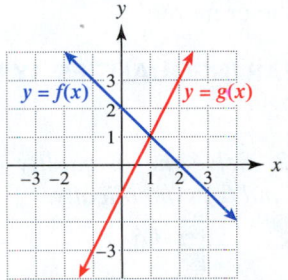

70. Use the following graph to solve each inequality.
(a) $f(x) > g(x)$ (b) $f(x) \leq g(x)$

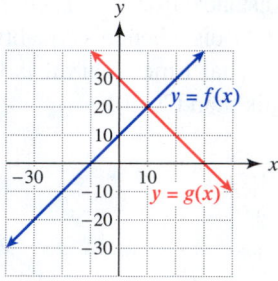

Exercises 71–82: Solve the inequality graphically.

71. $x - 1 < 0$

72. $4 - 2x \leq 0$

73. $2x \geq 0$

74. $3x + 6 > 0$

75. $4 - 2x \leq 8$

76. $-2x + 3 < -3$

77. $x - (2x + 4) > 0$

78. $2x + 1 \geq 3x - 4$

79. $2(x + 2) + 5 < -x$

80. $\dfrac{x + 5}{3} \geq \dfrac{1 - x}{2}$

81. $25(x - 1995) + 100 \leq 0$

82. $5(x - 1980) - 20 > 50$

Exercises 83 and 84: Solve the inequality graphically. Approximate boundary numbers to the nearest thousandth.

83. $\pi(\sqrt{2}x - 1.2) + \sqrt{3}x > \dfrac{\pi + 1}{3}$

84. $3.1x - 0.5(\pi - x) \leq \sqrt{5} - 0.4x$

CONCEPTS

85. Thinking Generally Let y_1 be a (nonconstant) linear function. If $y_1 < b$ when $x > a$, then solve $y_1 > b$.

86. Thinking Generally Let y_1 be a (nonconstant) linear function and $y_1 = 0$ when $x = a$. What are the possible solution sets to $y_1 \leq 0$?

87. Online Exploration Determine when the world average life expectancy exceeded 60 years.

88. Online Exploration Determine when the national debt was \$9 trillion or more.

SOLVING LINEAR INEQUALITIES BY MORE THAN ONE METHOD

Exercises 89–92: Solve the inequality (a) numerically, (b) graphically, and (c) symbolically.

89. $5x - 2 < 8$

90. $2x - 4 - 3(x + 3) \leq 0$

91. $2x \geq 3x - 3$

92. $x - 1 > \dfrac{4 - x}{2}$

INEQUALITIES AND FUNCTIONS

Exercises 93–98: Solve each equation or inequality for the given $f(x)$.

93. $f(x) = x - 5$
(a) $f(x) = 7$
(b) $f(x) > 7$

94. $f(x) = 4x$
(a) $f(x) = -2$
(b) $f(x) < -2$

95. $f(x) = 4 - 3x$
(a) $f(x) \geq 11$
(b) $f(x) \leq 11$

96. $f(x) = 9 - 3x$
(a) $f(x) < 0$
(b) $f(x) > 0$

97. $f(x) = 3(x + 7) + 1$
(a) $f(x) \leq -5$
(b) $f(x) \geq -5$

98. $f(x) = -2(4 - 3x)$
(a) $f(x) > 16$
(b) $f(x) < 16$

APPLICATIONS

99. *Profit* A band buys 500 blank DVDs and records a music video on them. The band sells the DVDs to earn a profit. Refer to the graph to answer the following questions.

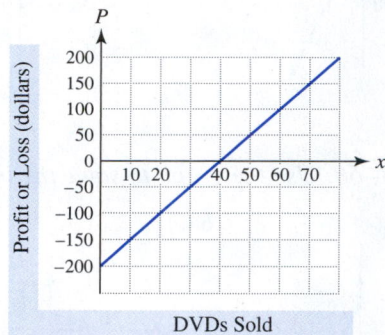

(a) How much did the band pay for the blank DVDs?
(b) How much did the band charge for their DVDs?
(c) Write an equation that gives the profit P from selling x DVDs.
(d) Determine the break-even point.
(e) When will the band make a profit?

100. *Average Temperature* On March 1 the average high temperature in Minnesota is about 21°F. On March 30 this average increases to about 46°F. Use a linear function to estimate the days in March when the average high temperature is 31°F or less.

101. *Geometry* Find values for x so that the perimeter of the figure is less than 50 feet.

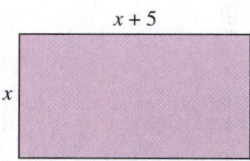

102. *Geometry* A rectangle is twice as long as it is wide. If the rectangle is to have a perimeter of 36 inches or less, what values for the width are possible?

103. *Distance and Time* Two cars are traveling in the same direction along a freeway. After x hours, the first car's distance in miles from a rest stop is given by $y_1 = 70x$ and the second car's distance in miles is given by $y_2 = 60x + 35$.
(a) What is the speed of each car?
(b) When are the cars the same distance from the rest stop?
(c) At what times is the first car farther from the rest stop than the second? Assume that $x \geq 0$.

104. *Sales of CDs and LP Records* From 1985 to 1990 sales of CDs, in millions, in the United States can be modeled by

$$y_1 = 51.6(x - 1985) + 9.1,$$

and sales of vinyl LP records, in millions, can be modeled by

$$y_2 = -31.9(x - 1985) + 167.7.$$

(a) Graph y_1 and y_2 in the viewing rectangle given by [1984, 1991, 1] by [0, 350, 50].
(b) Estimate the years when CD sales were greater than or equal to LP record sales.
(c) Solve part (b) symbolically.

105. *Altitude and Temperature* (Refer to Example 6.) If the temperature on the ground is $60°$ F, the air temperature x miles high is given by $T(x) = 60 - 19x$. Determine the altitudes at which the air temperature is greater than $0°$ F. Support your answer graphically or numerically. (*Source:* A. Miller.)

106. *Altitude and Dew Point* If the dew point on the ground is $70°$ F, then the dew point x miles high is given by $D(x) = 70 - 5.8x$. (*Source:* A. Miller.)
(a) For each 1-mile increase in altitude, how much does the dew point change?
(b) Determine the altitudes at which the dew point is greater than $30°$ F.
(c) Interpret the y-intercept on the graph of D.

107. *Body Mass Index* Body mass index, BMI, gives a recommended relationship between a person's height H in inches and weight W in pounds and is given by

$$\text{BMI} = \frac{705W}{H^2}.$$

If a person is 70 inches tall, approximate the range of weights W that result in a BMI between 19 and 25, inclusive.

108. *Grades* A student scores 75, 86, and 72 on three 100-point exams. If the final is worth 150 points, find the range of final exam scores S that gives the student an overall percentage of 80% or higher.

109. *Amazon versus eBay* From 2006 to 2012, the formula $A(x) = 4124x + 45,000$ gives the number of monthly *unique* visitors to Amazon x years after 2006. Similarly, $E(x) = -5176x + 75,000$ calculates the number of monthly unique visitors to eBay. Use each method to find the years when the number of visitors to Amazon is greater than the number of visitors to eBay. (*Source:* comScore.)
(a) Graphical   (b) Numerical   (c) Symbolical

110. *Computers Sold* From 2006 to 2012, the formula $A(x) = 7.2x + 10$ calculates the number of personal computers sold in millions by Acer x years after 2006. Similarly, $D(x) = -1.6x + 43$ calculates the number of personal computers sold in millions by Dell. Use each method to find the years when the number of computers sold by Dell was more than the number sold by Acer. (*Source:* Gartner.)
(a) Graphical   (b) Numerical   (c) Symbolical

WRITING ABOUT MATHEMATICS

111. Explain the following terms and give examples.
(a) Linear function   (b) Linear equation
(c) Linear inequality

112. Suppose that a student says that a linear equation and a linear inequality can be solved symbolically in exactly the same way. How would you respond?

Group Activity   Working with Real Data

Directions: Form a group of 2 to 4 people. Select someone to record the group's responses for this activity. All members of the group should work cooperatively to answer the questions. If your instructor asks for your results, each member of the group should be prepared to respond.

1. *Born Outside the United States* The foreign-born portion of the population increased from 5% in 1970 to 12% in 2010. This increase could be modeled by a linear function. Use these data to estimate the year when this percentage was 8.5%. (*Source:* U.S. Census Bureau.)

2. When do you expect the foreign-born population to be 15%?

3. What does your model say about the foreign-born population in 1940 and in 2600? Are your answers reasonable? What do these results say about your model?

3.4   Compound Inequalities

**Basic Concepts • Symbolic Solutions and Number Lines •
Numerical and Graphical Solutions • Interval Notation**

A LOOK INTO MATH ▶

A person weighing 143 pounds and needing to purchase a life vest for white-water rafting is not likely to find one designed exactly for this weight. Life vests are manufactured to support a range of body weights. A vest approved for weights between 100 and 160 pounds might be appropriate for this person. In other words, if a person's weight is w, this life vest is safe if $w \geq 100$ *and* $w \leq 160$. This example illustrates the concept of a *compound inequality*.

Basic Concepts

A **compound inequality** consists of two inequalities joined by the words *and* or *or*. The following are two examples of compound inequalities.

NEW VOCABULARY

☐ Compound inequality
☐ Intersection
☐ Three-part inequality
☐ Union
☐ Interval notation
☐ Infinity
☐ Negative infinity

1. $2x \geq -3$ **and** $2x < 5$ **2.** $x + 2 \geq 3$ **or** $x - 1 < -5$

First compound inequality   Second compound inequality

If a compound inequality contains the word *and*, a solution must satisfy *both* inequalities. For example, 1 is a solution to the first compound inequality because

$$2(\mathbf{1}) \geq -3 \quad \text{and} \quad 2(\mathbf{1}) < 5 \qquad \text{First compound inequality}$$
$$\textbf{True} \qquad\qquad\qquad \textbf{True} \qquad \text{with } x = 1$$

are both true statements.

If a compound inequality contains the word *or*, a solution must satisfy *at least one* of the two inequalities. Thus 5 is a solution to the second compound inequality, because the first statement is true.

$$\mathbf{5} + 2 \geq 3 \quad \text{or} \quad \mathbf{5} - 1 < -5 \qquad \text{Second compound inequality}$$
$$\textbf{True} \qquad\qquad\qquad \textbf{False} \qquad \text{with } x = 5$$

Note that 5 does not need to satisfy both statements for this compound inequality to be true.

EXAMPLE 1   **Determining solutions to compound inequalities**

Determine whether the given x-values are solutions to the compound inequalities.
(a) $x + 1 < 9$ and $2x - 1 > 8$ $x = 5, -5$
(b) $5 - 2x \leq -4$ or $5 - 2x \geq 4$ $x = 2, -3$

Solution
(a) Substitute $x = \mathbf{5}$ in the compound inequality $x + 1 < 9$ and $2\mathbf{x} - 1 > 8$.

$$\mathbf{5} + 1 < 9 \quad \text{and} \quad 2(\mathbf{5}) - 1 > 8$$
$$\textbf{True} \qquad\qquad\qquad \textbf{True}$$

Both inequalities are true, so 5 is a solution. Now substitute $x = \mathbf{-5}$.

$$\mathbf{-5} + 1 < 9 \quad \text{and} \quad 2(\mathbf{-5}) - 1 > 8$$
$$\textbf{True} \qquad\qquad\qquad \textbf{False}$$

To be a solution both inequalities must be true, so -5 is *not* a solution.

(b) Substitute $x = \mathbf{2}$ into the compound inequality $5 - 2x \le -4$ or $5 - 2x \ge 4$.

$$5 - 2(\mathbf{2}) \le -4 \quad \text{or} \quad 5 - 2(\mathbf{2}) \ge 4$$

False   **False**

Neither inequality is true, so 2 is *not* a solution. Now substitute $x = -\mathbf{3}$.

$$5 - 2(-\mathbf{3}) \le -4 \quad \text{or} \quad 5 - 2(-\mathbf{3}) \ge 4$$

False   **True**

At least one of the two inequalities is true, so -3 is a solution.

▍ **Now Try Exercises 7, 9**

CRITICAL THINKING

Graph the following inequalities and discuss your results.
1. $x < 2$ and $x > 5$
2. $x > 2$ or $x < 5$

Symbolic Solutions and Number Lines

We can use a number line to graph solutions to compound inequalities, such as

$$x \le 6 \quad \text{and} \quad x > -4.$$

The solution set for $x \le 6$ is shaded to the left of 6, with a bracket placed at $x = 6$, as shown in Figure 3.20. The solution set for $x > -4$ can be shown by shading a different number line to the right of -4 and placing a left parenthesis at -4. Because the inequalities are connected by *and*, the solution set consists of all numbers that are shaded on *both* number lines. The final number line represents the intersection of the two solution sets. That is, the solution set includes real numbers where the graphs "overlap." For any two sets A and B, the **intersection** of A and B, denoted $A \cap B$, is defined by

$$A \cap B = \{x \mid x \text{ is an element of } A \text{ and an element of } B\}.$$

Graphing a Compound Inequality

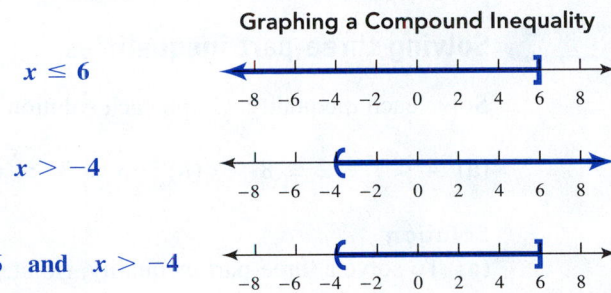

Figure 3.20

NOTE: A bracket, either [or], is used when an inequality contains $\le$ or $\ge$. A parenthesis, either (or), is used when an inequality contains $<$ or $>$. This notation makes clear whether an endpoint is included in the inequality.

EXAMPLE 2 **Solving a compound inequality containing "and"**

Solve $2x + 4 > 8$ and $5 - x < 9$. Graph the solution set.

Solution
First solve each linear inequality separately.

$$2x + 4 > 8 \quad \text{and} \quad 5 - x < 9$$
$$2x > 4 \quad \text{and} \quad -x < 4$$
$$x > 2 \quad \text{and} \quad x > -4$$

Graph $x > 2$ and $x > -4$ on two different number lines. On a third number line, shade solutions that appear on both of the first two number lines. As shown in Figure 3.21, the solution set is $\{x \mid x > 2\}$.

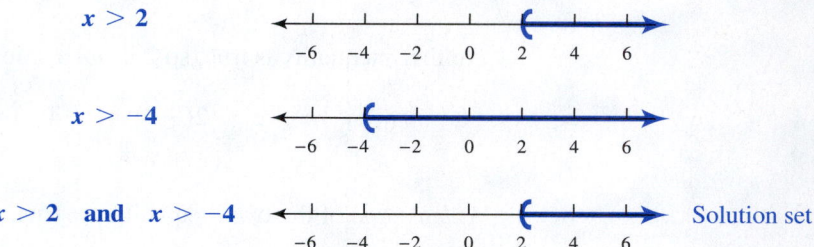

Figure 3.21

▌ **Now Try Exercise 43**

READING CHECK

What is a three-part inequality?

Sometimes a compound inequality containing the word *and* can be combined into a three-part inequality. For example, rather than writing

$$x > 5 \quad \text{and} \quad x \leq 10,$$

we could write the **three-part inequality**

$$5 < x \leq 10.$$

This three-part inequality is represented by the number line shown in Figure 3.22.

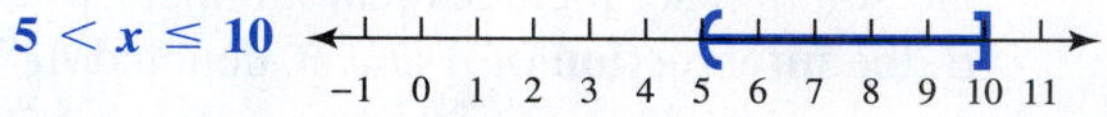

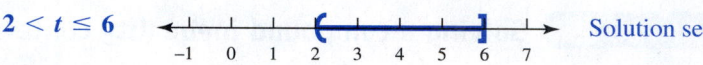

$5 < x \leq 10$

Figure 3.22

EXAMPLE 3 **Solving three-part inequalities**

Solve each inequality. Graph each solution set.

(a) $4 < t + 2 \leq 8$ **(b)** $-3 \leq 3z \leq 6$ **(c)** $-\dfrac{5}{2} < \dfrac{1 - m}{2} < 4$

Solution

(a) To solve a three-part inequality, isolate the variable by applying properties of inequalities to each part of the inequality.

$$4 < t + 2 \leq 8 \qquad \text{Given three-part inequality}$$
$$4 - 2 < t + 2 - 2 \leq 8 - 2 \qquad \text{Subtract 2 from each part.}$$
$$2 < t \leq 6 \qquad \text{Simplify each part.}$$

The solution set is $\{t \mid 2 < t \leq 6\}$. See Figure 3.23.

$2 < t \leq 6$ Solution set

Figure 3.23

(b) To simplify, divide each part by 3.

$$-3 \leq 3z \leq 6 \qquad \text{Given three-part inequality}$$
$$\frac{-3}{3} \leq \frac{3z}{3} \leq \frac{6}{3} \qquad \text{Divide each part by 3.}$$
$$-1 \leq z \leq 2 \qquad \text{Simplify each part.}$$

The solution set is $\{z \mid -1 \leq z \leq 2\}$. See Figure 3.24.

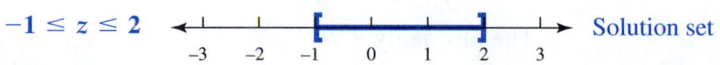

$-1 \leq z \leq 2$ Solution set

Figure 3.24

(c) Multiply each part by 2 to clear (eliminate) fractions.

$$-\frac{5}{2} < \frac{1-m}{2} < 4 \qquad \text{Given three-part inequality}$$

$$2 \cdot \left(-\frac{5}{2}\right) < 2 \cdot \left(\frac{1-m}{2}\right) < 2 \cdot 4 \qquad \text{Multiply each part by 2.}$$

$$-5 < 1 - m < 8 \qquad \text{Simplify each part.}$$

$$-5 - 1 < 1 - m - 1 < 8 - 1 \qquad \text{Subtract 1 from each part.}$$

$$-6 < -m < 7 \qquad \text{Simplify each part.}$$

$$-1 \cdot (-6) > -1 \cdot (-m) > -1 \cdot 7 \qquad \begin{array}{l}\text{Multiply each part by } -1; \\ \textit{reverse inequality symbols.}\end{array}$$

$$6 > m > -7 \qquad \text{Simplify each part.}$$

$$-7 < m < 6 \qquad \text{Rewrite inequality.}$$

The solution set is $\{m \mid -7 < m < 6\}$. See Figure 3.25.

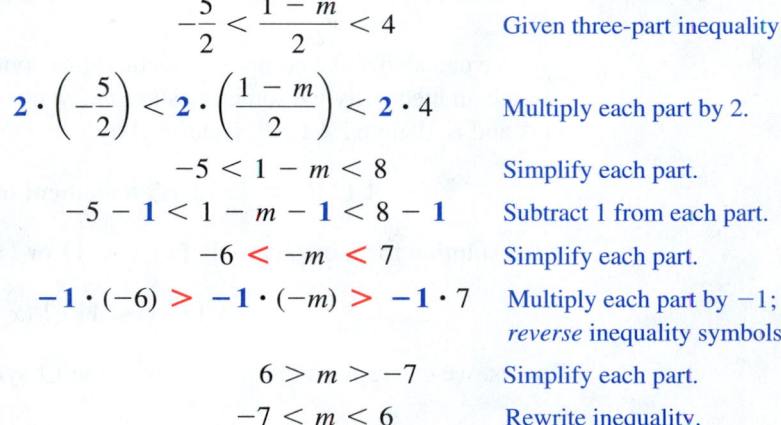

$-7 < m < 6$ Solution set

Figure 3.25

NOTE: Either $6 > m > -7$ or $-7 < m < 6$ is a correct way to write a three-part inequality. *However*, we usually write the smaller number on the left side and the larger number on the right side.

■ **Now Try Exercises 59, 63, 79**

STUDY TIP

When simplifying a three-part inequality, be sure to perform the same step on each of the three parts.

▶ **REAL-WORLD CONNECTION** Three-part inequalities occur frequently in applications. In the next example we find altitudes at which the air temperature is within a certain range.

EXAMPLE 4   **Solving a three-part inequality**

If the ground-level temperature is 80°F, the air temperature x miles above Earth's surface is cooler and can be modeled by $T(x) = 80 - 19x$. Find the altitudes at which the air temperature ranges from 42°F down to 23°F. (*Source:* A. Miller and R. Anthes, *Meteorology.*)

Solution
We write and solve the three-part inequality $23 \leq T(x) \leq 42$.

$$23 \leq 80 - 19x \leq 42 \qquad \text{Substitute for } T(x).$$

$$-57 \leq -19x \leq -38 \qquad \text{Subtract 80 from each part.}$$

$$\frac{-57}{-19} \geq x \geq \frac{-38}{-19} \qquad \text{Divide by } -19; \textit{ reverse inequality symbols.}$$

$$3 \geq x \geq 2 \qquad \text{Simplify.}$$

$$2 \leq x \leq 3 \qquad \text{Rewrite inequality.}$$

The air temperature ranges from 42°F to 23°F for altitudes between 2 and 3 miles.

■ **Now Try Exercise 119**

MAKING CONNECTIONS

Writing Three-Part Inequalities

The inequality $-2 < x < 1$ means that $x > -2$ *and* $x < 1$. A three-part inequality should *not* be used when *or* connects a compound inequality. Writing $x < -2$ or $x > 1$ as $1 < x < -2$ is **incorrect** because it states that x must be both greater than 1 *and* less than -2. It is impossible for any value of x to satisfy this statement.

We can also solve compound inequalities containing the word *or*. To write the solution to such an inequality we sometimes use *union* notation. For any two sets A and B, the **union** of A and B, denoted $A \cup B$, is defined by

$$A \cup B = \{x \mid x \text{ is an element of } A \textit{ or } \text{an element of } B\}.$$

If the solution to an inequality is $\{x \mid x < 1\}$ or $\{x \mid x \geq 3\}$, then it can also be written as

$$\{x \mid x < 1\} \cup \{x \mid x \geq 3\}.$$

That is, we can replace the word *or* with the $\cup$ symbol.

EXAMPLE 5   **Solving a compound inequality containing "or"**

Solve $x + 2 < -1$ or $x + 2 > 1$. Graph the solution set.

Solution

We first solve each linear inequality.

$$x + 2 < -1 \quad \text{or} \quad x + 2 > 1 \qquad \text{Given compound inequality}$$
$$x < -3 \quad \text{or} \quad x > -1 \qquad \text{Subtract 2.}$$

We can graph the simplified inequalities on different number lines, as shown in Figure 3.26. A solution must satisfy at least one of the two inequalities. Thus the solution set for the compound inequality results from taking the *union* of the first two number lines. We can write the solution, using set-builder notation, as $\{x \mid x < -3\} \cup \{x \mid x > -1\}$ or as $\{x \mid x < -3 \text{ or } x > -1\}$.

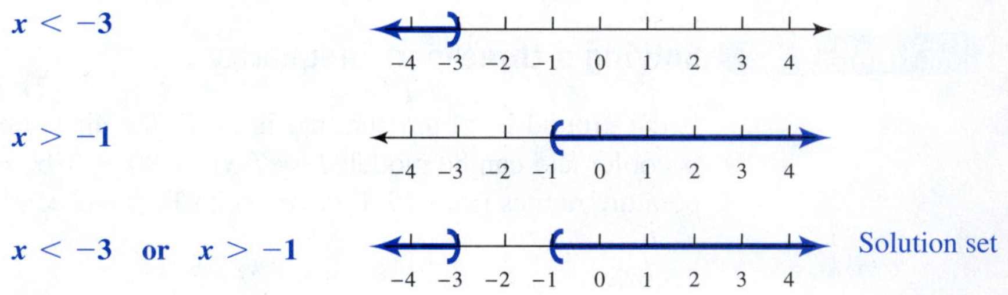

$x < -3$

$x > -1$

$x < -3$ or $x > -1$ — Solution set

Figure 3.26

Now Try Exercise 47

CRITICAL THINKING

Carbon dioxide is emitted when human beings breathe. In one study of college students, the amount of carbon dioxide exhaled in grams per hour was measured during both lectures and exams. The average amount exhaled during lectures L satisfied $25.33 \leq L \leq 28.17$, whereas the average amount exhaled during exams E satisfied $36.58 \leq E \leq 40.92$. What do these results indicate? Explain. (*Source:* T. Wang, *ASHRAE Trans.*)

Numerical and Graphical Solutions

Compound inequalities can also be solved graphically and numerically, as illustrated in the next example.

EXAMPLE 6 **Estimating numbers of Internet users**

The number of U.S. Internet users in millions during year x can be modeled by the formula $f(x) = 11.6(x - 2000) + 124$. Estimate the years when the number of users is expected to be between 240 and 275 million. (*Source:* The Nielsen Company.)

Solution

Numerical Solution Let $y_1 = 11.6(x - 2000) + 124$. Make a table of values, as shown in Figure 3.27(a). In 2010 the number of Internet users was 240 million, and in 2013 this number is about 275 million. Thus from 2010 to about 2013 the number of Internet users is expected to be between 240 million and 275 million.

Graphical Solution The graph of $y_1 = 11.6(x - 2000) + 124$ is shown between the graphs of $y_2 = 240$ and $y_3 = 275$ in Figures 3.27(b) and 3.27(c) from 2010 to about 2013, or when $2010 \le x \le 2013$.

CALCULATOR HELP

To find a point of intersection, see Appendix A (page AP-6).

[2006, 2015, 1] by [150, 350, 50]   [2006, 2015, 1] by [150, 300, 50]

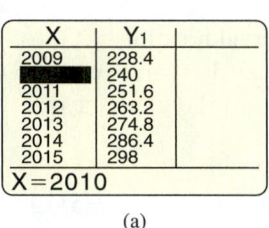

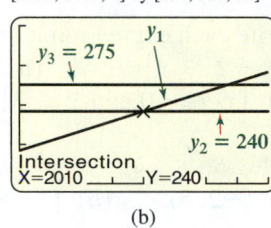

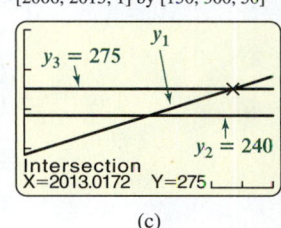

(a)   (b)   (c)

Figure 3.27

Now Try Exercise 113 (a) and (b)

Interval Notation

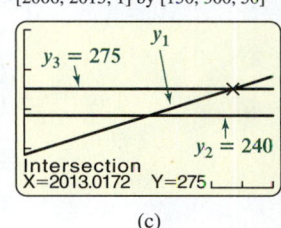

Figure 3.28

The solution set in Example 4 was $\{x \mid 2 \le x \le 3\}$. This solution set can be graphed on a number line, as shown in Figure 3.28.

A convenient notation for number line graphs is called **interval notation**. Instead of drawing the entire number line as in Figure 3.28, the solution set can be expressed as [2, 3] in interval notation. Because the solution set includes the endpoints 2 and 3, brackets are used. A solution set that includes all real numbers satisfying $-2 < x < 3$ can be expressed as $(-2, 3)$. Parentheses indicate that the endpoints are *not* included. The interval $0 \le x < 4$ is represented by $[0, 4)$.

MAKING CONNECTIONS

Points and Intervals

The expression (1, 2) may represent a point in the xy-plane or the interval $1 < x < 2$. To alleviate confusion, phrases such as "the point (1, 2)" or "the interval (1, 2)" are used.

Table 3.6 on the next page provides examples of interval notation. The symbol ∞ refers to **infinity**, and it does not represent a real number. The interval $(5, \infty)$ represents $x > 5$, which has no maximum x-value, so ∞ is used for the right endpoint. The symbol $-\infty$ may be used similarly and denotes **negative infinity**. Real numbers are denoted $(-\infty, \infty)$.

TABLE 3.6 Interval Notation

Inequality	Interval Notation	Number Line Graph
$-1 < x < 3$	$(-1, 3)$	
$-3 < x \leq 2$	$(-3, 2]$	
$-2 \leq x \leq 2$	$[-2, 2]$	
$x < -1$ or $x > 2$	$(-\infty, -1) \cup (2, \infty)$ ($\cup$ is the union symbol.)	
$x > -1$	$(-1, \infty)$	
$x \leq 2$	$(-\infty, 2]$	

EXAMPLE 7 **Writing inequalities in interval notation**

Write each expression in interval notation.
(a) $-2 \leq x < 5$ **(b)** $x \geq 3$ **(c)** $x < -5$ or $x \geq 2$
(d) $\{x \mid x > 0 \text{ and } x \leq 3\}$ **(e)** $\{x \mid x \leq 1 \text{ or } x \geq 3\}$

Solution
(a) $[-2, 5)$ **(b)** $[3, \infty)$ **(c)** $(-\infty, -5) \cup [2, \infty)$
(d) $(0, 3]$ **(e)** $(-\infty, 1] \cup [3, \infty)$

Now Try Exercises 13, 17, 23, 27, 37

EXAMPLE 8 **Solving an inequality**

Solve $2x + 1 \leq -1$ or $2x + 1 \geq 3$. Write the solution set in interval notation.

Solution
First solve each inequality.

$$2x + 1 \leq -1 \quad \text{or} \quad 2x + 1 \geq 3 \quad \text{Given compound inequality}$$
$$2x \leq -2 \quad \text{or} \quad 2x \geq 2 \quad \text{Subtract 1.}$$
$$x \leq -1 \quad \text{or} \quad x \geq 1 \quad \text{Divide by 2.}$$

The solution set may be written as $(-\infty, -1] \cup [1, \infty)$.

Now Try Exercise 55

EXAMPLE 9 **Writing domains in interval notation**

Write the domain of each function in interval notation.
(a) $f(x) = \sqrt{x - 1}$ **(b)** $g(x) = \frac{1}{2x + 1}$

Solution
(a) The square root $\sqrt{x - 1}$ is undefined when $x - 1$ is negative. Thus x is in the domain of f whenever $x - 1 \geq 0$, or $x \geq 1$. In interval notation the domain of f is $[1, \infty)$.

(b) The ratio $\frac{1}{2x+1}$ is undefined when $2x + 1 = 0$. The solution to this equation is $-\frac{1}{2}$. Thus the domain of f is all real numbers except $-\frac{1}{2}$. In interval notation the domain of f is $\left(-\infty, -\frac{1}{2}\right) \cup \left(-\frac{1}{2}, \infty\right)$.

Now Try Exercises 99, 101

3.4 Putting It All Together

CONCEPT	EXPLANATION	EXAMPLES
Compound Inequality	Two inequalities joined by *and* or *or*	$2x \geq 10$ and $x + 2 < 16$; $x < -1$ or $x > 2$
Three-Part Inequality	Can be used to write some types of compound inequalities involving *and*	$x > -2$ and $x \leq 3$ is equivalent to $-2 < x \leq 3$.
Interval Notation	Notation used to write sets of real numbers rather than using number lines or inequalities	$-2 \leq z \leq 4$ is equivalent to $[-2, 4]$. $x < 4$ is equivalent to $(-\infty, 4)$. $x \leq -2$ or $x > 0$ is equivalent to $(-\infty, -2] \cup (0, \infty)$.

TYPE OF INEQUALITY	METHOD TO SOLVE THE INEQUALITY
Solving a Compound Inequality Containing *and*	**STEP 1:** First solve each inequality individually. **STEP 2:** The solution set includes values that satisfy *both* inequalities from Step 1.
Solving a Compound Inequality Containing *or*	**STEP 1:** First solve each inequality individually. **STEP 2:** The solution set includes values that satisfy *at least one* of the inequalities from Step 1.
Solving a Three-Part Inequality	Work on all three parts at the same time. Be sure to perform the same step on each part. Continue until the variable is isolated in the middle part.

3.4 Exercises

MyMathLab 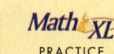 Math XL PRACTICE   WATCH   DOWNLOAD   READ   REVIEW

CONCEPTS AND VOCABULARY

1. Give an example of a compound inequality containing the word *and*.

2. Give an example of a compound inequality containing the word *or*.

3. Is 1 a solution to $x > 3$ and $x \leq 5$?

4. Is 1 a solution to $x < 3$ or $x \geq 5$?

5. Is the compound inequality $x \geq -5$ and $x \leq 5$ equivalent to $-5 \leq x \leq 5$?

6. Name three ways to solve a compound inequality.

Exercises 7–12: Determine whether the given values of x are solutions to the compound inequality.

7. $x - 1 < 5$ and $2x > 3$ $x = 2, x = 6$

8. $2x + 1 \geq 4$ and $1 - x \leq 3$ $x = -2, x = 3$

9. $3x < -5$ or $2x \geq 3$ $x = 0, x = 3$

10. $x + 1 \leq -4$ or $x + 1 \geq 4$ $x = -5, x = 2$

11. $2 - x > -5$ and $2 - x \leq 4$ $x = -3, x = 0$

12. $x + 5 \geq 6$ or $3x \leq 3$ $x = -1, x = 1$

INTERVAL NOTATION

Exercises 13–38: Write the inequality in interval notation.

13. $2 \leq x \leq 10$ 14. $-1 < x < 5$

15. $5 < x \leq 8$ 16. $-\frac{1}{2} \leq x \leq \frac{5}{6}$

17. $x < 4$ 18. $x \leq -3$

19. $x > -2$ 20. $x \geq 6$

21. $x \geq -2$ and $x < 5$ 22. $x \leq 6$ and $x \geq 2$

23. $x \leq 8$ and $x > -8$ 24. $x \geq -4$ and $x < 3$

25. $x \geq 6$ or $x > 3$ 26. $x \leq -4$ or $x < -3$

27. $x \leq -2$ or $x \geq 4$ 28. $x \leq -1$ or $x > 6$

29. $x < 1$ or $x \geq 5$ 30. $x < -3$ or $x > 3$

31.

32.

33.

34.

35. $\{x \mid x < 4\}$ 36. $\{x \mid -1 \leq x < 4\}$

37. $\{x \mid x < 1$ or $x > 2\}$ 38. $\{x \mid -\infty < x < \infty\}$

SYMBOLIC SOLUTIONS

Exercises 39–48: Solve the compound inequality. Graph the solution set on a number line.

39. $x \leq 3$ and $x \geq -1$ 40. $x \geq 5$ and $x > 6$

41. $2x < 5$ and $2x > -4$

42. $2x + 1 < 3$ and $x - 1 \geq -5$

43. $x + 2 > 5$ and $3 - x < 10$

44. $x + 2 > 5$ or $3 - x < 10$

45. $x \leq -1$ or $x \geq 2$

46. $2x \leq -6$ or $x \geq 6$

47. $5 - x > 1$ or $x + 3 \geq -1$

48. $1 - 2x > 3$ or $2x - 4 \geq 4$

Exercises 49–58: Solve the compound inequality. Write your answer in interval notation.

49. $x - 3 \leq 4$ and $x + 5 \geq -1$

50. $2z \geq -10$ and $z < 8$

51. $3t - 1 > -1$ and $2t - \frac{1}{2} > 6$

52. $2(x + 1) < 8$ and $-2(x - 4) > -2$

53. $x - 4 \geq -3$ or $x - 4 \leq 3$

54. $1 - 3n \geq 6$ or $1 - 3n \leq -4$

55. $-x < 1$ or $5x + 1 < -10$

56. $7x - 6 > 0$ or $-\frac{1}{2}x \leq 6$

57. $1 - 7x < -48$ and $3x + 1 \leq -9$

58. $3x - 4 \leq 8$ or $4x - 1 \leq 13$

Exercises 59–80: Solve the three-part inequality. Write your answer in interval notation.

59. $-2 \leq t + 4 < 5$ 60. $5 < t - 7 < 10$

61. $-\frac{5}{8} \leq y - \frac{3}{8} < 1$ 62. $-\frac{1}{2} < y - \frac{3}{2} < \frac{1}{2}$

63. $-27 \leq 3x \leq 9$ 64. $-4 < 2y < 22$

65. $\frac{1}{2} < -2y \leq 8$ 66. $-16 \leq -4x \leq 8$

67. $-4 < 5z + 1 \leq 6$ 68. $-3 \leq 3z + 6 < 9$

69. $3 \leq 4 - n \leq 6$ 70. $-1 < 3 - n \leq 1$

71. $-1 < 2z - 1 < 3$ 72. $2 \leq 4z + 5 \leq 6$

73. $-2 \leq 5 - \frac{1}{3}m < 2$ 74. $-\frac{3}{2} < 4 - 2m < \frac{7}{2}$

75. $100 \leq 10(5x - 2) \leq 200$

76. $-15 < 5(x - 1990) < 30$

77. $-3 < \dfrac{3z + 1}{4} < 1$ 78. $-3 < \dfrac{z - 1}{2} < 5$

79. $-\dfrac{5}{2} \leq \dfrac{2 - m}{4} \leq \dfrac{1}{2}$

80. $\dfrac{4}{5} \leq \dfrac{4 - 2m}{10} \leq 2$

NUMERICAL AND GRAPHICAL SOLUTIONS

Exercises 81–84: Use the table to solve the three-part inequality. Write your answer in interval notation.

81. $-3 \le 3x \le 6$

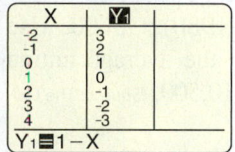

82. $-5 \le 2x - 1 \le 1$

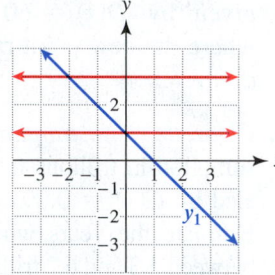

83. $-1 < 1 - x < 2$

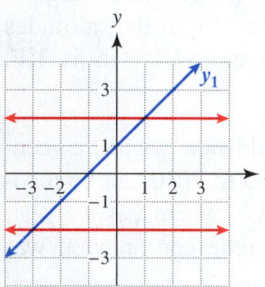

84. $-2 \le -2x < 4$

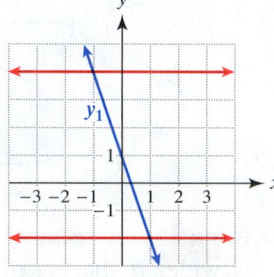

Exercises 85–88: Use the graph to solve the compound inequality. Write your answer in interval notation.

85. $-2 \le y_1 \le 2$

86. $1 \le y_1 < 3$

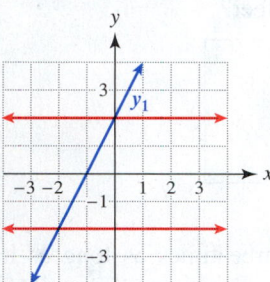

87. $y_1 < -2$ or $y_1 > 2$

88. $y_1 \le -2$ or $y_1 \ge 4$

89. *Distance* The function f, shown in the figure at the top of the next column, gives the distance y in miles between a car and Omaha, Nebraska, after x hours, where $0 \le x \le 6$.
(a) Is the car moving toward or away from Omaha? Explain.
(b) Determine the times when the car is 100 miles or 200 miles from Omaha.
(c) When is the car from 100 to 200 miles from Omaha?

(d) When is the car's distance from Omaha greater than or equal to 200 miles?

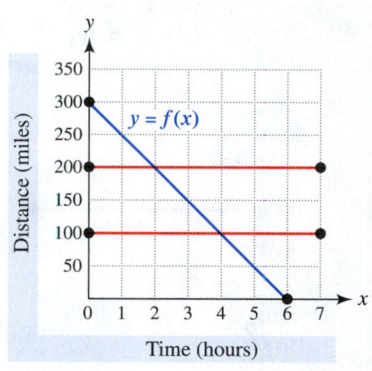

90. *Distance* The function g, shown in the figure, gives the distance y in miles between a train and Seattle after x hours, where $0 \le x \le 5$.
(a) Is the train moving toward or away from Seattle? Explain.
(b) Determine the times when the train is 150 miles or 300 miles from Seattle.
(c) When is the train from 150 to 300 miles from Seattle?
(d) When is the train's distance from Seattle less than or equal to 150 miles?

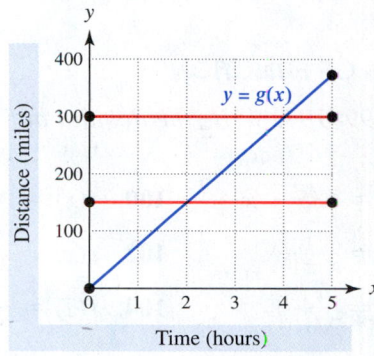

91. Use the figure to solve each equation or inequality. Let the domains of y_1, y_2, and y_3 be $0 \le x \le 8$.
(a) $y_1 = y_2$ (b) $y_2 = y_3$
(c) $y_1 \le y_2 \le y_3$ (d) $y_2 < y_1$

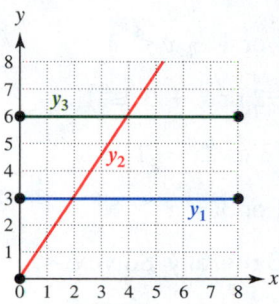

92. Use the figure to solve each equation or inequality. Let the domains of $y_1, y_2,$ and y_3 be $0 \leq x \leq 5$.
(a) $y_1 = y_2$ (b) $y_2 = y_3$
(c) $y_1 \leq y_2 \leq y_3$ (d) $y_2 < y_3$

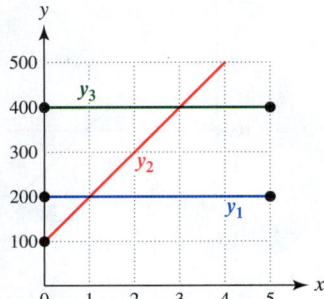

Exercises 93–98: Solve numerically or graphically. Write your answer in interval notation.

93. $-2 \leq 2x - 4 \leq 4$ **94.** $-1 \leq 1 - x \leq 3$

95. $x + 1 < -1$ or $x + 1 > 1$

96. $2x - 1 < -3$ or $2x - 1 > 5$

97. $95 \leq 25(x - 2000) + 45 \leq 295$

98. $42 \leq -13(x - 2005) + 120 \leq 94$

DOMAINS OF FUNCTIONS

Exercises 99–104: Write the domain of f in interval notation.

99. $f(x) = \sqrt{5 - x}$ **100.** $f(x) = \sqrt{3x + 6}$

101. $f(x) = \frac{1}{x + 1}$ **102.** $f(x) = \frac{x}{2x - 3}$

103. $f(x) = \frac{1}{\sqrt{x}}$ **104.** $f(x) = \frac{x}{\sqrt{x - 1}}$

USING MORE THAN ONE METHOD

Exercises 105–110: Solve symbolically, graphically, and numerically. Write the solution set in interval notation.

105. $4 \leq 5x - 1 \leq 14$ **106.** $-4 < 2x < 4$

107. $4 - x \geq 1$ or $4 - x < 3$

108. $x + 3 \geq -2$ or $x + 3 \leq 1$

109. $2x + 1 < 3$ or $2x + 1 \geq 7$

110. $3 - x \leq 4$ or $3 - x > 8$

111. Thinking Generally Solve $c < x + b \leq d$ for x.

112. Thinking Generally Solve $c \leq ax + b \leq d$ for x, if $a < 0$.

APPLICATIONS

113. *Online Betting* Global online betting losses in billions can be modeled by $L(x) = 2.5x - 5000$, where x is a year from 2006 to 2011. Use each method to estimate when losses ranged from $15 billion to $20 billion. (*Source:* Christiansen Capital Advisors.)
(a) Numerical
(b) Graphical
(c) Symbolic

114. *College Tuition* From 1980 to 2000, college tuition and fees at private colleges could be modeled by the linear function $f(x) = 575(x - 1980) + 3600$. Use each method to estimate when the average tuition and fees ranged from $8200 to $10,500. (*Source:* The College Board.)
(a) Numerical
(b) Graphical
(c) Symbolic

115. *Altitude and Dew Point* If the dew point D on the ground is $60°F$, then the dew point x miles high is given by $D(x) = 60 - 5.8x$. Find the altitudes where the dew point ranges from $57.1°F$ to $51.3°F$. (*Source:* A. Miller.)

116. *Cigarette Consumption* Worldwide cigarette consumption in trillions from 1950 to 2010 can be modeled by $C(x) = 0.09x - 173.8$, where x is the year. Estimate the years when cigarette consumption was between 5.3 and 6.2 trillion. (*Source:* Department of Agriculture.)

117. *Geometry* For what values of x is the perimeter of the rectangle from 40 to 60 feet?

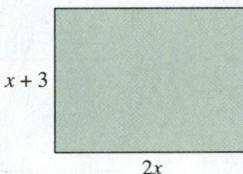

118. *Geometry* A rectangle is three times as long as it is wide. If the perimeter ranges from 100 to 160 inches, what values for the width are possible?

119. *Altitude and Temperature* If the air temperature at ground level is $70°F$, the air temperature x miles high is given by $T(x) = 70 - 19x$. Determine the altitudes at which the air temperature is from $41.5°F$ to $22.5°F$. (*Source:* A. Miller and R. Anthes, *Meteorology*.)

120. *Distance* A car's distance in miles from a rest stop after x hours is given by $f(x) = 70x + 50$.
(a) Make a table for $f(x)$ for $x = 4, 5, 6, \ldots, 10$ and use the table to solve the inequality $470 \leq f(x) \leq 680$. Explain your result.
(b) Solve the inequality in part (a) symbolically.

121. *Medicare Costs* In 2000 Medicare cost taxpayers $250 billion, and in 2010 it cost $500 billion. (*Source: Department of Health and Human Services.*)

(a) Find a linear function M that models these data x years after 2000

(b) Estimate when Medicare costs were from $300 billion to $400 billion.

122. *Temperature Conversion* Water freezes at 32°F, or 0°C, and boils at 212°F, or 100°C.

(a) Find a linear function $C(F)$ that converts Fahrenheit temperature to Celsius temperature.

(b) The greatest temperature ranges on Earth are recorded in Siberia, where temperature has varied from about –70°C to 35°C. Find this temperature range in Fahrenheit.

WRITING ABOUT MATHEMATICS

123. Suppose that the solution set for a compound inequality can be written as $x < -3$ or $x > 2$. A student writes it as $2 < x < -3$. Is the student's three-part inequality correct? Explain your answer.

124. How can you determine whether an x-value is a solution to a compound inequality containing the word *and*? Give an example. Repeat the question for a compound inequality containing the word *or*.

SECTIONS 3.3 and 3.4

Checking Basic Concepts

1. Solve the linear inequality $4 - 3x < \frac{1}{2}x$.

2. Use the graph to solve each equation or inequality.

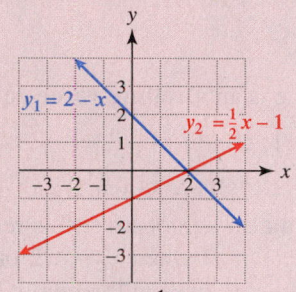

(a) $\frac{1}{2}x - 1 = 2 - x$ (b) $\frac{1}{2}x - 1 < 2 - x$

(c) $\frac{1}{2}x - 1 \geq 2 - x$

3. (a) Is 3 a solution to the compound inequality $x + 2 < 4$ or $2x - 1 \geq 3$?

(b) Is 3 a solution to the compound inequality $x + 2 < 4$ and $2x - 1 \geq 3$?

4. Solve each compound inequality. Write your answers in interval notation.

(a) $-5 \leq 2x + 1 \leq 3$

(b) $1 - x \leq -2$ or $1 - x > 2$

(c) $-2 < \dfrac{4 - 3x}{2} \leq 6$

3.5 Absolute Value Equations and Inequalities

Basic Concepts • Absolute Value Equations • Absolute Value Inequalities

A LOOK INTO MATH ▶

Monthly average temperatures can vary greatly from one month to another, whereas yearly average temperatures remain fairly constant from one year to the next. In Boston, Massachusetts, the yearly average temperature is 50°F, but monthly average temperatures can vary from 28°F to 72°F. Because 50°F − 28°F = 22°F and 72°F − 50°F = 22°F, the monthly average temperatures are always within 22°F of the yearly average temperature. If T represents a monthly average temperature, we can model this situation by using the *absolute value inequality*

$$|T - 50| \leq 22.$$

The absolute value is necessary because a monthly average temperature T can be either greater than or less than 50°F by as much as 22°F. In this section we discuss absolute value equations and inequalities. (*Source: A. Miller and J. Thompson, Elements of Meteorology.*)

NEW VOCABULARY

☐ Absolute value function
☐ Absolute value equation
☐ Absolute value inequality

Basic Concepts

In Chapter 1 we discussed the absolute value of a number. We can define the **absolute value function** by $f(x) = |x|$. To graph $y = |x|$, we begin by making a table of values, as shown in Table 3.7. Next we plot these points and then sketch the graph shown in Figure 3.29. Note that the graph is V-shaped.

TABLE 3.7

x	$|x|$
-2	2
-1	1
0	0
1	1
2	2

Absolute Value Function

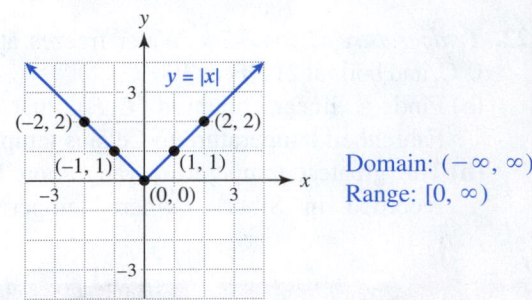

Domain: $(-\infty, \infty)$
Range: $[0, \infty)$

Figure 3.29

Because the input for $f(x) = |x|$ is any real number, the domain of f is all real numbers, or $(-\infty, \infty)$. The graph of the absolute value function shows that the output y (range) is any real number greater than or equal to 0. That is, the range is $[0, \infty)$.

Absolute Value Equations

An equation that contains an absolute value is an **absolute value equation**. Examples are

$$|x| = 2, \quad |2x - 1| = 5, \quad \text{and} \quad |5 - 3x| - 3 = 1.$$

Consider the absolute value equation $|x| = 2$. This equation has *two* solutions: -2 and 2 because $|-2| = 2$ and $|2| = 2$. We can also demonstrate this result with a table of values or a graph. Refer back to Table 3.7: $|x| = 2$ when $x = -2$ or $x = 2$. In Figure 3.30 the graph of $y_1 = |x|$ intersects the graph of $y_2 = 2$ at the points $(-2, 2)$ and $(2, 2)$. The x-values at these points of intersection correspond to the solutions -2 and 2.

Solving $|x| = 2$ Graphically

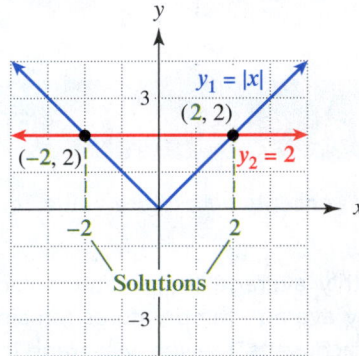

Figure 3.30

We generalize this discussion in the following manner.

SOLVING $|x| = k$

1. If $k > 0$, then $|x| = k$ is equivalent to $x = k$ or $x = -k$.

2. If $k = 0$, then $|x| = k$ is equivalent to $x = 0$.

3. If $k < 0$, then $|x| = k$ has no solutions.

EXAMPLE 1   Solving absolute value equations

Solve each equation.
(a) $|x| = 20$ **(b)** $|x| = -5$

Solution
(a) The solutions are -20 and 20.   **(b)** There are no solutions because $|x|$ is never negative.

Now Try Exercises 23, 25

EXAMPLE 2   Solving an absolute value equation

Solve $|2x - 5| = 3$ symbolically.

Solution
If $|2x - 5| = 3$, then either $2x - 5 = 3$ or $2x - 5 = -3$. Solve each equation separately.

$$2x - 5 = 3 \quad \text{or} \quad 2x - 5 = -3 \qquad \text{Equations to be solved}$$
$$2x = 8 \quad \text{or} \qquad 2x = 2 \qquad \text{Add 5.}$$
$$x = 4 \quad \text{or} \qquad x = 1 \qquad \text{Divide by 2.}$$

The solutions are **1** and **4**.

Now Try Exercise 31

A table of values can be used to solve the equation $|2x - 5| = 3$ from Example 2. Table 3.8 shows that $|2x - 5| = 3$ when $x = 1$ or $x = 4$.

Numerical Solution

TABLE 3.8 $|2x - 5| = 3$

x	0	1	2	3	4	5	6
$|2x - 5|$	5	3	1	1	3	5	7

Solutions are **1** and **4**.

Graphical Solution

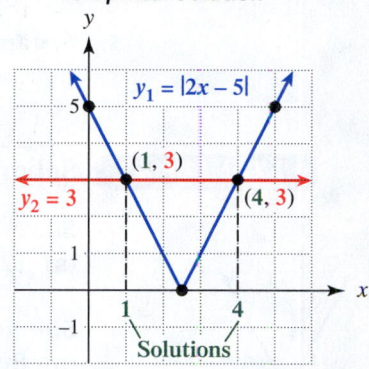

Figure 3.31

The equation from Example 2 can also be solved by graphing $y_1 = |2x - 5|$ and $y_2 = 3$. To graph y_1, first plot some of the points from Table 3.8. Its graph is V-shaped, as shown in Figure 3.31. Note that the x-coordinate of the "point" or vertex of the V can be found by solving the equation $2x - 5 = 0$ to obtain $\frac{5}{2}$. The graph of y_1 intersects the graph of y_2 at the points $(1, 3)$ and $(4, 3)$, giving the solutions **1** and **4**, so the graphical solutions agree with the numerical and symbolic solutions.

This discussion leads to the following result.

ABSOLUTE VALUE EQUATIONS

If $k > 0$, then

$$|ax + b| = k$$

is equivalent to

$$ax + b = k \quad \text{or} \quad ax + b = -k.$$

EXAMPLE 3   **Solving absolute value equations**

Solve.
(a) $|5 - x| - 2 = 8$ **(b)** $\left|\frac{1}{2}(x - 6)\right| = \frac{3}{4}$

Solution
(a) Start by adding 2 to each side to obtain $|5 - x| = 10$. This new equation is satisfied by the solution from either of the following equations.

$$5 - x = 10 \quad \text{or} \quad 5 - x = -10 \qquad \text{Equations to be solved}$$
$$-x = 5 \quad \text{or} \quad -x = -15 \qquad \text{Subtract 5.}$$
$$x = -5 \quad \text{or} \quad x = 15 \qquad \text{Multiply by } -1.$$

The solutions are -5 and 15.

(b) This equation is satisfied by the solution from either of the following equations.

$$\frac{1}{2}(x - 6) = \frac{3}{4} \quad \text{or} \quad \frac{1}{2}(x - 6) = -\frac{3}{4} \qquad \text{Equations to be solved}$$
$$4 \cdot \frac{1}{2}(x - 6) = 4 \cdot \frac{3}{4} \quad \text{or} \quad 4 \cdot \frac{1}{2}(x - 6) = 4\left(-\frac{3}{4}\right) \qquad \text{Multiply by 4 to clear fractions.}$$
$$2(x - 6) = 3 \quad \text{or} \quad 2(x - 6) = -3 \qquad \text{Simplify.}$$
$$2x - 12 = 3 \quad \text{or} \quad 2x - 12 = -3 \qquad \text{Distributive property}$$
$$2x = 15 \quad \text{or} \quad 2x = 9 \qquad \text{Add 12.}$$
$$x = \frac{15}{2} \quad \text{or} \quad x = \frac{9}{2} \qquad \text{Divide by 2.}$$

The solutions are $\frac{9}{2}$ and $\frac{15}{2}$.

Now Try Exercises 33, 37

EXAMPLE 4   **Solving absolute value equations with no solutions and one solution**

Solve.
(a) $|2x - 1| = -2$ **(b)** $|4 - 2x| = 0$

Solution
(a) Because an absolute value is never negative, there are no solutions. Figure 3.32 shows that the graph of $y_1 = |2x - 1|$ never intersects the graph of $y_2 = -2$.
(b) If $|y| = 0$, then $y = 0$. Thus $|4 - 2x| = 0$ when $4 - 2x = 0$ or when $x = 2$. The only solution is 2.

Now Try Exercises 35, 39

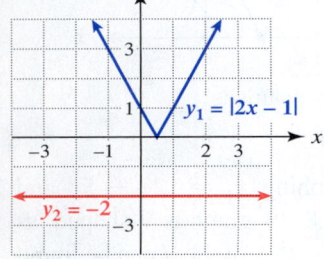

Figure 3.32

Sometimes an equation can have an absolute value on each side. An example would be $|2x| = |x - 3|$. In this situation either $2x = x - 3$ (the two expressions are equal) or $2x = -(x - 3)$ (the two expressions are opposites).

These concepts are summarized as follows.

SOLVING $|ax + b| = |cx + d|$

Let a, b, c, and d be constants. Then $|ax + b| = |cx + d|$ is equivalent to

$$ax + b = cx + d \quad \text{or} \quad ax + b = -(cx + d).$$

EXAMPLE 5   **Solving an absolute value equation**

Solve $|2x| = |x - 3|$.

Solution
Solve the following equalities.

$$2x = x - 3 \quad \text{or} \quad 2x = -(x - 3)$$
$$x = -3 \quad \text{or} \quad 2x = -x + 3$$
$$3x = 3$$
$$x = 1$$

The solutions are -3 and 1.

Now Try Exercise 43

Example 5 is solved graphically in Figure 3.33. The graphs of $y_1 = |2x|$ and $y_2 = |x - 3|$ are V-shaped and intersect at $(-3, 6)$ and $(1, 2)$. The solutions are -3 and 1.

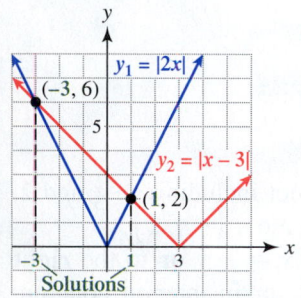

Figure 3.33

Absolute Value Inequalities

We can solve absolute value inequalities graphically. For example, to solve $|x| < 3$, let $y_1 = |x|$ and $y_2 = 3$ (see Figure 3.34). Their graphs intersect at $(-3, 3)$ and $(3, 3)$. The graph of y_1 is *below* the graph of y_2 for x-values *between*, but not including, $x = -3$ and $x = 3$. The solution set for $|x| < 3$ is $\{x \mid -3 < x < 3\}$ and is shaded on the x-axis.

Other absolute value inequalities can be solved graphically in a similar way. In Figure 3.35 the solutions to $|2x - 1| = 3$ are -1 and 2. The V-shaped graph of $y_1 = |2x - 1|$ is below the horizontal line $y_2 = 3$ when $-1 < x < 2$. Thus $|2x - 1| < 3$ whenever $-1 < x < 2$. The solution set is shaded on the x-axis.

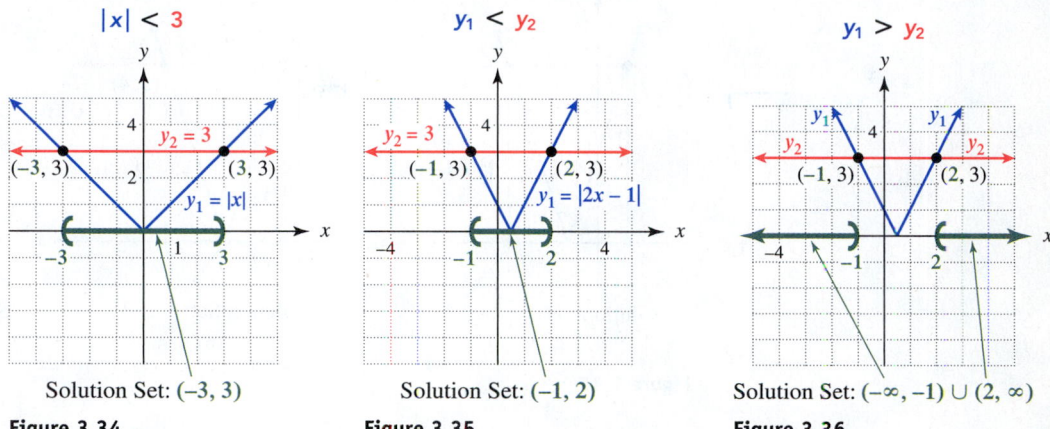

Solution Set: $(-3, 3)$ Solution Set: $(-1, 2)$ Solution Set: $(-\infty, -1) \cup (2, \infty)$

Figure 3.34   **Figure 3.35**   **Figure 3.36**

In Figure 3.36 the V-shaped graph of $y_1 = |2x - 1|$ is above the horizontal line $y_2 = 3$ both to the left of -1 and to the right of 2. That is, $|2x - 1| > 3$ whenever $x < -1$ or $x > 2$. The solution set is shaded on the x-axis.

ABSOLUTE VALUE INEQUALITIES

Let the solutions to $|ax + b| = k$ be c and d, where $c < d$ and $k > 0$.

1. $|ax + b| < k$ is equivalent to $c < x < d$.
2. $|ax + b| > k$ is equivalent to $x < c$ or $x > d$.

Similar statements can be made for inequalities involving $\leq$ or $\geq$.

EXAMPLE 6 **Solving absolute value equations and inequalities**

Solve each absolute value equation and inequality.

(a) $|2 - 3x| = 4$ **(b)** $|2 - 3x| < 4$ **(c)** $|2 - 3x| > 4$

Solution

(a) The given equation is equivalent to the following equations.

$$2 - 3x = 4 \quad \text{or} \quad 2 - 3x = -4 \qquad \text{Equations to be solved}$$

$$-3x = 2 \quad \text{or} \quad -3x = -6 \qquad \text{Subtract 2.}$$

$$x = -\frac{2}{3} \quad \text{or} \quad x = 2 \qquad \text{Divide by } -3.$$

The solutions are $-\frac{2}{3}$ and **2**.

STUDY TIP

Be sure you understand how to write the solution to Example 6(c).

(b) Solutions to $|2 - 3x| < 4$ include x-values **between**, but not including, $-\frac{2}{3}$ and **2**. Thus the solution set is $\left\{ x \mid -\frac{2}{3} < x < 2 \right\}$, or in interval notation, $\left(-\frac{2}{3}, 2 \right)$.

(c) Solutions to $|2 - 3x| > 4$ include x-values to the **left** of $x = -\frac{2}{3}$ **or** to the **right** of $x = 2$. Thus the solution set is $\left\{ x \mid x < -\frac{2}{3} \text{ or } x > 2 \right\}$, or in interval notation, $\left(-\infty, -\frac{2}{3} \right) \cup (2, \infty)$.

Now Try Exercise 51

VISUALIZING SOLUTIONS Figure 3.37(a) can be used to visualize the solution to $|2 - 3x| = 4$ in Example 6(a). Similarly, Figures 3.37(b) and 3.37(c) can be used to visualize the solutions to $|2 - 3x| < 4$ and $|2 - 3x| > 4$ in parts (b) and (c) of Example 6.

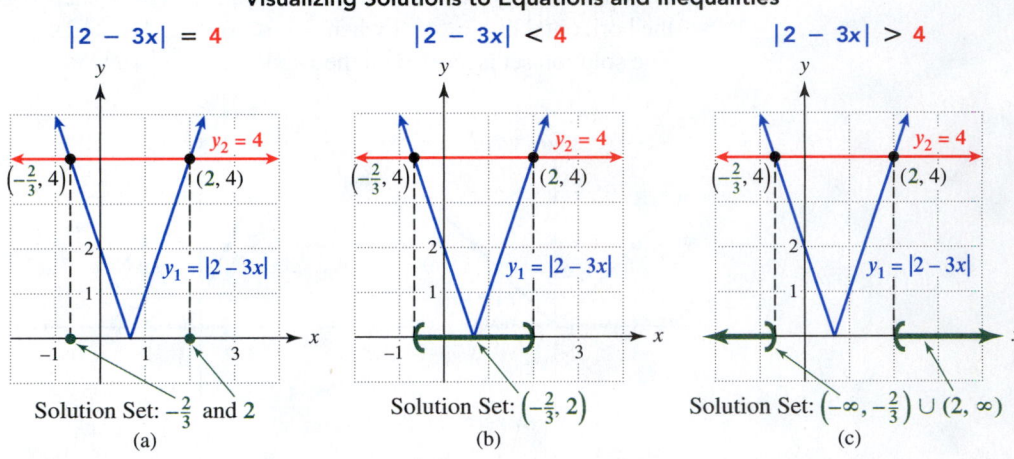

Visualizing Solutions to Equations and Inequalities

Figure 3.37

EXAMPLE 7 **Solving an absolute value inequality**

Solve $\left| \frac{2x - 5}{3} \right| > 3$. Write the solution set in interval notation.

Solution

Start by solving $\left| \frac{2x - 5}{3} \right| = 3$ as follows.

$$\frac{2x - 5}{3} = 3 \quad \text{or} \quad \frac{2x - 5}{3} = -3 \qquad \text{Equations to be solved}$$

$$2x - 5 = 9 \quad \text{or} \quad 2x - 5 = -9 \qquad \text{Multiply by 3.}$$

$$2x = 14 \quad \text{or} \quad 2x = -4 \qquad \text{Add 5.}$$

$$x = 7 \quad \text{or} \quad x = -2 \qquad \text{Divide by 2.}$$

Because the inequality symbol is $>$, the solution set is $x < -2$ or $x > 7$, or in interval notation, $(-\infty, -2) \cup (7, \infty)$.

Now Try Exercise 79

The results from Examples 6 and 7 can be generalized as follows.

CRITICAL THINKING

How many solutions are there to $|2x + 1| < 4$? How many solutions are there to $|2x + 1| < -4$? Explain.

INEQUALITIES AND ABSOLUTE VALUES

If $k > 0$ and $y = f(x)$, then

$$|y| < k \text{ is equivalent to } -k < y < k \text{ and}$$

$$|y| > k \text{ is equivalent to } y < -k \text{ or } y > k.$$

Similar statements can be made for inequalities involving $\leq$ and $\geq$.

In the next example, we use the fact that $-k \leq y \leq k$ is equivalent to $|y| \leq k$.

EXAMPLE 8 **Analyzing error**

An engineer is designing a circular cover for a container. The diameter d of the cover is to be 4.25 inches and must be accurate to within 0.01 inch. Write an absolute value inequality that gives acceptable values for d.

Solution

The diameter d must satisfy $4.24 \leq d \leq 4.26$. Subtracting 4.25 from each part gives

$$-0.01 \leq d - 4.25 \leq 0.01,$$

which is equivalent to $|d - 4.25| \leq 0.01$. The "distance" or difference between 4.25 and the diameter is less than or equal to 0.01.

CALCULATOR HELP

To graph an absolute value, see Appendix A (page AP-8).

Now Try Exercise 121

EXAMPLE 9 **Modeling temperature in Boston**

In A Look Into Math at the beginning of this section we discussed how the inequality $|T - 50| \leq 22$ models the range for the monthly average temperatures T in Boston.

(a) Solve this inequality and interpret the result.

(b) Give graphical support for part (a).

[0, 100, 10] by [0, 70, 10]

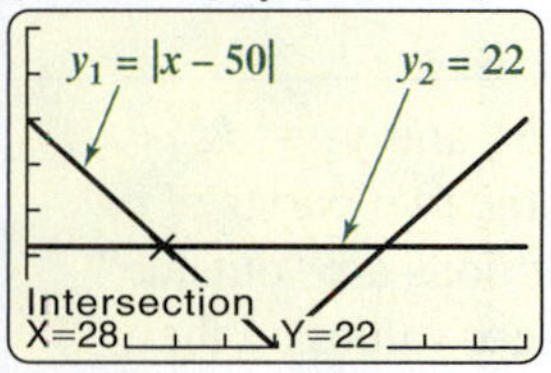

(a)

[0, 100, 10] by [0, 70, 10]

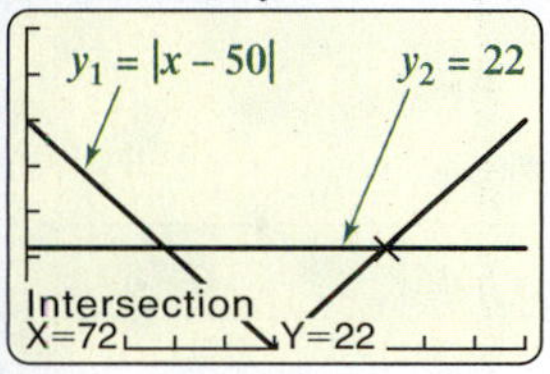

(b)

Figure 3.38

Solution

(a) ***Symbolic Solution*** Start by solving $|T - 50| = 22$.

$$T - 50 = 22 \quad \text{or} \quad T - 50 = -22 \qquad \text{Equations to be solved}$$

$$T = 72 \quad \text{or} \quad T = 28 \qquad \text{Add 50 to each side.}$$

Thus the solution set to $|T - 50| \leq 22$ is $\{T \mid 28 \leq T \leq 72\}$. Monthly average temperatures in Boston vary from 28°F to 72°F.

(b) ***Graphical Solution*** The graphs of $y_1 = |x - 50|$ and $y_2 = 22$ intersect at $(28, 22)$ and $(72, 22)$, as shown in Figures 3.38(a) and (b). The V-shaped graph of y_1 intersects the horizontal graph of y_2, or is below it, when $28 \leq x \leq 72$. Thus the solution set is $\{T \mid 28 \leq T \leq 72\}$. This result agrees with the symbolic result.

Now Try Exercise 113

Sometimes the solution set to an absolute value inequality can be either empty or the set of all real numbers. These two situations are illustrated in the next example.

EXAMPLE 10 **Solving absolute value inequalities**

Solve if possible.
(a) $|2x - 5| > -1$ **(b)** $|5x - 1| + 3 \leq 2$

Solution
(a) Because the absolute value of an expression cannot be negative, $|2x - 5|$ is greater than -1 for every x-value. The solution set is all real numbers, or $(-\infty, \infty)$.
(b) Subtracting 3 from each side results in $|5x - 1| \leq -1$. Because the absolute value is always greater than or equal to 0, no x-values satisfy this inequality. There are no solutions.

■ **Now Try Exercises 75, 77**

3.5 Putting It All Together

PROBLEM	SYMBOLIC SOLUTION	GRAPHICAL SOLUTION
$\|ax + b\| = k, k > 0$	Solve the equations $$ax + b = k$$ and $$ax + b = -k.$$	Graph $y_1 = \|ax + b\|$ and $y_2 = k$. Find the x-values of the two points of intersection.
$\|ax + b\| < k, k > 0$	If the solutions to $$\|ax + b\| = k$$ are c and d, $c < d$, then the solutions to $$\|ax + b\| < k$$ satisfy $$c < x < d.$$	Graph $y_1 = \|ax + b\|$ and $y_2 = k$. Find the x-values of the two points of intersection. The solutions are between these x-values on the number line, where the graph of y_1 lies below the graph of y_2.
$\|ax + b\| > k, k > 0$	If the solutions to $$\|ax + b\| = k$$ are c and d, $c < d$, then the solutions to $$\|ax + b\| > k$$ satisfy $$x < c \quad \text{or} \quad x > d.$$	Graph $y_1 = \|ax + b\|$ and $y_2 = k$. Find the x-values of the two points of intersection. The solutions are "outside" (left or right of) these x-values on the number line, where the graph of y_1 is above the graph of y_2.

3.5 Exercises

CONCEPTS AND VOCABULARY

1. Give an example of an absolute value equation.

2. Give an example of an absolute value inequality.

3. Is -3 a solution to $|x| = 3$?

4. Is -4 a solution to $|x| > 3$?

5. Is the solution set to $|x| = 5$ equal to $\{-5, 5\}$?

6. Is $|x| < 3$ equivalent to $x < -3$ or $x > 3$? Explain.

7. How many times does the graph of $y = |2x - 1|$ intersect the graph of $y = 5$?

8. How many times does the graph of $y = |2x - 1|$ intersect the graph of $y = -5$?

Exercises 9–14: Determine whether the given values of x are solutions to the absolute value equation or inequality.

9. $|2x - 5| = 1$ $x = -3, x = 3$

10. $|5 - 6x| = 1$ $x = 1, x = 0$

11. $|7 - 4x| \le 5$ $x = -2, x = 2$

12. $|2 + x| < 2$ $x = -4, x = -1$

13. $|7x + 4| > -1$ $x = -\frac{4}{7}, x = 2$

14. $|12x + 3| \ge 3$ $x = -\frac{1}{4}, x = 2$

Exercises 15 and 16: Use the graph to solve the equation.

15. $|x - 2| = 2$ 16. $|2x + 1| = 3$

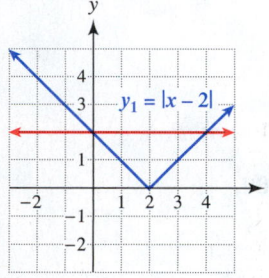 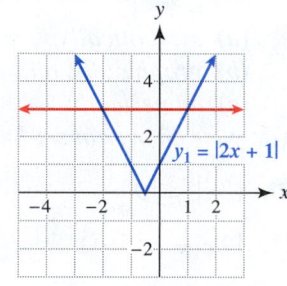

Exercises 17–22: Solve each equation or inequality.

17. $|x| = 3$ 18. $|x| = 5$

19. $|x| < 3$ 20. $|x| < 5$

21. $|x| > 3$ 22. $|x| > 5$

SYMBOLIC SOLUTIONS

Exercises 23–48: Solve the absolute value equation.

23. $|x| = 7$ 24. $|x| = 4$

25. $|x| = -6$ 26. $|x| = 0$

27. $|4x| = 9$ 28. $|-3x| = 7$

29. $|-2x| - 6 = 2$ 30. $|5x| + 1 = 5$

31. $|2x + 1| = 11$ 32. $|1 - 3x| = 4$

33. $|-2x + 3| + 3 = 4$

34. $|6x + 2| - 2 = 6$

35. $|3 - 4x| = 0$ 36. $|5x - 3| = -1$

37. $\left|\frac{1}{2}x - 1\right| = 5$ 38. $\left|6 - \frac{3}{4}x\right| = 3$

39. $|2x - 6| = -7$ 40. $\left|1 - \frac{2}{3}x\right| = 0$

41. $\left|\frac{2}{3}z - 1\right| - 3 = 8$ 42. $|1 - 2z| + 5 = 10$

43. $|z - 1| = |2z|$ 44. $|2z + 3| = |2 - z|$

45. $|3t + 1| = |2t - 4|$

46. $\left|\frac{1}{2}t - 1\right| = \left|3 - \frac{3}{2}t\right|$ 47. $\left|\frac{1}{4}x\right| = \left|3 + \frac{1}{4}x\right|$

48. $|2x - 1| = |2x + 2|$

Exercises 49–52: Solve each equation or inequality.

49. (a) $|2x| = 8$ (b) $|2x| < 8$
    (c) $|2x| > 8$

50. (a) $|3x - 9| = 6$
    (b) $|3x - 9| \le 6$
    (c) $|3x - 9| \ge 6$

51. (a) $|5 - 4x| = 3$
    (b) $|5 - 4x| \le 3$
    (c) $|5 - 4x| \ge 3$

52. (a) $\left|\frac{x - 5}{2}\right| = 2$ (b) $\left|\frac{x - 5}{2}\right| < 2$

    (c) $\left|\frac{x - 5}{2}\right| > 2$

Exercises 53–84: Solve the absolute value inequality. Write your answer in interval notation.

53. $|x| \le 3$ 54. $|x| < 2$

55. $|k| > 4$ 56. $|k| \ge 5$

57. $|t| \le -3$

58. $|t| < -1$

59. $|z| > 0$

60. $|2z| \ge 0$

61. $|2x| > 7$

62. $|-12x| < 30$

63. $|-4x + 4| < 16$

64. $|-5x - 8| > 2$

65. $2|x + 5| \ge 8$

66. $-3|x - 1| \ge -9$

67. $|8 - 6x| - 1 \le 2$

68. $4 - \left|\dfrac{2x}{3}\right| < -7$

69. $5 + \left|\dfrac{2 - x}{3}\right| \le 9$

70. $\left|\dfrac{x + 3}{5}\right| \le 12$

71. $|2x - 1| \le -3$

72. $|x + 6| \ge -5$

73. $|x + 1| - 1 > -3$

74. $-2|1 - 7x| \ge 2$

75. $|2z - 4| + 2 \le 1$

76. $|4 - z| \le 0$

77. $|3z - 1| > -3$

78. $|2z| \ge -2$

79. $\left|\dfrac{2 - t}{3}\right| \ge 5$

80. $\left|\dfrac{2t + 3}{5}\right| \ge 7$

81. $|t - 1| \le 0.1$

82. $|t - 2| \le 0.01$

83. $|b - 10| > 0.5$

84. $|b - 25| \ge 1$

NUMERICAL AND GRAPHICAL SOLUTIONS

Exercises 85 and 86: Use the table of $y = |ax + b|$ to solve each equation or inequality. Write your answers in interval notation for parts (b) and (c).

85. (a) $y = 2$ (b) $y < 2$ (c) $y > 2$

x	-2	-1	0	1	2	3	4
y	3	2	1	0	1	2	3

86. (a) $y = 6$ (b) $y \le 6$ (c) $y \ge 6$

x	-12	-6	0	6	12	18	24
y	9	6	3	0	3	6	9

Exercises 87 and 88: Use the graph of y_1 at the top of the next column to solve each equation or inequality. Write your answers in interval notation for parts (b) and (c).

87. (a) $|2x + 1| = 1$

   (b) $|2x + 1| \le 1$

   (c) $|2x + 1| \ge 1$

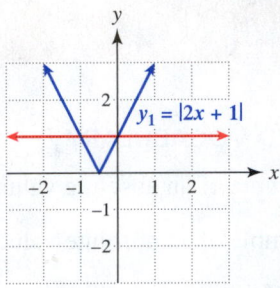

88. (a) $|x - 1| = 3$ (b) $|x - 1| < 3$

   (c) $|x - 1| > 3$

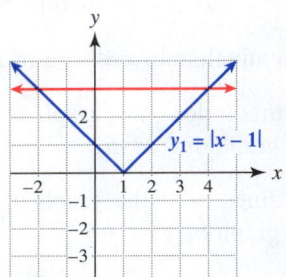

Exercises 89–98: Solve the inequality graphically. Write your answer in interval notation.

89. $|x| \ge 1$

90. $|x| < 2$

91. $|x - 1| \le 3$

92. $|x + 5| \ge 2$

93. $|4 - 2x| > 2$

94. $|1.5x - 3| \ge 6$

95. $|10 - 3x| < 4$

96. $|7 - 4x| \le 2.5$

97. $|8.1 - x| > -2$

98. $\left|\dfrac{5x - 9}{2}\right| \le -1$

USING MORE THAN ONE METHOD

Exercises 99–102: Solve the absolute value inequality

   (a) *symbolically,*
   (b) *graphically, and*
   (c) *numerically.*

Write your answer in set-builder notation.

99. $|3x| \le 9$

100. $|5 - x| \ge 3$

101. $|2x - 5| > 1$

102. $|-8 - 4x| < 6$

WRITING ABSOLUTE VALUE INEQUALITIES

Exercises 103–110: Write each compound inequality as an absolute value inequality. Do not simplify Exercises 107–110.

103. $-4 \le x \le 4$

104. $-0.1 < y < 0.1$

105. $y < -2$ or $y > 2$

106. $-0.1 \le x \le 0.1$

107. $-0.3 \le 2x + 1 \le 0.3$

108. $4x < -5$ or $4x > 5$ **109.** $\pi x \le -7$ or $\pi x \ge 7$

110. $-0.9 \le x - \sqrt{2} \le 0.9$

111. Thinking Generally If $a \ne 0$ and $k > 0$, then the graph of $y = |ax + b|$ intersects the graph of $y = k$ at _____ points.

112. Thinking Generally If a and k are positive, then the solution set to $|ax + b| < k$ is _____.

APPLICATIONS

Exercises 113–116: Average Temperatures (Refer to Example 9.) The given inequality models the range for the monthly average temperatures T in degrees Fahrenheit at the location specified.

(a) *Solve the inequality.*
(b) *Give a possible interpretation of the inequality.*

113. $|T - 43| \le 24$, Marquette, Michigan

114. $|T - 62| \le 19$, Memphis, Tennessee

115. $|T - 10| \le 36$, Chesterfield, Canada

116. $|T - 61.5| \le 12.5$, Buenos Aires, Argentina

117. Highest Elevations The table lists the highest elevation in each continent.

Continent	Elevation (feet)
Asia	29,028
S. America	22,834
N. America	20,320
Africa	19,340
Europe	18,510
Antarctica	16,066
Australia	7,310

Source: National Geographic.

(a) Calculate the average A of these elevations.
(b) Which continents have their highest elevations within 1000 feet of A?
(c) Which continents have their highest elevations within 5000 feet of A?
(d) Let E be an elevation. Write an absolute value inequality that says that E is within 5000 feet of A.

118. Distance Suppose that two cars, both traveling at a constant speed of 60 miles per hour, approach each other on a straight highway.

(a) If the two cars are initially 4 miles apart, sketch a graph of the distance between the two cars after x minutes, where $0 \le x \le 4$. (*Hint:* 60 miles per hour = 1 mile per minute.)
(b) Write an absolute value equation whose solution gives the times when the cars are 2 miles apart.
(c) Solve your equation from part (b).

119. Error in Measurements (Refer to Example 8.) The maximum error in the diameter of the can is restricted to 0.002 inch, so an acceptable diameter d must satisfy the absolute value inequality

$$|d - 2.5| \le 0.002.$$

Solve this inequality for d and interpret the result.

120. Error in Measurements Suppose that a person can operate a stopwatch accurately to within 0.02 second. If Byron Dyce's time in the 800-meter race is recorded as 105.30 seconds, write an absolute value inequality that gives the possible values for the actual time t.

121. Error in Measurements A circular lid is being designed for a container. The diameter d of the lid is to be 3.8 inches and must be accurate to within 0.03 inch. Write an absolute value inequality that gives acceptable values for d.

122. Manufacturing a Tire An engineer is designing a tire for a truck. The diameter d of the tire is to be 36 inches and the *circumference* must be accurate to within 0.1 inch. Write an absolute value inequality that gives acceptable values for d.

123. Relative Error If a quantity is measured to be x and the true value is t, then the relative error in the measurement is $\left|\frac{x - t}{t}\right|$. If the true measurement is $t = 20$ and you want the relative error to be less than 0.05 (5%), what values for x are possible?

124. Relative Error (Refer to the preceding exercise.) The volume V of a box is 50 cubic inches. How accurately must you measure the volume of the box for the relative error to be less than 3%?

WRITING ABOUT MATHEMATICS

125. If $a \neq 0$, how many solutions are there to the equation $|ax + b| = k$ when
(a) $k > 0$,   (b) $k = 0$, and   (c) $k < 0$?
Explain each answer.

126. Suppose that you know two solutions to the equation $|ax + b| = k$. How can you use these solutions to solve the inequalities $|ax + b| < k$ and $|ax + b| > k$? Give an example.

Checking Basic Concepts

Write answers in interval notation whenever possible.

1. Solve $\left|\frac{3}{4}x - 1\right| - 3 = 5$.

2. Solve the absolute value equation and inequalities.
   (a) $|3x - 6| = 8$

   (b) $|3x - 6| < 8$

   (c) $|3x - 6| > 8$

3. Solve the inequality $|-2(3 - x)| < 6$, and then solve $|-2(3 - x)| \geq 6$.

4. Use the graph to solve the equation and inequalities.

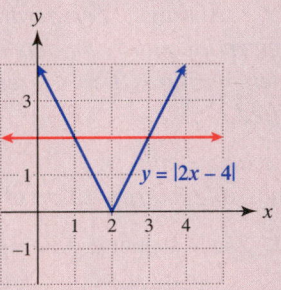

   (a) $|2x - 4| = 2$ **(b)** $|2x - 4| \leq 2$

   (c) $|2x - 4| \geq 2$

Summary

SECTION 3.1 ■ LINEAR EQUATIONS

Linear Equations in One Variable
Can be written as $ax + b = 0$, where $a \neq 0$. Linear equations have *one* solution.

Examples: $3x - 5 = 0$ and $x + 2 = 1 - 3x$

Symbolic, Graphical, and Numerical Solutions

Example: Solve $3x - 1 = 2$.

Symbolic Solution

$3x - 1 = 2$

$3x = 3$

$x = 1$

The solution is 1.

Numerical Solution

x	$3x - 1$
0	-1
1	2
2	5
3	8

$3x - 1 = 2$ when $x = 1$.

Graphical Solution

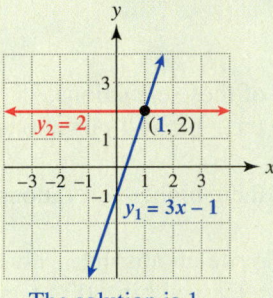

The solution is 1.

Lines

Standard Form	$ax + by = c$, where a, b, and c are constants with a and b not both 0.
Finding x-intercepts	Let $y = 0$ and solve for x.
Finding y-intercepts	Let $x = 0$ and solve for y.

Example: $3x - 4y = 24$

$3x - 4(0) = 24$ implies $x = 8$; the x-intercept is 8.

$3(0) - 4y = 24$ implies $y = -6$; the y-intercept is -6.

Identities and Contradictions

An *identity* is an equation that is always true, regardless of the values of any variables. A *contradiction* is an equation that is always false, regardless of the values of any variables.

Examples: $x + 3x = 4x$ (identity); $4x + 5 = 4x + 1$ (contradiction)

SECTION 3.2 ■ INTRODUCTION TO PROBLEM SOLVING

Solving a Formula for a Variable Use properties of algebra to solve for a variable.

Example: Solve $N = \frac{a + b}{2}$ for b.

$$2N = a + b \qquad \text{Multiply by 2.}$$
$$2N - a = b \qquad \text{Subtract } a.$$

Steps for Solving a Problem

STEP 1: Read the problem carefully and be sure that you understand it. (You may need to read the problem more than once.) Assign a variable to what you are being asked to find. If necessary, write other quantities in terms of this variable.

STEP 2: Write an equation that relates the quantities described in the problem. You may need to sketch a diagram, make a table, or refer to known formulas.

STEP 3: Solve the equation and determine the solution.

STEP 4: Look back and check your answer. Does it seem reasonable? Did you find the required information?

SECTION 3.3 ■ LINEAR INEQUALITIES

Linear Inequalities in One Variable Can be written as $ax + b > 0$, where $a \neq 0$; the symbol $>$ can be replaced with $<$, $\leq$, or $\geq$. Linear inequalities have *infinitely many* solutions.

Examples: $3x - 5 < 0$, $x + 2 \geq 1 - 3x$

Symbolic, Numerical, and Graphical Solutions

Example: Solve $4 - 2x \geq 0$.

Symbolic Solution

$$4 - 2x \geq 0$$
$$-2x \geq -4$$
$$x \leq 2$$

Reverse inequality.

Numerical Solution

x	$4 - 2x$
0	4
1	2
2	0
3	-2
4	-4

$4 - 2x \geq 0$ when $x \leq 2$.

Graphical Solution

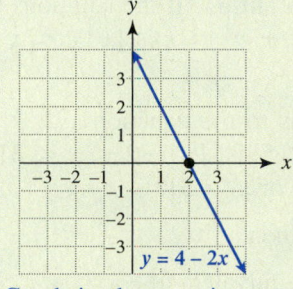

Graph is above or intersects the x-axis for $x \leq 2$.

SECTION 3.4 ■ COMPOUND INEQUALITIES

Compound Inequality Two inequalities connected by *and* or *or*.

Examples: For $x + 1 < 3$ *or* $x + 1 > 6$, a solution satisfies *at least* one of the inequalities.

For $2x + 1 < 3$ *and* $1 - x > 6$, a solution satisfies *both* inequalities.

Three-Part Inequality A compound inequality in the form $x > a$ *and* $x < b$ can be written as the three-part inequality $a < x < b$.

Example: $1 \leq x < 7$ means $x \geq 1$ *and* $x < 7$.

Interval Notation Can be used to identify intervals on the real number line

Examples: $-2 < x \leq 3$ is equivalent to $(-2, 3]$.

$x < 5$ is equivalent to $(-\infty, 5)$.

All real numbers are denoted $(-\infty, \infty)$.

SECTION 3.5 ■ ABSOLUTE VALUE EQUATIONS AND INEQUALITIES

Absolute Value Equations The graph of $y = |ax + b|$, $a \neq 0$, is V-shaped and intersects the horizontal line $y = k$ twice if $k > 0$. In this case there are two solutions to the equation $|ax + b| = k$ determined by $ax + b = k$ or $ax + b = -k$.

Example: The equation $|2x - 1| = 5$ has two solutions.

Symbolic Solution

$$2x - 1 = 5 \quad \text{or} \quad 2x - 1 = -5$$
$$2x = 6 \quad \text{or} \quad 2x = -4 \qquad \text{Add 1.}$$
$$x = \mathbf{3} \quad \text{or} \quad x = \mathbf{-2} \qquad \text{Divide by 2.}$$

Graphical Solution

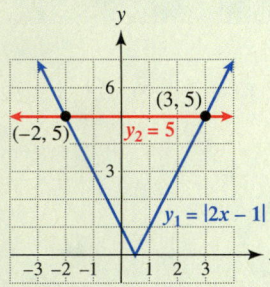

The solutions are -2 and 3.

Numerical Solution

x	-3	-2	-1	0	1	2	3
$|2x - 1|$	7	5	3	1	1	3	5

The solutions are -2 and 3.

Absolute Value Inequalities If the solutions to $|ax + b| = k$ are c and d with $c < d$, then the solution set for $|ax + b| < k$ is $\{x \mid c < x < d\}$, and the solution set for $|ax + b| > k$ is $\{x \mid x < c \text{ or } x > d\}$.

Example: The solutions to the equation $|2x - 1| = 5$ are -2 and 3, so the solution set for $|2x - 1| < 5$ is $\{x \mid -2 < x < 3\}$, and the solution set for $|2x - 1| > 5$ is $\{x \mid x < -2 \text{ or } x > 3\}$.

If $k > 0$ and $y = f(x)$, then

$$|y| < k \text{ is equivalent to } -k < y < k \text{ and}$$

$$|y| > k \text{ is equivalent to } y < -k \text{ or } y > k.$$

Examples: $|3 - x| < 5$ is equivalent to $-5 < 3 - x < 5$ and

$|3 - x| > 5$ is equivalent to $3 - x < -5$ or $3 - x > 5$.

CHAPTER 3   Review Exercises

SECTION 3.1

Exercises 1 and 2: Complete the table and then use the table to solve the equation.

1. $3x - 6 = 0$

x	0	1	2	3	4
$3x - 6$	-6				6

2. $5 - 2x = 3$

x	-1	0	1	2	3
$5 - 2x$					

3. Decide whether $-\frac{2}{3}$ is a solution to the linear equation $\frac{1}{2}z - 2(2z + 3) = 4z - 1$.

4. Use the graph to solve the equation $y_1 = y_2$.

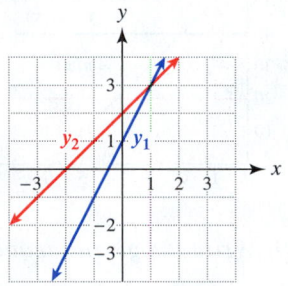

5. Solve $5 - 2x = -1 + x$ graphically.

6. Solve $4x - 3 = 5 + 2x$ graphically or numerically.

Exercises 7–16: Solve the equation.

7. $2x - 7 = 21$

8. $1 - 7x = -\frac{5}{2}$

9. $-2(4x - 1) = 1 - x$

10. $-\frac{3}{4}(x - 1) + 5 = 6$

11. $\frac{x - 4}{3} = 2$

12. $\frac{2x - 3}{2} = \frac{x + 3}{5}$

13. $-2(z - 1960) + 32 = 8$

14. $5 - (3 - 2z) + 4 = 5(2z - 3) - (3z - 2)$

15. $\frac{2}{3}\left(\frac{t - 3}{2}\right) + 4 = \frac{1}{3}t - (1 - t)$

16. $0.3r - 0.12(r - 1) = 0.4r + 2.1$

Exercises 17 and 18: Determine whether the equation is an identity or a contradiction.

17. $-4(5 - 3x) = 12x - 20$

18. $4x - (x - 3) = 3x - 3$

Exercises 19 and 20: Find the x- and y-intercepts. Then write the equation in slope–intercept form.

19. $4x - 5y = 20$

20. $2x + \frac{1}{2}y = -6$

SECTION 3.2

Exercises 21–26: Solve the equation for the given variable.

21. $5x - 4y = 20$ for y **22.** $-\frac{1}{3}x + \frac{1}{2}y = 1$ for y

23. $2a + 3b = a$ for a

24. $4m - 5n = 6m + 2n$ for n

25. $A = \frac{1}{2}h(a + b)$ for b **26.** $V = \frac{1}{3}\pi r^2 h$ for h

Exercises 27–30: Solve the equation for y. Let $y = f(x)$ and write a formula for $f(x)$.

27. $\frac{1}{2}x - \frac{3}{4}y = 2$ **28.** $-7(x - 4y) = y + 1$

29. $3x + 4y = 10 - y$ **30.** $\frac{y}{x} = 3$

Exercises 31 and 32: Translate the sentence into an equation and then solve the equation.

31. The sum of twice x and 25 is 19.

32. If 5 is subtracted from twice x, it equals x plus 1.

SECTION 3.3

33. Use the table to solve $f(x) < 5$, where $f(x)$ represents a linear function.

x	-2	-1	0	1	2
$f(x)$	7	5	3	1	-1

34. Solve $2(3 - x) + 4 < 0$.

35. Solve the inequality $y_1 \geq y_2$, using the graph.

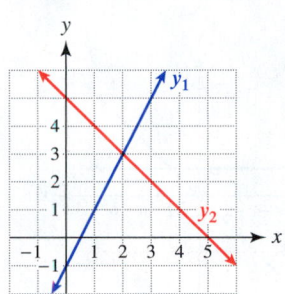

36. Solve each equation or inequality.
   (a) $5 - 4x = -2$ (b) $5 - 4x < -2$
   (c) $5 - 4x > -2$

Exercises 37–44: Solve the inequality.

37. $-2x + 1 \leq 3$

38. $x - 5 \geq 2x + 3$

39. $\dfrac{3x - 1}{4} > \dfrac{1}{2}$

40. $-3.2(x - 2) < 1.6x$

41. $\dfrac{2}{5}t - (5 - t) + 3 > 2\left(\dfrac{3 - t}{5}\right)$

42. $\dfrac{2(3x - 5)}{5} - 3 \geq \dfrac{2 - x}{3} + 2$

43. $0.05t - 0.15 \leq 0.03 - 0.75(t - 2)$

44. $0.6z - 1.55(z + 5) < 1 - 0.35(2z - 1)$

Exercises 45 and 46: Solve the inequality graphically. Approximate boundary numbers to the nearest thousandth.

45. $0.72(\pi - 1.3x) \geq 0.54$

46. $\sqrt{2} - (4 - \pi x) > \sqrt{7}$

SECTION 3.4

Use interval notation whenever possible for the remaining exercises.

Exercises 47–50: Solve the compound inequality. Graph the solution set on a number line.

47. $x + 1 \leq 3$ and $x + 1 \geq -1$

48. $2x + 7 < 5$ and $-2x \geq 6$

49. $5x - 1 \leq 3$ or $1 - x < -1$

50. $3x + 1 > -1$ or $3x + 1 < 10$

51. Use the table to solve $-2 \leq 2x + 2 \leq 4$.

x	-3	-2	-1	0	1	2	3
$2x + 2$	-4	-2	0	2	4	6	8

52. Use the following figure to solve each equation and inequality.
(a) $y_1 = y_2$ (b) $y_2 = y_3$
(c) $y_1 \leq y_2 \leq y_3$ (d) $y_2 < y_3$

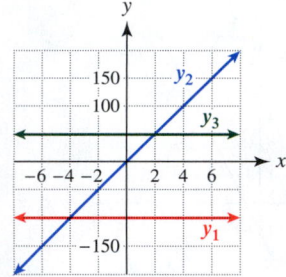

53. The graphs of y_1 and y_2 are shown at the top of the next column. Solve each equation and inequality.

(a) $y_1 = y_2$ (b) $y_1 < y_2$
(c) $y_1 > y_2$

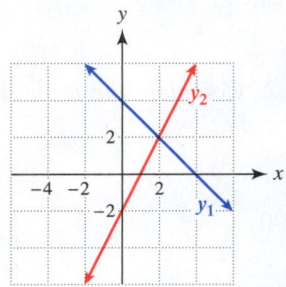

54. The graphs of three linear functions f, g, and h are shown in the following figure. Solve each equation and inequality.
(a) $f(x) = g(x)$ (b) $g(x) = h(x)$
(c) $f(x) < g(x) < h(x)$

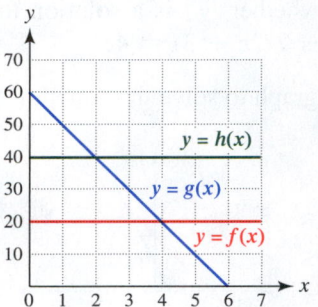

Exercises 55–60: Write the given inequality in interval notation.

55. $-3 \leq x \leq \frac{2}{3}$

56. $-6 < x \leq 45$

57. $x < \frac{7}{2}$

58. $x \geq 1.8$

59. $x > -3$ and $x < 4$

60. $x < 4$ or $x > 10$

Exercises 61–66: Solve the three-part inequality. Write the solution set in interval notation.

61. $-4 < x + 1 < 6$

62. $20 \leq 2x + 4 \leq 60$

63. $-3 < 4 - \frac{1}{3}x < 7$

64. $2 \leq \frac{1}{2}x - 2 \leq 12$

65. $-3 \leq \dfrac{4 - 5x}{3} - 2 < 3$

66. $30 \leq \dfrac{2x - 6}{5} - 4 < 50$

SECTION 3.5

Exercises 67–70: Determine whether the given values of x are solutions to the absolute value equation or inequality.

67. $|12x - 24| = 24$ $x = -3; x = 2$

68. $|5 - 3x| > 3$ $x = \frac{4}{3}; x = 0$

69. $|3x - 6| \leq 6$ $x = -3; x = 4$

70. $|2 + 3x| + 4 < 11$ $x = -3; x = \frac{2}{3}$

71. Use the accompanying table to solve each equation or inequality.
(a) $y_1 = 2$ (b) $y_1 < 2$ (c) $y_1 > 2$

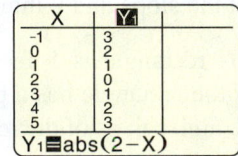

72. Use the graph of $y = |2x + 2|$ to solve each equation or inequality.
(a) $|2x + 2| = 4$ (b) $|2x + 2| \le 4$
(c) $|2x + 2| \ge 4$

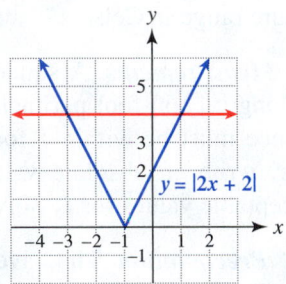

Exercises 73–78: Solve the absolute value equation.

73. $|x| = 22$ **74.** $|2x - 9| = 7$

75. $\left|4 - \frac{1}{2}x\right| = 17$ **76.** $\frac{1}{3}|3x - 1| + 1 = 9$

77. $|2x - 5| = |5 - 3x|$

78. $|-3 + 3x| = |-2x + 6|$

Exercises 79 and 80: Solve each absolute value equation or inequality.

79. (a) $|x + 1| = 7$ (b) $|x + 1| \le 7$

    (c) $|x + 1| \ge 7$

80. (a) $|1 - 2x| = 6$ (b) $|1 - 2x| \le 6$

    (c) $|1 - 2x| \ge 6$

Exercises 81–88: Solve the absolute value inequality.

81. $|x| > 3$ **82.** $|-5x| < 20$

83. $|4x - 2| \le 14$ **84.** $\left|1 - \frac{4}{5}x\right| \ge 3$

85. $|t - 4.5| \le 0.1$ **86.** $-2|13t - 5| \ge -4$

87. $|5 - 4x| > -5$ **88.** $|2t - 3| \le 0$

Exercises 89 and 90: Solve the inequality graphically.

89. $|2x| \ge 3$ **90.** $\left|\frac{1}{2}x - 1\right| \le 2$

Exercises 91 and 92: Write each compound inequality as an absolute value inequality.

91. $-0.05 \le x \le 0.05$

92. $5x - 1 < -4$ or $5x - 1 > 4$

APPLICATIONS

93. *Loan Interest* A student takes out two loans, one at 5% and the other at 7% annual interest. The total amount for both loans is $7700, and the total interest after one year is $469. Find the amount of each loan.

94. *Distance and Time* At first an athlete runs at 8 miles per hour and then runs at 10 miles per hour, traveling 12.8 miles in 1.4 hours. How long did the athlete run at each speed?

95. *Expensive Homes* A typical 2200-square-foot house in Honolulu costs $415,000, which is only 54% of the cost of the same house in San Francisco. (San Francisco has the most expensive housing in the United States.) How much would this Honolulu house cost in San Francisco? (*Source:* Runzheimer International.)

96. *Violent Crimes in the U.S.* The number of violent crimes reported has dropped from 10 million in 1992 to 5 million in 2008.
(a) Find a linear function $f(x) = mx + b$ that models the data x years after 1992.
(b) Use $f(x)$ to estimate the number of violent crimes in 2005.

97. *Distance Between Bicyclists* The following graph shows the distance between two bicyclists traveling toward each other along a straight road after x hours.
(a) After how long did the bicycle riders meet?
(b) When were they 20 miles apart?
(c) Find the times when they were less than 20 miles apart.
(d) Estimate the sum of the speeds of the bicyclists.

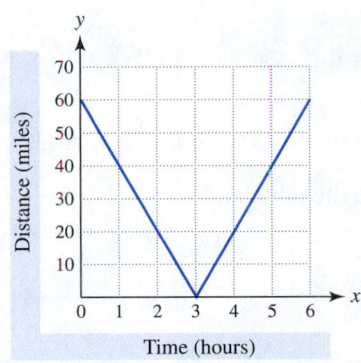

98. *Distance* Two cars, Car 1 and Car 2, are traveling in the same direction on a straight highway. Their distances in miles north of Austin, Texas, after x hours are shown in the following graph.

(a) Which car is traveling faster? Explain.

(b) How many hours elapse before the two cars are the same distance from Austin? How far are they from Austin?

(c) During what time interval is Car 1 closer to Austin than Car 2?

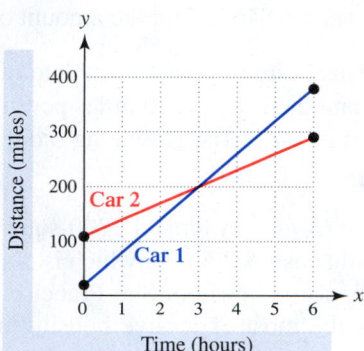

99. *Weight of a Fish* If a walleye has a length of x inches, where $30 \leq x \leq 35$, its weight W in pounds can be *estimated* by $W(x) = 1.11x - 23.3$. (*Source: Minnesota Department of Natural Resources.*)

(a) What length of walleye weighs 12 pounds?

(b) What lengths of walleye weigh less than 12 pounds?

100. *Air Temperature* Suppose that the air temperature at ground level is $60°$F and that the air cools $19°$F for each one-mile increase in altitude.

(a) Write $T(x) = ax + b$ so that T gives the temperature at an altitude of x miles.

(b) Estimate the altitudes at which the air temperature is from $40°$F to $20°$F.

101. *Age in the United States* The median age from 1970 to 2010 can be approximated by $f(x) = 0.238x + 27.9$, where x is the number of years after 1970.

(a) Estimate the year(s) when this number was 30 years.

(b) Interpret the slope and y-intercept.

102. *Geometry* A rectangle is 5 feet longer than twice its width. If the rectangle has a perimeter of 88 feet, what are the dimensions of the rectangle?

103. *Temperature* The formula $F = \frac{9}{5}C + 32$ may be used to convert Fahrenheit temperature to Celsius temperature. The temperature range at Houghton Lake, Michigan, has varied between $-48°$F and $107°$F. Solve a three-part inequality to find this temperature range in Celsius.

104. *Error in Measurements* A square garden is being fenced along its 160-foot perimeter. If the length L of the fence must be within 1 foot of the garden's perimeter, write an absolute value inequality that gives acceptable values for L. Solve your inequality.

105. *Average Precipitation* The average rainfall in Houston, Texas, is 3.9 inches per month. Each month's average A is within 1.7 inches of 3.9 inches. (*Source:* J. Williams, *The Weather Almanac 1995.*)

(a) Write an absolute value inequality that models this situation.

(b) Solve the inequality.

106. *Relative Error* If a quantity is measured to be T and the actual value is A, then the relative error in this measurement is $\left|\frac{T-A}{A}\right|$. If $A = 35$ and the relative error is to be less than 0.08 (8%), what values for T are possible?

CHAPTER 3 Test

CHAPTER Test Prep VIDEOS

Step-by-step test solutions are found on the Chapter Test Prep Videos available in *MyMathLab* and on **You Tube** (search "RockswoldInterAlg" and click on "Channels").

1. Is $\frac{1}{6}$ a solution to $3 - 4(1 - 2x) = 2x$?

2. Complete the table and then solve $2 - 3x \leq -1$.

x	-2	-1	0	1	2
$2 - 3x$					

3. Solve $3 - 5x = 18$. Check your answer.

4. Use the graph to the right to solve each equation and inequality. Write your answers for parts (b) and (c) in interval notation.

(a) $y_1 = y_2$ (b) $y_1 \geq y_2$
(c) $y_1 \leq y_2$

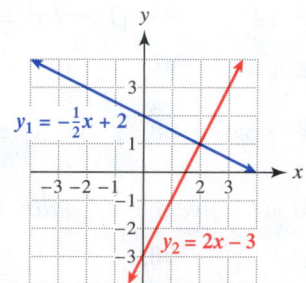

5. Solve $4 - 2x = 1 + x$ graphically.

6. Solve each equation.
(a) $-\frac{2}{3}(3x - 2) + 1 = x$

(b) $\dfrac{2z + 1}{3} = \dfrac{3(1 - z)}{2}$

(c) $0.4(2 - 3x) = 0.5x - 0.2$

7. Find the x- and y-intercepts for the graph of the equation $-6x + 3y = 9$. Then write the equation in slope–intercept form.

8. Translate the sentence into an equation and then solve. "If 2 is added to 5 times x, it equals x minus 4."

9. Solve $3x - 2y = 6$ for y. Let $y = f(x)$ and write a formula for $f(x)$.

Exercises 10 and 11: Solve the inequality. Write your answer in interval notation.

10. $-5x + 1 \le 4$ **11.** $3.1(3 - x) < 2.9x$

12. Graph the solution set to $2x + 6 < 2$ and $-3x \ge 3$ on a number line.

13. Use the table to solve the compound inequality $-3x < -3$ or $-3x > 6$. Write your answer in interval notation.

x	-3	-2	-1	0	1	2	3
$-3x$	9	6	3	0	-3	-6	-9

14. Use the following figure to solve the equations and inequalities. Write your answers for parts (c) and (d) in interval notation.
(a) $y_1 = y_2$ (b) $y_2 = y_3$
(c) $y_1 \le y_2 \le y_3$ (d) $y_2 < y_3$

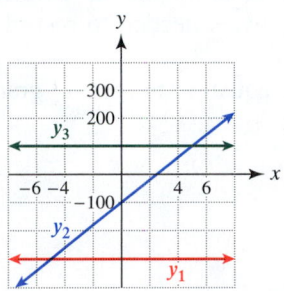

15. Solve $-2 < 2 + \frac{1}{2}x < 2$ and write the solution set in interval notation.

16. Solve the equation $\left| 2 - \frac{1}{3}x \right| = 6$.

17. Solve each inequality. Write your answer in interval notation.
(a) $|x| \le 5$
(b) $|x| > 0$

(c) $|2x + 7| \le -2$
(d) $|5x| > 10$
(e) $-2|2 - 5x| + 1 \le 5$
(f) $|3 - x| < 5$

18. Solve each equation or inequality. Write your answers for parts (b) and (c) in interval notation.
(a) $|1 - 5x| = 3$ (b) $|1 - 5x| \le 3$
(c) $|1 - 5x| \ge 3$

19. *Sport Drinks* The following graph shows a relationship between the calories in an 8-ounce serving of a sport drink and the corresponding grams of carbohydrates. Estimate the number of calories x in 8 ounces of a sport drink with the following grams of carbohydrates. (*Source: Runner's World.*)
(a) 16 grams of carbohydrates
(b) More than 16 grams of carbohydrates
(c) Less than 16 grams of carbohydrates
(d) Interpret the slope of the graph.

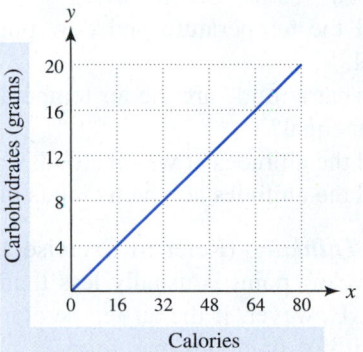

20. *Drinking Fluids and Exercise* To determine the number of ounces of fluid that a person should drink in a day, divide his or her weight in pounds by 2 and then add 0.4 ounce for every minute of exercise.
(a) Write a function that gives the fluid requirements for a person weighing 150 pounds and exercising x minutes a day.
(b) If a 150-pound runner needs 89 ounces of fluid each day, determine the runner's daily minutes of exercise.

21. *Solving a Formula* Solve the formula $d = \frac{1}{2}gt^2$ for g.

22. *Loan Interest* A student takes out two loans, one at 4% and the other at 6% annual interest. The total amount for both loans is $5000, and the total interest after one year is $230. Find the amount of each loan.

23. *Angles in a Triangle* The measure of the largest angle in a triangle is twice the measure of the smallest angle. The third angle is 16° more than the smallest angle. Find the measure of each angle.

24. *Prices* The sale price of a jacket is $81. If the retail price had been reduced by 25%, find the retail price of the jacket.

25. *Women Marines* The number of women in the Marine Corps from 1980 to 2010 can be approximated by $f(x) = 190x + 6000$, where x is the number of years after 1980.
   (a) Estimate the years when this number was from 7900 to 9800.
   (b) Interpret the slope and y-intercept.

CHAPTER 3   Extended and Discovery Exercises

1. *Clouds and Temperature* If the air temperature is greater than the dew point, clouds do not form. If the air temperature cools to the dew point, fog or clouds may appear. Suppose that the Fahrenheit temperature x miles high is given by $T(x) = 90 - 19x$ and that the dew point x miles high is given by $D(x) = 70 - 5.8x$. (*Source:* A. Miller.)
   (a) Find the temperature and dew point at ground level.
   (b) At what altitude are the air temperature and dew point equal?
   (c) Find the altitudes at which clouds do not form.
   (d) Find the altitudes at which clouds may form.

2. *Critical Thinking* (Refer to Exercise 1.) At ground level the dew point is usually less than the air temperature. However, if the air temperature reaches the dew point, fog or clouds may form.
   (a) Discuss why clouds are likely to form at some altitude in the sky.
   (b) Suppose that the dew point is slightly less than the air temperature near the ground. How does that affect the altitude at which clouds form?

3. *Recording Music* A compact disc (CD) can hold approximately 700 million bytes. One *byte* is capable of storing one letter of the alphabet. For example, the word "function" requires 8 bytes to store in computer memory. One million bytes is commonly referred to as a *megabyte* (MB). Recording music requires an enormous amount of memory. The accompanying table lists the megabytes x needed to record y seconds (sec) of music.

x (MB)	0.129	0.231	0.415	0.491
y (sec)	6.010	10.74	19.27	22.83

x (MB)	0.667	1.030	1.160	1.260
y (sec)	31.00	49.00	55.25	60.18

Source: Gateway 2000 System CD.

   (a) Make a scatterplot of the data.
   (b) What relationship seems to exist between x and y? Why does this relationship seem reasonable?
   (c) Find the slope–intercept form of a line that models the data. Interpret the slope of this line as a rate of change. Answers may vary.
   (d) Check your answer in part (c) by graphing the line and data in the same graph.
   (e) Write a linear equation whose solution gives the megabytes needed to record 120 seconds of music.
   (f) Solve the equation in part (e) graphically or symbolically.

CHAPTERS 1–3   Cumulative Review Exercises

Exercises 1 and 2: Classify each real number as one or more of the following: natural number, whole number, integer, rational number, or irrational number.

1. $-7, -\frac{3}{5}, 0, \sqrt{5}, \frac{8}{2}, 5.\overline{12}$

2. $-\frac{6}{3}, \frac{0}{9}, \sqrt{9}, \pi, 4.\overline{6}$

Exercises 3 and 4: State whether the equation illustrates an identity, commutative, associative, or distributive property.

3. $8x - 2x = 6x$

4. $(6 + z) + 3 = 6 + (z + 3)$

Exercises 5 and 6: Use properties of real numbers to evaluate the expression mentally.

5. $11 + 26 + (-1) + 14$

6. $7 \cdot 99$

7. Plot the numbers -4, $-\frac{5}{2}$, 0, and 3 on a number line.

8. Find the opposite of $-5y - 4$.

Exercises 9 and 10: Simplify the expression. Write the result using positive exponents.

9. $\dfrac{24x^{-4}y^2}{8xy^{-5}}$ **10.** $\left(\dfrac{3a^2}{4b^3}\right)^{-3}$

11. Evaluate the expression $15 - 2^3 \div 4$.

12. Write the number 0.000059 in scientific notation.

Exercises 13 and 14: Evaluate the formula for the given value of the variable.

13. $J = \sqrt{38 - t}$ $t = 13$

14. $r = 3 - z^3$ $z = -3$

15. Select the formula that best models the data in the table.

x	-2	-1	0	1	2
y	7	5	3	1	-1

   (i) $y = 2x + 11$ (ii) $y = -2x + 3$
   (iii) $y = x - 5$

16. Identify the domain and range of the given relation $S = \{(-2, 4), (0, 2), (1, 4), (3, 0)\}$.

Exercises 17 and 18: Evaluate the given formula for $x = -2, -1, 0, 1,$ and 2. Plot these ordered pairs.

17. $y = -2x + 1$ **18.** $y = \dfrac{x^3 + 4}{2}$

19. Sketch the graph of $f(x) = x^2 - 3$ by hand.

20. Find the domain of $f(x) = \dfrac{4}{x}$.

Exercises 21 and 22: Use the graph or table to evaluate $f(0)$ and $f(-2)$.

21.

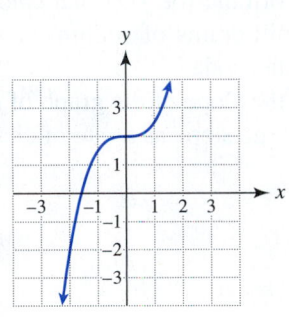

22.

x	-2	-1	0	1	2
$f(x)$	-4	-1	2	5	8

Exercises 23 and 24: Determine whether f is a linear function. If f is linear, give values for m and b so that f may be expressed as $f(x) = mx + b$.

23. $f(x) = 11 - 3x$

24. $f(x) = \sqrt{x} + 2$

25. Use the table to write the formula for $f(x) = mx + b$.

x	-2	-1	0	1	2
$f(x)$	-5	-1	3	7	11

26. Sketch a graph of $f(x) = 3$.

27. Find the slope and the y-intercept of the graph of $f(x) = 5x - 4$.

28. Calculate the slope of the line passing through the points $(3, -2)$ and $(-1, 6)$.

29. Sketch a line passing through the point $(1, 3)$ with slope $m = 2$.

30. Use the graph to express the equation of the line in slope–intercept form.

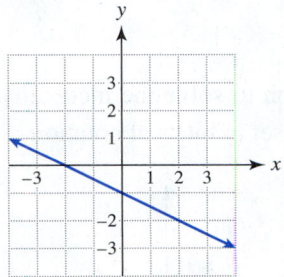

31. Let f be a linear function. Find the slope, x-intercept, and y-intercept of the graph of f.

x	-2	-1	0	1	2
$f(x)$	-3	0	3	6	9

32. Write the equation of the vertical line that passes through the point $(5, 7)$.

Exercises 33 and 34: Write the slope–intercept equation for a line satisfying the given conditions.

33. Parallel to $y = 3x - 4$, passing through $(2, 1)$

34. Perpendicular to $y = -x - 5$, passing through the point $(-3, 0)$

35. Determine whether $\frac{8}{5}$ is a solution to the equation $\frac{1}{2}x - 4(x - 1) = \frac{1}{4}x - 2$.

36. Use the graph to solve the equation $y_1 = y_2$.

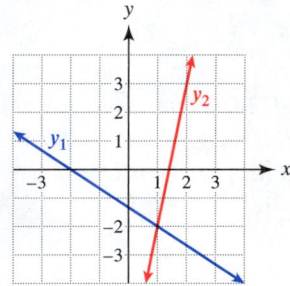

Exercises 37 and 38: Solve the equation.

37. $\frac{3}{4}(x - 2) + 4 = 2$

38. $\frac{2}{3}\left(\frac{t - 7}{4}\right) - 2 = \frac{1}{3}t - (2t + 3)$

Exercises 39 and 40: Find the x- and y-intercepts. Write the equation in slope–intercept form.

39. $x - \frac{1}{3}y = 2$

40. $5x - 4y = -10$

41. Use the table to solve $y < 2$, where y represents a linear function. Write the solution set in set-builder notation.

x	-2	-1	0	1	2
y	12	7	2	-3	-8

42. Use the graph to solve the inequality $y_1 \le y_2$. Write the solution set in interval notation.

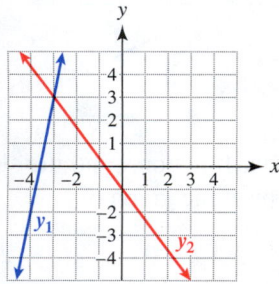

Exercises 43 and 44: Solve the inequality symbolically. Write the solution set in interval notation.

43. $\dfrac{4x - 9}{6} > \dfrac{1}{2}$

44. $\frac{2}{3}z - 2 \le \frac{1}{4}z - (2z + 2)$

Exercises 45 and 46: Solve the compound inequality. Graph the solution set on a number line.

45. $x + 2 > 1$ and $2x - 1 \le 9$

46. $4x + 7 < 1$ or $3x + 2 \ge 11$

Exercises 47 and 48: Solve the three-part inequality. Write the solution set in interval notation.

47. $-7 \le 2x - 3 \le 5$

48. $-8 \le -\frac{1}{2}x - 3 \le 5$

Exercises 49 and 50: Solve the absolute value inequality. Write the solution set in interval notation.

49. $|3x + 5| > 13$

50. $-3|2t - 11| \ge -9$

51. Use the graph to solve the equation and inequalities.
(a) $y_1 = 2$ (b) $y_1 \le 2$
(c) $y_1 \ge 2$

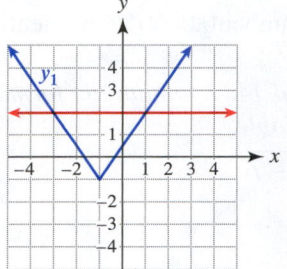

52. Solve the absolute value equation $\left|\frac{2}{3}x - 4\right| = 8$.

APPLICATIONS

53. *Calculating Interest* If P dollars are deposited in a savings account paying 5% annual interest, then the amount A in the account after t years is given by the formula $A = P(1.05)^t$. Find A for the given values of P and t. Round your answer to the nearest cent.
(a) $P = \$4300$, $t = 10$ years
(b) $P = \$11,000$, $t = 6$ years

54. *Modeling Motion* The table lists the distance d in miles traveled by a car for various elapsed times t in hours. Find an equation that models these data.

t (hours)	3	5	7	9
d (miles)	195	325	455	585

55. *Sodium Content* Some types of diet soda contain 120 milligrams of sodium per 12-ounce can.
(a) Give a formula for $f(x)$ that calculates the number of milligrams of sodium in x ounces of this type of diet soda.
(b) What is the slope of the graph of f?
(c) Interpret the slope as a rate of change.

56. *Graphical Model* A 150-liter aquarium is initially empty. A small hose attached to a faucet begins to fill the aquarium at a constant rate of 5 liters per minute. After 10 minutes, the faucet is turned off for 15 minutes and the small hose is replaced by a larger hose that can fill the aquarium at a rate of 10 liters per minute. The faucet is turned back on until the aquarium is full. Sketch a graph that depicts the amount of water in the aquarium after x minutes.

57. *Distance* A motorcyclist and a bicyclist are traveling toward each other on a straight road. Their distances in miles south of Euclid, Ohio, after x hours are shown in the graph to the right.

(a) Is the bicyclist traveling toward or away from Euclid? Explain.

(b) How many hours elapse before they are the same distance from Euclid?

(c) During what time period is the motorcyclist farther away from Euclid than the bicyclist?

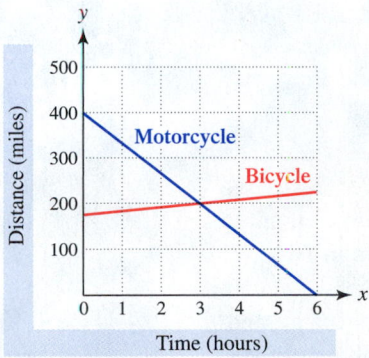

58. *Women Officers* The number of women officers in the Marine Corps from 1960 to 2010 may be modeled by $f(x) = 17(x - 1960) + 56$, where x represents the year. Estimate the years in which the number of women officers was between 600 and 800. (*Source:* Department of Defense.)

4 Systems of Linear Equations

The essence of mathematics is not to make simple things complicated, but to make complicated things simple.

—STANLEY GUDDER

In 1940, a physicist named John Atanasoff at Iowa State University needed to solve 29 equations with 29 variables simultaneously. This task was too difficult to do by hand, so he and a graduate student invented the first fully electronic digital computer. Thus the desire to solve a mathematical problem led to one of the most important inventions of the twentieth century. Today people can solve thousands of equations with thousands of variables. Solutions to such equations have resulted in better airplanes, cars, electronic devices, weather forecasts, and medical equipment.

Equations are even used in biology. The following table contains the weight W, neck size N, and chest size C for three black bears. Suppose that park rangers find a bear with a neck size of 22 inches and a chest size of 38 inches. Can they use the data in the table to estimate the bear's weight? Using systems of linear equations, they can estimate the bear's weight. See Example 8 and Exercise 61 in Section 4.6.

Black Bear Measurements

W (pounds)	N (inches)	C (inches)
80	16	26
344	28	45
416	31	54
?	22	38

Sources: A. Tucker, *Fundamentals of Computing*; M. Triola, *Elementary Statistics*; Minitab, Inc.

4.1 Systems of Linear Equations in Two Variables

Basic Concepts • Graphical and Numerical Solutions • Types of Linear Systems • Applications

A LOOK INTO MATH ▶

With a land-line telephone, information is typically sent along a single wire, or path. However, if a person is talking on a cell phone in a car, any *one* path could fail at any time because the location of the person's cell phone is always changing. As a result, information must be sent simultaneously along several paths so that this information can be reconstructed even when one or more paths fail during transmission. Mathematics, and *systems of equations* in particular, play an important role in determining how information is sent. See Example 10 and the discussion preceding it. (*Source:* Effros, M., and A. Goldsmith, *Scientific American.*)

NEW VOCABULARY

☐ System of equations
☐ Solution
☐ Standard form
☐ Consistent system
☐ Inconsistent system
☐ Independent equations
☐ Dependent equations

Basic Concepts

▶ **REAL-WORLD CONNECTION** Each year, more people are buying music albums through the Internet. A combined total of 120 million albums were sold online in 2007 and 2008, with a 20 million album increase from 2007 to 2008. (*Source:* Jupiter Research.)

To determine the number bought each year, we can let x be the number bought in 2008 and let y be the number bought in 2007, where both amounts are in millions of albums. Then the given information is described by the following *system of equations.*

$$x + y = 120 \qquad \text{The total is 120 million albums.}$$
$$x - y = 20 \qquad \text{The difference is 20 million albums.}$$

Each equation contains two variables, x and y. These two equations form a **system of two linear equations in two variables**. An ordered pair (x, y) is a **solution** to a system of equations if the values for x and y satisfy *both* equations. (We *solve* the above system of equations in Example 4.) Any system of two linear equations in two variables can be written in **standard form** as

$$ax + by = c$$
$$dx + ey = k,$$

where a, b, c, d, e, and k are constants.

EXAMPLE 1 **Testing for solutions**

Determine which ordered pair is a solution to the system of equations: $(0, 3)$ or $(-1, 2)$.

$$-x + 4y = 9$$
$$3x - 3y = -9$$

Solution
For $(\mathbf{0}, \mathbf{3})$ to be a solution, the values of $x = \mathbf{0}$ and $y = \mathbf{3}$ must satisfy *both* equations.

$$-\mathbf{0} + 4(\mathbf{3}) \stackrel{?}{=} 9 \quad ✗ \quad \text{False}$$
$$3(\mathbf{0}) - 3(\mathbf{3}) \stackrel{?}{=} -9 \quad ✓ \quad \text{True}$$

READING CHECK

How can you determine whether a point is a solution to a system of equations?

Because $(0, 3)$ does not satisfy *both* equations, $(0, 3)$ is *not* a solution. To test $(\mathbf{-1}, \mathbf{2})$, substitute $x = \mathbf{-1}$ and $y = \mathbf{2}$ in each equation.

$$-(\mathbf{-1}) + 4(\mathbf{2}) \stackrel{?}{=} 9 \quad ✓ \quad \text{True}$$
$$3(\mathbf{-1}) - 3(\mathbf{2}) \stackrel{?}{=} -9 \quad ✓ \quad \text{True}$$

Both equations are true, so $(-1, 2)$ *is* a solution.

Now Try Exercise 9

Graphical and Numerical Solutions

Graphical, numerical, and symbolic techniques can be used to solve systems of equations. In this section we focus on graphical and numerical techniques and delay discussion of symbolic techniques until the next section.

In the next example we solve a system of linear equations graphically. Note that sometimes it may be difficult to read a graph precisely, so it is important to check our solutions by using the technique described in Example 1.

EXAMPLE 2 **Solving a system of equations graphically**

Solve the following system of equations graphically. Check your answer.

$$y = x - 2$$
$$y = 4 - x$$

Solution

Both equations are in slope–intercept form, so we can graph

$$y = x - 2 \text{ and } y = 4 - x$$

immediately, as shown in Figure 4.1. Their graphs intersect at the point (3, 1). To check that (**3, 1**) is the solution, substitute $x = 3$ and $y = 1$ into each equation.

$$1 \overset{?}{=} 3 - 2 \checkmark \quad \text{True}$$
$$1 \overset{?}{=} 4 - 3 \checkmark \quad \text{True}$$

The answer checks.

Now Try Exercise 13

Solving a System

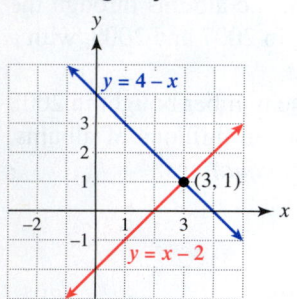

Figure 4.1

NOTE: In Figure 4.1 the blue line represents all points (x, y) that satisfy the equation $y = 4 - x$, and the red line represents all points (x, y) that satisfy the equation $y = x - 2$. We are looking for a point that lies on *both* lines because the solution must satisfy *both* equations. The only point on *both* lines is (**3, 1**).

In the next example we demonstrate two methods for graphing linear equations that are written in the standard form: $ax + by = c$. The first method uses the slope–intercept form to graph each line, and the second method uses the x- and y-intercepts to graph each line. The *same* solution is found with either method.

EXAMPLE 3 **Solving a system of equations graphically**

Solve the system of equations graphically.

$$x + 2y = 4$$
$$2x - y = 3$$

Solution

Method I: Finding Slope–Intercept Form Solve each equation for y and then use the slope–intercept form to graph the line.

$x + 2y = 4$	First equation	$2x - y = 3$	Second equation
$2y = -x + 4$	Subtract x.	$-y = -2x + 3$	Subtract $2x$.
$y = -\frac{1}{2}x + 2$	Multiply by $\frac{1}{2}$.	$y = 2x - 3$	Multiply by -1.

Now graph $y = -\frac{1}{2}x + 2$ and $y = 2x - 3$. See Figure 4.2. Because the graphs intersect at (2, 1), the solution is (2, 1).

Method II: Finding x- and y-Intercepts To find the *x*-intercepts, let $y = 0$ in each equation.

$x + 2y = 4$	First equation		$2x - y = 3$	Second equation
$x + 2(0) = 4$	Let $y = 0$.		$2x - 0 = 3$	Let $y = 0$.
$x = 4$	Solve for x.		$x = \frac{3}{2}$	Solve for x.

The *x*-intercept is **4**. The *x*-intercept is $\frac{3}{2}$.

To find the *y*-intercepts, let $x = 0$ in each equation.

$x + 2y = 4$	First equation		$2x - y = 3$	Second equation
$0 + 2y = 4$	Let $x = 0$.		$2(0) - y = 3$	Let $x = 0$.
$y = 2$	Solve for y.		$y = -3$	Solve for y.

The *y*-intercept is **2**. The *y*-intercept is **−3**.

Graph $x + 2y = 4$ by drawing a line that passes through (**4**, 0) and (0, **2**). The graph of $2x - y = 3$ passes through $\left(\frac{3}{2}, 0\right)$ and (0, **−3**). See Figure 4.3. Their graphs intersect at (2, 1).

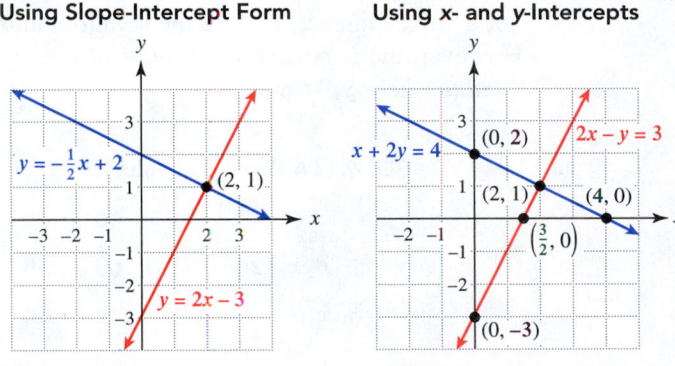

Using Slope-Intercept Form   **Using x- and y-Intercepts**

Figure 4.2   **Figure 4.3**

▮ **Now Try Exercises 17, 47**

In the next example we solve the system of equations presented earlier, which models sales of digital music albums.

EXAMPLE 4 ### Solving a system of equations graphically and numerically

Solve the following system of equations graphically and numerically. Interpret the solution.

$$x + y = 120$$
$$x - y = 20$$

Solution

Start by solving each equation for *y*.

$x + y = 120$	First equation	$x - y = 20$	Second equation
$y = -x + 120$	Subtract x.	$-y = -x + 20$	Subtract x.
		$y = x - 20$	Multiply by -1.

Graphical Solution The graphs of $y_1 = -x + 120$ and $y_2 = x - 20$ are shown in Figure 4.4 on the next page. Because *x* and *y* represent sales, which are never negative, we graph these lines only in the first quadrant. Their graphs intersect at the point (**70**, **50**). Thus **70** million albums were bought in 2008, and **50** million albums were bought in 2007. Check this solution.

Albums Sold over Internet

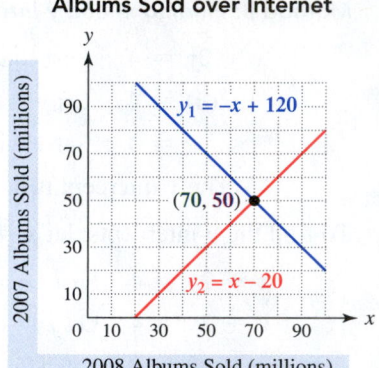

Figure 4.4 A Graphical Solution

Numerical Solution Make a table of values for $y_1 = -x + 120$ and $y_2 = x - 20$. Table 4.1 shows that, when $x = 70$, both of the expressions equal **50**. Thus the solution is (**70**, **50**). This type of numerical solution is based on trial and error. If the solution is not a "nice" integer, finding the solution numerically may be difficult or even *impossible*. However, the *important mathematical concept* to remember is that *you are looking for an x-value where* $y_1 = y_2$.

TABLE 4.1 A Numerical Solution

	x	50	60	**70**	80
$y_1 = -x + 120$		70	60	**50**	40
$y_2 = x - 20$		30	40	**50**	60

X	Y₁	Y₂
40	80	20
50	70	30
60	60	40
70	50	50
80	40	60
90	30	70
100	20	80
X=70		

Figure 4.5

In Figure 4.5 a calculator was used to create a table similar to Table 4.1.

Now Try Exercise 81

EXAMPLE 5   **Solving a system of equations graphically**

Solve the following system of equations graphically. Check your answer.

$$2x - 3y = 6$$
$$4x + y = 5$$

Solution
Solve the first equation for y.

$2x - 3y = 6$	First equation
$-3y = -2x + 6$	Subtract $2x$ from each side.
$\dfrac{-3y}{-3} = \dfrac{-2x}{-3} + \dfrac{6}{-3}$	Divide each term by -3.
$y = \dfrac{2}{3}x - 2$	Simplify.

Next, solve the second equation for y.

$4x + y = 5$	Second equation
$y = -4x + 5$	Subtract $4x$ from each side.

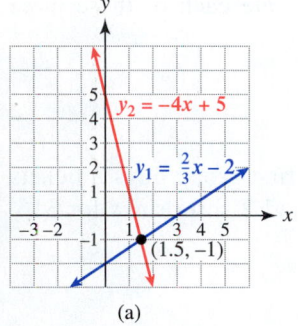

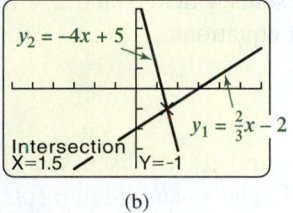

Figure 4.6

READING CHECK

Explain what a consistent system is and what an inconsistent system is.

Graph $y_1 = \frac{2}{3}x - 2$ and $y_2 = -4x + 5$, as shown in Figure 4.6(a). (A similar calculator graph is shown in Figure 4.6(b).) Their graphs appear to intersect at $(1.5, -1)$. To be *certain*, check this answer in the *given* equations.

$$2x - 3y = 6 \qquad \text{First equation}$$
$$2(\mathbf{1.5}) - 3(\mathbf{-1}) \overset{?}{=} 6 \qquad \text{Let } x = 1.5 \text{ and } y = -1.$$
$$6 = 6 \checkmark \qquad \text{The answer checks.}$$

Now substitute these values in the second equation. (It is essential to check *both* equations.)

$$4x + y = 5 \qquad \text{Second equation}$$
$$4(\mathbf{1.5}) + (\mathbf{-1}) \overset{?}{=} 5 \qquad \text{Let } x = 1.5 \text{ and } y = -1.$$
$$5 = 5 \checkmark \qquad \text{The answer checks.}$$

■ **Now Try Exercise 19**

Types of Linear Systems

A system of linear equations that has at least one solution is a **consistent system**; otherwise, it is an **inconsistent system**. A system of linear equations in two variables can be represented graphically by two lines in the xy-plane. Three different situations involving two lines are illustrated in Figure 4.7. In Figure 4.7(a) the lines intersect at a single point, which represents a *unique solution*. In this case the equations of the lines are called **independent equations**. In Figure 4.7(b) the two lines are identical, which occurs when the two equations are *equivalent*. For example, the equations $x + y = 1$ and $2x + 2y = 2$ are equivalent. If we divide each side of the second equation by 2 we obtain the first equation. As a result, their graphs are identical and every point on the line represents a solution to the system of linear equations. Thus there are *infinitely many solutions*, and the equations are called **dependent equations**. Finally, in Figure 4.7(c) the lines are parallel and do not intersect. There are *no solutions*, so the system is inconsistent.

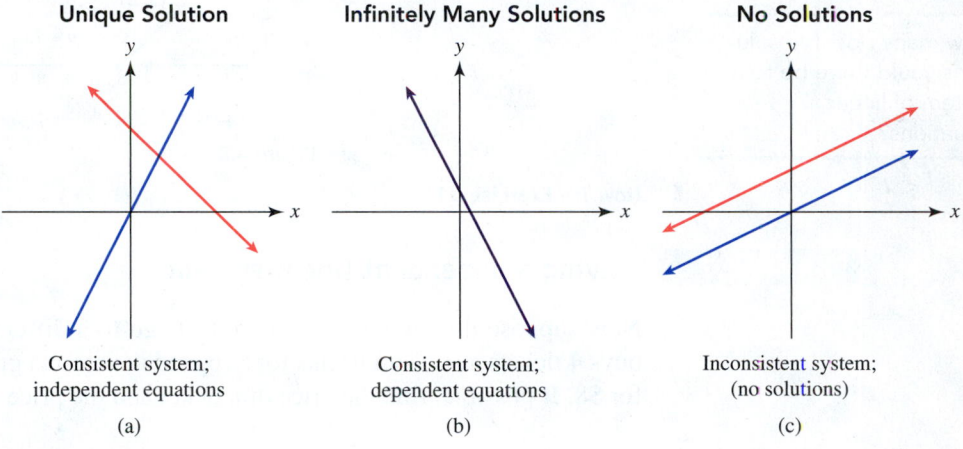

Figure 4.7

MAKING CONNECTIONS

Solutions and Points of Intersection

In Figure 4.7(a) there is only one point that lies on *both* lines. Thus there is only one solution, and it corresponds to this point. In Figure 4.7(b) the lines are identical. Any point that lies on the first line also lies on the second line. Thus all such points lie on *both* lines and each corresponds to a solution. There are infinitely many solutions. In Figure 4.7(c) the lines are parallel and do not intersect. There are no points that lie on *both* lines. Thus there are no solutions.

▶ **REAL-WORLD CONNECTION**   In the next three examples we illustrate each of these three types of equations with real-world applications.

EXAMPLE 6

Solving a linear system with a unique solution

Suppose that two groups of students go to a football game. The first group buys 3 tickets and 3 soft drinks for $15, and the second group buys 4 tickets and 2 soft drinks for $16. Find the price of a ticket and the price of a soft drink graphically.

Solution

Let x be the cost of a ticket and y be the cost of a soft drink. Then $3x + 3y = 15$ represents 3 tickets and 3 soft drinks costing $15, and $4x + 2y = 16$ represents 4 tickets and 2 soft drinks costing $16. This information can be written as a system of equations.

$$3x + 3y = 15 \quad \text{First group}$$
$$4x + 2y = 16 \quad \text{Second group}$$

To graph each equation, we will use the x- and y-intercepts (Method II), as discussed in Example 3. The x- and y-intercepts for $3x + 3y = 15$ are both 5. The x- and y-intercepts for $4x + 2y = 16$ are 4 and 8, respectively. Plotting the points $(5, 0)$, $(0, 5)$, $(4, 0)$, and $(0, 8)$ and sketching the corresponding lines in the first quadrant results in Figure 4.8. These graphs intersect at $(3, 2)$. Thus the cost of a ticket is $3 and the cost of a soft drink is $2. Note that 3 tickets and 3 soft drinks cost $3(3) + 3(2) = \$15$ and that 4 tickets and 2 soft drinks cost $4(3) + 2(2) = \$16$, so our solution is correct.

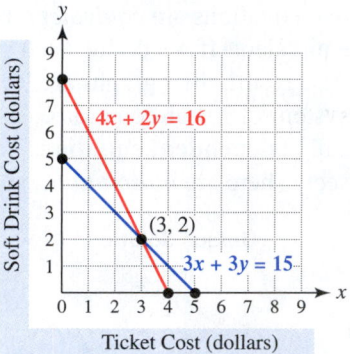

Figure 4.8

READING CHECK

How many possible solutions could there be to a system of linear equations?

Now Try Exercise 77

EXAMPLE 7

Solving a dependent linear system

Now suppose that two groups of students go to a different football game. The first group buys 4 tickets and 2 soft drinks for $16, and the second group buys 2 tickets and 1 soft drink for $8. If possible, find the price of a ticket and the price of a soft drink graphically.

Solution

Let x be the cost of a ticket and y be the cost of a soft drink. Then this situation can be modeled by the following system of equations.

$$4x + 2y = 16$$
$$2x + y = 8$$

Solve each equation for y.

$4x + 2y = 16$	First equation	$2x + y = 8$	Second equation
$2y = -4x + 16$	Subtract 4x.	$y = -2x + 8$	Subtract 2x.
$y = -2x + 8$	Divide by 2.		

Both equations simplify to the *same* slope–intercept form, $y = -2x + 8$, and thus the lines are identical, as shown in Figure 4.9.

CRITICAL THINKING

Suppose that a system of two linear equations with two variables is dependent. If you try to solve this system numerically by using a table, how could you recognize that the equations are indeed dependent? Explain your answer.

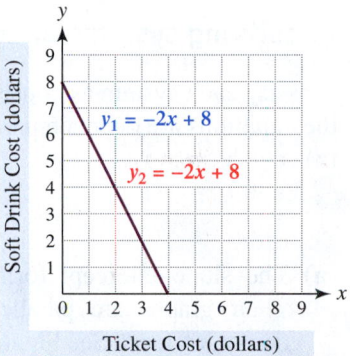

Figure 4.9

The system is consistent because there is at least one solution. The equations are dependent because not enough information is available to determine a unique solution. Note that the second group bought half what the first group bought and paid half as much. If 4 tickets and 2 soft drinks cost \$16, we would expect that 2 tickets and 1 soft drink cost \$8. These two purchases contain the same information about prices. As a result, the two equations contain the *same information* and are *equivalent*. Thus there are infinitely many solutions because every point on the line is a solution. For example, a ticket could cost \$3 and a soft drink could cost \$2, or a ticket could cost \$2 and a soft drink could cost \$4. The solution set can be expressed in set-builder notation as $\{(x, y) \mid 2x + y = 8\}$.

Now Try Exercise 29

EXAMPLE 8   **Recognizing an inconsistent linear system**

Now suppose that two groups of students go to a concert. The first group buys 4 tickets and 2 soft drinks for \$20, and the second group buys 2 tickets and 1 soft drink for \$12. If possible, find the price of a ticket and the price of a soft drink graphically.

Solution
Let x be the cost of a ticket and y be the cost of a soft drink. Then the system

$$4x + 2y = 20$$
$$2x + y = 12$$

models the data. Solving for y, we can write each equation in slope–intercept form.

$$y = -2x + 10$$
$$y = -2x + 12$$

Their graphs are parallel lines with slope -2 and different y-intercepts. Thus they do not intersect (see Figure 4.10). The linear system is *inconsistent* because there are no solutions. Note that the second group purchased half of what the first group purchased. If pricing had been *consistent*, we would expect the second group to have paid half, or \$10, instead of \$12. *Inconsistent pricing* resulted in an *inconsistent linear system*.

CRITICAL THINKING

What would a table of values look like if you were solving an inconsistent system of equations?

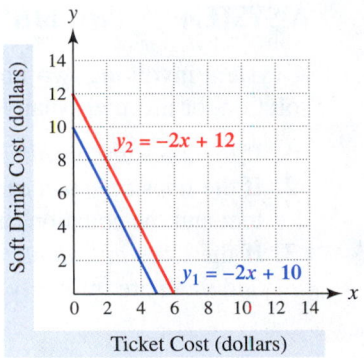

Figure 4.10

Now Try Exercise 33

EXAMPLE 9 **Classifying systems of equations**

Classify each system as consistent or inconsistent. If the system is consistent, state whether the equations are dependent or independent.

(a) $x + y = 1$
$x + y = -1$

(b) $x + 2y = 4$
$2x + 4y = 8$

(c) $x + y = 4$
$x - y = 2$

Solution

(a) The slope–intercept forms for these equations are $y = -x + 1$ and $y = -x - 1$. Their graphs are parallel lines with slope -1 and different y-intercepts, so they do not intersect. See Figure 4.11(a). There are *no solutions* and the system is inconsistent.

NOTE: The sum of two numbers, x and y, cannot equal both 1 and -1 at the same time, so it is reasonable that there are no solutions.

(b) Because the equations both have slope–intercept form $y = -\frac{1}{2}x + 2$, their graphs are identical. The system is consistent and the equations are dependent. See Figure 4.11(b). There are *infinitely many solutions* of the form $\{(x, y) \mid x + 2y = 4\}$.

NOTE: The second equation is exactly double the first equation. When one equation is a *nonzero* multiple of the other, the equations are dependent.

(c) The first equation has slope–intercept form $y = -x + 4$, and the second equation has slope–intercept form $y = x - 2$. These lines have different slopes and intersect at one point: $(3, 1)$, so the solution is *unique*. The system is consistent and the equations are independent. See Figure 4.11(c).

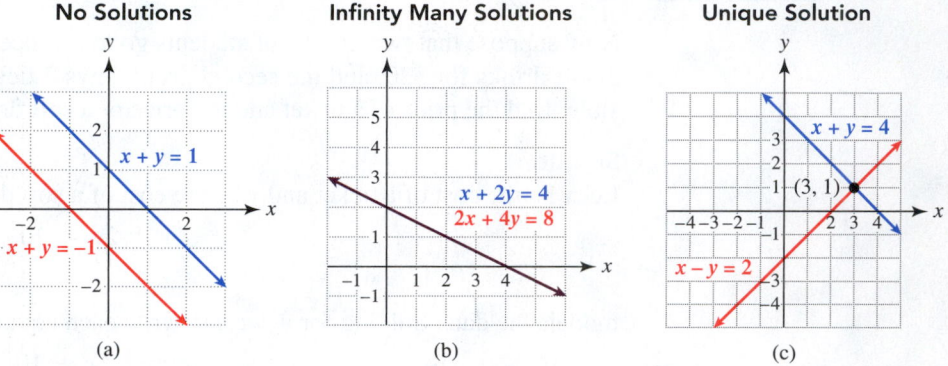

Figure 4.11

Now Try Exercises 25, 37, 39

A SYSTEM OF TWO LINEAR EQUATIONS IN TWO VARIABLES

A system involving two linear equations in two variables can have no solutions, one solution, or infinitely many solutions. Its graph consists of two lines.

1. If the lines *are parallel*, the system is inconsistent and there are no solutions.
2. If the lines *intersect at a single point*, there is one solution. The system is consistent and the equations are independent.
3. If the lines *are identical*, the system is consistent. The equations are dependent and there are infinitely many solutions.

Applications

▶ **REAL-WORLD CONNECTION** In A Look into Math at the beginning of this section, we discussed how systems of equations are used to transmit information reliably over wireless networks. To better understand how this is done, consider the following simple example.

Suppose we want to send the number 35 from one cell phone to another. Begin by letting $x = 3$ and $y = 5$. Instead of sending both 3 and 5 over one path, which could fail, we will use four paths. Over the first path we send $x = 3$, over the second $y = 5$, over the third $x + y = 8$, and over the fourth $2x + y = 11$. See Figure 4.12.

Systems of Equations in Wireless Networks

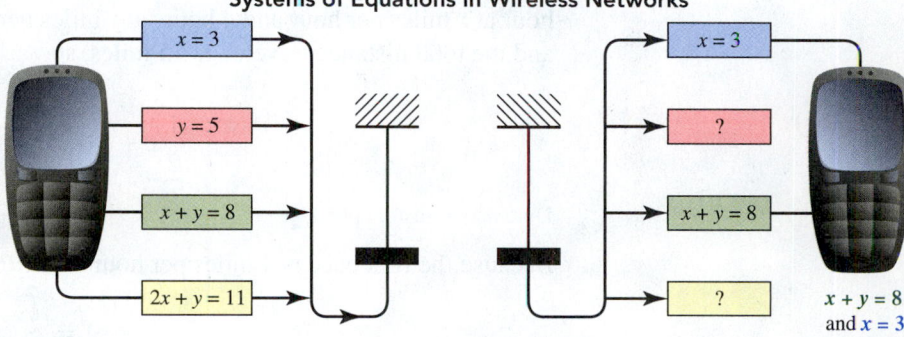

Figure 4.12 Sending the Number 35

CRITICAL THINKING

Why don't they send the equations $x = 3$ and $y = 5$ twice, rather than sending the other two equations?

In Figure 4.12 any two paths can fail and the receiving phone can still determine the two numbers, 3 and 5. For example, suppose that the second and fourth paths in Figure 4.12 fail. Then the receiving phone gets the information only from the first and third paths, $x = 3$ and $x + y = 8$. However, it can still determine the values x and y by solving the following system of linear equations.

$$x = 3 \quad \text{First path information}$$
$$x + y = 8 \quad \text{Third path information}$$

The solution is given by $x = 3$ and $y = 5$, so the number is 35.

NOTE: Wireless network systems are often designed so that any pair of equations is consistent and independent, and thus the solution (information) is unique.

EXAMPLE 10 | **Sending cell phone messages**

A person's age is being transmitted over a wireless network in a manner similar to that in Figure 4.12. Let x be the tens digit and y be the ones digit. Suppose that the first and second paths fail, resulting in the following system that must be solved.

$$x + y = 3 \quad \text{Third path information}$$
$$2x + y = 5 \quad \text{Fourth path information}$$

Solve this system graphically and find the person's age.

Solution

Begin by solving each equation for y.

$$y = 3 - x \quad \text{Subtract } x \text{ in third path equation.}$$
$$y = 5 - 2x \quad \text{Subtract } 2x \text{ in fourth path equation.}$$

Graph $y_1 = 3 - x$ and $y_2 = 5 - 2x$, as in Figure 4.13. Their graphs intersect at $(2, 1)$. It follows that $x = 2$, $y = 1$, and the person's age is 21.

Now Try Exercise 83

Decoding Cell Phone
Message

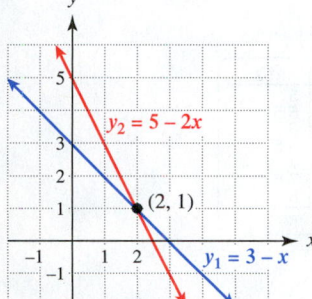

Figure 4.13 Person's Age: 21.

EXAMPLE 11 **Finding an athlete's running speeds**

An athlete jogs at a faster pace for 30 minutes and then jogs at a slower pace for 90 minutes. The first pace is 4 miles per hour faster than the second pace, and the athlete covers a total distance of 15 miles.

(a) Write a linear system whose solution gives the athlete's running speeds.

(b) Solve the resulting system graphically and numerically.

Solution

(a) Let x be the faster speed of the runner and y be the slower speed. The athlete runs $\frac{1}{2}$ hour at x miles per hour and $\frac{3}{2}$ hours at y miles per hour. *Rate times time equals distance* and the total distance traveled is 15 miles, so

$$\frac{1}{2}x + \frac{3}{2}y = 15.$$

Total distance

Distance at faster speed — Distance at slower speed

Because the first pace is 4 miles per hour faster than the second pace,

$$x - y = 4.$$

Difference in speeds

Faster speed — Slower speed

Thus we need to solve the following system of linear equations.

$$\frac{1}{2}x + \frac{3}{2}y = 15 \quad \text{First equation}$$

$$x - y = 4 \quad \text{Second equation}$$

(b) Multiply the first equation by 2 to clear fractions and then solve for y.

$$\mathbf{2}\left(\tfrac{1}{2}x\right) + \mathbf{2}\left(\tfrac{3}{2}y\right) = \mathbf{2}(15) \quad \text{Multiply first equation by 2.}$$

$$x + 3y = 30 \quad \text{Simplify.}$$

$$3y = -x + 30 \quad \text{Subtract } x.$$

$$y = -\frac{x}{3} + 10 \quad \text{Divide by 3.}$$

Solving the second equation for y gives

$$y = x - 4.$$

Graphical Solution We graph $Y_1 = -X/3 + 10$ and $Y_2 = X - 4$, as shown in Figure 4.14(a). The graphs intersect at the point (10.5, 6.5). Thus the athlete ran at 10.5 miles per hour and then slowed to 6.5 miles per hour.

Numerical Solution A numerical solution is shown in Figure 4.14(b). To find this solution, we set the increment for the x-values to 0.5.

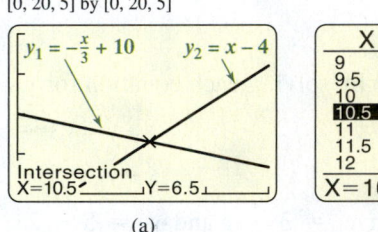

[0, 20, 5] by [0, 20, 5]

(a)

X	Y₁	Y₂
9	7	5
9.5	6.8333	5.5
10	6.6667	6
10.5	6.5	6.5
11	6.3333	7
11.5	6.1667	7.5
12	6	8

X=10.5

(b)

Figure 4.14

Now Try Exercise 73

4.1 Putting It All Together

SYSTEM	SOLUTIONS	GRAPH
Consistent, with a Unique Solution *Example:* $x + y = 3$ $x - y = 1$ Equations Are Independent	There is one solution given by $x = 2$ and $y = 1$. The solution is the ordered pair $(2, 1)$. *Check:* $2 + 1 = 3$ ✓ True $2 - 1 = 1$ ✓ True	Graph $y_1 = 3 - x$ and $y_2 = x - 1$. Their graphs intersect at $(2, 1)$.
Consistent, with Infinitely Many Solutions *Example:* $x + y = 1$ $3x + 3y = 3$ Equations Are Dependent	There are infinitely many solutions, such as $(2, -1)$ and $(0, 1)$. *Solution Set:* $\{(x, y) \mid x + y = 1\}$ Note that multiplying the first equation by 3 results in the second equation. Also, both equations have the same x- and y-intercepts: 1 and 1.	Graph $y_1 = 1 - x$ and $y_2 = (-3x + 3)/3$, or $y_2 = 1 - x$. The graphs are identical.
Inconsistent *Example:* $x + y = 1$ $x + y = 2$	There are no solutions. The sum of two variables, x and y, cannot equal both 1 and 2 at the same time.	Graph $y_1 = 1 - x$ and $y_2 = 2 - x$. The lines are parallel with slope -1 and do not intersect.

4.1 Exercises

MyMathLab | Math XL PRACTICE | WATCH | DOWNLOAD | READ | REVIEW

CONCEPTS AND VOCABULARY

1. Can a system of linear equations have exactly two solutions? Explain.

2. Give an example of a system of linear equations.

3. Name two ways to solve a system of linear equations.

4. If a graphical solution consists of two distinct, intersecting lines, what does this result indicate about the number of solutions?

5. If a graphical solution consists of two parallel lines, what does this result indicate about the number of solutions?

6. How many solutions does a dependent linear system have? How can you recognize a dependent linear system graphically?

7. The equations $x + y = 4$ and $x + y = 6$ determine what type of system of equations?

8. The equations $x + y = 5$ and $y + x = 5$ determine what type of system of equations?

Exercises 9–12: (Refer to Example 1.) Decide which of the ordered pairs is a solution for the linear system of equations.

9. $(1, -2), (4, 4)$
$2x - y = 4$
$3x + y = 1$

10. $(-3, -1), (3, 1)$
$x - 3y = 0$
$3x + y = 10$

11. $(4, 6), \left(-1, \frac{13}{3}\right)$
$x - 3y = -14$
$4x + 3y = 9$

12. $(4, 0), (3, 5)$
$5x - 4y = 20$
$-x + 4y = -4$

GRAPHICAL SOLUTIONS

Exercises 13–20: Solve the system graphically. Check your answers.

13. $y = 2x - 1$
$y = 2 - x$

14. $y = -2x$
$y = x + 3$

15. $y = \frac{1}{2}x + 1$
$y = 1 - \frac{1}{2}x$

16. $y = -x + 1$
$y = x + 3$

17. $2x + y = 4$
$x - y = 5$

18. $2x + y = -1$
$3x + 2y = -1$

19. $x + y = 3$
$3x - y = 3$

20. $3x + y = 2$
$x + 3y = 2$

Exercises 21–24: A system of two linear equations is represented graphically. Use the graph to find any possible solutions.

21.

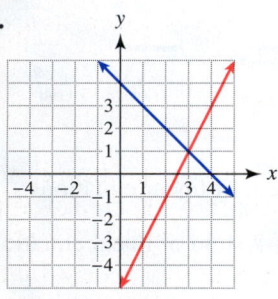

22.

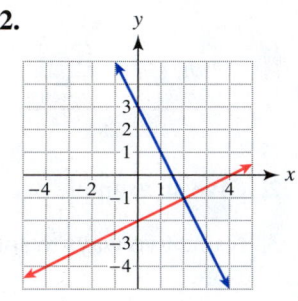

23.

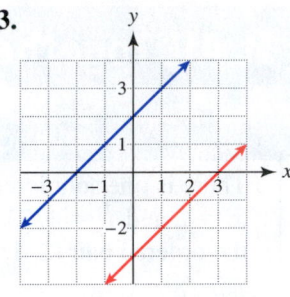

24.

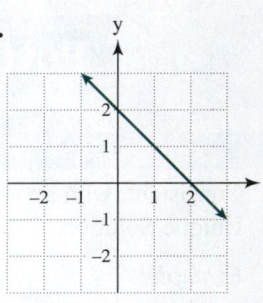

Exercises 25–42: Solve the system of equations. Determine whether the system is consistent or inconsistent. If the system is consistent, state whether the equations are dependent or independent.

25. $-x + y = 1$
$x + y = 3$

26. $x + y = 2$
$x - y = 2$

27. $2x + y = 5$
$-2x + y = -3$

28. $x + 2y = 3$
$2x - y = 1$

29. $x + y = 3$
$2x + 2y = 6$

30. $x - y = 1$
$-x + y = 3$

31. $3x - y = 0$
$2x + y = 5$

32. $-2x - y = -3$
$x + y = 2$

33. $-2x + y = 3$
$4x - 2y = 2$

34. $2x + 4y = 2$
$-x - 2y = -1$

35. $x + y = 6$
$x - y = 2$

36. $x + y = 9$
$x - y = 3$

37. $x - y = 4$
$2x - 2y = 4$

38. $2x + y = 5$
$4x + 2y = 10$

39. $6x - 4y = -2$
$-3x + 2y = 1$

40. $4x - 3y = 3$
$-8x + 6y = 1$

41. $4x + 3y = 2$
$5x + 2y = 6$

42. $-3x + 2y = 4$
$4x - y = 3$

Exercises 43–46: Solve the system of equations. Check your answer.

43. $2x + 2y = 4$
$x - 3y = -2$

44. $6x - y = 3$
$x - 2y = -5$

45. $-\frac{1}{2}x - \frac{1}{2}y = \frac{3}{2}$
$x - \frac{1}{2}y = 3$

46. $\frac{1}{2}x + \frac{1}{8}y = 1$
$-\frac{1}{2}x + \frac{5}{6}y = -1$

Exercises 47–50: Do the following.

(a) *Find the x- and y-intercepts for each equation.*
(b) *Use the intercepts to solve the system graphically.*

47. $x + y = 4$
$x - y = -2$

48. $2x + 3y = 12$
$2x - y = 4$

49. $-2x + 3y = 12$
$\quad\ -2x + \ y = \ \ 8$

50. $-x + y = 2$
$\quad\ \ x + y = 4$

 Exercises 51–58: Use technology to solve the system.

51. $\frac{1}{4}x + \frac{1}{2}y = \ \ \frac{3}{20}$
$\quad\ \frac{1}{8}x - \ \ y = -\frac{3}{10}$

52. $\frac{1}{2}x - \frac{3}{8}y = \ \ \frac{1}{8}$
$\quad\ \frac{1}{3}x - \frac{1}{2}y = -\frac{1}{12}$

53. $0.1x + 0.2y = 0.25$
$\quad\ 0.7x - 0.3y = 0.9$

54. $2.3x + 4.3y = 5.63$
$\quad\ 1.1x - 3.6y = 0.43$

55. $0.1x + 0.2y = 50$
$\quad\ 0.3x - 0.1y = 10$

56. $\ \ 0.5x + 0.2y = 14$
$\quad -0.1x + 0.4y = \ \ 6$

57. $\ \ \ \ x - 2y = \ \ 5$
$\quad -2x + 4y = -2$

58. $3x + 4y = \ \ 5$
$\quad\ 6x + 8y = 10$

NUMERICAL SOLUTIONS

Exercises 59–62: A system of two linear equations has been solved numerically. Find any possible solutions.

59. $y_1 = 5 - x$
$\quad\ y_2 = 2x - 1$

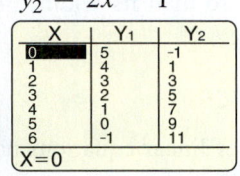

60. $y_1 = 1 - 2x$
$\quad\ y_2 = -1 - 3x$

61. $y_1 = 2 - x$
$\quad\ y_2 = x$

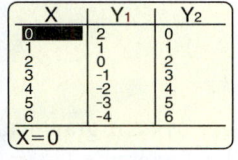

62. $y_1 = 2x - 1$
$\quad\ y_2 = -3 + 2x$

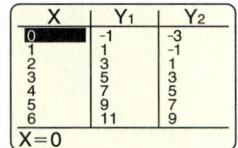

Exercises 63–68: Solve the system numerically.

63. $x + y = 3$
$\quad\ x - y = 7$

64. $2x + y = 3$
$\quad\ 3x - y = 7$

65. $\ \ 3x + 2y = \ \ 5$
$\quad\ -x - \ y = -5$

66. $2x + 3y = 3.5$
$\quad\ 3x + 2y = 6.5$

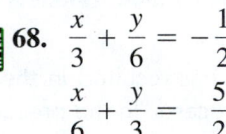 **67.** $0.5x - 0.1y = \ \ \ 0.1$
$\quad\ 0.1x - 0.3y = -0.4$

68. $\dfrac{x}{3} + \dfrac{y}{6} = -\dfrac{1}{2}$
$\quad\ \dfrac{x}{6} + \dfrac{y}{3} = \ \ \dfrac{5}{2}$

Exercises 69 and 70: **Thinking Generally** *Solve the system of linear equations, if $a \neq 0$.*

69. $x + \ ay = 1$
$\quad\ 2x + 2ay = 4$

70. $-ax + y = 4$
$\quad\ \ \ ax + y = 4$

WRITING AND SOLVING EQUATIONS

Exercises 71–80: Do the following.

   (a) *Write a system of linear equations that models the situation.*

   (b) *Solve the resulting system.*

71. The sum of two numbers is 18, and their difference is 6. Find the two numbers.

72. Twice a number minus a second number equals 5. The sum of the two numbers is 16. Find the two numbers.

73. *Time and Distance* An athlete ran for part of an hour at 6 miles per hour and for the rest of the hour at 8 miles per hour. The total distance traveled was 7 miles. How long did the athlete run at each speed?

74. *Time and Distance* A car was driven for 2 hours, part of the time at 40 miles per hour and the rest of the time at 60 miles per hour. The total distance was 90 miles. How long did the car travel at each speed?

75. *Rectangle* The perimeter of a rectangle is 76 inches. The rectangle's length is 4 inches longer than its width. Find the dimensions of this rectangle.

76. *Rectangle* The perimeter of a rectangle is 40 feet. The rectangle's length is three times as long as its width. Find the dimensions of the rectangle.

77. *Price* If 2 boxes of popcorn and 3 soft drinks cost $7 and 3 boxes of popcorn and 2 soft drinks cost $8, find the price of a box of popcorn and the price of a soft drink.

78. *Triangle* An isosceles triangle has a perimeter of 100 inches. The triangle's longest side is 10 inches longer than either of the other two sides. Find the length of each side of the triangle. (*Hint:* An isosceles triangle has two sides with equal measure.)

79. *Triangle* The largest angle in an isosceles triangle is 60° larger than either of the other two angles. Find the measure of each angle. (*Hint:* An isosceles triangle has two angles with equal measure.)

80. *Rectangle* The perimeter of a rectangle is 30 centimeters. The rectangle's length is twice as long as its width. Find the dimensions of the rectangle.

APPLICATIONS

81. *Lady Gaga Tickets* In 2010 concerts, where a minimum of 1000 tickets were sold, Lady Gaga had the highest average-priced ticket at Radio City Music Hall and Owl City had the lowest average-priced ticket at Terminal 5. The combined price of their average-priced tickets was $436 and the difference was $378. Determine both graphically and numerically the average-priced ticket for Lady Gaga and for Owl City. (*Source: New York Times.*)

82. *Nuclear Power* The United States, Europe, and Canada will produce about 670,000 megawatts of electricity from nuclear power by 2050. Europe and Canada combined will produce about 67.5% of the amount that the United States will produce. Determine both graphically and numerically how much electricity from nuclear power the United States will produce and how much Europe and Canada will produce together. (*Sources: Scientific American; MIT.*)

Exercises 83–86: Wireless Networks (Refer to Example 10.) The system of four equations represents a wireless transmission of a person's age.

$$x = 2 \qquad (1)$$
$$y = 3 \qquad (2)$$
$$x + y = 5 \qquad (3)$$
$$3x + y = 9 \qquad (4)$$

Solve the given pair of equations and find the age of the person. Do you get the same answer in each case?

83. Equations (1) and (3)
84. Equations (2) and (3)
85. Equations (2) and (4)
86. Equations (3) and (4)

87. *Winning Speeds* Speeds of the winners in the Tour de France peaked in 2005 due to blood doping. Because of massive disqualifications, these winning speeds decreased in 2007. The average of the 2005 and 2007 winning speeds was 25 mph, and the 2007 speed was 2 miles per hour slower than the 2005 speed. Find the 2005 and 2007 speeds. (*Source: Christiansen, J., Scientific American.*)

88. *Numbers* The average of two numbers is 105, and their difference is 24. Find the two numbers.

89. *Home Runs* In 1998, Mark McGwire and Sammy Sosa hit a total of 136 home runs. McGwire hit 4 more home runs than Sosa. How many home runs did each player hit?

90. *Student Loans* A student takes out two loans totaling $4000 to help pay for college expenses. One loan is at 10% annual interest, and the other is at 5% annual interest. The first-year interest is $250. Find the amount of each loan.

91. *Investment* A total of $600 is invested in two accounts at 4% and 5%. One hundred dollars more is invested in the 5% account. Determine the amount invested in each account.

92. *Dimensions* A small building has a perimeter of 14 yards. Its length is 1 yard more than its width. Find the dimensions of the building.

WRITING ABOUT MATHEMATICS

93. Discuss the types of systems of linear equations having two variables. Explain how you can recognize each type graphically.

94. Given

$$ax + by = c$$
$$dx + ey = k,$$

explain how to solve this linear system graphically. Demonstrate your method with an example.

4.2 The Substitution and Elimination Methods

The Substitution Method • The Elimination Method • Models and Applications

A LOOK INTO MATH ▶ The ability to solve systems of equations has resulted in the development of CAT scans, satellites, DVDs, and accurate weather forecasts. In the preceding section we showed how to solve a system of two linear equations in two variables using graphical and numerical methods. In this section we demonstrate how to solve these systems symbolically.

The Substitution Method

The **substitution method** is one way to solve a system symbolically. To apply the substitution method, we begin by solving an equation for one of its variables. Then we substitute the result into the other equation. For example, consider the following system of equations.

$$2x + y = 5 \qquad \text{First equation}$$
$$3x - 2y = 4 \qquad \text{Second equation}$$

NEW VOCABULARY

☐ Substitution method
☐ Elimination method

It is convenient to solve the first equation, $2x + y = 5$, for y to obtain $y = 5 - 2x$. Now substitute $(5 - 2x)$ for y in the second equation,

$$3x - 2(y) = 4 \quad \text{Second equation}$$
$$3x - 2(5 - 2x) = 4, \quad \text{Substitute.}$$

Be sure to insert parentheses.

to obtain a linear equation in one variable. Now solve for x.

$$3x - 2(5 - 2x) = 4 \quad \text{Equation to solve}$$
$$3x - 10 + 4x = 4 \quad \text{Distributive property}$$
$$7x - 10 = 4 \quad \text{Combine like terms.}$$
$$7x = 14 \quad \text{Add 10 to each side.}$$
$$x = 2 \quad \text{Divide each side by 7.}$$

To determine y, substitute 2 for x in $y = 5 - 2x$ to obtain

$$y = 5 - 2(2) = 1.$$

The solution is $(2, 1)$. To check this solution, let $x = 2$ and $y = 1$ in each equation.

$2x + y = 5$	First equation	$3x - 2y = 4$	Second equation
$2(2) + 1 \overset{?}{=} 5$	Substitute.	$3(2) - 2(1) \overset{?}{=} 4$	Substitute.
$5 = 5$ ✓	It checks.	$4 = 4$ ✓	It checks.

Because $(2, 1)$ satisfies *both* equations, it is the solution to the system.

Do not stop after solving for the first variable. You *must* also solve for the second variable. Remember that a solution to a system of equations in two variables consists of an *ordered pair*, not a single number.

READING CHECK

When using substitution to solve a system of equations, does it matter which variable you solve for? Why is it helpful to solve for a variable that has a coefficient of 1?

NOTE: When using substitution, you may begin by solving for *either variable in either equation*. The same result is obtained regardless of which equation is used. However, it is often simpler to solve for a variable with a coefficient of 1 because there is less likelihood of encountering fractions.

EXAMPLE 1   **Applying the substitution method**

Solve each system of equations.

(a) $\quad y = 2x$ $\qquad$ (b) $\quad 2x + y = -1$ $\qquad$ (c) $\quad -3x + 2y = 3$
$\quad\ \ x + y = 21$ $\qquad\qquad\ \ 2x - y = -3$ $\qquad\qquad\quad\ \ 2x - 4y = -6$

Solution
(a) The first equation is $y = 2x$, so we substitute $(2x)$ for y in the second equation.

$$x + y = 21 \quad \text{Second equation}$$
$$x + (2x) = 21 \quad \text{Let } y = (2x).$$
$$3x = 21 \quad \text{Add like terms.}$$
$$x = 7 \quad \text{Divide by 3.}$$

Substituting 7 for x in $y = 2x$ gives $y = 14$. The solution is $(7, 14)$.
(b) We start by solving for y in the first equation because its coefficient is 1.

$$2x + y = -1 \quad \text{First equation}$$
$$y = -2x - 1 \quad \text{Subtract } 2x.$$

Now we substitute $(-2x - 1)$ for y in the second equation.

Be sure to include parentheses.

$$2x - y = -3 \quad \text{Second equation}$$
$$2x - (-2x - 1) = -3 \quad \text{Let } y = (-2x - 1).$$
$$2x + 2x + 1 = -3 \quad \text{Distributive property}$$
$$4x = -4 \quad \text{Subtract 1; combine like terms.}$$
$$x = -1 \quad \text{Divide by 4.}$$

Substituting -1 for x in $y = -2x - 1$ gives $y = 1$. The solution is $(-1, 1)$.

(c) We start by solving for x in the second equation, but we could solve for y.

$$2x - 4y = -6 \quad \text{Second equation}$$
$$2x = 4y - 6 \quad \text{Add } 4y.$$
$$x = 2y - 3 \quad \text{Divide by 2.}$$

Substitute $(2y - 3)$ for x in the first equation.

$$-3x + 2y = 3 \quad \text{First equation}$$
$$-3(2y - 3) + 2y = 3 \quad \text{Let } x = (2y - 3).$$
$$-6y + 9 + 2y = 3 \quad \text{Distributive property}$$
$$-4y + 9 = 3 \quad \text{Combine like terms.}$$

Be sure to include parentheses.

$$-4y = -6 \quad \text{Subtract 9.}$$
$$y = \tfrac{3}{2} \quad \text{Divide by } -4.$$

Substituting $\tfrac{3}{2}$ for y in $x = 2y - 3$ gives $x = 2\left(\tfrac{3}{2}\right) - 3 = 0$. The solution is $\left(0, \tfrac{3}{2}\right)$.

▌ **Now Try Exercises 9, 17, 25**

EXAMPLE 2 **Finding per capita income**

In 2009, the *average* of the per capita (or per person) incomes for Massachusetts and Maine was \$43,000. The per capita income in Massachusetts exceeded the per capita income in Maine by \$13,000. Find the 2009 per capita income for each state.

Solution

Let x be the per capita income in Massachusetts and y be the per capita income in Maine. The following system of equations models the data.

$$\frac{x + y}{2} = 43{,}000 \quad \text{Their average is \$43,000.}$$

$$x - y = 13{,}000 \quad \text{Their difference is \$13,000.}$$

Begin by solving the second equation for x.

$$x = y + 13{,}000$$

Substitute $(y + 13{,}000)$ for x in the first equation and solve for y.

$$\frac{(y + 13{,}000) + y}{2} = 43{,}000 \quad \text{Let } x = (y + 13{,}000).$$

$$(y + 13{,}000) + y = 86{,}000 \quad \text{Multiply each side by 2.}$$

$$2y + 13{,}000 = 86{,}000 \quad \text{Combine like terms.}$$

$$2y = 73{,}000 \quad \text{Subtract 13,000 from each side.}$$

$$y = 36{,}500 \quad \text{Divide each side by 2.}$$

Substituting **36,500** for y in $x = y + 13{,}000$ yields $x = 49{,}500$. In 2009, the per capita income in Massachusetts was \$49,500, and in Maine it was \$36,500.

▌ **Now Try Exercise 97**

NOTE: A step-by-step procedure for the substitution method is given in Putting It All Together for this section.

The Elimination Method

The **elimination** (or addition) **method** is a second way to solve linear systems symbolically. This method is based on the property that *equals added to equals are equal*. That is, if

$$a = b \quad \text{and} \quad c = d, \text{ then}$$
$$a + c = b + d.$$

The goal of this method is to get an equation where one of the two variables has been eliminated. This task is sometimes accomplished by adding two equations. This method is demonstrated in the next example.

EXAMPLE 3 **Applying the elimination method**

Solve each system of equations.

(a) $x + y = 3$ (b) $4x + 3y = \quad 0$
    $x - y = 1$ $4x - 2y = -20$

Solution

(a) If we add the two equations, the y-variable will be eliminated.

$x + \ y = 3$	First equation
$\underline{x - \ y = 1}$	Second equation
$2x + 0y = 4 \quad \text{or} \quad x = 2$	Add equations and solve.

To find the value of y, we substitute $x = 2$ in either equation.

$x + y = 3$	First equation
$2 + y = 3$	Let $x = 2$.
$y = 1$	Subtract 2.

The solution is $(2, 1)$.

(b) If we multiply the first equation by -1 and then add the two equations, the x-variable will be eliminated. That is, the coefficients of x become additive inverses.

$-4x - 3y = \quad 0$	First equation times -1
$\underline{4x - 2y = -20}$	Second equation
$0x - 5y = -20 \quad \text{or} \quad y = 4$	Add equations and solve.

To find the value of x, we substitute $y = 4$ in either equation.

$4x + 3y = \quad 0$	First equation
$4x + 3(4) = \quad 0$	Let $y = 4$.
$4x = -12$	Subtract 12.
$x = -3$	Divide by 4.

The solution is $(-3, 4)$.

Now Try Exercises 49, 51

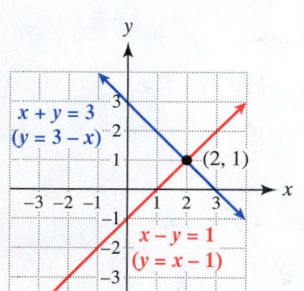

$x + y = 3$
$(y = 3 - x)$

$(2, 1)$

$x - y = 1$
$(y = x - 1)$

Figure 4.15

Solutions to systems of equations can be *supported graphically*. For example, if we graph the equations in Example 3(a), they intersect at the point (2, 1), as shown in Figure 4.15.

EXAMPLE 4 **Applying the elimination method**

Solve the following system by using elimination.

$$2x - y = 4$$
$$x + y = 1$$

Solution
Adding the two equations eliminates the variable y.

$$
\begin{aligned}
2x - y &= 4 \\
x + y &= 1 \\
\hline
3x + 0y &= 5
\end{aligned}
\quad \text{or} \quad x = \frac{5}{3}
$$
Add the two equations and solve for x.

Substituting $x = \frac{5}{3}$ in the second equation gives

$$\frac{5}{3} + y = 1 \quad \text{or} \quad y = -\frac{2}{3}.$$

The solution is $\left(\frac{5}{3}, -\frac{2}{3} \right)$.

▌ **Now Try Exercise 53**

NOTE: Example 4 illustrates that both the x- and y-values for a solution can be fractions and *not* integers.

In the next example, we use multiplication before we add the two equations.

EXAMPLE 5 **Multiplying before applying elimination**

Solve the system of equations.

$$x + \frac{1}{2}y = 1$$
$$-3x + 2y = 11$$

Solution
Neither variable can be eliminated by simply adding the given equations. However, if we multiply each side of the first equation by -4, we eliminate fractions and then addition of the two equations eliminates the y-variable.

$$
\begin{aligned}
-4x - 2y &= -4 \\
-3x + 2y &= 11 \\
\hline
-7x + 0y &= 7
\end{aligned}
\quad \text{or} \quad x = -1
$$

Multiply first equation by -4.
Second equation
Add equations and solve.

To find the value of y, we substitute $x = -1$ in the second equation.

$$
\begin{aligned}
-3x + 2y &= 11 & &\text{Second equation} \\
-3(-1) + 2y &= 11 & &\text{Let } x = -1. \\
2y &= 8 & &\text{Subtract 3 from each side.} \\
y &= 4 & &\text{Divide each side by 2.}
\end{aligned}
$$

The solution is $(-1, 4)$.

▌ **Now Try Exercise 55**

MAKING CONNECTIONS

Substitution and Elimination

Substitution and elimination are two symbolic methods that accomplish the *same* task: solving a system of linear equations. Be aware that one method may be easier to perform than the other, depending on the system of equations to be solved.

EXAMPLE 6

Multiplying before applying elimination

Solve the following system by using elimination. Support your answer graphically and numerically.

$$3x - 2y = 11$$
$$2x + 3y = 3$$

Solution

Symbolic Solution If we add (or subtract) these equations, neither variable will be eliminated. However, if we multiply the first equation by 3 and multiply the second equation by 2, we can eliminate y by adding the resulting equations.

$9x - 6y = 33$	Multiply first equation by 3.
$4x + 6y = 6$	Multiply second equation by 2.
$13x + 0y = 39$ or $x = 3$	Add the equations and solve.

Substituting $x = 3$ in the first equation, $3x - 2y = 11$, gives

$$3(3) - 2y = 11 \quad \text{or} \quad y = -1.$$

The solution is $(3, -1)$.

NOTE: We could have multiplied the first equation by 2 and the second equation by -3. Adding the resulting equations would have eliminated the variable x.

Graphical Solution If we use a calculator, we solve each equation for y.

$3x - 2y = 11$	First equation	$2x + 3y = 3$	Second equation
$-2y = 11 - 3x$	Subtract 3x.	$3y = 3 - 2x$	Subtract 2x.
$y = \dfrac{11 - 3x}{-2}$	Divide by −2.	$y = \dfrac{3 - 2x}{3}$	Divide by 3.

In Figure 4.16(a), the graphs of $Y_1 = (11 - 3X)/(-2)$ and $Y_2 = (3 - 2X)/3$ intersect at the point $(3, -1)$.

Numerical Solution In Figure 4.16(b), $y_1 = y_2 = -1$, when $x = 3$.

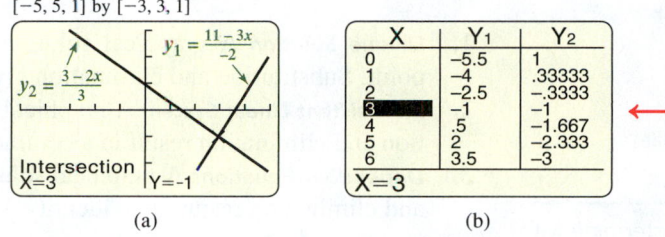

$[-5, 5, 1]$ by $[-3, 3, 1]$

(a)   (b)

Figure 4.16 Solution is $(3, -1)$.

Now Try Exercise 73

| EXAMPLE 7 | **Recognizing an inconsistent system** |

Use elimination to solve the following system.

$$3x - 4y = 5$$
$$-6x + 8y = 9$$

No Solutions

$[-6, 6, 1]$ by $[-4, 4, 1]$

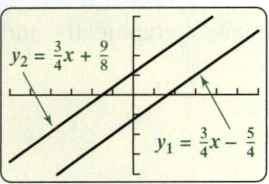

$y_2 = \frac{3}{4}x + \frac{9}{8}$

$y_1 = \frac{3}{4}x - \frac{5}{4}$

Figure 4.17

Solution
If we multiply the first equation by 2 and add, we obtain the following result.

$6x - 8y = 10$	Multiply first equation by 2.
$-6x + 8y = \;\;9$	Second equation
$\mathbf{0 = 19}$ ✗	Adding the two equations gives a false result.

The statement $0 = 19$ is a *contradiction*, which tells us that the system has no solutions. If we solve each equation for y and graph, we obtain two parallel lines with slope $\frac{3}{4}$ that never intersect. This result is shown in Figure 4.17, where the equations are graphed as $y_1 = \frac{3}{4}x - \frac{5}{4}$ and $y_2 = \frac{3}{4}x + \frac{9}{8}$.

▌ **Now Try Exercise 57**

| EXAMPLE 8 | **Recognizing dependent equations** |

Use elimination to solve the following system.

$$3x - 6y = 3$$
$$x - 2y = 1$$

Infinitely Many Solutions

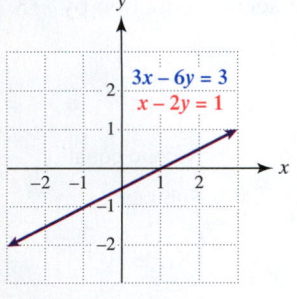

$3x - 6y = 3$
$x - 2y = 1$

Figure 4.18

Solution
If we multiply the second equation by -3 and add, we obtain the following.

$3x - 6y = \;\;3$	First equation
$-3x + 6y = -3$	Multiply second equation by -3.
$0 = \;\;0$	Adding the two equations gives a true result.

The statement $0 = 0$ is an *identity*, so the two equations are equivalent. That is, if an ordered pair (x, y) satisfies the first equation, it also satisfies the second equation. Thus the solution set may be expressed as $\{(x, y) \,|\, x - 2y = 1\}$. For example, $(1, 0)$ and $(5, 2)$ are both solutions because they both satisfy $x - 2y = 1$.

The graphs of these equations are identical lines, as shown in Figure 4.18.

▌ **Now Try Exercise 59**

MAKING CONNECTIONS

Graphical Solutions, Substitution, and Elimination

1. *Unique Solution* A graphical solution results in two lines intersecting at a unique point. Substitution and elimination give unique values for x and y.
2. *Inconsistent Linear System* A graphical solution results in two parallel lines. Substitution and elimination result in a contradiction, such as $0 = 1$.
3. *Dependent Equations* A graphical solution results in two identical lines. Substitution and elimination result in an identity, such as $0 = 0$.

NOTE: Conditions 2 and 3 occur when *both* variables are eliminated, and the resulting equation is either an identity or a contradiction.

READING CHECK

How can you determine with elimination that a system is inconsistent?

Models and Applications

In Section 3.2, we learned a four-step process for solving problems. We will apply this process to solving applications that involve two variables. A *brief summary* of these four steps is as follows. (For a full discussion, see page 166.)

STEPS FOR SOLVING A PROBLEM INVOLVING TWO VARIABLES

STEP 1: Read the problem. Assign two variables to what you are being asked to find.

STEP 2: Write two equations involving these variables.

STEP 3: Solve the system of equations.

STEP 4: Check your solution.

EXAMPLE 9   **Modeling tuition**

A student is attempting to graduate on schedule by taking 14 credits of day classes at one college and 4 credits of night classes at a different college. A credit for day classes costs $20 more than a credit for night classes. If the student's total tuition is $2440, how much does each type of credit cost?

Solution

STEP 1: Let x be the cost of a credit for day classes and y be the cost of a credit for night classes.

STEP 2: The data can be represented by the following system of equations.

$$x - y = 20 \qquad \text{Day credits cost \$20 more than night credits.}$$
$$14x + 4y = 2440 \qquad \text{The total cost is \$2440 for 14 credits during the day and 4 credits at night.}$$

STEP 3: To solve this system we can multiply the first equation by 4 and then add the two equations.

$$
\begin{array}{ll}
4x - 4y = 80 & \text{Multiply first equation by 4.} \\
14x + 4y = 2440 & \text{Second equation} \\
\hline
18x + 0y = 2520 \ \text{ or } \ x = 140 & \text{Add and solve for } x.
\end{array}
$$

Because $x = 140$ and $x - y = 20$, it follows that $y = 120$. Thus a credit for day classes costs $140 and a credit for night classes costs $120.

STEP 4: Check the solution. The difference is $140 - \$120 = \$20$ and

$$
\begin{array}{ll}
14 \times \$140 = \$1960 & \text{Cost of day classes} \\
4 \times \$120 = \$480 & \text{Cost of night classes} \\
\hline
\text{Total cost} = \$2440. \ \checkmark & \text{The answer checks.}
\end{array}
$$

Now Try Exercise 105

In the next example we make use of the formula $d = rt$. (*Distance equals rate times time.*) If we solve this formula for r to obtain $r = \frac{d}{t}$, we can find the average speed (rate) of a boat traveling on a river. This method can also be used to determine the speed of an airplane when there is a wind. See Exercise 99.

EXAMPLE 10

Modeling river travel

A boat travels 150 miles upstream in 10 hours, and the return trip takes 6 hours. Find the speed of the boat in still water and the speed of the current.

Solution

STEP 1: Let x be the speed of the boat and y be the speed of the river current.

STEP 2: The boat travels 150 miles upstream in 10 hours. Thus the speed of the boat **against** the current is $\frac{150}{10} = 15$ miles per hour, or $x - y = 15$. Similarly, the boat travels 150 miles downstream in 6 hours. Thus the speed of the boat **with** the current is $\frac{150}{6} = 25$ miles per hour, or $x + y = 25$.

STEP 3: We represent these data with the following equations.

$$
\begin{array}{ll}
x + \ y = 25 & \text{Speed with current} \\
\underline{x - \ y = 15} & \text{Speed against current} \\
2x + 0y = 40 \quad \text{or} \quad x = 20 & \text{Add equations and solve.}
\end{array}
$$

Substituting $x = 20$ in $x + y = 25$ gives $y = 5$. The boat can travel 20 miles per hour in still water, and the speed of the current is 5 miles per hour.

STEP 4: Check the solution. The speed of the boat *against* the current is $20 - 5 = 15$ miles per hour, and the speed of the boat *with* the current is $20 + 5 = 25$ miles per hour.

$$t = \frac{d}{r} = \frac{150}{15} = 10 \text{ hours} \qquad \text{Time to travel upstream}$$

$$t = \frac{d}{r} = \frac{150}{25} = 6 \text{ hours} \qquad \text{Time to travel downstream}$$

The answer checks. ✓

▮ **Now Try Exercise 101**

EXAMPLE 11

Burning calories while exercising

During strenuous exercise, an athlete can burn 12 calories per minute running and 10 calories per minute on a bicycle. In a 60-minute workout, an athlete burns 644 calories. How long did the athlete perform each activity?

Solution

STEP 1: Let x be the time spent running and y be the time spent bicycling.

STEP 2: Because the workout is 60 minutes long, we write $x + y = 60$. The number of calories burned running equals $12x$, and the number of calories burned bicycling equals $10y$. The total calories burned equals 644, so the system is as follows.

$$
\begin{array}{ll}
x + \quad y = \ 60 & \text{Total workout is 60 minutes.} \\
12x + 10y = 644 & \text{Total calories equals 644.}
\end{array}
$$

STEP 3: To solve this system we can multiply the first equation by -10 and add the equations to eliminate the y-variable.

$$
\begin{array}{ll}
-10x - 10y = -600 & \text{Multiply first equation by } -10. \\
\underline{12x + 10y = \quad 644} & \\
2x + \ 0y = \quad 44 \quad \text{or} \quad x = 22 & \text{Add equations and solve.}
\end{array}
$$

Substituting $x = 22$ in $x + y = 60$ gives $y = 38$. The athlete spent 22 minutes running and 38 minutes bicycling.

STEP 4: Check the solution. The total time is $22 + 38 = 60$ minutes and

$$22 \text{ minutes} \times 12 \text{ calories/minute} = 264 \text{ calories} \qquad \text{Calories burned running}$$

$$38 \text{ minutes} \times 10 \text{ calories/minute} = 380 \text{ calories} \qquad \text{Calories burned bicycling}$$

$$\text{Total calories burned} = 644 \text{ calories.} \checkmark \qquad \text{The answer checks.}$$

Now Try Exercise 89

EXAMPLE 12   **Mixing antifreeze**

A mixture of water and antifreeze in a car is currently 20% antifreeze. If the radiator holds 5 gallons of fluid and this mixture should be 40% antifreeze, how many gallons of radiator fluid should be drained and replaced with a mixture containing 90% antifreeze?

Solution

STEP 1: Let x be the gallons of 20% antifreeze that should remain in the radiator and y be the gallons of 90% antifreeze that should be added to the radiator.

STEP 2: The radiator holds 5 gallons of fluid, so $x + y = 5$. The final solution in the radiator should contain 5 gallons of 40% solution or $5(0.4) = 2$ gallons of (pure) antifreeze. Thus the antifreeze in the 20% solution plus the antifreeze in the 90% solution must equal 2 gallons. That is, $0.2x + 0.9y = 2$. Table 4.2 summarizes this situation.

TABLE 4.2 Mixing Antifreeze

	20% Solution	90% Solution	40% Solution
Radiator Fluid (gallons)	x	y	5
Pure Antifreeze (gallons)	$0.2x$	$0.9y$	2

The resulting system to be solved is

$$x + y = 5 \qquad \text{First row in table}$$

$$0.2x + 0.9y = 2. \qquad \text{Second row in table}$$

STEP 3: Multiply the first equation by 2 and the second equation by -10.

$$2x + 2y = 10 \qquad \text{Multiply first equation by 2.}$$

$$\underline{-2x - 9y = -20} \qquad \text{Multiply second equation by } -10.$$

$$0x - 7y = -10 \quad \text{or} \quad y = \tfrac{10}{7} \qquad \text{Add and solve.}$$

Thus $\frac{10}{7}$ gallons of 90% antifreeze solution should be added.

STEP 4: Check the solution. If $\frac{10}{7}$ gallons are drained from the 5-gallon radiator, then there are $5 - \frac{10}{7} = \frac{35}{7} - \frac{10}{7} = \frac{25}{7}$ gallons of 20% antifreeze remaining and $\frac{10}{7}$ gallons of 90% antifreeze added.

$$20\% \text{ of } \frac{25}{7} = \frac{2}{10} \times \frac{25}{7} = \frac{5}{7} \text{ gallons of pure antifreeze}$$

$$90\% \text{ of } \frac{10}{7} \text{ gallons} = \frac{9}{10} \times \frac{10}{7} = \frac{9}{7} \text{ gallons of pure antifreeze}$$

There are $\frac{5}{7} + \frac{9}{7} = \frac{14}{7} = \mathbf{2}$ gallons of pure antifreeze in the **5**-gallon radiator, so the new concentration is $\frac{2}{5} = 0.4$ or 40%. The answer checks. $\checkmark$

Now Try Exercise 91

4.2 Putting It All Together

CONCEPT	EXPLANATION
Substitution Method	**1.** Solve for a convenient variable such as y in the first equation. $$x + y = 3 \quad \text{or} \quad y = 3 - x$$ $$2x - y = 0$$ **2.** Substitute $(3 - x)$ in the second equation for y. Solve the equation for x. $$2x - (3 - x) = 0 \quad \text{or} \quad x = 1$$ **3.** Substitute $x = 1$ in either of the given equations and find y. $$1 + y = 3 \quad \text{or} \quad y = 2$$ **4.** The solution is $(1, 2)$. Check your answer.
Elimination Method	**1.** Multiply the first equation by -2 so that the coefficients of x in the two equations are additive inverses. $$x + 2y = 1 \qquad -2x - 4y = -2$$ $$2x - 3y = 9 \quad \rightarrow \quad 2x - 3y = 9$$ **2.** Eliminate x by adding the two equations. $$-2x - 4y = -2$$ $$\underline{2x - 3y = 9}$$ $$0x - 7y = 7 \quad \text{or} \quad y = -1$$ **3.** Substitute $y = -1$ in one of the given equations and solve for x. $$x + 2(-1) = 1 \quad \text{or} \quad x = 3$$ **4.** The solution is $(3, -1)$. Check your answer.

4.2 Exercises

CONCEPTS AND VOCABULARY

1. Name two symbolic methods for solving a linear system.

2. Are different solutions obtained when the same system of equations is solved symbolically, graphically, and numerically? Explain.

3. When using elimination, how can you recognize an inconsistent system of equations?

4. When using elimination, how can you recognize a system of dependent equations?

5. When you are using substitution to solve

$$3x + y = 4$$
$$5x - 7y = -2,$$

what is a good first step?

6. When you are using elimination to solve

$$2x - 5y = 4$$
$$x + 3y = 11,$$

what is a good first step?

7. When you are using elimination to solve

$$3x - y = 2$$
$$5x + y = 4,$$

what is a good first step?

8. When you are using elimination to solve

$$2x - 3y = 3$$
$$3x + 2y = 11,$$

what is a good first step?

SUBSTITUTION METHOD

Exercises 9–12: Solve the system of linear equations by using substitution.

9. $y = 2x$
$3x + y = 5$

10. $y = x + 1$
$x + 2y = 8$

11. $x = 2y - 1$
$x + 5y = 20$

12. $x = 3y$
$-x + 2y = 4$

Exercises 13 and 14: Use substitution to solve the system of equations. Then use the graph to support your answer.

13. $x - 2y = 0$
$3x + y = 7$

14. $2x + 3y = 8$
$3x - 2y = -14$

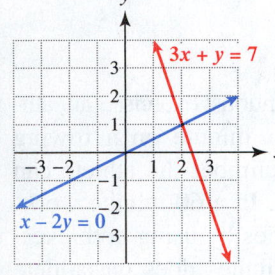
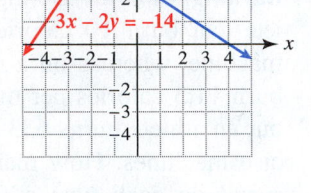

Exercises 15–20: Solve the system of equations by using substitution. Then solve the system graphically.

15. $y = 3x$
$x + y = 4$

16. $x = \frac{1}{2}y$
$2x + y = 4$

17. $x + y = 2$
$2x - y = 1$

18. $x - y = 2$
$x + 2y = -1$

19. $2x + 3y = 6$
$x - 2y = -4$

20. $2x - y = 2$
$3x + 2y = 10$

Exercises 21–36: Solve the system of equations by using substitution.

21. $4x + y = 3$
$2x - 3y = -2$

22. $3x + 4y = 32$
$2x - y = 3$

23. $x - 2y = 0$
$-3x + 2y = -1$

24. $-x + 3y = 3$
$2x + 5y = -\frac{1}{2}$

25. $2x - 5y = -1$
$-x + 3y = 0$

26. $5x + 10y = -5$
$-10x + 5y = 60$

27. $x - y = 1$
$2x + 6y = -2$

28. $4x - y = -4$
$-2x + 5y = 29$

29. $\frac{1}{2}x - \frac{1}{2}y = 1$
$2x - y = 6$

30. $\frac{3}{4}x + \frac{1}{2}y = 5$
$x - 2y = 12$

31. $\frac{1}{6}x - \frac{1}{3}y = -1$
$\frac{1}{3}x + \frac{5}{6}y = 7$

32. $\frac{3}{5}x - \frac{1}{10}y = 4$
$\frac{2}{5}x + \frac{1}{10}y = 6$

33. $\frac{1}{2}x + \frac{2}{3}y = -2$
$\frac{1}{4}x - \frac{1}{3}y = 3$

34. $\frac{1}{5}x - \frac{1}{10}y = \frac{7}{40}$
$\frac{1}{4}x - \frac{1}{5}y = \frac{11}{40}$

35. $0.1x + 0.4y = 1.3$
$0.3x - 0.2y = 1.1$

36. $1.5x - 4.1y = -1.6$
$2.7x - 0.1y = 0.76$

37. The following is a system of dependent equations.

$$x - y = 5$$
$$2x - 2y = 10$$

Solve this system by using substitution. Explain how you can recognize a system of dependent equations when you are using substitution.

38. The following system is inconsistent.

$$-x + 2y = 5$$
$$2x - 4y = 10$$

Solve this system by using substitution. Explain how you can recognize an inconsistent system when you are using substitution.

Exercises 39–42: (Refer to Exercises 37 and 38.) Solve the system of equations by using substitution, if possible.

39. $5x + 3y = 6$
$-\frac{5}{3}x - y = 2$

40. $2x - 4y = 5$
$-x + 2y = 1$

41. $x + 3y = -2$
$-\frac{1}{2}x - \frac{3}{2}y = 1$

42. $2x - 4y = 4$
$-3x + 6y = -6$

ELIMINATION METHOD

Exercises 43 and 44: Use elimination to solve the system of equations. Then use the graph to support your answer.

43. $x - y = 5$
$x + y = 9$

44. $3x - 2y = 9$
$5x + 2y = 7$

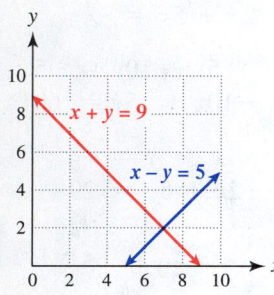

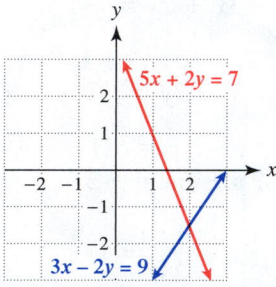

Exercises 45–48: Use elimination to solve the system of equations. Then solve the system graphically.

45. $x + y = 3$
$x - y = 1$

46. $2x - y = 3$
$x + 2y = -1$

47. $4x - y = 4$
$x + 2y = 1$

48. $x - y = 1$
$x + 4y = 6$

Exercises 49–64: Use elimination to solve the system.

49. $x + y = 4$
$x - y = 2$

50. $x - 3y = 5$
$-x + 5y = -7$

51. $-3x - y = 5$
$-3x + 2y = -1$

52. $-2x + 3y = 13$
$x + 3y = 7$

53. $2x + y = 4$
$2x - y = -2$

54. $2x + 3y = -6$
$2x - y = 4$

55. $6x - 4y = 12$
$3x + 5y = -6$

56. $-5x - 4y = -8$
$x - 4y = 10$

57. $2x - 4y = 5$
$-x + 2y = 9$

58. $x - 3y = 5$
$2x - 6y = 1$

59. $2x + y = 2$
$4x + 2y = 4$

60. $\frac{1}{2}x + \frac{1}{7}y = 2$
$7x + 2y = 28$

61. $2x + 4y = -22$
$75x + 15y = -120$

62. $-3x + 20y = 67$
$2x + 5y = 47$

63. $0.3x + 0.2y = 0.8$
$0.4x + 0.3y = 1.1$

64. $1.2x + 4.3y = 1.7$
$2.4x - 1.5y = 1.38$

Exercises 65–80: Solve the system of equations.

65. $2x + 3y = 7$
$2x - y = -13$

66. $2x - 3y = -25$
$x + 3y = 10$

67. $3u - 5v = 4$
$5u + v = 2$

68. $4u - 3v = -2$
$-8u + 9v = 5$

69. $2r - 3t = 7$
$-4r + 6t = -14$

70. $r - t = 10$
$-2r + 2t = -20$

71. $m - n = 5$
$m - n = 7$

72. $6m + 9n = 4$
$-4m - 6n = 2$

73. $2x - 3y = 2$
$3x - 5y = 4$

74. $x + 3y = \frac{5}{4}$
$-2x - 7y = -\frac{11}{4}$

75. $0.1x - 0.3y = -5$
$0.5x + 1.1y = 27$

76. $0.6x - 0.2y = 1.8$
$-0.1x - 0.5y = 1.3$

77. $\frac{1}{2}y - \frac{1}{2}z = -1$
$\frac{3}{4}y - \frac{1}{2}z = 1$

78. $\frac{3}{4}y + \frac{1}{4}z = 13$
$-\frac{1}{4}y + \frac{1}{4}z = -3$

79. $\frac{1}{5}x - \frac{2}{5}y = -\frac{3}{5}$
$x - y = 1$

80. $\frac{1}{10}x - \frac{1}{5}y = 2$
$-\frac{1}{5}x + \frac{1}{10}y = -1.9$

USING MORE THAN ONE METHOD

Exercises 81–86: Solve the system of equations (a) graphically, (b) numerically, and (c) symbolically.

81. $x - y = 4$
$x + y = 6$

82. $2x - y = 7$
$3x + y = 3$

83. $5x + 2y = 9$
$3x - y = 1$

84. $-x + 4y = 17$
$3x + 6y = 21$

85. $3x - 2y = 8.5$
$2x + 4y = 3$

86. $-x - 3y = 1.8$
$2x - 5y = 0.8$

*Exercises 87 and 88: **Thinking Generally** Solve the system of equations, where a is a constant and $a \neq 1$.*

87. $ax + y = 4$
$x + y = 4$

88. $2x + y = 1$
$2x + ay = a$

APPLICATIONS

89. *Burning Calories* During strenuous exercise, an athlete can burn 10 calories per minute on a rowing machine, whereas on a stair climber the athlete can burn 11.5 calories per minute. In a 60-minute workout an athlete burns 633 calories by using both exercise machines. How many minutes does the athlete spend on each type of workout equipment? (*Source: Runner's World.*)

90. *Burning Fat Calories* Two athletes engage in a strenuous 40-minute workout on a stair climber.

The heavier athlete burns 58 more fat calories than the lighter athlete. Together they burn a total of 290 fat calories. How many fat calories does each athlete burn? If 1 fat gram equals 9 fat calories, how many fat grams does each athlete burn? (*Source: Runner's World.*)

91. *Mixing Antifreeze* A mixture of water and antifreeze in a car is 10% antifreeze. In colder climates this mixture should contain 50% antifreeze. If the radiator contains 4 gallons of fluid, how many gallons of radiator fluid should be drained and replaced with a mixture containing 80% antifreeze?

92. *Mixing Acid* Determine the milliliters of 10% sulfuric acid and the milliliters of 25% sulfuric acid that should be mixed to obtain 20 milliliters of 18% sulfuric acid.

93. *Hotel Rooms* The revenue from renting 50 hotel rooms is $4945. If premium rooms cost $115 and regular rooms cost $80, find the number of each type of room rented.

94. *Coins* A sample of dimes and quarters totals $18. If there are 111 coins in all, how many of each type of coin are there?

95. *Supplementary Angles* The larger of two supplementary angles is 30° more than twice the smaller angle. Find the angles. (*Hint:* Supplementary angles sum to 180°.)

96. *Complementary Angles* The larger of two complementary angles is 6° less than twice the smaller angle. Find the angles. (*Hint:* Complementary angles sum to 90°.)

97. *Population* In 2010, there were 308 million people in the United States, and females outnumbered males by 5 million. Determine the number of males and the number of females in 2010. (*Source:* U.S. Census Bureau.)

98. *Per Capita Income* In 2009, the average of the per capita incomes for Alaska and Arizona was $38,000. The per capita income in Alaska exceeded the per capita income in Arizona by $10,000. Find the per capita income for each state. (*Source:* Bureau of Economic Analysis.)

99. *Airplane Speed* A plane flies 2400 miles from Orlando, Florida, to Los Angeles, California, against the wind in 4 hours and 10 minutes. Then the plane flies back to Orlando with the wind in 3 hours and 45 minutes. Find the speed of the plane with no wind and the speed of the wind.

100. *Airplane Speed* An airplane flies with the wind and travels 3000 miles in 5 hours. The return trip into the wind requires 6 hours. Find the speed of the wind and of the airplane in no wind.

101. *Boat Speed* A boat travels downstream 150 miles in 5 hours. The return trip takes 7 hours and 30 minutes. Find the speed of the boat without a current and the speed of the current.

102. *River Current* A tugboat can push a barge 165 miles upstream in 33 hours. The same tugboat and barge can make the return trip downstream in 15 hours. Determine the speed of the current and the speed of the tugboat when there is no current.

103. *Student Loans* A student takes out two loans to help pay for college. One loan is at 8% interest, and the other is at 9% interest. The total amount borrowed is $3500, and the interest after 1 year for both loans is $294. Find the amount of each loan.

104. *Student Loans* A student takes out two loans totaling $5000 at 4% and 6% annual interest. The total interest after one year is $254. Find the amount of each loan.

105. *College Tuition* A student takes 12 credits of day classes and 6 credits of night classes. A credit for day classes costs $30 more than a credit for night classes. If the student's total tuition is $1800, how much does each type of credit cost?

106. *Internet Gambling* Internet gambling has increased dramatically since 2000. The combined revenues for 2009 and 2010 were $47 billion, with a 4% increase from 2009 to 2010. Estimate the Internet gambling revenues in 2009 and in 2010 to the nearest billion dollars. (*Source:* Christiansen Capital Advisors, LLC.)

107. *Roof Trusses* Linear systems are used in the design of roof trusses for buildings (see the accompanying figures). One of the simplest types of roof trusses is an equilateral triangle. If a 200-pound force is applied to the peak of the truss, the weights W_1 and W_2 exerted on each rafter are determined by the following system of equations. Solve this system.

$$W_1 - W_2 = 0$$
$$\frac{\sqrt{3}}{2}(W_1 + W_2) = 200$$

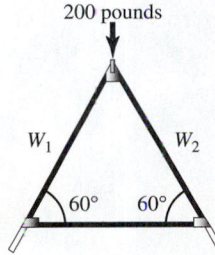

108. *Roof Trusses* (Refer to Exercise 107.) The weights W_1 and W_2 exerted on each rafter for the roof truss shown in the accompanying figure are determined by the following system of equations. Solve the system and interpret the result.

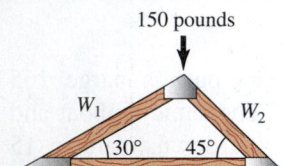

150 pounds

$$W_1 + \sqrt{2}W_2 = 300$$
$$\sqrt{3}W_1 - \sqrt{2}W_2 = 0$$

109. *Geometry* A basketball court has a perimeter of 296 feet. Its length is 44 feet greater than its width. Set up a system of two linear equations whose solution gives the length and width of the court. Solve the system (a) symbolically, (b) graphically, and (c) numerically.

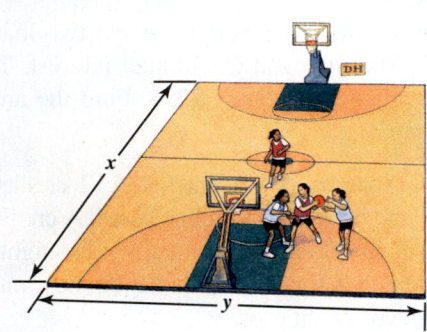

110. *Rectangle* The perimeter of a rectangle is 127 feet. The rectangle's length is 11 feet longer than its width. Find the dimensions of the rectangle.

111. *Rectangle* The perimeter of a rectangle is 41 meters. The rectangle's length is 5.5 meters longer than its width. Find the dimensions of the rectangle.

112. *Concert Tickets* Concert tickets cost $55 and $40. If 2000 tickets were sold for total receipts of $90,500, find how many of each type of ticket were sold (a) symbolically, (b) graphically, and 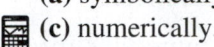 (c) numerically.

WRITING ABOUT MATHEMATICS

113. Explain what it means for a linear system to be inconsistent. Give an example.

114. Describe two possible ways that you could use elimination to solve the following system of linear equations. (Do not actually solve the system.)

$$ax - y = c$$
$$x + dy = e$$

SECTIONS 4.1 and 4.2 — Checking Basic Concepts

1. Solve the system of linear equations
$$2x - y = 5$$
$$-x + 3y = 0$$
   (a) graphically and
   (b) numerically.
   (c) Do your answers agree?

2. Use substitution to solve the system of equations.
$$4x - 3y = -1$$
$$x + 4y = 14$$

3. Use elimination to solve the system of equations. Are the equations dependent? Is the system inconsistent?
$$4x - 3y = -17$$
$$-6x + 2y = 23$$

4. *Complementary Angles* The smaller of two complementary angles is 40° less than the larger angle.
   (a) Write a system of linear equations whose solution gives the measure of each angle. (*Hint:* Complementary angles sum to 90°.)
   (b) Solve the system and interpret the result.

4.3 Systems of Linear Inequalities

Solving Linear Inequalities in Two Variables ●
Solving Systems of Linear Inequalities

A LOOK INTO MATH ▶ People often walk or jog in an effort to increase their heart rates and get in better shape. During strenuous exercise, older people should maintain lower heart rates than younger people. A person cannot maintain precisely one heart rate, so a range of heart rates is recommended by health professionals. For aerobic fitness, a 50-year-old's heart rate might be between 120 and 140 beats per minute, whereas a 20-year-old's heart rate might be between 140 and 160 beats per minute. Systems of linear inequalities can be used to model these situations. See Example 4.

NEW VOCABULARY

☐ Linear inequality in two variables
☐ Test point
☐ System of linear inequalities in two variables

Solving Linear Inequalities in Two Variables

▶ **REAL-WORLD CONNECTION** Suppose that candy costs $2 per pound and that peanuts cost $1 per pound. The total cost C of buying x pounds of candy and y pounds of peanuts is given by

$$C = 2x + y. \quad \text{Cost of candy and peanuts}$$

If we have at most **$5** to spend, the inequality

$$2x + y \le 5 \quad \text{Cost less than or equal to \$5}$$

must be satisfied. To determine the different weight combinations of candy and peanuts that could be bought, we could use the graph shown in Figure 4.19. Points located on the line $2x + y = 5$ represent weight combinations resulting in *exactly* a $5 purchase. For example, the point (**2**, **1**) satisfies the equation $2x + y = 5$ and represents buying **2** pounds of candy at $2 per pound and **1** pound of peanuts at $1 per pound for $5. Ordered pairs (x, y) located in the shaded region *below* the line represent purchases of *less than* $5. The point (**1**, **2**) lies in the shaded region and represents buying **1** pound of candy and **2** pounds of peanuts for $4, which is (clearly) less than $5. Note that if we substitute $x = 1$ and $y = 2$ in the inequality, we obtain

$$2(\mathbf{1}) + (\mathbf{2}) \le 5,$$

which is a true statement. (Note that points *above* the line represent purchases of more than $5.)

Purchases of Candy and Peanuts

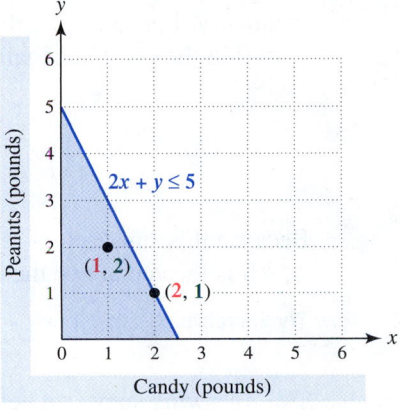

Figure 4.19

When the equals sign in a linear equation in two variables is replaced with one of the symbols $<$, $\leq$, $>$, or $\geq$, a **linear inequality in two variables** results. Examples include

$$2x + y \leq 5, \quad y \geq x - 5, \quad \text{and} \quad \frac{1}{2}x - \frac{3}{5}y < 8. \qquad \text{Linear inequalities in two variables}$$

EXAMPLE 1 **Solving linear inequalities**

Shade the solution set for each inequality.
(a) $x > 1$ **(b)** $y \leq 2x - 1$ **(c)** $x - 2y < 4$

Solution
(a) Begin by graphing the vertical line $x = 1$ with a *dashed* line because equality is *not* included. The solution set includes all points with x-values greater than 1, so shade the region to the right of this line, as shown in Figure 4.20(a).

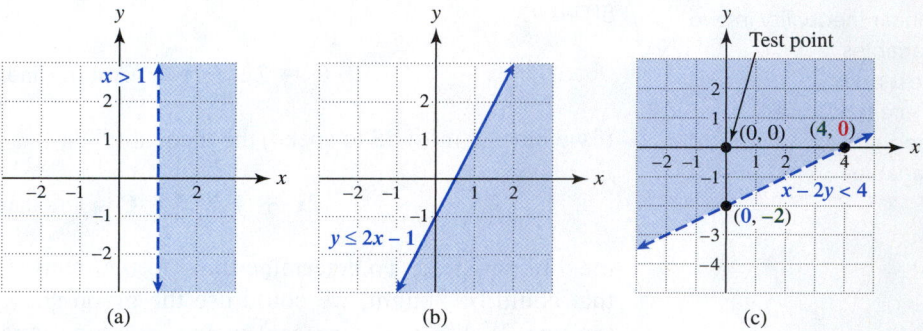

(a)   (b)   (c)

Figure 4.20

(b) Start by graphing $y = 2x - 1$ with a *solid* line because equality *is* included. The inequality sign is $\leq$, so the solution set includes all points on or below this line, as shown in Figure 4.20(b).

(c) Rather than solving for y, we start by finding the intercepts for the line $x - 2y = 4$.

x-intercept: $y = \mathbf{0}$ implies $x - 2(\mathbf{0}) = 4$ or $x = \mathbf{4}$

y-intercept: $x = \mathbf{0}$ implies $\mathbf{0} - 2y = 4$ or $y = -\mathbf{2}$

Plot the points $(\mathbf{4}, \mathbf{0})$ and $(\mathbf{0}, -\mathbf{2})$ and sketch a dashed line, as shown in Figure 4.20(c). To determine whether to shade above or below this line, substitute a *test point* in the inequality. For example, if we pick the *test point* $(0, 0)$ and then substitute $x = 0$ and $y = 0$ in the given inequality, we find the following result.

$$x - 2y < 4 \qquad \text{Given inequality}$$
$$0 - 2(0) < 4 \qquad \text{Let } x = 0 \text{ and } y = 0.$$
$$0 < 4 \checkmark \qquad \text{A true statement}$$

Because this substitution results in a true statement, shade the region *containing* $(0, 0)$, which is located above the dashed line.

Now Try Exercises 7, 11, 19

The following steps can be used to graph a linear inequality in two variables.

SOLVING A LINEAR INEQUALITY GRAPHICALLY

1. Replace the inequality symbol with an equals sign.
2. Graph the resulting line. You may want to plot the x- and y-intercepts first. Use a solid line if the inequality symbol is $\leq$ or $\geq$ and a dashed line if it is $<$ or $>$.
3. (a) If the inequality is in the form $x \leq k$ or $x < k$ (where k is a constant), shade to the *left* of the vertical line. If the inequality is in the form $x \geq k$ or $x > k$, shade to the *right* of the vertical line.
   (b) If the inequality is in the form $y \leq mx + b$ or $y < mx + b$, shade *below* this line. If the inequality is in the form $y \geq mx + b$ or $y > mx + b$, shade *above* this line.
   (c) If you are uncertain as to which region to shade, choose a **test point** that is *not* on the line. Substitute it in the given inequality. If the test point makes the inequality true, then shade the region containing the test point. Otherwise, shade on the other side of the line.

| EXAMPLE 2 | **Solving a linear inequality graphically** |

Shade the solution set for the linear inequality $4x - 3y < 12$.

Solution

Start graphing the line $4x - 3y = 12$ by finding its intercepts.

$$x\text{-intercept: } y = 0 \text{ implies } 4x - 3(0) = 12 \text{ or } x = 3.$$
$$y\text{-intercept: } x = 0 \text{ implies } 4(0) - 3y = 12 \text{ or } y = -4.$$

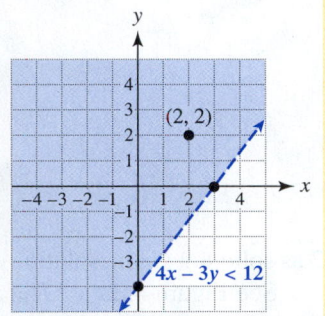

Figure 4.21

Plot the points $(3, 0)$ and $(0, -4)$ and sketch a dashed line because the inequality symbol is $<$. Any point that is not on this line can be used for a test point. We use the test point $(2, 2)$.

$4x - 3y < 12$	Given inequality
$4(2) - 3(2) < 12$	Let $x = 2$ and $y = 2$.
$2 < 12$ ✓	A true statement

Thus shade the region containing the point $(2, 2)$, as shown in Figure 4.21.

Now Try Exercise 21

Solving Systems of Linear Inequalities

A **system of linear inequalities** results when the equals signs in a system of linear equations are replaced with $<$, $\leq$, $>$, or $\geq$. For example, the system of linear equations

$$x + y = 4 \quad \text{System of linear equations}$$
$$y = x$$

STUDY TIP

A test point is helpful when locating the region that should be shaded.

becomes a system of linear inequalities when it is written as

$$x + y \leq 4 \quad \text{System of linear inequalities}$$
$$y \geq x.$$

A solution to a system of inequalities must satisfy *both* inequalities. For example, the ordered pair $(1, 2)$ is a solution to this system because substituting $x = 1$ and $y = 2$ makes both inequalities true.

$$1 + 2 \leq 4 \ ✓ \quad \text{True}$$
$$2 \geq 1 \ ✓ \quad \text{True}$$

To see *graphically* that $(1, 2)$ is a solution, consider the following. The solution set for $x + y \leq 4$ consists of points lying on the line $x + y = 4$ and all points below the line. This region is shaded in Figure 4.22(a). Similarly, the solutions to $y \geq x$ include the line $y = x$ and all points above it. This region is shaded in Figure 4.22(b).

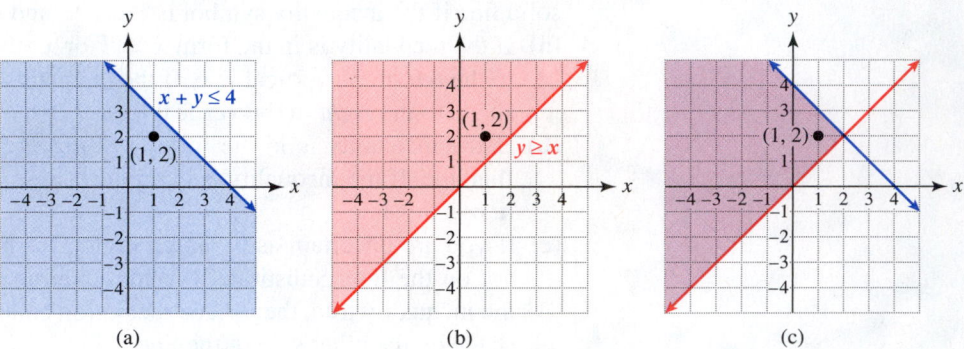

Figure 4.22

For a point (x, y) to be a solution to the *system* of linear inequalities, it must be located in *both* shaded regions shown in Figures 4.22(a) and 4.22(b). The *intersection* of the shaded regions is shown in Figure 4.22(c). Note that the point $(1, 2)$ is located in each shaded region shown in Figures 4.22(a) and 4.22(b). Therefore the point $(1, 2)$ is located in the shaded region shown in Figure 4.22(c) and is a solution of the system of linear inequalities.

EXAMPLE 3 **Solving a system of linear inequalities**

Shade the solution set for each system of inequalities.
(a) $x \leq -1$ **(b)** $y < 2x$ **(c)** $2x - y < 2$
    $y \geq 2$ $x + y > 3$ $x + 2y \geq 6$

Solution
(a) Graph the vertical line $x = -1$ and the horizontal line $y = 2$ as solid lines. The solution set is to the left of the line $x = -1$ and above the line $y = 2$. See Figure 4.23(a). The test point $(-2, 3)$ satisfies both inequalities and lies in the shaded region. Note that the boundary to the region is included because both inequalities include equality $(\leq, \geq)$.

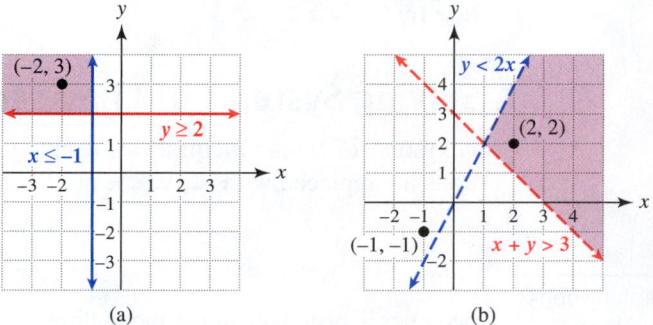

Figure 4.23

(b) Graph $y = 2x$ and $x + y = 3$ as dashed lines. These two lines divide the xy-plane into four regions. The test point $(-1, -1)$ does not satisfy the inequalities, so do not shade the region containing it. However, the point $(2, 2)$ does satisfy *both* inequalities. To verify this fact, substitute $x = 2$ and $y = 2$ in each inequality.

$$2 < 2(2) \checkmark \quad \text{True}$$

$$2 + 2 > 3 \quad \checkmark \quad \text{True}$$

Thus shade the region shown in Figure 4.23(b).

CRITICAL THINKING

Is the point where the two lines intersect in Figure 4.22(c) (on page 256) part of the solution set? Why or why not? Repeat this question for Figure 4.24.

(c) Graph $2x - y = 2$ and $x + 2y = 6$ as dashed and solid lines, respectively. The test point $(2, 3)$ satisfies *both* inequalities, so shade the region containing it, as shown in Figure 4.24.

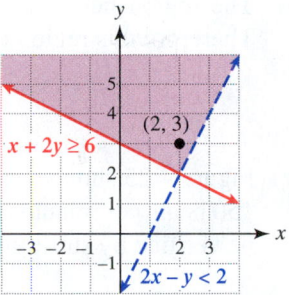

Figure 4.24

■ **Now Try Exercises 27, 33, 37**

NOTE: Finding a test point that satisfies a system of linear inequalities may require trial and error. You may need to pick one test point from each region determined by the two lines.

▶ **REAL-WORLD CONNECTION** In the next example we demonstrate how systems of inequalities can model the situation discussed in A Look Into Math for this section.

EXAMPLE 4   **Modeling target heart rates**

When exercising, people often try to maintain target heart rates that are percentages of their maximum heart rates. A person's maximum heart rate (MHR) is $MHR = 220 - A$, where A represents age and the MHR is in beats per minute. The shaded region shown in Figure 4.25 represents target heart rates for aerobic fitness for various ages. (*Source*: Hebb Industries, Inc.)

(a) Estimate the range R of heart rates that are acceptable for someone 40 years old.

(b) By choosing two points on each line in Figure 4.25 and applying the point–slope form, we can show that the equations for these lines are approximately

$$T = -0.8A + 196 \quad \text{Upper line in target region}$$

and

$$T = -0.7A + 154, \quad \text{Lower line in target region}$$

where A represents age and T represents the target heart rate. Write a system of inequalities whose solution set is the shaded region, including the two lines.

(c) Use Figure 4.25 to determine whether $(30, 150)$ is a solution. Then verify your answer by using the system of inequalities.

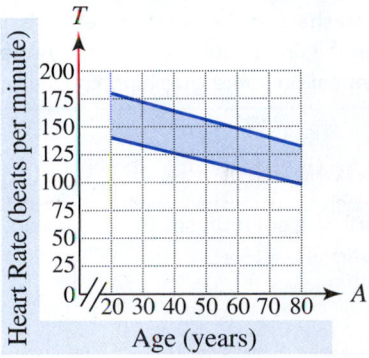

Figure 4.25

Target Heart Rates

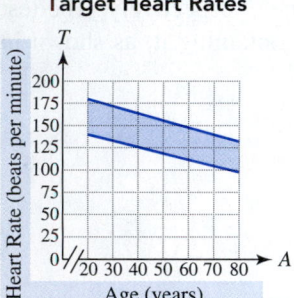

Figure 4.25 (Repeated)

Solution

(a) Figure 4.25 reveals that when $A = 40$, the target heart rates T are *approximately* $125 \leq T \leq 165$ beats per minute.

(b) The region lies **below** the upper line, **above** the lower line, and includes both lines. Therefore this region is modeled by

$$T \leq -0.8A + 196$$
$$T \geq -0.7A + 154.$$

(c) Figure 4.25 shows that the point representing a 30-year-old person with a heart rate of 150 beats per minute lies in the shaded region, so (30, 150) is a solution. This result can be verified by substituting $A = 30$ and $T = 150$ in the system.

$$150 \leq -0.8(30) + 196 = 172 \checkmark \quad \text{True}$$
$$150 \geq -0.7(30) + 154 = 133 \checkmark \quad \text{True}$$

∎ **Now Try Exercise 67**

EXAMPLE 5 **Solving a system of linear inequalities with technology**

Shade the solution set for the system of inequalities, using a graphing calculator.

$$2x + y \leq 5$$
$$-2x + y \geq 1$$

Solution

Begin by solving each inequality for y to obtain $y \leq 5 - 2x$ and $y \geq 2x + 1$. The graphs of $Y_1 = 5 - 2X$ and $Y_2 = 2X + 1$ are shown in Figure 4.26(a). The solution set lies below y_1 and above y_2. Figure 4.26(b) shows how to shade this region. The solution set is the region comprising small squares in Figure 4.26(c).

$[-15, 15, 5]$ by $[-10, 10, 5]$ $\qquad\qquad\qquad\qquad\qquad$ $[-15, 15, 5]$ by $[-10, 10, 5]$

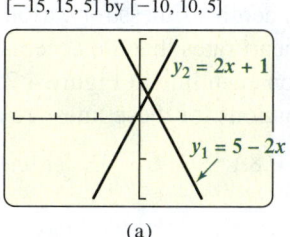

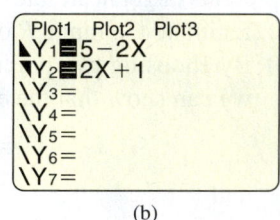

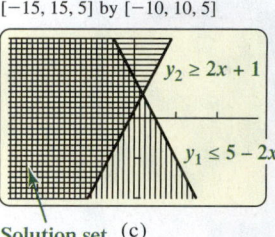

(a) $\qquad\qquad\qquad\qquad$ (b) $\qquad\qquad\qquad\qquad$ Solution set (c)

CALCULATOR HELP

To shade a graph, see Appendix A (pages AP-6 and AP-7).

Figure 4.26

∎ **Now Try Exercise 51**

TECHNOLOGY NOTE

Shading of Linear Inequalities
When shading the solution set for a linear inequality, graphing calculators often show solid lines even if a line should be dashed. For example, the graphs for $y < 5 - 2x$ and $y \leq 5 - 2x$ are *identical* on some graphing calculators.

CRITICAL THINKING

Graph the solution set to each inequality in the same *xy*-plane. Does the system have any solutions?

$$3x + y \geq 6$$
$$3x + y \leq 2$$

4.3 Putting It All Together

CONCEPT	EXPLANATION	EXAMPLES
Linear Inequality in Two Variables	$ax + by > c$, where a, b, and c are constants. ($>$ may be replaced with $<$, $\leq$, or $\geq$.)	$y \geq 5$, $x - y < -10$, and $6x - 7y \leq 22$
System of Linear Inequalities in Two Variables	Two or more linear inequalities where a solution must satisfy *all* inequalities	$x + y < 11$ $x - y \geq 4$ $(8, 2)$ is a solution because $x = 8$ and $y = 2$ make *both* inequalities true.
Solution Set to a System of Linear Inequalities in Two Variables	Set of all solutions; typically a region in the xy-plane (infinitely many solutions) Two intersecting lines divide the xy-plane into four regions. To determine which region should be shaded, choose one test point from each region. The region that contains the test point satisfying both inequalities should be shaded.	$y \leq 5 - x$ (below $y = 5 - x$) and $y < 2x - 3$ (below $y = 2x - 3$) Use a dashed line for $<$ or $>$. Use a solid line for $\leq$ or $\geq$. The test point $(3, -2)$ satisfies both inequalities.

4.3 Exercises

MyMathLab   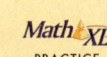 Math XL PRACTICE   WATCH   DOWNLOAD   READ   REVIEW

CONCEPTS AND VOCABULARY

1. If a system of inequalities contains two inequalities in two variables, how many of the inequalities does a solution satisfy?

2. Can a system of linear inequalities have infinitely many solutions? Explain.

3. Does $(3, 1)$ satisfy $5x - 2y > 8$? Does $(2, 1)$ satisfy this inequality?

4. Does $(3, 4)$ satisfy the following system of inequalities?

$$x - 2y \geq -8$$
$$2x - y < 1$$

Is $(-2, 0)$ a solution to the system?

5. When you are graphing the solution set to $y \leq x$, the boundary should be a (dashed/solid) line.

6. When you are graphing the solution set to $2x - y > 2$, the boundary should be a (dashed/solid) line.

LINEAR INEQUALITIES

Exercises 7–24: Shade the solution set in the xy-plane.

7. $x > 2$

8. $x \leq -1$

9. $y \leq 1$

10. $y > 0$

11. $y < -2x$

12. $y < 2x$

13. $y \geq x + 1$

14. $y \geq x - 2$

15. $y \geq 3x$

16. $y \leq 1 - x$

17. $y < 3x - 2$

18. $x + y > 3$

19. $2x - y \leq 4$

20. $4x + 2y < 8$

21. $-x + 3y > 3$

22. $-2x + y \leq 0$

23. $5x - 2y \leq -10$

24. $-3x + 2y < 6$

Exercises 25 and 26: **Thinking Generally** *Shade the solution set.*

25. $x > a$ if $a < 0$

26. $x + y \geq a$ if $a > 0$

SYSTEMS OF INEQUALITIES

Exercises 27–48: Shade the solution set in the xy-plane.

27. $x \geq 2$
$y \leq 1$

28. $x < 3$
$y > -2$

29. $x \leq 2$
$y > -x$

30. $y \geq 2$
$y \geq x$

31. $x \geq y$
$y \leq 3$

32. $y > x + 1$
$y < 2$

33. $y \geq x$
$x \leq -2y$

34. $y > x - 1$
$x > y - 2$

35. $x + y < 3$
$x - y < 3$

36. $2x + y > -2$
$x + y \leq -1$

37. $2x + 3y \geq 6$
$x + 2y < 3$

38. $-x + y \leq 0$
$3x + y \leq 3$

39. $\frac{1}{2}x + y \leq 1$
$x - \frac{1}{3}y < 1$

40. $2x - y > 3$
$y \leq x + 2$

41. $x \geq 1 + 2y$
$y \leq 1 - x$

42. $4x + y \leq 4$
$x - 3y \leq 3$

43. $\frac{1}{2}x - y \geq 1$
$x - \frac{1}{3}y \geq 1$

44. $2x + y > 4$
$x - y > 1$

45. $3x < 6 - 2y$
$2y \geq 2 - 2x$

46. $-4x + 6y > 12$
$2x - 4y < 8$

47. $\frac{1}{2}x - \frac{1}{2}y > 4$
$-x - \frac{1}{3}y < 1$

48. $-\frac{2}{3}x + \frac{1}{6}y \geq \frac{4}{3}$
$-\frac{1}{4}x - \frac{1}{3}y \geq 1$

 Exercises 49–54: (Refer to Example 5.) Use a graphing calculator to shade the solution set for the system of inequalities.

49. $y \leq 5$
$y \geq -3$

50. $y \geq -x$
$y \leq x + 1$

51. $x + 2y \geq 8$
$6x - 3y \geq 10$

52. $y \geq 2.1x - 3.5$
$y \leq 2.1x - 1.7$

53. $0.9x + 1.7y \leq 3.2$
$1.9x - 0.7y \leq 1.3$

54. $21x \geq 31y - 51$
$5x - 17y \leq 18$

Exercises 55–58: Match the given inequality or system of inequalities with its graph (a.–d.).

55. $x + y \geq 2$

56. $y \leq x + 1$
$y \geq x - 3$

57. $y \leq \frac{1}{2}x$

58. $y \geq x + 1$
$y \leq 5$

a.
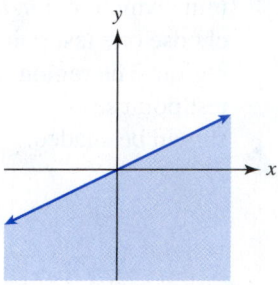

b.
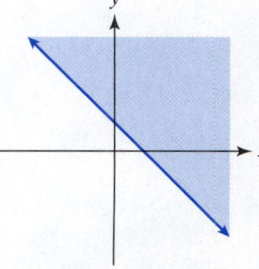

c.
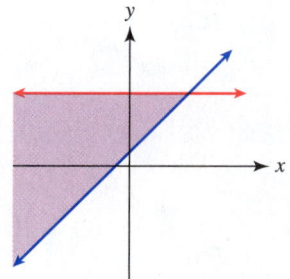

d.
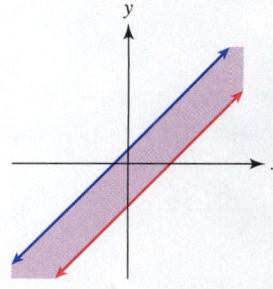

Exercises 59–62: Use the graph to write the inequality or system of inequalities.

59.
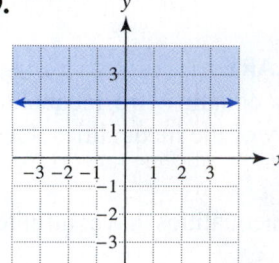

60.
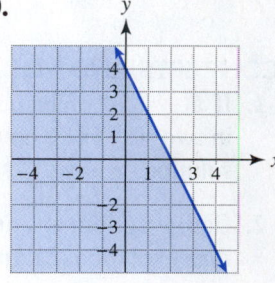

61.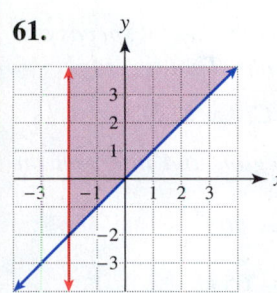

62.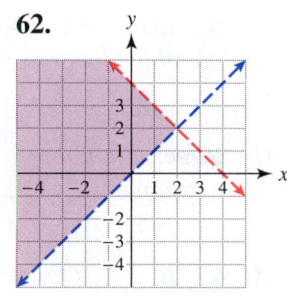

(b) The upper line has a slope of -0.6 and passes through the point $(20, 140)$, and the lower line has a slope of -0.5 and passes through the point $(20, 110)$. Write a system of inequalities whose solution set lies in the shaded region in the graph.

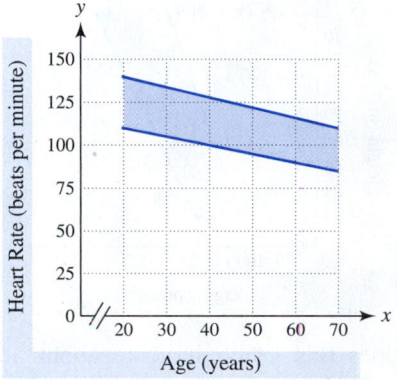

APPLICATIONS

63. *Sales* Suppose that candy costs \$3 per pound and cashews cost \$5 per pound. Let x be the number of pounds of candy bought and y be the number of pounds of cashews bought. Graph the region in the xy-plane that represents all possible weight combinations that cost *less than* \$15.

64. *Tickets and Popcorn* At a movie theater, tickets cost \$8 and a bag of popcorn costs \$4. Let x be the number of tickets bought and y be the number of bags of popcorn bought. Graph the region in the xy-plane that represents all possible combinations of tickets and bags of popcorn that cost \$32 or less.

65. *Retail Sales* A small business manufactures compact disc players and radios. Because every CD player contains a radio, the business must make at least as many radios as CD players. The business can make at most a total of 40 compact disc players and radios per day. Let x be the number of compact disc players manufactured and y be the number of radios manufactured. Graph the region that represents all possible combinations of CD players and radios that can be manufactured in one day.

66. *Retail Sales* A small business can manufacture at most 40 large crates and 30 small crates per day. Let x be the number of large crates manufactured per day and y be the number of small crates manufactured per day. Graph the region in the xy-plane that represents all possible combinations of large and small crates that can be manufactured.

67. *Target Heart Rate* (Refer to Example 4.) The graph at the top of the next column shows target heart rates for general health and weight loss for a person x years old. These heart rates should be maintained for longer periods of time than the times specified for aerobic fitness.

(a) What range of heart rates is recommended for someone 30 years old?

68. *Deserts, Grasslands, and Forests* Two factors that have a critical effect on plant growth are temperature and precipitation. If a region has too little precipitation, it will be a desert. Forests tend to grow in regions where trees can exist at relatively low temperatures and rainfall is sufficient. At other levels of precipitation and temperature, grasslands may prevail. The following figure illustrates the relationship among forests, grasslands, and deserts as suggested by average annual temperature T in degrees Fahrenheit and precipitation P in inches. (*Source:* A. Miller and J. Thompson, *Elements of Meteorology.*)

(a) Determine a system of linear inequalities that describes where grasslands are likely to occur.

(b) Bismarck, North Dakota, has an average annual temperature of 40°F and precipitation of 15 inches. According to the graph, what type of plant growth would you expect near Bismarck? Do these values satisfy the system of inequalities from part (a)?

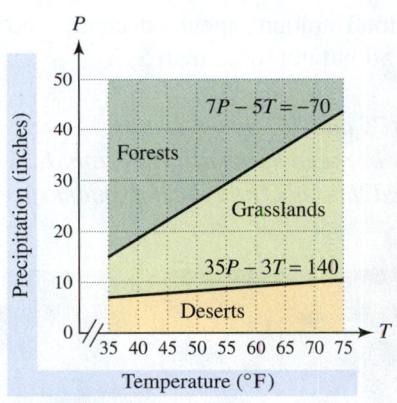

Exercises 69–72: **Weight and Height** *The following figure shows a weight and height graph. The weight w is listed in pounds, and the height h in inches. The shaded area is the recommended region.* (**Source:** Department of Agriculture.)

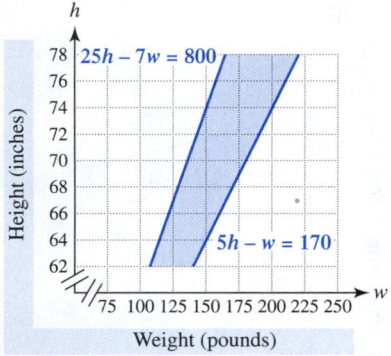

Weight (pounds)

69. What does this graph indicate about an individual weighing 125 pounds with a height of 70 inches?

70. For a person 74 inches tall, use the graph to estimate the recommended weight range.

71. Use the graph to find a system of linear inequalities that describes the recommended region.

72. Explain why inequalities, rather than equalities, are more appropriate for describing recommended weight and height combinations.

GRAPH SKETCHING

Exercises 73 and 74: Suppose that the cost of candy is $3 per pound and the cost of peanuts is $2 per pound. Let x be the number of pounds of candy purchased and let y be the number of pounds of peanuts purchased. Sketch a graph that shows all possible weights of candy and peanuts that could be purchased for each situation.

73. The amount spent on candy is $6 or less and the amount spent on peanuts is $8 or less.

74. The total amount spent on candy and peanuts is at least $6 but not more than $12.

Exercises 75 and 76: **Speed Limits** *A car is driving along a freeway between mile markers A and B, where the minimum speed limit is C and the maximum speed limit is D.*

(a) *Shade the region that gives the lawful speeds and possible locations for the car. That is, graph*

$$x \geq A, \quad x \leq B, \quad y \geq C, \quad y \leq D.$$

(b) *Select one point in the region and interpret the coordinates.*

75. $A = 5$, $B = 9$, $C = 40$, $D = 70$

76. $A = 3$, $B = 7$, $C = 45$, $D = 65$

77. *Climatic Boundaries* Scientists have set boundaries for key environmental processes that, if crossed, could threaten Earth's habitability. The boundary for carbon dioxide (CO_2) concentrations is less than or equal to 350 parts per million (ppm), and the boundary for ozone thickness is greater than or equal to 275 Dobson units. (*Source: Scientific American*, April 2010.)

(a) Graph CO_2 units on the x-axis and ozone units on the y-axis. Shade the region where habitability of Earth is within the boundaries.

(b) Plot the point $(387, 283)$. This represents the current state of CO_2 and ozone on Earth. Is this point within the boundaries?

78. *Climatic Boundaries* (Refer to Exercise 77.) The boundary for percent of land used as cropland is 15% or less, and the boundary for human freshwater consumption in cubic kilometers per year is 4000.

(a) Graph cropland use on the x-axis and freshwater consumption units on the y-axis. Shade the region where habitability of Earth is within the boundaries.

(b) Plot the point $(11.5, 2600)$. This represents the current state of cropland use and freshwater consumption on Earth. Is this point within the boundaries?

WRITING ABOUT MATHEMATICS

79. Explain how a system of linear inequalities in two variables can have no solutions. Give an example to illustrate your answer.

80. A student changes the system of linear inequalities

$$
\begin{array}{c}
2x - y \geq 6 \\
x - y \leq 4
\end{array}
\quad \text{to} \quad
\begin{array}{c}
y \geq 2x - 6 \\
y \leq \quad x - 4.
\end{array}
$$

The student tests the point $(-2, -8)$, which satisfies the second system but not the first. Explain the student's errors.

Group Activity   Working with Real Data

Directions: Form a group of 2 to 4 people. Select someone to record the group's responses for this activity. All members of the group should work cooperatively to answer the questions. If your instructor asks for your results, each member of the group should be prepared to respond.

1. *Netflix Subscribers* In 2006 Netflix had about 6 million subscribers. By 2009 this number had increased to about 12 million. (*Source:* Netflix.)
   (a) Set up a linear system to find values for a and b so that the graph of $y_1 = ax + b$ passes through the points (2006, 6) and (2009, 12). (*Hint:* To obtain the first equation, let $x = 2006$ and $y_1 = 6$ in $y_1 = ax + b$. To obtain the second equation, let $x = 2009$ and $y_1 = 12$.)
   (b) Have one member of the group solve the system symbolically, a second member solve it graphically, and a third member solve it numerically. If the group consists of only two members, solve it symbolically and graphically. How do your answers compare?

2. *Blockbuster Stores* In 2006 Blockbuster had about 5300 stores. By 2009 this number had decreased to 4400. (*Source:* Blockbuster.)
   (a) Set up a linear system to find values for c and d so that the graph of $y_2 = cx + d$ passes through the points (2006, 5300) and (2009, 4400).
   (b) Have one member of the group solve the system symbolically, a second member solve it graphically, and a third member solve it numerically. If the group consists of only two members, solve it symbolically and graphically. How do your answers compare?

3. *Rented Movie Sales* Discuss how the data in Exercises 1 and 2 are related. Do you think this trend will continue or change?

4.4   Introduction to Linear Programming

Basic Concepts • Region of Feasible Solutions • Solving Linear Programming Problems

A LOOK INTO MATH ▶

During World War II, large numbers of troops were on the front. Keeping these soldiers supplied with equipment and food was an essential but complex task. To solve this problem, a new type of mathematics called *linear programming* was invented. Today, linear programming is important to business and the social sciences because it is a procedure that can be used to optimize quantities such as cost and profit. Linear programming applications frequently contain thousands of variables. In this section, we focus on problems involving only two variables. However, the concepts discussed in this section are important to your understanding of larger problems.

NEW VOCABULARY

☐ Objective function
☐ Constraint
☐ Feasible solutions
☐ Linear programming problem
☐ Optimal value
☐ Vertex

Basic Concepts

▶ **REAL-WORLD CONNECTION** Suppose that a small business sells candy for $3 per pound and freshly ground coffee for $5 per pound. All inventory is sold by the end of the day. The revenue R collected in dollars is given by

$$R = 3x + 5y,$$

where x is the pounds of candy sold and y is the pounds of coffee sold. For example, if the business sells **80** pounds of candy and **40** pounds of coffee during a day, then its revenue is

$$R = 3(80) + 5(40) = \$440.$$

The function $R = 3x + 5y$ is called an **objective function**.

Suppose also that the company cannot package more than 150 pounds of candy and coffee per day. Then the inequality

$$x + y \le 150$$

represents a **constraint** on the objective function, which limits the company's revenue for any one day. A goal of this business might be to maximize the objective function

$$R = 3x + 5y, \qquad \text{Revenue equation}$$

subject to the constraints

$$x + y \le 150 \qquad \text{Constraint on production}$$
$$x \ge 0, y \ge 0. \qquad \text{Variables cannot be negative.}$$

Note that the constraints $x \ge 0$ and $y \ge 0$ are included because the number of pounds of candy or coffee cannot be negative. The problem that we have described is called a *linear programming problem*. Before learning how to solve a linear programming problem, we need to discuss the region of *feasible solutions*.

Region of Feasible Solutions

The constraints for a linear programming problem consist of linear inequalities. These inequalities are satisfied by some points in the xy-plane but not by others. The set of solutions to these constraints is called the **feasible solutions**. For example, the *region of feasible solutions* to the constraints for the business just described is shaded in Figure 4.27.

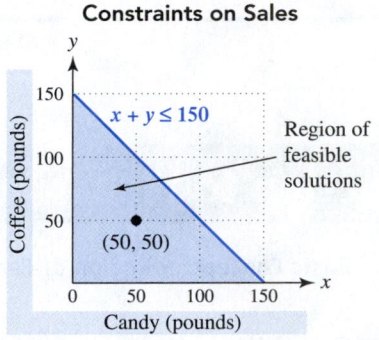

Constraints on Sales

Figure 4.27

The point (50, 50) lies in the shaded region and represents the business selling 50 pounds of candy and 50 pounds of coffee. In the next example, we shade the region of feasible solutions to a set of constraints.

EXAMPLE 1 | **Finding the region of feasible solutions**

Shade the region of feasible solutions to the following constraints.

$$x + 2y \le 30$$
$$2x + y \le 30$$
$$x \ge 0, y \ge 0$$

Solution

The feasible solutions are the ordered pairs (x, y) that *satisfy all four* inequalities. They lie on or below the lines $x + 2y = 30$ and $2x + y = 30$, and on or above the line $y = 0$ and on or to the right of $x = 0$, as shown in Figure 4.28. Note that the inequalities $x \ge 0$ and $y \ge 0$ restrict the feasible solutions to quadrant I and the x- and y-axes.

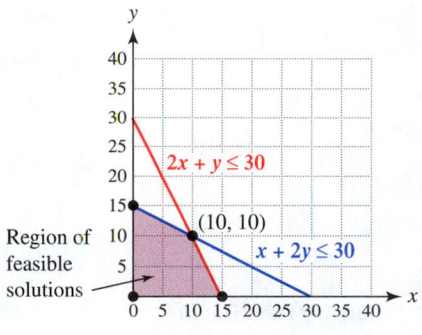

Figure 4.28

■ **Now Try Exercise 15**

STUDY TIP

Be sure that you can graph inequalities (see Example 1) before you attempt to solve a linear programming problem.

Solving Linear Programming Problems

A **linear programming problem** consists of an *objective function* and a system of linear inequalities called *constraints*. The solution set for the system of linear inequalities is called the *region of feasible solutions*. The objective function describes a quantity that is to be optimized. The **optimal value** to a linear programming problem often results in maximum revenue or minimum cost.

When the system of constraints has only two variables, the boundary of the region of feasible solutions often consists of line segments intersecting at points called *vertices* (plural of **vertex**). To solve a linear programming problem we use the *fundamental theorem of linear programming*.

FUNDAMENTAL THEOREM OF LINEAR PROGRAMMING

If the optimal value for a linear programming problem exists, then it occurs at a vertex of the region of feasible solutions.

The fundamental theorem of linear programming is used to solve the following linear programming problem. A justification of this theorem is given after Example 2.

EXAMPLE 2   **Maximizing an objective function**

Maximize the objective function $R = 2x + 3y$ subject to

$$x + 2y \leq 30$$
$$2x + y \leq 30$$
$$x \geq 0, y \geq 0.$$

Solution

The region of feasible solutions is shaded in Figure 4.28 from Example 1. Note that the vertices on the boundary of feasible solutions are $(0, 0)$, $(15, 0)$, $(10, 10)$, and $(0, 15)$. To find the maximum value of R, substitute each vertex in the formula for R. The maximum value of R is **50** when $x = $ **10** and $y = $ **10**. See Table 4.3 on the next page.

TABLE 4.3

Vertex	$R = 2x + 3y$	
(0, 0)	$2(0) + 3(0) = 0$	
(15, 0)	$2(15) + 3(0) = 30$	
(10, 10)	$2(10) + 3(10) = 50$	← **Maximum R is 50.**
(0, 15)	$2(0) + 3(15) = 45$	

■ **Now Try Exercise 33**

The following steps are helpful in solving linear programming word problems.

> ### STEPS FOR SOLVING A LINEAR PROGRAMMING WORD PROBLEM
>
> **STEP 1:** Read the problem carefully. Consider making a table.
>
> **STEP 2:** Write the objective function and all the constraints.
>
> **STEP 3:** Sketch a graph of the region of feasible solutions. Identify all vertices.
>
> **STEP 4:** Evaluate the objective function at each vertex. A maximum (or a minimum) occurs at a vertex.
>
> **NOTE:** If the region is unbounded, a maximum (or minimum) may not exist.

JUSTIFICATION OF THE FUNDAMENTAL THEOREM To better understand the fundamental theorem of linear programming, consider the following example. Suppose that we want to maximize profit $P = 3x + 7y$ subject to the following four constraints:

$$x \geq 1, \qquad x \leq 5, \qquad y \geq x, \qquad \text{and} \qquad y \leq 6.$$

The corresponding region of feasible solutions is shown in Figure 4.29.

Each value of P determines a unique line. For example, if $P = 70$, then the equation for P becomes $3x + 7y = 70$. The resulting line, shown in Figure 4.30, does not intersect the region of feasible solutions. Thus there are no values for x and y that lie in this region and result in a profit of 70. Figure 4.30 also shows the lines that result from letting $P = 0$, 10, and 30. If $P = 10$, then the line intersects the region of feasible solutions only at the vertex $(1, 1)$. This means that if $x = 1$ and $y = 1$, then $P = 3(1) + 7(1) = 10$. If $P = 30$, then the line $3x + 7y = 30$ intersects the region of feasibility infinitely many times. However, it appears that values greater than 30 are possible for P. In Figure 4.31 lines are drawn for $P = 57$, 63, and 70. Notice that there are no points of intersection for $P = 63$ or $P = 70$, but there is one vertex in the region of feasible solutions at $(5, 6)$ that gives $P = 57$. Thus the maximum value of P is 57 and this maximum occurs at a vertex of the region of feasible solutions. The fundamental theorem of linear programming generalizes this result.

CRITICAL THINKING

What is the minimum value for P subject to the given constraints?

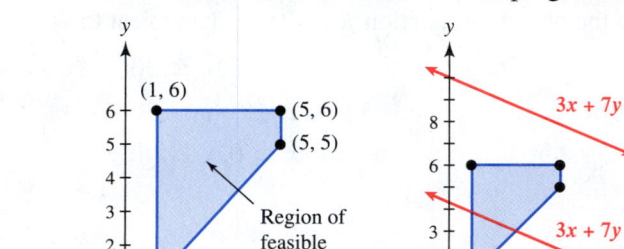

Figure 4.29

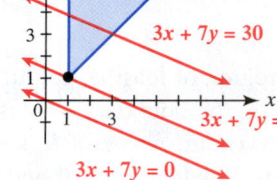

Figure 4.30

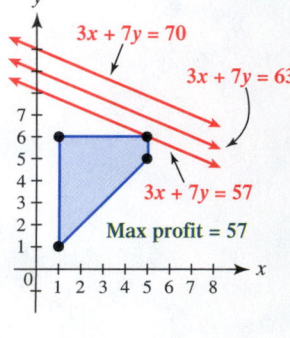

Figure 4.31

EXAMPLE 3 | **Minimizing the cost of vitamins**

A breeder is mixing two different vitamins, Brand X and Brand Y, into pet food. Each serving of pet food should contain at least 60 units of vitamin A and 30 units of vitamin C. Brand X costs 80 cents per ounce and Brand Y costs 50 cents per ounce. Each ounce of Brand X contains 15 units of vitamin A and 10 units of vitamin C, whereas each ounce of Brand Y contains 20 units of vitamin A and 5 units of vitamin C. Determine how much of each brand of vitamin should be mixed to produce a minimum cost per serving.

Solution

STEP 1: Begin by listing the information, as illustrated in Table 4.4.

TABLE 4.4

Brand	Amount	Vitamin A	Vitamin C	Cost
X	x	15	10	80 cents
Y	y	20	5	50 cents
Minimum		60	30	

STEP 2: If x ounces of Brand X are purchased at 80 cents per ounce and if y ounces of Brand Y are purchased at 50 cents per ounce, then the total cost C is given by $C = 80x + 50y$. Because each ounce of Brand X contains 15 units of vitamin A and each ounce of Brand Y contains 20 units of vitamin A, the total number of units of vitamin A is $15x + 20y$. If each serving of pet food must contain at least 60 units of vitamin A, the constraint is $15x + 20y \geq 60$. Similarly, because each serving requires at least 30 units of vitamin C, $10x + 5y \geq 30$. The linear programming problem then becomes the following.

Minimize:	$C = 80x + 50y$	Cost (in cents)
Subject to:	$15x + 20y \geq 60$	Vitamin A constraint
	$10x + 5y \geq 30$	Vitamin C constraint
	$x \geq 0, y \geq 0$	Amounts cannot be negative.

STEP 3: The region containing the feasible solutions is shown in Figure 4.32.

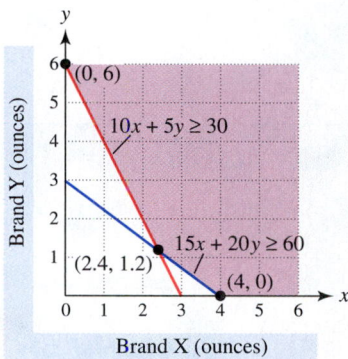

Figure 4.32

NOTE: To determine the vertex (2.4, 1.2), solve the system of equations

$$15x + 20y = 60$$
$$10x + 5y = 30$$

by using elimination.

STEP 4: In Figure 4.32 on the previous page, the vertices are $(0, 6)$, $(2.4, 1.2)$, and $(4, 0)$. Evaluate the objective function $C = 80x + 50y$ at each vertex, as shown in Table 4.5.

TABLE 4.5

Vertex	$C = 80x + 50y$
$(0, 6)$	$80(0) + 50(6) = 300$
$(2.4, 1.2)$	$80(2.4) + 50(1.2) = 252$ ← **Minimum cost (cents)**
$(4, 0)$	$80(4) + 50(0) = 320$

The minimum cost occurs when **2.4** ounces of Brand X and **1.2** ounces of Brand Y are mixed, at a cost of **$2.52** per serving.

Now Try Exercise 47

4.4 Putting It All Together

The following table summarizes the basic concept of linear programming.

CONCEPT	COMMENTS	EXAMPLE
Linear Programming	In a linear programming problem, the maximum or minimum of an objective function is found, subject to a set of constraints. If a solution exists, it occurs at a vertex in the region of feasible solutions.	Maximize $R = x + 2y$, subject to $3x + 2y \leq 15$, $2x + 3y \leq 15$, $x \geq 0$, and $y \geq 0$. The vertices are $(0, 0)$, $(5, 0)$, $(3, 3)$, and $(0, 5)$. The maximum, $R = 10$, occurs at vertex $(0, 5)$.

4.4 Exercises

MyMathLab

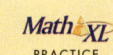 Math XL PRACTICE   WATCH   DOWNLOAD   READ   REVIEW

CONCEPTS AND VOCABULARY

1. A procedure used in business to optimize quantities such as cost and profit is called _____.

2. In linear programming, the function to be optimized is called the _____ function.

3. The region in the xy-plane that satisfies the constraints is called the region of _____.

4. In linear programming, constraints typically consist of a system of _____.

5. If the optimal value for a linear programming problem exists, then it occurs at a(n) _____ of the region of feasible solutions.

6. To find the optimal value in a linear programming problem, substitute each vertex in the _____ function.

REGIONS OF FEASIBLE SOLUTIONS

Exercises 7–20: Shade the region of feasible solutions for the following constraints.

7. $x + y \leq 3$
   $x \geq 0, y \geq 0$

8. $2x + y \leq 4$
   $x \geq 0, y \geq 0$

9. $4x + 3y \leq 12$
   $x \geq 0, y \geq 0$

10. $5x + 3y \leq 15$
   $x \geq 0, y \geq 0$

11. $x \leq 5$
   $y \leq 2$
   $x \geq 0, y \geq 0$

12. $x \leq 3$
   $y \leq 4$
   $x \geq 1, y \geq 1$

13. $x + y \leq 5$
   $x + y \geq 2$
   $x \geq 0, y \geq 0$

14. $2x + y \leq 6$
   $x + y \geq 3$
   $x \geq 0, y \geq 0$

15. $3x + 2y \leq 6$
   $2x + 3y \leq 6$
   $x \geq 0, y \geq 0$

16. $5x + 3y \leq 30$
   $3x + 5y \leq 30$
   $x \geq 0, y \geq 0$

17. $x + y \leq 3$
   $x + 3y \geq 3$
   $x \geq 0, y \geq 0$

18. $3x + y \geq 6$
   $x + 2y \geq 6$
   $x \geq 0, y \geq 0$

19. $x + 2y \geq 4$
   $3x + 2y \geq 6$
   $x \geq 0, y \geq 0$

20. $4x + 3y \geq 12$
   $3x + 4y \geq 12$
   $x \geq 0, y \geq 0$

LINEAR PROGRAMMING

Exercises 21–24: Find the maximum of R on the region of feasible solutions shown in the figure.

21. $R = 4x + 5y$

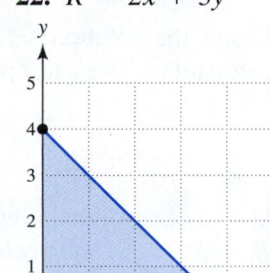

22. $R = 2x + 3y$

23. $R = x + 3y$

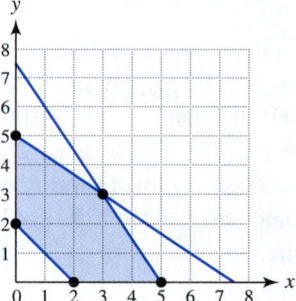

24. $R = 12x + 9y$

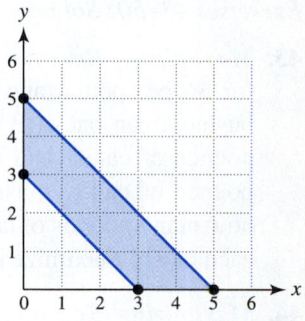

Exercises 25–28: Use the figures in Exercises 21–24 to complete Exercises 25–28, respectively, to minimize C.

25. $C = 2x + 3y$

26. $C = 3x + y$

27. $C = 5x + y$

28. $C = 2x + 7y$

Exercises 29–36: Maximize the objective function R, subject to the given constraints.

29. $R = 3x + 5y$
   $x + y \leq 150$
   $x \geq 0, y \geq 0$

30. $R = 6x + 5y$
   $x + y \leq 8$
   $x \geq 0, y \geq 0$

31. $R = 3x + 2y$
   $2x + y \leq 6$
   $x \geq 0, y \geq 0$

32. $R = x + 3y$
   $x + 2y \leq 4$
   $x \geq 0, y \geq 0$

33. $R = 12x + 9y$
   $3x + y \leq 6$
   $x + 3y \leq 6$
   $x \geq 0, y \geq 0$

34. $R = 10x + 30y$
   $x + 3y \leq 12$
   $3x + y \leq 12$
   $x \geq 0, y \geq 0$

35. $R = 4x + 5y$
   $x + y \geq 2$
   $x + 2y \leq 4$
   $x \geq 0, y \geq 0$

36. $R = 3x + 7y$
   $x + y \geq 1$
   $3x + y \leq 3$
   $x \geq 0, y \geq 0$

Exercises 37–42: Minimize the objective function C, subject to the given constraints.

37. $C = x + 2y$
   $x \leq 3, y \leq 2$
   $x \geq 0, y \geq 0$

38. $C = 3x + y$
   $x \leq 5, y \leq 3$
   $x \geq 1, y \geq 1$

39. $C = 8x + 15y$
   $x + y \geq 4$
   $x \geq 0, y \geq 0$

40. $C = x + 2y$
   $3x + 4y \geq 12$
   $x \geq 0, y \geq 0$

41. $C = 30x + 40y$
   $2x + y \leq 6$
   $x + y \geq 2$
   $x \geq 0, y \geq 0$

42. $C = 50x + 70y$
   $2x + 3y \leq 6$
   $x + y \geq 1$
   $x \geq 0, y \geq 0$

APPLICATIONS

Exercises 43–50: Solve the linear programming problem.

43. *Maximizing Revenue* A small business sells candy for $4 per pound and coffee for $6 per pound. The business can package and sell at most a total of 100 pounds of candy and coffee per day, but at least 20 pounds of candy must be sold each day. Determine how many pounds of candy and coffee need to be sold each day to maximize revenue.

44. *Maximizing Revenue* An organic foods store sells almonds for $10 per pound and flax seed for $5 per pound. The business can package and sell, at most, a total of 50 pounds of almonds and flax seed per day, but it must sell at least 15 pounds of flax seed per day. Determine the maximum revenue.

45. *Minimizing Cost* It costs a business $50 to make a graphing calculator and $20 to make a scientific calculator. Each week the company must produce at least 90 calculators. At least twice as many scientific calculators must be made as graphing calculators. Determine the minimum cost.

46. *Minimizing Cost* It costs a business $20 to make one compact disc player and $10 to make one radio. Each week the company must make a combined total of at least 50 compact disc players and radios. At least as many compact disc players as radios must be manufactured. Determine how many compact disc players and radios should be made to minimize weekly costs.

47. *Vitamin Cost* A pet owner is mixing two different brands of vitamins, Brand X and Brand Y, into pet food. Brand X costs 90 cents per ounce and Brand Y costs 60 cents per ounce. Each serving is a mixture of the two brands and should contain at least 40 units of vitamin A and 30 units of vitamin C. Each ounce of Brand X contains 20 units of vitamin A and 10 units

of vitamin C, whereas each ounce of Brand Y contains 10 units of vitamin A and 10 units of vitamin C. Determine how much of each brand of vitamin should be mixed to produce a minimum cost per serving.

48. *Pet Food Cost* A pet owner is buying two brands of food, X and Y, for his animals. Each serving of the mixture of the two foods should contain at least 60 grams of protein and 40 grams of fat. Brand X costs 75 cents per unit and Brand Y costs 50 cents per unit. Each unit of Brand X contains 20 grams of protein and 10 grams of fat, whereas each unit of Brand Y contains 10 grams of protein and 10 grams of fat. Determine how much of each brand should be bought to obtain a minimum cost per serving.

49. *Raising Animals* A breeder can raise no more than 50 hamsters and mice but no more than 20 hamsters. If she sells the hamsters for $15 each and the mice for $10 each, find the maximum revenue.

50. *Maximizing Profit* A business manufactures two parts, X and Y. Machines A and B are needed to make each part. To make part X, machine A is needed 3 hours and machine B is needed 1 hour. To make part Y, machine A is needed 1 hour and machine B is needed 2 hours. Machine A is available 60 hours per week and machine B is available 50 hours per week. The profit from part X is $300 and the profit from part Y is $250. How many parts of each type should be made to maximize weekly profit?

WRITING ABOUT MATHEMATICS

51. Give the steps for solving a linear programming word problem.

52. Is the vertex that gives the optimal solution always unique? Could there be more than one vertex that gives the optimal solution? Explain your reasoning.

SECTIONS 4.3 and 4.4	**Checking Basic Concepts**

1. Write an inequality that describes the shaded region in the graph.

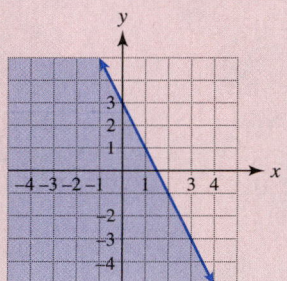

2. Graph the solution set for the linear system of inequalities. Use a test point to check your graph.

$$2x + 3y \le 6$$
$$x + 2y \ge 4$$

3. Use linear programming to find the maximum of $R = 3x + 2y$, subject to the constraints

$$3x + 2y \le 12$$
$$4x + 4y \le 20$$
$$x \ge 0, y \ge 0.$$

4.5 | Systems of Linear Equations in Three Variables

Basic Concepts • Solving Linear Systems with Substitution and Elimination •
Modeling Data • Systems of Equations with No Solutions •
Systems of Equations with Infinitely Many Solutions

A LOOK INTO MATH ▶

In the Chapter 4 opening discussion on page 224, we saw how systems of linear equations are used to predict the weight of black bears. In Section 4.1 we saw how systems of linear equations are used to transmit information reliably in wireless networks. In this section we will see how systems of linear equations are used to predict antelope populations in Wyoming. (See Examples 3 and 7.) The same mathematical concept, systems of linear equations, is used to model a variety of real-world situations. This is an amazing power of mathematics: *one simple but profound concept solves many complicated problems.*

Basic Concepts

NEW VOCABULARY

☐ Linear system in three
   variables
☐ Ordered triple

When we solve a linear system in two variables, we can express a solution as an ordered pair (x, y). A linear equation in two variables can be represented graphically by a line. A system of two linear equations with a unique solution can be represented graphically by two lines intersecting at a point, as shown in Figure 4.33.

When solving **linear systems in three variables,** we often use the variables x, y, and z. A solution is expressed as an **ordered triple** (x, y, z), rather than an ordered pair (x, y). For example, if the ordered triple $(1, 2, 3)$ is a solution, $x = 1$, $y = 2$, and $z = 3$ satisfy each equation. A linear equation in three variables can be represented by a flat plane in space. If the solution is unique, we can represent a linear system of three equations in three variables graphically by three planes intersecting at a single point, as illustrated in Figure 4.34.

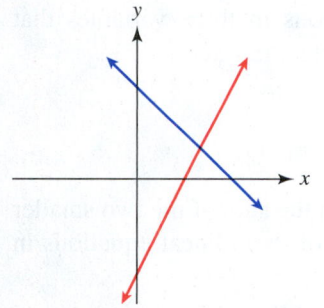

Figure 4.33

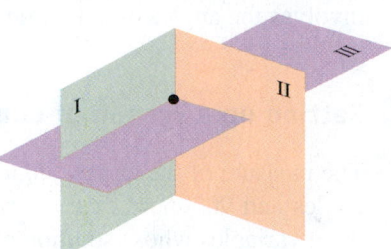

Figure 4.34

NOTE: The three planes in Figure 4.34 all intersect at right angles. In general, three planes can intersect at a point even if they are not at right angles to each other.

CRITICAL THINKING

Each of the following depictions of three planes in space represents a system of three linear equations in three variables. How many solutions are there in each case? Explain.

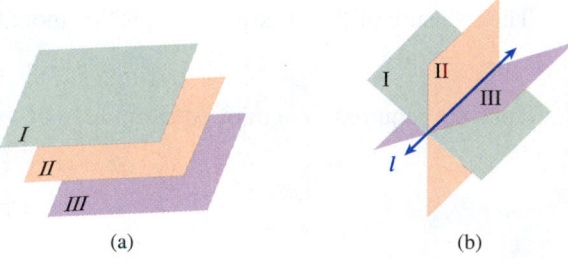

(a)   (b)

The next example shows how to check whether an ordered triple is a solution to a system of linear equations in three variables.

> **EXAMPLE 1** **Checking for solutions to a system of three equations**
>
> Determine whether $(4, 2, -1)$ or $(-1, 0, 3)$ is a solution to the system.
>
> $$2x - 3y + z = 1$$
> $$x - 2y + 2z = 5$$
> $$2y + z = 3$$
>
> **Solution**
> To check $(4, 2, -1)$, substitute $x = 4$, $y = 2$, and $z = -1$ in each equation.
>
> $$2(4) - 3(2) + (-1) \overset{?}{=} 1 \checkmark \quad \text{True}$$
> $$4 - 2(2) + 2(-1) \overset{?}{=} 5 \text{ X} \quad \text{False}$$
> $$2(2) + (-1) \overset{?}{=} 3 \checkmark \quad \text{True}$$
>
> The ordered triple $(4, 2, -1)$ *does not satisfy all three equations, so it is not a solution.* Next, substitute $x = -1$, $y = 0$, and $z = 3$ in each equation.
>
> $$2(-1) - 3(0) + 3 \overset{?}{=} 1 \checkmark \quad \text{True}$$
> $$-1 - 2(0) + 2(3) \overset{?}{=} 5 \checkmark \quad \text{True}$$
> $$2(0) + 3 \overset{?}{=} 3 \checkmark \quad \text{True}$$
>
> The ordered triple $(-1, 0, 3)$ *satisfies all three equations, so it is a solution.*
>
> **Now Try Exercise 9**

In the next example we set up a system of three equations in three variables that involves the angles of a triangle. You are asked to solve this system in Exercise 51.

> **EXAMPLE 2** **Setting up a system of equations**
>
> The measure of the largest angle in a triangle is 40° greater than the sum of the two smaller angles and 90° more than the smallest angle. Set up a system of three linear equations in three variables whose solution gives the measure of each angle.
>
> **Solution**
> Let x, y, and z be the measures of the three angles from largest to smallest. Because the sum of the measures of the angles in a triangle equals 180°, we have
>
> $$x + y + z = 180.$$
>
> The measure of the largest angle x is 40° greater than the sum of the measures of the two smaller angles $y + z$, so
>
> $$x - (y + z) = 40 \quad \text{or} \quad x - y - z = 40.$$
>
> The measure of the largest angle x is 90° more than the measure of the smallest angle z, so
>
> $$x - z = 90.$$
>
> Thus the required system of equations can be written as follows.
>
> $$x + y + z = 180$$
> $$x - y - z = 40$$
> $$x - z = 90$$
>
> **Now Try Exercise 49(a)**

▶ **REAL-WORLD CONNECTION**   In the next example we show how a linear system involving three equations and three variables can be used to model a real-world situation. We solve this system of equations in Example 7.

EXAMPLE 3   **Modeling real data with a linear system**

The Bureau of Land Management studies antelope populations in Wyoming. It monitors the number of adult antelope, the number of fawns each spring, and the severity of the winter. The first two columns of Table 4.6 contain counts of fawns and adults for three representative winters. The third column shows the severity of each winter. The severity of the winter is measured from 1 to 5, with 1 being mild and 5 being severe.

TABLE 4.6

Fawns (F)	Adults (A)	Winter (W)
405	870	3
414	848	2
272	684	5
?	750	4

We want to use the data in the first three rows of the table to estimate the number of fawns F in the fourth row when the number of adults is 750 and the severity of the winter is 4. To do so, we use the formula

$$F = a + bA + cW,$$

where a, b, and c are constants. Write a system of linear equations whose solution gives appropriate values for a, b, and c.

Solution
From the first row in the table, when $F = \mathbf{405}$, $A = \mathbf{870}$, and $W = \mathbf{3}$, the formula

$$F = a + bA + cW$$

becomes

$$\mathbf{405} = a + b(\mathbf{870}) + c(\mathbf{3}).$$

Similarly, $F = \mathbf{414}$, $A = \mathbf{848}$, and $W = \mathbf{2}$ gives

$$\mathbf{414} = a + b(\mathbf{848}) + c(\mathbf{2}),$$

and $F = \mathbf{272}$, $A = \mathbf{684}$, and $W = \mathbf{5}$ yields

$$\mathbf{272} = a + b(\mathbf{684}) + c(\mathbf{5}).$$

To find values for a, b, and c we can solve the following system of linear equations.

$$\mathbf{405} = a + \mathbf{870}b + \mathbf{3}c$$
$$\mathbf{414} = a + \mathbf{848}b + \mathbf{2}c$$
$$\mathbf{272} = a + \mathbf{684}b + \mathbf{5}c$$

We can also write these equations as a linear system in the following form.

$$a + 870b + 3c = 405$$
$$a + 848b + 2c = 414$$
$$a + 684b + 5c = 272$$

Finding values for a, b, and c will allow us to use the formula $F = a + bA + cW$ to predict the number of fawns F when the number of adults A is 750 and the severity of the winter W is 4 (see Example 7).

Now Try Exercise 53(a)

NOTE: Linear systems of two equations can have no solutions, one solution, or infinitely many solutions. The same is true for larger linear systems. In the following subsection we focus on linear systems having one solution.

Solving Linear Systems with Substitution and Elimination

When solving systems of linear equations with more than two variables, we usually use both substitution and elimination. However, in the next example we use only substitution to solve a particular type of linear system in three variables.

EXAMPLE 4   **Using substitution to solve a linear system of equations**

Solve the following system.

$$2x - y + z = 7$$
$$3y - z = 1$$
$$z = 2$$

Solution

The last equation gives us the value of z immediately. We can substitute $z = 2$ in the second equation and determine y.

$$3y - z = 1 \quad \text{Second equation}$$
$$3y - 2 = 1 \quad \text{Substitute } z = 2.$$
$$3y = 3 \quad \text{Add 2 to each side.}$$
$$y = 1 \quad \text{Divide each side by 3.}$$

Knowing that $y = 1$ and $z = 2$ allows us to find x by using the first equation.

$$2x - y + z = 7 \quad \text{First equation}$$
$$2x - 1 + 2 = 7 \quad \text{Let } y = 1 \text{ and } z = 2.$$
$$2x = 6 \quad \text{Simplify and subtract 1.}$$
$$x = 3 \quad \text{Divide each side by 2.}$$

Thus $x = 3$, $y = 1$, and $z = 2$, and the solution is $(3, 1, 2)$.

Now Try Exercise 13

In the next example we use elimination and substitution to solve a system of linear equations. This four-step method is summarized by the following.

SOLVING A LINEAR SYSTEM IN THREE VARIABLES

STEP 1: Eliminate one variable, such as x, from two of the equations.

STEP 2: Use the two resulting equations in two variables to eliminate one of the variables, such as y. Solve for the remaining variable, z.

STEP 3: Substitute z in one of the two equations from Step 2. Solve for the unknown variable y.

STEP 4: Substitute values for y and z in one of the given equations and find x. The solution is (x, y, z).

EXAMPLE 5 **Solving a linear system in three variables**

Solve the following system.

$$
\begin{aligned}
x - y + 2z &= 6 \\
2x + y - 2z &= -3 \\
-x - 2y + 3z &= 7
\end{aligned}
$$

Solution

STEP 1: We begin by eliminating the variable x from the second and third equations. To eliminate x from the second equation, we multiply the first equation by -2 and then add it to the second equation. To eliminate x from the third equation, we add the first and third equations.

$$
\begin{array}{ll}
-2x + 2y - 4z = -12 & \text{First equation times } -2 \\
\underline{2x + y - 2z = \ \ -3} & \text{Second equation} \\
3y - 6z = -15 & \text{Add.}
\end{array}
\qquad
\begin{array}{ll}
x - y + 2z = \ \ 6 & \text{First equation} \\
\underline{-x - 2y + 3z = \ \ 7} & \text{Third equation} \\
-3y + 5z = 13 & \text{Add.}
\end{array}
$$

STEP 2: Take the two resulting equations from Step 1 and eliminate either variable. Here we add the two equations to eliminate the variable y.

$$
\begin{array}{ll}
3y - 6z = -15 & \\
\underline{-3y + 5z = \ \ 13} & \\
-z = \ \ -2 & \text{Add the equations.} \\
z = \ \ \ \ 2 & \text{Multiply by } -1.
\end{array}
$$

STEP 3: Now we can use substitution to find the values of x and y. We let $z = \mathbf{2}$ in either equation used in Step 2 to find y.

$$
\begin{array}{ll}
3y - 6z = -15 & \text{From Step 2} \\
3y - 6(\mathbf{2}) = -15 & \text{Substitute } z = 2. \\
3y - 12 = -15 & \text{Multiply.} \\
3y = \ \ -3 & \text{Add 12.} \\
y = \ \ \mathbf{-1} & \text{Divide by 3.}
\end{array}
$$

STEP 4: Finally, we substitute $y = \mathbf{-1}$ and $z = \mathbf{2}$ in one of the given equations to find x.

$$
\begin{array}{ll}
x - \mathbf{y} + 2z = 6 & \text{First given equation} \\
x - (\mathbf{-1}) + 2(\mathbf{2}) = 6 & \text{Let } y = -1 \text{ and } z = 2. \\
x + 1 + 4 = 6 & \text{Simplify.} \\
x = \mathbf{1} & \text{Subtract 5.}
\end{array}
$$

The solution is $(\mathbf{1}, \mathbf{-1}, \mathbf{2})$. Check this solution.

▮ Now Try Exercise 25

▶ **REAL-WORLD CONNECTION** In the next example we solve a system of linear equations to determine the number of tickets sold at a play.

EXAMPLE 6 **Finding the number of tickets sold**

One thousand tickets were sold for a play, which generated $3800 in revenue. The prices of the tickets were $3 for children, $4 for students, and $5 for adults. There were 100 fewer student tickets sold than adult tickets. Find the number of each type of ticket sold.

Solution

Let x be the number of tickets sold to children, y be the number of tickets sold to students, and z be the number of tickets sold to adults. The total number of tickets sold was 1000, so

$$x + y + z = 1000. \quad (1)$$

Each child's ticket cost \$3, so the revenue generated from selling x tickets was $3x$. Similarly, the revenue generated from students was $4y$, and the revenue from adults was $5z$. Total ticket sales were \$3800, so

$$3x + 4y + 5z = 3800. \quad (2)$$

The equation $z - y = 100$, or

$$y - z = -100, \quad (3)$$

must also be satisfied, because 100 fewer tickets were sold to students than to adults.

To find the number of each type of ticket sold, we need to solve the following system of linear equations.

$$\begin{aligned} x + \ y + \ z &= \ 1000 \\ 3x + 4y + 5z &= \ 3800 \\ y - \ z &= -100 \end{aligned}$$

STEP 1: We begin by eliminating the variable x from the second equation. To do so, we multiply the first equation by -3 and add the second equation.

$$\begin{array}{ll} -3x - 3y - 3z = -3000 & \text{First given equation times } -3 \\ \underline{3x + 4y + 5z = \ \ 3800} & \text{Second equation} \\ y + 2z = \ \ \ 800 & \text{Add.} \end{array}$$

STEP 2: We then multiply the resulting equation from Step 1 by -1 and add the third given equation to eliminate y.

$$\begin{array}{ll} -y - 2z = -800 & \text{Equation from Step 1 times } -1 \\ \underline{y - \ z = -100} & \text{Third given equation} \\ -3z = -900 & \text{Add the equations.} \\ z = \ \ \ 300 & \text{Divide by } -3. \end{array}$$

STEP 3: To find y we can substitute $z = \mathbf{300}$ in one of the equations from Step 2.

$$\begin{array}{ll} y - z = -100 & \text{Step 2 equation} \\ y - \mathbf{300} = -100 & \text{Let } z = 300. \\ y = \ \ \mathbf{200} & \text{Add 300.} \end{array}$$

STEP 4: Finally, we substitute $y = \mathbf{200}$ and $z = \mathbf{300}$ in the first given equation.

$$\begin{array}{ll} x + y + z = 1000 & \text{First given equation} \\ x + \mathbf{200} + \mathbf{300} = 1000 & \text{Let } y = 200 \text{ and } z = 300. \\ x = \ \ \mathbf{500} & \text{Subtract 500.} \end{array}$$

Thus **500** tickets were sold to children, **200** to students, and **300** to adults.

■ **Now Try Exercise 47**

Modeling Data

In the next example we solve the system of equations that we discussed in Example 3.

EXAMPLE 7   **Predicting fawns in the spring**

Solve the following linear system for a, b, and c.

$$a + 870b + 3c = 405$$
$$a + 848b + 2c = 414$$
$$a + 684b + 5c = 272$$

Then use $F = a + bA + cW$ to predict the number of fawns when there are 750 adults and the severity of the winter is 4.

Solution

STEP 1: We begin by eliminating the variable a from the second and third equations. To do so, we add the second and third equations to the first equation times -1.

$$
\begin{array}{ll}
-a - 870b - 3c = -405 & \quad -a - 870b - 3c = -405 \quad \text{First times } -1 \\
\underline{a + 848b + 2c = 414} & \quad \underline{a + 684b + 5c = 272} \quad \text{Second/third equation} \\
-22b - c = 9 & \quad -186b + 2c = -133 \quad \text{Add.}
\end{array}
$$

STEP 2: We use the two resulting equations from Step 1 to eliminate c. To do so we multiply the equation $-22b - c = 9$ by 2 and add it to the other equation.

$$
\begin{array}{ll}
-44b - 2c = 18 & \quad (-22b - c = 9) \text{ times } 2 \\
\underline{-186b + 2c = -133} & \\
-230b = -115 & \quad \text{Add the equations.} \\
b = 0.5 & \quad \text{Divide by } -230.
\end{array}
$$

STEP 3: To find c we substitute $b = \mathbf{0.5}$ in either equation used in Step 2.

$$
\begin{array}{ll}
-44\mathbf{b} - 2c = 18 & \quad \text{First equation in Step 2} \\
-44(\mathbf{0.5}) - 2c = 18 & \quad \text{Let } b = 0.5. \\
-22 - 2c = 18 & \quad \text{Multiply.} \\
-2c = 40 & \quad \text{Add 22.} \\
c = \mathbf{-20} & \quad \text{Divide by } -2.
\end{array}
$$

STEP 4: Finally, we substitute $b = \mathbf{0.5}$ and $c = \mathbf{-20}$ in one of the given equations to find a.

$$
\begin{array}{ll}
a + 870\mathbf{b} + 3\mathbf{c} = 405 & \quad \text{First given equation} \\
a + 870(\mathbf{0.5}) + 3(\mathbf{-20}) = 405 & \quad \text{Let } b = 0.5 \text{ and } c = -20. \\
a + 435 - 60 = 405 & \quad \text{Multiply.} \\
a = \mathbf{30} & \quad \text{Solve for } a.
\end{array}
$$

CRITICAL THINKING

Give reasons why the coefficient for A is positive and the coefficient for W is negative in the formula

$$F = 30 + 0.5A - 20W.$$

The solution is $a = \mathbf{30}$, $b = \mathbf{0.5}$, and $c = \mathbf{-20}$. Thus we may write

$$F = a + bA + cW$$
$$= \mathbf{30} + \mathbf{0.5}A - \mathbf{20}W. \quad \text{Modeling equation}$$

If there are 750 adults and the winter has a severity of 4, this model predicts

$$F = 30 + 0.5(750) - 20(4)$$
$$= 325 \text{ fawns.}$$

Now Try Exercise 53

Systems of Equations with No Solutions

It is possible for a system of three linear equations in three variables to be inconsistent and have no solutions. If we apply substitution and elimination to this type of system, we arrive at a contradiction. This case is demonstrated in the next example.

EXAMPLE 8 **Recognizing an inconsistent system**

Solve the system, if possible.

$$x + y + z = 4$$
$$-x + y + z = 2$$
$$y + z = 1$$

Solution

STEP 1: If we add the first two equations, we can eliminate x. The variable x is already eliminated from the third equation.

$x + y + z = 4$	First equation
$-x + y + z = 2$	Second equation
$2y + 2z = 6$	Add.

STEP 2: If we multiply the third *given* equation by -2 and add it to the resulting equation in Step 1, we arrive at a contradiction.

$-2y - 2z = -2$	Third equation times -2
$2y + 2z = 6$	Equation from Step 1
$0 = 4$ ✗	Add. (Contradiction)

Because $0 = 4$ is a contradiction, there are no solutions to the given system of equations.

Now Try Exercise 31

Systems of Equations with Infinitely Many Solutions

It is possible for a system of linear equations in three variables to have infinitely many solutions. If we apply substitution and elimination to this type of system, we arrive at an identity. This case is demonstrated in the next example.

EXAMPLE 9 **Solving a system with infinitely many solutions**

Solve the system.

$$x + y + z = 2$$
$$x - y + z = 4$$
$$3x - y + 3z = 10$$

Solution

STEP 1: To eliminate y from the second equation, add the first equation to the second. To eliminate y from the third equation, add the first equation to the third equation.

$x + y + z = 2$	First equation		$x + y + z = 2$	First equation
$x - y + z = 4$	Second equation		$3x - y + 3z = 10$	Third equation
$2x + 2z = 6$	Add.		$4x + 4z = 12$	Add.

STEP 2: If we multiply the first resulting equation in Step 1 by -2 and add it to the second resulting equation in Step 1, we arrive at an identity.

$$-4x - 4z = -12 \quad \text{($2x + 2z = 6$) times -2}$$
$$\underline{4x + 4z = 12} \quad \text{Second equation from Step 1}$$
$$0 = 0 \quad \text{Add. (Identity)}$$

The variable x can be written in terms of z by solving $2x + 2z = 6$ for x.

$$2x + 2z = 6 \qquad \text{Equation from Step 1}$$
$$2x = 6 - 2z \quad \text{Subtract $2z$.}$$
$$x = 3 - z \quad \text{Divide by 2.}$$

STEP 3: To find y in terms of z, substitute $3 - z$ for x in the first *given* equation.

$$x + y + z = 2 \quad \text{First equation}$$
$$(3 - z) + y + z = 2 \quad \text{Let $x = 3 - z$.}$$
$$y = -1 \quad \text{Solve for y.}$$

All solutions have the form $(3 - z, -1, z)$, where z can be any real number. For example, if $z = 1$, then $(2, -1, 1)$ is one of infinitely many solutions to the system of equations. Note that in this particular system, y must always equal -1.

■ **Now Try Exercise 33**

READING CHECK

What is the possible number of solutions to a system of linear equations in three variables?

4.5 Putting It All Together

In this section we discussed how to solve a system of three linear equations in three variables. Systems of linear equations can have no solutions, one solution, or infinitely many solutions.

CONCEPT	EXPLANATION/EXAMPLE
System of Linear Equations in Three Variables	The following is a system of three linear equations in three variables. Each equation is linear. $$x - 2y + z = 0$$ $$-x + y + z = 4$$ $$-y + 4z = 10$$
Solution to a Linear System in Three Variables	The solution to a linear system in three variables is an ordered triple, expressed as (x, y, z). The solution to the preceding system is $(1, 2, 3)$ because substituting $x = 1$, $y = 2$, and $z = 3$ in each equation results in a true statement. We can check solutions this way. $$(1) - 2(2) + (3) = 0 \checkmark \quad \text{True}$$ $$-(1) + (2) + (3) = 4 \checkmark \quad \text{True}$$ $$-(2) + 4(3) = 10 \checkmark \quad \text{True}$$

continued on next page

continued from previous page

CONCEPT	EXPLANATION
Solving a Linear System with Substitution and Elimination	**STEP 1:** Eliminate one variable, such as x, from two of the equations.
	STEP 2: Use the two resulting equations in two variables to eliminate one of the variables, such as y. Solve for the remaining variable z.
	STEP 3: Substitute z in one of the two equations from Step 2. Solve for the unknown variable y.
	STEP 4: Substitute values for y and z in one of the given equations and find x. The solution is (x, y, z).

4.5   Exercises

CONCEPTS AND VOCABULARY

1. Can a system of three linear equations in three variables have exactly two solutions? Explain.

2. Give an example of a system of three linear equations in three variables.

3. Does the ordered triple $(1, 2, 3)$ satisfy the equation $x + y + z = 6$?

4. Does $(3, 4)$ represent a solution to the equation $x + y + z = 7$? Explain.

5. To solve uniquely for two variables, how many equations do you usually need?

6. To solve uniquely for three variables, how many equations do you usually need?

7. If a contradiction occurs while solving a system of linear equations, then there is/are _____ solution(s).

8. If an identity occurs while solving a linear system, how many solutions are there?

SOLVING LINEAR SYSTEMS

Exercises 9–12: Determine which ordered triple is a solution to the linear system.

9. $(1, 2, 3), (0, 2, 4)$
$$x + y + z = 6$$
$$x - y - z = -4$$
$$-x - y + z = 0$$

10. $(-1, 0, 2), (0, 4, 4)$
$$2x + y - 3z = -8$$
$$x - 3y + 2z = -4$$
$$3x - 2y + z = -4$$

11. $(1, 0, 3), (-1, 1, 2)$
$$3x - 2y + z = -3$$
$$-x + 3y - 2z = 0$$
$$x + 4y + 2z = 7$$

12. $\left(\frac{1}{2}, \frac{3}{2}, -\frac{1}{2}\right), (-1, 0, -2)$
$$x + 3y - 4z = 7$$
$$-x + 5y + 3z = \frac{11}{2}$$
$$3x - 2y - 7z = 2$$

Exercises 13–18: (Refer to Example 4.) Use substitution to solve the system of linear equations. Check your solution.

13. $\begin{aligned} x + y - z &= 1 \\ 2y + z &= -1 \\ z &= 1 \end{aligned}$

14. $\begin{aligned} 2x + y - 3z &= 1 \\ y + 4z &= 0 \\ z &= -1 \end{aligned}$

15. $\begin{aligned} -x - 3y + z &= -2 \\ 2y + 3z &= 3 \\ z &= 2 \end{aligned}$

16. $\begin{aligned} 3x + 2y - 3z &= -4 \\ -y + 2z &= 4 \\ z &= 0 \end{aligned}$

17. $\begin{aligned} a - b + 2c &= 3 \\ -3b + c &= 4 \\ c &= -2 \end{aligned}$

18. $\begin{aligned} 5a + 2b - 3c &= 10 \\ 5b - 2c &= -4 \\ c &= 3 \end{aligned}$

Exercises 19–44: Solve the system, if possible.

19. $\begin{aligned} x + y - z &= 11 \\ -x + 2y + 3z &= -1 \\ 2z &= 4 \end{aligned}$

20. $\begin{aligned} x + 2y - 3z &= -7 \\ -2x + y + z &= -1 \\ 3z &= 9 \end{aligned}$

21. $\begin{aligned} x + y - z &= -2 \\ -x + z &= 1 \\ y + 2z &= 3 \end{aligned}$

22. $\begin{aligned} x + y - 3z &= 11 \\ -2x + y + 2z &= 1 \\ -3y + 3z &= -21 \end{aligned}$

23. $x + y - 2z = -7$
$y + z = -1$
$-y + 3z = 9$

24. $2x + 3y + z = 5$
$y + 2z = 4$
$-2y + z = 2$

25. $x + 2y + 2z = 1$
$x + y + z = 0$
$-x - 2y + 3z = -11$

26. $x + y - z = 0$
$x - 3y + z = -2$
$x - y + 3z = 8$

27. $x + y + z = 5$
$y + z = 6$
$x + z = 3$

28. $x + y + z = 0$
$x - y - z = 6$
$-x + y - z = 4$

29. $x + 2y + 3z = 24$
$-x + y + 2z = 1$
$x + y - 2z = 9$

30. $5x - 15y + z = 22$
$-10x + 12y - 2z = -8$
$4x - 2y - 3z = 9$

31. $x + y + z = 2$
$x - y + z = 1$
$x + z = 3$

32. $4x - y + 3z = 3$
$2x + y + z = 2$
$x - y + z = 1$

33. $x + y + z = 6$
$x - y + z = 2$
$-x + 5y - z = 6$

34. $x - y + z = 3$
$2x - y + z = 2$
$-x - y + z = 5$

35. $2x + y + z = 3$
$2x - y - z = 9$
$x + y - z = 0$

36. $x + 3y + z = -8$
$x - 2y = 11$
$2y - z = -16$

37. $2x + 6y - 2z = 47$
$2x + y + 3z = -28$
$-x + y + z = -\frac{7}{2}$

38. $x + y + 2z = 23$
$3x - y + 3z = 8$
$2x + 2y + z = 13$

39. $x + 3y - 4z = \frac{13}{2}$
$-2x + 3y - z = \frac{1}{2}$
$3x + z = 4$

40. $x - 2y + z = \frac{9}{2}$
$4x - y + 3z = 9$
$x + 2y = -\frac{3}{2}$

41. $2x - 2y + z = 4$
$x - y + z = 1$
$x - y + z = 3$

42. $x - 2y + 3z = 1$
$x + 3y + z = 2$
$3x + 4y + 5z = 7$

43. $x + y + z = 5$
$x - y + z = 3$
$2x + y + 2z = 9$

44. $x + y - 2z = 0$
$x + 2y - 4z = -1$
$y - 2z = -1$

Exercises 45 and 46: **Thinking Generally** *Solve the system of linear equations, if a ≠ 0.*

45. $x + y + z = a$
$x + y + z = 2a$
$-x + y + z = 0$

46. $x + y + z = a$
$-x - y - z = -a$
$y - z = 0$

APPLICATIONS

47. *Finding Costs* The accompanying table shows the costs of purchasing different combinations of hamburgers, fries, and soft drinks.

Hamburgers	Fries	Soft Drinks	Total Cost
1	2	4	$10
1	4	6	$15
0	3	2	$6

(a) Let x be the cost of a hamburger, y the cost of fries, and z the cost of a soft drink. Write a system of three linear equations that represents the data in the table.

(b) Solve the system and interpret your answer.

48. *Cost of CDs* The accompanying table shows the total cost of purchasing combinations of differently priced CDs. The types of CDs are labeled A, B, and C.

A	B	C	Total Cost
1	1	1	$37
3	2	1	$69
1	1	4	$82

(a) Let x be the cost of a CD of type A, y the cost of a CD of type B, and z the cost of a CD of type C. Write a system of three linear equations that represents the data in the table.

(b) Solve the system and interpret your answer.

49. *Geometry* The largest angle in a triangle is 55° more than the smallest angle. The sum of the measures of the two smaller angles is 10° more than the measure of the largest angle.

(a) Let x, y, and z be the measures of the three angles from largest to smallest. Write a system of three linear equations whose solution gives the measure of each angle.

(b) Solve the system.

(c) Check your solution.

50. *Geometry* The perimeter of a triangle is 90 inches. The longest side is 20 inches longer than the shortest side and 10 inches longer than the remaining side.

(a) Let x, y, and z be the lengths of the three sides from largest to smallest. Write a system of three linear equations whose solution gives the lengths of each side.

(b) Solve the system.

(c) Check your solution.

51. *Geometry* (Refer to Example 2.) Solve the system to find the measure of each angle.

$$x + y + z = 180$$
$$x - y - z = 40$$
$$x \qquad - z = 90$$

52. *Loan Mixture* A student takes out a total of $5000 in three loans: one subsidized, one unsubsidized, and one from the parents of the student. The subsidized loan is $200 more than the combined total of the unsubsidized and parent loans. The unsubsidized loan is twice the amount of the parent loan. Find the amount of each loan.

53. *Predicting Fawns* (Refer to Examples 3 and 7.) The accompanying table shows counts for fawns and adult deer and the severity of the winter. These data may be modeled by the equation $F = a + bA + cW$.

Fawns (F)	Adults (A)	Winter (W)
525	600	4
365	400	2
805	900	5
?	500	3

(a) Use the first three rows to write a system of three linear equations in three variables whose solution gives values for a, b, and c.
(b) Solve the system.
(c) Predict the number of fawns when there are 500 adults and the winter has severity 3.

54. *Business Production* A business has three machines that manufacture containers. Together they make 100 containers per day, whereas the two fastest machines can make 80 containers per day. The fastest machine makes 34 more containers per day than the slowest machine.
(a) Let x, y, and z be the number of containers that the machines make from fastest to slowest. Write a system of three equations whose solution gives

the number of containers that each machine can make.
(b) Solve the system.

55. *Mixture Problem* One type of lawn fertilizer consists of a mixture of nitrogen, N, phosphorus, P, and potassium, K. An 80-pound sample contains 8 more pounds of nitrogen and phosphorus than potassium. There is 9 times as much potassium as phosphorus.
(a) Write a system of three equations whose solution gives the amount of nitrogen, phosphorus, and potassium in this sample.
(b) Solve the system.

56. *Predicting Home Prices* Selling prices of homes can depend on several factors such as size and age. The accompanying table shows the selling price for three homes. In this table, price P is given in thousands of dollars, age A in years, and home size S in thousands of square feet. These data may be modeled by the equation $P = a + bA + cS$.

Price (P)	Age (A)	Size (S)
190	20	2
320	5	3
50	40	1
?	10	2.5

(a) Write a system of linear equations whose solution gives a, b, and c.
(b) Solve the system.
(c) Predict the price of a home that is 10 years old and has 2500 square feet.

57. *Investment Mixture* A sum of $30,000 was invested in three mutual funds. In one year, the first fund grew by 4%, the second by 5%, and the third by 7.5%. Total earnings were $1775. The amount invested in the third fund was $2000 less than the combined amount invested in the other two funds. Use a linear system of equations to determine the amount invested in each fund.

58. *Football Tickets* A total of 2500 tickets to a football game were sold. Prices were $2 for children, $3 for students, and $5 for adults. Twice as many tickets were sold to students as to children, and ticket revenues were $7250. Use a system of linear equations to determine how many of each type of ticket were sold.

*Exercises 59–62: **Wireless Networks** (Refer to Section 4.1, Example 10.) The system of six equations (1–6) represents a wireless transmission over six paths of a three-digit PIN given by x, y, and z.*

$$x = 2 \qquad (1)$$
$$y = 3 \qquad (2)$$
$$z = 1 \qquad (3)$$
$$x + y + z = 6 \qquad (4)$$
$$2x + y + z = 8 \qquad (5)$$
$$x + 2y + z = 9 \qquad (6)$$

Solve each set of three equations and find the PIN. Do you get the same answer in each case?

59. (1), (2), and (4)   **60.** (2), (3), and (5)

61. (1), (5), and (6)   **62.** (2), (5), and (6)

WRITING ABOUT MATHEMATICS

63. In the previous section we solved problems with two variables; to obtain a unique solution we needed two linear equations. In this section we solved problems with three variables; to obtain a unique solution we needed three equations. Try to generalize these results. In the design of aircraft, problems commonly involve 100,000 variables. How many equations are required to solve such problems? Can such problems be solved by hand? If not, how are such problems solved? Explain your answers.

64. In Exercise 56 the price of a home was estimated by its age and size. What other factors might affect the price of a home? Explain how these factors might affect the number of variables and equations in the linear system.

Group Activity   Working with Real Data

Directions: Form a group of 2 to 4 people. Select someone to record the group's responses for this activity. All of the members of the group should work cooperatively to answer the questions. If your instructor asks for your results, each member of the group should be prepared to respond.

CEO Salaries In 2010, hourly wages for the top three CEOs (chief executive officers) of U.S. corporations were calculated based on their working 14 hours per day for 365 days. Together the three CEOs made $75,000 per hour. The first CEO earned $3,000 more per hour than the hourly wage of the second CEO, and the first CEO earned $6,000 more per hour than the third CEO. (*Sources:* Forbes.com; Department of Labor.)

(a) Let x, y, and z be the hourly wages in *thousands* of dollars for the top CEOs from greatest to least. Write a system of equations whose solution gives these hourly wages.

(b) Solve the system. Interpret the answer.

(c) The average private sector American worker earned $22.78 per hour in 2010. How many hours did the average private sector American work to earn an amount equal to one hour of work by the first CEO in 2010?

4.6 Matrix Solutions of Linear Systems

Representing Systems of Linear Equations with Matrices •
Matrices and Social Networks • Gauss–Jordan Elimination •
Using Technology to Solve Systems of Linear Equations (Optional)

A LOOK INTO MATH ▶

Suppose that the size of a bear's head and its overall length are known. Can its weight be estimated from these variables? Can a bear's weight be estimated if its neck size and chest size are known? In this section we show that systems of linear equations can be used to make such estimates.

NEW VOCABULARY

☐ Matrix
☐ Element
☐ Dimension of a matrix
☐ Square matrix
☐ Augmented matrix
☐ Main diagonal
☐ Reduced row–echelon form
☐ Gauss–Jordan elimination
☐ Matrix row transformations

Representing Systems of Linear Equations with Matrices

Arrays of numbers are used frequently in many different real-world situations. Spreadsheets often make use of arrays. A **matrix** is a rectangular array of numbers. Each number in a matrix is called an **element**. The following are examples of *matrices* (plural of matrix), with their dimensions written below them.

Matrices and Their Dimensions

$$\begin{bmatrix} 2 & 0 \\ 3 & 1 \end{bmatrix} \quad \begin{bmatrix} -1.2 & 5 & 0 \\ 1 & 0 & 1 \\ 4 & -5 & 7 \end{bmatrix} \quad \begin{bmatrix} 3 & -6 & 0 & \sqrt{3} \\ 1 & 4 & 0 & 9 \\ -3 & 1 & 1 & 18 \\ -10 & -4 & 5 & -1 \end{bmatrix} \quad \begin{bmatrix} 4 & 2 \\ 0 & 1 \\ 1 & 0 \end{bmatrix} \quad \begin{bmatrix} 1 & 5 & -1 \\ 3 & 4 & 2 \end{bmatrix}$$

$\boxed{2 \times 2}$ $\quad$ $\boxed{3 \times 3}$ $\qquad$ $\boxed{4 \times 4}$ $\qquad$ $\boxed{3 \times 2}$ $\quad$ $\boxed{2 \times 3}$

Rows × Columns

▶ **REAL-WORLD CONNECTION** The dimension of a matrix is stated much like the dimensions of a rectangular room. We might say that a room is m feet long and n feet wide. Similarly, the **dimension of a matrix** is $m \times n$ (m by n) if it has m rows and n columns. For example, the last matrix in the preceding group has a dimension of 2×3 because it has 2 rows and 3 columns. If the number of rows equals the number of columns, the matrix is a **square matrix**. The first three matrices in that group are square matrices.

Matrices can be used to represent a system of linear equations. For example, if we have the system of equations

$$3x - y + 2z = 7$$
$$x - 2y + z = 0$$
$$2x + 5y - 7z = -9,$$

we can represent the system with the following **augmented matrix**. Note how the coefficients of the variables are placed in the matrix. A vertical line is positioned in the matrix where the equals signs occur in the system. The rows and columns are labeled, and the elements of the **main diagonal** of the augmented matrix are circled. The matrix has dimension 3×4.

Augmented Matrix

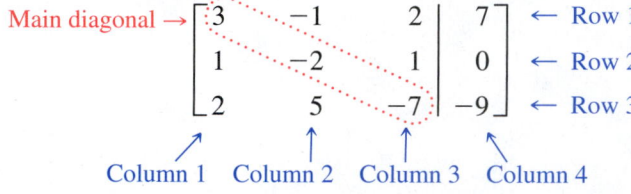

EXAMPLE 1   **Representing a linear system**

Represent each linear system with an augmented matrix. State the dimension of the matrix.

(a) $x - 2y = 9$
$\quad\ \ 6x + 7y = 16$

(b) $x - 3y + 7z = 4$
$\quad\ \ 2x + 5y - z = 15$
$\quad\ \ 2x + y \quad\quad = 8$

Solution

(a) This system can be represented by the following 2×3 augmented matrix.

$$\left[\begin{array}{cc|c} 1 & -2 & 9 \\ 6 & 7 & 16 \end{array}\right]$$

(b) This system can be represented by the following 3×4 augmented matrix.

$$\left[\begin{array}{ccc|c} 1 & -3 & 7 & 4 \\ 2 & 5 & -1 & 15 \\ 2 & 1 & 0 & 8 \end{array}\right]$$

Now Try Exercises 11, 13

Matrices and Social Networks

Today there are several different types of online social networks, such as Facebook and Twitter. Mathematics is essential to the success of these social networks. Consider Figure 4.35, which represents a simple social network of four people.

Simple Social Network

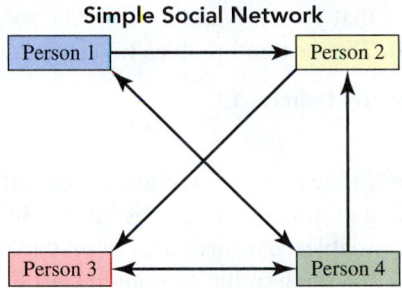

Figure 4.35

In Figure 4.35 the arrow from Person 1 to Person 2 indicates that Person 1 likes Person 2, but Person 2 does not like Person 1 because there is no arrow from Person 2 to Person 1. On the other hand, Person 3 and Person 4 like each other because there is a double arrow between them. In the next example we can see how a matrix can be used to represent this simple social network.

EXAMPLE 2   **Representing social networks**

Use a matrix to model the social network shown in Figure 4.35.

Solution

To model a social network with four people, we will use a square 4×4 matrix. Because Person 1 likes Person 2, we put a **1** in row 1 column 2. Because Person 1 also likes Person 4, we put a **1** in row 1 column 4. Continuing in the same manner, we arrive at the following matrix.

A Social Network

$$\left[\begin{array}{cccc} 0 & 1 & 0 & 1 \\ 0 & 0 & 1 & 0 \\ 0 & 0 & 0 & 1 \\ 1 & 1 & 1 & 0 \end{array}\right]$$

Now Try Exercise 21

Gauss–Jordan Elimination

A convenient matrix form for representing a system of linear equations is **reduced row–echelon form**. The following augmented matrices are examples of reduced row–echelon form. Note that there are 1s on the main diagonal with 0s above and below the 1s.

$$\left[\begin{array}{cc|c} 1 & 0 & 3 \\ 0 & 1 & -2 \end{array}\right] \quad \left[\begin{array}{ccc|c} 1 & 0 & 0 & 3 \\ 0 & 1 & 0 & 1 \\ 0 & 0 & 1 & -1 \end{array}\right] \quad \left[\begin{array}{ccc|c} 1 & 0 & 0 & 8 \\ 0 & 1 & 0 & 2 \\ 0 & 0 & 1 & 3 \end{array}\right]$$

If an augmented matrix representing a linear system is in reduced row–echelon form, we can usually determine the solution easily.

EXAMPLE 3 **Determining a solution from reduced row–echelon form**

Each matrix represents a system of linear equations. Find the solution.

(a) $\left[\begin{array}{ccc|c} 1 & 0 & 0 & 2 \\ 0 & 1 & 0 & -3 \\ 0 & 0 & 1 & 5 \end{array}\right]$ **(b)** $\left[\begin{array}{cc|c} 1 & 0 & 10 \\ 0 & 1 & -4 \end{array}\right]$

Solution
(a) The top row represents $1x + 0y + 0z = 2$ or $x = 2$. The second and third rows tell us that $y = -3$ and $z = 5$. The solution is $(2, -3, 5)$.
(b) The system involves two equations in two variables. The solution is $(10, -4)$.

▮ **Now Try Exercise 19**

In the previous section we solved systems of three linear equations in three variables by using elimination and substitution. In real life, systems of equations often contain thousands of variables. To solve a large system of equations, we need an efficient method. Long before the invention of the computer, Carl Friedrich Gauss (1777–1855) developed a method called *Gaussian elimination* to solve systems of linear equations. Even though it was developed more than 150 years ago, it is still used today in modern computers and calculators.

We will use a numerical method called **Gauss–Jordan elimination** to solve a linear system. It makes use of the following **matrix row transformations.**

MATRIX ROW TRANSFORMATIONS

For any augmented matrix representing a system of linear equations, the following row transformations result in an equivalent system of linear equations.

1. Any two rows may be interchanged.
2. The elements of any row may be multiplied by a nonzero constant.
3. Any row may be changed by adding to (or subtracting from) its elements a non-zero multiple of the corresponding elements of another row.

Gauss–Jordan elimination can be used to transform an augmented matrix into reduced row–echelon form. Its objective is to use these matrix row transformations to obtain a matrix that has the following reduced row–echelon form, where (a, b) represents the solution.

$$\left[\begin{array}{cc|c} 1 & 0 & a \\ 0 & 1 & b \end{array}\right]$$

This method is illustrated in the next example.

EXAMPLE 4 **Transforming a matrix into reduced row–echelon form**

Use Gauss–Jordan elimination to transform the augmented matrix of the linear system into reduced row–echelon form. Find the solution.

$$x + y = 5$$
$$-x + y = 1$$

Solution
Both the linear system and the augmented matrix are shown.

Linear System	Augmented Matrix
$x + y = 5$	$\begin{bmatrix} 1 & 1 & \vert & 5 \\ -1 & 1 & \vert & 1 \end{bmatrix}$
$-x + y = 1$	

First, we want to obtain a 0 in the second row, where the -1 is highlighted. To do so, we add row 1 to row 2 and place the result in row 2. This step is denoted $R_2 + R_1$ and eliminates the x-variable from the second equation.

$$x + y = 5 \qquad \begin{bmatrix} 1 & 1 & \vert & 5 \\ 0 & 2 & \vert & 6 \end{bmatrix}$$
$$2y = 6 \qquad R_2 + R_1 \rightarrow$$

To obtain a 1 where the 2 in the second row is located, we multiply the second row by $\frac{1}{2}$, denoted $\frac{1}{2}R_2$.

$$x + y = 5 \qquad \begin{bmatrix} 1 & 1 & \vert & 5 \\ 0 & 1 & \vert & 3 \end{bmatrix}$$
$$y = 3 \qquad \frac{1}{2}R_2 \rightarrow$$

Next, we need to obtain a 0 where the 1 is highlighted. We do so by subtracting row 2 from row 1 and placing the result in row 1, denoted $R_1 - R_2$.

$$x = 2 \qquad R_1 - R_2 \rightarrow \begin{bmatrix} 1 & 0 & \vert & 2 \\ 0 & 1 & \vert & 3 \end{bmatrix}$$
$$y = 3$$

This matrix is in reduced row–echelon form. The solution is (2, 3).

Now Try Exercise 23

In the next example, we use Gauss–Jordan elimination to solve a system with three linear equations and three variables. To do so we transform the matrix into the following reduced row–echelon form, where (a, b, c) represents the solution.

$$\begin{bmatrix} 1 & 0 & 0 & \vert & a \\ 0 & 1 & 0 & \vert & b \\ 0 & 0 & 1 & \vert & c \end{bmatrix}$$

EXAMPLE 5 **Transforming a matrix to reduced row–echelon form**

Use Gauss–Jordan elimination to transform the augmented matrix of the linear system into reduced row–echelon form. Find the solution.

$$x + y + 2z = 1$$
$$-x + z = -2$$
$$2x + y + 5z = -1$$

Solution

The linear system and the augmented matrix are both shown.

Linear System **Augmented Matrix**

$$\begin{aligned} x + y + 2z &= 1 \\ -x + z &= -2 \\ 2x + y + 5z &= -1 \end{aligned} \qquad \left[\begin{array}{ccc|c} 1 & 1 & 2 & 1 \\ -1 & 0 & 1 & -2 \\ 2 & 1 & 5 & -1 \end{array}\right]$$

First, we want to put 0s in the second and third rows, where the -1 and 2 are highlighted. To obtain a 0 in the first position of the second row we add row 1 to row 2 and place the result in row 2, denoted $R_2 + R_1$. To obtain a 0 in the first position of the third row we subtract 2 times row 1 from row 3 and place the result in row 3, denoted $R_3 - 2R_1$. Row 1 does not change. These steps eliminate the x-variable from the second and third equations.

$$\begin{aligned} x + y + 2z &= 1 \\ y + 3z &= -1 \\ -y + z &= -3 \end{aligned} \qquad \begin{aligned} \\ R_2 + R_1 &\rightarrow \\ R_3 - 2R_1 &\rightarrow \end{aligned} \left[\begin{array}{ccc|c} 1 & 1 & 2 & 1 \\ 0 & 1 & 3 & -1 \\ 0 & -1 & 1 & -3 \end{array}\right]$$

To eliminate the y-variable in row 1, we subtract row 2 from row 1. To eliminate the y-variable from row 3, we add row 2 to row 3.

$$\begin{aligned} x - z &= 2 \\ y + 3z &= -1 \\ 4z &= -4 \end{aligned} \qquad \begin{aligned} R_1 - R_2 &\rightarrow \\ \\ R_3 + R_2 &\rightarrow \end{aligned} \left[\begin{array}{ccc|c} 1 & 0 & -1 & 2 \\ 0 & 1 & 3 & -1 \\ 0 & 0 & 4 & -4 \end{array}\right]$$

To obtain a 1 in row 3, where the highlighted 4 is located, we multiply row 3 by $\frac{1}{4}$.

$$\begin{aligned} x - z &= 2 \\ y + 3z &= -1 \\ z &= -1 \end{aligned} \qquad \begin{aligned} \\ \\ \tfrac{1}{4}R_3 &\rightarrow \end{aligned} \left[\begin{array}{ccc|c} 1 & 0 & -1 & 2 \\ 0 & 1 & 3 & -1 \\ 0 & 0 & 1 & -1 \end{array}\right]$$

For the matrix to be in reduced row–echelon form, we need 0s in the highlighted locations. We first add row 3 to row 1 and then subtract 3 times row 3 from row 2.

$$\begin{aligned} x &= 1 \\ y &= 2 \\ z &= -1 \end{aligned} \qquad \begin{aligned} R_1 + R_3 &\rightarrow \\ R_2 - 3R_3 &\rightarrow \\ \end{aligned} \left[\begin{array}{ccc|c} 1 & 0 & 0 & 1 \\ 0 & 1 & 0 & 2 \\ 0 & 0 & 1 & -1 \end{array}\right]$$

This matrix is now in reduced row–echelon form. The solution is $(1, 2, -1)$.

Now Try Exercise 33

CRITICAL THINKING

An *inconsistent* system of linear equations has no solutions, and a system of *dependent* linear equations has infinitely many solutions. Suppose that an augmented matrix row reduces to either of the following matrices. Explain what each matrix indicates about the given system of linear equations.

$$\left[\begin{array}{ccc|c} 1 & 0 & 0 & 2 \\ 0 & 1 & 0 & 3 \\ 0 & 0 & 0 & 1 \end{array}\right] \qquad \left[\begin{array}{ccc|c} 1 & 0 & 1 & 2 \\ 0 & 1 & 2 & 3 \\ 0 & 0 & 0 & 0 \end{array}\right]$$

In the next example we find the amounts invested in three mutual funds.

EXAMPLE 6

Determining investment amounts in mutual funds

A total of $8000 was invested in three funds that grew at a rate of 5%, 10%, and 20% over 1 year. After 1 year, the combined value of the three funds had grown by $1200. Five times as much money was invested at 20% as at 10%. Find the amount invested in each fund.

Solution

Let x be the amount invested at 5%, y be the amount invested at 10%, and z be the amount invested at 20%. The total amount invested was $8000, so

$$x + y + z = 8000.$$

The growth in the first mutual fund, paying 5% of x, is given by $0.05x$. Similarly, the growths in the other mutual funds are given by $0.10y$ and $0.20z$. As the total growth was $1200, we can write

$$0.05x + 0.10y + 0.20z = 1200.$$

Multiplying each side of this equation by 20 to eliminate decimals results in

$$x + 2y + 4z = 24{,}000.$$

Five times as much was invested at 20% as at 10%, so $z = 5y$, or $5y - z = 0$.

These three equations can be written as a system of linear equations and as an augmented matrix.

Linear System

$$\begin{aligned} x + y + z &= 8{,}000 \\ x + 2y + 4z &= 24{,}000 \\ 5y - z &= 0 \end{aligned}$$

Augmented Matrix

$$\left[\begin{array}{ccc|c} 1 & 1 & 1 & 8{,}000 \\ 1 & 2 & 4 & 24{,}000 \\ 0 & 5 & -1 & 0 \end{array}\right]$$

A 0 can be obtained in the highlighted position by subtracting row 1 from row 2.

$$\begin{aligned} x + y + z &= 8{,}000 \\ y + 3z &= 16{,}000 \\ 5y - z &= 0 \end{aligned} \quad R_2 - R_1 \to \left[\begin{array}{ccc|c} 1 & 1 & 1 & 8{,}000 \\ 0 & 1 & 3 & 16{,}000 \\ 0 & 5 & -1 & 0 \end{array}\right]$$

Zeros can be obtained in the highlighted positions by subtracting row 2 from row 1 and by subtracting 5 times row 2 from row 3.

$$\begin{aligned} x - 2z &= -8{,}000 \\ y + 3z &= 16{,}000 \\ -16z &= -80{,}000 \end{aligned} \quad \begin{aligned} R_1 - R_2 \to \\ \\ R_3 - 5R_2 \to \end{aligned} \left[\begin{array}{ccc|c} 1 & 0 & -2 & -8{,}000 \\ 0 & 1 & 3 & 16{,}000 \\ 0 & 0 & -16 & -80{,}000 \end{array}\right]$$

To obtain a 1 in the highlighted position, multiply row 3 by $-\frac{1}{16}$.

$$\begin{aligned} x - 2z &= -8{,}000 \\ y + 3z &= 16{,}000 \\ z &= 5{,}000 \end{aligned} \quad -\frac{1}{16}R_3 \to \left[\begin{array}{ccc|c} 1 & 0 & -2 & -8{,}000 \\ 0 & 1 & 3 & 16{,}000 \\ 0 & 0 & 1 & 5{,}000 \end{array}\right]$$

To obtain a 0 in each of the highlighted positions, add twice row 3 to row 1 and subtract three times row 3 from row 2.

$$\begin{aligned} x &= 2{,}000 \\ y &= 1{,}000 \\ z &= 5{,}000 \end{aligned} \quad \begin{aligned} R_1 + 2R_3 \to \\ R_2 - 3R_3 \to \end{aligned} \left[\begin{array}{ccc|c} 1 & 0 & 0 & 2{,}000 \\ 0 & 1 & 0 & 1{,}000 \\ 0 & 0 & 1 & 5{,}000 \end{array}\right]$$

Thus $2000 was invested at 5%, $1000 at 10%, and $5000 at 20%.

Now Try Exercise 65

Using Technology to Solve Systems of Linear Equations (Optional)

▶ **REAL-WORLD CONNECTION** Examples 5 and 6 involve a lot of arithmetic. Trying to solve a large system of equations by hand is an enormous—if not impossible—task. In the real world, people use technology to solve large systems. In the next example we solve the linear systems from Examples 4 and 5 with a graphing calculator.

EXAMPLE 7 **Using technology**

Use a graphing calculator to solve the following systems of equations.

(a)
$$\begin{aligned} x + y &= 5 \\ -x + y &= 1 \end{aligned}$$

(b)
$$\begin{aligned} x + y + 2z &= 1 \\ -x + + z &= -2 \\ 2x + y + 5z &= -1 \end{aligned}$$

Solution

(a) Enter the 2×3 augmented matrix from Example 4 in a graphing calculator, as shown in Figure 4.36(a). Then transform the matrix into reduced row–echelon form (rref), as shown in Figure 4.36(b). The solution is $(2, 3)$.

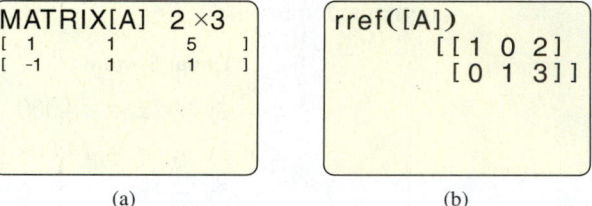

(a)   (b)

Figure 4.36

(b) Enter the 3×4 augmented matrix from Example 5, as shown in Figure 4.37(a). (The fourth column of A can be seen by scrolling right.) Then transform the matrix into reduced row–echelon form (rref), as shown in Figure 4.37(b). The solution is $(1, 2, -1)$.

CALCULATOR HELP

To enter a matrix and put it in reduced row–echelon form, see Appendix A (pages AP-8 and AP-9).

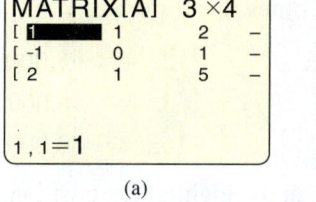
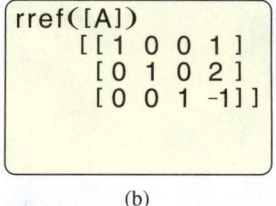

(a)   (b)

Figure 4.37

▌ Now Try Exercises 41, 43

▶ **REAL-WORLD CONNECTION** In the next example we use technology to solve a type of application that was first presented in the chapter introduction.

EXAMPLE 8 **Modeling the weight of male bears**

The data shown in Table 4.7 give the weight W, head length H, and overall length L of three bears. These data can be modeled with the equation $W = a + bH + cL$, where a, b, and c are constants that we need to determine. (*Sources:* M. Triola, *Elementary Statistics*; Minitab, Inc.)

TABLE 4.7

W (pounds)	H (inches)	L (inches)
362	16	72
300	14	68
147	11	52
?	13	65

(a) Set up a system of equations whose solution gives values for constants a, b, and c.
(b) Solve the system.
(c) Predict the weight of a bear with $H = 13$ inches and $L = 65$ inches.

Solution
(a) Substitute each row of Table 4.7 in the equation $W = a + bH + cL$.

$$362 = a + b(16) + c(72)$$
$$300 = a + b(14) + c(68)$$
$$147 = a + b(11) + c(52)$$

Rewrite this system as

$$a + 16b + 72c = 362$$
$$a + 14b + 68c = 300$$
$$a + 11b + 52c = 147$$

and represent it as the augmented matrix

$$A = \begin{bmatrix} 1 & 16 & 72 & | & 362 \\ 1 & 14 & 68 & | & 300 \\ 1 & 11 & 52 & | & 147 \end{bmatrix}.$$

(b) Enter A and put it in reduced row–echelon form, as shown in Figures 4.38(a) and (b), respectively. The solution is $a = -374$, $b = 19$, and $c = 6$.

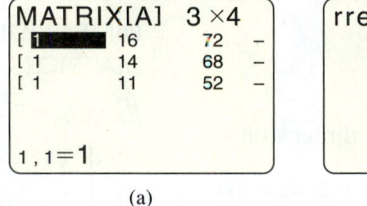

(a)   (b)

Figure 4.38

(c) For $W = a + bH + cL$, use

$$W = -374 + 19H + 6L$$

to predict the weight of a bear with head length $H = 13$ and overall length $L = 65$.

$$W = -374 + 19(13) + 6(65) = 263 \text{ pounds}$$

■ **Now Try Exercise 61**

4.6   Putting It All Together

A matrix is a rectangular array of numbers. An augmented matrix may be used to represent any system of linear equations. One common method for solving a system of linear equations is Gauss–Jordan elimination. Matrix row operations may be used to transform an augmented matrix to reduced row–echelon form.

CONCEPT	EXPLANATION/EXAMPLE
Augmented Matrix	A linear system can be represented by an augmented matrix. The following matrix has dimension 3×4. **Linear System** $$\begin{aligned} x + 2y - z &= 6 \\ -2x + y - z &= 7 \\ 2x \quad\quad + 3z &= -11 \end{aligned}$$ **Augmented Matrix** $$\begin{bmatrix} 1 & 2 & -1 & 6 \\ -2 & 1 & -1 & 7 \\ 2 & 0 & 3 & -11 \end{bmatrix}$$
Reduced Row–Echelon Form	The following augmented matrix is in reduced row–echelon form, which results from transforming the preceding system to reduced row–echelon form. There are 1s along the main diagonal and 0s elsewhere in the first three columns. The solution to the linear system is $(-1, 2, -3)$. $$\begin{bmatrix} 1 & 0 & 0 & -1 \\ 0 & 1 & 0 & 2 \\ 0 & 0 & 1 & -3 \end{bmatrix}$$

4.6   Exercises

CONCEPTS AND VOCABULARY

1. What is a matrix?

2. Give an example of a matrix and state its dimension.

3. Give an example of an augmented matrix and state its dimension.

4. If an augmented matrix is used to solve a system of three linear equations in three variables, what will be its dimension?

5. Give an example of a matrix that is in reduced row–echelon form.

6. Identify the elements on the main diagonal in the augmented matrix.

$$\begin{bmatrix} 4 & -6 & -1 & 3 \\ 6 & 2 & -2 & 9 \\ 7 & 5 & -3 & 1 \end{bmatrix}$$

DIMENSIONS OF MATRICES AND AUGMENTED MATRICES

Exercises 7–10: State the dimension of the matrix.

7. $\begin{bmatrix} 3 & -3 & 7 \\ 2 & 6 & -2 \\ 4 & 2 & 5 \end{bmatrix}$

8. $\begin{bmatrix} -2 & 3 & 0 \\ 1 & -8 & 4 \end{bmatrix}$

9. $\begin{bmatrix} 1 & 7 \\ 0 & 2 \\ 2 & -5 \end{bmatrix}$

10. $\begin{bmatrix} 4 & 2 & -3 & -1 \\ 4 & -3 & 2 & -7 \\ 14 & 6 & 4 & 0 \end{bmatrix}$

Exercises 11–14: Represent the linear system as an augmented matrix.

11. $\begin{aligned} x - 3y &= 1 \\ -x + 3y &= -1 \end{aligned}$

12. $\begin{aligned} 4x + 2y &= -5 \\ 5x + 8y &= 2 \end{aligned}$

13. $2x - y + 2z = -4$
$x - 2y = 2$
$-x + y - 2z = -6$

14. $3x - 2y + z = 5$
$-x + 2z = -4$
$x - 2y + z = -1$

Exercises 15–20: Write the system of linear equations that the augmented matrix represents. Use the variables x, y, and z.

15. $\begin{bmatrix} 1 & 2 & | & -6 \\ 5 & -1 & | & 4 \end{bmatrix}$

16. $\begin{bmatrix} 1 & -5 & | & 7 \\ 0 & -3 & | & 6 \end{bmatrix}$

17. $\begin{bmatrix} 1 & -1 & 2 & | & 6 \\ 2 & 1 & -2 & | & 1 \\ -1 & 2 & -1 & | & 3 \end{bmatrix}$

18. $\begin{bmatrix} 3 & -1 & 2 & | & -1 \\ 2 & -2 & 2 & | & 4 \\ 1 & 7 & -2 & | & 2 \end{bmatrix}$

19. $\begin{bmatrix} 1 & 0 & 0 & | & 4 \\ 0 & 1 & 0 & | & -2 \\ 0 & 0 & 1 & | & 7 \end{bmatrix}$

20. $\begin{bmatrix} 1 & 0 & 0 & | & 6 \\ 0 & 1 & 0 & | & -2 \\ 0 & 0 & 1 & | & 4 \end{bmatrix}$

SOCIAL NETWORKS

Exercises 21 and 22: (Refer to Example 2.) Use a 4 × 4 matrix to represent the social network.

21.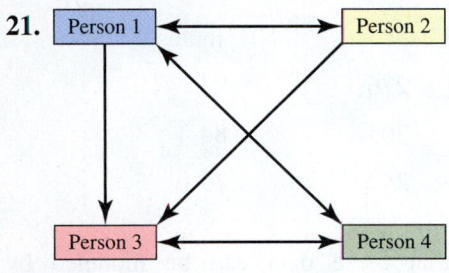

22.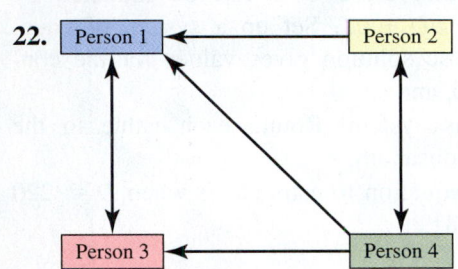

GAUSS–JORDAN ELIMINATION

Exercises 23–40: Use Gauss–Jordan elimination to find the solution. Write the solution as an ordered pair or ordered triple and check the solution.

23. $x + y = 4$
$x + 3y = 10$

24. $x - 3y = -7$
$2x + y = 0$

25. $2x + 3y = 3$
$-2x + 2y = 7$

26. $x + 3y = -14$
$2x + 5y = -24$

27. $x - y = 5$
$x + 3y = -1$

28. $x + 4y = 1$
$3x - 2y = 10$

29. $4x - 8y = -10$
$x + y = 2$

30. $x - 7y = -16$
$4x + 10y = 50$

31. $x + y + z = 6$
$2y - z = 1$
$y + z = 5$

32. $x + y + z = 3$
$x + y - z = 2$
$y + z = 2$

33. $x + 2y + 3z = 6$
$-x + 3y + 4z = 0$
$x + y - 2z = -6$

34. $2x - 4y + 2z = 10$
$-x + 3y - 4z = -19$
$2x - y - 6z = -28$

35. $x + y + z = 0$
$2x + y + 2z = -1$
$x + y = 0$

36. $x + y - 2z = 5$
$x + 2y - 2z = 4$
$-x - y + z = -4$

37. $x + y + z = 3$
$-x - z = -2$
$x + y + 2z = 4$

38. $x + 2y - z = 3$
$-x - y + z = 0$
$x + 2y = 5$

39. $x + 2y + z = 3$
$2x + y - z = -6$
$-x - y + 2z = 5$

40. $x + y + z = -3$
$x - y - z = -1$
$-2x + y + 4z = 4$

*Exercises 41–50: **Technology** Use a graphing calculator to solve the system of linear equations.*

41. $x + 4y = 13$
$5x - 3y = -50$

42. $9x - 11y = 7$
$5x + 6y = 16$

43. $2x - y + 3z = 9$
$-4x + 5y + 2z = 12$
$2x + 7z = 23$

44. $3x - 2y + 4z = 29$
$2x + 3y - 7z = -14$
$5x - y + 11z = 59$

45. $6x + 2y + z = 4$
$-2x + 4y + z = -3$
$2x - 8y = -2$

46. $-x - 9y + 2z = -28.5$
$2x - y + 4z = -17$
$x - y + 8z = -9$

47. $4x + 3y + 12z = -9.25$
$15y + 8z = -4.75 + x$
$7z = -5.5 - 6y$

48. $5x + 4y = 13.3 + z$
$7y + 9z = 16.9 - x$
$x - 3y + 4z = -4.1$

49. $1.2x - 0.9y + 2.7z = 5.37$
$3.1x - 5.1y + 7.2z = 14.81$
$1.8y + 6.38 = 3.6z - 0.2x$

50. $11x + 13y - 17z = 380$
$5x - 14y - 19z = 24$
$-21y + 46z = -676 + 7x$

Exercises 51–56: (Refer to the Critical Thinking box in this section.) Row-reduce the matrix associated with the given system to determine whether the system of linear equations is inconsistent or has dependent equations.

51. $\begin{aligned} x + 2y &= 4 \\ -2x - 4y &= -8 \end{aligned}$

52. $\begin{aligned} x - 5y &= 4 \\ -2x + 10y &= 8 \end{aligned}$

53. $\begin{aligned} x + y + z &= 3 \\ x + y - z &= 1 \\ x + y \quad\;\; &= 3 \end{aligned}$

54. $\begin{aligned} x + y + z &= 5 \\ x - y - z &= 8 \\ 2x + 2y + 2z &= 6 \end{aligned}$

55. $\begin{aligned} x + 2y + 3z &= 14 \\ 2x - 3y - 2z &= -10 \\ 3x - y + z &= 4 \end{aligned}$

56. $\begin{aligned} x + 2y + 3z &= 6 \\ -x + 3y + 4z &= 6 \\ 5y + 7z &= 12 \end{aligned}$

Exercises 57 and 58: **Thinking Generally** *The matrix represents a linear system of equations. Solve the system, if a and b are nonzero constants.*

57. $\begin{bmatrix} a & 0 & 0 & | & 1 \\ 0 & b & 0 & | & 1 \\ 0 & 0 & ab & | & 2 \end{bmatrix}$

58. $\begin{bmatrix} a & 0 & 0 & | & 1 \\ 0 & b & 0 & | & 2 \\ 0 & 0 & 0 & | & a \end{bmatrix}$

APPLICATIONS

59. *Weight of a Bear* Use the results of Example 8 to estimate the weight of a bear with a head length of 12 inches and an overall length of 60 inches.

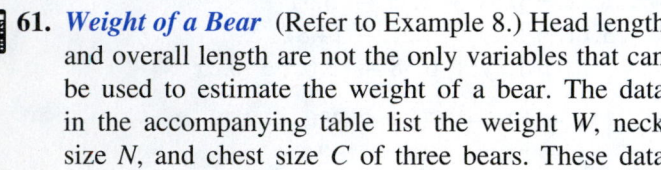 **60.** *Garbage and Household Size* A larger household produces more garbage, on average, than a smaller household. If we know the amount of metal *M* and plastic *P* waste produced each week, we can estimate the household size *H* from $H = a + bM + cP$. The table contains representative data for three households. (*Source:* M. Triola, *Elementary Statistics.*)

H (people)	*M* (pounds)	*P* (pounds)
3	2.00	1.40
2	1.50	0.65
6	4.00	3.40

(a) Set up a system of equations whose solution gives values for the constants *a*, *b*, and *c*.
(b) Solve this system.
(c) Predict the size of a household that produces 3 pounds of metal waste and 2 pounds of plastic waste each week.

61. *Weight of a Bear* (Refer to Example 8.) Head length and overall length are not the only variables that can be used to estimate the weight of a bear. The data in the accompanying table list the weight *W*, neck size *N*, and chest size *C* of three bears. These data

can be modeled by $W = a + bN + cC$. (*Sources:* M. Triola, *Elementary Statistics;* Minitab, Inc.)

W (pounds)	*N* (inches)	*C* (inches)
80	16	26
344	28	45
416	31	54

(a) Set up a system of equations whose solution gives values for the constants *a*, *b*, and *c*.
(b) Solve this system. Round each value to the nearest tenth.
(c) Predict the weight of a bear with neck size *N* = 22 inches and chest size *C* = 38 inches.

62. *Old Faithful Geyser* In Yellowstone National Park, Old Faithful Geyser has been a favorite attraction for decades. Although this geyser erupts about every 80 minutes, this time interval varies, as do the duration and height of the eruptions. The accompanying table shows the height *H*, duration *D*, and time interval *T* for three representative eruptions. (*Source:* National Park Service.)

H (feet)	*D* (seconds)	*T* (minutes)
160	276	94
125	203	84
140	245	79

(a) Assume that these data can be modeled by $H = a + bD + cT$. Set up a system of equations whose solution gives values for the constants *a*, *b*, and *c*.
(b) Solve this system. Round each value to the nearest thousandth.
(c) Use this equation to estimate *H* when *D* = 220 and *T* = 81.

63. *Jogging Speeds* A runner in preparation for a marathon jogs at 5, 6, and 8 miles per hour. The runner travels a total distance of 12.5 miles in 2 hours and jogs the same length of time at 5 miles per hour and at 8 miles per hour. How long does the runner jog at each speed?

64. *Mixture Problem* Three types of candy that cost $2, $3, and $4 per pound are to be mixed to produce a 5-pound bag of candy that costs $14.50. If there are to be equal amounts of the $3-per-pound candy and the $4-per-pound candy, how much of each type of candy should be included in the mixture?

65. *Interest and Investments* (Refer to Example 6.) A total of $3000 is invested at 3%, 4%, and 6% annual interest. The interest earned after 1 year equals $145. The amount invested at 6% is triple the amount invested at 3%. Find the amount invested at each rate.

66. *Geometry* The measure of the largest angle in a triangle is twice the measure of the smallest angle. The remaining angle is 10° less than the largest angle. Find the measure of each angle.

67. **Online Exploration** Go to your Facebook profile and select three or four of your friends plus yourself. Answers will vary.
  (a) Draw a social network that shows the friendships between each pair of people. Use a double arrow to represent this. (See Figure 4.35.)
  (b) Use a matrix to represent your social network.

68. **Online Exploration** Go online and look up a diagram of the molecular structure of water (H_2O).
  (a) Draw double arrows wherever the hydrogen and oxygen are connected.
  (b) Use a matrix to represent the structure of water. Let the first row represent oxygen.

WRITING ABOUT MATHEMATICS

69. Explain what the dimension of a matrix means. What is the difference between a matrix that has dimension 3×4 and one that has dimension 4×3?

70. Discuss the advantages of using technology to transform an augmented matrix to reduced row–echelon form. Are there any disadvantages? Explain.

SECTIONS 4.5 and 4.6 — Checking Basic Concepts

1. Determine which ordered triple is the solution to the system of equations: $(5, -4, 0)$ or $(1, 3, -1)$.

$$x - y + 7z = -9$$
$$2x - 2y + 5z = -9$$
$$-x + 3y - 2z = 10$$

2. Solve the system of equations by using elimination and substitution.

$$x - y + z = 2$$
$$2x - 3y + z = -1$$
$$-x + y + z = 4$$

3. Use an augmented matrix to represent the system of equations. Solve the system by using
  (a) Gauss–Jordan elimination and
   (b) technology.

$$x + 2y + z = 1$$
$$x + y + z = -1$$
$$y + z = 1$$

4. A total of $1500 is invested at 1%, 2%, and 4% annual interest. The interest after 1 year equals $46. The amount invested at 2% is double the amount invested at 1%. Find the amount invested at each rate.

4.7   Determinants

Calculation of Determinants • Area of Regions • Cramer's Rule

A LOOK INTO MATH ▶

Surveyors commonly calculate the areas of parcels of land. To do so, they frequently divide the land into triangular regions. When the coordinates of the vertices of a triangle are known, determinants may be used to find the area of the triangle. *A determinant is a real number that can be calculated for any square matrix.* In this section we use determinants to find areas and to solve systems of linear equations.

Calculation of Determinants

NEW VOCABULARY

☐ Determinant
☐ Expansion by minors
☐ Minors
☐ Cramer's rule

▶ **REAL-WORLD CONNECTION** The concept of determinants originated with the Japanese mathematician Seki Kowa (1642–1708), who used them to solve systems of linear equations. Later, Gottfried Leibniz (1646–1716) formally described determinants and also used them to solve systems of linear equations. (*Source: Historical Topics for the Mathematical Classroom, NCTM.*)

We begin by defining a determinant of a 2×2 matrix.

DETERMINANT OF A 2×2 MATRIX

The **determinant** of

$$A = \begin{bmatrix} a & b \\ c & d \end{bmatrix}$$

is a *real number* defined by

$$\det A = ad - cb.$$

EXAMPLE 1   **Calculating determinants**

Find det A for each 2×2 matrix.

(a) $A = \begin{bmatrix} 1 & 2 \\ 3 & 4 \end{bmatrix}$ **(b)** $A = \begin{bmatrix} -1 & -3 \\ 2 & -8 \end{bmatrix}$

Solution

(a) The determinant is calculated as follows.

$$\det A = \det \begin{bmatrix} 1 & 2 \\ 3 & 4 \end{bmatrix} = (1)(4) - (3)(2) = -2$$

(b) Similarly,

$$\det A = \det \begin{bmatrix} -1 & -3 \\ 2 & -8 \end{bmatrix} = (-1)(-8) - (2)(-3) = 14.$$

▮ **Now Try Exercise 5**

READING CHECK

How do you calculate the determinant of a 2×2 matrix?

We can use determinants of 2×2 matrices to find determinants of 3×3 matrices. This method is called **expansion of a determinant by minors**.

DETERMINANT OF A 3 × 3 MATRIX

$$\det A = \det \begin{bmatrix} a_1 & b_1 & c_1 \\ a_2 & b_2 & c_2 \\ a_3 & b_3 & c_3 \end{bmatrix}$$

$$= a_1 \cdot \det \begin{bmatrix} b_2 & c_2 \\ b_3 & c_3 \end{bmatrix} - a_2 \cdot \det \begin{bmatrix} b_1 & c_1 \\ b_3 & c_3 \end{bmatrix} + a_3 \cdot \det \begin{bmatrix} b_1 & c_1 \\ b_2 & c_2 \end{bmatrix}$$

The 2 × 2 matrices in this equation are called **minors**.

EXAMPLE 2   **Calculating 3 × 3 determinants**

Evaluate det A.

(a) $A = \begin{bmatrix} 2 & 1 & -1 \\ -1 & 3 & 2 \\ 4 & -3 & -5 \end{bmatrix}$ (b) $A = \begin{bmatrix} 5 & -2 & 4 \\ 0 & 2 & 1 \\ -1 & 4 & -4 \end{bmatrix}$

Solution

(a) We evaluate the determinant as follows.

$$\det \begin{bmatrix} \mathbf{2} & 1 & -1 \\ \mathbf{-1} & 3 & 2 \\ \mathbf{4} & -3 & -5 \end{bmatrix} = \mathbf{2} \cdot \det \begin{bmatrix} 3 & 2 \\ -3 & -5 \end{bmatrix} - (\mathbf{-1}) \cdot \det \begin{bmatrix} 1 & -1 \\ -3 & -5 \end{bmatrix}$$

$$+ \, \mathbf{4} \cdot \det \begin{bmatrix} 1 & -1 \\ 3 & 2 \end{bmatrix}$$

$$= 2(-9) + 1(-8) + 4(5)$$

$$= -6$$

(b) We evaluate the determinant as follows.

$$\det \begin{bmatrix} \mathbf{5} & -2 & 4 \\ \mathbf{0} & 2 & 1 \\ \mathbf{-1} & 4 & -4 \end{bmatrix} = \mathbf{5} \cdot \det \begin{bmatrix} 2 & 1 \\ 4 & -4 \end{bmatrix} - (\mathbf{0}) \cdot \det \begin{bmatrix} -2 & 4 \\ 4 & -4 \end{bmatrix}$$

$$+ \, (\mathbf{-1}) \cdot \det \begin{bmatrix} -2 & 4 \\ 2 & 1 \end{bmatrix}$$

$$= 5(-12) - 0(-8) + (-1)(-10)$$

$$= -50$$

Now Try Exercises 13, 15

Many graphing calculators can evaluate the determinant of a matrix, as illustrated in the next example, where we evaluate the determinants from Example 2.

EXAMPLE 3   **Using technology to find determinants**

Find each determinant of A, using a graphing calculator.

(a) $A = \begin{bmatrix} 2 & 1 & -1 \\ -1 & 3 & 2 \\ 4 & -3 & -5 \end{bmatrix}$ (b) $A = \begin{bmatrix} 5 & -2 & 4 \\ 0 & 2 & 1 \\ -1 & 4 & -4 \end{bmatrix}$

Solution

(a) Begin by entering the matrix and then evaluate the determinant, as shown in Figure 4.39. The result is det $A = -6$, which agrees with our earlier calculation.

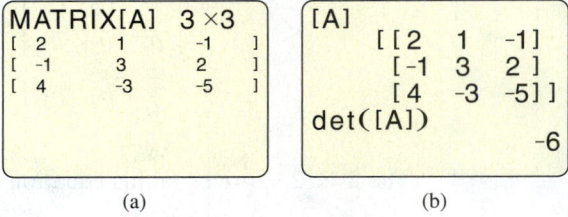

(a)   (b)

Figure 4.39

(b) The determinant of A evaluates to -50 (see Figure 4.40).

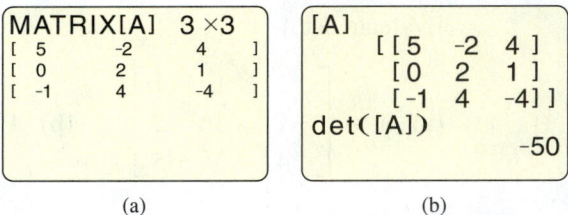

(a)   (b)

Figure 4.40

▌ **Now Try Exercise 21**

CALCULATOR HELP

To find a determinant, see Appendix A (pages AP-9 and AP-10).

Area of Regions

CRITICAL THINKING

Suppose that you are given three distinct points in the xy-plane and $D = 0$. What must be true about the three points?

▶ **REAL-WORLD CONNECTION** A determinant may be used to find the area of a triangle. For example, if a triangle has vertices (a_1, a_2), (b_1, b_2), and (c_1, c_2), its area equals the absolute value of D, where

$$D = \frac{1}{2}\det\begin{bmatrix} a_1 & b_1 & c_1 \\ a_2 & b_2 & c_2 \\ 1 & 1 & 1 \end{bmatrix}.$$

If the vertices are entered in the columns of D counterclockwise as they appear in the xy-plane, D will be positive. (*Source:* W. Taylor, *The Geometry of Computer Graphics.*)

EXAMPLE 4 **Computing the area of a triangular parcel of land**

A triangular parcel of land is shown in Figure 4.41. If all units are miles, find the area of the parcel of land by using a determinant.

Solution

The vertices of the triangular parcel of land are $(2, 2)$, $(5, 4)$, and $(3, 8)$. The area of the triangle is

$$D = \frac{1}{2}\det\begin{bmatrix} 2 & 5 & 3 \\ 2 & 4 & 8 \\ 1 & 1 & 1 \end{bmatrix} = \frac{1}{2}\cdot 16 = 8.$$

The area of the triangle is 8 square miles.

▌ **Now Try Exercise 27**

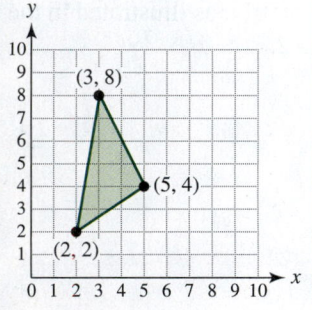

Figure 4.41

Cramer's Rule

▶ **REAL-WORLD CONNECTION** Determinants were developed by Gabriel Cramer (1704–1752). His work, published in 1750, provided a method called **Cramer's rule** for solving systems of linear equations.

CRAMER'S RULE FOR LINEAR SYSTEMS IN TWO VARIABLES

The solution to the system of linear equations

$$a_1x + b_1y = c_1$$
$$a_2x + b_2y = c_2$$

is given by $x = \frac{E}{D}$ and $y = \frac{F}{D}$, where

$$E = \det\begin{bmatrix} c_1 & b_1 \\ c_2 & b_2 \end{bmatrix}, \quad F = \det\begin{bmatrix} a_1 & c_1 \\ a_2 & c_2 \end{bmatrix}, \quad \text{and} \quad D = \det\begin{bmatrix} a_1 & b_1 \\ a_2 & b_2 \end{bmatrix} \neq 0.$$

NOTE: If $D = 0$, the system has either no solutions or infinitely many solutions.

EXAMPLE 5 | **Using Cramer's rule**

Use Cramer's rule to solve the following linear systems.
(a) $3x - 4y = 18$ **(b)** $-4x + 9y = -24$
      $7x + 5y = -1$ $6x + 17y = -25$

Solution

(a) $E = \det\begin{bmatrix} c_1 & b_1 \\ c_2 & b_2 \end{bmatrix} = \det\begin{bmatrix} 18 & -4 \\ -1 & 5 \end{bmatrix} = (18)(5) - (-1)(-4) = \mathbf{86}$

$F = \det\begin{bmatrix} a_1 & c_1 \\ a_2 & c_2 \end{bmatrix} = \det\begin{bmatrix} 3 & 18 \\ 7 & -1 \end{bmatrix} = (3)(-1) - (7)(18) = \mathbf{-129}$

$D = \det\begin{bmatrix} a_1 & b_1 \\ a_2 & b_2 \end{bmatrix} = \det\begin{bmatrix} 3 & -4 \\ 7 & 5 \end{bmatrix} = (3)(5) - (7)(-4) = \mathbf{43}$

Because $x = \frac{E}{D} = \frac{86}{43} = 2$ and $y = \frac{F}{D} = \frac{-129}{43} = -3$, the solution is $(2, -3)$.

(b) $E = \det\begin{bmatrix} c_1 & b_1 \\ c_2 & b_2 \end{bmatrix} = \det\begin{bmatrix} -24 & 9 \\ -25 & 17 \end{bmatrix} = (-24)(17) - (-25)(9) = \mathbf{-183}$

$F = \det\begin{bmatrix} a_1 & c_1 \\ a_2 & c_2 \end{bmatrix} = \det\begin{bmatrix} -4 & -24 \\ 6 & -25 \end{bmatrix} = (-4)(-25) - (6)(-24) = \mathbf{244}$

$D = \det\begin{bmatrix} a_1 & b_1 \\ a_2 & b_2 \end{bmatrix} = \det\begin{bmatrix} -4 & 9 \\ 6 & 17 \end{bmatrix} = (-4)(17) - (6)(9) = \mathbf{-122}$

Because $x = \frac{E}{D} = \frac{-183}{-122} = 1.5$ and $y = \frac{F}{D} = \frac{244}{-122} = -2$, the solution is $(1.5, -2)$.

Now Try Exercises 33, 35

Cramer's rule can be applied to systems that have any number of linear equations. Cramer's rule for three linear equations is discussed in the Extended and Discovery Exercises at the end of this chapter.

4.7 Putting It All Together

CONCEPT	EXPLANATION
Determinant of a 2×2 Matrix	The determinant of a 2×2 matrix A is given by $$\det A = \det \begin{bmatrix} a & b \\ c & d \end{bmatrix} = ad - cb.$$
Determinant of a 3×3 Matrix	The determinant of a 3×3 matrix A is given by $$\det A = \det \begin{bmatrix} a_1 & b_1 & c_1 \\ a_2 & b_2 & c_2 \\ a_3 & b_3 & c_3 \end{bmatrix}$$ $$= a_1 \cdot \det \begin{bmatrix} b_2 & c_2 \\ b_3 & c_3 \end{bmatrix} - a_2 \cdot \det \begin{bmatrix} b_1 & c_1 \\ b_3 & c_3 \end{bmatrix}$$ $$+ a_3 \cdot \det \begin{bmatrix} b_1 & c_1 \\ b_2 & c_2 \end{bmatrix}.$$
Area of a Triangle	If a triangle has vertices (a_1, a_2), (b_1, b_2), and (c_1, c_2), its area equals the absolute value of D, where $$D = \frac{1}{2} \det \begin{bmatrix} a_1 & b_1 & c_1 \\ a_2 & b_2 & c_2 \\ 1 & 1 & 1 \end{bmatrix}.$$
Cramer's Rule for Linear Systems in Two Variables	The solution to the linear system $$a_1 x + b_1 y = c_1$$ $$a_2 x + b_2 y = c_2$$ is given by $x = \frac{E}{D}$ and $y = \frac{F}{D}$, where $$E = \det \begin{bmatrix} c_1 & b_1 \\ c_2 & b_2 \end{bmatrix}, \quad F = \det \begin{bmatrix} a_1 & c_1 \\ a_2 & c_2 \end{bmatrix}, \quad \text{and}$$ $$D = \det \begin{bmatrix} a_1 & b_1 \\ a_2 & b_2 \end{bmatrix} \neq 0.$$ **NOTE:** If $D = 0$, then the system has either no solutions or infinitely many solutions.

4.7 Exercises

MyMathLab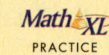

CONCEPTS AND VOCABULARY

1. We can find the determinant of a(n) _____ matrix.

2. If we find a determinant, the answer is a(n) _____.

3. Cramer's rule can be used to solve a(n) _____.

4. If the first column of a matrix is all 0s, then its determinant equals _____.

CALCULATING DETERMINANTS

Exercises 5–20: Evaluate det A by hand where A is the given matrix.

5. $\begin{bmatrix} 1 & -2 \\ 3 & -8 \end{bmatrix}$ 6. $\begin{bmatrix} 5 & -1 \\ 3 & 7 \end{bmatrix}$

7. $\begin{bmatrix} -3 & 7 \\ 8 & -1 \end{bmatrix}$ 8. $\begin{bmatrix} 0 & -7 \\ -3 & 1 \end{bmatrix}$

9. $\begin{bmatrix} 23 & 4 \\ 6 & -13 \end{bmatrix}$ 10. $\begin{bmatrix} 44 & -51 \\ -9 & 32 \end{bmatrix}$

11. $\begin{bmatrix} 1 & -1 & 2 \\ 0 & 1 & -3 \\ 0 & -4 & 7 \end{bmatrix}$ 12. $\begin{bmatrix} 2 & -1 & -5 \\ -1 & 4 & -2 \\ 0 & 1 & 4 \end{bmatrix}$

13. $\begin{bmatrix} 2 & -1 & 0 \\ 1 & -2 & 6 \\ 0 & 1 & 8 \end{bmatrix}$ 14. $\begin{bmatrix} 0 & 1 & -4 \\ 3 & -6 & 10 \\ 4 & -2 & 7 \end{bmatrix}$

15. $\begin{bmatrix} -1 & 3 & 5 \\ 3 & -3 & 5 \\ 2 & -3 & 7 \end{bmatrix}$ 16. $\begin{bmatrix} 6 & -1 & 9 \\ 7 & 0 & -3 \\ 2 & 5 & -1 \end{bmatrix}$

17. $\begin{bmatrix} 5 & 0 & 0 \\ 0 & -2 & 0 \\ 0 & 0 & 5 \end{bmatrix}$ 18. $\begin{bmatrix} 1 & 2 & 3 \\ 2 & 4 & 6 \\ 3 & 6 & 9 \end{bmatrix}$

19. $\begin{bmatrix} 0 & 2 & -3 \\ 0 & 3 & -9 \\ 0 & 5 & 9 \end{bmatrix}$ 20. $\begin{bmatrix} 3 & -1 & 2 \\ 0 & 5 & 7 \\ 0 & 0 & -1 \end{bmatrix}$

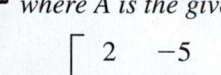 *Exercises 21–24: Use technology to calculate det A, where A is the given matrix.*

21. $\begin{bmatrix} 2 & -5 & 13 \\ 10 & 15 & -10 \\ 17 & -19 & 22 \end{bmatrix}$ 22. $\begin{bmatrix} 1.6 & 3.1 & 5.7 \\ 2.1 & 6.7 & 8.1 \\ -0.4 & -0.8 & -3.1 \end{bmatrix}$

23. $\begin{bmatrix} 17 & 0 & 4 \\ -9 & 14 & 1.5 \\ 13 & 67 & -11 \end{bmatrix}$ 24. $\begin{bmatrix} 121 & 45 & -56 \\ -45 & 87 & 32 \\ -14 & -34 & 67 \end{bmatrix}$

Exercises 25 and 26: **Thinking Generally** *Find det A, if a, b, and c are nonzero constants.*

25. $A = \begin{bmatrix} a & 0 & 0 \\ 0 & b & 0 \\ 0 & 0 & c \end{bmatrix}$ 26. $A = \begin{bmatrix} 0 & 0 & 0 \\ a & b & 0 \\ 0 & 0 & c \end{bmatrix}$

CALCULATING AREA

Exercises 27–32: Find the area of the figure by using a determinant. Assume that units are feet.

27.
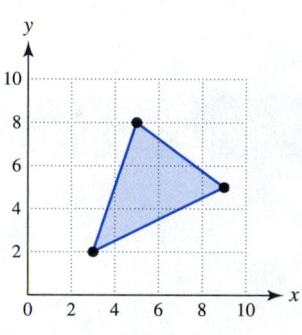

28.
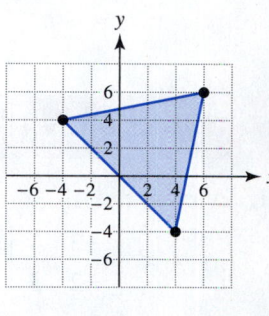

29.
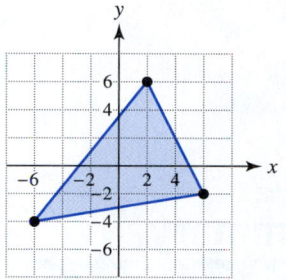

30.
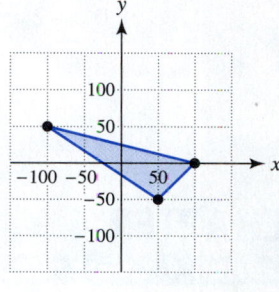

31.
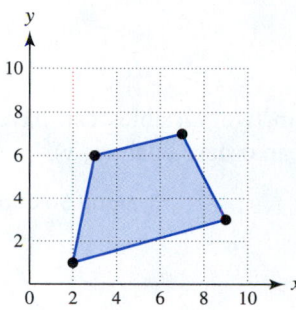

32.
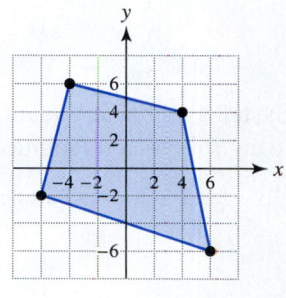

CRAMER'S RULE

Exercises 33–38: Solve the system of equations by using Cramer's rule.

33. $5x + 3y = 4$
    $6x - 4y = 20$

34. $-5x + 4y = -5$
    $4x + 4y = -32$

35. $7x - 5y = -3$
    $-4x + 6y = -8$

36. $-4x - 9y = -17$
    $8x + 4y = 9$

37. $8x = 3y - 61$
    $-x = 4y - 23$

38. $15y = -188 - 22x$
    $23y = -173 - 16x$

WRITING ABOUT MATHEMATICS

39. Suppose that one row of a 3×3 matrix A is all 0s. What is the value of det A? Give an example and explain your answer.

40. Explain how to evaluate a determinant of a 2×2 matrix. Can you find the determinant of a 2×3 matrix? Explain your answer.

Checking Basic Concepts

1. Evaluate det A.

(a) $A = \begin{bmatrix} -3 & 4 \\ -2 & 3 \end{bmatrix}$

(b) $A = \begin{bmatrix} 1 & -2 & 3 \\ 5 & 1 & 1 \\ 0 & 2 & -1 \end{bmatrix}$

2. Use Cramer's rule to solve the system of linear equations.

$$2x - y = -14$$
$$3x - 4y = -36$$

3. Find the area of a triangle that has the vertices $(-1, 2)$, $(5, 6)$, and $(2, -3)$.

CHAPTER 4   Summary

SECTION 4.1 ■ SYSTEMS OF LINEAR EQUATIONS IN TWO VARIABLES

Systems of Linear Equations in Two Variables

$$ax + by = c$$
$$dx + ey = k$$

Systems of linear equations in two variables can have no solutions, one solution, or infinitely many solutions. A solution is an ordered pair (x, y).

No Solutions	Unique Solution	Infinitely Many Solutions
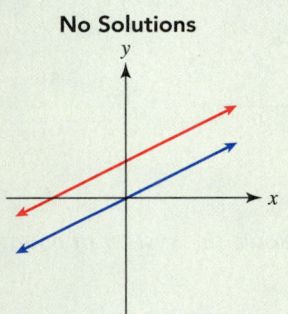	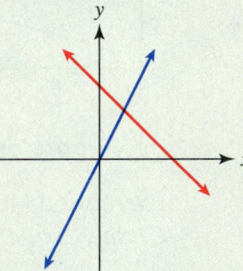	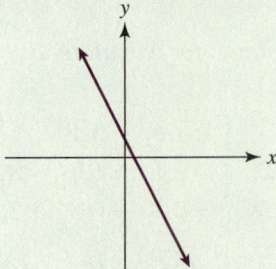
Inconsistent System; (No Solutions)	Consistent System; Independent Equations	Consistent System; Dependent Equations

Examples:

$x + y = 4$	$x + y = 4$	$x + y = 1$
$x + y = 0$	$x - y = 2$	$2x + 2y = 2$
No solutions	The solution is $(3, 1)$.	$\{(x, y) \mid x + y = 1\}$

Graphical and Numerical Solutions These methods use graphs and tables to solve systems.

Example: $2x + y = 4$
   $x + y = 3$

Graph and make a table for $y_1 = 4 - 2x$ and $y_2 = 3 - x$. The solution is $(1, 2)$.

Graphical Solution

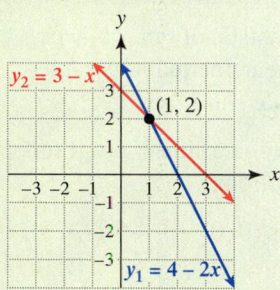

Numerical Solution

x	$4 - 2x$	$3 - x$
0	4	3
1	2	2
2	0	1
3	-2	0

SECTION 4.2 ■ THE SUBSTITUTION AND ELIMINATION METHODS

Two symbolic methods for solving systems of linear equations are *substitution* and *elimination*.

Substitution

$$x + y = 5 \quad \text{or} \quad y = 5 - x$$
$$x - y = -3$$

Substituting in the second equation, $x - (5 - x) = -3$, yields $x = 1$ and $y = 4$. The solution is $(1, 4)$.

Elimination

$x +$	$y = 5$	First equation
$x -$	$y = -3$	Second equation
$2x + 0y =$	2	Add the equations.

Thus $x = 1$ and $y = 4$. The solution is $(1, 4)$.

SECTION 4.3 ■ SYSTEMS OF LINEAR INEQUALITIES

Linear Inequalities in Two Variables

$$ax + by > c,$$

where $>$ can be replaced by $<$, $\leq$, or $\geq$. The solution set is typically a region in the xy-plane.

Example: $x + y \leq 4$

Any point in the shaded region must satisfy the given inequality. The *test point*, $(0, 0)$, lies in the shaded region and satisfies the inequality $x + y \leq 4$ because $0 + 0 \leq 4$.

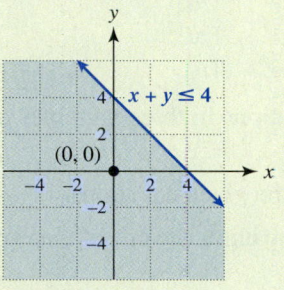

Systems of Linear Inequalities in Two Variables

$$ax + by > c$$
$$dx + ey > k,$$

where $>$ can be replaced by $<$, $\leq$, or $\geq$.

Example: $x + y \leq 2$
$$y \geq x$$

The test point $(-2, 1)$ satisfies both inequalities, so shade the region containing $(-2, 1)$.

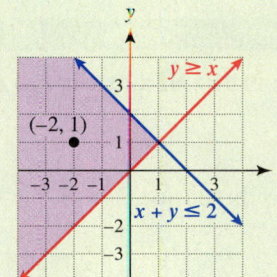

Linear Programming Problem   A linear programming problem consists of an *objective function* to be minimized or maximized (optimized) and a system of linear inequalities called *constraints*. The solution set to the constraints is called the *region of feasible solutions*.

Fundamental Theorem of Linear Programming   If the optimal value for a linear programming problem exists, then it occurs at a vertex of the region of feasible solutions.

Example: The maximum of $R = 2x + 3y$ must occur at one of the vertices $(0, 0)$, $(0, 4.5)$, $(3, 3)$, or $(4.5, 0)$.

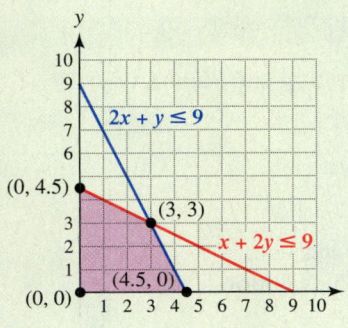

Vertex	$R = 2x + 3y$	
$(0, 0)$	$2(0) + 3(0) = 0$	
$(0, 4.5)$	$2(0) + 3(4.5) = 13.5$	
$(3, 3)$	$2(3) + 3(3) = $ **15**	← **Maximum**
$(4.5, 0)$	$2(4.5) + 3(0) = 9$	

Maximum of $R = 15$ occurs at vertex $(3, 3)$.

Solution to a System of Linear Equations in Three Variables   An ordered triple (x, y, z) that satisfies all three equations

Example:
$$x - y + 2z = 3$$
$$2x - y + z = 5$$
$$x + y + z = 6$$

The solution is $(\mathbf{3}, \mathbf{2}, \mathbf{1})$ because these values for (x, y, z) satisfy all three equations.

$$\mathbf{3} - \mathbf{2} + 2(\mathbf{1}) = 3 \checkmark \quad \text{True}$$
$$2(\mathbf{3}) - \mathbf{2} + \mathbf{1} = 5 \checkmark \quad \text{True}$$
$$\mathbf{3} + \mathbf{2} + \mathbf{1} = 6 \checkmark \quad \text{True}$$

Elimination and Substitution   Systems of linear equations in three variables can be solved by elimination and substitution, using the following steps.

STEP 1:   Eliminate one variable, such as x, from two of the given equations.

STEP 2:   Use the two resulting equations in two variables to eliminate one of the variables, such as y. Solve for the remaining variable z.

STEP 3:   Substitute z in one of the two equations from Step 2. Solve for the unknown variable y.

STEP 4:   Substitute values for y and z in one of the given equations. Then find x. The solution is the ordered triple (x, y, z).

Matrix   A rectangular array of numbers is a matrix. If a matrix has m rows and n columns, it has dimension $m \times n$.

Example: Matrix $A = \begin{bmatrix} 3 & -1 & 7 \\ 0 & 6 & -2 \end{bmatrix}$ has dimension 2×3.

Augmented Matrix Any linear system can be represented with an augmented matrix.

Linear System	Augmented Matrix
$4x - 3y = 5$	$\begin{bmatrix} 4 & -3 & 5 \\ 1 & 2 & 4 \end{bmatrix}$
$x + 2y = 4$	

Gauss–Jordan Elimination A numerical method that uses matrix row transformations to tranform a matrix into reduced row–echelon form

Example: The matrix $\begin{bmatrix} 4 & -3 & 5 \\ 1 & 2 & 4 \end{bmatrix}$ reduces to $\begin{bmatrix} 1 & 0 & 2 \\ 0 & 1 & 1 \end{bmatrix}$.

The solution to the system is $(2, 1)$.

SECTION 4.7 ■ DETERMINANTS

Determinant for a 2 × 2 Matrix A determinant is a *real number*. The determinant of a 2 × 2 matrix is

$$\det A = \det \begin{bmatrix} a & b \\ c & d \end{bmatrix} = ad - cb.$$

Example: $\det \begin{bmatrix} 2 & 3 \\ 4 & 5 \end{bmatrix} = (2)(5) - (4)(3) = -2$

Determinant for a 3 × 3 Matrix

$$\det A = \det \begin{bmatrix} a_1 & b_1 & c_1 \\ a_2 & b_2 & c_2 \\ a_3 & b_3 & c_3 \end{bmatrix}$$

$$= a_1 \cdot \det \begin{bmatrix} b_2 & c_2 \\ b_3 & c_3 \end{bmatrix} - a_2 \cdot \det \begin{bmatrix} b_1 & c_1 \\ b_3 & c_3 \end{bmatrix} + a_3 \cdot \det \begin{bmatrix} b_1 & c_1 \\ b_2 & c_2 \end{bmatrix}$$

Example: $\det \begin{bmatrix} 2 & 3 & 2 \\ 3 & 7 & -3 \\ 0 & 0 & -1 \end{bmatrix} = 2 \det \begin{bmatrix} 7 & -3 \\ 0 & -1 \end{bmatrix} - 3 \det \begin{bmatrix} 3 & 2 \\ 0 & -1 \end{bmatrix} + 0$

$$= 2(-7) - 3(-3) = -5$$

Cramer's rule uses determinants to solve linear systems of equations. Determinants can also be used to find areas of triangles. See Putting It All Together for Section 4.7.

CHAPTER 4 Review Exercises

SECTION 4.1

Exercises 1 and 2: Which ordered pair is a solution?

1. $(2, -1), (3, 2)$
$$3x - 2y = 5$$
$$-2x + 4y = 2$$

2. $(4, -3), (-1, 2)$
$$x - 5y = 19$$
$$4x + 3y = 7$$

3. A system of linear equations is represented graphically. Find any solutions. Check your answer.

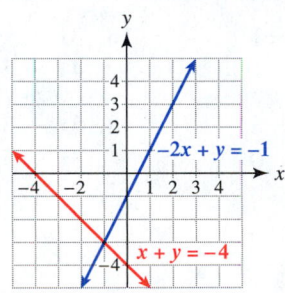

4. Use the accompanying table to solve

$$3x - y = 2$$
$$-x + y = 5.$$

Check your answer.

x	2	2.5	3	3.5	4
$3x - 2$	4	5.5	7	8.5	10
$x + 5$	7	7.5	8	8.5	9

Exercises 5–8: Solve the system of equations graphically. Determine whether the system is consistent or inconsistent. If the system is consistent, state whether the equations are dependent or independent.

5. $x + y = 6$
   $x - y = -4$

6. $x - y = -2$
   $-2x + 2y = 4$

7. $4x + 2y = 1$
   $2x + y = 5$

8. $x - 3y = 5$
   $x + 5y = -3$

Exercises 9 and 10: Solve the system numerically.

9. $3x + y = 7$
   $6x + y = 16$

10. $4x + 2y = -6$
   $3x - y = -7$

Exercises 11 and 12: Complete the following.

(a) *Write a system that models the situation.*
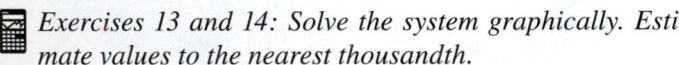
(b) *Solve the resulting system graphically.*

11. The sum of two numbers is 25 and their difference is 10. Find the two numbers.

12. Three times a number minus two times another number equals 19. The sum of the two numbers is 18. Find the two numbers.

Exercises 13 and 14: Solve the system graphically. Estimate values to the nearest thousandth.

13. $\pi x - 2.1y = \sqrt{2}$
   $\sqrt{3}x + y = \frac{5}{6}$

14. $\sqrt{5}x - \pi y = \frac{2}{7}$
   $x + 0.3y = \pi$

SECTION 4.2

Exercises 15–18: Use substitution to solve the system.

15. $2x + 5y = -1$
   $x + 2y = -1$

16. $3x + y = 6$
   $4x + 5y = 8$

17. $2x - 3y = -8$
   $4x + 2y = 0$

18. $5x + 3y = -1$
   $3x - 5y = -21$

Exercises 19–22: Use elimination to solve the system of equations, if possible.

19. $3x + y = 4$
   $2x - y = -2$

20. $2x + 3y = -13$
   $3x - 2y = 0$

21. $3x - y = 5$
   $-6x + 2y = -10$

22. $8x - 6y = 7$
   $-4x + 3y = 11$

SECTION 4.3

Exercises 23–30: Shade the solution set in the xy-plane.

23. $y \geq 2$

24. $y < 2x - 3$

25. $2x - y \leq 4$

26. $-x + 3y > 3$

27. $y - x \geq 1$
   $y \leq 2$

28. $-x + y \leq 3$
   $3x + 2y \geq 6$

29. $y > x - 1$
   $y < 4 - 3x$

30. $x + y \geq 5$
   $2x - 3 < 6$

Exercises 31 and 32: Use the graph to write the system of inequalities.

31.
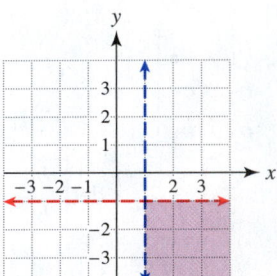

32.
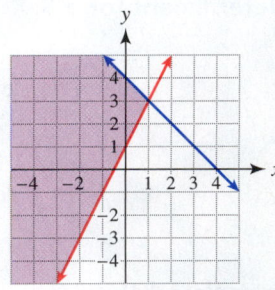

SECTION 4.4

33. Find the maximum of $R = 7x + 8y$ in the shaded region of feasible solutions.

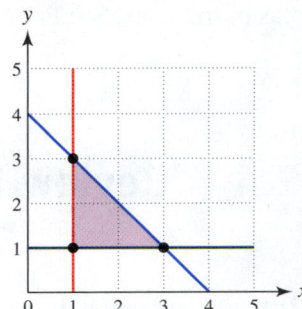

34. Use the figure from Exercise 33 to find the minimum of $C = x + 2y$.

Exercises 35 and 36: Maximize the objective function R subject to the given constraints.

35. $R = 2x + y$

$$x + y \leq 3$$
$$x + y \geq 1$$
$$x \geq 0, y \geq 0$$

36. $R = 6x + 9y$

$$3x + y \leq 12$$
$$x + 3y \leq 12$$
$$x \geq 0, y \geq 0$$

SECTION 4.5

37. Is $(3, -4, 5)$ a solution for $x + y + z = 4$?

38. Decide whether either ordered triple is a solution: $(1, -1, 2)$ or $(1, 0, 5)$.

$$2x - 3y + z = 7$$
$$-x - y + 3z = 6$$
$$3x - 2y + z = 7$$

Exercises 39–44: Use elimination and substitution to solve the system of linear equations, if possible.

39.
$$x - y - 2z = -11$$
$$-x + 2y + 3z = 16$$
$$3z = 6$$

40.
$$x + y = 4$$
$$-2x + y + 3z = -2$$
$$x - 2y + 5z = -26$$

41.
$$2x - y = -5$$
$$x + 2y + z = 7$$
$$-2x + y + z = 7$$

42.
$$2x + 3y + z = 6$$
$$-x + 2y + 2z = 3$$
$$x + y + 2z = 4$$

43.
$$x - y + 3z = 2$$
$$2x + y + 4z = 3$$
$$x + 2y + z = 5$$

44.
$$x - y + 3z = 3$$
$$x + y - z = 1$$
$$x + z = 2$$

SECTION 4.6

Exercises 45–48: Write the system of linear equations as an augmented matrix. Then use Gauss–Jordan elimination to solve the system, writing the solution as an ordered triple. Check your solution.

45.
$$x + y + z = -6$$
$$x + 2y + z = -8$$
$$y + z = -5$$

46.
$$x + y + z = -3$$
$$-x + y = 5$$
$$y + z = -1$$

47.
$$x + 2y - z = 1$$
$$-x + y - 2z = 5$$
$$2y + z = 10$$

48.
$$2x + 2y - 2z = -14$$
$$-2x - 3y + 2z = 12$$
$$x + y - 4z = -22$$

Exercises 49 and 50: **Technology** *Use a graphing calculator to solve the system of linear equations.*

49.
$$3x - 2y + 6z = -17$$
$$-2x - y + 5z = 20$$
$$4y + 7z = 30$$

50.
$$19x - 13y - 7z = 7.4$$
$$22x + 33y - 8z = 110.5$$
$$10x - 56y + 9z = 23.7$$

SECTION 4.7

Exercises 51–54: Evaluate det A, if A is the given matrix.

51. $\begin{bmatrix} 6 & -5 \\ -4 & 2 \end{bmatrix}$

52. $\begin{bmatrix} 0 & -6 \\ 5 & 9 \end{bmatrix}$

53. $\begin{bmatrix} 3 & -5 & -3 \\ 1 & 4 & 7 \\ 0 & -3 & 1 \end{bmatrix}$

54. $\begin{bmatrix} -2 & -1 & -7 \\ 2 & 1 & -3 \\ 3 & -5 & 8 \end{bmatrix}$

Exercises 55 and 56: Use technology to calculate det A, where A is the given matrix.

55. $\begin{bmatrix} 22 & -45 & 3 \\ 15 & -12 & -93 \\ 5 & 81 & -21 \end{bmatrix}$

56. $\begin{bmatrix} 0.5 & -7.3 & 9.6 \\ 0.1 & 3.1 & 9.2 \\ -0.5 & -1.9 & 5.4 \end{bmatrix}$

Exercises 57 and 58: Use a determinant to find the area of the triangle. Assume that the units are feet.

57.

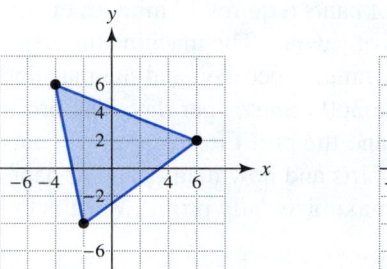

58.

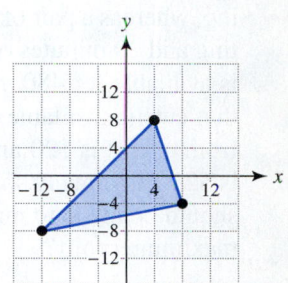

Exercises 59–62: Use Cramer's rule to solve the system.

59. $7x + 6y = 8$
$\quad\; 5x - 8y = 18$

60. $-2x + 5y = 25$
$\quad\;\; 3x + 4y = -3$

61. $3x - 6y = 1.5$
$\quad\; 7x - 5y = 8$

62. $-5x + 4y = -47$
$\quad\;\; 6x - 7y = 63$

APPLICATIONS

63. *Pedestrian Fatalities* Forty-seven percent of pedestrian fatalities occur on Friday and Saturday nights. The combined total of pedestrian fatalities in 1994 and 2004 was 10,130. There were 848 more fatalities in 1994 than in 2004. Find the number of pedestrian fatalities during each year. (*Source:* National Highway Traffic Safety Administration.)

64. *Burning Calories* During strenuous exercise, an athlete burns 11.5 calories per minute on a stair climber and 9 calories per minute on a stationary bicycle. In a 30-minute workout, the athlete burns 290 calories. How many minutes are spent on each type of workout equipment? (*Source:* Runner's World.)

65. *Candy Sales* Suppose that candy costs $4 per pound and cashews cost $5 per pound. Let x be the number of pounds of candy bought and y be the number of pounds of cashews bought. Graph the region in the xy-plane that represents all possible weight combinations of candy and cashews that cost less than or equal to $20.

66. *Loan Mixture* A student takes out two loans for a total of $4000. The unsubsidized loan is $500 less than the subsidized loan. Find the amount of each loan.

67. *Linear Programming* A business makes shirts and pants, which require both cutting and sewing. A shirt requires 20 minutes of cutting and 10 minutes of sewing, whereas a pair of pants requires 10 minutes of cutting and 20 minutes of sewing. The machine that sews is available for 480 minutes per day, and the machine that cuts is available 360 minutes per day. The profit from a shirt is $20 and the profit from a pair of pants is $25. How many shirts and how many pairs of pants should be made to maximize daily profit? What is the maximum profit?

68. *Mixing Antifreeze* A car radiator should contain 4 gallons of fluid that is 40% antifreeze. An auto mechanic has a 30% solution of antifreeze and a 55% solution of antifreeze. If the car radiator is empty, how many gallons of each solution should be added?

69. *Boat Speed* A boat travels 18 miles down a river in 1 hour. The return trip against the current takes 1.5 hours. Find the average speed of the boat and the average speed of the current.

70. *Tickets* Tickets for a football game cost $8 and $12. If 480 tickets were sold for total receipts of $4620, how many of each type of ticket were sold?

71. *Determining Costs* The accompanying table shows the costs for purchasing different combinations of malts, cones, and ice cream bars.

Malts	Cones	Bars	Total Cost
1	3	5	$14
1	2	4	$11
0	1	3	$5

  (a) Let m be the cost of a malt, c the cost of a cone, and b the cost of an ice cream bar. Write a system of equations that represents the data in the table.
  (b) Solve this system.

72. *Geometry* The largest angle in a triangle is 20° more than the sum of the two smaller angles. The measure of the largest angle is 85° more than the smallest angle. Find the measure of each angle in the triangle.

73. *Mixture Problem* Three types of candy that cost $1.50, $2.00, and $2.50 per pound are to be mixed to produce 12 pounds of candy worth $26.00. If there is to be 2 pounds more of the $2.50 candy than the $2.00 candy, how much of each type of candy should be used in the mixture?

74. *Estimating the Chest Size of a Bear* The accompanying table shows the chest size C, weight W, and overall length L of three bears. These data can be modeled with the formula $C = a + bW + cL$. (*Sources:* M. Triola, *Elementary Statistics*; Minitab, Inc.)

C (inches)	W (pounds)	L (inches)
40	202	63
50	365	70
55	446	77
?	300	68

  (a) Set up a system of linear equations whose solution gives values for the constants a, b, and c.
  (b) Solve this system. Round each value to the nearest thousandth.
  (c) Predict the chest size of a bear weighing 300 pounds and having a length of 68 inches.

Exercises 1–4: Solve the system of equations graphically and by using substitution. Determine whether the system is consistent or inconsistent. If the system is consistent, state whether the equations are dependent or independent.

1. $2x + y = 7$
$3x - 2y = 7$

2. $8x - 4y = 3$
$-4x + 2y = 6$

3. $2x - 6y = 2$
$-3x + 9y = -3$

4. $2x - 5y = 36$
$-4x + 3y = -23$

5. Solve the system of equations using elimination.

$$2x + 5y = -1$$
$$3x + 2y = -7$$

6. The difference of two numbers is 34. The first number is twice the second number.
   (a) Write a system of linear equations that models the situation.
   (b) Solve the system using elimination.

7. Solve the system graphically. Estimate values to the nearest thousandth.

$$-\pi x + \sqrt{3}y = 3.3$$
$$\sqrt{5}x + (1 + \sqrt{2})y = 2.1$$

8. Solve the system numerically.

$$3x + y = 17$$
$$2x - 3y = 37$$

9. Shade the solution set for $y \le 2$ in the xy-plane.

10. Shade the solution set for $x - 2y > 3$ in the xy-plane.

11. Shade the solution set in the xy-plane. Use a test point to check your graph.

$$-2x + y \ge 3$$
$$x - 2y < -3$$

Exercises 12–14: Use substitution and elimination to solve the system, if possible.

12. $x + 3y = 2$
$-2x + y + z = 5$
$y + z = -3$

13. $x + y - z = 1$
$2x - 3y + z = 0$
$x - 4y + 2z = 2$

14. $x - y + z = 1$
$x + y - z = 1$
$3x + y - z = 3$

15. Use the graph to write the system of inequalities.

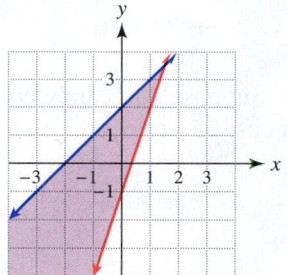

16. Consider the system of linear equations.

$$x + y + z = 2$$
$$x - y - z = 3$$
$$2x + 2y + z = 6$$

   (a) Write the system as an augmented matrix.
   (b) Use Gauss–Jordan elimination to solve the system, writing the solution as an ordered triple.

17. Evaluate det A if $A = \begin{bmatrix} 3 & 2 & -1 \\ 6 & 2 & -6 \\ 0 & 8 & -3 \end{bmatrix}$.

18. Solve the system of equations with Cramer's rule.

$$5x - 3y = 7$$
$$-4x + 2y = 11$$

19. *College Tuition* In 2010, the average cost of private tuition, including room and board, was \$19,700 more than public tuition. Private tuition was 3.6 times higher than public tuition. (*Source:* The College Board.)
   (a) Write a system of linear equations that models the situation.
   (b) Solve the system of linear equations and interpret the solution.

20. *Burning Calories* An athlete burns 12 calories per minute while running and 9 calories per minute on a rowing machine. During a one-hour workout, the athlete burns 669 calories. How many minutes are spent on each type of exercise?

21. *Airplane Speed* An airplane travels 600 miles into the wind in 2.5 hours. The return trip with the wind takes 2 hours. Find the average speed of the airplane with no wind and the average wind speed.

22. *Geometry* The largest angle in a triangle is 50° more than the smallest angle. The sum of the measures of the smaller two angles is 10° more than the largest angle. Find the measure of each angle.

23. *Linear Programming* Maximize $R = x + 2y$ subject to the constraints

$$x \geq 0, y \geq 0$$
$$2x + 3y \leq 6.$$

24. *Candy Sales* Suppose that candy costs $3 per pound and cashews cost $5 per pound. Let x be the number of pounds of candy bought and y be the number of pounds of cashews bought. Graph the region in the xy-plane that represents all possible weight combinations of candy and cashews that cost less than $30.

25. *Loan Mixture* A student takes out two loans, one that is subsidized and one that is not, for a total of $4400. The unsubsidized loan is $700 more than the subsidized loan. Find the amount of each loan.

CHAPTER 4 Extended and Discovery Exercises

CRAMER'S RULE

Exercises 1–6: Cramer's rule can be applied to systems of three equations in three variables. For the system of equations

$$a_1x + b_1y + c_1z = d_1$$
$$a_2x + b_2y + c_2z = d_2$$
$$a_3x + b_3y + c_3z = d_3,$$

the solution can be written as follows.

$$D = \det\begin{bmatrix} a_1 & b_1 & c_1 \\ a_2 & b_2 & c_2 \\ a_3 & b_3 & c_3 \end{bmatrix}, \quad E = \det\begin{bmatrix} d_1 & b_1 & c_1 \\ d_2 & b_2 & c_2 \\ d_3 & b_3 & c_3 \end{bmatrix}$$

$$F = \det\begin{bmatrix} a_1 & d_1 & c_1 \\ a_2 & d_2 & c_2 \\ a_3 & d_3 & c_3 \end{bmatrix}, \quad G = \det\begin{bmatrix} a_1 & b_1 & d_1 \\ a_2 & b_2 & d_2 \\ a_3 & b_3 & d_3 \end{bmatrix}$$

If $D \neq 0$, a unique solution exists and is given by

$$x = \frac{E}{D}, \quad y = \frac{F}{D}, \quad z = \frac{G}{D}.$$

Use Cramer's rule to solve the equations.

1. $\quad x + y + z = 6$
$\quad 2x + y + 2z = 9$
$\quad\quad\quad y + 3z = 9$

2. $\quad\quad\quad y + z = 1$
$\quad 2x - y - z = -1$
$\quad x + y - z = 3$

3. $\quad x + \quad\quad z = 2$
$\quad x + y \quad\quad = 0$
$\quad\quad\quad y + 2z = 1$

4. $\quad x + y + 2z = 1$
$\quad -x - 2y - 3z = -2$
$\quad\quad\quad y - 3z = 5$

5. $\quad x + \quad\quad 2z = 7$
$\quad -x + y + z = 5$
$\quad 2x - y + 2z = 6$

6. $\quad x + 2y + 3z = -1$
$\quad 2x - 3y - z = 12$
$\quad x + 4y - 2z = -12$

MATRICES AND ROAD MAPS

Exercises 7–12: Adjacency Matrix A matrix A can be used to represent a map showing distances between cities. Let a_{ij} denote the number in row i and column j of a matrix A. Now consider the following map illustrating freeway distances in miles between four cities. Each city has been assigned a number. For example, there is a direct route from Denver, Colorado (city 1), to Colorado Springs, Colorado (city 2), of approximately 60 miles. Therefore $a_{12} = 60$ in the accompanying matrix A. (Note that a_{12} is the number in row 1 and column 2.) The distance from Colorado Springs to Denver is also 60 miles, so $a_{21} = 60$. As there is no direct freeway connection between Las Vegas, Nevada (city 4), and Colorado Springs (city 2), we let $a_{24} = a_{42} = *$. The matrix A is called an adjacency matrix. (Source: S. Baase, Computer Algorithms: Introduction to Design and Analysis.)

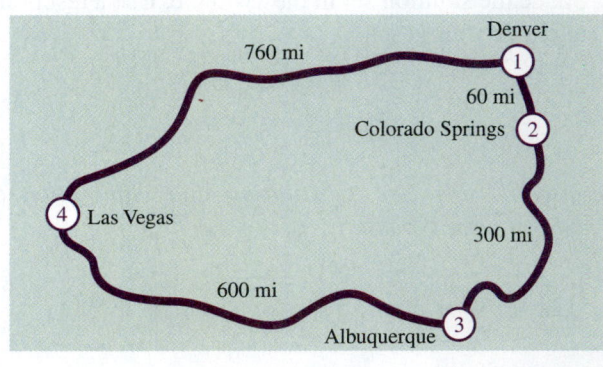

$$A = \begin{bmatrix} 0 & 60 & * & 760 \\ 60 & 0 & 300 & * \\ * & 300 & 0 & 600 \\ 760 & * & 600 & 0 \end{bmatrix}$$

7. Explain how to use A to find the freeway distance from Denver to Las Vegas.

8. Explain how to use A to find the freeway distance from Denver to Albuquerque.

9. If a map shows 20 cities, what would be the dimension of the adjacency matrix? How many elements would there be in this matrix?

10. Why are there only zeros on the main diagonal of A?

11. What does $a_{14} + a_{41}$ equal?

12. What does $a_{11} + a_{44}$ equal?

Exercises 13 and 14: (Refer to Exercises 7–12.) Determine an adjacency matrix A for the road map.

13.

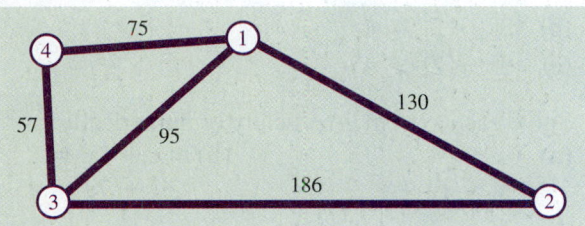

14.

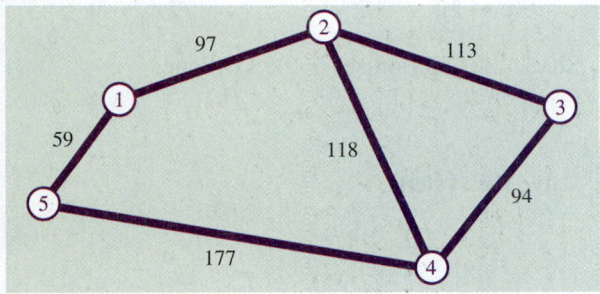

Exercises 15 and 16: (Refer to Exercises 7–12.) Sketch a road map represented by the adjacency matrix A. Is your answer unique?

15. $A = \begin{bmatrix} 0 & 30 & 20 & 5 \\ 30 & 0 & 15 & * \\ 20 & 15 & 0 & 25 \\ 5 & * & 25 & 0 \end{bmatrix}$

16. $A = \begin{bmatrix} 0 & 5 & * & 13 & 20 \\ 5 & 0 & 5 & * & * \\ * & 5 & 0 & 13 & * \\ 13 & * & 13 & 0 & 10 \\ 20 & * & * & 10 & 0 \end{bmatrix}$

SOLVING AN EQUATION IN FOUR VARIABLES

17. ***Weight of a Bear*** In Section 4.6 we estimated the weight of a bear by using two variables. We may be able to make more accurate estimates by using four variables. The accompanying table shows the weight W, neck size N, overall length L, and chest size C for four bears. (*Sources:* M. Triola, *Elementary Statistics;* Minitab, Inc.)

W (pounds)	N (inches)	L (inches)	C (inches)
125	19	57.5	32
316	26	65	42
436	30	72	48
514	30.5	75	54
?	24	63	39

(a) We can model the data in the table with the equation $W = a + bN + cL + dC$, where a, b, c, and d are constants. To do so, represent a system of linear equations by a 4×5 augmented matrix whose solution gives values for a, b, c, and d.

(b) Solve the system with a graphing calculator. Round each value to the nearest thousandth.

(c) Predict the weight of a bear with $N = 24$, $L = 63$, and $C = 39$. Interpret the result.

CHAPTERS 1–4   Cumulative Review Exercises

Exercises 1 and 2: State whether the equation illustrates an identity, commutative, associative, or distributive property.

1. $x + 0 = x$

2. $1 + (b + 3) = (1 + b) + 3$

3. Evaluate $\frac{1+2}{3-1} - 2^2 \cdot 4$ by hand.

4. Evaluate $-1 + 8 - 19 + 12$ mentally.

5. Find the opposite of $\frac{a}{b}$.

6. Simplify each expression. Write the result using positive exponents.
   (a) $\left(\frac{4^2}{3}\right)^{-2}$
   (b) $\frac{(2ab^{-4})^2}{a^7(b^2)^{-3}}$
   (c) $(x^2y^3)^{-2}(xy^2)^4$

7. Write 0.0056 in scientific notation.

8. Find $f(-2)$, if $f(x) = 5 - 4x$.

9. Evaluate $V = \frac{1}{3}\pi r h^2$ for $r = 3$ and $h = 2$.

10. Identify the domain of $f(x) = \frac{x}{2x + 1}$.

Exercises 11 and 12: Graph f by hand.

11. $f(x) = 2 - 4x$

12. $f(x) = 2x^2$

13. Use the graph to answer the following.
   (a) Does the graph represent a function?
   (b) Identify the domain and range.
   (c) Evaluate $f(-1)$ and $f(2)$.
   (d) Identify the x- and y-intercepts.
   (e) Find the slope of the line.
   (f) Find a formula for $f(x)$.

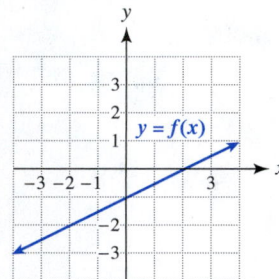

14. Find the slope–intercept form of the line that passes through $(-6, 3)$ and is perpendicular to $y = -\frac{3}{4}x$.

15. Let f be a linear function. Find the slope, x-intercept, and y-intercept of the graph of f.

x	-2	-1	0	2
$f(x)$	-6	-2	2	10

16. Solve $\frac{1}{2}(1 - x) + 2x = 3x + 1$.

17. Solve $2(1 - x) > 4x - (2 + x)$.

18. Solve $x + 1 < 2$ or $x + 1 > 4$.

19. Solve $-3 \le 2 - 4x < 5$.

20. Solve each equation or inequality.
   (a) $|3x - 2| = 4$
   (b) $|3x - 2| < 4$
   (c) $|3x - 2| > 4$

21. Solve each system graphically or numerically.
   (a) $2x + y = 3$
       $x - 2y = -1$
   (b) $4x + 3y = 8$
       $2x + 7y = 4$

22. Solve each system, if possible.
   (a) $x + 2y = 4$
       $3x - 2y = -2$
   (b) $-x + y = 2$
       $2x - 2y = 5$

23. Shade the solution set in the xy-plane.
   (a) $x + 2y < 3$
   (b) $x + y \le 2$
       $3x - y > 6$

24. Solve the system.
$$-x + 2y + z = 2$$
$$3x + y + z = 12$$
$$x - y - 2z = -1$$

25. Use Gauss–Jordan elimination to solve the system.
$$x + y + z = -3$$
$$2x - y + z = -11$$
$$-x - y + z = -5$$

26. Calculate det $\begin{bmatrix} 1 & 2 & 0 \\ -1 & 0 & 4 \\ 2 & 4 & 3 \end{bmatrix}$.

APPLICATIONS

27. *Investment* Suppose x dollars are deposited in an account paying 7% annual interest.
   (a) Write a formula that calculates the interest I after 1 year.
   (b) Calculate $I(500)$ and interpret the result.

28. *Tuition and Fees*   Average tuition and fees in dollars at private colleges from 1980 to 2000 are modeled by $T(x) = 572.3(x - 1980) + 3617$.
(a) Evaluate $T(1990)$ and interpret the result.
(b) Interpret the slope of the graph of T.

29. *Geometry*   A rectangle is 7 inches longer than it is wide. If the perimeter is 74 inches, find the dimensions of the rectangle.

30. *Error in Measurements*   The perimeter of a square is measured to be 12.4 feet. If the measurement of the perimeter P is accurate to within 0.2 foot, write an absolute value inequality that describes this situation.

31. *Investment*   Five thousand dollars are invested in three accounts paying 5%, 6%, and 8% interest. The amount invested at 8% is $500 more than the amount invested at 6%, and the amount invested at 6% is $1000 less than the total amount invested at 5% and 8%. Find the amount invested at each interest rate.

32. *Population*   In 2010, the combined population of Florida and Texas was 43 million. The population of Texas was 7 million greater than the population of Florida. Solve a linear system to determine the population of each state.

5 Polynomial Expressions and Functions

The book of nature is written in the language of mathematics.

—GALILEO

Video games have been available to consumers for many years. However, most of these older games lacked the graphic capability to make the action appear real on a screen. During the past three decades the computing power of processors has *doubled every 2 years*. Now video game systems, such as PlayStation 3, Wii, and Xbox 360, are taking advantage of the new generation of powerful processors to make action scenes look incredibly realistic. Thousands of objects can be tracked and manipulated simultaneously, making the experience more lifelike to the user.

Video games are mathematically intensive because separate equations must be formed and solved in real time to determine the position of each person and object on the screen. Whether it is an athlete, a football, or an exploding object, *mathematics is essential* to making the movements look real.

Because flying objects do not follow straight-line paths, we cannot use linear functions and equations to model their paths. Instead, we need functions whose graphs are curves. Polynomial functions are frequently used in modeling *nonlinear motion* of objects and are excellent at describing flying objects. In this chapter you will begin to learn about polynomials and how they can be used to model the motion of objects like golf balls in flight. (See Section 5.3, Example 6.)

Source: Microsoft.

5.1 Polynomial Functions

Monomials and Polynomials • Addition and Subtraction of Polynomials •
Polynomial Functions • Evaluating Polynomials • Operations on Functions •
Applications and Models

A LOOK INTO MATH ▶

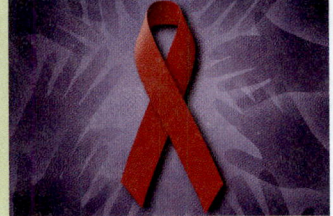

Many quantities in applications cannot be modeled with linear functions and equations. If data points do not lie on a line, we say that the data are *nonlinear*. For example, a scatterplot of the *cumulative* number of AIDS deaths from 1981 through 2007 is nonlinear, as shown in Figure 5.1. Monomials and polynomials are often used to model nonlinear data such as these. (*Source:* U.S. Department of Health and Human Services.)

NEW VOCABULARY

☐ Term
☐ Monomial
☐ Degree of monomial
☐ Coefficient
☐ Leading coefficient
☐ Degree of a polynomial
☐ Linear polynomial
☐ Quadratic polynomial
☐ Cubic polynomial
☐ Like terms
☐ Opposite of a polynomial
☐ Linear function
☐ Quadratic function
☐ Cubic function

U.S. AIDS Deaths

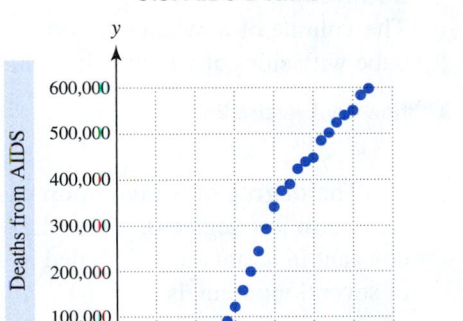

Figure 5.1

Monomials and Polynomials

A **term** is a number, a variable, or a *product* of numbers and variables raised to powers. Examples of terms include

Terms: -15, y, $4x^2$, $3x^{-2}$, $-xy^2$, x^3y^{-2}, $\dfrac{5}{x}$,   and $\dfrac{x}{2}$.

If the variables in a term have only *nonnegative integer* exponents, the term is called a **monomial**. Examples of monomials include

Monomials: -15, y, $4x^2$, $3x^{-2}$, $-xy^2$, x^3y^{-2}, $\dfrac{5}{x}$ and $\dfrac{x}{2}$.

Not monomials

NOTE: Although a monomial can have a negative sign, it cannot contain any addition or subtraction signs. Also, a monomial cannot have division by a variable.

EXAMPLE 1   **Identifying monomials**

Determine whether the expression is a monomial.

(a) $-8x^3y^5$ **(b)** xy^{-1} **(c)** $9 + x^2$ **(d)** $\dfrac{2}{x}$

Solution

(a) The expression $-8x^3y^5$ represents a monomial because it is a product of the number -8 and the variables x and y, which have the nonnegative integer exponents 3 and 5.

(b) The expression xy^{-1} is not a monomial because y has a negative exponent.

(c) The expression $9 + x^2$ is not a monomial; it is the sum of two monomials, 9 and x^2.

(d) The expression $\frac{2}{x}$ is not a monomial because it involves division by a variable. Note also that $\frac{2}{x} = 2x^{-1}$, which has a negative exponent.

▌ **Now Try Exercises 15, 17, 19, 21**

Monomials occur in applications involving geometry, as illustrated in the next example.

EXAMPLE 2 **Writing monomials**

Write a monomial that represents the volume of a cube with sides of length x.

Volume: x^3

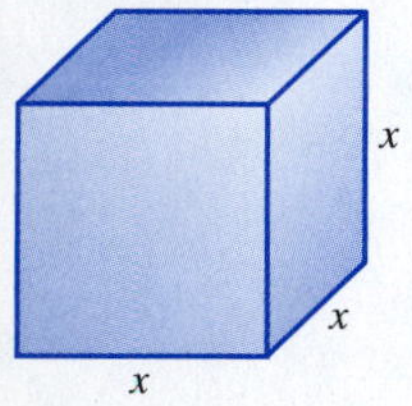

Figure 5.2

Solution

The volume of a rectangular box equals the product of its length, width, and height. For a cube with sides of length x, its volume is given by $x \cdot x \cdot x = x^3$. See Figure 5.2.

▌ **Now Try Exercise 23**

The **degree of a monomial** equals the sum of the exponents of the variables. A constant term has degree 0, unless the term is 0 (which has an undefined degree). The numeric constant in a monomial is called its **coefficient**. Table 5.1 shows the degree and coefficient of several monomials.

TABLE 5.1

Monomial	64	$4x^2y^3$	$-5x^2$	xy^3
Degree	0	5	2	4
Coefficient	64	4	-5	1

A **polynomial** is either a monomial or a sum of monomials. Examples include

$$5x^4z^2, \quad 9x^4 - 5, \quad 4x^2 + 5xy - y^2, \quad \text{and} \quad 4 - y^2 + 5y^4 + y^5.$$

1 term   2 terms   3 terms   4 terms

Polynomials containing one variable are called **polynomials of one variable**. The second and fourth polynomials shown are examples of polynomials of one variable. The **leading coefficient** of a polynomial of one variable is the coefficient of the monomial with greatest degree. The **degree of a polynomial** equals the degree of the monomial with greatest degree. Table 5.2 shows several polynomials of one variable along with their degrees and leading coefficients. A polynomial of degree 1 is a **linear polynomial**, a polynomial of degree 2 is a **quadratic polynomial**, and a polynomial of degree 3 is a **cubic polynomial**.

STUDY TIP

Study Table 5.2 carefully. It will help you understand the vocabulary of this section.

TABLE 5.2

Polynomial	Degree	Leading Coefficient	Type
-98	0	-98	Constant
$2x - 7$	1	2	Linear
$-5z + 9z^2 + 7$	2	9	Quadratic
$-2x^3 + 4x^2 + x - 1$	3	-2	Cubic
$7 - x + 4x^2 + x^5$	5	1	Fifth degree

Addition and Subtraction of Polynomials

Suppose that we have 3 rectangles of the same dimension having length x and width y, as shown in Figure 5.3. The total area is given by

$$xy + xy + xy.$$

This area is equivalent to 3 times xy, which can be expressed as $3xy$. In symbols we write

$$xy + xy + xy = 3xy.$$

Total Area: 3xy

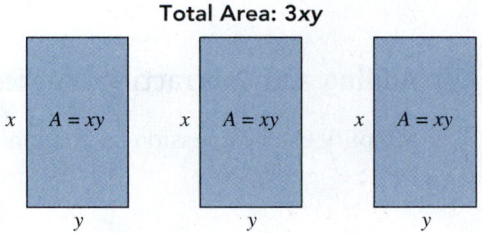

Figure 5.3

We can add these three terms because they are *like terms*. If two or more terms contain the *same variables raised to the same powers*, we call them **like terms**. We can add or subtract *like* terms but not *unlike* terms. For example, if one cube has sides of length x and another cube has sides of length y, their respective volumes are x^3 and y^3. The total volume equals

$$x^3 + y^3,$$

but we cannot combine these terms into one term because they are unlike terms. See Figure 5.4.

Unlike Terms

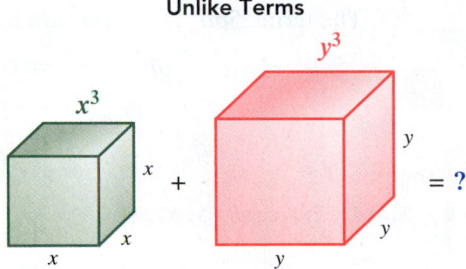

Figure 5.4

However, the distributive property can be used to simplify the expression $2x^3 + 3x^3$ as

$$2x^3 + 3x^3 = (2 + 3)x^3 = 5x^3$$

because $2x^3$ and $3x^3$ are like terms. See Figure 5.5.

Like Terms

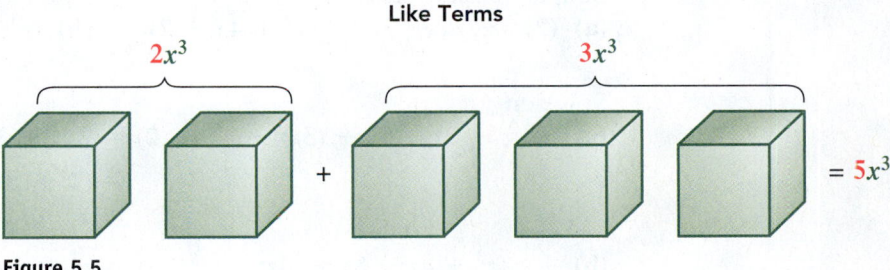

Figure 5.5

READING CHECK

What are like terms? Give examples.

Table 5.3 lists examples of like terms and unlike terms.

TABLE 5.3

Like terms	$-x^4, 2x^4$	$-2x^2y, 4x^2y$	$a^2b^4, \frac{1}{2}a^2b^4$
Unlike terms	$-x^3, 2x^4$	$-2xy^2, 4x^2y$	$ab^5, \frac{1}{2}a^2b^4$

To add or subtract monomials we simply apply the distributive property.

EXAMPLE 3 **Adding and subtracting monomials**

Simplify each expression by combining like terms.
(a) $8x^2 - 4x^2 + x^3$ (b) $9x - 6xy^2 + 2xy^2 + 4x$ (c) $5ab^2 - 4a^2 - ab^2 + a^2$

Solution
(a) The terms $8x^2$ and $-4x^2$ are like terms, so they can be combined.

$$8x^2 - 4x^2 + x^3 = (8 - 4)x^2 + x^3 \qquad \text{Combine like terms.}$$
$$= 4x^2 + x^3 \qquad \text{Simplify.}$$

However, $4x^2$ and x^3 are unlike terms and cannot be combined.
(b) The terms $9x$ and $4x$ can be combined, as can $-6xy^2$ and $2xy^2$.

$$9x - 6xy^2 + 2xy^2 + 4x = 9x + 4x - 6xy^2 + 2xy^2 \qquad \text{Commutative property}$$
$$= (9 + 4)x + (-6 + 2)xy^2 \qquad \text{Combine like terms.}$$
$$= 13x - 4xy^2 \qquad \text{Simplify.}$$

(c) The terms $5ab^2$ and $-ab^2$ are like terms, as are $-4a^2$ and a^2, so they can be combined.

$$5ab^2 - 4a^2 - ab^2 + a^2 = 5ab^2 - 1ab^2 - 4a^2 + 1a^2 \qquad \text{Commutative property}$$
$$= (5 - 1)ab^2 + (-4 + 1)a^2 \qquad \text{Combine like terms.}$$
$$= 4ab^2 - 3a^2 \qquad \text{Simplify.}$$

Now Try Exercises 41, 47, 49

To add two polynomials we combine like terms, as in the next example.

EXAMPLE 4 **Adding polynomials**

Simplify each expression.
(a) $(2x^2 - 3x + 7) + (3x^2 + 4x - 2)$ (b) $(z^3 + 4z + 8) + (4z^2 - z + 6)$

Solution
(a) $(2x^2 - 3x + 7) + (3x^2 + 4x - 2) = 2x^2 + 3x^2 - 3x + 4x + 7 - 2$
$$= (2 + 3)x^2 + (-3 + 4)x + (7 - 2)$$
$$= 5x^2 + x + 5$$

(b) $(z^3 + 4z + 8) + (4z^2 - z + 6) = z^3 + 4z^2 + 4z - z + 8 + 6$
$$= z^3 + 4z^2 + (4 - 1)z + (8 + 6)$$
$$= z^3 + 4z^2 + 3z + 14$$

Now Try Exercises 55, 57

We can also add polynomials vertically, as in the next example.

EXAMPLE 5   **Adding polynomials vertically**

Find the sum.
(a) $(3x^2 - 5xy - 7y^2) + (xy + 4y^2 - x^2)$
(b) $(3x^3 - 2x + 7) + (x^3 + 5x^2 - 9)$

Solution

(a) Polynomials can be added vertically by placing like terms in the same columns and then adding each column.

$$
\begin{array}{l}
3x^2 - 5xy - 7y^2 \\
\underline{-x^2 + xy + 4y^2} \\
2x^2 - 4xy - 3y^2
\end{array}
\qquad \text{Add each column.}
$$

(b) Note that the first polynomial does not contain an x^2-term and that the second polynomial does not contain an x-term. When you are adding vertically, leave a blank for a missing term.

$$
\begin{array}{l}
3x^3 - 2x + 7 \\
\underline{x^3 + 5x^2 - 9} \\
4x^3 + 5x^2 - 2x - 2
\end{array}
\qquad \text{Add each column.}
$$

❙ **Now Try Exercises 59, 63**

Recall that to subtract integers we add the first integer and the *additive inverse* or *opposite* of the second integer. For example, to evaluate $3 - 5$ we perform the following operations.

$$
\begin{aligned}
3 - 5 &= 3 + (-5) \qquad \text{Add the opposite.} \\
&= -2 \qquad\qquad\ \text{Simplify.}
\end{aligned}
$$

Similarly, to subtract two polynomials we add the first polynomial and the opposite of the second polynomial. To find the **opposite of a polynomial**, we negate each term. For example, the opposite of $2x - 1$ can be found as follows.

Opposite of **2x − 1**

$$-(2x - 1) = -1(2x - 1) = -2x + 1$$

Multiplying by -1 is equivalent to negating an expression.

Thus to find the opposite of $2x - 1$ we *negate* each term and get $-2x + 1$. Table 5.4 shows three polynomials and their opposites.

TABLE 5.4

Polynomial	Opposite
$9 - x$	$-9 + x$
$5x^2 + 4x - 1$	$-5x^2 - 4x + 1$
$-x^4 + 5x^3 - x^2 + 5x - 1$	$x^4 - 5x^3 + x^2 - 5x + 1$

EXAMPLE 6 **Subtracting polynomials**

Simplify.
(a) $(y^5 + 3y^3) - (-y^4 + 2y^3)$ **(b)** $(5x^3 + 9x^2 - 6) - (5x^3 - 4x^2 - 7)$

Solution

(a) The opposite of the polynomial $(-y^4 + 2y^3)$ is $(y^4 - 2y^3)$. Rather than subtracting, we add the opposite.

$$(y^5 + 3y^3) - (-y^4 + 2y^3) = (y^5 + 3y^3) + (y^4 - 2y^3)$$
$$= y^5 + y^4 + (3 - 2)y^3$$
$$= y^5 + y^4 + y^3$$

(b) The opposite of the polynomial $(5x^3 - 4x^2 - 7)$ is $(-5x^3 + 4x^2 + 7)$.

$$(5x^3 + 9x^2 - 6) - (5x^3 - 4x^2 - 7) = (5x^3 + 9x^2 - 6) + (-5x^3 + 4x^2 + 7)$$
$$= (5 - 5)x^3 + (9 + 4)x^2 + (-6 + 7)$$
$$= 0x^3 + 13x^2 + 1$$
$$= 13x^2 + 1$$

Now Try Exercises 71, 73

The following summarizes addition and subtraction of polynomials.

> **ADDITION AND SUBTRACTION OF POLYNOMIALS**
>
> **1.** To *add* two polynomials, combine like terms.
>
> **2.** To *subtract* two polynomials, add the first polynomial and the opposite of the second polynomial.
>
> **NOTE:** The *opposite* of a polynomial is found by changing the sign of every term.

CRITICAL THINKING

Is the sum of two quadratic polynomials always a quadratic polynomial? Explain.

Polynomial Functions

The following expressions are examples of polynomials of one variable.

$$1 - 5x, \quad 3x^2 - 5x + 1, \quad \text{and} \quad x^3 + 5$$

As a result, we say that the following are *symbolic representations* of polynomial functions of one variable.

$$f(x) = 1 - 5x, \quad g(x) = 3x^2 - 5x + 1, \quad \text{and} \quad h(x) = x^3 + 5$$

Function f is a **linear function** because it has degree 1, function g is a **quadratic function** because it has degree 2, and function h is a **cubic function** because it has degree 3.

NOTE: The domain of *every* polynomial function is *all* real numbers.

| EXAMPLE 7 | **Identifying polynomial functions** |

Determine whether $f(x)$ represents a polynomial function. If possible, identify the type of polynomial function and its degree.

(a) $f(x) = 5x^3 - x + 10$

(b) $f(x) = x^{-2.5} + 1$

(c) $f(x) = 1 - 2x$

(d) $f(x) = \dfrac{3}{x - 1}$

Solution

(a) The expression $5x^3 - x + 10$ is a cubic polynomial, so $f(x)$ represents a cubic polynomial function. It has degree 3.

(b) $f(x)$ does not represent a polynomial function because the variables in a polynomial must have *nonnegative integer* exponents.

(c) $f(x) = 1 - 2x$ represents a polynomial function that is linear. It has degree 1.

(d) $f(x)$ does not represent a polynomial function because $\frac{3}{x - 1}$ is not a polynomial.

Now Try Exercises 79, 81, 83, 85

Evaluating Polynomials

Frequently, monomials and polynomials represent functions or formulas that can be evaluated. This situation is illustrated in the next three examples.

| EXAMPLE 8 | **Evaluating a polynomial function graphically and symbolically** |

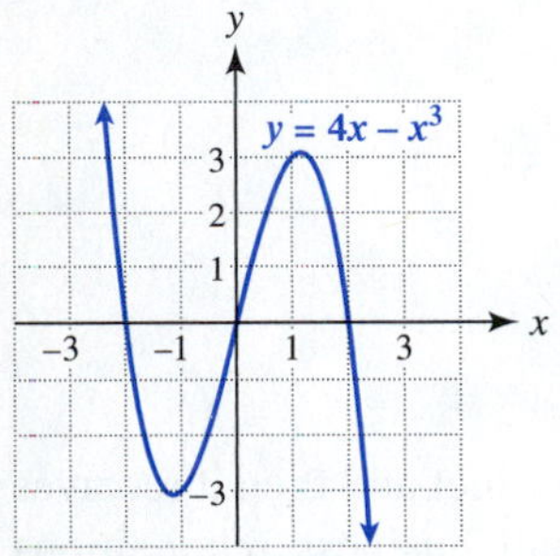

A graph of $f(x) = 4x - x^3$ is shown in Figure 5.6. Evaluate $f(-1)$ graphically and check your result symbolically.

Solution

Graphical Evaluation To evaluate $f(-1)$ graphically, find -1 on the x-axis and move down until the graph of f is reached. Then move horizontally to the y-axis, as shown in Figure 5.7. Thus when $x = -1$, $y = -3$ and $f(-1) = -3$.

Symbolic Evaluation When $x = -1$, evaluation of $f(x) = 4x - x^3$ is performed as follows.

$$f(-1) = 4(-1) - (-1)^3 = -4 - (-1) = -3$$

$$f(-1) = -3$$

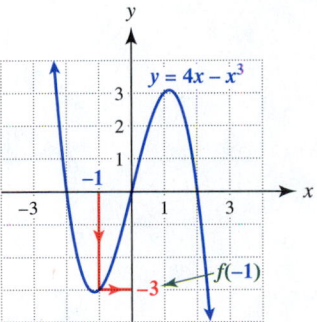

Figure 5.7

Figure 5.6

Now Try Exercises 93, 97

EXAMPLE 9

Writing and evaluating a polynomial

Write a polynomial that represents the total volume of two identical boxes having square bases. Make a sketch to illustrate your formula. Find the total volume of the boxes if each base is 11 inches on a side and the height of each box is 5 inches.

Solution

Let b be the length (width) of the square base and h be the height of a side. The volume of one box is length times width times height, or b^2h, as illustrated in Figure 5.8, and the volume of two boxes is $2b^2h$. To calculate their total volumes, let $b = \mathbf{11}$ and $h = \mathbf{5}$ in the expression $2b^2h$. Then

$$2 \cdot \mathbf{11}^2 \cdot \mathbf{5} = 1210 \text{ cubic inches.}$$

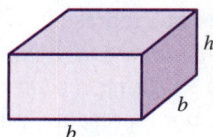

Figure 5.8 Volume: b^2h

Now Try Exercise 125

EXAMPLE 10

Evaluating a polynomial function symbolically

Evaluate $f(x)$ at the given value of x.
(a) $f(x) = -3x^4 - 2, \quad x = 2$ **(b)** $f(x) = -2x^3 - 4x^2 + 5, \quad x = -3$

Solution
(a) Be sure to evaluate exponents before multiplying.

$$f(\mathbf{2}) = -3(\mathbf{2})^4 - 2 = -3 \cdot 16 - 2 = -50$$

(b) $f(\mathbf{-3}) = -2(\mathbf{-3})^3 - 4(\mathbf{-3})^2 + 5 = -2(-27) - 4(9) + 5 = 23$

Now Try Exercises 101, 105

Operations on Functions

▶ **REAL-WORLD CONNECTION** A business incurs a *cost* to make its product and then it receives *revenue* from selling this product. For example, suppose a small business reconditions motorcycles. The graphs of its cost and of its revenue for reconditioning and selling x motorcycles are shown in Figure 5.9.

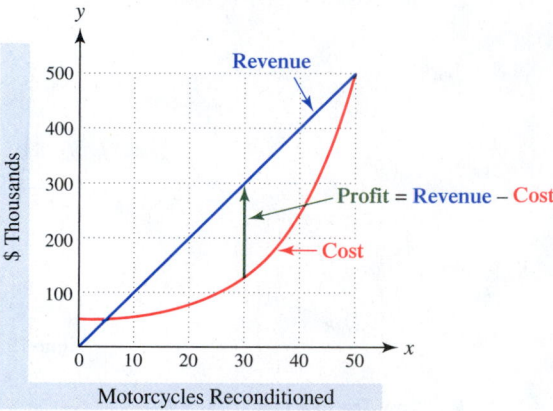

Profit, Revenue, and Cost

Figure 5.9

In general, *profit equals revenue minus cost*. In Figure 5.9, **profit** is shown visually as the length of the vertical green arrow between the graphs of **revenue** and **cost**. For any x-value, the distance by which revenue is above cost is called the profit for reconditioning and selling x motorcycles. *Maximum profit for the company occurs at the x-value where the length of the vertical green arrow is greatest.*

If we let $C(x)$, $R(x)$, and $P(x)$ be functions that calculate the cost, revenue, and profit, respectively, for reconditioning and selling x motorcycles, then

$$P(x) = R(x) - C(x).$$

Profit equals Revenue minus Cost.

This example helps explain why we subtract functions in the real world. Functions can be added in a similar manner.

Given two functions, f and g, we define the sum, $f + g$, and difference, $f - g$, as follows.

SUM AND DIFFERENCE OF TWO FUNCTIONS

The sum and difference of two functions f and g are defined as

$$(f + g)(x) = f(x) + g(x)$$

$$(f - g)(x) = f(x) - g(x)$$

provided both $f(x)$ and $g(x)$ are defined.

EXAMPLE 11   **Calculating sums and differences of functions**

Let $f(x) = 2x^2 + 1$ and $g(x) = 5 - x^2$. Find each sum or difference.
(a) $(f + g)(1)$ **(b)** $(f - g)(-2)$
(c) $(f + g)(x)$ **(d)** $(f - g)(x)$

Solution
(a) Because $f(1) = 2 \cdot 1^2 + 1 = 3$ and $g(1) = 5 - 1^2 = 4$, it follows that

$$(f + g)(1) = f(1) + g(1) = 3 + 4 = 7.$$

(b) Because $f(-2) = 2 \cdot (-2)^2 + 1 = 9$ and $g(-2) = 5 - (-2)^2 = 1$, it follows that

$$(f - g)(-2) = f(-2) - g(-2) = 9 - 1 = 8.$$

(c) $(f + g)(x) = f(x) + g(x)$ Addition of functions
                $= (2x^2 + 1) + (5 - x^2)$ Substitute for $f(x)$ and $g(x)$.
                $= x^2 + 6$ Combine like terms.

(d) $(f - g)(x) = f(x) - g(x)$ Subtraction of functions
                $= (2x^2 + 1) - (5 - x^2)$ Substitute for $f(x)$ and $g(x)$.
                $= (2x^2 + 1) + (-5 + x^2)$ Add the opposite.
                $= 3x^2 - 4$ Combine like terms.

Now Try Exercise 115

Applications and Models

▶ **REAL-WORLD CONNECTION** Polynomials are used to model a wide variety of data. A scatterplot of the cumulative number of reported AIDS cases in thousands from 1984 to 1994 is shown in Figure 5.10. In this graph $x = 4$ corresponds to 1984, $x = 5$ to 1985, and so on until $x = 14$ represents 1994. These data can be modeled with a quadratic function, as shown in Figure 5.11, where $f(x) = 4.1x^2 - 25x + 46$ is graphed with the data. Note that $f(x)$ was found by using a graphing calculator. (*Source:* U.S. Department of Health and Human Services.)

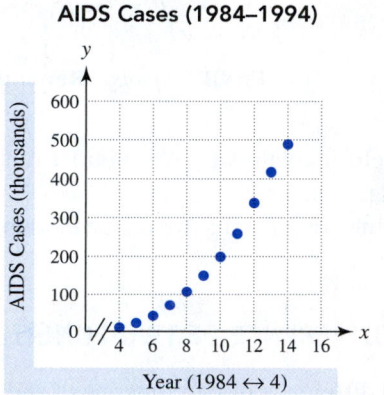

Figure 5.10   **Figure 5.11**

Limits on Modeling: After 1994, effective drugs and education greatly slowed the spread of AIDS in the United States. As a result, the modeling function f does *not* accurately describe the data after 1994. In general, we have to be careful about using a modeling function to predict too far into the future.

EXAMPLE 12 **Modeling AIDS cases in the United States**

Use $f(x) = 4.1x^2 - 25x + 46$ to model the number of AIDS cases.
(a) Estimate the number of AIDS cases reported by 1987. Compare it to the actual value of 71.4 thousand.
(b) By 1997, the total number of AIDS cases reported was 631 thousand. What estimate does $f(x)$ give? Discuss your result.

Solution
(a) The value $x = 7$ corresponds to 1987. Let $f(x) = 4.1x^2 - 25x + 46$ and evaluate $f(7)$, which gives

$$f(7) = 4.1(7)^2 - 25(7) + 46 = 71.9.$$

This model estimates that a cumulative total of 71.9 thousand AIDS cases were reported by 1987. This result compares favorably to the actual value of 71.4 thousand cases.
(b) To estimate the number in 1997, we evaluate $f(17)$ because $x = 17$ corresponds to 1997, obtaining

$$f(17) = 4.1(17)^2 - 25(17) + 46 = 805.9.$$

This result is considerably more than the actual value of 631 thousand. The reason is that f models data only from 1984 through 1994. After 1994, f gives estimates that are too large because the growth in AIDS cases has slowed more in recent years.

Now Try Exercise 123

▶ **REAL-WORLD CONNECTION** A well-conditioned athlete's heart rate can reach 200 beats per minute during strenuous physical activity. Upon quitting, a typical heart rate decreases rapidly at first and then more gradually after a few minutes, as illustrated in the next example.

EXAMPLE 13   **Modeling heart rate of an athlete**

Let $P(t) = 1.875t^2 - 30t + 200$ model an athlete's heart rate (or pulse P) in beats per minute (bpm) t minutes after strenuous exercise has stopped, where $0 \le t \le 8$. (*Source: V. Thomas, Science and Sport.*)

(a) What is the initial heart rate when the athlete stops exercising?
(b) What is the heart rate after 8 minutes?
(c) A graph of P is shown in Figure 5.12. Interpret this graph.

Athlete's Heart Rate

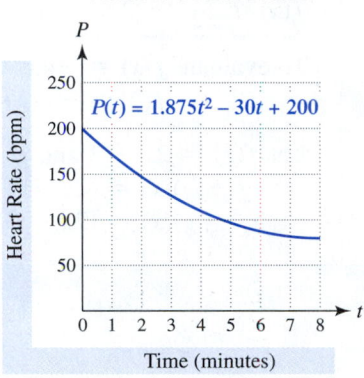

Figure 5.12

Solution
(a) To find the initial heart rate, evaluate $P(t)$ at $t = 0$, or

$$P(0) = 1.875(0)^2 - 30(0) + 200 = 200.$$

When the athlete stops exercising, the heart rate is 200 beats per minute. (This result agrees with the graph.)
(b) $P(8) = 1.875(8)^2 - 30(8) + 200 = 80$ beats per minute.
(c) The heart rate does not drop at a constant rate; rather, it drops rapidly at first and then gradually begins to level off.

Now Try Exercise 121

5.1   Putting It All Together

CONCEPT	EXAMPLES
Addition of Polynomials	$(-3x^2 + 2x - 7) + (4x^2 - x + 1) = -3x^2 + 4x^2 + 2x - x - 7 + 1$ $\qquad\qquad\qquad\qquad = (-3 + 4)x^2 + (2 - 1)x + (-7 + 1)$ $\qquad\qquad\qquad\qquad = x^2 + x - 6$ $(x^4 + 8x^3 - 7x) + (5x^4 - 3x) = x^4 + 5x^4 + 8x^3 - 7x - 3x$ $\qquad\qquad\qquad\qquad = (1 + 5)x^4 + 8x^3 + (-7 - 3)x$ $\qquad\qquad\qquad\qquad = 6x^4 + 8x^3 - 10x$
Opposite of a Polynomial	The opposite of $-5x^2 + 3x - 5$ is $5x^2 - 3x + 5$. The opposite of $-6x^6 + 4x^4 - 8x^2 - 17$ is $6x^6 - 4x^4 + 8x^2 + 17$.

continued on next page

continued from previous page

CONCEPT	EXAMPLES
Subtraction of Polynomials (Add the Opposite.)	$(5x^2 - 6x + 1) - (-5x^2 + 3x - 5) = (5x^2 - 6x + 1) + (5x^2 - 3x + 5)$ $= (5 + 5)x^2 + (-6 - 3)x + (1 + 5)$ $= 10x^2 - 9x + 6$ $(x^4 - 6x^2 + 5x) - (x^4 - 5x + 7) = (x^4 - 6x^2 + 5x) + (-x^4 + 5x - 7)$ $= (1 - 1)x^4 - 6x^2 + (5 + 5)x - 7$ $= -6x^2 + 10x - 7$
Polynomial Functions	The following represent polynomial functions. $f(x) = 3$ Degree = 0 Constant $f(x) = 5x - 3$ Degree = 1 Linear $f(x) = x^2 - 2x - 1$ Degree = 2 Quadratic $f(x) = 3x^3 + 2x^2 - 6$ Degree = 3 Cubic
Evaluating a Polynomial Function	To evaluate $f(x) = -4x^2 + 3x - 1$ at $x = \mathbf{2}$, substitute $\mathbf{2}$ for x. $f(\mathbf{2}) = -4(\mathbf{2})^2 + 3(\mathbf{2}) - 1 = -16 + 6 - 1 = -11$
Sums and Differences of Functions	Let $f(x) = 2x + 3$ and $g(x) = x + 2$. $(f + g)(x) = (2x + 3) + (x + 2) = 3x + 5$ $(f - g)(x) = (2x + 3) - (x + 2) = x + 1$

5.1 Exercises

CONCEPTS AND VOCABULARY

1. Give an example of a monomial.

2. What are the degree and leading coefficient of the polynomial $3x^2 - x^3 + 1$?

3. Are $-5x^3y$ and $6xy^3$ like terms? Explain.

4. Give an example of a polynomial that has 3 terms and is degree 4.

5. Does the opposite of $x^2 + 1$ equal $-x^2 + 1$? Explain.

6. Evaluate $4x^3y$ when $x = 2$ and $y = 3$.

7. When you evaluate a polynomial function at one value of x, can you calculate two answers? Explain.

8. Could the graph of a polynomial function be a line? Explain.

9. If $f(x) = x^2 + x$, then $f(2) =$ _____.

10. If $f(x) = x^2$ and $g(x) = 2x^2$, then it follows that $(f + g)(x) =$ _____.

11. If $f(x) = 3x^2 + 7$ and $g(x) = 2x^2 - 5$, then it follows that $(f - g)(x) =$ _____.

12. Identify which entries in the following list are terms.
$$7, \ 3 - x^2, \ -5x^6, \ xy^3, \ \frac{1}{2x}, \ x^{-3}$$

MONOMIALS AND POLYNOMIALS

Exercises 13–22: Determine whether the expression is a monomial.

13. x^4

14. x^{-4}

15. $2x^2y + y^2$

16. $5 - \sqrt{x}$

17. $-4x^3y^3$

18. xy

19. $\dfrac{3}{x - 2}$

20. $\pi x^4 y^2 z$

21. $x^{-3}y$

22. x^{-2}

Exercises 23–28: Write a monomial that represents the described quantity.

23. The area of a square with sides equal to x

24. The circumference of a circle with diameter d

25. The area of a circle with radius r

26. The area of three congruent triangles with base b and height h

27. The number of members in a marching band that has x rows with y people in each row

28. The revenue from selling w items for z dollars each

Exercises 29–34: Identify the degree and coefficient.

29. $3x^7$

30. $-5y^3$

31. $-3x^2y^5$

32. xy^5

33. $-x^3y^3$

34. $\sqrt{2}xy$

Exercises 35–40: Identify the degree and leading coefficient of the polynomial.

35. $5x^2 - 4x + \frac{3}{4}$

36. $-9y^4 + y^2 + 5$

37. $5 - x + 3x^2 - \frac{2}{5}x^3$

38. $7x + 4x^4 - \frac{4}{3}x^3$

39. $8x^4 + 3x^3 - 4x + x^5$

40. $5x^2 - x^3 + 7x^4 + 10$

Exercises 41–52: Combine like terms whenever possible.

41. $x^2 + 4x^2$

42. $-3z + 5z$

43. $6y^4 - 3y^4$

44. $9xy - 7xy$

45. $5x^2y + 8xy^2$

46. $5x + 4y$

47. $9x^2 - x + 4x - 6x^2$

48. $-xy^2 - \frac{1}{2}xy^2$

49. $x^2 + 9xy - y^2 + 4x^2 + y^2$

50. $6xy + 4x - 6xy$

51. $4x + 7x^3y^7 - \frac{1}{2}x^3y^7 + 9x - \frac{3}{2}x^3y^7$

52. $19x^3 + x^2 - 3x^3 + x - 4x^2 + 1$

Exercises 53–64: Add the polynomials.

53. $(3x + 1) + (-x + 1)$

54. $(5y^3 + y) + (12y^3 - 5y)$

55. $(x^2 - 2x + 15) + (-3x^2 + 5x - 7)$

56. $(4x) + (1 - 4.5x)$

57. $(3x^3 - 4x + 3) + (5x^2 + 4x + 12)$

58. $(y^5 + y) + \left(5 - y + \frac{1}{3}y^2\right)$

59. $(x^4 - 3x^2 - 4) + \left(-8x^4 + x - \frac{1}{2}\right)$

60. $(3z + z^4 + 2) + (-3z^4 - 5 + z^2)$

61. $(4r^4 - r + 2) + (r^3 - 5r)$

62. $(5t^3 + 3t) + (4t^4 - 3t^3 + 1)$

63. $(4xy - x^2 + y^2) + (4y^2 - 8xy - x^2)$

64. $(3x^2 + 6xy - y^2) + (8y^2 - xy - 2x^2)$

Exercises 65–70: Find the opposite of the polynomial.

65. $6x^5$

66. $-5y^7$

67. $19x^5 - 5x^3 + 3x$

68. $-x^2 - x - 5$

69. $-7z^4 + z^2 - 8$

70. $6 - 4x + 5x^2 - \frac{1}{10}x^3$

Exercises 71–78: Subtract the polynomials.

71. $(5x - 3) - (2x + 4)$

72. $(10x + 5) - (-6x - 4)$

73. $(x^2 - 3x + 1) - (-5x^2 + 2x - 4)$

74. $(-x^2 + x - 5) - (x^2 - x + 5)$

75. $3(4x^4 + 2x^2 - 9) - 4(x^4 - 2x^2 - 5)$

76. $2(8x^3 + 5x^2 - 3x + 1) - 5(-5x^3 + 6x - 11)$

77. $4(x^4 - 1) - (4x^4 + 3x + 7)$

78. $(5x^4 - 6x^3 + x^2 + 5) - (x^3 + 11x^2 + 9x - 3)$

Exercises 79–86: Determine whether $f(x)$ represents a polynomial function. If possible, identify the degree and type of polynomial function.

79. $f(x) = x^4 + 5x^2$

80. $f(x) = 5x^3 + 1$

81. $f(x) = x^{-2}$

82. $f(x) = |x|$

83. $f(x) = \dfrac{1}{x^2 + 1}$

84. $f(x) = 5x - 7$

85. $f(x) = 3x - x^2$

86. $f(x) = x + 3x^{-2}$

EVALUATING POLYNOMIALS

Exercises 87–92: Evaluate the polynomial at the given value(s) of the variable(s).

87. $-4x^2$ $x = 2$

88. $-2y^3$ $y = -3$

89. $2x^2y$ $x = 2, y = 3$

90. $-xy^3$ $x = 4, y = -1$

91. $a^2b - ab^2$ $a = -3, b = 4$

92. $3a^3 + 2b^3$ $a = -2, b = 3$

Exercises 93–96: Evaluate the expressions graphically.

93. $f(1)$ and $f(-2)$ **94.** $f(0)$ and $f(2)$

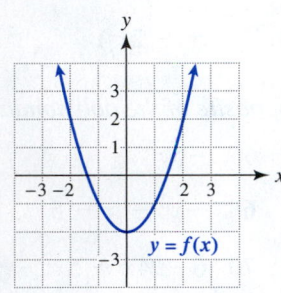
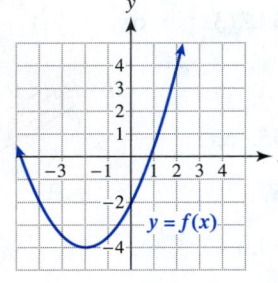

95. $f(-1)$ and $f(2)$ **96.** $f(-1)$ and $f(0)$

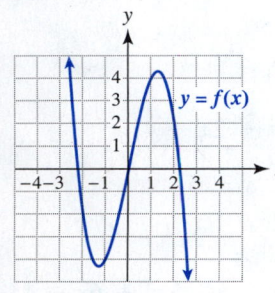
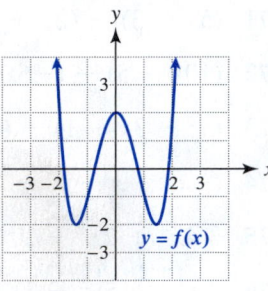

Exercises 97–110: Evaluate $f(x)$ at the given x-value.

97. $f(x) = 3x^2$ $x = -2$

98. $f(x) = x^2 - 2x$ $x = 3$

99. $f(x) = 5 - 4x$ $x = -\frac{1}{2}$

100. $f(x) = x^2 + 4x - 5$ $x = -3$

101. $f(x) = 0.5x^4 - 0.3x^3 + 5$ $x = -1$

102. $f(x) = 6 - 2x + x^3$ $x = 0$

103. $f(x) = -x^3$ $x = -1$

104. $f(x) = 3 - 2x$ $x = \frac{3}{10}$

105. $f(x) = -x^2 - 3x$ $x = -3$

106. $f(x) = 2x^3 - 4x + 1$ $x = 1$

107. $f(x) = 1 - 2x + x^2$ $x = 2.4$

108. $f(x) = 4x^2 - 20x + 25$ $x = 1.8$

109. $f(x) = x^5 - 5$ $x = -1$

110. $f(x) = 1.2x^4 - 5.7x + 3$ $x = \frac{3}{2}$

111. Thinking Generally If $f(x) = x^2 - 2x$, then it follows that $f(a) = $ _____.

112. Thinking Generally If $f(x) = 2x - 1$, then it follows that $f(a + 2) = $ _____.

Exercises 113–120: For the given $f(x)$ and $g(x)$, find the following and simplify.

    (a) $(f + g)(2)$ **(b)** $(f - g)(-1)$
    (c) $(f + g)(x)$ **(d)** $(f - g)(x)$

113. $f(x) = 3x - 1, g(x) = 5 - x$

114. $f(x) = -4x + 3, g(x) = 5x$

115. $f(x) = -3x^2, g(x) = x^2 + 1$

116. $f(x) = 2x^2 - 2, g(x) = 3 - x^2$

117. $f(x) = -x^3, g(x) = 3x^3$

118. $f(x) = x^3 + x, g(x) = 2x - 3x^3$

119. $f(x) = x^2 - 2x + 1, g(x) = 4x^2 + 3x$

120. $f(x) = 3 + x - 2x^2, g(x) = 5x + 8$

APPLICATIONS

121. *Women on the Run* The number of women participating in the New York City Marathon has increased dramatically in recent years. The polynomial function $f(x) = 3.56x^2 + 360x + 1300$ models the number of women running each year from 1980 through 2010, where $x = 0$ corresponds to 1980, $x = 10$ to 1990, and $x = 30$ to 2010. See the accompanying figure. (*Source: Runner's World.*)

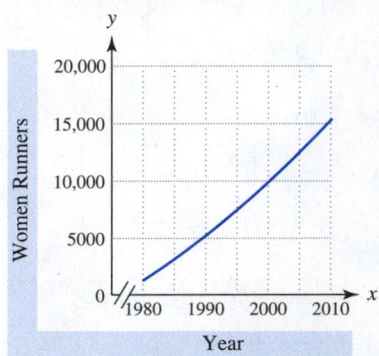

(a) Use the graph to estimate the number of women running in 1980 and in 2000.

(b) Answer part (a) using the polynomial. How do your answers compare with your answers in part (a)?

(c) Use the polynomial to find the increase in women runners from 1980 to 2000.

122. *A PC for All?* Worldwide sales of computers have climbed as prices have continued to drop. The function $f(x) = 0.29x^2 + 8x + 19$ models the number of personal computers sold in millions during year x, where $x = 0$ corresponds to 1990, $x = 1$ to 1991, and so on until $x = 25$ corresponds to 2015.

Estimate the number of personal computers sold in 2010, using both the graph and the polynomial. (*Source:* eTForcasts.)

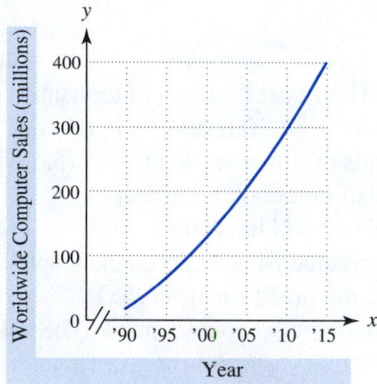

123. *U.S. AIDS Deaths* The following scatterplot shows the cumulative number of reported AIDS deaths. The data may be modeled x years after 1980 by $f(x) = 2.4x^2 - 14x + 23$, where the output is in thousands of deaths.

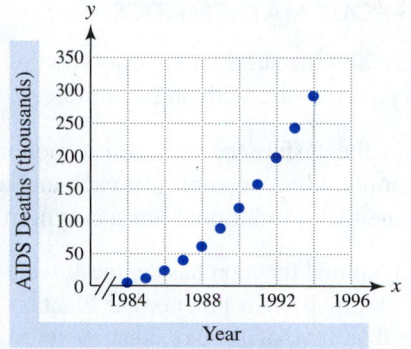

(a) Use $f(x)$ to estimate the cumulative total of AIDS deaths in 1990. Compare it with the actual value of 121.6 thousand.

(b) In 1997 the cumulative number of AIDS deaths was 390 thousand. What estimate does $f(x)$ give? Discuss your result.

124. *Geometry* A house has 8-foot-high ceilings. Three rooms are x feet by y feet and two rooms are x feet by z feet. Write a polynomial that gives the total volume of the five rooms. Find their total volume if $x = 10$ feet, $y = 12$ feet, and $z = 7$ feet.

125. Write a monomial that represents the volume of five identical cubes with sides having length s.

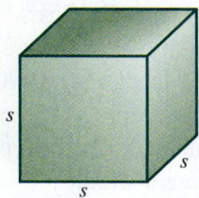

126. Write a monomial that represents the area of six identical rectangles with length L and width W.

127. *Squares and Circles* Write a polynomial that gives the sum of the areas of a square with sides of length x and a circle with radius x. Find the combined area when $x = 10$ inches.

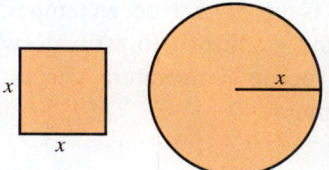

128. *Spheres* Write a monomial that gives the volume of 9 spheres with radius y. Find the combined volumes when $y = 3$ feet. (*Hint:* $V = \frac{4}{3}\pi r^3$.)

129. *Heart Rate* (Refer to Example 13.) Make a table of $f(t) = 1.875t^2 - 30t + 200$, starting at $t = 3$ and incrementing by 1. When was the athlete's heart rate between 80 and 110, inclusive?

130. *Life Expectancy* The life expectancy of a female born in 1970 is 75 years, and in 2015 it is expected to be 84 years. (*Source:* Bureau of the Census.)

(a) Find values for a and b so that $f(x) = ax + b$ models the data. (*Hint:* The graph of f should pass through the point (1970, 75) and the point (2015, 84).)

(b) Estimate the life expectancy in 2010.

131. *Microsoft Stocks* If a person had bought $2500 of Microsoft stock in 1986, it would have been worth about $1.2 million in 1999. The following table shows the cost of this stock (adjusted for splits) in various years.

Year	1986	1994	1999
Cost ($)	0.19	20	90

Source: USA Today.

Which of the following polynomials models the data best, where $x = 6$ corresponds to 1986, $x = 14$ to 1994, and $x = 19$ to 1999? Explain how you made your decision.

(i) $f(x) = 0.886x^2 - 15.25x + 59.79$

(ii) $g(x) = 0.882x^2 - 15.4x + 60.15$

132. *Ocean Temperatures* The cubic polynomial

$$f(x) = -0.064x^3 + 0.56x^2 + 2.9x + 61$$

models the ocean temperature in degrees Fahrenheit at Naples, Florida. In this formula $x = 1$ corresponds to January, $x = 2$ to February, and so on. (*Source:* J. Williams.)
(a) What is the average ocean temperature in April?
(b) Use the graph of f to estimate when the maximum ocean temperature occurs. What is this maximum?

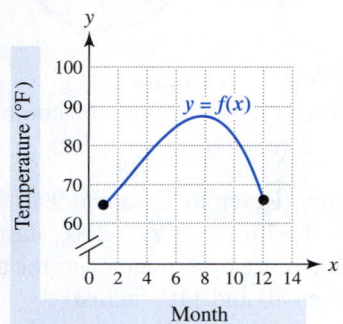

Month

133. *Profit from DVDs* Suppose that the cost in dollars for a band to produce x DVDs is given by the function $C(x) = 4x + 2000$. If the band sells DVDs for $16 each, complete the following.
(a) Write a function R that gives the revenue for selling x DVDs.
(b) If profit equals revenue minus cost, write a formula $P(x)$ for the profit.
(c) Evaluate $P(3000)$ and interpret the answer.

134. *Profit* A company makes and sells sailboats. The company's cost function in thousands of dollars is $C(x) = 2x + 20$, and the revenue function in thousands of dollars is $R(x) = 4x$, where x is the number of sailboats.
(a) Evaluate and interpret $C(5)$.

(b) Interpret the y-intercepts on the graphs of C and R.
(c) Give the profit function $P(x)$.
(d) How many sailboats need to be sold to break even?

135. *Profit* A company makes and sells notebook computers. Their cost function in thousands of dollars is $C(x) = 0.3x + 100$, and their revenue function in thousands of dollars is $R(x) = 0.75x$, where x is the number of notebook computers.
(a) Evaluate and interpret $C(100)$.
(b) Interpret the y-intercepts on the graphs of C and R.
(c) Give the profit function $P(x)$.
(d) How many computers need to be sold to make a profit?

136. *Income Tax* Let the state tax be 7% and the federal tax be 15%.
(a) Find a function T that calculates the total tax on a taxable income of x dollars.
(b) Evaluate $T(4000)$ and interpret the result.

WRITING ABOUT MATHEMATICS

137. Discuss how to subtract two polynomials. Demonstrate your method with an example.

138. Explain the difference between a monomial and a polynomial. Give examples of each and an example that is neither a monomial nor a polynomial.

139. If a polynomial function has degree 1, what can be said about its graph? If a polynomial function has degree greater than 1, what can be said about its graph?

140. Explain how to evaluate a function graphically. Use your method to evaluate $f(x) = x^2$ at $x = -2$.

Group Activity   Working with Real Data

Directions: Form a group of 2 to 4 people. Select someone to record the group's responses for this activity. All members of the group should work cooperatively to answer the questions. If your instructor asks for the results, each member of the group should be prepared to respond.

Female Graduates The table shows the number of bachelor degrees awarded to women for various years

Year	1992	2002	2012
Degrees (thousands)	533	632	968

Source: National Center for Education Statistics.

(a) Verify that $f(x) = 1.185x^2 - 6.69x + 541.64$ models the data by evaluating $f(2)$, $f(12)$, and $f(22)$, where x is years *after* 1990.
(b) Interpret the y-intercept.
(c) Evaluate $f(15)$ and $f(50)$. Interpret each answer. Which one is more accurate?

5.2 Multiplication of Polynomials

Review of Basic Properties • Multiplying Polynomials • Some Special Products •
Multiplying Functions

A LOOK INTO MATH ▶ In 2010 Serena Williams claimed her thirteenth Grand Slam title at Wimbledon. Her serves reached speeds of 122 miles per hour, and her first serves had a 66% chance of being in bounds. More and more, mathematics is being used to evaluate an athlete's performance. In Exercise 133 we use a polynomial to estimate the likelihood that either the first serve or the second serve will be in bounds. (*Source: The Associated Press.*)

NEW VOCABULARY

☐ Monomial
☐ Binomial
☐ Trinomial

Review of Basic Properties

Distributive properties are used frequently in the multiplication of polynomials. For all real numbers a, b, and c

$$a(b + c) = ab + ac \quad \text{and} \qquad \text{Distributive properties}$$
$$a(b - c) = ab - ac.$$

In the next example we use these distributive properties to multiply expressions.

EXAMPLE 1 **Using distributive properties**

Multiply.
(a) $4(5 + x)$ **(b)** $-3(x - 4y)$ **(c)** $(2x - 5)(6)$

Solution
(a) $4(5 + x) = 4 \cdot 5 + 4 \cdot x = 20 + 4x$
(b) $-3(x - 4y) = -3 \cdot x - (-3) \cdot (4y) = -3x + 12y$
(c) $(2x - 5)(6) = 2x \cdot 6 - 5 \cdot 6 = 12x - 30$

Now Try Exercises 21, 23, 27

VISUALIZING THE DISTRIBUTIVE PROPERTY We can visualize the distributive property in Example 1(a) by using rectangles. In Figure 5.13(a) there is a large rectangle having width 4 and length $5 + x$, so its area must be $4(5 + x)$. In Figure 5.13(b) the large rectangle has been split into two smaller rectangles having equivalent total areas of $4(5)$ and $4(x)$. Finally, in Figure 5.13(c), small squares and rectangles have been sketched to show that the total area is 20 squares, each having area 1, plus 4 rectangles with area x.

The Distributive Property

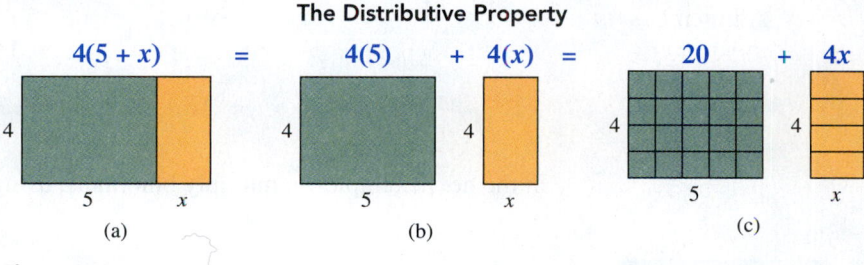

Figure 5.13

In Section 1.3 we discussed several properties of exponents. The following three properties of exponents are frequently used in the multiplication of polynomials.

PROPERTIES OF EXPONENTS

For any numbers a and b and integers m and n,

$$a^m \cdot a^n = a^{m+n}, \quad (a^m)^n = a^{mn}, \quad \text{and} \quad (ab)^n = a^n b^n.$$

We use these three properties to simplify expressions in the next two examples.

EXAMPLE 2 **Multiplying powers of variables**

Multiply each expression.
(a) $-2x^3 \cdot 4x^5$ **(b)** $(3xy^3)(4x^2y^2)$ **(c)** $6y^3(2y - y^2)$ **(d)** $-mn(m^2 - n)$

Solution
(a) $-2x^3 \cdot 4x^5 = (-2)(4)x^3 x^5 = -8x^{3+5} = -8x^8$
(b) $(3xy^3)(4x^2y^2) = (3)(4)xx^2y^3y^2 = 12x^{1+2}y^{3+2} = 12x^3y^5$

NOTE: We cannot simplify $12x^3y^5$ further because x^3 and y^5 have different bases.

(c) $6y^3(2y - y^2) = 6y^3 \cdot 2y - 6y^3 \cdot y^2 = 12y^4 - 6y^5$
(d) $-mn(m^2 - n) = -mn \cdot m^2 + mn \cdot n = -m^3n + mn^2$

Now Try Exercises 13, 17, 29, 31

EXAMPLE 3 **Using properties of exponents**

Simplify.
(a) $(x^2)^5$ **(b)** $(2x)^3$ **(c)** $(5x^3)^2$ **(d)** $(-mn^3)^2$

Solution
(a) $(x^2)^5 = x^{2 \cdot 5} = x^{10}$ **(b)** $(2x)^3 = 2^3x^3 = 8x^3$
(c) $(5x^3)^2 = 5^2(x^3)^2 = 25x^6$ **(d)** $(-mn^3)^2 = (-m)^2(n^3)^2 = m^2n^6$

Now Try Exercises 83, 85, 87

Multiplying Polynomials

A polynomial with one term is a **monomial**, with two terms a **binomial**, and with three terms a **trinomial**. Examples are shown in Table 5.5.

TABLE 5.5

Monomials	$2x^2$	$-3x^4y$	9
Binomials	$3x - 1$	$2x^3 - x$	$x^2 + 5$
Trinomials	$x^2 - 3x + 5$	$5x^4 - 2x + 10$	$2x^3 - x^2 - 2$

In the next example we multiply binomials, using geometric and symbolic techniques.

EXAMPLE 4 **Multiplying binomials**

Multiply $(x + 1)(x + 3)$
(a) geometrically and   **(b)** symbolically.

Area: $(x + 1)(x + 3)$

$x + 1$

$x + 3$

(a)

Area: $x^2 + 4x + 3$

1   x 3

x x^2 $3x$

x 3

(b)

Figure 5.14

Solution

(a) To multiply $(x + 1)(x + 3)$ geometrically, draw a rectangle $x + 1$ wide and $x + 3$ long, as shown in Figure 5.14(a). The area of the rectangle equals the product $(x + 1)(x + 3)$. This large rectangle can be divided into four smaller rectangles as shown in Figure 5.14(b). The sum of the areas of these four smaller rectangles equals the area of the large rectangle. The smaller rectangles have areas of x^2, x, $3x$, and 3. Thus

$$(x + 1)(x + 3) = x^2 + x + 3x + 3$$
$$= x^2 + 4x + 3.$$

(b) To multiply $(x + 1)(x + 3)$ symbolically we apply the distributive property.

$$(x + 1)(x + 3) = (x + 1)(x) + (x + 1)(3)$$
$$= x \cdot x + 1 \cdot x + x \cdot 3 + 1 \cdot 3$$
$$= x^2 + x + 3x + 3$$
$$= x^2 + 4x + 3$$

Now Try Exercise 37

GRAPHICAL AND NUMERICAL SUPPORT It is possible to give graphical and numerical support to our result in Example 4 by letting $y_1 = (x + 1)(x + 3)$ and by letting $y_2 = x^2 + 4x + 3$. The graphs of y_1 and y_2 appear to be identical in Figures 5.15(a) and 5.15(b). Figure 5.15(c) shows that $y_1 = y_2$ for each value of x in the table.

$[-6, 6, 1]$ by $[-4, 4, 1]$

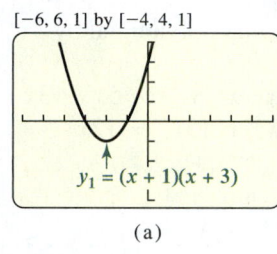

$y_1 = (x + 1)(x + 3)$

(a)

$[-6, 6, 1]$ by $[-4, 4, 1]$

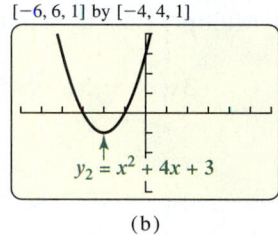

$y_2 = x^2 + 4x + 3$

(b)

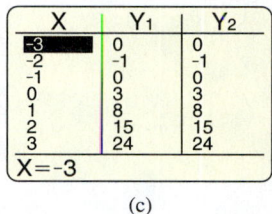

X	Y1	Y2
-3	0	0
-2	-1	-1
-1	0	0
0	3	3
1	8	8
2	15	15
3	24	24

X=-3

(c)

Figure 5.15

FOIL One way to multiply $(x + 1)$ by $(x + 3)$ is to multiply every term in $x + 1$ by every term in $x + 3$. That is,

$$(x + 1)(x + 3) = x^2 + 3x + x + 3$$
$$= x^2 + 4x + 3.$$

NOTE: This process of multiplying binomials is sometimes called *FOIL*, which reminds us to multiply the first terms (F), outside terms (O), inside terms (I), and last terms (L).

Multiply the *First terms* to obtain x^2. $(x + 1)(x + 3)$

Multiply the *Outside terms* to obtain $3x$. $(x + 1)(x + 3)$

Multiply the *Inside terms* to obtain x. $(x + 1)(x + 3)$

Multiply the *Last terms* to obtain 3. $(x + 1)(x + 3)$

The following method summarizes how to multiply two polynomials in general.

MULTIPLICATION OF POLYNOMIALS

The product of two polynomials may be found by multiplying every term in the first polynomial by every term in the second polynomial.

EXAMPLE 5 **Multiplying polynomials**

Multiply each binomial.
(a) $(2x - 1)(x + 2)$ **(b)** $(1 - 3x)(2 - 4x)$ **(c)** $(x^2 + 1)(5x - 3)$

Solution

(a) $(2x - 1)(x + 2) = 2x \cdot x + 2x \cdot 2 - 1 \cdot x - 1 \cdot 2$
$= 2x^2 + 4x - x - 2$
$= 2x^2 + 3x - 2$

(b) $(1 - 3x)(2 - 4x) = 1 \cdot 2 - 1 \cdot 4x - 3x \cdot 2 + 3x \cdot 4x$
$= 2 - 4x - 6x + 12x^2$
$= 2 - 10x + 12x^2$

(c) $(x^2 + 1)(5x - 3) = x^2 \cdot 5x - x^2 \cdot 3 + 1 \cdot 5x - 1 \cdot 3$
$= 5x^3 - 3x^2 + 5x - 3$

Now Try Exercises 53, 57, 63

EXAMPLE 6 **Multiplying polynomials**

Multiply each expression.
(a) $3x(x^2 + 5x - 4)$ **(b)** $-x^2(x^4 - 2x + 5)$ **(c)** $(x + 2)(x^2 + 4x - 3)$

Solution
(a) Multiply each term in the second polynomial by $3x$.

$3x(x^2 + 5x - 4) = 3x \cdot x^2 + 3x \cdot 5x - 3x \cdot 4$
$= 3x^3 + 15x^2 - 12x$

(b) $-x^2(x^4 - 2x + 5) = -x^2 \cdot x^4 + x^2 \cdot 2x - x^2 \cdot 5$
$= -x^6 + 2x^3 - 5x^2$

(c) $(x + 2)(x^2 + 4x - 3) = x \cdot x^2 + x \cdot 4x - x \cdot 3 + 2 \cdot x^2 + 2 \cdot 4x - 2 \cdot 3$
$= x^3 + 4x^2 - 3x + 2x^2 + 8x - 6$
$= x^3 + 6x^2 + 5x - 6$

Now Try Exercises 67, 69, 73

EXAMPLE 7 **Multiplying polynomials**

Multiply each expression.
(a) $2ab^3(a^2 - 2ab + 3b^2)$ **(b)** $4m(mn^2 + 3m)(m^2n - 4n)$

Solution
(a) Multiply each term in the second polynomial by $2ab^3$.

$2ab^3(a^2 - 2ab + 3b^2) = 2ab^3 \cdot a^2 - 2ab^3 \cdot 2ab + 2ab^3 \cdot 3b^2$
$= 2a^3b^3 - 4a^2b^4 + 6ab^5$

(b) Start by multiplying every term in $(mn^2 + 3m)$ by $4m$.

$4m(mn^2 + 3m)(m^2n - 4n) = (4m \cdot mn^2 + 4m \cdot 3m)(m^2n - 4n)$

$= (4m^2n^2 + 12m^2)(m^2n - 4n)$

$= 4m^2n^2 \cdot m^2n - 4m^2n^2 \cdot 4n + 12m^2 \cdot m^2n - 12m^2 \cdot 4n$
$= 4m^4n^3 - 16m^2n^3 + 12m^4n - 48m^2n$

Now Try Exercises 75, 77

$$
\begin{array}{r}
212 \\
\times\ 32 \\
\hline
424 \\
636 \\
\hline
6784
\end{array}
$$

MULTIPLYING POLYNOMIALS VERTICALLY Before we demonstrate a second way to multiply polynomials in Example 8, we will review how to multiply 212 and 32 vertically, as shown in the margin. First, we multiply 2 and 212 to get 424. Second, we multiply 3 and 212 to get 636, but we must move 636 one place to the left so that the ones, tens, and hundreds digits are aligned correctly. Finally, we add to get 6784.

EXAMPLE 8 | **Multiplying polynomials vertically**

Multiply $(3x - 4y)(2x^2 + xy - 4y^2)$.

Solution
Start by stacking the polynomials as follows. Note that unlike terms can be multiplied but not added.

$$
\begin{array}{r}
2x^2 + xy - 4y^2 \qquad \text{First row} \\
3x - 4y \\
\hline
-8x^2y - 4xy^2 + 16y^3 \qquad \text{Multiply first row by } -4y. \\
6x^3 + 3x^2y - 12xy^2 \qquad\qquad\quad \text{Multiply first row by } 3x. \\
\hline
6x^3 - 5x^2y - 16xy^2 + 16y^3 \qquad \text{Add columns.}
\end{array}
$$

■ **Now Try Exercise 73**

▶ **REAL-WORLD CONNECTION** Polynomials frequently occur in business applications. For example, the **demand** D for buying a new video game might vary with its price. If the price is high, fewer games are sold, and if the price is low, more games are sold. The revenue R from selling D games at price p is given by $R = pD$. This concept is used in the next example.

EXAMPLE 9 | **Using polynomials in a business application**

Let the demand D, or number of games sold in thousands, be given by $D = 30 - \frac{1}{5}p$, where $p \geq 10$ is the price of the game in dollars.
(a) Find the demand when $p = \$30$ and when $p = \$60$.
(b) Write an expression for the revenue R. Multiply your expression.
(c) Find the revenue when $p = \$40$.

Solution
(a) When $p = \mathbf{30}, D = 30 - \frac{1}{5}(\mathbf{30}) = 24$ thousand games.

When $p = \mathbf{60}, D = 30 - \frac{1}{5}(\mathbf{60}) = 18$ thousand games.

(b) $R = pD = p(30 - \frac{1}{5}p) = 30p - \frac{1}{5}p^2$

(c) Let $R = 30p - \frac{1}{5}p^2$. When $p = \mathbf{40}, R = 30(\mathbf{40}) - \frac{1}{5}(\mathbf{40})^2 = \$880$ thousand.

■ **Now Try Exercise 131**

Some Special Products

The product of a sum and a difference often occurs in mathematics.

$$
\begin{aligned}
(a + b)(a - b) &= a \cdot a - a \cdot b + b \cdot a - b \cdot b \\
&= a^2 - ab + ba - b^2 \\
&= a^2 - b^2
\end{aligned}
$$

That is, the product of a sum and difference of two numbers equals the difference of their squares.

PRODUCT OF A SUM AND DIFFERENCE

For any real numbers a and b,

$$(a + b)(a - b) = a^2 - b^2.$$

EXAMPLE 10   Finding the product of a sum and difference

Multiply.
(a) $(x + 3)(x - 3)$ **(b)** $(5 - 4x^2)(5 + 4x^2)$

Solution
(a) If we let $a = x$ and $b = 3$, we can apply the rule

$$(a + b)(a - b) = a^2 - b^2.$$

Thus

$$(x + 3)(x - 3) = (x)^2 - (3)^2$$
$$= x^2 - 9.$$

(b) Because $(a - b)(a + b) = a^2 - b^2$, we can multiply as follows.

$$(5 - 4x^2)(5 + 4x^2) = (5)^2 - (4x^2)^2$$
$$= 25 - 16x^4$$

Now Try Exercises 91, 93

EXAMPLE 11   Finding the product of a sum and difference

Multiply each expression.
(a) $4rt(r - 4t)(r + 4t)$ **(b)** $(2z + 5k^4)(2z - 5k^4)$

Solution
(a) Start by finding the product of the difference and sum, then simplify.

$$4rt(r - 4t)(r + 4t) = 4rt(r^2 - 16t^2)$$
$$= 4rt \cdot r^2 - 4rt \cdot 16t^2$$
$$= 4r^3t - 64rt^3$$

(b) $(2z + 5k^4)(2z - 5k^4) = (2z)^2 - (5k^4)^2$
$$= 4z^2 - 25k^8$$

Now Try Exercises 107, 109

Two other special products involve *squaring a binomial*.

$$(a + b)^2 = (a + b)(a + b) \qquad\qquad (a - b)^2 = (a - b)(a - b)$$
$$= a^2 + ab + ba + b^2 \qquad\qquad = a^2 - ab - ba + b^2$$
$$= a^2 + 2ab + b^2 \qquad\qquad = a^2 - 2ab + b^2$$

The first product is illustrated geometrically in Figure 5.16, where each side of a square has length $(a + b)$. The area of the square is

$$(a + b)(a + b) = (a + b)^2.$$

$(a + b)^2 = a^2 + 2ab + b^2$

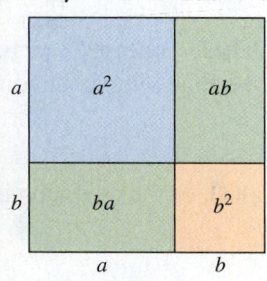

Figure 5.16

This area can also be computed by adding the areas of the four small rectangles.

$$a^2 + ab + ba + b^2 = a^2 + 2ab + b^2$$

Thus $(a + b)^2 = a^2 + 2ab + b^2$. *To obtain the middle term, multiply the two terms in the binomial and double the result.*

SQUARE OF A BINOMIAL

For any real numbers a and b,

$$(a + b)^2 = a^2 + 2ab + b^2 \quad \text{and} \quad (a - b)^2 = a^2 - 2ab + b^2.$$

NOTE: $(a + b)^2 \neq a^2 + b^2$ and $(a - b)^2 \neq a^2 - b^2$

EXAMPLE 12 **Squaring a binomial**

Multiply.
(a) $(x + 5)^2$ **(b)** $(3 - 2x)^2$ **(c)** $(2m^3 + 4n)^2$ **(d)** $\big((2z - 3) + k\big)\big((2z - 3) - k\big)$

Solution
(a) If we let $a = x$ and $b = 5$, we can apply $(a + b)^2 = a^2 + 2ab + b^2$. Thus

$$(x + 5)^2 = (x)^2 + 2(x)(5) + (5)^2$$

> To find the middle term, multiply x and 5 and double the result.

$$= x^2 + 10x + 25.$$

(b) Applying the formula $(a - b)^2 = a^2 - 2ab + b^2$, we find

$$(3 - 2x)^2 = (3)^2 - 2(3)(2x) + (2x)^2$$
$$= 9 - 12x + 4x^2.$$

(c) $(2m^3 + 4n)^2 = (2m^3)^2 + 2(2m^3)(4n) + (4n)^2$
$$= 4m^6 + 16m^3n + 16n^2$$

(d) If $a = 2z - 3$ and $b = k$, we can use $(a + b)(a - b) = a^2 - b^2$.

$$\big((2z - 3) + k\big)\big((2z - 3) - k\big) = (2z - 3)^2 - k^2$$
$$= (2z)^2 - 2(2z)(3) + (3)^2 - k^2$$
$$= 4z^2 - 12z + 9 - k^2$$

Now Try Exercises 97, 99, 111, 117

NOTE: If you forget these special products, you can still use techniques learned earlier to multiply the polynomials in Examples 10–12. Consider Example 12(b).

$$(3 - 2x)^2 = (3 - 2x)(3 - 2x)$$

$$= 3 \cdot 3 - 3 \cdot 2x - 2x \cdot 3 + 2x \cdot 2x$$
$$= 9 - 6x - 6x + 4x^2$$
$$= 9 - 12x + 4x^2 \checkmark$$

CRITICAL THINKING

Suppose that a friend is convinced that the expressions

$$(x + 3)^2 \quad \text{and} \quad x^2 + 9$$

are equivalent. How could you convince your friend that $(x + 3)^2 \neq x^2 + 9$?

Multiplying Functions

In Section 5.1 we discussed how to add and subtract functions. These concepts can be extended to include multiplication of two functions, f and g, by using the following definition.

$$(fg)(x) = f(x)g(x) \qquad \text{Multiplying functions}$$

NOTE: There is no operation symbol written between f and g in $(fg)(x)$, so the operation is assumed to be multiplication, just as in the expression xy.

STUDY TIP

Examples like this are sometimes useful for helping to understand why we multiply two functions in real life.

▶ **REAL-WORLD CONNECTION** When would someone ever need to multiply two functions? Consider the following example. From 1980 to 2010, the average person used $G(x) = -24x + 1900$ gallons of water per day, where x is years after 1980. (Note that G includes all types of water usage in the United States, not just residential usage.) During the same time period, the population P of the United States in millions was $P(x) = 2.6x + 230$. If we multiply the gallons G of water each person used by the number of people P, we get the total amount of water used per day (in millions of gallons). That is, we multiply $G(x)$ and $P(x)$ to obtain the total water consumption $T(x)$.

$$T(x) = G(x)P(x)$$
$$= (-24x + 1900)(2.6x + 230)$$

You are asked to multiply these binomials and interpret your answer in Exercise 135.

EXAMPLE 13 **Multiplying polynomial functions**

Let $f(x) = x^2 - x$ and $g(x) = 2x^2 + 4$. Find $(fg)(3)$ and $(fg)(x)$.

Solution
To find $(fg)(3)$, evaluate $f(3)$ and $g(3)$. Then multiply the results.

$$(fg)(3) = f(3)g(3) = (3^2 - 3)(2 \cdot 3^2 + 4) = (6)(22) = 132$$

To find $(fg)(x)$, multiply the formulas for $f(x)$ and $g(x)$.

$$(fg)(x) = f(x)g(x) = (x^2 - x)(2x^2 + 4) = 2x^4 - 2x^3 + 4x^2 - 4x$$

Now Try Exercise 121

5.2 Putting It All Together

CONCEPT	EXPLANATION	EXAMPLES
Distributive Properties	For all real numbers a, b, and c, $a(b + c) = ab + ac$ and $a(b - c) = ab - ac$.	$4(3 + a) = 12 + 4a$, $5(x - 1) = 5x - 5$, and $-(b - 5) = -1(b - 5) = -b + 5$
Multiplying Polynomials	The product of two polynomials may be found by multiplying every term in the first polynomial by every term in the second polynomial.	$(2x + 3)(x - 7) = 2x \cdot x - 2x \cdot 7 + 3 \cdot x - 3 \cdot 7$ $= 2x^2 - 14x + 3x - 21$ $= 2x^2 - 11x - 21$

CONCEPT	EXPLANATION	EXAMPLES
Properties of Exponents	For numbers a and b and integers m and n, $$a^m \cdot a^n = a^{m+n},$$ $$(a^m)^n = a^{mn}, \quad \text{and}$$ $$(ab)^n = a^n b^n.$$	$5^2 \cdot 5^6 = 5^8$, $(2^3)^2 = 2^6$, $(5y^4)^2 = 25y^8$, $(a^2 b^3)^4 = a^8 b^{12}$, and $(-4rt^3)^2 = 16r^2 t^6$
Special Products of Binomials	Product of a sum and difference $$(a + b)(a - b) = a^2 - b^2$$ Squares of binomials $$(a + b)^2 = a^2 + 2ab + b^2$$ $$(a - b)^2 = a^2 - 2ab + b^2$$	$(x + 2)(x - 2) = x^2 - 4$, $(7y - 6z^2)(7y + 6z^2) = 49y^2 - 36z^4$, $(x + 4)^2 = x^2 + 8x + 16$, $(x - 4)^2 = x^2 - 8x + 16$, and $(2m - 3n^2)^2 = 4m^2 - 12mn^2 + 9n^4$
Multiplication of Functions	$(fg)(x) = f(x)g(x)$	If $f(x) = x^2$ and $g(x) = x + 2$, then $(fg)(x) = x^2(x + 2) = x^3 + 2x^2$.

5.2 Exercises

MyMathLab  WATCH DOWNLOAD READ REVIEW

CONCEPTS AND VOCABULARY

1. $5(x - 4) = 5x - 20$ illustrates what property?

2. Give an example of a monomial, a binomial, and a trinomial.

3. Simplify the expression $x^3 \cdot x^5$.

4. Simplify $(2x)^3$ and $(x^2)^3$.

5. $(a + b)(a - b) = $ _____

6. $(a + b)^2 = $ _____

7. $x(x - 1) = $ _____

8. Does $(x + 1)^2$ equal $x^2 + 1$? Explain.

9. (True or False?) $x^3(x^2 - x^4)$ equals $x^6 - x^{12}$.

10. (True or False?) $b^3 b^5$ equals b^8.

11. Which of the following (a. – d.) simplifies to $4x^2 y^6$?
  a. $-(2xy^3)^2$ **b.** $-2^2(xy^3)^2$
  c. $4x^2(y^3)^3$ **d.** $(-2xy^3)^2$

12. Which of the following (a. – d.) simplifies to $4x^3 - 4x$?
  a. $4x(x + 1)(x - 1)$ **b.** $4(x - 1)(x + 1)$
  c. $x(x + 2)(x - 2)$ **d.** $4(x^2 - x)$

MULTIPLYING POLYNOMIALS

Exercises 13–20: Multiply the monomials.

13. $x^4 \cdot x^8$ **14.** $2x \cdot 4x^3$

15. $-5y^7 \cdot 4y$ **16.** $3xy^2 \cdot 6x^3 y^2$

17. $(-xy)(4x^3 y^5)$

18. $(4z^3)(-5z^2)$

19. $(5y^2 z)(4x^2 yz^5)$

20. $x^2(-xy^2)$

Exercises 21–32: Multiply.

21. $5(y + 2)$ **22.** $4(y - 7)$

23. $-2(5x + 9)$ **24.** $-3x(5 + x)$

25. $-6y(y - 3)$ **26.** $(2y - 5)8y^3$

27. $(9 - 4x)3$ **28.** $-(5 - x^2)$

29. $-ab(a^2 - b^2)$ **30.** $a^2 b^2(1 - 4ab)$

31. $-5m(n^3 + m)$ **32.** $7n(3n - 2m^2)$

Exercises 33–36: Use the figure to write the product. Find the area of the rectangle if x = 5 inches.

33. $(x + 1)(x + 2)$

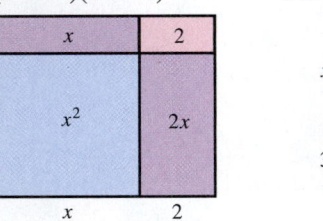

34. $(x + 3)(x + 4)$

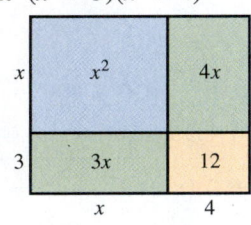

35. $(2x + 1)(x + 1)$

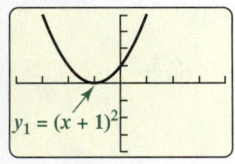

36. $(2x + 4)(3x + 2)$

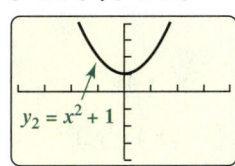

Exercises 37–40: (Refer to Example 4.) Multiply geometrically and symbolically. Support your answer graphically or numerically.

37. $(x + 5)(x + 6)$

38. $(x + 1)(x + 4)$

39. $(2x + 1)(2x + 1)$

40. $(x + 3)(2x + 4)$

Exercises 41 and 42: The pair of graphing calculator screens show graphs of y_1 and y_2. Does $y_1 = y_2$? Interpret this result in general.

41. $y_1 = (x + 1)^2, y_2 = x^2 + 1$

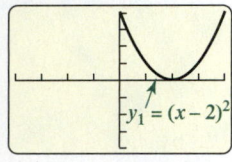

42. $y_1 = (x - 2)^2, y_2 = x^2 - 4$

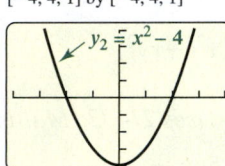

Exercises 43–48: Determine whether the equation is an identity. (Hint: If it is an identity, the equation must be true for all x-values.)

43. $(x - 2)^2 = x^2 - 4$

44. $(x + 4)^2 = x^2 + 16$

45. $(x + 3)(x - 2) = x^2 + x - 6$

46. $(5x - 1)^2 = 25x^2 - 10x + 1$

47. $2z(3z + 1) = 6z^2 + 1$

48. $-(z^2 + 4z) = -z^2 + 4z$

Exercises 49–66: Multiply the binomials.

49. $(x + 5)(x + 10)$

50. $(x - 5)(x - 10)$

51. $(x - 3)(x - 4)$

52. $(x + 3)(x + 4)$

53. $(2z - 1)(z + 2)$

54. $(2z + 1)(z - 2)$

55. $(y + 3)(y - 4)$

56. $(2y + 1)(5y + 1)$

57. $(4x - 3)(4 - 9x)$

58. $(1 - x)(1 + 2x)$

59. $(-2z + 3)(z - 2)$

60. $(z - 2)(4z + 3)$

61. $\left(z - \frac{1}{2}\right)\left(z + \frac{1}{4}\right)$

62. $\left(z - \frac{1}{3}\right)\left(z - \frac{1}{6}\right)$

63. $(x^2 + 1)(2x^2 - 1)$

64. $(x^2 - 2)(x^2 + 4)$

65. $(x + y)(x - 2y)$

66. $(x^2 + y^2)(x - y)$

Exercises 67–80: Multiply the polynomials.

67. $4x(x^2 - 2x - 3)$

68. $2x(3 - x + x^2)$

69. $-x(x^4 - 3x^2 + 1)$

70. $-3m^2(4m^3 + m^2 - 2m)$

71. $(2n^2 - 4n + 1)(3n^2)$

72. $(x - y + 5)(xy)$

73. $(x + 1)(x^2 + 2x - 3)$

74. $(2x - 1)(3x^2 - x + 6)$

75. $z(2 + z)(1 - z - z^2)$

76. $z^2(1 - z)(2 + z)$

77. $2ab^2(2a^2 - ab + 3b^2)$

78. $2n(mn^2 + 2n)(3m^2n - 3n)$

79. $(2r - 4t)(3r^2 + rt - t^2)$

80. $-2(x - y)(x^2 + xy + y^2)$

Exercises 81–90: Simplify the expression.

81. $(2^3)^2$

82. $(x^3)^5$

83. $2(z^3)^6$

84. $(5y)^3$

85. $(-5x)^2$

86. $(2y)^4$

87. $(-2xy^2)^3$

88. $(3x^2y^3)^4$

89. $(-4a^2b^3)^2$

90. $-(5r^3t)^2$

Exercises 91–118: Multiply the expression.

91. $(x - 4)(x + 4)$

92. $(x + 5)(x - 5)$

93. $(3 - 2x)(3 + 2x)$

94. $(4 - 5x)(4 + 5x)$

95. $(x - y)(x + y)$ **96.** $(2x + 2y)(2x - 2y)$

97. $(x + 2)^2$ **98.** $(y + 5)^2$

99. $(2x + 1)^2$ **100.** $(3x + 5)^2$

101. $(x - 1)^2$

102. $(x - 7)^2$

103. $(3x - 2)^2$

104. $(6x - 5)^2$

105. $3x(x + 1)(x - 1)$

106. $-4x(3x - 5)^2$

107. $3rt(t - 2r)(t + 2r)$

108. $5r^2t^2(t - 4)(t + 4)$

109. $(a^2 + 2b^2)(a^2 - 2b^2)$

110. $(2a + 5b^4)(2a - 5b^4)$

111. $(3m^3 + 5n^2)^2$

112. $(6m + 4n^2)^2$

113. $(x^3 - 2y^3)^2$

114. $(6m - n^4)^2$

115. $\big((x - 3) + y\big)\big((x - 3) - y\big)$

116. $\big((2m - 1) + n\big)\big((2m - 1) - n\big)$

117. $\big(r - (t + 2)\big)\big(r + (t + 2)\big)$

118. $\big(y - (z + 1)\big)\big(y + (z + 1)\big)$

Exercises 119 and 120: **Thinking Generally** *Use*

$$(a + b)(a - b) = a^2 - b^2$$

to multiply the pair of integers mentally.

119. $102, 98$

120. $51, 49$

Exercises 121–126: Find $(fg)(2)$ *and* $(fg)(x)$.

121. $f(x) = x + 1, g(x) = x - 2$

122. $f(x) = 2x + 3, g(x) = 4 - x$

123. $f(x) = x^2, g(x) = 3 - 5x$

124. $f(x) = x^2 - 4x, g(x) = x^2 - 5$

125. $f(x) = 2x^2 + 1, g(x) = x^2 + x - 3$

126. $f(x) = x^3 - 3x^2, g(x) = 2x^2 + 4x$

AREA

Exercises 127–130: Express the shaded area of the given figure in terms of x. Find the area if x = 20 feet.

127.

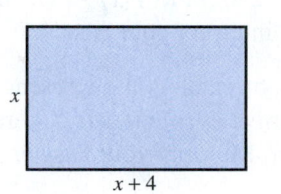

128.

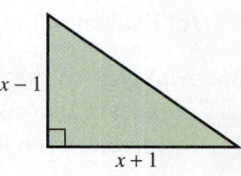

129.

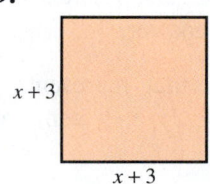

130.

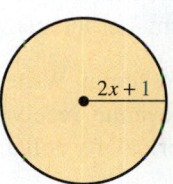

APPLICATIONS

Exercises 131 and 132: **Price and Demand** *(Refer to Example 9.) Let D be the demand in thousands of units and p be the price in dollars of one unit.*

  (a) *Find D when* $p = 20$.

  (b) *Write an expression for the revenue R. Multiply your expression.*

  (c) *Find the revenue when* $p = 30$.

131. $D = 40 - \frac{1}{2}p$ **132.** $D = 20 - \frac{1}{10}p$

133. **Tennis Serves** Serena Williams hit 66% of her first serves *in* bounds at Wimbledon 2010, or, equivalently, 34% of these serves were *out* of bounds. If x represents the percentage of a player's serves that are *out* of bounds, then the percent chance that two consecutive serves will be *in* bounds is given by $100\left(1 - \frac{x}{100}\right)^2$.

  (a) Multiply this expression.

  (b) Evaluate both the given expression and the expression you found in part (a) with $x = 34$. Are your answers the same?

  (c) Interpret your answer from part (b).

134. **Interest** If the annual interest rate r is expressed as a decimal and N dollars are deposited into an account, then the amount in the account after two years is $N(1 + r)^2$.

  (a) Multiply this expression.

  (b) Let $r = 0.10$ and $N = 200$. Use both the given expression and the expression you obtained in part (a) to find the amount of money in the account after 2 years. Do your answers agree?

135. *Water Usage* (Refer to the Real-World Connection preceding Example 13.)
(a) Interpret the numbers -24 and 1900 in the formula $G(x) = -24x + 1900$.
(b) Let $P(x) = 2.6x + 230$. Find $T(x) = G(x)P(x)$.
(c) Evaluate $T(30)$ and interpret your answer.

136. *Numbers* Write a polynomial that represents the product of two consecutive even integers, where x is the greater even integer. Multiply your answer.

137. *Numbers* Write a polynomial that represents the product of two consecutive integers, where x is the smaller integer. Multiply your answer.

138. *Revenue* Write a polynomial that represents the revenue received from selling $2y + 1$ items at a price of x dollars each.

139. *Rectangular Pen* A rectangular pen has a perimeter of 100 feet. If x represents its width, write a polynomial that gives the area of the pen in terms of x.

140. *Numbers* Find an expression that represents the product of the sum and difference of two numbers. Let the two numbers be x and y.

WRITING ABOUT MATHEMATICS

141. Suppose that a student insists that
$$(x + 5)(x + 5)$$
equals $x^2 + 25$. Explain how you could convince the student otherwise.

142. Explain how to multiply two polynomials. Give an example of how your method works. Does your method give the correct result for any two polynomials?

SECTIONS 5.1 and 5.2 | **Checking Basic Concepts**

1. Combine like terms.
(a) $8x^2 + 4x - 5x^2 + 3x$
(b) $(5x^2 - 3x + 2) - (3x^2 - 5x^3 + 1)$

2. Write a polynomial that represents the revenue from ticket sales if $x + 120$ tickets are sold for x dollars each. Evaluate the polynomial for $x = 10$ and interpret the result.

3. An athlete's heart rate t minutes after exercise is stopped is modeled by
$$f(t) = 2t^2 - 25t + 160,$$
where $0 \le t \le 6$.
(a) What was the initial heart rate?
(b) What was the heart rate after 4 minutes?

4. Multiply the expressions.
(a) $-5(x - 6)$
(b) $4x^3(3x^2 - 5x)$
(c) $(2x - 1)(x + 3)$

5. Multiply the following special products.
(a) $(5x - 6)(5x + 6)$
(b) $(3x - 4)^2$

6. Make a sketch of a rectangle whose area illustrates that $(x + 2)(x + 4) = x^2 + 6x + 8$.

5.3 Factoring Polynomials

Common Factors • Factoring and Equations • Factoring by Grouping

A LOOK INTO MATH ▶

Video games involving sports frequently have to show a ball traveling through the air. To make the motion look realistic, the game must solve an equation in real time to determine when the ball hits the ground. For example, suppose that a golf ball is hit with an *upward* velocity of 88 feet per second, or 60 miles per hour. Then its height h in feet above the ground after t seconds is modeled by the *quadratic function* $h(t) = 88t - 16t^2$. To estimate when the ball strikes the ground, we can solve the *quadratic equation*

$$88t - 16t^2 = 0. \qquad \text{At ground level, } h = 0.$$

One method for solving this equation is factoring (see Example 6). In this section we discuss factoring and how it is used to solve equations.

NEW VOCABULARY

☐ Greatest common
   factor (GCF)
☐ Zero-product property
☐ Zero
☐ Expressions/equations
☐ Parabola

Common Factors

When factoring a polynomial, we first look for factors that are common to each term. By applying a distributive property, we can write a polynomial as two factors. For example, each term in $2x^2 + 4x$ contains a factor of $2x$.

$$2x^2 = 2x \cdot x$$
$$4x = 2x \cdot 2$$

Thus the polynomial $2x^2 + 4x$ can be factored as follows.

$$2x^2 + 4x = 2x(x + 2)$$

Common factor

EXAMPLE 1 **Finding common factors**

Factor.
(a) $4x^2 + 5x$ **(b)** $12x^3 - 4x^2$ **(c)** $6z^3 - 2z^2 + 4z$ **(d)** $4x^3y^2 + x^2y^3$

Solution
(a) Both $4x^2$ and $5x$ contain a common factor of x. That is,

$$4x^2 = x \cdot 4x \quad \text{and} \quad 5x = x \cdot 5.$$

Thus $4x^2 + 5x = x(4x + 5)$.
(b) Both $12x^3$ and $4x^2$ contain a common factor of $4x^2$. That is,

$$12x^3 = 4x^2 \cdot 3x \quad \text{and} \quad 4x^2 = 4x^2 \cdot 1.$$

Thus $12x^3 - 4x^2 = 4x^2(3x - 1)$.
(c) Each of the terms $6z^3, 2z^2$, and $4z$ contains a common factor of $2z$. That is,

$$6z^3 = 2z \cdot 3z^2, \quad 2z^2 = 2z \cdot z, \quad \text{and} \quad 4z = 2z \cdot 2.$$

Thus $6z^3 - 2z^2 + 4z = 2z(3z^2 - z + 2)$.
(d) Both $4x^3y^2$ and x^2y^3 contain a common factor of x^2y^2. That is,

$$4x^3y^2 = x^2y^2 \cdot 4x \quad \text{and} \quad x^2y^3 = x^2y^2 \cdot y.$$

Thus $4x^3y^2 + x^2y^3 = x^2y^2(4x + y)$.

Now Try Exercises 9, 17, 21, 27

Many times we factor out the *greatest common factor*. For example, the polynomial $15x^4 - 5x^2$ has a common factor of **$5x$**. We could write this polynomial as

$$15x^4 - 5x^2 = 5x(3x^3 - x).$$

However, we can also factor out **$5x^2$** to obtain

$$15x^4 - 5x^2 = 5x^2(3x^2 - 1).$$

Because $5x^2$ is the common factor with the greatest degree and largest (integer) coefficient, we say that $5x^2$ is the **greatest common factor** (GCF) of $15x^4 - 5x^2$. In Example 1, we factored out the GCF in each case.

EXAMPLE 2 **Factoring greatest common factors**

Factor.
(a) $24x^5 + 12x^3 - 6x^2$ **(b)** $6m^3n^2 - 3mn^2 + 9m$ **(c)** $-9x^3 + 6x^2 - 3x$

Solution
(a) The GCF of $24x^5$, $12x^3$, and $6x^2$ is $6x^2$.

$$24x^5 = 6x^2 \cdot 4x^3, \quad 12x^3 = 6x^2 \cdot 2x, \quad \text{and} \quad 6x^2 = 6x^2 \cdot 1$$

Thus $24x^5 + 12x^3 - 6x^2 = 6x^2(4x^3 + 2x - 1)$.
(b) The GCF of $6m^3n^2$, $3mn^2$, and $9m$ is $3m$.

$$6m^3n^2 = 3m \cdot 2m^2n^2, \quad 3mn^2 = 3m \cdot n^2, \quad \text{and} \quad 9m = 3m \cdot 3$$

Thus $6m^3n^2 - 3mn^2 + 9m = 3m(2m^2n^2 - n^2 + 3)$.
(c) Rather than factoring out $3x$, we can also factor out $-3x$ and make the leading coefficient of the remaining expression positive.

$$-9x^3 = -3x \cdot 3x^2, \quad 6x^2 = -3x \cdot -2x, \quad \text{and} \quad -3x = -3x \cdot 1$$

Thus $-9x^3 + 6x^2 - 3x = -3x(3x^2 - 2x + 1)$.

Now Try Exercises 25, 33, 39

Factoring and Equations

To solve equations by using factoring, we use the **zero-product property**. It states that if the product of two numbers is 0, then at least one of the numbers must equal 0.

ZERO-PRODUCT PROPERTY

For all real numbers a and b, if $ab = 0$, then $a = 0$ or $b = 0$ (or both).

NOTE: The zero-product property works only for 0. If $ab = 1$, then it does *not* follow that $a = 1$ or $b = 1$. For example, $a = \frac{1}{3}$ and $b = 3$ also satisfy the equation $ab = 1$.

Sometimes factoring needs to be performed on an equation before the zero-product property can be applied. For example, the left side of the equation

$$2x^2 + 4x = 0$$

may be factored to obtain

$$2x(x + 2) = 0.$$

Note that $2x$ times $(x + 2)$ equals 0. By the zero-product property, we must have either

$$2x = 0 \quad \text{or} \quad x + 2 = 0.$$

Solving each equation for x gives

$$x = \mathbf{0} \quad \text{or} \quad x = -\mathbf{2}.$$

The x-values of $\mathbf{0}$ and $-\mathbf{2}$ are called **zeros** of the polynomial $2x^2 + 4x$ because when they are substituted in $2x^2 + 4x$, the result is $\mathbf{0}$. That is, $2(\mathbf{0})^2 + 4(\mathbf{0}) = \mathbf{0}$ and $2(-\mathbf{2})^2 + 4(-\mathbf{2}) = \mathbf{0}$.

READING CHECK

What are the differences between expressions and equations?

MAKING CONNECTIONS

Expressions and Equations

There are important distinctions between expressions and equations. We often *factor expressions* and *solve equations*. For example, when we solved the *equation* $2x^2 + 4x = 0$, we started by factoring the *expression* $2x^2 + 4x$, as $2x(x + 2)$. The equivalent equation, $2x(x + 2) = 0$, resulted. Then the zero-product property was used to obtain the *solutions to the equation*: 0 and -2.

An equation is a statement that two expressions are equal. Equations *always* have an equals sign. We often solve equations, but we do *not* solve expressions.

EXAMPLE 3 **Applying the zero-product property**

Solve.
(a) $x(x - 1) = 0$ **(b)** $2x(x + 3) = 0$ **(c)** $(2x - 1)(3x + 2) = 0$

Solution
(a) The expression $x(x - 1)$ is the product of x and $(x - 1)$. By the zero-product property, either $x = 0$ or $x - 1 = 0$. The solutions are 0 and 1.

(b)

$2x(x + 3) = 0$		Given equation
$2x = 0$ or $x + 3 = 0$		Zero-product property
$x = 0$ or $x = -3$		Solve each equation.

The solutions are 0 and -3.

(c)

$(2x - 1)(3x + 2) = 0$		Given equation
$2x - 1 = 0$ or $3x + 2 = 0$		Zero-product property
$2x = 1$ or $3x = -2$		Add 1; subtract 2.
$x = \dfrac{1}{2}$ or $x = -\dfrac{2}{3}$		Divide by 2; divide by 3.

The solutions are $\frac{1}{2}$ and $-\frac{2}{3}$.

▐ **Now Try Exercises 45, 49**

EXAMPLE 4 **Solving polynomial equations with factoring**

Solve each polynomial equation.
(a) $x^2 + 3x = 0$ **(b)** $4x^2 = 16x$ **(c)** $2x^3 + 2x^2 = 0$

Solution
(a) We begin by factoring out the greatest common factor x.

$x^2 + 3x = 0$	Given equation
$x(x + 3) = 0$	Factor out x.
$x = 0$ or $x + 3 = 0$	Zero-product property
$x = 0$ or $x = -3$	Solve.

The solutions are 0 and -3.

(b) Write the equation $4x^2 = 16x$ so that there is a 0 on its right side before applying the zero-product property.

$$4x^2 = 16x \qquad \text{Given equation}$$
$$4x^2 - 16x = 0 \qquad \text{Subtract } 16x \text{ from each side.}$$
$$4x(x - 4) = 0 \qquad \text{Factor out } 4x.$$
$$4x = 0 \quad \text{or} \quad x - 4 = 0 \qquad \text{Zero-product property}$$
$$x = 0 \quad \text{or} \quad x = 4 \qquad \text{Solve.}$$

The solutions are 0 and 4.

(c) Start by factoring out the GCF of $2x^2$.

$$2x^3 + 2x^2 = 0 \qquad \text{Given equation}$$
$$2x^2(x + 1) = 0 \qquad \text{Factor out } 2x^2.$$
$$2x^2 = 0 \quad \text{or} \quad x + 1 = 0 \qquad \text{Zero-product property}$$
$$x = 0 \quad \text{or} \quad x = -1 \qquad \text{Solve.}$$

The solutions are -1 and 0.

Now Try Exercises 67, 71, 75

Polynomial equations can also be solved numerically and graphically. For example, to find a graphical solution to $x^2 - 2x = 0$, graph $y = x^2 - 2x$ by making a table of values, as shown in Table 5.6. When the points are plotted and connected, a $\cup$-shaped curve results, with x-intercepts 0 and 2. This curve shown in Figure 5.17 is called a *parabola*.

NOTE: The zeros, **0** and **2**, of $x^2 - 2x$ in Table 5.6; the x-intercepts, **0** and **2**, in Figure 5.17; and the solutions, **0** and **2**, to the equation $x^2 - 2x = 0$ are *identical*.

Numerical Solution

TABLE 5.6

$y = x^2 - 2x$

x	y
-1	3
0	**0**
1	-1
2	**0**
3	3

Zeros: 0, 2

Graphical Solution

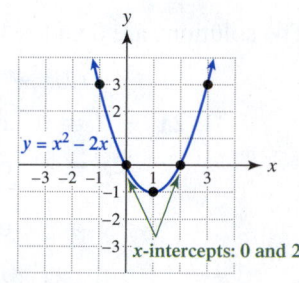

$y = x^2 - 2x$

x-intercepts: 0 and 2

Figure 5.17

Symbolic Solution

$$x^2 - 2x = 0$$
$$x(x - 2) = 0$$
$$x = 0 \quad \text{or} \quad x - 2 = 0$$
$$x = 0 \quad \text{or} \quad x = 2$$

Solutions: 0, 2

These concepts are generalized in the following.

MAKING CONNECTIONS

Zeros, x-Intercepts, and Solutions

The following statements are *equivalent*, where k is a real number and $P(x)$ is a polynomial.

1. A *zero* of a polynomial $P(x)$ is k.
2. An *x-intercept* on the graph of $y = P(x)$ is k.
3. A *solution* to the equation $P(x) = 0$ is k.

See the Making Connections after the solution to Example 5.

In the next example we solve an equation numerically, graphically, and symbolically.

EXAMPLE 5 **Solving an equation**

Solve the equation $4x - x^2 = 0$ numerically, graphically, and symbolically.

Solution

Numerical Solution Make a table of values for $y = 4x - x^2$, as shown in Table 5.7. Note that 0 and 4 are zeros of $4x - x^2$ and solutions to $4x - x^2 = 0$, because $4(0) - 0^2 = 0$ and $4(4) - 4^2 = 0$.

Graphical Solution Plot the points given in Table 5.7 and sketch a curve through them, as shown in Figure 5.18. The curve is a $\cap$-shaped graph, or parabola, with x-intercepts 0 and 4, which are the solutions to the given equation.

Numerical Solution   Graphical Solution

TABLE 5.7

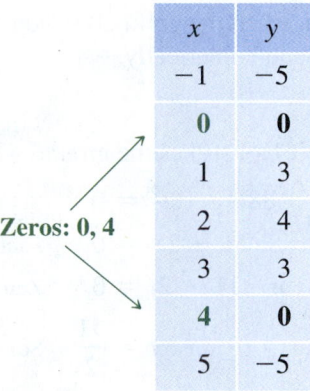

x	y
-1	-5
0	**0**
1	3
2	4
3	3
4	0
5	-5

Zeros: 0, 4

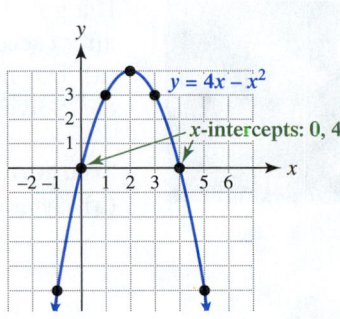

Figure 5.18

Symbolic Solution Start by factoring the left side of the equation.

$$4x - x^2 = 0 \qquad \text{Given equation}$$
$$x(4 - x) = 0 \qquad \text{Factor out } x.$$
$$x = 0 \quad \text{or} \quad 4 - x = 0 \qquad \text{Zero-product property}$$
$$x = 0 \quad \text{or} \quad x = 4 \qquad \text{Solve each equation.}$$

Solutions: 0, 4

NOTE: The numerical and graphical solutions agree with the symbolic solutions.

Now Try Exercise 77

MAKING CONNECTIONS

1. The *zeros* of $4x - x^2$ are 0 and 4, as shown in Table 5.7.
2. The *x-intercepts* for $y = 4x - x^2$ are 0 and 4, as shown in Figure 5.18.
3. The *solutions* to $4x - x^2 = 0$ are 0 and 4, as shown in the *Symbolic Solution*.

GRAPH OF $y = ax^2 + bx + c$

The graph of $y = ax^2 + bx + c$ is a **parabola**. It is a $\cup$-shaped graph if $a > 0$ and a $\cap$-shaped graph if $a < 0$. This graph can have zero, one, or two x-intercepts.

CALCULATOR HELP

To find an x-intercept or zero, see Appendix A (page AP-10).

TECHNOLOGY NOTE:

Locating x-intercepts
Many calculators have the capability to locate an x-intercept or zero. This method is illustrated in the accompanying figures, where the solution 2 to the equation $x^2 - 4 = 0$ is found.

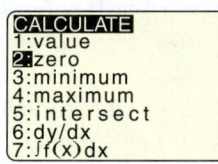 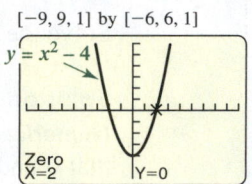

In the next example we use factoring to estimate when a golf ball strikes the ground.

EXAMPLE 6

Modeling the flight of a golf ball

If a golf ball is hit upward at 88 feet per second, or 60 miles per hour, its height h in feet after t seconds is modeled by $h(t) = 88t - 16t^2$.
(a) Use factoring to determine when the golf ball strikes the ground.
(b) Solve part (a) graphically and numerically.

Solution
(a) *Symbolic Solution* The golf ball strikes the ground when its height is 0.

$$88t - 16t^2 = 0 \qquad \text{Set the formula equal to 0.}$$

$$8t(11 - 2t) = 0 \qquad \text{Factor out } 8t.$$

$$8t = 0 \quad \text{or} \quad 11 - 2t = 0 \qquad \text{Zero-product property}$$

$$t = 0 \quad \text{or} \quad t = \frac{11}{2} \qquad \text{Solve for } t.$$

The ball strikes the ground after $\frac{11}{2}$, or 5.5 seconds. The solution of $t = 0$ is not used in this problem because it corresponds to when the ball is initially hit from the ground.

(b) *Graphical and Numerical Solutions* Graph $y_1 = 88x - 16x^2$. Using a graphing calculator to find an x-intercept (or zero) yields the solution, 5.5. See Figure 5.19(a). Numerical support is shown in Figure 5.19(b), where $y_1 = 0$ when $x = 5.5$.

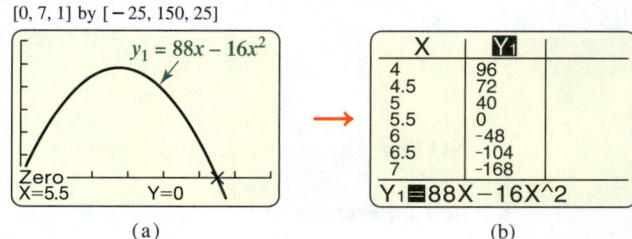

Figure 5.19

Now Try Exercise 101

Factoring by Grouping

Factoring by grouping is a technique that makes use of the associative and distributive properties. The next example illustrates the first step in this factoring technique.

EXAMPLE 7

Factoring out binomials

Factor.
(a) $2x(x + 1) + 3(x + 1)$ **(b)** $2x^2(3x - 2) - x(3x - 2)$

Solution

(a) Both terms in the expression $2x(x + 1) + 3(x + 1)$ contain the binomial $(x + 1)$. Therefore the distributive property can be used to factor out this expression.

$$2x(x + 1) + 3(x + 1) = (2x + 3)(x + 1)$$

(b) Both terms in the expression $2x^2(3x - 2) - x(3x - 2)$ contain the binomial $(3x - 2)$. Therefore the distributive property can be used to factor this expression.

$$2x^2(3x - 2) - x(3x - 2) = (2x^2 - x)(3x - 2)$$

Now Try Exercises 81, 83

Now consider the polynomial

$$3t^3 + 6t^2 + 2t + 4.$$

We can factor this polynomial by first grouping it into two binomials.

$(3t^3 + 6t^2) + (2t + 4)$ Associative property

$3t^2(t + 2) + 2(t + 2)$ Factor out common factors.

$(3t^2 + 2)(t + 2)$ Factor out $(t + 2)$.

The following steps summarize factoring a polynomial with four terms by grouping.

FACTORING BY GROUPING

STEP 1: Use parentheses to group the terms into binomials with common factors. Begin by writing the expression with a plus sign between the binomials.

STEP 2: Factor out the common factor in each binomial.

STEP 3: Factor out the common binomial. If there is no common binomial, try a different grouping.

These steps are used in Example 8.

EXAMPLE 8   **Factoring by grouping**

Factor each polynomial.
(a) $x^3 - 2x^2 + 3x - 6$ (b) $12x^3 - 9x^2 - 8x + 6$ (c) $2x - 2y + ax - ay$

Solution

(a) $x^3 - 2x^2 + 3x - 6 = (x^3 - 2x^2) + (3x - 6)$ Group terms.

$= x^2(x - 2) + 3(x - 2)$ Factor out x^2 and 3.

$= (x^2 + 3)(x - 2)$ Factor out $x - 2$.

(b) $12x^3 - 9x^2 - 8x + 6 = (12x^3 - 9x^2) + (-8x + 6)$ Write with a plus sign between binomials.

$= 3x^2(4x - 3) - 2(4x - 3)$ Factor out $3x^2$ and -2.

$= (3x^2 - 2)(4x - 3)$ Factor out $4x - 3$.

(c) $2x - 2y + ax - ay = (2x - 2y) + (ax - ay)$ Group terms.

$= 2(x - y) + a(x - y)$ Factor out 2 and a.

$= (2 + a)(x - y)$ Factor out $x - y$.

Now Try Exercises 85, 93, 95

CRITICAL THINKING

Solve
$xy - 4x - 3y + 12 = 0.$

5.3 Putting It All Together

CONCEPT	EXPLANATION	EXAMPLES
Greatest Common Factor (GCF)	We can sometimes factor the greatest common factor out of a polynomial. The GCF has the largest (integer) coefficient and greatest degree possible.	$3x^4 - 9x^3 + 12x^2 = 3x^2(x^2 - 3x + 4)$ The terms in the expression $(x^2 - 3x + 4)$ have no common factor, so $3x^2$ is called the *greatest common factor* of $3x^4 - 9x^3 + 12x^2$.
Zero-Product Property	If the product of two numbers is 0, then at least one of the numbers equals 0.	$xy = 0$ implies that $x = 0$ or $y = 0$. $x(2x + 1) = 0$ implies that $x = 0$ or $2x + 1 = 0$.
Factoring and Equations	Factoring may be used to solve equations.	$$6x^2 - 9x = 0$$ $$3x(2x - 3) = 0 \quad \text{Common factor, } 3x$$ $$3x = 0 \quad \text{or} \quad 2x - 3 = 0 \quad \text{Zero-product property}$$ $$x = 0 \quad \text{or} \quad x = \tfrac{3}{2} \quad \text{Solve for } x.$$
Factoring by Grouping	Factoring by grouping is a method that can be used to factor four terms into a product of two binomials. It involves the associative and distributive properties.	$$4x^3 + 6x^2 + 10x + 15 = (4x^3 + 6x^2) + (10x + 15)$$ $$= 2x^2(2x + 3) + 5(2x + 3)$$ $$= (2x^2 + 5)(2x + 3)$$
Zeros, x-Intercepts, and Solutions	An x-intercept for the graph of $y = P(x)$ corresponds to a zero of $P(x)$ and to a solution to the equation $P(x) = 0$. These three concepts are closely connected.	*x*-Intercepts: **0, 1** Zeros of $x^2 - x$: **0, 1** Solutions to $x^2 - x = 0$: **0, 1**

5.3 Exercises

MyMathLab   Math XL PRACTICE    WATCH   DOWNLOAD   READ   REVIEW

CONCEPTS AND VOCABULARY

1. Give one reason for factoring expressions.

2. If $x(x - 3) = 0$, what can be said about either x or $(x - 3)$? What property did you use?

3. Is $2x$ a common factor of $(4x^3 - 12x^2)$? Explain.

4. Is $2x$ the GCF of $(4x^3 - 12x^2)$? Explain.

5. If $ab = 2$, must either $a = 2$ or $b = 2$? Explain.

6. If $xy = 0$, must either $x = 0$ or $y = 0$? Explain.

7. Which of the following (a.–d.) is a factorization of $4x^3 - 6x^2$?
  a. $2x(2x - 3)$ **b.** $4x(x^2 - 2x)$
  c. $2x^2(2x + 3)$ **d.** $2x^2(2x - 3)$

8. Which of the following (a.–d.) is a factorization of $36a^2b^3 - 24a^3b^2$?
  a. $12ab(3b - 2a)$ **b.** $12a^2b^2(3b - 2a)$
  c. $6a^2b(6b - 4a)$ **d.** $12a^2b^2(3b + 2a)$

FACTORING AND EQUATIONS

Exercises 9–36: Factor out the greatest common factor.

9. $10x - 15$ **10.** $32 - 16x$

11. $4x + 6y$ **12.** $50a + 20b$

13. $9r - 15t$ **14.** $16m - 24n$

15. $2x^3 - 5x$ **16.** $3y - 9y^2$

17. $8a^3 + 10a$ **18.** $20b^3 + 25b^2$

19. $6r^3 - 18r^5$ **20.** $7n^2 - 21n^4$

21. $8x^3 - 4x^2 + 16x$ **22.** $5x^3 - x^2 + 4x$

23. $9n^4 - 6n^2 + 3n$ **24.** $5n^4 + 10n^2 - 25n$

25. $6t^6 - 4t^4 + 2t^2$ **26.** $15t^6 + 25t^4 - 20t^2$

27. $5x^2y^2 - 15x^2y^3$ **28.** $21xy + 14x^3y^3$

29. $6a^3b^2 - 15a^2b^3$ **30.** $45a^2b + 30a^3b^2$

31. $18mn^2 + 12m^2n^3$ **32.** $24m^2n^3 - 36m^3n^2$

33. $15x^2y + 10xy - 25x^2y^2$

34. $14a^3b^2 - 21a^2b^2 + 35a^2b$

35. $4a^2 - 2ab + 6ab^2$

36. $5a^2 + 10a^2b^2 - 15ab$

Exercises 37–42: (Refer to Example 2(c).) Factor out the negative of the greatest common factor.

37. $-2x^2 + 4x - 6$ **38.** $-7x^5 - 21x^3 - 14x^2$

39. $-8z^4 - 16z^3$ **40.** $-8z^5 - 24z^4$

41. $-4m^2n^3 - 6mn^2 - 8mn$

42. $-13m^4n^4 - 13m^3n^3 + 26m^2n^2$

Exercises 43–52: Use the zero-product property to solve the equation.

43. $mn = 0$ **44.** $xyz = 0$

45. $3z(z + 4) = 0$ **46.** $2z(z - 1) = 0$

47. $(r - 1)(r + 3) = 0$ **48.** $(2r + 3)(r - 5) = 0$

49. $(x + 2)(3x - 1) = 0$ **50.** $(4y - 3)(2y + 1) = 0$

51. $3x(y - 6) = 0$ **52.** $7m(3n + 1) = 0$

Exercises 53–58: Use the graph of $y = P(x)$ to do the following.

  (a) *Find the x-intercepts.*
  (b) *Solve the equation $P(x) = 0$.*
  (c) *Find the zeros of $P(x)$.*

53. 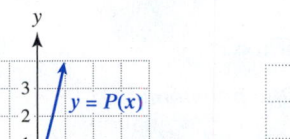   **54.**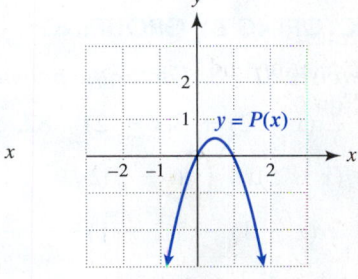

55. 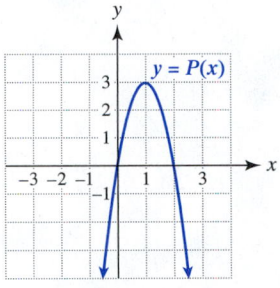   **56.**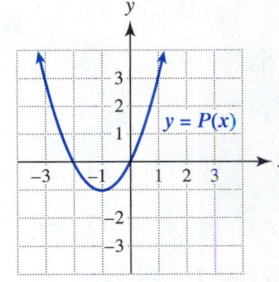

57. 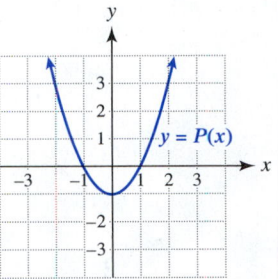   **58.**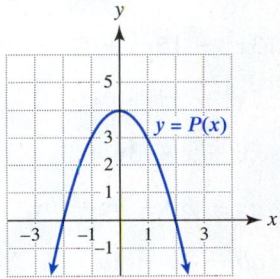

Exercises 59–64: Solve the equation graphically. Check your answers.

59. $x^2 - 2x = 0$ **60.** $x^2 + x = 0$

61. $x - x^2 = 0$ **62.** $3x - x^2 = 0$

63. $x^2 + 4x = 0$ **64.** $x^2 - 4x = 0$

Exercises 65–76: Solve the equation.

65. $x^2 - x = 0$ **66.** $4x - 2x^2 = 0$

67. $5x^2 - x = 0$ **68.** $4x^2 + 3x = 0$

69. $10x^2 + 5x = 0$

70. $6x - 12x^2 = 0$

71. $15x^2 = 10x$

72. $4x = 8x^2$

73. $25x = 10x^2$

74. $34x^2 = 51x$

75. $32x^4 - 16x^3 = 0$

76. $45x^4 - 30x^3 = 0$

 Exercises 77–80: Solve the equation numerically, graphically, and symbolically.

77. $x^2 + 2x = 0$

78. $x^2 + 3x = 0$

79. $2x^2 - 3x = 0$

80. $2x - 3x^2 = 0$

FACTORING BY GROUPING

Exercises 81–98: Factor the polynomial.

81. $2x(x + 2) + 3(x + 2)$ **82.** $5(x - 1) - 2x(x - 1)$

83. $(x - 5)x^2 - (x - 5)2$

84. $7x(x - 1) - 3(x - 1)$

85. $x^3 + 3x^2 + 2x + 6$

86. $4x^3 + 3x^2 + 8x + 6$

87. $6x^3 - 4x^2 + 9x - 6$

88. $x^3 - 3x^2 - 5x + 15$

89. $2x^3 - 3x^2 + 2x - 3$

90. $8x^3 - 2x^2 + 12x - 3$

91. $x^3 - 7x^2 - 3x + 21$

92. $6x^3 - 15x^2 - 4x + 10$

93. $3x^3 - 15x^2 + 5x - 25$

94. $2x^4 - x^3 + 4x - 2$

95. $xy + x + 3y + 3$

96. $ax + bx - ay - by$

97. $ab - 3a + 2b - 6$

98. $2ax - 6bx - ay + 3by$

99. *Graphical Interpretation* Which graph might represent the height h of a golf ball in flight after t seconds? Which graph might represent the height h of a yo-yo above the ground after t seconds?

a. h

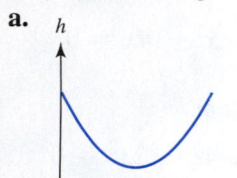

b. h

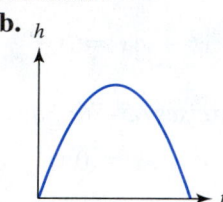

100. **Thinking Generally** The zero-product property states $ab = 0$ implies that either $a = 0$ or $b = 0$.
(a) Why isn't there a "five-product" property?
(b) Solve $ax^2 - bx = 0$, where $a \neq 0$.

APPLICATIONS

101. *Flight of a Golf Ball* (Refer to Example 6.) If a golf ball is hit with an upward velocity of 128 feet per second (about 87 mph), its height h in feet above the ground can be modeled by

$$h(t) = -16t^2 + 128t,$$

where t is in seconds.
(a) Use factoring to determine when the golf ball strikes the ground.
(b) Use the graph of $y = h(t)$ to estimate when the golf ball strikes the ground.

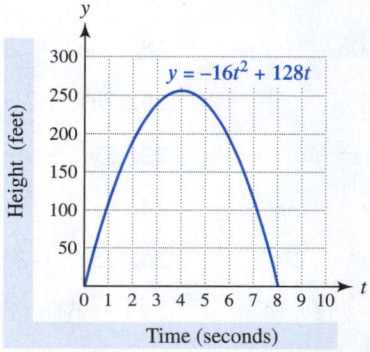

(c) Make a table of h, starting at $t = 3$ and incrementing by 1. Does the table support your answer in part (a)?
(d) Use the table and graph of h to determine the maximum height of the golf ball. At what value of t does this occur?

102. *Flight of a Golf Ball* The height h in feet reached by a golf ball after t seconds is given by

$$h(t) = -16t^2 + 96t.$$

After how many seconds does the golf ball strike the ground?

 Exercises 103 and 104: *Climate Change* *Atmospheric carbon dioxide (CO_2) in parts per million (ppm) from 1958 to 2010 is modeled by $f(x) = 0.0148x^2 + 0.686x + 315$, where x represents years after 1958. Do the following.* (Source: The Honolulu Advertiser.)

103. Evaluate $f(52)$ and interpret the result.

104. Solve the equation $f(x) = 340$ graphically. Interpret the result.

105. *Geometry* The area of a rectangle with width y and length 8 is $A = 8y$. A circle with radius y has area $A = \pi y^2$. For what positive value of y are these two areas equal?

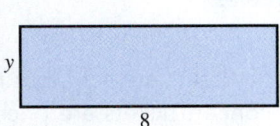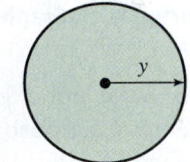

106. *Mobile Data Subscribers* The number of mobile data subscribers S in millions x years after 2000 is approximated by $S(x) = 1.368x^2 + 6x + 19$. *(Source: Business Insider.)*

(a) Evaluate $S(9)$ and interpret your answer.

(b) Estimate graphically when there were 100 million subscribers.

107. *Space Shuttle Altitude* Approximate altitudes y in feet reached by the space shuttle *Endeavour*, t seconds after liftoff, are shown in the table.

t (seconds)	20	30	40
y (feet)	4500	10,000	18,000

t (seconds)	50	60
y (feet)	28,000	40,000

Source: NASA.

(a) The function $f(t) = 11t^2 + 6t$ models the data, where $y = f(t)$. Make a table of f, starting at $t = 20$ and incrementing by 10. Comment on how well f models the data.

(b) Predict the altitude reached by the shuttle 70 seconds after liftoff. Round your answer to the nearest thousand feet.

(c) Solve the equation $11t^2 + 6t = 0$. Do your results have any physical meaning? Explain.

108. *Average Temperature* Suppose that the monthly average temperatures T in degrees Fahrenheit for a city are given by $T(x) = 24x - 2x^2$, where $x = 1$ corresponds to January, $x = 2$ to February, and so on.

(a) What is the average temperature in October?

(b) Use factoring to solve the quadratic equation $24x - 2x^2 = 0$. Do your results have any physical meaning?

(c) Support your results in part (b) either graphically or numerically.

109. *Area* Suppose that the area of a rectangular room is 216 square feet. Can you find the dimensions of the room from this information? Explain.

110. *Area* Suppose that the area of a rectangular room is $4x^2 + 8x$ square feet. If the length of the room is $4x$, is it possible to determine the width of the room in terms of x? Explain.

111. **Online Exploration** Go online and search for images of parabolas. Using these images, discuss two real-world applications of parabolas.

112. **Online Exploration** Go online and search for images of cubic polynomials. Using these images, discuss one real-world situation in which cubic polynomials occur.

WRITING ABOUT MATHEMATICS

113. A student solves the following equation *incorrectly*.

$$x^2 - 5x \overset{?}{=} 1$$
$$x(x - 5) \overset{?}{=} 1$$
$$x \overset{?}{=} 1 \quad \text{or} \quad x - 5 \overset{?}{=} 1$$
$$x \overset{?}{=} 1 \quad \text{or} \quad x \overset{?}{=} 6$$

What is the student's mistake? Explain.

114. Explain how you would solve $ax^2 + bx = 0$ symbolically and graphically.

5.4 Factoring Trinomials

Factoring $x^2 + bx + c$ • **Factoring Trinomials by Grouping** • **Factoring Trinomials with FOIL** • **Factoring with Graphs and Tables**

A LOOK INTO MATH ▶ Items usually sell better at a lower price. At a higher price fewer items are sold, but more money is made on each item. For example, suppose that if concert tickets are priced at $100 no one will buy a ticket, but for each $1 reduction in price 100 additional tickets will be sold. If the promoters of the concert need to gross $240,000 from ticket sales, what ticket price accomplishes this goal? To solve this problem we need to set up and solve a polynomial equation. In this section we describe how to solve polynomial equations by factoring trinomials in the form $x^2 + bx + c$ and $ax^2 + bx + c$. See Example 3.

Factoring $x^2 + bx + c$

NEW VOCABULARY

☐ Symbolic factoring
☐ Numerical factoring
☐ Graphical factoring

The product $(x + 3)(x + 4)$ can be found as follows.

$$(x + 3)(x + 4) = x^2 + 4x + 3x + 12 \quad \text{Multiplication}$$
$$= x^2 + 7x + 12$$

The middle term $7x$ is found by calculating the sum $4x + 3x$, and the last term is found by calculating the product $3 \cdot 4 = 12$.

When we factor polynomials, we are *reversing* the process of multiplication. To factor $x^2 + 7x + 12$ we must find m and n that satisfy

$$x^2 + 7x + 12 = (x + m)(x + n).$$

Because

$$(x + m)(x + n) = x^2 + (m + n)x + mn,$$

it follows that $mn = 12$ and $m + n = 7$. To determine m and n we list factors of 12 and their sum, as shown in Table 5.8.

Because $3 \cdot 4 = 12$ and $3 + 4 = 7$, we can write the factored form as

$$x^2 + 7x + 12 = (x + 3)(x + 4). \quad \text{Factoring}$$

This result can always be checked by multiplying the two binomials.

$$(x + 3)(x + 4) = x^2 + 7x + 12$$

$$7x \longleftarrow \text{The middle term checks. ✓}$$

TABLE 5.8
Factor Pairs for 12

Factors	Sum
1, 12	13
2, 6	8
3, 4	7

FACTORING $x^2 + bx + c$

To factor the trinomial $x^2 + bx + c$, find numbers m and n that satisfy

$$m \cdot n = c \quad \text{and} \quad m + n = b.$$

Then $x^2 + bx + c = (x + m)(x + n)$. Note that the coefficient of x^2 is 1.

Factoring ⟶

NOTE: $x^2 + 7x + 12 = (x + 3)(x + 4)$ Multiplying and factoring
⟵ Multiplying   are reverse operations.

EXAMPLE 1

Factoring the form $x^2 + bx + c$

Factor each trinomial.
(a) $x^2 + 10x + 16$ **(b)** $x^2 - 5x - 24$ **(c)** $x^2 + 7x - 30$

TABLE 5.9
Factor Pairs for 16

Factors	Sum
1, 16	17
2, 8	**10**
4, 4	8

Solution
(a) We need to find a factor pair for 16 whose sum is 10. From Table 5.9 the required factor pair is $m = \mathbf{2}$ and $n = \mathbf{8}$. Thus

$$x^2 + 10x + 16 = (x + \mathbf{2})(x + \mathbf{8}).$$

(b) Factors of -24 whose sum equals -5 are $\mathbf{3}$ and $\mathbf{-8}$. Thus

$$x^2 - 5x - 24 = (x + \mathbf{3})(x - \mathbf{8}).$$

(c) Factors of -30 whose sum equals 7 are $\mathbf{-3}$ and $\mathbf{10}$. Thus

$$x^2 + 7x - 30 = (x - \mathbf{3})(x + \mathbf{10}).$$

Now Try Exercises 17, 23, 25

EXAMPLE 2

Removing common factors first

Factor completely.
(a) $3x^2 + 15x + 18$ **(b)** $5x^3 + 5x^2 - 60x$

Solution
(a) If we first factor out the common factor of 3, the resulting trinomial is easier to factor.

$$3x^2 + 15x + 18 = 3(x^2 + \mathbf{5}x + \mathbf{6})$$

Now we find m and n such that $mn = \mathbf{6}$ and $m + n = \mathbf{5}$. These numbers are $\mathbf{2}$ and $\mathbf{3}$.

$$3x^2 + 15x + 18 = 3(x^2 + 5x + 6)$$
$$= 3(x + \mathbf{2})(x + \mathbf{3})$$

(b) First, we factor out the common factor of $5x$. Then we factor the resulting trinomial.

$$5x^3 + 5x^2 - 60x = 5x(x^2 + x - 12)$$
$$= 5x(x - 3)(x + 4)$$

Now Try Exercises 55, 67

▶ **REAL-WORLD CONNECTION** Factoring may be used to solve the problem presented in A Look Into Math for this section. To do so we let x be the number of dollars that the price of a ticket is reduced below \$100. Then $100 - x$ represents the price of a ticket. For each \$1 reduction in price, 100 additional tickets will be sold, so the number of tickets sold is given by $100x$. Because the number of tickets sold times the price of each ticket equals the total revenue, we need to solve

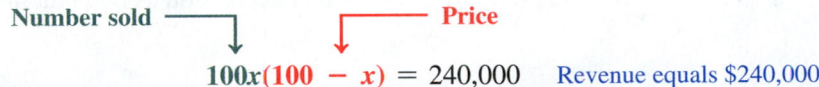

Number sold   Price

$$100x(\mathbf{100 - x}) = 240{,}000 \quad \text{Revenue equals \$240,000.}$$

to determine when gross sales will reach \$240,000. We cannot immediately apply the zero-product property because the right side of the equation is *not* 0.

EXAMPLE 3

Determining ticket price

Solve the equation $100x(100 - x) = 240{,}000$ symbolically. What should be the price of the tickets and how many tickets will be sold? Support your answer graphically.

CRITICAL THINKING

In Example 3 what price will maximize the revenue from ticket sales? (*Hint:* See Figure 5.20.)

Solution

Symbolic Solution Begin by applying the distributive property.

$$100x(100 - x) = 240,000 \qquad \text{Given equation}$$
$$10,000x - 100x^2 = 240,000 \qquad \text{Distributive property}$$
$$-100x^2 + 10,000x - 240,000 = 0 \qquad \text{Subtract 240,000; rewrite.}$$
$$-100(x^2 - 100x + 2400) = 0 \qquad \text{Factor out } -100.$$
$$-100(x - 40)(x - 60) = 0 \qquad \text{Factor.}$$
$$x = 40 \quad \text{or} \quad x = 60 \qquad \text{Solve.}$$

If $x = 40$, the price is $100 - 40 = \$60$ per ticket and $100 \cdot 40 = 4000$ tickets are sold at $60 each. If $x = 60$, the price is $100 - 60 = \$40$ and $100 \cdot 60 = 6000$ tickets are sold at $40 each. Either price gives the desired ticket revenue of $240,000.

Graphical Solution The graphs of $y_1 = 100x(100 - x)$ and $y_2 = 240,000$ intersect in Figure 5.20 at $x = 40$ and $x = 60$, which agrees with the symbolic solution.

CALCULATOR HELP

To find a point of intersection, see Appendix A (page AP-6).

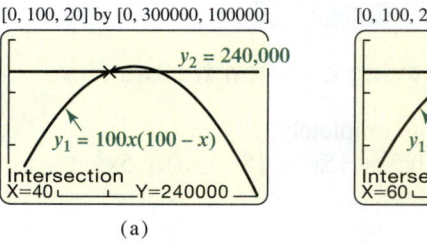
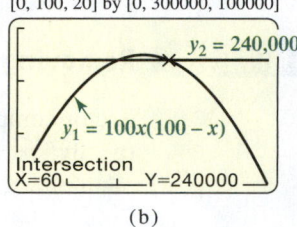

(a)   (b)

Figure 5.20

NOTE: These graphs also illustrate how revenue increases to a maximum value and then decreases as the price of the tickets decreases from $100 to $0.

▌ **Now Try Exercise 103**

Factoring Trinomials by Grouping

In this subsection we use grouping to factor trinomials $ax^2 + bx + c$ with $a \neq 1$.

> ### FACTORING $ax^2 + bx + c$ BY GROUPING
>
> To factor $ax^2 + bx + c$ perform the following steps. (Assume that a, b, and c have no factor in common.)
>
> 1. Find numbers m and n such that $mn = ac$ and $m + n = b$. (This step may require trial and error.)
> 2. Write the trinomial as $ax^2 + mx + nx + c$.
> 3. Use grouping to factor this expression as two binomials.

For example, one way to factor $3x^2 + 14x + 8$ is to find two numbers m and n such that $mn = 3 \cdot 8 = 24$ and $m + n = 14$. Because $2 \cdot 12 = 24$ and $2 + 12 = 14$, $m = 2$ and $n = 12$. Using grouping, we can now factor this trinomial.

$$3x^2 + 14x + 8 = 3x^2 + 2x + 12x + 8 \qquad \text{Write } 14x \text{ as } 2x + 12x.$$
$$= (3x^2 + 2x) + (12x + 8) \qquad \text{Associative property}$$
$$= x(3x + 2) + 4(3x + 2) \qquad \text{Factor out } x \text{ and } 4.$$
$$= (x + 4)(3x + 2) \qquad \text{Factor out } 3x + 2.$$

EXAMPLE 4 **Factoring $ax^2 + bx + c$ by grouping**

Factor each trinomial.
(a) $3x^2 + 17x + 10$ **(b)** $12y^2 + 5y - 3$ **(c)** $6r^2 - 19r + 10$

Solution

(a) In this trinomial $a = 3$, $b = 17$, and $c = 10$. Because $mn = ac$ and $m + n = b$, the numbers m and n satisfy $mn = 30$ and $m + n = 17$. Thus $m = \mathbf{2}$ and $n = \mathbf{15}$.

$$3x^2 + \mathbf{17}x + 10 = 3x^2 + \mathbf{2}x + \mathbf{15}x + 10 \qquad \text{Write } 17x \text{ as } 2x + 15x.$$
$$= (3x^2 + 2x) + (15x + 10) \qquad \text{Associative property}$$
$$= \mathbf{\textit{x}(3x + 2)} + \mathbf{5(3x + 2)} \qquad \text{Factor out } x \text{ and } 5.$$
$$= \mathbf{(\textit{x} + 5)(3x + 2)} \qquad \text{Distributive property}$$

(b) In this trinomial $a = 12$, $b = 5$, and $c = -3$. Because $mn = ac$ and $m + n = b$, the numbers m and n satisfy $mn = -36$ and $m + n = 5$. Thus $m = \mathbf{9}$ and $n = \mathbf{-4}$.

$$12y^2 + \mathbf{5}y - 3 = 12y^2 + \mathbf{9}y - \mathbf{4}y - 3 \qquad \text{Write } 5y \text{ as } 9y - 4y.$$
$$= (12y^2 + 9y) + (-4y - 3) \qquad \text{Associative property}$$

Be sure to factor out the GCF from each quantity. →
$$= \mathbf{3\textit{y}(4y + 3)} - \mathbf{1(4y + 3)} \qquad \text{Factor out } 3y \text{ and } -1.$$
$$= \mathbf{(3\textit{y} - 1)(4y + 3)} \qquad \text{Distributive property}$$

(c) In this trinomial $a = 6$, $b = -19$, and $c = 10$. Because $mn = ac$ and $m + n = b$, the numbers m and n satisfy $mn = 60$ and $m + n = -19$. Thus $m = \mathbf{-4}$ and $n = \mathbf{-15}$.

$$6r^2 - \mathbf{19}r + 10 = 6r^2 - \mathbf{4}r - \mathbf{15}r + 10 \qquad \text{Write } -19r \text{ as } -4r - 15r.$$
$$= (6r^2 - 4r) + (-15r + 10) \qquad \text{Associative property}$$
$$= \mathbf{2\textit{r}(3r - 2)} - \mathbf{5(3r - 2)} \qquad \text{Factor out } 2r \text{ and } -5.$$
$$= \mathbf{(2\textit{r} - 5)(3r - 2)} \qquad \text{Distributive property}$$

Now Try Exercises **35, 41, 43**

Factoring Trinomials with FOIL

An alternative to factoring trinomials by grouping is to use FOIL in reverse. For example, the factors of $3x^2 + 7x + 2$ are two binomials.

$$3x^2 + 7x + 2 \overset{?}{=} (\_\_\_ + \_\_\_)(\_\_\_ + \_\_\_)$$

The expressions to be placed in the four blanks are yet to be found. By the FOIL method, we know that the product of the first terms is $3x^2$. Because $3x^2 = 3x \cdot x$, we can write

$$3x^2 + 7x + 2 \overset{?}{=} (\,\underline{3x}\, + \_\_\_)(\,\underline{x}\, + \_\_\_).$$

The product of the last terms in each binomial must equal 2. Because $2 = \mathbf{1 \cdot 2}$, we can put the 1 and 2 in the blanks, but we must be sure to place them correctly so that the product of the *outside terms* plus the product of the *inside terms* equals $7x$.

$$(3x + 1)(x + 2) = 3x^2 + 7x + 2$$

$\lfloor 1x \rfloor$

$+6x$

$7x$ ← Middle term checks. ✔

If we had interchanged the 1 and 2, we would have obtained an incorrect result.

$$(3x + 2)(x + 1) = 3x^2 + 5x + 2$$

$\lfloor 2x \rfloor$

$+3x$

$5x$ ← Middle term is *not* $7x$. ✗

In the next example we factor expressions of the form $ax^2 + bx + c$, where $a \neq 1$. In this situation, we may need to *guess and check* or use *trial and error* a few times before finding the correct factors.

EXAMPLE 5 **Factoring the form $ax^2 + bx + c$**

Factor each trinomial.
(a) $2x^2 + 9x + 4$ (b) $6x^2 - x - 2$ (c) $4x^3 - 14x^2 + 6x$

Solution
(a) The factors of $2x^2$ are $2x$ and x, so we begin by writing

$$2x^2 + 9x + 4 \stackrel{?}{=} (2x + \_\_)(x + \_\_).$$

The factors of the last term, 4, are either 1 and 4 or 2 and 2. Selecting the factors 2 and 2 results in a middle term of $6x$ rather than $9x$.

$$(2x + 2)(x + 2) = 2x^2 + 6x + 4$$

Middle term is *not* $9x$. ✗

Next we try the factors 1 and 4.

$$(2x + 4)(x + 1) = 2x^2 + 6x + 4$$

Middle term is *not* $9x$. ✗

Again we obtain the wrong middle term. By interchanging the 1 and 4, we find the correct factorization.

$$(2x + 1)(x + 4) = 2x^2 + 9x + 4$$

Middle term is $9x$. ✓

(b) The factors of $6x^2$ are either $2x$ and $3x$ or $6x$ and x. The factors of -2 are either -1 and 2 or 1 and -2. To obtain a middle term of $-x$ we use the following factors.

$$(3x - 2)(2x + 1) = 6x^2 - x - 2$$

It checks. ✓

To find the correct factorization we may need to guess and check a few times.
(c) Each term contains a common factor of $2x$, so we do the following step first.

$$4x^3 - 14x^2 + 6x = 2x(2x^2 - 7x + 3)$$

Next we factor $2x^2 - 7x + 3$. The factors of $2x^2$ are $2x$ and x. Because the middle term is negative, we use -1 and -3 for factors of 3.

$$4x^3 - 14x^2 + 6x = 2x(2x^2 - 7x + 3)$$
$$= 2x(2x - 1)(x - 3)$$

Now Try Exercises 37, 39, 69

NOTE: If a trinomial does not have any common factors, then its binomial factors will not have any common factors either. Thus in Example 5(a), $(2x + 2)$ and $(2x + 4)$ cannot be factors of $2x^2 + 9x + 4$. These factors can be eliminated without checking.

EXAMPLE 6 **Estimating passing distance**

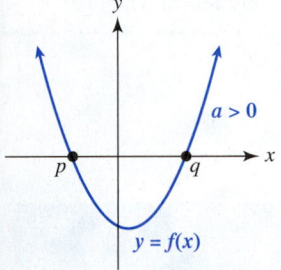

A car traveling 48 miles per hour accelerates at a constant rate to pass a car in front of it. A no-passing zone begins 2000 feet away. The distance d traveled in feet by the car after t seconds is $d(t) = 3t^2 + 70t$. How long does it take the car to reach the no-passing zone?

Solution
We must determine the time when $3t^2 + 70t = 2000$.

$$3t^2 + 70t = 2000 \qquad \text{Equation to be solved}$$
$$3t^2 + 70t - 2000 = 0 \qquad \text{Subtract 2000.}$$
$$(3t - 50)(t + 40) = 0 \qquad \text{Factor.}$$
$$3t - 50 = 0 \quad \text{or} \quad t + 40 = 0 \qquad \text{Zero-product property}$$
$$t = \frac{50}{3} \quad \text{or} \quad t = -40 \qquad \text{Solve.}$$

The car reaches the no-passing zone after $\frac{50}{3} \approx 16.7$ seconds. (The solution $t = -40$ has no physical meaning in this problem.)

■ **Now Try Exercise 107**

CRITICAL THINKING

Can every trinomial be factored with the methods discussed in this section? Try to factor the following trinomials and then make a conjecture.

$$x^2 + 2x + 2, \quad x^2 - x + 1, \quad \text{and} \quad 2x^2 + x + 2.$$

Factoring with Graphs and Tables

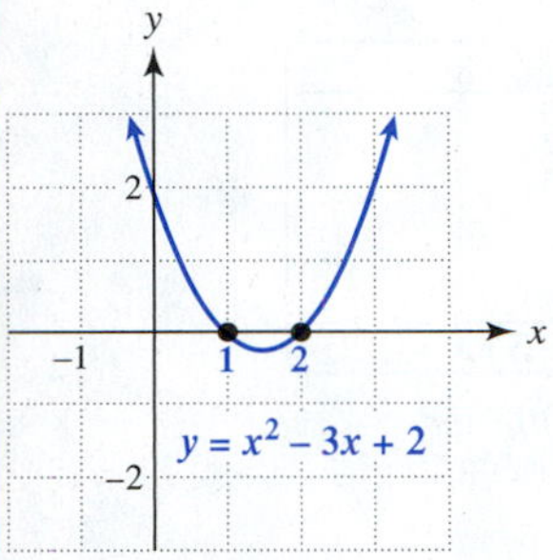

Polynomials can also be factored graphically. One way to do so is to use a graph of the polynomial to find its zeros. A number p is a *zero* if a value of **0** results when p is substituted in the polynomial. For example, both -2 and 2 are zeros of $x^2 - 4$ because

$$(-2)^2 - 4 = 0 \quad \text{and} \quad (2)^2 - 4 = 0.$$

Now consider the trinomial $x^2 - 3x + 2$, whose graph is shown in Figure 5.21. Its x-intercepts or zeros are **1** and **2**, and this trinomial factors as

$$x^2 - 3x + 2 = (x - 1)(x - 2).$$

We generalize these concepts as follows.

Figure 5.21

$y = x^2 - 3x + 2$

$f(x) = a(x - p)(x - q)$

Figure 5.22

$y = f(x)$

$a > 0$

FACTORING WITH A GRAPH OR A TABLE OF VALUES

To factor the trinomial $x^2 + bx + c$, either graph or make a table of $y = x^2 + bx + c$. If the zeros of the trinomial are p and q, then the trinomial can be factored as

$$x^2 + bx + c = (x - p)(x - q).$$

If the trinomial $ax^2 + bx + c$ has zeros p and q, then it may be factored as

$$ax^2 + bx + c = a(x - p)(x - q).$$

See Figure 5.22.

NOTE: If the coefficient a of x^2 is not 1, then a must be included in the factoring. For example, the zeros of $2x^2 - 8x + 6$ are **1** and **3**, so it can be factored as $2(x - 1)(x - 3)$. Note that the leading coefficient **2** must be included as part of the factorization.

EXAMPLE 7 | **Factoring with technology**

Factor each trinomial graphically or numerically.
(a) $x^2 - 2x - 24$ **(b)** $2x^2 - 51x + 220$

Solution

(a) Graph $y_1 = x^2 - 2x - 24$. When the trace feature is used, the zeros of the trinomial are **−4** and **6**, as shown in Figure 5.23. Thus the trinomial factors as follows.

$$x^2 - 2x - 24 = \left(x - (-4)\right)(x - 6)$$
$$= (x + 4)(x - 6)$$

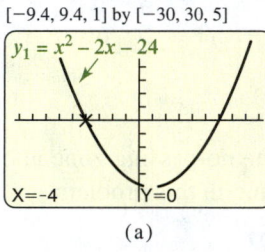
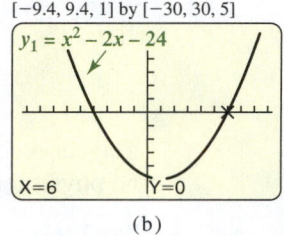

Figure 5.23

(b) Construct a table for $y_1 = 2x^2 - 51x + 220$. Figure 5.24 reveals that the zeros are **5.5** and **20**. The leading coefficient of $2x^2 - 51x + 220$ is **2**, so we factor this expression as follows.

$$2x^2 - 51x + 220 = 2(x - 5.5)(x - 20)$$
$$= (2x - 11)(x - 20)$$

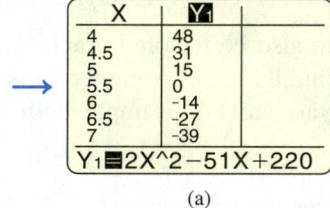
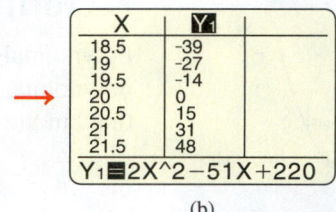

Figure 5.24

Now Try Exercises 95, 99

READING CHECK

If you know the zeros and leading coefficient of a polynomial, how can you factor it?

5.4 Putting It All Together

Equations of the form $ax^2 + bx + c = 0$ occur in applications. One way to solve such equations is to factor the trinomial $ax^2 + bx + c$. This factorization may be done symbolically, graphically, or numerically. The following table includes an explanation and examples of each technique. To factor by grouping, see the box on page 356.

TECHNIQUE	EXPLANATION	EXAMPLE
Symbolic Factoring	To factor $x^2 + bx + c$ find two factors of c whose sum is b. To factor $ax^2 + bx + c$ find factors of ax^2 and c so that the middle term is bx. Grouping can also be used. With grouping we start by finding two numbers whose product is ac and whose sum is b.	To factor $x^2 - 3x - 4$, find two factors of -4 whose sum is -3. These factors are 1 and -4. $$x^2 - 3x - 4 = (x + 1)(x - 4)$$ It checks. $\longrightarrow -3x$ ✓ To factor $2x^2 + 3x - 2$ find factors of $2x^2$ and -2 so that the middle term is $3x$. $$2x^2 + 3x - 2 = (2x - 1)(x + 2)$$ It checks. $\longrightarrow 3x$ ✓
Graphical Factoring	If p and q are zeros of $x^2 + bx + c$, then the expression factors as $$(x - p)(x - q).$$ If p and q are zeros of $ax^2 + bx + c$, then the expression factors as $$a(x - p)(x - q).$$	To factor $x^2 - 3x - 4$ $(a = 1)$, graph $y = x^2 - 3x - 4$. The zeros (x-intercepts) are -1 and 4, and it follows that $$x^2 - 3x - 4 = (x + 1)(x - 4).$$ $y = x^2 - 3x - 4$
Numerical Factoring	If p and q are zeros of $x^2 + bx + c$, then the expression factors as $$(x - p)(x - q).$$ If p and q are zeros of $ax^2 + bx + c$, then the expression factors as $$a(x - p)(x - q).$$	To factor $2x^2 + 3x - 2$, make a table for $y_1 = 2x^2 + 3x - 2$. The zeros are 0.5 and -2. $$2x^2 + 3x - 2 = 2(x - 0.5)(x + 2)$$ $$= (2x - 1)(x + 2)$$ X Y1 -2 0 -1.5 -2 -1 -3 -.5 -3 0 -2 .5 0 1 3 Y1=2X^2+3X-2

5.4 Exercises

 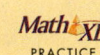

CONCEPTS AND VOCABULARY

1. What is a trinomial? Give an example.

2. Name two methods for factoring $ax^2 + bx + c$.

3. The trinomial $3x^2 - x - 3$ is written in the form $ax^2 + bx + c$. Identify a, b, and c.

4. If you factor the trinomial $x^2 + 3x + 2$, do you obtain $(x + 1)(x + 2)$? Explain.

5. Suppose that the graph of $y = x^2 + bx + c$ has x-intercepts -3 and 1. Factor $x^2 + bx + c$.

6. The graph of $y = 6x^2 - 15x + 6$ has x-intercepts $\frac{1}{2}$ and 2. Factor $6x^2 - 15x + 6$.

7. Which of the following (a.–d.) is a factorization of $x^2 - 3x - 18$?
   a. $(x - 2)(x + 9)$ **b.** $(x + 6)(x + 3)$
   c. $(x + 3)(x - 6)$ **d.** $(x - 9)(4x + 2)$

8. Which of the following (a.–d.) is a factorization of $12x^2 - 7x - 12$?
   a. $(6x - 3)(2x + 4)$ **b.** $(x - 1)(12x + 6)$
   c. $(3x + 4)(4x - 3)$ **d.** $(3x - 4)(4x + 3)$

FACTORING TRINOMIALS

Exercises 9–16: Determine whether the factors of the trinomial are the given binomials.

9. $x^2 + 5x + 6$ $(x + 2)(x + 3)$

10. $x^2 - 13x + 30$ $(x - 10)(x - 3)$

11. $x^2 - x - 20$ $(x + 5)(x - 4)$

12. $x^2 + 3x - 28$ $(x + 4)(x - 7)$

13. $6z^2 - 11z + 4$ $(2z - 1)(3z - 4)$

14. $4z^2 - 19z + 12$ $(2z - 4)(2z - 3)$

15. $10m^2 - 21m + 10$ $(5m + 2)(2m - 5)$

16. $12n^2 + 5n - 2$ $(3n + 2)(4n - 1)$

Exercises 17–72: Factor completely.

17. $x^2 + 7x + 10$ 18. $x^2 + 3x - 10$

19. $x^2 + 8x + 12$ 20. $x^2 - 8x + 12$

21. $x^2 - 13x + 36$ 22. $x^2 + 11x + 24$

23. $x^2 - 7x - 8$ 24. $x^2 - 21x - 100$

25. $z^2 + z - 72$ 26. $z^2 + 6z - 55$

27. $t^2 - 15t + 56$ 28. $t^2 - 14t + 40$

29. $y^2 - 18y + 72$ 30. $y^2 - 15y + 54$

31. $m^2 - 18m - 40$ 32. $m^2 - 22m - 75$

33. $n^2 - 20n - 300$ 34. $n^2 - 13n - 30$

35. $2x^2 + 7x + 3$ 36. $2x^2 - 5x - 3$

37. $6x^2 - 7x - 5$ 38. $10x^2 + 3x - 1$

39. $4z^2 + 19z + 12$ 40. $4z^2 + 17z + 4$

41. $6t^2 - 17t + 12$ 42. $6t^2 - 13t + 6$

43. $10y^2 + 13y - 3$ 44. $10y^2 + 23y - 5$

45. $6m^2 - m - 12$ 46. $20m^2 - m - 12$

47. $42n^2 + 5n - 25$ 48. $42n^2 + 65n + 25$

49. $1 + x - 2x^2$ 50. $3 - 5x - 2x^2$

51. $20 + 7x - 6x^2$ 52. $4 + 13x - 12x^2$

53. $5y^2 + 5y - 30$ 54. $3y^2 - 27y + 24$

55. $2z^2 + 12z + 16$ 56. $4z^2 + 32z + 60$

57. $z^3 + 9z^2 + 14z$ 58. $z^3 + 7z^2 + 12z$

59. $t^3 - 10t^2 + 21t$ 60. $t^3 - 11t^2 + 24t$

61. $m^4 + 6m^3 + 5m^2$ 62. $m^4 - m^3 - 2m^2$

63. $5x^3 + x^2 - 6x$ 64. $2x^3 + 8x^2 - 24x$

65. $6x^3 + 21x^2 + 9x$ 66. $12x^3 - 8x^2 - 20x$

67. $2x^3 - 14x^2 + 20x$ 68. $7x^3 + 35x^2 + 42x$

69. $60z^3 + 230z^2 - 40z$ 70. $24z^3 + 8z^2 - 80z$

71. $4x^4 + 10x^3 - 6x^2$ 72. $30x^4 + 3x^3 - 9x^2$

Exercises 73–76: **Thinking Generally** *Factor the given expression completely.*

73. $x^2 + (2 + 3)x + 2 \cdot 3$

74. $x^2 - (3 + 4)x + 3 \cdot 4$

75. $x^2 + (a + b)x + ab$

76. $x^2 - (a + b)x + ab$

77. **Thinking Generally** If $x^2 + bx + c = 0$ for $x = 1$ and $x = 2$, find b and c.

78. **Thinking Generally** If $x^2 + bx = 0$ for $x = -4$, find b.

Exercises 79–82: **Multiplying Functions** *Use factoring to find $f(x)$ and $g(x)$ so that $h(x) = f(x) g(x)$.*

79. $h(x) = x^2 + 3x + 2$

80. $h(x) = x^2 - 4x + 3$

81. $h(x) = x^2 - 2x - 8$

82. $h(x) = x^2 + 3x - 10$

FACTORING WITH GRAPHS AND TABLES

Exercises 83–88: Factor the expression by using the graph of $y = f(x)$, where $f(x)$ is the given expression. Check your answer by multiplying.

83. $x^2 - 6x + 8$

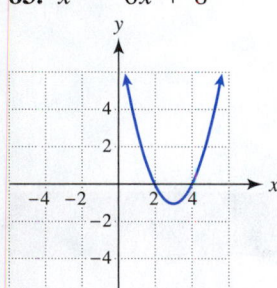

84. $x^2 - x - 6$

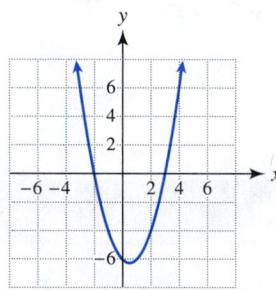

85. $2x^2 - 2x - 4$

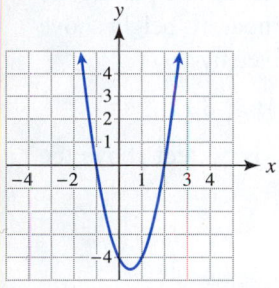

86. $3x^2 + 3x - 18$

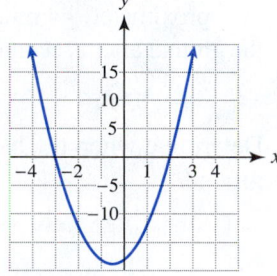

87. $2 + x - x^2$

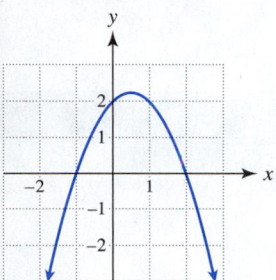

88. $3 - 2x - x^2$

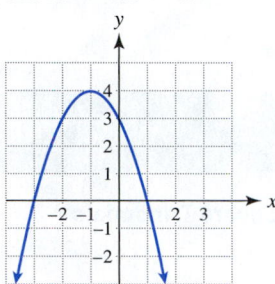

Exercises 89–92: Use the table to factor the expression. Check your answer by multiplying.

89. $x^2 - 3x - 4$

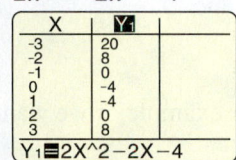

90. $x^2 - 40x + 300$

91. $2x^2 - 2x - 4$

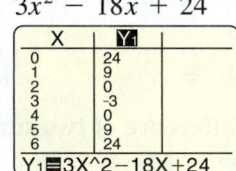

92. $3x^2 - 18x + 24$

Exercises 93–102: Factor graphically or numerically.

93. $x^2 + 3x - 10$

94. $x^2 + 7x + 12$

95. $x^2 - 3x - 28$

96. $x^2 - 25x + 100$

97. $2x^2 - 14x + 20$

98. $12x^2 - 6x - 6$

99. $5x^2 - 30x - 200$

100. $12x^2 - 30x + 12$

101. $8x^2 - 44x + 20$

102. $20x^2 + 3x - 9$

APPLICATIONS

103. *Ticket Prices* (Refer to Example 3.) If tickets are sold for $35, 1200 tickets are expected to be sold. For each $1 reduction in ticket price, an additional 100 tickets will be sold.
(a) Find the revenue from ticket sales when ticket prices are reduced by x dollars.
(b) Find the ticket prices that give sales of $54,000.

104. *Ticket Prices* One airline ticket costs $295. For each additional ticket sold to a group of up to 30 people, the price of every ticket is reduced by $5. For example, 2 tickets cost $2 \cdot \$290 = \$580$ and 3 tickets cost $3 \cdot \$285 = \$855$.
(a) Find the cost of x tickets.
(b) What is the cost of 20 tickets?
(c) How many tickets are sold if the total cost is $2500?

105. *Area of a Rectangle* A rectangle has an area of 91 square feet. Its length is 6 feet more than its width. Find the dimensions of the rectangle.

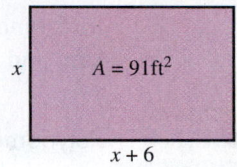

106. *Area of a Triangle* The base of a triangle is 2 inches less than its height. If the area of the triangle is 60 square inches, find its height.

107. *Passing Distances* (Refer to Example 6.) The distance traveled in feet by a car passing another car after t seconds is given by $2t^2 + 88t$.
(a) Use factoring to determine how long it takes for the car to travel 600 feet.
(b) Solve part (a) either graphically or numerically.

108. *Numbers* Find the two consecutive positive integers whose product is 132.

WRITING ABOUT MATHEMATICS

109. A student factors each trinomial into the given pair of binomials.

(i) $x^2 - 5x - 6$ $(x - 1)(x + 6)$
(ii) $10x^2 - 11x - 6$ $(5x - 3)(2x + 2)$
(iii) $12x^2 + 13x + 3$ $(6x + 1)(2x + 3)$

Explain the error that the student is making.

110. Suppose that a graph of $y = x^2 + bx + c$ intersects the x-axis at p and at q. Explain how you could use the graph to factor $x^2 + bx + c$.

SECTIONS 5.3 and 5.4

Checking Basic Concepts

1. Factor out the greatest common factor.
(a) $3x^2 - 6x$ (b) $16x^3 - 8x^2 + 4x$

2. Use factoring to solve each equation.
(a) $x^2 - 2x = 0$ (b) $9x^2 = 81x$

3. Factor each trinomial.
(a) $x^2 + 3x - 10$ (b) $x^2 - 3x - 10$
(c) $8x^2 + 14x + 3$

4. Solve $x^2 + 3x + 2 = 0$.

5. A baseball is thrown upward at 64 feet per second or approximately 44 miles per hour. Its height above the ground after t seconds is given by

$$h(t) = -16t^2 + 64t + 8.$$

Determine when the baseball is 56 feet above the ground.

5.5 Special Types of Factoring

Difference of Two Squares • Perfect Square Trinomials • Sum and Difference of Two Cubes

A LOOK INTO MATH ▶

Polynomials are frequently used to estimate and model things in real life, such as the curved shape of a car's exterior. Because of this fact, people have spent a lot of time developing factoring methods for polynomials. Two primary reasons for factoring are to *solve equations* and to *simplify expressions*. Although not all polynomials can be factored easily, many polynomials can be factored. In this section we focus on some special types of factoring.

Difference of Two Squares

NEW VOCABULARY

☐ Difference of two squares
☐ Perfect square trinomials
☐ Sum/difference of cubes

When we factor polynomials, we are *reversing* the process of multiplying polynomials. In Section 5.2 we discussed the equation

$$(a - b)(a + b) = a^2 - b^2. \qquad \text{Multiplying}$$

We can use this equation to factor a **difference of two squares**. For example, if we want to factor $x^2 - 25$, we can substitute x for a and 5 for b in

$$a^2 - b^2 = (a - b)(a + b) \qquad \text{Factoring}$$

to get

$$x^2 - 5^2 = (x - 5)(x + 5).$$

DIFFERENCE OF TWO SQUARES

For any real numbers a and b,

$$a^2 - b^2 = (a - b)(a + b).$$

NOTE: The sum of two squares *cannot* be factored completely using real numbers. For example, $x^2 + y^2$ cannot be factored, whereas $x^2 - y^2$ can be factored. *It is important to remember that* $x^2 + y^2 \neq (x + y)^2$.

EXAMPLE 1   **Factoring the difference of two squares**

Factor each polynomial, if possible.
(a) $9x^2 - 64$ **(b)** $4x^2 + 9y^2$ **(c)** $9x^4 - y^6$ **(d)** $4a^3 - 4a$

Solution
(a) Note that $9x^2 = (3x)^2$ and $64 = 8^2$.

$$\begin{aligned} 9x^2 - 64 &= (3x)^2 - (8)^2 & \text{Write as } a^2 - b^2. \\ &= (3x - 8)(3x + 8) & \text{Factor as } (a - b)(a + b). \end{aligned}$$

(b) Because $4x^2 + 9y^2$ is the *sum* of two squares, it *cannot* be factored.

(c) If we let $a^2 = 9x^4$ and $b^2 = y^6$, then $a = 3x^2$ and $b = y^3$. Thus

$$\begin{aligned} 9x^4 - y^6 &= (3x^2)^2 - (y^3)^2 & \text{Write as } a^2 - b^2. \\ &= (3x^2 - y^3)(3x^2 + y^3). & \text{Factor as } (a - b)(a + b). \end{aligned}$$

(d) Start by factoring out the common factor of $4a$.

$$\begin{aligned} 4a^3 - 4a &= 4a(a^2 - 1) \\ &= 4a(a - 1)(a + 1) \end{aligned}$$

Now Try Exercises 17, 19, 25, 29

EXAMPLE 2   **Applying the difference of two squares**

Factor each expression.
(a) $(n + 1)^2 - 9$ **(b)** $x^4 - y^4$ **(c)** $6r^2 - 24t^4$ **(d)** $x^3 + 3x^2 - 4x - 12$

Solution
(a) Use $a^2 - b^2 = (a - b)(a + b)$, with $a = n + 1$ and $b = 3$.

$$\begin{aligned} (n + 1)^2 - 9 &= (n + 1)^2 - 3^2 & 9 = 3^2 \\ &= \big((n + 1) - 3\big)\big((n + 1) + 3\big) & \text{Difference of squares} \\ &= (n - 2)(n + 4) & \text{Combine terms.} \end{aligned}$$

(b) Use $a^2 - b^2 = (a - b)(a + b)$, with $a = x^2$ and $b = y^2$.

$$\begin{aligned} x^4 - y^4 &= (x^2)^2 - (y^2)^2 & \text{Write as squares.} \\ &= (x^2 - y^2)(x^2 + y^2) & \text{Difference of squares} \\ &= (x - y)(x + y)(x^2 + y^2) & \text{Difference of squares again} \end{aligned}$$

(c) Start by factoring out the common factor of 6.

$$6r^2 - 24t^4 = 6(r^2 - 4t^4) \qquad \text{Factor out 6.}$$
$$= 6\left(r^2 - (2t^2)^2\right) \qquad \text{Write as squares.}$$
$$= 6(r - 2t^2)(r + 2t^2) \qquad \text{Difference of squares}$$

(d) Start factoring by using *grouping* and then factor the difference of squares.

$$x^3 + 3x^2 - 4x - 12 = (x^3 + 3x^2) + (-4x - 12) \qquad \text{Associative property}$$
$$= \mathbf{\color{red}x^2(x + 3)} - \mathbf{\color{red}4(x + 3)} \qquad \text{Factor out } x^2 \text{ and } -4.$$
$$= \mathbf{\color{red}(x^2 - 4)(x + 3)} \qquad \text{Factor out } (x + 3).$$
$$= (x - 2)(x + 2)(x + 3) \qquad \text{Difference of squares}$$

Now Try Exercises 23, 31, 37, 39

Perfect Square Trinomials

In Section 5.2 we also showed how to multiply $(a + b)^2$ and $(a - b)^2$.

$$(a + b)^2 = a^2 + 2ab + b^2 \quad \text{and} \quad (a - b)^2 = a^2 - 2ab + b^2$$

STUDY TIP

It takes practice to recognize
perfect square trinomials.

The expressions $a^2 + 2ab + b^2$ and $a^2 - 2ab + b^2$ are **perfect square trinomials**. If we can recognize a perfect square trinomial, we can use these formulas to factor it.

PERFECT SQUARE TRINOMIALS

For any real numbers a and b,

$$a^2 + 2ab + b^2 = (a + b)^2 \quad \text{and}$$
$$a^2 - 2ab + b^2 = (a - b)^2.$$

EXAMPLE 3   **Factoring perfect square trinomials**

Factor.
(a) $x^2 + 6x + 9$ **(b)** $81x^2 - 72x + 16$

Solution
(a) Let $\mathbf{\color{red}a^2 = x^2}$ and $\mathbf{\color{red}b^2 = 3^2}$. For a *perfect square trinomial*, the middle term must be $2ab$.

$$\mathbf{\color{blue}2ab = 2(x)(3) = 6x,} \checkmark$$

which equals the given middle term. Thus $a^2 + 2ab + b^2 = (a + b)^2$ implies

$$x^2 + \mathbf{\color{red}6x} + 9 = (\mathbf{\color{red}x + 3})^2.$$

(b) Let $\mathbf{\color{red}a^2 = (9x)^2}$ and $\mathbf{\color{red}b^2 = 4^2}$. Again, the middle term must be $2ab$.

$$\mathbf{\color{blue}2ab = 2(9x)(4) = 72x,} \checkmark$$

which equals the given middle term. Thus $a^2 - 2ab + b^2 = (a - b)^2$ implies

$$81x^2 - 72x + 16 = (\mathbf{\color{red}9x - 4})^2.$$

Now Try Exercises 51, 57

EXAMPLE 4 **Factoring perfect square trinomials**

Factor each expression.

(a) $4x^2 + 4xy + y^2$ (b) $9r^2 - 12rt + 4t^2$ (c) $25a^3 + 10a^2b + ab^2$

Solution

(a) Let $a^2 = (2x)^2$ and $b^2 = y^2$. The middle term must be $2ab$.

$$2ab = 2(2x)(y) = 4xy, \checkmark$$

which equals the given middle term. Thus $a^2 + 2ab + b^2 = (a + b)^2$ implies

$$4x^2 + 4xy + y^2 = (2x + y)^2.$$

(b) Let $a^2 = (3r)^2$ and $b^2 = (2t)^2$. The middle term must be $2ab$.

$$2ab = 2(3r)(2t) = 12rt, \checkmark$$

which equals the given middle term. Thus $a^2 - 2ab + b^2 = (a - b)^2$ implies

$$9r^2 - 12rt + 4t^2 = (3r - 2t)^2.$$

(c) Factor out the common factor of a. Then factor the resulting perfect square trinomial.

$$25a^3 + 10a^2b + ab^2 = a(25a^2 + 10ab + b^2)$$
$$= a(5a + b)^2$$

Now Try Exercises 63, 65, 67

MAKING CONNECTIONS

Special Factoring and General Techniques

If you do *not* recognize a polynomial as being either the difference of two squares or a perfect square trinomial, then you can factor the polynomial by using the methods discussed in earlier sections.

Sum and Difference of Two Cubes

The **sum or difference of two cubes** may be factored. This fact is justified by the following two equations.

$$(a + b)(a^2 - ab + b^2) = a^3 + b^3 \quad \text{and} \qquad \text{Sum of cubes}$$
$$(a - b)(a^2 + ab + b^2) = a^3 - b^3 \qquad \text{Difference of cubes}$$

These equations can be verified by multiplying the left side to obtain the right side. For example,

$$(a + b)(a^2 - ab + b^2) = a \cdot a^2 - a \cdot ab + a \cdot b^2 + b \cdot a^2 - b \cdot ab + b \cdot b^2$$
$$= a^3 - a^2b + ab^2 + a^2b - ab^2 + b^3$$
$$= a^3 + b^3.$$

SUM AND DIFFERENCE OF TWO CUBES

For any real numbers a and b,

$$a^3 + b^3 = (a + b)(a^2 - ab + b^2) \quad \text{and}$$
$$a^3 - b^3 = (a - b)(a^2 + ab + b^2).$$

EXAMPLE 5 **Factoring a sum or difference of two cubes**

Factor each polynomial.
(a) $x^3 + 8$ **(b)** $27x^3 - 64y^3$

Solution
(a) Because $x^3 = (x)^3$ and $8 = 2^3$, we let $a = x, b = 2$, and factor. Substituting in

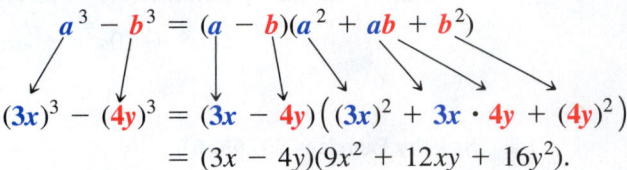

gives

$$x^3 + 2^3 = (x + 2)(x^2 - x \cdot 2 + 2^2)$$
$$= (x + 2)(x^2 - 2x + 4).$$

Note that $x^2 - 2x + 4$ does not factor further.
(b) Here, $27x^3 = (3x)^3$ and $64y^3 = (4y)^3$, so

$$27x^3 - 64y^3 = (3x)^3 - (4y)^3.$$

Substituting $a = 3x$ and $b = 4y$ in

$$a^3 - b^3 = (a - b)(a^2 + ab + b^2)$$

gives

$$(3x)^3 - (4y)^3 = (3x - 4y)\big((3x)^2 + 3x \cdot 4y + (4y)^2\big)$$
$$= (3x - 4y)(9x^2 + 12xy + 16y^2).$$

▌ **Now Try Exercises 69, 75**

EXAMPLE 6 **Factoring a sum or difference of two cubes**

Factor each expression.
(a) $x^6 + 8y^3$ **(b)** $27p^9 - 8q^6$

Solution
(a) Let $a^3 = (x^2)^3$ and $b^3 = (2y)^3$. Then $a^3 + b^3 = (a + b)(a^2 - ab + b^2)$ implies

$$x^6 + 8y^3 = (x^2 + 2y)(x^4 - 2x^2y + 4y^2).$$

(b) Let $a^3 = (3p^3)^3$ and $b^3 = (2q^2)^3$. Then $a^3 - b^3 = (a - b)(a^2 + ab + b^2)$ implies

$$27p^9 - 8q^6 = (3p^3 - 2q^2)(9p^6 + 6p^3q^2 + 4q^4).$$

▌ **Now Try Exercises 81, 83**

5.5 Putting It All Together

In this section we presented three special types of factoring, which are summarized in the following table.

TYPE OF FACTORING	DESCRIPTION	EXAMPLE
The Difference of Two Squares	To factor the difference of two squares, use $$a^2 - b^2 = (a - b)(a + b).$$ **NOTE:** $a^2 + b^2$ cannot be factored.	$25x^2 - 16 = (5x - 4)(5x + 4)$ $a = 5x, b = 4$ $4x^2 + y^2$ cannot be factored.
A Perfect Square Trinomial	To factor a perfect square trinomial, use $$a^2 + 2ab + b^2 = (a + b)^2 \text{ or }$$ $$a^2 - 2ab + b^2 = (a - b)^2.$$	$x^2 + 4x + 4 = (x + 2)^2$ $a = x, b = 2$ $16x^2 - 24x + 9 = (4x - 3)^2$ $a = 4x, b = 3$ Be sure to check the middle term.
The Sum and Difference of Two Cubes	The sum and difference of two cubes can be factored as $$a^3 + b^3 = (a + b)$$ $$\cdot (a^2 - ab + b^2) \text{ or }$$ $$a^3 - b^3 = (a - b)$$ $$\cdot (a^2 + ab + b^2).$$	$x^3 + 8y^3 = (x + 2y)\left(x^2 - x \cdot 2y + (2y)^2\right)$ $\qquad = (x + 2y)(x^2 - 2xy + 4y^2)$ $a = x, b = 2y$ $125x^3 - 64 = (5x - 4)\left((5x)^2 + 5x \cdot 4 + 4^2\right)$ $\qquad = (5x - 4)(25x^2 + 20x + 16)$ $a = 5x, b = 4$

5.5 Exercises

CONCEPTS AND VOCABULARY

1. Give an example of a difference of two squares.

2. Give an example of a perfect square trinomial.

3. Give an example of the sum of two cubes.

4. Factor $a^2 - b^2$.

5. Factor $a^2 + 2ab + b^2$.

6. Factor $a^3 - b^3$.

7. Which of the following (a.–d.) is a factorization of $25x^2 - 16$?
   a. $(5x - 8)(5x + 8)$ **b.** $(5x - 4)(5x - 4)$
   c. $(5x - 4)(5x + 4)$ **d.** $(25x - 1)(x - 16)$

8. Which of the following (a.–d.) is a factorization of $36x^2 - 48x + 16$?
   a. $(9x - 8)(2x - 2)$ **b.** $4(3x - 2)^2$
   c. $(12x + 4)(3x + 4)$ **d.** $4(3x + 2)^2$

DIFFERENCE OF TWO SQUARES

Exercises 9–12: Determine whether the expression represents a difference of two squares. Factor the expression.

9. $x^2 - 25$

10. $16x^2 - 100$

11. $x^3 + y^3$

12. $9x^2 + 36y^2$

Exercises 13–38: Factor the expression, if possible.

13. $x^2 - 36$

14. $y^2 - 144$

15. $25 - z^2$

16. $36 - y^2$

17. $4z^2 + 25$

18. $9z^2 + 64$

19. $36x^2 - 100$

20. $4y^2 - 64$

21. $49a^2 - 64b^2$

22. $9x^2 - 4y^2$

23. $64z^2 - 25z^4$

24. $64z^4 - 49z^2$

25. $5x^3 - 125x$

26. $100x^3 - x$

27. $4r^4 + t^6$

28. $9t^4 - 25r^6$

29. $16t^4 - r^2$

30. $t^4 + 9r^2$

31. $(x + 1)^2 - 25$

32. $(x - 2)^2 - 9$

33. $100 - (n - 4)^2$

34. $81 - (n + 3)^2$

35. $y^4 - 16$

36. $16z^4 - 1$

37. $16x^4 - y^4$

38. $r^4 - 81t^4$

Exercises 39–42: Use grouping to help factor the expression completely.

39. $x^3 + x^2 - x - 1$

40. $x^3 + x^2 - 9x - 9$

41. $4x^3 - 8x^2 - x + 2$

42. $9x^3 - 18x^2 - 16x + 32$

PERFECT SQUARE TRINOMIALS

Exercises 43–50: Determine whether the trinomial is a perfect square. If it is, factor it.

43. $x^2 + x + 16$

44. $4x^2 - 2x + 25$

45. $x^2 + 8x + 16$

46. $x^2 - 4x + 4$

47. $4z^2 - 4z + 1$

48. $9z^2 + 6z + 4$

49. $16t^2 - 12t + 9$

50. $4t^2 + 12t + 9$

Exercises 51–68: Factor the expression.

51. $x^2 + 2x + 1$

52. $x^2 - 6x + 9$

53. $4x^2 + 20x + 25$

54. $x^2 + 10x + 25$

55. $x^2 - 12x + 36$

56. $x^2 + 20x + 100$

57. $9z^2 - 24z + 16$

58. $36z^2 + 12z + 1$

59. $4y^4 + 4y^3 + y^2$

60. $16z^4 - 24z^3 + 9z^2$

61. $9z^3 - 6z^2 + z$

62. $49y^2 + 42y + 9$

63. $9x^2 + 6xy + y^2$

64. $25x^2 + 30xy + 9y^2$

65. $49a^2 - 28ab + 4b^2$

66. $64a^2 - 16ab + b^2$

67. $4x^4 - 4x^3y + x^2y^2$

68. $4x^3y + 12x^2y^2 + 9xy^3$

SUM AND DIFFERENCE OF TWO CUBES

Exercises 69–86: Factor the expression.

69. $x^3 - 8$

70. $x^3 + 8$

71. $y^3 + z^3$

72. $y^3 - z^3$

73. $27x^3 - 8$

74. $64 - y^3$

75. $64z^3 + 27t^3$

76. $125t^3 - 64r^3$

77. $8x^4 + 125x$

78. $x^3y^2 + 8y^5$

79. $27y - 8x^3y$

80. $x^3y^3 + 1$

81. $z^6 - 27y^3$

82. $z^6 + 27y^3$

83. $125z^6 + 8y^9$

84. $125z^6 - 8y^9$

85. $5m^6 + 40n^3$

86. $10m^9 - 270n^6$

GENERAL FACTORING

Exercises 87–120: Factor the expression completely.

87. $25x^2 - 64$

88. $25x^2 - 30x + 9$

89. $x^3 + 27$

90. $4 - 16y^2$

91. $64x^2 + 16x + 1$

92. $2x^2 - 5x + 3$

93. $3x^2 + 14x + 8$

94. $125x^3 - 1$

95. $x^4 + 8x$

96. $2x^3 - 12x^2 + 18x$

97. $64x^3 + 8y^3$

98. $54 - 16x^3$

99. $2r^2 - 8t^2$

100. $a^3 - ab^2$

101. $a^3 + 4a^2b + 4ab^2$

102. $8r^4 + rt^3$

103. $x^2 - 3x + 2$

104. $x^2 + 4x - 5$

105. $4z^2 - 25$

106. $(z + 1)^2 - 49$

107. $x^4 + 16x^3 + 64x^2$

108. $4x^2 - 12xy + 9y^2$

109. $z^3 - 1$

110. $8z^3 + 1$

111. $3t^2 - 5t - 8$

112. $15t^2 - 11t + 2$

113. $7a^3 + 20a^2 - 3a$

114. $b^3 - b^2 - 2b$

115. $x^6 - y^6$

116. $a^8 - b^8$

117. $100x^2 - 1$

118. $4x^2 + 28x + 49$

119. $p^3q^3 - 27$

120. $a^2b^2 - c^2d^2$

121. ***Difference of Two Squares*** Write the difference in the areas between the large square with sides of length x and a smaller square with sides of length 5, using factored form.

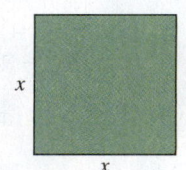
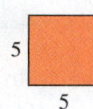

122. *Difference of Two Cubes* Write the difference in the volumes between a large cube with sides of length 2 and a smaller cube with sides of length x, using factored form.

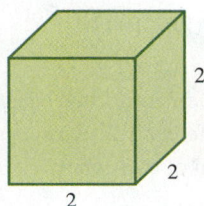 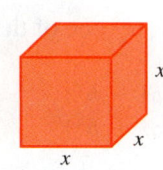

WRITING ABOUT MATHEMATICS

123. Explain how factoring $x^3 + y^3$ is similar to factoring $x^3 - y^3$.

124. Can $x^2 - y^2$ be factored? Can $x^2 + y^2$ be factored? Explain your answers.

Group Activity   Working with Real Data

Directions: Form a group of 2 to 4 people. Select someone to record the group's responses for this activity. All members of the group should work cooperatively to answer the questions. If your instructor asks for your results, each member of the group should be prepared to respond.

Exercises 1–6: *Geometric Factoring* The polynomial $x^2 + 4x + 3$ can be factored by grouping.

$$x^2 + 4x + 3 = x^2 + x + 3x + 3$$
$$= (x^2 + x) + (3x + 3)$$
$$= x(x + 1) + 3(x + 1)$$
$$= (x + 3)(x + 1)$$

This factorization is represented geometrically in the accompanying figure. Note that the outside dimensions of the large rectangle are $x + 3$ by $x + 1$ and therefore its area is $(x + 3)(x + 1)$. This area equals the sum of the four smaller rectangles inside the large rectangle. Thus

$$(x + 3)(x + 1) = x^2 + x + 3x + 3 = x^2 + 4x + 3.$$

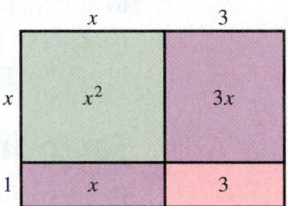

Factor the following expressions geometrically. If the rectangle turns out to be a square, what can you say about the trinomial?

1. $x^2 + 3x + 2$ **2.** $x^2 + 6x + 5$

3. $4x^2 + 4x + 1$ **4.** $x^2 + 6x + 9$

5. $6x^2 + 29x + 20$ **6.** $8x^2 + 59x + 21$

5.6   Summary of Factoring

Guidelines for Factoring Polynomials • **Factoring Polynomials**

A LOOK INTO MATH ▶   Mathematics is being used more and more to model all types of situations. Modeling is not only used in science; it is also used to describe consumer behavior and to search the Internet. To model the world around us, mathematical equations need to be solved. One important technique for solving equations is to set an expression equal to 0 and then factor. Factoring is an important method for solving equations and for simplifying complicated expressions. In this section, we discuss general guidelines that can be used to factor polynomials.

Guidelines for Factoring Polynomials

The following guidelines can be used to factor polynomials in general.

FACTORING POLYNOMIALS

STEP 1: Factor out the greatest common factor, if possible.

STEP 2: **A.** If the polynomial has *four terms*, try factoring by grouping.

B. If the polynomial is a *binomial*, try one of the following.
1. $a^2 - b^2 = (a - b)(a + b)$ Difference of two squares
2. $a^3 - b^3 = (a - b)(a^2 + ab + b^2)$ Difference of two cubes
3. $a^3 + b^3 = (a + b)(a^2 - ab + b^2)$ Sum of two cubes

C. If the polynomial is a *trinomial*, check for a perfect square.
1. $a^2 + 2ab + b^2 = (a + b)^2$ Perfect square trinomial
2. $a^2 - 2ab + b^2 = (a - b)^2$ Perfect square trinomial

Otherwise, apply FOIL or grouping, as described in Section 5.4.

STEP 3: Check to make sure that the polynomial is *completely* factored.

These guidelines give a general strategy for factoring a polynomial by hand. It is important to always perform Step 1 first. Factoring out the greatest common factor first usually makes it easier to factor the resulting polynomial. In the next subsection we apply these guidelines to several polynomials.

Factoring Polynomials

In the first example, we apply Step 1 to a polynomial with a common factor.

EXAMPLE 1 **Factoring out a common factor**

Factor $3x^3 - 9x^2$.

Solution
STEP 1: The greatest common factor is $\mathbf{3x^2}$.

$$3x^3 - 9x^2 = \mathbf{3x^2}(\mathbf{x - 3})$$

STEP 2B: The binomial, $\mathbf{x - 3}$, cannot be factored further.

STEP 3: The completely factored polynomial is $3x^2(x - 3)$.

Now Try Exercise 19

In several of the next examples we apply more than one factoring technique.

EXAMPLE 2 **Factoring a difference of squares**

Factor $5x^3 - 20x$.

Solution
STEP 1: The greatest common factor is $\mathbf{5x}$.

$$5x^3 - 20x = \mathbf{5x}(\mathbf{x^2 - 4})$$

STEP 2B: We can factor the binomial $\mathbf{x^2 - 4}$ as a difference of squares.

$$\mathbf{5x}(\mathbf{x^2 - 4}) = \mathbf{5x}(\mathbf{x - 2})(\mathbf{x + 2})$$

STEP 3: The completely factored polynomial is $5x(x - 2)(x + 2)$.

Now Try Exercise 21

EXAMPLE 3 **Factoring a difference of squares**

Factor $5x^4 - 80$.

Solution

STEP 1: The greatest common factor is 5.

$$5x^4 - 80 = 5(x^4 - 16)$$

STEP 2B: We can factor $x^4 - 16$ as a difference of squares *twice*.

$$5(x^4 - 16) = 5(x^2 - 4)(x^2 + 4) = 5(x - 2)(x + 2)(\mathbf{x^2 + 4})$$

The sum of squares, $\mathbf{x^2 + 4}$, cannot be factored further.

STEP 3: The completely factored polynomial is $5(x - 2)(x + 2)(x^2 + 4)$.

■ **Now Try Exercise 23**

EXAMPLE 4 **Factoring a perfect square trinomial**

Factor $12x^3 + 36x^2 + 27x$.

Solution

STEP 1: The greatest common factor is $3x$.

$$12x^3 + 36x^2 + 27x = 3x(4x^2 + 12x + 9)$$

STEP 2C: The trinomial $4x^2 + 12x + 9$ is a perfect square trinomial.

$$3x(\mathbf{4x^2 + 12x + 9}) = 3x(\mathbf{2x + 3})(\mathbf{2x + 3})$$

STEP 3: The completely factored polynomial is $3x(2x + 3)^2$.

■ **Now Try Exercise 37**

EXAMPLE 5 **Factoring a sum of cubes**

Factor $-8x^4 - 27x$.

Solution

STEP 1: The negative of the greatest common factor is $-x$.

$$-8x^4 - 27x = -x(8x^3 + 27)$$

STEP 2B: The binomial $8x^3 + 27$ can be factored as a sum of cubes.

$$-x(8x^3 + 27) = -x(2x + 3)(\mathbf{4x^2 - 6x + 9})$$

The trinomial, $\mathbf{4x^2 - 6x + 9}$, cannot be factored further.

STEP 3: The completely factored polynomial is $-x(2x + 3)(4x^2 - 6x + 9)$.

■ **Now Try Exercise 43**

EXAMPLE 6 **Factoring by grouping**

Factor $x^3 - 2x^2 + x - 2$.

Solution

STEP 1: There are no common factors.

STEP 2A: Because there are four terms, we apply grouping.

$$x^3 - 2x^2 + x - 2 = (x^3 - 2x^2) + (x - 2) \qquad \text{Associative property}$$
$$= x^2(x - 2) + 1(x - 2) \qquad \text{Factor out } x^2 \text{ and 1.}$$
$$= (x^2 + 1)(x - 2) \qquad \text{Distributive property}$$

The sum of squares, $x^2 + 1$, cannot be factored further.

STEP 3: The completely factored polynomial is $(x^2 + 1)(x - 2)$.

Now Try Exercise 31

EXAMPLE 7 **Factoring a trinomial**

Factor $2x^4 - 7x^3 + 6x^2$.

Solution

STEP 1: The greatest common factor is x^2.

$$2x^4 - 7x^3 + 6x^2 = x^2(2x^2 - 7x + 6)$$

STEP 2C: We can factor the resulting trinomial by using FOIL.

$$x^2(2x^2 - 7x + 6) = x^2(x - 2)(2x - 3)$$

STEP 3: The completely factored polynomial is $x^2(x - 2)(2x - 3)$.

Now Try Exercise 27

EXAMPLE 8 **Factoring with two variables**

Factor $12x^2 - 27a^2$.

Solution

STEP 1: The greatest common factor is 3.

$$12x^2 - 27a^2 = 3(4x^2 - 9a^2)$$

STEP 2B: We can factor the binomial as the difference of squares.

$$3(4x^2 - 9a^2) = 3(2x - 3a)(2x + 3a)$$

STEP 3: The completely factored polynomial is $3(2x - 3a)(2x + 3a)$.

Now Try Exercise 49

In the next example we use common factors, grouping, the difference of two cubes, and the difference of two squares to factor a polynomial.

EXAMPLE 9 **Applying several techniques**

Factor $2x^5 - 8x^3 - 2x^2 + 8$.

Solution

STEP 1: The expression has a greatest common factor of 2.

$$2x^5 - 8x^3 - 2x^2 + 8 = 2(x^5 - 4x^3 - x^2 + 4)$$

STEP 2A: The resulting four terms can be factored by grouping.

$$2(x^5 - 4x^3 - x^2 + 4) = 2\big((x^5 - 4x^3) + (-x^2 + 4)\big) \qquad \text{Associative property}$$
$$= 2\big(x^3(x^2 - 4) - 1(x^2 - 4)\big) \qquad \text{Factor out } x^3 \text{ and } -1.$$
$$= 2(x^3 - 1)(x^2 - 4) \qquad \text{Distributive property}$$

STEP 2B: Both of the resulting binomials can be factored further, one as the difference of two cubes and the other as the difference of two squares.

$$2(x^3 - 1)(x^2 - 4) = 2(x - 1)(x^2 + x + 1)(x - 2)(x + 2)$$

The trinomial, $x^2 + x + 1$, cannot be factored further.

STEP 3: The completely factored polynomial is $2(x - 1)(x^2 + x + 1)(x - 2)(x + 2)$.

▌ **Now Try Exercise 57**

5.6 Putting It All Together

When using the guidelines presented in this section for general factoring strategy, the following rules can be helpful.

CONCEPT	EXPLANATION	EXAMPLES
Greatest Common Factor	Factor out the greatest common factor, or monomial, that occurs in each term.	$4x^2 - 8x = 4x(x - 2)$ $2x^2 - 4x + 8 = 2(x^2 - 2x + 4)$ $x^5 + x^3 = x^3(x^2 + 1)$
Factoring by Grouping	Use the associative and distributive properties to factor a polynomial with four terms.	$x^3 - 2x^2 + 3x - 6$ $= (x^3 - 2x^2) + (3x - 6)$ $= x^2(x - 2) + 3(x - 2)$ $= (x^2 + 3)(x - 2)$
Factoring Binomials	Use the difference of squares, the difference of cubes, or the sum of cubes.	$4x^2 - 9 = (2x - 3)(2x + 3)$ $x^3 - 8 = (x - 2)(x^2 + 2x + 4)$ $x^3 + 8 = (x + 2)(x^2 - 2x + 4)$
Factoring Trinomials	Use FOIL or grouping to factor a trinomial after factoring out the greatest common factor.	$x^2 - 3x + 2 = (x - 1)(x - 2)$ $6x^2 + 13x + 5 \qquad \text{(grouping)}$ $= (6x^2 + 3x) + (10x + 5)$ $= 3x(2x + 1) + 5(2x + 1)$ $= (3x + 5)(2x + 1)$

5.6   Exercises

MyMathLab

CONCEPTS AND VOCABULARY

1. When factoring a polynomial, what is a good first step?

2. What is the greatest common factor for $4x^2 + 2x$?

3. Can you factor $x^2 + 4$? Explain.

4. What method might you use to factor a polynomial with four terms?

WARM UP

Exercises 5–18: Factor completely, if possible.

5. $a^2 - a$

6. $2x - 4$

7. $a^2 - 9$

8. $4b^2 - 1$

9. $x^2 - 2x + 1$

10. $a^2 + 2ab + b^2$

11. $x^3 - a^3$

12. $x^3 + a^3$

13. $a^2 + 4$

14. $36x^2 + y^2$

15. $x(x + 2) - 3(x + 2)$

16. $2x(x - 3) + (x - 3)$

17. $x^3 + 2x^2 + x + 2$

18. $x^3 - x^2 + 2x - 2$

FACTORING

Exercises 19–58: Factor completely.

19. $6x^2 - 14x$

20. $-27x^3 - 15x$

21. $2x^3 - 18x$

22. $8x^2 - 18$

23. $4a^4 - 64$

24. $b^4 - 81$

25. $6x^3 - 13x^2 - 15x$

26. $10x^4 + 13x^3 - 3x^2$

27. $2x^4 - 5x^3 - 25x^2$

28. $10x^3 + 28x^2 - 6x$

29. $2x^4 + 5x^2 + 3$

30. $2x^4 + 2x^2 - 4$

31. $x^3 + 3x^2 + x + 3$

32. $x^3 + 5x^2 + 4x + 20$

33. $5x^3 - 5x^2 + 10x - 10$

34. $5x^4 - 20x^3 + 10x - 40$

35. $ax + bx - ay - by$

36. $ax - bx - ay + by$

37. $18x^2 + 12x + 2$

38. $-3x^2 + 30x - 75$

39. $-4x^3 + 24x^2 - 36x$

40. $18x^3 - 60x^2 + 50x$

41. $8x^3 - 27$

42. $27x^3 + 8$

43. $-x^4 - 8x$

44. $x^5 - 27x^2$

45. $x^4 - 2x^3 - x + 2$

46. $x^4 + 3x^3 + x + 3$

47. $r^4 - 16$

48. $r^4 - 81$

49. $25x^2 - 4a^2$

50. $9y^2 - 16z^2$

51. $2x^4 - 2y^4$

52. $a^4 - b^4$

53. $9x^3 + 6x^2 - 3x$

54. $8x^3 + 28x^2 - 16x$

55. $(z - 2)^2 - 9$

56. $(y + 2)^2 - 4$

57. $3x^5 - 27x^3 + 3x^2 - 27$

58. $2x^5 - 8x^3 - 16x^2 + 64$

APPLICATIONS

59. *Greenhouse Gas Puzzle* Worldwide atmospheric carbon dioxide levels continue to increase and are suspected to contribute to climate change. The solution to the polynomial equation

$$x^3 - 390x^2 + 2x - 780 = 0$$

gives the record high amount of carbon dioxide in parts per million (ppm) in 2010, as measured at Mauna Loa Observatory. (*Source: The Honolulu Advertiser.*)
(a) Factor the expression on the left side of the equation.
(b) Solve the equation and determine this record amount.

60. *Greenhouse Gas Puzzle* (Refer to the preceding exercise.) The solution to the polynomial equation

$$2x^3 - 620x^2 + x - 310 = 0$$

gives the amount of carbon dioxide in parts per million (ppm) in 1958, as measured at Mauna Loa Observatory. (*Source: The Honolulu Advertiser.*)
(a) Factor the expression on the left side of the equation.
(b) Solve the equation and determine this amount.

61. *Dimensions of a Box* A box has a volume of 6 cubic feet. The height of the box is 1 foot more than its length, and its length is 1 foot more than its width. Let x be the height.
- **(a)** Write a polynomial that gives the volume of the box in terms of x.
- **(b)** Write an equation whose solution gives the height of the box.
- **(c)** Solve the equation by using factoring.

62. *Dimensions of a Rectangle* The length of a rectangle is 2 inches more than its width, and the area is 24 square inches. Let x be the width.
- **(a)** Write a polynomial that gives the area of the rectangle in terms of x.
- **(b)** Write an equation whose solution gives the width of the rectangle.
- **(c)** Solve the equation by using factoring.

WRITING ABOUT MATHEMATICS

63. Give the basic guidelines for factoring a polynomial.

64. What are some ways to recognize that a polynomial is completely factored?

SECTIONS 5.5 and 5.6

Checking Basic Concepts

1. Factor each special type of polynomial.
- **(a)** $25x^2 - 16$
- **(b)** $x^2 + 12x + 36$
- **(c)** $9x^2 - 30x + 25$
- **(d)** $x^3 - 27$
- **(e)** $81x^4 - 16$

2. Factor each expression.
- **(a)** $x^3 - 2x^2 + 3x - 6$
- **(b)** $x^2 - 4x - 21$
- **(c)** $12x^3 + 2x^2 - 4x$

5.7 Polynomial Equations

Quadratic Equations • Higher Degree Equations • Equations in Quadratic Form • Applications

A LOOK INTO MATH ▶ Polynomials are used to solve important problems. One example is found in highway construction where the elevations of hills and valleys are modeled by quadratic polynomials. (*Source:* F. Mannering and W. Kilareski, *Principles of Highway Engineering and Traffic Analysis.*)

The quadratic polynomial

$$h(x) = -0.0001x^2 + 100$$

could model the height of a hill, where $x = 0$ corresponds to the peak or crest of the hill, as illustrated in Figure 5.25. To determine any locations where the height of the hill is **75** feet, we can solve the *quadratic equation*

$$-0.0001x^2 + 100 = 75.$$

See Example 3.

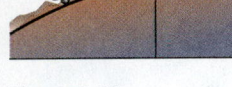

Figure 5.25

NEW VOCABULARY

☐ Quadratic equation
☐ Quadratic form

Quadratic Equations

Any **quadratic equation** can be written as $ax^2 + bx + c = 0$, where a, b, and c are constants, with $a \neq 0$. Quadratic equations can sometimes be solved by factoring and then applying the zero-product property. They can also be solved graphically. These techniques are demonstrated in the next two examples.

EXAMPLE 1 **Solving a quadratic equation**

Solve $x^2 - 4 = 0$ symbolically and graphically.

Solution

Symbolic Solution Start by factoring the left side of the equation.

$$x^2 - 4 = 0 \qquad \text{Given equation}$$
$$(x - 2)(x + 2) = 0 \qquad \text{Difference of squares}$$
$$x - 2 = 0 \quad \text{or} \quad x + 2 = 0 \qquad \text{Zero-product property}$$
$$x = 2 \quad \text{or} \quad x = -2 \qquad \text{Solve each equation.}$$

The solutions are -2 and 2.

Graphical Solution Start by making Table 5.10. Plot these points and connect them with a smooth curve called a parabola, as shown in Figure 5.26. The x-intercepts, or zeros, are -2 and 2, which are also the solutions to the equation.

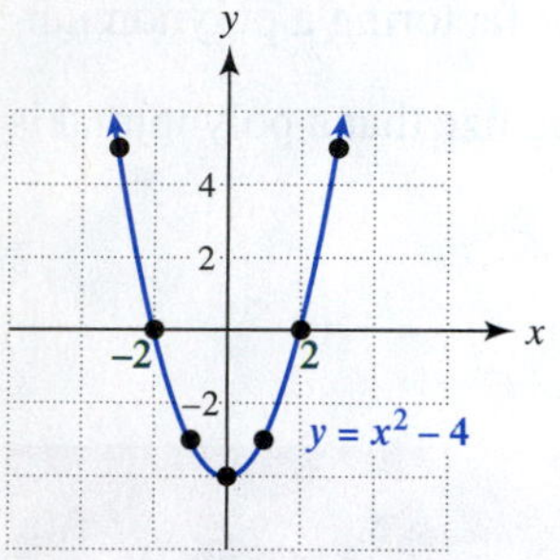

Zeros: $-2, 2$

Figure 5.26

TABLE 5.10 $y = x^2 - 4$

x	-3	-2	-1	0	1	2	3
y	5	0	-3	-4	-3	0	5

Zeros are -2 and 2.

Now Try Exercise 7

EXAMPLE 2 **Solving quadratic equations**

Solve each equation.
(a) $4x^2 - 20x + 25 = 0$ **(b)** $3x^2 + 11x = 20$ **(c)** $x^2 + 25 = 0$

Solution
(a) Start by factoring the left side, which is a perfect square trinomial.

$$4x^2 - 20x + 25 = 0 \qquad \text{Given equation}$$
$$(2x - 5)(2x - 5) = 0 \qquad \text{Factor.}$$
$$2x - 5 = 0 \qquad \text{Zero-product property}$$
$$x = \frac{5}{2} \qquad \text{Solve.}$$

The only solution is $\frac{5}{2}$.

(b) Start by subtracting 20 from each side to obtain a 0 on the right side.

$$3x^2 + 11x = 20 \qquad \text{Given equation}$$
$$3x^2 + 11x - 20 = 0 \qquad \text{Subtract 20.}$$
$$(x + 5)(3x - 4) = 0 \qquad \text{Factor.}$$
$$x + 5 = 0 \quad \text{or} \quad 3x - 4 = 0 \qquad \text{Zero-product property}$$
$$x = -5 \quad \text{or} \quad x = \frac{4}{3} \qquad \text{Solve.}$$

The solutions are -5 and $\frac{4}{3}$.

(c) $x^2 + 25 = 0$ implies that $x^2 = -25$, which has no solutions because $x^2 \geq 0$ for all real numbers.

Now Try Exercises 25, 29, 33

MAKING CONNECTIONS

Quadratic Equations and Solutions

Example 2 demonstrates that a quadratic equation can have no solutions, one solution, or two solutions. The following graphs illustrate this concept.

No solutions
$x^2 + 25 = 0$

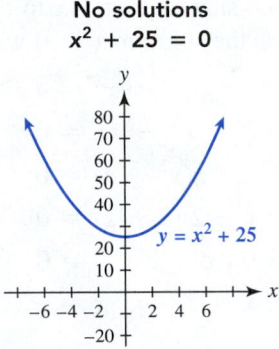

One solution
$4x^2 - 20x + 25 = 0$

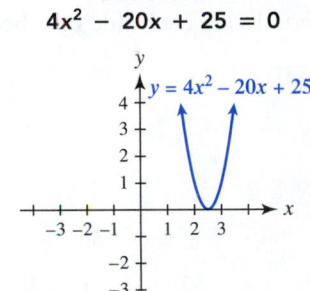

Two solutions
$3x^2 + 11x + 20$

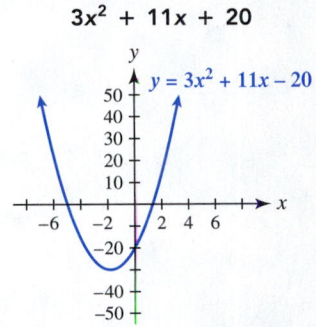

READING CHECK

How many solutions can a quadratic equation have? Explain your answer graphically.

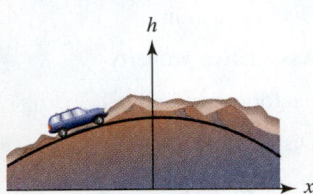

Figure 5.25 (Repeated)

EXAMPLE 3 **Determining highway elevations**

Solve the equation $-0.0001x^2 + 100 = 75$, which is discussed in A Look Into Math for this section. See Figure 5.25 repeated in the margin.

Solution

Start by subtracting 75 from each side of the equation. Then multiply each side by **−10,000** to clear decimals.

$$-0.0001x^2 + 100 = 75 \qquad \text{Given equation}$$
$$-0.0001x^2 + 25 = 0 \qquad \text{Subtract 75.}$$
$$\mathbf{-10{,}000}(-0.0001x^2 + 25) = \mathbf{-10{,}000}(0) \qquad \text{Multiply by } -10{,}000.$$
$$x^2 - 250{,}000 = 0 \qquad \text{Distributive property}$$
$$(x - 500)(x + 500) = 0 \qquad \sqrt{250{,}000} = 500$$
$$x - 500 = 0 \quad \text{or} \quad x + 500 = 0 \qquad \text{Zero-product property}$$
$$x = 500 \quad \text{or} \quad x = -500 \qquad \text{Solve each equation.}$$

Thus the highway elevation is 75 feet at a distance of 500 feet on either side of the crest of the hill.

Now Try Exercise 75

Higher Degree Equations

Using the techniques of factoring that we discussed in earlier sections, we can solve equations that involve higher degree polynomials. These higher degree equations can have more than two solutions.

EXAMPLE 4 | **Solving higher degree equations**

Solve each equation.
(a) $x^3 = 4x$ **(b)** $2x^3 + 2x^2 - 12x = 0$ **(c)** $x^3 - 5x^2 - x + 5 = 0$

Solution
(a) Start by subtracting $4x$ from each side of $x^3 = 4x$ to obtain a 0 on the right side. (Do *not* divide each side by x because the solution $x = 0$ will be lost.)

$$x^3 = 4x \qquad \text{Given equation}$$
$$x^3 - 4x = 0 \qquad \text{Subtract } 4x.$$
$$x(x^2 - 4) = 0 \qquad \text{Factor out } x.$$
$$x(x - 2)(x + 2) = 0 \qquad \text{Difference of squares}$$
$$x = 0 \quad \text{or} \quad x - 2 = 0 \quad \text{or} \quad x + 2 = 0 \qquad \text{Zero-product property}$$
$$x = 0 \quad \text{or} \quad x = 2 \quad \text{or} \quad x = -2 \qquad \text{Solve each equation.}$$

The solutions are $-2, 0,$ and 2.

(b) Start by factoring out the common factor $2x$.

$$2x^3 + 2x^2 - 12x = 0 \qquad \text{Given equation}$$
$$2x(x^2 + x - 6) = 0 \qquad \text{Factor out } 2x.$$
$$2x(x + 3)(x - 2) = 0 \qquad \text{Factor the trinomial.}$$
$$2x = 0 \quad \text{or} \quad x + 3 = 0 \quad \text{or} \quad x - 2 = 0 \qquad \text{Zero-product property}$$
$$x = 0 \quad \text{or} \quad x = -3 \quad \text{or} \quad x = 2 \qquad \text{Solve each equation.}$$

The solutions are $-3, 0,$ and 2.

(c) To solve this equation use *grouping* to factor the left side.

$$x^3 - 5x^2 - x + 5 = 0 \qquad \text{Given equation}$$
$$(x^3 - 5x^2) + (-x + 5) = 0 \qquad \text{Associative property}$$
$$x^2(x - 5) - 1(x - 5) = 0 \qquad \text{Factor out } x^2 \text{ and } -1.$$
$$(x^2 - 1)(x - 5) = 0 \qquad \text{Distributive property}$$
$$(x - 1)(x + 1)(x - 5) = 0 \qquad \text{Difference of squares}$$
$$x - 1 = 0 \quad \text{or} \quad x + 1 = 0 \quad \text{or} \quad x - 5 = 0 \qquad \text{Zero-product property}$$
$$x = 1 \quad \text{or} \quad x = -1 \quad \text{or} \quad x = 5 \qquad \text{Solve each equation.}$$

The solutions are $-1, 1,$ and 5.

Now Try Exercises 43, 47, 51

In the next example, we demonstrate how an equation can be solved symbolically, graphically, and numerically with the aid of a graphing calculator.

EXAMPLE 5 | **Solving a polynomial equation**

Solve $16x^4 - 64x^3 + 64x^2 = 0$ symbolically, graphically, and numerically.

Solution
Symbolic Solution Begin by factoring out the common factor $16x^2$.

$$16x^4 - 64x^3 + 64x^2 = 0 \qquad \text{Given equation}$$
$$16x^2(x^2 - 4x + 4) = 0 \qquad \text{Factor out } 16x^2.$$
$$16x^2(x - 2)^2 = 0 \qquad \text{Perfect square trinomial}$$

Solving results in

$$16x^2 = 0 \quad \text{or} \quad (x - 2)^2 = 0 \qquad \text{Zero-product property}$$

$$x = 0 \quad \text{or} \quad x = 2. \qquad \text{Solve.}$$

Graphical Solution Graph $y_1 = 16x^4 - 64x^3 + 64x^2$, as shown in Figures 5.27(a) and 5.27(b). When the trace feature is used, solutions to $y_1 = 0$ occur when $x = 0$ and $x = 2$.

Numerical Solution Construct the table for $y_1 = 16x^4 - 64x^3 + 64x^2$, as shown in Figure 5.27(c). Solutions occur at $x = 0$ and $x = 2$.

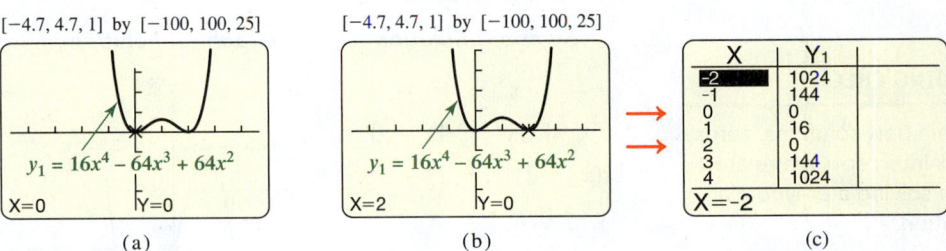

$[-4.7, 4.7, 1]$ by $[-100, 100, 25]$ $[-4.7, 4.7, 1]$ by $[-100, 100, 25]$

(a)   (b)   (c)

Figure 5.27

Now Try Exercise 61

Equations in Quadratic Form

Sometimes equations do not appear to be quadratic, but they can be solved by using factoring techniques that we have already applied to quadratic equations. Such equations have **quadratic form** and sometimes need to be factored more than once, as demonstrated in the next example.

EXAMPLE 6   **Solving equations in quadratic form**

Solve each equation.
(a) $x^4 - 81 = 0$ **(b)** $x^4 - 5x^2 + 4 = 0$

Solution
(a) To solve this equation use the difference of squares twice.

$$(x^2)^2 - 9^2 = 0 \qquad \text{Rewrite given equation.}$$

$$(x^2 - 9)(x^2 + 9) = 0 \qquad \text{Difference of squares}$$

$$(x - 3)(x + 3)(x^2 + 9) = 0 \qquad \text{Difference of squares again}$$

$$x - 3 = 0 \quad \text{or} \quad x + 3 = 0 \quad \text{or} \quad x^2 + 9 = 0 \qquad \text{Zero-product property}$$

$$x = 3 \quad \text{or} \quad x = -3 \quad \text{or} \quad x^2 = -9 \qquad \text{Solve each equation.}$$

The solutions are -3 and 3 because $x^2 = -9$ has no real number solutions.
(b) To solve this equation start by factoring the trinomial.

$$x^4 - 5x^2 + 4 = 0 \qquad \text{Given equation}$$

$$(x^2 - 4)(x^2 - 1) = 0 \qquad \text{Factor.}$$

$$(x - 2)(x + 2)(x - 1)(x + 1) = 0 \qquad \text{Difference of squares twice}$$

$$x = 2 \quad \text{or} \quad x = -2 \quad \text{or} \quad x = 1 \quad \text{or} \quad x = -1 \qquad \text{Solve.}$$

The solutions are $-2, -1, 1,$ and 2.

Now Try Exercises 65, 69

NOTE: To factor the equation in Example 6(a), it may be helpful to let $z = x^2$ and $z^2 = x^4$. Then the given equation becomes $z^2 - 81 = 0$, or $(z - 9)(z + 9) = 0$ after factoring. Substituting $z = x^2$ gives the factored equation $(x^2 - 9)(x^2 + 9) = 0$. Similar comments can be made for Example 6(b).

MAKING CONNECTIONS

Symbolic, Graphical, and Numerical Solutions

The following illustrates how to solve $x^3 - 4x = 0$ with each method. Note that the solutions to $x^3 - 4x = 0$, the x-intercepts of the graph of $y = x^3 - 4x$, and the zeros of $x^3 - 4x$ are *all identical*.

READING CHECK

Explain how solutions, zeros, and x-intercepts are related when solving a polynomial equation.

Symbolic Solution

$$x^3 - 4x = 0$$
$$x(x^2 - 4) = 0$$
$$x(x - 2)(x + 2) = 0$$
$$x = 0, x = 2, x = -2$$

Solutions: $-2, 0, 2$

Graphical Solution

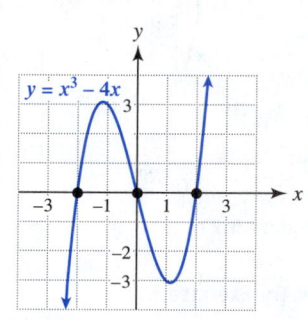

x-intercepts: $-2, 0, 2$

Numerical Solution

x	$x^3 - 4x$
-3	-15
-2	0
-1	3
0	0
1	-3
2	0
3	15

Zeros: $-2, 0, 2$

Applications

▶ **REAL-WORLD CONNECTION** Many times applications involving geometry make use of quadratic equations. In the next example we solve a quadratic equation to find the dimensions of a picture.

EXAMPLE 7 **Finding the dimensions of a picture**

A frame surrounding a picture is 3 inches wide. The picture inside the frame is 5 inches wider than it is high. If the overall area of the picture and frame is 336 square inches, find the dimensions of the picture inside the frame.

Solution

Let x be the height of the picture and $x + 5$ be its width. The dimensions of the picture and frame, in inches, are illustrated in Figure 5.28.

The overall area is given by $(x + 6)(x + 11)$, which equals 336 square inches.

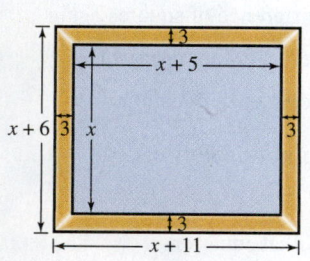

Figure 5.28

$$(x + 6)(x + 11) = 336 \quad \text{Equation to solve}$$
$$x^2 + 17x + 66 = 336 \quad \text{Multiply binomials.}$$
$$x^2 + 17x - 270 = 0 \quad \text{Subtract 336.}$$
$$(x - 10)(x + 27) = 0 \quad \text{Factor.}$$
$$x = 10 \quad \text{or} \quad x = -27 \quad \text{Zero-product property; solve.}$$

The only valid solution for x is 10 inches. Because the width is 5 inches more than the height, the dimensions are 10 inches by 15 inches.

Now Try Exercise 73

▶ **REAL-WORLD CONNECTION** Highway engineers often use quadratic functions to calculate stopping distances for cars. The faster a car is traveling, the farther it takes for the car to stop. The stopping distance for a car can be used to estimate its speed, as demonstrated in the next example. (*Source:* F. Mannering.)

EXAMPLE 8 **Finding the speed of a car**

For a car traveling at x miles per hour on wet, level pavement, highway engineers sometimes estimate stopping distance d in feet by

$$d(x) = \frac{1}{9}x^2 + \frac{11}{3}x.$$

Use this formula to approximate the speed of a car that takes 180 feet to stop.

Solution
We need to solve the equation $\frac{1}{9}x^2 + \frac{11}{3}x = 180$. Start by multiplying by **9** to eliminate fractions.

$\frac{1}{9}x^2 + \frac{11}{3}x = 180$	Equation to be solved
$9\left(\frac{1}{9}x^2 + \frac{11}{3}x\right) = 9 \cdot 180$	Multiply by 9.
$x^2 + 33x = 1620$	Distributive property
$x^2 + 33x - 1620 = 0$	Subtract 1620.
$(x - 27)(x + 60) = 0$	Factor.
$x = 27 \quad \text{or} \quad x = -60$	Zero-product property; solve.

The speed of the car is 27 miles per hour.

■ **Now Try Exercise 81**

NOTE: To factor $x^2 + 33x - 1620$ in Example 8, we find values for m and n such that $mn = -1620$ and $m + n = 33$, which may require some trial and error.

EXAMPLE 9 **Finding the dimensions of an iPod Mini**

The width of an iPod Mini is 1.5 inches more than its thickness, and its length is 3 inches more than its thickness. If the volume of an iPod Mini is 3.5 cubic inches, find its dimensions graphically.

Solution
Let x be its thickness, $x + 1.5$ be its width, and $x + 3$ be its length. Because volume equals thickness times width times length, we have

$$x(x + 1.5)(x + 3) = 3.5.$$

The graphs of $y_1 = x(x + 1.5)(x + 3)$ and $y_2 = 3.5$ intersect at $(0.5, 3.5)$, as shown in Figure 5.29. Thus the iPod Mini is 0.5 inch thick, 2 inches wide, and 3.5 inches long.

[−6, 6, 1] by [−2, 6, 1]

$y_2 = 3.5$

$\leftarrow y_1$

Intersection
X=.5   Y=3.5

Figure 5.29

■ **Now Try Exercise 83**

5.7 Putting It All Together

The following is a typical strategy for solving a polynomial equation. When you have finished, *be sure to check your answers*.

STEP 1: If necessary, rewrite the equation so that a 0 appears on one side of the equation.

STEP 2: Factor out common factors.

STEP 3: Factor the remaining polynomial using the techniques presented in this chapter.

STEP 4: Apply the *zero-product* property.

STEP 5: Solve each resulting equation.

The following table summarizes how to solve three types of higher degree polynomial equations.

CONCEPT	EXPLANATION	EXAMPLE
Higher Degree Equations	Factor out common factors; then use the techniques for factoring.	$2x^3 - 32x = 0$ $2x(x^2 - 16) = 0$ $2x(x - 4)(x + 4) = 0$ $x = 0$ or $x = 4$ or $x = -4$
Equations in Quadratic Form	Factor these forms as $ax^2 + bx + c$.	$x^4 - 13x^2 + 36 = 0$ $(x^2 - 4)(x^2 - 9) = 0$ $(x - 2)(x + 2)(x - 3)(x + 3) = 0$ $x = 2$ or $x = -2$ or $x = 3$ or $x = -3$
Solving by Grouping	Use grouping to help factor a cubic expression having four terms. Apply the distributive property.	$x^3 - 7x^2 - 16x + 112 = 0$ $(x^3 - 7x^2) + (-16x + 112) = 0$ $x^2(x - 7) - 16(x - 7) = 0$ $(x^2 - 16)(x - 7) = 0$ $(x - 4)(x + 4)(x - 7) = 0$ $x = 4$ or $x = -4$ or $x = 7$

5.7 Exercises

MyMathLab

CONCEPTS AND VOCABULARY

1. Explain why factoring is important in mathematics.

2. Factoring a polynomial is the *reverse* of _____ polynomials.

3. When you are solving $x^2 = 16$ by factoring, what is a good first step?

4. To solve $x^2 - 3x - 3 = 1$ could you start by factoring the left side of the equation? Explain.

5. What is the solution to $(x - 1)^2 = 0$?

6. Does the equation $x^2 + 4 = 0$ have any real number solutions? Explain.

SOLVING QUADRATIC EQUATIONS

Exercises 7–12: Solve the equation graphically. Then solve the equation symbolically.

7. $x^2 - 1 = 0$

8. $x^2 - 9 = 0$

9. $\frac{1}{4}x^2 - 1 = 0$

10. $\frac{1}{16}x^2 - 1 = 0$

11. $x^2 - x - 2 = 0$

12. $x^2 - x - 6 = 0$

Exercises 13–34: Solve the given quadratic equation, if possible.

13. $z^2 - 64 = 0$

14. $z^2 - 100 = 0$

15. $4y^2 - 1 = 0$

16. $9y^2 - 36 = 0$

17. $x^2 - 3x - 4 = 0$

18. $x^2 - 8x + 7 = 0$

19. $x^2 + 4x - 12 = 0$

20. $x^2 - 3x - 28 = 0$

21. $2x^2 + 5x - 3 = 0$

22. $3x^2 + 8x + 4 = 0$

23. $2x^2 = 32$

24. $4x^2 = 64$

25. $z^2 + 14z + 49 = 0$

26. $z^2 + 64 = 16z$

27. $9t^2 + 1 = 6t$

28. $49t^2 + 28t + 4 = 0$

29. $15n^2 = 7n + 2$

30. $7n^2 + 57n + 8 = 0$

31. $24m^2 + 23m = 12$

32. $11m^2 = 31m + 6$

33. $x^2 + 12 = 0$

34. $2x^2 + 3 = 0$

Exercises 35–38: Use the graph to solve the equation. Then solve the equation by factoring.

35. $x^2 + x - 6 = 0$

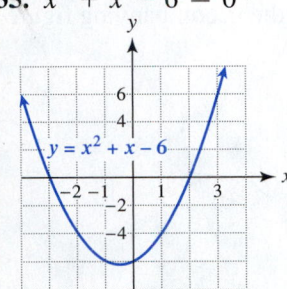

$y = x^2 + x - 6$

36. $x^2 - 2x - 3 = 0$

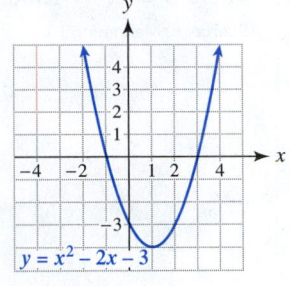

$y = x^2 - 2x - 3$

37. $2x^2 - 4x - 16 = 0$

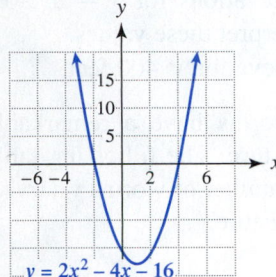

$y = 2x^2 - 4x - 16$

38. $2x^2 + 3x - 2 = 0$

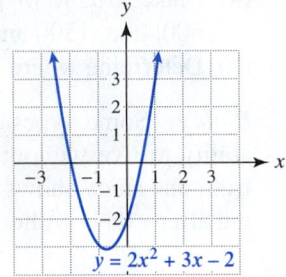

$y = 2x^2 + 3x - 2$

Exercises 39–42: The graph of the quadratic expression $y = ax^2 + bx + c$ is $\cap$-shaped when $a < 0$. If possible, sketch a graph of a quadratic expression with $a < 0$ that satisfies the condition.

39. No x-intercepts

40. One x-intercept

41. Two x-intercepts

42. Three x-intercepts

HIGHER DEGREE EQUATIONS

Exercises 43–54: Solve the equation.

43. $z^3 = 9z$

44. $2z^3 + 8z = 0$

45. $x^3 + x = 0$

46. $x^3 = x$

47. $2x^3 - 6x^2 = 20x$

48. $3x^3 + 15x^2 + 12x = 0$

49. $t^4 - 4t^3 - 5t^2 = 0$

50. $2t^4 - 8t^3 + 6t^2 = 0$

51. $x^3 + 7x^2 - 4x - 28 = 0$

52. $x^3 + 5x^2 = 9x + 45$

53. $n^3 - 2n^2 - 36n + 72 = 0$

54. $n^3 - 10n^2 - 4n + 40 = 0$

Exercises 55–62: (Refer to Example 5.) Solve the equation

(a) *symbolically,*

(b) *graphically, and*

(c) *numerically.*

55. $x^2 - 2x - 15 = 0$

56. $x^2 + 7x + 10 = 0$

57. $2x^2 - 3x = 2$

58. $4t^2 + 25 = 20t$

59. $3t^2 + 18t + 15 = 0$

60. $4z^2 = 16$

61. $4x^4 + 16x^2 = 16x^3$

62. $2x^3 + 12x^2 + 18x = 0$

EQUATIONS IN QUADRATIC FORM

Exercises 63–72: Solve the equation.

63. $x^4 - 2x^2 - 8 = 0$

64. $x^4 - 8x^2 - 9 = 0$

65. $x^4 - 26x^2 + 25 = 0$

66. $x^4 - 21x^2 - 100 = 0$

67. $x^4 - 13x^2 + 36 = 0$

68. $x^4 - 18x^2 + 81 = 0$

69. $x^4 - 16 = 0$

70. $4x^4 = x^2 + 18$

71. $9x^4 - 13x^2 + 4 = 0$

72. $64x^4 - 180x^2 + 81 = 0$

APPLICATIONS

73. *Picture Frame* (Refer to Example 7.) A rectangular frame surrounding a picture is made from boards that are 2 inches wide. The picture inside the frame is 4 inches wider than it is high. If the overall area of the picture and frame is 525 square inches, find the dimensions of the picture.

74. *Picture Frame* The outside measurements of a rectangular frame for a picture are 30 inches by 40 inches. The area of the rectangular picture inside the frame is 936 square inches. Find the width of the frame that surrounds the picture.

75. *Modeling a Hill* (Refer to Example 3.) The elevation E in feet of a highway x feet along a hill is modeled by

$$E(x) = -0.0001x^2 + 500,$$

where $-2000 \le x \le 2000$. Determine the x-values where the elevation is 400 feet.

76. *Numbers* If a positive number n is increased by 3, its square equals 121. Find n.

77. *Sidewalk Around a Pool* A 5-foot-wide sidewalk around a rectangular swimming pool has a total area of 900 square feet. Find the dimensions of the swimming pool if the pool is 20 feet longer than it is wide.

78. *Height Reached by a Baseball* The height in feet reached by a batted baseball after t seconds is given by

$$h(t) = -16t^2 + 88t + 4.$$

Determine when the baseball is 100 feet in the air.

79. *Height Reached by a Baseball* The height in feet reached by a batted baseball after t seconds is given by

$$h(t) = -16t^2 + 66t + 2.$$

A graph of $y = h(t)$ is shown.
(a) Interpret the y-intercept.
(b) Use the graph to estimate when the baseball is 70 feet in the air.
(c) Solve part (b) symbolically.
(d) Why are symbolic solutions important?

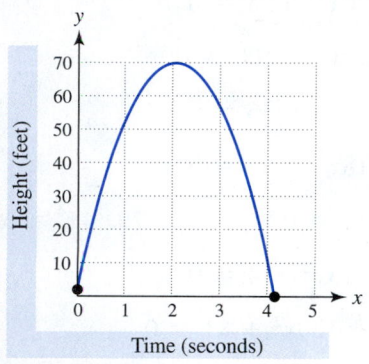

80. *Geometry* A rectangle is 2 inches longer than it is wide. If each side is increased by 3 inches, the area increases by 183 square inches. Find the dimensions of the rectangle.

Exercises 81 and 82: (Refer to Example 8.) On dry, level pavement, the stopping distance d in feet for a car traveling at x miles per hour can be estimated by

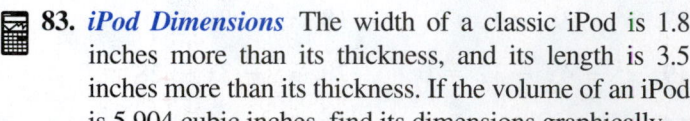

$$d(x) = \frac{1}{11}x^2 + \frac{11}{3}x.$$

Use this formula to approximate the speed of a car that takes d feet to stop.

81. $d = 220$ feet   **82.** $d = 638$ feet

83. *iPod Dimensions* The width of a classic iPod is 1.8 inches more than its thickness, and its length is 3.5 inches more than its thickness. If the volume of an iPod is 5.904 cubic inches, find its dimensions graphically.

84. *Box Dimensions* The width of a rectangular mailing box is 9 inches more than its height, and its length is 11 inches more than its height. If the volume of the box is 286 cubic inches, find its dimensions graphically.

85. *Modeling a Highway* In highway design, quadratic polynomials are used not only to model hills but also to model valleys. Valleys or sags in the road are sometimes referred to as *sag curves*. Suppose that the elevation E in feet of the sag curve x (horizontal) feet along a proposed route is modeled by

$$E(x) = 0.0002x^2 - 0.3x + 500,$$

where $0 \le x \le 1500$. See the accompanying figure.
(*Source:* F. Mannering.)

(a) Make a table of the elevations for $x = 0$, 300, 600, . . . , 1500 and interpret these values.
(b) Determine where the elevation is 400 feet.

86. *Biology* Some types of worms have a remarkable ability to live without moisture. The following table from one study shows the number of worms y surviving after x days without moisture.

x (days)	0	20	40
y (worms)	50	48	45

x (days)	80	120	160
y (worms)	36	20	3

Source: D. Brown and P. Rothery, *Models in Biology.*

(a) Plot the data and graph

$$y = -0.00138x^2 - 0.076x + 50.1$$

in the same viewing rectangle.

(b) Discuss how well the equation models the data.

(c) Estimate the day x when 10 worms remained.

WRITING ABOUT MATHEMATICS

87. Explain why the graph of $y = ax^2 + bx + c$ is used to model a dip or valley in a highway rather than the graph of $y = |ax + b|$. Make a sketch to support your reasoning. Assume that $a > 0$.

88. Explain the basic steps that you would use to solve the quadratic equation $ax^2 + bx + c = 0$ graphically and numerically. Assume that the equation has two solutions.

SECTION 5.7

Checking Basic Concepts

1. Use the graph to solve $x^2 + 4x + 3 = 0$. Then solve the equation symbolically.

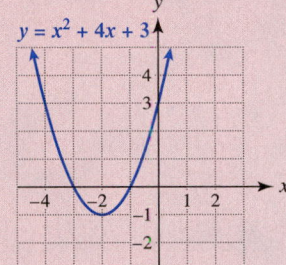

$y = x^2 + 4x + 3$

2. Solve $x^2 - 9 = 0$ graphically. Then solve the equation by factoring.

3. If possible, solve each equation.
    (a) $x^2 + 9 = 6x$
    (b) $x^2 + 9 = 0$
    (c) $12x^2 + 7x - 10 = 0$

4. Solve each equation.
    (a) $x^3 + 5x = 0$
    (b) $x^4 - 81 = 0$
    (c) $z^4 + z^2 = 20$
    (d) $x^3 + 2x^2 - 49x = 98$

CHAPTER 5   Summary

SECTION 5.1 • POLYNOMIAL FUNCTIONS

Monomials and Polynomials A monomial is a term with only *nonnegative integer* exponents. A polynomial is either a monomial or a sum of monomials.

Examples: Monomial: $-4x^3y^2$ has degree 5 and coefficient -4

           Polynomial: $3x^2 - 4x + 2$ has degree 2 and leading coefficient 3

Addition and Subtraction of Polynomials To add two polynomials, combine like terms. When subtracting two polynomials, add the first polynomial and the opposite of the second polynomial. Combine like terms.

Examples: $(3x^3 - 4x) + (-5x^3 + 7x) = (3 - 5)x^3 + (-4 + 7)x$

                                  $= -2x^3 + 3x$

      $(3xy^2 + 4x^2) - (4xy^2 - 6x) = (3xy^2 + 4x^2) + (-4xy^2 + 6x)$

                                    $= 3xy^2 - 4xy^2 + 4x^2 + 6x$

                                    $= -xy^2 + 4x^2 + 6x$

Polynomial Functions If the formula for a function f is a polynomial, then f is a polynomial function. The following are examples of polynomial functions.

$$f(x) = 2x - 7 \qquad \text{Linear} \qquad \text{Degree 1}$$
$$f(x) = 4x^2 - 5x + 1 \qquad \text{Quadratic} \qquad \text{Degree 2}$$
$$f(x) = 9x^3 + 6x \qquad \text{Cubic} \qquad \text{Degree 3}$$

Example: $f(x) = 4x^3 - 2x^2 + 4$ represents a polynomial function of one variable.

$$f(2) = 4(2)^3 - 2(2)^2 + 4 = 32 - 8 + 4 = 28$$

SECTION 5.2 • MULTIPLICATION OF POLYNOMIALS

Product of Two Polynomials The product of two polynomials may be found by multiplying every term in the first polynomial by every term in the second polynomial.

Example: $(2x + 3)(3x - 4) = 6x^2 - 8x + 9x - 12 = 6x^2 + x - 12$

Special Products

Product of a Sum and Difference $\qquad (a + b)(a - b) = a^2 - b^2$

$\qquad\qquad\qquad\qquad\qquad\qquad$ **Example:** $(x + 3)(x - 3) = x^2 - 9$

Square of a Binomial $\qquad\qquad\qquad (a + b)^2 = a^2 + 2ab + b^2$

$\qquad\qquad\qquad\qquad\qquad\qquad\quad (a - b)^2 = a^2 - 2ab + b^2$

$\qquad\qquad\qquad\qquad\qquad\qquad$ **Examples:** $(x + 3)^2 = x^2 + 6x + 9$

$\qquad\qquad\qquad\qquad\qquad\qquad\qquad\qquad\quad (x - 3)^2 = x^2 - 6x + 9$

SECTION 5.3 • FACTORING POLYNOMIALS

Common Factors When factoring a polynomial, start by factoring out the greatest common factor (GCF).

Examples: $8z^3 - 12z^2 + 16z = 4z(2z^2 - 3z + 4)$

$\qquad\qquad\quad 18x^2y^3 - 12x^3y^2 = 6x^2y^2(3y - 2x)$

Zero-Product Property For all real numbers a and b, if $ab = 0$, then $a = 0$ or $b = 0$ (or both).

Example: $(x - 3)(x + 2) = 0$ implies that either $x - 3 = 0$ or $x + 2 = 0$.

Solving Polynomial Equations by Factoring Obtain a zero on one side of the equation. Factor the other side of the equation and apply the zero-product property.

Example: $\qquad\quad x^2 - 7x = 0$

$\qquad\qquad\quad x(x - 7) = 0 \qquad$ Factor out common factor.

$\qquad\quad x = 0 \quad \text{or} \quad x = 7 \qquad$ Zero-product property; solve.

Factoring by Grouping Grouping can be used to factor some polynomial expressions with four terms by applying the distributive and associative properties.

Example: $x^3 + x^2 + 5x + 5 = (x^3 + x^2) + (5x + 5) \qquad$ Associative property

$\qquad\qquad\qquad\qquad\quad = x^2(x + 1) + 5(x + 1) \qquad$ Distributive property

$\qquad\qquad\qquad\qquad\quad = (x^2 + 5)(x + 1) \qquad$ Distributive property

SECTION 5.4 • FACTORING TRINOMIALS

Factoring $x^2 + bx + c$ Find integers m and n that satisfy $mn = c$ and $m + n = b$. Then $x^2 + bx + c = (x + m)(x + n)$.

Example: $x^2 + 3x + 2 = (x + 1)(x + 2)$ **$m = 1, n = 2$**

Factoring $ax^2 + bx + c$ Either grouping or FOIL (in reverse) can be used to factor $ax^2 + bx + c$. Trinomials can also be factored graphically or numerically.

Examples: *Grouping* Find m and n so that $mn = ac = 30$ and $m + n = b = 17$.

$$6x^2 + 17x + 5 = (6x^2 + 2x) + (15x + 5) \qquad m = 2, n = 15$$
$$a \qquad b \qquad c$$
$$= 2x(3x + 1) + 5(3x + 1)$$
$$= (2x + 5)(3x + 1)$$

FOIL Correctly factoring a trinomial requires checking the middle term.

$$2x^2 + 3x - 14 = (2x + 7)(x - 2)$$
$$7x$$
$$-4x$$
It checks. $\longrightarrow$ $3x$ ✓

Graphically or Numerically The x-intercepts on the graph of $f(x) = 2x^2 + x - 1$ are $x = 0.5$ and $x = -1$. The zeros of f are 0.5 and -1. Thus $f(x)$ can be factored as

$$2x^2 + x - 1 = 2(x - 0.5)(x + 1)$$
$$= (2x - 1)(x + 1).$$

SECTION 5.5 • SPECIAL TYPES OF FACTORING

Difference of Two Squares $a^2 - b^2 = (a - b)(a + b)$

Example: $49x^2 - 36y^2 = (7x)^2 - (6y)^2 = (7x - 6y)(7x + 6y)$ **$a = 7x, b = 6y$**

Perfect Square Trinomials

$$a^2 + 2ab + b^2 = (a + b)^2 \quad \text{and} \quad a^2 - 2ab + b^2 = (a - b)^2$$

Examples: $x^2 + 2xy + y^2 = (x + y)^2$ and $9r^2 - 12r + 4 = (3r - 2)^2$

Sum and Difference of Two Cubes

$$a^3 + b^3 = (a + b)(a^2 - ab + b^2) \quad \text{and}$$
$$a^3 - b^3 = (a - b)(a^2 + ab + b^2)$$

Examples: $x^3 + y^3 = (x + y)(x^2 - xy + y^2)$ **$a = x, b = y$**
$8t^3 - 27r^6 = (2t - 3r^2)(4t^2 + 6tr^2 + 9r^4)$ **$a = 2t, b = 3r^2$**

SECTION 5.6 • SUMMARY OF FACTORING

Factoring Polynomials The following steps can be used as general guidelines for factoring a polynomial.

STEP 1: Factor out the greatest common factor, if possible.

STEP 2: **A.** If the polynomial has four terms, try factoring by grouping.

B. If the polynomial is a binomial, consider the difference of two squares, difference of two cubes, or sum of two cubes.

C. If the polynomial is a trinomial, check for a perfect square. Otherwise, apply FOIL or grouping, as described in Section 5.4.

STEP 3: Check to make sure that the polynomial is *completely* factored.

SECTION 5.7 • POLYNOMIAL EQUATIONS

Solving Polynomial Equations Follow the strategy presented in Putting It All Together for Section 5.7.

Example:

$x^4 = 2x^2 + 8$	Given equation
$x^4 - 2x^2 - 8 = 0$	Subtract $2x^2 + 8$ from each side.
$(x^2 - 4)(x^2 + 2) = 0$	Factor trinomial.
$(x - 2)(x + 2)(x^2 + 2) = 0$	Difference of squares
$x = 2$ or $x = -2$	Zero-product property; solve.

CHAPTER 5   Review Exercises

SECTION 5.1

1. Give an example of a binomial and of a trinomial.

2. Write a monomial that represents the volume of 10 identical boxes x inches tall with square bases y inches on a side. What is the total volume of the boxes if $x = 5$ and $y = 8$?

Exercises 3–6: Identify the degree and coefficient.

3. $-4x^5$
4. y^3

5. $5xy^6$
6. $-9y^{10}$

Exercises 7–9: Combine like terms.

7. $5x - 4x + 10x$
8. $9x^3 - 5x^3 + x^2$

9. $6x^3y - 4x^3 + 8x^3y + 5x^3$

10. Evaluate $2x^3y^2 - 3y$ when $x = 2$ and $y = -1$.

Exercises 11 and 12: Identify the degree and leading coefficient of the polynomial.

11. $5x^2 - 3x + 5$

12. $x^3 - 5x + 5 + 2x^4$

Exercises 13–16: Combine the polynomials.

13. $(3x^2 - x + 7) + (5x^2 + 4x - 8)$

14. $(6z^3 + z) + (17z^3 - 4z^2)$

15. $(-4x^2 - 6x + 1) - (-3x^2 - 7x + 1)$

16. $(3x^3 - 5x + 7) - (8x^3 + x^2 - 2x + 1)$

Exercises 17 and 18: Evaluate $f(x)$ at the value of x.

17. $f(x) = 2x^2 - 3x + 2$ $x = -1$

18. $f(x) = 1 - x - 4x^3$ $x = 4$

19. Use the graph to evaluate $f(2)$.

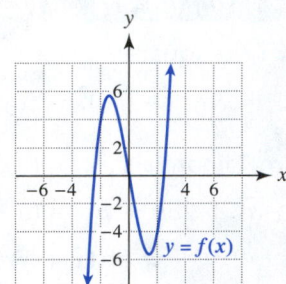

20. If $f(x) = x^2 - 1$ and $g(x) = x - 3x^2$, then find $(f + g)(-2)$ and $(f - g)(x)$.

SECTION 5.2

Exercises 21 and 22: Multiply.

21. $5(3x - 4)$
22. $-2x(1 + x - 4x^2)$

Exercises 23–26: Multiply the monomials.

23. $x^3 \cdot x^5$
24. $-2x^3 \cdot 3x$

25. $(-7xy^7)(6xy)$
26. $(12xy^4)(5x^2y)$

Exercises 27–38: Multiply the expressions.

27. $(x + 4)(x + 5)$
28. $(x - 7)(x - 8)$

29. $(6x + 3)(2x - 9)$
30. $\left(y - \frac{1}{3}\right)\left(y + \frac{1}{3}\right)$

31. $4x^2(2x^2 - 3x - 1)$ **32.** $-x(4 + 5x - 7x^2)$

33. $(4x + y)(4x - y)$ **34.** $(x + 3)^2$

35. $(2y - 5)^2$

36. $(a - b)(a^2 + ab + b^2)$

37. $(5m - 2n^4)^2$

38. $\big((r - 1) + t\big)\big((r - 1) - t\big)$

SECTION 5.3

Exercises 39 and 40: Factor out the greatest common factor.

39. $25x^2 - 30x$ **40.** $12x^4 + 8x^3 - 16x^2$

Exercises 41–44: Use factoring to solve the equation.

41. $x^2 + 3x = 0$ **42.** $7x^4 = 28x^2$

43. $2t^2 - 3t + 1 = 0$

44. $4z(z - 3) + 4(z - 3) = 0$

Exercises 45–47: Use grouping to factor the polynomial.

45. $2x^3 + 2x^2 - 3x - 3$

46. $z^3 + z^2 + z + 1$ **47.** $ax - bx + ay - by$

48. If $f(x) = x + 1$ and $g(x) = x - 3$, find $(fg)(4)$ and $(fg)(x)$.

Exercises 49 and 50: Use the graph to do the following.

(a) Find the x-intercepts.
(b) Solve the equation $P(x) = 0$.
(c) Find the zeros of $P(x)$.

49. **50.**

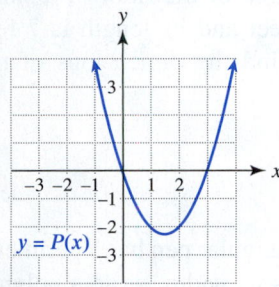
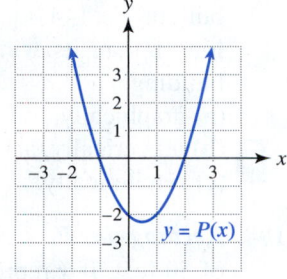

SECTION 5.4

Exercises 51–58: Factor completely.

51. $x^2 + 8x + 12$ **52.** $x^2 - 5x - 50$

53. $9x^2 + 25x - 6$ **54.** $4x^2 - 22x + 10$

55. $x^3 - 4x^2 + 3x$ **56.** $2x^4 + 14x^3 + 20x^2$

57. $5x^4 + 15x^3 - 90x^2$ **58.** $10x^3 - 90x^2 + 200x$

Exercises 59 and 60: Use the table to factor the expression. Check your answer by multiplying.

59. $x^2 - 2x - 15$ **60.** $x^2 - 24x + 143$

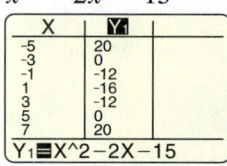
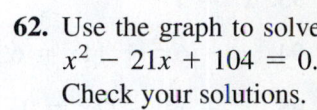

61. Use the graph to factor $x^2 + 3x - 28$. Check your answer by multiplying.

62. Use the graph to solve $x^2 - 21x + 104 = 0$. Check your solutions.

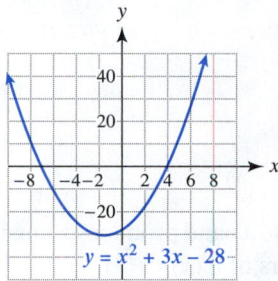
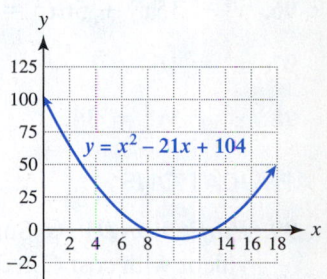

SECTION 5.5

Exercises 63–76: Factor completely.

63. $t^2 - 49$ **64.** $4y^2 - 9x^2$

65. $x^2 + 4x + 4$ **66.** $16x^2 - 8x + 1$

67. $x^3 - 27$ **68.** $64x^3 + 27y^3$

69. $10y^3 - 10y$ **70.** $4r^4 - t^6$

71. $m^4 - 16n^4$ **72.** $n^3 - 2n^2 - n + 2$

73. $25a^2 - 30ab + 9b^2$ **74.** $2r^3 - 12r^2t + 18rt^2$

75. $a^6 + 27b^3$ **76.** $8p^6 - q^3$

SECTION 5.6

Exercises 77–86: Factor completely.

77. $5x^3 - 10x^2$ **78.** $-2x^3 + 32x$

79. $x^4 - 16y^4$ **80.** $4x^3 + 8x^2 - 12x$

81. $-2x^3 + 11x^2 - 12x$ **82.** $x^4 - 8x^2 - 9$

83. $64a^3 + b^3$ **84.** $8 - y^3$

85. $(z + 3)^2 - 16$

86. $x^4 - 5x^3 - 4x^2 + 20x$

SECTION 5.7

Exercises 87 and 88: Solve the equation symbolically and graphically.

87. $x^2 - 16 = 0$ **88.** $x^2 - 2x - 3 = 0$

Exercises 89–98: Solve the equation.

89. $4x^2 - 28x + 49 = 0$

90. $x^2 + 8 = 0$

91. $3x^2 = 2x + 5$

92. $4x^2 + 5x = 6$

93. $x^3 = x$

94. $x^3 - 6x^2 + 11x = 6$

95. $x^3 + x^2 - 72x = 0$

96. $x^4 - 15x^3 + 56x^2 = 0$

97. $x^4 = 16$

98. $x^4 + 5x^2 = 36$

APPLICATIONS

99. *Profit from DVDs* Suppose a band produces x music videos with cost C in dollars of $C(x) = 3x + 9000$. If they sell each DVD for \$15, do the following.
(a) Write a revenue function R that gives the amount the band makes for selling x DVDs.
(b) If profit equals revenue minus cost, write a formula $P(x)$ for the profit function.
(c) Evaluate $P(4000)$ and interpret the answer.

100. *Dimensions of a Box* The length of a box is 5 inches more than its width, its width is 5 inches more than its height, and its volume is 168 cubic inches. Let x be the height.
(a) Write an expression that gives the volume of the box in terms of x.
(b) Write an equation whose solution gives the height of the box.
(c) Find the dimensions of the box.

101. *Picture Frame* A rectangular frame surrounding a picture is made from boards that are 3 inches wide. The picture inside the frame is 2 inches wider than it is high. If the overall area of the picture and frame is 224 square inches, find the dimensions of the picture.

102. *Geometry* Suppose that a triangle has base $x + 2$ and height $x - 3$. Find a polynomial that gives the area of the triangle.

103. *High Temperatures* The formula

$$f(x) = -1.466x^2 + 20.25x + 9$$

models the monthly average high temperatures in degrees Fahrenheit at Columbus, Ohio. In this formula, let $x = 1$ be January, $x = 2$ be February, and so on. (*Source:* J. Williams, *The Weather Almanac.*)

(a) What is the average high temperature in May?
(b) Make a table of $f(x)$, starting at $x = 1$ and incrementing by 1. During what month is the average temperature the greatest?
(c) Graph f in [1, 12, 1] by [30, 90, 10] and interpret the graph.

104. *Softball Pitching* Suppose that the likelihood, or chance, that a pitch in softball will be a strike is x percent. Then the likelihood, as a percentage, that two consecutive pitches will *not* be strikes is given by $100\left(1 - \frac{x}{100}\right)^2$.
(a) Multiply this expression.
(b) Let $x = 70$. Use both the given expression and the expression you found in part (a) to obtain the likelihood that two consecutive pitches are not strikes.

105. *Numbers* Write a polynomial that represents the product of three consecutive integers, where x corresponds to the largest integer.

106. *Area* Use the area of a rectangle to illustrate that $(x + 2)(x + 7) = x^2 + 9x + 14$.

107. *Area* Suppose that the area of the floor of a small building is 144 square feet and its length is 7 feet longer than its width. Find the dimensions of the building
(a) graphically,
(b) numerically, and
(c) symbolically.

108. *Rectangular Pen* A rectangular pen has a perimeter of 50 feet. If x represents its width, write a polynomial that gives the area of the pen in terms of x.

109. *Flight of a Golf Ball* If a golf ball is hit upward with a velocity of 66 feet per second (45 mph), its height h in feet above the ground after t seconds can be modeled by

$$h(t) = -16t^2 + 66t.$$

(a) Determine when the ball strikes the ground.
(b) When is the height of the ball 50 feet?

110. *Ticket Prices* If tickets are sold for \$50, then 600 tickets are expected to be sold. For each \$1 reduction in price, an additional 20 tickets will be sold.
(a) Write an equation that gives the revenue R from ticket sales when ticket prices are reduced by x dollars.
(b) Determine the ticket price that results in sales of \$32,000.

 111. *Modeling a Hill* The elevation E of a hill x feet along a highway can be modeled by

$$E(x) = -0.0001x^2 + 500,$$

where E is in feet. The crest of the hill corresponds to $x = 0$. Find values of x where E is 400 feet.

CHAPTER 5   Test

 Step-by-step test solutions are found on the Chapter Test Prep Videos available in *MyMathLab* and on YouTube (search "RockswoldInterAlg" and click on "Channels").

Exercises 1 and 2: Simplify by combining like terms.

1. $x^2y^2 - 4x + 9x - 5x^2y^2$

2. $(-2x^3 - 6x + 1) - (5x^3 - x^2 + x - 10)$

3. Evaluate $f(x) = 2x^3 - x^2 - 5x + 2$ at $x = -2$.

4. Use the graph to evaluate $f(2)$.

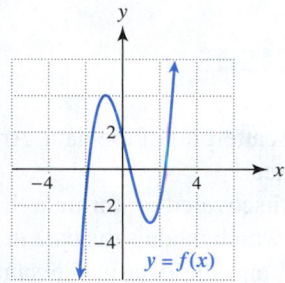

$y = f(x)$

5. Give the degree and coefficient of $3x^3y$.

6. Write a monomial that gives the volume of 8 identical boxes with width x, length y, and height z.

7. Evaluate $-x^2y + 3xy^2$ when $x = -2$ and $y = 3$.

8. If $f(x) = 2x - 5$ and $g(x) = 1 - x^3$, then find $(f - g)(4)$ and $(f + g)(x)$.

Exercises 9–16: Multiply and simplify.

9. $-\frac{2}{5}x^2(10x - 5)$

10. $2xy^7 \cdot 7xy$

11. $(2x + 1)(5x - 7)$

12. $(5 - 3x)^2$

13. $(5x - 4y)(5x + 4y)$

14. $-2x^2(x^2 - 3x + 2)$

15. $(x - 2y)(x^2 + 2xy + 4y^2)$

16. $2x^2(x - 1)(x + 1)$

Exercises 17–25: Factor completely.

17. $x^2 - 3x - 10$

18. $2x^3 + 6x$

19. $3x^2 + 7x - 20$

20. $5x^4 - 5x^2$

21. $2x^3 + x^2 - 10x - 5$

22. $49x^2 - 14x + 1$

23. $x^3 + 8$

24. $4x^2y^4 + 8x^4y^2$

25. $a^2 - 3ab + 2b^2$

26. Identify the degree and leading coefficient of the polynomial $5 - 4x + x^3$.

27. When we multiply $(2m^3 - 4n^2)^2$ we get which of the following (a.–d.)?
a. $4m^6 - 8m^3n^2 + 16n^4$ b. $4m^6 - 16n^4$
c. $4m^6 - 16m^3n^2 + 16n^4$ d. $4m^6 + 16n^4$

28. Use the graph to factor $x^2 + 2x - 48$.

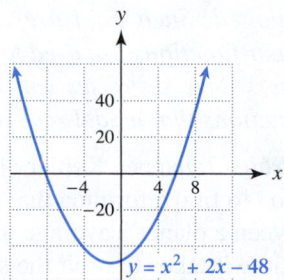

$y = x^2 + 2x - 48$

29. *Numbers* Write a polynomial that represents the product of two consecutive even integers, where x is the smaller integer. Multiply the expression.

Exercises 30–34: Use factoring to solve the polynomial equation.

30. $5x^2 = 15x$

31. $4t^2 + 19t - 5 = 0$

32. $2z^4 - 8z^2 = 0$

33. $x^4 - 2x^2 + 1 = 0$

34. $x(x - 3) + (x - 3) = 0$

35. *Area* A rectangular picture frame is 4 inches wider than it is high and has an area of 221 square inches.
(a) Write an equation whose solution gives the height of the picture frame.
(b) Solve the equation in part (a).
(c) Find the perimeter of the picture frame.

36. *Dew Point* The formula

$$f(x) = -0.091x^3 + 0.66x^2 + 5.78x + 23.5$$

models the monthly average dew point in degrees Fahrenheit in Birmingham, Alabama. In this formula $x = 1$ corresponds to January, $x = 2$ to February, and so on. (*Source:* J. Williams, *The Weather Almanac.*)

(a) What is the average dew point in May?

(b) Use the graph of f to discuss how the dew point changes during the year.

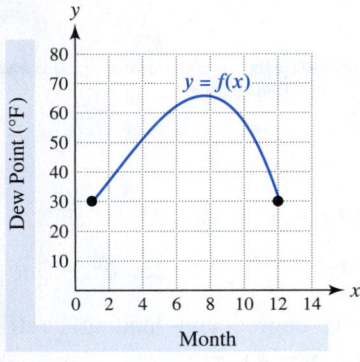

Month

37. *Geometry* Multiply $(x + 2)(x + 3)$ geometrically by drawing a rectangle.

38. *Height Reached by a Baseball* After a baseball is hit, the height h in feet reached after t seconds is given by

$$h(t) = -16t^2 + 96t + 3.$$

Determine when the baseball is 131 feet in the air.

CHAPTER 5   Extended and Discovery Exercises

MODELING NONLINEAR DATA

Exercises 1–3: Real-world data often do not lie on a line when they are plotted. Such data are called nonlinear data, and nonlinear functions are used to model this type of data. In the next three exercises you will be asked to find your own functions that model real-world data.

 1. *Planetary Orbits* Johannes Kepler (1571–1630) was the first person to find a formula that models the relationship between a planet's average distance from the sun and the time it takes to orbit the sun. The following table lists a planet's average distance x from the sun and the number of years y it takes to orbit the sun. In this table distances have been normalized so that Earth's distance is exactly 1. For example, Saturn is 9.54 times farther from the sun than Earth is and requires 29.5 years to orbit the sun.

Planet	x (distance)	y (period)
Mercury	0.387	0.241
Venus	0.723	0.615
Earth	1.00	1.00
Mars	1.52	1.88
Jupiter	5.20	11.9
Saturn	9.54	29.5

Source: C. Ronan, *The Natural History of the Universe.*

(a) Make a scatterplot of the data. Are the data linear or nonlinear?

(b) Kepler discovered a formula having the form $y = x^k$, which models the data in the table. Graph $y = x^{2.5}$ and the data in the same viewing rectangle. Does y model the data accurately? Explain.

(c) Try to discover a value for k so that $y = x^k$ models the data.

(d) **Online Exploration** The average distances of Neptune and Pluto (no longer a planet) from the sun are 30.1 and 39.4, respectively. Estimate how long it takes for each of them to orbit the sun. Go online to see whether your answers are accurate.

2. *Aging of America* Americans are living longer. The table shows the number of Americans who were more than 100 years old for various years.

Year	1994	1996	1998
Number (thousands)	50	56	65

Year	2000	2002	2004
Number (thousands)	75	94	110

Source: U.S. Census Bureau.

(a) Make a scatterplot of the data. Choose an appropriate viewing rectangle.

(b) Use trial and error to find a value for k so that

$$y = k(x - 1994)^2 + 50$$

models the data.

(c) Use y to estimate the number of Americans that were more than 100 years old in 2006.

3. *Charitable Giving* Average charitable giving varies according to income. The average amount given for three different incomes is listed in the table.

Income	$10,000	$40,000	$100,000
Giving	$412	$843	$2550

Source: Gallup for Independent Sector.

(a) Let x be income and y be charitable giving. Set up an augmented matrix whose solution gives values for a, b, and c so that the graph of $y = ax^2 + bx + c$ passes through the points $(10000, 412)$, $(40000, 843)$, and $(100000, 2550)$.

(b) Use a graphing calculator to solve this system of linear equations.

(c) Estimate the average charitable giving for someone earning $20,000. Compare your answer to the actual value of $525.

CHAPTERS 1–5   Cumulative Review Exercises

1. Evaluate $A = \frac{1}{2}ab$ when $a = 5$ and $b = \frac{3}{2}$.

2. Find the reciprocal of $\frac{a}{b-a}$.

3. Simplify each expression. Write the result using positive exponents.

(a) $\left(\dfrac{x^{-3}}{y}\right)^2$ **(b)** $\dfrac{(3r^{-1}t)^{-4}}{r^2(t^2)^{-2}}$ **(c)** $(ab^{-2})^4(ab^3)^{-1}$

4. Write 5.859×10^4 in standard (decimal) form.

5. Find $f(-3)$, if $f(x) = -2x^2 + 3x$.

6. Find the domain of $f(x) = \sqrt{x - 4}$.

Exercises 7 and 8: Graph f by hand.

7. $f(x) = 3 - 2x$ **8.** $f(x) = x^2 - x - 2$

9. Use the graph to answer the following.
   (a) Does the graph represent a function?
   (b) Identify the domain and range.
   (c) Evaluate $f(1)$ and $f(2)$.
   (d) Identify the x-intercepts.
   (e) Solve the equation $f(x) = 0$.
   (f) Write $f(x)$ completely factored, if $f(x)$ is quadratic with leading coefficient 1.

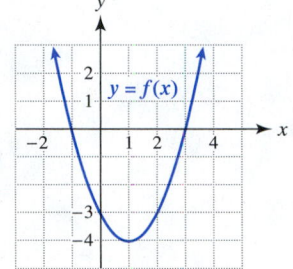

10. Find the slope–intercept form of the line that passes through $(-2, 5)$ and is parallel to $2x - y = 4$.

11. Let f be a linear function. Find a formula for $f(x)$.

x	-2	-1	1	2
$f(x)$	12	7	-3	-8

12. Solve $-2(5x - 1) = 1 - (5 - x)$.

13. Solve $3x + 4 \leq x - 1$.

14. Solve $|5x - 10| > 5$.

15. Solve $-2 < 2 - 7x \leq 2$.

16. Solve the system graphically or numerically.

$$4x + 2y = 10$$
$$-x + 5y = 3$$

17. Solve each system symbolically.

(a) $\begin{aligned} -x - 2y &= 5 \\ 2x + 4y &= -10 \end{aligned}$ **(b)** $\begin{aligned} 3x + 2y &= 7 \\ 2x - 2y &= 3 \end{aligned}$

18. Shade the solution set in the xy-plane.

$$2x + y \leq 4$$
$$x - 2y < -2$$

19. Solve the system.

$$x + y - z = -2$$
$$x - y + z = -6$$
$$x - y - 2z = 3$$

20. Use Gauss–Jordan elimination to solve the system.

$$x - y - z = -2$$
$$2x + y + z = 8$$
$$-x - y + 2z = 0$$

21. Calculate $\det \begin{bmatrix} 0 & 1 & -3 \\ 1 & 0 & -1 \\ 3 & 1 & 0 \end{bmatrix}$

22. Solve $x^2 + 4 = 0$, if possible.

Exercises 23–28: Multiply the expression.

23. $-2x(x^2 - 2x + 5)$ **24.** $(2a + b)(2a - b)$

25. $(x - 1)(x + 1)(x + 3)$ **26.** $(4x + 9)(2x - 1)$

27. $(x^2 + 3y^3)^2$ **28.** $-2x(1 - x^2)$

Exercises 29–34: Factor the expression.

29. $x^2 - 8x - 33$ **30.** $10x^3 + 65x^2 - 35x$

31. $4x^2 - 100$ **32.** $49x^2 - 70x + 25$

33. $r^4 - r$ **34.** $x^3 + 2x^2 + x + 2$

Exercises 35–38: Solve the equation.

35. $4x^2 - 1 = 0$ **36.** $3x^2 + 14x - 5 = 0$

37. $x^3 + 4x = 4x^2$ **38.** $x^4 = x^2$

 Exercises 39 and 40: Solve each equation graphically or numerically to the nearest tenth.

39. $\sqrt{2}x - 1.1(x - \pi) = 1 - 2x$

40. $(\pi - 1)x^2 - \sqrt{3} = 5 - 1.3x$

APPLICATIONS

41. *Investment* Suppose $4000 is deposited in two accounts paying 6% and 7% annual interest. If the interest after 1 year is $257, how much is invested at each rate?

42. *Cost of a Car* Suppose the cost C of an antique car from 1990 to 2010 is modeled by

$$C(x) = 750(x - 1990) + 6800.$$

(a) Evaluate $C(1995)$ and interpret the result.
(b) Interpret the slope of the graph of C.

43. *Geometry* A rectangular box with a square base has a volume of 1008 cubic inches. If its height is 5 inches less than the length of an edge of the base, find the dimensions of the box.

44. *Angles in a Triangle* The measure of the largest angle in a triangle is 20° less than the sum of the measures of the two smaller angles. The sum of the measures of the two larger angles is 100° greater than the measure of the smaller angle. Find the measure of each angle.

6 Rational Expressions and Functions

The future belongs to those who believe in the beauty of their dreams.

—ELEANOR ROOSEVELT

On average, a person spends an estimated 45 to 60 minutes waiting each day. This waiting could be at the doctor's office, at the grocery store, in airports, or stuck in traffic. In a 70-year life span, a person might spend as much as 3 years waiting!

Motorists in the nation's top 68 urban areas spend about 50 percent more time stuck in traffic than they did 10 years ago. Traffic congestion delays travelers 3.7 billion hours and wastes 2.3 billion gallons of fuel at a cost of $63 billion each year. Many urban areas are not adding enough roads and highways to keep traffic moving.

The time spent waiting in lines is subject to a *nonlinear effect*. For example, you can put more cars on the road up to a point. Then, if the traffic intensity increases even slightly, congestion and long lines increase dramatically. See the graph below. This phenomenon cannot be modeled by linear or polynomial functions. Instead we need a new function, called a rational function, to describe this type of behavior. (See Example 6 in Section 6.4.)

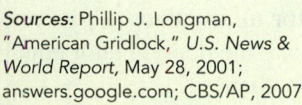

Sources: Phillip J. Longman, "American Gridlock," *U.S. News & World Report*, May 28, 2001; answers.google.com; CBS/AP, 2007.

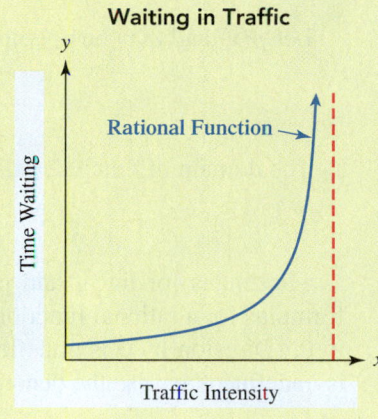

Waiting in Traffic

Rational Function

Time Waiting

Traffic Intensity

6.1   Introduction to Rational Functions and Equations

Recognizing and Using Rational Functions • Solving Rational Equations • Operations on Functions

A LOOK INTO MATH ▶

Suppose that you are trying to exit a parking ramp. If the parking attendant can wait on 5 cars per minute and cars are randomly leaving the parking ramp at a rate of 4 cars per minute, how long is your wait? What happens to your wait if cars start arriving randomly at nearly the parking attendant's rate of 5 cars per minute? Using a new function, called a *rational function*, we can answer these questions and also describe such things as how people participate in social networks. (See Example 8 and Exercise 105.)

Recognizing and Using Rational Functions

NEW VOCABULARY

☐ Rational expression
☐ Rational function
☐ Vertical asymptote
☐ Rational equation
☐ Extraneous solution

In Chapter 5 we discussed polynomials. Examples of polynomials include

$$4, \quad x, \quad x^2, \quad 2x - 5, \quad 3x^2 - 6x + 1, \quad \text{and} \quad 3x^4 - 7.$$

Rational expressions result when a polynomial is divided by a nonzero polynomial. Three examples of rational expressions are

$$\frac{4}{x}, \quad \frac{x^2}{2x - 5}, \quad \text{and} \quad \frac{3x^2 - 6x + 1}{3x^4 - 7}. \quad \text{Rational expressions}$$

EXAMPLE 1   **Recognizing a rational expression**

Determine whether each expression is rational.

(a) $\dfrac{4 - x}{5x^2 + 3}$ (b) $\dfrac{x^2 - 1}{\sqrt{x} + 2x}$ (c) $\dfrac{6}{x - 1}$

Solution
(a) This expression is rational because both $4 - x$ and $5x^2 + 3$ are polynomials.
(b) This expression is not rational because $\sqrt{x} + 2x$ is not a polynomial.
(c) This expression is rational because both 6 and $x - 1$ are polynomials.

Now Try Exercises 11, 13, 15

Rational expressions are used to define *rational functions*. For example, $\frac{6}{x - 1}$ is a rational *expression*, so $f(x) = \frac{6}{x - 1}$ represents a rational *function*.

> ### RATIONAL FUNCTION
>
> Let $p(x)$ and $q(x)$ be polynomials. Then a **rational function** is given by
>
> $$f(x) = \frac{p(x)}{q(x)}.$$
>
> The domain of f includes all x-values such that $q(x) \neq 0$.

A Rational Function

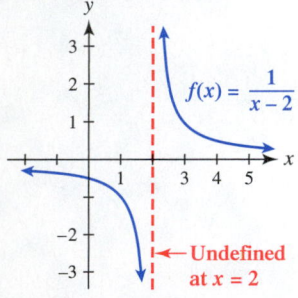

$f(x) = \frac{1}{x - 2}$

← Undefined at $x = 2$

Figure 6.1

Formulas for linear and polynomial functions are *defined* for all x-values. However, formulas for a rational function are *undefined* for x-values that make the denominator equal to 0. (Division by 0 is undefined.) For example, if $f(x) = \frac{1}{x - 2}$, then $f(2) = \frac{1}{2 - 2} = \frac{1}{0}$ is undefined because the denominator equals 0. A graph of $y = f(x)$ is shown in Figure 6.1.

The graph does not cross the dashed vertical line $x = 2$, because $f(x)$ is *undefined* at $x = 2$. The red vertical dashed line $x = 2$ is called a *vertical asymptote*, and is used as an aid for sketching a graph of f. It is not actually part of the graph of the function.

EXAMPLE 2   **Identifying the domain of rational functions**

Identify the domain of each function.

(a) $f(x) = \dfrac{1}{x + 2}$ (b) $g(x) = \dfrac{2x}{x^2 - 3x + 2}$ (c) $h(t) = \dfrac{4}{t^3 - t}$

Solution

(a) The domain of f includes all x-values for which the expression $f(x) = \dfrac{1}{x+2}$ is defined. Thus we must *exclude* values of x for which the denominator equals 0.

$$x + 2 = 0 \quad \text{Set denominator equal to 0.}$$
$$x = -2 \quad \text{Subtract 2.}$$

In set-builder notation, $D = \{x \mid x \neq -2\}$.

(b) The domain of g includes all values of x except where the denominator equals 0.

$$x^2 - 3x + 2 = 0 \quad \text{Set denominator equal to 0.}$$
$$(x - 1)(x - 2) = 0 \quad \text{Factor.}$$
$$x = 1 \quad \text{or} \quad x = 2 \quad \text{Zero-product property; solve.}$$

Thus $D = \{x \mid x \neq 1, x \neq 2\}$.

(c) The domain of h includes all values of t except where the denominator equals 0.

$$t^3 - t = 0 \quad \text{Set denominator equal to 0.}$$
$$t(t^2 - 1) = 0 \quad \text{Factor out } t.$$
$$t(t - 1)(t + 1) = 0 \quad \text{Factor difference of squares.}$$
$$t = 0 \quad \text{or} \quad t = 1 \quad \text{or} \quad t = -1 \quad \text{Zero-product property; solve.}$$

Thus $D = \{t \mid t \neq 0, t \neq 1, t \neq -1\}$.

Now Try Exercises 21, 27, 29

To graph a rational function by hand, we usually start by making a table of values, as demonstrated in the next example. Because the graphs of rational functions are typically nonlinear, it is a good idea to plot at least 3 points on each side of an x-value where the formula is undefined, that is, where the denominator equals 0.

EXAMPLE 3   **Graphing a rational function**

Graph $f(x) = \dfrac{1}{x}$. State the domain of f.

Solution

Make a table of values for $f(x) = \frac{1}{x}$, as shown in Table 6.1. Notice that $x = 0$ is not in the domain of f, and a dash can be used to denote this undefined value. The domain of f is all real numbers such that $x \neq 0$. Start by picking three x-values on each side of 0.

TABLE 6.1

x	-2	-1	$-\frac{1}{2}$	0	$\frac{1}{2}$	1	2
$\frac{1}{x}$	$-\frac{1}{2}$	-1	-2	—	2	1	$\frac{1}{2}$

↑ —— **Undefined**

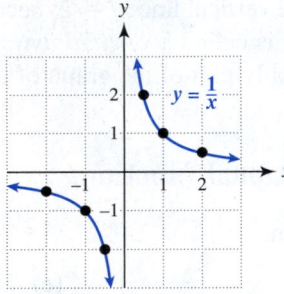

Figure 6.2

Plot the points shown in Table 6.1 and then connect the points with a smooth curve, as shown in Figure 6.2. Because $f(0)$ is undefined, the graph of $f(x) = \frac{1}{x}$ does not cross the line $x = 0$ (the y-axis). The line $x = 0$ is a vertical asymptote.

TABLE 6.1 **(Repeated)**

x	-2	-1	$-\frac{1}{2}$	0	$\frac{1}{2}$	1	2
$\frac{1}{x}$	$-\frac{1}{2}$	-1	-2	—	2	1	$\frac{1}{2}$

↑——— **Undefined**

▌ **Now Try Exercise 31**

In the next example we evaluate a rational function in three ways.

EXAMPLE 4 **Evaluating a rational function**

Use Table 6.2, the formula for $f(x)$, and Figure 6.3 to evaluate $f(-1)$, $f(1)$, and $f(2)$.

(a) **TABLE 6.2**

x	$f(x)$
-3	$\frac{3}{2}$
-2	$\frac{4}{3}$
-1	1
0	0
1	—
2	4
3	3

(b) $f(x) = \dfrac{2x}{x - 1}$

(c)

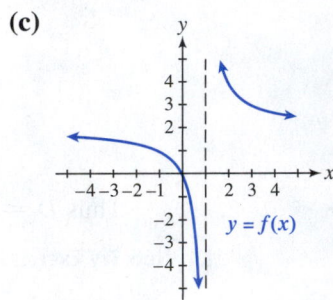

Figure 6.3

Solution

(a) *Numerical Evaluation* Table 6.2 shows that

$$f(-1) = 1, \quad f(1) \text{ is undefined}, \quad \text{and} \quad f(2) = 4.$$

(b) *Symbolic Evaluation*

$$f(-1) = \frac{2(-1)}{-1 - 1} = 1$$

$$f(1) = \frac{2(1)}{1 - 1} = \frac{2}{0}, \text{ which is undefined. Input 1 is } \textit{not} \text{ in the domain of } f.$$

$$f(2) = \frac{2(2)}{2 - 1} = 4$$

(c) *Graphical Evaluation* To evaluate $f(-1)$ graphically, find $x = -1$ on the x-axis and move upward to the graph of f. The y-value is **1** at the point of intersection, so $f(-1) = 1$, as shown in Figure 6.4(a). In Figure 6.4(b) the red, dashed vertical line $x = 1$ is called a vertical asymptote. Because the graph of f does not intersect this line, $f(1)$ is undefined. Figure 6.4(c) reveals that $f(2) = 4$.

Evaluating a Rational Function Graphically

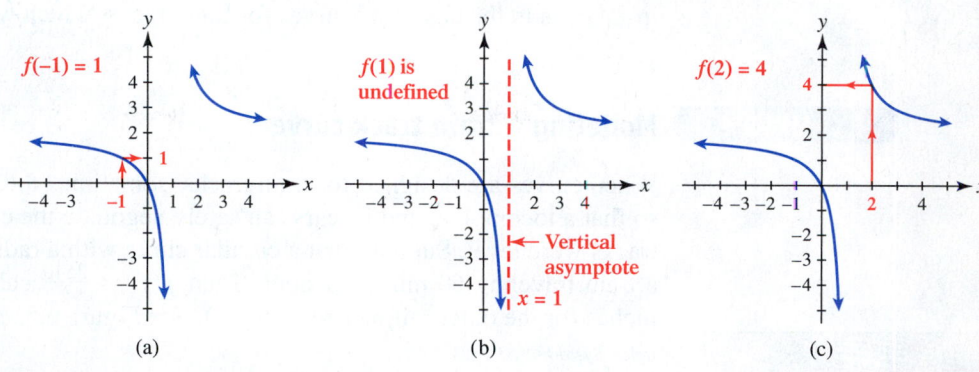

Figure 6.4

■ **Now Try Exercises 43, 49, 53**

MAKING CONNECTIONS

Asymptotes and Graphs of Rational Functions

A **vertical asymptote** often occurs at x-values in the graph of a rational function $y = f(x)$ where the denominator equals 0. A vertical asymptote can be used as an aid to sketch the graph of a rational function. However, the graph of a rational function *never* crosses a vertical asymptote, so a vertical asymptote is *not* part of the graph of f.

On either side of a vertical asymptote, the y-values on the graph of a rational function typically become either very large (approach ∞) or very small (approach $-\infty$). See the figure below.

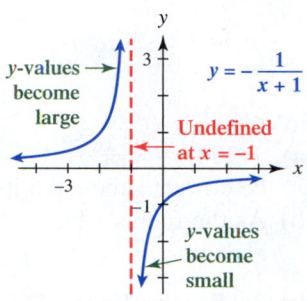

TECHNOLOGY NOTE

Asymptotes, Dot Mode, and Decimal Windows

When a rational function is graphed on a graphing calculator in connected mode, pseudo-asymptotes often occur because the calculator is simply connecting dots to draw a graph. The accompanying figures show the graph of $y = \frac{2}{x - 2}$ in connected mode, in dot mode, and with a *decimal*, or *friendly*, window. In dot mode, pixels in the calculator screen are not connected. With dot mode (and sometimes with a decimal window) pseudo-asymptotes do not appear. To learn more about these features, consult your owner's manual.

CALCULATOR HELP

To set a calculator in dot mode or to set a decimal window, see Appendix A (pages AP-10 and AP-11).

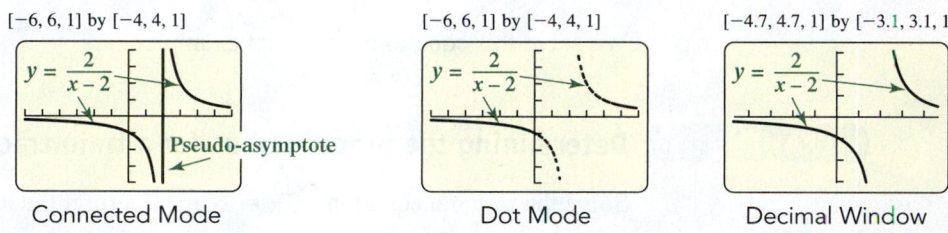

▶ **REAL-WORLD CONNECTION** Applications involving rational functions are numerous. One instance is in the design of curves for train tracks, which we discuss in the next example.

EXAMPLE 5 Modeling a train track curve

When curves are designed for train tracks, sometimes the outer rail is elevated, or banked, so that a locomotive and its cars can safely negotiate the curve at a higher speed than if the tracks were level. Suppose that a circular curve with a radius of r feet is being designed for a train traveling 60 miles per hour. Then $f(r) = \frac{2540}{r}$ calculates the proper elevation y in inches for the outer rail, where $y = f(r)$. See Figure 6.5. (*Source:* L. Haefner, *Introduction to Transportation Systems.*)

(a) Evaluate $f(300)$ and interpret the result.
(b) A graph of f is shown in Figure 6.6. Discuss how the elevation of the outer rail changes as the radius r increases.

Figure 6.5

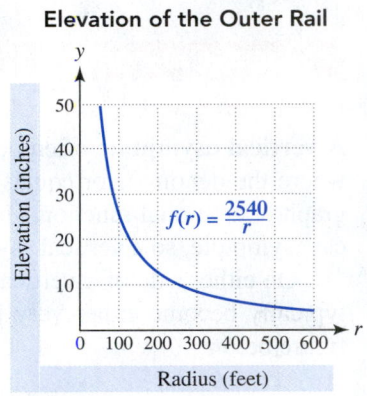

Elevation of the Outer Rail

$$f(r) = \frac{2540}{r}$$

Figure 6.6

Solution
(a) $f(300) = \frac{2540}{300} \approx 8.5$. Thus the outer rail on a curve with a radius of 300 feet should be elevated about 8.5 inches for a train to safely travel through it at 60 miles per hour.
(b) As the radius r increases (and the curve becomes less sharp), the outer rail needs less elevation.

▌ **Now Try Exercise 101(a)**

Solving Rational Equations

▶ **REAL-WORLD CONNECTION** When working with rational functions, we commonly encounter rational equations. In Example 5, we used $f(r) = \frac{2540}{r}$ to calculate the elevation of the outer rail in a train track curve. Now suppose that the outer rail for a curve is elevated **6** inches. What should be the radius of the curve? To answer this question, we need to solve the **rational equation**

$$\frac{2540}{r} = 6. \quad \text{Rational equation}$$

We solve this equation in the next example.

EXAMPLE 6 Determining the proper radius for a train track curve

Solve the rational equation $\frac{2540}{r} = 6$ and interpret the result.

Solution
We begin by multiplying each side of the equation by r.

$$r \cdot \frac{2540}{r} = 6 \cdot r \qquad \text{Multiply by } r.$$

$$2540 = 6r \qquad \text{Simplify.}$$

$$\frac{2540}{6} = r \qquad \text{Divide by 6.}$$

$$r = 423.\overline{3} \qquad \text{Rewrite.}$$

A train track curve designed for 60 miles per hour and banked 6 inches should have a radius of about 423 feet.

▊ **Now Try Exercise 101(d)**

Multiplying each side of the rational equation in the form $\frac{N}{D} = C$ by the denominator D is often a good step for solving this equation (see Example 6), because

$$D \cdot \frac{N}{D} = C \cdot D \quad \text{simplifies to} \quad N = CD,$$

provided that $D \neq 0$. For example, to solve

$$\frac{x}{12} = 5,$$

we multiply each side of the equation by **12**, because

$$12 \cdot \frac{x}{12} = 5 \cdot 12 \quad \text{simplifies to} \quad x = 5 \cdot 12 = 60.$$

We use this technique in Example 7. Be sure to check your answer.

EXAMPLE 7 ▶ **Solving rational equations**

Solve each rational equation and check your answer.

(a) $\dfrac{3x}{2x - 1} = 3$ **(b)** $\dfrac{x + 1}{x - 2} = \dfrac{3}{x - 2}$ **(c)** $\dfrac{6}{x + 1} = x$

Solution
(a) First note that $\frac{1}{2}$ cannot be a solution to this equation because $x = \frac{1}{2}$ results in the left side of the equation being undefined; division by 0 is not possible. Multiply each side of the equation by the denominator, $2x - 1$.

$$(2x - 1) \cdot \frac{3x}{2x - 1} = 3 \cdot (2x - 1) \qquad \text{Multiply by } (2x - 1).$$

$$3x = 6x - 3 \qquad \text{Simplify.}$$

$$-3x = -3 \qquad \text{Subtract } 6x.$$

$$x = 1 \qquad \text{Divide by } -3.$$

To check your answer, substitute **1** in the given equation for x.

$$\frac{3(\mathbf{1})}{2(\mathbf{1}) - 1} = 3 \ \checkmark \qquad \text{The answer checks.}$$

(b) Each side of the equation is undefined when $x = 2$. Thus 2 cannot be a solution. Begin by multiplying each side of the equation by the denominator, $x - 2$.

$$(x - 2) \cdot \frac{x + 1}{x - 2} = \frac{3}{x - 2} \cdot (x - 2) \qquad \text{Multiply by } x - 2.$$

$$x + 1 = 3 \qquad \text{Simplify.}$$

$$x = 2 \qquad \text{Subtract 1.}$$

In this case 2 is not a valid solution. There are no solutions. Instead, 2 is called an **extraneous solution**, which cannot be used. It is *important* to check your answers, as follows.

$$\frac{2 + 1}{2 - 2} \stackrel{?}{=} \frac{3}{2 - 2} \qquad \text{Let } x = 2 \text{ in given equation.}$$

$$\frac{3}{0} \stackrel{?}{=} \frac{3}{0} \; ✗ \qquad \text{Division by 0 is undefined.}$$

READING CHECK

Why is it important to check your answer when solving a rational equation?

(c) The left side of the equation is undefined when $x = -1$. Thus -1 cannot be a solution. To solve this equation start by multiplying each side by the denominator, $x + 1$.

$$(x + 1) \cdot \frac{6}{x + 1} = x \cdot (x + 1) \qquad \text{Multiply by } x + 1.$$

$$6 = x^2 + x \qquad \text{Simplify.}$$

$$x^2 + x - 6 = 0 \qquad \text{Rewrite the equation.}$$

$$(x + 3)(x - 2) = 0 \qquad \text{Factor.}$$

$$x = -3 \quad \text{or} \quad x = 2 \qquad \text{Zero-product property}$$

Checking these results confirms that both -3 and 2 are solutions.

■ **Now Try Exercises 57, 65, 69**

Like other equations, rational equations can be solved symbolically, graphically, and numerically, as shown in this Making Connections.

MAKING CONNECTIONS

Symbolic, Numerical, and Graphical Solutions to a Rational Equation

Symbolic Solution

$$\frac{x + 3}{2x} = 2$$

$$2x \cdot \frac{x + 3}{2x} = 2 \cdot 2x$$

$$x + 3 = 4x$$

$$3 = 3x$$

$$x = 1$$

The solution is **1**.

Numerical Solution

x	$(x + 3)/(2x)$
-2	-0.25
-1	-1
0	—
1	2
2	1.25

When $x = 1$, $\frac{x + 3}{2x} = 2$.

Graphical Solution

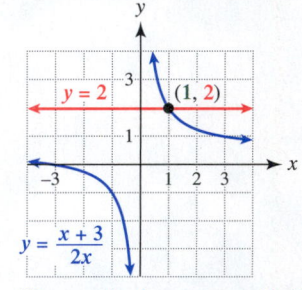

The graphs intersect at $(1, 2)$.

STUDY TIP

It can be helpful to talk and study with other students. However, make sure that others are not "doing the work for you."

▶ **REAL-WORLD CONNECTION** You may have noticed that a relatively small percentage of people do the vast majority of postings on social networks, such as Facebook and Twitter. This phenomenon is called *participation inequality*. That is, a vast majority of the population falls under the category of "lurkers," who are on the network but are not posting material. This characteristic of a social network can be modeled approximately by a rational function, as illustrated in the next example. (*Source*: Wu, Michael, *The Economics of 90–9–1*.)

EXAMPLE 8   **Modeling social network participation**

The rational function given by

$$f(x) = \frac{100}{101 - x}, \ 5 \le x \le 100,$$

models participation inequality in a social network. In this formula, $f(x)$ outputs the percentage of the postings done by the least active (bottom) x percent of the population.
(a) Evaluate $f(95)$. Interpret your answer.
(b) A graph of $y = f(x)$ is shown in Figure 6.7. Interpret the graph.

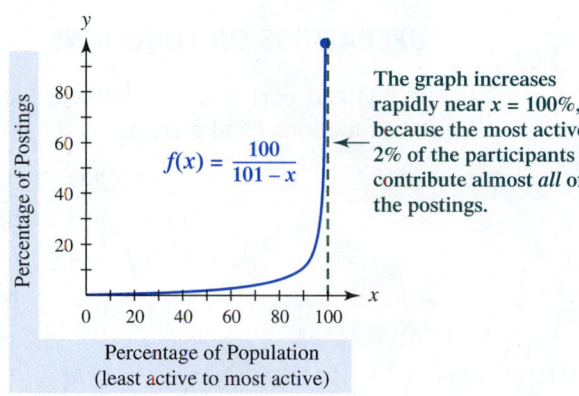

Participation in a Social Network

The graph increases rapidly near $x = 100\%$, because the most active 2% of the participants contribute almost *all* of the postings.

$$f(x) = \frac{100}{101 - x}$$

Percentage of Postings

Percentage of Population
(least active to most active)

Figure 6.7 Participation Inequality

(c) Solve the rational equation $\frac{100}{101 - x} = 9$. Interpret your answer.

Solution
(a) $f(\mathbf{95}) = \dfrac{100}{101 - \mathbf{95}} = \dfrac{100}{6} \approx 16.7\%$ Let $x = 95$.

This means that the least active 95% of the population contributes only 16.7% of the postings, so the most active 5% of the population is responsible for the remaining 83.3% of the postings.
(b) The graph shows participation inequality visually. The graph remains at a relatively low percentage until $x = 90\%$. This means that the bottom 90% of the population does very few postings. For $x \ge 90\%$ the graph rises rapidly because the top 10% contributes a vast majority of the postings.
(c) To solve this equation, we begin by multiplying each side by $(\mathbf{101 - x})$.

$\dfrac{100}{101 - x} = 9$	Given equation
$(\mathbf{101 - x}) \cdot \dfrac{100}{\mathbf{101 - x}} = 9(\mathbf{101 - x})$	Multiply by $(101 - x)$.
$100 = 9(101 - x)$	Simplify left side.
$100 = 909 - 9x$	Distributive property
$\mathbf{9x} + 100 - \mathbf{100} = 909 - \mathbf{100} + \mathbf{9x} - 9x$	Add $9x$. Subtract 100.
$9x = 809$	Simplify.
$x = \dfrac{809}{9} \approx 90\%$	Simplify.

This result indicates that the least active 90% of the population contributes only 9% of the postings.

Now Try Exercise 105

CRITICAL THINKING

Suppose that a social network had *participation equality*, in which every member contributed an equal number of postings. Sketch a graph like Figure 6.7 that describes this social network.

Operations on Functions

▶ **REAL-WORLD CONNECTION** The ratio of two functions can give us information. For example, suppose that $G(x)$ gives the revenue for Google x years after its startup, and that $F(x)$ gives the revenue for Facebook x years after its startup. The function of their ratios, $R(x) = \frac{G(x)}{F(x)}$, gives a measure of how their revenues compare in size. (See Exercises 111 and 112.)

We have performed addition, subtraction, multiplication, and division on numbers and variables. In Chapter 5, we performed addition, subtraction, and multiplication on functions. In this subsection we will review these operations and also perform division on functions.

OPERATIONS ON FUNCTIONS

If $f(x)$ and $g(x)$ are both defined, then the sum, difference, product, and quotient of two functions f and g are defined by

$$(f + g)(x) = f(x) + g(x) \qquad \text{Sum}$$
$$(f - g)(x) = f(x) - g(x) \qquad \text{Difference}$$
$$(fg)(x) = f(x) \cdot g(x) \qquad \text{Product}$$
$$\left(\frac{f}{g}\right)(x) = \frac{f(x)}{g(x)}, \text{ where } g(x) \neq 0. \quad \text{Quotient}$$

EXAMPLE 9   **Performing arithmetic on functions**

Use $f(x) = x^2$ and $g(x) = 2x - 4$ to evaluate each of the following.

(a) $(f + g)(3)$ **(b)** $(fg)(-1)$ **(c)** $\left(\dfrac{f}{g}\right)(0)$ **(d)** $(f/g)(2)$

Solution
(a) $(f + g)(3) = f(3) + g(3) = 3^2 + (2 \cdot 3 - 4) = 9 + 2 = 11$
(b) $(fg)(-1) = f(-1) \cdot g(-1) = (-1)^2 \cdot (2 \cdot (-1) - 4) = 1 \cdot (-6) = -6$
(c) $\left(\dfrac{f}{g}\right)(0) = \dfrac{f(0)}{g(0)} = \dfrac{0^2}{2 \cdot 0 - 4} = \dfrac{0}{-4} = 0$

(d) Note that $(f/g)(2)$ is equivalent to $\left(\frac{f}{g}\right)(2)$.

$$(f/g)(2) = \frac{f(2)}{g(2)} = \frac{2^2}{2 \cdot 2 - 4} = \frac{4}{0},$$

which is not possible because division by 0 is undefined. Thus $(f/g)(2)$ is *undefined*.

Now Try Exercise 89

In the next example, we find the sum, difference, product, and quotient of two functions for a general x.

EXAMPLE 10   **Performing arithmetic on functions**

Use $f(x) = 4x - 5$ and $g(x) = 3x + 1$ to evaluate each of the following.

(a) $(f + g)(x)$ **(b)** $(f - g)(x)$ **(c)** $(fg)(x)$ **(d)** $\left(\dfrac{f}{g}\right)(x)$

Solution

(a) $(f + g)(x) = f(x) + g(x) = (4x - 5) + (3x + 1) = 7x - 4$

(b) $(f - g)(x) = f(x) - g(x) = (4x - 5) - (3x + 1) = x - 6$

(c) $(fg)(x) = f(x) \cdot g(x) = (4x - 5)(3x + 1) = 12x^2 - 11x - 5$

(d) $\left(\dfrac{f}{g}\right)(x) = \dfrac{f(x)}{g(x)} = \dfrac{4x - 5}{3x + 1}$

▮ **Now Try Exercise 93**

6.1   Putting It All Together

CONCEPT	EXPLANATION	EXAMPLE
Rational Function	Let $p(x)$ and $q(x)$ be polynomials with $q(x) \neq 0$. Then a *rational function* is given by $$f(x) = \frac{p(x)}{q(x)}.$$	$$f(x) = \frac{3x}{x + 2}$$ The domain of f includes all real numbers except $x = -2$. A vertical asymptote occurs at $x = -2$ in the graph of f.
Rational Equation	An equation that contains rational expressions is a *rational equation*. When you are solving a rational equation, multiplication is often a good first step. *Be sure to check your results.*	To solve $\frac{3x}{x + 1} = 6$, begin by multiplying each side by $x + 1$. $$(x + 1) \cdot \frac{3x}{x + 1} = 6 \cdot (x + 1)$$ $$3x = 6x + 6$$ $$-3x = 6$$ $$x = -2$$
Division of Functions	$$(f/g)(x) = \frac{f(x)}{g(x)}, \; g(x) \neq 0$$	If $f(x) = x^2$ and $g(x) = x + 1$, then $$(f/g)(x) = \frac{x^2}{x + 1}.$$

6.1   Exercises

CONCEPTS AND VOCABULARY

1. What is a rational expression? Give an example.

2. If $f(x) = \frac{6}{x - 4}$, for what input x is $f(x)$ undefined?

3. What would be a good first step in solving the rational equation $\frac{3}{x + 7} = x$?

4. Is 5 a solution to $\frac{x + 5}{x - 5} = \frac{10}{x - 5}$? Explain.

5. Does $x \cdot \frac{5 + x}{x}$ simplify to $5x$? Explain.

6. If $x \neq 3$, then $\frac{x - 3}{x - 3} = $ _____.

7. If $b \neq 0$, then $b \cdot \frac{a}{b} = $ _____.

8. The domain of $f(x) = \frac{p(x)}{q(x)}$ includes all x-values such that $q(x) \neq$ _____.

9. Which of the following expressions (a.–d.) is not a rational function?

   a. $f(x) = \frac{1}{x}$ **b.** $f(x) = x^2 + 1$

   c. $f(x) = \sqrt{x}$ **d.** $f(x) = \frac{2x^2}{x - 1}$

10. Which (a.–d.) is the domain of $f(x) = \frac{2x}{2x - 1}$?

   a. $\{x \mid x \neq \frac{1}{2}\}$ **b.** $\{x \mid x \neq 1\}$
   c. $\{x \mid x \neq 0\}$ **d.** $\{x \mid x = 1\}$

Exercises 11–16: Determine if the expression is rational.

11. $\frac{2}{x}$ **12.** $\frac{5 - 2x}{x^2}$

13. $\frac{3x - 5}{2x + 1}$ **14.** $|x| + \frac{4}{|x|}$

15. $\frac{\sqrt{x}}{x^2 - 1}$ **16.** $\frac{4x^2 - 3x + 1}{x^3 - 4x}$

RATIONAL FUNCTIONS

Exercises 17–20: Write a symbolic representation (or formula) for the rational function f described.

17. Divide x by the quantity x plus 1.

18. Add 2 to x and then divide the result by the quantity x plus 5.

19. Divide x squared by the quantity x minus 2.

20. Compute the reciprocal of twice x.

Exercises 21–30: Identify the domain of f. Write your answer in set-builder notation.

21. $f(x) = \frac{x + 1}{x + 2}$ **22.** $f(x) = \frac{2x}{x - 4}$

23. $f(x) = \frac{1 - x}{3x - 1}$ **24.** $f(x) = \frac{x}{5x + 4}$

25. $f(t) = \frac{2}{t^2 - 4}$ **26.** $f(t) = \frac{5t}{9 - t^2}$

27. $f(x) = \frac{x^2 + 1}{x^2 - 3x + 2}$ **28.** $f(x) = \frac{2x^2 + x + 5}{2x^2 - x - 15}$

29. $f(x) = \frac{5}{x^3 - 4x}$ **30.** $f(x) = \frac{x^2 + 4}{9x - x^3}$

Exercises 31–42: Graph $y = f(x)$. Be sure to include any vertical asymptotes as dashed lines. State the domain of f in set-builder notation.

31. $f(x) = \frac{1}{x - 1}$ **32.** $f(x) = \frac{1}{x + 3}$

33. $f(x) = \frac{1}{2x}$ **34.** $f(x) = \frac{2}{x}$

35. $f(x) = \frac{1}{x + 2}$ **36.** $f(x) = \frac{1}{x - 2}$

37. $f(x) = \frac{4}{x^2 + 1}$ **38.** $f(x) = \frac{6}{x^2 + 2}$

39. $f(x) = \frac{3}{2x - 3}$ **40.** $f(x) = \frac{1}{3x + 2}$

41. $f(x) = \frac{1}{x^2 - 1}$ **42.** $f(x) = \frac{4}{4 - x^2}$

Exercises 43–48: Evaluate $f(x)$ at the given value of x.

43. $f(x) = \frac{1}{x - 1}$ $x = -2$

44. $f(x) = \frac{3x}{x^2 - 1}$ $x = 2$

45. $f(x) = \frac{x + 1}{x - 1}$ $x = -3$

46. $f(x) = \frac{2x + 1}{3x - 1}$ $x = 0$

47. $f(x) = \frac{x^2 - 3x + 5}{x^2 + 1}$ $x = -2$

48. $f(x) = \frac{5}{x^2 - x}$ $x = -1$

Exercises 49–52: Use the graph to evaluate each expression. Give the equation of any vertical asymptotes.

49. $f(-3)$ and $f(1)$ **50.** $f(-3)$ and $f(2)$

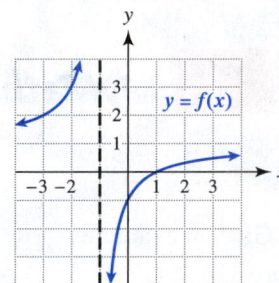
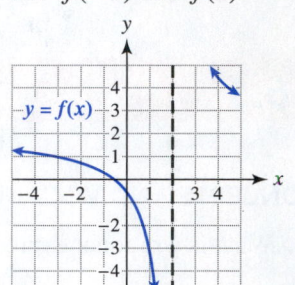

51. $f(-1)$ and $f(2)$ **52.** $f(-1)$ and $f(1)$

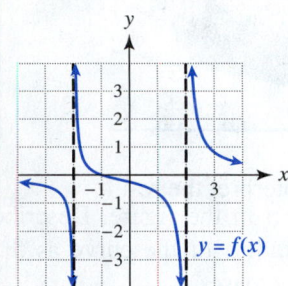 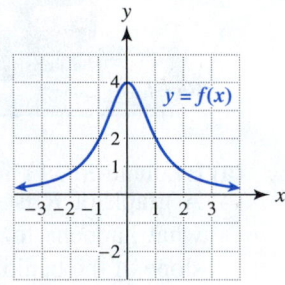

Exercises 53 and 54: Complete the table. Then evaluate $f(1)$.

53.

x	-2	-1	0	1	2
$f(x) = \frac{1}{x-1}$					

54.

x	-2	-1	0	1	2
$f(x) = \frac{2x}{x+2}$					

RATIONAL EQUATIONS

Exercises 55–74: Solve. Check your result.

55. $\dfrac{3}{x} = 5$

56. $\dfrac{5}{x} = 7$

57. $\dfrac{1}{x-2} = -1$

58. $\dfrac{-5}{x+4} = -5$

59. $\dfrac{x}{x+1} = 2$

60. $\dfrac{3x}{x+3} = -6$

61. $\dfrac{2x+1}{3x-2} = 1$

62. $\dfrac{x+3}{4x+1} = \dfrac{3}{5}$

63. $\dfrac{3}{x+2} = x$

64. $\dfrac{6}{2x+1} = x$

65. $\dfrac{6}{x+1} = 3x$

66. $\dfrac{4}{x-3} = x$

67. $\dfrac{1}{x^2-1} = -1$

68. $\dfrac{5x}{x^2-5} = \dfrac{x+2}{x^2-5}$

69. $\dfrac{x}{x-5} = \dfrac{2x-5}{x-5}$

70. $\dfrac{x}{x+1} = \dfrac{2x+1}{x+1}$

71. $\dfrac{4x}{x+2} = \dfrac{-8}{2x+4}$

72. $\dfrac{6x}{3x-2} = \dfrac{4}{4-6x}$

73. $\dfrac{2x}{x+2} = \dfrac{x-4}{x+2}$

74. $\dfrac{x+3}{x+1} = \dfrac{3x+4}{x+1}$

Exercises 75–80: Solve the rational equation either graphically or numerically. Check your answer.

75. $\dfrac{4+x}{2x} = -0.5$

76. $\dfrac{2x-1}{x-5} = -\dfrac{5}{2}$

77. $\dfrac{2x}{x^2-4} = -\dfrac{2}{3}$

78. $\dfrac{x-1}{x+2} = x$

79. $\dfrac{1}{x-1} = x-1$

80. $\dfrac{2}{x+2} = x+1$

USING MORE THAN ONE METHOD

*Exercises 81–84: Solve the equation (**a**) symbolically, (**b**) graphically, and (**c**) numerically.*

81. $\dfrac{1}{x+2} = 1$

82. $\dfrac{3}{x-1} = 2$

83. $\dfrac{x}{2x+1} = \dfrac{2}{5}$

84. $\dfrac{4x-2}{x+1} = \dfrac{3}{2}$

Exercises 85 and 86: Solve to the nearest hundredth.

85. $\dfrac{x^3}{x-1} = \dfrac{-1}{x+1}$

86. $\dfrac{x^2-1}{2x^2} = \dfrac{x+2}{1-x}$

87. Thinking Generally If $a = \frac{x}{b}$, then $x =$ _____.

88. Thinking Generally If $\frac{a}{b} = \frac{c}{d}$, then $ad =$ _____.

OPERATIONS ON FUNCTIONS

Exercises 89–92: Use $f(x)$ and $g(x)$ to evaluate each of the following.

(**a**) $(f+g)(3)$ (**b**) $(f-g)(-2)$
(**c**) $(fg)(5)$ (**d**) $(f/g)(0)$

89. $f(x) = 5x,\ g(x) = x+1$

90. $f(x) = x^2+2,\ g(x) = -2x$

91. $f(x) = 2x-1,\ g(x) = 4x^2$

92. $f(x) = x^2-1,\ g(x) = x+2$

Exercises 93–96: Use $f(x)$ and $g(x)$ to find each of the following.

(**a**) $(f+g)(x)$ (**b**) $(f-g)(x)$
(**c**) $(fg)(x)$ (**d**) $(f/g)(x)$

93. $f(x) = x+1,\ g(x) = x+2$

94. $f(x) = -3x,\ g(x) = x-1$

95. $f(x) = 1-x,\ g(x) = x^2$

96. $f(x) = x^2+4,\ g(x) = 6x$

APPLICATIONS

*Exercises 97–100: **Graphical Interpretation** Match the physical situation with the graph (a.–d.) of the rational function that models it best.*

97. A population of fish that increases and then levels off

98. An insect population that dies out

99. The length of a ticket line as the rate at which people arrive in line increases

100. The wind speed during a day that is initially calm, becomes windy, and then is calm again

a.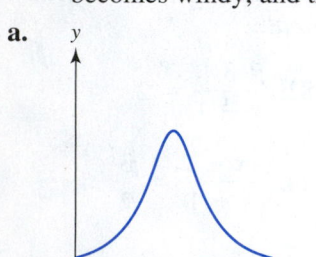

b.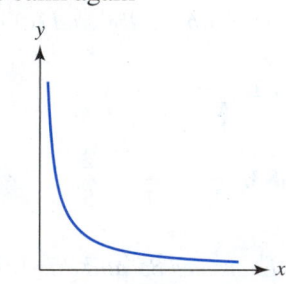

c.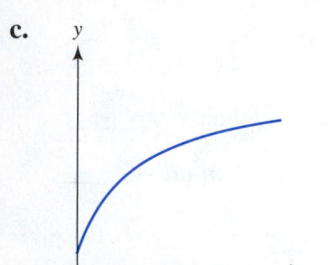

d.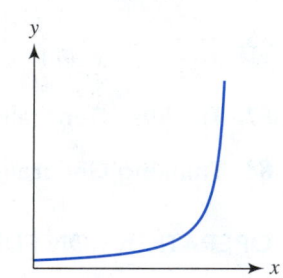

101. ***Train Track Curves*** (Refer to Example 5.) Let $f(r) = \frac{2540}{r}$ compute the elevation of the outer rail in inches for a curve designed for 60 miles per hour with a radius of r feet. (*Source*: L. Haefner.)
(a) Evaluate $f(400)$ and interpret the result.
(b) Construct a table of f, starting at $r = 100$ and incrementing by 50.
(c) From the table, if the radius of the curve doubles, what happens to the elevation of the outer rail?
(d) If the outer rail is elevated 5 inches, what should be the radius of the curve?

102. ***Highway Curves*** Engineers need to calculate a minimum safe radius for highway curves. To make a sharp curve safer, the road can be banked, or elevated, as illustrated in the figure at the top of the next column. If a curve is designed for a speed of 40 miles per hour and is banked with slope m, then a minimum radius R is computed by

$$R(m) = \frac{1600}{15m + 2}.$$

(*Source*: N. Garber.)

(a) Evaluate $R(0.1)$ and interpret the result.
(b) A graph of R is shown in the figure. Describe what happens to the radius of the curve as the slope of the banking increases.

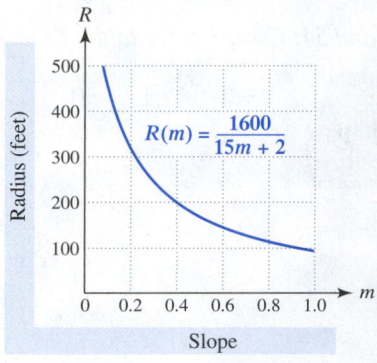

(c) If the curve has a radius of 320 feet, what is m?

103. ***Uphill Highway Grade*** The *grade x* of a hill is a measure of its steepness and corresponds to the slope of the road. For example, if a road rises 10 feet for every 100 feet of horizontal distance, it has an uphill grade of $x = \frac{10}{100}$, or 10%, as illustrated in the figure.

The braking distance for a car traveling 30 miles per hour on a wet, *uphill* grade x is given by

$$D(x) = \frac{900}{10.5 + 30x}.$$

(*Source*: N. Garber.)

(a) Evaluate $D(0.05)$ and interpret the result.
(b) If the braking distance for this car is 60 feet, find the uphill grade x.

104. ***Downhill Highway Grade*** (See Exercise 103.) The braking distance for a car traveling 30 miles per hour on a wet, *downhill* grade x is given by

$$S(x) = \frac{900}{10.5 - 30x}.$$

(a) Evaluate $S(0.05)$ and interpret the result.

(b) Make a table for $D(x)$ from Exercise 103 and $S(x)$, starting at $x = 0$ and incrementing by 0.05.

(c) How do the braking distances for uphill and downhill grades compare? Does this result agree with your driving experience?

105. *Time Spent in Line* If a parking lot attendant can wait on 5 vehicles per minute and vehicles are leaving the lot randomly at an average rate of x vehicles per minute, then the average time T in minutes spent waiting in line *and* paying the attendant is given by

$$T(x) = \frac{1}{5 - x},$$

where $x < 5$. (*Source:* N. Garber.)

(a) Evaluate $T(4)$ and interpret the result.

(b) A graph of T is shown in the figure. Interpret the graph as x increases from 0 to 5. Does this result agree with your intuition?

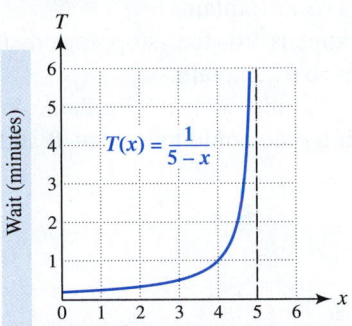

Traffic Rate (vehicles per minute)

(c) Find x if the waiting time is 3 minutes.

106. *People Waiting in Line* At a post office, workers can wait on 50 people per hour. If people arrive randomly at an average rate of x per hour, then the average number of people N waiting in line is given by

$$N(x) = \frac{x^2}{2500 - 50x},$$

where $x < 50$. (*Source:* N. Garber.)

(a) Evaluate $N(30)$ and interpret the result.

(b) A graph of N is shown in the figure. Interpret the graph as x increases from 0 to 50. Does this result agree with your intuition?

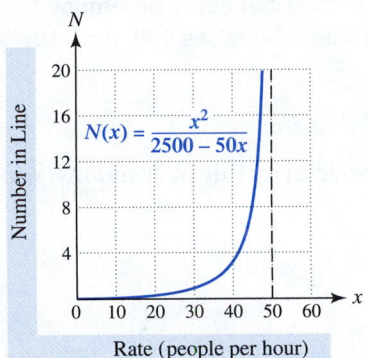

Rate (people per hour)

(c) Find x if $N = 8$.

107. *Winning Number* A jar contains x balls. Each ball has a unique number written on it and only one ball has the winning number. The likelihood, or probability P, of *not* drawing the winning ball is given by

$$P(x) = \frac{x - 1}{x},$$

where $x > 0$. For example, if $P = 0.99$, then there is a 99% chance of not drawing the winning ball.

(a) Evaluate $P(1)$ and $P(50)$ and interpret the result.

(b) Graph P in $[0, 100, 10]$ by $[0, 1, 0.1]$.

(c) What happens to the probability of not winning as the number of balls increases? Does this result agree with your intuition? Explain.

(d) How many balls are in the jar if the probability of not winning is 0.975?

108. *Insect Population* Suppose that an insect population in thousands per acre is modeled by

$$P(x) = \frac{5x + 2}{x + 1},$$

where $x \geq 0$ is time in months.

(a) Evaluate $P(10)$ and interpret the result.

(b) Graph P in $[0, 50, 10]$ by $[0, 6, 1]$.

(c) What happens to the insect population after several years?

(d) After how many months is the insect population 4.8 thousand per acre?

*Exercises 109 and 110: **Remembering What You Learn*** After a test students often forget what they learned. The rational function

$$R(x) = \frac{100}{1.2x + 1}, \quad 0 \leq x \leq 5,$$

gives an estimate of the percentage of the material a student remembers x days after a test.

109. Evaluate $R(1)$ and $R(3)$. Interpret your results.

110. If a student takes notes in class, these percentages increase by 30% for $1 \leq x \leq 5$. Write another function $N(x)$ that models this result. Evaluate $N(3)$.

111. *Google and Facebook Revenues* (Refer to the Real-World Connection that is discussed before Example 9.) The function $G(x) = 350(x - 3)^2 + 50$ gives the revenue for Google x years after its startup, and $F(x) = 206(x - 3)^2 + 150$ gives the revenue for Facebook x years after its startup. Both revenues are in millions of dollars and $x \geq 3$.

(a) Write a formula for $R(x) = \frac{G(x)}{F(x)}$.

(b) Evaluate $R(6)$ and interpret your result.

112. *Google and Yahoo Revenues* (Refer to Exercise 111.) The function $G(x) = 350(x - 3)^2 + 50$ gives the revenue for Google x years after its startup, and $Y(x) = 117(x - 3)^2 + 50$ gives the revenue for Yahoo x years after its startup. Both revenues are in millions of dollars and $x \geq 3$.
(a) Write a formula for $R(x) = \frac{G(x)}{Y(x)}$.
(b) Evaluate $R(5)$ and interpret your result.

WRITING ABOUT MATHEMATICS

113. Is every polynomial function a rational function? Explain your answer.

114. The domain of a polynomial function includes all real numbers. Does the domain of a rational function include all real numbers? Explain your answer and give an example.

Group Activity   Working with Real Data

Directions: Form a group of 2 to 4 people. Select someone to record the group's responses for this activity. All members of the group should work cooperatively to answer the questions. If your instructor asks for your results, each member of the group should be prepared to respond.

Slippery Roads If a car is moving on a level highway, its stopping distance depends on road conditions. If the road is slippery, stopping may take longer. A measure of the slipperiness of a road is the coefficient of friction x between the tire and the road, where x satisfies $0 < x \leq 1$. A smaller value for x indicates that the road is slipperier. The stopping distance D in feet for a car traveling at 50 miles per hour on a road with a coefficient of friction x is given by

$$D(x) = \frac{250}{3x}.$$

(*Source:* L. Haefner.)

(a) Graph D. Identify any vertical asymptotes.
(b) What happens to the stopping distance as x increases to 1? Explain.
(c) What happens to the stopping distance as x decreases to 0? Explain.
(d) In reality, could $x = 0$? Explain. What would happen if a road could have $x = 0$ and a car tried to stop?

6.2   Multiplication and Division of Rational Expressions

Simplifying Rational Expressions • Review of Multiplication and Division of Fractions • Multiplication of Rational Expressions • Division of Rational Expressions

A LOOK INTO MATH ▶ Rational expressions are sometimes used by businesses to describe the average cost to produce a particular item. If a business knows the average cost to produce each item and also knows the number of items produced, then they can easily determine the total cost by multiplying the two expressions. See Example 5 and Exercises 109 and 110.

Simplifying Rational Expressions

NEW VOCABULARY

☐ Basic principle of fractions
☐ Lowest terms
☐ Reciprocal of a polynomial
☐ Reciprocal of a rational expression

When simplifying fractions, we often use the **basic principle of fractions**, which states that

$$\frac{a \cdot c}{b \cdot c} = \frac{a}{b}.$$

This principle holds because $\frac{c}{c} = 1$ and $\frac{a}{b} \cdot 1 = \frac{a}{b}$. That is,

$$\frac{a \cdot c}{b \cdot c} = \frac{a}{b} \cdot \frac{c}{c} = \frac{a}{b} \cdot 1 = \frac{a}{b}.$$

For example, this principle can be used to simplify a fraction.

$$\frac{6}{44} = \frac{3 \cdot \mathbf{2}}{22 \cdot \mathbf{2}} = \frac{3}{22}$$

This same principle can also be used to simplify rational expressions. For example,

$$\frac{(z + 1)(z + 3)}{z(z + 3)} = \frac{z + 1}{z}, \quad \text{provided } z \neq -3.$$

SIMPLIFYING RATIONAL EXPRESSIONS

The following principle can be used to simplify rational expressions, where A, B, and C are polynomials.

$$\frac{A \cdot C}{B \cdot C} = \frac{A}{B} \qquad B \text{ and } C \text{ not zero}$$

To *simplify* a rational expression, we *completely factor* the numerator and the denominator and then apply the principle for simplifying rational expressions until the expression is in **lowest terms**. These concepts are illustrated in the next example.

EXAMPLE 1   **Simplifying rational expressions**

Simplify each expression.

(a) $\dfrac{9x}{3x^2}$ **(b)** $\dfrac{5y - 10}{10y - 20}$ **(c)** $\dfrac{2z^2 - 3z - 9}{z^2 + 2z - 15}$ **(d)** $\dfrac{a^2 - b^2}{a + b}$

Solution

(a) First factor out the greatest common factor, $3x$, in the numerator and denominator.

$$\frac{9x}{3x^2} = \frac{3 \cdot \mathbf{3x}}{x \cdot \mathbf{3x}} = \frac{3}{x}$$

(b) Use the distributive property and then simplify the rational expressions.

$$\frac{5y - 10}{10y - 20} = \frac{5(\mathbf{y - 2})}{10(\mathbf{y - 2})} = \frac{5}{10} = \frac{1}{2} \qquad \text{Provided } y \neq 2$$

(c) Start by factoring the numerator and denominator.

$$\frac{2z^2 - 3z - 9}{z^2 + 2z - 15} = \frac{(2z + 3)(\mathbf{z - 3})}{(z + 5)(\mathbf{z - 3})} = \frac{2z + 3}{z + 5} \qquad \text{Provided } z \neq 3$$

(d) Start by factoring the numerator as the difference of squares.

$$\frac{a^2 - b^2}{a + b} = \frac{(a - b)(\mathbf{a + b})}{\mathbf{a + b}} = a - b \qquad \text{Provided } a + b \neq 0$$

▌ **Now Try Exercises 23, 25, 39, 43**

NOTE: In Example 1(b) the expressions $\frac{5y - 10}{10y - 20}$ and $\frac{1}{2}$ are equal, *provided* that $y \neq 2$. If $y = 2$, then $\frac{5y - 10}{10y - 20}$ is not defined but $\frac{1}{2}$ is defined. When we simplify rational expressions, we will usually not list these excluded values; nonetheless, be aware that the given expression may *not* be defined for some values of the variable, whereas the simplified expression may be defined for these same values.

STUDY TIP

If you tried to solve a problem but need help, ask a question in class. Other students will likely have the same question.

There are a number of ways that a negative sign can be placed in a fraction. For example,

$$-\frac{2}{3} = \frac{-2}{3} = \frac{2}{-3}$$

illustrates three fractions that are equal.

To simplify some types of rational expressions, it may be helpful to factor out -1.

$$2 - x = -1(-2 + x) = -(x - 2) \qquad \text{Factor out } -1.$$

Thus $2 - x = -(x - 2)$. As a result, the following rational expression simplifies to -1.

$$\frac{2 - x}{x - 2} = \frac{-(x - 2)}{x - 2} = -\frac{x - 2}{x - 2} = -1$$

Some examples of this result are the following.

$$\frac{4 - y}{y - 4} = -1, \quad \frac{z - 3}{3 - z} = -1, \quad \text{and} \quad \frac{x - y}{y - x} = -1$$

These concepts are applied in the next example.

EXAMPLE 2 | **Distributing or factoring out a negative sign**

Simplify each expression.

(a) $-\dfrac{1 - z}{z - 1}$ **(b)** $\dfrac{-y - 2}{4y + 8}$ **(c)** $\dfrac{3 - x}{x - 3}$

Solution
(a) Start by distributing the negative sign over the numerator.

$$-\frac{1 - z}{z - 1} = \frac{-(1 - z)}{z - 1} = \frac{-1 + z}{z - 1} = \frac{z - 1}{z - 1} = 1$$

Note that the negative sign could also be distributed over the denominator.

(b) Factor -1 out of the numerator and 4 out of the denominator.

$$\frac{-y - 2}{4y + 8} = \frac{-1(y + 2)}{4(y + 2)} = -\frac{1}{4}$$

(c) Start by factoring -1 out of the numerator.

$$\frac{3 - x}{x - 3} = \frac{-1(-3 + x)}{x - 3} = \frac{-1(x - 3)}{x - 3} = -1$$

Now Try Exercises 47, 49, 51

MAKING CONNECTIONS

Negative Signs and Rational Expressions

In general, $(b - a)$ equals $-1(a - b)$. As a result, if $a \neq b$, then

$$\frac{b - a}{a - b} = \frac{-1(a - b)}{a - b} = -1.$$

See Example 2(c).

Review of Multiplication and Division of Fractions

Recall that to multiply two fractions we use the property

$$\frac{a}{b} \cdot \frac{c}{d} = \frac{ac}{bd}.$$

For example, $\dfrac{2}{5} \cdot \dfrac{3}{7} = \dfrac{2 \cdot 3}{5 \cdot 7} = \dfrac{6}{35}.$

EXAMPLE 3 **Multiplying fractions**

Multiply and simplify the product.

(a) $\dfrac{4}{9} \cdot \dfrac{3}{8}$ (b) $\dfrac{2}{3} \cdot \dfrac{3}{4} \cdot \dfrac{5}{6}$

Solution

(a) $\dfrac{4}{9} \cdot \dfrac{3}{8} = \dfrac{4 \cdot 3}{9 \cdot 8} = \dfrac{12}{72} = \dfrac{1 \cdot 12}{6 \cdot 12} = \dfrac{1}{6}$ (b) $\dfrac{2}{3} \cdot \dfrac{3}{4} \cdot \dfrac{5}{6} = \dfrac{6 \cdot 5}{12 \cdot 6} = \dfrac{5}{12}$

Now Try Exercises 11, 13

Recall that to divide two fractions we "invert and multiply." That is, we change a division problem to a multiplication problem, using the property

$$\frac{a}{b} \div \frac{c}{d} = \frac{a}{b} \cdot \frac{d}{c}.\quad\text{"Invert and multiply."}$$

For example, $\dfrac{3}{4} \div \dfrac{5}{4} = \dfrac{3}{4} \cdot \dfrac{4}{5} = \dfrac{3 \cdot 4}{5 \cdot 4} = \dfrac{3}{5}.$

↑———— "Invert and multiply."

Multiplication of Rational Expressions

Multiplying rational expressions is similar to multiplying fractions.

READING CHECK

How do you multiply rational expressions?

PRODUCTS OF RATIONAL EXPRESSIONS

To multiply two rational expressions, multiply numerators and multiply denominators.

$$\frac{A}{B} \cdot \frac{C}{D} = \frac{AC}{BD}\qquad B \text{ and } D \text{ not zero}$$

EXAMPLE 4 **Multiplying rational expressions**

Multiply. Leave your answer in factored form when appropriate.

(a) $\dfrac{1}{x} \cdot \dfrac{x+1}{2x}$ (b) $\dfrac{x-1}{x} \cdot \dfrac{x-1}{x+2}$

Solution

(a) $\dfrac{1}{x} \cdot \dfrac{x+1}{2x} = \dfrac{1 \cdot (x+1)}{x \cdot 2x} = \dfrac{x+1}{2x^2}$

(b) $\dfrac{x-1}{x} \cdot \dfrac{x-1}{x+2} = \dfrac{(x-1)(x-1)}{x(x+2)} = \dfrac{(x-1)^2}{x(x+2)}$

Now Try Exercises 61, 63

In the next example we see how rational expressions are sometimes multiplied to help businesses calculate costs.

EXAMPLE 5 **Finding total cost of producing earbuds**

The average cost (per item) in dollars to produce x pairs of earbuds is given by

$$C(x) = \frac{12x + 6000}{x}.$$

(a) A company expects to sell 2000 pairs. Evaluate $C(2000)$ and interpret the result.
(b) Write an expression $T(x)$ that gives the total cost of producing x pairs. Simplify it.
(c) Evaluate $T(2000)$ and interpret the result.

Solution

(a) Because $C(x) = \frac{12x + 6000}{x}$, $C(2000) = \frac{12(2000) + 6000}{2000} = \$15$. When 2000 earbuds are made, the average cost is \$15 per pair.

(b) To find the total cost we must multiply the average cost $C(x)$ by the number of pairs x of earbuds sold.

$$T(x) = C(x) \cdot x$$

$$= \frac{12x + 6000}{x} \cdot \frac{x}{1}$$

$$= 12x + 6000$$

(c) $T(2000) = 12(2000) + 6000 = \$30{,}000$. It costs \$30,000 to make 2000 pairs of earbuds.

■ **Now Try Exercise 109**

Many times the product of two rational expressions can be simplified. This process is demonstrated in the next two examples.

EXAMPLE 6 | **Multiplying rational expressions**

Multiply and simplify. Leave your answer in factored form when appropriate.

(a) $\dfrac{5x}{8} \cdot \dfrac{4}{10x^2}$ **(b)** $\dfrac{x - 2}{2x - 1} \cdot \dfrac{x + 1}{2x - 4}$ **(c)** $\dfrac{x^2 - 1}{x^2 - 4} \cdot \dfrac{x + 2}{x - 1}$

Solution

(a)

$$\frac{5x}{8} \cdot \frac{4}{10x^2} = \frac{20x}{80x^2} \qquad \text{Multiply rational expressions.}$$

$$= \frac{1}{4x} \qquad \text{Simplify.}$$

(b)

$$\frac{x - 2}{2x - 1} \cdot \frac{x + 1}{2x - 4} = \frac{(x - 2)(x + 1)}{(2x - 1)(2x - 4)} \qquad \text{Multiply rational expressions.}$$

$$= \frac{(x - 2)(x + 1)}{(2x - 1)2(x - 2)} \qquad \text{Factor } 2x - 4.$$

$$= \frac{(x + 1)(x - 2)}{2(2x - 1)(x - 2)} \qquad \text{Commutative property}$$

$$= \frac{x + 1}{2(2x - 1)} \qquad \text{Simplify.}$$

(c)

$$\frac{x^2 - 1}{x^2 - 4} \cdot \frac{x + 2}{x - 1} = \frac{(x^2 - 1)(x + 2)}{(x^2 - 4)(x - 1)} \qquad \text{Multiply rational expressions.}$$

$$= \frac{(x - 1)(x + 1)(x + 2)}{(x - 2)(x + 2)(x - 1)} \qquad \text{Difference of squares}$$

$$= \frac{(x + 1)(x - 1)(x + 2)}{(x - 2)(x - 1)(x + 2)} \qquad \text{Commutative property}$$

$$= \frac{x + 1}{x - 2} \qquad \text{Simplify.}$$

■ **Now Try Exercises 65, 71, 73**

EXAMPLE 7 **Multiplying rational expressions**

Multiply and simplify. Leave your answer in factored form when appropriate.

(a) $\dfrac{2x^2y^3}{3xy^2} \cdot \dfrac{(2x^3y)^2}{2(xy)^3}$ **(b)** $\dfrac{x^3 - x}{x - 1} \cdot \dfrac{x + 1}{x}$ **(c)** $\dfrac{a}{b} \cdot \dfrac{a^2 - b^2}{2} \cdot \dfrac{b}{a + b}$

Solution

(a)
$$\dfrac{2x^2y^3}{3xy^2} \cdot \dfrac{(2x^3y)^2}{2(xy)^3} = \dfrac{2x^2y^3 \cdot 4x^6y^2}{3xy^2 \cdot 2x^3y^3}$$ Multiply; properties of exponents.

$$= \dfrac{8x^8y^5}{6x^4y^5}$$ Multiply; properties of exponents.

$$= \dfrac{4x^4}{3}$$ Simplify.

(b)
$$\dfrac{x^3 - x}{x - 1} \cdot \dfrac{x + 1}{x} = \dfrac{(x^3 - x)(x + 1)}{x(x - 1)}$$ Multiply.

$$= \dfrac{x(x^2 - 1)(x + 1)}{x(x - 1)}$$ Factor out x.

$$= \dfrac{x(x - 1)(x + 1)(x + 1)}{x(x - 1)}$$ Difference of squares

$$= (x + 1)(x + 1)$$ Simplify.
$$= (x + 1)^2$$ Rewrite.

(c)
$$\dfrac{a}{b} \cdot \dfrac{a^2 - b^2}{2} \cdot \dfrac{b}{a + b} = \dfrac{a(a^2 - b^2)b}{b(2)(a + b)}$$ Multiply.

$$= \dfrac{a(a - b)(a + b)b}{2(a + b)b}$$ Difference of squares

$$= \dfrac{a(a - b)}{2}$$ Simplify.

Now Try Exercises 69, 75, 83

Division of Rational Expressions

When dividing two rational expressions, *multiply* the first expression by the reciprocal of the second expression. This technique is similar to the division of fractions.

READING CHECK

How do you divide rational expressions?

QUOTIENTS OF RATIONAL EXPRESSIONS

To divide two rational expressions, multiply by the reciprocal of the divisor.

$$\dfrac{A}{B} \div \dfrac{C}{D} = \dfrac{A}{B} \cdot \dfrac{D}{C} \qquad B, C, \text{ and } D \text{ not zero}$$

NOTE: This technique of multiplying the first rational expression and the reciprocal of the second rational expression is sometimes summarized as "invert and multiply."

The **reciprocal of a polynomial** $p(x)$ is $\frac{1}{p(x)}$, and the **reciprocal of a rational expression** $\frac{p(x)}{q(x)}$ is $\frac{q(x)}{p(x)}$. The next example demonstrates how to find reciprocals.

EXAMPLE 8 | **Finding reciprocals**

Write the reciprocal of each expression.

(a) $3x + 4$　(b) $\dfrac{5}{x^2 + 1}$　(c) $\dfrac{x - 7}{x + 7}$

Solution

(a) The reciprocal of $3x + 4$ is $\dfrac{1}{3x + 4}$.　(b) The reciprocal of $\dfrac{5}{x^2 + 1}$ is $\dfrac{x^2 + 1}{5}$.

(c) The reciprocal of $\dfrac{x - 7}{x + 7}$ is $\dfrac{x + 7}{x - 7}$.

■ **Now Try Exercises 53, 57, 59**

EXAMPLE 9 | **Dividing two rational expressions**

Divide and simplify.

(a) $\dfrac{2}{x} \div \dfrac{2x - 1}{4x}$　(b) $\dfrac{x^2 - 1}{x^2 + x - 6} \div \dfrac{x - 1}{x + 3}$

Solution

(a)
$$\dfrac{2}{x} \div \dfrac{2x - 1}{4x} = \dfrac{2}{x} \cdot \dfrac{4x}{2x - 1} \qquad \text{“Invert and multiply.”}$$

$$= \dfrac{8x}{x(2x - 1)} \qquad \text{Multiply.}$$

$$= \dfrac{8}{2x - 1} \qquad \text{Simplify.}$$

(b)
$$\dfrac{x^2 - 1}{x^2 + x - 6} \div \dfrac{x - 1}{x + 3} = \dfrac{x^2 - 1}{x^2 + x - 6} \cdot \dfrac{x + 3}{x - 1} \qquad \text{“Invert and multiply.”}$$

$$= \dfrac{(x + 1)(x - 1)}{(x - 2)(x + 3)} \cdot \dfrac{x + 3}{x - 1} \qquad \text{Factor completely.}$$

$$= \dfrac{(x + 1)(x - 1)(x + 3)}{(x - 2)(x - 1)(x + 3)} \qquad \text{Commutative property}$$

$$= \dfrac{x + 1}{x - 2} \qquad \text{Simplify.}$$

■ **Now Try Exercises 85, 99**

EXAMPLE 10 | **Dividing rational expressions**

Divide and simplify. Leave your answer in factored form when appropriate.

(a) $\dfrac{7a^2}{4b^3} \div \dfrac{21a}{8b^4}$　(b) $\dfrac{2x + 2}{x - 1} \div (x + 1)$　(c) $\dfrac{x^2 - 25}{x^2 + 5x + 4} \div \dfrac{x^2 - 10x + 25}{2x^2 + 8x}$

Solution

(a)
$$\dfrac{7a^2}{4b^3} \div \dfrac{21a}{8b^4} = \dfrac{7a^2}{4b^3} \cdot \dfrac{8b^4}{21a} \qquad \text{“Invert and multiply.”}$$

$$= \dfrac{56a^2b^4}{84ab^3} \qquad \text{Multiply rational expressions.}$$

$$= \dfrac{2ab}{3} \qquad \text{Simplify.}$$

(b) $\dfrac{2x + 2}{x - 1} \div (x + 1) = \dfrac{2x + 2}{x - 1} \cdot \dfrac{1}{x + 1}$ "Invert and multiply."

$= \dfrac{2(x + 1)}{x - 1} \cdot \dfrac{1}{x + 1}$ Factor.

$= \dfrac{2(x + 1)}{(x - 1)(x + 1)}$ Multiply rational expressions.

$= \dfrac{2}{x - 1}$ Simplify.

(c) $\dfrac{x^2 - 25}{x^2 + 5x + 4} \div \dfrac{x^2 - 10x + 25}{2x^2 + 8x} = \dfrac{x^2 - 25}{x^2 + 5x + 4} \cdot \dfrac{2x^2 + 8x}{x^2 - 10x + 25}$

$= \dfrac{(x - 5)(x + 5)}{(x + 1)(x + 4)} \cdot \dfrac{2x(x + 4)}{(x - 5)(x - 5)}$

$= \dfrac{2x(x + 5)(x - 5)(x + 4)}{(x + 1)(x - 5)(x - 5)(x + 4)}$

$= \dfrac{2x(x + 5)}{(x + 1)(x - 5)}$

▌ **Now Try Exercises 87, 89, 97**

In the next example, we divide three rational expressions. It is important to realize that *the associative property does not apply to division*. For example,

$$(24 \div 6) \div 2 = 4 \div 2 = 2 \quad \text{but} \quad 24 \div (6 \div 2) = 24 \div 3 = 8.$$

The order in which we do division *does matter*. As a result, we perform division from *left to right*, so

$$24 \div 6 \div 2 = 2.$$

EXAMPLE 11   **Dividing three fractions**

Divide and simplify $\dfrac{4x}{y} \div \dfrac{x^2}{y} \div \dfrac{3y}{2x}$.

Solution

$\dfrac{4x}{y} \div \dfrac{x^2}{y} \div \dfrac{3y}{2x} = \left(\dfrac{4x}{y} \div \dfrac{x^2}{y} \right) \div \dfrac{3y}{2x}$ Divide left to right.

$= \left(\dfrac{4x}{y} \cdot \dfrac{y}{x^2} \right) \div \dfrac{3y}{2x}$ "Invert and multiply."

$= \dfrac{4xy}{x^2 y} \div \dfrac{3y}{2x}$ Multiply fractions.

$= \dfrac{4xy}{x^2 y} \cdot \dfrac{2x}{3y}$ "Invert and multiply."

$= \dfrac{8x^2 y}{3x^2 y^2}$ Multiply fractions.

$= \dfrac{8}{3y}$ Simplify.

▌ **Now Try Exercise 101**

In the next example, we find the length of a rectangle when its area and width are expressed as polynomials.

EXAMPLE 12 **Finding the dimensions of a rectangle**

The area A of a rectangle is $3x^2 + 14x + 15$ and its width W is $x + 3$. See Figure 6.8.
(a) Find the length L of the rectangle.
(b) Find the length if the width is 12 inches.

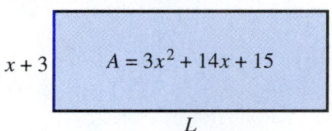

$$x + 3 \quad\quad A = 3x^2 + 14x + 15$$

$$L$$

Figure 6.8

Solution
(a) Because the area equals length times width, $A = LW$, length equals area divided by width, or $L = \frac{A}{W}$. To determine L, factor the expression for A and then simplify.

$$L = \frac{3x^2 + 14x + 15}{x + 3} = \frac{(3x + 5)(x + 3)}{x + 3} = 3x + 5.$$

The length of the rectangle is $3x + 5$.
(b) If the width is 12 inches, then $x + 3 = 12$ or $x = 9$. The length is

$$L = 3x + 5 = 3(9) + 5 = 32 \text{ inches.}$$

▌ **Now Try Exercise 111**

6.2 Putting It All Together

CONCEPT	EXPLANATION	EXAMPLES
Simplifying Rational Expressions	Use the principle $$\frac{A \cdot C}{B \cdot C} = \frac{A}{B}.$$ B and C not zero	$$\frac{5a(a + b)}{7b(a + b)} = \frac{5a}{7b} \text{ and}$$ $$\frac{x^2 - 1}{x - 1} = \frac{(x + 1)(x - 1)}{x - 1} = x + 1$$
Multiplying Rational Expressions	To multiply two rational expressions, multiply numerators and multiply denominators. $$\frac{A}{B} \cdot \frac{C}{D} = \frac{AC}{BD}$$ B and D not zero	$$\frac{5a}{4b^2} \cdot \frac{2b^3}{10a^3} = \frac{10ab^3}{40a^3b^2} = \frac{b}{4a^2}$$
Dividing Rational Expressions	To divide two rational expressions, multiply by the reciprocal of the divisor. ("Invert and multiply.") $$\frac{A}{B} \div \frac{C}{D} = \frac{A}{B} \cdot \frac{D}{C}$$ B, C, and D not zero └── "Invert and multiply."	$$\frac{x}{x + 1} \div \frac{x + 2}{x + 1} = \frac{x}{x + 1} \cdot \frac{x + 1}{x + 2}$$ $$= \frac{x(x + 1)}{(x + 2)(x + 1)}$$ $$= \frac{x}{x + 2}$$

6.2   Exercises

MyMathLab

CONCEPTS AND VOCABULARY

1. Simplify $\frac{2x + 3}{2x + 3}$.

2. Simplify $\frac{x - 3}{3 - x}$.

3. Is $\frac{x^2 - 1}{x - 1}$ equal to x? Explain.

4. Is $\frac{3}{3 + x}$ equal to $\frac{1}{x}$? Is it equal to $1 + \frac{3}{x}$?

5. To divide $\frac{2}{3}$ by $\frac{5}{7}$ multiply _____ by _____.

6. $\frac{a}{b} \cdot \frac{c}{d} = $ _____.

7. $\frac{a}{b} \div \frac{c}{d} = $ _____.

8. $\frac{ac}{bc} = $ _____.

9. Which of the following expressions (a.–d.) is equal to the quotient $\frac{x^2 - 1}{x^2 + 2x + 1}$?

   a. $\frac{1}{2x + 1}$

   b. $\frac{x + 1}{x - 1}$

   c. $\frac{x - 1}{x + 1}$

   d. $x - 1$

10. Which of the following expressions (a.–d.) is equal to the quotient $\frac{5x}{y} \div \frac{15x - 15}{3y^2}$?

    a. $\frac{75x(x - 1)}{3y^3}$

    b. $\frac{xy}{x - 1}$

    c. $\frac{3xy}{x - 1}$

    d. $\frac{y}{x(x - 1)}$

REVIEW OF FRACTIONS

Exercises 11–22: Simplify.

11. $\frac{1}{2} \cdot \frac{4}{5}$

12. $\frac{5}{6} \cdot \frac{3}{10}$

13. $\frac{7}{8} \cdot \frac{4}{3} \cdot (-3)$

14. $4 \cdot \frac{7}{4} \cdot \frac{1}{2}$

15. $\frac{3}{8} \cdot 2$

16. $-5 \cdot \frac{2}{7}$

17. $-\frac{7}{11} \div 14$

18. $6 \div \frac{6}{5}$

19. $\frac{5}{7} \div \frac{15}{14}$

20. $-\frac{2}{3} \div 2$

21. $6 \div \left(-\frac{1}{3}\right)$

22. $\frac{10}{9} \div \frac{5}{3}$

SIMPLIFYING RATIONAL EXPRESSIONS

Exercises 23–52: Simplify the rational expression.

23. $\frac{5x}{x^2}$

24. $\frac{18t^3}{6t}$

25. $\frac{3z + 6}{z + 2}$

26. $\frac{6}{12x + 6}$

27. $\frac{2z + 2}{3z + 3}$

28. $\frac{4x - 12}{2x - 6}$

29. $\frac{(x - 1)(x + 1)}{x - 1}$

30. $(x + 2) \cdot \frac{x - 6}{x + 2}$

31. $\frac{x^2 - 4}{x + 2}$

32. $\frac{(x + 2)(x - 5)}{(x + 10)(x + 2)}$

33. $\frac{x(x - 1)}{(x + 1)(x - 1)}$

34. $\frac{(2x - 1)(x - 3)}{(x - 3)(2x + 5)}$

35. $\frac{(3x + 1)(x + 2)}{(x + 2)(5x - 2)}$

36. $\frac{(3x + 2)(x - 7)}{x(x - 7)}$

37. $\frac{x + 5}{x^2 + 2x - 15}$

38. $\frac{x^2 - 9}{x^2 + 6x + 9} \cdot (x + 3)$

39. $\frac{x^2 + 2x}{x^2 + 3x + 2}$

40. $\frac{x^2 - 3x - 10}{x^2 - 6x + 5}$

41. $\frac{6x^2 + 7x - 5}{2x^2 - 11x + 5}$

42. $\frac{30x^2 - 7x - 15}{6x^2 + 7x - 10}$

43. $\frac{a^2 - b^2}{a - b}$

44. $\frac{a^2 + 2ab + b^2}{a + b}$

45. $\frac{m^3 + n^3}{m + n}$

46. $\frac{m^3 - n^3}{m - n}$

47. $-\frac{4 - t}{t - 4}$

48. $-\frac{t - r}{r - t}$

49. $\frac{4m - n}{-4m + n}$

50. $\frac{2n + 10m}{-n - 5m}$

51. $\frac{5 - y}{y - 5}$

52. $\frac{a - b}{b - a}$

RECIPROCALS

Exercises 53–60: Write the reciprocal of the expression.

53. $4x$

54. $5x - 2$

55. $\frac{2a}{5b}$

56. $\frac{5b^3}{7a^2}$

57. $\frac{3 - x}{5 - x}$

58. $\frac{x^2 - 4}{3x + 1}$

59. $\frac{1}{x^2 + 1}$

60. $\frac{-1}{5 - x}$

MULTIPLICATION AND DIVISION OF RATIONAL EXPRESSIONS

Exercises 61–104: Multiply or divide. Leave your answer in factored form when appropriate.

61. $\dfrac{2}{x} \cdot \dfrac{x-1}{3x}$

62. $\dfrac{2}{x-3} \cdot \dfrac{x}{x+1}$

63. $\dfrac{x-2}{x} \cdot \dfrac{x-3}{x+4}$

64. $\dfrac{x}{x+5} \cdot \dfrac{x-6}{x+4}$

65. $\dfrac{1}{2x} \cdot \dfrac{4x}{2}$

66. $\dfrac{5a^2}{7} \cdot \dfrac{7}{10a}$

67. $\dfrac{5a}{4} \cdot \dfrac{12}{5a}$

68. $\dfrac{a^2b}{4c} \cdot \dfrac{8c^2}{3ab^2}$

69. $\dfrac{9x^2y^4}{8xy^6} \cdot \dfrac{(2xy^2)^3}{3(xy)^4}$

70. $\dfrac{(7rt)^2}{8} \cdot \dfrac{8r}{49(rt^2)^3}$

71. $\dfrac{x+1}{2x-5} \cdot \dfrac{2x-5}{x}$

72. $\dfrac{x+1}{x} \cdot \dfrac{x}{x+2}$

73. $\dfrac{b^2+1}{b^2-1} \cdot \dfrac{b-1}{b+1}$

74. $\dfrac{(x-5)(x+3)}{3x-1} \cdot \dfrac{x(3x-1)}{(x-5)}$

75. $\dfrac{x^2-2x-35}{2x^3-3x^2} \cdot \dfrac{x^3-x^2}{2x-14}$

76. $\dfrac{2x+4}{x+1} \cdot \dfrac{x^2+3x+2}{4x+2}$

77. $\dfrac{3n-9}{n^2-9} \cdot (n^3+27)$

78. $(10n+15) \cdot \dfrac{n+1}{6n+9}$

79. $\dfrac{3n-9}{n^2-9} \cdot \dfrac{n^3+27}{12}$

80. $\dfrac{10n+15}{n^2-1} \cdot \dfrac{n+1}{6n+9}$

81. $\dfrac{x}{y} \cdot \dfrac{2y}{x} \cdot \dfrac{2}{xy}$

82. $\dfrac{4z^2}{r^3} \cdot \dfrac{5r}{2z} \cdot \dfrac{r}{z}$

83. $\dfrac{x-1}{y} \cdot \dfrac{y(x+y)}{2} \cdot \dfrac{y}{x+y}$

84. $\dfrac{x}{y^2} \cdot \dfrac{y^3-y^2}{x+y} \cdot \dfrac{x^2+xy}{5x^2}$

85. $\dfrac{3x}{2} \div \dfrac{2x}{5}$

86. $\dfrac{x^2+x}{2x+6} \div \dfrac{x}{x+3}$

87. $\dfrac{8a^4}{3b} \div \dfrac{a^5}{9b^2}$

88. $\dfrac{5m^4}{n^2} \div 5m$

89. $(2n+4) \div \dfrac{n+2}{n-1}$

90. $\dfrac{n+1}{n+3} \div \dfrac{n+1}{n+3}$

91. $\dfrac{6b}{b+2} \div \dfrac{3b^4}{2b+4}$

92. $\dfrac{5x^5}{x-2} \div \dfrac{10x^3}{5x-10}$

93. $\dfrac{3a+1}{a^7} \div \dfrac{a+1}{3a^8}$

94. $\dfrac{16-x^2}{x+3} \div \dfrac{x+4}{9-x^2}$

95. $\dfrac{x+5}{x-x^3} \div \dfrac{25-x^2}{x^3}$

96. $\dfrac{x^2+x-12}{2x^2-9x-5} \div \dfrac{x^2+7x+12}{2x^2-7x-4}$

97. $\dfrac{x^2-3x+2}{x^2+5x+6} \div \dfrac{x^2+x-2}{x^2+2x-3}$

98. $\dfrac{2x^2+x-1}{6x^2+x-2} \div \dfrac{2x^2+5x+3}{6x^2+13x+6}$

99. $\dfrac{x^2-4}{x^2+x-2} \div \dfrac{x-2}{x-1}$

100. $\dfrac{x^2+2x+1}{x-2} \div \dfrac{x+1}{2x-4}$

101. $\dfrac{3y}{x^2} \div \dfrac{y^2}{x} \div \dfrac{y}{5x}$

102. $\dfrac{x+1}{y-2} \div \dfrac{2x+2}{y-2} \div \dfrac{x}{y}$

103. $\dfrac{x-3}{x-1} \div \dfrac{x^2}{x-1} \div \dfrac{x-3}{x}$

104. $\dfrac{2x}{x-2} \div \dfrac{x+2}{x} \div \dfrac{7x}{x^2-4}$

OPERATIONS ON FUNCTIONS

105. If $f(x) = \dfrac{x+2}{x^2-3x+2}$ and $g(x) = \dfrac{x-1}{x+2}$, then find $(fg)(x)$.

106. If $f(x) = \dfrac{2}{x^2 - 5x + 4}$ and $g(x) = \dfrac{x-4}{x+1}$, then find $(fg)(x)$.

107. Find $(f/g)(x)$ if $f(x) = x^2 - 1$ and $g(x) = x + 1$.

108. Find $(f/g)(x)$ if $f(x) = \dfrac{2x-4}{5}$ and $g(x) = \dfrac{4-2x}{15}$.

APPLICATIONS

109. *Total Production Cost* (Refer to Example 5.) The average cost per item in dollars for a company to produce x digital cameras is given by

$$C(x) = \frac{50x + 20{,}000}{x}.$$

(a) The company expects to sell 5000 cameras. Evaluate $C(5000)$ and interpret the result.
(b) Write an expression $T(x)$ that gives the total cost of producing x cameras. Simplify it.
(c) Evaluate $T(5000)$ and interpret the result.

110. *Total Production Cost* (Refer to Example 5.) The average cost per item in dollars for a company to produce x flash drives is given by

$$C(x) = \frac{18x + 36{,}000}{x}.$$

(a) The company expects to sell 9000 flash drives. Evaluate $C(9000)$ and interpret the result.
(b) Write an expression for $T(x)$ that gives the total cost of producing x flash drives. Simplify this expression.
(c) Evaluate $T(9000)$ and interpret the result.

GEOMETRY

111. *Area of a Rectangle* The area A of a rectangle is $5x^2 + 12x + 4$ and its width W is $x + 2$, as shown in the figure.
(a) Find the length L of the rectangle.
(b) Find the length if the width is 8 feet.

112. *Area of a Rectangle* The area A of a rectangle is given by $x^2 - 1$.
(a) Find its length if its width is $x - 1$.
(b) Find the dimensions and area of the rectangle when $x = 15$.

113. *Volume of a Box* The volume V of a box with a square bottom is $4x^3 + 4x^2 + x$.
(a) If its height is x, find the area of the bottom.
(b) Find the dimensions of the box when $x = 10$.

114. *Area of a Triangle* The area A of a triangle is $6x^2 - x - 15$. Find its height if the base of the triangle is $2x + 3$.

WRITING ABOUT MATHEMATICS

115. A student does the following to simplify a rational expression. Is the work correct? Explain any errors and how you would correct them.

$$\frac{3x + x^2}{3x} \stackrel{?}{=} 1 + x^2$$

116. Explain how to multiply two rational expressions and how to divide two rational expressions.

SECTIONS 6.1 and 6.2	**Checking Basic Concepts**

1. Let $f(x) = \dfrac{x}{x-1}$.

(a) Evaluate $f(2)$.
(b) Find the domain of f.
(c) Graph f. Identify any vertical asymptotes.

2. Solve each equation. Check your answers.

(a) $\dfrac{6}{2x+3} = 3$ (b) $\dfrac{2}{x-1} = x$

3. Simplify $\dfrac{x^2 - 6x - 7}{x^2 - 1}$.

4. Multiply or divide each expression.

(a) $\dfrac{2x^2}{x^2 - 1} \cdot \dfrac{x+1}{4x}$

(b) $\dfrac{1}{x-2} \div \dfrac{3}{(x-2)(x+3)}$

6.3 Addition and Subtraction of Rational Expressions

Least Common Multiples • Review of Addition and Subtraction of Fractions •
Addition of Rational Expressions • Subtraction of Rational Expressions

A LOOK INTO MATH ▶

NEW VOCABULARY

☐ Least common multiple
☐ Common denominator
☐ Least common denominator

Rational expressions occur in many real-world applications. For example, addition of rational expressions is used in the design of electrical circuits. (See Example 8.) They are also used in business to analyze average cost. (See Exercises 91 and 92.) In this section we demonstrate how to add and subtract rational expressions. These techniques are similar to techniques used to add and subtract fractions. We begin by discussing least common multiples, which are used to find least common denominators.

Least Common Multiples

▶ **REAL-WORLD CONNECTION** Two friends work part-time at a store. The first person works every fourth day and the second person works every sixth day. If they are both working today, how many days pass before they both work on the same day again?

We can answer this question by listing the days that each person works.

First person: 4, 8, **12**, 16, 20, **24**, 28, 32, **36**, 40,…
Second person: 6, **12**, 18, **24**, 30, **36**, 42,…

After 12 days, the two friends work on the same day and the next time is after 24 days. The numbers 12 and 24 are *common multiples* of 4 and 6. (Find two more.) However, **12** is the **least common multiple** (LCM) of 4 and 6. Note that 12 is the smallest positive number that has both 4 and 6 as factors.

Another way to find the least common multiple for 4 and 6 is to first factor each number into prime numbers:

$$4 = \mathbf{2 \cdot 2} \quad \text{and} \quad 6 = \mathbf{2} \cdot \mathbf{3}.$$

To find the least common multiple, list each factor the *greatest* number of times that it occurs in either factorization. Then find the product of the listed numbers. For our example, the factor 2 occurs two times in the factorization of 4 and only once in the factorization of 6, so list 2 two times. The factor 3 appears only once in the factorization of 6 and not at all in the factorization of 4, so list it once:

$$\mathbf{2}, \mathbf{2}, \mathbf{3}.$$

The least common multiple is their product: $\mathbf{2} \cdot \mathbf{2} \cdot \mathbf{3} = 12$. This same procedure can also be used to find the least common multiple for two (or more) polynomials.

FINDING THE LEAST COMMON MULTIPLE

The least common multiple (LCM) of two polynomials can be found as follows.

STEP 1: Factor each polynomial completely.

STEP 2: List each factor the greatest number of times that it occurs in either factorization.

STEP 3: Find the product of this list of factors. The result is the LCM.

The next example illustrates how to use this procedure.

EXAMPLE 1 **Finding least common multiples**

Find the least common multiple for each pair of expressions.
(a) $4x$, $5x^3$ **(b)** $x^2 - 2x$, $(x - 2)^2$
(c) $x + 2$, $x - 1$ **(d)** $x^2 + 4x + 4$, $x^2 + 3x + 2$

Solution

(a) **STEP 1:** Factor each polynomial completely.

$$4x = \mathbf{2 \cdot 2 \cdot x} \quad \text{and} \quad 5x^3 = \mathbf{5 \cdot x \cdot x \cdot x}$$

STEP 2: The factor 2 occurs twice, the factor 5 occurs once, and the factor x occurs at most three times. The list is $\mathbf{2, 2, 5, x, x}$, and $\mathbf{x}$.
STEP 3: The LCM is the product $\mathbf{2 \cdot 2 \cdot 5 \cdot x \cdot x \cdot x}$, or $20x^3$.

(b) **STEP 1:** Factor each polynomial completely.

$$x^2 - 2x = \mathbf{x(x - 2)} \quad \text{and} \quad (x - 2)^2 = \mathbf{(x - 2)(x - 2)}$$

STEP 2: The factor x occurs once, and the factor $(x - 2)$ occurs at most twice. The list of factors is $\mathbf{x}$, $\mathbf{(x - 2)}$, and $\mathbf{(x - 2)}$.
STEP 3: The LCM is the product $x(x - 2)^2$, which is left in factored form.

(c) **STEP 1:** Neither polynomial can be factored.
STEP 2: The list of factors is $\mathbf{(x + 2)}$ and $\mathbf{(x - 1)}$.
STEP 3: The LCM is the product $(x + 2)(x - 1)$, or $x^2 + x - 2$.

(d) **STEP 1:** Factor each polynomial as follows.

$$x^2 + 4x + 4 = \mathbf{(x + 2)(x + 2)} \quad \text{and} \quad x^2 + 3x + 2 = \mathbf{(x + 1)(x + 2)}$$

STEP 2: The factor $(x + 1)$ occurs once and $(x + 2)$ occurs at most twice.
STEP 3: The LCM is the product $\mathbf{(x + 1)(x + 2)^2}$, which is left in factored form.

Now Try Exercises 15, 17, 19, 21

Review of Addition and Subtraction of Fractions

Recall that to add two fractions we use the property

$$\frac{a}{c} + \frac{b}{c} = \frac{a + b}{c}. \qquad \text{Like denominators}$$

This property requires that the fractions have like denominators. For example,

$$\frac{1}{5} + \frac{3}{5} = \frac{1 + 3}{5} = \frac{4}{5}. \qquad \text{Like denominators}$$

When the denominators are not alike, we must find a common denominator. Before adding two fractions, such as $\frac{1}{6}$ and $\frac{3}{4}$, we write them with 12 as their common denominator. That is, we multiply each fraction by 1, written in an appropriate form. For example, to write $\frac{1}{6}$ with a denominator of 12 we multiply $\frac{1}{6}$ by 1, written as $\frac{2}{2}$.

$$\frac{1}{6} = \frac{1}{6} \cdot \frac{\mathbf{2}}{\mathbf{2}} = \frac{\mathbf{2}}{12} \quad \text{and} \quad \frac{3}{4} = \frac{3}{4} \cdot \frac{\mathbf{3}}{\mathbf{3}} = \frac{\mathbf{9}}{12}$$

Once the fractions have a common denominator, we can add them, as in

$$\frac{1}{6} + \frac{3}{4} = \frac{\mathbf{2}}{\mathbf{12}} + \frac{\mathbf{9}}{\mathbf{12}} = \frac{\mathbf{11}}{\mathbf{12}}.$$

Note that a **common denominator** for $\frac{1}{6}$ and $\frac{3}{4}$ is also a *common multiple* of 6 and 4. The **least common denominator** (LCD) for $\frac{1}{6}$ and $\frac{3}{4}$ is equal to the *least common multiple* (LCM) of 6 and 4. Thus the least common denominator is **12**.

EXAMPLE 2 **Adding fractions**

Find the sum.

(a) $\dfrac{3}{4} + \dfrac{1}{8}$ (b) $\dfrac{3}{5} + \dfrac{2}{7}$

Solution

(a) The LCD is 8.

$$\frac{3}{4} + \frac{1}{8} = \frac{3}{4} \cdot \frac{2}{2} + \frac{1}{8} = \frac{6}{8} + \frac{1}{8} = \frac{7}{8}$$

(b) The LCD is 35.

$$\frac{3}{5} + \frac{2}{7} = \frac{3}{5} \cdot \frac{7}{7} + \frac{2}{7} \cdot \frac{5}{5} = \frac{21}{35} + \frac{10}{35} = \frac{31}{35}$$

Now Try Exercises 23, 25

Recall that subtraction of fractions is similar to addition of fractions. To subtract two fractions with like denominators we use the property

$$\frac{a}{c} - \frac{b}{c} = \frac{a - b}{c}. \quad \text{Like denominators}$$

For example, $\dfrac{3}{11} - \dfrac{7}{11} = \dfrac{3 - 7}{11} = -\dfrac{4}{11}$.

EXAMPLE 3 **Subtracting fractions**

Find the difference.

(a) $\dfrac{3}{10} - \dfrac{2}{15}$ (b) $\dfrac{3}{8} - \dfrac{5}{6}$

Solution

(a) The LCD is 30.

$$\frac{3}{10} - \frac{2}{15} = \frac{3}{10} \cdot \frac{3}{3} - \frac{2}{15} \cdot \frac{2}{2} = \frac{9}{30} - \frac{4}{30} = \frac{5}{30} = \frac{1}{6}$$

(b) The LCD is 24.

$$\frac{3}{8} - \frac{5}{6} = \frac{3}{8} \cdot \frac{3}{3} - \frac{5}{6} \cdot \frac{4}{4} = \frac{9}{24} - \frac{20}{24} = -\frac{11}{24}$$

Now Try Exercises 27, 29

Addition of Rational Expressions

Addition of rational expressions is similar to addition of fractions.

SUMS OF RATIONAL EXPRESSIONS

To add two rational expressions with like denominators, add their numerators. The denominator does not change.

$$\frac{A}{C} + \frac{B}{C} = \frac{A + B}{C} \qquad C \text{ not zero}$$

EXAMPLE 4 | **Adding rational expressions with like denominators**

Add and simplify.

(a) $\dfrac{x}{x + 2} + \dfrac{3x - 1}{x + 2}$ (b) $\dfrac{x}{3x^2 + 4x - 4} + \dfrac{2}{3x^2 + 4x - 4}$

Solution

(a) The expressions have like denominators, so add the numerators.

$$\dfrac{x}{x + 2} + \dfrac{3x - 1}{x + 2} = \dfrac{x + 3x - 1}{x + 2} \quad \text{Add numerators.}$$

$$= \dfrac{4x - 1}{x + 2} \quad \text{Combine like terms.}$$

(b) The expressions have like denominators, so add the numerators. The resulting sum can be simplified by factoring the denominator.

$$\dfrac{x}{3x^2 + 4x - 4} + \dfrac{2}{3x^2 + 4x - 4} = \dfrac{x + 2}{3x^2 + 4x - 4} \quad \text{Add numerators.}$$

$$= \dfrac{x + 2}{(3x - 2)(x + 2)} \quad \text{Factor denominator.}$$

$$= \dfrac{1}{3x - 2} \quad \text{Simplify.}$$

Now Try Exercises 31, 45

READING CHECK

How do you add rational expressions with unlike denominators?

To add rational expressions with *unlike denominators*, we must first write each expression so that it has the same common denominator. For example, the least common denominator for $\frac{1}{x + 2}$ and $\frac{2}{x - 1}$ equals the least common multiple of $x + 2$ and $x - 1$, which was shown to be their product $(x + 2)(x - 1)$ in Example 1(c). To rewrite $\frac{1}{x + 2}$ with the new denominator, we multiply it by 1, expressed as $\frac{x - 1}{x - 1}$.

$$\dfrac{1}{x + 2} \cdot 1 = \dfrac{1}{x + 2} \cdot \dfrac{x - 1}{x - 1} = \dfrac{x - 1}{(x + 2)(x - 1)}$$

Similarly, we multiply $\frac{2}{x - 1}$ by 1, expressed as $\frac{x + 2}{x + 2}$.

$$\dfrac{2}{x - 1} \cdot 1 = \dfrac{2}{x - 1} \cdot \dfrac{x + 2}{x + 2} = \dfrac{2x + 4}{(x - 1)(x + 2)}$$

Now the two rational expressions have like denominators and can be added.

$$\dfrac{1}{x + 2} + \dfrac{2}{x - 1} = \dfrac{x - 1}{(x + 2)(x - 1)} + \dfrac{2x + 4}{(x - 1)(x + 2)} \quad \text{Write with LCD.}$$

$$= \dfrac{x - 1 + 2x + 4}{(x + 2)(x - 1)} \quad \text{Add numerators.}$$

$$= \dfrac{3x + 3}{(x + 2)(x - 1)} \quad \text{Combine like terms.}$$

EXAMPLE 5 | **Adding rational expressions with unlike denominators**

Add and simplify. Leave your answer in factored form when appropriate.

(a) $\dfrac{1}{x} + \dfrac{2}{x^2}$ (b) $\dfrac{1}{x^2 - 9} + \dfrac{2}{x + 3}$

(c) $\dfrac{a}{a - b} + \dfrac{b}{a + b}$ (d) $\dfrac{1}{x - 1} + \dfrac{1}{1 - x}$

Solution

(a) The LCD for $\dfrac{1}{x} + \dfrac{2}{x^2}$ is x^2.

$$\frac{1}{x} \cdot \frac{x}{x} + \frac{2}{x^2} = \frac{x}{x^2} + \frac{2}{x^2} \qquad \text{Write with LCD.}$$

$$= \frac{x+2}{x^2} \qquad \text{Add numerators.}$$

(b) The LCD for $\dfrac{1}{x^2-9} + \dfrac{2}{x+3}$ is $x^2 - 9 = (x-3)(x+3)$.

$$\frac{1}{x^2-9} + \frac{2}{x+3} \cdot \frac{x-3}{x-3} = \frac{1}{(x+3)(x-3)} + \frac{2x-6}{(x+3)(x-3)} \qquad \text{Write with LCD.}$$

$$= \frac{2x-5}{(x+3)(x-3)} \qquad \text{Add numerators.}$$

(c) The LCD for $\dfrac{a}{a-b} + \dfrac{b}{a+b}$ is $(a-b)(a+b)$.

$$\frac{a}{a-b} \cdot \frac{a+b}{a+b} + \frac{b}{a+b} \cdot \frac{a-b}{a-b} = \frac{a^2+ab}{(a-b)(a+b)} + \frac{ab-b^2}{(a-b)(a+b)} \qquad \text{Write with LCD.}$$

$$= \frac{a^2 + 2ab - b^2}{(a-b)(a+b)} \qquad \text{Add numerators.}$$

(d) For $\dfrac{1}{x-1} + \dfrac{1}{1-x}$, multiply the second expression by $\dfrac{-1}{-1}$.

$$\frac{1}{x-1} + \frac{1}{1-x} \cdot \frac{-1}{-1} = \frac{1}{x-1} + \frac{-1}{(1-x)(-1)} \qquad \text{Write with LCD.}$$

$$= \frac{1}{x-1} + \frac{-1}{x-1} \qquad \text{Distributive property}$$

$$= \frac{0}{x-1} \qquad \text{Add numerators.}$$

$$= 0 \qquad \text{Simplify.}$$

▌ **Now Try Exercises 49, 55, 59**

NOTE: When one denominator is the opposite of the other, try multiplying one expression by 1 in the form $\frac{-1}{-1}$.

Subtraction of Rational Expressions

Subtraction of rational expressions is similar to subtraction of fractions.

DIFFERENCES OF RATIONAL EXPRESSIONS

To subtract two rational expressions with like denominators, subtract their numerators. The denominator does not change.

$$\frac{A}{C} - \frac{B}{C} = \frac{A-B}{C} \qquad C \text{ not zero}$$

When you are subtracting numerators of rational expressions, it is *essential* to apply the distributive property correctly: subtract every term in the numerator of the second rational expression. For example,

$$\frac{1}{x+1} - \frac{2x-1}{x+1} = \frac{1-(2x-1)}{x+1} = \frac{1-2x+1}{x+1} = \frac{2-2x}{x+1}.$$

Be sure to place parentheses around the second numerator before subtracting.

EXAMPLE 6 | **Subtracting rational expressions with like denominators**

Subtract and simplify.

(a) $\dfrac{3}{x^2} - \dfrac{x+3}{x^2}$ (b) $\dfrac{2x}{x^2-1} - \dfrac{x+1}{x^2-1}$

Solution

(a) The expressions have like denominators, so subtract the numerators.

$$\frac{3}{x^2} - \frac{x+3}{x^2} = \frac{3-(x+3)}{x^2} \qquad \text{Subtract numerators.}$$

$$= \frac{3-x-3}{x^2} \qquad \text{Distributive property}$$

$$= \frac{-x}{x^2} \qquad \text{Subtract.}$$

$$= -\frac{1}{x} \qquad \text{Simplify.}$$

(b) The expressions have like denominators, so subtract the numerators.

$$\frac{2x}{x^2-1} - \frac{x+1}{x^2-1} = \frac{2x-(x+1)}{x^2-1} \qquad \text{Subtract numerators.}$$

$$= \frac{2x-x-1}{x^2-1} \qquad \text{Distributive property}$$

$$= \frac{x-1}{(x+1)(x-1)} \qquad \text{Combine like terms; factor.}$$

$$= \frac{1}{x+1} \qquad \text{Simplify.}$$

Now Try Exercises 33, 41

READING CHECK

How do you subtract rational expressions with like denominators?

TECHNOLOGY NOTE

Graphical and Numerical Support
We can give support to our work in Example 6(a) by letting $Y_1 = 3/X^2 - (X+3)/X^2$, the given expression, and $Y_2 = -1/X$, the simplified expression. In Figures 6.9(a) and 6.9(b) the graphs of y_1 and y_2 appear to be identical. In Figure 6.9(c) numerical support is given, where $y_1 = y_2$ for each value of x.

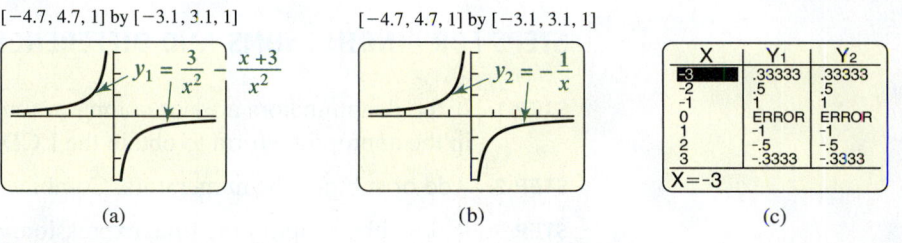

[−4.7, 4.7, 1] by [−3.1, 3.1, 1]   [−4.7, 4.7, 1] by [−3.1, 3.1, 1]

(a)   (b)   (c)

Figure 6.9

EXAMPLE 7 | **Subtracting rational expressions with unlike denominators**

Subtract and simplify. Leave your answer in factored form when appropriate.

(a) $\dfrac{5a}{b^2} - \dfrac{4b}{a^2}$ **(b)** $\dfrac{x-1}{x} - \dfrac{5}{x+5}$ **(c)** $\dfrac{1}{x^2-3x+2} - \dfrac{1}{x^2-x-2}$

Solution

(a) The LCD is a^2b^2.

$$\frac{5a}{b^2} \cdot \frac{a^2}{a^2} - \frac{4b}{a^2} \cdot \frac{b^2}{b^2} = \frac{5a^3}{b^2a^2} - \frac{4b^3}{a^2b^2} \qquad \text{Write with LCD.}$$

$$= \frac{5a^3 - 4b^3}{a^2b^2} \qquad \text{Subtract numerators.}$$

(b) The LCD is $x(x+5)$.

$$\frac{x-1}{x} \cdot \frac{x+5}{x+5} - \frac{5}{x+5} \cdot \frac{x}{x} \qquad \text{Write with LCD.}$$

$$= \frac{(x-1)(x+5)}{x(x+5)} - \frac{5x}{x(x+5)} \qquad \text{Multiply.}$$

$$= \frac{(x-1)(x+5) - 5x}{x(x+5)} \qquad \text{Subtract numerators.}$$

$$= \frac{x^2 + 4x - 5 - 5x}{x(x+5)} \qquad \text{Multiply binomials.}$$

$$= \frac{x^2 - x - 5}{x(x+5)} \qquad \text{Combine like terms.}$$

(c) Because $x^2 - 3x + 2 = (x-2)(x-1)$ and $x^2 - x - 2 = (x-2)(x+1)$, the LCD is $(x-1)(x+1)(x-2)$.

$$\frac{1}{(x-2)(x-1)} \cdot \frac{(x+1)}{(x+1)} - \frac{1}{(x-2)(x+1)} \cdot \frac{(x-1)}{(x-1)} \qquad \text{Write with LCD.}$$

$$= \frac{(x+1)}{(x-2)(x-1)(x+1)} - \frac{(x-1)}{(x-2)(x+1)(x-1)} \qquad \text{Multiply.}$$

$$= \frac{(x+1) - (x-1)}{(x-2)(x-1)(x+1)} \qquad \text{Subtract numerators.}$$

$$= \frac{x+1 - x+1}{(x-2)(x-1)(x+1)} \qquad \text{Distributive property}$$

$$= \frac{2}{(x-2)(x-1)(x+1)} \qquad \text{Simplify numerator.}$$

Now Try Exercises 47, 51, 69

The following procedure summarizes how to add or subtract rational expressions.

STEPS FOR FINDING SUMS AND DIFFERENCES OF RATIONAL EXPRESSIONS

STEP 1: If the denominators are not common, multiply each expression by 1 written in the appropriate form to obtain the LCD.

STEP 2: Add or subtract the numerators. Combine like terms.

STEP 3: If possible, simplify the final expression.

Figure 6.10

▶ **REAL-WORLD CONNECTION** Sums of rational expressions occur in applications such as electrical circuits. Suppose that two light bulbs are wired in parallel so that electricity can flow through either light bulb, as depicted in Figure 6.10. If their resistances are R_1 and R_2, their combined resistance R can be computed by the equation

$$\frac{1}{R} = \frac{1}{R_1} + \frac{1}{R_2}.$$

Resistance is often measured in a unit called *ohms*. A standard 60-watt light bulb might have a resistance of about 200 ohms. (**Source:** R. Weidner and R. Sells, *Elementary Classical Physics, Vol. 2.*)

EXAMPLE 8 **Modeling electrical resistance**

A 100-watt light bulb with a resistance of $R_1 = 120$ ohms and a 75-watt light bulb with a resistance of $R_2 = 160$ ohms are placed in an electrical circuit, as shown in Figure 6.10. Find their combined resistance.

Solution
Let $R_1 = 120$ and $R_2 = 160$ in the given equation and solve for R.

$$\frac{1}{R} = \frac{1}{R_1} + \frac{1}{R_2} \qquad \text{Given equation}$$

$$= \frac{1}{120} + \frac{1}{160} \qquad \text{Substitute for } R_1 \text{ and } R_2.$$

$$= \frac{1}{120} \cdot \frac{4}{4} + \frac{1}{160} \cdot \frac{3}{3} \qquad \text{LCD is 480.}$$

$$= \frac{4}{480} + \frac{3}{480} \qquad \text{Multiply.}$$

$$= \frac{7}{480} \qquad \text{Add.}$$

Because $\frac{1}{R} = \frac{7}{480}$, $R = \frac{480}{7} \approx 69$ ohms.

▪ **Now Try Exercise 95**

CRITICAL THINKING

If $\frac{1}{R} = \frac{1}{R_1} + \frac{1}{R_2}$, does it follow that $R = R_1 + R_2$? Explain your answer.

6.3 Putting It All Together

CONCEPT	EXPLANATION	EXAMPLE
Least Common Multiple (LCM)	To find the LCM, follow the three-step procedure presented on page 424.	To find the LCM of $2x^2 - 2x$ and $8x^2$, factor each expression: $$2x^2 - 2x = \mathbf{2} \cdot \mathbf{x} \cdot (x - 1) \quad \text{and}$$ $$8x^2 = \mathbf{2} \cdot \mathbf{2} \cdot \mathbf{2} \cdot \mathbf{x} \cdot \mathbf{x}.$$ The LCM is $$\mathbf{2} \cdot \mathbf{2} \cdot \mathbf{2} \cdot \mathbf{x} \cdot \mathbf{x} \cdot (x - 1) = 8x^2(x - 1).$$
Least Common Denominator (LCD)	*The* LCD *equals the* LCM *of the denominators.*	The LCD for $$\frac{3x}{2x^2 - 2x} \text{ and } \frac{5}{8x^2}$$ is $8x^2(x - 1)$.

continued on next page

continued from previous page

CONCEPT	EXPLANATION	EXAMPLE
Addition and Subtraction of Rational Expressions	To add (or subtract) two rational expressions with like denominators, add (or subtract) their numerators. The denominator does not change. $$\frac{A}{C} + \frac{B}{C} = \frac{A + B}{C}$$ $$\frac{A}{C} - \frac{B}{C} = \frac{A - B}{C} \quad \textcolor{blue}{C \text{ not zero}}$$ If the denominators are *not alike*, write each term with the LCD first.	*Like denominators* $$\frac{x}{x + 1} + \frac{3x}{x + 1} = \frac{4x}{x + 1}$$ *Unlike denominators* $$\frac{1}{x} - \frac{2x}{x + 1} = \frac{1}{x} \cdot \frac{x + 1}{x + 1} - \frac{2x}{x + 1} \cdot \frac{x}{x}$$ $$= \frac{x + 1}{x(x + 1)} - \frac{2x^2}{x(x + 1)}$$ $$= \frac{-2x^2 + x + 1}{x(x + 1)}$$

6.3 Exercises

MyMathLab    PRACTICE   WATCH   DOWNLOAD   READ   REVIEW

CONCEPTS AND VOCABULARY

1. What is the LCM for 6 and 9?

2. What is the LCD for $\frac{1}{6}$ and $\frac{1}{9}$?

3. What is the LCM for $x^2 - 25$ and $x + 5$?

4. What is the LCD for $\frac{1}{x^2 - 25}$ and $\frac{1}{x + 5}$?

5. What do you need to find before you can add $\frac{1}{4}$ and $\frac{1}{3}$?

6. What do you need to find before you can add the expressions $\frac{2}{x}$ and $\frac{1}{2x - 1}$?

7. $\frac{a}{c} + \frac{b}{c} =$ _____.

8. $\frac{a}{c} - \frac{b}{c} =$ _____.

9. Which of the following (a.–d.) equals $\frac{1 - x}{2 + x} + \frac{x}{2 + x}$?

   a. $\dfrac{1}{4 + 2x}$ **b.** $\dfrac{1}{2 + x}$

   c. $\dfrac{1}{(2 + x)^2}$ **d.** $\dfrac{x - x^2}{2 + x}$

10. Which of the following (a.–d.) equals $\frac{1}{x} - \frac{2 - x}{2x}$?

    a. $\dfrac{1}{2}$ **b.** $\dfrac{x - 1}{2x}$

    c. $\dfrac{1 + x}{x}$ **d.** $\dfrac{x}{2}$

LEAST COMMON MULTIPLES

Exercises 11–22: Find the least common multiple.

11. 10, 15   12. 9, 12

13. 34, 51   14. 24, 36

15. $6a, 9a^2$ 16. $12ab, 6a^2$

17. $10x^2, 25(x^2 - x)$ 18. $x^2 - 4, x^2 + 2x$

19. $x^2 + 2x + 1, x^2 - 4x - 5$

20. $4x^2 + 20x + 25, 2x^2 + 5x$

21. $x + y, x - y$ 22. $(x - y)^2, x^2 - y^2$

REVIEW OF FRACTIONS

Exercises 23–30: Simplify.

23. $\frac{1}{7} + \frac{4}{7}$ 24. $\frac{2}{5} + \frac{1}{2}$

25. $\frac{2}{3} + \frac{5}{6} + \frac{1}{4}$ 26. $\frac{3}{11} + \frac{1}{2} + \frac{1}{6}$

27. $\frac{1}{10} - \frac{3}{10}$ 28. $\frac{2}{9} - \frac{1}{11}$

29. $\frac{3}{2} - \frac{1}{8}$ 30. $\frac{3}{12} - \frac{5}{16}$

ADDITION AND SUBTRACTION OF RATIONAL EXPRESSIONS

Exercises 31–84: Add or subtract. Leave your answer in factored form when appropriate.

31. $\frac{1}{x} + \frac{3}{x}$ 32. $\frac{2}{x - 1} + \frac{x}{x - 1}$

33. $\frac{2}{x^2 - 4} - \frac{x + 1}{x^2 - 4}$ 34. $\frac{2x - 1}{x^2 + 6} - \frac{2x + 1}{x^2 + 6}$

35. $\frac{4}{x^2} + \frac{5}{x^2}$ 36. $\frac{8}{b^3} - \frac{5}{b^3}$

37. $\frac{4}{xy} - \frac{7}{xy}$ 38. $\frac{9}{a^2 b} + \frac{1}{a^2 b}$

39. $\dfrac{x}{x+1} + \dfrac{1}{x+1}$

40. $\dfrac{1}{2x-1} - \dfrac{2x}{2x-1}$

41. $\dfrac{2z}{4-z} - \dfrac{3z-4}{4-z}$

42. $\dfrac{z}{z^2-9} - \dfrac{3}{z^2-9}$

43. $\dfrac{4r}{5t^2} + \dfrac{r}{5t^2}$

44. $\dfrac{1}{2ab} + \dfrac{1}{2ab}$

45. $\dfrac{3t}{t^2-t-6} + \dfrac{2-2t}{t^2-t-6}$

46. $\dfrac{t}{t^2+5t} + \dfrac{5}{t^2+5t}$

47. $\dfrac{5b}{3a} - \dfrac{7b}{5a}$

48. $\dfrac{a}{ab^2} - \dfrac{b}{a^2b}$

49. $\dfrac{4}{n-4} + \dfrac{3}{2-n}$

50. $\dfrac{3}{2n-1} - \dfrac{3}{1-2n}$

51. $\dfrac{x}{x+4} - \dfrac{x+1}{x}$

52. $\dfrac{4x}{x+2} + \dfrac{x-5}{x-2}$

53. $\dfrac{2}{x^2} - \dfrac{4x-1}{x}$

54. $\dfrac{2x}{x-5} - \dfrac{x}{x+5}$

55. $\dfrac{x+3}{x-5} + \dfrac{5}{x-3}$

56. $\dfrac{x}{2x-1} + \dfrac{1-x}{3x}$

57. $\dfrac{4n}{n^2-9} - \dfrac{8}{n-3}$

58. $\dfrac{3n}{(4n-3)^2} - \dfrac{1}{4n-3}$

59. $\dfrac{x}{x^2-9} + \dfrac{5x}{x-3}$

60. $\dfrac{a^2+1}{a^2-1} + \dfrac{a}{1-a^2}$

61. $\dfrac{b}{2b-4} - \dfrac{b-1}{b-2}$

62. $\dfrac{y^2}{2-y} - \dfrac{y}{y^2-4}$

63. $\dfrac{2x}{x-5} + \dfrac{2x-1}{3x^2-16x+5}$

64. $\dfrac{x+3}{2x-1} + \dfrac{3}{10x^2-5x}$

65. $\dfrac{4x}{x-y} - \dfrac{9}{x+y}$

66. $\dfrac{1}{a-b} - \dfrac{3a}{a^2-b^2}$

67. $\dfrac{3}{(x-1)(x-2)} + \dfrac{4x}{(x+1)(x-2)}$

68. $\dfrac{x}{(x-1)^2} - \dfrac{1}{(x-1)(x+3)}$

69. $\dfrac{3}{x^2-x-6} - \dfrac{2}{x^2+5x+6}$

70. $\dfrac{x}{x^2-5x+4} + \dfrac{2}{x^2-2x-8}$

71. $\dfrac{3}{x^2-2x+1} + \dfrac{1}{x^2-3x+2}$

72. $\dfrac{x}{x^2-4} - \dfrac{1}{x^2+4x+4}$

73. $\dfrac{3x}{x^2+2x-3} + \dfrac{1}{x^2-2x+1}$

74. $\dfrac{3x}{x-y} - \dfrac{3y}{x^2-2xy+y^2}$

75. $\dfrac{4c}{ab} + \dfrac{3b}{ac} - \dfrac{2a}{bc}$

76. $x + \dfrac{1}{x-1} - \dfrac{1}{x+1}$

77. $5 - \dfrac{6}{n^2-36} + \dfrac{3}{n-6}$

78. $\dfrac{6}{t-1} + \dfrac{2}{t-2} + \dfrac{1}{t}$

79. $\dfrac{3}{x-5} - \dfrac{1}{x-3} - \dfrac{2x}{x-5}$

80. $\dfrac{2x+1}{x-1} - \dfrac{3}{x+1} + \dfrac{x}{x-1}$

81. $\dfrac{5}{2x-3} + \dfrac{x}{x+1} - \dfrac{x}{2x-3}$

82. $\dfrac{1}{x-3} - \dfrac{2}{x+3} + \dfrac{x}{x^2-9}$

83. $\dfrac{1}{x-1} - \dfrac{2}{x+1} + \dfrac{x}{x^2-1}$

84. $\dfrac{2}{x-2} + \dfrac{1}{x+3} - \dfrac{1}{x-1}$

Exercises 85 and 86: **Thinking Generally** *Add or subtract.*

85. $\dfrac{a-b}{a+b} + \dfrac{a+b}{a-b}$

86. $\dfrac{1}{a^2-b^2} - \dfrac{1}{a^2+b^2}$

ADDITION AND SUBTRACTION OF FUNCTIONS

Exercises 87–90: Find $(f+g)(x)$ *and* $(f-g)(x)$.

87. $f(x) = \dfrac{1}{x+1},\ g(x) = \dfrac{x}{x+1}$

88. $f(x) = \dfrac{1}{2x},\ g(x) = \dfrac{3}{4x}$

89. $f(x) = \dfrac{1}{x+2},\ g(x) = \dfrac{1}{x-2}$

90. $f(x) = \dfrac{1}{x+1},\ g(x) = \dfrac{1}{x}$

APPLICATIONS

91. *Average Cost* The average cost per item in dollars to make x video games is given by

$$M(x) = \frac{5x + 1{,}000{,}000}{x},$$

and the average cost per item in dollars to package the same video game is given by

$$P(x) = \frac{2x + 500{,}000}{x}.$$

Write a function $C(x)$ that gives the total cost per item of making and packaging x video games.

92. *Breaking Even* Use the result from Exercise 91 to determine the number of video games the company must sell to break even, if it charges $37 per game.

93. *Planet Alignment* Suppose that Saturn and Jupiter are in alignment with respect to the sun. If Jupiter orbits the sun every 12 years and Saturn orbits the sun every 30 years, how many years will it be before these two planets are back in the same location with the same alignment again?

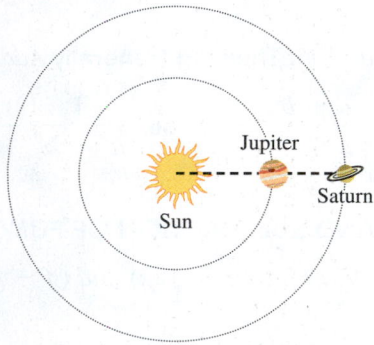

NOT TO SCALE

94. *Working Shifts* Two friends work part-time at a store. The first person works every sixth day and the second person works every tenth day. If they are both working today, how many days pass before they both work on the same day again?

95. *Electrical Resistance* (Refer to Example 8.) A 150-watt light bulb with a resistance of 80 ohms and a 40-watt light bulb with a resistance of 300 ohms are wired in parallel. Find the combined resistance.

96. *Electrical Resistance* Solve the formula

$$\frac{1}{R} = \frac{1}{R_1} + \frac{1}{R_2}$$

for R. Use your formula to solve Exercise 95.

97. *Photography* A lens in a camera has a focal length, which is important in focusing the camera. If an object is at distance D from a lens that has a focal length F, then to be in focus the distance S between the lens and the film should satisfy the equation

$$\frac{1}{S} = \frac{1}{F} - \frac{1}{D}.$$

(See the accompanying figure.) If the focal length is $F = 0.1$ foot and the object is $D = 10$ feet from the camera, find S.

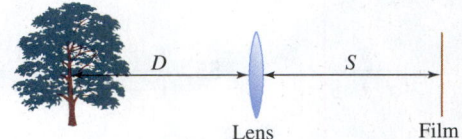

98. *Photography* (Continuation of Exercise 97.) Suppose that the object is closer to the lens than the focal length of the lens. That is, suppose $D < F$. According to the formula, is it possible to focus the image on the film? Explain your reasoning.

99. *Intensity of a Light Bulb* The intensity of a light bulb at a point varies according to its distance from the light bulb. The expression $\frac{8}{d^2}$ approximates the intensity of a 100-watt light bulb in watts per square meter (W/m^2) at a distance of d meters. Calculate the combined intensity at a point d meters from two 100-watt light bulbs.

100. *Intensity of a Light Bulb* (Refer to the preceding exercise.) Suppose that one 100-watt light bulb is at a distance d and another 100-watt light bulb is at a distance $\frac{1}{2}d$ from a point. Write an expression for the combined intensity of the two light bulbs.

WRITING ABOUT MATHEMATICS

101. A student does the following to add two rational expressions. Explain the error that the student is making and how it can be corrected.

$$\frac{x}{x + 1} + \frac{6x}{x} \stackrel{?}{=} \frac{7x}{2x + 1}$$

102. Explain in words how to subtract two rational expressions with like denominators.

6.4 Rational Equations

Solving Rational Equations • Solving an Equation for a Variable

Although thousands of mobile phone applications are being written each year, most users who download these applications use them only once. For example, only about 25% of people who downloaded an application are still using it on the second day. This aspect of mobile phone applications can be modeled by a rational function. See Example 4.

Solving Rational Equations

A rational expression is a ratio of two polynomials. In Sections 6.2 and 6.3, we *simplified* and *evaluated* rational expressions. Unlike a rational *expression*, a rational *equation* always contains an equals sign, and we *solve* this equation for values of the variables that make the equation a true statement. One way to solve a rational equation is to *clear fractions* by multiplying each side of the equation by the least common denominator (LCD). For example, to solve

$$\frac{1}{2x} + \frac{2}{3} = \frac{5}{2x}$$

multiply each side (or each term) by the LCD (LCM of 3 and $2x$), which is **$6x$**.

$$6x\left(\frac{1}{2x} + \frac{2}{3}\right) = 6x\left(\frac{5}{2x}\right) \qquad \text{Multiply each side by } 6x.$$

$$\frac{6x}{2x} + \frac{12x}{3} = \frac{30x}{2x} \qquad \text{Distributive property}$$

$$3 + 4x = 15 \qquad \text{Simplify.}$$

$$4x = 12 \qquad \text{Subtract 3 from each side.}$$

$$x = 3 \qquad \text{Divide each side by 4.}$$

Check this answer to verify that 3 is indeed a solution. *When solving a rational equation, always check your answer.* If one of your *possible answers* makes the denominator of an expression in the given equation equal to 0, it cannot be a solution.

EXAMPLE 1 **Solving rational equations**

Solve each equation and check your answer.

(a) $\dfrac{x+1}{2x} - \dfrac{x-1}{4x} = \dfrac{1}{3}$ (b) $\dfrac{3}{x-2} + \dfrac{5}{x+2} = \dfrac{12}{x^2-4}$

Solution

(a) Note that the left side of the equation is undefined when $x = 0$. Thus 0 cannot be a solution. Start by determining the LCD, which is **$12x$**.

$$12x\left(\frac{x+1}{2x} - \frac{x-1}{4x}\right) = 12x\left(\frac{1}{3}\right) \qquad \text{Multiply each side by } 12x.$$

$$\frac{12x(x+1)}{2x} - \frac{12x(x-1)}{4x} = \frac{12x}{3} \qquad \text{Distributive property}$$

$$6(x+1) - 3(x-1) = 4x \qquad \text{Simplify.}$$

$$6x + 6 - 3x + 3 = 4x \qquad \text{Distributive property}$$

$$3x + 9 = 4x \qquad \text{Combine like terms.}$$

$$x = 9 \qquad \text{Solve for } x.$$

Check:
$$\frac{9+1}{2(9)} - \frac{9-1}{4(9)} \stackrel{?}{=} \frac{1}{3}$$ Let $x = 9$ in $\frac{x+1}{2x} - \frac{x-1}{4x} = \frac{1}{3}$.

$$\frac{10}{18} - \frac{8}{36} \stackrel{?}{=} \frac{1}{3}$$ Simplify.

$$\frac{1}{3} = \frac{1}{3} \checkmark$$ It checks.

(b) Note that both sides of $\frac{3}{x-2} + \frac{5}{x+2} = \frac{12}{x^2-4}$ are undefined when $x = 2$ or $x = -2$. Thus neither 2 nor -2 can be a solution. The LCD is $x^2 - 4 = (x-2)(x+2)$.

$$\frac{3(x-2)(x+2)}{x-2} + \frac{5(x-2)(x+2)}{x+2} = \frac{12(x^2-4)}{x^2-4}$$ Multiply each term by the LCD.

$$3(x+2) + 5(x-2) = 12$$ Simplify.

$$3x + 6 + 5x - 10 = 12$$ Distributive property

$$8x - 4 = 12$$ Combine like terms.

$$8x = 16$$ Add 4.

$$x = 2$$ Divide by 8.

Check: We already noted that 2 cannot be a solution. However, checking this answer gives the following result.

$$\frac{3}{2-2} + \frac{5}{2+2} \stackrel{?}{=} \frac{12}{2^2-4}$$ Let $x = 2$ in $\frac{3}{x-2} + \frac{5}{x+2} = \frac{12}{x^2-4}$.

$$\frac{3}{0} + \frac{5}{4} \stackrel{?}{=} \frac{12}{0} ✗$$ Simplify fractions.

Both sides are undefined because division by 0 is not possible. There are no solutions.

Now Try Exercises 31, 39

READING CHECK

What is a good first step when solving a rational equation?

The next example illustrates how to solve a rational equation graphically. Because graphical solutions are only estimates, be sure to check them by substituting observed values in the given equation.

EXAMPLE 2   **Solving a rational equation graphically**

Solve the equation $\frac{4}{x-1} = 2x$ graphically. Check your results.

Solution

Start by graphing $y_1 = \frac{4}{x-1}$ and $y_2 = 2x$. The graph of y_2 is a line with slope 2 passing through the origin. To help graph y_1 make a table of values or use a graphing calculator. The two graphs intersect when $x = -1$ and $x = 2$, as shown in Figure 6.11.

Check:
$$\frac{4}{-1-1} \stackrel{?}{=} 2(-1)$$ Let $x = -1$. $$\frac{4}{2-1} \stackrel{?}{=} 2(2)$$ Let $x = 2$.

$$-2 = -2 \checkmark$$ It checks. $$4 = 4 \checkmark$$ It checks.

Now Try Exercises 53, 57

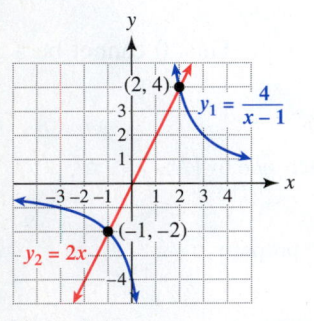

Figure 6.11

The next example illustrates how to solve rational equations and support the results with a graphing calculator.

| EXAMPLE 3 | **Solving rational equations** |

Solve each equation. Support your results either graphically or numerically.

(a) $\dfrac{1}{2} + \dfrac{x}{3} = \dfrac{x}{5}$ **(b)** $\dfrac{3}{x+3} - 2 = \dfrac{x}{x+3}$ **(c)** $\dfrac{3}{x-2} = \dfrac{5}{x+2}$

Solution

(a) The LCD is the product of 2, 3, and 5, which is **30**.

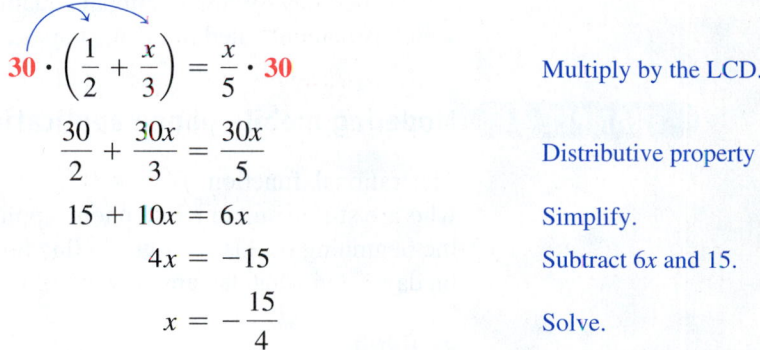

$$30 \cdot \left(\frac{1}{2} + \frac{x}{3} \right) = \frac{x}{5} \cdot 30 \qquad \text{Multiply by the LCD.}$$

$$\frac{30}{2} + \frac{30x}{3} = \frac{30x}{5} \qquad \text{Distributive property}$$

$$15 + 10x = 6x \qquad \text{Simplify.}$$

$$4x = -15 \qquad \text{Subtract } 6x \text{ and } 15.$$

$$x = -\frac{15}{4} \qquad \text{Solve.}$$

This result can be supported graphically by letting $y_1 = \frac{1}{2} + \frac{x}{3}$ and $y_2 = \frac{x}{5}$. Their graphs intersect at the point $(-3.75, -0.75)$, as shown in Figure 6.12. Therefore the solution to this equation is -3.75, or $-\frac{15}{4}$.

(b) Multiply each side by the LCD, which is $x + 3$.

$$(x + 3) \cdot \left(\frac{3}{x+3} - 2 \right) = \frac{x}{x+3} \cdot (x + 3) \qquad \text{Multiply by the LCD.}$$

$$3 - 2(x + 3) = x \qquad \text{Distributive property}$$

$$3 - 2x - 6 = x \qquad \text{Distributive property}$$

$$-3 = 3x \qquad \text{Add } 2x \text{ to each side.}$$

$$x = -1 \qquad \text{Divide by 3; rewrite.}$$

Figure 6.13 supports the finding that -1 is a solution by showing that the graphs of the equations $y_1 = \frac{3}{x+3} - 2$, and $y_2 = \frac{x}{x+3}$ intersect at $(-1, -0.5)$.

NOTE: Because graphs of rational expressions are often nonlinear, it may be difficult to locate graphical solutions.

(c) Multiply each side by the LCD, which is $(x - 2)(x + 2)$.

$$(x - 2)(x + 2) \cdot \frac{3}{x-2} = \frac{5}{x+2} \cdot (x - 2)(x + 2) \qquad \text{Multiply by the LCD.}$$

$$\frac{3(x - 2)(x + 2)}{x - 2} = \frac{5(x - 2)(x + 2)}{x + 2} \qquad \text{Multiply expressions.}$$

$$3(x + 2) = 5(x - 2) \qquad \text{Simplify.}$$

$$3x + 6 = 5x - 10 \qquad \text{Distributive property}$$

$$16 = 2x \qquad \text{Add 10 and subtract } 3x.$$

$$x = 8 \qquad \text{Divide by 2; rewrite.}$$

To support this result numerically, let $y_1 = \frac{3}{x-2}$ and $y_2 = \frac{5}{x+2}$. The table in Figure 6.14 shows that $y_1 = y_2 = 0.5$ when $x = 8$. The solution is 8.

Now Try Exercises 67, 69

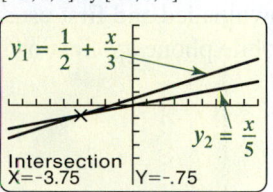

[−9, 9, 1] by [−6, 6, 1]

$y_1 = \frac{1}{2} + \frac{x}{3}$
$y_2 = \frac{x}{5}$
Intersection
X=-3.75   Y=-.75

Figure 6.12

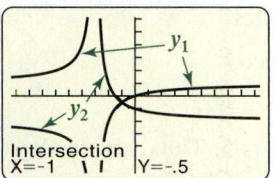

[−9.4, 9.4, 1] by [−6.2, 6.2, 1]

y_1
y_2
Intersection
X=-1   Y=-.5

Figure 6.13

X	Y₁	Y₂
5	1	.71429
6	.75	.625
7	.6	.55556
8	.5	.5
9	.42857	.45455
10	.375	.41667
11	.33333	.38462

X=8

Figure 6.14

TECHNOLOGY NOTE

Entering Rational Expressions
When entering a rational expression, use parentheses around the numerator and denominator, as in the following example. Enter

$$y = \frac{x + 1}{2x - 1} \quad \text{as} \quad Y_1 = (X + 1)/(2X - 1).$$

Sometimes solving a rational equation results in the need to solve a quadratic equation, which is demonstrated in the next application.

EXAMPLE 4

Modeling mobile phone applications

The rational function $f(x) = \frac{100}{x^2}$, $1 \le x \le 10$, approximates the percentage of people who are still using a mobile phone application after x days. (Refer to A Look Into Math at the beginning of this section.) In this formula, the application is downloaded and first used on day 1. On what day are only 4% of people still using a typical mobile phone application?

Solution
We must solve the rational equation $\frac{100}{x^2} = 4$.

$$\frac{100}{x^2} = 4 \qquad \text{Equation to be solved}$$

$$x^2 \cdot \frac{100}{x^2} = 4 \cdot x^2 \qquad \text{Multiply each side by } x^2.$$

$$100 = 4x^2 \qquad \text{Simplify.}$$

$$4x^2 - 100 = 0 \qquad \text{Rewrite equation.}$$

$$(2x - 10)(2x + 10) = 0 \qquad \text{Difference of squares}$$

$$2x - 10 = 0 \quad \text{or} \quad 2x + 10 = 0 \qquad \text{Zero-product property}$$

$$x = 5 \quad \text{or} \quad x = -5 \qquad \text{Solve.}$$

Since we do not allow a negative number of days, the only solution is 5. Thus, after 5 days about 4% of people are still using a typical mobile phone application. (Check this.)

Now Try Exercise 93

EXAMPLE 5

Solving rational equations

Solve the equation.

(a) $\dfrac{1}{x} + \dfrac{1}{x^2} = \dfrac{3}{4}$ **(b)** $\dfrac{2x}{x + 2} + \dfrac{3x}{x - 1} = 7$

Solution
(a) The LCD is $4x^2$. Note that 0 cannot be a solution.

$$4x^2 \cdot \left(\frac{1}{x} + \frac{1}{x^2} \right) = \frac{3}{4} \cdot 4x^2 \qquad \text{Multiply by } 4x^2.$$

$$\frac{4x^2}{x} + \frac{4x^2}{x^2} = \frac{12x^2}{4} \qquad \text{Distributive property}$$

$$4x + 4 = 3x^2 \qquad \text{Simplify.}$$

$$0 = 3x^2 - 4x - 4 \qquad \text{Subtract } 4x \text{ and } 4.$$

$$0 = (3x + 2)(x - 2) \qquad \text{Factor.}$$

$$x = -\frac{2}{3} \quad \text{or} \quad x = 2 \qquad \text{Solve.}$$

Both solutions check.

(b) The LCD is $(x + 2)(x - 1)$. Note that -2 and 1 cannot be solutions.

$$(x + 2)(x - 1)\left(\frac{2x}{x + 2} + \frac{3x}{x - 1}\right) = 7(x + 2)(x - 1) \quad \text{Multiply by the LCD.}$$

$$\frac{2x(x + 2)(x - 1)}{x + 2} + \frac{3x(x + 2)(x - 1)}{x - 1} = 7(x + 2)(x - 1) \quad \text{Distributive property}$$

$$2x(x - 1) + 3x(x + 2) = 7(x + 2)(x - 1) \quad \text{Simplify.}$$

$$2x^2 - 2x + 3x^2 + 6x = 7(x^2 + x - 2) \quad \text{Multiply.}$$

$$5x^2 + 4x = 7x^2 + 7x - 14 \quad \text{Simplify.}$$

$$0 = 2x^2 + 3x - 14 \quad \text{Subtract } 5x^2 \text{ and } 4x.$$

$$0 = (2x + 7)(x - 2) \quad \text{Factor.}$$

$$x = -\frac{7}{2} \quad \text{or} \quad x = 2 \quad \text{Solve.}$$

Both solutions check.

❚ Now Try Exercises 35, 41

▶ **REAL-WORLD CONNECTION** A customer often has to wait to check out at a store, and a commuter often has to wait in a traffic jam. During a lifetime, a typical person spends thousands of hours waiting in line. Rational equations are frequently used to estimate the time spent waiting in line whenever arrivals are random. This concept is illustrated in the next example.

EXAMPLE 6 **Modeling waiting time**

Suppose that cars are arriving randomly at a construction zone and that the flag person can instruct x drivers per minute. If cars arrive at an average rate of 7 cars per minute, the average time T in minutes for each driver to wait and talk to the flag person is

$$T(x) = \frac{1}{x - 7},$$

where $x > 7$. How many drivers per minute should the flag person be able to instruct to keep the average wait to 1 minute? (*Source:* N. Garber.)

Solution
We must determine x such that $\dfrac{1}{x - 7} = 1$.

$$\frac{1}{x - 7} = 1 \quad \text{Equation to be solved}$$

$$(x - 7) \cdot \frac{1}{x - 7} = 1 \cdot (x - 7) \quad \text{Multiply by the LCD.}$$

$$1 = x - 7 \quad \text{Simplify.}$$

$$8 = x \quad \text{Add 7 to each side.}$$

The flag person should be able to instruct 8 drivers per minute to limit the average wait to 1 minute.

❚ Now Try Exercise 87(a)

Rational equations occur in applications involving time and rate, as demonstrated in the next example.

EXAMPLE 7 **Determining the time required to empty a pool**

A pump can empty a swimming pool in 50 hours. To speed up the process, a second pump that can empty the pool in 80 hours is used. How long will it take for both pumps working together to empty the pool?

Solution

Because the first pump can empty the entire pool in 50 hours, it can empty $\frac{1}{50}$ of the pool in 1 hour, $\frac{2}{50}$ of the pool in 2 hours, $\frac{3}{50}$ of the pool in 3 hours, and in general, it can empty $\frac{t}{50}$ of the pool in t hours. That is, the amount of water pumped equals the rate $\frac{1}{50}$ times the time t. The second pump can empty the pool in 80 hours, so (using similar reasoning) it can empty $\frac{t}{80}$ of the pool in t hours.

Together the pumps can empty

$$\frac{t}{50} + \frac{t}{80}$$

of the pool in t hours. The job will be complete when the fraction of the pool that is empty equals 1. Thus we must solve the equation

$$\frac{t}{50} + \frac{t}{80} = 1.$$

To clear fractions we can multiply each side by **(50)(80)**. (We could also use the LCD of 400.)

$$\mathbf{(50)(80)}\left(\frac{t}{50} + \frac{t}{80}\right) = 1\mathbf{(50)(80)} \qquad \text{Multiply by common denominator.}$$

$$\frac{\mathbf{50 \cdot 80} \cdot t}{\mathbf{50}} + \frac{50 \cdot \mathbf{80} \cdot t}{\mathbf{80}} = 4000 \qquad \text{Distributive property}$$

$$80t + 50t = 4000 \qquad \text{Simplify.}$$

$$130t = 4000 \qquad \text{Combine like terms.}$$

$$t = \frac{4000}{130} \approx 30.8 \text{ hours} \qquad \text{Solve.}$$

The two pumps can empty the pool in about 30.8 hours.

Now Try Exercise 95

CRITICAL THINKING

A bicyclist rides uphill at 6 miles per hour for 1 mile and then rides downhill at 12 miles per hour for 1 mile. What is the average speed of the bicyclist?

▶ **REAL-WORLD CONNECTION** If a person drives a car 60 miles per hour for 4 hours, the total distance d traveled is

$$d = 60 \cdot 4 = 240 \text{ miles.}$$

That is, distance equals the product of the rate (speed) and the elapsed time. This may be written as $d = rt$ and expressed verbally as "distance equals rate times time." This formula is used in Example 8.

EXAMPLE 8 **Solving an application**

The winner of a 3-mile race finishes 3 minutes ahead of another runner. If the winner runs 2 miles per hour faster than the slower runner, find the average speed of each runner.

Solution

Let x be the speed of the slower runner. Then $x + 2$ represents the speed of the winner. See Figure 6.15. To determine the time for each runner to finish the race, divide each side of $d = rt$ by r to obtain

$$t = \frac{d}{r}.$$

Figure 6.15

The slower runner runs 3 miles at x miles per hour, so the time is $\frac{3}{x}$; the winner runs 3 miles at $x + 2$ miles per hour, so the winning time is $\frac{3}{x+2}$. We summarize this in Table 6.3.

TABLE 6.3

	Distance (mi)	Rate (mi/hr)	Time (hr)
Winning Runner	3	$x + 2$	$\frac{3}{x+2}$
Slower Runner	3	x	$\frac{3}{x}$

If you add 3 minutes (or, equivalently, $\frac{3}{60} = \frac{1}{20}$ hour) to the winner's time, it equals the slower runner's time, or

$$\frac{3}{x+2} + \frac{1}{20} = \frac{3}{x}.$$

NOTE: The runners' speeds are in miles per *hour*, so you need to keep time in hours.

Multiply each side by the LCD, which is **$20x(x + 2)$**.

$$\mathbf{20x(x + 2)}\left(\frac{3}{x+2} + \frac{1}{20}\right) = \mathbf{20x(x + 2)}\left(\frac{3}{x}\right) \qquad \text{Multiply by the LCD.}$$

$$60x + x(x + 2) = 60(x + 2) \qquad \text{Distributive property}$$

$$60x + x^2 + 2x = 60x + 120 \qquad \text{Distributive property}$$

$$x^2 + 2x - 120 = 0 \qquad \text{Rewrite the equation.}$$

$$(x + 12)(x - 10) = 0 \qquad \text{Factor.}$$

$$x = -12 \quad \text{or} \quad x = 10 \qquad \text{Solve.}$$

The only valid solution is 10 because a running speed cannot be negative. Thus the slower runner is running at 10 miles per hour, and the faster runner is running at 12 miles per hour.

▮ **Now Try Exercise 97**

Solving an Equation for a Variable

▶ **REAL-WORLD CONNECTION** Sometimes in science and other disciplines, it is necessary to solve a formula for a variable. These formulas are often rational equations. This technique is demonstrated in the next two examples.

EXAMPLE 9   **Solving an equation for a variable**

Solve the equation $P = \frac{nrT}{V}$ for V. (This formula is used to calculate the pressure of a gas.)

Solution
Start by multiplying each side of the equation by V.

$$V \cdot P = V\left(\frac{nrT}{V}\right) \qquad \text{Multiply by } V.$$

$$PV = nrT \qquad \text{Simplify.}$$

$$\frac{PV}{P} = \frac{nrT}{P} \qquad \text{Divide each side by } P.$$

$$V = \frac{nrT}{P} \qquad \text{Simplify.}$$

▮ **Now Try Exercise 73**

EXAMPLE 10   **Solving an equation involving electricity**

Solve the equation $\dfrac{1}{R} = \dfrac{1}{R_1} + \dfrac{1}{R_2}$ for R_2.

Solution

Start by multiplying each side of the equation by the LCD: $R R_1 R_2$.

$$(R R_1 R_2)\frac{1}{R} = (R R_1 R_2)\left(\frac{1}{R_1} + \frac{1}{R_2}\right) \qquad \text{Multiply by the LCD.}$$

$$R_1 R_2 = R R_2 + R R_1 \qquad \text{Distributive property}$$

$$R_1 R_2 - R R_2 = R R_1 \qquad \text{Subtract } R R_2.$$

$$R_2(R_1 - R) = R R_1 \qquad \text{Factor out } R_2.$$

$$R_2 = \frac{R R_1}{R_1 - R} \qquad \text{Divide each side by } R_1 - R.$$

Now Try Exercise 79

6.4   Putting It All Together

CONCEPT	EXPLANATION	EXAMPLE
Rational Equations	To solve a rational equation, first multiply each side of the equation by the LCD. Then solve the resulting polynomial equation. Be sure to check your answer(s). The LCD is the LCM of the denominators.	Multiply each term by $4x^2$. $\dfrac{1}{4x} + \dfrac{2}{x^2} = \dfrac{1}{x}$ $\dfrac{4x^2}{4x} + \dfrac{8x^2}{x^2} = \dfrac{4x^2}{x}$ $x + 8 = 4x$ $8 = 3x$ $x = \dfrac{8}{3}$

The following example illustrates how to solve the rational equation $\frac{2}{x} + 2 = 3$ symbolically, graphically, and numerically with the aid of a graphing calculator, where we let $y_1 = \frac{2}{x} + 2$ and $y_2 = 3$.

Symbolic Solution

$$\frac{2}{x} + 2 = 3$$

$$\frac{2}{x} = 1$$

$$2 = x$$

The solution is **2**.

Numerical Solution

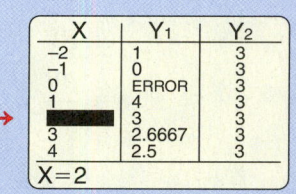

$y_1 = 3$ when $x = $ **2**.

Graphical Solution

$[-4.7, 4.7, 1]$ by $[-1, 5, 1]$

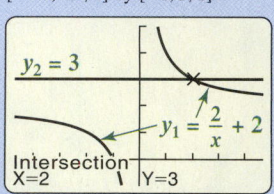

The graphs intersect at (**2**, 3).

6.4 Exercises

CONCEPTS AND VOCABULARY

1. When solving the equation $\frac{4}{x + 2} = 2$, what is a good first step?

2. In general, what is a good first step when solving a rational equation?

3. If $\frac{1}{x} = 2$, then $x =$ _____.

4. If $\frac{a}{b} = c$, then $a =$ _____.

5. To clear an equation of fractions, multiply each side of the equation by the _____.

6. To solve a rational equation graphically, graph each side of the equation and determine x-coordinates of the point(s) where the graphs _____.

7. (True or False?) A solution to $\frac{x}{x + 2} = \frac{2}{2 - x}$ is 2.

8. (True or False?) A solution to $\frac{2x}{x - 2} + \frac{x}{x + 2} = 3$ is -6.

SOLVING RATIONAL EQUATIONS

Exercises 9–14: Find the LCD.

9. $\frac{1}{x}, \frac{1}{5}$

10. $\frac{1}{x - 1}, \frac{1}{x + 5}$

11. $\frac{1}{x - 1}, \frac{1}{x^2 - 1}$

12. $\frac{1}{x^2 - x}, \frac{1}{2x}$

13. $\frac{3}{2}, \frac{x}{2x + 1}, \frac{x}{2x - 4}$

14. $\frac{1}{x}, \frac{1}{x^2 - 4x}, \frac{1}{2x}$

Exercises 15–48: Solve. Check your result.

15. $\frac{x}{3} + \frac{1}{2} = \frac{5}{6}$

16. $\frac{7}{8} - \frac{x}{4} = -\frac{11}{8}$

17. $\frac{2}{x} - \frac{7}{3} = -\frac{29}{15}$

18. $\frac{1}{3x} + \frac{1}{x} = \frac{4}{15}$

19. $\frac{x}{x - 1} = \frac{4}{3}$

20. $\frac{4}{5x} = -\frac{1}{x + 1}$

21. $\frac{1}{2x} - \frac{5}{3x} = 1$

22. $\frac{3}{4x} + \frac{5}{6x} = -\frac{19}{12}$

23. $\frac{1}{x + 1} - 1 = \frac{3}{x + 1}$

24. $\frac{2}{x - 2} + 2 = \frac{4}{x - 2}$

25. $\frac{1}{x - 3} + \frac{x}{x - 3} = \frac{2x}{x - 3}$

26. $\frac{6}{x + 2} + \frac{3x}{x + 2} = \frac{3}{x + 2}$

27. $\frac{3}{x - 1} = \frac{6}{x + 4}$

28. $\frac{1}{x + 5} = \frac{2}{2x + 1}$

29. $\frac{6}{3z + 4} = \frac{4}{2z - 5}$

30. $\frac{2}{z - 1} - \frac{5}{z + 1} = \frac{4}{z^2 - 1}$

31. $\frac{5}{t - 1} + \frac{2}{t + 2} = \frac{15}{t^2 + t - 2}$

32. $\frac{3}{5t} - \frac{1}{t + 1} = \frac{6}{5t^2 + 5t}$

33. $\frac{1}{3n} - 2 = \frac{1}{2n}$

34. $\frac{4}{n} - 3 = \frac{1}{3n}$

35. $\frac{1}{x} + \frac{1}{x^2} = 2$

36. $\frac{1}{x} - \frac{1}{x + 1} = \frac{1}{56}$

37. $\frac{x}{x + 2} = \frac{4}{x - 3}$

38. $\frac{2x}{1 - x} - \frac{1}{x} = 0$

39. $\frac{2w + 1}{3w} - \frac{4w - 3}{w} = 0$

40. $\frac{1}{2w} - \frac{2}{1 - w} = -\frac{3}{2}$

41. $\frac{3}{2y} + \frac{2y}{y - 4} = -\frac{11}{2}$

42. $\frac{3}{y - 2} + \frac{6y}{y + 1} = 2$

43. $\frac{1}{(x - 1)^2} + \frac{3}{x^2 - 1} = \frac{5}{x^2 - 1}$

44. $\frac{1}{x^2 - 4} + \frac{1}{(x - 2)^2} = \frac{2}{(x + 2)^2}$

45. $\frac{x^2 + x - 2}{x - 2} - \frac{4}{x - 2} = 0$

46. $\frac{x^2 - 4x - 5}{x - 1} + \frac{8}{x - 1} = 0$

47. $\frac{4}{x + 3} - \frac{x}{3 - x} = \frac{18}{x^2 - 9}$

48. $\frac{12}{x^2 - 9} + \frac{2}{x + 3} = \frac{x}{3 - x}$

Exercises 49 and 50: Solve $f(x) = 0$ for the given $f(x)$.

49. $f(x) = \dfrac{1}{x} - 4x$

50. $f(x) = \dfrac{2 - x}{x} + 2$

Exercises 51 and 52: **Thinking Generally** *Solve for x.*

51. $\dfrac{a}{x} - \dfrac{b}{x} = c$

52. $\dfrac{ab}{x} - c = \dfrac{1}{x}$

Exercises 53–56: Use the graph to solve the rational equation. Then check your answer.

53. $\dfrac{2}{x} + 1 = 3$

54. $\dfrac{1}{x - 1} = \dfrac{2}{x}$

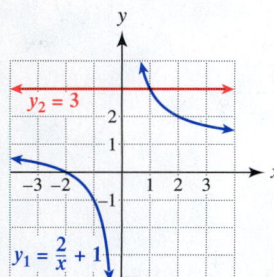
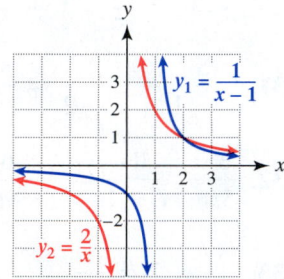

55. $\dfrac{3}{x^2 - 1} = \dfrac{1}{2}x$

56. $\dfrac{1}{x^2} = -x$

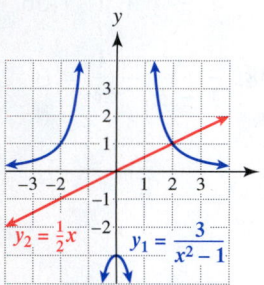
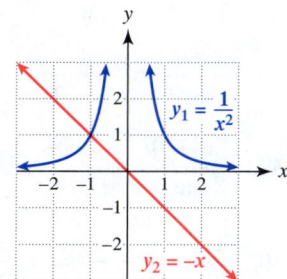

Exercises 57–66: Solve the rational equation graphically. Approximate your answer(s) to the nearest hundredth when appropriate.

57. $\dfrac{1}{x - 1} = \dfrac{1}{2}$

58. $\dfrac{1}{x + 1} = 2$

59. $\dfrac{3}{x + 2} = x$

60. $\dfrac{3}{x - 2} = x$

61. $\dfrac{1}{x - 2} = 2$

62. $\dfrac{2x}{x + 1} = 3$

63. $\dfrac{1}{x} + \dfrac{1}{x^2} = \dfrac{15}{4}$

64. $\dfrac{1}{x} + \dfrac{1}{2x} = 2x$

65. $\dfrac{1}{x + 2} - \dfrac{1}{x - 2} = \dfrac{4}{3}$

66. $\dfrac{x}{2x + 4} - \dfrac{2x}{x^2 - 4} = 2.7$

USING MORE THAN ONE METHOD

*Exercises 67–70: Solve the equation (**a**) symbolically, (**b**) graphically, and (**c**) numerically.*

67. $\dfrac{1}{x} + \dfrac{2}{3} = 1$

68. $\dfrac{1}{x} - \dfrac{1}{x^2} = -2$

69. $\dfrac{1}{x} + \dfrac{1}{x + 2} = \dfrac{4}{3}$

70. $\dfrac{1}{x - 2} + \dfrac{1}{x + 2} = -\dfrac{2}{3}$

Exercises 71 and 72: Solve the equation graphically to the nearest hundredth.

71. $\dfrac{x}{x^2 - 1} - \dfrac{3}{x + 2} = 2$

72. $\dfrac{2.2}{x - 0.1} + \dfrac{x}{0.4x - 1} = 1$

SOLVING AN EQUATION FOR A VARIABLE

Exercises 73–86: Solve for the specified variable.

73. $t = \dfrac{d}{r}$ for r

74. $r = \dfrac{C}{2\pi}$ for C

75. $h = \dfrac{2A}{b}$ for b

76. $P = VT$ for T

77. $\dfrac{1}{2a} = \dfrac{1}{b}$ for a

78. $\dfrac{a}{b} = \dfrac{c}{d}$ for d

79. $\dfrac{1}{R} = \dfrac{1}{R_1} + \dfrac{1}{R_2}$ for R_1

80. $S = \dfrac{5}{8H}$ for H

81. $T = \dfrac{1}{15 - x}$ for x

82. $h = \dfrac{V}{\pi r^2}$ for V

83. $\dfrac{1}{r} = \dfrac{1}{t + 1}$ for t

84. $R = \dfrac{R_1 R_2}{R_1 + R_2}$ for R_2

85. $\dfrac{1}{r} = \dfrac{a}{a + b}$ for b

86. $\dfrac{1}{t + 1} = \dfrac{a}{a - b}$ for b

APPLICATIONS

87. *Waiting in Traffic* Suppose that cars are arriving randomly at a construction zone and the flag person can instruct x drivers per hour. If cars arrive at an average rate of 80 cars per hour, the average time T in minutes for each driver to wait and talk to the flag person is given by the formula

$$T(x) = \dfrac{1}{x - 80},$$

where $x > 80$. (*Source:* N. Garber.)

(a) How many drivers per hour should the flag person be able to instruct to keep the average wait to 0.5 minute?

(b) What happens to T if x is only slightly more than 80?

88. *Waiting in Line* At a parking lot, the attendant can wait on 60 people per hour. If people arrive randomly at an average rate of x cars per hour, the average number of cars N waiting to exit the lot is given by

$$N(x) = \frac{x^2}{3600 - 60x},$$

where $x < 60$. (*Source: N. Garber.*)

(a) Approximate x when $N(x) = 8$.

(b) Interpret your answer.

89. *Mowing the Lawn* Suppose that a person with a push mower can mow a large lawn in 5 hours, whereas the same lawn can be mowed with a riding mower in 2 hours.

(a) Write an equation whose solution gives the time needed to mow the lawn if both mowers are used at the same time.

(b) Solve the equation in part (a) symbolically.

(c) Solve the equation in part (a) either graphically or numerically.

90. *Working Together* It takes a more experienced person 1 hour less to complete a task than it does a less experienced person. If they work together, they can complete the task in 1.2 hours. How long does it take each person to complete the task working alone?

91. *Finding Glitches in New Products* Before a new product such as a mobile phone is released, there is usually a time period when people test it to find any problems. After several people have tested a product, most problems are found and fixed. The rational function

$$f(x) = \frac{100x}{x + 1}, \quad 0 \le x \le 20$$

approximates the percentage of the problems that are found after x testers have used a new product.

(a) Calculate $f(3)$. Interpret your answer.

(b) How many testers are necessary to find 95% of glitches?

(c) Graph $y = f(x)$. Interpret the graph for large values of x.

92. *Cost–Benefit Function* A cost–benefit function C computes the cost in millions of dollars to implement a city recycling project aiming at getting x percent of its citizens to participate, where

$$C(x) = \frac{2x}{100 - x}, \quad 0 \le x < 100.$$

(a) Evaluate $C(90)$ and interpret the result.

(b) Evaluate $C(95) - C(90)$. Comment about the additional cost to get an extra 5% participation.

(c) Graph $y = C(x)$. Use the graph to explain why expecting 100% participation is not practical.

(d) If the city has $8 million to spend, what participation rate can be expected?

93. *Economics of the NFL Draft* The top picks in the NFL draft are often paid high salaries. However, compared to the top draft picks, lower picks often receive considerably lower salaries. The following function S approximates the percentage of salary that draft pick x receives compared to the number one draft pick's salary.

$$S(x) = \frac{500}{x + 4}$$

(a) Evaluate $S(36)$ and interpret your answer.

(b) Determine the pick that receives 20% of the number one pick's salary.

94. *Cost of Making Cookies* Because of startup expenses, suppose that it costs $100 to make the first specialty cookie and $1 for each cookie thereafter. The average cost for making x cookies is given by

$$C(x) = \frac{x + 99}{x}.$$

(a) What is the average cost to make 10 cookies?

(b) How many cookies should be made to have an average cost of $1.19?

(c) What happens to the average cost as the number of cookies made becomes very large?

95. *Pumping Water* (Refer to Example 7.) Suppose that a large pump can empty a swimming pool in 40 hours and that a small pump can empty the same pool in 70 hours.

(a) Write an equation whose solution gives the time needed for both pumps to empty the pool.

(b) Solve the equation in part (a) symbolically.

(c) Solve the equation in part (a) either graphically or numerically.

96. *Filling a Water Tank* A large pump can fill a 10,000-gallon tank 5 hours faster than a small pump. The large pump outputs water 100 gallons per hour faster than the small pump. Find the number of gallons pumped in 1 hour by each pump.

97. *Running a Race* (Refer to Example 8.) The winner of a 5-mile race finishes 7.5 minutes ahead of the second-place runner. On average, the winner ran 2 miles per hour faster than the second-place runner. Find the average running speed for each runner.

98. *Waiting in Line* If a clerk can wait on 30 people per hour and people arrive randomly at x people per hour, then the average amount of time spent waiting and checking out, *in hours,* is given by

$$T(x) = \frac{1}{30 - x}.$$

If the average wait is 3 minutes, how many people are arriving per hour, on average?

99. *Speed of a Current* In still water a tugboat can travel 15 miles per hour. It travels 36 miles upstream and then 36 miles downstream in a total time of 5 hours. Find the speed of the current.

100. *Wind Speed* Without any wind an airplane flies at 200 miles per hour. The plane travels 800 miles into the wind and then returns with the wind in a total time of 8 hours and 20 minutes. Find the average speed of the wind.

101. *Speed of an Airplane* When there is a 50-mile-per-hour wind, an airplane can fly 675 miles with the wind in the same time that it can fly 450 miles against the wind. Find the speed of the plane when there is no wind.

102. *Waiting in Line* If a ticket agent can wait on 25 people per hour and people are arriving randomly at an average rate of x people per hour, the average time T in hours for people waiting in line *and* paying the agent is

$$T(x) = \frac{1}{25 - x},$$

where $x < 25$. (*Source:* N. Garber.)
(a) Solve the equation $T(x) = 5$ symbolically and interpret the result.
(b) Solve part (a) either graphically or numerically.

103. *Working Alone* It takes one employee 3 hours longer to mow a football field than it does a more experienced employee. Together they can mow the grass in 2 hours. How long does it take for each person to mow the football field working alone?

104. *Painting a House* It takes a less experienced painter 20 hours longer to paint a house than it does a more

experienced painter. Together they can paint the house in 24 hours. How long does it take for each painter to paint the house working alone?

105. *Walking Speed* One person can walk 1 mile per hour faster than another person. The faster person can walk 12 miles in the time it takes the slower person to walk 9 miles. What is the walking speed of each person?

106. *Freeway Speed* On a freeway, a car passes another car that is traveling in the same direction but is going 5 miles per hour slower. The faster car travels 340 miles in the time it takes the slower car to travel 315 miles. What is the speed of each car?

107. *Emptying a Pool* An inlet pipe can fill a pool in 60 hours, whereas an outlet pipe can empty the pool in 40 hours. If both pipes are left open, how long will it take to empty the pool if the pool is full initially?

108. *Emptying a Boat* A small leak will fill an empty boat in 5 hours. The boat's bilge pump will empty a full boat in 3 hours. If the boat were half full and leaking, how long would it take the pump to empty the boat?

Exercises 109 and 110: Graphical Interpretation Use the graph of the rational function f to decide if there is a solution to the given equation. If there is a solution, discuss its value.

109. $f(x) = 100$ **110.** $f(x) = -100$

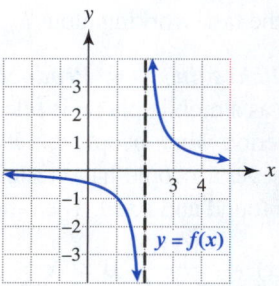

WRITING ABOUT MATHEMATICS

111. A student does the following to solve a rational equation. Is the work correct? Explain any errors and how you might correct them.

$$\frac{1}{x + 1} + \frac{2}{x - 1} = -1$$

$$(x - 1) + 2(x + 1) \stackrel{?}{=} -1$$

$$3x + 1 \stackrel{?}{=} -1$$

$$x \stackrel{?}{=} -\frac{2}{3}$$

112. Explain the difference between simplifying the rational expression $\frac{x + 1}{x^2 - 1}$ and solving the rational equation $\frac{x + 1}{x^2 - 1} = \frac{1}{2}$.

Checking Basic Concepts

1. Simplify each expression.

   (a) $\dfrac{x}{x^2 - 1} + \dfrac{1}{x^2 - 1}$

   (b) $\dfrac{1}{x - 2} - \dfrac{3}{x}$

   (c) $\dfrac{1}{x(x - 1)} + \dfrac{1}{x^2 - 1} - \dfrac{2}{x(x + 1)}$

2. Solve each equation.

   (a) $\dfrac{6}{x} - \dfrac{1}{2} = 1$ (b) $\dfrac{3}{2x - 1} = \dfrac{2}{x + 1}$

   (c) $\dfrac{2}{x - 1} + \dfrac{3}{x + 2} = \dfrac{x}{x^2 + x - 2}$

3. If a gas pump can fill a 200-gallon tank in 12 minutes and a smaller gas pump can fill the same tank in 28 minutes, how long will it take for the pumps to fill the tank together?

4. Solve $\dfrac{1}{S} = \dfrac{1}{F} - \dfrac{1}{D}$ for S.

6.5   Complex Fractions

Basic Concepts • Simplifying Complex Fractions

A LOOK INTO MATH ▶

If two and a half pizzas are cut into fourths, then there are 10 pieces of pizza. This problem can be written as

$$\frac{2 + \dfrac{1}{2}}{\dfrac{1}{4}} = 10.$$

The expression on the left side of the equation is called a *complex fraction*. Typically, we want to simplify a complex fraction to a fraction with the standard form $\frac{a}{b}$. In this section we discuss how to simplify complex fractions.

NEW VOCABULARY

☐ Complex fraction

Basic Concepts

A **complex fraction** is a rational expression that contains fractions in its numerator, denominator, or both. Examples of complex fractions include

$$\frac{1 + \dfrac{1}{x}}{1 - \dfrac{1}{x}}, \quad \frac{2x}{\dfrac{4}{x} + \dfrac{3}{y}}, \quad \text{and} \quad \frac{\dfrac{a}{3} + \dfrac{a}{4}}{a - \dfrac{1}{a - 1}}. \qquad \text{Complex fractions}$$

Complex fractions involve division of fractions. When dividing two fractions, we can multiply the first fraction by the reciprocal of the second fraction.

SIMPLIFYING COMPLEX FRACTIONS

For any real numbers a, b, c, and d,

$$\frac{\dfrac{a}{b}}{\dfrac{c}{d}} = \frac{a}{b} \cdot \frac{d}{c},$$

where b, c, and d are not zero.

EXAMPLE 1 **Simplifying basic types of complex fractions**

Simplify.

(a) $\dfrac{\dfrac{7}{8}}{\dfrac{3}{4}}$ **(b)** $\dfrac{\dfrac{a}{4b}}{\dfrac{a}{6b^2}}$ **(c)** $\dfrac{2 + \dfrac{1}{2}}{\dfrac{1}{4}}$

Solution

(a) Rather than dividing $\frac{7}{8}$ by $\frac{3}{4}$, multiply $\frac{7}{8}$ by $\frac{4}{3}$.

$$\frac{\dfrac{7}{8}}{\dfrac{3}{4}} = \frac{7}{8} \cdot \frac{4}{3} \qquad \text{“Invert and multiply.”}$$

$$= \frac{7}{6} \qquad \text{Multiply and simplify.}$$

(b) Rather than dividing $\frac{a}{4b}$ by $\frac{a}{6b^2}$, multiply $\frac{a}{4b}$ by $\frac{6b^2}{a}$.

$$\frac{\dfrac{a}{4b}}{\dfrac{a}{6b^2}} = \frac{a}{4b} \cdot \frac{6b^2}{a} \qquad \text{“Invert and multiply.”}$$

$$= \frac{6ab^2}{4ab} \qquad \text{Multiply fractions.}$$

$$= \frac{3b}{2} \qquad \text{Simplify.}$$

(c) Start by writing $2 + \frac{1}{2}$ as one fraction.

$$\frac{2 + \dfrac{1}{2}}{\dfrac{1}{4}} = \frac{\dfrac{5}{2}}{\dfrac{1}{4}} \qquad \text{Write numerator as an improper fraction.}$$

$$= \frac{5}{2} \cdot \frac{4}{1} \qquad \text{“Invert and multiply.”}$$

$$= 10 \qquad \text{Simplify.}$$

READING CHECK

What does the phrase "invert and multiply" mean?

Now Try Exercises 7, 11, 13

Simplifying Complex Fractions

There are two basic strategies for simplifying a complex fraction. The first is to simplify both the numerator and the denominator and then divide the resulting fractions as was done in Example 1(c). The second is to multiply the numerator and denominator by the least common denominator of the fractions in the numerator and denominator.

SIMPLIFYING THE NUMERATOR AND DENOMINATOR The next example illustrates the method whereby the numerator and denominator are simplified first.

EXAMPLE 2   **Simplifying complex fractions**

Simplify. Leave your answer in factored form when appropriate.

(a) $\dfrac{1 + \dfrac{1}{x}}{1 - \dfrac{1}{x}}$
   (b) $\dfrac{2t - \dfrac{1}{2t}}{2t + \dfrac{1}{2t}}$
   (c) $\dfrac{\dfrac{1}{z+1} - \dfrac{1}{z-1}}{\dfrac{2}{z+2} - \dfrac{2}{z-2}}$

Solution
(a) First, simplify the numerator by writing it as one term. The LCD is x.

$$1 + \frac{1}{x} = \frac{x}{x} + \frac{1}{x} = \frac{x+1}{x}$$

Second, simplify the denominator by writing it as one term.

$$1 - \frac{1}{x} = \frac{x}{x} - \frac{1}{x} = \frac{x-1}{x}$$

Finally, simplify the complex fraction.

$$\frac{1 + \dfrac{1}{x}}{1 - \dfrac{1}{x}} = \frac{\dfrac{x+1}{x}}{\dfrac{x-1}{x}} \qquad \text{Simplify.}$$

$$= \frac{x+1}{x} \cdot \frac{x}{x-1} \qquad \text{"Invert and multiply."}$$

$$= \frac{x+1}{x-1} \qquad \text{Multiply and simplify.}$$

(b) Combine the terms in the numerator and in the denominator. The LCD is $2t$.

$$\frac{2t - \dfrac{1}{2t}}{2t + \dfrac{1}{2t}} = \frac{\dfrac{4t^2}{2t} - \dfrac{1}{2t}}{\dfrac{4t^2}{2t} + \dfrac{1}{2t}} \qquad \text{Write with the LCD.}$$

$$= \frac{\dfrac{4t^2 - 1}{2t}}{\dfrac{4t^2 + 1}{2t}} \qquad \text{Combine terms.}$$

$$= \frac{4t^2 - 1}{2t} \cdot \frac{2t}{4t^2 + 1} \qquad \text{"Invert and multiply."}$$

$$= \frac{4t^2 - 1}{4t^2 + 1} \qquad \text{Multiply and simplify.}$$

(c) The LCD is $(z + 1)(z - 1)$ for the numerator and $(z + 2)(z - 2)$ for the denominator.

$$\frac{\dfrac{1}{z+1} - \dfrac{1}{z-1}}{\dfrac{2}{z+2} - \dfrac{2}{z-2}} = \frac{\dfrac{z-1}{(z+1)(z-1)} - \dfrac{z+1}{(z+1)(z-1)}}{\dfrac{2(z-2)}{(z+2)(z-2)} - \dfrac{2(z+2)}{(z+2)(z-2)}}$$

$$= \frac{\dfrac{-2}{(z+1)(z-1)}}{\dfrac{-8}{(z+2)(z-2)}}$$

$$= \frac{-2}{(z+1)(z-1)} \cdot \frac{(z+2)(z-2)}{-8}$$

$$= \frac{(z+2)(z-2)}{4(z+1)(z-1)}$$

Now Try Exercises 27, 29, 31

MULTIPLYING BY THE LEAST COMMON DENOMINATOR A second strategy for simplifying a complex fraction is to multiply by **1** in the form $\frac{\text{LCD}}{\text{LCD}}$. To do this, we multiply the numerator and denominator by the LCD of the fractions in the numerator *and* denominator. For example, the LCD for the complex fraction

$$\frac{1 - \dfrac{1}{x}}{1 + \dfrac{1}{2x}}$$

is $2x$. To simplify, multiply the complex fraction by **1**, expressed in the form $\frac{2x}{2x}$.

$$\frac{\left(1 - \dfrac{1}{x}\right) \cdot 2x}{\left(1 + \dfrac{1}{2x}\right) \cdot 2x} = \frac{2x - \dfrac{2x}{x}}{2x + \dfrac{2x}{2x}} \qquad \text{Distributive property}$$

$$= \frac{2x - 2}{2x + 1} \qquad \text{Simplify fractions.}$$

In the next example we use this method to simplify other complex fractions.

CRITICAL THINKING

Does the expression $\dfrac{\frac{x}{y}}{2 + \frac{x}{y}}$ equal $\frac{1}{3}$? Explain.

Does the expression $\dfrac{2 + \frac{x}{y}}{\frac{x}{y}}$ equal 3? Explain.

EXAMPLE 3 **Simplifying complex fractions**

Simplify.

(a) $\dfrac{1 + \dfrac{1}{x}}{1 + \dfrac{1}{y}}$ **(b)** $\dfrac{\dfrac{1}{x} - \dfrac{1}{y}}{x - y}$ **(c)** $\dfrac{\dfrac{3}{x-1} - \dfrac{2}{x}}{\dfrac{1}{x-1} + \dfrac{3}{x}}$ **(d)** $\dfrac{n^{-2} + m^{-2}}{1 + (nm)^{-2}}$

Solution

(a) The LCD for the numerator *and* the denominator is xy.

$$\dfrac{1 + \dfrac{1}{x}}{1 + \dfrac{1}{y}} = \dfrac{\left(1 + \dfrac{1}{x}\right) \cdot xy}{\left(1 + \dfrac{1}{y}\right) \cdot xy} \qquad \text{Multiply the numerator and the denominator by the LCD.}$$

$$= \dfrac{xy + y}{xy + x} \qquad \text{Distributive property}$$

(b) The LCD is xy. Multiply the given expression by $\dfrac{xy}{xy}$.

$$\dfrac{\left(\dfrac{1}{x} - \dfrac{1}{y}\right) \cdot xy}{(x - y) \cdot xy} = \dfrac{\dfrac{xy}{x} - \dfrac{xy}{y}}{xy(x - y)} \qquad \text{Distributive and commutative properties}$$

$$= \dfrac{y - x}{xy(x - y)} \qquad \text{Simplify fractions in numerator.}$$

$$= \dfrac{-1(x - y)}{xy(x - y)} \qquad \text{Factor out } -1\text{; rewrite numerator.}$$

$$= -\dfrac{1}{xy} \qquad \text{Simplify.}$$

(c) The LCD is $x(x - 1)$. Multiply the given expression by $\dfrac{x(x - 1)}{x(x - 1)}$.

$$\dfrac{\left(\dfrac{3}{x - 1} - \dfrac{2}{x}\right) \cdot x(x - 1)}{\left(\dfrac{1}{x - 1} + \dfrac{3}{x}\right) \cdot x(x - 1)} = \dfrac{\dfrac{3x(x - 1)}{x - 1} - \dfrac{2x(x - 1)}{x}}{\dfrac{x(x - 1)}{x - 1} + \dfrac{3x(x - 1)}{x}} \qquad \text{Distributive property}$$

$$= \dfrac{3x - 2(x - 1)}{x + 3(x - 1)} \qquad \text{Simplify.}$$

$$= \dfrac{3x - 2x + 2}{x + 3x - 3} \qquad \text{Distributive property}$$

$$= \dfrac{x + 2}{4x - 3} \qquad \text{Combine like terms.}$$

(d) Start by rewriting the ratio with positive exponents.

$$\dfrac{n^{-2} + m^{-2}}{1 + (nm)^{-2}} = \dfrac{\dfrac{1}{n^2} + \dfrac{1}{m^2}}{1 + \dfrac{1}{n^2 m^2}}$$

The LCD for the numerator *and* the denominator is $n^2 m^2$.

$$\dfrac{\dfrac{1}{n^2} + \dfrac{1}{m^2}}{1 + \dfrac{1}{n^2 m^2}} = \dfrac{\left(\dfrac{1}{n^2} + \dfrac{1}{m^2}\right) \cdot n^2 m^2}{\left(1 + \dfrac{1}{n^2 m^2}\right) \cdot n^2 m^2} \qquad \text{Multiply by LCD.}$$

$$= \dfrac{m^2 + n^2}{n^2 m^2 + 1} \qquad \text{Distributive property}$$

Now Try Exercises 25, 37, 39, 43

6.5   Putting It All Together

CONCEPT	EXPLANATION	EXAMPLES
Complex Fraction	A rational expression that contains fractions in its numerator, denominator, or both	$\dfrac{\dfrac{1}{x} + \dfrac{1}{2x+1}}{1 - \dfrac{1}{2x+1}}$ and $\dfrac{\dfrac{a}{2b} - \dfrac{b}{2a}}{\dfrac{a}{2b} + \dfrac{b}{2a}}$
Simplifying Basic Complex Fractions	$\dfrac{\dfrac{a}{b}}{\dfrac{c}{d}} = \dfrac{a}{b} \cdot \dfrac{d}{c}$ $b, c,$ and d not zero	$\dfrac{\dfrac{2y}{z}}{\dfrac{y}{z-1}} = \dfrac{2y}{z} \cdot \dfrac{z-1}{y} = \dfrac{2(z-1)}{z}$
Method I: Simplifying the Numerator and Denominator First and then Dividing	Combine the terms in the numerator, combine the terms in the denominator, and then invert and multiply.	$\dfrac{\dfrac{1}{2b} + \dfrac{1}{2a}}{\dfrac{1}{2b} - \dfrac{1}{2a}} = \dfrac{\dfrac{a+b}{2ab}}{\dfrac{a-b}{2ab}}$ $= \dfrac{a+b}{2ab} \cdot \dfrac{2ab}{a-b}$ $= \dfrac{a+b}{a-b}$
Method II: Multiplying the Numerator and Denominator by the LCD	Start by multiplying the numerator and denominator by the LCD of the numerator *and* the denominator.	$\dfrac{\dfrac{1}{2b} + \dfrac{1}{2a}}{\dfrac{1}{2b} - \dfrac{1}{2a}} = \dfrac{\left(\dfrac{1}{2b} + \dfrac{1}{2a}\right) \cdot 2ab}{\left(\dfrac{1}{2b} - \dfrac{1}{2a}\right) \cdot 2ab}$ $= \dfrac{a+b}{a-b}$ Note that the LCD is $2ab$.

6.5   Exercises

CONCEPTS AND VOCABULARY

1. $\dfrac{\dfrac{5}{7}}{\dfrac{3}{11}} = $ _____.

2. $\dfrac{\dfrac{a}{b}}{\dfrac{c}{d}} = $ _____.

3. What is a good first step when simplifying

$$\dfrac{2 + \dfrac{1}{x-1}}{2 - \dfrac{1}{x-1}}?$$

4. Explain what a complex fraction is.

5. Write the phrase "the quantity z plus three-fourths divided by the quantity z minus three-fourths" as a complex fraction.

6. Write the expression $\frac{a}{b} \div \frac{a-b}{a+b}$ as a complex fraction.

SIMPLIFYING COMPLEX FRACTIONS

Exercises 7–48: Simplify the complex fraction. Leave your answer in factored form when appropriate.

7. $\dfrac{\dfrac{1}{5}}{\dfrac{4}{7}}$

8. $\dfrac{\dfrac{3}{7}}{\dfrac{2}{9}}$

9. $\dfrac{1 + \dfrac{1}{3}}{1 - \dfrac{1}{3}}$

10. $\dfrac{\dfrac{1}{2} - 3}{\dfrac{1}{2} + 3}$

11. $\dfrac{2 + \dfrac{2}{3}}{2 - \dfrac{1}{4}}$

12. $\dfrac{\dfrac{1}{2} + \dfrac{3}{4}}{\dfrac{1}{2} - \dfrac{3}{4}}$

13. $\dfrac{\dfrac{a}{b}}{\dfrac{3a}{2b^2}}$

14. $\dfrac{\dfrac{7}{y}}{\dfrac{14}{y}}$

15. $\dfrac{\dfrac{x}{2y}}{\dfrac{2x}{3y}}$

16. $\dfrac{\dfrac{ab^2}{2c}}{\dfrac{a}{4bc}}$

17. $\dfrac{\dfrac{8}{n+1}}{\dfrac{4}{n-1}}$

18. $\dfrac{\dfrac{n}{m-2}}{\dfrac{3n}{m-2}}$

19. $\dfrac{\dfrac{2k+3}{k}}{\dfrac{k-4}{k}}$

20. $\dfrac{\dfrac{2k}{k-7}}{\dfrac{1}{(k-7)^2}}$

21. $\dfrac{\dfrac{3}{z^2-4}}{\dfrac{z}{z^2-4}}$

22. $\dfrac{\dfrac{z}{(z-2)^2}}{\dfrac{2z}{z^2-4}}$

23. $\dfrac{\dfrac{x}{x^2-16}}{\dfrac{1}{x-4}}$

24. $\dfrac{\dfrac{4y}{x-y}}{\dfrac{1}{x^2-y^2}}$

25. $\dfrac{1 + \dfrac{1}{x}}{x+1}$

26. $\dfrac{2-x}{\dfrac{1}{x} - \dfrac{1}{2}}$

27. $\dfrac{\dfrac{1}{x-3}}{\dfrac{1}{x} - \dfrac{3}{x-3}}$

28. $\dfrac{5 + \dfrac{1}{x-1}}{\dfrac{1}{x-1} - \dfrac{1}{4}}$

29. $\dfrac{\dfrac{1}{x} + \dfrac{2}{x^2}}{\dfrac{3}{x} - \dfrac{1}{x^2}}$

30. $\dfrac{\dfrac{1}{x-1} + \dfrac{2}{x}}{2 - \dfrac{1}{x}}$

31. $\dfrac{\dfrac{1}{x+3} + \dfrac{2}{x-3}}{2 - \dfrac{1}{x-3}}$

32. $\dfrac{\dfrac{1}{x} + \dfrac{2}{x}}{\dfrac{1}{x-1} + \dfrac{x}{2}}$

33. $\dfrac{\dfrac{4}{x-5}}{\dfrac{1}{x+5} + \dfrac{1}{x}}$

34. $\dfrac{\dfrac{1}{x-4} + \dfrac{1}{x-4}}{1 - \dfrac{1}{x+4}}$

35. $\dfrac{\dfrac{1}{p^2q} + \dfrac{1}{pq^2}}{\dfrac{1}{p^2q} - \dfrac{1}{pq^2}}$

36. $\dfrac{\dfrac{1}{p-1}}{\dfrac{1}{p-1} + 2}$

37. $\dfrac{\dfrac{1}{a} + \dfrac{1}{b}}{\dfrac{1}{b} - \dfrac{1}{a}}$

38. $\dfrac{\dfrac{1}{a} + \dfrac{1}{b} + \dfrac{1}{c}}{\dfrac{1}{ab} + \dfrac{1}{bc}}$

39. $\dfrac{\dfrac{1}{x} + \dfrac{1}{x+1}}{\dfrac{2}{x+1} - \dfrac{1}{x+1}}$

40. $\dfrac{\dfrac{2}{x-1} - 4}{\dfrac{1}{x-1} + \dfrac{1}{x-2}}$

41. $\dfrac{3^{-1} - 4^{-1}}{5^{-1} + 4^{-1}}$

42. $\dfrac{1 + 2^{-3}}{2^{-3} - 1}$

43. $\dfrac{m^{-1} - 2n^{-2}}{1 + (mn)^{-2}}$

44. $\dfrac{1 + p^{-2}}{1 - p^{-2}}$

45. $\dfrac{1 - (2n + 1)^{-1}}{1 + (2n + 1)^{-1}}$

46. $\dfrac{(a + b)^{-1} - (a - b)^{-1}}{1 + (a - b)^{-1}}$

47. $\dfrac{\dfrac{x}{x^2 - 4} - \dfrac{1}{x^2 - 4}}{\dfrac{1}{x + 4}}$

48. $\dfrac{\dfrac{1}{x^2 + 2x + 1} - \dfrac{1}{x^2 - 2x + 1}}{(x + 1)(x - 1)}$

APPLICATIONS

49. *Annuity* If P dollars are deposited every month in an account paying an annual interest rate r expressed as a decimal, then the amount A in the account after 2 years can be approximated by

$$\left(P \left(1 + \dfrac{r}{12} \right)^{24} - P \right) \div \dfrac{r}{12}.$$

Write this expression as a complex fraction.

50. *Annuity* (Continuation of the preceding exercise.) Use a calculator to evaluate the expression when $r = 0.06$ (6%) and $P = \$250$. Interpret the result.

51. *Resistance in Electricity* Light bulbs are often wired so that electricity can flow through either bulb as illustrated in the accompanying figure.

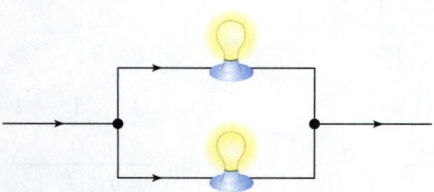

In this way, if one bulb burns out, the other bulb still works. If two light bulbs have resistances R_1 and R_2, then their combined resistance R is given by the complex fraction

$$R = \dfrac{1}{\dfrac{1}{R_1} + \dfrac{1}{R_2}}.$$

Simplify this formula.

52. *Resistance in Electricity* (Refer to the preceding exercise.) Evaluate the formula

$$R = \dfrac{1}{\dfrac{1}{R_1} + \dfrac{1}{R_2}}$$

when $R_1 = 75$ ohms and $R_2 = 100$ ohms.

53. *The Power of Practice* If a person practices doing a task, he or she usually gets faster at completing the task. This observation can be modeled by

$$T(x) = \dfrac{10}{\dfrac{1}{10}x + \dfrac{9}{10}},$$

where $T(x)$ gives the time in seconds to perform a simple task after practicing x times.
(a) Evaluate $T(11)$ and interpret the result.
(b) Rewrite $T(x)$ as a fraction in standard form $\frac{a}{b}$, without using decimals.

54. *Doppler Effect* Have you ever noticed how a siren sounds higher when it is approaching you and then its pitch drops when it passes you? This observation, called the Doppler effect, is described by the following equation.

$$F = \dfrac{f}{1 - \frac{v}{c}}$$

Simplify the complex fraction on the right side of the equation.

WRITING ABOUT MATHEMATICS

55. Describe a method to simplify a complex fraction.

56. A student simplifies a complex fraction as follows. Explain what the student's mistake is and how to work it correctly.

$$\dfrac{\dfrac{1}{x} + \dfrac{1}{y}}{\dfrac{1}{x} + \dfrac{1}{y}} \stackrel{?}{=} \dfrac{\dfrac{1}{x}}{\dfrac{1}{x}} + \dfrac{\dfrac{1}{y}}{\dfrac{1}{y}} \stackrel{?}{=} 1 + 1 = 2$$

6.6   Modeling with Proportions and Variation

Proportions • Direct Variation • Inverse Variation • Joint Variation

A LOOK INTO MATH ▶

Proportions are used frequently to solve problems. The following are examples.

- If someone earns $100 per day, that person can earn $500 in 5 days.
- If 1 out of 3 adults believes in ghosts, 77 million out of 231 million adults believe in ghosts.
- If a person walks a mile in 16 minutes, that person can walk a half mile in 8 minutes.

Many applications involve proportions or variation. In this section we discuss some of them.

Proportions

NEW VOCABULARY

☐ Proportion
☐ Similar figures
☐ Similar triangles
☐ Directly proportional
☐ Varies directly
☐ Constant of proportionality
☐ Constant of variation
☐ Inversely proportional
☐ Varies inversely
☐ Varies jointly

REAL-WORLD CONNECTION A 650-megabyte compact disc (CD) can store about 74 minutes of music. Suppose that we have already selected some music to record on the CD and 256 megabytes are still available. Using proportions we can determine how many more minutes of music could be recorded. A **proportion** is a statement (equation) that two ratios (fractions) are equal.

Let x be the number of minutes available on the CD. Then **74** minutes *are to* **650** megabytes *as* **x** minutes *are to* **256** megabytes, which can be written as the proportion

$$\frac{74}{650} = \frac{x}{256}. \qquad \frac{\text{Minutes}}{\text{Megabytes}} = \frac{\text{Minutes}}{\text{Megabytes}}$$

Solving this equation for x gives

$$x = \frac{74 \cdot 256}{650} \approx 29.1 \text{ minutes.} \qquad \text{Multiply by 256; simplify.}$$

About 29 minutes are still available on the CD.

MAKING CONNECTIONS

Proportions and Fractional Parts

We could have solved the preceding problem by noting that the fraction of the CD still available for recording music is $\frac{256}{650}$. So $\frac{256}{650}$ of 74 minutes equals

$$\frac{256}{650} \cdot 74 \approx 29.1 \text{ minutes.}$$

The following property is a convenient way to solve proportions:

$$\frac{a}{b} = \frac{c}{d} \quad \text{is equivalent to} \quad ad = bc,$$

provided $b \neq 0$ and $d \neq 0$. This is a result of multiplying each side of the equation by a common denominator bd, and is sometimes referred to as *clearing fractions*.

$$bd \cdot \frac{a}{b} = \frac{c}{d} \cdot bd \qquad \text{Multiply by } bd.$$

$$\frac{bda}{b} = \frac{cbd}{d} \qquad \text{Property of multiplying fractions}$$

$$ad = bc \qquad \text{Simplify.}$$

For example, the proportion

is equivalent to

$$6x = 5 \cdot 8 \quad \text{or} \quad x = \frac{40}{6} = \frac{20}{3}.$$

READING CHECK

Explain how to solve the proportion $\frac{4}{x} = \frac{6}{7}$. Solve the equation for x.

▶ **REAL-WORLD CONNECTION** Flooding in Fargo, North Dakota, has been an annual problem in recent years. With snow drifts reaching 10 to 15 feet high in some areas, an accurate estimate of the water content in the snow is essential for predicting spring flood levels. In the next example we estimate the water content in snow.

EXAMPLE 1 **Calculating the water content in snow**

Six inches of light, fluffy snow is equivalent to about half an inch of rain in terms of water content. If 21 inches of such snow fall, use proportions to estimate the water content.

Solution
Let x be the equivalent amount of rain. Then 6 inches of snow is to $\frac{1}{2}$ inch of rain as 21 inches of snow is to x inches of rain, which can be written as the proportion

$$\frac{6}{\frac{1}{2}} = \frac{21}{x}. \qquad \frac{\text{Snow}}{\text{Rain}} = \frac{\text{Snow}}{\text{Rain}}$$

Solving this equation gives

$$6x = \frac{21}{2} \quad \text{or} \quad x = \frac{21}{12} = 1.75.$$

Thus 21 inches of light, fluffy snow is equivalent to about 1.75 inches of rain.

Now Try Exercise 67

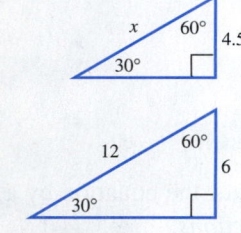

Figure 6.16

▶ **REAL-WORLD CONNECTION** Proportions frequently occur in geometry when we work with **similar figures**. Two triangles are similar if the measures of their corresponding angles are equal. Corresponding sides of **similar triangles** are proportional. Figure 6.16 shows two right triangles that are similar because each has angles of 30°, 60°, and 90°.

We can find the length of side x by using proportions. Side x is to 12 as 4.5 is to 6, which can be written as the proportion

$$\frac{x}{12} = \frac{4.5}{6}. \qquad \frac{\text{Hypotenuse}}{\text{Hypotenuse}} = \frac{\text{Shorter leg}}{\text{Shorter leg}}$$

Solving yields the equation

$$6x = 4.5(12) \qquad \text{Clear fractions.}$$
$$x = 9. \qquad \text{Divide by 6.}$$

EXAMPLE 2   **Calculating the height of a tree**

A 6-foot-tall person casts a 4-foot-long shadow. If a nearby tree casts a 44-foot-long shadow, estimate the height of the tree. See Figure 6.17.

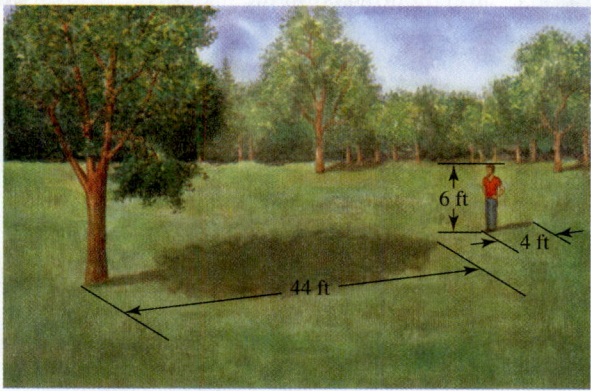

Figure 6.17

Solution

The triangles in Figure 6.18 are similar because the measures of corresponding angles are equal. Therefore corresponding sides are proportional. Let h be the height of the tree.

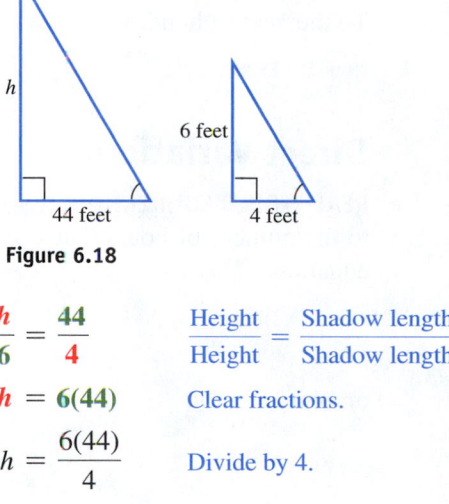

Figure 6.18

$$\frac{h}{6} = \frac{44}{4} \qquad \frac{\text{Height}}{\text{Height}} = \frac{\text{Shadow length}}{\text{Shadow length}}$$

$$4h = 6(44) \qquad \text{Clear fractions.}$$

$$h = \frac{6(44)}{4} \qquad \text{Divide by 4.}$$

$$h = 66 \qquad \text{Simplify.}$$

The tree is 66 feet tall.

Now Try Exercise 65

▶ **REAL-WORLD CONNECTION**   Surveys are based on the concept of proportions and are used in a wide variety of areas because it is usually impossible to ask *every* individual the same question. For example, in 2011 Nielsen polled 2131 people about what 4G meant. At that time, only 482 people knew that 4G meant superfast wireless. By using proportions, we might conclude that only about 52 million out of 231 million adults in the United States knew what 4G meant. See Exercise 70.

Biologists also use proportions to determine fish populations in a lake. They tag a small number of a particular species of fish and then release them. It is assumed that over time these tagged fish will distribute themselves evenly throughout the lake. Later, a sample of fish is taken and the number of tagged fish is compared to the total number of fish in the sample. These numbers can be used to estimate the population of a particular species of fish in the lake. (*Source:* Minnesota Department of Natural Resources.)

Estimating fish population

In May, 500 tagged largemouth bass are released into a lake. Later in the summer, a sample of 194 largemouth bass from the lake contains 28 tagged fish. Estimate the population of largemouth bass in the lake to the nearest hundred.

Solution
The sample contains 28 tagged bass out of a total of 194 bass. The ratio of tagged bass to the total number of bass in this sample is $\frac{28}{194}$. Let B be the total number of largemouth bass in the lake. Then the ratio of tagged bass to the total number of bass in the lake is $\frac{500}{B}$. If the bass are evenly distributed throughout the lake, then these two ratios are equal and form the following proportion.

Tagged bass in sample ⟍　　　⟋ Total number of tagged bass

$$\frac{28}{194} = \frac{500}{B}$$

Total number of bass in sample ⟋　　　⟍ Total number of bass in the lake

To solve this proportion we clear fractions to obtain the following equation.

$$28B = 194 \cdot 500 \qquad \text{Clear fractions.}$$

$$B = \frac{194 \cdot 500}{28} \approx 3464 \qquad \text{Divide by 28 and approximate.}$$

To the nearest hundred there are approximately 3500 largemouth bass in the lake.

Now Try Exercise 71

Direct Variation

▶ **REAL-WORLD CONNECTION** If your wage is $9 per hour, the amount you earn is proportional to the number of hours that you work. If you work H hours, your total pay P satisfies the equation

$$\frac{P}{H} = \frac{9}{1}, \qquad \frac{\text{Pay}}{\text{Hours}}$$

or, equivalently,

$$P = 9H.$$

We say that your pay P is *directly proportional* to the number of hours H worked, and the *constant of proportionality* is 9.

Direct Variation

Figure 6.19

DIRECT VARIATION

Let x and y denote two quantities. Then y is **directly proportional** to x, or y **varies directly** with x, if there is a nonzero number k such that

$$y = kx.$$

The number k is called the **constant of proportionality**, or the **constant of variation**.

The graph of $y = kx$ for $x > 0$ is a line passing through the origin, as illustrated in Figure 6.19. Sometimes data in a scatterplot indicate that two quantities are directly proportional. The constant of proportionality k corresponds to the *slope m* of the line in Figure 6.19.

EXAMPLE 4 | **Modeling college tuition**

Table 6.4 lists the tuition for taking various numbers of credits.

TABLE 6.4

Credits	3	5	8	11	17
Tuition	$189	$315	$504	$693	$1071

(a) A scatterplot of the data is shown in Figure 6.20. Could the data be modeled using a line?

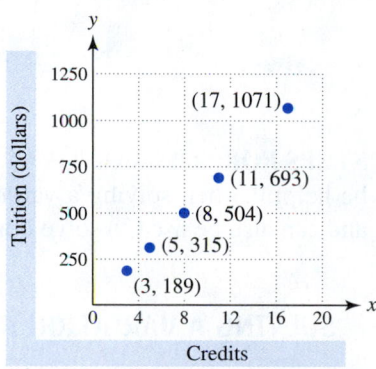

Figure 6.20

(b) Explain why tuition is directly proportional to the number of credits taken.
(c) Find the constant of proportionality. Interpret your result.
(d) Predict the cost of taking 15 credits.

Solution
(a) The data are linear and suggest a line passing through the origin.
(b) Because the data can be modeled by a line passing through the origin, tuition is directly proportional to the number of credits taken. Hence doubling the credits will double the tuition and tripling the credits will triple the tuition. Zero credits cost $0.
(c) The slope of the line equals the constant of proportionality k. If we use the first and last data points (**3**, **189**) and (**17**, **1071**), the slope is

$$k = \frac{\mathbf{1071} - \mathbf{189}}{\mathbf{17} - \mathbf{3}} = \mathbf{63}.$$

That is, tuition is **$63** per credit. If we graph the line $y = \mathbf{63}x$, it models the data as shown in Figure 6.21. This graph can also be created with a graphing calculator.

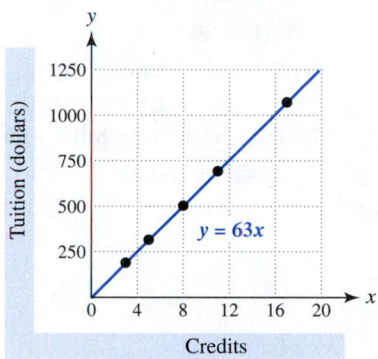

Figure 6.21

READING CHECK

If y varies directly with x, what happens to y if x triples?

(d) If y represents tuition and x represents the credits taken, 15 credits would cost

$$y = 63(15) = \$945.$$

▌ **Now Try Exercise 81**

> **MAKING CONNECTIONS**

Ratios and the Constant of Proportionality

The constant of proportionality in Example 4 can also be found by calculating the ratios $\frac{y}{x}$, where y is the tuition and x is the credits taken. Note that each ratio in the table equals 63 because the equation $y = 63x$ is equivalent to the equation $\frac{y}{x} = 63$.

x	3	5	8	11	17
y	189	315	504	693	1071
$\frac{y}{x}$	63	63	63	63	63

← Ratios are constant for direct variation.

STEPS FOR SOLVING A VARIATION PROBLEM The following four-step process can be helpful when solving a variation application. This process is used to solve Example 5 and can also be used to solve other types of variation problems.

> **SOLVING A VARIATION APPLICATION**
>
> When solving a variation problem, the following steps can be used.
>
> **STEP 1:** Write the general equation for the type of variation problem that you are solving.
>
> **STEP 2:** Substitute given values in this equation so the constant of variation k is the only unknown value in the equation. Solve for k.
>
> **STEP 3:** Substitute the value of k in the general equation in Step 1.
>
> **STEP 4:** Use this equation to find the requested quantity.

▶ **REAL-WORLD CONNECTION** If you have ever dived into deep water, you may have experienced pressure on your eardrums. The water pressure exerted on a diver *varies directly* with the depth, as illustrated in the next example.

EXAMPLE 5 **Modeling pressure on a diver**

At a depth of 18 feet, a diver experiences water pressure equal to 8 pounds per square inch. Find the pressure at 63 feet.

Solution

STEP 1: The general equation for direct variation is $y = kx$, where, in this case, y is the pressure and x is the depth.

STEP 2: Substitute 8 for y and 18 for x in $y = kx$. Solve for k.

$$8 = k(18) \qquad y = kx$$

$$\frac{8}{18} = k \qquad \text{Divide each side by 18.}$$

$$k = \frac{4}{9} \qquad \text{Simplify; rewrite equation.}$$

STEP 3: Replace k with $\frac{4}{9}$ in $y = kx$ to obtain $y = \frac{4}{9}x$.

STEP 4: Find the pressure by letting $x = 63$ in $y = \frac{4}{9}x$.

$$y = \tfrac{4}{9}(63) = 28 \text{ pounds per square inch}$$

■ **Now Try Exercise 83**

NOTE: In Example 5 the water depth increased by a factor of $\frac{63}{18} = 3.5$, so with direct variation the pressure also increased by a factor of 3.5, or $3.5(8) = 28$ pounds per square inch.

Inverse Variation

▶ **REAL-WORLD CONNECTION** When two quantities vary inversely, an *increase* in one quantity results in a *decrease* in the second quantity. For example, at 25 miles per hour a car travels 100 miles in 4 hours, whereas at 50 miles per hour the car travels 100 miles in 2 hours. Doubling the speed (or rate) decreases the travel time by half. Distance equals rate times time, so $d = rt$. Thus

$$100 = rt, \quad \text{or equivalently,} \quad t = \frac{100}{r}.$$

We say that the time t to travel 100 miles is *inversely proportional* to the speed or rate r. The constant of proportionality or constant of variation is 100 in this example.

Inverse Variation

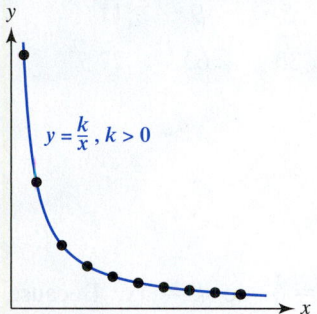

Figure 6.22

> ### INVERSE VARIATION
>
> Let x and y denote two quantities. Then y is **inversely proportional** to x, or y **varies inversely** with x, if there is a nonzero number k such that
>
> $$y = \frac{k}{x}.$$

NOTE: We assume that the constant k is positive.

The data shown in Figure 6.22 represent inverse variation and are modeled by $y = \frac{k}{x}$. Note that as x increases, y decreases.

▶ **REAL-WORLD CONNECTION** A wrench is commonly used to loosen a nut on a bolt. See Figure 6.23. If the nut is difficult to loosen, a wrench with a longer handle is often helpful because a longer wrench requires less force. The length of the wrench and the force needed to loosen a nut on a bolt are *inversely* proportional, as illustrated in the next example.

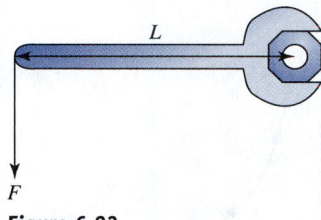

Figure 6.23

EXAMPLE 6 **Loosening a nut on a bolt**

It takes a force F of 10 pounds to loosen a nut on a bolt with an 8-inch wrench. Find the force necessary to loosen the same nut if a 16-inch wrench is used.

Solution

STEP 1: The equation for inverse variation is $F = \frac{k}{L}$, where L is the length of the wrench.

STEP 2: Substitute 10 for F and 8 for L in $F = \frac{k}{L}$. Solve for k.

$$10 = \frac{k}{8} \qquad F = \frac{k}{L}$$

$$8 \cdot 10 = \frac{k}{8} \cdot 8 \qquad \text{Multiply each side by 8.}$$

$$80 = k \qquad \text{Simplify.}$$

STEP 3: Replace k with 80 in $F = \frac{k}{L}$ to obtain $F = \frac{80}{L}$.

STEP 4: Find the necessary force by letting $L = 16$ in $F = \frac{80}{L}$.

$$F = \frac{80}{16} = 5 \text{ pounds}$$

■ **Now Try Exercise 79**

READING CHECK

If y varies inversely with x, what happens to y if x doubles?

NOTE: In Example 6 the length of the wrench doubled, so with inverse variation the necessary force was half as much, or $\frac{1}{2}(10) = 5$ pounds.

EXAMPLE 7   **Analyzing data**

Determine whether the data in each table represent direct variation, inverse variation, or neither. For direct and inverse variation, find an equation. Graph the data and equation.

(a)

x	4	5	10	20
y	50	40	20	10

(b)

x	2	5	9	11
y	14	35	63	77

(c)

x	2	4	6	8
y	10	16	24	48

Solution

(a) As x increases, y decreases. With inverse variation $y = \frac{k}{x}$, so $k = xy$. Because $xy = 200$ for each data point, $k = 200$ and the equation $y = \frac{200}{x}$ models the data. The data represent inverse variation. The data and equation are graphed in Figure 6.24(a).

(b) Because $\frac{y}{x} = 7$ for each data point, the equation $y = 7x$ models the data. These data represent direct variation. The data and equation are graphed in Figure 6.24(b).

(c) Neither the product xy nor the ratio $\frac{y}{x}$ is constant for the data in the table. Therefore these data represent neither direct variation nor inverse variation. The data are plotted in Figure 6.24(c). Note that the data values increase and are nonlinear.

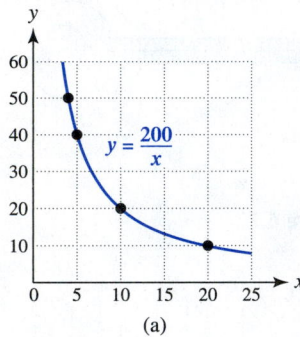
(a)

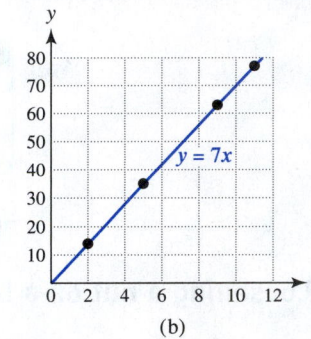
(b)

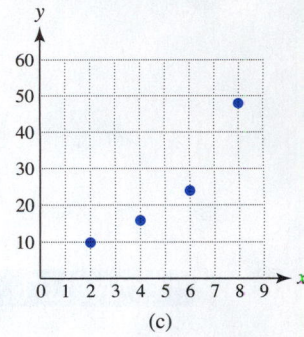
(c)

Figure 6.24

■ **Now Try Exercises 49, 51, 53**

Joint Variation

▶ **REAL-WORLD CONNECTION** In many applications a quantity depends on more than one variable. In *joint variation* a quantity varies with the product of more than one variable. For example, the formula for the area A of a rectangle is given by

$$A = WL,$$

where W and L are the width and length, respectively. Thus the area of a rectangle varies jointly with the width and length.

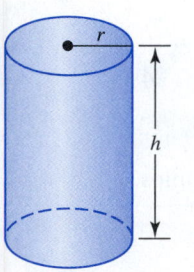

Figure 6.25 $V = \pi r^2 h$

JOINT VARIATION

Let x, y, and z denote three quantities. Then z **varies jointly** with x and y if there is a nonzero number k such that

$$z = kxy.$$

Sometimes joint variation can involve a power of a variable. For example, the volume V of a cylinder is given by $V = \pi r^2 h$, where r is its radius and h is its height, as illustrated in Figure 6.25. In this case we say that the volume varies jointly with the height and the *square* of the radius. The constant of variation is $k = \pi$.

EXAMPLE 8 ### Finding the strength of a rectangular beam

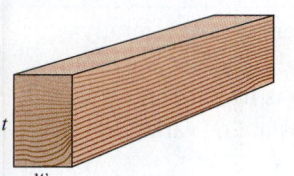

Figure 6.26

The strength S of a rectangular beam varies jointly with its width w and the square of its thickness t. See Figure 6.26. If a beam 3 inches wide and 5 inches thick supports 750 pounds, how much can a similar beam 2 inches wide and 6 inches thick support?

Solution
The strength of a beam is modeled by

$$S = kwt^2, \qquad \text{S varies jointly with w and the square of t.}$$

where k is a constant of variation. We can find k by substituting $S = \mathbf{750}$, $w = \mathbf{3}$, and $t = \mathbf{5}$ in the formula.

$$\mathbf{750} = k \cdot \mathbf{3} \cdot \mathbf{5}^2 \qquad \text{Substitute in } S = kwt^2.$$

$$k = \frac{750}{3 \cdot 5^2} \qquad \text{Solve for } k; \text{ rewrite.}$$

$$= 10 \qquad \text{Simplify.}$$

Thus $S = 10wt^2$ models the strength of this type of beam. When $w = \mathbf{2}$ and $t = \mathbf{6}$, the beam can support

$$S = 10 \cdot \mathbf{2} \cdot \mathbf{6}^2 = \mathbf{720} \text{ pounds.}$$

Now Try Exercise 91

CRITICAL THINKING

Compare the increased strength of a beam if the width doubles and if the thickness doubles. What happens to the strength of a beam if both the width and thickness triple?

6.6   Putting It All Together

CONCEPT	EXPLANATION	EXAMPLES
Proportion	A statement (equation) that two ratios (fractions) are equal The proportion $\frac{a}{b} = \frac{c}{d}$ can be expressed verbally as "a is to b as c is to d."	10 is to 13 as 32 is to x: $\frac{10}{13} = \frac{32}{x}$. x is to 7 as 2 is to 5: $\frac{x}{7} = \frac{2}{5}$.

continued on next page

continued from previous page

CONCEPT	EXPLANATION	EXAMPLES
Direct Variation	Two quantities x and y vary according to the equation $y = kx$, where k is a nonzero constant of proportionality (or variation). The ratios $\frac{y}{x}$ all equal k.	$y = 3x$ or $\frac{y}{x} = 3$; $k = 3$ 	x	1	2	4
---	---	---	---
y	3	6	12	 If x doubles, then y also doubles.
Inverse Variation	Two quantities x and y vary according to the equation $y = \frac{k}{x}$, where k is a nonzero constant of proportionality (or variation). The products xy all equal k.	$y = \frac{2}{x}$ or $xy = 2$; $k = 2$ 	x	1	2	4
---	---	---	---
y	2	1	$\frac{1}{2}$	 If x doubles, then y decreases by half.
Joint Variation	Three quantities x, y, and z vary according to the equation $z = kxy$, where k is a constant.	The area A of a triangle varies jointly with b and h according to the equation $A = \frac{1}{2}bh$, where b is its base and h is its height. The constant of variation is $k = \frac{1}{2}$.

6.6 Exercises

MyMathLab

CONCEPTS AND VOCABULARY

1. What is a proportion?

2. If 5 is to 6 as x is to 7, write a proportion that allows you to find x.

3. Suppose that y is directly proportional to x. If x doubles, what happens to y?

4. Suppose that y is inversely proportional to x. If x doubles, what happens to y?

5. If y varies inversely with x, then xy equals a(n) _____.

6. If y varies directly with x, then $\frac{y}{x}$ equals a(n) _____.

7. If z varies jointly with x and y, then $z =$ _____.

8. If z varies jointly with the square of x and the cube of y, then $z =$ _____.

9. Would a food bill B generally vary directly or vary inversely with the number of people N being fed? Explain your reasoning.

10. Would the time T needed to paint a building vary directly or inversely with the number of painters N working on the job? Explain your reasoning.

PROPORTIONS

Exercises 11–22: Solve the proportion.

11. $\dfrac{x}{14} = \dfrac{5}{7}$

12. $\dfrac{x}{5} = \dfrac{4}{9}$

13. $\dfrac{8}{x} = \dfrac{2}{3}$

14. $\dfrac{5}{11} = \dfrac{9}{x}$

15. $\dfrac{6}{13} = \dfrac{h}{156}$

16. $\dfrac{25}{a} = \dfrac{15}{8}$

17. $\dfrac{7}{4z} = \dfrac{5}{3}$

18. $\dfrac{3}{2} = \dfrac{2x}{9}$

19. $\dfrac{2}{3x+1} = \dfrac{5}{x}$

20. $\dfrac{7}{x-1} = -\dfrac{5}{3x}$

21. $\dfrac{4}{x} = \dfrac{x}{9}$

22. $\dfrac{2x}{5} = \dfrac{40}{x}$

Exercises 23–30: Complete the following.
  (a) Write a proportion that models the situation.
  (b) Solve the proportion for x.

23. 7 is to 9 as 10 is to x

24. x is to 11 as 9 is to 2

25. A triangle has sides of 3, 4, and 6. In a similar triangle the shortest side is 5 and the longest side is x.

26. A rectangle has sides of 9 and 14. In a similar rectangle the longer side is 8 and the shorter side is x.

27. If you earn $78 in 6 hours, then you earn x dollars in 8 hours.

28. If 12 gallons of gas contain 1.2 gallons of ethanol, 18 gallons of gas contain x gallons of ethanol.

29. If 2 compact discs can record 120 minutes of music, 5 compact discs can record x minutes.

30. If a gas pump fills a 30-gallon tank in 8 minutes, it can fill a 17-gallon tank in x minutes.

VARIATION

Exercises 31–36: Let y be directly proportional to x.
  (a) Find the constant of proportionality k.
  (b) Use $y = kx$ to find y when $x = 7$.

31. $y = 6$ when $x = 3$ **32.** $y = 7$ when $x = 14$

33. $y = 5$ when $x = 2$ **34.** $y = 11$ when $x = 22$

35. $y = -120$ when $x = 16$

36. $y = -34$ when $x = 17$

Exercises 37–42: Let y be inversely proportional to x.
  (a) Find the constant of proportionality k.
  (b) Use $y = \frac{k}{x}$ to find y when $x = 10$.

37. $y = 5$ when $x = 4$ **38.** $y = 2$ when $x = 30$

39. $y = 100$ when $x = \frac{1}{2}$ **40.** $y = \frac{1}{4}$ when $x = 40$

41. $y = 20$ when $x = 20$ **42.** $y = \frac{45}{4}$ when $x = 8$

Exercises 43–48: Let z vary jointly with x and y.
  (a) Find the constant of variation k.
  (b) Use $z = kxy$ to find z when $x = 5$ and $y = 7$.

43. $z = 6$ when $x = 3$ and $y = 8$

44. $z = 135$ when $x = 2.5$ and $y = 9$

45. $z = 5775$ when $x = 25$ and $y = 21$

46. $z = 1530$ when $x = 22.5$ and $y = 4$

47. $z = 25$ when $x = \frac{1}{2}$ and $y = 5$

48. $z = 12$ when $x = \frac{1}{4}$ and $y = 12$

Exercises 49–56: (Refer to Example 7.)
  (a) Determine whether the data represent direct variation, inverse variation, or neither.
  (b) If the data represent either direct or inverse variation, find an equation that models the data.
  (c) Graph the equation and the data.

49.

x	2	3	4	5
y	3	4.5	6	7.5

50.

x	3	6	9	12
y	12	6	4	3

51.

x	10	20	30	40
y	12	6	5	4

52.

x	4	6	12	20
y	10	20	30	40

53.

x	2	6	10	14
y	105	35	21	15

54.

x	1	5	9	15
y	6	30	54	90

55. Thinking Generally (Do not graph.)

x	-2	-1	1	2
y	$4b$	$2b$	$-2b$	$-4b$

56. Thinking Generally (Do not graph.)

x	$-2a$	$-a$	a	$2a$
y	$-b$	$-2b$	$2b$	b

Exercises 57–62: Use the graph to determine whether the data represent direct variation, inverse variation, or neither. Find the constant of variation whenever possible.

57.

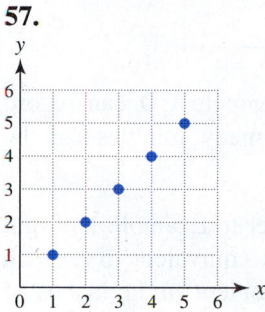

58.

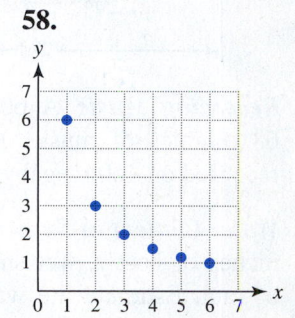

59.

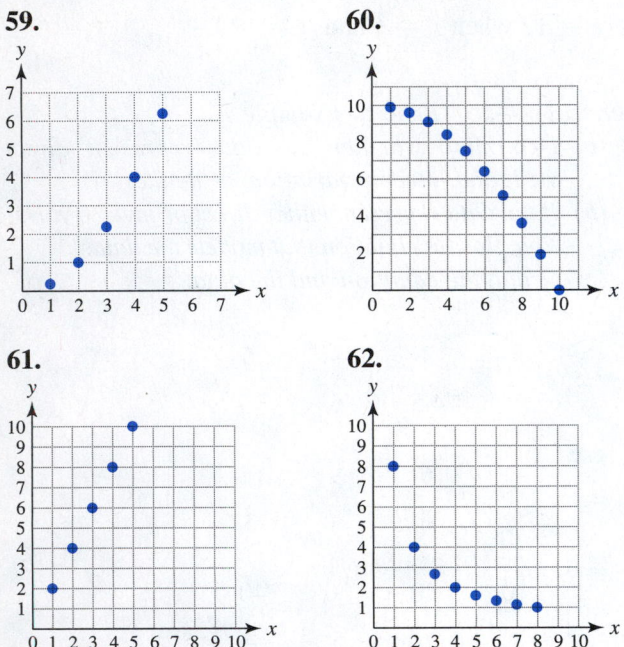

60.

61.

62.

63. Online Exploration Go online and find a study that gives its sample size. Use proportions to generalize the study to the larger U.S. population. You may need to also look up the population of the United States.

64. Online Exploration Go online and find the dimensions of a famous building or structure. Use proportions to draw a view of the structure to scale. Record both the actual dimensions and your scale dimensions.

APPLICATIONS

65. *Height of a Tree* (Refer to Example 2.) A 6-foot-tall person casts a 7-foot shadow, and a nearby tree casts a 27-foot shadow. Estimate the height of the tree.

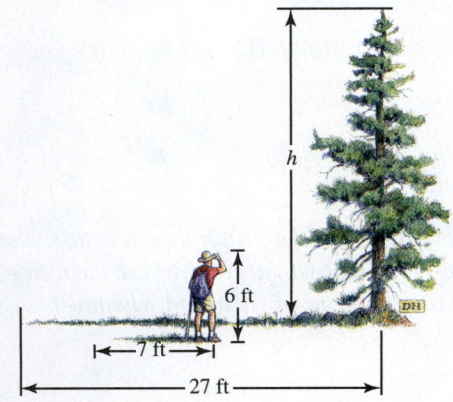

66. *Recording Music* A 600-megabyte CD can record 68 minutes of music. How many minutes can be recorded on 360 megabytes?

67. *Water Content in Snow* (Refer to Example 1.) Eight inches of heavy, wet snow is equivalent to an inch of rain. Estimate the water content in 11 inches of heavy, wet snow.

68. *Wages* If a person earns $143 in 11 hours, how much will that person earn in 17 hours?

69. *Buying 4G Phones* In a 2011 Nielsen poll with a sample size of 2131 respondents, 618 were planning to buy a 4G phone in 2011. Of the 231 million adults in the United States, about how many were planning to buy a 4G phone in 2011?

70. *What Does 4G Mean?* In a 2011 Nielsen poll with a sample size of 2131 respondents, only 482 knew that 4G meant superfast wireless. Of the 231 million adults in the United States, about how many knew what 4G meant?

71. *Largemouth Bass Population* (Refer to Example 3.) Three hundred largemouth bass are tagged and released into a lake. Later, a sample of 112 largemouth bass contains 17 tagged bass. Estimate the largemouth bass population to the nearest hundred.

72. *Eagle Population* Twenty-two bald eagles are tagged and released into the wilderness. Later, an observed sample of 56 bald eagles contains 7 eagles that are tagged. Estimate the bald eagle population in this wilderness area.

73. *Braking Distance* The distance d required for a car to stop after the brakes have been applied is directly proportional to the square of the car's speed x. If a car moving at 60 miles per hour requires 300 feet to stop on dry, level pavement, how many feet would be required to stop the car at 30 miles per hour?

74. *Points in Basketball* If a player scores 132 points in 11 games, use proportions to estimate the number of points the player might score in a 20-game season.

75. *Rolling Resistance of Cars* If you were to try to push a car, you would experience *rolling resistance*. This resistance equals the force necessary to keep the car moving slowly in neutral gear. The following table shows the rolling resistance R for passenger cars of different gross weights W. (*Source*: N. Garber and L. Hoel, *Traffic and Highway Engineering.*)

W (pounds)	2000	2500	3000	3500
R (pounds)	24	30	36	42

(a) Find an equation that models the data. Graph the equation with the data.

(b) Find the rolling resistance of a 3200-pound car.

76. *Transportation Costs* The use of a toll bridge varies inversely according to the toll. When the toll is $0.75, 6000 vehicles use the bridge. Estimate the number of users if the toll is $0.40. (*Source:* N. Garber.)

77. *Hooke's Law* The table shows the distance D that a spring stretches when a weight W is hung on it.

W (pounds)	2	6	9	15
D (inches)	1.5	4.5	6.75	11.25

(a) Do the data represent direct or inverse variation? Explain.

(b) Find an equation that models the data.

(c) How far will the spring stretch if an 11-pound weight is hung on it, as depicted in the figure?

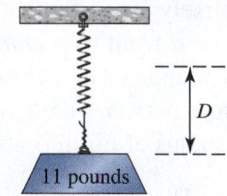

11 pounds

78. *Flow of Water* The gallons of water G flowing in 1 minute through a hose with a cross-sectional area A are shown in the table.

A (square inch)	0.2	0.3	0.4	0.5
G (gallons)	5.4	8.1	10.8	13.5

(a) Do the data represent direct or inverse variation? Explain.

(b) Find an equation that models the data. Graph the equation with the data.

(c) Interpret the constant of variation k.

79. *Tightening Lug Nuts* (Refer to Example 6.) When a tire is mounted on a car, the lug nuts should not be over-tightened. The following table shows the maximum force used with wrenches of different lengths.

L (inches)	8	10	16
F (pounds)	150	120	75

Source: Tires Plus.

(a) Model the data, using the equation $F = \frac{k}{L}$.

(b) How much force should be used with a wrench 20 inches long?

80. *Loosening a Nut on a Bolt* (Refer to Example 6.) It takes a force F of 20 pounds to loosen a nut on a bolt with a wrench having a handle length L of 12 inches. Find the force necessary to loosen the nut if a 10-inch wrench is used.

81. *Cost of Tuition* (Refer to Example 4.) The cost of tuition is directly proportional to the number of credits taken. If 6 credits cost $435, find the cost of 11 credits. What does the constant of proportionality represent?

82. *Diving and Pressure* (Refer to Example 5.) At a depth of 45 feet a diver experiences water pressure equal to 20 pounds per square inch. Find the pressure at 72 feet.

83. *Whales and Diving* (Refer to Example 5.) At a depth of 1800 feet a sperm whale experiences water pressure equal to 800 pounds per square inch. Find the pressure at 2700 feet.

84. *Stretching a Spring* A 10-pound weight causes a spring to stretch 2 inches. Use direct variation to predict how much an 18-pound weight will stretch the same spring.

85. *Air Temperature and Altitude* In the first 6 miles of Earth's atmosphere, air cools as the altitude increases. The following graph shows the temperature change y in degrees Fahrenheit at an altitude of x miles. (*Source:* A. Miller and R. Anthes, *Meteorology.*)

(a) Does this graph represent direct variation or inverse variation?

(b) Find an equation for the line in the graph.

(c) Is the constant of proportionality k positive or negative? Interpret k.

(d) Estimate the temperature change 3.5 miles high.

86. *Ozone and UV Radiation* Depletion of the ozone layer has caused an increase in the amount of UV radiation reaching Earth's surface. An increase in UV radiation is associated with skin cancer. The graph on the next page shows the percentage increase y in UV radiation for a decrease in the ozone layer of x percent. (*Source:* R. Turner, D. Pearce, and I. Bateman, *Environmental Economics.*)

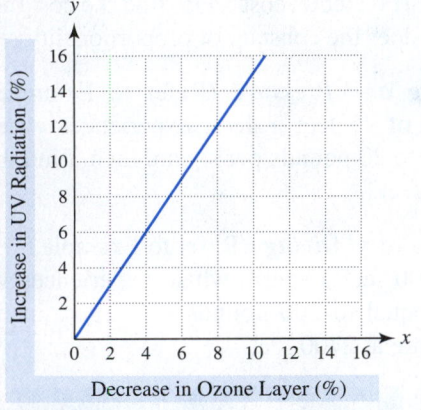

(a) Does this graph represent direct variation or inverse variation?
(b) Find an equation for the line in the graph.
(c) Estimate the percentage increase in UV radiation if the ozone layer decreases by 7%.

87. *Electrical Resistance* The electrical resistance of a wire is directly proportional to its length. If a 35-foot-long wire has a resistance of 2 ohms, find the resistance of a 25-foot-long wire.

88. *Resistance and Current* The current that flows through an electrical circuit is inversely proportional to the resistance. When the resistance R is 150 ohms, the current I is 0.8 amp. Find the current when the resistance is 40 ohms.

89. *Joint Variation* The variable z varies jointly with the second power of x and the third power of y. Write a formula for z if $z = 31.9$ when $x = 2$ and $y = 2.5$.

90. *Wind Power* The electrical power generated by a windmill varies jointly with the square of the diameter of the area swept out by the blades and the cube of the wind velocity. If a windmill with a 10-foot diameter in a 16-mile-per-hour wind generates 15,392 watts, how much power would be generated if the blades swept out an area 12 feet in diameter and the wind speed was 15 miles per hour?

91. *Strength of a Beam* (Refer to Example 8.) If a wood beam 5 inches wide and 3 inches thick supports 300 pounds, how much can a similar beam 5 inches wide and 2 inches thick support?

92. *Carpeting* The cost of carpet for a rectangular room varies jointly with its width and length. If a room 11 feet wide and 14 feet long costs $539 to carpet, find the cost to carpet a room 17 feet by 19 feet. Interpret the constant of variation k.

93. *Weight on the Moon* The weight of a person on the moon is directly proportional to the weight of the person on Earth. If a 175-pound person weighs 28 pounds on the moon, how much will a 220-pound person weigh on the moon?

94. *Weight Near Earth* The weight W of a person near Earth is inversely proportional to the square of the person's distance d from the *center* of Earth. If a person weighs 200 pounds when $d = 4000$ miles, how much does the same person weigh when $d = 7000$ miles? (*Note:* The radius of Earth is about 4000 miles.)

95. *Ohm's Law* The voltage V in an electrical circuit varies jointly with the amperage I and resistance R. If $V = 220$ when $I = 10$ and $R = 22$, find V when $I = 15$ and $R = 50$.

96. *Revenue* The revenue R from selling x items at price p varies jointly with x and p. If $R = \$24{,}000$ when $x = 3000$ and $p = \$8$, find the number of items x sold when $R = \$30{,}000$ and $p = \$6$.

WRITING ABOUT MATHEMATICS

97. Explain in words what it means for a quantity y to be directly proportional to a quantity x.

98. Explain in words what it means for a quantity y to be inversely proportional to a quantity x.

SECTIONS 6.5 AND 6.6 **Checking Basic Concepts**

1. Simplify each complex fraction.

(a) $\dfrac{3 - \dfrac{1}{x^2}}{3 + \dfrac{1}{x^2}}$ (b) $\dfrac{\dfrac{2}{x-1} - \dfrac{2}{x+1}}{\dfrac{4}{x^2-1}}$

2. Suppose that y is directly proportional to x and that $y = 6$ when $x = 8$.
   (a) Find the constant of proportionality k.
   (b) Use $y = kx$ to find y when $x = 11$.

3. Determine whether the data represent direct variation or inverse variation. Find an equation that models the data.

(a)

x	10	15	25	40
y	4	6	10	16

(b)

x	5	10	15	20
y	24	12	8	6

6.7 Division of Polynomials

Division by a Monomial • Division by a Polynomial • Synthetic Division

Girolamo Cardano (1501–1576)

NEW VOCABULARY

☐ Divisor
☐ Quotient
☐ Remainder
☐ Synthetic division

The study of polynomials has occupied the minds of mathematicians for centuries. During the sixteenth century, Italian mathematicians, such as Cardano, discovered how to solve higher degree polynomial equations. In this section we demonstrate symbolic methods for dividing polynomials. Division is often needed to factor higher degree polynomials and to solve polynomial equations. (*Source:* H. Eves, *An Introduction to the History of Mathematics.*)

Division by a Monomial

To divide a polynomial by a monomial we use the two properties

$$\frac{a+b}{c} = \frac{a}{c} + \frac{b}{c} \quad \text{and} \quad \frac{a-b}{c} = \frac{a}{c} - \frac{b}{c}.$$

For example,

$$\frac{3x^2 + x}{x} = \frac{3x^2}{x} + \frac{x}{x} = 3x + 1. \quad \text{Divide each term by } x.$$

Factoring and the basic principle of fractions can also be applied to simplify fractions, as follows.

$$\frac{3x^2 + x}{x} = \frac{x(3x + 1)}{x} = 3x + 1$$

When dividing natural numbers, we can check our work by using multiplication. Because

$$\frac{6}{2} = 3,$$

it follows that $2 \cdot 3 = 6$. Similarly, to check whether

$$\frac{3x^2 + x}{x} = 3x + 1,$$

we multiply x and $3x + 1$.

$$x(3x + 1) = x \cdot 3x + x \cdot 1 \quad \text{Distributive property}$$
$$= 3x^2 + x \checkmark \quad \text{It checks.}$$

Graphical support for this division problem is shown in Figures 6.27(a) and 6.27(b), where the graphs of $y_1 = \frac{3x^2 + x}{x}$ and $y_2 = 3x + 1$ appear to be identical. In Figure 6.27(c) numerical support is given, where $y_1 = y_2$ for all x-values except 0. Note that we assumed that $x \neq 0$ in $\frac{3x^2 + x}{x}$ because 0 would result in this expression being undefined.

Graphical and Numerical Support for $\dfrac{3x^2 + x}{x} = 3x + 1$

[−4.7, 4.7, 1] by [−3.1, 3.1, 1]   [−4.7, 4.7, 1] by [−3.1, 3.1, 1]

$y_1 = \dfrac{3x^2 + x}{x}$ $y_2 = 3x + 1$

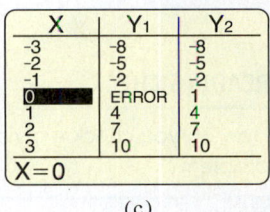

X	Y₁	Y₂
-3	-8	-8
-2	-5	-5
-1	-2	-2
0	ERROR	1
1	4	4
2	7	7
3	10	10

X=0

(a)   (b)   (c)

Figure 6.27

EXAMPLE 1 **Dividing by a monomial**

Divide and check.

(a) $\dfrac{5x^3 - 15x^2}{15x}$ **(b)** $\dfrac{4x^2 + 8x - 12}{4x^2}$ **(c)** $\dfrac{5x^2y + 10xy^2}{5xy}$

Solution

(a) $\dfrac{5x^3 - 15x^2}{15x} = \dfrac{5x^3}{15x} - \dfrac{15x^2}{15x} = \dfrac{x^2}{3} - x$

Check: $15x\left(\dfrac{x^2}{3} - x\right) = \dfrac{15x \cdot x^2}{3} - 15x \cdot x$

$= 5x^3 - 15x^2$ ✓

(b) $\dfrac{4x^2 + 8x - 12}{4x^2} = \dfrac{4x^2}{4x^2} + \dfrac{8x}{4x^2} - \dfrac{12}{4x^2} = 1 + \dfrac{2}{x} - \dfrac{3}{x^2}$

Check: $4x^2\left(1 + \dfrac{2}{x} - \dfrac{3}{x^2}\right) = 4x^2 \cdot 1 + \dfrac{4x^2 \cdot 2}{x} - \dfrac{4x^2 \cdot 3}{x^2}$

$= 4x^2 + 8x - 12$ ✓

(c) $\dfrac{5x^2y + 10xy^2}{5xy} = \dfrac{5x^2y}{5xy} + \dfrac{10xy^2}{5xy} = x + 2y$

Check: $5xy(x + 2y) = 5xy \cdot x + 5xy \cdot 2y$

$= 5x^2y + 10xy^2$ ✓

▪ **Now Try Exercises 9, 11, 19**

MAKING CONNECTIONS

Division and "Canceling" Incorrectly

When dividing expressions, students commonly "cancel" incorrectly. For example,

$$\dfrac{x - 3x^3}{x} \neq 1 - 3x^3 \quad \text{and} \quad \dfrac{x - 3x^3}{x} \neq -3x^2.$$

Rather, $\qquad \dfrac{x - 3x^3}{x} = \dfrac{x}{x} - \dfrac{3x^3}{x} = 1 - 3x^2.$

Divide the monomial into *every* term in the numerator.

Division by a Polynomial

First, we briefly review division of natural numbers.

Quotient ⟶ 58 R 1 ⟵ Remainder
Divisor ⟶ 3)175 ⟵ Dividend
$\underline{15}$
25
$\underline{24}$
1

READING CHECK

How do you check a division problem?

Because $3 \cdot 58 + 1 = 175$, the answer checks. The quotient and remainder also can be expressed as $58\frac{1}{3}$. Division of polynomials is similar to long division of natural numbers.

EXAMPLE 2   **Dividing polynomials**

Divide $4x^2 - 2x + 1$ by $2x + 1$ and check.

Solution
Begin by dividing $2x$ into $4x^2$.

$$
\begin{array}{r}
2x \\
2x + 1\overline{)4x^2 - 2x + 1} \\
\underline{4x^2 + 2x} \\
-4x + 1
\end{array}
$$

$\dfrac{4x^2}{2x} = 2x$

$2x(2x + 1) = 4x^2 + 2x$
Subtract $-2x - 2x = -4x$.
Bring down the 1.

Next, divide $2x$ into $-4x$.

$$
\begin{array}{r}
2x - 2 \\
2x + 1\overline{)4x^2 - 2x + 1} \\
\underline{4x^2 + 2x} \\
-4x + 1 \\
\underline{-4x - 2} \\
3
\end{array}
$$

$\dfrac{-4x}{2x} = -2$

$-2(2x + 1) = -4x - 2$
Subtract $1 - (-2) = 3$.
Remainder is 3.

The quotient is $2x - 2$ with remainder 3, which can also be written as

$$2x - 2 + \frac{3}{2x + 1}, \qquad \text{Quotient} + \frac{\text{Remainder}}{\text{Divisor}}$$

in the same manner as 58 R 1 was expressed earlier as $58\frac{1}{3}$. Polynomial division is checked by multiplying the divisor and the quotient and adding the remainder.

$$(2x + 1)(2x - 2) + 3 = 4x^2 - 4x + 2x - 2 + 3$$
$$ = 4x^2 - 2x + 1 \longleftarrow \text{Dividend} \ \checkmark$$

Divisor   Quotient   Remainder

Now Try Exercise 25

EXAMPLE 3   **Dividing polynomials**

Divide $2x^3 - 5x^2 + 4x - 1$ by $x - 1$.

Solution
Begin by dividing x into $2x^3$.

$$
\begin{array}{r}
2x^2 \\
x - 1\overline{)2x^3 - 5x^2 + 4x - 1} \\
\underline{2x^3 - 2x^2} \\
-3x^2 + 4x
\end{array}
$$

$\dfrac{2x^3}{x} = 2x^2$

$2x^2(x - 1) = 2x^3 - 2x^2$
Subtract $-5x^2 - (-2x^2) = -3x^2$.
Bring down $4x$.

Next, divide x into $-3x^2$.

$$
\begin{array}{r}
2x^2 - 3x \\
x - 1\overline{)2x^3 - 5x^2 + 4x - 1} \\
\underline{2x^3 - 2x^2} \\
-3x^2 + 4x \\
\underline{-3x^2 + 3x} \\
x - 1
\end{array}
$$

$\dfrac{-3x^2}{x} = -3x$

$-3x(x - 1) = -3x^2 + 3x$
Subtract $4x - 3x = x$.
Bring down -1.

Finally, divide x into x.

$$
\begin{array}{r}
2x^2 - 3x + 1 \\
x - 1 \overline{)2x^3 - 5x^2 + 4x - 1} \\
\underline{2x^3 - 2x^2} \\
-3x^2 + 4x \\
\underline{-3x^2 + 3x} \\
x - 1 \\
\underline{x - 1} \\
0
\end{array}
$$

$\dfrac{x}{x} = 1$

$1(x - 1) = x - 1$

Subtract. Remainder is 0.

The quotient is $2x^2 - 3x + 1$, and the remainder is 0.

■ **Now Try Exercise 29**

EXAMPLE 4 **Dividing by a quadratic divisor**

Divide $3x^3 - 2x^2 - 4x + 4$ by $x^2 - 1$.

Solution
Begin by writing $x^2 - 1$ as $x^2 + \mathbf{0x} - 1$.

$$
\begin{array}{r}
3x - 2 \\
x^2 + \mathbf{0x} - 1 \overline{)3x^3 - 2x^2 - 4x + 4} \\
\underline{3x^3 + 0x^2 - 3x} \\
-2x^2 - x + 4 \\
\underline{-2x^2 - 0x + 2} \\
-x + 2
\end{array}
$$

The quotient is $3x - 2$ with remainder $-x + 2$, or $2 - x$, which can be written as

$$
3x - 2 + \frac{2 - x}{x^2 - 1}.
$$

■ **Now Try Exercise 37**

EXAMPLE 5 **Dividing into a polynomial with missing terms**

Divide $2x^3 - 3$ by $x + 1$.

Solution
Begin by writing $2x^3 - 3$ as $2x^3 + \mathbf{0x^2} + \mathbf{0x} - 3$.

$$
\begin{array}{r}
2x^2 - 2x + 2 \\
x + 1 \overline{)2x^3 + \mathbf{0x^2} + \mathbf{0x} - 3} \\
\underline{2x^3 + 2x^2} \\
-2x^2 + 0x \\
\underline{-2x^2 - 2x} \\
2x - 3 \\
\underline{2x + 2} \\
-5
\end{array}
$$

The quotient is $2x^2 - 2x + 2$ with remainder -5, which can be written as

$$
2x^2 - 2x + 2 - \frac{5}{x + 1}.
$$

■ **Now Try Exercise 31**

Synthetic Division

A shortcut called **synthetic division** can be used to divide $x - k$, where k is a constant, into a polynomial. For example, to divide $x - 2$ into $3x^3 - 8x^2 + 7x - 6$, we do the following (with the equivalent long division shown at the right).

Synthetic Division

$$\begin{array}{r|rrrr} 2 & 3 & -8 & 7 & -6 \\ & & 6 & -4 & 6 \\ \hline & 3 & -2 & 3 & 0 \end{array}$$

Add to find row 3.

Long Division of Polynomials

$$\begin{array}{r} 3x^2 - 2x + 3 \\ x - 2\overline{)3x^3 - 8x^2 + 7x - 6} \\ \underline{3x^3 - 6x^2} \\ -2x^2 + 7x \\ \underline{-2x^2 + 4x} \\ 3x - 6 \\ \underline{3x - 6} \\ 0 \end{array}$$

Note that the blue numbers in the expression for long division correspond to the third row in synthetic division. The remainder is 0, which is the last number in the third row. The quotient is $3x^2 - 2x + 3$. Its coefficients are 3, -2, and 3 and are located in the third row. To divide $x - 2$ into $3x^3 - 8x^2 + 7x - 6$ with synthetic division use the following steps.

STEP 1: In the top row write **2** (the value of k) on the left and then write the coefficients of the dividend $3x^3 - 8x^2 + 7x - 6$.

STEP 2: **(a)** Copy the leading coefficient **3** of $3x^3 - 8x^2 + 7x - 6$ into the third row and multiply it by **2** (the value of k). Write the result 6 in the second row below -8. Add -8 and 6 in the second column to obtain the -2 in the third row.

(b) Repeat the process by multiplying -2 by **2** and place the result -4 below **7**. Then add **7** and -4 to obtain **3**.

(c) Multiply **3** by **2** and place the result 6 below the -6. Adding 6 and -6 gives **0**.

STEP 3: The last number in the third row is **0**, which is the remainder. The other numbers in the third row are the coefficients of the quotient, which is $3x^2 - 2x + 3$.

EXAMPLE 6 **Performing synthetic division**

Use synthetic division to divide $x^4 - 5x^3 + 9x^2 - 10x + 3$ by $x - 3$.

Solution
Because the divisor is $x - 3$, the value of k is **3**.

$$\begin{array}{r|rrrrr} 3 & 1 & -5 & 9 & -10 & 3 \\ & & 3 & -6 & 9 & -3 \\ \hline & 1 & -2 & 3 & -1 & 0 \end{array}$$

The quotient is $x^3 - 2x^2 + 3x - 1$ and the remainder is **0**. This result is expressed by

$$\frac{x^4 - 5x^3 + 9x^2 - 10x + 3}{x - 3} = x^3 - 2x^2 + 3x - 1 + \frac{0}{x - 3}, \quad \text{or}$$

$$\frac{x^4 - 5x^3 + 9x^2 - 10x + 3}{x - 3} = x^3 - 2x^2 + 3x - 1.$$

Now Try Exercise 49

MAKING CONNECTIONS

Factors and Remainders

Multiplying the last equation in the solution to Example 6 by $x - 3$ gives

$$x^4 - 5x^3 + 9x^2 - 10x + 3 = (x - 3)(x^3 - 2x^2 + 3x - 1).$$

That is, $x - 3$ is a *factor* of $x^4 - 5x^3 + 9x^2 - 10x + 3$ because the remainder is 0.

This concept is true in general: If a polynomial $p(x)$ is divided by $x - k$ and the remainder is 0, then $x - k$ is a factor of $p(x)$. See Exercises 61 and 62.

EXAMPLE 7 | **Performing synthetic division**

Use synthetic division to divide $2x^3 - x + 5$ by $x + 1$.

Solution

Write $2x^3 - x + 5$ as $\mathbf{2x^3 + 0x^2 - x + 5}$. The divisor $x + 1$ can be written as

$$x + 1 = x - (\mathbf{-1}),$$

so we let $k = \mathbf{-1}$.

$$
\begin{array}{r|rrrr}
-1 & 2 & 0 & -1 & 5 \\
   &   & -2 & 2 & -1 \\
\hline
   & 2 & -2 & 1 & 4
\end{array}
$$

The remainder is **4**, and the quotient is $\mathbf{2x^2 - 2x + 1}$. This result can also be expressed as

$$\frac{2x^3 - x + 5}{x + 1} = 2x^2 - 2x + 1 + \frac{4}{x + 1}.$$

▪ **Now Try Exercise 53**

CRITICAL THINKING

Suppose that $p(x)$ is a polynomial and that $(x - k)$ divides into $p(x)$ with remainder 0.

1. Evaluate $p(k)$.
2. Give one x-intercept on the graph of $p(x)$.

6.7 | Putting It All Together

TYPE OF DIVISION	EXPLANATION	EXAMPLE
By a Monomial	Use the property $\frac{a \pm b}{c} = \frac{a}{c} \pm \frac{b}{c}$ to divide a polynomial by a monomial. Be sure to divide the denominator into every term of the numerator.	$\dfrac{8a^3 - 4a^2}{2a} = \dfrac{8a^3}{2a} - \dfrac{4a^2}{2a} = 4a^2 - 2a$

TYPE OF DIVISION	EXPLANATION	EXAMPLE
By a Polynomial	Division by a polynomial may be done in a manner similar to long division of natural numbers. See Examples 2–5.	When $2x^2 - 7x + 4$ is divided by $2x - 1$, the quotient is $x - 3$ and the remainder is 1. This result may be expressed as $$\frac{2x^2 - 7x + 4}{2x - 1} = x - 3 + \frac{1}{2x - 1}.$$
Synthetic	Synthetic division is a fast way to divide a polynomial by a divisor in the form $x - k$.	Divide $4x^2 + 7x - 14$ by $x + 3$. $$\begin{array}{r} -3\,|\ \ 4\quad 7\quad -14 \\ \underline{\quad -12\quad 15} \\ 4\quad -5\quad\ 1 \end{array}$$ The quotient is $4x - 5$ with remainder 1.

6.7 Exercises

CONCEPTS AND VOCABULARY

1. To divide a monomial into a polynomial, divide the monomial into every _____ in the numerator.

2. $\frac{a + b}{c} =$ _____.

3. Because $\frac{21}{5} = 4$ with remainder 1, it follows that $21 =$ _____ · _____ + _____.

4. Because $\frac{3x^2 - 11x + 8}{3x - 5} = x - 2$ with remainder -2, $3x^2 - 11x + 8 = ($_____$) \cdot ($_____$) + ($_____$)$.

5. Is it possible to use synthetic division to divide $3x^3 - 2x^2 + x - 3$ by $x^2 - 2x$? Explain.

6. Where is the remainder located when you finish with synthetic division?

DIVISION BY A MONOMIAL

Exercises 7–12: Divide and check. Give graphical or numerical support for your result.

7. $\frac{4x - 6}{2}$

8. $\frac{4x^2 - 8x + 12}{4}$

9. $\frac{6x^3 - 9x}{3x}$

10. $\frac{3x^4 + x^2}{6x^2}$

11. $(4x^2 - x + 1) \div 2x^2$

12. $(5x^3 - 4x + 2) \div 20x$

Exercises 13–22: Divide.

13. $\frac{9x^2 - 12x - 3}{3}$

14. $\frac{10x^3 - 15x^2 + 5x}{5}$

15. $\frac{12a^3 - 18a}{6a}$

16. $\frac{50x^4 + 25x^2 + 100x}{25x}$

17. $(16x^3 - 24x) \div (12x)$

18. $(2y^4 - 4y^2 + 16) \div (2y^2)$

19. $(a^2b^2 - 4ab + ab^2) \div (ab)$

20. $(10x^3y^2 + 5x^2y^3) \div (5x^2y^2)$

21. $\frac{6m^4n^4 + 3m^2n^2 - 12}{3m^2n^2}$

22. $\frac{6p^2q^4 - 9p^6q^2}{-3pq}$

DIVISION BY A POLYNOMIAL

Exercises 23–28: Divide and check.

23. $\frac{3x^2 - 16x + 21}{x - 3}$

24. $\frac{4x^2 - x - 18}{x + 2}$

25. $\dfrac{2x^3 + 3x^2 - 2x - 2}{2x + 3}$

26. $\dfrac{2x^3 - 7x^2 + 8}{x - 2}$

27. $(10x^2 - 5) \div (x - 1)$

28. $(6x^2 - 11x + 4) \div (2x - 3)$

Exercises 29–42: Divide.

29. $\dfrac{4x^3 + 8x^2 - x - 2}{x + 2}$ **30.** $\dfrac{3x^3 - 4x^2 + 3x - 2}{x - 1}$

31. $\dfrac{x^3 + 3x - 4}{x + 4}$ **32.** $\dfrac{2x^3 - x^2 + 5}{x - 2}$

33. $(3x^3 + 8x^2 - 21x + 7) \div (3x - 1)$

34. $(14x^3 + 3x^2 - 9x + 3) \div (7x - 2)$

35. $(2a^4 + 5a^3 - 2a^2 - 5a) \div (2a + 5)$

36. $(4b^4 + 10b^3 - 2b^2 - 3b + 5) \div (2b + 1)$

37. $\dfrac{3x^3 + 4x^2 - 12x - 16}{x^2 - 4}$

38. $\dfrac{2x^4 - x^3 - 5x^2 + 4x - 12}{2x^2 - x + 3}$

39. $\dfrac{x^3 + x^2 - x}{x^2 - 1}$ **40.** $\dfrac{x^3 + x^2 - 6x}{x^2 - 2x}$

41. $(2a^4 - 3a^3 + 14a^2 - 8a + 10) \div (a^2 - a + 5)$

42. $(2z^3 + 3z^2 - 9z - 12) \div (z^2 - 1)$

43. Thinking Generally If the divisor is $a - b$, the quotient is $a^2 + ab + b^2$, and the remainder is 1, find the dividend.

44. Thinking Generally If the dividend is given by $x^3 + x^2 - 3x - 2$, the quotient is $x^2 - 3$, and the remainder is 1, find the divisor.

SYNTHETIC DIVISION

Exercises 45–56: Use synthetic division to divide.

45. $\dfrac{x^2 + 3x - 1}{x - 1}$ **46.** $\dfrac{2x^2 + x - 1}{x - 3}$

47. $(3x^2 - 22x + 7) \div (x - 7)$

48. $(5x^2 + 29x - 6) \div (x + 6)$

49. $\dfrac{x^3 + 7x^2 + 14x + 8}{x + 4}$ **50.** $\dfrac{2x^3 + 3x^2 + 2x + 4}{x + 1}$

51. $\dfrac{2x^3 + x^2 - 1}{x - 2}$ **52.** $\dfrac{x^3 + x - 2}{x + 3}$

53. $(2x^4 + 3x^2 - 4) \div (x + 2)$

54. $(x^3 - 2x^2 - 2x + 4) \div (x - 4)$

55. $(b^4 - 1) \div (b - 1)$

56. $(a^2 + a) \div (a + 2.5)$

GEOMETRY

57. *Area of a Rectangle* Use the figure to find the length L of the rectangle from its width and area A. Determine the length when $x = 8$ feet.

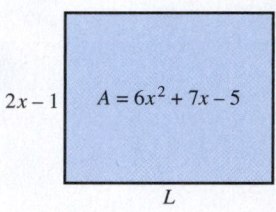

58. *Area of a Rectangle* Use the figure to find the width W of the rectangle from its length and area A. Determine the width when $x = 20$ inches.

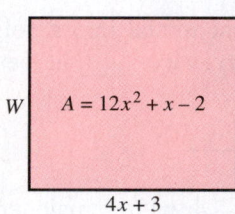

59. *Volume of a Box* The volume of a rectangular box is $x^3 + 7x^2 + 14x + 8$. If the width of the box is $x + 1$ and its length is $x + 4$, find the height of the box.

60. *Area of a Triangle* The area A of a triangle is $12x^2 - 5x - 3$. If the height of the triangle is $3x + 1$, find the length of its base.

THEOREMS ABOUT POLYNOMIALS

61. *Remainder Theorem* Divide $p(x)$ by $x - k$ and determine the remainder. Then evaluate $p(k)$.
  (a) $p(x) = x^2 - 3x + 2$ $\qquad k = 1$
  (b) $p(x) = 3x^2 + 5x - 2$ $\qquad k = -2$
  (c) $p(x) = 3x^3 - 4x^2 - 5x + 3$ $\qquad k = 2$
  (d) $p(x) = x^3 - 2x^2 - x + 2$ $\qquad k = -1$
  (e) $p(x) = x^4 - 5x^2 - 1$ $\qquad k = 3$

62. *Remainder Theorem* Try to generalize the results from Exercise 61. It can be done, and the result is called the **remainder theorem**.

63. *Factor Theorem* Evaluate $p(k)$. Determine whether $x - k$ is a factor of $p(x)$.
  (a) $p(x) = x^2 + x - 6$ $k = 2$
  (b) $p(x) = x^2 + 4x - 5$ $k = 1$
  (c) $p(x) = x^2 + 8x + 11$ $k = -2$
  (d) $p(x) = x^3 + x^2 + x + 1$ $k = -1$
  (e) $p(x) = x^3 - 3x^2 - x - 3$ $k = 2$

64. *Factor Theorem* Try to generalize the results from Exercise 63. It can be done, and the result is called the **factor theorem**.

WRITING ABOUT MATHEMATICS

65. A student dividing $2x^3 + 5x^2 - 13x + 5$ by $2x - 1$ does the following.

$$
\begin{array}{r}
1\rfloor\ \ 2 \quad 5 \quad -13 \quad 5 \\
\underline{\quad 2 \quad 7 \quad -6} \\
2 \quad 7 \quad -6 \quad -1
\end{array}
$$

What would you tell the student?

66. If you add, subtract, or multiply two polynomials, is the result a polynomial? If you divide two polynomials, is the result always a polynomial? Explain.

SECTION 6.7   **Checking Basic Concepts**

1. Divide and simplify.
  (a) $\dfrac{2x - x^2}{x^2}$ **(b)** $\dfrac{6a^2 - 9a + 15}{3a}$

2. Divide $\dfrac{10x^2 - x + 4}{2x + 1}$.

3. Use synthetic division to divide $2x^3 - 5x^2 - 1$ by each expression.
  (a) $x - 1$ **(b)** $x + 2$

CHAPTER 6   Summary

SECTION 6.1 ■ INTRODUCTION TO RATIONAL FUNCTIONS AND EQUATIONS

Rational Functions A rational function is given by $f(x) = \dfrac{p(x)}{q(x)}$, where $p(x)$ and $q(x)$ are polynomials. The domain of f includes all x-values such that $q(x) \neq 0$.

Example: $f(x) = \dfrac{4}{x - 2}$ defines a rational function with domain $\{x \mid x \neq 2\}$.

Graphs of Rational Functions Graphs of rational functions are usually curves. A vertical asymptote typically occurs at an x-value where the denominator equals zero but the numerator does not. The graph of a rational function never crosses a vertical asymptote. Graphing a rational function by hand may require plotting several points on each side of a vertical asymptote.

Example: $f(x) = \dfrac{2x}{x - 2}$

A vertical asymptote occurs at $x = 2$. The graph does not cross this vertical asymptote.

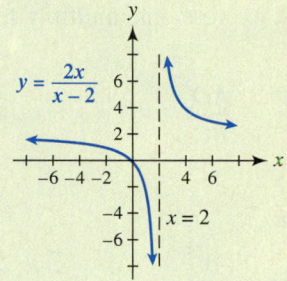

Rational Equations A rational equation contains one or more rational expressions. To solve a rational equation in the form $\frac{A}{B} = C$, start by multiplying each side of the equation by B to obtain $A = BC$. This step clears fractions from the equation.

Example: To solve $\frac{2x}{x-3} = 8$, start by multiplying each side by $x - 3$.

$$(x - 3)\frac{2x}{x - 3} = (x - 3)8$$

$$2x = 8x - 24$$

$$x = 4 \qquad\qquad \text{Check your answer.}$$

Operations on Functions If $f(x)$ and $g(x)$ are both defined, then the sum, difference, product, and quotient of two functions f and g are defined by

$$(f + g)(x) = f(x) + g(x) \qquad\qquad \text{Sum}$$

$$(f - g)(x) = f(x) - g(x) \qquad\qquad \text{Difference}$$

$$(fg)(x) = f(x) \cdot g(x) \qquad\qquad \text{Product}$$

$$\left(\frac{f}{g}\right)(x) = \frac{f(x)}{g(x)}, \text{ where } g(x) \neq 0. \quad \text{Quotient}$$

Example: Let $f(x) = x^2 - 1$ and $g(x) = x^2 + 1$.

$$(f + g)(x) = f(x) + g(x) = (x^2 - 1) + (x^2 + 1) = 2x^2$$

$$(f - g)(x) = f(x) - g(x) = (x^2 - 1) - (x^2 + 1) = -2$$

$$(fg)(x) = f(x) \cdot g(x) = (x^2 - 1)(x^2 + 1) = x^4 - 1$$

$$\left(\frac{f}{g}\right)(x) = \frac{f(x)}{g(x)} = \frac{x^2 - 1}{x^2 + 1}$$

SECTION 6.2 ■ MULTIPLICATION AND DIVISION OF RATIONAL EXPRESSIONS

Simplifying a Rational Expression To simplify a rational expression, factor the numerator and the denominator. Then apply the following principle.

$$\frac{PR}{QR} = \frac{P}{Q}$$

Example: $\dfrac{x^2 - 1}{x^2 + 2x + 1} = \dfrac{(x - 1)(x + 1)}{(x + 1)(x + 1)} = \dfrac{x - 1}{x + 1}$

Multiplying Rational Expressions To multiply two rational expressions, multiply numerators and multiply denominators. Simplify the result, if possible.

$$\frac{A}{B} \cdot \frac{C}{D} = \frac{AC}{BD} \qquad B \text{ and } D \text{ not zero}$$

Example: $\dfrac{3}{x - 1} \cdot \dfrac{x - 1}{x + 1} = \dfrac{3(x - 1)}{(x - 1)(x + 1)} = \dfrac{3}{x + 1}$

Dividing Rational Expressions To divide two rational expressions, multiply by the reciprocal of the divisor. Simplify the result, if possible.

$$\frac{A}{B} \div \frac{C}{D} = \frac{A}{B} \cdot \frac{D}{C} = \frac{AD}{BC} \qquad B, C, \text{ and } D \text{ not zero}$$

Example: $\dfrac{x + 1}{x^2 - 4} \div \dfrac{x + 1}{x - 2} = \dfrac{x + 1}{(x - 2)(x + 2)} \cdot \dfrac{x - 2}{x + 1}$

$$= \dfrac{(x + 1)(x - 2)}{(x - 2)(x + 2)(x + 1)} = \dfrac{1}{x + 2}$$

SECTION 6.3 ■ ADDITION AND SUBTRACTION OF RATIONAL EXPRESSIONS

Addition and Subtraction of Rational Expressions with Like Denominators To add (subtract) two rational expressions with like denominators, add (subtract) their numerators. The denominator does not change.

$$\frac{A}{C} + \frac{B}{C} = \frac{A + B}{C} \quad \text{and} \quad \frac{A}{C} - \frac{B}{C} = \frac{A - B}{C}$$

Examples: $\dfrac{2x}{2x + 1} + \dfrac{3x}{2x + 1} = \dfrac{5x}{2x + 1}$ and

$$\frac{x}{x^2 - 1} - \frac{1}{x^2 - 1} = \frac{x - 1}{x^2 - 1} = \frac{x - 1}{(x - 1)(x + 1)} = \frac{1}{x + 1}$$

Finding the Least Common Denominator The least common denominator (LCD) is the least common multiple (LCM) of the denominators.

Example: The LCD for $\dfrac{1}{x(x + 1)}$ and $\dfrac{1}{x^2}$ is $x^2(x + 1)$.

Addition and Subtraction of Rational Expressions with Unlike Denominators First write each rational expression with the LCD. Then add or subtract the numerators.

Examples: $\dfrac{2}{x + 1} - \dfrac{1}{x} = \dfrac{2x}{x(x + 1)} - \dfrac{x + 1}{x(x + 1)} = \dfrac{2x - (x + 1)}{x(x + 1)} = \dfrac{x - 1}{x(x + 1)}$

$$\frac{1}{x + 1} + \frac{2}{x - 1} = \frac{1(x - 1)}{(x + 1)(x - 1)} + \frac{2(x + 1)}{(x - 1)(x + 1)} = \frac{3x + 1}{(x - 1)(x + 1)}$$

SECTION 6.4 ■ RATIONAL EQUATIONS

Solving Rational Equations A first step in solving a rational equation is to multiply each side by the LCD of the rational expressions to clear fractions from the equation. *Be sure to check all answers.*

Example: The LCD for the equation $\frac{2}{x} + 1 = \frac{4 - x}{x}$ is x.

$$\frac{2}{x} + 1 = \frac{4 - x}{x} \qquad \text{Given equation}$$

$$x\left(\frac{2}{x} + 1\right) = \left(\frac{4 - x}{x}\right)x \qquad \text{Multiply by } x.$$

$$2 + x = 4 - x \qquad \text{Clear fractions.}$$

$$2x = 2 \qquad \text{Add } x \text{ and subtract 2.}$$

$$x = 1 \qquad \text{Divide by 2. Check your answer.}$$

SECTION 6.5 ■ COMPLEX FRACTIONS

Complex Fractions A complex fraction is a rational expression that contains fractions in its numerator, denominator, or both. The following equation can be used to help simplify basic complex fractions.

$$\frac{\dfrac{a}{b}}{\dfrac{c}{d}} = \frac{a}{b} \cdot \frac{d}{c} \qquad \text{"Invert and multiply."}$$

Example: $\dfrac{\dfrac{1 - x}{2}}{\dfrac{4}{1 + x}} = \dfrac{1 - x}{2} \cdot \dfrac{1 + x}{4} = \dfrac{(1 - x)(1 + x)}{(2)(4)} = \dfrac{1 - x^2}{8}$

Simplifying Complex Fractions

Method I: Combine terms in the numerator, combine terms in the denominator, and simplify the resulting expression.

Method II: Multiply the numerator and the denominator by the LCD and simplify the resulting expression.

Example: *Method I:*
$$\dfrac{2 - \dfrac{1}{b}}{2 + \dfrac{1}{b}} = \dfrac{\dfrac{2b - 1}{b}}{\dfrac{2b + 1}{b}} = \dfrac{2b - 1}{b} \cdot \dfrac{b}{2b + 1} = \dfrac{2b - 1}{2b + 1}$$

           Method II: The LCD is b.

$$\dfrac{\left(2 - \dfrac{1}{b}\right)b}{\left(2 + \dfrac{1}{b}\right)b} = \dfrac{2b - \dfrac{b}{b}}{2b + \dfrac{b}{b}} = \dfrac{2b - 1}{2b + 1}$$

SECTION 6.6 ■ MODELING WITH PROPORTIONS AND VARIATION

Proportions   A proportion is a statement (equation) that two ratios (fractions) are equal.

Example: $\frac{5}{x} = \frac{4}{7}$ (in words, 5 *is to x as* 4 *is to* 7.)

Similar Triangles   Two triangles are similar if the measures of their corresponding angles are equal. Corresponding sides of similar triangles are proportional.

Example: A right triangle has legs with lengths 3 and 4. A similar right triangle has a shorter leg with length 6. Its longer leg can be found by solving the proportion $\frac{3}{6} = \frac{4}{x}$, or $x = 8$.

Direct Variation   A quantity y is *directly proportional* to a quantity x, or y *varies directly* with x, if there is a nonzero constant k such that $y = kx$. The number k is called the *constant of proportionality* or the *constant of variation*.

Example: If y varies directly with x, then the ratio $\frac{y}{x}$ always equals k. The following data satisfy $\frac{y}{x} = 4$, so the constant of variation is 4. Thus $y = 4x$.

Direct Variation

x	1	2	3	4
y	4	8	12	16

$\frac{4}{1} = \frac{8}{2} = \frac{12}{3} = \frac{16}{4} = 4$, so $k = 4$.

Inverse Variation   A quantity y is *inversely proportional* to a quantity x, or y *varies inversely* with x, if there is a nonzero constant k such that $y = \frac{k}{x}$. (We assume that $k > 0$.)

Example: If y varies inversely with x, then $xy = k$. The following data satisfy $xy = 12$, so the constant of variation is 12. Thus $y = \frac{12}{x}$.

Inverse Variation

x	1	2	4	6
y	12	6	3	2

$1 \cdot 12 = 2 \cdot 6 = 4 \cdot 3 = 6 \cdot 2 = 12$, so $k = 12$.

Joint Variation   The quantity z *varies jointly* with x and y if $z = kxy$, $k \neq 0$.

Example: The area A of a rectangle varies jointly with the width W and length L because $A = LW$. Note that $k = 1$ in this example.

Division of a Polynomial by a Monomial Divide the monomial in the denominator into *every* term of the numerator.

Example: $\dfrac{4x^4 - 8x^3 + 16x^2}{8x^2} = \dfrac{4x^4}{8x^2} - \dfrac{8x^3}{8x^2} + \dfrac{16x^2}{8x^2} = \dfrac{x^2}{2} - x + 2$

Division of a Polynomial by a Polynomial Division of polynomials is similar to long division of natural numbers.

Example: Divide $2x^3 - 3x + 3$ by $x - 1$. (Be sure to include $\mathbf{0x^2}$.)

$$
\begin{array}{r}
2x^2 + 2x - 1 \\
x - 1 \overline{)2x^3 + \mathbf{0x^2} - 3x + 3} \\
\underline{2x^3 - 2x^2} \\
2x^2 - 3x \\
\underline{2x^2 - 2x} \\
-x + 3 \\
\underline{-x + 1} \\
2
\end{array}
$$

The quotient is $2x^2 + 2x - 1$ with remainder 2, which can be written as

$$2x^2 + 2x - 1 + \frac{2}{x - 1}.$$

Synthetic Division Synthetic division is a fast way to divide a polynomial by an expression in the form $x - k$, where k is a constant.

Example: Divide $3x^3 + 4x^2 - 7x - 1$ by $x + 2$.

$$
\begin{array}{r|rrrr}
-2 & 3 & 4 & -7 & -1 \\
 & & -6 & 4 & 6 \\
\hline
 & 3 & -2 & -3 & 5
\end{array}
$$

The quotient is $3x^2 - 2x - 3$ with remainder 5.

CHAPTER 6   Review Exercises

SECTION 6.1

Exercises 1 and 2: Write a symbolic representation (or formula) for the rational function f described.

1. Divide 1 by the quantity x minus 1.

2. Subtract 3 from x and then divide the result by x.

3. Sketch a graph of $f(x) = \frac{1}{2x + 4}$. Show any vertical asymptotes as dashed lines. State the domain of f in set-builder notation.

4. Evaluate $f(x) = \frac{1}{x^2 - 1}$ at $x = 3$. Find any x-values that are not in the domain of f.

Exercises 5 and 6: If possible, evaluate $f(x)$ for each x.

5. $f(x) = \dfrac{3}{x + 2}$ $\quad x = -3, x = 2$

6. $f(x) = \dfrac{2x}{x^2 - 4}$ $\quad x = -2, x = 3$

Exercises 7 and 8: Use the graph of f to evaluate the expressions. Give the equation of any vertical asymptotes.

7. $f(0)$ and $f(2)$

8. $f(-3)$ and $f(-2)$

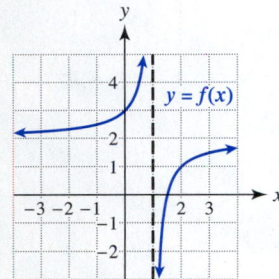

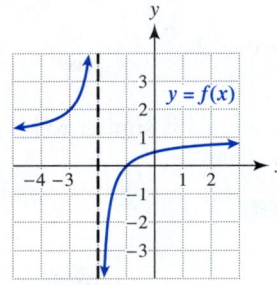

Exercises 9–12: Solve. Check your answer.

9. $\dfrac{3-x}{x} = 2$

10. $\dfrac{1}{x+3} = \dfrac{1}{5}$

11. $\dfrac{x}{x-2} = \dfrac{2x-2}{x-2}$

12. $\dfrac{4}{2-3x} = -1$

13. Use $f(x) = 2x^2 - 3x$ and $g(x) = 2x - 3$ to find each of the following.
   (a) $(f+g)(3)$ **(b)** $(fg)(3)$

14. Use $f(x) = x^2 - 1$ and $g(x) = x - 1$ to find each of the following.
   (a) $(f-g)(x)$ **(b)** $(f/g)(x)$

SECTION 6.2

Exercises 15–20: Simplify the expression.

15. $\dfrac{4a}{6a^4}$

16. $\dfrac{(x-3)(x+2)}{(x+1)(x-3)}$

17. $\dfrac{x^2-4}{x-2}$

18. $\dfrac{x^2-6x-7}{2x^2-x-3}$

19. $\dfrac{8-x}{x-8}$

20. $-\dfrac{3-2x}{2x-3}$

Exercises 21–30: Simplify the expression.

21. $\dfrac{1}{2y} \cdot \dfrac{4y^2}{8}$

22. $\dfrac{x+2}{x-5} \cdot \dfrac{x-5}{x+1}$

23. $\dfrac{x^2+1}{x^2-1} \cdot \dfrac{x-1}{x+1}$

24. $\dfrac{x^2+2x+1}{x^2-9} \cdot \dfrac{x+3}{x+1}$

25. $\dfrac{1}{3y} \div \dfrac{1}{9y^4}$

26. $\dfrac{2x+2}{3x-3} \div \dfrac{x+1}{x-1}$

27. $\dfrac{x^2+2x}{x^2-25} \div \dfrac{x+2}{x+5}$

28. $\dfrac{x^2+2x-15}{x^2+4x+3} \div \dfrac{x-3}{x+1}$

29. $\dfrac{x^2+x-2}{2x^2-7x+3} \cdot \dfrac{x^2-3x}{x^2+2x-3}$

30. $\dfrac{x^2+3x+2}{x^2+7x+12} \div \dfrac{x^2+4x+4}{x^2+4x+3}$

SECTION 6.3

Exercises 31–36: Find the least common multiple.

31. $36, 24, 16$

32. $4ab, a^2b$

33. $9x^2y, 6xy^3$

34. $(x-1), (x+2)$

35. $x^2-9, x(x+3)$

36. $x^2, x-3, x^2-6x+9$

Exercises 37–50: Simplify.

37. $\dfrac{1}{x+4} + \dfrac{3}{x+4}$

38. $\dfrac{2}{x} + \dfrac{x-3}{x}$

39. $\dfrac{1}{x+1} - \dfrac{x}{x+1}$

40. $\dfrac{2x}{x-2} - \dfrac{2}{x-2}$

41. $\dfrac{4}{1-t} + \dfrac{t}{t-1}$

42. $\dfrac{2}{y-2} - \dfrac{2}{y+2}$

43. $\dfrac{4b}{a^2c} - \dfrac{3a}{b^2c}$

44. $\dfrac{r}{5t^2} + \dfrac{t}{5r^2}$

45. $\dfrac{4}{a^2-b^2} - \dfrac{2}{a+b}$

46. $\dfrac{a}{a-b} + \dfrac{b}{a+b}$

47. $\dfrac{1}{x^2-3x+2} + \dfrac{1}{x^2+x-2}$

48. $\dfrac{1}{x^2-5x+6} - \dfrac{1}{x^2+x-6}$

49. $\dfrac{1}{x-2} + \dfrac{2}{x+2} - \dfrac{x}{x-2}$

50. $\dfrac{2x}{2x-1} - \dfrac{3}{2x+1} - \dfrac{1}{2x-1}$

SECTION 6.4

Exercises 51–60: Solve. Check your result.

51. $\dfrac{4}{x} - \dfrac{5}{2x} = \dfrac{1}{2}$

52. $\dfrac{1}{x-4} = \dfrac{3}{2x-1}$

53. $\dfrac{2}{x^2} - \dfrac{1}{x} = 1$

54. $\dfrac{1}{x+4} - \dfrac{1}{x} = 1$

55. $\dfrac{1}{x^2 - 1} - \dfrac{1}{x-1} = \dfrac{2}{3}$

56. $\dfrac{1}{x-3} + \dfrac{1}{x+3} = \dfrac{-5}{x^2 - 9}$

57. $\dfrac{1}{(x-2)^2} - \dfrac{1}{x^2 - 4} = \dfrac{2}{(x-2)^2}$

58. $\dfrac{1}{x^2 + 3x + 2} + \dfrac{x}{x+2} = \dfrac{1}{2}$

59. $\dfrac{-3x - 3}{x-3} + \dfrac{x^2 + x}{x-3} = 7x + 7$

60. $\dfrac{1}{x-4} + \dfrac{x}{x+4} = \dfrac{8}{x^2 - 16}$

Exercises 61 and 62: Use the graph to solve the rational equation. Then check your answer.

61. $\dfrac{1}{x} = \dfrac{2}{x+2}$

62. $\dfrac{3}{x-1} = \dfrac{1}{2}x$

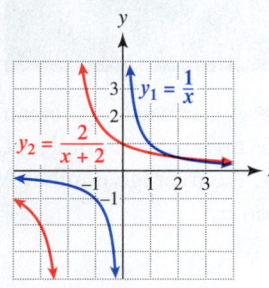
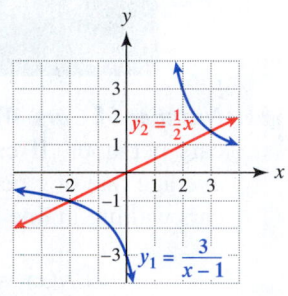

Exercises 63–66: Solve for the specified variable.

63. $m = \dfrac{y_2 - y_1}{x_2 - x_1}$ for y_2

64. $T = \dfrac{a}{b+2}$ for a

65. $\dfrac{1}{f} = \dfrac{1}{p} + \dfrac{1}{q}$ for p

66. $I = \dfrac{2a + 3b}{ab}$ for b

SECTION 6.5

Exercises 67–74: Simplify the complex fraction.

67. $\dfrac{\frac{3}{5}}{\frac{10}{13}}$

68. $\dfrac{\frac{4}{ab}}{\frac{2}{bc}}$

69. $\dfrac{\frac{2n-1}{n}}{\frac{3n}{2n+1}}$

70. $\dfrac{\frac{1}{x-y}}{\frac{1}{x^2 - y^2}}$

71. $\dfrac{2 + \frac{3}{x}}{2 - \frac{3}{x}}$

72. $\dfrac{\frac{1}{x} + \frac{1}{2}}{\frac{1}{4} - \frac{2}{x}}$

73. $\dfrac{\frac{4}{x} + \frac{1}{x-1}}{\frac{1}{x} - \frac{2}{x-1}}$

74. $\dfrac{\frac{1}{x+3} - \frac{1}{x-3}}{\frac{4}{x+3} - \frac{2}{x-3}}$

SECTION 6.6

Exercises 75–78: Solve the proportion.

75. $\dfrac{x}{6} = \dfrac{6}{20}$

76. $\dfrac{11}{x} = \dfrac{5}{7}$

77. $\dfrac{x+1}{5} = \dfrac{x}{3}$

78. $\dfrac{3}{7} = \dfrac{4}{x-1}$

79. If $x + 1$ is to 5 as 10 is to 15, use a proportion to find the value of x.

80. A rectangle has sides with lengths 7 and 8. Find the longer side of a similar rectangle whose shorter side has length 11.

81. Suppose that y varies directly with x. If $y = 8$ when $x = 2$, find y when $x = 7$.

82. Suppose that y varies inversely with x. If $y = 5$ when $x = 10$, find y when $x = 25$.

83. Suppose that z varies jointly with x and y. If $z = 483$ when $x = 23$ and $y = 7$, find the constant of variation k.

84. Suppose that z varies jointly with x and the square of y. If $z = 891$ when $x = 22$ and $y = 3$, find z when $x = 10$ and $y = 4$.

Exercises 85 and 86: Use the table to determine whether the data represent direct or inverse variation. Find an equation that models the data.

85.

x	2	4	5	8
y	100	50	40	25

86.

x	3	7	8	11
y	9	21	24	33

Exercises 87 and 88: Decide if the graph represents direct or inverse variation. Find the constant of variation k.

87.

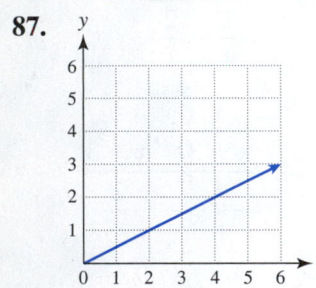

88.

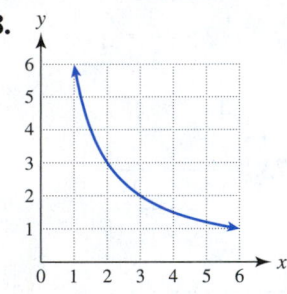

SECTION 6.7

Exercises 89–96: Divide.

89. $\dfrac{10x + 15}{5}$

90. $\dfrac{2x^2 + x}{x}$

91. $(4x^3 - x^2 + 2x) \div (2x^2)$

92. $(4a^3b - 6ab^2) \div (2a^2b^2)$

93. $\dfrac{2x^2 - x - 2}{x + 1}$

94. $(x^3 + x - 1) \div (x + 1)$

95. $(6x^3 - 7x^2 + 5x - 1) \div (3x - 2)$

96. $\dfrac{2x^3 - 9x^2 + 21x - 21}{x^2 - 2}$

Exercises 97 and 98: Use synthetic division to divide.

97. $\dfrac{2x^2 - 11x + 13}{x - 3}$

98. $(3x^3 + 10x^2 - 4x + 19) \div (x + 4)$

APPLICATIONS

99. *Time Spent in Line* Suppose that amusement park attendants can wait on 10 vehicles per minute and vehicles are arriving at the park randomly at an average rate of x vehicles per minute. Then the average time T in minutes spent waiting in line *and* paying the attendant is given by

$$T(x) = \frac{1}{10 - x},$$

where $x < 10$. (*Source:* N. Garber.)

(a) Construct a table of T, starting at $x = 9$ and incrementing by 0.1.

(b) What happens to the waiting time as the value of x approaches 10? Interpret this result.

100. *Downhill Highway Grade* The braking distance D in feet for a car traveling downhill at 40 miles per hour on a wet grade x is given by

$$D(x) = \frac{1600}{9.6 - 30x}.$$

(*Source:* N. Garber.)

(a) In this formula a level road is represented by $x = 0$ and a 10% downhill grade is represented by $x = 0.1$. Evaluate $D(0)$ and $D(0.1)$. How much does the downhill grade add to the braking distance compared to a level road?

(b) If the braking distance is 200 feet, estimate the downhill grade x.

101. *Number of Cars Waiting* A car wash can clean 15 cars per hour. If cars are arriving randomly at an average rate of x per hour, the average number N of cars waiting in line is given by

$$N(x) = \frac{x^2}{225 - 15x},$$

where $x < 15$. (*Source:* N. Garber.)

(a) Estimate the average length of a line when 14 cars per hour are arriving.

(b) A graph of $y = N(x)$ is shown in the figure. Interpret the graph as x increases. Does it agree with your intuition?

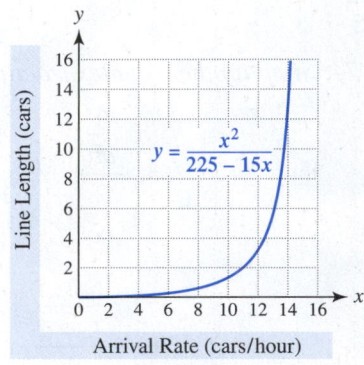

102. *Fish Population* Suppose that a fish population P in thousands in a lake after x years is modeled by

$$P(x) = \frac{5}{x^2 + 1}.$$

(a) Evaluate $P(0)$ and interpret the result.
(b) Graph P in $[0, 6, 1]$ by $[0, 6, 1]$.
(c) What happens to the population over this 6-year period?
(d) When was the population 1000?

103. *Working Together* Suppose that two students are working collaboratively to solve a large number of quadratic equations. The first student can solve all the problems in 2 hours, whereas the second student can solve them in 3 hours.

(a) Write an equation whose solution gives the time needed to solve all the problems if the two students work together and share answers.
(b) Solve the equation in part (a) symbolically.
(c) Solve the equation in part (a) either graphically or numerically.

104. *Speed of a Boat* In still water a riverboat can travel 12 miles per hour. It travels 48 miles upstream and then 48 miles downstream in a total time of 9 hours. Find the speed of the current.

105. *Recording Music* A 650-megabyte CD can record 74 minutes of music. Estimate the number of minutes that can be recorded on 387 megabytes.

106. *Height of a Building* A 5-foot-tall person casts a 3-foot-long shadow, while a nearby building casts a 26-foot-long shadow. Find the height of the building.

107. *Light Bulbs* The approximate resistances R for light bulbs of wattage W are measured and recorded in the following table. (*Source:* D. Horn, *Basic Electronics Theory.*)

W (watts)	50	100	200	250
R (ohms)	242	121	60.5	48.4

(a) Make a scatterplot of the data. Do the data represent direct or inverse variation?
(b) Find an equation that models the data. What is the constant of proportionality k?
(c) Find R for a 55-watt light bulb.

108. *Scales* The distance D that the spring in a produce scale stretches is directly proportional to the weight W of the fruits and vegetables placed in the pan.

(a) If 5 pounds of apples stretch the spring 0.3 inch, find an equation that relates D and W. What is the constant of proportionality k?
(b) How far will 7 pounds of oranges stretch the spring?

109. *Air Temperature and Altitude* In the first 30,000 feet of Earth's atmosphere the *change* in air temperature is directly proportional to the altitude. If the temperature is $80°F$ on the ground and $62°F$ 4000 feet above the ground, find the air temperature at 6000 feet. (*Source:* L. Battan, *Weather in Your Life.*)

110. *Skin Cancer and UV Radiation* Depletion of the ozone layer has caused an increase in the amount of UV radiation reaching Earth's surface. The following table shows the estimated percent increase in skin cancer y for an x percent increase in the amount of UV radiation reaching Earth's surface. (*Source:* R. Turner, D. Pearce, and I. Bateman, *Environmental Economics.*)

x (%)	0	1	2	3	4
y (%)	0	3.5	7	10.5	14

(a) Do these data represent direct variation, inverse variation, or neither?
(b) Find an equation that models the data.
(c) Estimate the percent increase in skin cancer if UV radiation increases by 2.3%.

CHAPTER 6   Test

 Step-by-step test solutions are found on the Chapter Test Prep Videos available in **MyMathLab** and on YouTube (search "RockswoldInterAlg" and click on "Channels").

1. Write a symbolic representation (formula) for $f(x)$ that divides x by the quantity x plus 2.

2. Let $f(x) = \frac{1}{4x^2 - 1}$.
(a) Evaluate $f(-2)$.
(b) Write the domain of f in set-builder notation.

3. Graph $y = \frac{1}{x + 2}$. Show any vertical asymptotes as dashed lines.

Exercises 4–6: Simplify the expression.

4. $\dfrac{2a^3}{4a^2}$

5. $\dfrac{1 - 2t}{2t - 1}$

6. $\dfrac{x^2 - 2x - 15}{2x^2 - x - 21}$

Exercises 7–15: Simplify.

7. $\dfrac{x^2 + 4}{x^2 - 4} \cdot \dfrac{x - 2}{x + 2}$

8. $\dfrac{1}{4y^2} \div \dfrac{1}{8y^4}$

9. $\dfrac{x}{x + 5} + \dfrac{1 - x}{x + 5}$

10. $\dfrac{2x}{x - 2} - \dfrac{1}{x + 2}$

11. $\dfrac{a^2}{3b} - \dfrac{2b^3}{5a}$

12. $\dfrac{x}{x - 2} - \dfrac{4}{x^2} - \dfrac{2}{x}$

13. $\dfrac{3 + \frac{3}{x}}{3 - \frac{3}{x}}$

14. $\dfrac{1}{z + 4} - \dfrac{z}{(z + 4)^2}$

15. $\dfrac{\dfrac{1}{x - 2} + \dfrac{x}{x - 2}}{\dfrac{1}{3} - \dfrac{5}{x - 2}}$

Exercises 16–21: Solve. Check your result.

16. $\dfrac{t}{5t + 1} = \dfrac{2}{7}$

17. $\dfrac{x}{2x - 1} = \dfrac{x + 2}{x + 4}$

18. $\dfrac{x + 4}{2 - x} - \dfrac{2x - 1}{2 - x} = 0$

19. $\dfrac{5}{x + 1} = 4$

20. $\dfrac{1}{x + 2} + \dfrac{1}{x - 2} = \dfrac{4}{x^2 - 4}$

21. $\dfrac{1}{x^2 - 4} - \dfrac{1}{x - 2} = \dfrac{1}{x + 2}$

22. Solve $F = \dfrac{Gm}{r^2}$ for m.

23. 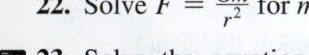 Solve the equation $\dfrac{\sqrt{2}}{x^2 + 2} - \dfrac{1.7x}{x - 1.2} = \sqrt{3}$ to the nearest hundredth.

24. Let $f(x) = x - 1$ and $g(x) = x^2 - 2x + 1$. Find $(f - g)(2)$ and $(f/g)(x)$.

25. A triangle has sides with lengths 12, 15, and 20. Find the longest side of a similar triangle with a shortest side of length 7.

26. Suppose that y varies directly with x. If $y = 8$ when $x = 23$, find y when $x = 10$.

27. Use the table to determine whether the data represent direct or inverse variation. Find, an equation that models the data.

x	2	4	5	10
y	50	25	20	10

28. Determine whether the data represent direct or inverse variation. Find an equation that models the data.

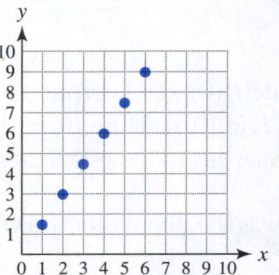

Exercises 29 and 30: Divide.

29. $\dfrac{4a^3 + 10a}{2a}$

30. $\dfrac{3x^3 + 5x^2 - 2}{x + 2}$

31. *Height of a Building* A person 73 inches tall casts a shadow 50 inches long while a nearby building casts a shadow 15 feet long. Find the height of the building.

32. *Time Spent in Line* Suppose that parking lot attendants can wait on 25 vehicles per minute and vehicles are arriving at the lot randomly at an average rate of x vehicles per minute. Then the average time T in minutes spent waiting in line *and* paying the attendant is given by

$$T(x) = \dfrac{1}{25 - x},$$

where $x < 25$. (*Source:* N. Garber.)

(a) Graph T in $[0, 25, 5]$ by $[0, 2, 0.5]$. Identify any vertical asymptotes.

(b) If the wait is 1 minute, how many vehicles are arriving on average?

33. *Working Together* Suppose that one pump can empty a pool in 24 hours, and a second pump can empty the pool in 30 hours.

(a) Write an equation whose solution gives the time needed for the pumps working together to empty the pool.

(b) Solve the equation in part (a).

34. *Dew Point and Altitude* In the first 30,000 feet of Earth's atmosphere the change in the dew point is directly proportional to the altitude. If the dew point is 50°F on the ground and 39°F 10,000 feet above the ground, find the dew point at 7500 feet. (*Source:* L. Battan, *Weather in Your Life*.)

CHAPTER 6   Extended and Discovery Exercises

OTHER TYPES OF DIRECT VARIATION

Exercises 1–4: Sometimes a quantity y varies directly with a power of x. For example, the area A of a circle varies directly with the second power of the radius r, since $A = \pi r^2$ and π is a constant. Let x and y denote two quantities and n be a positive number. Then y is directly proportional to the nth power of x, or y varies directly with the nth power of x, if there exists a nonzero number k such that

$$y = kx^n.$$

1. Let y be directly proportional to the second power of x. If x doubles, what happens to y?

2. Let y be directly proportional to the $\frac{1}{2}$ power of x. If x quadruples, what happens to y?

3. *Allometric Growth* If x is the weight of a fiddler crab and y is the weight of its claws, then y is directly proportional to the 1.25 power of x; that is, $y = kx^{1.25}$. Suppose that a typical crab with a body weight of 2.1 grams has claws weighing 1.125 grams. (*Source:* D. Brown and P. Rothery, *Models in Biology.*)

    (a) Find the constant of proportionality k. Round your answer to the nearest thousandth.

    (b) Estimate the weight of a fiddler crab with claws weighing 0.8 gram.

4. *Volume* The volume V of a cylinder is directly proportional to the square of its radius r. If a cylinder with a radius of 5 inches has a volume of 150 cubic inches, what is the volume of a cylinder with the same height and a radius of 7 inches?

OTHER TYPES OF INVERSE VARIATION

Exercises 5–8: Sometimes a quantity varies inversely with a power of x. Let x and y denote two quantities and n be a positive number. Then y is inversely proportional to the nth power of x, or y varies inversely with the nth power of x, if there exists a nonzero number k such that

$$y = \frac{k}{x^n}.$$

5. Let y be inversely proportional to the second power of x. If x doubles, what happens to y?

6. Let y be inversely proportional to the third power of x. If x doubles, what happens to y?

7. *Earth's Gravity* The weight W of an object can be modeled by $W = \frac{k}{d^2}$, where d is the distance that the object is from Earth's center and k is a constant. Earth's radius is about 4000 miles.

    (a) Find k for a person who weighs 200 pounds on Earth's surface.

    (b) Graph W in a convenient viewing rectangle. At what distance from Earth's center is this person's weight 50 pounds?

    (c) How far from the center of Earth would an object be if its weight were 1% of its weight on the surface of Earth?

8. *Modeling Brightness* Inverse variation occurs when the intensity of a light is measured. If you increase your distance from a light bulb, the intensity of the light decreases. Intensity I is inversely proportional to the second power of the distance d. The equation $I = \frac{k}{d^2}$ models this phenomenon. The following table gives the intensity of a 100-watt light bulb at various distances. (*Source:* R. Weidner and R. Sells, *Elementary Classical Physics, Vol. 2.*)

d (meters)	0.5	2	3	4
I (watts/square meter)	31.68	1.98	0.88	0.495

    (a) Find the constant of proportionality k.

    (b) Graph I in [0, 5, 1] by [0, 30, 5]. What happens to the intensity as the distances increase?

    (c) If the distance from the light bulb doubles, what happens to the intensity?

    (d) Determine d when $I = 1$ watt per square meter.

CHAPTERS 1–6   Cumulative Review Exercises

Exercises 1 and 2: Evaluate the formula for the given value of the variable.

1. $y = \sqrt{74 + t}$ $t = 7$

2. $r = 16 - w^2$ $w = -4$

3. Classify each number as one or more of the following: natural number, whole number, integer, rational number, or irrational number.

$$-\frac{12}{4}, 0, \sqrt{3}, 1, 2.\overline{11}, \frac{13}{2}$$

4. Select the formula that best models the data.
   a. $y = 2x - 5$ **b.** $y = x - 7$ **c.** $y = 2x + 3$

x	−2	−1	0	1	2
y	−9	−7	−5	−3	−1

Exercises 5 and 6: Simplify the expression. Write the result using positive exponents.

5. $\dfrac{18x^{-2}y^3}{3x^2y^{-3}}$

6. $\left(\dfrac{2c^2}{3d^3}\right)^{-2}$

7. Write 67,300,000,000 in scientific notation.

8. Identify the domain and range of the relation given by $S = \{(-3, 5), (0, 1), (-1, -2), (4, 0)\}$.

Exercises 9 and 10: Evaluate the given formula for $x = -2, -1, 0, 1,$ and 2. Plot the resulting ordered pairs.

9. $y = 3x - 1$

10. $y = \dfrac{4 - x^2}{2}$

11. Sketch the graph of $f(x) = x^2 + 2$ by hand.

12. Find the domain of $f(x) = \dfrac{5}{x - 1}$.

13. Use the table to write the formula for $f(x) = ax + b$.

x	−2	−1	0	1	2
$f(x)$	−5	−3	−1	1	3

14. Sketch a graph of $f(x) = -2$.

15. Find the slope and the y-intercept of the graph of $f(x) = -3x + 2$.

16. Calculate the slope of the line passing through the points $(6, -3)$ and $(2, 9)$.

17. Sketch the graph of a line passing through the point $(3, -1)$ with slope $m = -2$.

18. Use the graph to express the equation of the line in slope–intercept form.

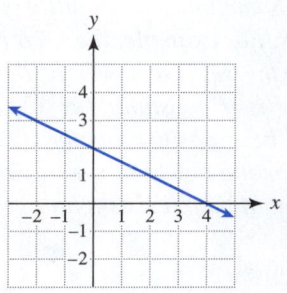

Exercises 19 and 20: Write the slope–intercept form for a line satisfying the given conditions.

19. Parallel to $y = 2x + 3$, passing through $(1, 4)$

20. Perpendicular to $y = \frac{2}{3}x - 2$, passing through $(2, 1)$

Exercises 21 and 22: Solve the equation.

21. $\dfrac{2}{5}(x + 1) - 6 = -4$

22. $\dfrac{1}{4}\left(\dfrac{t - 5}{3}\right) - 6 = \dfrac{2}{3}t - (3t + 7)$

23. Use the graph to solve the equation $y_1 = y_2$.

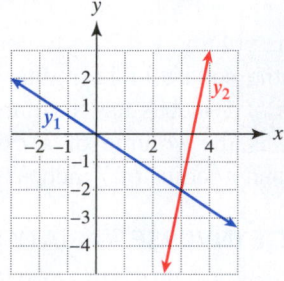

Exercises 24 and 25: Solve the inequality symbolically. Write the solution set in interval notation.

24. $\dfrac{2x + 3}{4} \le \dfrac{1}{3}$

25. $\dfrac{1}{2}z - 5 > \dfrac{3}{4}z - (2z + 5)$

Exercises 26 and 27: Solve the compound inequality. Graph the solution set on a number line.

26. $-6 \le -\dfrac{2}{3}x - 4 < -2$

27. $3x - 1 \le 5$ or $2x + 5 > 13$

28. Solve the absolute value equation $\left|\frac{1}{3}x + 6\right| = 4$.

Exercises 29 and 30: Solve the absolute value inequality. Write the solution set in interval notation.

29. $|2x - 3| < 11$ **30.** $-3|t - 5| \leq -18$

31. Determine which ordered pair is a solution to the system of equations: $(2, -6)$, $(-8, -1)$.
$$x + 2y = -10$$
$$3x - 10y = -14$$

32. Shade the solution set in the xy-plane.
$$x + y < 4$$
$$x - 2y \geq 1$$

Exercises 33 and 34: Solve the system of equations.

33. $2x - 8y = 5$ **34.** $2x - 3y = 12$
  $4x + 2y = 1$ $-x + 2y = -6$

35. Maximize the objective function R subject to the given constraints.
$$R = 2x + 3y$$
$$2x + y \leq 6$$
$$x + 2y \leq 6$$
$$x \geq 0, y \geq 0$$

36. Use elimination and substitution to solve the system of linear equations.
$$2x + 3y - z = 3$$
$$3x - y + 4z = 10$$
$$2x + y - 2z = -1$$

37. Write the system of linear equations as an augmented matrix. Then use Gauss–Jordan elimination to solve the system. Write the solution as an ordered triple.
$$x + y - z = 4$$
$$-x - y - z = 0$$
$$x - 2y + z = -9$$

38. Evaluate $\det A$ if $A = \begin{bmatrix} 4 & -2 \\ 1 & 3 \end{bmatrix}$.

39. Find the area of the triangle by using a determinant. Assume that the units are inches.

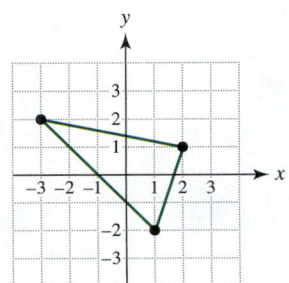

40. Use Cramer's rule to solve the system of equations.
$$6x + 7y = 8$$
$$-8x + 5y = 18$$

Exercises 41 and 42: Multiply the expressions.

41. $3x^2(x^3 + 5x - 2)$ **42.** $(2x - 5)(x + 3)$

43. Use factoring to solve $6x + 3x^2 = 0$.

44. Use grouping to factor $3a^3 - a^2 + 15a - 5$.

Exercises 45–48: Factor completely.

45. $4x^2 + 5x - 6$ **46.** $6x^3 - 9x^2 - 6x$

47. $9a^2 - 4b^2$ **48.** $64t^3 + 27$

Exercises 49 and 50: Solve the equation.

49. $3x^2 - 11x = 4$ **50.** $x^4 - x^3 - 30x^2 = 0$

Exercises 51–54: Simplify the expressions.

51. $\dfrac{x^2 + 3x - 10}{x^2 - 4} \cdot \dfrac{x - 2}{x + 5}$

52. $\dfrac{x^2 + 2x - 24}{x^2 + 3x - 18} \div \dfrac{x + 3}{x^2 - 9}$

53. $\dfrac{2}{t + 2} - \dfrac{t}{t^2 - 4}$ **54.** $\dfrac{4a}{3ab^2} + \dfrac{b}{a^2c}$

Exercises 55 and 56: Solve. Check your result.

55. $\dfrac{1}{x + 4} + \dfrac{1}{x - 4} = \dfrac{-7}{x^2 - 16}$

56. $\dfrac{1}{y^2 + 3y - 4} - \dfrac{y}{y - 1} = -1$

57. Solve $J = \dfrac{y + z}{z}$ for z.

58. Simplify the complex fraction.
$$\dfrac{\dfrac{4}{x^2} + \dfrac{1}{x}}{\dfrac{4}{x^2} - \dfrac{1}{x}}$$

59. Suppose that y varies inversely with x. If $y = 2$ when $x = 4$, find y when x is 16.

60. Simplify $(x^3 + 2x + 11) \div (x + 2)$.

Exercises 61 and 62: Let $f(x) = x^2 - 3x + 2$ and let $g(x) = x - 2$. Simplify the expression.

61. $(f + g)(3)$ **62.** $(f/g)(x)$

APPLICATIONS

63. *Temperature Scales* The accompanying table shows equivalent temperatures in degrees Fahrenheit and degrees Celsius.

°F	−40	32	59	95	212
°C	−40	0	15	35	100

(a) Plot the data. Let the *x*-axis correspond to the Fahrenheit temperature and the *y*-axis correspond to the Celsius temperature. What type of relation exists between the data?

(b) Find $f(x) = m(x - h) + k$ so that *f* receives the Fahrenheit temperature as input and outputs the corresponding Celsius temperature.

(c) If the temperature is 104° F, what is the equivalent temperature in degrees Celsius?

64. *Geometry* An isosceles triangle has two shorter sides of the same length and a longer side that is 7 inches longer than half of the length of either of the shorter sides. If the perimeter of the triangle is 22 inches, what are the lengths of the three sides?

65. *Burning Calories* During strenuous exercise, an athlete burns 690 calories per hour on a stair climber and 540 calories per hour on a stationary bicycle. During a 90-minute workout the athlete burns 885 calories. How much time (in hours) is spent on each type of exercise equipment? (*Source: Runner's World.*)

66. *Tickets* The price of admission to a county fair is $2 for children and $5 for adults. If a group of 30 people pays $78 to enter the fairgrounds, find the number of children and the number of adults in the group.

67. *Area of a Rectangle* A rectangle has an area of 165 square feet. Its length is 4 feet more than its width. Find the dimensions of the rectangle.

68. *Flight of a Ball* If a ball is thrown upward with a velocity of 44 feet per second (30 miles per hour), its height *h* in feet above the ground after *t* seconds can be modeled by

$$h(t) = -16t^2 + 44t.$$

(a) Determine when the ball strikes the ground.

(b) When does the ball reach a height of 18 feet?

69. *Working Together* An amateur painter can paint a room in 15 hours. A professional painter can paint the same room in 10 hours.

(a) Write an equation whose solution gives the time to paint the room if the painters work together.

(b) Solve the equation in part (a).

70. *Height of a Tree* A 6-foot-tall person casts a 4-foot-long shadow while a nearby tree casts a 32-foot-long shadow. Find the height of the tree.

7 Radical Expressions and Functions

Bear in mind that the wonderful things you learn in schools are the work of many generations, produced by enthusiastic effort and infinite labor in every country.

—ALBERT EINSTEIN

Throughout history, people have created (or discovered) new numbers. Often these new numbers were met with resistance and regarded as being imaginary or unreal. The number 0 was not invented at the same time as the natural numbers. There was no Roman numeral for 0, which is one reason why our calendar started with A.D. 1 and, as a result, the twenty-first century began in 2001. No doubt there were skeptics during the time of the Roman Empire who questioned why anyone needed a number to represent nothing. Negative numbers also met strong resistance. After all, how could anyone possibly have −6 apples?

In this chapter we describe a new number system called *complex numbers*, which involve square roots of negative numbers. The Italian mathematician Cardano (1501–1576) was one of the first mathematicians to work with complex numbers and called them useless. René Descartes (1596–1650) originated the term *imaginary number*, which is associated with complex numbers. Today complex numbers are used in many applications, such as electricity, fiber optics, and the design of airplanes. We are privileged to study in a period of days what took people centuries to discover.

Source: Historical Topics for the Mathematics Classroom, Thirty-first Yearbook, NCTM.

7.1   Radical Expressions and Functions

Radical Notation • The Square Root Function • The Cube Root Function

A LOOK INTO MATH ▶

Suppose you start a summer painting business, and initially you are the only employee. Because business is good, you hire another employee and productivity goes up, more than enough to pay for the new employee. As time goes on, you hire more employees. Eventually, productivity starts to level off because there are too many employees to keep busy. This situation is common in business and can be modeled by using a function involving a square root. (See Example 12.)

Radical Notation

SQUARE ROOTS Recall the definition of the square root of a number a.

NEW VOCABULARY

☐ Radical sign
☐ Radicand
☐ Radical expression
☐ *nth* root
☐ Index
☐ Odd root
☐ Even root
☐ Principal *nth* root
☐ Square root function
☐ Cube root function

> **SQUARE ROOT**
>
> The number b is a *square root* of a if $b^2 = a$.

EXAMPLE 1   **Finding square roots**

Find the square roots of 100.

Solution

The square roots of 100 are **10** *and* **−10** because **10**^{2} = 100 and (**−10**)^2 = 100.

Now Try Exercise 1

Every positive number a has two square roots: one positive and one negative. Recall that the *positive square root* is called the *principal square root* and is denoted $\sqrt{a}$. The *negative square root* is denoted $-\sqrt{a}$. To identify both square roots, we write $\pm\sqrt{a}$. The symbol $\pm$ is read "plus or minus." The symbol $\sqrt{}$ is called the **radical sign**. The expression under the radical sign is called the **radicand**, and an expression containing a radical sign is called a **radical expression**. Examples of radical expressions include

$$\sqrt{6}, \quad 5 + \sqrt{x + 1}, \quad \text{and} \quad \sqrt{\frac{3x}{2x - 1}}.$$

> **MAKING CONNECTIONS**
>
> **Expressions and Equations**
>
> Expressions and equations are *different* mathematical concepts. An expression does not contain an equals sign, whereas an equation *always* contains an equals sign. *An equation is a statement that two expressions are equal.* For example,
>
> $$\sqrt{x + 1} \quad \text{and} \quad \sqrt{5 - x} \qquad \text{Expressions}$$
>
> are two different expressions, and
>
> $$\sqrt{x + 1} = \sqrt{5 - x} \qquad \text{Equation}$$
>
> is an equation. We often *solve an equation*, but we *do not solve an expression*. Instead, we *simplify* and *evaluate* expressions.

In the next example we show how to find the principal square root of an expression.

EXAMPLE 2   **Finding principal square roots**

Evaluate each square root.

(a) $\sqrt{25}$ **(b)** $\sqrt{0.49}$ **(c)** $\sqrt{\frac{4}{9}}$ **(d)** $\sqrt{c^2}, c > 0$

Solution

(a) Because $5 \cdot 5 = 25$, the principal, or *positive*, square root of 25 is $\sqrt{25} = 5$.

(b) Because $(0.7)(0.7) = 0.49$, the principal square root of 0.49 is $\sqrt{0.49} = 0.7$.

(c) Because $\frac{2}{3} \cdot \frac{2}{3} = \frac{4}{9}$, the principal square root of $\frac{4}{9}$ is $\sqrt{\frac{4}{9}} = \frac{2}{3}$.

(d) The principal square root of c^2 is $\sqrt{c^2} = c$, as c is positive.

Now Try Exercises 15, 17, 19, 21

The square roots of many real numbers, such as $\sqrt{17}$, $\sqrt{1.2}$, and $\sqrt{\frac{5}{7}}$, cannot be conveniently evaluated (or approximated) by hand. In these cases we sometimes use a calculator to give a decimal *approximation*, as demonstrated in the next example.

EXAMPLE 3   **Approximating a square root**

Approximate $\sqrt{17}$ to the nearest thousandth.

Solution

Figure 7.1 shows that $\sqrt{17} \approx 4.123$, rounded to the nearest thousandth. This result means that $4.123 \times 4.123 \approx 17$.

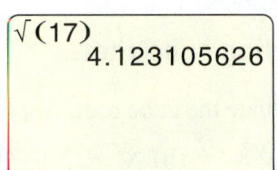

```
√(17)
          4.123105626
```

Figure 7.1

Now Try Exercise 39

SQUARE ROOTS OF NEGATIVE NUMBERS   The square root of a *negative* number is *not* a real number. For example, $\sqrt{-4} \neq 2$ because $2 \cdot 2 \neq -4$ and $\sqrt{-4} \neq -2$ because $(-2)(-2) \neq -4$. (Later in this chapter, we will use the complex numbers to identify square roots of negative numbers.)

READING CHECK

Are square roots of negative numbers real numbers? Explain.

▶ **REAL-WORLD CONNECTION**   Have you ever noticed that if you climb up a hill or a tower you can see farther to the horizon? This phenomenon can be described by a formula containing a square root, as demonstrated in the next example.

EXAMPLE 4 **Seeing the horizon**

A formula for calculating the distance d in miles that one can see to the horizon on a clear day is approximated by $d = 1.22\sqrt{x}$, where x is the elevation, in feet, of a person.
(a) Approximate how far a 6-foot-tall person can see to the horizon.
(b) Approximate how far a person can see from an 8000-foot mountain.

Solution
(a) Let $x = 6$ in the formula $d = 1.22\sqrt{x}$.

$$d = 1.22\sqrt{6} \approx 3, \text{ or about 3 miles.}$$

(b) If $x = 8000$, then $d = 1.22\sqrt{8000} \approx 109$, or about 109 miles.

Now Try Exercise 107

CUBE ROOTS The cube root of a number a is denoted $\sqrt[3]{a}$.

CALCULATOR HELP

To calculate a cube root, see Appendix A (page AP-1).

CUBE ROOT

The number b is a *cube root* of a if $b^3 = a$.

Although the square root of a negative number is *not* a real number, the cube root of a negative number is a negative real number. *Every real number has one real cube root.*

EXAMPLE 5 **Finding cube roots**

Evaluate the cube root. Approximate your answer to the nearest hundredth when appropriate.
(a) $\sqrt[3]{8}$ **(b)** $\sqrt[3]{-27}$ **(c)** $\sqrt[3]{\frac{1}{64}}$ **(d)** $\sqrt[3]{d^6}$ **(e)** $\sqrt[3]{16}$

Solution
(a) $\sqrt[3]{8} = 2$ because $2^3 = 2 \cdot 2 \cdot 2 = 8$.
(b) $\sqrt[3]{-27} = -3$ because $(-3)^3 = (-3)(-3)(-3) = -27$.
(c) $\sqrt[3]{\frac{1}{64}} = \frac{1}{4}$ because $\left(\frac{1}{4}\right)^3 = \frac{1}{4} \cdot \frac{1}{4} \cdot \frac{1}{4} = \frac{1}{64}$.
(d) $\sqrt[3]{d^6} = d^2$ because $(d^2)^3 = d^2 \cdot d^2 \cdot d^2 = d^{2+2+2} = d^6$.
(e) $\sqrt[3]{16}$ is not an integer. Figure 7.2 shows that $\sqrt[3]{16} \approx 2.52$.

Now Try Exercises 23, 25, 27, 41

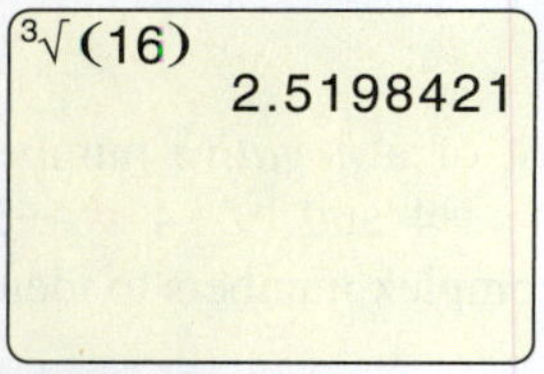

```
³√(16)
        2.5198421
```

Figure 7.2

NOTE: $\sqrt[3]{-b} = -\sqrt[3]{b}$ for any real number b. That is, the cube root of a negative is the negative of the cube root. For example, $\sqrt[3]{-8} = -\sqrt[3]{8} = -2$.

Table 7.1 illustrates how to evaluate both square roots and cube roots. If the radical expression is undefined, a dash is used.

TABLE 7.1 Radicals

Expression	$\sqrt{64}$	$-\sqrt{64}$	$\sqrt{-64}$	$\sqrt[3]{64}$	$-\sqrt[3]{64}$	$\sqrt[3]{-64}$
Evaluated	8	-8	—	4	-4	-4

*N*th ROOTS We can generalize square roots and cube roots to include *n*th roots of a number *a*. The number *b* is an **nth root** of *a* if $b^n = a$, where *n* is a positive integer. For example, $2^5 = 32$, so the 5th root of 32 is 2 and can be written as $\sqrt[5]{32} = 2$.

THE NOTATION $\sqrt[n]{a}$

The equation $\sqrt[n]{a} = b$ means that $b^n = a$, where *n* is a natural number called the **index**. If *n* is odd, we are finding an **odd root** and if *n* is even, we are finding an **even root**.

1. If $a > 0$, then $\sqrt[n]{a}$ is a positive number. $\sqrt[4]{16} = 2$ and is positive.
2. If $a < 0$ and
   (a) *n* is odd, then $\sqrt[n]{a}$ is a negative number. $\sqrt[3]{-8} = -2$ and is negative.
   (b) *n* is even, then $\sqrt[n]{a}$ is *not* a real number. $\sqrt[4]{-8}$ is undefined.

If $a > 0$ and *n* is even, then *a* has two real *n*th roots: one positive and one negative. The positive root is denoted $\sqrt[n]{a}$ and called the **principal *n*th root** of *a*. For example, $(-3)^4 = 81$ *and* $3^4 = 81$, but $\sqrt[4]{81} = 3$ in the same way *principal square roots* are calculated.

EXAMPLE 6   **Finding *n*th roots**

Find each root, if possible.

(a) $\sqrt[4]{16}$ **(b)** $\sqrt[5]{-32}$ **(c)** $\sqrt[4]{-81}$ **(d)** $-\sqrt[4]{81}$

Solution
(a) $\sqrt[4]{16} = 2$ because $2^4 = 2 \cdot 2 \cdot 2 \cdot 2 = 16$.
(b) $\sqrt[5]{-32} = -2$ because $(-2)^5 = (-2)(-2)(-2)(-2)(-2) = -32$.
(c) An *even* root of a *negative* number is *not* a real number.
(d) $-\sqrt[4]{81} = -3$ because $\sqrt[4]{81} = 3$.

Now Try Exercises 33, 35, 37

ABSOLUTE VALUE Consider the calculations

$$\sqrt{3^2} = \sqrt{9} = \mathbf{3}, \quad \sqrt{(-4)^2} = \sqrt{16} = \mathbf{4}, \quad \text{and} \quad \sqrt{(-6)^2} = \sqrt{36} = \mathbf{6}.$$

In general, the expression $\sqrt{x^2}$ equals $|x|$. Graphical support is shown in Figure 7.3, where the graphs of $Y_1 = \sqrt{(X\char`^2)}$ and $Y_2 = \text{abs}(X)$ appear to be identical.

CRITICAL THINKING

Evaluate $\sqrt[6]{(-2)^6}$ and $\sqrt[3]{(-2)^3}$. Now simplify $\sqrt[n]{x^n}$ when *n* is even and when *n* is odd.

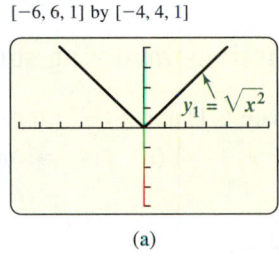

[−6, 6, 1] by [−4, 4, 1]   [−6, 6, 1] by [−4, 4, 1]

(a)   (b)

Figure 7.3

THE EXPRESSION $\sqrt{x^2}$

For every real number *x*, $\sqrt{x^2} = |x|$.

EXAMPLE 7 | **Simplifying expressions**

Write each expression in terms of an absolute value.

(a) $\sqrt{(-3)^2}$ (b) $\sqrt{(x+1)^2}$ (c) $\sqrt{z^2 - 4z + 4}$

Solution

(a) $\sqrt{x^2} = |x|$, so $\sqrt{(-3)^2} = |-3|$

(b) $\sqrt{(x+1)^2} = |x+1|$

(c) $\sqrt{z^2 - 4z + 4} = \sqrt{(z-2)^2} = |z-2|$

Now Try Exercises 45, 49, 51

The Square Root Function

The **square root function** is given by $f(x) = \sqrt{x}$. The domain of the square root function is all nonnegative real numbers because we have *not* defined the square root of a negative number. Table 7.2 lists three points that lie on the graph of $f(x) = \sqrt{x}$. In Figure 7.4 these points are plotted and the graph of $y = \sqrt{x}$ has been sketched. The graph does not appear to the left of the origin because $f(x) = \sqrt{x}$ is undefined for negative inputs.

TABLE 7.2

x	$\sqrt{x}$
0	0
1	1
4	2

Square Root Function

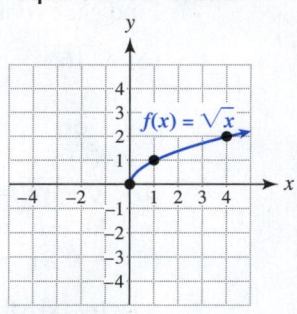

Figure 7.4

TECHNOLOGY NOTE

Square Roots of Negative Numbers

If a table of values for $y_1 = \sqrt{x}$ includes both negative and positive values for x, then many calculators give error messages when x is negative, as shown in the figure in the margin.

X	Y1
-9	ERROR
-4	ERROR
-1	ERROR
0	0
1	1
4	2
9	3

$Y_1 \blacksquare \sqrt{(X)}$

EXAMPLE 8 | **Evaluating functions involving square roots**

If possible, evaluate $f(1)$ and $f(-2)$ for each $f(x)$.

(a) $f(x) = \sqrt{2x - 1}$ (b) $f(x) = \sqrt{4 - x^2}$

Solution

(a) $f(1) = \sqrt{2(1) - 1} = \sqrt{1} = 1$

$f(-2) = \sqrt{2(-2) - 1} = \sqrt{-5}$, which does not equal a real number.

(b) $f(1) = \sqrt{4 - (1)^2} = \sqrt{3}$

$f(-2) = \sqrt{4 - (-2)^2} = \sqrt{0} = 0$

Now Try Exercises 61, 63

▶ **REAL-WORLD CONNECTION** A good punter can kick a football so that the ball has a long *hang time*. Hang time is the length of time that the ball is in the air, and a long hang time gives the kicking team time to run down the field and stop the punt return. By using a function involving a square root, we can estimate hang time.

EXAMPLE 9

Calculating hang time

If a football is kicked x feet high, then the time T in seconds that the ball is in the air is given by the function

$$T(x) = \frac{1}{2}\sqrt{x}.$$

(a) Find the hang time if the ball is kicked 50 feet into the air.
(b) Does the hang time double if the ball is kicked 100 feet in the air?

Solution
(a) The hang time is $T(50) = \frac{1}{2}\sqrt{50} \approx 3.5$ seconds.
(b) The hang time is $T(100) = \frac{1}{2}\sqrt{100} = 5$ seconds. The time does *not* double when the height doubles.

■ **Now Try Exercise 105**

CRITICAL THINKING

How high would a football have to be kicked to have twice the hang time of a football kicked 50 feet into the air?

EXAMPLE 10

Finding the domain of a square root function

Let $f(x) = \sqrt{x - 1}$.
(a) Find the domain of f. Write your answer in interval notation.
(b) Graph $y = f(x)$ and compare it to the graph of $y = \sqrt{x}$.

Solution
(a) For $f(x)$ to be defined, $x - 1$ cannot be negative. Thus valid inputs for x must satisfy

$$x - 1 \geq 0 \quad \text{or} \quad x \geq 1.$$

The domain is $[1, \infty)$.
(b) Table 7.3 lists points that lie on the graph of $y = \sqrt{x - 1}$. Note in Figure 7.5 that the graph appears only when $x \geq 1$. This graph is similar to $y = \sqrt{x}$ (see Figure 7.4) except that it is shifted one unit to the right.

TABLE 7.3

x	$\sqrt{x - 1}$
1	0
2	1
5	2

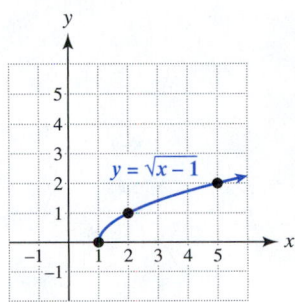

Figure 7.5

■ **Now Try Exercises 75, 89**

EXAMPLE 11

Finding the domains of square root functions

Find the domain of each function. Write your answer in interval notation.

(a) $f(x) = \sqrt{4 - 2x}$ **(b)** $g(x) = \sqrt{x^2 + 1}$

Solution

(a) To determine when $f(x) = \sqrt{4 - 2x}$ is defined, we must solve $4 - 2x \geq 0$.

$$4 - 2x \geq 0 \qquad \text{Inequality to be solved}$$
$$4 \geq 2x \qquad \text{Add } 2x \text{ to each side.}$$
$$2 \geq x \qquad \text{Divide each side by 2.}$$

The domain is $(-\infty, 2]$.

(b) Regardless of the value of x in $g(x) = \sqrt{x^2 + 1}$, the expression $x^2 + 1$ is always positive because $x^2 \geq 0$. Thus $g(x)$ is defined for all real numbers, and its domain is $(-\infty, \infty)$.

▮ **Now Try Exercises 83, 85**

MAKING CONNECTIONS

Domains of Functions and Their Graphs

In Example 11, the domains of f and g were found *symbolically*. Notice that the graph of f does not appear to the right of $x = 2$ because the domain of f is $(-\infty, 2]$, whereas the graph of g appears for all values of x because the domain of g is $(-\infty, \infty)$.

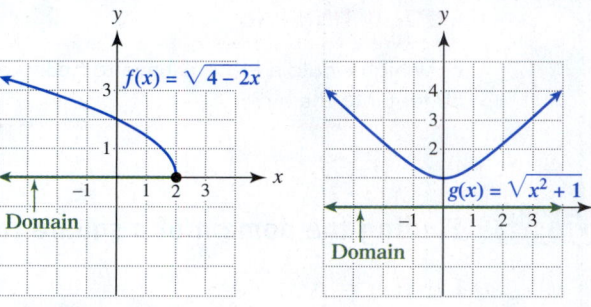

▶ **REAL-WORLD CONNECTION** (Refer to A Look Into Math at the beginning of this section.) In the next example, we analyze the benefit of adding employees to a small business.

EXAMPLE 12

Increasing productivity with more employees

The function $R(x) = 108\sqrt{x}$ gives the total revenue per year in thousands of dollars generated by a small business having x employees. A graph of $y = R(x)$ is shown in Figure 7.6.

Revenue as a Function of Employees

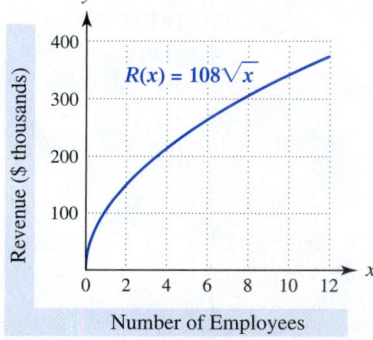

Figure 7.6

(a) Approximate values for $R(4)$, $R(8)$, and $R(12)$ by using both the formula and the graph.
(b) Evaluate $R(12) - R(8)$ and $R(8) - R(4)$ using the formula. Interpret your answer.

Solution

(a) $R(4) = 108\sqrt{4} = \$216$ thousand, $R(8) = 108\sqrt{8} \approx \$305$ thousand, and finally $R(12) = 108\sqrt{12} \approx \$374$ thousand.

We can graphically evaluate $R(4) \approx \$215$ thousand, $R(8) \approx \$305$ thousand, and $R(12) \approx \$375$ thousand, as shown in Figure 7.7.

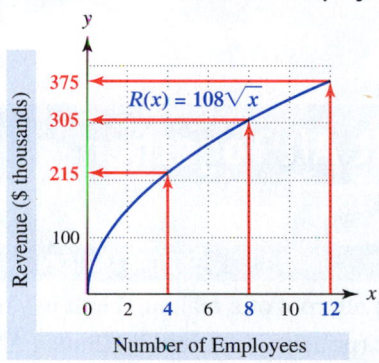

Revenue as a Function of Employees

Figure 7.7

(b) $R(8) - R(4) \approx 305 - 216 = \$89$ thousand
$R(12) - R(8) \approx 374 - 305 = \$69$ thousand

There is more revenue gained from 4 to 8 employees than there is from 8 to 12 employees. Because the graph of $R(x) = 108\sqrt{x}$ starts to level off, there is a limited benefit to adding employees.

Now Try Exercise 109

Cube Root Function

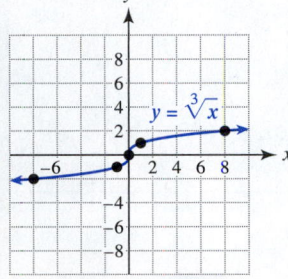

Figure 7.8

The Cube Root Function

We can define the **cube root function** by $f(x) = \sqrt[3]{x}$. Cube roots are defined for both positive and negative numbers, so *the domain of the cube root function includes all real numbers*. Table 7.4 lists points that lie on the graph of the cube root function. Figure 7.8 shows a graph of $y = \sqrt[3]{x}$.

TABLE 7.4 Cube Root Function

x	-27	-8	-1	0	1	8	27
$\sqrt[3]{x}$	-3	-2	-1	0	1	2	3

MAKING CONNECTIONS

Domains of the Square Root and Cube Root Functions

$f(x) = \sqrt{x}$ equals a real number for any nonnegative x. Thus $D = [0, \infty)$.
$f(x) = \sqrt[3]{x}$ equals a real number for any x. Thus $D = (-\infty, \infty)$.

Examples: $\sqrt{4} = 2$ but $\sqrt{-4}$ is *not* a real number. $\sqrt[3]{8} = 2$ and $\sqrt[3]{-8} = -2$.

EXAMPLE 13   **Evaluating functions involving cube roots**

Evaluate $f(1)$ and $f(-3)$ for each $f(x)$.

(a) $f(x) = \sqrt[3]{x^2 - 1}$ **(b)** $f(x) = \sqrt[3]{2 - x^2}$

Solution

(a) $f(\mathbf{1}) = \sqrt[3]{1^2 - 1} = \sqrt[3]{0} = 0;\ f(\mathbf{-3}) = \sqrt[3]{(-3)^2 - 1} = \sqrt[3]{8} = 2$

(b) $f(\mathbf{1}) = \sqrt[3]{2 - 1^2} = \sqrt[3]{1} = 1;\ f(\mathbf{-3}) = \sqrt[3]{2 - (-3)^2} = \sqrt[3]{-7}$ or $-\sqrt[3]{7}$

■ **Now Try Exercises 65, 69**

7.1   Putting It All Together

CONCEPT	EXPLANATION	EXAMPLES
nth Root of a Real Number	An nth root of a real number a is b if $b^n = a$, and the (principal) nth root is denoted $\sqrt[n]{a}$. If $a < 0$ and n is even, $\sqrt[n]{a}$ is not a real number.	The square roots of 25 are 5 and -5. The principal square root is $\sqrt{25} = 5$. $\sqrt[3]{-125} = -5$ because $(-5)^3 = -125$. $\sqrt[4]{-9}$ is not a real number.
Square Root and Cube Root Functions	$f(x) = \sqrt{x}$ and $g(x) = \sqrt[3]{x}$ The cube root function g is defined for all inputs, whereas the square root function f is defined only for nonnegative inputs. **Square Root Function** **Cube Root Function** Domain: $[0, \infty)$ Domain: $(-\infty, \infty)$	$f(64) = \sqrt{64} = 8$ $f(-64) = \sqrt{-64}$ is *not* a real number. $g(64) = \sqrt[3]{64} = 4$ $g(-64) = \sqrt[3]{-64} = -4$

7.1   Exercises

 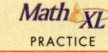

CONCEPTS AND VOCABULARY

1. What are the square roots of 9?

2. What is the principal square root of 9?

3. What is the cube root of 8?

4. Does every real number have a cube root?

5. If $b^n = a$ and $b > 0$, then $\sqrt[n]{a} = $ _____.

6. What is $\sqrt{x^2}$ equal to?

7. Evaluate $\sqrt{-25}$, if possible.

8. Evaluate $\sqrt[3]{-27}$, if possible.

9. Which of the following (a.–d.) equals -4?
   a. $\sqrt{16}$
   b. $\sqrt{-16}$
   c. $-\sqrt[3]{16}$
   d. $\sqrt[3]{-64}$

10. Which of the following (a.–d.) equals $|2x + 1|$?
    a. $\sqrt{(2x + 1)^2}$
    b. $\sqrt[3]{(2x + 1)^3}$
    c. $\left(\sqrt{2x + 1}\right)^2$
    d. $\sqrt{(2x)^2 + 1}$

11. Sketch a graph of the square root function.

12. Sketch a graph of the cube root function.

13. What is the domain of the square root function?

14. What is the domain of the cube root function?

RADICAL EXPRESSIONS

Exercises 15–38: Evaluate the expression by hand, if possible. Variables represent any real number.

15. $\sqrt{9}$

16. $\sqrt{121}$

17. $\sqrt{0.36}$

18. $\sqrt{0.64}$

19. $\sqrt{\frac{16}{25}}$

20. $\sqrt{\frac{9}{49}}$

21. $\sqrt{x^2}, x > 0$

22. $\sqrt{(x - 1)^2}, x > 1$

23. $\sqrt[3]{27}$

24. $\sqrt[3]{64}$

25. $\sqrt[3]{-64}$

26. $-\sqrt[3]{-1}$

27. $\sqrt[3]{\frac{8}{27}}$

28. $\sqrt[3]{-\frac{1}{125}}$

29. $-\sqrt[3]{x^9}$

30. $\sqrt[3]{(x + 1)^6}$

31. $\sqrt[3]{(2x)^6}$

32. $\sqrt[3]{27x^3}$

33. $\sqrt[4]{81}$

34. $\sqrt[5]{-1}$

35. $\sqrt[5]{-243}$

36. $\sqrt[4]{625}$

37. $\sqrt[4]{-16}$

38. $\sqrt[6]{-64}$

Exercises 39–44: Approximate to the nearest hundredth.

39. $-\sqrt{5}$

40. $\sqrt{11}$

41. $\sqrt[3]{5}$

42. $\sqrt[3]{-13}$

43. $\sqrt[5]{-7}$

44. $\sqrt[4]{6}$

Exercises 45–58: Simplify the expression. Assume that all variables are real numbers.

45. $\sqrt{(-4)^2}$

46. $\sqrt{9^2}$

47. $\sqrt{y^2}$

48. $\sqrt{z^4}$

49. $\sqrt{(x - 5)^2}$

50. $\sqrt{(2x - 1)^2}$

51. $\sqrt{x^2 - 2x + 1}$

52. $\sqrt{4x^2 + 4x + 1}$

53. $\sqrt[4]{y^4}$

54. $\sqrt[4]{x^8 z^4}$

55. $\sqrt[4]{x^{12}}$

56. $\sqrt[6]{x^6}$

57. $\sqrt[5]{x^5}$

58. $\sqrt[5]{32(x + 4)^5}$

SQUARE AND CUBE ROOT FUNCTIONS

Exercises 59–74: If possible, evaluate the function at the given value(s) of the variable.

59. $f(x) = \sqrt{x - 1}$ $x = 10, 0$

60. $f(x) = \sqrt{4 - 3x}$ $x = -4, 1$

61. $f(x) = \sqrt{3 - 3x}$ $x = -1, 5$

62. $f(x) = \sqrt{x - 5}$ $x = -1, 5$

63. $f(x) = \sqrt{x^2 - x}$ $x = -4, 3$

64. $f(x) = \sqrt{2x^2 - 3}$ $x = -1, 2$

65. $f(x) = \sqrt[3]{x^2 - 8}$ $x = -3, 4$

66. $f(x) = \sqrt[3]{2x^2}$ $x = -2, 2$

67. $f(x) = \sqrt[3]{x - 9}$ $x = 1, 10$

68. $f(x) = \sqrt[3]{5x - 2}$ $x = -5, 2$

69. $f(x) = \sqrt[3]{3 - x^2}$ $x = -2, 3$

70. $f(x) = \sqrt[3]{-1 - x^2}$ $x = 0, 3$

71. $T(h) = \frac{1}{2}\sqrt{h}$ $h = 64$

72. $L(k) = 2\sqrt{k + 2}$ $k = 23$

73. $f(x) = \sqrt{x + 5} + \sqrt{x}$ $x = 4$

74. $f(x) = \dfrac{\sqrt{x - 5} - \sqrt{x}}{2}$ $x = 9$

Exercises 75–88: Find the domain of f. Write your answer in interval notation.

75. $f(x) = \sqrt{x + 2}$

76. $f(x) = \sqrt{x - 1}$

77. $f(x) = \sqrt{x - 2}$

78. $f(x) = \sqrt{x + 1}$

79. $f(x) = \sqrt{2x - 4}$

80. $f(x) = \sqrt{4x + 2}$

81. $f(x) = \sqrt{1 - x}$

82. $f(x) = \sqrt{6 - 3x}$

83. $f(x) = \sqrt{8 - 5x}$

84. $f(x) = \sqrt{3 - 2x}$

85. $f(x) = \sqrt{3x^2 + 4}$

86. $f(x) = \sqrt{1 + 2x^2}$

87. $f(x) = \dfrac{1}{\sqrt{2x + 1}}$

88. $f(x) = \dfrac{1}{\sqrt{x - 1}}$

Exercises 89–94: Graph the equation. Compare the graph to either $y = \sqrt{x}$ or $y = \sqrt[3]{x}$.

89. $y = \sqrt{x} + 2$ **90.** $y = \sqrt{x} - 1$

91. $y = \sqrt{x + 2}$ **92.** $y = \sqrt[3]{x} + 2$

93. $y = \sqrt[3]{x + 2}$ **94.** $y = \sqrt[3]{x} - 1$

REPRESENTATIONS OF ROOT FUNCTIONS

Exercises 95–102: Give symbolic, numerical, and graphical representations for the function f.

95. Function f takes the square root of x and then adds 1 to the result.

96. Function f takes the square root of x and then subtracts 2 from the result.

97. Function f takes the square root of the quantity three times x.

98. Function f takes the square root of the quantity x plus 1.

99. Function f takes the cube root of x and then multiplies the result by 2.

100. Function f takes the cube root of the quantity 4 times x.

101. Function f takes the cube root of the quantity x minus 1.

102. Function f takes the cube root of x and then adds 1.

AREA OF A TRIANGLE

*Exercises 103 and 104: **Heron's Formula** Suppose the lengths of the sides of a triangle are a, b, and c as illustrated in the figure.*

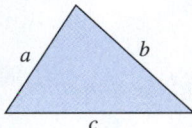

*If the **semiperimeter** (half of the perimeter) of the triangle is $s = \frac{1}{2}(a + b + c)$, then the area of the triangle is*

$$A = \sqrt{s(s - a)(s - b)(s - c)}.$$

Find the area A of the triangle with the given sides.

103. $a = 3, b = 4, c = 5$

104. $a = 5, b = 9, c = 10$

APPLICATIONS

105. *Jumping* (Refer to Example 9.) If a person jumps 4 feet off the ground, estimate how long the person is in the air.

106. *Hang Time* (Refer to Example 9.) Find the hang time for a golf ball hit 80 feet into the air.

107. *Distance to the Horizon* (Refer to Example 4.) Use the formula $d = 1.22\sqrt{x}$ to estimate how many miles a person can see from a jet airliner at 10,000 feet.

108. *Distance to the Horizon* (Refer to Example 4.) Use the formula $d = 1.22\sqrt{x}$ to estimate how many miles a 5-foot-tall person can see standing on the deck of a ship that is 50 feet above the ocean.

109. *Increasing Productivity* (Refer to Example 12.) Use $R(x) = 108\sqrt{x}$ to evaluate $R(16) - R(15)$. If the salary for the sixteenth employee is \$25,000, is it a good decision to hire the sixteenth employee?

110. *Increasing Productivity* (Refer to Example 12.) Use $R(x) = 108\sqrt{x}$ to evaluate $R(2) - R(1)$. If the salary for the second employee is \$25,000, is it a good decision to hire the second employee?

111. *Productivity of Workers* If workers are given more equipment, or *physical capital*, they are often more productive. For example, a carpenter definitely needs a hammer to be productive, but probably does not need 20 hammers. There is a leveling off in a worker's productivity as more is spent on equipment. The function $P(x) = 400\sqrt{x} + 8000$ approximates the worth of the goods produced by a typical U.S. worker in dollars when x dollars are spent on equipment per worker.
(a) Evaluate $P(25,000)$ and interpret the result.
(b) Sketch a graph of $y = P(x)$.
(c) Use the graph from (b) and the formula to evaluate $P(50,000) - P(25,000)$ and also to evaluate $P(75,000) - P(50,000)$. Interpret the result.

112. *Design of Open Channels* To protect cities from flooding during heavy rains, open channels are sometimes constructed to handle runoff. The rate R at which water flows through the channel is modeled by $R = k\sqrt{m}$, where m is the slope of the channel and k is a constant determined by the shape of the channel. (*Source:* N. Garber and L. Hoel, *Traffic and Highway Design.*)

(a) Suppose that a channel has a slope of $m = 0.01$ (or 1%) and a runoff rate of $R = 340$ cubic feet per second (cfs). Find k.

(b) If the slope of the channel increases to $m = 0.04$ (or 4%), what happens to R? Be specific.

WRITING ABOUT MATHEMATICS

113. Try to calculate $\sqrt{-7}$, $\sqrt[4]{-56}$, and $\sqrt[6]{-10}$ with a calculator. Describe what happens when you evaluate an even root of a negative number. Does the same difficulty occur when you evaluate an odd root of a negative number? Try to evaluate $\sqrt[3]{-7}$, $\sqrt[5]{-56}$, and $\sqrt[7]{-10}$. Explain.

114. Explain the difference between a root and a positive integer power of a number. Give examples.

7.2 Rational Exponents

Basic Concepts • Properties of Rational Exponents

A LOOK INTO MATH ▶

In today's society, people are used to having a lot of choices when it comes to online videos. In fact, if the average online video has 1,000,000 viewers, then about 200,000 of them, or 20%, have abandoned the video in 10 seconds or less. This average viewer abandonment rate can be modeled by a new function A, given by

$$A(x) = 7.3x^{7/16},$$

where A computes the percentage of viewers who abandon an online video after x seconds. But what does the exponent $\frac{7}{16}$ mean? To answer this question we need to discuss rational exponents and their properties. See Example 3. (*Source:* Business Insider.)

Basic Concepts

When m and n are integers, the product rule states that $a^m a^n = a^{m+n}$. This rule can be extended to include exponents that are fractions. For example,

$$4^{1/2} \cdot 4^{1/2} = 4^{1/2 + 1/2} = 4^1 = 4.$$

That is, if we multiply the expression $4^{1/2}$ by itself, the result is 4. Because we also know that $\sqrt{4} \cdot \sqrt{4} = 4$, this discussion suggests that $4^{1/2} = \sqrt{4}$ and leads to the following definition.

THE EXPRESSION $a^{1/n}$

If n is an integer greater than 1 and a is a real number, then
$$a^{1/n} = \sqrt[n]{a}.$$

$7^{1/5} = \sqrt[5]{7}$

NOTE: If $a < 0$ and n is an even positive integer, then $a^{1/n}$ is not a real number.

In the next two examples, we show how to interpret rational exponents.

EXAMPLE 1 | **Interpreting rational exponents**

Write each expression in radical notation. Then evaluate the expression and round to the nearest hundredth when appropriate.

(a) $36^{1/2}$ **(b)** $23^{1/5}$ **(c)** $x^{1/3}$ **(d)** $(5x)^{1/2}$

Solution

(a) The exponent $\frac{1}{2}$ indicates a square root. Thus $36^{1/2} = \sqrt{36}$, which evaluates to 6.

(b) The exponent $\frac{1}{5}$ indicates a fifth root. Thus $23^{1/5} = \sqrt[5]{23}$, which is not an integer. Figure 7.9 shows this expression approximated in both exponential and radical notation. In either case $23^{1/5} \approx 1.87$.

(c) The exponent $\frac{1}{3}$ indicates a cube root, so $x^{1/3} = \sqrt[3]{x}$.

(d) The exponent $\frac{1}{2}$ indicates a square root, so $(5x)^{1/2} = \sqrt{5x}$.

Now Try Exercises 37, 45, 59, 63

```
23^(1/5)
        1.872171231
5ˣ√(23)
        1.872171231
```

Figure 7.9

CALCULATOR HELP

To calculate other roots, see Appendix A (pages AP-7 and AP-8).

Suppose that we want to define the expression $8^{2/3}$. On the one hand, using properties of exponents we have

$$8^{1/3} \cdot 8^{1/3} = 8^{1/3 + 1/3} = 8^{2/3}.$$

On the other hand, we have

$$8^{1/3} \cdot 8^{1/3} = \sqrt[3]{8} \cdot \sqrt[3]{8} = 2 \cdot 2 = 4.$$

Thus $8^{2/3} = 4$, and that value is obtained whether we interpret $8^{2/3}$ as either

$$8^{2/3} = (8^{1/3})^2 = (\sqrt[3]{8})^2 = 2^2 = 4$$

or

$$8^{2/3} = (8^2)^{1/3} = \sqrt[3]{8^2} = \sqrt[3]{64} = 4.$$

This result suggests the following definition.

STUDY TIP

Be sure that you understand how rational exponents relate to radical notation.

THE EXPRESSION $a^{m/n}$

If m and n are positive integers with $\frac{m}{n}$ in lowest terms, then

$$a^{m/n} = \sqrt[n]{a^m} = (\sqrt[n]{a})^m. \qquad 2^{3/4} = \sqrt[4]{2^3} = (\sqrt[4]{2})^3$$

NOTE: If $a < 0$ and n is an even integer, then $a^{m/n}$ is *not* a real number.

The exponent $\frac{m}{n}$ indicates that we either take the nth root and then calculate the mth power of the result or calculate the mth power and then take the nth root. For example, $4^{3/2}$ means that we can either take the **square root** of 4 and then **cube** the result or we can **cube** 4 and then take the **square root** of the result. In either case the result is the same: $4^{3/2} = 8$. This concept is illustrated in the next example.

EXAMPLE 2 | **Interpreting rational exponents**

Write each expression in radical notation. Evaluate the expression by hand when possible.
(a) $(-27)^{2/3}$ **(b)** $12^{3/5}$

Solution

(a) The exponent $\frac{2}{3}$ indicates either that we take the **cube root** of -27 and then **square** it or that we **square** -27 and then take the **cube root**. Thus

$$(-27)^{2/3} = (\sqrt[3]{-27})^2 = (-3)^2 = 9$$

or

$$(-27)^{2/3} = \sqrt[3]{(-27)^2} = \sqrt[3]{729} = 9.$$

Same result

(b) The exponent $\frac{3}{5}$ indicates either that we take the **fifth root** of 12 and then **cube** it or that we **cube** 12 and then take the **fifth root**. Thus

$$12^{3/5} = \left(\sqrt[5]{12}\right)^3 \quad \text{or} \quad 12^{3/5} = \sqrt[5]{12^3}.$$

This result cannot be evaluated by hand.

▮ **Now Try Exercises 47, 51**

TECHNOLOGY NOTE

Rational Exponents
When evaluating expressions with rational (fractional) exponents, be sure to put parentheses around the fraction. For example, most calculators will evaluate 8^(2/3) and 8^2/3 differently. The accompanying figure shows that evaluating $8^{2/3}$ correctly as 8^(2/3) results in 4 but evaluating $8^{2/3}$ incorrectly as 8^2/3 results in $\frac{8^2}{3} = 21.\overline{3}$.

Correct $\longrightarrow$ | 8^(2/3)
| 4
Incorrect $\longrightarrow$ | 8^2/3
| 21.33333333

▶ **REAL-WORLD CONNECTION** In the next example, we evaluate the function A presented in A Look Into Math at the beginning of this section. This function calculates the average viewer abandonment rate for online videos.

EXAMPLE 3   Calculating abandonment rates for online videos

The function A, given by $A(x) = 7.3x^{7/16}$, computes the percentage of viewers who abandon an online video after x seconds.
(a) Find $A(10)$ and $A(90)$. Interpret your answers.
(b) Use radical notation to write $A(x)$.

Solution

(a) To approximate these expressions we will use a calculator, as shown in Figure 7.10.

$$A(\mathbf{10}) = 7.3(\mathbf{10})^{7/16} \approx \mathbf{20\%}$$
$$A(\mathbf{90}) = 7.3(\mathbf{90})^{7/16} \approx \mathbf{52\%}$$

After **10** seconds about **20%** of the viewers have abandoned the online video, and after **90** seconds over half, or **52%**, of the viewers have abandoned the online video.

(b) The exponent $\frac{7}{16}$ means we raise x to the **seventh** power and then take the **sixteenth** root. That is, $A(x) = 7.3\sqrt[16]{x^7}$.

▮ **Now Try Exercise 99**

7.3*10^(7/16)
    19.99046333
7.3*90^(7/16)
    52.27619364

Figure 7.10

From properties of exponents we know that $a^{-n} = \frac{1}{a^n}$, where n is a positive integer. We now define this property for negative rational exponents.

THE EXPRESSION $a^{-m/n}$

If m and n are positive integers with $\frac{m}{n}$ in lowest terms, then

$$a^{-m/n} = \frac{1}{a^{m/n}}, \qquad a \neq 0.$$

$$2^{-3/4} = \frac{1}{2^{3/4}}$$

Examples of changing rational exponents to radical expressions are shown in Table 7.5.

TABLE 7.5 Converting Forms

Expression	$a^{3/4}$	$5^{1/7}$	$x^{-5/3}$
Radical Form	$\sqrt[4]{a^3}$	$\sqrt[7]{5}$	$\dfrac{1}{\sqrt[3]{x^5}}$

EXAMPLE 4 **Interpreting negative rational exponents**

Write each expression in radical notation and then evaluate.
(a) $64^{-1/3}$ **(b)** $81^{-3/4}$

Solution

(a) $64^{-1/3} = \dfrac{1}{64^{1/3}} = \dfrac{1}{\sqrt[3]{64}} = \dfrac{1}{4}.$

(b) $81^{-3/4} = \dfrac{1}{81^{3/4}} = \dfrac{1}{(\sqrt[4]{81})^3} = \dfrac{1}{3^3} = \dfrac{1}{27}.$

▌ **Now Try Exercises 53, 55**

EXAMPLE 5 **Converting to rational exponents**

Use rational exponents to write each radical expression.
(a) $\sqrt[5]{x^4}$ **(b)** $\dfrac{1}{\sqrt{b^5}}$ **(c)** $\sqrt[3]{(x-1)^2}$ **(d)** $\sqrt[3]{a^2 + b^2}$

Solution
(a) $\sqrt[5]{x^4} = x^{4/5}$

(b) $\dfrac{1}{\sqrt{b^5}} = b^{-5/2}$

(c) $\sqrt[3]{(x-1)^2} = (x-1)^{2/3}$

(d) $\sqrt[3]{a^2 + b^2} = (a^2 + b^2)^{1/3}$ Note that $\sqrt[3]{a^2 + b^2} \neq a^{2/3} + b^{2/3}$.

▌ **Now Try Exercises 27, 29, 33, 35**

▶ **REAL-WORLD CONNECTION** In the next example, we use a formula from biology that involves a rational exponent.

EXAMPLE 6   **Analyzing stepping frequency**

When smaller (four-legged) animals walk, they tend to take faster, shorter steps, whereas larger animals tend to take slower, longer steps. If an animal is h feet high at the shoulder, then the number N of steps per second that the animal takes *while walking* can be estimated by $N(h) = 1.6h^{-1/2}$.   (*Source:* C. Pennycuick, *Newton Rules Biology.*)

(a) Use radical notation to write $N(h)$.

(b) Estimate the stepping frequency for an adult African elephant that is 10 feet high at the shoulders and a baby elephant that is only 33 inches high.

Solution

(a) We can rewrite the formula as $N(h) = \dfrac{1.6}{\sqrt{h}}$.

(b) We can evaluate N as follows. Note that 33 inches is $\dfrac{33}{12} = 2.75$ feet.

$$N(10) = \frac{1.6}{\sqrt{10}} \approx 0.51 \quad \text{and} \quad N(2.75) = \frac{1.6}{\sqrt{2.75}} \approx 0.96$$

While walking, the adult elephant takes about 1 step every 2 seconds, and the baby elephant takes about 1 step every second.

Now Try Exercise 101

Properties of Rational Exponents

Any rational number can be written as a ratio of two integers. That is, if p is a rational number, then $p = \frac{m}{n}$, where m and n are integers. Properties for integer exponents also apply to rational exponents, with one exception. If n is even in the expression $a^{m/n}$ and $\frac{m}{n}$ is written in lowest terms, then a must be nonnegative for the result to be a real number.

For example, the expression $(-8)^{3/2}$ is not a real number because -8 is negative and 2 is even. That is, $(-8)^{3/2} = (\sqrt{-8})^3$, which is undefined because the square root of a negative number is not a real number.

PROPERTIES OF EXPONENTS

Let p and q be rational numbers written in lowest terms. For all real numbers a and b for which the expressions are real numbers, the following properties hold.

1. $a^p \cdot a^q = a^{p+q}$ Product rule for exponents $3^{1/7} \cdot 3^{2/7} = 3^{3/7}$

2. $a^{-p} = \dfrac{1}{a^p}, \ \dfrac{1}{a^{-p}} = a^p$ Negative exponents $7^{-1/4} = \dfrac{1}{7^{1/4}}, \ \dfrac{1}{7^{-1/4}} = 7^{1/4}$

3. $\left(\dfrac{a}{b}\right)^{-p} = \left(\dfrac{b}{a}\right)^{p}$ Negative exponents for quotients $\left(\dfrac{3}{7}\right)^{-1/3} = \left(\dfrac{7}{3}\right)^{1/3}$

4. $\dfrac{a^p}{a^q} = a^{p-q}$ Quotient rule for exponents $\dfrac{5^{4/7}}{5^{1/7}} = 5^{3/7}$

5. $(a^p)^q = a^{pq}$ Power rule for exponents $(2^{1/3})^{1/2} = 2^{1/6}$

6. $(ab)^p = a^p b^p$ Power rule for products $(5x)^{2/3} = 5^{2/3} x^{2/3}$

7. $\left(\dfrac{a}{b}\right)^p = \dfrac{a^p}{b^p}$ Power rule for quotients $\left(\dfrac{5}{3}\right)^{3/4} = \dfrac{5^{3/4}}{3^{3/4}}$

In the next two examples, we apply these properties.

EXAMPLE 7 **Applying properties of exponents**

Write each expression using rational exponents and simplify. Write the answer with a positive exponent. Assume that all variables are positive numbers.

(a) $\sqrt{x} \cdot \sqrt[3]{x}$ **(b)** $\sqrt[3]{27x^2}$ **(c)** $\dfrac{\sqrt[4]{16x}}{\sqrt[3]{x}}$ **(d)** $\left(\dfrac{x^2}{81}\right)^{-1/2}$

Solution

(a) $\sqrt{x} \cdot \sqrt[3]{x} = x^{1/2} \cdot x^{1/3}$ Use rational exponents.

$= x^{1/2 + 1/3}$ Product rule for exponents

$= x^{5/6}$ Simplify: $\frac{1}{2} + \frac{1}{3} = \frac{3}{6} + \frac{2}{6} = \frac{5}{6}$.

(b) $\sqrt[3]{27x^2} = (27x^2)^{1/3}$ Use rational exponents.

$= 27^{1/3}(x^2)^{1/3}$ Power rule for products

$= 3x^{2/3}$ Simplify; power rule for exponents.

(c) $\dfrac{\sqrt[4]{16x}}{\sqrt[3]{x}} = \dfrac{(16x)^{1/4}}{x^{1/3}}$ Use rational exponents.

$= \dfrac{16^{1/4}x^{1/4}}{x^{1/3}}$ Power rule for products

$= 16^{1/4}x^{1/4 - 1/3}$ Quotient rule for exponents

$= 2x^{-1/12}$ Simplify: $\frac{1}{4} - \frac{1}{3} = \frac{3}{12} - \frac{4}{12} = -\frac{1}{12}$.

$= \dfrac{2}{x^{1/12}}$ Negative exponents

(d) $\left(\dfrac{x^2}{81}\right)^{-1/2} = \left(\dfrac{81}{x^2}\right)^{1/2}$ Negative exponents for quotients

$= \dfrac{(81)^{1/2}}{(x^2)^{1/2}}$ Power rule for quotients

$= \dfrac{9}{x}$ Simplify; power rule for exponents.

▌ **Now Try Exercises 77, 83, 91, 97**

EXAMPLE 8 **Applying properties of exponents**

Write each expression with positive rational exponents and simplify, if possible.

(a) $\sqrt[3]{\sqrt{x+1}}$ **(b)** $\sqrt[5]{c^{15}}$ **(c)** $\dfrac{y^{-1/2}}{x^{-1/3}}$ **(d)** $\sqrt{x}(\sqrt{x} - 1)$

Solution

(a) $\sqrt[3]{\sqrt{x+1}} = \left((x+1)^{1/2}\right)^{1/3} = (x+1)^{1/6}$

(b) $\sqrt[5]{c^{15}} = c^{15/5} = c^3$

(c) $\dfrac{y^{-1/2}}{x^{-1/3}} = \dfrac{x^{1/3}}{y^{1/2}}$

(d) $\sqrt{x}(\sqrt{x} - 1) = x^{1/2}(x^{1/2} - 1) = x^{1/2}x^{1/2} - x^{1/2} = x - x^{1/2}$

▌ **Now Try Exercises 85, 89, 95**

7.2 Putting It All Together

CONCEPT	EXPLANATION	EXAMPLES
Rational Exponents	If m and n are positive integers with $\frac{m}{n}$ in lowest terms, $$a^{m/n} = \sqrt[n]{a^m} = \left(\sqrt[n]{a}\right)^m.$$ If $a < 0$ and n is even, $a^{m/n}$ is not a real number.	$8^{4/3} = \left(\sqrt[3]{8}\right)^4 = 2^4 = 16$ $(-27)^{3/4} = \left(\sqrt[4]{-27}\right)^3$ is *not* a real number.
Properties of Exponents	Let p and q be rational numbers. **1.** $a^p \cdot a^q = a^{p+q}$ **2.** $a^{-p} = \dfrac{1}{a^p}, \dfrac{1}{a^{-p}} = a^p$ **3.** $\left(\dfrac{a}{b}\right)^{-p} = \left(\dfrac{b}{a}\right)^p$ **4.** $\dfrac{a^p}{a^q} = a^{p-q}$ **5.** $(a^p)^q = a^{pq}$ **6.** $(ab)^p = a^p b^p$ **7.** $\left(\dfrac{a}{b}\right)^p = \dfrac{a^p}{b^p}$	**1.** $2^{1/3} \cdot 2^{2/3} = 2^{1/3+2/3} = 2^1 = 2$ **2.** $2^{-1/2} = \dfrac{1}{2^{1/2}}, \dfrac{1}{3^{-1/4}} = 3^{1/4}$ **3.** $\left(\dfrac{3}{4}\right)^{-4/5} = \left(\dfrac{4}{3}\right)^{4/5}$ **4.** $\dfrac{7^{2/3}}{7^{1/3}} = 7^{2/3-1/3} = 7^{1/3}$ **5.** $\left(8^{2/3}\right)^{1/2} = 8^{(2/3)\cdot(1/2)} = 8^{1/3} = 2$ **6.** $(2x)^{1/3} = 2^{1/3} x^{1/3}$ **7.** $\left(\dfrac{x}{y}\right)^{1/6} = \dfrac{x^{1/6}}{y^{1/6}}$

7.2 Exercises

MyMathLab  Math XL PRACTICE   WATCH   DOWNLOAD   READ   REVIEW

CONCEPTS AND VOCABULARY

1. Simplify $4^{1/2}$.

2. Simplify $8^{1/3}$.

3. Simplify $4^{-1/2}$.

4. Simplify $8^{-1/3}$.

5. Use a rational exponent to write $\sqrt{x}$.

6. Use a rational exponent to write $\sqrt[3]{a^4}$.

7. Write $a^{1/n}$ in radical notation.

8. Write $a^{m/n}$ in radical notation.

Exercises 9–16: Match the given expression to the expression (a.–h.) that it equals.

9. $\sqrt{x^3}$

10. $\sqrt[3]{8^2}$

11. $25^{-1/2}$

12. $27^{-2/3}$

13. $x^{1/5}$

14. $(x^{1/3})^2$

15. $\sqrt{x} \cdot \sqrt[3]{x}$

16. $\dfrac{x^{1/2}}{x^{1/3}}$

a. $\sqrt[6]{x^5}$

b. $\sqrt[6]{x}$

c. $x^{3/2}$

d. $\sqrt[5]{x}$

e. $\sqrt[3]{x^2}$

f. 4

g. $\dfrac{1}{5}$

h. $\dfrac{1}{9}$

CONVERTING FORMS

Exercises 17–26: Use radical notation to write each expression.

17. $7^{1/2}$

18. $10^{1/2}$

19. $a^{1/3}$

20. $y^{1/4}$

21. $x^{5/6}$

22. $x^{3/2}$

23. $(x + 5)^{1/2}$

24. $(x + y)^{2/3}$

25. $b^{-2/3}$

26. $b^{-3/4}$

Exercise 27–36: Use rational exponents to write each expression.

27. $\sqrt{t}$

28. $\sqrt[3]{t}$

29. $\sqrt[3]{(x + 1)}$

30. $\sqrt{y - 3}$

31. $\dfrac{1}{\sqrt{x + 1}}$

32. $\dfrac{1}{\sqrt[3]{x - 3}}$

33. $\sqrt{a^2 - b^2}$

34. $\sqrt[3]{a^3 + b^3}$

35. $\dfrac{1}{\sqrt[3]{x^7}}$

36. $\dfrac{1}{\sqrt[3]{y^4}}$

RATIONAL EXPONENTS

Exercises 37–44: Approximate to the nearest hundredth.

37. $16^{1/5}$

38. $7^{1/4}$

39. $5^{1/3}$

40. $11^{1/2}$

41. $9^{3/5}$

42. $13^{5/4}$

43. $4^{-3/7}$

44. $2^{-3/4}$

Exercises 45–64: Write each expression in radical notation. Evaluate the expression by hand when possible.

45. $9^{1/2}$

46. $100^{1/2}$

47. $8^{1/3}$

48. $(-8)^{1/3}$

49. $\left(\frac{4}{9}\right)^{1/2}$

50. $\left(\frac{64}{27}\right)^{1/3}$

51. $(-8)^{2/3}$

52. $(16)^{3/2}$

53. $\left(\frac{1}{8}\right)^{-1/3}$

54. $\left(\frac{1}{81}\right)^{-1/4}$

55. $16^{-3/4}$

56. $(32)^{-2/5}$

57. $(4^{1/2})^{-3}$

58. $(8^{1/3})^{-2}$

59. $z^{1/4}$

60. $b^{1/3}$

61. $y^{-2/5}$

62. $z^{-3/4}$

63. $(3x)^{1/3}$

64. $(5x)^{1/2}$

Exercises 65–72: Use a positive rational exponent to write the expression.

65. $\sqrt{y}$

66. $\sqrt{z + 1}$

67. $\sqrt{x} \cdot \sqrt{x}$

68. $\sqrt[3]{x} \cdot \sqrt[3]{x}$

69. $\sqrt[3]{8x^2}$

70. $\sqrt[3]{27z}$

71. $\dfrac{\sqrt{49x}}{\sqrt[3]{x^2}}$

72. $\dfrac{\sqrt[3]{8x}}{\sqrt[4]{x^3}}$

Exercises 73 and 74: **Thinking Generally** *Simplify the expression. Assume that a and b are positive integers.*

73. $(b^{1/a}b^{2/a})^a$

74. $b^{(b-1)/b} \cdot \sqrt[b]{b}$

Exercises 75–98: Use positive rational exponents to simplify the expression. Assume that all variables are positive.

75. $(x^2)^{3/2}$

76. $(y^4)^{1/2}$

77. $\sqrt[3]{x^3y^6}$

78. $\sqrt{16x^4}$

79. $\sqrt{y^3} \cdot \sqrt[3]{y^2}$

80. $\left(\dfrac{x^6}{81}\right)^{1/4}$

81. $\left(\dfrac{x^6}{27}\right)^{2/3}$

82. $\left(\dfrac{1}{x^8}\right)^{-1/4}$

83. $\left(\dfrac{x^2}{y^6}\right)^{-1/2}$

84. $\dfrac{\sqrt{x}}{\sqrt[3]{27x^6}}$

85. $\sqrt{\sqrt{y}}$

86. $\sqrt{\sqrt[3]{(3x)^2}}$

87. $(a^{-1/2})^{4/3}$

88. $(x^{-3/2})^{2/3}$

89. $\dfrac{(k^{1/2})^{-3}}{(k^2)^{1/4}}$

90. $\dfrac{(b^{3/4})^4}{(b^{4/5})^{-5}}$

91. $\sqrt{b} \cdot \sqrt[4]{b}$

92. $\sqrt[3]{t} \cdot \sqrt[5]{t}$

93. $p^{1/2}\left(p^{3/2} + p^{1/2}\right)$

94. $d^{3/4}(d^{1/4} - d^{-1/4})$

95. $\sqrt[3]{x}\left(\sqrt{x} - \sqrt[3]{x^2}\right)$

96. $\frac{1}{2}\sqrt{x}\left(\sqrt{x} + \sqrt[4]{x^2}\right)$

97. $\dfrac{\sqrt[3]{27x}}{\sqrt{x}}$

98. $\dfrac{\sqrt[5]{32x^2}}{\sqrt[3]{8x}}$

APPLICATIONS

99. *Online Videos* (Refer to Example 3.) The function given by $A(x) = 7.3x^{7/16}$ computes the percentage of viewers who abandon an online video after x seconds for $0 \leq x \leq 120$.
   (a) Make a table of values for $x = 0, 20, 40, 60,$ and 80. Round values to the nearest percent.
   (b) Interpret the table in terms of how people watch online videos.

100. *Online Videos* (Refer to Example 3.) Suppose that the function given by $A(x) = 7.3x^{4/3}$ computes the percentage of viewers who abandon an online video after x seconds. Evaluate $A(10)$. Is this function realistic?

101. *Animal Stepping Frequency* (Refer to Example 6.) Use the formula $N(h) = 1.6h^{-1/2}$ to estimate the stepping frequency of a dog that is 2.5 feet high at the shoulders. (*Source:* C. Pennycuick.)

102. *Animal Pulse Rate* According to one model, an animal's heart rate varies according to its weight. The formula $N(w) = 885w^{-1/2}$ gives an estimate for the average number N of beats per minute for an animal that weighs w pounds. Use the formula to estimate the heart rate for a horse that weighs 800 pounds. (*Source:* C. Pennycuick.)

103. *Bird Wings* Heavier birds tend to have larger wings than lighter birds do. For some birds the relationship between the surface area A of the bird's wings in square inches and its weight W in pounds can be modeled by $A = 100\sqrt[3]{W^2}$. (*Source:* C. Pennycuick, *Newton Rules Biology*.)
  (a) Find the area of the wings when the weight is 8 pounds.
  (b) Write this formula with rational exponents.

104. *Planet Orbits and Distance* Johannes Kepler (1571–1630) discovered a relationship between a planet's distance D from the sun and the time T it takes to orbit the sun. This formula is $T = \sqrt{D^3}$, where T is in Earth years and $D = 1$ corresponds to the distance between Earth and the sun, or 93,000,000 miles.
  (a) The planet Neptune is 30 times farther from the sun than Earth ($D = 30$). Estimate the number of years required for Neptune to orbit the sun.
  (b) Write this formula with rational exponents.

105. *Baby's Head Size* If a female baby's head circumference is 32.5 centimeters at birth, then the function given by $H(x) = 35.2x^{3/40}$ gives the infant's head circumference at x months for $1 \le x \le 36$.
  (a) Evaluate $H(12)$ and $H(24)$.
  (b) Does an infant's head circumference increase the most in the first year or the second year?

106. *Musical Tones* One octave on a piano contains 12 keys (including both the black and white keys). The frequency of each successive key increases by a factor of $2^{1/12}$. For example, middle C is two keys below the first D above it. Therefore the frequency of this D is

$$2^{1/12} \cdot 2^{1/12} = 2^{1/6} \approx 1.12$$

times the frequency of middle C.
  (a) If two tones are one octave apart, how do their frequencies compare?
  (b) The A tone below middle C on a piano has a frequency of 220 cycles per second. Middle C is 3 keys above this A note. Estimate the frequency of middle C.

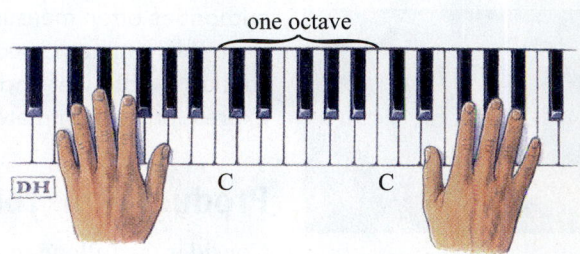

one octave

107. **Online Exploration** Go online and look up the shoulder height of three animals of different sizes. Use the formula in Exercise 101 to calculate their stepping frequencies while walking.

108. **Online Exploration** Go online and look up the weight of three animals of different sizes. Use the formula in Exercise 102 to calculate their heart rates.

WRITING ABOUT MATHEMATICS

109. Explain the meaning of the rational exponent in $x^{3/5}$.

110. Explain the meaning of each expression: $\sqrt[3]{x^5}$ and $(\sqrt[3]{x})^5$. Are these two expressions equal?

SECTIONS
7.1 and 7.2 **Checking Basic Concepts**

1. Find the following.
   (a) The square roots of 49
   (b) The principal square root of 49

2. Evaluate.
   (a) $\sqrt[3]{-8}$ (b) $-\sqrt[4]{81}$

3. Write the expression in radical notation.
   (a) $x^{3/2}$ (b) $x^{2/3}$ (c) $x^{-2/5}$

4. Simplify $\sqrt{(x-1)^2}$ for any real number x.

5. Evaluate each function at the given value of x.
   (a) $f(x) = \sqrt{x}$, $x = 9$
   (b) $g(x) = \sqrt[3]{x}$, $x = 125$
   (c) $h(x) = x^{7/12}$, $x = \frac{12}{7}$

7.3 Simplifying Radical Expressions

Product Rule for Radical Expressions • Quotient Rule for Radical Expressions

A LOOK INTO MATH ▶ When a car stops suddenly, it often leaves skid marks. If the car is involved in an accident, authorities often measure the length of the skid marks M and use this information to determine the minimum speed S of the car. One formula that is sometimes used when the road surface is both wet and level is given by $S = \sqrt{9M}$. In Example 4 we simplify this radical expression and estimate the speed of a car based on the length of its skid marks.

Product Rule for Radical Expressions

Consider the following examples of multiplying radical expressions. The equations

NEW VOCABULARY

☐ Perfect nth power
☐ Perfect square
☐ Perfect cube

$$\sqrt{4} \cdot \sqrt{25} = 2 \cdot 5 = 10 \quad \text{and} \quad \sqrt{4 \cdot 25} = \sqrt{100} = 10$$

imply that

$$\sqrt{4} \cdot \sqrt{25} = \sqrt{4 \cdot 25} \quad \text{(see Figure 7.11(a)).}$$

Similarly, the equations

$$\sqrt[3]{8} \cdot \sqrt[3]{27} = 2 \cdot 3 = 6 \quad \text{and} \quad \sqrt[3]{8 \cdot 27} = \sqrt[3]{216} = 6$$

imply that

$$\sqrt[3]{8} \cdot \sqrt[3]{27} = \sqrt[3]{8 \cdot 27} \quad \text{(see Figure 7.11(b)).}$$

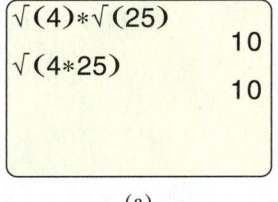
(a)

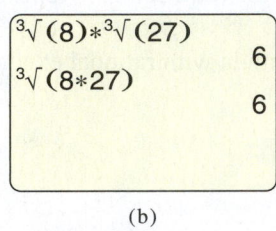
(b)

Figure 7.11

These examples suggest that *the product of the roots is equal to the root of the product.*

PRODUCT RULE FOR RADICAL EXPRESSIONS

Let a and b be real numbers, where $\sqrt[n]{a}$ and $\sqrt[n]{b}$ are both defined. Then

$$\sqrt[n]{a} \cdot \sqrt[n]{b} = \sqrt[n]{a \cdot b}.$$
$$\sqrt[4]{2} \cdot \sqrt[4]{8} = \sqrt[4]{2 \cdot 8}$$

NOTE: The product rule works only when the radicals have the *same index*. For example, the product $\sqrt{2} \cdot \sqrt[3]{4}$ cannot be simplified because the indexes are 2 and 3. (However, by using rational exponents, we can simplify this product. See Example 6(b).)

We apply the product rule in the next two examples.

EXAMPLE 1 **Multiplying radical expressions**

Multiply each radical expression.

(a) $\sqrt{5} \cdot \sqrt{20}$ (b) $\sqrt[3]{-3} \cdot \sqrt[3]{9}$ (c) $\sqrt[4]{\frac{1}{3}} \cdot \sqrt[4]{\frac{1}{9}} \cdot \sqrt[4]{\frac{1}{3}}$

Solution

(a) $\sqrt{5} \cdot \sqrt{20} = \sqrt{5 \cdot 20} = \sqrt{100} = 10$

(b) $\sqrt[3]{-3} \cdot \sqrt[3]{9} = \sqrt[3]{-3 \cdot 9} = \sqrt[3]{-27} = -3$

(c) The product rule can also be applied to three or more factors. Thus

$$\sqrt[4]{\frac{1}{3}} \cdot \sqrt[4]{\frac{1}{9}} \cdot \sqrt[4]{\frac{1}{3}} = \sqrt[4]{\frac{1}{3} \cdot \frac{1}{9} \cdot \frac{1}{3}} = \sqrt[4]{\frac{1}{81}} = \frac{1}{3},$$

because $\frac{1}{3} \cdot \frac{1}{3} \cdot \frac{1}{3} \cdot \frac{1}{3} = \frac{1}{81}$.

▪ **Now Try Exercises 13, 15, 21**

EXAMPLE 2 **Multiplying radical expressions containing variables**

Multiply each radical expression. Assume that all variables are positive.

(a) $\sqrt{x} \cdot \sqrt{x^3}$ (b) $\sqrt[3]{2a} \cdot \sqrt[3]{5a}$ (c) $\sqrt{11} \cdot \sqrt{xy}$ (d) $\sqrt[5]{\frac{2x}{y}} \cdot \sqrt[5]{\frac{16y}{x}}$

Solution

(a) $\sqrt{x} \cdot \sqrt{x^3} = \sqrt{x \cdot x^3} = \sqrt{x^4} = x^2$

(b) $\sqrt[3]{2a} \cdot \sqrt[3]{5a} = \sqrt[3]{2a \cdot 5a} = \sqrt[3]{10a^2}$

(c) $\sqrt{11} \cdot \sqrt{xy} = \sqrt{11xy}$

(d) $\sqrt[5]{\frac{2x}{y}} \cdot \sqrt[5]{\frac{16y}{x}} = \sqrt[5]{\frac{2x}{y} \cdot \frac{16y}{x}}$ Product rule

$\qquad\qquad\qquad = \sqrt[5]{\frac{32xy}{xy}}$ Multiply fractions.

$\qquad\qquad\qquad = \sqrt[5]{32}$ Simplify.

$\qquad\qquad\qquad = 2$ $2^5 = 32$

▪ **Now Try Exercises 23, 51, 57, 61**

An integer a is a **perfect nth power** if there exists an integer b such that $b^n = a$. Thus 36 is a **perfect square** because $6^2 = 36$. Similarly, 8 is a **perfect cube** because $2^3 = 8$, and 81 is a *perfect fourth power* because $3^4 = 81$. Table 7.6 lists examples of perfect squares and perfect cubes.

TABLE 7.6

Perfect Squares	1	4	9	16	25	36
Perfect Cubes	1	8	27	64	125	216

To simplify a square root, we sometimes need to recognize the *largest* perfect square factor of the radicand. For example, 50 has several factors.

$$1, \ 2, \ 5, \ 10, \ 25, \ 50 \quad \text{Factors of 50}$$

From Table 7.6, the perfect square factors of 50 are **1** and **25**. The *largest* perfect square factor of 50 is **25**.

The product rule for radicals can be used to simplify radical expressions. For example, because the largest perfect square factor of 50 is 25, the expression $\sqrt{50}$ can be simplified as

$$\sqrt{50} = \sqrt{25 \cdot 2} = \sqrt{25} \cdot \sqrt{2} = 5\sqrt{2}.$$

This procedure is generalized as follows.

READING CHECK

Explain how to simplify $\sqrt{40}$.

SIMPLIFYING RADICALS (nth ROOTS)

STEP 1: Determine the largest perfect nth power factor of the radicand.

STEP 2: Use the product rule to factor out and simplify this perfect nth power.

EXAMPLE 3 **Simplifying radical expressions**

Simplify each expression.
(a) $\sqrt{300}$ (b) $\sqrt[3]{16}$ (c) $\sqrt{54}$ (d) $\sqrt[4]{512}$

Solution
(a) First note that $300 = 100 \cdot 3$ and that 100 is the largest perfect square factor of 300.

$$\sqrt{300} = \sqrt{100 \cdot 3} = \sqrt{100} \cdot \sqrt{3} = 10\sqrt{3}$$

(b) The largest perfect cube factor of 16 is 8. (See Table 7.6.) Thus

$$\sqrt[3]{16} = \sqrt[3]{8} \cdot \sqrt[3]{2} = 2\sqrt[3]{2}.$$

(c) $\sqrt{54} = \sqrt{9} \cdot \sqrt{6} = 3\sqrt{6}$

(d) $\sqrt[4]{512} = \sqrt[4]{256} \cdot \sqrt[4]{2} = 4\sqrt[4]{2}$ because $4^4 = 256$.

Now Try Exercises 73, 75, 77, 79

▶ **REAL-WORLD CONNECTION** In A Look Into Math at the beginning of this section, we discussed a formula for estimating the minimum speed of a car by using its skid marks. In the next example, we simplify and use this formula.

EXAMPLE 4 **Calculating minimum speed**

A car is in an accident and leaves skid marks on wet, level pavement that are 121 feet long. The formula $S = \sqrt{9M}$ calculates the minimum speed S in miles per hour for skid marks M feet long.
(a) Simplify this formula.
(b) Determine the minimum speed of the car.

Solution
(a) Because 9 is a perfect square, we can factor it out.

$$\sqrt{9M} = \sqrt{9} \cdot \sqrt{M} = 3\sqrt{M}$$

Thus $S = 3\sqrt{M}$.
(b) $S = 3\sqrt{121} = 3 \cdot 11 = 33$ miles per hour.

Now Try Exercise 111

NOTE: To simplify a cube root of a negative number we factor out the negative of the largest perfect cube factor. For example, $-16 = -8 \cdot 2$, so $\sqrt[3]{-16} = \sqrt[3]{-8} \cdot \sqrt[3]{2} = -2\sqrt[3]{2}$. This procedure can be used with any odd root of a negative number. (See Example 5(c).)

EXAMPLE 5 **Simplifying radical expressions**

Simplify each expression. Assume that all variables are positive.
(a) $\sqrt{25x^4}$ **(b)** $\sqrt{32n^3}$ **(c)** $\sqrt[3]{-16x^3y^5}$ **(d)** $\sqrt[3]{2a} \cdot \sqrt[3]{4a^2b}$

Solution
(a) $\sqrt{25x^4} = 5x^2$ Perfect square: $(5x^2)^2 = 25x^4$
(b) $\sqrt{32n^3} = \sqrt{(16n^2)2n}$ $16n^2$ is the largest perfect square factor.
            $= \sqrt{16n^2} \cdot \sqrt{2n}$ Product rule
            $= 4n\sqrt{2n}$ $(4n)^2 = 16n^2$
(c) $\sqrt[3]{-16x^3y^5} = \sqrt[3]{(-8x^3y^3)2y^2}$ $8x^3y^3$ is the largest perfect cube factor.
            $= \sqrt[3]{-8x^3y^3} \cdot \sqrt[3]{2y^2}$ Product rule
            $= -2xy\sqrt[3]{2y^2}$ $(-2xy)^3 = -8x^3y^3$
(d) $\sqrt[3]{2a} \cdot \sqrt[3]{4a^2b} = \sqrt[3]{(2a)(4a^2b)}$ Product rule
            $= \sqrt[3]{(8a^3)b}$ $8a^3$ is the largest perfect cube factor.
            $= \sqrt[3]{8a^3} \cdot \sqrt[3]{b}$ Product rule
            $= 2a\sqrt[3]{b}$ $(2a)^3 = 8a^3$

Now Try Exercises 45, 85, 89, 91

The product rule for radical expressions cannot be used if the radicals do not have the same indexes. In this case we use rational exponents, as illustrated in the next example.

EXAMPLE 6 **Multiplying radicals with different indexes**

Simplify each expression. Write your answer in radical notation.

(a) $\sqrt{5} \cdot \sqrt[4]{5}$ **(b)** $\sqrt{2} \cdot \sqrt[3]{4}$ **(c)** $\sqrt[3]{x} \cdot \sqrt[4]{x}$

Solution

(a) Because $\sqrt{5} = 5^{1/2}$ and $\sqrt[4]{5} = 5^{1/4}$,

$$\sqrt{5} \cdot \sqrt[4]{5} = 5^{1/2} \cdot 5^{1/4} = 5^{1/2 + 1/4} = 5^{3/4}.$$

In radical notation, $5^{3/4} = \sqrt[4]{5^3} = \sqrt[4]{125}$.

(b) Because $\sqrt[3]{4} = \sqrt[3]{2^2} = 2^{2/3}$,

$$\sqrt{2} \cdot \sqrt[3]{4} = 2^{1/2} \cdot 2^{2/3} = 2^{1/2 + 2/3} = 2^{7/6}.$$

In radical notation, $2^{7/6} = \sqrt[6]{2^7} = \sqrt[6]{2^6 \cdot 2^1} = \sqrt[6]{2^6} \cdot \sqrt[6]{2} = 2\sqrt[6]{2}$.

(c) $\sqrt[3]{x} \cdot \sqrt[4]{x} = x^{1/3} \cdot x^{1/4} = x^{7/12} = \sqrt[12]{x^7}$

Now Try Exercises 101, 103, 107

Quotient Rule for Radical Expressions

$\sqrt{(4/9)}\blacktriangleright$Frac
                    2/3
$\sqrt{(4)}/\sqrt{(9)}\blacktriangleright$Frac
                    2/3

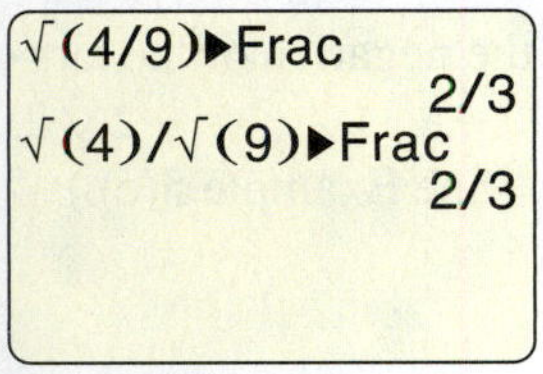

Figure 7.12

Consider the following examples of dividing radical expressions. The equations

$$\sqrt{\frac{4}{9}} = \sqrt{\frac{2}{3} \cdot \frac{2}{3}} = \frac{2}{3} \quad \text{and} \quad \frac{\sqrt{4}}{\sqrt{9}} = \frac{2}{3}$$

imply that

$$\sqrt{\frac{4}{9}} = \frac{\sqrt{4}}{\sqrt{9}} \quad \text{(see Figure 7.12)}.$$

These examples suggest that *the root of a quotient is equal to the quotient of the roots.*

CALCULATOR HELP

To use the Frac feature, see Appendix A (pages AP-1 and AP-2).

> **QUOTIENT RULE FOR RADICAL EXPRESSIONS**
>
> Let a and b be real numbers, where $\sqrt[n]{a}$ and $\sqrt[n]{b}$ are both defined and $b \neq 0$. Then
>
> $$\sqrt[n]{\frac{a}{b}} = \frac{\sqrt[n]{a}}{\sqrt[n]{b}}.$$
>
> $$\sqrt[3]{\frac{23}{5}} = \frac{\sqrt[3]{23}}{\sqrt[3]{5}}$$

EXAMPLE 7 **Simplifying quotients**

Simplify each radical expression. Assume that all variables are positive.

(a) $\sqrt[3]{\dfrac{5}{8}}$ **(b)** $\sqrt[4]{\dfrac{x}{16}}$ **(c)** $\sqrt{\dfrac{16}{y^2}}$

Solution

(a) $\sqrt[3]{\dfrac{5}{8}} = \dfrac{\sqrt[3]{5}}{\sqrt[3]{8}} = \dfrac{\sqrt[3]{5}}{2}$ **(b)** $\sqrt[4]{\dfrac{x}{16}} = \dfrac{\sqrt[4]{x}}{\sqrt[4]{16}} = \dfrac{\sqrt[4]{x}}{2}$

(c) $\sqrt{\dfrac{16}{y^2}} = \dfrac{\sqrt{16}}{\sqrt{y^2}} = \dfrac{4}{y}$ because $y > 0$.

Now Try Exercises 25, 27, 29

EXAMPLE 8	**Simplifying quotients**

Simplify each radical expression. Assume that all variables are positive.

(a) $\dfrac{\sqrt{40}}{\sqrt{10}}$ **(b)** $\dfrac{\sqrt[3]{2}}{\sqrt[3]{16}}$ **(c)** $\dfrac{\sqrt{x^2 y}}{\sqrt{y}}$

Solution

(a) $\dfrac{\sqrt{40}}{\sqrt{10}} = \sqrt{\dfrac{40}{10}} = \sqrt{4} = 2$

(b) $\dfrac{\sqrt[3]{2}}{\sqrt[3]{16}} = \sqrt[3]{\dfrac{2}{16}} = \sqrt[3]{\dfrac{1}{8}} = \dfrac{1}{2}$ because $\dfrac{1}{2} \cdot \dfrac{1}{2} \cdot \dfrac{1}{2} = \dfrac{1}{8}$.

(c) $\dfrac{\sqrt{x^2 y}}{\sqrt{y}} = \sqrt{\dfrac{x^2 y}{y}} = \sqrt{x^2} = x$ because $x > 0$.

Now Try Exercises 33, 39, 41

MAKING CONNECTIONS

Rules for Radical Expressions and Rational Exponents

The rules for radical expressions are a result of the properties of rational exponents.

$$\sqrt[n]{a \cdot b} = \sqrt[n]{a} \cdot \sqrt[n]{b} \quad \text{is equivalent to} \quad (a \cdot b)^{1/n} = a^{1/n} \cdot b^{1/n}.$$

$$\sqrt[n]{\dfrac{a}{b}} = \dfrac{\sqrt[n]{a}}{\sqrt[n]{b}} \quad \text{is equivalent to} \quad \left(\dfrac{a}{b}\right)^{1/n} = \dfrac{a^{1/n}}{b^{1/n}}.$$

READING CHECK

Use rational exponents to explain why $\sqrt[3]{x} \cdot \sqrt[3]{x} \cdot \sqrt[3]{x} = x$.

EXAMPLE 9	**Simplifying radical expressions**

Simplify each radical expression. Assume that all variables are positive.

(a) $\sqrt[4]{\dfrac{16x^3}{y^4}}$ **(b)** $\sqrt{\dfrac{5a^2}{8}} \cdot \sqrt{\dfrac{5a^3}{2}}$

Solution

(a) To simplify this expression, we first use the quotient rule for radical expressions and then apply the product rule for radical expressions.

$$\sqrt[4]{\dfrac{16x^3}{y^4}} = \dfrac{\sqrt[4]{16x^3}}{\sqrt[4]{y^4}} \qquad \text{Quotient rule}$$

$$= \dfrac{\sqrt[4]{16}\,\sqrt[4]{x^3}}{\sqrt[4]{y^4}} \qquad \text{Product rule}$$

$$= \dfrac{2\sqrt[4]{x^3}}{y} \qquad \text{Evaluate 4th roots.}$$

(b) To simplify this expression, we use both the product and quotient rules.

$$\sqrt{\frac{5a^2}{8}} \cdot \sqrt{\frac{5a^3}{2}} = \sqrt{\frac{5a^2 \cdot 5a^3}{8 \cdot 2}} \qquad \text{Product rule}$$

$$= \sqrt{\frac{25a^5}{16}} \qquad \text{Multiply.}$$

$$= \frac{\sqrt{25a^5}}{\sqrt{16}} \qquad \text{Quotient rule}$$

$$= \frac{\sqrt{25a^4 \cdot a}}{\sqrt{16}} \qquad \text{Factor out largest perfect square.}$$

$$= \frac{\sqrt{25a^4} \cdot \sqrt{a}}{\sqrt{16}} \qquad \text{Product rule}$$

$$= \frac{5a^2\sqrt{a}}{4} \qquad (5a^2)^2 = 25a^4$$

Now Try Exercises 95, 97

EXAMPLE 10 | **Simplifying products and quotients of roots**

Simplify each expression. Assume all radicands are positive.

(a) $\sqrt{x - 3} \cdot \sqrt{x + 3}$ **(b)** $\dfrac{\sqrt[3]{x^2 + 3x + 2}}{\sqrt[3]{x + 1}}$

Solution

(a) Start by applying the product rule for radical expressions.

$$\sqrt{x - 3} \cdot \sqrt{x + 3} = \sqrt{(x - 3)(x + 3)} \qquad \text{Product rule}$$

$$= \sqrt{x^2 - 9} \qquad \text{Multiply binomials.}$$

NOTE: The expression $\sqrt{x^2 - 9}$ does *not* simplify further. It is important to realize that

$$\sqrt{x^2 - 9} \neq \sqrt{x^2} - \sqrt{9} = x - 3.$$

For example, $\sqrt{5^2 - 3^2} = \sqrt{16} = 4$, but $\sqrt{5^2} - \sqrt{3^2} = 5 - 3 = 2$.

(b) Start by applying the quotient rule for radical expressions.

$$\frac{\sqrt[3]{x^2 + 3x + 2}}{\sqrt[3]{x + 1}} = \sqrt[3]{\frac{x^2 + 3x + 2}{x + 1}} \qquad \text{Quotient rule}$$

$$= \sqrt[3]{\frac{(x + 1)(x + 2)}{x + 1}} \qquad \text{Factor trinomial.}$$

$$= \sqrt[3]{x + 2} \qquad \text{Simplify quotient.}$$

Now Try Exercises 63, 67

7.3 Putting It All Together

CONCEPT	EXPLANATION	EXAMPLES
Product Rule for Radical Expressions	Let a and b be real numbers, where $\sqrt[n]{a}$ and $\sqrt[n]{b}$ are both defined. Then $$\sqrt[n]{a} \cdot \sqrt[n]{b} = \sqrt[n]{a \cdot b}.$$	$$\sqrt{2} \cdot \sqrt{32} = \sqrt{64} = 8$$ $$\sqrt{500} = \sqrt{100} \cdot \sqrt{5} = 10\sqrt{5}$$
Simplifying Radicals (nth roots)	**STEP 1:** Find the largest perfect nth power factor of the radicand. **STEP 2:** Factor out and simplify this perfect nth power.	$$\sqrt{12} = \sqrt{4 \cdot 3} = \sqrt{4} \cdot \sqrt{3} = 2\sqrt{3}$$ $$\sqrt[3]{81} = \sqrt[3]{27 \cdot 3} = \sqrt[3]{27} \cdot \sqrt[3]{3} = 3\sqrt[3]{3}$$ $$\sqrt[4]{x^5} = \sqrt[4]{x^4 \cdot x} = \sqrt[4]{x^4} \cdot \sqrt[4]{x} = x\sqrt[4]{x},$$ provided $x \geq 0$.
Quotient Rule for Radical Expressions	Let a and b be real numbers, where $\sqrt[n]{a}$ and $\sqrt[n]{b}$ are both defined and $b \neq 0$. Then $$\sqrt[n]{\frac{a}{b}} = \frac{\sqrt[n]{a}}{\sqrt[n]{b}}.$$	$$\frac{\sqrt{60}}{\sqrt{15}} = \sqrt{\frac{60}{15}} = \sqrt{4} = 2$$ $$\sqrt[3]{\frac{x^2}{-27}} = \frac{\sqrt[3]{x^2}}{\sqrt[3]{-27}} = \frac{\sqrt[3]{x^2}}{-3} = -\frac{\sqrt[3]{x^2}}{3}$$

7.3 Exercises

MyMathLab

CONCEPTS AND VOCABULARY

1. Does $\sqrt{2} \cdot \sqrt{3}$ equal $\sqrt{6}$?

2. Does $\sqrt{5} \cdot \sqrt[3]{5}$ equal 5?

3. $\sqrt[3]{a} \cdot \sqrt[3]{b} = $ _____   **4.** $\dfrac{\sqrt{a}}{\sqrt{b}} = \sqrt{?}$

5. $\dfrac{\sqrt[n]{a}}{\sqrt[n]{b}} = \sqrt[n]{?}$ **6.** $\dfrac{\sqrt{3}}{\sqrt{27}} = $ _____

7. Does $\sqrt{50} = \sqrt{25} + \sqrt{25}$?

8. Does $\sqrt{50} = 5\sqrt{2}$?

9. Is $\sqrt[3]{3}$ equal to 1? Explain.

10. Is 64 a perfect cube? Explain.

MULTIPLYING AND DIVIDING

Exercises 11–62: Simplify the expression. Assume that all variables are positive.

11. $\sqrt{3} \cdot \sqrt{3}$ **12.** $\sqrt{2} \cdot \sqrt{18}$

13. $\sqrt{2} \cdot \sqrt{50}$ **14.** $\sqrt[3]{-2} \cdot \sqrt[3]{-4}$

15. $\sqrt[3]{4} \cdot \sqrt[3]{16}$ **16.** $\sqrt[3]{x} \cdot \sqrt[3]{x^2}$

17. $\sqrt{\dfrac{9}{25}}$ **18.** $\sqrt[3]{\dfrac{x}{8}}$

19. $\sqrt{\dfrac{1}{2}} \cdot \sqrt{\dfrac{1}{8}}$ **20.** $\sqrt{\dfrac{5}{3}} \cdot \sqrt{\dfrac{1}{3}}$

21. $\sqrt[3]{\dfrac{2}{3}} \cdot \sqrt[3]{\dfrac{4}{3}} \cdot \sqrt[3]{\dfrac{1}{3}}$ **22.** $\sqrt[4]{\dfrac{8}{3}} \cdot \sqrt[4]{\dfrac{4}{9}} \cdot \sqrt[4]{\dfrac{8}{3}}$

23. $\sqrt{x^3} \cdot \sqrt{x^3}$ **24.** $\sqrt{z} \cdot \sqrt{z^7}$

25. $\sqrt[3]{\dfrac{7}{27}}$ **26.** $\sqrt[3]{\dfrac{9}{64}}$

27. $\sqrt[4]{\dfrac{x}{81}}$ **28.** $\sqrt[4]{\dfrac{16x}{81}}$

29. $\sqrt{\dfrac{9}{z^2}}$ **30.** $\sqrt{\dfrac{x^2}{81}}$

31. $\sqrt{\dfrac{x}{2}} \cdot \sqrt{\dfrac{x}{8}}$ **32.** $\sqrt{\dfrac{4}{y}} \cdot \sqrt{\dfrac{y}{5}}$

33. $\dfrac{\sqrt{45}}{\sqrt{5}}$ **34.** $\dfrac{\sqrt{7}}{\sqrt{28}}$

35. $\sqrt[3]{-4} \cdot \sqrt[3]{-16}$ **36.** $\sqrt[3]{9} \cdot \sqrt[3]{3}$

37. $\sqrt[4]{9} \cdot \sqrt[4]{9}$ **38.** $\sqrt[5]{16} \cdot \sqrt[5]{-2}$

39. $\dfrac{\sqrt[5]{64}}{\sqrt[5]{-2}}$ **40.** $\dfrac{\sqrt[4]{324}}{\sqrt[4]{4}}$

41. $\dfrac{\sqrt{a^2 b}}{\sqrt{b}}$ **42.** $\dfrac{\sqrt{4xy^2}}{\sqrt{x}}$

43. $\dfrac{\sqrt[3]{54}}{\sqrt[3]{2}}$ **44.** $\dfrac{\sqrt[3]{x^3 y^7}}{\sqrt[3]{y^4}}$

45. $\sqrt{4x^4}$

46. $\sqrt[3]{-8y^3}$

47. $\sqrt[3]{-5a^6}$

48. $\sqrt{9x^2y}$

49. $\sqrt[4]{16x^4y}$

50. $\sqrt[3]{8xy^3}$

51. $\sqrt{3x} \cdot \sqrt{12x}$

52. $\sqrt{6x^5} \cdot \sqrt{6x}$

53. $\sqrt[3]{8x^6y^3z^9}$

54. $\sqrt{16x^4y^6}$

55. $\sqrt[4]{\dfrac{3}{4}} \cdot \sqrt[4]{\dfrac{27}{4}}$

56. $\sqrt[5]{\dfrac{4}{-9}} \cdot \sqrt[5]{\dfrac{8}{-27}}$

57. $\sqrt[3]{12} \cdot \sqrt[3]{ab}$

58. $\sqrt{5x} \cdot \sqrt{5z}$

59. $\sqrt[4]{25z} \cdot \sqrt[4]{25z}$

60. $\sqrt[5]{3z^2} \cdot \sqrt[5]{7z}$

61. $\sqrt[5]{\dfrac{7a}{b^2}} \cdot \sqrt[5]{\dfrac{b^2}{7a^6}}$

62. $\sqrt[3]{\dfrac{8m}{n}} \cdot \sqrt[3]{\dfrac{n^4}{m^2}}$

Exercises 63–68: Use properties of polynomials to simplify the expression. Assume all radicands are positive.

63. $\sqrt{x+4} \cdot \sqrt{x-4}$

64. $\sqrt[3]{x-1} \cdot \sqrt[3]{x^2+x+1}$

65. $\sqrt[3]{a+1} \cdot \sqrt[3]{a^2-a+1}$

66. $\sqrt{b-1} \cdot \sqrt{b+1}$

67. $\dfrac{\sqrt{x^2+2x+1}}{\sqrt{x+1}}$

68. $\dfrac{\sqrt{x^2-4x+4}}{\sqrt{x-2}}$

Exercises 69–74: Complete the equation.

69. $\sqrt{500} = \underline{\quad} \sqrt{5}$ **70.** $\sqrt{28} = \underline{\quad} \sqrt{7}$

71. $\sqrt{8} = \underline{\quad} \sqrt{2}$ **72.** $\sqrt{99} = \underline{\quad} \sqrt{11}$

73. $\sqrt{45} = \underline{\quad} \sqrt{5}$ **74.** $\sqrt{243} = \underline{\quad} \sqrt{3}$

Exercises 75–98: Simplify the radical expression by factoring out the largest perfect nth power. Assume that all variables are positive.

75. $\sqrt{200}$

76. $\sqrt{72}$

77. $\sqrt[3]{81}$

78. $\sqrt[3]{256}$

79. $\sqrt[4]{64}$

80. $\sqrt[5]{27 \cdot 81}$

81. $\sqrt[5]{-64}$

82. $\sqrt[3]{-81}$

83. $\sqrt{b^5}$

84. $\sqrt{t^3}$

85. $\sqrt{8n^3}$

86. $\sqrt{32a^2}$

87. $\sqrt{12a^2b^5}$

88. $\sqrt{20a^3b^2}$

89. $\sqrt[3]{-125x^4y^5}$

90. $\sqrt[3]{-81a^5b^2}$

91. $\sqrt[3]{5t} \cdot \sqrt[3]{125t}$

92. $\sqrt[4]{4bc^3} \cdot \sqrt[4]{64ab^3c^2}$

93. $\sqrt[4]{\dfrac{9t^5}{r^8}} \cdot \sqrt[4]{\dfrac{9r}{5t}}$

94. $\sqrt[5]{\dfrac{4t^6}{r}} \cdot \sqrt[5]{\dfrac{8t}{r^6}}$

95. $\sqrt[3]{\dfrac{27x^2}{y^3}}$

96. $\sqrt[4]{\dfrac{32x^8}{z^4}}$

97. $\sqrt{\dfrac{7a^2}{27}} \cdot \sqrt{\dfrac{7a}{3}}$

98. $\sqrt{\dfrac{8a}{125}} \cdot \sqrt{\dfrac{2a}{5}}$

Exercises 99 and 100: **Thinking Generally** *Let m and n be positive integers and assume each expression exists.*

99. Simplify $\left(\sqrt[mn]{a^m b^m} \right)^n$ so that it equals ab.

100. Simplify $\left(\sqrt[mn]{a^n b^m} \right) \cdot \left(\sqrt[mn]{a^m b^n} \right)$ so that it equals $\sqrt[m]{ab} \cdot \sqrt[n]{ab}$.

Exercises 101–110: Simplify the expression. Let all variables be positive and write your answer in radical notation.

101. $\sqrt{3} \cdot \sqrt[3]{3}$

102. $\sqrt{5} \cdot \sqrt[3]{5}$

103. $\sqrt[4]{8} \cdot \sqrt[3]{4}$

104. $\sqrt[5]{16} \cdot \sqrt{2}$

105. $\sqrt[4]{27} \cdot \sqrt[3]{9} \cdot \sqrt{3}$

106. $\sqrt[5]{16} \cdot \sqrt[3]{16}$

107. $\sqrt[4]{x^3} \cdot \sqrt[3]{x}$

108. $\sqrt[4]{x^3} \cdot \sqrt{x}$

109. $\sqrt[4]{rt} \cdot \sqrt[3]{r^2t}$

110. $\sqrt[3]{a^3b^2} \cdot \sqrt{a^2b}$

APPLICATIONS

111. *Minimum Speed* (Refer to Example 4.) If the pavement is dry cement, then $S = \sqrt{25M}$ is an estimate for a car's speed S in miles per hour when it leaves skid marks M feet long.
  (a) Simplify this formula.
  (b) Determine the minimum speed of the car if the skid marks are 100 feet.

112. *Minimum Speed* (Refer to Example 4.) If the road surface is gravel, then $S = \sqrt{15M}$ is an estimate for a car's speed S in miles per hour when it leaves skid marks M feet long in the gravel. Determine the minimum speed of the car if the skid marks are 30 feet.

WRITING ABOUT MATHEMATICS

113. Explain what it means for a positive integer to be a *perfect square* or a *perfect cube*. List the positive integers that are perfect squares and less than 101. List the positive integers that are perfect cubes and less than 220.

114. Explain how the product and quotient rules for radical expressions are the result of properties of rational exponents.

Group Activity   Working with Real Data

Directions: Form a group of 2 to 4 people. Select someone to record the group's responses for this activity. All members of the group should work cooperatively to answer the questions. If your instructor asks for your results, each member of the group should be prepared to respond.

Designing a Paper Cup A paper drinking cup is being designed in the shape shown in the accompanying figure. The amount of paper needed to manufacture the cup is determined by the surface area S of the cup, which is given by

$$S = \pi r \sqrt{r^2 + h^2},$$

where r is the radius and h is the height.

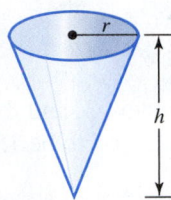

(a) Approximate S to the nearest hundredth when $r = 1.5$ inches and $h = 4$ inches.

(b) Could the formula for S be simplified as follows?

$$\pi r \sqrt{r^2 + h^2} \stackrel{?}{=} \pi r (\sqrt{r^2} + \sqrt{h^2})$$
$$\stackrel{?}{=} \pi r (r + h)$$

Try evaluating this second formula with $r = 1.5$ inches and $h = 4$ inches. Is your answer the same as in part (a)? Explain.

(c) Discuss why evaluating real-world formulas correctly is important.

(d) In general, does $\sqrt{a + b}$ equal $\sqrt{a} + \sqrt{b}$? Justify your answer by completing the following table. Approximate answers to the nearest hundredth when appropriate.

a	b	$\sqrt{a + b}$	$\sqrt{a} + \sqrt{b}$
0	4		
4	0		
5	4		
9	7		
4	16		
25	100		

7.4   Operations on Radical Expressions

Addition and Subtraction • Multiplication • Rationalizing the Denominator

A LOOK INTO MATH ▶

Our productivity as a nation has increased significantly during the past 70 years because of new technology. For example, if a typical business bought $10,000 in equipment per worker in 2005, it would have generated about $48,000 per worker in revenue. This same investment by a business in 1935 would have generated only $23,000 in revenue (adjusted to 2005 dollars). This situation can be described by the following two functions.

$$N(x) = 400\sqrt{x} + 8000 \quad \text{and} \quad O(x) = 195\sqrt{x} + 3500$$

2005 technology (new)   1935 technology (old)

What does the difference of these two functions represent? In this section we learn how to subtract expressions that contain radicals. See Example 7.

NEW VOCABULARY

☐ Like radicals
☐ Rationalizing the denominator
☐ Conjugate

Addition and Subtraction

We can add $2x^2$ and $5x^2$ to obtain $7x^2$ because they are *like* terms. That is,

$$2x^2 + 5x^2 = (2 + 5)x^2 = 7x^2. \quad \text{Like terms}$$

We can also add and subtract *like radicals*. **Like radicals** *have the same index and the same radicand*. For example, we can add $3\sqrt{2}$ and $5\sqrt{2}$ because they are like radicals.

$$3\sqrt{2} + 5\sqrt{2} = (3 + 5)\sqrt{2} = 8\sqrt{2} \quad \text{Like radicals}$$

EXAMPLE 1 Adding like radicals

If possible, add the expressions and simplify.
(a) $10\sqrt{11} + 4\sqrt{11}$ **(b)** $5\sqrt[3]{6} + \sqrt[3]{6}$
(c) $4 + 5\sqrt{3}$ **(d)** $\sqrt{7} + \sqrt{11}$

Solution
(a) These terms are like radicals because they have the same index, 2, and the same radicand, 11.

$$10\sqrt{11} + 4\sqrt{11} = (10 + 4)\sqrt{11} = 14\sqrt{11}$$

(b) These terms are like radicals because they have the same index, 3, and the same radicand, 6. Note that the coefficient on the second term is understood to be 1.

$$5\sqrt[3]{6} + 1\sqrt[3]{6} = (5 + 1)\sqrt[3]{6} = 6\sqrt[3]{6}$$

(c) The expression $4 + 5\sqrt{3}$ can be written as $4\sqrt{1} + 5\sqrt{3}$. These terms cannot be added because they are not like radicals.

NOTE: $4 + 5\sqrt{3} \neq 9\sqrt{3}$ because multiplication is performed before addition.

(d) The expression $\sqrt{7} + \sqrt{11}$ contains unlike radicals that *cannot* be added.

■ **Now Try Exercises 19, 21, 23, 25**

Sometimes two radical expressions that are not alike can be added by changing them to like radicals. For example, $\sqrt{20}$ and $\sqrt{5}$ are unlike radicals. However,

because $\qquad \sqrt{20} = \sqrt{4 \cdot 5} = \sqrt{4} \cdot \sqrt{5} = 2\sqrt{5},$

we have $\qquad \sqrt{20} + \sqrt{5} = 2\sqrt{5} + 1\sqrt{5} = 3\sqrt{5}.$

We cannot combine $x + x^2$ because they are unlike terms. Similarly, we cannot combine $\sqrt{2} + \sqrt{5}$ because they are unlike radicals. When combining radicals, the first step is to see if we can write pairs of terms as like radicals, as demonstrated in the next example.

EXAMPLE 2 Finding like radicals

Write each pair of terms as like radicals, if possible.
(a) $\sqrt{45}, \sqrt{20}$ **(b)** $\sqrt{27}, \sqrt{5}$ **(c)** $5\sqrt[3]{16}, 4\sqrt[3]{54}$

Solution
(a) The expressions $\sqrt{45}$ and $\sqrt{20}$ are unlike radicals. However, they can be changed to like radicals, as follows.

$$\sqrt{45} = \sqrt{9 \cdot 5} = \sqrt{9} \cdot \sqrt{5} = 3\sqrt{5} \quad \text{and}$$

$$\sqrt{20} = \sqrt{4 \cdot 5} = \sqrt{4} \cdot \sqrt{5} = 2\sqrt{5}$$

The expressions $3\sqrt{5}$ and $2\sqrt{5}$ are like radicals.

(b) Because $\sqrt{27} = \sqrt{9 \cdot 3} = \sqrt{9} \cdot \sqrt{3} = 3\sqrt{3}$, the given expressions $\sqrt{27}$ and $\sqrt{5}$ are unlike radicals and cannot be written as like radicals.

(c) $5\sqrt[3]{16} = 5\sqrt[3]{8 \cdot 2} = 5\sqrt[3]{8} \cdot \sqrt[3]{2} = 5 \cdot 2 \cdot \sqrt[3]{2} = 10\sqrt[3]{2}$ and

$4\sqrt[3]{54} = 4\sqrt[3]{27 \cdot 2} = 4\sqrt[3]{27} \cdot \sqrt[3]{2} = 4 \cdot 3 \cdot \sqrt[3]{2} = 12\sqrt[3]{2}$

The expressions $10\sqrt[3]{2}$ and $12\sqrt[3]{2}$ are like radicals.

Now Try Exercises 9, 11, 13

We use these techniques to add radical expressions in the next two examples.

EXAMPLE 3   **Adding radical expressions**

Add the expressions and simplify.
(a) $\sqrt{12} + 7\sqrt{3}$ **(b)** $\sqrt[3]{16} + \sqrt[3]{2}$ **(c)** $3\sqrt{2} + \sqrt{8} + \sqrt{18}$

Solution
(a)
$$
\begin{aligned}
\sqrt{12} + 7\sqrt{3} &= \sqrt{4 \cdot 3} + 7\sqrt{3} \\
&= \sqrt{4} \cdot \sqrt{3} + 7\sqrt{3} \\
&= 2\sqrt{3} + 7\sqrt{3} \\
&= 9\sqrt{3}
\end{aligned}
$$

(b)
$$
\begin{aligned}
\sqrt[3]{16} + \sqrt[3]{2} &= \sqrt[3]{8 \cdot 2} + \sqrt[3]{2} \\
&= \sqrt[3]{8} \cdot \sqrt[3]{2} + \sqrt[3]{2} \\
&= 2\sqrt[3]{2} + 1\sqrt[3]{2} \\
&= 3\sqrt[3]{2}
\end{aligned}
$$

(c)
$$
\begin{aligned}
3\sqrt{2} + \sqrt{8} + \sqrt{18} &= 3\sqrt{2} + \sqrt{4 \cdot 2} + \sqrt{9 \cdot 2} \\
&= 3\sqrt{2} + \sqrt{4} \cdot \sqrt{2} + \sqrt{9} \cdot \sqrt{2} \\
&= 3\sqrt{2} + 2\sqrt{2} + 3\sqrt{2} \\
&= 8\sqrt{2}
\end{aligned}
$$

Now Try Exercises 29, 31, 43

EXAMPLE 4   **Adding radical expressions**

Add the expressions and simplify. Assume that all variables are positive.
(a) $\sqrt[4]{32} + 3\sqrt[4]{2}$ **(b)** $-2\sqrt{4x} + \sqrt{x}$ **(c)** $3\sqrt{3k} + 5\sqrt{12k} + 9\sqrt{48k}$

Solution
(a) Because $\sqrt[4]{32} = \sqrt[4]{16 \cdot 2} = \sqrt[4]{16} \cdot \sqrt[4]{2} = 2\sqrt[4]{2}$, we can add and simplify as follows.
$$\sqrt[4]{32} + 3\sqrt[4]{2} = 2\sqrt[4]{2} + 3\sqrt[4]{2} = 5\sqrt[4]{2}$$
(b) Note that $\sqrt{4x} = \sqrt{4} \cdot \sqrt{x} = 2\sqrt{x}$.
$$-2\sqrt{4x} + \sqrt{x} = -2(2\sqrt{x}) + \sqrt{x} = -4\sqrt{x} + 1\sqrt{x} = -3\sqrt{x}$$
(c) Note that $\sqrt{12k} = \sqrt{4} \cdot \sqrt{3k} = 2\sqrt{3k}$ and that $\sqrt{48k} = \sqrt{16} \cdot \sqrt{3k} = 4\sqrt{3k}$.
$$
\begin{aligned}
3\sqrt{3k} + 5\sqrt{12k} + 9\sqrt{48k} &= 3\sqrt{3k} + 5(2\sqrt{3k}) + 9(4\sqrt{3k}) \\
&= (3 + 10 + 36)\sqrt{3k} \\
&= 49\sqrt{3k}
\end{aligned}
$$

Now Try Exercises 45, 47, 49

Subtraction of radical expressions is similar to addition, as illustrated in the next three examples.

EXAMPLE 5 | **Subtracting like radicals**

Simplify the expressions.

(a) $5\sqrt{7} - 3\sqrt{7}$ (b) $8\sqrt[3]{5} - 3\sqrt[3]{5} + \sqrt[3]{11}$ (c) $5\sqrt{z} + \sqrt[3]{z} - 2\sqrt{z}$

Solution

(a) $5\sqrt{7} - 3\sqrt{7} = (5 - 3)\sqrt{7} = 2\sqrt{7}$

(b) $8\sqrt[3]{5} - 3\sqrt[3]{5} + \sqrt[3]{11} = (8 - 3)\sqrt[3]{5} + \sqrt[3]{11} = 5\sqrt[3]{5} + \sqrt[3]{11}$

(c) $5\sqrt{z} + \sqrt[3]{z} - 2\sqrt{z} = 5\sqrt{z} - 2\sqrt{z} + \sqrt[3]{z}$ Commutative property

$\qquad\qquad\qquad = (5 - 2)\sqrt{z} + \sqrt[3]{z}$ Distributive property

$\qquad\qquad\qquad = 3\sqrt{z} + \sqrt[3]{z}$ Subtract.

NOTE: We cannot combine $3\sqrt{z} + \sqrt[3]{z}$ because their indexes are different. That is, one term is a square root and the other is a cube root.

Now Try Exercises 33, 35, 39

EXAMPLE 6 | **Subtracting radical expressions**

Subtract and simplify. Assume that all variables are positive.

(a) $3\sqrt[3]{xy^2} - 2\sqrt[3]{xy^2}$ (b) $\sqrt{16x^3} - \sqrt{x^3}$ (c) $\sqrt[3]{\dfrac{5x}{27}} - \dfrac{\sqrt[3]{5x}}{6}$

Solution

(a) $3\sqrt[3]{xy^2} - 2\sqrt[3]{xy^2} = (3 - 2)\sqrt[3]{xy^2} = \sqrt[3]{xy^2}$

(b) $\sqrt{16x^3} - \sqrt{x^3} = \sqrt{16x^2} \cdot \sqrt{x} - \sqrt{x^2} \cdot \sqrt{x}$ Factor out perfect squares.

$\qquad\qquad\qquad = 4x\sqrt{x} - x\sqrt{x}$ Simplify.

$\qquad\qquad\qquad = (4x - x)\sqrt{x}$ Distributive property

$\qquad\qquad\qquad = 3x\sqrt{x}$ Subtract.

(c) $\sqrt[3]{\dfrac{5x}{27}} - \dfrac{\sqrt[3]{5x}}{6} = \dfrac{\sqrt[3]{5x}}{\sqrt[3]{27}} - \dfrac{\sqrt[3]{5x}}{6}$ Quotient rule for radical expressions

$\qquad\qquad\qquad = \dfrac{\sqrt[3]{5x}}{3} - \dfrac{\sqrt[3]{5x}}{6}$ Evaluate $\sqrt[3]{27} = 3$.

$\qquad\qquad\qquad = \dfrac{2\sqrt[3]{5x}}{6} - \dfrac{\sqrt[3]{5x}}{6}$ Find a common denominator.

$\qquad\qquad\qquad = \dfrac{2\sqrt[3]{5x} - \sqrt[3]{5x}}{6}$ Subtract numerators.

$\qquad\qquad\qquad = \dfrac{\sqrt[3]{5x}}{6}$ Simplify.

Now Try Exercises 55, 63, 65

▶ **REAL-WORLD CONNECTION** In A Look Into Math at the beginning of this section, we discussed how functions with radicals could describe an increase in productivity due to technology. In the next example, we calculate the difference of these two functions and interpret the result.

EXAMPLE 7 Increasing productivity with new technology

The functions given by

$$N(x) = 400\sqrt{x} + 8000 \quad \text{and} \quad O(x) = 195\sqrt{x} + 3500$$

2005 technology (new)　　　　　1935 technology (old)

approximate the increase in revenue resulting from investing x dollars in equipment (per worker).
(a) Find their difference $D(x) = N(x) - O(x)$. Simplify your answer.
(b) Evaluate $D(40,000)$ and interpret the result.

Solution
(a) $D(x) = N(x) - O(x)$

$$
\begin{aligned}
&= (400\sqrt{x} + 8000) - (195\sqrt{x} + 3500) && \text{Substitute.}\\
&= 400\sqrt{x} - 195\sqrt{x} + 8000 - 3500 && \text{Rewrite terms.}\\
&= (400 - 195)\sqrt{x} + 4500 && \text{Distributive property}\\
&= 205\sqrt{x} + 4500 && \text{Simplify.}
\end{aligned}
$$

The difference is given by $D(x) = 205\sqrt{x} + 4500$.
(b) $D(40,000) = 205\sqrt{40,000} + 4500 = \$45,500$. An investment of \$40,000 per worker resulted in \$45,500 *more* revenue per worker in 2005 than in 1935 (adjusted to 2005 dollars).

■ **Now Try Exercise 81**

EXAMPLE 8 Subtracting radical expressions

Subtract and simplify. Assume that all variables are positive.

(a) $\dfrac{5\sqrt{2}}{3} - \dfrac{2\sqrt{2}}{4}$　**(b)** $\sqrt[4]{81a^5b^6} - \sqrt[4]{16ab^2}$　**(c)** $3\sqrt[3]{\dfrac{n^5}{27}} - 2\sqrt[3]{n^2}$

Solution

(a)
$$
\begin{aligned}
\frac{5\sqrt{2}}{3} - \frac{2\sqrt{2}}{4} &= \frac{5\sqrt{2}}{3} \cdot \frac{4}{4} - \frac{2\sqrt{2}}{4} \cdot \frac{3}{3} && \text{LCD is 12.}\\[2mm]
&= \frac{20\sqrt{2}}{12} - \frac{6\sqrt{2}}{12} && \text{Multiply fractions.}\\[2mm]
&= \frac{14\sqrt{2}}{12} && \text{Subtract numerators.}\\[2mm]
&= \frac{7\sqrt{2}}{6} && \text{Simplify.}
\end{aligned}
$$

(b)
$$
\begin{aligned}
\sqrt[4]{81a^5b^6} - \sqrt[4]{16ab^2} &= \sqrt[4]{81a^4b^4} \cdot \sqrt[4]{ab^2} - \sqrt[4]{16} \cdot \sqrt[4]{ab^2} && \text{Factor out perfect powers.}\\[2mm]
&= 3ab\sqrt[4]{ab^2} - 2\sqrt[4]{ab^2} && \text{Simplify.}\\[2mm]
&= (3ab - 2)\sqrt[4]{ab^2} && \text{Distributive property}
\end{aligned}
$$

(c) $3\sqrt[3]{\dfrac{n^5}{27}} - 2\sqrt[3]{n^2} = 3\sqrt[3]{\dfrac{n^3}{27}} \cdot \sqrt[3]{n^2} - 2\sqrt[3]{n^2}$ Factor out perfect cube.

$\qquad\qquad\qquad\qquad = \dfrac{3\sqrt[3]{n^3}}{\sqrt[3]{27}} \cdot \sqrt[3]{n^2} - 2\sqrt[3]{n^2}$ Quotient rule

$\qquad\qquad\qquad\qquad = n\sqrt[3]{n^2} - 2\sqrt[3]{n^2}$ Simplify: $\sqrt[3]{n^3} = n$ and $\sqrt[3]{27} = 3$.

$\qquad\qquad\qquad\qquad = (n - 2)\sqrt[3]{n^2}$ Distributive property

Now Try Exercises 69, 77, 79

Radicals often occur in geometry. In the next example, we find the perimeter of a triangle by adding radical expressions.

EXAMPLE 9 **Finding the perimeter of a triangle**

Find the *exact* perimeter of the right triangle shown in Figure 7.13. Then approximate your answer to the nearest hundredth of a foot.

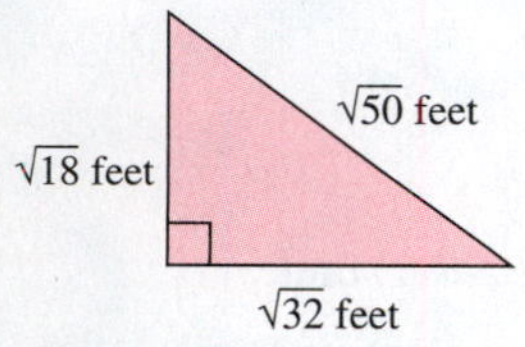

$\sqrt{50}$ feet

$\sqrt{18}$ feet

$\sqrt{32}$ feet

Figure 7.13

Solution

The sum of the lengths of the sides of the triangle is

$$\color{blue}\sqrt{18}\color{black} + \color{red}\sqrt{32}\color{black} + \color{blue}\sqrt{50}\color{black} = 3\sqrt{2} + 4\sqrt{2} + 5\sqrt{2} = 12\sqrt{2}.$$

The perimeter is $12\sqrt{2} \approx 16.97$ feet.

Now Try Exercise 125

Multiplication

Some types of radical expressions can be multiplied like binomials. For example, because

$$(x + 1)(x + 2) = x^2 + 3x + 2,$$

we have $(\sqrt{x} + 1)(\sqrt{x} + 2) = (\sqrt{x})^2 + 2\sqrt{x} + 1\sqrt{x} + 2 = x + 3\sqrt{x} + 2,$

provided that $x \geq 0$. The next example demonstrates this technique.

EXAMPLE 10 **Multiplying radical expressions**

Multiply and simplify.
(a) $(\sqrt{b} - 4)(\sqrt{b} + 5)$ **(b)** $(4 + \sqrt{3})(4 - \sqrt{3})$

Solution
(a) This expression can be multiplied and then simplified.

$$(\sqrt{b} - 4)(\sqrt{b} + 5) = \sqrt{b} \cdot \sqrt{b} + 5\sqrt{b} - 4\sqrt{b} - 4 \cdot 5$$

$$= b + \sqrt{b} - 20$$

Compare this product with $(b - 4)(b + 5) = b^2 + b - 20$.

(b) This expression is in the form $(a + b)(a - b)$, which equals $a^2 - b^2$.

$$(4 + \sqrt{3})(4 - \sqrt{3}) = (4)^2 - (\sqrt{3})^2 \quad a = 4, b = \sqrt{3}$$

$$= 16 - 3$$

$$= 13$$

Now Try Exercises 85, 87

NOTE: Example 10(b) illustrates a special case for multiplying radicals. In general,

$$(\sqrt{a} + \sqrt{b})(\sqrt{a} - \sqrt{b}) = (\sqrt{a})^2 - (\sqrt{b})^2 = a - b,$$

provided a and b are nonnegative.

Rationalizing the Denominator

In mathematics it is common to write expressions without radicals in the denominator. Quotients containing radical expressions in the numerator or denominator can appear to be different but actually be equal. For example, $\frac{1}{\sqrt{3}}$ and $\frac{\sqrt{3}}{3}$ represent the same real number even though they look as though they are unequal. To show that they are equal, we multiply the first quotient by 1 in the form $\frac{\sqrt{3}}{\sqrt{3}}$.

$$\frac{1}{\sqrt{3}} \cdot \frac{\sqrt{3}}{\sqrt{3}} = \frac{1 \cdot \sqrt{3}}{\sqrt{3} \cdot \sqrt{3}} = \frac{\sqrt{3}}{3}$$

If the denominator of a quotient contains only one term with one square root, then we can rationalize the denominator by multiplying the numerator and denominator by this square root. For example, the denominator of $\frac{1}{\sqrt{3}}$ contains one term, which is $\sqrt{3}$. Therefore, we multiplied $\frac{1}{\sqrt{3}}$ by $\frac{\sqrt{3}}{\sqrt{3}}$ to rationalize the denominator.

NOTE: $\sqrt{b} \cdot \sqrt{b} = \sqrt{b^2} = b$ for any *positive* number b.

One way to *standardize* radical expressions is to remove any radical expressions from the denominator. This process is called **rationalizing the denominator**. Exercise 133 suggests one reason why people rationalized denominators before calculators were invented. The next example demonstrates how to rationalize the denominator of several quotients.

EXAMPLE 11   **Rationalizing the denominator**

Rationalize each denominator. Assume that all variables are positive.

(a) $\dfrac{1}{\sqrt{2}}$ **(b)** $\dfrac{3}{5\sqrt{3}}$ **(c)** $\sqrt{\dfrac{x}{24}}$ **(d)** $\dfrac{xy}{\sqrt{y^3}}$

Solution
(a) We start by multiplying this expression by 1 in the form $\frac{\sqrt{2}}{\sqrt{2}}$.

$$\frac{1}{\sqrt{2}} \cdot \frac{\sqrt{2}}{\sqrt{2}} = \frac{\sqrt{2}}{\sqrt{4}} = \frac{\sqrt{2}}{2}$$

Note that the expression $\frac{\sqrt{2}}{2}$ does not have a radical in the denominator.

(b) We multiply $\dfrac{3}{5\sqrt{3}}$ by 1 in the form $\dfrac{\sqrt{3}}{\sqrt{3}}$.

$$\frac{3}{5\sqrt{3}} \cdot \frac{\sqrt{3}}{\sqrt{3}} = \frac{3\sqrt{3}}{5\sqrt{9}} = \frac{3\sqrt{3}}{5 \cdot 3} = \frac{\sqrt{3}}{5}$$

(c) Because $\sqrt{24} = \sqrt{4} \cdot \sqrt{6} = 2\sqrt{6}$, we start by simplifying $\sqrt{\dfrac{x}{24}}$.

$$\sqrt{\frac{x}{24}} = \frac{\sqrt{x}}{\sqrt{24}} = \frac{\sqrt{x}}{2\sqrt{6}}$$

To rationalize the denominator we multiply this expression by 1 in the form $\dfrac{\sqrt{6}}{\sqrt{6}}$.

$$\frac{\sqrt{x}}{2\sqrt{6}} = \frac{\sqrt{x}}{2\sqrt{6}} \cdot \frac{\sqrt{6}}{\sqrt{6}} = \frac{\sqrt{6x}}{12}$$

(d) Because $\sqrt{y^3} = \sqrt{y^2} \cdot \sqrt{y} = y\sqrt{y}$, we start by simplifying $\dfrac{xy}{\sqrt{y^3}}$.

$$\frac{xy}{\sqrt{y^3}} = \frac{xy}{y\sqrt{y}} = \frac{x}{\sqrt{y}}$$

To rationalize the denominator we multiply by 1 in the form $\dfrac{\sqrt{y}}{\sqrt{y}}$.

$$\frac{x}{\sqrt{y}} \cdot \frac{\sqrt{y}}{\sqrt{y}} = \frac{x\sqrt{y}}{y}$$

▮ **Now Try Exercises 97, 101, 103, 105**

EXAMPLE 12   **Rationalizing a denominator in geometry**

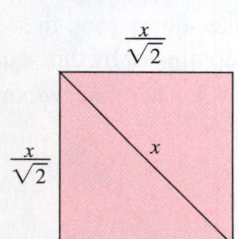

$\dfrac{x}{\sqrt{2}}$

$\dfrac{x}{\sqrt{2}}$ x

Figure 7.14

A square with a diagonal of length x has sides of length $\dfrac{x}{\sqrt{2}}$. Find the perimeter and rationalize the denominator. See Figure 7.14.

Solution
The perimeter is 4 times the length of one side.

$$4 \cdot \frac{x}{\sqrt{2}} = \frac{4x}{\sqrt{2}}$$

To rationalize the denominator we multiply the ratio by 1 in the form $\dfrac{\sqrt{2}}{\sqrt{2}}$.

$$\frac{4x}{\sqrt{2}} \cdot \frac{\sqrt{2}}{\sqrt{2}} = \frac{4x\sqrt{2}}{2} = 2x\sqrt{2}$$

The perimeter is $2x\sqrt{2}$.

▮ **Now Try Exercise 127**

When the denominator is either a sum or difference containing a square root, we rationalize the denominator by multiplying the numerator and denominator by the *conjugate* of the denominator. In this case, the **conjugate** of the denominator is found by changing a $+$ sign to a $-$ sign or vice versa. Table 7.7 lists examples of conjugates.

TABLE 7.7

Expression	$1 + \sqrt{2}$	$\sqrt{3} - 2$	$\sqrt{x} + 7$	$\sqrt{a} - \sqrt{b}$
Conjugate	$1 - \sqrt{2}$	$\sqrt{3} + 2$	$\sqrt{x} - 7$	$\sqrt{a} + \sqrt{b}$

In the next two examples we use this method to rationalize a denominator.

EXAMPLE 13 **Using a conjugate to rationalize the denominator**

Rationalize the denominator of $\dfrac{1}{1 + \sqrt{2}}$.

Solution
From Table 7.7, the conjugate of $1 + \sqrt{2}$ is $\mathbf{1 - \sqrt{2}}$.

$$\frac{1}{1 + \sqrt{2}} = \frac{1}{1 + \sqrt{2}} \cdot \frac{\mathbf{(1 - \sqrt{2})}}{\mathbf{(1 - \sqrt{2})}}$$ Multiply numerator and denominator by the conjugate.

$$= \frac{1 - \sqrt{2}}{(1)^2 - (\sqrt{2})^2}$$ $(a + b)(a - b) = a^2 - b^2$

$$= \frac{1 - \sqrt{2}}{1 - 2}$$ Simplify.

READING CHECK

How do you rationalize a denominator?

$$= \frac{1 - \sqrt{2}}{-1}$$ Subtract.

$$= \frac{1}{-1} - \frac{\sqrt{2}}{-1}$$ $\frac{a - b}{c} = \frac{a}{c} - \frac{b}{c}$

$$= -1 + \sqrt{2}$$ Simplify.

▌ **Now Try Exercise 107**

EXAMPLE 14 **Rationalizing the denominator**

Rationalize the denominator.
(a) $\dfrac{3 + \sqrt{5}}{2 - \sqrt{5}}$ **(b)** $\dfrac{\sqrt{x}}{\sqrt{x} - 2}$

Solution
(a) The conjugate of the denominator is $\mathbf{2 + \sqrt{5}}$.

$$\frac{3 + \sqrt{5}}{2 - \sqrt{5}} = \frac{(3 + \sqrt{5})}{(2 - \sqrt{5})} \cdot \frac{\mathbf{(2 + \sqrt{5})}}{\mathbf{(2 + \sqrt{5})}}$$ Multiply by 1.

$$= \frac{6 + 3\sqrt{5} + 2\sqrt{5} + (\sqrt{5})^2}{(2)^2 - (\sqrt{5})^2}$$ Multiply.

$$= \frac{11 + 5\sqrt{5}}{4 - 5}$$ Combine terms.

$$= -11 - 5\sqrt{5}$$ Simplify.

(b) The conjugate of the denominator is $\mathbf{\sqrt{x} + 2}$.

$$\frac{\sqrt{x}}{\sqrt{x} - 2} = \frac{\sqrt{x}}{(\sqrt{x} - 2)} \cdot \frac{\mathbf{(\sqrt{x} + 2)}}{\mathbf{(\sqrt{x} + 2)}}$$ Multiply by 1.

$$= \frac{x + 2\sqrt{x}}{x - 4}$$ Multiply.

▌ **Now Try Exercises 111, 115**

EXAMPLE 15   **Rationalizing a denominator having a cube root**

Rationalize the denominator of $\dfrac{5}{\sqrt[3]{x}}$.

Solution

The expression $\dfrac{5}{\sqrt[3]{x}}$ is equal to $\dfrac{5}{x^{1/3}}$. To rationalize the denominator, $x^{1/3}$, we can multiply it by $x^{2/3}$ because $x^{1/3} \cdot x^{2/3} = x^{1/3+2/3} = x^1$.

$$\dfrac{5}{x^{1/3}} = \dfrac{5}{x^{1/3}} \cdot \dfrac{x^{2/3}}{x^{2/3}} \qquad \text{Multiply by 1.}$$

$$= \dfrac{5x^{2/3}}{x^{1/3+2/3}} \qquad \text{Product rule}$$

$$= \dfrac{5\sqrt[3]{x^2}}{x} \qquad \text{Add; write in radical notation.}$$

Now Try Exercise 121

7.4   Putting It All Together

CONCEPT	EXPLANATION	EXAMPLES
Like Radicals	Like radicals have the same index and the same radicand.	$7\sqrt{5}$ and $3\sqrt{5}$ are like radicals. $5\sqrt[3]{ab}$ and $\sqrt[3]{ab}$ are like radicals. $\sqrt[3]{5}$ and $\sqrt[3]{4}$ are unlike radicals. $\sqrt[3]{7}$ and $\sqrt{7}$ are unlike radicals.
Adding and Subtracting Radical Expressions	Combine like radicals when adding or subtracting. We cannot combine unlike radicals such as $\sqrt{2}$ and $\sqrt{5}$. But sometimes we can rewrite radicals and then combine.	$6\sqrt{13} + \sqrt{13} = (6+1)\sqrt{13} = 7\sqrt{13}$ $\sqrt{40} - \sqrt{10} = \sqrt{4} \cdot \sqrt{10} - \sqrt{10}$ $\qquad\qquad = 2\sqrt{10} - \sqrt{10}$ $\qquad\qquad = \sqrt{10}$
Multiplying Radical Expressions	Radical expressions can sometimes be multiplied like binomials.	$(\sqrt{a} - 5)(\sqrt{a} + 5) = a - 25$ and $(\sqrt{x} - 3)(\sqrt{x} + 1) = x - 2\sqrt{x} - 3$
Conjugate	The conjugate is found by changing a $+$ sign to a $-$ sign or vice versa.	*Expression* *Conjugate* $\sqrt{x} + 7$ $\sqrt{x} - 7$ $\sqrt{a} - 2\sqrt{b}$ $\sqrt{a} + 2\sqrt{b}$
Rationalizing a Denominator Having One Term	Write the quotient without a radical expression in the denominator.	To rationalize $\dfrac{5}{\sqrt{7}}$, multiply the expression by 1 in the form $\dfrac{\sqrt{7}}{\sqrt{7}}$. $\dfrac{5}{\sqrt{7}} \cdot \dfrac{\sqrt{7}}{\sqrt{7}} = \dfrac{5\sqrt{7}}{\sqrt{49}} = \dfrac{5\sqrt{7}}{7}$
Rationalizing a Denominator Having Two Terms	Multiply the numerator and denominator by the conjugate of the denominator.	$\dfrac{1}{2 - \sqrt{3}} = \dfrac{1}{2 - \sqrt{3}} \cdot \dfrac{(2 + \sqrt{3})}{(2 + \sqrt{3})}$ $= \dfrac{2 + \sqrt{3}}{4 - 3} = 2 + \sqrt{3}$

7.4 Exercises

CONCEPTS AND VOCABULARY

1. $\sqrt{a} + \sqrt{a} = $ _____

2. $\sqrt[3]{b} + \sqrt[3]{b} + \sqrt[3]{b} = $ _____

3. You cannot simplify $\sqrt[3]{4} + \sqrt[3]{7}$ because they are not _____ radicals.

4. Can you simplify $4\sqrt{15} - 3\sqrt{15}$? Explain.

5. Does $6 + 3\sqrt{5}$ equal $9\sqrt{5}$?

6. To rationalize the denominator of $\frac{2}{\sqrt{7}}$, multiply this expression by _____.

7. What is the conjugate of $\sqrt{t} - 5$?

8. To rationalize the denominator of $\frac{1}{5 - \sqrt{2}}$, multiply this expression by _____.

LIKE RADICALS

Exercises 9–18: (Refer to Example 2.) Write the terms as like radicals, if possible. Assume that all variables are positive.

9. $\sqrt{12}, \sqrt{24}$

10. $\sqrt{18}, \sqrt{27}$

11. $\sqrt{7}, \sqrt{28}, \sqrt{63}$

12. $\sqrt{200}, \sqrt{300}, \sqrt{500}$

13. $\sqrt[3]{16}, \sqrt[3]{-54}$

14. $\sqrt[3]{80}, \sqrt[3]{10}$

15. $\sqrt{x^2 y}, \sqrt{4y^2}$

16. $\sqrt{x^5 y^3}, \sqrt{9xy}$

17. $\sqrt[3]{8xy}, \sqrt[3]{x^4 y^4}$

18. $\sqrt[3]{64x^4}, \sqrt[3]{-8x}$

ADDITION AND SUBTRACTION OF RADICALS

Exercises 19–80: If possible, simplify the expression. Assume that all variables are positive.

19. $2\sqrt{3} + 7\sqrt{3}$

20. $8\sqrt{7} + 2\sqrt{7}$

21. $4\sqrt[3]{5} + 2\sqrt[3]{5}$

22. $\sqrt[3]{13} + 3\sqrt[3]{13}$

23. $7 + 4\sqrt{7}$

24. $8 - 4\sqrt{3}$

25. $2\sqrt{3} + 3\sqrt{2}$

26. $\sqrt{6} + \sqrt{17}$

27. $\sqrt{3} + \sqrt[3]{3}$

28. $\sqrt{6} + \sqrt[4]{6}$

29. $\sqrt[3]{16} + 3\sqrt[3]{2}$

30. $\sqrt[3]{24} + \sqrt[3]{81}$

31. $\sqrt{2} + \sqrt{18} + \sqrt{32}$

32. $2\sqrt{3} + \sqrt{12} + \sqrt{27}$

33. $11\sqrt{11} - 5\sqrt{11}$

34. $9\sqrt{5} + \sqrt{2} - \sqrt{5}$

35. $\sqrt{x} + \sqrt{x} - \sqrt{y}$

36. $\sqrt{xy^2} - \sqrt{x}$

37. $\sqrt[3]{z} + \sqrt[3]{z}$

38. $\sqrt[3]{y} - \sqrt[3]{y}$

39. $2\sqrt[3]{6} - 7\sqrt[3]{6}$

40. $18\sqrt[3]{3} + 3\sqrt[3]{3}$

41. $\sqrt[3]{y^6} - \sqrt[3]{y^3}$

42. $2\sqrt{20} + 7\sqrt{5} + 3\sqrt{2}$

43. $3\sqrt{28} + 3\sqrt{7}$

44. $9\sqrt{18} - 2\sqrt{8}$

45. $\sqrt[4]{48} + 4\sqrt[4]{3}$

46. $\sqrt[4]{32} + \sqrt[4]{16}$

47. $\sqrt{9x} + \sqrt{16x}$

48. $-3\sqrt{x} + 5\sqrt{x}$

49. $3\sqrt{2k} + \sqrt{8k} + \sqrt{18k}$

50. $3\sqrt{k} + 2\sqrt{4k} + \sqrt{9k}$

51. $\sqrt{44} - 4\sqrt{11}$

52. $\sqrt[4]{5} + 2\sqrt[4]{5}$

53. $2\sqrt[3]{16} + \sqrt[3]{2} - \sqrt{2}$

54. $5\sqrt[3]{x} - 3\sqrt[3]{x}$

55. $\sqrt[3]{xy} - 2\sqrt[3]{xy}$

56. $3\sqrt{x^3} - \sqrt{x}$

57. $\sqrt{4x + 8} + \sqrt{x + 2}$

58. $\sqrt{2a + 1} + \sqrt{8a + 4}$

59. $\sqrt{9x + 18} - \sqrt{4x + 8}$

60. $\sqrt{25x - 25} - \sqrt{4x - 4}$

61. $\sqrt{x^3 + x^2} - \sqrt{x + 1}, x \geq 0$

62. $\sqrt{x^3 - x^2} - \sqrt{4x - 4}, x \geq 1$

63. $\sqrt{25x^3} - \sqrt{x^3}$

64. $\sqrt{36x^5} - \sqrt{25x^5}$

65. $\sqrt[3]{\frac{7x}{8}} - \frac{\sqrt[3]{7x}}{3}$

66. $\sqrt[3]{\frac{8x^2}{27}} - \sqrt[3]{\frac{x^2}{8}}$

67. $\frac{4\sqrt{3}}{3} + \frac{\sqrt{3}}{6}$

68. $\frac{8\sqrt{5}}{7} + \frac{4\sqrt{5}}{2}$

69. $\frac{15\sqrt{8}}{4} - \frac{2\sqrt{2}}{5}$

70. $\frac{23\sqrt{11}}{2} - \frac{\sqrt{44}}{8}$

71. $2\sqrt[4]{64} - \sqrt[4]{324} + \sqrt[4]{4}$

72. $2\sqrt[3]{16} - 5\sqrt[3]{54} + 10\sqrt[3]{2}$

73. $5\sqrt[4]{x^5} - \sqrt[4]{x}$

74. $20\sqrt[3]{b^4} - 4\sqrt[3]{b}$

75. $\sqrt{64x^3} - \sqrt{x} + 3\sqrt{x}$

76. $2\sqrt{3z} + 3\sqrt{12z} + 3\sqrt{48z}$

77. $\sqrt[4]{81a^5 b^5} - \sqrt[4]{ab}$

78. $\sqrt[4]{xy^5} - \sqrt[4]{x^5 y}$

79. $5\sqrt[3]{\frac{n^4}{125}} - 2\sqrt[3]{n}$

80. $\sqrt[3]{\frac{8x}{27}} - \frac{2\sqrt[3]{x}}{3}$

OPERATIONS ON FUNCTIONS

Exercises 81–84: Find $(f + g)(x)$ and $(f - g)(x)$.

81. $f(x) = 5\sqrt{x} - 2$, $g(x) = -2\sqrt{x} + 3$

82. $f(x) = 3\sqrt{4x - 4}$, $g(x) = 5\sqrt{x - 1}$

83. $f(x) = \sqrt[3]{8x} + 1$, $g(x) = 2\sqrt[3]{x} - 1$

84. $f(x) = \sqrt[3]{27x}$, $g(x) = \sqrt[3]{64x}$

MULTIPLYING BINOMIALS CONTAINING RADICALS

Exercises 85–96: Multiply and simplify.

85. $(\sqrt{x} - 3)(\sqrt{x} + 2)$ **86.** $(2\sqrt{x} + 1)(\sqrt{x} + 4)$

87. $(3 + \sqrt{7})(3 - \sqrt{7})$ **88.** $(5 - \sqrt{5})(5 + \sqrt{5})$

89. $(11 - \sqrt{2})(11 + \sqrt{2})$

90. $(6 + \sqrt{3})(6 - \sqrt{3})$

91. $(\sqrt{x} + 8)(\sqrt{x} - 8)$

92. $(\sqrt{ab} - 3)(\sqrt{ab} + 3)$

93. $(\sqrt{ab} - \sqrt{c})(\sqrt{ab} + \sqrt{c})$

94. $(\sqrt{2x} + \sqrt{3y})(\sqrt{2x} - \sqrt{3y})$

95. $(\sqrt{x} - 7)(\sqrt{x} + 8)$

96. $(\sqrt{ab} - 1)(\sqrt{ab} - 2)$

RATIONALIZING THE DENOMINATOR

Exercises 97–124: Rationalize the denominator.

97. $\dfrac{1}{\sqrt{7}}$ **98.** $\dfrac{1}{\sqrt{23}}$

99. $\dfrac{4}{\sqrt{3}}$ **100.** $\dfrac{8}{\sqrt{2}}$

101. $\dfrac{5}{3\sqrt{5}}$ **102.** $\dfrac{6}{11\sqrt{3}}$

103. $\sqrt{\dfrac{b}{12}}$ **104.** $\sqrt{\dfrac{5b}{72}}$

105. $\dfrac{rt}{2\sqrt{r^3}}$ **106.** $\dfrac{m^2 n}{2\sqrt{m^5}}$

107. $\dfrac{1}{3 - \sqrt{2}}$ **108.** $\dfrac{1}{\sqrt{3} - 2}$

109. $\dfrac{\sqrt{2}}{\sqrt{5} + 2}$ **110.** $\dfrac{\sqrt{3}}{\sqrt{3} + 2}$

111. $\dfrac{\sqrt{7} - 2}{\sqrt{7} + 2}$ **112.** $\dfrac{\sqrt{3} - 1}{\sqrt{3} + 1}$

113. $\dfrac{1}{\sqrt{7} - \sqrt{6}}$ **114.** $\dfrac{1}{\sqrt{8} - \sqrt{7}}$

115. $\dfrac{\sqrt{z}}{\sqrt{z} - 3}$ **116.** $\dfrac{2\sqrt{z}}{2 - \sqrt{z}}$

117. $\dfrac{\sqrt{a} + \sqrt{b}}{\sqrt{a} - \sqrt{b}}$ **118.** $\dfrac{\sqrt{x} - 2\sqrt{y}}{\sqrt{x} + 2\sqrt{y}}$

119. $\dfrac{1}{\sqrt{x + 1} - \sqrt{x}}$ **120.** $\dfrac{1}{\sqrt{a + 1} + \sqrt{a}}$

121. $\dfrac{3}{\sqrt[3]{x}}$ **122.** $\dfrac{6}{5\sqrt[3]{x}}$

123. $\dfrac{1}{\sqrt[3]{x^2}}$ **124.** $\dfrac{2}{\sqrt[3]{(x - 2)^2}}$

GEOMETRY

125. *Perimeter* (Refer to Example 9.) Find the exact perimeter of the right triangle. Then approximate your answer.

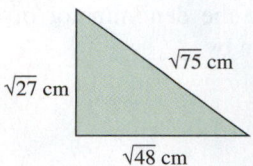

126. *Perimeter* Find the exact perimeter of the rectangle. Then approximate your answer.

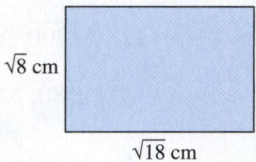

127. *Geometry* (Refer to Example 12.) A square has a diagonal with length $\sqrt{3}$. Find the perimeter and rationalize the denominator.

128. *Geometry* A rectangle has sides of length $\sqrt{8}$ and $\sqrt{32}$. Find the perimeter of the rectangle.

129. *Geometry* (Refer to Example 12.) A square has a diagonal that is 60 feet long. Find the exact perimeter of the rectangle and simplify your answer.

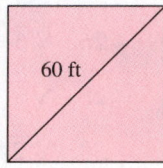

130. *Geometry* A square has a diagonal that is 10 feet long. Find the exact perimeter of the square and simplify your answer.

131. *Geometry* A square has an area of x square feet. Find the length of the diagonal of the square.

132. *Geometry* A square has an area of 16 square feet. Find the length of the diagonal of the square.

WRITING ABOUT MATHEMATICS

133. Suppose that a student knows that $\sqrt{3} \approx 1.73205$ and does not have a calculator. Which of the following expressions would be easier to evaluate by hand? Why?

$$\frac{1}{\sqrt{3}} \quad \text{or} \quad \frac{\sqrt{3}}{3}$$

134. A student simplifies an expression *incorrectly:*

$$\sqrt{8} + \sqrt[3]{16} \stackrel{?}{=} \sqrt{4 \cdot 2} + \sqrt[3]{8 \cdot 2}$$
$$\stackrel{?}{=} \sqrt{4} \cdot \sqrt{2} + \sqrt[3]{8} \cdot \sqrt[3]{2}$$
$$\stackrel{?}{=} 2\sqrt{2} + 2\sqrt[3]{2}$$
$$\stackrel{?}{=} 4\sqrt{4}$$
$$\stackrel{?}{=} 8.$$

Explain any errors that the student made. What would you do differently?

| SECTIONS 7.3 and 7.4 | **Checking Basic Concepts** |

1. Simplify each expression. Assume that all variables are positive.
   (a) $(64^{-3/2})^{1/3}$ **(b)** $\sqrt{5} \cdot \sqrt{20}$

   (c) $\sqrt[3]{-8x^4y}$ **(d)** $\sqrt{\dfrac{4b}{5}} \cdot \sqrt{\dfrac{4b^3}{5}}$

2. Simplify $\sqrt[3]{7} \cdot \sqrt{7}$

3. Simplify each expression.
   (a) $\sqrt{3} \cdot \sqrt{12}$ **(b)** $\dfrac{\sqrt[3]{81}}{\sqrt[3]{3}}$
   (c) $\sqrt{36x^6}, x > 0$

4. Simplify each expression.
   (a) $5\sqrt{6} + 2\sqrt{6} + \sqrt{7}$
   (b) $8\sqrt[3]{x} - 3\sqrt[3]{x}$
   (c) $\sqrt{9x} - \sqrt{4x}$

5. Simplify each expression.
   (a) $\sqrt[3]{xy^4} - \sqrt[3]{x^4y}$
   (b) $(4 - \sqrt{2})(4 + \sqrt{2})$

6. Rationalize the denominator of $\dfrac{6}{2\sqrt{6}}$.

7. Rationalize the denominator of $\dfrac{2}{\sqrt{5} - 1}$.

7.5   More Radical Functions

Root Functions • Power Functions • Modeling with Power Functions (Optional)

A LOOK INTO MATH ▶

Radical expressions often occur in biology. In Example 5, we discuss how heavier birds tend to have larger wings. This relationship between weight and wing size can be modeled by a radical function. Another application of radical functions in biology is discussed in Example 3, where a function is used to predict the surface area of a person, given their weight.

Root Functions

NEW VOCABULARY

☐ Root function
☐ Power function

Earlier in this chapter we discussed the square root and cube root functions. There are also functions for higher roots. Figures 7.15–7.18 on the next page show graphs of four common root functions. Note that the higher the root, the more slowly the graph increases for $x \geq 1$. Also, note that the domains for functions of even roots include only the nonnegative real numbers, whereas the domains of functions of odd roots include all real numbers.

Square Root Function

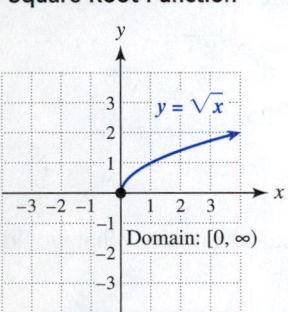

Figure 7.15

Cube Root Function

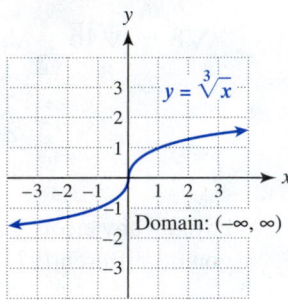

Figure 7.16

Fourth Root Function

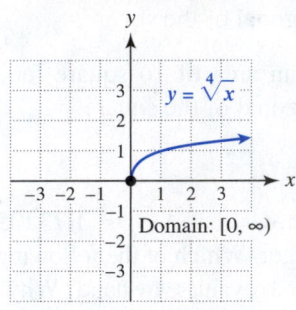

Figure 7.17

Fifth Root Function

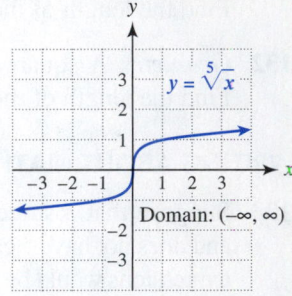

Figure 7.18

EXAMPLE 1 **Evaluating root functions**

If possible, evaluate each root function f at the given x-values.

(a) $f(x) = \sqrt{x + 1}, \quad x = -5, x = 3$

(b) $f(x) = \sqrt[3]{1 - x}, \quad x = -7, x = 28$

(c) $f(x) = \sqrt[4]{x - 7}, \quad x = 5, x = 88$

(d) $f(x) = \sqrt[5]{x}, \qquad x = -1, x = 32$

Solution

(a) $f(-5) = \sqrt{-5 + 1} = \sqrt{-4}$, which is *not* a real number.

$f(3) = \sqrt{3 + 1} = \sqrt{4} = 2$ because $2^2 = 4$.

(b) $f(-7) = \sqrt[3]{1 - (-7)} = \sqrt[3]{8} = 2$ because $2^3 = 8$.

$f(28) = \sqrt[3]{1 - 28} = \sqrt[3]{-27} = -3$ because $(-3)^3 = -27$.

(c) $f(5) = \sqrt[4]{5 - 7} = \sqrt[4]{-2}$, which is not a real number.

$f(88) = \sqrt[4]{88 - 7} = \sqrt[4]{81} = 3$ because $3^4 = 81$.

(d) $f(-1) = \sqrt[5]{-1} = -1$ because $(-1)^5 = -1$.

$f(32) = \sqrt[5]{32} = 2$ because $2^5 = 32$.

▮ **Now Try Exercises 9, 11, 13, 15**

Power Functions

Power functions are a generalization of root functions. Examples of power functions include

$$f(x) = x^{1/2}, \; g(x) = x^{2/3}, \quad \text{and} \quad h(x) = x^{-3/5}.$$

The exponent for a power function is frequently a rational number p, written in lowest terms. That is, we usually write $f(x) = x^{1/2}$, rather than $f(x) = x^{3/6}$. The power function $f(x) = x^{m/n}$ can also be written as $f(x) = \sqrt[n]{x^m}$. A power function f is defined for all real numbers when n is odd and defined for only nonnegative numbers when n is even.

POWER FUNCTION

If a function f can be represented by

$$f(x) = x^p,$$

where p is a rational number, then f is a **power function**. If $p = \frac{1}{n}$, where $n \geq 2$ is an integer, then f is also a **root function**, which is given by

$$f(x) = \sqrt[n]{x}.$$

EXAMPLE 2 | **Evaluating power functions**

If possible, evaluate $f(x)$ at the given value of x.
(a) $f(x) = x^{0.75}$ at $x = 16$ **(b)** $f(x) = x^{1/4}$ at $x = -81$

Solution
(a) $0.75 = \frac{3}{4}$, so $f(x) = x^{3/4}$. Thus $f(\mathbf{16}) = \mathbf{16}^{3/4} = (16^{1/4})^3 = 2^3 = 8$.

NOTE: $16^{1/4} = \sqrt[4]{16} = 2$.

(b) $f(-81) = (-81)^{1/4} = \sqrt[4]{-81}$, which is undefined. There is no real number a such that $a^4 = -81$ because a^4 is never negative.

∎ **Now Try Exercises 23, 27**

▶ **REAL-WORLD CONNECTION** The surface area of the skin covering the human body is influenced by both the height and weight of a person. A taller person tends to have a larger surface area, as does a heavier person. See Extended and Discovery Exercise 2 at the end of this chapter. In the next example, we use a power function from biology to model this situation.

EXAMPLE 3 | **Modeling surface area of the human body**

The surface area of a person who is 66 inches tall and weighs w pounds can be estimated by $S(w) = 327w^{0.425}$, where S is in square inches. (*Source:* H. Lancaster, *Quantitative Methods in Biological and Medical Sciences.*)
(a) Find S if this person weighs 130 pounds.
(b) If the person gains 20 pounds, by how much does the person's surface area increase?
(c) A graph of $y = S(w)$ is shown in Figure 7.19. Suppose that a person's weight is more than 50 pounds and it doubles. Use the graph to determine if the person's surface area also doubles.

Surface Area of Human Body

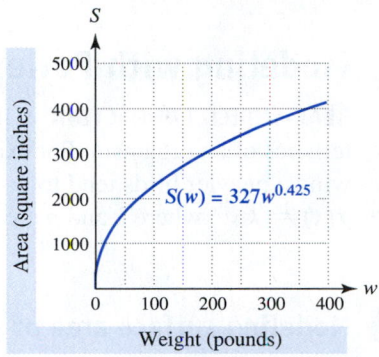

$S(w) = 327w^{0.425}$

Area (square inches)

Weight (pounds)

Figure 7.19

Solution
(a) $S(130) = 327(130)^{0.425} \approx 2588$ square inches
(b) Because $S(150) = 327(150)^{0.425} \approx 2750$ square inches, the surface area of the person increases by about $2750 - 2588 = 162$ square inches.
(c) For $w \geq 50$ the graph increases more slowly. This means that the person's surface area (S-values) will not double even if his or her weight doubles (w-values). For example, $A(100) \approx 2300$ square inches and $A(200) \approx 3100$ square inches. The surface area did not double.

∎ **Now Try Exercise 51**

In the next example, we investigate the graph of $y = x^p$ for different values of p.

EXAMPLE 4 **Graphing power functions**

The graphs of three power functions,

$$f(x) = x^{1/3}, \quad g(x) = x^{0.75}, \quad \text{and} \quad h(x) = x^{1.4},$$

are shown in Figure 7.20, where the two decimal exponents have been written as fractions. Discuss how the value of p affects the graph of $y = x^p$ when $x > 1$ and when $0 < x < 1$.

Power Functions

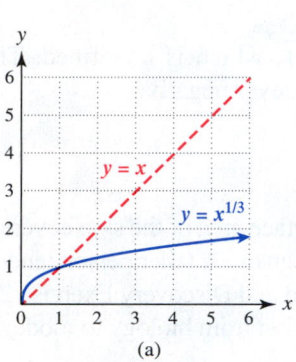

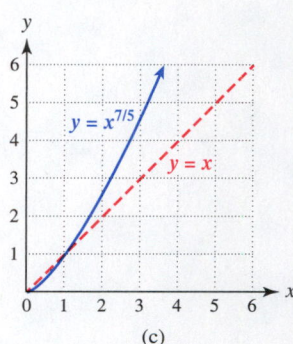

(a) (b) (c)

Figure 7.20

Solution
The graphs of these power functions have the following characteristics.
1. If $x > 1$, then $x^{1/3} < x^{3/4} < x^{7/5}$. Larger exponents result in graphs that increase (rise) *faster* for $x > 1$.
2. If $0 < x < 1$, then $x^{7/5} < x^{3/4} < x^{1/3}$. Smaller exponents result in graphs that have larger y-values for $0 < x < 1$.
3. All graphs pass through the points $(0, 0)$ and $(1, 1)$.

Now Try Exercise 35

Modeling with Power Functions (Optional)

▶ **REAL-WORLD CONNECTION** Allometry is the study of the relative sizes of different characteristics of an organism. For example, the weight of a bird is related to the surface area of its wings; heavier birds tend to have larger wings. Allometric relations are often modeled with $f(x) = kx^p$, where k and p are constants. (*Source:* C. Pennycuick, *Newton Rules Biology.*)

EXAMPLE 5 **Modeling surface area of wings**

Weights w of various birds and the corresponding surface area of their wings A are shown in Table 7.8.

TABLE 7.8 Weight and Wing Size

w (kilograms)	0.5	2.0	3.5	5.0
A (square meters)	0.069	0.175	0.254	0.325

(a) Make a scatterplot of the data. Discuss any trends in the data.
(b) Biologists modeled the data with $A(w) = kw^{2/3}$, where k is a constant. Find k.
(c) Graph A and the data in the same viewing rectangle.
(d) Estimate the area of the wings of a 3-kilogram bird.

CALCULATOR HELP

To make a scatterplot, see Appendix A (pages AP-3 and AP-4).

[0, 6, 1] by [0, 0.4, 0.1]

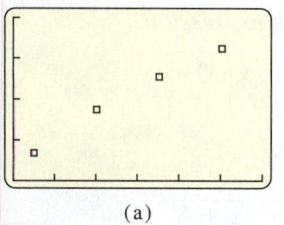

(a)

[0, 6, 1] by [0, 0.4, 0.1]

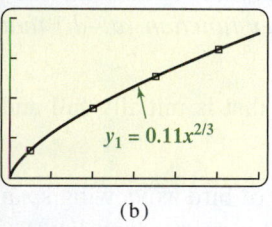

$y_1 = 0.11x^{2/3}$

(b)

Figure 7.21

Solution

(a) A scatterplot of the data is shown in Figure 7.21(a). As the weight of a bird increases so does the surface area of its wings.

(b) To determine k, substitute one of the data points into $A(w)$. We use (**2.0**, **0.175**).

$$A(w) = kw^{2/3} \qquad \text{Given formula}$$

$$\mathbf{0.175} = k(\mathbf{2})^{2/3} \qquad \text{Let } w = 2 \text{ and } A(w) = 0.175$$
$$\text{(any data point could be used).}$$

$$k = \frac{\mathbf{0.175}}{2^{2/3}} \qquad \text{Solve for } k; \text{ rewrite equation.}$$

$$k \approx 0.11 \qquad \text{Approximate } k.$$

Thus $A(w) = 0.11w^{2/3}$.

(c) The data and graph of $Y_1 = 0.11X^\wedge(2/3)$ are shown in Figure 7.21(b). Note that the graph appears to pass through each data point.

(d) $A(\mathbf{3}) = 0.11(\mathbf{3})^{2/3} \approx 0.23$ square meter

Now Try Exercise 59

7.5 Putting It All Together

FUNCTION	EXPLANATION
Root	$f(x) = \sqrt[3]{x}, \quad g(x) = \sqrt[4]{x}, \quad \text{and} \quad h(x) = \sqrt[5]{x}$ Odd root functions are defined for all real numbers, and even root functions are defined for nonnegative real numbers.
Power	$f(x) = x^p$, where p is a rational number, is a power function. If $p = \frac{1}{n}$, where $n \geq 2$ is an integer, f is also a root function, given by $f(x) = \sqrt[n]{x}$. ***Examples:*** $f(x) = x^{5/3}$ Power function $\qquad\qquad\quad g(x) = x^{1/4} = \sqrt[4]{x}$ Root and power function

---

7.5 Exercises

CONCEPTS AND VOCABULARY

1. Sketch a graph of $y = x^{1/2}$.

2. Sketch a graph of $y = x^{1/3}$.

3. What is the domain of $f(x) = x^{1/2}$?

4. What is the domain of $f(x) = x^{1/3}$?

5. Give a symbolic representation for a power function.

6. Give a symbolic representation for a root function.

7. What is the domain of $f(x) = \sqrt[4]{x}$?

8. What is the domain of $f(x) = \sqrt[5]{x}$?

ROOT FUNCTIONS

Exercises 9–16: If possible, evaluate each root function f at the given x-values. When the result is not an integer, approximate it to the nearest hundredth.

9. $f(x) = \sqrt{x^2 - 1}, \quad x = -2, x = 0$

10. $f(x) = \sqrt{2x - 5}, \quad x = 0, x = 3$

11. $f(x) = \sqrt[4]{1 - x}$, $x = -5$, $x = 2$

12. $f(x) = \sqrt[4]{x + 1}$, $x = -15$, $x = 15$

13. $f(x) = \sqrt[5]{4 - 3x}$, $x = -3$, $x = 1$

14. $f(x) = \sqrt[5]{1 - x^2}$, $x = 3$, $x = 1$

15. $f(x) = \sqrt[3]{1 - x}$, $x = -5$, $x = 2$

16. $f(x) = \sqrt[3]{(x + 1)^2}$, $x = -2$, $x = 7$

POWER FUNCTIONS

Exercises 17–22: Use radical notation to write f(x).

17. $f(x) = x^{1/2}$ **18.** $f(x) = x^{1/3}$

19. $f(x) = x^{2/3}$ **20.** $f(x) = x^{3/4}$

21. $f(x) = x^{-1/5}$ **22.** $f(x) = x^{-2/5}$

Exercises 23–30: If possible, evaluate f(x) at the given values of x. When appropriate, approximate the answer to the nearest hundredth.

23. $f(x) = x^{5/2}$ $x = 4, x = 5$

24. $f(x) = x^{-3/4}$ $x = 1, x = 3$

25. $f(x) = x^{-7/5}$ $x = -32, x = 10$

26. $f(x) = x^{4/3}$ $x = -8, x = 27$

27. $f(x) = x^{1/4}$ $x = 256, x = -10$

28. $f(x) = x^{3/4}$ $x = 16, x = -1$

29. $f(x) = x^{2/5}$ $x = 32, x = -32$

30. $f(x) = x^{5/6}$ $x = -5, x = 64$

Exercises 31–34: Graph y = f (x). Write the domain of f in interval notation.

31. $f(x) = x^{1/4}$ **32.** $f(x) = x^{1/5}$

33. $f(x) = x^{2/3}$ **34.** $f(x) = (x + 1)^{1/4}$

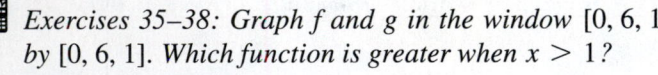

Exercises 35–38: Graph f and g in the window [0, 6, 1] by [0, 6, 1]. Which function is greater when x > 1?

35. $f(x) = x^{1/5}, g(x) = x^{1/3}$

36. $f(x) = x^{4/5}, g(x) = x^{5/4}$

37. $f(x) = x^{1.2}, g(x) = x^{0.45}$

38. $f(x) = x^{-1.4}, g(x) = x^{1.4}$

Exercises 39 and 40: **Thinking Generally** *Let* $0 < q < p$, *where p and q are rational numbers.*

39. For $x > 1$, is $x^p > x^q$ or $x^p < x^q$?

40. For $0 < x < 1$, is $x^p > x^q$ or $x^p < x^q$?

Exercises 41 and 42: **Operations on Functions** *For the given f(x) and g(x), evaluate each expression.*

(a) $(f + g)(2)$ **(b)** $(f - g)(x)$
(c) $(fg)(x)$ **(d)** $(f/g)(x)$

41. $f(x) = \sqrt{8x}, g(x) = \sqrt{2x}$

42. $f(x) = \sqrt{9x + 18}, g(x) = \sqrt{4x + 8}$

Exercises 43–46: **Graphical Interpretation** *Match the situation with the graph of the power function (a.–d.) that models it best.*

43. Amount of water in a barrel that is initially full and has a hole near the bottom

44. The average weight of a type of bird as its wing span increases (*Hint:* The exponent is greater than 1.)

45. Money made after x hours by a person who works for a fixed hourly wage

46. Rapid growth of an insect population that eventually slows down

a. 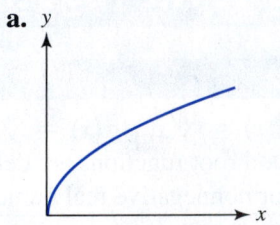   **b.**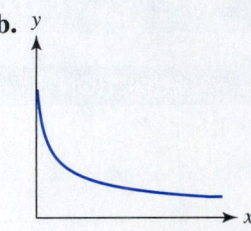

c. 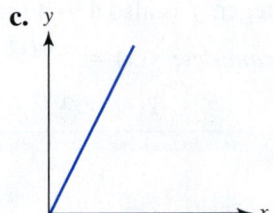   **d.**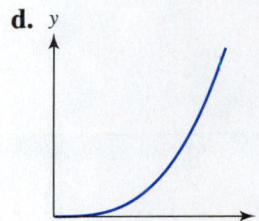

Exercises 47–50 **Graphical Recognition** *Match the equation with its graph (a.–d.). Do not use a calculator.*

47. $y = x^{7/4}$ **48.** $y = x^{4/3}$

49. $y = x^{1/3}$ **50.** $y = x^{1/2}$

a. 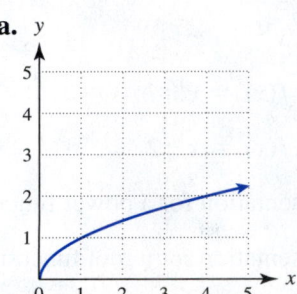   **b.**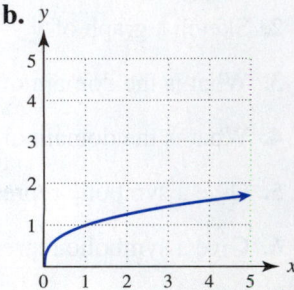

c. 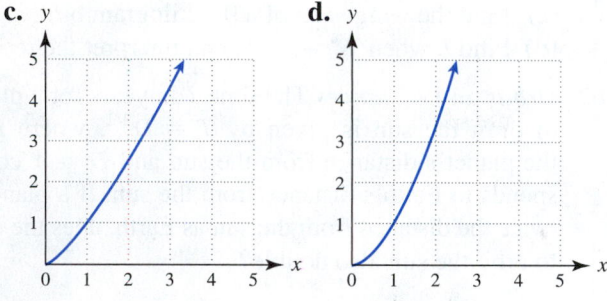   **d.**

APPLICATIONS

Exercises 51 and 52: **Surface Area** *(Refer to Example 3.) The surface area of the skin of a person who is 70 inches tall and weighs w pounds can be estimated by* $S(w) = 342w^{0.425}$, *where S is in square inches. Evaluate each of the following.*

51. $S(150)$ **52.** $S(200)$

Exercises 53–56 **Learning Curves** *The graph shows the percentage P of times that individuals completed a simple task correctly after practicing it x times.*

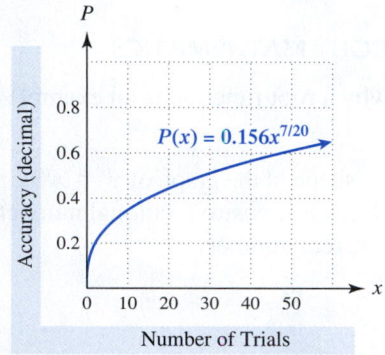

53. If a group of people does 10 practice trials each, what percentage of the time is the task done correctly?

54. How many trials need to be performed to reach 50% accuracy?

55. If a person doubles the number of practice trials, does the accuracy always double? Explain.

56. What happens to a person's accuracy after a large number of trials?

57. *Aging More Slowly* In his theory of relativity, Albert Einstein showed that if a person travels at nearly the speed of light, then time slows down significantly. Suppose that there are twins; one remains on Earth and the other leaves in a very fast spaceship having velocity v. If the twin on Earth ages T_0 years, then according to Einstein the twin in the spaceship ages T years, where

$$T(v) = T_0\sqrt{1 - (v/c)^2}.$$

In this formula c represents the speed of light, which is 186,000 miles per second.
   (a) Evaluate T when $v = 0.8c$ (eight-tenths the speed of light) and $T_0 = 10$ years. (*Hint:* Simplify $\frac{v}{c}$ without using 186,000 miles per second.)
   (b) Interpret your result.

58. *Increasing Your Weight* (Refer to Exercise 57.) Albert Einstein also showed that the weight (mass) of an object increases when traveling near the speed of light. If a person's weight on Earth is W_0, then the same person's weight W in a spaceship traveling at velocity v is

$$W(v) = \frac{W_0}{\sqrt{1 - (v/c)^2}}.$$

   (a) Evaluate W when $v = 0.6c$ (six-tenths the speed of light) and $W_0 = 220$ pounds (100 kilograms).
   (b) Interpret your result.

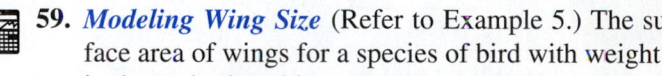 **59.** *Modeling Wing Size* (Refer to Example 5.) The surface area of wings for a species of bird with weight w is shown in the table.

w (kilograms)	2	3	4
A (square meters)	0.254	0.333	0.403

   (a) Make a scatterplot of the data.
   (b) The data can be modeled with $A(w) = kw^{2/3}$. Find k.
   (c) Graph A and the data in the same viewing rectangle. Does the graph of A pass through the data points?
   (d) Estimate the area of the wings of a 2.5-kilogram bird.

 60. *Pulse Rate in Animals* The following table lists typical pulse rates R in beats per minute (bpm) for animals with various weights W in pounds.
(*Source:* C. Pennycuick.)

W (pounds)	20	150	500	1500
R (beats per minute)	198	72	40	23

   (a) Describe what happens to the pulse rate as the weight of the animal increases.
   (b) Plot the data in [0, 1600, 400] by [0, 220, 20].
   (c) If $R = kW^{-1/2}$, find k.
   (d) Find R when $W = 700$, and interpret the result.

61. *Modeling Wing Span* (Refer to Example 5.) Biologists have found that the weight W of a bird and the length L of its wing span are related by $L = kW^{1/3}$, where k is a constant. The following table lists L and W for one species of bird. (*Source:* C. Pennycuick.)

W (kilograms)	0.1	0.4	0.8	1.1
L (meters)	0.422	0.670	0.844	0.938

(a) Use the data to approximate the value of k.
(b) Graph L and the data in the same viewing rectangle. What happens to L as W increases?

(c) Find the wing span of a 0.7-kilogram bird.
(d) Find L when $W = 0.65$, and interpret the result.

62. *Orbits and Distances* The time T in years for a planet to orbit the sun is given by $T = D^{3/2}$, where D is the planet's distance from the sun and $D = 1$ corresponds to Earth's distance from the sun. If a planet is twice the distance from the sun as Earth, does the time to orbit the sun also double?

WRITING ABOUT MATHEMATICS

63. Explain why a root function is an example of a power function.

64. Discuss the shape of the graph of $y = x^p$ as p increases. Assume that p is a positive rational number and that x is a positive real number.

Group Activity   Working with Real Data

Directions: Form a group of 2 to 4 people. Select someone to record the group's responses for this activity. All members of the group should work cooperatively to answer the questions. If your instructor asks for your results, each member of the group should be prepared to respond.

Simple Pendulum Gravity is responsible for an object falling toward Earth. The farther the object falls, the faster it is moving when it hits the ground. For each second that an object falls, its speed increases by a constant amount, called the *acceleration due to gravity*, denoted g. One way to calculate the value of g is to use a simple pendulum. See the accompanying figure.

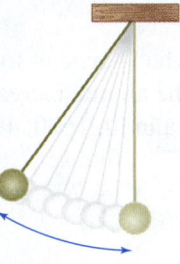

The time T for a pendulum to swing back and forth once is called its *period* and is given by

$$T = 2\pi\sqrt{\frac{L}{g}},$$

where L equals the length of the pendulum. The table lists the periods of pendulums with different lengths.

L (feet)	0.5	1.0	1.5
T (seconds)	0.78	1.11	1.36

(a) Solve the formula for g.
(b) Use the table to determine the value of g. (*Note:* The units for g are feet per second per second.)
(c) Interpret your result.

7.6 Equations Involving Radical Expressions

Solving Radical Equations • **The Distance Formula** • **Solving the Equation $x^n = k$**

A LOOK INTO MATH ▶

In a fast-paced society, people do not spend time watching videos that they find boring. For example, earlier in this chapter we learned how the function

$$A(x) = 7.3\sqrt[16]{x^7}$$

computes the percentage of viewers who abandon an online video after x seconds. If we want to find the number of seconds that it takes for **50%** of viewers to abandon a typical online video, we could *solve* the *radical equation*

$$7.3\sqrt[16]{x^7} = 50$$

(See Exercise 119.) In this section we discuss how to solve radical equations and equations with rational exponents.

Solving Radical Equations

NEW VOCABULARY

☐ Extraneous solutions
☐ Pythagorean theorem
☐ Distance

Many times, equations contain either radical expressions or rational exponents. Examples include

$$\sqrt{x} = 6, \quad 5x^{1/2} = 1, \quad \text{and} \quad \sqrt[3]{x-1} = 3. \quad \text{Radical equations}$$

One strategy for solving an equation containing a square root is to isolate the square root (if necessary) and then square each side of the equation. This technique is an example of the *power rule for solving equations*.

POWER RULE FOR SOLVING EQUATIONS

If each side of an equation is raised to the same positive integer power, then any solutions to the given equation are among the solutions to the new equation. That is, the solutions to the equation $a = b$ are among the solutions to $a^n = b^n$.

NOTE: After applying the power rule, the new equation can have solutions that do not satisfy the given equation, which are sometimes called **extraneous solutions**. We must *always* check our answers in the *given* equation after applying the power rule.

READING CHECK

When you apply the power rule to solve an equation, what must you do with your solutions? Why?

We illustrate the power rule in the next example.

EXAMPLE 1 **Solving a radical equation symbolically**

Solve $\sqrt{2x - 1} = 3$. Check your solution.

Solution
Begin by squaring each side of the equation. That is, apply the power rule with $n = 2$.

$$\sqrt{2x - 1} = 3 \qquad \text{Given equation}$$
$$(\sqrt{2x - 1})^2 = 3^2 \qquad \text{Square each side.}$$
$$2x - 1 = 9 \qquad \text{Simplify.}$$
$$2x = 10 \qquad \text{Add 1.}$$
$$x = 5 \qquad \text{Divide by 2.}$$

To check this answer we substitute $x = 5$ in the given equation.

$$\sqrt{2(5) - 1} \overset{?}{=} 3$$
$$3 = 3 \checkmark \text{ It checks.}$$

Now Try Exercise 21

NOTE: To simplify $(\sqrt{2x - 1})^2$ in Example 1, we used the fact that

$$\left(\sqrt{a}\right)^2 = \sqrt{a} \cdot \sqrt{a} = a.$$

The following steps can be used to solve a radical equation.

SOLVING A RADICAL EQUATION

STEP 1: Isolate a radical term on one side of the equation.

STEP 2: Apply the power rule by raising each side of the equation to the power equal to the index of the isolated radical term.

STEP 3: Solve the equation. If it still contains a radical, repeat Steps 1 and 2.

STEP 4: Check your answers by substituting each result in the *given* equation.

NOTE: In earlier sections, we *simplified expressions*. In this section, we *solve equations*. Equations contain equals signs and when we solve an equation, we try to find values of the variable that make the equation a true statement.

In the next example, we apply these steps to a radical equation.

EXAMPLE 2 **Isolating the radical term**

Solve $\sqrt{4 - x} + 5 = 8$.

Solution
STEP 1: To isolate the radical term, we subtract 5 from each side of the equation.

$$\sqrt{4 - x} + 5 = 8 \qquad \text{Given equation}$$
$$\sqrt{4 - x} = 3 \qquad \text{Subtract 5.}$$

STEP 2: The isolated term involves a square root, so we must **square** each side.

$$(\sqrt{4-x})^2 = (3)^2 \quad \text{Square each side.}$$

STEP 3: Next we solve the resulting equation. (It is not necessary to repeat Steps 1 and 2 because the resulting equation does not contain any radical expressions.)

$$4 - x = 9 \qquad \text{Simplify.}$$
$$-x = 5 \qquad \text{Subtract 4.}$$
$$x = -5 \qquad \text{Multiply by } -1.$$

STEP 4: To check this answer we substitute $x = -5$ in the given equation.

$$\sqrt{4 - (-5)} + 5 \stackrel{?}{=} 8$$
$$\sqrt{9} + 5 \stackrel{?}{=} 8$$
$$8 = 8 \checkmark \quad \text{It checks.}$$

■ **Now Try Exercise 23**

The next example shows that we must check our answers to identify extraneous solutions when squaring each side of an equation.

EXAMPLE 3 **Solving a radical equation**

Solve $\sqrt{3x + 3} = 2x - 1$. Check your results and then solve the equation graphically.

Solution
Symbolic Solution Begin by squaring each side of the equation.

$$\sqrt{3x + 3} = 2x - 1 \qquad \text{Given equation}$$
$$(\sqrt{3x + 3})^2 = (2x - 1)^2 \qquad \text{Square each side.}$$
$$3x + 3 = 4x^2 - 4x + 1 \qquad \text{Multiply.}$$
$$0 = 4x^2 - 7x - 2 \qquad \text{Subtract } 3x + 3.$$
$$0 = (4x + 1)(x - 2) \qquad \text{Factor.}$$
$$x = -\frac{1}{4} \quad \text{or} \quad x = 2 \qquad \text{Solve for } x.$$

To check these values substitute $x = -\frac{1}{4}$ and $x = 2$ in the given equation.

$$\sqrt{3\left(-\frac{1}{4}\right) + 3} \stackrel{?}{=} 2\left(-\frac{1}{4}\right) - 1$$
$$\sqrt{2.25} \stackrel{?}{=} -1.5$$
$$1.5 \neq -1.5 \; \text{✗} \qquad \text{It does not check.}$$

Thus $-\frac{1}{4}$ is an *extraneous solution*. Next substitute $x = 2$ in the given equation.

$$\sqrt{3 \cdot 2 + 3} \stackrel{?}{=} 2 \cdot 2 - 1$$
$$\sqrt{9} \stackrel{?}{=} 3$$
$$3 = 3 \checkmark \qquad \text{It checks.}$$

The only solution is 2. Next we check with a graphical solution.

[-5, 5, 1] by [-5, 5, 1]

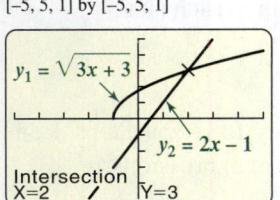

Figure 7.22

CALCULATOR HELP

To find a point of intersection, see Appendix A (page AP-6).

Graphical Solution The solution **2** to the equation $\sqrt{3x + 3} = 2x - 1$ is supported graphically in Figure 7.22, where the graphs of $y_1 = \sqrt{3x + 3}$ and $y_2 = 2x - 1$ intersect at the point (**2**, 3). *Note that the graphical solution does not give an extraneous solution.*

Now Try Exercises 27, 79(a), (b)

Example 3 demonstrates that *checking a solution is essential when you are squaring each side of an equation.* Squaring may introduce extraneous solutions, which are solutions to the resulting equation but are not solutions to the given equation.

CRITICAL THINKING

Will a numerical solution give extraneous solutions?

When an equation contains two or more terms with square roots, it may be necessary to square each side of the equation more than once. In these situations, isolate one of the square roots and then square each side of the equation. If a radical term remains after simplifying, repeat these steps. We apply this technique in the next example.

EXAMPLE 4 **Squaring twice**

Solve $\sqrt{2x} - 1 = \sqrt{x + 1}$.

Solution
Begin by squaring each side of the equation.

$\sqrt{2x} - 1 = \sqrt{x + 1}$	Given equation
$(\sqrt{2x} - 1)^2 = (\sqrt{x + 1})^2$	Square each side.
$(\sqrt{2x})^2 - 2(\sqrt{2x})(1) + 1^2 = x + 1$	$(a - b)^2 = a^2 - 2ab + b^2$
$2x - 2\sqrt{2x} + 1 = x + 1$	Simplify.
$2x - 2\sqrt{2x} = x$	Subtract 1.
$x = 2\sqrt{2x}$	Subtract x and add $2\sqrt{2x}$.
$x^2 = 4(2x)$	Square each side again.
$x^2 - 8x = 0$	Subtract $8x$.
$x(x - 8) = 0$	Factor.
$x = 0 \quad \text{or} \quad x = 8$	Solve.

To check these answers substitute $x = 0$ and $x = 8$ in the given equation.

$$\sqrt{2 \cdot 0} - 1 \stackrel{?}{=} \sqrt{0 + 1} \qquad\qquad \sqrt{2 \cdot 8} - 1 \stackrel{?}{=} \sqrt{8 + 1}$$

$$-1 \neq 1 \text{ ✗ } \text{ It does not check.} \qquad\qquad 3 = 3 \text{ ✓ } \text{ It checks.}$$

The only solution is 8.

Now Try Exercise 43

In the next example we apply the power rule to an equation that contains a cube root.

EXAMPLE 5

Solving an equation containing a cube root

Solve $\sqrt[3]{4x - 7} = 4$.

Solution

STEP 1: The cube root term is already isolated, so we proceed to Step 2.

STEP 2: Because the index is 3, we **cube** each side of the equation.

$$\sqrt[3]{4x - 7} = 4 \qquad \text{Given equation}$$
$$(\sqrt[3]{4x - 7})^3 = (4)^3 \qquad \text{Cube each side.}$$

STEP 3: We solve the resulting equation.

$$4x - 7 = 64 \qquad \text{Simplify.}$$
$$4x = 71 \qquad \text{Add 7 to each side.}$$
$$x = \frac{71}{4} \qquad \text{Divide each side by 4.}$$

STEP 4: To check this answer we substitute $x = \frac{71}{4}$ in the given equation.

$$\sqrt[3]{4\left(\frac{71}{4}\right) - 7} \overset{?}{=} 4$$
$$\sqrt[3]{64} \overset{?}{=} 4$$
$$4 = 4 \checkmark \quad \text{It checks.}$$

Now Try Exercise 31

▶ **REAL-WORLD CONNECTION** Larger birds tend to have larger wings. This relationship can sometimes be modeled by $A(W) = 100\sqrt[3]{W^2}$, where W is the weight in pounds of the bird and A is the area of the wings in square inches. In the next example, we use the formula to determine the weight of a bird that has wings with a surface area of 600 square inches. (*Source:* C. Pennycuick, *Newton Rules Biology.*)

EXAMPLE 6

Finding the weight of a bird

Solve the equation $600 = 100\sqrt[3]{W^2}$ to determine the weight in pounds of a bird having wings with an area of 600 square inches.

Solution

Begin by dividing each side of the equation $600 = 100\sqrt[3]{W^2}$ by 100 to isolate the radical term on the right side of the equation.

$$\frac{600}{100} = \sqrt[3]{W^2} \qquad \text{Divide each side by 100.}$$
$$(6)^3 = (\sqrt[3]{W^2})^3 \qquad \text{Cube each side.}$$
$$216 = W^2 \qquad \text{Simplify.}$$
$$\sqrt{216} = W \qquad \text{Take principal square root, } W > 0.$$
$$W \approx 14.7 \qquad \text{Approximate.}$$

The weight of the bird is approximately 14.7 pounds.

Now Try Exercise 117

TECHNOLOGY NOTE

Graphing Radical Expressions
The equation in Example 6 can be solved graphically. Sometimes it is more convenient to use rational exponents than radical notation. Thus $y = 100\sqrt[3]{W^2}$ can be entered as $Y_1 = 100X^{(2/3)}$. (Be sure to include parentheses around the 2/3.) The accompanying figure shows y_1 intersecting the line $y_2 = 600$ near the point (14.7, 600), which supports our symbolic result.

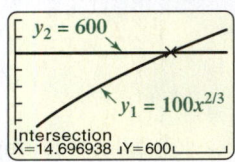

[0, 20, 5] by [0, 800, 100]

In the next example we solve an equation that would be difficult to solve symbolically, but an *approximate* solution can be found graphically.

EXAMPLE 7 **Solving an equation with rational exponents graphically**

Solve $x^{2/3} = 3 - x^2$ graphically.

Solution
Graph $Y_1 = X^{(2/3)}$ and $Y_2 = 3 - X^2$. The graphs intersect near $(-1.34, 1.21)$ and $(1.34, 1.21)$, as shown in Figure 7.23. Thus the solutions are given by $x \approx \pm 1.34$.

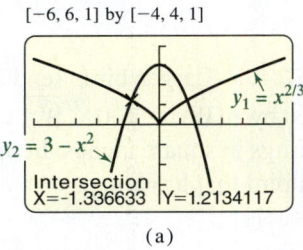

[−6, 6, 1] by [−4, 4, 1]

(a)

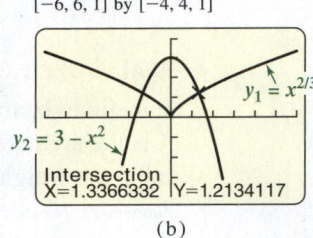

[−6, 6, 1] by [−4, 4, 1]

(b)

Figure 7.23

Now Try Exercise 73

The Distance Formula

$a^2 + b^2 = c^2$

Figure 7.24

▶ **REAL-WORLD CONNECTION** One of the most famous theorems in mathematics is the **Pythagorean theorem**. It states that if a right triangle has legs a and b with hypotenuse c (see Figure 7.24), then

$$a^2 + b^2 = c^2. \quad \text{Pythagorean Theorem}$$

For example, if the legs of a right triangle are $a = 3$ and $b = 4$, the hypotenuse is $c = 5$ because $3^2 + 4^2 = 5^2$. Also, if the sides of a triangle satisfy $a^2 + b^2 = c^2$, it is a right triangle.

EXAMPLE 8 **Applying the Pythagorean theorem**

A rectangular television screen has a width of 20 inches and a height of 15 inches. Find the diagonal of the television. Why is it called a 25-inch television?

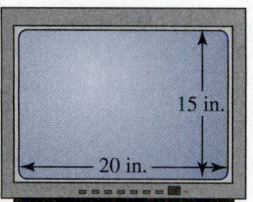

Figure 7.25

Solution

In Figure 7.25, let $a = 20$ and $b = 15$. Then the diagonal of the television corresponds to the hypotenuse of a right triangle with legs of 20 inches and 15 inches.

$$
\begin{aligned}
c^2 &= a^2 + b^2 && \text{Pythagorean theorem} \\
c &= \sqrt{a^2 + b^2} && \text{Take the principal square root, } c > 0. \\
c &= \sqrt{20^2 + 15^2} && \text{Substitute } a = 20 \text{ and } b = 15. \\
c &= 25 && \text{Simplify.}
\end{aligned}
$$

A 25-inch television has a diagonal of 25 inches.

▌ **Now Try Exercise 125**

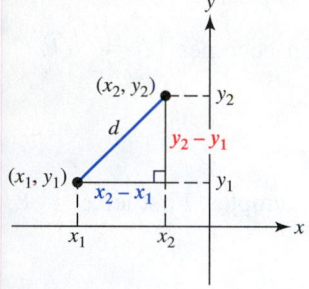

Figure 7.26

The Pythagorean theorem can be used to determine the distance between two points. Suppose that a line segment has endpoints (x_1, y_1) and (x_2, y_2), as illustrated in Figure 7.26. The lengths of the legs of the right triangle are $x_2 - x_1$ and $y_2 - y_1$. The distance d is the hypotenuse of a right triangle. Applying the Pythagorean theorem, we have

$$d^2 = (x_2 - x_1)^2 + (y_2 - y_1)^2.$$

Distance is nonnegative, so we let d be the principal square root and obtain

$$d = \sqrt{(x_2 - x_1)^2 + (y_2 - y_1)^2}.$$

DISTANCE FORMULA

The **distance** d between the points (x_1, y_1) and (x_2, y_2) in the xy-plane is

$$d = \sqrt{(x_2 - x_1)^2 + (y_2 - y_1)^2}.$$

EXAMPLE 9 **Finding distance between points**

Find the distance between the points $(-2, 3)$ and $(1, -4)$.

Solution

Start by letting $(x_1, y_1) = (-2, 3)$ and $(x_2, y_2) = (1, -4)$. Then substitute these values into the distance formula.

$$
\begin{aligned}
d &= \sqrt{(x_2 - x_1)^2 + (y_2 - y_1)^2} && \text{Distance formula} \\
&= \sqrt{(1 - (-2))^2 + (-4 - 3)^2} && \text{Substitute.} \\
&= \sqrt{9 + 49} && \text{Simplify.} \\
&= \sqrt{58} && \text{Add.} \\
&\approx 7.62 && \text{Approximate.}
\end{aligned}
$$

The distance between the points, as shown in Figure 7.27, is exactly $\sqrt{58}$ units, or about 7.62 units. Note that we would obtain the same result if we let $(x_1, y_1) = (1, -4)$ and $(x_2, y_2) = (-2, 3)$.

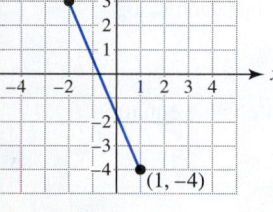

Figure 7.27

▌ **Now Try Exercise 109**

NOTE: In Example 9, $\sqrt{9 + 49} \neq \sqrt{9} + \sqrt{49} = 3 + 7 = 10$. In general, for any a and b, $\sqrt{a^2 + b^2} \neq a + b$.

Solving the Equation $x^n = k$

The equation $x^n = k$, where n is a *positive integer*, can be solved by taking the nth root of each side of the equation. The following technique allows us to find all *real* solutions to this equation.

> ### SOLVING THE EQUATION $x^n = k$
>
> Take the nth root of each side of $x^n = k$ to obtain $\sqrt[n]{x^n} = \sqrt[n]{k}$.
> 1. If n is odd, then $\sqrt[n]{x^n} = x$ and the equation becomes $x = \sqrt[n]{k}$.
> 2. If n is *even* and $k > 0$, then $\sqrt[n]{x^n} = |x|$ and the equation becomes $|x| = \sqrt[n]{k}$. (If $k < 0$, there are no real solutions.)

To understand this technique better, consider the following examples. First let $x^3 = 8$, so that n is odd. Taking the cube root of each side gives

$$\sqrt[3]{x^3} = \sqrt[3]{8}, \text{ which is equivalent to } x = \sqrt[3]{8}, \text{ or } x = 2.$$

Next let $x^2 = 4$, so that n is even. Taking the square root of each side gives

$$\sqrt{x^2} = \sqrt{4}, \text{ which is equivalent to } |x| = \sqrt{4}, \text{ or } |x| = 2.$$

The solutions to $|x| = 2$ are -2 or 2, which can be written as ± 2.

EXAMPLE 10 **Solving the equation $x^n = k$**

Solve each equation.
(a) $x^3 = -64$ **(b)** $x^2 = 12$ **(c)** $2(x - 1)^4 = 32$

Solution
(a) Taking the cube root of each side of $x^3 = -64$ gives

$$\sqrt[3]{x^3} = \sqrt[3]{-64} \text{ or } x = -4.$$

(b) Taking the square root of each side of $x^2 = 12$ gives

$$\sqrt{x^2} = \sqrt{12} \text{ or } |x| = \sqrt{12}.$$

The equation $|x| = \sqrt{12}$ is equivalent to $x = \pm\sqrt{12}$.
(c) First divide each side of $2(x - 1)^4 = 32$ by 2 to isolate the power of $x - 1$.

$$(x - 1)^4 = 16 \qquad \text{Divide each side by 2.}$$
$$\sqrt[4]{(x - 1)^4} = \sqrt[4]{16} \qquad \text{Take the 4th root of each side.}$$
$$|x - 1| = 2 \qquad \text{Simplify: } \sqrt[4]{y^4} = |y|.$$
$$x - 1 = -2 \text{ or } x - 1 = 2 \qquad \text{Solve the absolute value equation.}$$
$$x = -1 \text{ or } x = 3 \qquad \text{Add 1 to each side.}$$

Now Try Exercises 47, 55, 67

▶ **REAL-WORLD CONNECTION** In some parts of the United States, wind power is used to generate electricity. Suppose that the diameter of the circular path created by the blades of a wind-powered generator is 8 feet. Then the wattage W generated by a wind velocity of v miles per hour is modeled by

$$W(v) = 2.4v^3.$$

If the wind blows at 10 miles per hour, the generator can produce about

$$W(10) = 2.4 \cdot 10^3 = 2400 \text{ watts}.$$

(**Source:** *Conquering the Sciences,* Sharp Electronics.)

EXAMPLE 11 **Modeling a wind generator**

The formula $W(v) = 2.4v^3$ is used to calculate the watts generated when there is a wind velocity of v miles per hour.
(a) Find a function f that calculates the wind velocity when W watts are being produced.
(b) If the wattage doubles, has the wind velocity also doubled? Explain.

Solution
(a) Given W we need a formula to find v, so solve $W = 2.4v^3$ for v.

$$W = 2.4v^3 \qquad \text{Given formula}$$

$$\frac{W}{2.4} = v^3 \qquad \text{Divide by 2.4.}$$

$$\sqrt[3]{\frac{W}{2.4}} = \sqrt[3]{v^3} \qquad \text{Take the cube root of each side.}$$

$$v = \sqrt[3]{\frac{W}{2.4}} \qquad \text{Simplify and rewrite equation.}$$

Thus $f(W) = \sqrt[3]{\frac{W}{2.4}}$.
(b) Suppose that the power generated is 1000 watts. Then the wind speed is

$$f(1000) = \sqrt[3]{\frac{1000}{2.4}} \approx 7.5 \text{ miles per hour.}$$

If the power doubles to 2000 watts, then the wind speed is

$$f(2000) = \sqrt[3]{\frac{2000}{2.4}} \approx 9.4 \text{ miles per hour.}$$

Thus for the wattage to double, the wind speed does not need to double.

Now Try Exercise 133

7.6 Putting It All Together

CONCEPT	DESCRIPTION	EXAMPLE
Power Rule for Solving Equations	If each side of an equation is raised to the same positive integer power, any solutions to the given equation are among the solutions to the new equation.	$\sqrt{2x} = x$ $2x = x^2$ Square each side. $x^2 - 2x = 0$ Rewrite equation. $x = 0$ or $x = 2$ Factor and solve. *Be sure to check any solutions.*

continued on next page

continued from previous page

CONCEPT	DESCRIPTION	EXAMPLE
Pythagorean Theorem	If c is the hypotenuse of a right triangle and a and b are its legs, then $a^2 + b^2 = c^2$. 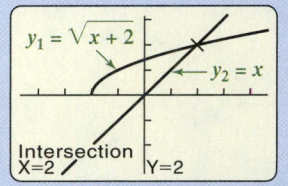	If the sides of the right triangle are $a = 5$, $b = 12$, and $c = 13$, then they satisfy $a^2 + b^2 = c^2$ or $5^2 + 12^2 = 13^2$.
Distance Formula	The distance d between the points (x_1, y_1) and (x_2, y_2) is $$d = \sqrt{(x_2 - x_1)^2 + (y_2 - y_1)^2}.$$	The distance between the points $(2, 3)$ and $(-3, 4)$ is $$d = \sqrt{(-3 - 2)^2 + (4 - 3)^2}$$ $$= \sqrt{(-5)^2 + (1)^2} = \sqrt{26}.$$
Solving the Equation $x^n = k$, Where n Is a Positive Integer	Take the nth root of each side to obtain $\sqrt[n]{x^n} = \sqrt[n]{k}$. Then, **1.** $x = \sqrt[n]{k}$, if n is odd. **2.** $x = \pm \sqrt[n]{k}$, if n is even and $k \geq 0$.	**1.** n odd: If $x^5 = 32$, then $x = \sqrt[5]{32} = 2$. **2.** n even: If $x^4 = 81$, then $x = \pm \sqrt[4]{81} = \pm 3$.

Equations involving radical expressions can be solved symbolically, numerically, and graphically. All symbolic solutions must be checked; *numerical and graphical methods do not give extraneous solutions.* These concepts are illustrated for the equation $\sqrt{x + 2} = x$, where $y_1 = \sqrt{x + 2}$ and $y_2 = x$.

Symbolic Solution

$$\sqrt{x + 2} = x$$
$$x + 2 = x^2$$
$$x^2 - x - 2 = 0$$
$$(x - 2)(x + 1) = 0$$
$$x = 2 \quad \text{or} \quad x = -1$$

Check: $\sqrt{2 + 2} = 2$ ✓

$\sqrt{-1 + 2} \neq -1$ ✗

The only solution is **2**.

Numerical Solution

X	Y₁	Y₂
-2	0	-2
-1	1	-1
0	1.4142	0
1	1.7321	1
2	2	2
3	2.2361	3
4	2.4495	4

X=2

$y_1 = y_2$ when $x = $ **2**, so **2** is a solution. Note that -1 is *not* a solution.

Graphical Solution

$[-4.7, 4.7, 1]$ by $[-3.1, 3.1, 1]$

$y_1 = \sqrt{x + 2}$

$y_2 = x$

Intersection X=2 Y=2

The graphs intersect at $(2, 2)$. Note that there is no point of intersection when $x = -1$.

7.6 Exercises

MyMathLab Math XL PRACTICE WATCH DOWNLOAD READ REVIEW

CONCEPTS AND VOCABULARY

1. What is a good first step for solving $\sqrt{4x - 1} = 5$?

2. What is a good first step for solving $\sqrt[3]{x + 1} = 6$?

3. Can an equation involving rational exponents have more than one solution?

4. When you square each side of an equation to solve for an unknown, what must you do with any answers?

5. What is the Pythagorean theorem used for?

6. If the legs of a right triangle are 3 and 4, what is the length of the hypotenuse?

7. What formula can you use to find the distance d between two points?

8. Write the equation $\sqrt{x} + \sqrt[4]{x^3} = 2$ with rational exponents.

SIMPLIFYING RADICALS

Exercises 9–16: Simplify. Assume radicands of square roots are positive.

9. $\sqrt{2} \cdot \sqrt{2}$ **10.** $\sqrt{5} \cdot \sqrt{5}$

11. $\sqrt{x} \cdot \sqrt{x}$ **12.** $\sqrt{2x} \cdot \sqrt{2x}$

13. $(\sqrt{2x + 1})^2$ **14.** $(\sqrt{7x})^2$

15. $(\sqrt[3]{5x^2})^3$ **16.** $(\sqrt[3]{2x - 5})^3$

SYMBOLIC SOLUTIONS

Exercises 17–46: Solve the equation symbolically. Check your results.

17. $\sqrt{x} = 8$ **18.** $\sqrt{3z} = 6$

19. $\sqrt[4]{x} = 3$ **20.** $\sqrt[3]{x - 4} = 2$

21. $\sqrt{2t + 4} = 4$ **22.** $\sqrt{y + 4} = 3$

23. $\sqrt{x + 1} - 3 = 4$ **24.** $\sqrt{2x + 5} + 2 = 5$

25. $2\sqrt{x - 2} + 1 = 5$ **26.** $-\sqrt{x + 7} - 1 = -7$

27. $\sqrt{x + 6} = x$ **28.** $\sqrt{z + 6} = z$

29. $\sqrt[3]{x} = 3$ **30.** $\sqrt[3]{x + 10} = 4$

31. $\sqrt[3]{2z - 4} = -2$ **32.** $\sqrt[3]{z - 1} = -3$

33. $\sqrt[4]{t + 1} = 2$ **34.** $\sqrt[4]{5t} = 5$

35. $\sqrt{5z - 1} = \sqrt{z + 1}$ **36.** $y = \sqrt{y + 1} + 1$

37. $\sqrt{1 - x} = 1 - x$ **38.** $\sqrt[3]{4x} = x$

39. $\sqrt{b^2 - 4} = b - 2$ **40.** $\sqrt{b^2 - 2b + 1} = b$

41. $\sqrt{1 - 2x} = x + 7$ **42.** $\sqrt{4 - y} = y - 2$

43. $\sqrt{x} = \sqrt{x - 5} + 1$ **44.** $\sqrt{x - 1} = \sqrt{x + 4} - 1$

45. $\sqrt{2t - 2} + \sqrt{t} = 7$ **46.** $\sqrt{x + 1} - \sqrt{x - 6} = 1$

SOLVING THE EQUATION $x^n = k$

Exercises 47–68: (Refer to Example 10.) Solve.

47. $x^2 = 49$ **48.** $x^2 = 9$

49. $2z^2 = 200$ **50.** $3z^2 = 48$

51. $(t + 1)^2 = 16$ **52.** $(t - 5)^2 = 81$

53. $(4 - 2x)^2 = 100$ **54.** $(3x - 6)^2 = 25$

55. $b^3 = 64$ **56.** $a^3 = 1000$

57. $2t^3 = -128$ **58.** $3t^3 = -81$

59. $(x + 1)^3 = 8$ **60.** $(4 - x)^3 = -1$

61. $(2 - 5z)^3 = -125$ **62.** $(2x + 4)^3 = 125$

63. $x^4 = 16$ **64.** $x^4 = 7$

65. $x^5 = 12$ **66.** $x^5 = -32$

67. $2(x + 2)^4 = 162$ **68.** $\frac{1}{2}(x - 1)^5 = 16$

GRAPHICAL SOLUTIONS

Exercises 69–78: Solve graphically. Approximate solutions to the nearest hundredth when appropriate.

69. $\sqrt[3]{x + 5} = 2$ **70.** $\sqrt[3]{x} + \sqrt{x} = 3.43$

71. $\sqrt{2x - 3} = \sqrt{x} - \frac{1}{2}$ **72.** $x^{4/3} - 1 = 2$

73. $x^{5/3} = 2 - 3x^2$ **74.** $x^{3/2} = \sqrt{x + 2} - 2$

75. $z^{1/3} - 1 = 2 - z$ **76.** $z^{3/2} - 2z^{1/2} - 1 = 0$

77. $\sqrt{y + 2} + \sqrt{3y + 2} = 2$

78. $\sqrt{x + 1} - \sqrt{x - 1} = 4$

USING MORE THAN ONE METHOD

Exercises 79–82: Solve the equation
   (a) *symbolically,*
   (b) *graphically, and*
   (c) *numerically.*

79. $2\sqrt{x} = 8$ **80.** $\sqrt[3]{5 - x} = 2$

81. $\sqrt{6z - 2} = 8$ **82.** $\sqrt{y + 4} = \frac{y}{3}$

SOLVING AN EQUATION FOR A VARIABLE

Exercises 83–86: Solve for the indicated variable.

83. $T = 2\pi\sqrt{\dfrac{L}{32}}$ for L

84. $Z = \sqrt{L^2 + R^2}$ for R

85. $r = \sqrt{\dfrac{A}{\pi}}$ for A

86. $F = \dfrac{1}{2\pi\sqrt{LC}}$ for C

PYTHAGOREAN THEOREM

Exercises 87–94: If the sides of a triangle are a, b, and c and they satisfy $a^2 + b^2 = c^2$, the triangle is a right triangle. Determine whether the triangle with the given sides is a right triangle.

87. $a = 6$ $b = 8$ $c = 10$

88. $a = 5$ $b = 12$ $c = 13$

89. $a = \sqrt{5}$ $b = \sqrt{9}$ $c = \sqrt{14}$

90. $a = 4$ $b = 5$ $c = 7$

91. $a = 7$ $b = 24$ $c = 25$

92. $a = 1$ $b = \sqrt{3}$ $c = 2$

93. $a = 8$ $b = 8$ $c = 16$

94. $a = 11$ $b = 60$ $c = 61$

Exercises 95–98: Find the length of the missing side in the right triangle.

95.   **96.**

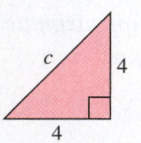   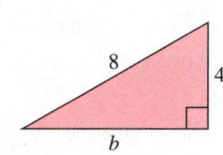

97.   **98.**

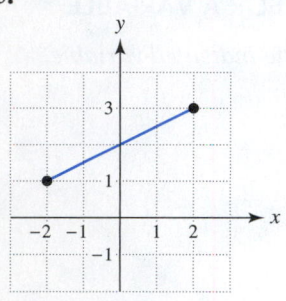   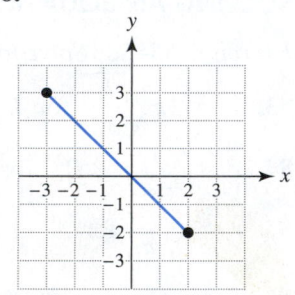

Exercises 99–104: A right triangle has legs a and b with hypotenuse c. Find the length of the missing side.

99. $a = 3, b = 4$ **100.** $a = 4, b = 7$

101. $a = \sqrt{3}, c = 8$ **102.** $a = \sqrt{6}, c = \sqrt{10}$

103. $b = 48, c = 50$ **104.** $b = 10, c = 26$

DISTANCE FORMULA

Exercises 105–108: Find the length of the line segment.

105.   **106.**

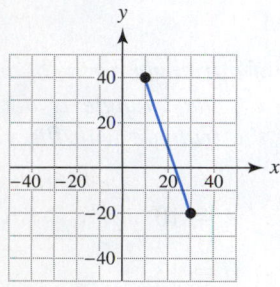   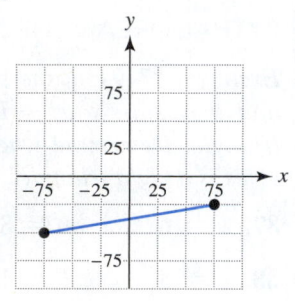

107.   **108.**

Exercises 109–112: Find the distance between the points.

109. $(-1, 2), (4, 10)$ **110.** $(5, -40), (-6, 20)$

111. $(0, -3), (4, 0)$ **112.** $(3, 9), (-4, 2)$

Exercises 113–116: Find x if the distance between the points is d. Assume that $x \geq 0$.

113. $(x, 3), (0, 6)$ $d = 5$

114. $(x, -1), (6, 11)$ $d = 13$

115. $(x, -5), (62, 6)$ $d = 61$

116. $(x, 3), (12, -4)$ $d = 25$

APPLICATIONS

Exercises 117 and 118: **Weight of a Bird** *(Refer to Example 6.) Estimate the weight W of a bird having wings of area A. Let $A = 100\sqrt[3]{W^2}$.*

117. $A = 400$ square inches

118. $A = 1000$ square inches

Exercises 119 and 120: **Abandonment Rate** *(Refer to A Look Into Math for this section.) Use $A(x) = 7.3\sqrt[16]{x^7}$ to find the average number of seconds x that it takes for A percent of the viewers to abandon an online movie.*

119. $A = 50$ **120.** $A = 20$

Exercises 121–124: **Distance to the Horizon** *Because of Earth's curvature, a person can see a limited distance to the horizon. The higher the location of the person, the farther that person can see. The distance D in miles to the horizon can be estimated by $D(h) = 1.22\sqrt{h}$, where h is the height of the person above the ground in feet.*

121. Find D for a 6-foot-tall person standing on level ground.

122. Find D for a person on top of Mount Everest with a height of 29,028 feet.

123. What height allows a person to see 20 miles?

124. How high does a plane need to fly for the pilot to be able to see 100 miles?

125. *Diagonal of a Television* (Refer to Example 8.) A rectangular television screen is 11.4 inches by 15.2 inches. Find the diagonal of the television screen.

126. *Dimensions of a Television* The height of a television with a 13-inch diagonal is $\frac{3}{4}$ of its width. Find the width and height of the television set.

127. *DVD and Picture Dimensions* If the picture shown on a television set is h units high and w units wide, the *aspect ratio* of the picture is $\frac{w}{h}$ (see the accompanying figure). Digital video discs support the newer aspect ratio of $\frac{16}{9}$ rather than the older ratio of $\frac{4}{3}$. If the width of a picture with an aspect ratio of $\frac{16}{9}$ is 29 inches, approximate the height and diagonal of the rectangular picture. (*Source:* J. Taylor, *DVD Demystified.*)

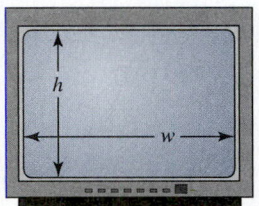

128. *Flood Control* The spillway capacity of a dam is important in flood control. Spillway capacity Q in cubic feet of water per second flowing over the spillway depends on the width W and the depth D of the spillway, as illustrated in the accompanying figure. If W and D are measured in feet, capacity can be modeled by $Q = 3.32WD^{3/2}$. (*Source:* D. Callas, Project Director, *Snapshots of Applications in Mathematics.*)
   (a) Find the capacity of a spillway with $W = 20$ feet and $D = 5$ feet.
   (b) A spillway with a width of 30 feet is to have a capacity of $Q = 2690$ cubic feet per second. Estimate to the nearest foot the appropriate depth of the spillway.

129. *Sky Diving* When sky divers initially fall from an airplane, their velocity v in miles per hour after free falling d feet can be approximated by $v = \frac{60}{11}\sqrt{d}$. (Because of air resistance, they will eventually reach a terminal velocity.) How far do sky divers need to fall to attain the following velocities? (These values for d represent minimum distances.)
   (a) 60 miles per hour   **(b)** 100 miles per hour

130. *Guy Wire* A guy wire attached to the top of a 30-foot-long pole is anchored 10 feet from the base of the pole, as illustrated in the figure. Find the length of the guy wire to the nearest tenth of a foot.

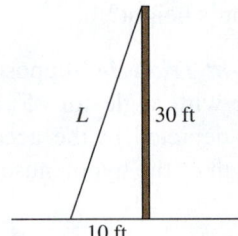

131. *Skid Marks* Vehicles involved in accidents often leave skid marks, which can be used to determine how fast a vehicle was traveling. To determine this speed, officials often use a test vehicle to compare skid marks on the same section of road. Suppose that a vehicle in a crash left skid marks D feet long and that a test vehicle traveling at v miles per hour leaves skid marks d feet long. Then the speed V of the vehicle involved in the crash is given by

$$V = v\sqrt{\frac{D}{d}}.$$

(*Source:* N. Garber and L. Hoel, *Traffic and Highway Engineering.*)
   (a) Find V if $v = 30$ mph, $D = 285$ feet, and $d = 178$ feet. Interpret your result.
   (b) A test vehicle traveling at 45 mph leaves skid marks 255 feet long. How long would the skid marks be for a vehicle traveling 60 miles per hour?

132. *Highway Curves* If a circular highway curve without any banking has a radius of R feet, the speed limit L in miles per hour for the curve is $L = 1.5\sqrt{R}$. (*Source:* N. Garber.)

  (a) Find the speed limit for a curve having a radius of 400 feet.

  (b) If the radius of a curve doubles, what happens to the speed limit?

  (c) A curve with a 40-mile-per-hour speed limit is being designed. What should be its radius?

133. *Wind Power* (Refer to Example 11.) If a wind-powered generator has blades that create a circular path with a diameter of 10 feet, then the wattage W generated by a wind velocity of v miles per hour is modeled by $W(v) = 3.8v^3$.

  (a) If the wind velocity doubles, what happens to the wattage generated?

  (b) Solve $W = 3.8v^3$ for v.

  (c) If the wind generator is producing 30,400 watts, find the wind speed.

134. *Height and Weight* Suppose that the weight of a person is directly proportional to the cube of the person's height. If one person weighs twice as much as a (similarly proportioned) second person, by what factor is the heavier person's height greater than the shorter person's height?

135. *45°–45° Right Triangle* Suppose that the legs of a right triangle with angles of 45° and 45° both have length a, as depicted in the accompanying figure. Find the length of the hypotenuse.

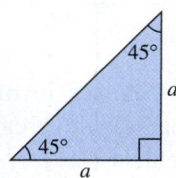

136. *30°–60° Right Triangle* In a right triangle with angles of 30° and 60°, the shortest side is half the length of the hypotenuse (see the accompanying figure). If the hypotenuse has length c, find the length of the other two sides, a and b, in terms of c.

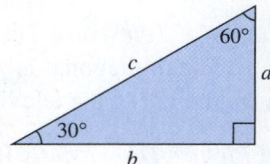

WRITING ABOUT MATHEMATICS

137. A student solves an equation *incorrectly* as follows.

$$\sqrt{3 - x} = \sqrt{x} - 1$$
$$(\sqrt{3 - x})^2 \overset{?}{=} (\sqrt{x})^2 - (1)^2$$
$$3 - x \overset{?}{=} x - 1$$
$$-2x \overset{?}{=} -4$$
$$x \overset{?}{=} 2$$

  (a) How could you convince the student that the answer is wrong?

  (b) Discuss where any errors were made.

138. When each side of an equation is squared, you must check your results. Explain why.

SECTIONS
7.5 and 7.6 **Checking Basic Concepts**

1. Sketch a graph of each function and then evaluate $f(-1)$, if possible.

  (a) $f(x) = \sqrt{x}$

  (b) $f(x) = \sqrt[3]{x}$

  (c) $f(x) = \sqrt{x^2}$

2. Evaluate $f(x) = 0.2x^{2/3}$ when $x = 64$.

3. Find the domain of $f(x) = \sqrt{x - 4}$. Write your answer in interval notation.

4. Solve each equation. Check your answers.

  (a) $\sqrt{2x - 4} = 2$

  (b) $\sqrt[3]{x - 1} = 3$

  (c) $\sqrt{3x} = 1 + \sqrt{x + 1}$

5. Find the distance between $(-3, 5)$ and $(2, -7)$.

6. A 16-inch diagonal television set has a rectangular picture with a width of 12.8 inches. Find the height of the picture.

7. Solve $(x + 1)^4 = 16$.

7.7 Complex Numbers

Basic Concepts • Addition, Subtraction, and Multiplication • Powers of i • Complex Conjugates and Division

A LOOK INTO MATH ▶ Mathematics is both applied and theoretical. A common misconception is that abstract or theoretical mathematics is unimportant in today's world. Many new ideas with great practical importance were first developed as abstract concepts with no particular application in mind. For example, complex numbers, which are related to square roots of negative numbers, started as an abstract concept to solve equations. Today, complex numbers are used in many sophisticated applications, such as the design of electrical circuits, ships, and airplanes. The *fractal image* shown to the left would not have been discovered without complex numbers.

NEW VOCABULARY

- ☐ Imaginary unit
- ☐ Complex number
- ☐ Standard form
- ☐ Real part
- ☐ Imaginary part
- ☐ Imaginary number
- ☐ Pure imaginary number
- ☐ Complex conjugate

Basic Concepts

A graph of $y = x^2 + 1$ is shown in Figure 7.28. There are no x-intercepts, so the quadratic equation $x^2 + 1 = 0$ has no real number solutions.

If we try to solve $x^2 + 1 = 0$ by subtracting 1 from each side, the result is $x^2 = -1$. Because $x^2 \geq 0$ for any real number x, there are no real solutions. However, mathematicians have invented (or discovered) solutions.

$$x^2 = -1$$
$$x = \pm\sqrt{-1} \quad \text{Solve for } x.$$

We now define a number called the **imaginary unit**, denoted i.

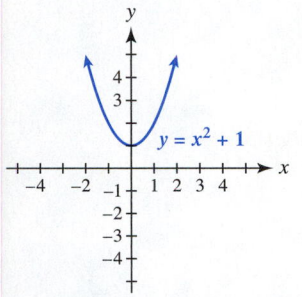

Figure 7.28

> #### PROPERTIES OF THE IMAGINARY UNIT i
>
> $$i = \sqrt{-1} \quad \text{and} \quad i^2 = -1$$

By creating the number i, the solutions to the equation $x^2 + 1 = 0$ are i and $-i$. Using the real numbers and the imaginary unit i, we can define a new set of numbers called the *complex numbers*. A **complex number** can be written in **standard form**, as $a + bi$, where a and b are real numbers. The **real part** is a and the **imaginary part** is b. Every real number a is also a complex number because it can be written $a + 0i$. A complex number $a + bi$ with $b \neq 0$ is an **imaginary number**. A complex number $a + bi$ with $a = 0$ and $b \neq 0$ is sometimes called a **pure imaginary number**. Examples of pure imaginary numbers include $4i$ and $-2i$. Table 7.9 lists several complex numbers with their real and imaginary parts.

TABLE 7.9

Complex Number: $a + bi$	$-3 + 2i$	5	$-3i$	$-1 + 7i$	$-5 - 2i$	$4 + 6i$
Real Part: a	-3	5	0	-1	-5	4
Imaginary Part: b	2	0	-3	7	-2	6

Figure 7.29 on the next page shows how different sets of numbers are related. Note that *the set of complex numbers contains the set of real numbers*.

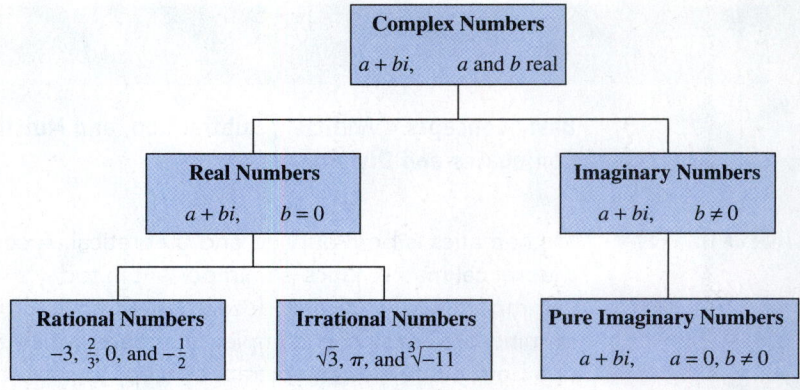

Figure 7.29

Using the imaginary unit i, we may write the square root of a negative number as a complex number. For example, $\sqrt{-2} = i\sqrt{2}$, and $\sqrt{-4} = i\sqrt{4} = 2i$. This method is summarized as follows.

CALCULATOR HELP

To set your calculator in $a + bi$ mode, see Appendix A (page AP-7).

THE EXPRESSION $\sqrt{-a}$

If $a > 0$, then $\sqrt{-a} = i\sqrt{a}$.

NOTE: Although it is standard for a complex number to be expressed as $a + bi$, we often write $\sqrt{2}i$ as $i\sqrt{2}$ so that it is clear that i is not under the square root. Similarly $\frac{1}{2}\sqrt{2}i$ is sometimes written as $\frac{1}{2}i\sqrt{2}$ or $\frac{i\sqrt{2}}{2}$.

EXAMPLE 1 **Writing the square root of a negative number**

Write each square root using the imaginary unit i.

(a) $\sqrt{-25}$ **(b)** $\sqrt{-7}$ **(c)** $\sqrt{-20}$

Solution
(a) $\sqrt{-25} = i\sqrt{25} = 5i$
(b) $\sqrt{-7} = i\sqrt{7}$
(c) $\sqrt{-20} = i\sqrt{20} = i\sqrt{4}\sqrt{5} = 2i\sqrt{5}$

Now Try Exercises 13, 15, 19

Addition, Subtraction, and Multiplication

Addition can be defined for complex numbers in a manner similar to how we add binomials. For example,

$$(-3 + 2x) + (2 - x) = (-3 + 2) + (2x - x) = -1 + x.$$

ADDITION AND SUBTRACTION To add the complex numbers $(-3 + 2i)$ and $(2 - i)$, add the real parts and then add the imaginary parts.

$$(-3 + 2i) + (2 - i) = (-3 + 2) + (2i - i)$$
$$= (-3 + 2) + (2 - 1)i$$
$$= -1 + i$$

This same process works for subtraction.

$$(\mathbf{6 - 3}i) - (\mathbf{2 + 5}i) = (\mathbf{6 - 2}) + (\mathbf{-3}i - \mathbf{5}i)$$
$$= (6 - 2) + (-3 - 5)i$$
$$= 4 - 8i$$

This method is summarized as follows.

SUM OR DIFFERENCE OF COMPLEX NUMBERS

Let $a + bi$ and $c + di$ be two complex numbers. Then

$$(a + bi) + (c + di) = (a + c) + (b + d)i \qquad \text{Sum}$$

and
$$(a + bi) - (c + di) = (a - c) + (b - d)i. \qquad \text{Difference}$$

READING CHECK

How do you add and subtract complex numbers?

EXAMPLE 2 **Adding and subtracting complex numbers**

Write each sum or difference in standard form.
(a) $(-7 + 2i) + (3 - 4i)$ **(b)** $3i - (5 - i)$

Solution
(a) $(\mathbf{-7 + 2}i) + (\mathbf{3 - 4}i) = (\mathbf{-7 + 3}) + (\mathbf{2 - 4})i = \mathbf{-4 - 2}i$
(b) $3i - (5 - i) = 3i - 5 + i = -5 + (3 + 1)i = -5 + 4i$

Now Try Exercises 23, 29

TECHNOLOGY NOTE

CALCULATOR HELP

To access the imaginary unit i, see Appendix A (page AP-7).

Complex Numbers
Many calculators can perform arithmetic with complex numbers. The figure shows a calculator display for the results in Example 2.

```
(-7+2i)+(3-4i)
                -4-2i
3i-(5-i)
                -5+4i
```

MULTIPLICATION We multiply two complex numbers in the same way that we multiply binomials and then we apply the property $i^2 = -1$. For example,

$$(2 - 3x)(1 + 4x) = 2 + 8x - 3x - 12x^2 = 2 + 5x - 12x^2.$$

In the next example we find the product of $2 - 3i$ and $1 + 4i$ in a similar manner.

EXAMPLE 3 **Multiplying complex numbers**

Write each product in standard form.
(a) $(2 - 3i)(1 + 4i)$ **(b)** $(5 - 2i)(5 + 2i)$

Solution
(a) Multiply the complex numbers like binomials.

$$
\begin{aligned}
(2 - 3i)(1 + 4i) &= (2)(1) + (2)(4i) - (3i)(1) - (3i)(4i) \\
&= 2 + 8i - 3i - 12i^2 \\
&= 2 + 5i - 12(\mathbf{-1}) \qquad i^2 = -1 \\
&= 14 + 5i
\end{aligned}
$$

(b)
$$
\begin{aligned}
(5 - 2i)(5 + 2i) &= (5)(5) + (5)(2i) - (2i)(5) - (2i)(2i) \\
&= 25 + 10i - 10i - 4i^2 \\
&= 25 - 4(\mathbf{-1}) \qquad i^2 = -1 \\
&= 29
\end{aligned}
$$

These results are supported in Figure 7.30.

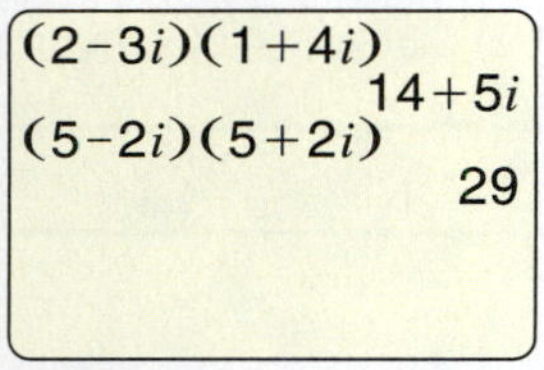

Now Try Exercises 31, 35

```
(2-3i)(1+4i)
           14+5i
(5-2i)(5+2i)
              29
```

Figure 7.30

Powers of i

An interesting pattern appears when powers of i are calculated.

$$
\begin{aligned}
i^1 &= \mathbf{\textit{i}} \\
i^2 &= \mathbf{-1} \\
i^3 &= i^2 \cdot i = -1 \cdot i = \mathbf{-\textit{i}} \\
i^4 &= i^2 \cdot i^2 = (-1)(-1) = \mathbf{1} \\
i^5 &= i^4 \cdot i = (1)i = \mathbf{\textit{i}} \\
i^6 &= i^4 \cdot i^2 = (1)(-1) = \mathbf{-1} \\
i^7 &= i^4 \cdot i^3 = (1)(-i) = \mathbf{-\textit{i}} \\
i^8 &= i^4 \cdot i^4 = (1)(1) = \mathbf{1}
\end{aligned}
$$

The powers of i cycle with the pattern i, -1, $-i$, and 1. These examples suggest the following method for calculating powers of i.

POWERS OF i

The value of i^n can be found by dividing n (a positive integer) by 4. If the remainder is r, then

$$i^n = i^r.$$

Note that $i^0 = 1$, $i^1 = i$, $i^2 = -1$, and $i^3 = -i$.

EXAMPLE 4   **Calculating powers of *i***

Evaluate each expression.

(a) i^9 (b) i^{19} (c) i^{40}

Solution

(a) When 9 is divided by 4, the result is 2 with remainder **1**. Thus $i^9 = i^1 = i$.
(b) When 19 is divided by 4, the result is 4 with remainder **3**. Thus $i^{19} = i^3 = -i$.
(c) When 40 is divided by 4, the result is 10 with remainder **0**. Thus $i^{40} = i^0 = 1$.

■ **Now Try Exercises 49, 51, 55**

Complex Conjugates and Division

The **complex conjugate** of $a + bi$ is $a - bi$. To find the conjugate, we change the sign of the imaginary part b. Table 7.10 contains some complex numbers and their conjugates.

TABLE 7.10   **Complex Conjugates**

Number	$2 + 5i$	$6 - 3i$	$-2 + 7i$	$-1 - i$	5	$-5i$
Conjugate	$2 - 5i$	$6 + 3i$	$-2 - 7i$	$-1 + i$	5	$5i$

EXAMPLE 5   **Finding complex conjugates**

Find the complex conjugate of each number.
(a) $5 + 3i$ (b) $-2 - i$ (c) $-4i$ (d) -6

Solution

(a) The conjugate is found by changing the imaginary part of $5 + 3i$ to its opposite. The conjugate is $5 - 3i$.

(b) The conjugate is $-2 + i$.

(c) $-4i$ can be written as $0 - 4i$. The conjugate is $0 + 4i$, or $4i$.

(d) -6 can be written as $-6 + 0i$. The conjugate is $-6 - 0i$, or -6. The conjugate of a real number is simply the same real number.

■ **Now Try Exercises 57, 59, 61, 63**

NOTE: The product of a complex number and its conjugate is a real number. That is,

$$(a + bi)(a - bi) = a^2 + b^2.$$

For example, $(3 + 4i)(3 - 4i) = 3^2 + 4^2 = 25$, which is a real number.

This property of complex conjugates is used to divide two complex numbers. To convert the quotient $\frac{2 + 3i}{3 - i}$ into standard form $a + bi$, we multiply the numerator and the denominator by the complex conjugate of the *denominator*, which is $3 + i$. The next example illustrates this method.

EXAMPLE 6   **Dividing complex numbers**

Write each quotient in standard form.

(a) $\dfrac{2 + 3i}{3 - i}$ (b) $\dfrac{4}{2i}$

Solution

(a) Multiply the numerator and denominator by **3 + i**.

$$\frac{2 + 3i}{3 - i} = \frac{(2 + 3i)(\mathbf{3 + i})}{(3 - i)(\mathbf{3 + i})} \qquad \text{Multiply by 1.}$$

$$= \frac{2(3) + (2)(i) + (3i)(3) + (3i)(i)}{(3)(3) + (3)(i) - (i)(3) - (i)(i)} \qquad \text{Multiply.}$$

$$= \frac{6 + 2i + 9i + 3i^2}{9 + 3i - 3i - i^2} \qquad \text{Simplify.}$$

$$= \frac{6 + 11i + 3(\mathbf{-1})}{9 - (\mathbf{-1})} \qquad i^2 = \mathbf{-1}$$

$$= \frac{3 + 11i}{10} \qquad \text{Simplify.}$$

$$= \frac{3}{10} + \frac{11}{10}i \qquad \frac{a + bi}{c} = \frac{a}{c} + \frac{b}{c}i$$

(b) Multiply the numerator and denominator by **− 2i**.

$$\frac{4}{2i} = \frac{(4)(\mathbf{-2i})}{(2i)(\mathbf{-2i})} \qquad \text{Multiply by 1.}$$

$$= \frac{-8i}{-4i^2} \qquad \text{Simplify.}$$

$$= \frac{-8i}{-4(\mathbf{-1})} \qquad i^2 = \mathbf{-1}$$

$$= \frac{-8i}{4} \qquad \text{Simplify.}$$

$$= -2i \qquad \text{Divide.}$$

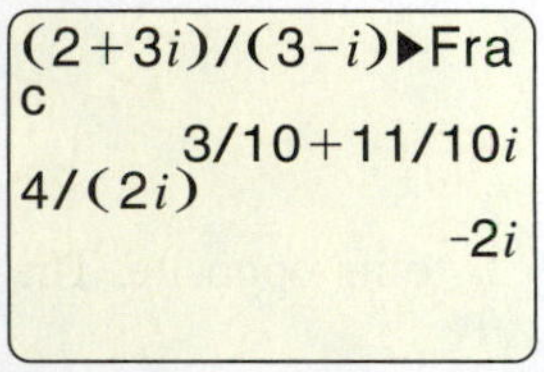

```
(2+3i)/(3-i)▶Fra
c
        3/10+11/10i
4/(2i)
              -2i
```

Figure 7.31

These results are supported in Figure 7.31.

▪ **Now Try Exercises 71, 73**

7.7   Putting It All Together

CONCEPT	EXPLANATION	EXAMPLES
Complex Numbers	A complex number can be expressed as $a + bi$, where a and b are real numbers.	The real part of $5 - 3i$ is 5 and the imaginary part is -3.
	The imaginary unit i satisfies $i = \sqrt{-1}$ and $i^2 = -1$. As a result, we can write $\sqrt{-a} = i\sqrt{a}$ if $a > 0$.	$\sqrt{-13} = i\sqrt{13}$ and $\sqrt{-9} = 3i$

CONCEPT	EXPLANATION	EXAMPLES
Addition, Subtraction, and Multiplication	To add (subtract) complex numbers, add (subtract) the real parts and then add (subtract) the imaginary parts.	$(3 + 6i) + (-1 + 2i)$ Sum $= (3 + -1) + (6 + 2)i$ $= 2 + 8i$ $(2 - 5i) - (1 + 4i)$ Difference $= (2 - 1) + (-5 - 4)i$ $= 1 - 9i$
	Multiply complex numbers in a manner similar to the way *FOIL* is used to multiply binomials. Then apply the property $i^2 = -1$.	$(-1 + 2i)(3 + i)$ Product $= (-1)(3) + (-1)(i) + (2i)(3) + (2i)(i)$ $= -3 - i + 6i + 2i^2$ $= -3 + 5i + 2(-1)$ $= -5 + 5i$
Complex Conjugates	The conjugate of $a + bi$ is $a - bi$.	The conjugate of $3 - 5i$ is $3 + 5i$. The conjugate of $2i$ is $-2i$.
Division	To simplify a quotient, multiply the numerator and denominator by the complex conjugate of the *denominator*. Then simplify the expression and write it in standard form as $a + bi$.	$\dfrac{10}{1 + 2i} = \dfrac{10(1 - 2i)}{(1 + 2i)(1 - 2i)}$ Quotient $= \dfrac{10 - 20i}{5}$ $= 2 - 4i$

7.7 Exercises

MyMathLab   Math XL PRACTICE   WATCH   DOWNLOAD   READ   REVIEW

CONCEPTS AND VOCABULARY

1. Give an example of a complex number that is not a real number.

2. Can you give an example of a real number that is not a complex number? Explain.

3. $\sqrt{-1} =$ _____   4. $i^2 =$ _____

5. $\sqrt{-a} =$ _____, if $a > 0$.

6. The complex conjugate of $10 + 7i$ is _____.

7. The standard form for a complex number is _____.

8. Write $\dfrac{2 + 4i}{2}$ in standard form.

9. The real part of $4 - 5i$ is _____.

10. The imaginary part of $4 - 5i$ is _____.

11. The imaginary part of -7 is _____.

12. The number $7i$ is a(n) _____ imaginary number.

COMPLEX NUMBERS

Exercises 13–22: Use i to write the expression.

13. $\sqrt{-5}$ 14. $\sqrt{-21}$

15. $\sqrt{-100}$ 16. $\sqrt{-49}$

17. $\sqrt{-144}$ 18. $\sqrt{-64}$

19. $\sqrt{-12}$ 20. $\sqrt{-8}$

21. $\sqrt{-18}$ 22. $\sqrt{-48}$

Exercises 23–46: Write the expression in standard form.

23. $(5 + 3i) + (-2 - 3i)$

24. $(1 - i) + (5 - 7i)$

25. $2i + (-8 + 5i)$

26. $-3i + 5i$

27. $(2 - 7i) - (1 + 2i)$

28. $(1 + 8i) - (3 + 9i)$ **29.** $5i - (10 - 2i)$

30. $-3(4 - 3i)$ **31.** $(3 + 2i)(-1 + 5i)$

32. $(1 + i) - (1 - i)$ **33.** $4(5 - 3i)$

34. $(1 + 2i)(-6 - i)$ **35.** $(5 + 4i)(5 - 4i)$

36. $(3 + 5i)(3 - 5i)$ **37.** $(-4i)(5i)$

38. $(-6i)(-4i)$

39. $3i + (2 - 3i) - (1 - 5i)$

40. $4 - (5 - 7i) + (3 + 7i)$

41. $(2 + i)^2$ **42.** $(-1 + 2i)^2$

43. $2i(-3 + i)$ **44.** $5i(1 - 9i)$

45. $i(1 + i)^2$ **46.** $2i(1 - i)^2$

Exercises 47 and 48: **Thinking Generally** *Write in standard form.*

47. $(a + 3bi)(a - 3bi)$ **48.** $(a + bi) - (a - bi)$

Exercises 49–56: (Refer to Example 4.) Simplify.

49. i^{11} **50.** i^{50}

51. i^{21} **52.** i^{103}

53. i^{58} **54.** i^{61}

55. i^{64} **56.** i^{28}

Exercises 57–64: Write the complex conjugate.

57. $3 + 4i$ **58.** $1 - 4i$

59. $-6i$ **60.** -10

61. $5 - 4i$ **62.** $7 + 2i$

63. -1 **64.** $19i$

Exercises 65–78: Write the expression in standard form.

65. $\dfrac{2}{1 + i}$ **66.** $\dfrac{-6}{2 - i}$

67. $\dfrac{3i}{5 - 2i}$ **68.** $\dfrac{-8}{2i}$

69. $\dfrac{8 + 9i}{5 + 2i}$ **70.** $\dfrac{3 - 2i}{1 + 4i}$

71. $\dfrac{5 + 7i}{1 - i}$ **72.** $\dfrac{-7 + 4i}{3 - 2i}$

73. $\dfrac{2 - i}{i}$ **74.** $\dfrac{3 + 2i}{-i}$

75. $\dfrac{1}{i} + \dfrac{1}{2i}$ **76.** $\dfrac{3}{4i} + \dfrac{2}{i}$

77. $\dfrac{1}{-1 + i} - \dfrac{2}{i}$ **78.** $-\dfrac{3}{2i} - \dfrac{2}{1 + i}$

APPLICATIONS

Exercises 79 and 80: ***Corrosion in Airplanes*** *Corrosion in the metal surface of an airplane can be difficult to detect visually. One test used to locate it involves passing an alternating current through a small area on the plane's surface. If the current varies from one region to another, it may indicate that corrosion is occurring. The impedance Z (or opposition to the flow of electricity) of the metal is related to the voltage V and current I by the equation $Z = \frac{V}{I}$, where Z, V, and I are complex numbers. Calculate Z for the given values of V and I.* (**Source:** Society for Industrial and Applied Mathematics.)

79. $V = 40 + 70i, I = 2 + 3i$

80. $V = 10 + 20i, I = 3 + 7i$

81. **Online Exploration** We graph real numbers on a number line. Can we graph complex numbers on a number line? Go online and find out how to graph complex numbers.

82. **Online Exploration** Go online and find out how to calculate the absolute value of a complex number. If a complex number is graphed, what does the absolute value tell us?

WRITING ABOUT MATHEMATICS

83. A student multiplies $(2 + 3i)(4 - 5i)$ *incorrectly* to obtain $8 - 15i$. What is the student's mistake?

84. A student divides the ratio $\frac{6 - 10i}{3 + 2i}$ *incorrectly* to obtain $2 - 5i$. What is the student's mistake?

CHAPTER 7 SUMMARY   563

SECTION 7.7

Checking Basic Concepts

1. Use i to write each expression.

   (a) $\sqrt{-64}$

   (b) $\sqrt{-17}$

2. Simplify each expression.

   (a) $(2 - 3i) + (1 - i)$ (b) $4i - (2 + i)$

   (c) $(3 - 2i)(1 + i)$ (d) $\dfrac{3}{2 - 2i}$

CHAPTER 7   Summary

SECTION 7.1 ▪ RADICAL EXPRESSIONS AND FUNCTIONS

Radicals and Radical Notation

Square Root	b is a square root of a if $b^2 = a$.
Principal Square Root	$\sqrt{a} = b$ if $b^2 = a$ and $b \geq 0$.

Examples: $\sqrt{16} = 4$, $-\sqrt{9} = -3$,

and $\pm\sqrt{36} = \pm 6$

Cube Root b is a cube root of a if $b^3 = a$.

Examples: $\sqrt[3]{27} = 3$, $\sqrt[3]{-8} = -2$

nth Root b is an nth root of a if $b^n = a$.

Example: $\sqrt[4]{16} = 2$ because $2^4 = 16$.

NOTE: An *even* root of a *negative* number is not a real number. Also, $\sqrt[n]{a}$ denotes the *principal* nth root.

Absolute Value   The expressions $|x|$ and $\sqrt{x^2}$ are equivalent.

Example: $\sqrt{(x + y)^2} = |x + y|$

The Square Root Function The square root function is denoted $f(x) = \sqrt{x}$. Its domain is $\{x \mid x \geq 0\}$ and its graph is shown in the figure.

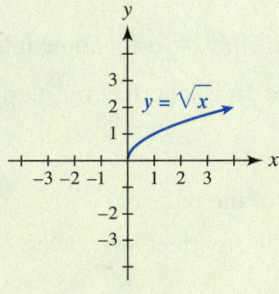

The Cube Root Function The cube root function is denoted $f(x) = \sqrt[3]{x}$. Its domain is all real numbers and its graph is shown in the figure.

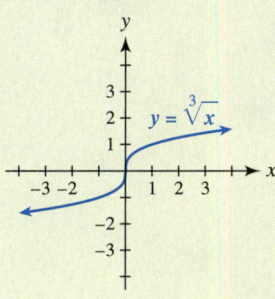

SECTION 7.2 ▪ RATIONAL EXPONENTS

The Expression $a^{1/n}$ $a^{1/n} = \sqrt[n]{a}$ if n is an integer greater than 1.

Examples: $5^{1/2} = \sqrt{5}$ and $64^{1/3} = \sqrt[3]{64} = 4$

The Expression $a^{m/n}$ $a^{m/n} = \sqrt[n]{a^m}$ or $a^{m/n} = (\sqrt[n]{a})^m$

Examples: $8^{2/3} = \sqrt[3]{8^2} = \sqrt[3]{64} = 4$ and
$$8^{2/3} = (\sqrt[3]{8})^2 = (2)^2 = 4$$

Properties of Exponents

Product Rule $a^p a^q = a^{p+q}$

Negative Exponents $a^{-p} = \dfrac{1}{a^p}, \dfrac{1}{a^{-p}} = a^p$

Negative Exponents for Quotients $\left(\dfrac{a}{b}\right)^{-p} = \left(\dfrac{b}{a}\right)^p$

Quotient Rule for Exponents $\dfrac{a^p}{a^q} = a^{p-q}$

Power Rule for Exponents $(a^p)^q = a^{pq}$

Power Rule for Products $(ab)^p = a^p b^p$

Power Rule for Quotients $\left(\dfrac{a}{b}\right)^p = \dfrac{a^p}{b^p}$

SECTION 7.3 ▪ SIMPLIFYING RADICAL EXPRESSIONS

Product Rule for Radical Expressions Provided each expression is defined,
$$\sqrt[n]{a} \cdot \sqrt[n]{b} = \sqrt[n]{a \cdot b}.$$

Example: $\sqrt[3]{3} \cdot \sqrt[3]{9} = \sqrt[3]{27} = 3$

Perfect nth Power An integer a is a perfect nth power if $b^n = a$ for some integer b.

Examples: 25 is a perfect square, 8 is a perfect cube, and 16 is a perfect fourth power.

Simplifying Radicals (nth roots)

STEP 1: Determine the largest perfect nth power factor of the radicand.

STEP 2: Use the product rule to factor out and simplify this perfect nth power.

Examples: $\sqrt{32} = \sqrt{16 \cdot 2} = \sqrt{16} \cdot \sqrt{2} = 4\sqrt{2}$
$$\sqrt[3]{32} = \sqrt[3]{8 \cdot 4} = \sqrt[3]{8} \cdot \sqrt[3]{4} = 2\sqrt[3]{4}$$

Quotient Rule for Radical Expressions Provided each expression is defined,

$$\sqrt[n]{\frac{a}{b}} = \frac{\sqrt[n]{a}}{\sqrt[n]{b}}.$$

Example: $\dfrac{\sqrt[3]{24}}{\sqrt[3]{3}} = \sqrt[3]{\dfrac{24}{3}} = \sqrt[3]{8} = 2$

SECTION 7.4 ■ OPERATIONS ON RADICAL EXPRESSIONS

Addition and Subtraction Combine like radicals.

Examples: $2\sqrt[3]{4} + 3\sqrt[3]{4} = 5\sqrt[3]{4}$ and $\sqrt{5} - 2\sqrt{5} = -\sqrt{5}$

Multiplication Sometimes radical expressions can be multiplied like binomials.

Examples: $(4 - \sqrt{2})(2 + \sqrt{2}) = 8 + 4\sqrt{2} - 2\sqrt{2} - 2 = 6 + 2\sqrt{2}$
$(5 - \sqrt{3})(5 + \sqrt{3}) = (5)^2 - (\sqrt{3})^2 = 25 - 3 = 22$ because
$(a - b)(a + b) = a^2 - b^2.$

Rationalizing the Denominator One technique is to multiply the numerator and denominator by the conjugate of the denominator if the denominator is a binomial containing one or more square roots.

Examples: $\dfrac{1}{4 + \sqrt{2}} = \dfrac{1}{(4 + \sqrt{2})} \cdot \dfrac{(4 - \sqrt{2})}{(4 - \sqrt{2})} = \dfrac{4 - \sqrt{2}}{(4)^2 - (\sqrt{2})^2} = \dfrac{4 - \sqrt{2}}{14}$

$\dfrac{4}{\sqrt{7}} = \dfrac{4}{\sqrt{7}} \cdot \dfrac{\sqrt{7}}{\sqrt{7}} = \dfrac{4\sqrt{7}}{7}$

SECTION 7.5 ■ MORE RADICAL FUNCTIONS

Root Functions $f(x) = x^{1/n} = \sqrt[n]{x}$

Odd root functions are defined for all real numbers, and even root functions are defined for nonnegative real numbers.

Examples: $f(x) = x^{1/3} = \sqrt[3]{x}$, $g(x) = x^{1/4} = \sqrt[4]{x}$,   and $h(x) = x^{1/5} = \sqrt[5]{x}$

Power Functions If a function can be defined by $f(x) = x^p$, where p is a rational number, then it is a power function.

Examples: $f(x) = x^{4/5}$ and $g(x) = x^{2.3}$

SECTION 7.6 ■ EQUATIONS INVOLVING RADICAL EXPRESSIONS

Power Rule for Solving Radical Equations The solutions to $a = b$ are among the solutions to $a^n = b^n$, where n is a positive integer.

Example: The solutions to the equation $\sqrt{3x + 3} = 2x - 1$ are among the solutions to the equation $3x + 3 = (2x - 1)^2$.

Solving Radical Equations

STEP 1: Isolate a radical term on one side of the equation.

STEP 2: Apply the power rule by raising each side of the equation to the power equal to the index of the isolated radical term.

STEP 3: Solve the equation. If it still contains a radical, repeat Steps 1 and 2.

STEP 4: Check your answers by substituting each result in the *given* equation.

Example: To isolate the radical in $\sqrt{x + 1} + 4 = 6$ subtract 4 from each side to obtain $\sqrt{x + 1} = 2$. Next square each side, which gives $x + 1 = 4$ or $x = 3$. Checking verifies that 3 is a solution.

Pythagorean Theorem If a right triangle has legs a and b with hypotenuse c, then

$$a^2 + b^2 = c^2.$$

Example: If a right triangle has legs 8 and 15, then the hypotenuse equals

$$c = \sqrt{8^2 + 15^2} = \sqrt{289} = 17.$$

The Distance Formula The distance d between (x_1, y_1) and (x_2, y_2) is

$$d = \sqrt{(x_2 - x_1)^2 + (y_2 - y_1)^2}.$$

Example: The distance between $(-1, 3)$ and $(4, 5)$ is

$$d = \sqrt{\left(4 - (-1)\right)^2 + (5 - 3)^2} = \sqrt{25 + 4} = \sqrt{29}.$$

Solving the Equation $x^n = k$ Let n be a positive integer.
Take the nth root of each side of $x^n = k$ to obtain $\sqrt[n]{x^n} = \sqrt[n]{k}$.

1. If n is *odd*, then $\sqrt[n]{x^n} = x$ and the equation becomes $x = \sqrt[n]{k}$.

2. If n is *even* and $k \geq 0$, then $\sqrt[n]{x^n} = |x|$ and the equation becomes $|x| = \sqrt[n]{k}$.

Examples: $x^3 = -27$ implies that $x = \sqrt[3]{-27} = -3$.

$x^4 = 81$ implies that $|x| = \sqrt[4]{81} = 3$ or $x = \pm 3$.

SECTION 7.7 ■ COMPLEX NUMBERS

Complex Numbers

Imaginary Unit	$i = \sqrt{-1}$ and $i^2 = -1$
Standard Form	$a + bi$, where a and b are real numbers
	Examples: $4 + 3i$, $5 - 6i$, 8, and $-2i$
Real Part	The real part of $a + bi$ is a.
	Example: The real part of $3 - 2i$ is 3.
Imaginary Part	The imaginary part of $a + bi$ is b.
	Example: The imaginary part of $2 - i$ is -1.
Arithmetic Operations	Arithmetic operations are similar to arithmetic operations on binomials.

Examples:

$$(2 + 2i) + (3 - i) = 5 + i \qquad \text{Sum}$$

$$(1 - i) - (1 - 2i) = i \qquad \text{Difference}$$

$$(1 - i)(1 + i) = 1^2 - i^2 = 1 - (-1) = 2 \quad \text{Product}$$

$$\frac{2}{1 - i} = \frac{2}{1 - i} \cdot \frac{1 + i}{1 + i} = \frac{2 + 2i}{2} = 1 + i \quad \text{Quotient}$$

Powers of i   The value of i^n equals i^r, where r is the remainder when n is divided by 4. Note that $i^0 = 1$, $i^1 = i$, $i^2 = -1$, and $i^3 = -i$.

Example: $i^{21} = i^1 = i$ because when 21 is divided by 4 the remainder is 1.

CHAPTER 7   Review Exercises

SECTION 7.1

Exercises 1–12: Simplify the expression.

1. $\sqrt{4}$

2. $\sqrt{36}$

3. $\sqrt{9x^2}$

4. $\sqrt{(x-1)^2}$

5. $\sqrt[3]{-64}$

6. $\sqrt[3]{-125}$

7. $\sqrt[3]{x^6}$

8. $\sqrt[3]{27x^3}$

9. $\sqrt[4]{16}$

10. $\sqrt[5]{-1}$

11. $\sqrt[4]{x^8}$

12. $\sqrt[5]{(x+1)^5}$

SECTION 7.2

Exercises 13–16: Write the expression in radical notation.

13. $14^{1/2}$

14. $(-5)^{1/3}$

15. $\left(\dfrac{x}{y}\right)^{3/2}$

16. $(xy)^{-2/3}$

Exercises 17–20: Evaluate the expression.

17. $(-27)^{2/3}$

18. $16^{1/4}$

19. $16^{3/2}$

20. $81^{3/4}$

Exercises 21–24: Simplify the expression. Assume that all variables are positive.

21. $(z^3)^{2/3}$

22. $(x^2y^4)^{1/2}$

23. $\left(\dfrac{x^2}{y^6}\right)^{3/2}$

24. $\left(\dfrac{x^3}{y^6}\right)^{-1/3}$

SECTION 7.3

Exercises 25–40: Simplify the expression. Assume that all variables are positive.

25. $\sqrt{2} \cdot \sqrt{32}$

26. $\sqrt[3]{-4} \cdot \sqrt[3]{2}$

27. $\sqrt[3]{x^4} \cdot \sqrt[3]{x^2}$

28. $\dfrac{\sqrt{80}}{\sqrt{20}}$

29. $\sqrt[3]{-\dfrac{x}{8}}$

30. $\sqrt{\dfrac{1}{3}} \cdot \sqrt{\dfrac{1}{3}}$

31. $\sqrt{48}$

32. $\sqrt{54}$

33. $\sqrt[3]{\dfrac{3}{x}} \cdot \sqrt[3]{\dfrac{9}{x^2}}$

34. $\sqrt{32a^3b^2}$

35. $\sqrt{3xy} \cdot \sqrt{27xy}$

36. $\sqrt[3]{-25z^2} \cdot \sqrt[3]{-5z^2}$

37. $\sqrt{x^2 + 2x + 1}$

38. $\sqrt[4]{\dfrac{2a^2}{b}} \cdot \sqrt[4]{\dfrac{8a^3}{b^3}}$

39. $2\sqrt{x} \cdot \sqrt[3]{x}$

40. $\sqrt[3]{rt} \cdot \sqrt[4]{r^2t^4}$

SECTION 7.4

Exercises 41–50: Simplify the expression. Assume that all variables are positive.

41. $3\sqrt{3} + \sqrt{3}$

42. $\sqrt[3]{x} + 2\sqrt[3]{x}$

43. $3\sqrt[3]{5} - 6\sqrt[3]{5}$

44. $\sqrt[4]{y} - 2\sqrt[4]{y}$

45. $2\sqrt{12} + 7\sqrt{3}$

46. $3\sqrt{18} - 2\sqrt{2}$

47. $7\sqrt[3]{16} - \sqrt[3]{2}$

48. $\sqrt{4x+4} + \sqrt{x+1}$

49. $\sqrt{4x^3} - \sqrt{x}$

50. $\sqrt[3]{ab^4} + 2\sqrt[3]{a^4b}$

Exercises 51–56: Multiply and simplify.

51. $(1 + \sqrt{2})(3 + \sqrt{2})$

52. $(7 - \sqrt{5})(1 + \sqrt{3})$

53. $(3 + \sqrt{6})(3 - \sqrt{6})$

54. $(10 - \sqrt{5})(10 + \sqrt{5})$

55. $(\sqrt{a} + \sqrt{2b})(\sqrt{a} - \sqrt{2b})$

56. $(\sqrt{xy} - 1)(\sqrt{xy} + 2)$

Exercises 57–62: Rationalize the denominator.

57. $\dfrac{4}{\sqrt{5}}$

58. $\dfrac{r}{2\sqrt{t}}$

59. $\dfrac{1}{\sqrt{2} + 3}$

60. $\dfrac{2}{5 - \sqrt{7}}$

61. $\dfrac{1}{\sqrt{8} - \sqrt{7}}$

62. $\dfrac{\sqrt{a} - \sqrt{b}}{\sqrt{a} + \sqrt{b}}$

SECTION 7.5

Exercises 63 and 64: Graph the equation.

63. $y = \sqrt[4]{x}$

64. $y = \sqrt[3]{x}$

Exercises 65 and 66: Write f(x) in radical notation and evaluate f(4).

65. $f(x) = x^{1/2}$

66. $f(x) = x^{2/7}$

Exercises 67 and 68: Graph the equation. Compare the graph to either $y = \sqrt{x}$ or $y = \sqrt[3]{x}$.

67. $y = \sqrt{x} - 2$

68. $y = \sqrt[3]{x} - 1$

Exercises 69–72: Find the domain of f. Write your answer in interval notation.

69. $f(x) = \sqrt{x-1}$ **70.** $f(x) = \sqrt{6-2x}$

71. $f(x) = \sqrt{x^2+1}$ **72.** $f(x) = \dfrac{1}{\sqrt{x+2}}$

SECTION 7.6

Exercises 73–78: Solve. Check your answer.

73. $\sqrt{x+2} = x$ **74.** $\sqrt{2x-1} = \sqrt{x+3}$

75. $\sqrt[3]{x-1} = 2$ **76.** $\sqrt[3]{3x} = 3$

77. $\sqrt{2x} = x - 4$ **78.** $\sqrt{x+1} = \sqrt{x+2}$

 Exercises 79 and 80: Solve graphically. Approximate solutions to the nearest hundredth when appropriate.

79. $\sqrt[3]{2x-1} = 2$ **80.** $x^{2/3} = 3 - x$

Exercises 81 and 82: A right triangle has legs a and b with hypotenuse c. Find the length of the missing side.

81. $a = 4, b = 7$ **82.** $a = 5, c = 8$

Exercises 83 and 84: Find the exact distance between the given points.

83. $(-2, 3), (2, -2)$ **84.** $(2, -3), (-4, 1)$

Exercises 85–94: Solve.

85. $x^2 = 121$ **86.** $2z^2 = 32$

87. $(x-1)^2 = 16$ **88.** $x^3 = 64$

89. $(x-1)^3 = 8$ **90.** $(2x-1)^3 = 27$

91. $x^4 = 256$ **92.** $x^5 = -1$

93. $(x-3)^5 = -32$ **94.** $3(x+1)^4 = 3$

SECTION 7.7

Exercises 95–100: Write the complex expression in standard form.

95. $(1-2i) + (-3+2i)$

96. $(1+3i) - (3-i)$

97. $(1-i)(2+3i)$ **98.** $\dfrac{3+i}{1-i}$

99. $\dfrac{i(4+i)}{2-3i}$ **100.** $(1-i)^2(1+i)$

APPLICATIONS

101. *Hang Time* A football is punted and has a hang time T of 4.6 seconds. Use the formula $T(h) = \frac{1}{2}\sqrt{h}$ to estimate to the nearest foot the height h to which it was kicked.

102. *Baseball Diamond* The four bases of a baseball diamond form a square that is 90 feet on a side. Find the distance from home plate to second base.

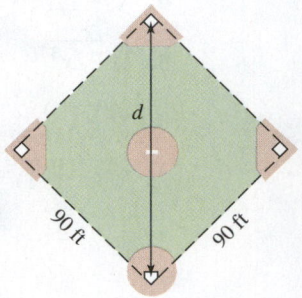

103. *Falling Time* The time T in seconds for an object to fall from a height of h feet is given by $T = \frac{1}{4}\sqrt{h}$. If a person steps off a 10-foot-high board into a swimming pool, how long is the person in the air?

104. *Geometry* A cube has sides of length $\sqrt{5}$.
 (a) Find the area of one side of the cube.
 (b) Find the volume of the cube.
 (c) Find the length of the diagonal of one of the sides.
 (d) Find the distance from A to B in the figure.

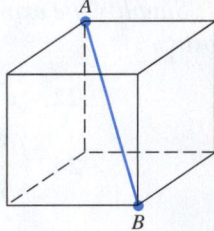

105. *Pendulum* The time required for a pendulum to swing back and forth is called its *period* (see the figure). The period T of a pendulum does not depend on its weight, only on its length L and gravity. It is given by $T = 2\pi\sqrt{\frac{L}{32.2}}$, where T is in seconds and L is in feet. Estimate the length of a pendulum with a period of 1 second.

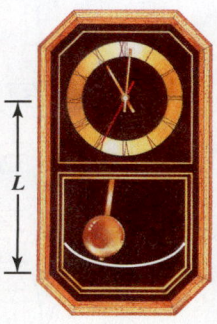

106. *Pendulum* (Refer to Exercise 105.) If a pendulum were on the moon, its period could be calculated by $T = 2\pi\sqrt{\frac{L}{5.1}}$. Estimate the length of a pendulum with a period of 1 second on the moon. Compare your answer to that for Exercise 105.

107. *Population Growth* In 1790 the population of the United States was 4 million, and by 2000 it had grown to 281 million. The *average* annual percentage growth in the population r (expressed as a decimal) can be determined by the polynomial equation $281 = 4(1 + r)^{210}$. Solve this equation for r and interpret the result.

108. *Highway Curves* If a circular highway curve is banked with a slope of $m = \frac{1}{10}$ (see the accompanying figure) and has a radius of R feet, then the speed limit L in miles per hour for the curve is given by

$$L = \sqrt{3.75R}.$$

(*Source:* N. Garber and L. Hoel, *Traffic and Highway Engineering.*)

(a) Find the speed limit if R is 500 feet.

(b) With no banking, the speed limit is given by $L = 1.5\sqrt{R}$. Find the speed limit for a curve with no banking and a radius of 500 feet. How does banking affect the speed limit? Does this result agree with your intuition?

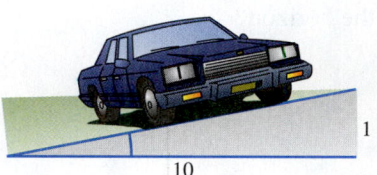

109. *Geometry* Find the length of a side of a square if the square has an area of 7 square feet.

CHAPTER 7   Test

 Step-by-step test solutions are found on the Chapter Test Prep Videos available in **MyMathLab** and on YouTube™ (search "RockswoldInterAlg" and click on "Channels").

Exercises 1–6: Simplify the expression.

1. $\sqrt[3]{-27}$

2. $\sqrt{(z + 1)^2}$

3. $\sqrt{25x^4}$

4. $\sqrt[3]{8z^6}$

5. $\sqrt[4]{16x^4y^5}$; $x > 0, y > 0$

6. $(\sqrt{3} - \sqrt{2})(\sqrt{3} + \sqrt{2})$

Exercises 7 and 8: Write the expression in radical notation.

7. $7^{2/5}$

8. $\left(\dfrac{x}{y}\right)^{-2/3}$

Exercises 9 and 10: Evaluate the expression by hand.

9. $(-8)^{4/3}$

10. $36^{-3/2}$

Exercises 11 and 12: Use rational exponents to write the expression.

11. $\sqrt[3]{x^4}$

12. $\sqrt{x} \cdot \sqrt[5]{x}$

13. Find the domain of $f(x) = \sqrt{4 - x}$. Write your answer in interval notation.

14. Sketch a graph of $y = \sqrt{x + 3}$.

Exercises 15–22: Simplify the expression. Assume that all variables are positive.

15. $(2z^{1/2})^3$

16. $\left(\dfrac{y^2}{z^3}\right)^{-1/3}$

17. $\sqrt{3} \cdot \sqrt{27}$

18. $\dfrac{\sqrt{y^3}}{\sqrt{4y}}$

19. $7\sqrt{7} - 3\sqrt{7} + \sqrt{5}$

20. $7\sqrt[3]{x} - \sqrt[3]{x}$

21. $4\sqrt{18} + \sqrt{8}$

22. $\dfrac{\sqrt[3]{32}}{\sqrt[3]{4}}$

23. Solve each equation.
   (a) $\sqrt{x - 2} = 5$
   (b) $\sqrt[3]{x + 1} = 2$
   (c) $(x - 1)^3 = 8$
   (d) $\sqrt{2x + 2} = x - 11$

24. Rationalize the denominator.
   (a) $\dfrac{2}{3\sqrt{7}}$
   (b) $\dfrac{1}{1 + \sqrt{5}}$

25. Solve $\sqrt{3x} - x + 1 = \sqrt[3]{x - 1}$ graphically. Round solutions to the nearest hundredth.

26. One leg of a right triangle has length 7 and the hypotenuse has length 13. Find the length of the third side.

27. Find the distance between $(-3, 5)$ and $(-1, 7)$.

Exercises 28–31: Write the complex expression in standard form.

28. $(-5 + i) + (7 - 20i)$

29. $(3i) - (6 - 5i)$

30. $\left(\dfrac{1}{2} - i\right)\left(\dfrac{1}{2} + i\right)$

31. $\dfrac{2i}{5 + 2i}$

32. *Distance to the Horizon* A formula for calculating the distance d in miles that one can see to the horizon on a clear day is approximated by the formula $d = 1.22\sqrt{x}$, where x is the elevation, in feet, of a person.

(a) If a person climbs a cliff and stops at an elevation of 200 feet, how far can the person see to the horizon?

(b) How high should the person climb to see 25 miles to the horizon?

33. *Wing Span of a Bird* The wing span L of a bird with weight W can sometimes be modeled by $L = 27.4W^{1/3}$, where L is in inches and W is in pounds. Use this formula to estimate the weight of a bird that has a wing span of 30 inches. (*Source:* C. Pennycuick, *Newton Rules Biology.*)

CHAPTER 7 Extended and Discovery Exercises

1. *Modeling Wood in a Tree* Forestry services estimate the volume of timber in a given area of forest. To make such estimates, scientists have developed formulas to find the amount of wood contained in a tree with height h in feet and diameter d in inches. One study concluded that the volume V of wood in cubic feet in a tree is given by $V = kh^{1.12}d^{1.98}$, where k is a constant. Note that the diameter is measured 4.5 feet above the ground. (*Source:* B. Ryan, B. Joiner, and T. Ryan, *Minitab Handbook.*)

(a) A tree with an 11-inch diameter and a 47-foot height has a volume of 11.4 cubic feet. Approximate the constant k.

(b) Estimate the volume of wood in the same type of tree with $d = 20$ inches and $h = 105$ feet.

2. *Area of Skin* The surface area of the skin covering the human body is a function of more than one variable. Both height and weight influence the surface area of a person's body. Hence a taller person tends to have a larger surface area, as does a heavier person. A formula to determine the area of a person's skin in square inches is $S = 15.7w^{0.425}h^{0.725}$, where w is weight in pounds and h is height in inches. (*Source:* H. Lancaster, *Quantitative Methods in Biological and Medical Sciences.*)

(a) Use S to estimate the area of a person's skin who is 65 inches tall and weighs 154 pounds.

(b) If a person's weight doubles, what happens to the area of the person's skin? Explain.

(c) If a person's height doubles, what happens to the area of the person's skin? Explain.

3. *Minimizing Cost* A natural gas line running along a river is to be connected from point A to a cabin on the other bank located at point D, as illustrated in the figure. The width of the river is 500 feet, and the distance from point A to point C is 1000 feet. The cost of running the pipe along the shoreline is $30 per foot, and the cost of running it underwater is $50 per foot. The cost of connecting the gas line from A to D is to be minimized.

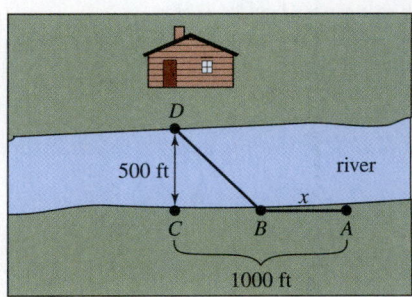

(a) Write an expression that gives the cost of running the line from A to B if the distance between these points is x feet.

(b) Find the distance from B to D in terms of x.

(c) Write an expression that gives the cost of running the line from B to D.

(d) Use your answer from parts (a) and (c) to write an expression that gives the cost of running the line from A to B to D.

(e) Graph your expression from part (d) in the window [0, 1000, 100] by [40000, 60000, 5000] to determine the value of x that minimizes the cost of the line going from A to D. What is the minimum cost?

1. Evaluate $S = 4\pi r^2$ when $r = 3$.

2. Identify the domain and range of the relation given by $S = \{(-1, 2), (0, 4), (1, 2)\}$.

3. Simplify each expression. Write the result using positive exponents.

(a) $\left(\dfrac{ab^2}{b^{-1}}\right)^{-3}$ (b) $\dfrac{(x^2y)^3}{x^2(y^2)^{-3}}$

(c) $(rt)^2(r^2t)^3$

4. Write 0.00043 in scientific notation.

5. Find $f(3)$ if $f(x) = \dfrac{x}{x - 2}$. What is the domain of f?

6. Find the domain of $f(x) = \sqrt[3]{2x}$.

Exercises 7 and 8: Graph f by hand.

7. $f(x) = 1 - 2x$

8. $f(x) = \sqrt{x + 1}$

9. Sketch a graph of $f(x) = x^2 - 4$.
  (a) Identify the domain and range.
  (b) Evaluate $f(-2)$.
  (c) Identify the x-intercepts.
  (d) Solve the equation $f(x) = 0$.

10. Find the slope–intercept form of the line that passes through $(-1, 2)$ and is perpendicular to $y = -2x$.

11. Let f be a linear function. Find a formula for $f(x)$.

x	-2	-1	1	2
$f(x)$	7	4	-2	-5

12. Sketch a graph of a line passing through $(-1, 1)$ with slope $-\frac{1}{2}$.

13. Solve $5x - (3 - x) = \frac{1}{2}x$.

14. Solve $2x - 5 \le 4 - x$.

15. Solve $|x - 2| \le 3$.

16. Solve $-1 \le 1 - 2x \le 6$.

17. Solve each system symbolically, if possible.
  (a) $2x - y = 4$ (b) $3x - 4y = 2$
  $\quad\;\; x + y = 8$ $\quad\;\; x - \frac{4}{3}y = 1$

18. Shade the solution set in the xy-plane.
$$x + 2y \le 2$$
$$-x + 3y \ge 3$$

19. Solve the system.
$$\begin{aligned} x + 2y - z &= 6 \\ x - 3y + z &= -2 \\ x + y + z &= 6 \end{aligned}$$

20. Calculate $\det \begin{bmatrix} 4 & 2 & -1 \\ 2 & 1 & 0 \\ 0 & -2 & 1 \end{bmatrix}$.

Exercises 21–24: Multiply the expression.

21. $4x(4 - x^3)$ **22.** $(x - 4)(x + 4)$

23. $(5x + 3)(x - 2)$ **24.** $(4x + 9)^2$

Exercises 25–30: Factor the expression.

25. $9x^2 - 16$ **26.** $x^2 - 4x + 4$

27. $15x^3 - 9x^2$ **28.** $12x^2 - 5x - 3$

29. $r^3 - 1$ **30.** $x^3 - 3x^2 + 5x - 15$

Exercises 31 and 32: Solve each equation.

31. $x^2 - 3x + 2 = 0$ **32.** $x^3 = 4x$

Exercises 33 and 34: Simplify the expression.

33. $\dfrac{x^2 + 3x + 2}{x - 3} \div \dfrac{x + 1}{2x - 6}$

34. $\dfrac{2}{x - 1} + \dfrac{5}{x}$

Exercises 35–42: Simplify the expression. Assume all variables are positive.

35. $\sqrt{36x^2}$ **36.** $\sqrt[3]{64}$

37. $16^{-3/2}$ **38.** $\sqrt[4]{625}$

39. $\sqrt{2x} \cdot \sqrt{8x}$ **40.** $\sqrt{x} \cdot \sqrt[4]{x}$

41. $\dfrac{\sqrt[3]{16x^4}}{\sqrt[3]{2x}}$ **42.** $4\sqrt{12x} - 2\sqrt{3x}$

43. Multiply $(2x + \sqrt{3})(x - \sqrt{3})$.

44. Find the domain of $f(x) = \sqrt{1 - x}$.

45. Graph $f(x) = 3\sqrt[3]{x}$.

46. Find the distance between $(-2, 3)$ and $(1, 2)$.

47. Write $(1 - i)(2 + 3i)$ in standard form.

48. The lengths of the legs of a right triangle are 5 and 12. What is the length of the hypotenuse?

Exercises 49–52: Solve symbolically.

49. $2\sqrt{x + 3} = x$ **50.** $\sqrt[3]{x - 1} = 3$

51. $\sqrt{x + 4} = 2\sqrt{x + 5}$ **52.** $\frac{1}{3}x^4 = 27$

Exercises 53 and 54: Solve graphically and approximate your answer to the nearest hundredth.

53. $\sqrt[3]{x^2 - 2} + x = \sqrt{x}$ **54.** $x - \sqrt[3]{x} = \sqrt{x + 2}$

APPLICATIONS

55. *Calculating Water Flow* The gallons G of water in a tank after t minutes are given by $G(t) = 300 - 15t$. Interpret the y-intercept and the slope of the graph of G as a rate of change.

56. *Distance* A person jogs away from home at 8 miles per hour for 1 hour, rests in a park for 2 hours, and then walks back home at 4 miles per hour. Sketch a graph that models the distance y that the person is from home after x hours.

57. *Investment* Suppose $2000 is deposited in two accounts paying 5% and 4% annual interest. If the total interest after 1 year is $93, how much is invested at each rate?

58. *Geometry* A rectangular box with a square base has a volume of 256 cubic inches. If its height is 4 inches less than the length of an edge of the base, find the dimensions of the box.

59. *Height of a Building* A 5-foot 3-inch person casts a shadow that is 7.5 feet while a nearby building casts a 32-foot shadow. Find the height of the building.

60. *Angles in a Triangle* The measure of the largest angle in a triangle is 20° less than the sum of the measures of the two smaller angles. The sum of the measures of the two larger angles is 90° greater than the measure of the smaller angle. Find the measure of each angle.

8

Quadratic Functions and Equations

There is no branch of mathematics, however abstract, which may not some day be applied to the real world.

—NIKOLAI LOBACHEVSKY

W hat size television should you buy? Should you buy a 32-inch screen or a 50-inch screen? According to a home entertainment article in *Money* magazine, the answer depends on how far you sit from your television. The farther you sit from your television, the larger it should be. For example, if you sit only 6 feet from the screen, then a 32-inch television would be adequate; if you sit 10 feet from the screen, then a 50-inch television is more appropriate.

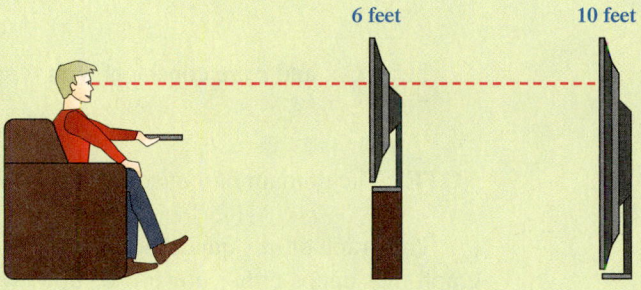

We can use a *quadratic function* to calculate the size S of the television screen needed for a person who sits x feet from the screen. You might want to use S to determine the size of screen that is recommended for you. (See Exercises 97 and 98 in Section 8.1.) Quadratic functions are a special type of polynomial function that occur frequently in applications involving economics, road construction, falling objects, geometry, and modeling real-world data. In this chapter we discuss quadratic functions and equations.

Source: Money, January 2007, p. 107; hdguru.com.

8.1 Quadratic Functions and Their Graphs

Graphs of Quadratic Functions ● **Min–Max Applications** ● **Basic Transformations of**
$y = ax^2$ ● **Transformations of** $y = ax^2 + bx + c$ **(Optional)**

A LOOK INTO MATH ▶

Suppose that a hotel is considering giving a group discount on room rates. The regular price is $80, but for each room rented the price decreases by $2. On the one hand, if the hotel rents one room, it makes only $78. On the other hand, if the hotel rents 40 rooms, the rooms are all free and the hotel makes nothing. Is there an optimal number of rooms between 1 and 40 that should be rented to maximize the revenue from the group?

In Figure 8.1 the hotel's revenue is graphed. From the graph it is apparent that "peak" revenue occurs when 20 rooms are rented. Obviously, this graph is not described by a linear function; rather, in Example 7 we see that it is described by a *quadratic function*.

NEW VOCABULARY

☐ Quadratic function
☐ Vertex
☐ Axis of symmetry
☐ Reflection
☐ Vertical shift

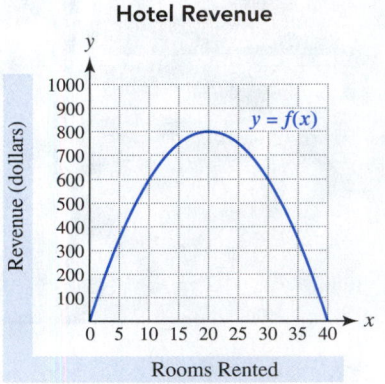

Figure 8.1

Graphs of Quadratic Functions

In Chapter 5 we discussed how a quadratic function could be represented by a polynomial of degree 2. We now give an alternative definition of a quadratic function.

> **QUADRATIC FUNCTION**
>
> A **quadratic function** can be written in the form
> $$f(x) = ax^2 + bx + c,$$
> where a, b, and c are constants with $a \neq 0$.

NOTE: The domain of a quadratic function is all real numbers.

READING CHECK

How can you identify the vertex and the axis of symmetry on the graph of a parabola?

The graph of *any* quadratic function is a *parabola*. Recall that a parabola is a ∪-shaped graph that opens either upward or downward. The graph of the simple quadratic function $y = x^2$ is a parabola that opens upward, with its *vertex* located at the origin, as shown in Figure 8.2(a). The **vertex** is the *lowest* point on the graph of a parabola that opens upward and the *highest* point on the graph of a parabola that opens downward. A parabola opening downward is shown in Figure 8.2(b). Its vertex is the point (0, 2) and is the highest point on the graph. If we were to fold the *xy*-plane along the *y*-axis, or the line $x = 0$, the left and right sides of the graph would match. That is, the graph is symmetric with respect to the line $x = 0$. In this case the line $x = 0$ is the **axis of symmetry** for the graph. Figure 8.2(c) shows a parabola that opens upward with vertex (2, −1) and axis of symmetry $x = 2$.

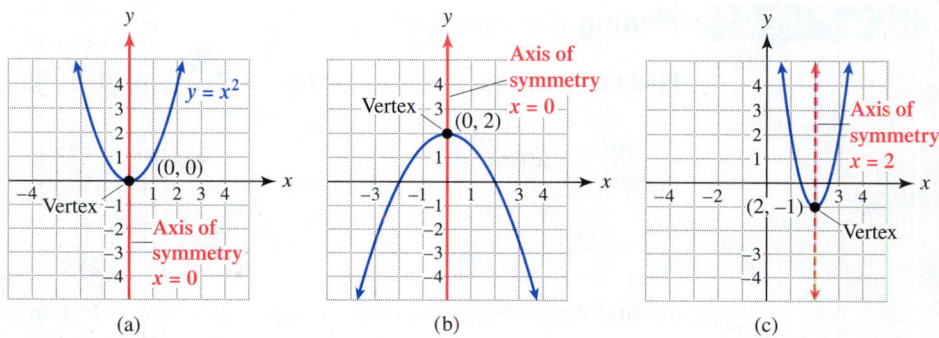

Figure 8.2

EXAMPLE 1

Identifying the vertex and the axis of symmetry

Use the graph of the quadratic function to identify the vertex, the axis of symmetry, and whether the parabola opens upward or downward.

(a)   **(b)**

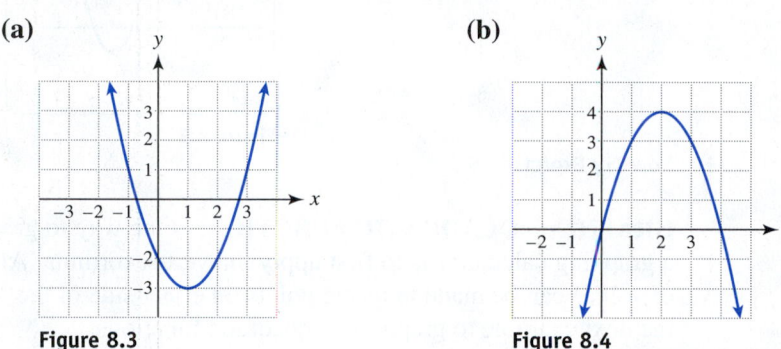

Figure 8.3   Figure 8.4

Solution

(a) The vertex is the lowest point on the graph shown in Figure 8.3, and its coordinates are $(1, -3)$. The axis of symmetry is the vertical line passing through the vertex, so its equation is $x = 1$. The parabola opens upward.

(b) The vertex is the highest point on the graph shown in Figure 8.4, and its coordinates are $(2, 4)$. The axis of symmetry is the vertical line passing through the vertex, so its equation is $x = 2$. The parabola opens downward.

Now Try Exercises 29, 31

THE VERTEX FORMULA In order to graph a parabola by hand, it is helpful to know the location of the vertex. The following formula can be used to find the coordinates of the vertex for *any* parabola. This formula is derived by *completing the square*, a technique discussed later in the next section.

READING CHECK

How do you use the vertex formula to find the coordinates of the vertex?

> ### VERTEX FORMULA
>
> The x-coordinate of the vertex of the graph of $y = ax^2 + bx + c$, $a \neq 0$, is given by
> $$x = -\frac{b}{2a}.$$
> To find the y-coordinate of the vertex, substitute this x-value in $y = ax^2 + bx + c$.

NOTE: The equation of the axis of symmetry for $f(x) = ax^2 + bx + c$ is $x = -\frac{b}{2a}$, and the vertex is the point $\left(-\frac{b}{2a}, f\left(-\frac{b}{2a}\right)\right)$.

EXAMPLE 2 **Finding the vertex of a parabola**

Find the vertex for the graph of $f(x) = 2x^2 - 4x + 1$. Support your answer graphically.

Solution
For $f(x) = \mathbf{2}x^2 - \mathbf{4}x + 1$, $a = \mathbf{2}$ and $b = \mathbf{-4}$. The x-coordinate of the vertex is

$$x = -\frac{b}{2a} = -\frac{(\mathbf{-4})}{2(\mathbf{2})} = \mathbf{1}. \qquad \text{\textit{x}-coordinate of vertex}$$

To find the y-coordinate of the vertex, substitute $x = \mathbf{1}$ in the given formula.

$$f(\mathbf{1}) = 2(\mathbf{1})^2 - 4(\mathbf{1}) + 1 = \mathbf{-1} \qquad \text{\textit{y}-coordinate of vertex}$$

Thus the vertex is located at $(\mathbf{1}, \mathbf{-1})$, which is supported by Figure 8.5.

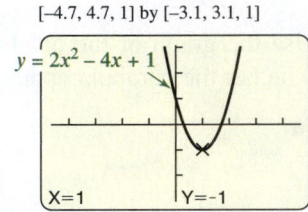

[−4.7, 4.7, 1] by [−3.1, 3.1, 1]

Figure 8.5

Now Try Exercise 15

GRAPHING QUADRATIC FUNCTIONS One way to graph a quadratic function without a graphing calculator is to first apply the vertex formula. After the vertex is located, a table of values can be made to locate points on either side of the vertex. This technique is used in the next example to graph three quadratic functions.

EXAMPLE 3 **Graphing quadratic functions**

Identify the vertex and axis of symmetry on the graph of $y = f(x)$. Graph $y = f(x)$.
(a) $f(x) = x^2 - 1$ **(b)** $f(x) = -(x + 1)^2$ **(c)** $f(x) = x^2 + 4x + 3$

Solution
(a) Begin by applying the vertex formula with $a = 1$ and $b = 0$ to locate the vertex.

$$x = -\frac{b}{2a} = -\frac{0}{2(1)} = \mathbf{0} \qquad \text{\textit{x}-coordinate of vertex}$$

The y-coordinate of the vertex is found by evaluating $f(\mathbf{0})$.

$$y = f(\mathbf{0}) = \mathbf{0}^2 - 1 = \mathbf{-1} \qquad \text{\textit{y}-coordinate of vertex}$$

Thus the coordinates of the vertex are $(\mathbf{0}, \mathbf{-1})$. Table 8.1 is made by finding points on either side of the vertex. Plotting these points and connecting a smooth ∪-shaped graph results in Figure 8.6. The axis of symmetry is the vertical line passing through the vertex, so its equation is $x = \mathbf{0}$.

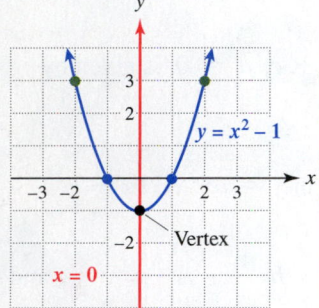

Figure 8.6

TABLE 8.1

	x	$y = x^2 - 1$	
	-2	3	
	-1	0	
Vertex →	$\mathbf{0}$	$\mathbf{-1}$	Equal
	1	0	
	2	3	

(b) Before we can apply the vertex formula, we need to determine values for a and b by multiplying the expression for $f(x) = -(x + 1)^2$.

$$-(x + 1)^2 = -(x^2 + 2x + 1) \qquad \text{Multiply.}$$
$$= -x^2 - 2x - 1 \qquad \text{Distribute the negative sign.}$$

Substitute $a = -1$ and $b = -2$ into the vertex formula.

$$x = -\frac{b}{2a} = -\frac{(-2)}{2(-1)} = \mathbf{-1} \qquad x\text{-coordinate of vertex}$$

The y-coordinate of the vertex is found by evaluating $f(\mathbf{-1})$.

$$y = f(\mathbf{-1}) = -(\mathbf{-1} + 1)^2 = \mathbf{0} \qquad y\text{-coordinate of vertex}$$

Thus the coordinates of the vertex are $(-1, 0)$. Table 8.2 is made by finding points on either side of the vertex. Plotting these points and connecting a smooth ∩-shaped graph results in Figure 8.7. The axis of symmetry is the vertical line passing through the vertex, so its equation is $x = -1$.

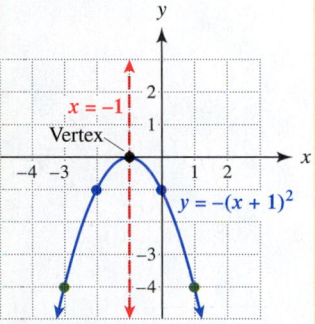

Figure 8.7

TABLE 8.2

	x	$y = -(x + 1)^2$	
	-3	-4	
	-2	-1	
Vertex →	-1	0	Equal
	0	-1	
	1	-4	

(c) The graph of $f(x) = x^2 + 4x + 3$ can be found in a manner similar to that used for the graphs in (a) and (b). See Table 8.3 and Figure 8.8. With $a = 1$ and $b = 4$, the vertex formula can be used to show that the vertex is located at $(-2, -1)$. The equation of the axis of symmetry is $x = -2$.

TABLE 8.3

	x	$y = x^2 + 4x + 3$	
	-5	8	
	-4	3	
	-3	0	
Vertex →	-2	-1	Equal
	-1	0	
	0	3	
	1	8	

Figure 8.8

Now Try Exercises 35, 41, 43

STUDY TIP

The vertex formula is important to memorize because it is often used when graphing parabolas.

Min–Max Applications

▶ **REAL-WORLD CONNECTION** Have you ever noticed that the first piece of pizza tastes really great and gives a person a lot of satisfaction? The second piece may be almost as good. After a few more pieces, there is often a point where the satisfaction starts to level off, and it can even go down as a person starts to eat too much. Eventually, if a person overeats, he or she might regret eating any pizza at all.

In Figure 8.9 a parabola models the total satisfaction received from eating x slices of pizza. Notice that the satisfaction level increases and then decreases. Although maximum satisfaction would vary with the individual, it always occurs at the vertex. (This is a seemingly simple model, but it is nonetheless used frequently in economics to describe consumer satisfaction from buying x identical items.)

Satisfaction from Eating Pizza

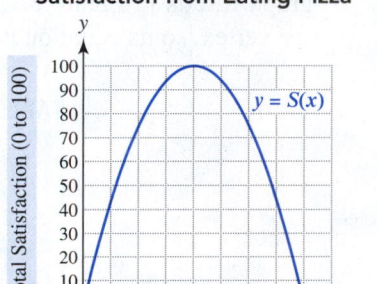

Figure 8.9

EXAMPLE 4 | **Finding maximum satisfaction from pizza**

The graph of $S(x) = -6.25x^2 + 50x$ is shown in Figure 8.9. Use the vertex formula to determine the number of slices of pizza that give maximum satisfaction. Does this maximum agree with what is shown in the graph of $y = S(x)$?

Solution
The parabola in Figure 8.9 opens downward, so the greatest y-value occurs at the vertex. To find the x-coordinate of the vertex, we can apply the vertex formula. Substitute $a = -6.25$ and $b = 50$ into the vertex formula.

$$x = -\frac{b}{2a} = -\frac{50}{2(-6.25)} = 4 \qquad \text{x-coordinate of vertex}$$

Maximum satisfaction on a scale of 1 to 100 occurs when **4** slices of pizza are eaten, which agrees with Figure 8.9. (This number obviously varies from individual to individual.)

▎**Now Try Exercise 81**

READING CHECK

How do you determine if a parabola has a maximum y-value or a minimum y-value?

FINDING MIN–MAX The graph in Figure 8.9 is a parabola that opens downward and whose maximum y-value of 100 occurs at the vertex. When a quadratic function f is used to model real data, the y-coordinate of the vertex represents either a maximum value of $f(x)$ or a minimum value of $f(x)$. For example, Figure 8.10(a) shows a parabola that opens upward. The minimum y-value on this graph is **1** and occurs at the vertex (2, **1**). Similarly, Figure 8.10(b) shows a parabola that opens downward. The maximum y-value on this graph is **3** and occurs at the vertex (−3, **3**).

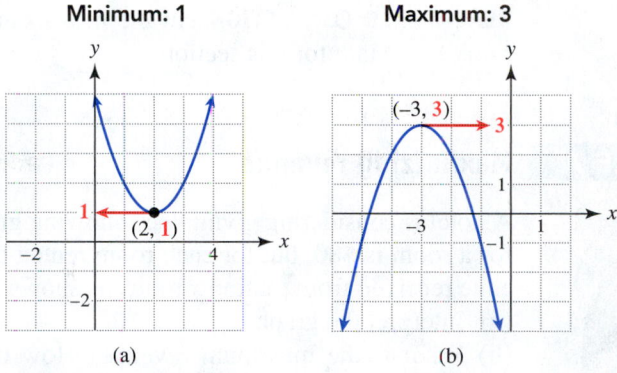

Figure 8.10

EXAMPLE 5   **Finding a minimum *y*-value**

Find the minimum *y*-value on the graph of $f(x) = x^2 - 4x + 3$.

Locating a Minimum *y*-Value

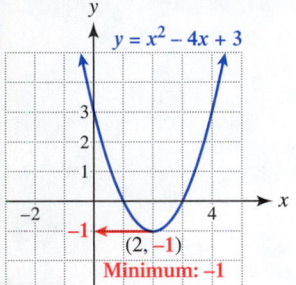

Figure 8.11

Solution

To locate the minimum *y*-value on a parabola, first apply the vertex formula with $a = 1$ and $b = -4$ to find the *x*-coordinate of the vertex.

$$x = -\frac{b}{2a} = -\frac{(-4)}{2(1)} = \mathbf{2} \qquad \text{\textit{x}-coordinate of vertex}$$

The minimum *y*-value on the graph is found by evaluating $f(2)$.

$$y = f(\mathbf{2}) = \mathbf{2}^2 - 4(\mathbf{2}) + 3 = \mathbf{-1} \qquad \text{\textit{y}-coordinate of vertex}$$

Thus the minimum *y*-coordinate on the graph of $y = f(x)$ is $\mathbf{-1}$. This result is supported by Figure 8.11.

Now Try Exercise 53

▶ **REAL-WORLD CONNECTION**   In the next example, we demonstrate finding a maximum height reached by a baseball.

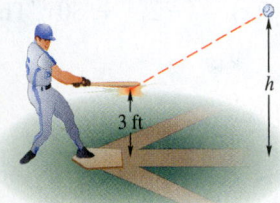

EXAMPLE 6   **Finding maximum height**

A baseball is hit into the air, and its height *h* in feet after *t* seconds can be calculated by $h(t) = -16t^2 + 96t + 3$.
(a) What is the height of the baseball when it is hit?
(b) Determine the maximum height of the baseball.

Solution
(a) The baseball is hit when $t = \mathbf{0}$, so $h(\mathbf{0}) = -16(\mathbf{0})^2 + 96(\mathbf{0}) + 3 = 3$ feet.
(b) The graph of *h* opens downward because $a = -16 < 0$. Thus the maximum height of the baseball occurs at the vertex. To find the vertex, we apply the vertex formula with $a = \mathbf{-16}$ and $b = \mathbf{96}$ because $h(t) = \mathbf{-16}t^2 + \mathbf{96}t + 3$.

$$t = -\frac{b}{2a} = -\frac{\mathbf{96}}{2(\mathbf{-16})} = \mathbf{3} \text{ seconds}$$

The maximum height of the baseball occurs at $t = \mathbf{3}$ seconds and is

$$h(\mathbf{3}) = -16(\mathbf{3})^2 + 96(\mathbf{3}) + 3 = 147 \text{ feet.}$$

Now Try Exercise 87

▶ **REAL-WORLD CONNECTION** In the next example, we answer the question presented in A Look Into Math for this section.

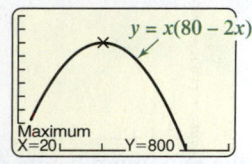

| EXAMPLE 7 | **Maximizing revenue** |

A hotel is considering giving the following group discount on room rates. The regular price for a room is $80, but for each room rented the price decreases by $2. A graph of the revenue received from renting x rooms is shown in Figure 8.12.

(a) Interpret the graph.
(b) What is the maximum revenue? How many rooms should be rented to receive the maximum revenue?
(c) Write a formula for $f(x)$ whose graph is shown in Figure 8.12.
(d) Use $f(x)$ to determine symbolically the maximum revenue and the number of rooms that should be rented.

Hotel Revenue

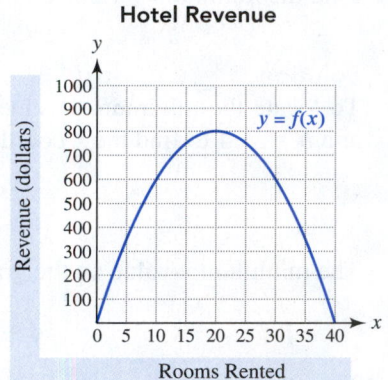

Figure 8.12

TECHNOLOGY NOTE

Locating a Vertex
Graphing calculators can locate a vertex with the MAXIMUM or MINIMUM utility. The maximum in Example 7 is found in the figure.

[0, 50, 10] by [0, 1000, 100]

Solution
(a) The revenue increases at first, reaches a maximum (which corresponds to the vertex), and then decreases.
(b) In Figure 8.12 the vertex is (**20**, **800**). Thus the maximum revenue of **$800** occurs when **20** rooms are rented.
(c) If x rooms are rented, the price for each room is $80 - 2x$. The revenue equals the number of rooms rented times the price of each room. Thus $f(x) = x(80 - 2x)$.
(d) First, multiply $x(80 - 2x)$ to obtain $80x - 2x^2$ and then let $f(x) = -2x^2 + 80x$. The x-coordinate of the vertex is

$$x = -\frac{b}{2a} = -\frac{80}{2(-2)} = \mathbf{20}.$$

The y-coordinate is $f(\mathbf{20}) = -2(\mathbf{20})^2 + 80(\mathbf{20}) = \mathbf{800}$. These calculations verify our results in part (b).

Now Try Exercise 91

CALCULATOR HELP

To find a minimum or maximum, see Appendix A (page AP-11).

Basic Transformations of $y = ax^2$

THE GRAPH OF $y = ax^2$, $a > 0$ First, graph $y_1 = \frac{1}{2}x^2$, $y_2 = x^2$, and $y_3 = 2x^2$, as shown in Figure 8.13(a). Note that $a = \frac{1}{2}$, $a = 1$, and $a = 2$, respectively, and that as a increases, the resulting parabola becomes narrower. The graph of $y_1 = \frac{1}{2}x^2$ is wider than the graph of $y_2 = x^2$, and the graph of $y_3 = 2x^2$ is narrower than the graph of $y_2 = x^2$. In general, the graph of $y = ax^2$ is wider than the graph of $y = x^2$ when $0 < a < 1$ and narrower than the graph of $y = x^2$ when $a > 1$. When $a > 0$, the graph of $y = ax^2$ opens upward and never lies *below* the x-axis.

THE GRAPH OF $y = ax^2$, $a < 0$ When $a < 0$, the graph of $y = ax^2$ never lies *above* the x-axis because, for any input x, the product $ax^2 \leq 0$. The graphs of $y_4 = -\frac{1}{2}x^2$, $y_5 = -x^2$, and $y_6 = -2x^2$ are shown in Figure 8.13(b) and open downward. The graph of $y_4 = -\frac{1}{2}x^2$ is wider than the graph of $y_5 = -x^2$ and the graph of $y_6 = -2x^2$ is narrower than the graph of $y_5 = -x^2$.

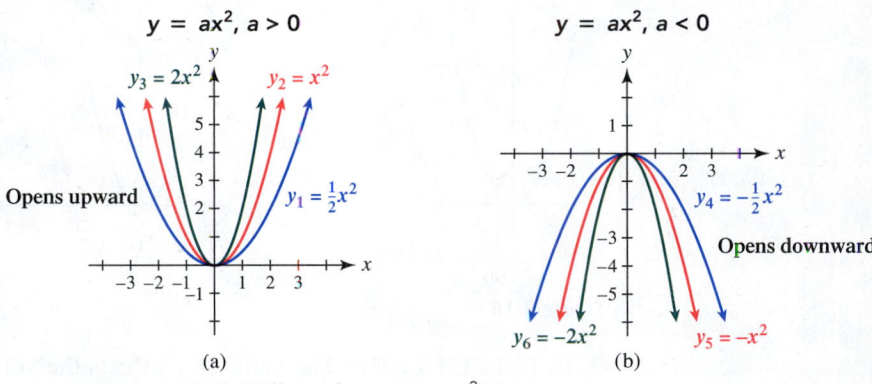

(a)   (b)

Figure 8.13 The Effect of a on $y = ax^2$

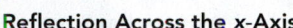
Reflection Across the x-Axis

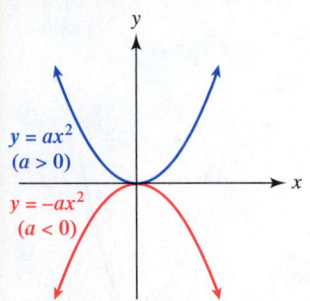

Figure 8.14

THE GRAPH OF $y = ax^2$

The graph of $y = ax^2$ is a parabola with the following characteristics.

1. The vertex is $(0, 0)$, and the axis of symmetry is given by $x = 0$.
2. It opens upward if $a > 0$ and opens downward if $a < 0$.
3. It is wider than the graph of $y = x^2$, if $0 < |a| < 1$. It is narrower than the graph of $y = x^2$, if $|a| > 1$.

REFLECTIONS OF $y = ax^2$ In Figure 8.13 the graph of $y_1 = \frac{1}{2}x^2$ can be *transformed* into the graph of $y_4 = -\frac{1}{2}x^2$ by *reflecting* it across the x-axis. The graph of y_4 is a **reflection** of the graph of y_1 across the x-axis. In general, *the graph of $y = -ax^2$ is a reflection of the graph of $y = ax^2$ across the x-axis*, as shown in Figure 8.14. That is, if we folded the xy-plane along the x-axis the two graphs would match.

EXAMPLE 8   **Graphing $y = ax^2$**

Compare the graph of $g(x) = -3x^2$ to the graph of $f(x) = x^2$. Then graph both functions on the same coordinate axes.

Solution
Both graphs are parabolas. However, the graph of g opens downward and is narrower than the graph of f. These graphs are shown in Figure 8.15.

Now Try Exercise 67

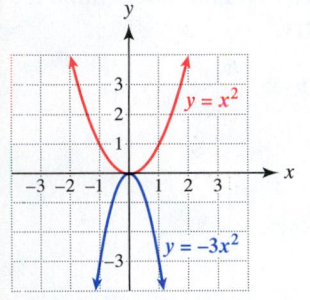

Figure 8.15

Transformations of $y = ax^2 + bx + c$ (Optional)

The graph of a quadratic function is a parabola, and any quadratic function f can be written as $f(x) = ax^2 + bx + c$, where a, b, and c are constants with $a \neq 0$. As a result, the values of a, b, and c determine both the shape and position of the parabola in the xy-plane. In this subsection, we summarize some of the effects that these constants have on the graph of f.

THE EFFECTS OF a The effects of a on the graph of $y = ax^2$ were already discussed, and these effects can be generalized to include the graph of $y = ax^2 + bx + c$.

1. **Width:** The graph of $f(x) = ax^2 + bx + c$ is wider than the graph of $y = x^2$ if $0 < |a| < 1$ and narrower if $|a| > 1$. See Figures 8.16(a) and (b).

2. **Opening:** The graph of $f(x) = ax^2 + bx + c$ opens upward if $a > 0$ and downward if $a < 0$. See Figure 8.16(c).

The Effects of a on $y = ax^2 + bx + c$

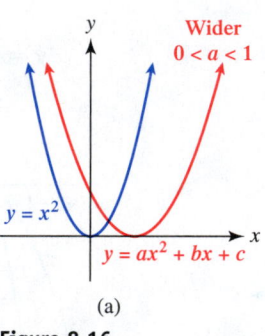
(a)

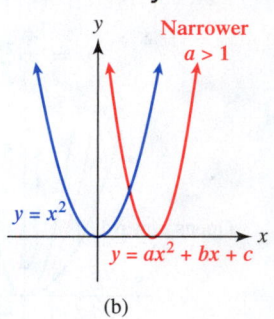
(b)

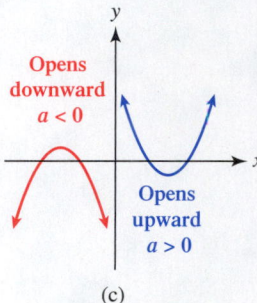
(c)

Figure 8.16

THE EFFECTS OF c The value of c affects the vertical placement of the parabola in the xy-plane. This placement is often called **vertical shift**.

1. **y-Intercept:** Because $f(0) = a(0)^2 + b(0) + c = c$, it follows that the y-intercept for the graph of $y = f(x)$ is c. See Figure 8.17(a).

2. **Vertical Shift:** The graph of $f(x) = ax^2 + bx + c$ is shifted vertically c units compared to the graph of $y = ax^2 + bx$. If $c < 0$, the shift is downward; if $c > 0$, the shift is upward. See Figures 8.17(b) and (c). The parabolas in both figures have identical shapes.

The Effects of c on $y = ax^2 + bx + c$

The Combined Effects of a and b

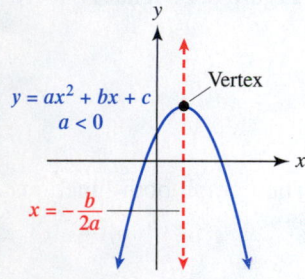

Figure 8.18

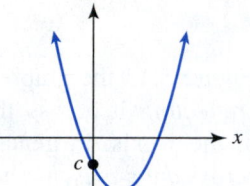
(a) y-intercept: c

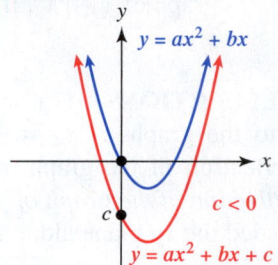
(b) Shifted downward: $c < 0$

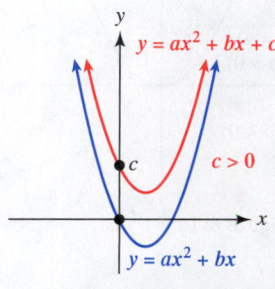
(c) Shifted upward: $c > 0$

Figure 8.17

THE COMBINED EFFECTS OF a AND b The combined values of a and b determine the x-coordinate of the vertex and the equation of the axis of symmetry.

1. **Vertex:** The x-coordinate of the vertex is $-\frac{b}{2a}$.

2. **Axis of Symmetry:** The axis of symmetry is given by $x = -\frac{b}{2a}$.

Figure 8.18 illustrates these concepts.

EXAMPLE 9 **Analyzing the graph of $y = ax^2 + bx + c$**

Let $f(x) = -\frac{1}{2}x^2 + x + \frac{3}{2}$.

(a) Does the graph of f open upward or downward? Is this graph wider or narrower than the graph of $y = x^2$?

(b) Find the axis of symmetry and the vertex.

(c) Find the y-intercept and any x-intercepts.

(d) Sketch a graph of f.

Solution

(a) If $f(x) = -\frac{1}{2}x^2 + x + \frac{3}{2}$, then $a = -\frac{1}{2}$, $b = 1$, and $c = \frac{3}{2}$. Because $a = -\frac{1}{2} < 0$, the parabola opens downward. Also, because $0 < |a| < 1$, the graph is wider than the graph of $y = x^2$.

(b) The axis of symmetry is $x = -\dfrac{b}{2a} = -\dfrac{1}{2\left(-\frac{1}{2}\right)} = 1$, or $x = 1$. Because

$$f(1) = -\frac{1}{2}(1)^2 + (1) + \frac{3}{2} = -\frac{1}{2} + 1 + \frac{3}{2} = 2,$$

the vertex is (**1**, **2**).

(c) The y-intercept equals c, or $\frac{3}{2}$. To find x-intercepts we let $y = 0$ and solve for x.

$$-\frac{1}{2}x^2 + x + \frac{3}{2} = 0 \quad \text{Equation to be solved}$$

$$x^2 - 2x - 3 = 0 \quad \text{Multiply by } -2; \text{ clear fractions.}$$

$$(x + 1)(x - 3) = 0 \quad \text{Factor.}$$

$$x + 1 = 0 \quad \text{or} \quad x - 3 = 0 \quad \text{Zero-product property}$$

$$x = -1 \quad \text{or} \quad x = 3 \quad \text{Solve.}$$

The x-intercepts are -1 and 3.

(d) Start by plotting the vertex and intercepts, as shown in Figure 8.19. Then sketch a smooth, ∩-shaped graph that connects these points.

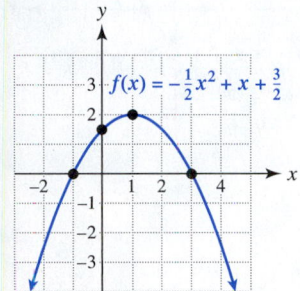

Figure 8.19

▌ **Now Try Exercise 75**

8.1   Putting It All Together

CONCEPT	EXPLANATION	EXAMPLES
Quadratic Function	Can be written as $f(x) = ax^2 + bx + c$, $a \neq 0$	$f(x) = x^2 + x - 2$ and $g(x) = -2x^2 + 4$ $(b = 0)$
Vertex of a Parabola	The x-coordinate of the vertex for the function $f(x) = ax^2 + bx + c$ with $a \neq 0$ is given by $$x = -\frac{b}{2a}.$$ The y-coordinate of the vertex is found by substituting this x-value in the equation. Hence the vertex is $\left(-\frac{b}{2a}, f\left(-\frac{b}{2a}\right)\right)$.	If $f(x) = -2x^2 + 8x - 7$, then $$x = -\frac{8}{2(-2)} = 2$$ and $$f(2) = -2(2)^2 + 8(2) - 7 = 1.$$ The vertex is (**2**, **1**). The graph of f opens downward because $a < 0$.
Graph of a Quadratic Function	Its graph is a parabola that opens upward if $a > 0$ and downward if $a < 0$. The value of $|a|$ affects the width of the parabola. The vertex can be used to determine the maximum or minimum output of a quadratic function.	The graph of $y = -\frac{1}{4}x^2$ opens downward and is wider than the graph of $y = x^2$, as shown in the figure. Each graph has its vertex at $(0, 0)$.

continued on next page

continued from previous page

The following gives symbolic, numerical, and graphical representations for the quadratic function f that *squares the input x and then subtracts* 1.

Symbolic Representation

$$f(x) = x^2 - 1$$

Numerical Representation

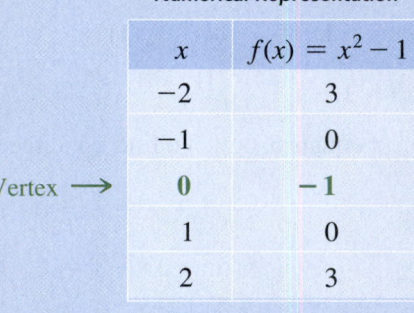

x	$f(x) = x^2 - 1$
-2	3
-1	0
0 (Vertex →)	**−1**
1	0
2	3

Equal

Graphical Representation

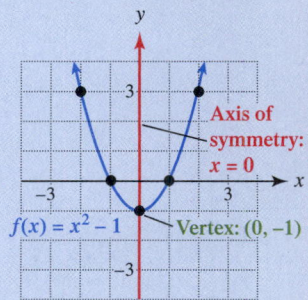

Axis of symmetry: $x = 0$

$f(x) = x^2 - 1$ Vertex: $(0, -1)$

8.1 Exercises

CONCEPTS AND VOCABULARY

1. The graph of a quadratic function is called a(n) _____.

2. If a parabola opens upward, what is the lowest point on the parabola called?

3. If a parabola is symmetric with respect to the y-axis, the y-axis is called the _____.

4. The vertex on the graph of $y = x^2$ is _____.

5. Sketch a parabola that opens downward with a vertex of $(1, 2)$.

6. If $y = ax^2 + bx + c$, the x-coordinate of the vertex is given by $x =$ _____.

7. Compared to the graph of $y = x^2$, the graph of $y = 2x^2$ is (wider/narrower).

8. The graph of $y = -x^2$ is similar to the graph of $y = x^2$ except that it is _____ across the x-axis.

9. Any quadratic function can be written in the form $f(x) =$ _____.

10. If a parabola opens downward, the point with the largest y-value is called the _____.

11. (True or False?) The axis of symmetry for the graph of $y = ax^2 + bx + c$ is given by $x = -\frac{b}{2a}$.

12. (True or False?) If the vertex of a parabola is located at (a, b), then the axis of symmetry is given by $x = b$.

13. (True or False?) If a parabola opens downward and its vertex is (a, b), then the minimum y-value on the parabola is b.

14. (True or False?) The graph of $y = -ax^2$ with $a > 0$ opens downward.

VERTEX FORMULA

Exercises 15–24: Find the vertex of the parabola.

15. $f(x) = x^2 - 4x - 2$

16. $f(x) = 2x^2 + 6x - 3$

17. $f(x) = -\frac{1}{3}x^2 - 2x + 1$

18. $f(x) = 5 - 4x + x^2$

19. $f(x) = 3 - 2x^2$

20. $f(x) = \frac{1}{4}x^2 - 3x - 2$

21. $f(x) = -0.3x^2 + 0.6x + 1.1$

22. $f(x) = 25 - 10x + 20x^2$

23. $f(x) = 6x - x^2$

24. $f(x) = x - \frac{1}{2}x^2$

GRAPHS OF QUADRATIC FUNCTIONS

Exercises 25–28: Use the given graph of f to evaluate the expressions.

25. $f(-2)$ and $f(0)$ **26.** $f(-2)$ and $f(2)$

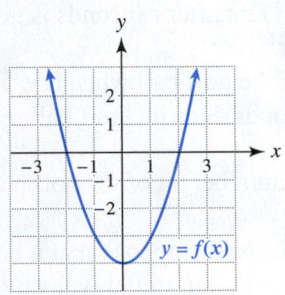
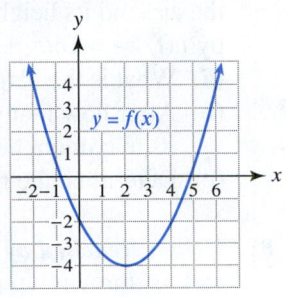

27. $f(-3)$ and $f(1)$ **28.** $f(-1)$ and $f(2)$

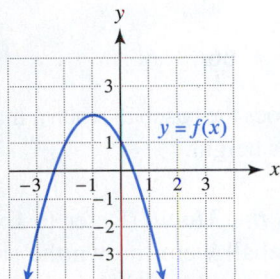
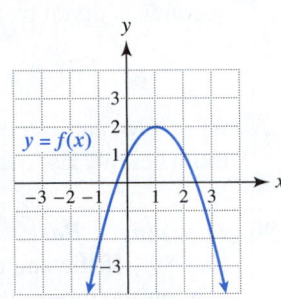

Exercises 29–32: Identify the vertex, axis of symmetry, and whether the parabola opens upward or downward.

29.   **30.**

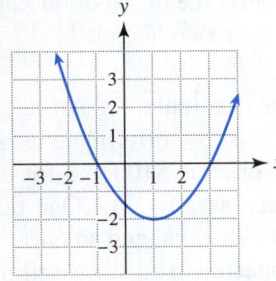
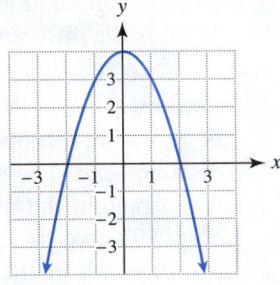

31.   **32.**

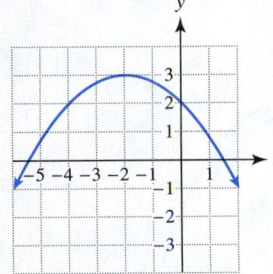
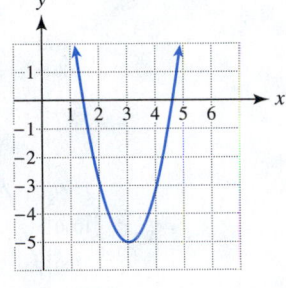

Exercises 33–52: Do the following for the given f(x).
(a) Identify the vertex and axis of symmetry on the graph of y = f(x).
(b) Graph y = f(x).
(c) Evaluate f(−2) and f(3).

33. $f(x) = \frac{1}{2}x^2$ **34.** $f(x) = -3x^2$

35. $f(x) = x^2 - 2$ **36.** $f(x) = x^2 - 1$

37. $f(x) = -3x^2 + 1$ **38.** $f(x) = \frac{1}{2}x^2 + 2$

39. $f(x) = (x - 1)^2$ **40.** $f(x) = (x + 2)^2$

41. $f(x) = -(x + 2)^2$ **42.** $f(x) = -(x - 1)^2$

43. $f(x) = x^2 + x - 2$ **44.** $f(x) = x^2 - 2x + 2$

45. $f(x) = 2x^2 - 3$ **46.** $f(x) = 1 - 2x^2$

47. $f(x) = 2x - x^2$ **48.** $f(x) = x^2 + 2x - 8$

49. $f(x) = -2x^2 + 4x - 1$

50. $f(x) = -\frac{1}{2}x^2 + 2x - 3$

51. $f(x) = \frac{1}{4}x^2 - x + 5$ **52.** $f(x) = 3 - 6x - 4x^2$

MIN–MAX

Exercises 53–58: Find the minimum y-value on the graph of y = f(x).

53. $f(x) = x^2 + 2x - 1$ **54.** $f(x) = x^2 + 6x + 2$

55. $f(x) = x^2 - 5x$ **56.** $f(x) = x^2 - 3x$

57. $f(x) = 2x^2 + 2x - 3$ **58.** $f(x) = 3x^2 - 3x + 7$

Exercises 59–64: Find the maximum y-value on the graph of y = f(x).

59. $f(x) = -x^2 + 2x + 5$

60. $f(x) = -x^2 + 4x - 3$

61. $f(x) = 4x - x^2$

62. $f(x) = 6x - x^2$

63. $f(x) = -2x^2 + x - 5$

64. $f(x) = -5x^2 + 15x - 2$

65. *Numbers* Find two positive numbers whose sum is 20 and whose product is maximum.

66. **Thinking Generally** Find two positive numbers whose sum is k and whose product is maximum.

TRANSFORMATIONS OF GRAPHS

Exercises 67–74: Graph f(x). Compare the graph of $y = f(x)$ to the graph of $y = x^2$.

67. $f(x) = -x^2$ **68.** $f(x) = -2x^2$

69. $f(x) = 2x^2$ **70.** $f(x) = 3x^2$

71. $f(x) = \frac{1}{4}x^2$ **72.** $f(x) = \frac{1}{2}x^2$

73. $f(x) = -\frac{1}{2}x^2$ **74.** $f(x) = -\frac{3}{2}x^2$

Exercises 75–80: (Refer to Example 9.) Use the given f(x) to complete the following.

(a) *Does the graph of f open upward or downward? Is this graph wider, narrower, or the same as the graph of $y = x^2$?*

(b) *Find the axis of symmetry and the vertex.*

(c) *Find the y-intercept and any x-intercepts.*

(d) *Sketch a graph of f.*

75. $f(x) = \frac{1}{2}x^2 + x - \frac{3}{2}$ **76.** $f(x) = -x^2 + 4x + 5$

77. $f(x) = 2x - x^2$ **78.** $f(x) = x - 2x^2$

79. $f(x) = 2x^2 + 2x - 4$ **80.** $f(x) = \frac{1}{2}x^2 - \frac{1}{2}x - 1$

APPLICATIONS

Exercises 81 and 82: **Eating Pizza** *(Refer to Example 4.) Let S(x) denote the satisfaction, on a scale from 0 to 100, from eating x pieces of pizza. Find the number of pieces that gives the maximum satisfaction.*

81. $S(x) = -\frac{100}{9}x^2 + \frac{200}{3}x$

82. $S(x) = -16x^2 + 80x$

Exercises 83–86: **Quadratic Models** *Match the physical situation with the graph (a.–d.) that models it best.*

83. The height y of a stone thrown from ground level after x seconds

84. The number of people attending a popular movie x weeks after its opening

85. The temperature after x hours in a house when the furnace quits and then a repair person fixes it

86. U.S. population from 1800 to the present

a.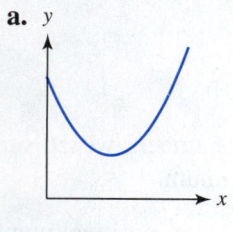

b.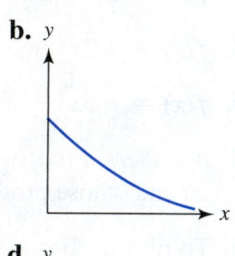

c.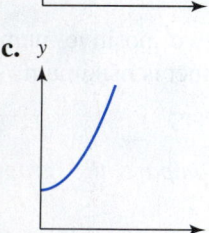

d.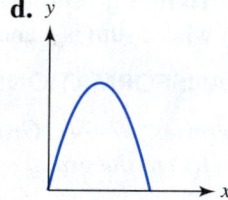

87. **Height Reached by a Baseball** (Refer to Example 6.) A baseball is hit into the air, and its height h in feet after t seconds is given by $h(t) = -16t^2 + 64t + 2$.

(a) What is the height of the baseball when it is hit?

(b) After how many seconds does the baseball reach its maximum height?

(c) Determine the maximum height of the baseball.

88. **Height Reached by a Golf Ball** A golf ball is hit into the air, and its height h in feet after t seconds is given by $h(t) = -16t^2 + 128t$.

(a) What is the height of the golf ball when it is hit?

(b) After how many seconds does the golf ball reach its maximum height?

(c) Determine the maximum height of the golf ball.

89. **Height Reached by a Baseball** Suppose that a baseball is thrown upward with an initial velocity of 66 feet per second (45 miles per hour) and it is released 6 feet above the ground. Its height h after t seconds is given by

$$h(t) = -16t^2 + 66t + 6.$$

After how many seconds does the baseball reach a maximum height? Estimate this height.

90. **Throwing a Baseball on the Moon** (Refer to Exercise 89.) If the same baseball were thrown the same way on the moon, its height h above the moon's surface after t seconds would be

$$h(t) = -2.55t^2 + 66t + 6.$$

Does the baseball go higher on the moon or on Earth? What is the difference in these two heights?

91. **Concert Tickets** (Refer to Example 7.) An agency is promoting concert tickets by offering a group-discount rate. The regular price is $100 and for each ticket bought the price decreases by $1. (One ticket costs $99, two tickets cost $98 *each*, and so on.)

(a) A graph of the revenue received from selling x tickets is shown in the figure. Interpret the graph.

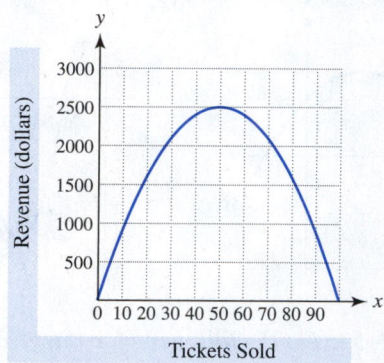

(b) What is the maximum revenue? How many tickets should be sold to maximize revenue?

(c) Write a formula for $y = f(x)$ whose graph is shown in the figure.

(d) Use $f(x)$ to determine symbolically the maximum revenue and the number of tickets that should be sold to maximize revenue.

92. *Maximizing Revenue* The regular price for a round-trip ticket to Las Vegas, Nevada, charged by an airline charter company is $300. For a group rate the company will reduce the price of each ticket by $1.50 for every passenger on the flight.

(a) Write a formula $f(x)$ that gives the revenue from selling x tickets.

(b) Determine how many tickets should be sold to maximize the revenue. What is the maximum revenue?

93. *Monthly Facebook Visitors* The number of *unique* monthly Facebook visitors in millions x years after 2006 can be modeled by

$$V(x) = 10.75x^2 - 24x + 35.$$

(a) Evaluate $V(1)$, $V(2)$, $V(3)$, and $V(4)$. Interpret your answer.

(b) Explain why numbers of Facebook visitors can not be modeled by a linear function.

94. *Cell Phone Complexity* Consumers often enjoy having certain features on their cell phones, such as e-mail and the ability to surf the Web. However, as phones become more and more complicated to operate, the benefits from the additional complexity start to decrease. The following parabola models this general situation. Interpret why a parabola is appropriate to model this consumer experience.

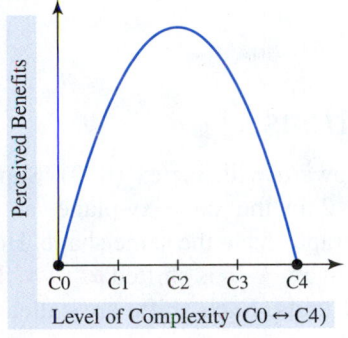

95. *Maximizing Area* A farmer is fencing a rectangular area for cattle and uses a straight portion of a river as one side of the rectangle, as illustrated in the figure. Note that there is no fence along the river. If the farmer has 1200 feet of fence, find the dimensions for the rectangular area that give a maximum area for the cattle.

96. *Maximizing Area* A rectangular pen being constructed for a pet requires 60 feet of fence.

(a) Write a formula $f(x)$ that gives the area of the pen if one side of the pen has length x.

(b) Find the dimensions of the pen that give the largest area. What is the largest area?

Exercises 97 and 98: Large-Screen Televisions (Refer to the introduction to this chapter.) Use the formula

$$S(x) = -0.227x^2 + 8.155x - 8.8,$$

where $6 \leq x \leq 16$, to estimate the recommended screen size in inches when viewers sit x feet from the screen.

97. $x = 8$ feet

98. $x = 12$ feet

99. *Carbon Emissions* Past and future carbon emissions in billions of metric tons during year x can be modeled by

$$C(x) = \frac{1}{300}x^2 - \frac{199}{15}x + \frac{39{,}619}{3},$$

where $1990 \leq x \leq 2020$. (*Source:* U.S. Department of Energy.)

(a) Evaluate $C(1990)$. Interpret your answer.

(b) Find the expected *increase* in carbon emissions from 1990 to 2020.

100. *Seedling Growth* In a study of the effect of temperature on the growth of melon seedlings, the seedlings were grown at different temperatures, and their heights were measured after a fixed period of time. The findings of this study can be modeled by

$$f(x) = -0.095x^2 + 5.4x - 52.2,$$

where x is the temperature in degrees Celsius and the output $f(x)$ gives the resulting average height in centimeters. (*Source:* R. Pearl, "The growth of *Cucumis melo* seedlings at different temperatures.")

 (a) Graph f in [20, 40, 5] by [0, 30, 5].
(b) Estimate graphically the temperature that resulted in the greatest height for the melon seedlings.
(c) Solve part (b) symbolically.

WRITING ABOUT MATHEMATICS

101. If $f(x) = ax^2 + bx + c$, explain how the values of a and c affect the graph of f.

102. Suppose that a quantity Q is modeled by the formula $Q(x) = ax^2 + bx + c$ with $a < 0$. Explain how to find the x-value that maximizes $Q(x)$. How do you find the maximum value of $Q(x)$?

8.2 Parabolas and Modeling

Vertical and Horizontal Translations • Vertex Form • Modeling with Quadratic Functions (Optional)

A LOOK INTO MATH ▶

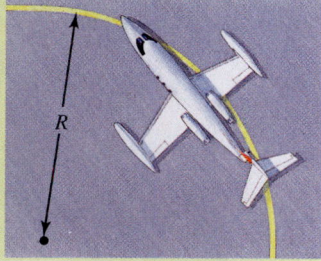

A taxiway used by an airplane to exit a runway often contains curves. A curve that is too sharp for the speed of the plane is a safety hazard. The scatterplot shown in Figure 8.20 gives an appropriate radius R of a curve designed for an airplane taxiing at x miles per hour. The data are nonlinear because they do not lie on a line. In this section we explain how a quadratic function may be used to model such data. First, we discuss translations of parabolas. (*Source:* FAA.)

NEW VOCABULARY

☐ Translations
☐ Vertex form
☐ Completing the square

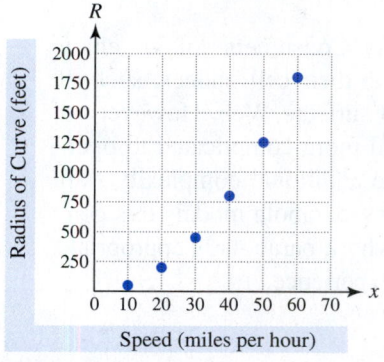

Figure 8.20

Vertical and Horizontal Translations

The graph of $y = x^2$ is a parabola opening upward with vertex $(0, 0)$. Suppose that we graph $y_1 = x^2$, $y_2 = x^2 + 1$, and $y_3 = x^2 - 2$ in the same xy-plane, as calculated in Table 8.4 and shown in Figure 8.21. All three graphs have the same shape. However, compared to the graph of $y_1 = x^2$, the graph of $y_2 = x^2$ **+ 1** is shifted *upward* **1** unit and the graph of $y_3 = x^2$ **− 2** is shifted *downward* **2** units. Such shifts are called **translations** because they do not change the shape of a graph—only its position.

Vertical Shifts of y = x²

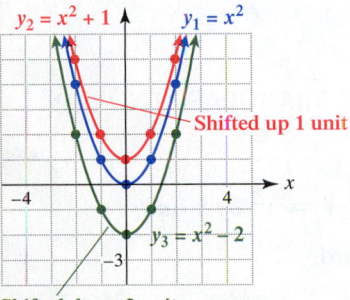

Figure 8.21

TABLE 8.4

x	$y_2 = x^2 + 1$	$y_1 = x^2$	$y_3 = x^2 - 2$
−2	5	4	2
−1	2	1	−1
0	1	0	−2
1	2	1	−1
2	5	4	2

↑ The x-values do NOT change.   ↑ Add 1 to find the y-values.   Subtract 2 to find the y-values.

Next, suppose that we graph $y_1 = x^2$ and $y_2 = (x - 1)^2$ in the same xy-plane. Compare Tables 8.5 and 8.6. Note that the y-values are equal when the x-value for y_2 is 1 unit *larger* than the x-value for y_1. For example, $y_1 = 4$ when $x = -2$ and $y_2 = 4$ when $x = -1$. Thus the graph of $y_2 = (x - 1)^2$ has the same shape as the graph of $y_1 = x^2$ except that it is translated *horizontally to the right* 1 unit, as illustrated in Figure 8.22.

Horizontal Shift of y = x²

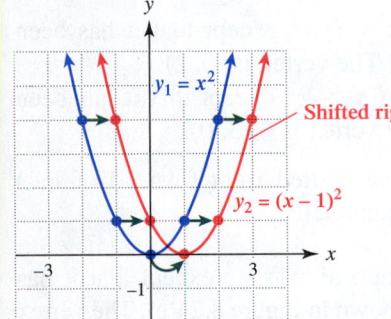

Figure 8.22

TABLE 8.5

x	$y_1 = x^2$
−2	4
−1	1
0	0
1	1
2	4

TABLE 8.6

x	$y_2 = (x - 1)^2$
−1	4
0	1
1	0
2	1
3	4

Add 1 to find the x-values.   The y-values do NOT change.

The graphs $y_1 = x^2$ and $y_2 = (x + 2)^2$ are shown in Figure 8.23. Note that Tables 8.7 and 8.8 show their y-values to be equal when the x-value for y_2 is 2 units *smaller* than the x-value for y_1. As a result, the graph of $y_2 = (x + 2)^2$ has the same shape as the graph of $y_1 = x^2$ except that it is translated *horizontally to the left* 2 units.

Horizontal Shift of y = x²

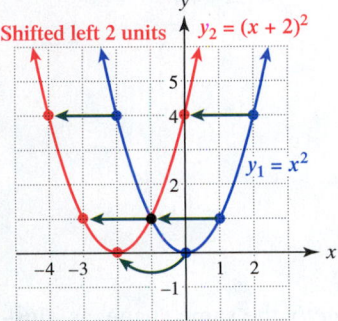

Figure 8.23

TABLE 8.7

x	$y_1 = x^2$
−2	4
−1	1
0	0
1	1
2	4

TABLE 8.8

x	$y_2 = (x + 2)^2$
−4	4
−3	1
−2	0
−1	1
0	4

Subtract 2 to find the x-values.   The y-values do NOT change.

These results are summarized as follows.

VERTICAL AND HORIZONTAL TRANSLATIONS OF PARABOLAS

Let h and k be positive numbers.

To graph	*shift the graph of $y = x^2$ by k units*
$y = x^2 + k$	upward.
$y = x^2 - k$	downward.

To graph	*shift the graph of $y = x^2$ by h units*
$y = (x - h)^2$	right.
$y = (x + h)^2$	left.

EXAMPLE 1   **Translating the graph $y = x^2$**

Sketch the graph of the equation and identify the vertex.

(a) $y = x^2 + 2$ **(b)** $y = (x + 3)^2$ **(c)** $y = (x - 2)^2 - 3$

Solution

(a) The graph of $y = x^2 + 2$ is similar to the graph of $y = x^2$ except that it has been translated *upward* 2 units, as shown in Figure 8.24(a). The vertex is $(0, 2)$.

(b) The graph of $y = (x + 3)^2$ is similar to the graph of $y = x^2$ except that it has been translated *left* 3 units, as shown in Figure 8.24(b). The vertex is $(-3, 0)$.

NOTE: If you are thinking that the graph should be shifted right (instead of left) 3 units, try graphing $y = (x + 3)^2$ on a graphing calculator.

(c) The graph of $y = (x - 2)^2 - 3$ is similar to the graph of $y = x^2$ except that it has been translated downward 3 units *and* right 2 units, as shown in Figure 8.24(c). The vertex is $(2, -3)$.

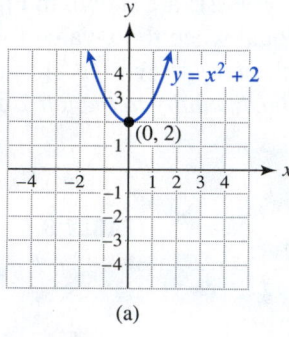
(a)

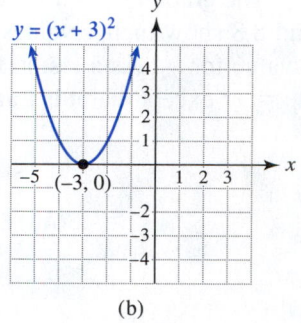
(b)

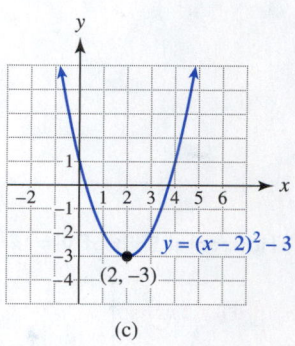
(c)

Figure 8.24

▌ **Now Try Exercises 15, 19, 27**

Vertex Form

The graphs of $y = 2x^2$ and $y = 2x^2 - 12x + 20$ have *exactly* the same shape, as illustrated in Figures 8.25 and 8.26. However, the vertex for $y = 2x^2$ is $(0, 0)$, whereas the vertex for $y = 2x^2 - 12x + 20$ is $(3, 2)$.

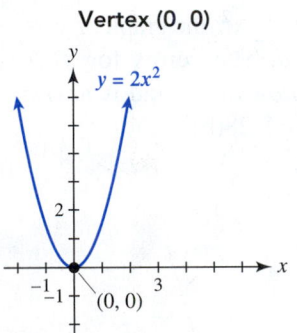

Figure 8.25   **Figure 8.26**   **Figure 8.27**

When the graph of $y = 2x^2$ (in Figure 8.25) is shifted **3** units right and **2** units upward by graphing the equation

Shift 2 units upward

$$y = 2(x - 3)^2 + 2, \quad \text{Vertex form}$$

Shift 3 units right

we get the graph in Figure 8.27, which is identical to the graph of $y = 2x^2 - 12x + 20$ in Figure 8.26. Thus $y = 2(x - 3)^2 + 2$ and $y = 2x^2 - 12x + 20$ are equivalent equations for the same parabola.

In general, the equation for any parabola can be written as either $y = ax^2 + bx + c$ or $y = a(x - h)^2 + k$, where the vertex is (h, k). The second form is sometimes called *vertex form*.

CRITICAL THINKING

Expand $y = 2(x - 3)^2 + 2$ and combine like terms. What does it equal?

READING CHECK

Give the vertex form. Explain it with an example.

VERTEX FORM

The **vertex form** of the equation of a parabola with vertex (h, k) is

$$y = a(x - h)^2 + k,$$

where $a \neq 0$ is a constant. If $a > 0$, the parabola opens upward; if $a < 0$, the parabola opens downward.

NOTE: Vertex form is sometimes called **standard form for a parabola with a vertical axis**.

In the next three examples, we demonstrate graphing parabolas in vertex form, finding their equations, and writing vertex forms of equations.

EXAMPLE 2   **Graphing parabolas in vertex form**

Compare the graph of $y = f(x)$ to the graph of $y = x^2$. Then sketch a graph of $y = f(x)$ and $y = x^2$ in the same xy-plane.

(a) $f(x) = \frac{1}{2}(x - 5)^2 + 2$ **(b)** $f(x) = -3(x + 5)^2 - 3$

Solution

(a) Compared to the graph of $y = x^2$, the graph of $y = f(x)$ is *translated 5 units right* and *2 units upward*. The vertex for $f(x)$ is $(5, 2)$, whereas the vertex of $y = x^2$ is $(0, 0)$. Because $a = \frac{1}{2}$, the graph of $y = f(x)$ *opens upward* and is *wider* than the graph of $y = x^2$. These graphs are shown in Figure 8.28(a) on the next page.

(b) Compared to the graph of $y = x^2$, the graph of $y = -3(x + 5)^2 - 3$ is *translated 5 units left* and *3 units downward*. The vertex for $f(x)$ is $(-5, -3)$. Because $a = -3$, the graph of $y = f(x)$ *opens downward* and is *narrower* than the graph of $y = x^2$. These graphs are shown in Figure 8.28(b).

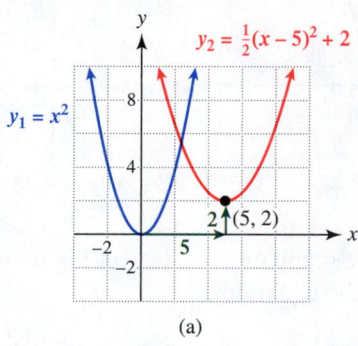

(a)

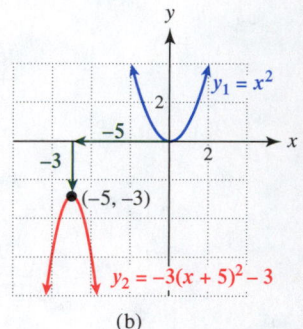

(b)

Figure 8.28

Now Try Exercises 35, 37

EXAMPLE 3 **Finding equations of parabolas**

Write the vertex form of a parabola with $a = 2$ and vertex $(-2, 1)$. Then express this equation in the form $y = ax^2 + bx + c$.

Solution
The vertex form of a parabola is given by $y = a(x - h)^2 + k$, where the vertex is (h, k). For $a = 2, h = -2$, and $k = 1$, the equation becomes

$$y = 2(x - (-2))^2 + 1 \quad \text{or} \quad y = 2(x + 2)^2 + 1.$$

To write $y = 2(x + 2)^2 + 1$ in the form $y = ax^2 + bx + c$, do the following.

$$
\begin{aligned}
y &= 2(x^2 + 4x + 4) + 1 && \text{Multiply } (x + 2)^2 \text{ as } x^2 + 4x + 4. \\
&= 2x^2 + 8x + 8 + 1 && \text{Distributive property} \\
&= 2x^2 + 8x + 9 && \text{Add.}
\end{aligned}
$$

The equivalent equation is $y = 2x^2 + 8x + 9$.

Now Try Exercise 41

In the next example, the vertex formula is used to write the equation of a parabola in vertex form.

EXAMPLE 4 **Using the vertex formula to write vertex form**

Find the vertex on the graph of $y = 3x^2 + 6x + 1$. Write this equation in vertex form.

Solution
We can use the vertex formula, $x = -\frac{b}{2a}$, to find the x-coordinate of the vertex with $a = 3$ and $b = 6$.

$$x = -\frac{b}{2a} = -\frac{6}{2(3)} = -1 \quad \text{x-coordinate of vertex}$$

To find the y-coordinate, let $x = -1$ in $y = 3x^2 + 6x + 1$.

$$y = 3(-1)^2 + 6(-1) + 1 = -2 \quad \text{Let } x = -1.$$

The vertex is $(-1, -2)$. We can now find the vertex form with $a = 3$, $h = -1$, and $k = -2$.

$$
\begin{array}{ll}
y = a(x - h)^2 + k & \text{Vertex form} \\
y = 3(x - (-1))^2 - 2 & \text{Substitute.} \\
y = 3(x + 1)^2 - 2 & \text{Simplify.}
\end{array}
$$

The equations $y = 3x^2 + 6x + 1$ and $y = 3(x + 1)^2 - 2$ are equivalent, and their graphs represent the same parabola with vertex $(-1, -2)$.

▌ **Now Try Exercise 53**

Completing the Square

$x^2 + 4x + 4 = (x + 2)^2$

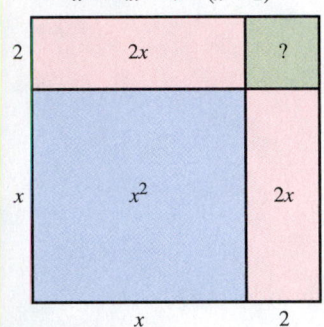

Figure 8.29

COMPLETING THE SQUARE TO FIND THE VERTEX The vertex of a parabola can be found by using a technique called **completing the square**. To complete the square for the expression $x^2 + 4x$, consider Figure 8.29.

Note that the blue and pink areas sum to

$$x^2 + 2x + 2x = x^2 + 4x.$$

The area of the small green square must be $2 \cdot 2 = 4$. Thus to *complete the square* for $x^2 + 4x$, we add 4. That is,

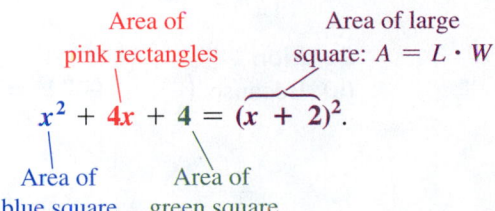

In the next example, we use the above result to complete the square and find the vertex on the graph of $y = x^2 + 4x$.

▌ EXAMPLE 5   **Writing vertex form by completing the square**

Write the equation $y = x^2 + 4x$ in vertex form by completing the square. Identify the vertex.

Solution

As discussed, we must add 4 to $x^2 + 4x$ to complete the square. Because we are given an equation, we will add 4 *and* subtract 4 from the right side of the equation in order to keep the equation "balanced."

$$
\begin{array}{ll}
y = x^2 + 4x. & \text{Given equation} \\
y = x^2 + 4x + 4 - 4 & \text{Add and subtract 4 on the right.} \\
y = (x^2 + 4x + 4) - 4 & \text{Associative property} \\
y = (x + 2)^2 - 4 & \text{Perfect square trinomial}
\end{array}
$$

The vertex form is $y = (x + 2)^2 - 4$, and the vertex is $(-2, -4)$.

▌ **Now Try Exercise 61**

NOTE: Adding 4 and subtracting 4 from the right side of the equation in Example 5 is equivalent to adding 0 to the right side, which does not change the equation.

In general, to complete the square for $x^2 + bx$ we must add $\left(\frac{b}{2}\right)^2$, as illustrated in Figure 8.30. This technique is shown in Example 6.

READING CHECK

Use the figure to *complete the square* for $x^2 + 8x$.

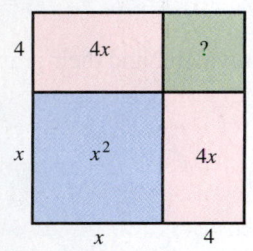

Completing the Square

$$x^2 + bx + \left(\frac{b}{2}\right)^2 = \left(x + \frac{b}{2}\right)^2$$

Figure 8.30

EXAMPLE 6   Writing vertex form

Write each equation in vertex form. Identify the vertex.
(a) $y = x^2 - 6x - 1$ **(b)** $y = x^2 + 3x + 4$ **(c)** $y = 2x^2 + 4x - 1$

Solution
(a) Because $\left(\frac{b}{2}\right)^2 = \left(\frac{-6}{2}\right)^2 = 9$, add *and* subtract 9 on the right side.

$$
\begin{aligned}
y &= x^2 - 6x - 1 && \text{Given equation} \\
&= (x^2 - 6x + 9) - 9 - 1 && \text{Add and subtract 9.} \\
&= (x - 3)^2 - 10 && \text{Perfect square trinomial}
\end{aligned}
$$

The vertex is $(3, -10)$.

(b) Because $\left(\frac{b}{2}\right)^2 = \left(\frac{3}{2}\right)^2 = \frac{9}{4}$, add *and* subtract $\frac{9}{4}$ on the right side.

$$
\begin{aligned}
y &= x^2 + 3x + 4 && \text{Given equation} \\
&= \left(x^2 + 3x + \frac{9}{4}\right) - \frac{9}{4} + 4 && \text{Add and subtract } \tfrac{9}{4}. \\
&= \left(x + \frac{3}{2}\right)^2 + \frac{7}{4} && \text{Perfect square trinomial}
\end{aligned}
$$

The vertex is $\left(-\frac{3}{2}, \frac{7}{4}\right)$.

(c) This equation is slightly different because the leading coefficient is 2 rather than 1. Start by factoring 2 from the first two terms on the right side.

$$
\begin{aligned}
y &= 2x^2 + 4x - 1 && \text{Given equation} \\
&= 2(x^2 + 2x) - 1 && \text{Factor out 2.} \\
&= 2(x^2 + 2x + 1 - 1) - 1 && \left(\tfrac{b}{2}\right)^2 = \left(\tfrac{2}{2}\right)^2 = 1 \\
&= 2(x^2 + 2x + 1) - 2 - 1 && \text{Distributive property: } 2 \cdot (-1) \\
&= 2(x + 1)^2 - 3 && \text{Perfect square trinomial}
\end{aligned}
$$

The vertex is $(-1, -3)$.

Now Try Exercises 65, 69, 73

Modeling with Quadratic Functions (Optional)

▶ **REAL-WORLD CONNECTION** In A Look Into Math for this section, we discussed airport taxiway curves designed for airplanes. The data previously shown in Figure 8.20 are listed in Table 8.9.

TABLE 8.9 Safe Taxiway Speed

x (mph)	10	20	30	40	50	60
R (ft)	50	200	450	800	1250	1800

Source: Federal Aviation Administration.

A second scatterplot of the data is shown in Figure 8.31. The data may be modeled by $R(x) = ax^2$ for some value a. To illustrate this relation, graph R for different values of a. In Figures 8.32–8.34, R has been graphed for $a = 2$, -1, and $\frac{1}{2}$, respectively. When $a > 0$ the parabola opens upward, and when $a < 0$ the parabola opens downward. Larger values of $|a|$ make a parabola narrower, whereas smaller values of $|a|$ make the parabola wider. Through trial and error, $a = \frac{1}{2}$ gives a good fit to the data, so $R(x) = \frac{1}{2}x^2$ models the data.

[−70, 70, 10] by [−2000, 2000, 500]

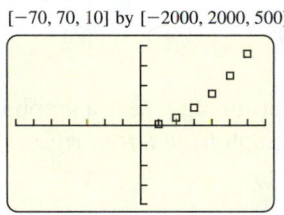

Figure 8.31

[−70, 70, 10] by [−2000, 2000, 500]

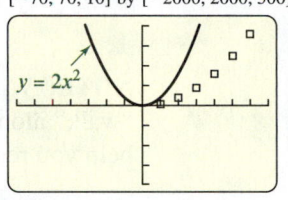

Figure 8.32 $a = 2$

[−70, 70, 10] by [−2000, 2000, 500]

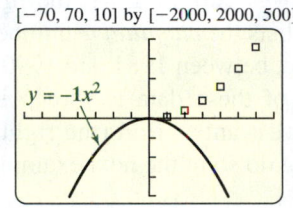

Figure 8.33 $a = -1$

[−70, 70, 10] by [−2000, 2000, 500]

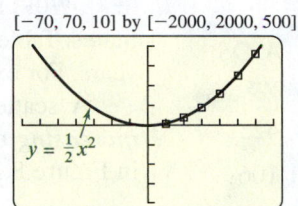

Figure 8.34 $a = \frac{1}{2}$

CALCULATOR HELP

To make a scatterplot, see Appendix A (pages AP-3 and AP-4).

This value of a can also be found *symbolically*, as demonstrated in the next example.

EXAMPLE 7 **Modeling safe taxiway speed**

Find a value for the constant a so that $R(x) = ax^2$ models the data in Table 8.9. Check your result by making a table of values for $R(x)$.

Solution

From Table 8.9, when $x = \mathbf{10}$ miles per hour, the curve radius is **50** feet. Therefore

$$R(\mathbf{10}) = \mathbf{50} \quad \text{or} \quad a(\mathbf{10})^2 = \mathbf{50}. \qquad R(x) = ax^2$$

Solving for a gives

$$a = \frac{\mathbf{50}}{\mathbf{10}^2} = \frac{1}{2}.$$

X	Y1
0	0
10	50
20	200
30	450
40	800
50	1250
60	1800

Y1■.5X²

Figure 8.35

To be sure that $R(x) = \frac{1}{2}x^2$ is correct, make a table, as shown in Figure 8.35. Its values agree with those in Table 8.9.

Now Try Exercise 87

INCREASING AND DECREASING The concept of increasing and decreasing is frequently used when modeling data with functions. Suppose that the graph of the equation $y = x^2$ shown in Figure 8.36 represents a valley. If we walk from *left to right*, the valley "goes down" and then "goes up." Mathematically, we say that the graph of $y = x^2$ is *decreasing* when $x \leq 0$ and *increasing* when $x \geq 0$. In Figure 8.33, the graph of $y = -1x^2$ increases when $x \leq 0$ and decreases when $x \geq 0$, and in Figure 8.24(c) on page 590 the graph decreases when $x \leq 2$ and increases when $x \geq 2$.

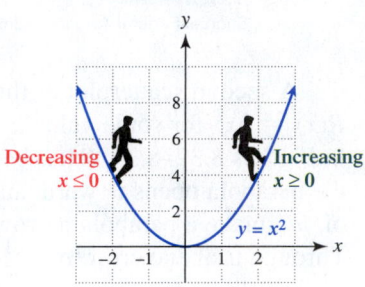

Figure 8.36

NOTE: When determining where a graph is increasing and where it is decreasing, we must "walk" along the graph *from left to right*. (We read English from left to right, which might help you remember.)

TABLE 8.10

Year	U.S. AIDS Cases
1981	425
1984	11,106
1987	71,414
1990	199,608
1993	417,835
1996	609,933

Source: Department of Health and Human Services.

▶ **REAL-WORLD CONNECTION** In 1981, the first cases of AIDS were reported in the United States. Table 8.10 lists the *cumulative* number of AIDS cases in the United States for various years. For example, between 1981 and 1990, a total of 199,608 AIDS cases were reported.

A scatterplot of these data is shown in Figure 8.37. To model these nonlinear and *increasing* data, we want to find (the right half of) a parabola with the shape illustrated in Figure 8.38. We do so in the next example.

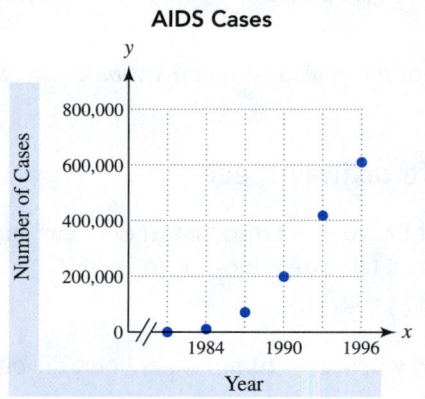

Figure 8.37

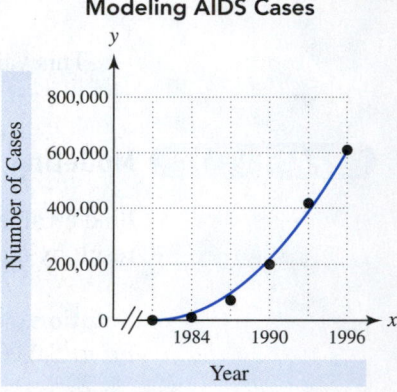

Figure 8.38

NOTE: From 1981 to 1996, the number of AIDS cases grew rapidly and can be modeled by a quadratic function. After 1996, the growth in AIDS cases slowed so a quadratic model is much less accurate. When modeling data with functions, it is important to remember that data can change character over time. As a result, the same modeling function may not be appropriate for every time period.

EXAMPLE 8   **Modeling AIDS cases**

Use the data in Table 8.10 to complete the following.
(a) Make a scatterplot of the data in [1980, 1997, 2] by [−10000, 800000, 100000].
(b) The lowest data point in Table 8.10 is (1981, 425). Let this point be the vertex of a parabola that opens upward. Graph $y = a(x - 1981)^2 + 425$ together with the data by first letting $a = 1000$.
(c) Use trial and error to adjust the value of a until the graph models the data.
(d) Use your final equation to estimate the number of AIDS cases in 1992. Compare it to the known value of 338,786.

Solution
(a) A scatterplot of the data is shown in Figure 8.39.

[1980, 1997, 2] by
[−10000, 800000, 100000]

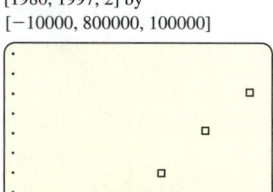

Figure 8.39

[1980, 1997, 2] by
[−10000, 800000, 100000]

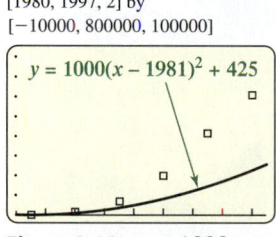

$y = 1000(x - 1981)^2 + 425$

Figure 8.40 $a = 1000$

[1980, 1997, 2] by
[−10000, 800000, 100000]

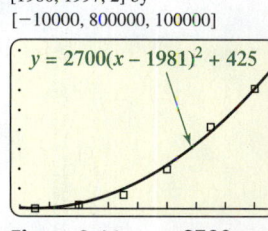

$y = 2700(x - 1981)^2 + 425$

Figure 8.41 $a = 2700$

(b) A graph of $y = 1000(x - 1981)^2 + 425$ is shown in Figure 8.40. To have a better fit of the data, a larger value for a is needed.
(c) Figure 8.41 shows the effect of adjusting the value of a to **2700**. This value provides a reasonably good fit. (Note that you may decide on a slightly different value for a.)
(d) If $a = 2700$, the modeling equation becomes

$$y = 2700(x - 1981)^2 + 425.$$

To estimate the number of AIDS cases in 1992, substitute $x = 1992$ to obtain

$$y = 2700(1992 - 1981)^2 + 425 = 327{,}125.$$

This number is about 12,000 less than the known value of **338,786**.

Now Try Exercise 91

8.2   Putting It All Together

CONCEPT	EXPLANATION	EXAMPLES
Translations of Parabolas	Compared to the graph $y = x^2$, the graph of $y = x^2 + k$ is shifted vertically k units and the graph of $y = (x - h)^2$ is shifted horizontally h units.	Compared to the graph of $y = x^2$, the graph of $y = x^2 - 4$ is shifted *downward* 4 units. Compared to the graph of $y = x^2$, the graph of $y = (x - 4)^2$ is shifted *right* 4 units and the graph of $y = (x + 4)^2$ is shifted *left* 4 units.
Vertex Form	The vertex form of the equation of a parabola with vertex (h, k) is $$y = a(x - h)^2 + k,$$ where $a \neq 0$ is a constant. If $a > 0$, the parabola opens upward; if $a < 0$, the parabola opens downward.	The graph of $y = 3(x + 2)^2 - 7$ has a vertex of $(-2, -7)$ and opens upward because $3 > 0$.

continued on next page

continued from previous page

CONCEPT	EXPLANATION	EXAMPLES
Completing the Square Method	To complete the square to obtain the vertex form, add *and* subtract $\left(\frac{b}{2}\right)^2$ on the right side of the equation $y = x^2 + bx + c$. Then factor the perfect square trinomial.	If $y = x^2 + 10x - 3$, then add *and* subtract $\left(\frac{b}{2}\right)^2 = \left(\frac{10}{2}\right)^2 = 25$ on the right side of this equation. $$y = (x^2 + 10x + 25) - 25 - 3$$ $$= (x + 5)^2 - 28$$ The vertex is $(-5, -28)$.

8.2 Exercises

MyMathLab 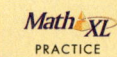 WATCH   DOWNLOAD   READ   REVIEW

CONCEPTS AND VOCABULARY

1. Compared to the graph of $y = x^2$, the graph of $y = $ _____ is shifted upward 2 units.

2. Compared to the graph of $y = x^2$, the graph of $y = $ _____ is shifted to the right 2 units.

3. The vertex of $y = (x - 1)^2 + 2$ is _____.

4. The vertex of $y = (x + 1)^2 - 2$ is _____.

5. A quadratic function f may be written either in the form _____ or _____.

6. The vertex form of a parabola is given by _____ and its vertex is _____.

7. The graph of the equation $y = -x^2$ is a parabola that opens _____.

8. The x-coordinate of the vertex of $y = ax^2 + bx + c$ is $x = $ _____.

9. Compared to the graph of $y = x^2$, the graph of $y = x^2 + k$ with $k > 0$ is shifted k units _____.
   a. upward   b. downward
   c. left   d. right

10. Compared to the graph of $y = x^2$, the graph of $y = (x - k)^2$ with $k > 0$ is shifted k units _____.
   a. upward   b. downward
   c. left   d. right

TABLES AND TRANSLATIONS

Exercises 11–14: Complete the table for each translation of $y = x^2$. State what the translation does.

11.

x	−2	−1	0	1	2
$y = x^2$					
$y = x^2 - 3$					

12.

x	−2	−1	0	1	2
$y = x^2$					
$y = x^2 + 5$					

13.

x					
$y = x^2$	4	1	0	1	4

x					
$y = (x - 3)^2$	4	1	0	1	4

14.

x					
$y = x^2$	16	4	0	4	16

x					
$y = (x + 4)^2$	16	4	0	4	16

GRAPHS OF PARABOLAS

Exercises 15–34: Do the following.
(a) Sketch a graph of the equation.
(b) Identify the vertex.
(c) Compare the graph of $y = f(x)$ to the graph of $y = x^2$. (State any transformations used.)

15. $f(x) = x^2 - 4$

16. $f(x) = x^2 - 1$

17. $f(x) = 2x^2 + 1$

18. $f(x) = \frac{1}{2}x^2 + 1$

19. $f(x) = (x - 3)^2$

20. $f(x) = (x + 1)^2$

21. $f(x) = -x^2$

22. $f(x) = -(x + 2)^2$

23. $f(x) = 2 - x^2$

24. $f(x) = (x - 1)^2$

25. $f(x) = (x + 2)^2$

26. $f(x) = (x - 2)^2 - 3$

27. $f(x) = (x + 1)^2 - 2$

28. $f(x) = (x - 3)^2 + 1$

29. $f(x) = (x - 1)^2 + 2$

30. $f(x) = \frac{1}{2}(x + 3)^2 - 3$

31. $f(x) = 2(x - 5)^2 - 4$

32. $f(x) = -3(x + 4)^2 + 5$

33. $f(x) = -\frac{1}{2}(x + 3)^2 + 1$

34. $f(x) = 2(x - 5)^2 + 10$

Exercises 35–38: Compare the graph of $y = f(x)$ to the graph of $y = x^2$. Then sketch a graph of $y = f(x)$ and $y = x^2$ in the same xy-plane.

35. $f(x) = \frac{1}{2}(x - 1)^2 - 2$

36. $f(x) = 2(x + 2)^2 - 1$

37. $f(x) = -2(x + 1)^2 + 3$

38. $f(x) = -\frac{1}{2}(x - 2)^2 + 2$

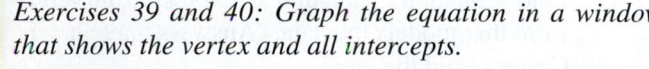

Exercises 39 and 40: Graph the equation in a window that shows the vertex and all intercepts.

39. $f(x) = -0.4x^2 + 6x - 10$

40. $f(x) = 3x^2 - 40x + 50$

VERTEX FORM

Exercises 41–44: (Refer to Example 3.) Write the vertex form of a parabola that satisfies the conditions given. Then write the equation in the form $y = ax^2 + bx + c$.

41. Vertex $(3, 4)$ and $a = 3$

42. Vertex $(-1, 3)$ and $a = -5$

43. Vertex $(5, -2)$ and $a = -\frac{1}{2}$

44. Vertex $(-2, -6)$ and $a = \frac{3}{4}$

Exercises 45–48: Write the vertex form of a parabola that satisfies the conditions given. Assume that $a = \pm 1$.

45. Opens upward, vertex $(1, 2)$

46. Opens downward, vertex $(-1, -2)$

47. Opens downward, vertex $(0, -3)$

48. Opens upward, vertex $(5, -4)$

Exercises 49–52: Write the vertex form of the parabola shown in the graph. Assume that $a = \pm 1$.

49.
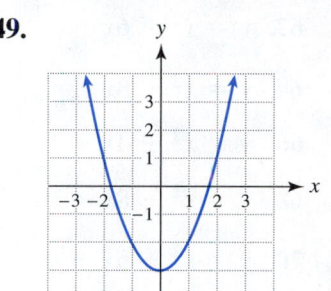

50.
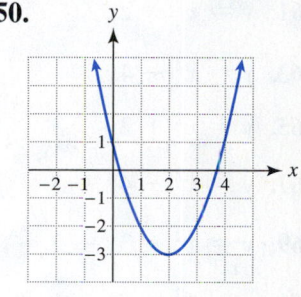

51.
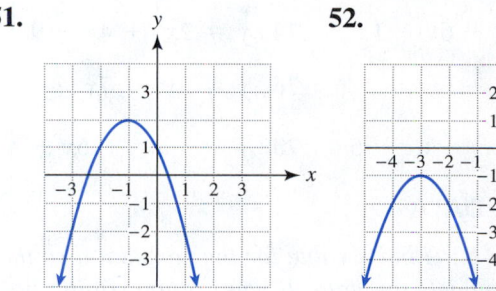

52.

Exercises 53–58: (Refer to Example 4.) Do the following.
(a) Find the vertex on the graph of the equation.
(b) Write the equation in vertex form.

53. $y = 4x^2 - 8x + 5$

54. $y = 3x^2 - 12x + 15$

55. $y = -x^2 - 2x - 3$

56. $y = -x^2 - 4x - 5$

57. $y = -2x^2 - 4x + 1$

58. $y = 2x^2 - 16x + 27$

COMPLETING THE SQUARE

Exercises 59 and 60: Use the given figure to determine what number should be added to the expression to complete the square.

59. $x^2 + 2x$ **60.** $x^2 + 6x$

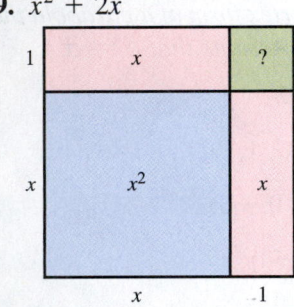
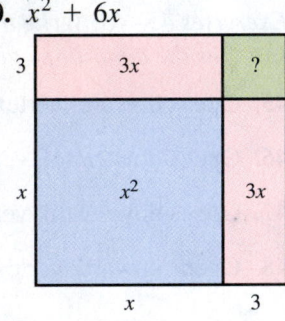

Exercises 61–78: (Refer to Example 6.) Write the equation in vertex form. Identify the vertex.

61. $y = x^2 + 2x$ **62.** $y = x^2 + 6x$

63. $y = x^2 - 4x$ **64.** $y = x^2 - 8x$

65. $y = x^2 + 2x - 3$ **66.** $y = x^2 + 4x + 1$

67. $y = x^2 - 4x + 5$ **68.** $y = x^2 - 8x + 10$

69. $y = x^2 + 3x - 2$ **70.** $y = x^2 + 5x - 4$

71. $y = x^2 - 7x + 1$ **72.** $y = x^2 - 3x + 5$

73. $y = 3x^2 + 6x - 1$ **74.** $y = 2x^2 + 4x - 9$

75. $y = 2x^2 - 3x$ **76.** $y = 3x^2 - 7x$

77. $y = -2x^2 - 8x + 5$ **78.** $y = -3x^2 + 6x + 1$

MODELING DATA

Exercises 79–82: Find a value for the constant a so that $f(x) = ax^2$ models the data. If you are uncertain about your value for a, check it by making a table of values.

79.

x	1	2	3
y	2	8	18

80.

x	-2	0	2
y	6	0	6

81.

x	2	4	6	8
y	1.2	4.8	10.8	19.2

82.

x	5	10	15	20
y	17.5	70	157.5	280

 Exercises 83–86: Modeling Quadratic Data (Refer to Example 8.) Find a quadratic function expressed in vertex form that models the data in the given table.

83.

x	1	2	3	4
y	-3	-1	5	15

84.

x	-2	-1	0	1	2
y	5	2	-7	-22	-43

85.

x	1980	1990	2000	2010
y	6	55	210	450

86.

x	1990	1995	2000	2005
y	10	60	205	470

87. *Braking Distance* The table lists approximate braking distances D in feet for cars traveling at x miles per hour on dry, level pavement.

x	12	24	36	48
D	12	48	108	192

 (a) Make a scatterplot of the data.
(b) Find a function given by $D(x) = ax^2$ that models these data.

88. *Health Care Costs* The table lists approximate *annual* percent increases in the cost of health insurance premiums between 1992 and 2000.

Year	1992	1994	1996	1998	2000
Increase	11%	4%	1%	4%	11%

Source: Kaiser Family Foundation.

(a) Describe what happened to health care costs from 1992 to 2000.
(b) Would a linear function model these data? Explain.
(c) What type of function might model these data? Explain.
(d) What might be a good choice for the vertex? Explain.
(e) Find a quadratic function C expressed in vertex form that models the data. (Answers may vary.)
(f) Graph C and the data.

89. *Sub-Saharan Africa* The table lists actual and projected real gross domestic product (GDP) per capita for selected years in Sub-Saharan Africa.

Year	1980	1995	2010	2025
Real GDP	$1000	$600	$1000	$2200

Source: IMF WEO, Standard Chartered Research.

(a) Describe what happens to the real GDP.
(b) Would a linear function model these data? Explain.
(c) What type of function might model these data? Explain.

(d) What might be a good choice for the vertex? Explain.
(e) Find a quadratic function C expressed in vertex form that models the data. (Answers may vary.)
(f) Graph C and the data.

90. *Tax Theory and Modeling* If a government wants to generate enough revenue from income taxes, then it is important to assess the correct tax rates. On the one hand, if tax rates are too low, then enough revenue may not be generated. On the other hand, if taxes are too high, people may work less and not earn as much money. Again, enough revenue may not be generated. The following parabolic graph illustrates this phenomenon.

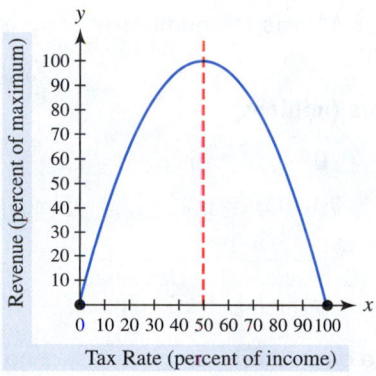

(a) The points $(0, 0)$ and $(100, 0)$ are on this graph. Explain why.
(b) Explain why a linear model is not suitable.
(c) According to *this* graph, what rate maximizes revenue?

91. *U.S. AIDS Deaths* (Refer to Example 8.) The table lists cumulative numbers of AIDS deaths *in thousands* for selected years.

Year	1982	1986	1990	1994
Deaths (thousands)	1	25	123	300

Source: Department of Health and Human Services.

(a) Determine $f(x) = a(x - h)^2 + k$, so that f models these data.
(b) Estimate the number of AIDS deaths in 1992 and compare it to the actual value of 202 thousand.

 92. *Head Start Enrollment* The table lists numbers of students *in thousands* enrolled in Head Start for selected years.

Year	1966	1980	1995
Students (thousands)	733	376	750

(a) Determine $f(x) = a(x - h)^2 + k$, so that f models these data.
(b) Estimate Head Start enrollment in 1990 and compare it to the actual value of 541 thousand.

INCREASING AND DECREASING

Exercises 93–96: Use the graph of $y = f(x)$ to determine the intervals where f is increasing and where f is decreasing.

93.

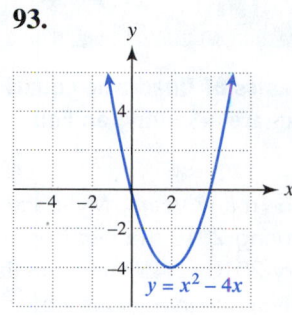

94.

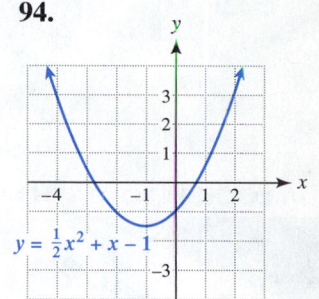

95.

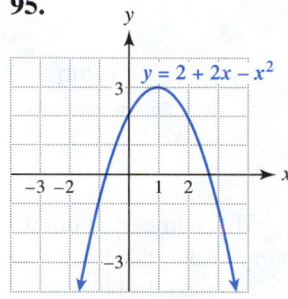

96.

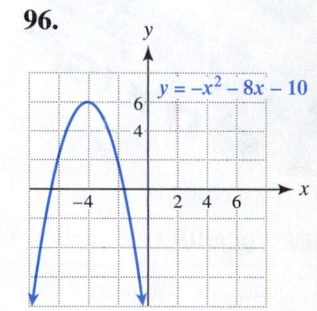

Exercises 97–102: Determine the intervals where $y = f(x)$ is increasing and where it is decreasing.

97. $y = 3x^2 - 4x + 1$ **98.** $y = 2x^2 - 5x - 2$

99. $y = -x^2 - 3x$ **100.** $y = -2x^2 + 3x - 4$

101. $y = 5 - x - 4x^2$ **102.** $y = 3 + x + x^2$

WRITING ABOUT MATHEMATICS

103. Explain how to find the vertex of $y = x^2 + bx + c$ by completing the square.

104. If $f(x) = a(x - h)^2 + k$, explain how the values of a, h, and k affect the graph of $y = f(x)$.

SECTIONS 8.1 and 8.2
Checking Basic Concepts

1. Graph each quadratic function. Identify the vertex and axis of symmetry.
   (a) $f(x) = x^2 - 2$
   (b) $f(x) = x^2 - 2x - 2$

2. Compare the graph of $y_1 = 2x^2$ to the graph of $y_2 = -\frac{1}{2}x^2$.

3. Find the maximum y-value on the graph of the equation $y = -3x^2 + 12x - 5$.

4. Sketch a graph of $y = f(x)$. Compare this graph to the graph of $y = x^2$.
   (a) $f(x) = (x - 1)^2 + 2$
   (b) $f(x) = -(x + 3)^2$

5. Write the vertex form for each equation.
   (a) $y = x^2 + 14x - 7$
   (b) $y = 4x^2 + 8x - 2$

8.3 Quadratic Equations

Basics of Quadratic Equations • The Square Root Property • Completing the Square • Solving an Equation for a Variable • Applications of Quadratic Equations

A LOOK INTO MATH ▶

NEW VOCABULARY

☐ Quadratic equation
☐ Square root property

For many years, MySpace was the largest social network in the United States. However, during 2010 the number of unique monthly visitors fell dramatically from 70 million in January 2010 to 45 million in January 2011. Table 8.11 lists the number of unique visitors V to MySpace in millions x months after January 2010.

TABLE 8.11 Unique MySpace Visitors (millions)

x (months after Jan. 2010)	0	6	12
V (unique monthly visitors)	70	63	45

Source: comScore.

Decreased 7 million   Decreased 18 million

The number of unique visitors to MySpace decreased faster in the second half of 2010 than it did in the first half, so a linear function will *not* model these data. Instead, these data can be modeled by the *quadratic function*

$$V(x) = -\frac{25}{144}x^2 + 70. \qquad \text{Quadratic function}$$

If we want to determine the month when MySpace had **56** million unique visitors, then we could solve the *quadratic equation*

$$-\frac{25}{144}x^2 + 70 = \mathbf{56}. \qquad \text{Quadratic equation}$$

In this section we learn how to solve this and other quadratic equations. See Exercise 125.

Basics of Quadratic Equations

Any quadratic function f can be represented by $f(x) = ax^2 + bx + c$ with $a \neq 0$. Examples of quadratic functions include

$$f(x) = \mathbf{2x^2 - 1},\, g(x) = \mathbf{-\tfrac{1}{3}x^2 + 2x},\, \text{and } h(x) = x^2 + 2x - 1. \qquad \text{Quadratic functions}$$

Quadratic functions can be used to write quadratic equations. Examples of quadratic equations include

$$2x^2 - 1 = 0, \quad -\tfrac{1}{3}x^2 + 2x = 0, \quad \text{and} \quad x^2 + 2x - 1 = 0.$$ Quadratic equations

READING CHECK

How can you identify a quadratic equation?

QUADRATIC EQUATION

A **quadratic equation** is an equation that can be written as

$$ax^2 + bx + c = 0,$$

where a, b, and c are constants with $a \neq 0$.

Solutions to the quadratic equation $ax^2 + bx + c = 0$ correspond to x-intercepts of the graph of $y = ax^2 + bx + c$. Because the graph of a quadratic function is either $\cup$-shaped or $\cap$-shaped, it can intersect the x-axis zero, one, or two times, as illustrated in Figure 8.42. Hence a quadratic equation can have zero, one, or two real solutions.

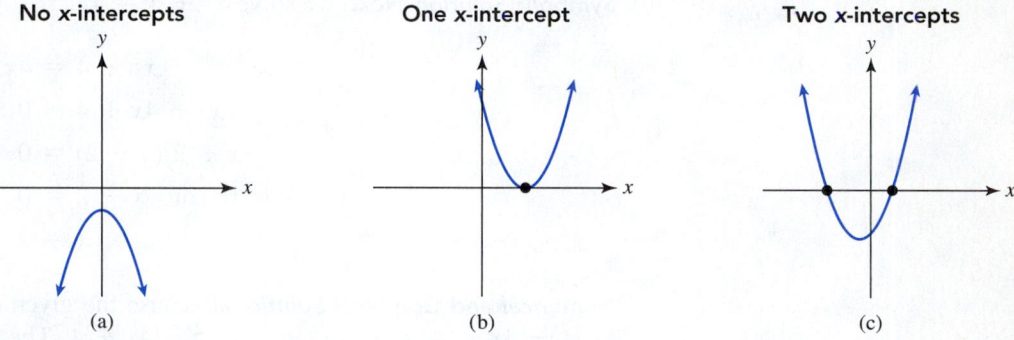

Figure 8.42

NOTE: Some quadratic equations may not have 0 on the right side of the equation, such as $4x^2 = 1$. To solve this quadratic equation graphically, we will sometimes rewrite it as $4x^2 - 1 = 0$ and graph $y = 4x^2 - 1$. Then solutions are x-intercepts, because $y = 0$ on the x-axis.

We have already solved quadratic equations by factoring, graphing, and constructing tables. In the next example we apply these three techniques to quadratic equations that have no real solutions, one real solution, and two real solutions.

EXAMPLE 1   **Solving quadratic equations**

Solve each quadratic equation. Support your results numerically and graphically.
(a) $2x^2 + 1 = 0$ (No real solutions)
(b) $x^2 + 4 = 4x$ (One real solution)
(c) $x^2 - 6x + 8 = 0$ (Two real solutions)

Solution
(a) *Symbolic Solution*

$$2x^2 + 1 = 0 \qquad \text{Given equation}$$
$$2x^2 = -1 \qquad \text{Subtract 1 from each side.}$$
$$x^2 = -\frac{1}{2} \qquad \text{Divide each side by 2.}$$

This equation has no real-number solutions because $x^2 \geq 0$ for all real numbers x.

Numerical and Graphical Solution The points in Table 8.12 for $y = 2x^2 + 1$ are plotted in Figure 8.43 and connected with a parabolic graph. The graph of $y = 2x^2 + 1$ has no x-intercepts, indicating that there are no real solutions.

TABLE 8.12

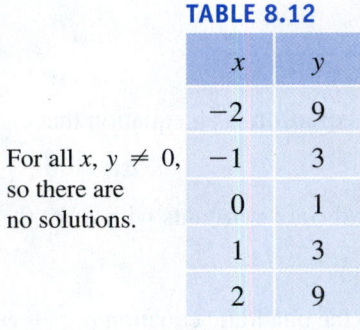

For all x, $y \neq 0$, so there are no solutions.

x	y
-2	9
-1	3
0	1
1	3
2	9

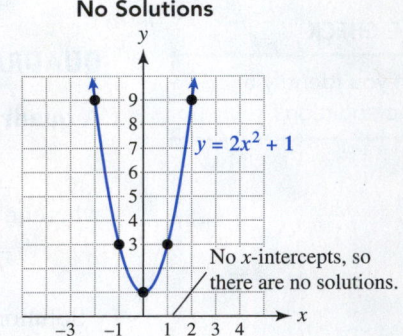

No Solutions

$y = 2x^2 + 1$

No x-intercepts, so there are no solutions.

Figure 8.43

(b) **Symbolic Solution** Next, we solve $x^2 + 4 = 4x$.

$$
\begin{array}{ll}
x^2 + 4 = 4x & \text{Given equation} \\
x^2 - 4x + 4 = 0 & \text{Subtract } 4x \text{ from each side.} \\
(x - 2)(x - 2) = 0 & \text{Factor.} \\
x - 2 = 0 \quad \text{or} \quad x - 2 = 0 & \text{Zero-product property} \\
x = \mathbf{2} & \text{There is one solution.}
\end{array}
$$

Numerical and Graphical Solution Because the given quadratic equation is equivalent to $x^2 - 4x + 4 = 0$, we let $y = x^2 - 4x + 4$. The points in Table 8.13 are plotted in Figure 8.44 and connected with a parabolic graph. The graph of $y = x^2 - 4x + 4$ has one x-intercept, **2**. Note that in Table 8.13, $y = \mathbf{0}$ when $x = \mathbf{2}$, indicating that the equation has one solution.

TABLE 8.13

x	y
0	4
1	1
$\mathbf{2}$	$\mathbf{0}$
3	1
4	4

One solution: **2**

One Solution

$y = x^2 - 4x + 4$

One x-intercept: **2**, so there is one solution: **2**.

Figure 8.44

(c) **Symbolic Solution** Next, we solve $x^2 - 6x + 8 = 0$.

$$
\begin{array}{ll}
x^2 - 6x + 8 = 0 & \text{Given equation} \\
(x - 2)(x - 4) = 0 & \text{Factor.} \\
x - 2 = 0 \quad \text{or} \quad x - 4 = 0 & \text{Zero-product property} \\
x = \mathbf{2} \quad \text{or} \quad x = \mathbf{4} & \text{There are two solutions.}
\end{array}
$$

Numerical and Graphical Solution The points in Table 8.14 for $y = x^2 - 6x + 8$ are plotted in Figure 8.45 and connected with a parabolic graph. The graph of

$y = x^2 - 6x + 8$ has two x-intercepts, **2** and **4**, indicating two solutions. Note that in Table 8.14 $y = 0$ when $x = 2$ or $x = 4$.

TABLE 8.14

x	y
0	8
1	3
2	**0**
3	−1
4	**0**
5	3
6	8

Two solutions: **2** and **4**

Two Solutions

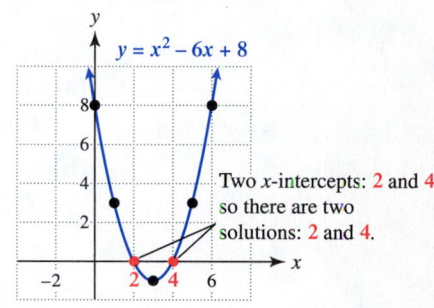

Two x-intercepts: **2** and **4**, so there are two solutions: **2** and **4**.

Figure 8.45

■ **Now Try Exercises 29, 37, 39**

READING CHECK

How many solutions can a quadratic equation have?

The Square Root Property

The **square root property** is used to solve quadratic equations that have no x-terms. The following is an example of the square root property.

$$x^2 = 25 \quad \text{is equivalent to} \quad x = \pm 5.$$

The equation $x = \pm 5$ (read "x equals plus or minus 5") indicates that either $x = 5$ or $x = -5$. Each value is a solution because $(5)^2 = 25$ and $(-5)^2 = 25$.

We can derive this result in general for $k \geq 0$.

$$x^2 = k \qquad \text{Given quadratic equation}$$
$$\sqrt{x^2} = \sqrt{k} \qquad \text{Take the square root of each side.}$$
$$|x| = \sqrt{k} \qquad \sqrt{x^2} = |x| \text{ for all } x.$$
$$x = \pm\sqrt{k} \qquad |x| = b \text{ implies } x = \pm b, b \geq 0.$$

This result is summarized by the *square root property*.

SQUARE ROOT PROPERTY

Let k be a nonnegative number. Then the solutions to the equation

$$x^2 = k$$

are given by $x = \pm\sqrt{k}$. If $k < 0$, then this equation has no real solutions.

Before applying the square root property in the next two examples, we review a quotient property of square roots. If a and b are positive numbers, then

$$\sqrt{\frac{a}{b}} = \frac{\sqrt{a}}{\sqrt{b}}.$$

For example, $\sqrt{\frac{25}{36}} = \frac{\sqrt{25}}{\sqrt{36}} = \frac{5}{6}$.

EXAMPLE 2 **Using the square root property**

Solve each equation.
(a) $x^2 = 7$ **(b)** $16x^2 - 9 = 0$ **(c)** $(x - 4)^2 = 25$

Solution
(a) $x^2 = 7$ is equivalent to $x = \pm\sqrt{7}$ by the square root property. The solutions are $\sqrt{7}$ and $-\sqrt{7}$.

(b)

$$16x^2 - 9 = 0 \qquad \text{Given equation}$$
$$16x^2 = 9 \qquad \text{Add 9 to each side.}$$
$$x^2 = \frac{9}{16} \qquad \text{Divide each side by 16.}$$
$$x = \pm\sqrt{\frac{9}{16}} \qquad \text{Square root property}$$
$$x = \pm\frac{3}{4} \qquad \text{Simplify.}$$

READING CHECK

When do you use the square root property to solve an equation?

The solutions are $\frac{3}{4}$ and $-\frac{3}{4}$.

(c)

$$(x - 4)^2 = 25 \qquad \text{Given equation}$$
$$(x - 4) = \pm\sqrt{25} \qquad \text{Square root property}$$
$$x - 4 = \pm 5 \qquad \text{Simplify.}$$
$$x = 4 \pm 5 \qquad \text{Add 4 to each side.}$$
$$x = 9 \quad \text{or} \quad x = -1 \qquad \text{Evaluate } 4 + 5 \text{ and } 4 - 5.$$

The solutions are 9 and -1.

Now Try Exercises 51, 53, 57

▶ **REAL-WORLD CONNECTION** If an object is dropped from a height of h feet, its distance d above the ground after t seconds is given by

$$d(t) = h - 16t^2.$$

This formula can be used to estimate the time it takes for a falling object to hit the ground.

EXAMPLE 3 **Modeling a falling object**

A toy falls 30 feet from a window. How long does the toy take to hit the ground?

Solution
The height h of the window above the ground is 30 feet, so let $d(t) = 30 - 16t^2$. The toy strikes the ground when the distance d above the ground equals 0.

$$30 - 16t^2 = 0 \qquad \text{Equation to solve for } t$$
$$-16t^2 = -30 \qquad \text{Subtract 30 from each side.}$$
$$t^2 = \frac{30}{16} \qquad \text{Divide each side by } -16.$$
$$t = \pm\sqrt{\frac{30}{16}} \qquad \text{Square root property}$$
$$t = \pm\frac{\sqrt{30}}{4} \qquad \text{Simplify.}$$

Time cannot be negative in this problem, so the appropriate solution is $t = \frac{\sqrt{30}}{4} \approx 1.4$. The toy hits the ground after about 1.4 seconds.

Now Try Exercise 115

Completing the Square

In Section 8.2 we used the *method of completing the square* to find the vertex of a parabola. This method can also be used to solve quadratic equations. Because

$$x^2 + bx + \left(\frac{b}{2}\right)^2 = \left(x + \frac{b}{2}\right)^2,$$

we can solve a quadratic equation in the form $x^2 + bx = d$, where b and d are constants, by adding $\left(\frac{b}{2}\right)^2$ to each side and then factoring the resulting perfect square trinomial.

In the equation $x^2 + 6x = 7$ we have $b = 6$, so we add $\left(\frac{6}{2}\right)^2 = 9$ to each side.

$$x^2 + 6x = 7 \qquad \text{Given equation}$$
$$x^2 + 6x + 9 = 7 + 9 \qquad \text{Add 9 to each side.}$$
$$(x + 3)^2 = 16 \qquad \text{Perfect square trinomial}$$
$$x + 3 = \pm 4 \qquad \text{Square root property}$$
$$x = -3 \pm 4 \qquad \text{Add } -3 \text{ to each side.}$$
$$x = 1 \quad \text{or} \quad x = -7 \qquad \text{Simplify } -3 + 4 \text{ and } -3 - 4.$$

The solutions are 1 and -7. Note that after completing the square, the left side of the equation is a perfect square trinomial. We show how to create one in the next example.

READING CHECK

Use the figure to *complete the square* for $x^2 + 6x$.

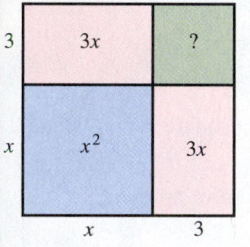

EXAMPLE 4 **Creating a perfect square trinomial**

Find the term that should be added to $x^2 - 10x$ to form a perfect square trinomial.

Solution

The coefficient of the x-term is -10, so we let $b = -10$. To complete the square we divide b by 2 and then square the result.

$$\left(\frac{b}{2}\right)^2 = \left(\frac{-10}{2}\right)^2 = 25$$

If we add 25 to $x^2 - 10x$, a perfect square trinomial is formed.

$$x^2 - 10x + 25 = (x - 5)^2$$

▎**Now Try Exercise 67**

EXAMPLE 5 **Completing the square when the leading coefficient is 1**

Solve the equation $x^2 - 4x + 2 = 0$.

Solution

Start by writing the equation in the form $x^2 + bx = d$.

$$x^2 - 4x + 2 = 0 \qquad \text{Given equation}$$
$$x^2 - 4x = -2 \qquad \text{Subtract 2.}$$
$$x^2 - 4x + 4 = -2 + 4 \qquad \text{Add } \left(\frac{b}{2}\right)^2 = \left(\frac{-4}{2}\right)^2 = 4.$$
$$(x - 2)^2 = 2 \qquad \text{Perfect square trinomial; add.}$$
$$x - 2 = \pm\sqrt{2} \qquad \text{Square root property}$$
$$x = 2 \pm \sqrt{2} \qquad \text{Add 2.}$$

The solutions are $2 + \sqrt{2} \approx 3.41$ and $2 - \sqrt{2} \approx 0.59$.

▎**Now Try Exercise 73**

EXAMPLE 6 | **Completing the square when the leading coefficient is not 1**

Solve the equation $2x^2 + 7x - 5 = 0$.

Solution
Start by writing the equation in the form $x^2 + bx = d$. That is, add 5 to each side and then divide each side by 2 so that the leading coefficient of the x^2-term becomes 1.

$$2x^2 + 7x - 5 = 0 \qquad \text{\color{blue}Given equation}$$

$$2x^2 + 7x = 5 \qquad \text{\color{blue}Add 5 to each side.}$$

$$x^2 + \frac{7}{2}x = \frac{5}{2} \qquad \text{\color{blue}Divide each side by 2.}$$

$$x^2 + \frac{7}{2}x + \frac{49}{16} = \frac{5}{2} + \frac{49}{16} \qquad \text{\color{blue}Add } \left(\frac{b}{2}\right)^2 = \left(\frac{7}{4}\right)^2 = \frac{49}{16}.$$

$$\left(x + \frac{7}{4}\right)^2 = \frac{89}{16} \qquad \text{\color{blue}Perfect square trinomial; add.}$$

$$x + \frac{7}{4} = \pm \frac{\sqrt{89}}{4} \qquad \text{\color{blue}Square root property}$$

$$x = -\frac{7}{4} \pm \frac{\sqrt{89}}{4} \qquad \text{\color{blue}Add } -\frac{7}{4}.$$

$$x = \frac{-7 \pm \sqrt{89}}{4} \qquad \text{\color{blue}Combine fractions.}$$

The solutions are $\frac{-7 + \sqrt{89}}{4} \approx 0.61$ and $\frac{-7 - \sqrt{89}}{4} \approx -4.1$.

▌ **Now Try Exercise 81**

CRITICAL THINKING

What happens if you try to solve

$$2x^2 - 13 = 1$$

by completing the square? What method could you use to solve this problem?

Solving an Equation for a Variable

We often need to solve an equation or formula for a variable. For example, the formula $V = \frac{1}{3}\pi r^2 h$ calculates the volume of the cone shown in Figure 8.46. Let's say that we know the volume V is 120 cubic inches and the height h is 15 inches. We can then find the radius of the cone by solving the equation for r.

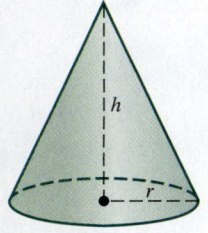

Figure 8.46

$$V = \frac{1}{3}\pi r^2 h \qquad \text{\color{blue}Solve the equation for } r.$$

$$3V = \pi r^2 h \qquad \text{\color{blue}Multiply by 3.}$$

$$\frac{3V}{\pi} = r^2 h \qquad \text{\color{blue}Divide by } \pi.$$

$$\frac{3V}{\pi h} = r^2 \qquad \text{\color{blue}Divide by } h.$$

$$r = \pm \sqrt{\frac{3V}{\pi h}} \qquad \text{\color{blue}Square root property; rewrite.}$$

Because $r \geq 0$, we use the positive or *principal square root*. Thus for $V = 120$ cubic inches and $h = 15$ inches,

$$r = \sqrt{\frac{3(120)}{\pi(15)}} = \sqrt{\frac{24}{\pi}} \approx 2.8 \text{ inches.}$$

EXAMPLE 7   **Solving equations for variables**

Solve each equation for the specified variable.
(a) $s = -\frac{1}{2}gt^2 + h$ for t **(b)** $d^2 = x^2 + y^2$ for y

Solution
(a) Begin by subtracting h from each side of the equation.

$$s = -\frac{1}{2}gt^2 + h \qquad \text{Solve the equation for } t.$$

$$s - h = -\frac{1}{2}gt^2 \qquad \text{Subtract } h.$$

$$-2(s - h) = gt^2 \qquad \text{Multiply by } -2.$$

$$\frac{2h - 2s}{g} = t^2 \qquad \text{Divide by } g; \text{ simplify.}$$

$$t = \pm\sqrt{\frac{2h - 2s}{g}} \qquad \text{Square root property; rewrite.}$$

(b) Begin by subtracting x^2 from each side of the equation.

$$d^2 = x^2 + y^2 \qquad \text{Solve the equation for } y.$$

$$d^2 - x^2 = y^2 \qquad \text{Subtract } x^2.$$

$$y = \pm\sqrt{d^2 - x^2} \qquad \text{Square root property; rewrite.}$$

Now Try Exercises 105, 107

Applications of Quadratic Equations

▶ **REAL-WORLD CONNECTION** In Section 8.2 we modeled curves on airport taxiways by using $R(x) = \frac{1}{2}x^2$. In this formula x represented the airplane's speed in miles per hour, and R represented the radius of the curve in feet. This formula may be used to determine the speed limit for a curve with a radius of **650** feet by solving the *quadratic equation*

$$\frac{1}{2}x^2 = 650.$$

In the next example, we solve this quadratic equation. (*Source:* FAA.)

EXAMPLE 8   **Finding a safe speed limit**

Solve $\frac{1}{2}x^2 = 650$ and interpret any solutions.

Solution
Use the square root property to solve this problem.

$$\frac{1}{2}x^2 = 650 \qquad \text{Given equation}$$

$$x^2 = 1300 \qquad \text{Multiply by 2.}$$

$$x = \pm\sqrt{1300} \qquad \text{Square root property}$$

The solutions are $\sqrt{1300} \approx 36$ and $-\sqrt{1300} \approx -36$. The solution of $x \approx 36$ indicates that a safe speed limit for a curve with a radius of 650 feet is about 36 miles per hour. (The negative solution has no physical meaning in this problem.)

Now Try Exercise 113

▶ **REAL-WORLD CONNECTION** From 1989 to 2009 there was a dramatic increase in compact disc and tape sales and then a dramatic decrease. Their global music sales S in billions of dollars x years after 1989 are modeled by

$$S(x) = -0.095x^2 + 1.85x + 6.$$

A graph of $S(x) = -0.095x^2 + 1.85x + 6$ is shown in Figure 8.47. (*Source:* RIAA, Bain Analysis.)

Global Music Sales from CDs and Tapes

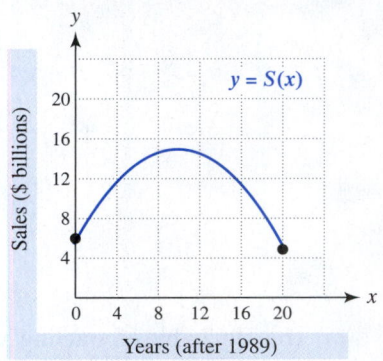

Figure 8.47

When were sales equal to $12 billion? Because the graph is parabolic, it appears that there are two answers to the question: about 4 years and about 15 years after 1989. In the next example, we use graphical and numerical methods to find these years. In Section 8.4 (Exercise 113), you are asked to answer this question symbolically.

EXAMPLE 9 ▸ **Modeling global CD and tape sales**

Use each method to determine the years when global sales of CDs and tapes were $12 billion.
(a) Graphical   **(b)** Numerical

Solution

(a) *Graphical Solution* Graph $y_1 = -0.095x^2 + 1.85x + 6$ and $y_2 = 12$, as shown in Figures 8.48(a) and (b). Their graphs intersect near the points (4.11, 12) and (15.36, 12). Because x is years *after* 1989, sales were $12 billion dollars in about 1993 (1989 + 4) and about 2004 (1989 + 15).

CALCULATOR HELP

To find a point of intersection, see Appendix A (page AP-6).

Determining When Sales Were $12 Billion

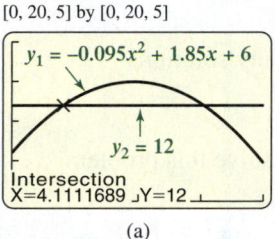

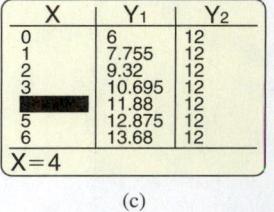

Figure 8.48

(b) *Numerical Solution* Make a table of $y_1 = -0.095x^2 + 1.85x + 6$ and $y_2 = 12$, as shown in Figure 8.48(c). When $x = 4$, y_1 is approximately 12. Thus sales were $12 billion around 1993. If we were to scroll down in the table further, we would see that sales also equal $12 billion when $x \approx 15$, or in 2004. The graphical and numerical solutions agree.

▌ **Now Try Exercise 117**

8.3 Putting It All Together

Quadratic equations can have no real solutions, one real solution, or two real solutions. Symbolic techniques for solving quadratic equations include factoring, the square root property, and completing the square. We discussed factoring in Chapter 5, so the following table summarizes only the square root property and the method of completing the square.

TECHNIQUE	DESCRIPTION	EXAMPLES
Square Root Property	If $k \geq 0$, the solutions to the equation $x^2 = k$ are $\pm\sqrt{k}$.	$x^2 = 100$ is equivalent to $x = \pm 10$ and $x^2 = 13$ is equivalent to $x = \pm\sqrt{13}$. $x^2 = -2$ has no real solutions.
Method of Completing the Square	To solve an equation in the form $x^2 + bx = d$, add $\left(\frac{b}{2}\right)^2$ to each side of the equation. Factor the resulting perfect square trinomial and solve for x by applying the square root property.	To solve $x^2 + 8x = 3$, add $\left(\frac{8}{2}\right)^2 = 16$ to each side. $x^2 + 8x + 16 = 3 + 16$ Add 16. $(x + 4)^2 = 19$ Perfect square trinomial $x + 4 = \pm\sqrt{19}$ Square root property $x = -4 \pm\sqrt{19}$ Add -4.

8.3 Exercises

MyMathLab   Math XL PRACTICE   WATCH   DOWNLOAD   READ   REVIEW

CONCEPTS AND VOCABULARY

1. Give an example of a quadratic equation. How many real solutions can a quadratic equation have?

2. Is a quadratic equation a linear equation or a nonlinear equation?

3. Name three symbolic methods that can be used to solve a quadratic equation.

4. Sketch a graph of a quadratic function that has two x-intercepts and opens downward.

5. Sketch a graph of a quadratic function that has no x-intercepts and opens upward.

6. If the graph of $y = ax^2 + bx + c$ intersects the x-axis twice, how many solutions does the equation $ax^2 + bx + c = 0$ have? Explain.

7. Solve $x^2 = 64$. What property did you use?

8. To solve $x^2 + bx = 6$ by completing the square, what value should be added to each side of the equation?

Exercises 9–16: Is the given equation quadratic?

9. $x^2 - 3x + 1 = 0$

10. $2x^2 - 3 = 0$

11. $3x + 1 = 0$

12. $x^3 - 3x^2 + x = 0$

13. $-3x^2 + x = 16$

14. $x^2 - 1 = 4x$

15. $x^2 = \sqrt{x} + 1$

16. $\frac{1}{x - 1} = 5$

SOLVING QUADRATIC EQUATIONS

Exercises 17–20: Approximate to the nearest hundredth.

17. (a) $1 \pm \sqrt{7}$ (b) $-2 \pm \sqrt{11}$

18. (a) $\pm\frac{\sqrt{3}}{2}$ (b) $\pm\frac{2\sqrt{5}}{7}$

19. (a) $\frac{3 \pm \sqrt{13}}{5}$ (b) $\frac{-5 \pm \sqrt{6}}{9}$

20. (a) $\frac{2}{5} \pm \frac{\sqrt{5}}{5}$ (b) $-\frac{3}{7} \pm \frac{\sqrt{3}}{7}$

Exercises 21–24: A graph of $y = ax^2 + bx + c$ is given. Use this graph to solve $ax^2 + bx + c = 0$, if possible.

21.

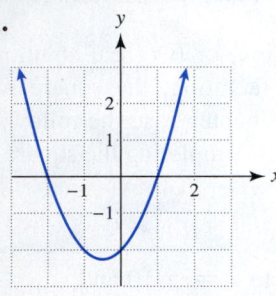

22.

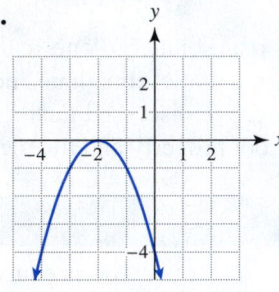

23.

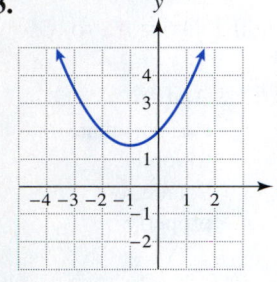

24.

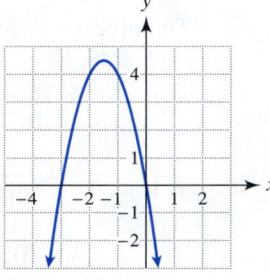

Exercises 25–28: A table of $y = ax^2 + bx + c$ is given. Use this table to solve $ax^2 + bx + c = 0$.

25.

X	Y₁
-3	6
-2	0
-1	-4
0	-6
1	-6
2	-4
3	0

Y₁■X^2−X−6

26.

X	Y₁
-6	0
-4	-16
-2	-24
0	-24
2	-16
4	0
6	24

Y₁■X^2+2X−24

27.

X	Y₁
-2	9
-1.5	4
-1	1
-.5	0
0	1
.5	4
1	9

Y₁■4X^2+4X+1

28.

X	Y₁
-4	-2.5
-3	0
-2	1.5
-1	2
0	1.5
1	0
2	-2.5

Y₁■−X^2/2−X+3/2

Exercises 29–40: Use each method to solve the equation.

(a) Symbolic **(b)** Graphical **(c)** Numerical

29. $x^2 - 4x - 5 = 0$

30. $x^2 - x - 6 = 0$

31. $x^2 + 2x = 3$

32. $x^2 + 4x = 5$

33. $x^2 = 9$

34. $x^2 = 4$

35. $4x^2 - 4x = 3$

36. $2x^2 + x = 1$

37. $x^2 + 2x = -1$

38. $4x^2 - 4x + 1 = 0$

39. $x^2 + 2 = 0$

40. $-4x^2 - 1 = 2$

Exercises 41–50: Solve by factoring.

41. $x^2 + 2x - 35 = 0$

42. $2x^2 - 7x + 3 = 0$

43. $6x^2 - x - 1 = 0$

44. $x^2 + 4x + 6 = -3x$

45. $4x^2 + 13x + 9 = x$

46. $9x^2 + 4 = 12x$

47. $25x^2 - 350 = 125x$

48. $20x^2 + 150 = 130x$

49. $2(5x^2 + 9) = 27x$

50. $15(3x^2 + x) = 10$

Exercises 51–62: Use the square root property to solve.

51. $x^2 = 144$

52. $4x^2 - 5 = 0$

53. $5x^2 - 64 = 0$

54. $3x^2 = 7$

55. $(x + 1)^2 = 25$

56. $(x + 4)^2 = 9$

57. $(x - 1)^2 = 64$

58. $(x - 3)^2 = 0$

59. $(2x - 1)^2 = 5$

60. $(5x + 3)^2 = 7$

61. $10(x - 5)^2 = 50$

62. $7(3x + 1)^2 = 14$

COMPLETING THE SQUARE

Exercises 63–66: To solve by completing the square, what value should you add to each side of the equation?

63. $x^2 + 4x = -3$

64. $x^2 - 6x = 4$

65. $x^2 - 5x = 4$

66. $x^2 + 3x = 1$

Exercises 67–70: (Refer to Example 4.) Find the term that should be added to the expression to form a perfect square trinomial. Write the resulting perfect square trinomial in factored form.

67. $x^2 - 8x$

68. $x^2 - 5x$

69. $x^2 + 9x$

70. $x^2 + x$

Exercises 71–86: Solve by completing the square.

71. $x^2 - 2x = 24$

72. $x^2 - 2x + \frac{1}{2} = 0$

73. $x^2 + 6x - 2 = 0$

74. $x^2 - 16x = 5$

75. $x^2 - 3x = 5$

76. $x^2 + 5x = 2$

77. $x^2 - 5x + 1 = 0$

78. $x^2 - 9x + 7 = 0$

79. $x^2 - 4 = 2x$

80. $x^2 + 1 = 7x$

81. $2x^2 - 3x = 4$

82. $3x^2 + 6x - 5 = 0$

83. $4x^2 - 8x - 7 = 0$

84. $25x^2 - 20x - 1 = 0$

85. $36x^2 + 18x + 1 = 0$ **86.** $12x^2 + 8x - 2 = 0$

Exercises 87–96: Solve by any method.

87. $3x^2 + 12x = 36$ **88.** $6x^2 + 9x = 27$

89. $x^2 + 4x = -2$ **90.** $x^2 + 6x + 3 = 0$

91. $3x^2 - 4 = 2$ **92.** $-2x^2 + 3 = 1$

93. $-6x^2 + 70 = 16x$

94. $-15x^2 + 25x + 10 = 0$

95. $-3x(x - 8) = 6$ **96.** $-2x(4 - x) = 8$

Exercises 97 and 98: **Thinking Generally** *Solve for x. Assume a and c are positive.*

97. $ax^2 - c = 0$ **98.** $ax^2 + bx = 0$

SOLVING EQUATIONS BY MORE THAN ONE METHOD

Exercises 99–104: Solve the quadratic equation

(a) *symbolically,*
(b) *graphically, and*
(c) *numerically.*

99. $x^2 - 3x - 18 = 0$ **100.** $\frac{1}{2}x^2 + 2x - 6 = 0$

101. $x^2 - 8x + 15 = 0$ **102.** $2x^2 + 3 = 7x$

103. $4(x^2 + 35) = 48x$ **104.** $4x(2 - x) = -5$

SOLVING AN EQUATION FOR A VARIABLE

Exercises 105–112: Solve for the specified variable.

105. $x = y^2 - 1$ for y

106. $x = 9y^2$ for y **107.** $K = \frac{1}{2}mv^2$ for v

108. $c^2 = a^2 + b^2$ for b

109. $E = \dfrac{k}{r^2}$ for r **110.** $W = I^2R$ for I

111. $LC = \dfrac{1}{(2\pi f)^2}$ for f

112. $F = \dfrac{KmM}{r^2}$ for r

APPLICATIONS

113. *Safe Curve Speed* (Refer to Example 8.) Find a safe speed limit x for an airport taxiway curve with the given radius R by using $R = \frac{1}{2}x^2$.
(a) $R = 450$ feet   (b) $R = 800$ feet

114. *Braking Distance* The braking distance y in feet that it takes for a car to stop on wet, level pavement can be estimated by $y = \frac{1}{9}x^2$, where x is the speed of the car in miles per hour. Find the speed associated with each braking distance. (*Source:* L. Haefner, *Introduction to Transportation Systems.*)
(a) 25 feet   (b) 361 feet   (c) 784 feet

115. *Falling Object* (Refer to Example 3.) How long does it take for a toy to hit the ground if it is dropped out of a window 60 feet above the ground? Does it take twice as long as it takes to fall from a window 30 feet above the ground?

116. *Falling Object* If a metal ball is thrown *downward* with an initial velocity of 22 feet per second (15 mph) from a 100-foot water tower, its height h in feet above the ground after t seconds is modeled by

$$h(t) = -16t^2 - 22t + 100.$$

(a) Determine symbolically when the height of the ball is 62 feet.
(b) Support your result in part (a) either graphically or numerically.
(c) If the ball is thrown *upward* at 22 feet per second, then its height is given by

$$h(t) = -16t^2 + 22t + 100.$$

Determine when the height of the ball is 80 feet.

Exercises 117 and 118: *Television Size* (*Refer to the introduction of this chapter.*) *The size S of the television screen recommended for a person who sits x feet from the screen ($6 \leq x \leq 15$) is given by*

$$S(x) = -0.227x^2 + 8.155x - 8.8.$$

If a person buys a television set with a size S screen, how far from the screen should the person sit?

117. $S = 42$ inches   **118.** $S = 50$ inches

119. *Distance* Two athletes start jogging at the same time. One jogs north at 6 miles per hour while the second jogs east at 8 miles per hour. After how long are the two athletes 20 miles apart?

120. *Geometry* A triangle has an area of 35 square inches, and its base is 3 inches more than its height. Find the base and height of the triangle.

121. *Construction* A rectangular plot of land has an area of 520 square feet and is 6 feet longer than it is wide.
(a) Write a quadratic equation in the form $ax^2 + bx + c = 0$, whose solution gives the width of the rectangular plot of land.
(b) Solve the equation.

122. *Modeling Motion* The height y in feet of a tennis ball after x seconds is shown in the graph. Estimate when the ball was 25 feet above the ground.

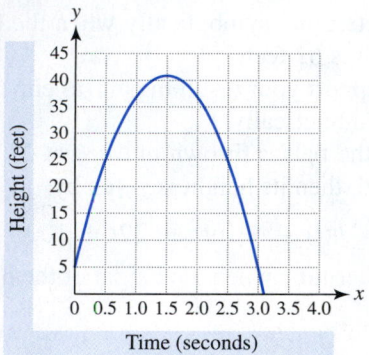

Time (seconds)

123. *Seedling Growth* (Refer to Exercise 100, Section 8.1.) The heights of melon seedlings grown at different temperatures are shown in the following graph. At what temperatures were the heights of the seedlings about 22 centimeters? (*Source:* R. Pearl, "The growth of *Cucumis melo* seedlings at different temperatures.")

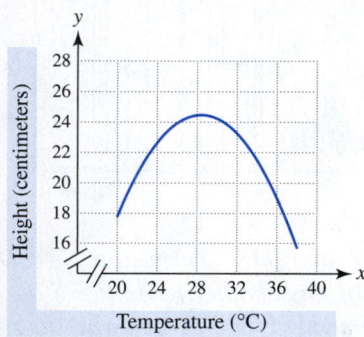

Temperature (°C)

124. *U.S. Population* The three tables show the population of the United States in millions from 1800 through 2010 for selected years.

Year	1800	1820	1840	1860
Population	5	10	17	31

Year	1880	1900	1920	1940
Population	50	76	106	132

Year	1960	1980	2000	2010
Population	179	226	269	308

Source: U.S. Census Bureau.

(a) Without plotting the data, how do you know that the data are nonlinear?
(b) These data are modeled (approximately) by

$$f(x) = 0.0066(x - 1800)^2 + 5.$$

Find the vertex of the graph of f and interpret it.
(c) Estimate when the U.S. population reached 85 million.

125. *MySpace Visitors* (Refer to A Look Into Math at the beginning of this section.) The number of unique monthly visitors V in millions to MySpace x months after January 2010 can be modeled by

$$V(x) = -\frac{25}{144}x^2 + 70.$$

Determine the month when this number of visitors was 56 million.

126. *Federal Debt* The federal debt D in trillions of dollars held by foreign and international investors x years after 1970 can be modeled by

$$D(x) = \frac{1}{320}x^2 - \frac{3}{80}x,$$

where $15 \leq x \leq 40$.
(a) Evaluate $D(40)$ and interpret the result.
 (b) Determine the year when the federal debt held by foreign and international investors first reached $500 billion ($0.5 trillion).

WRITING ABOUT MATHEMATICS

127. Suppose that you are asked to solve

$$ax^2 + bx + c = 0.$$

Explain how the graph of $y = ax^2 + bx + c$ can be used to find any real solutions to the equation.

128. Explain why a quadratic equation could not have more than two solutions. (*Hint:* Consider the graph of $y = ax^2 + bx + c$.)

Group Activity   Working With Real Data

Directions: Form a group of 2 to 4 people. Select someone to record the group's responses for this activity. All members of the group should work cooperatively to answer the questions. If your instructor asks for the results, each member of the group should be prepared to respond.

Personal Consumption   Although there was an economic downturn in 2008, personal consumption has grown overall. Table 8.15 lists (approximate) personal consumption C in dollars for selected years x.

TABLE 8.15 Personal Consumption (dollars)

x	1959	1982	1990	1998	2009
C	$300	$2000	$4000	$6000	$10,000

Source: BEA.

(a) Make a scatterplot of these data. Use the window [1955, 2010, 10] by [0, 12000, 2000].

(b) Find a quadratic function given by

$$C(x) = a(x - 1959)^2 + k$$

that models the data. Graph C and the data. (Answers may vary.)

(c) Estimate when personal consumption was $8000. Compare your answer to the actual value of 2005.

(d) Use your function C to predict consumption in 2020 to the nearest thousand dollars.

8.4   The Quadratic Formula

Solving Quadratic Equations • The Discriminant • Quadratic Equations Having Complex Solutions

A LOOK INTO MATH ▶

To model the stopping distance of a car, highway engineers compute two quantities. The first quantity is the *reaction distance*, which is the distance a car travels from the time a driver first recognizes a hazard until the brakes are applied. The second quantity is *braking distance*, which is the distance a car travels after a driver applies the brakes. *Stopping distance* equals the sum of the reaction distance and the braking distance. If a car is traveling x miles per hour, highway engineers estimate the reaction distance in feet as $\frac{11}{3}x$ and the braking distance in feet as $\frac{1}{9}x^2$. See Figure 8.49. (*Source:* L. Haefner, *Introduction to Transportation Systems.*)

NEW VOCABULARY

□ Quadratic formula
□ Discriminant

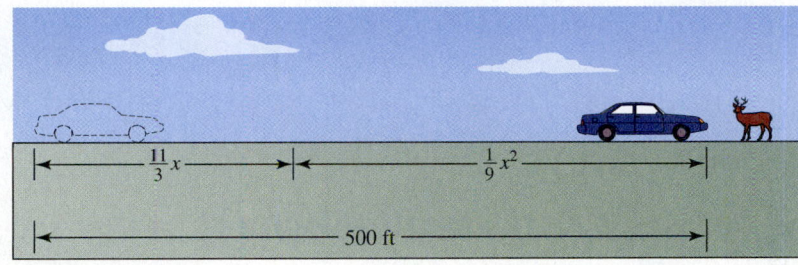

Figure 8.49 Stopping Distance

To estimate the total stopping distance d in feet, add the two expressions to obtain

$$d(x) = \frac{1}{9}x^2 + \frac{11}{3}x.$$

If a car's headlights don't illuminate the road beyond **500** feet, a safe nighttime speed limit x for the car can be determined by solving the quadratic equation

$$\frac{1}{9}x^2 \quad + \quad \frac{11}{3}x \quad = \quad 500. \qquad \text{Quadratic equation}$$

Braking Distance + Reaction Distance = Stopping Distance

In this section we learn how to solve this equation with the quadratic formula. (See Example 4 and Exercises 109–112.)

Solving Quadratic Equations

Thus far, we have solved quadratic equations by factoring, the square root property, and completing the square. In this subsection, we derive a formula that can be used to solve *any* quadratic equation. To do this, we solve the general quadratic equation $ax^2 + bx + c = 0$ for x by completing the square. The resulting formula is called the **quadratic formula**. We assume that $a > 0$ and derive this formula as follows.

$ax^2 + bx + c = 0$	Quadratic equation
$ax^2 + bx = -c$	Subtract c.
$x^2 + \dfrac{b}{a}x = -\dfrac{c}{a}$	Divide by a.
$x^2 + \dfrac{b}{a}x + \dfrac{b^2}{4a^2} = -\dfrac{c}{a} + \dfrac{b^2}{4a^2}$	Add $\left(\dfrac{b/a}{2}\right)^2 = \dfrac{b^2}{4a^2}$.
$\left(x + \dfrac{b}{2a}\right)^2 = -\dfrac{c}{a} + \dfrac{b^2}{4a^2}$	Perfect square trinomial
$\left(x + \dfrac{b}{2a}\right)^2 = -\dfrac{c \cdot 4a}{a \cdot 4a} + \dfrac{b^2}{4a^2}$	Multiply $-\dfrac{c}{a}$ by $\dfrac{4a}{4a}$.
$\left(x + \dfrac{b}{2a}\right)^2 = -\dfrac{4ac}{4a^2} + \dfrac{b^2}{4a^2}$	Simplify.
$\left(x + \dfrac{b}{2a}\right)^2 = \dfrac{-4ac + b^2}{4a^2}$	Add fractions.
$\left(x + \dfrac{b}{2a}\right)^2 = \dfrac{b^2 - 4ac}{4a^2}$	Rewrite.
$x + \dfrac{b}{2a} = \pm\sqrt{\dfrac{b^2 - 4ac}{4a^2}}$	Square root property
$x = -\dfrac{b}{2a} \pm \sqrt{\dfrac{b^2 - 4ac}{4a^2}}$	Add $-\dfrac{b}{2a}$.
$x = -\dfrac{b}{2a} \pm \dfrac{\sqrt{b^2 - 4ac}}{2a}$	Property of square roots
$x = \dfrac{-b \pm \sqrt{b^2 - 4ac}}{2a}$	Combine fractions.

STUDY TIP

Memorize the quadratic formula.

QUADRATIC FORMULA

The solutions to $ax^2 + bx + c = 0$ with $a \neq 0$ are given by

$$x = \frac{-b \pm \sqrt{b^2 - 4ac}}{2a}.$$

NOTE: The quadratic formula can be used to solve *any* quadratic equation. It always "works."

EXAMPLE 1   **Solving a quadratic equation having two solutions**

Solve the equation $2x^2 - 3x - 1 = 0$. Support your results graphically.

Solution

Symbolic Solution Let $a = 2$, $b = -3$, and $c = -1$ in $2x^2 - 3x - 1 = 0$.

$$x = \frac{-b \pm \sqrt{b^2 - 4ac}}{2a} \qquad \text{Quadratic formula}$$

$$x = \frac{-(-3) \pm \sqrt{(-3)^2 - 4(2)(-1)}}{2(2)} \qquad \text{Substitute for } a, b, \text{ and } c.$$

$$x = \frac{3 \pm \sqrt{17}}{4} \qquad \text{Simplify.}$$

The solutions are $\frac{3 + \sqrt{17}}{4} \approx 1.78$ and $\frac{3 - \sqrt{17}}{4} \approx -0.28$.

Graphical Solution The graph of $y = 2x^2 - 3x - 1$ is shown in Figure 8.50. Note that the two x-intercepts correspond to the two solutions for $2x^2 - 3x - 1 = 0$. Estimating from this graph, we see that the solutions are approximately -0.25 and 1.75, which supports our symbolic solution. (You could also use a graphing calculator to find the x-intercepts.)

Two Solutions

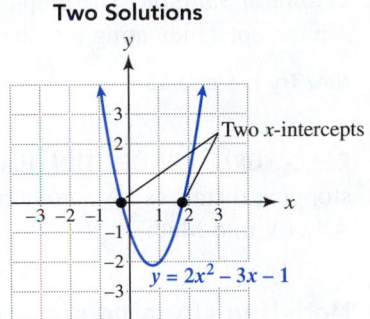

Figure 8.50

Now Try Exercise 9

CRITICAL THINKING

Use the equation and results from Example 1 to evaluate each expression mentally.

$$2\left(\frac{3 + \sqrt{17}}{4}\right)^2 - 3\left(\frac{3 + \sqrt{17}}{4}\right) - 1 \quad \text{and} \quad 2\left(\frac{3 - \sqrt{17}}{4}\right)^2 - 3\left(\frac{3 - \sqrt{17}}{4}\right) - 1$$

EXAMPLE 2 **Solving a quadratic equation having one solution**

Solve the equation $25x^2 + 20x + 4 = 0$. Support your result graphically.

Solution

Symbolic Solution Let $a = 25$, $b = 20$, and $c = 4$ in $25x^2 + 20x + 4 = 0$.

$$x = \frac{-b \pm \sqrt{b^2 - 4ac}}{2a} \qquad \text{Quadratic formula}$$

$$= \frac{-20 \pm \sqrt{20^2 - 4(25)(4)}}{2(25)} \qquad \text{Substitute for } a, b, \text{ and } c.$$

$$= \frac{-20 \pm \sqrt{0}}{50} \qquad \text{Simplify.}$$

$$= \frac{-20}{50} = -0.4 \qquad \sqrt{0} = 0$$

There is one solution, -0.4.

One Solution

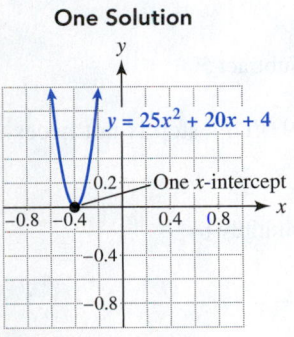

Figure 8.51

Graphical Solution The graph of $y = 25x^2 + 20x + 4$ is shown in Figure 8.51. Note that the one x-intercept, -0.4, corresponds to the solution to $25x^2 + 20x + 4 = 0$.

Now Try Exercise 11

EXAMPLE 3 **Recognizing a quadratic equation having no real solutions**

Solve the equation $5x^2 - x + 3 = 0$. Support your result graphically.

Solution

Symbolic Solution Let $a = \mathbf{5}, b = \mathbf{-1}$, and $c = \mathbf{3}$ in $\mathbf{5}x^2 - \mathbf{1}x + \mathbf{3} = 0$.

$$x = \frac{-b \pm \sqrt{b^2 - 4ac}}{2a} \qquad \text{Quadratic formula}$$

$$= \frac{-(\mathbf{-1}) \pm \sqrt{(\mathbf{-1})^2 - 4(\mathbf{5})(\mathbf{3})}}{2(\mathbf{5})} \qquad \text{Substitute for } a, b, \text{ and } c.$$

$$= \frac{1 \pm \sqrt{-59}}{10} \qquad \text{Simplify.}$$

There are *no real solutions* to this equation because $\sqrt{-59}$ *is not a real number*. (Later in this section we discuss how to find complex solutions to quadratic equations like this one.)

Graphical Solution The graph of $y = 5x^2 - x + 3$ is shown in Figure 8.52. There are no x-intercepts, indicating that the equation $5x^2 - x + 3 = 0$ has no real solutions.

No Real Solutions

[Graph showing parabola $y = 5x^2 - x + 3$ with No x-intercepts]

Figure 8.52

▌ **Now Try Exercise 13**

▶ **REAL-WORLD CONNECTION** Earlier in this section we discussed how engineers estimate safe stopping distances for automobiles. In the next example we solve the equation presented in A Look Into Math.

EXAMPLE 4 **Modeling stopping distance**

If a car's headlights do not illuminate the road beyond 500 feet, estimate a safe nighttime speed limit x for the car by solving $\frac{1}{9}x^2 + \frac{11}{3}x = 500$.

Solution

Begin by subtracting 500 from each side of the given equation.

$$\frac{1}{9}x^2 + \frac{11}{3}x - 500 = 0 \qquad \text{Subtract 500.}$$

To eliminate fractions, multiply each side by the LCD, which is 9. (This step is not necessary, but it makes the problem easier to work.)

$$x^2 + 33x - 4500 = 0 \qquad \text{Multiply by 9.}$$

STUDY TIP

The quadratic formula can be used to solve *any* quadratic equation.

Now let $a = \mathbf{1}, b = \mathbf{33}$, and $c = \mathbf{-4500}$ in the quadratic formula.

$$x = \frac{-b \pm \sqrt{b^2 - 4ac}}{2a} \qquad \text{Quadratic formula}$$

$$= \frac{-33 \pm \sqrt{33^2 - 4(1)(-4500)}}{2(1)} \qquad \text{Substitute for } a, b, \text{ and } c.$$

$$= \frac{-33 \pm \sqrt{19{,}089}}{2} \qquad \text{Simplify.}$$

The solutions are

$$\frac{-33 + \sqrt{19{,}089}}{2} \approx 52.6 \quad \text{and} \quad \frac{-33 - \sqrt{19{,}089}}{2} \approx -85.6.$$

The negative solution has no physical meaning because negative speeds are not possible. The other solution is 52.6, so an appropriate speed limit might be 50 miles per hour.

▌ **Now Try Exercise 109**

The Discriminant

The expression $b^2 - 4ac$ in the quadratic formula is called the **discriminant**. It provides information about the number of solutions to a quadratic equation.

READING CHECK

How can you use the discriminant to determine the number of solutions to a quadratic equation?

THE DISCRIMINANT AND QUADRATIC EQUATIONS

To determine the number of solutions to the quadratic equation $ax^2 + bx + c = 0$, evaluate the discriminant $b^2 - 4ac$.

1. If $b^2 - 4ac > 0$, there are two real solutions.
2. If $b^2 - 4ac = 0$, there is one real solution.
3. If $b^2 - 4ac < 0$, there are no real solutions; there are two complex solutions.

GRAPHS AND THE DISCRIMINANT The graph of $y = ax^2 + bx + c$ can be used to determine the sign of the discriminant, $b^2 - 4ac$. Figure 8.53 illustrates several possibilities. For example, in Figure 8.53(a) neither graph with $a > 0$ or $a < 0$ intersects the x-axis. Therefore these graphs indicate that the equation $ax^2 + bx + c = 0$ has no real solutions and that the discriminant must be negative: $b^2 - 4ac < 0$. Figures 8.53(b) and (c) can be done similarly.

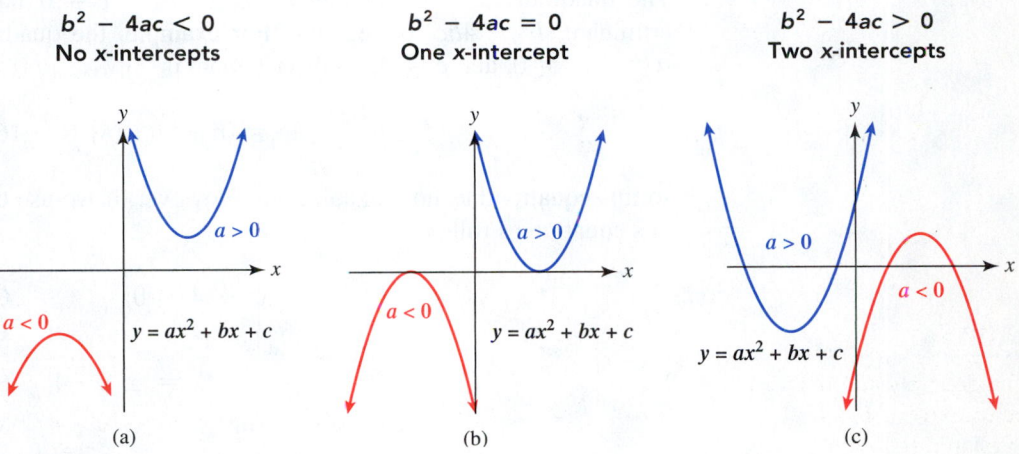

Figure 8.53

EXAMPLE 5

Analyzing graphs of quadratic functions

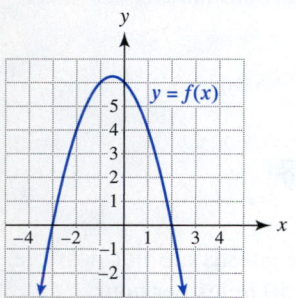

Figure 8.54

A graph of $f(x) = ax^2 + bx + c$ is shown in Figure 8.54.
(a) State whether $a > 0$ or $a < 0$.
(b) Solve the equation $ax^2 + bx + c = 0$.
(c) Determine whether the discriminant is positive, negative, or zero.

Solution
(a) The parabola opens downward, so $a < 0$.
(b) The graph of $f(x) = ax^2 + bx + c$ intersects the x-axis at -3 and 2. Therefore $f(-3) = 0$ and $f(2) = 0$. The solutions to $ax^2 + bx + c = 0$ are -3 and 2.
(c) There are two solutions, so the discriminant is positive.

▌ **Now Try Exercise 33**

EXAMPLE 6

Using the discriminant

Use the discriminant to determine the number of solutions to $4x^2 + 25 = 20x$. Then solve the equation, using the quadratic formula.

Solution
Write the equation as $4x^2 - 20x + 25 = 0$ so that $a = 4, b = -20$, and $c = 25$. The discriminant evaluates to

$$b^2 - 4ac = (-20)^2 - 4(4)(25) = 0.$$

Thus there is **one real solution**.

$$
\begin{aligned}
x &= \frac{-b \pm \sqrt{b^2 - 4ac}}{2a} \qquad &&\text{Quadratic formula} \\
&= \frac{-(-20) \pm \sqrt{0}}{2(4)} \qquad &&\text{Substitute.} \\
&= \frac{20}{8} = 2.5 \qquad &&\text{Simplify.}
\end{aligned}
$$

The only solution is 2.5.

▌ **Now Try Exercise 39(a), (b)**

Quadratic Equations Having Complex Solutions

The quadratic equation written as $ax^2 + bx + c = 0$ has no real solutions if the discriminant, $b^2 - 4ac$, is negative. For example, the quadratic equation $x^2 + 4 = 0$ has $a = 1, b = 0$, and $c = 4$. Its discriminant is

$$b^2 - 4ac = 0^2 - 4(1)(4) = -16 < 0,$$

so this equation has no real solutions. However, if we use complex numbers, we can solve this equation as follows.

$$
\begin{aligned}
x^2 + 4 &= 0 \qquad &&\text{Given equation} \\
x^2 &= -4 \qquad &&\text{Subtract 4.} \\
x &= \pm\sqrt{-4} \qquad &&\text{Square root property} \\
x = \sqrt{-4} \quad \text{or} \quad x &= -\sqrt{-4} \qquad &&\text{Meaning of } \pm \\
x = 2i \quad \text{or} \quad x &= -2i \qquad &&\text{The expression } \sqrt{-a}
\end{aligned}
$$

No x-intercepts

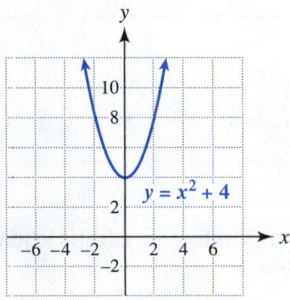

Figure 8.55

The solutions are $\pm 2i$. We check each solution to $x^2 + 4 = 0$ as follows.

$$(2i)^2 + 4 = (2)^2 i^2 + 4 = 4(-1) + 4 = 0 \checkmark \quad \text{It checks.}$$
$$(-2i)^2 + 4 = (-2)^2 i^2 + 4 = 4(-1) + 4 = 0 \checkmark \quad \text{It checks.}$$

The fact that the equation $x^2 + 4 = 0$ has only imaginary solutions is apparent from the graph of $y = x^2 + 4$, shown in Figure 8.55. This parabola does not intersect the x-axis, so the equation $x^2 + 4 = 0$ has no real solutions.

These results can be generalized as follows.

THE EQUATION $x^2 + k = 0$

If $k > 0$, the solutions to $x^2 + k = 0$ are given by $x = \pm i\sqrt{k}$.

NOTE: This result is a form of the *square root property* that includes complex solutions.

EXAMPLE 7 Solving a quadratic equation having complex solutions

Solve $x^2 + 5 = 0$.

Solution
The solutions are $\pm i\sqrt{5}$. That is, $x = i\sqrt{5}$ or $x = -i\sqrt{5}$.

■ **Now Try Exercise 57**

When $b \neq 0$, the preceding method cannot be used. Consider the quadratic equation $2x^2 + x + 3 = 0$, which has $a = 2$, $b = 1$, and $c = 3$. Its discriminant is negative.

$$b^2 - 4ac = 1^2 - 4(2)(3) = -23 < 0$$

This equation has two complex solutions, as demonstrated in the next example.

EXAMPLE 8 Solving a quadratic equation having complex solutions

Solve $2x^2 + x + 3 = 0$. Write your answer in standard form: $a + bi$.

Solution
Let $a = 2$, $b = 1$, and $c = 3$.

$$x = \frac{-b \pm \sqrt{b^2 - 4ac}}{2a} \qquad \text{Quadratic formula}$$

$$= \frac{-1 \pm \sqrt{1^2 - 4(2)(3)}}{2(2)} \qquad \text{Substitute for } a, b, \text{ and } c.$$

$$= \frac{-1 \pm \sqrt{-23}}{4} \qquad \text{Simplify.}$$

$$= \frac{-1 \pm i\sqrt{23}}{4} \qquad \sqrt{-23} = i\sqrt{23}$$

$$= -\frac{1}{4} \pm i\frac{\sqrt{23}}{4} \qquad \text{Property of fractions}$$

CALCULATOR HELP

To set your calculator in $a + bi$ mode or to access the imaginary unit i, see Appendix A (page AP-7).

The solutions are $-\frac{1}{4} + i\frac{\sqrt{23}}{4}$ and $-\frac{1}{4} - i\frac{\sqrt{23}}{4}$.

■ **Now Try Exercise 73**

CRITICAL THINKING

Use the results of Example 8 to evaluate each expression mentally.

$$2\left(-\tfrac{1}{4} + i\tfrac{\sqrt{23}}{4}\right)^2 + \left(-\tfrac{1}{4} + i\tfrac{\sqrt{23}}{4}\right) + 3 \quad \text{and} \quad 2\left(-\tfrac{1}{4} - i\tfrac{\sqrt{23}}{4}\right)^2 + \left(-\tfrac{1}{4} - i\tfrac{\sqrt{23}}{4}\right) + 3$$

Sometimes we can use properties of radicals to simplify a solution to a quadratic equation, as demonstrated in the next example.

EXAMPLE 9 **Solving a quadratic equation having complex solutions**

Solve $\tfrac{3}{4}x^2 + 1 = x$. Write your answer in standard form: $a + bi$.

Solution
Begin by subtracting x from each side of the equation and then multiply by 4 to clear fractions. The resulting equation is $3x^2 - 4x + 4 = 0$. Substitute $a = 3$, $b = -4$, and $c = 4$ in the quadratic formula.

$$x = \frac{-b \pm \sqrt{b^2 - 4ac}}{2a} \qquad \text{Quadratic formula}$$

$$= \frac{-(-4) \pm \sqrt{(-4)^2 - 4(3)(4)}}{2(3)} \qquad \text{Substitute.}$$

$$= \frac{4 \pm \sqrt{-32}}{6} \qquad \text{Simplify.}$$

$$= \frac{4 \pm 4i\sqrt{2}}{6} \qquad \sqrt{-32} = i\sqrt{32} = i\sqrt{16}\sqrt{2} = 4i\sqrt{2}$$

$$= \frac{2}{3} \pm \frac{2}{3}i\sqrt{2} \qquad \text{Property of fractions; simplify.}$$

▌ **Now Try Exercise 79**

In the next example, we use completing the square to obtain complex solutions.

EXAMPLE 10 **Completing the square to find complex solutions**

Solve $x(x + 2) = -2$ by completing the square.

Solution
After applying the distributive property, we obtain the equation $x^2 + 2x = -2$. Because $b = 2$, add $\left(\tfrac{b}{2}\right)^2 = \left(\tfrac{2}{2}\right)^2 = 1$ to each side of the equation.

$$x^2 + 2x = -2 \qquad \text{Equation to be solved}$$

$$x^2 + 2x + 1 = -2 + 1 \qquad \text{Add 1 to each side.}$$

$$(x + 1)^2 = -1 \qquad \text{Perfect square trinomial; add.}$$

$$x + 1 = \pm\sqrt{-1} \qquad \text{Square root property}$$

$$x + 1 = \pm i \qquad \sqrt{-1} = i, \text{ the imaginary unit}$$

$$x = -1 \pm i \qquad \text{Add } -1 \text{ to each side.}$$

The solutions are $-1 + i$ and $-1 - i$.

▌ **Now Try Exercise 91**

8.4 Putting It All Together

CONCEPT	EXPLANATION	EXAMPLES
Quadratic Formula	The quadratic formula can be used to solve *any* quadratic equation written as $ax^2 + bx + c = 0$. The solutions are given by $$x = \frac{-b \pm \sqrt{b^2 - 4ac}}{2a}.$$	For the equation $$2x^2 - 3x + 1 = 0$$ with $a = 2$, $b = -3$, and $c = 1$, the solutions are $$\frac{-(-3) \pm \sqrt{(-3)^2 - 4(2)(1)}}{2(2)} = \frac{3 \pm \sqrt{1}}{4} = 1, \frac{1}{2}.$$
The Discriminant	The expression $b^2 - 4ac$ is called the discriminant. 1. $b^2 - 4ac > 0$ indicates two real solutions. 2. $b^2 - 4ac = 0$ indicates one real solution. 3. $b^2 - 4ac < 0$ indicates no real solutions; rather, there are two complex solutions.	For the equation $$x^2 + 4x - 1 = 0$$ with $a = 1$, $b = 4$, and $c = -1$, the discriminant is $$b^2 - 4ac = 4^2 - 4(1)(-1) = 20 > 0,$$ indicating two real solutions.
Quadratic Formula and Complex Solutions	If the discriminant is negative $(b^2 - 4ac < 0)$, the two solutions are complex numbers that are not real numbers. If $k > 0$, the solutions to $x^2 + k = 0$ are given by $x = \pm i\sqrt{k}$.	Solve $2x^2 - x + 3 = 0$. $$x = \frac{-(-1) \pm \sqrt{(-1)^2 - 4(2)(3)}}{2(2)}$$ $$= \frac{1 \pm \sqrt{-23}}{4} = \frac{1}{4} \pm i\frac{\sqrt{23}}{4}$$ $x^2 + 7 = 0$ is equivalent to $x = \pm i\sqrt{7}$.

Three Ways to Solve $x^2 - x - 2 = 0$

Symbolic Solution

Solve $x^2 - x - 2 = 0$.

$$x = \frac{-(-1) \pm \sqrt{(-1)^2 - 4(1)(-2)}}{2(1)}$$

$$x = \frac{1 \pm 3}{2}$$

$$x = 2, -1$$

Solutions are -1 and 2.

Numerical Solution

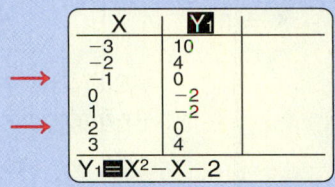

$x^2 - x - 2 = 0$ when $x = -1$ or 2.

Graphical Solution

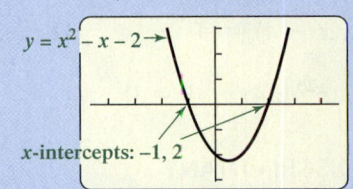

The x-intercepts are -1 and 2.

8.4 Exercises

MyMathLab   Math XL PRACTICE   WATCH   DOWNLOAD   READ   REVIEW

CONCEPTS AND VOCABULARY

1. What is the quadratic formula used for?

2. What basic algebraic technique is used to derive the quadratic formula?

3. Write the discriminant.

4. If the discriminant evaluates to 0, what does that indicate about the quadratic equation?

5. Name four symbolic techniques for solving a quadratic equation.

6. Does every quadratic equation have at least one real solution? Explain.

7. Solve $x^2 - k = 0$, if $k > 0$.

8. Solve $x^2 + k = 0$, if $k > 0$.

THE QUADRATIC FORMULA

Exercises 9–14: Use the quadratic formula to solve the equation. Support your result graphically. If there are no real solutions, say so.

9. $2x^2 + 11x - 6 = 0$ **10.** $x^2 + 2x - 24 = 0$

11. $-x^2 + 2x - 1 = 0$ **12.** $3x^2 - x + 1 = 0$

13. $-2x^2 + x - 1 = 0$ **14.** $-x^2 + 4x - 4 = 0$

Exercises 15–32: Solve by using the quadratic formula. If there are no real solutions, say so.

15. $x^2 - 6x - 16 = 0$ **16.** $2x^2 - 9x + 7 = 0$

17. $4x^2 - x - 1 = 0$ **18.** $-x^2 + 2x + 1 = 0$

19. $-3x^2 + 2x - 1 = 0$ **20.** $x^2 + x + 3 = 0$

21. $36x^2 - 36x + 9 = 0$ **22.** $4x^2 - 5.6x + 1.96 = 0$

23. $2x(x - 3) = 2$ **24.** $x(x + 1) + x = 5$

25. $(x - 1)(x + 1) + 2 = 4x$

26. $\frac{1}{2}(x - 6) = x^2 + 1$ **27.** $\frac{1}{2}x(x + 1) = 2x^2 - \frac{3}{2}$

28. $\frac{1}{2}x^2 - \frac{1}{4}x + \frac{1}{2} = x$ **29.** $2x(x - 1) = 7$

30. $3x(x - 4) = 4$ **31.** $-3x^2 + 10x - 5 = 0$

32. $-2x^2 + 4x - 1 = 0$

THE DISCRIMINANT

Exercises 33–38: A graph of $y = ax^2 + bx + c$ is shown.
  (a) State whether $a > 0$ or $a < 0$.
  (b) Solve $ax^2 + bx + c = 0$, if possible.
  (c) Determine whether the discriminant is positive, negative, or zero.

33.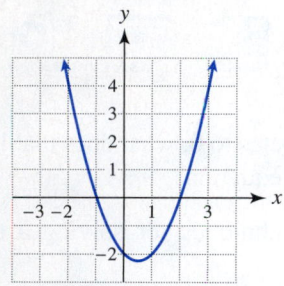

34.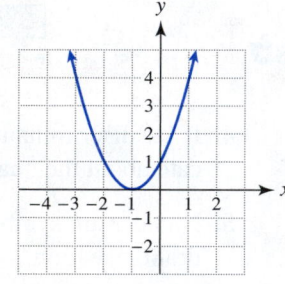

35.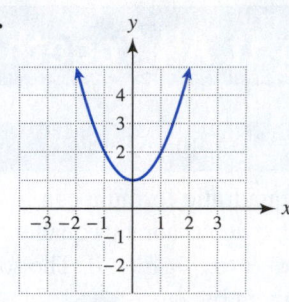

36.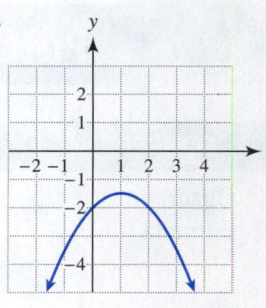

37.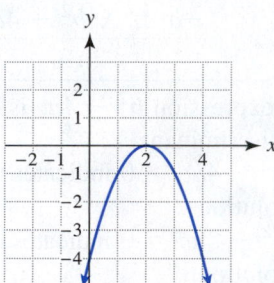

38.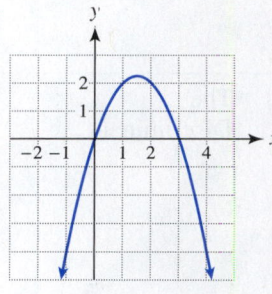

Exercises 39–46: Do the following for the given equation.
  (a) Evaluate the discriminant.
  (b) How many real solutions are there?
  (c) Support your answer for part (b) graphically.

39. $3x^2 + x - 2 = 0$ **40.** $5x^2 - 13x + 6 = 0$

41. $x^2 - 4x + 4 = 0$ **42.** $\frac{1}{4}x^2 + 4 = 2x$

43. $\frac{1}{2}x^2 + \frac{3}{2}x + 2 = 0$ **44.** $x - 3 = 2x^2$

45. $x(x + 3) = 3$

46. $(4x - 1)(x - 3) = -25$

Exercises 47–56: Use the quadratic formula to find any x-intercepts on the graph of the equation.

47. $y = x^2 - 2x - 1$ **48.** $y = x^2 + 3x + 1$

49. $y = -2x^2 - x + 3$ **50.** $y = -3x^2 - x + 4$

51. $y = x^2 + x + 5$ **52.** $y = 3x^2 - 2x + 5$

53. $y = x^2 + 9$ **54.** $y = x^2 + 11$

55. $y = 3x^2 + 4x - 2$ **56.** $y = 4x^2 - 2x - 3$

COMPLEX SOLUTIONS

Exercises 57–88: Solve the equation. Write complex solutions in standard form.

57. $x^2 + 9 = 0$ **58.** $x^2 + 16 = 0$

59. $x^2 + 80 = 0$ **60.** $x^2 + 20 = 0$

61. $x^2 + \frac{1}{4} = 0$ **62.** $x^2 + \frac{9}{4} = 0$

63. $16x^2 + 9 = 0$ **64.** $25x^2 + 36 = 0$

65. $x^2 = -6$ **66.** $x^2 = -75$

67. $x^2 - 3 = 0$ **68.** $x^2 - 8 = 0$

69. $x^2 + 2 = 0$ **70.** $x^2 + 4 = 0$

71. $x^2 - x + 2 = 0$ **72.** $x^2 + 2x + 3 = 0$

73. $2x^2 + 3x + 4 = 0$ **74.** $3x^2 - x = 1$

75. $x^2 + 1 = 4x$ **76.** $3x^2 + 2 = x$

77. $x(x + 1) = -2$ **78.** $x(x - 4) = -8$

79. $5x^2 + 2x + 4 = 0$ **80.** $7x^2 - 2x + 4 = 0$

81. $\frac{1}{2}x^2 + \frac{3}{4}x = -1$ **82.** $-\frac{1}{3}x^2 + x = 2$

83. $x(x + 2) = x - 4$ **84.** $x - 5 = 2x(2x + 1)$

85. $x(2x - 1) = 1 + x$ **86.** $2x = x(3 - 4x)$

87. $x^2 = x(1 - x) - 2$ **88.** $2x^2 = 2x(5 - x) - 8$

COMPLETING THE SQUARE

Exercises 89–94: Solve by completing the square.

89. $x^2 + 2x + 4 = 0$ **90.** $x^2 - 2x + 2 = 0$

91. $x(x + 4) = -5$ **92.** $x(8 - x) = 25$

93. $2x^2 - 4x + 6 = 0$ **94.** $2x^2 + 2x + 1 = 0$

YOU DECIDE THE METHOD

Exercises 95–108: Find exact solutions to the quadratic equation, using a method of your choice. Explain why you chose the method you did. Answers may vary.

95. $x^2 - 3x + 2 = 0$ **96.** $x^2 + 2x + 1 = 0$

97. $0.5x^2 - 1.75x = 1$ **98.** $\frac{3}{5}x^2 + \frac{9}{10}x = \frac{3}{5}$

99. $x^2 - 5x + 2 = 0$ **100.** $2x^2 - x - 4 = 0$

101. $2x^2 + x = -8$ **102.** $4x^2 = 2x - 3$

103. $4x^2 - 1 = 0$ **104.** $3x^2 = 9$

105. $3x^2 + 6 = 0$ **106.** $4x^2 + 7 = 0$

107. $9x^2 + 1 = 6x$ **108.** $10x^2 + 15x = 25$

APPLICATIONS

Exercises 109–112: **Modeling Stopping Distance** *(Refer to Example 4.) Use* $d = \frac{1}{9}x^2 + \frac{11}{3}x$ *to find a safe speed x for the following stopping distances d.*

109. 42 feet

110. 152 feet

111. 390 feet

112. 726 feet

113. *Global Music Sales* (Refer to Example 9, Section 8.3.) From 1989 to 2009, global music sales S of compact discs and tapes in billions of dollars can be modeled by $S(x) = -0.095x^2 + 1.85x + 6$, where x is years after 1989. Estimate symbolically when sales were $12 billion. (*Source: RIAA, Bain Analysis.*)

114. *Monthly Facebook Visitors* The number of *unique* monthly Facebook visitors in millions x years after 2006 can be modeled by

$$V(x) = 10.75x^2 - 24x + 35.$$

Assuming current trends continue, estimate symbolically when this number might reach 530 million.

115. *Groupon's Growth* Groupon negotiates coupons for items discounted 50–90% off by having thousands of subscribers. As a result, it experienced a dramatic increase in value from October 2010 to March 2011. The function

$$G(x) = 0.4x^2 + 1.8x + 6$$

approximates the company's value in billions of dollars x months after October 2010.
(a) Evaluate $G(0)$. Interpret the result.
(b) Determine when Groupon's value was about $15 billion.

116. *Foursquare Users* Foursquare provides a service that allows your friends to know your whereabouts by "checking in." From March 2010 to March 2011, it experienced amazing growth. The function

$$F(x) = \frac{1}{18}x^2 - \frac{1}{12}x + \frac{1}{2}$$

approximates Foursquare users in millions x months after March 2010.
(a) Evaluate $F(3)$. Interpret the result.
(b) Determine symbolically when Foursquare had 4.25 million users.

117. *U.S. AIDS Deaths* The cumulative numbers in thousands of AIDS deaths from 1984 through 1994 may be modeled by

$$f(x) = 2.39x^2 + 5.04x + 5.1,$$

where $x = 0$ corresponds to 1984, $x = 1$ corresponds to 1985, and so on until $x = 10$ corresponds to 1994. See the accompanying graph. Use the formula for $f(x)$ to estimate the year when the total number of AIDS deaths reached 200 thousand. Compare your result with that shown in the graph.

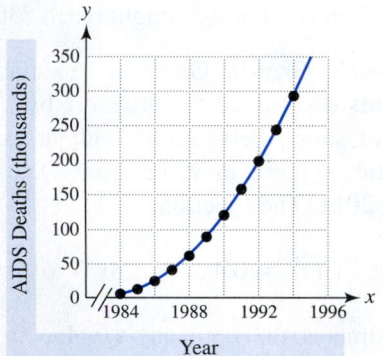

118. *Canoeing* A camper paddles a canoe 2 miles downstream in a river that has a 2-mile-per-hour current. To return to camp, the canoeist travels upstream on a different branch of the river. It is 4 miles long and has a 1-mile-per-hour current. The total trip (both ways) takes 3 hours. Find the average speed of the canoe in still water. (*Hint:* Time equals distance divided by rate.)

119. *Airplane Speed* A pilot flies 500 miles against a 20-mile-per-hour wind. On the next day, the pilot flies back home with a 10-mile-per-hour tail wind. The total trip (both ways) takes 4 hours. Find the speed of the airplane without a wind.

120. *Distance* Two cars leave an intersection, one traveling south and one traveling east, as shown in the figure at the top of the next column. After 1 hour the two cars are 50 miles apart and the car traveling east has traveled 10 miles farther than the car traveling south. How far did each car travel?

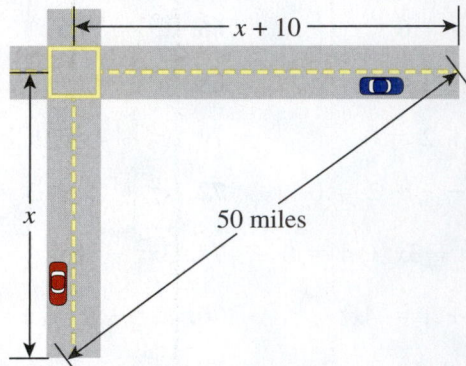

121. *Screen Dimensions* The width of a rectangular computer screen is 3 inches more than its height. If the area of the screen is 154 square inches, find its dimensions
(a) graphically,
(b) numerically, and
(c) symbolically.

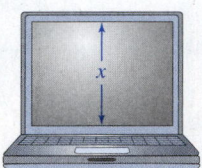

122. *Sidewalk Dimension* A rectangular flower garden in a park is 30 feet wide and 40 feet long. A sidewalk around the perimeter of the garden is being planned, as shown in the figure. The gardener has enough money to pour 624 square feet of cement sidewalk. Find the width of the sidewalk.

123. *Modeling Water Flow* When water runs out of a hole in a cylindrical container, the height of the water in the container can often be modeled by a quadratic function. The data in the table show the height y in centimeters of water at 30-second intervals in a metal can that has a small hole in it.

Time	0	30	60	90
Height	16	11.9	8.4	5.3

Time	120	150	180
Height	3.1	1.4	0.5

These data are modeled by

$$f(x) = 0.0004x^2 - 0.15x + 16.$$

(a) Explain why a linear function would not be appropriate for modeling these data.

(b) Use the table to estimate the time at which the height was 7 centimeters.

(c) Use the quadratic formula to solve part (b).

124. *Hospitals* The general trend in the number of hospitals in the United States from 1945 through 2000 is shown in the graph and can be modeled by

$$f(x) = -1.38x^2 + 84x + 5865,$$

where $x = 5$ corresponds to 1945, $x = 10$ corresponds to 1950, and so on until $x = 60$ represents 2000.

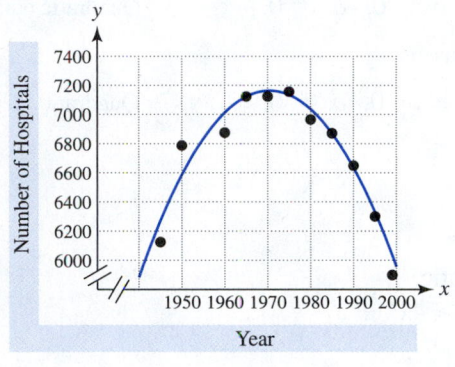

Year

(a) Describe any trends in the numbers of hospitals from 1945 to 2000.

(b) What information does the vertex give?

(c) Use the formula for $f(x)$ to estimate the number of hospitals in 1970. Compare your result with that shown in the graph.

(d) Use the formula for $f(x)$ to estimate the year (or years) when there were 6300 hospitals. Compare your result with that shown in the graph.

WRITING ABOUT MATHEMATICS

125. Explain how the discriminant $b^2 - 4ac$ can be used to determine the number of solutions to a quadratic equation.

126. Let $f(x) = ax^2 + bx + c$ be a quadratic function. If you know the value of $b^2 - 4ac$, what information does this give you about the graph of f? Explain your answer.

SECTIONS 8.3 and 8.4

Checking Basic Concepts

1. Solve the quadratic equation $2x^2 - 7x + 3 = 0$ symbolically and graphically.

2. Use the square root property to solve $x^2 = 5$.

3. Complete the square to solve $x^2 - 4x + 1 = 0$.

4. Solve the equation $x^2 + y^2 = 1$ for y.

5. Use the quadratic formula to solve each equation.
(a) $2x^2 = 3x + 1$
(b) $9x^2 - 24x + 16 = 0$
(c) $x^2 + x + 2 = 0$

6. Calculate the discriminant for each equation and give the number of *real* solutions.
(a) $x^2 - 5x + 5 = 0$
(b) $2x^2 - 5x + 4 = 0$
(c) $49x^2 - 56x + 16 = 0$

7. Solve each equation.
(a) $x^2 + 5 = 0$
(b) $x^2 + x + 3 = 0$

8.5   Quadratic Inequalities

Basic Concepts • Graphical and Numerical Solutions • Symbolic Solutions

A LOOK INTO MATH ▶

Parabolas are frequently used in highway design. Sometimes it is necessary for engineers to determine where a highway is below a particular elevation. To do this, they may need to solve a *quadratic inequality*. (See Example 5.) In this section we discuss how to solve quadratic inequalities graphically, numerically, and symbolically.

Basic Concepts

If the equals sign in a quadratic equation is replaced with $>$, $\geq$, $<$, or $\leq$, a **quadratic inequality** results. Examples of quadratic inequalities include

$$x^2 + 4x - 3 < 0, \quad 5x^2 \geq 5, \quad \text{and} \quad 1 - z \leq z^2. \qquad \text{Quadratic inequalities}$$

Any quadratic equation can be written as

$$ax^2 + bx + c = 0, \ a \neq 0, \qquad \text{Quadratic equation}$$

NEW VOCABULARY

☐ Quadratic inequality
☐ Test value

so any quadratic inequality can be written as

$$ax^2 + bx + c > 0, \ a \neq 0, \qquad \text{Quadratic inequality}$$

where $>$ may be replaced with $\geq$, $<$, or $\leq$.

EXAMPLE 1   **Identifying a quadratic inequality**

Determine whether the inequality is quadratic.
(a) $5x + x^2 - x^3 \leq 0$ **(b)** $4 + 5x^2 > 4x^2 + x$

Solution
(a) The inequality $5x + x^2 - x^3 \leq 0$ is not quadratic because it has an x^3-term.
(b) Write the inequality as follows.

$$\begin{aligned}
4 + 5x^2 &> 4x^2 + x && \text{Given inequality} \\
4 + 5x^2 - 4x^2 - x &> 0 && \text{Subtract } 4x^2 \text{ and } x. \\
4 + x^2 - x &> 0 && \text{Combine like terms.} \\
x^2 - x + 4 &> 0 && \text{Rewrite.}
\end{aligned}$$

Because the inequality can be written in the form $ax^2 + bx + c > 0$ with $a = 1$, $b = -1$, and $c = 4$, it is a quadratic inequality.

Now Try Exercises 7, 11

Graphical and Numerical Solutions

Equality often is the boundary between *greater than* and *less than*, so a first step in solving an inequality is to determine the x-values where equality occurs. We begin by using this concept with graphical techniques.

A graph of $y = x^2 - x - 2$ has x-intercepts **−1** and **2**. See Figure 8.56. The solutions to $x^2 - x - 2 = 0$ are given by $x = -1$ or $x = 2$. Between the x-intercepts the graph dips **below** the x-axis and the y-values are **negative**. Thus solutions to $x^2 - x - 2 < 0$ satisfy $-1 < x < 2$. To support this result we select a **test value**. For example, 0 lies between −1 and 2. If we substitute $x = 0$ in $x^2 - x - 2 < 0$, it results in a true statement.

$$0^2 - 0 - 2 < 0 \ ✓ \qquad \text{True}$$

Visualizing Quadratic Inequalities

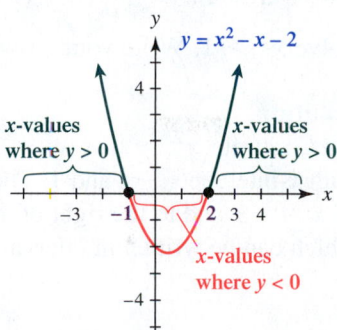

Figure 8.56

When $x < -1$ or $x > 2$, the graph lies **above** the x-axis and the y-values are **positive**. Thus the solutions to $x^2 - x - 2 > 0$ satisfy $x < -1$ or $x > 2$. For example, 3 is greater than 2 and -3 is less than -1. Therefore both 3 and -3 are solutions. We can verify this result by substituting **3** and -3 as test values in $x^2 - x - 2 > 0$.

$$3^2 - 3 - 2 > 0 \quad \checkmark \quad \text{True}$$
$$(-3)^2 - (-3) - 2 > 0 \quad \checkmark \quad \text{True}$$

In the next three examples, we use these concepts to solve quadratic inequalities.

EXAMPLE 2 **Solving a quadratic inequality**

Make a table of values for $y = x^2 - 3x - 4$ and then sketch the graph. Use the table and graph to solve $x^2 - 3x - 4 \leq 0$. Write your answer in interval notation.

Solution
The points calculated for Table 8.16 are plotted in Figure 8.57 and connected with a smooth ∪-shaped graph.

Numerical Solution Table 8.16 shows that $x^2 - 3x - 4$ equals 0 when $x = -1$ or $x = 4$. Between these values, $x^2 - 3x - 4$ is negative, so the solution set to $x^2 - 3x - 4 \leq 0$ is given by $-1 \leq x \leq 4$ or, in interval notation, $[-1, 4]$.

Graphical Solution In Figure 8.57 the graph of $y = x^2 - 3x - 4$ shows that the x-intercepts are -1 and **4**. Between these values, the graph dips **below** the x-axis. Thus the solution set is $[-1, 4]$.

STUDY TIP

Learn how to use a parabola to help solve a quadratic inequality.

TABLE 8.16

x	$y = x^2 - 3x - 4$
-2	6
-1	0
0	-4
1	-6
2	-6
3	-4
4	0
5	6

x-intercepts ← -1 ... 4

Less than or equal to 0

Solving $x^2 - 3x - 4 \leq 0$
x-Intercepts: -1, 4

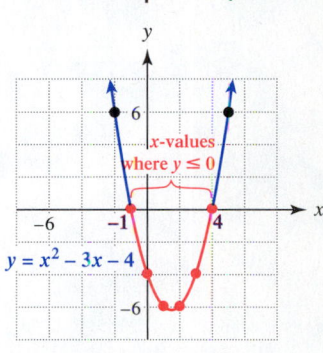

Figure 8.57

Now Try Exercises 19, 27

EXAMPLE 3 **Solving a quadratic inequality**

Solve $x^2 > 1$. Write your answer in interval notation.

Solution

First, rewrite $x^2 > 1$ as $x^2 - 1 > 0$. The graph of $y = x^2 - 1$ is shown in Figure 8.58 with x-intercepts -1 and 1. The graph lies **above** the x-axis and is shaded green to the left of $x = -1$ and to the right of $x = 1$. Thus the solution set is given by $x < -1$ or $x > 1$, which can be written in interval notation as $(-\infty, -1) \cup (1, \infty)$.

Solving $x^2 - 1 > 0$
x-Intercepts: $-1, 1$

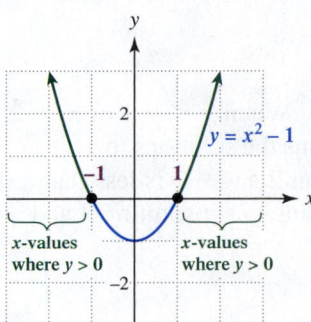

Figure 8.58

■ **Now Try Exercise 49**

EXAMPLE 4 **Solving some special cases**

Solve each of the inequalities graphically.
(a) $x^2 + 1 > 0$ **(b)** $x^2 + 1 < 0$ **(c)** $(x - 1)^2 \leq 0$

Solution
(a) Because the graph of $y = x^2 + 1$, shown in Figure 8.59, is always above the x-axis, $x^2 + 1$ is always greater than 0. The solution set includes all real numbers, or $(-\infty, \infty)$.
(b) Because the graph of $y = x^2 + 1$, shown in Figure 8.59, never goes below the x-axis, $x^2 + 1$ is never less than 0. Thus there are no real solutions.
(c) Because the graph of $y = (x - 1)^2$, shown in Figure 8.60, never goes below the x-axis, $(x - 1)^2$ is never less than 0. When $x = 1$, $y = 0$, so **1** is the only solution to the inequality $(x - 1)^2 \leq 0$.

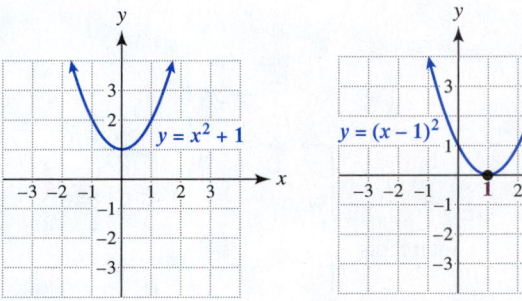

Figure 8.59   **Figure 8.60**

■ **Now Try Exercises 51, 55, 57**

The following chart summarizes visual solutions for several possibilities when solving quadratic equations and inequalities.

Visualizing Solutions to Quadratic Equations and Inequalities

Two x-Intercepts

Case 1: Upward Opening

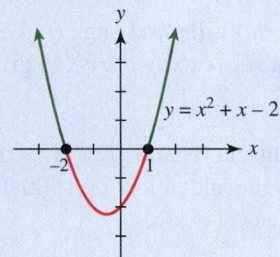

$y = x^2 + x - 2$

Equation/Inequality	Solutions
$x^2 + x - 2 = 0$	$x = -2, 1$
$x^2 + x - 2 < 0$	$-2 < x < 1$
$x^2 + x - 2 > 0$	$x < -2$ or $x > 1$

Case 2: Downward Opening

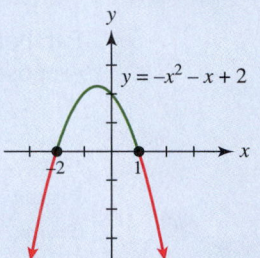

$y = -x^2 - x + 2$

Equation/Inequality	Solutions
$-x^2 - x + 2 = 0$	$x = -2, 1$
$-x^2 - x + 2 < 0$	$x < -2$ or $x > 1$
$-x^2 - x + 2 > 0$	$-2 < x < 1$

One x-Intercept

Case 1: Upward Opening

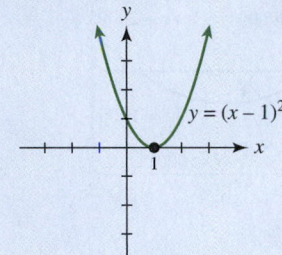

$y = (x - 1)^2$

Equation/Inequality	Solutions
$(x - 1)^2 = 0$	$x = 1$
$(x - 1)^2 < 0$	no solutions
$(x - 1)^2 > 0$	$x < 1$ or $x > 1$

Case 2: Downward Opening

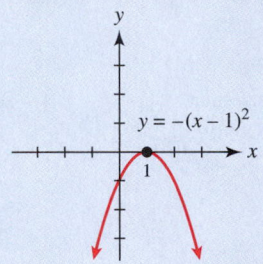

$y = -(x - 1)^2$

Equation/Inequality	Solutions
$-(x - 1)^2 = 0$	$x = 1$
$-(x - 1)^2 < 0$	$x < 1$ or $x > 1$
$-(x - 1)^2 > 0$	no solutions

No x-Intercepts

Case 1: Upward Opening

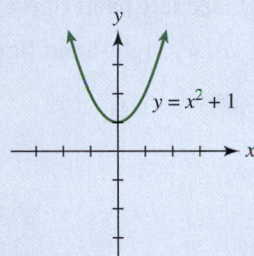

$y = x^2 + 1$

Equation/Inequality	Solutions
$x^2 + 1 = 0$	no solutions
$x^2 + 1 < 0$	no solutions
$x^2 + 1 > 0$	all real numbers

Case 2: Downward Opening

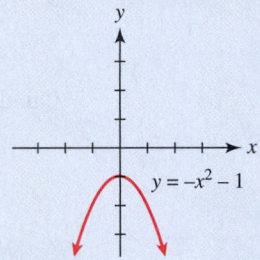

$y = -x^2 - 1$

Equation/Inequality	Solutions
$-x^2 - 1 = 0$	no solutions
$-x^2 - 1 < 0$	all real numbers
$-x^2 - 1 > 0$	no solutions

▶ **REAL-WORLD CONNECTION** In the next example we show how quadratic inequalities are used in highway design, as discussed in A Look Into Math for this section.

| EXAMPLE 5 | **Determining elevations on a sag curve** |

Parabolas are frequently used in highway design to model hills and sags (valleys) along a proposed route. Suppose that the elevation E in feet of a sag, or *sag curve*, is given by

$$E(x) = 0.00004x^2 - 0.4x + 2000,$$

where x is the horizontal distance in feet along the sag curve and $0 \leq x \leq 10{,}000$. See Figure 8.61. Estimate graphically the x-values where the elevation is 1500 feet or less. (***Source:*** F. Mannering and W. Kilareski, *Principles of Highway Engineering and Traffic Analysis.*)

E

Figure 8.61

Solution
Graphical Solution We must solve the quadratic inequality

$$0.00004x^2 - 0.4x + 2000 \leq 1500.$$

Let $y_1 = 0.00004x^2 - 0.4x + 2000$ be the sag curve and $y_2 = 1500$ be a line with an elevation of 1500 feet. Their graphs intersect at $x \approx 1464$ and $x \approx 8536$, as shown in Figure 8.62. The elevation of the road is less than 1500 feet between these x-values. Therefore the elevation is 1500 feet or less when $1464 \leq x \leq 8536$ (approximately).

Determining Where the Elevation is 1500 Feet

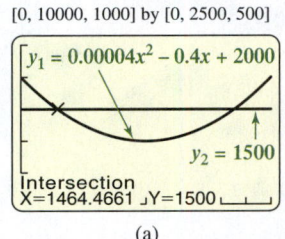
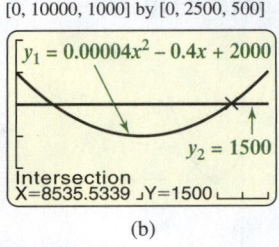

(a)   (b)

Figure 8.62

CALCULATOR HELP

To find a point of intersection, see Appendix A (page AP-6).

▌ **Now Try Exercise 63**

Symbolic Solutions

To solve a quadratic inequality symbolically, we first solve the corresponding equality. We can then write the solution to the inequality, using the following method.

SOLUTIONS TO QUADRATIC INEQUALITIES

Let $ax^2 + bx + c = 0$, $a > 0$, have two real solutions p and q, where $p < q$.

$ax^2 + bx + c < 0$ is equivalent to $p < x < q$ (see left-hand figure).

$ax^2 + bx + c > 0$ is equivalent to $x < p$ or $x > q$ (see right-hand figure).

Quadratic inequalities involving $\leq$ or $\geq$ can be solved similarly.

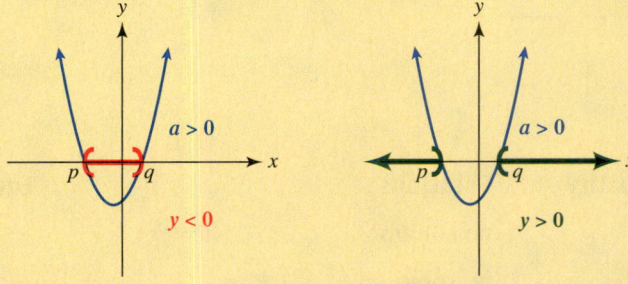

Solutions lie between p and q.   Solutions lie "outside" p and q.

READING CHECK

How can you distinguish a linear inequality from a quadratic inequality?

One way to handle the situation where $a < 0$ is to multiply each side of the inequality by -1, in which case we must be sure to *reverse* the inequality symbol. For example, the inequality $-2x^2 + 8 \leq 0$ has $a = -2$, which is negative. If we multiply each side of this inequality by -1, we obtain $2x^2 - 8 \geq 0$ and now $a = 2$, which is positive.

EXAMPLE 6   **Solving quadratic inequalities**

Solve each inequality symbolically. Write your answer in interval notation.
(a) $6x^2 - 7x - 5 \geq 0$ **(b)** $x(3 - x) > -18$

Solution

(a) Begin by solving $6x^2 - 7x - 5 = 0$.

$$6x^2 - 7x - 5 = 0 \qquad \text{Quadratic equation}$$
$$(2x + 1)(3x - 5) = 0 \qquad \text{Factor.}$$
$$2x + 1 = 0 \quad \text{or} \quad 3x - 5 = 0 \qquad \text{Zero-product property}$$
$$x = -\frac{1}{2} \quad \text{or} \quad x = \frac{5}{3} \qquad \text{Solve.}$$

Therefore solutions to $6x^2 - 7x - 5 \geq 0$ lie "**outside**" these two values and satisfy $x \leq -\frac{1}{2}$ or $x \geq \frac{5}{3}$. In interval notation the solution set is $\left(-\infty, -\frac{1}{2}\right] \cup \left[\frac{5}{3}, \infty\right)$.

(b) First, rewrite the inequality as follows.

$$x(3 - x) > -18 \qquad \text{Given inequality}$$
$$3x - x^2 > -18 \qquad \text{Distributive property}$$
$$3x - x^2 + 18 > 0 \qquad \text{Add 18.}$$
$$-x^2 + 3x + 18 > 0 \qquad \text{Rewrite.}$$
$$x^2 - 3x - 18 < 0 \qquad \text{Multiply by } -1 \text{ because } a < 0;$$
$$\text{reverse the inequality symbol.}$$

Next, solve $x^2 - 3x - 18 = 0$.

$$(x + 3)(x - 6) = 0 \qquad \text{Factor.}$$
$$x = -3 \quad \text{or} \quad x = 6 \qquad \text{Solve.}$$

Solutions to $x^2 - 3x - 18 < 0$ lie **between** these two values and satisfy $-3 < x < 6$. In interval notation the solution set is $(-3, 6)$.

Now Try Exercises 31, 37

CRITICAL THINKING

The graph of $y = -x^2 + x + 12$ is a parabola opening downward with x-intercepts -3 and 4. Solve each inequality.
(a) $-x^2 + x + 12 > 0$
(b) $-x^2 + x + 12 < 0$

EXAMPLE 7   **Finding the dimensions of a building**

A rectangular building needs to be 7 feet longer than it is wide, as illustrated in Figure 8.63. The area of the building must be at least 450 square feet. What widths x are possible for this building? Support your results with a table of values.

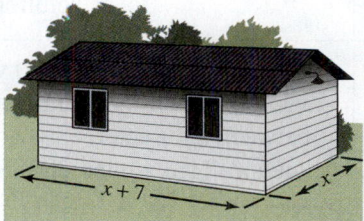

Figure 8.63

Solution

Symbolic Solution If x is the width of the building, $x + 7$ is the length of the building and its area is $x(x + 7)$. The area must be at least 450 square feet, so the inequality $x(x + 7) \geq 450$ must be satisfied.

First solve the following quadratic equation.

$$x(x + 7) = 450 \qquad \text{Quadratic equation}$$

$$x^2 + 7x = 450 \qquad \text{Distributive property}$$

$$x^2 + 7x - 450 = 0 \qquad \text{Subtract 450.}$$

$$x = \frac{-7 \pm \sqrt{7^2 - 4(1)(-450)}}{2(1)} \qquad \begin{array}{l}\text{Quadratic formula: } a = 1, \\ b = 7, \text{ and } c = -450\end{array}$$

$$= \frac{-7 \pm \sqrt{1849}}{2} \qquad \text{Simplify.}$$

$$= \frac{-7 \pm 43}{2} \qquad \sqrt{1849} = 43$$

$$= 18, -25 \qquad \text{Evaluate.}$$

Thus the solutions to $x(x + 7) \geq 450$ are $x \leq -25$ or $x \geq 18$. The width is positive, so the building width must be 18 feet or more.

Numerical Solution A table of values is shown in Figure 8.64, where $y_1 = x(x + 7)$ equals 450 when $x = 18$. For $x \geq 18$ the area is *at least* 450 square feet.

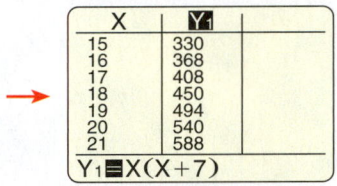

X	Y1
15	330
16	368
17	408
18	450
19	494
20	540
21	588

Y1■X(X+7)

Figure 8.64

Now Try Exercise 67

8.5 Putting It All Together

METHOD	EXPLANATION
Solving a Quadratic Inequality Symbolically	Let $ax^2 + bx + c = 0$, $a > 0$, have two real solutions p and q, where $p < q$. $ax^2 + bx + c < 0$ is equivalent to $p < x < q$. $ax^2 + bx + c > 0$ is equivalent to $x < p$ or $x > q$. *Examples:* The solutions to $x^2 - 3x + 2 = 0$ are 1 and 2. The solutions to $x^2 - 3x + 2 < 0$ satisfy $1 < x < 2$. The solutions to $x^2 - 3x + 2 > 0$ satisfy $x < 1$ or $x > 2$.
Solving a Quadratic Inequality Graphically	Given $ax^2 + bx + c < 0$ with $a > 0$, graph $y = ax^2 + bx + c$ and locate any x-intercepts. If there are two x-intercepts, then solutions correspond to x-values between the x-intercepts. Solutions to $ax^2 + bx + c > 0$ correspond to x-values "outside" the x-intercepts. See the box on page 631.
Solving a Quadratic Inequality Numerically	If a quadratic inequality is expressed as $ax^2 + bx + c < 0$ with $a > 0$, then we can solve $y = ax^2 + bx + c = 0$ with a table. If there are two solutions, then the solutions to the given inequality lie between these values. Solutions to $ax^2 + bx + c > 0$ lie before or after these values in the table.

8.5 Exercises

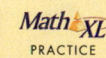 

CONCEPTS AND VOCABULARY

1. How is a quadratic inequality different from a quadratic equation?

2. Do quadratic inequalities typically have two solutions? Explain.

3. Is 3 a solution to $x^2 < 7$?

4. Is 5 a solution to $x^2 \geq 25$?

5. The solutions to $x^2 - 2x - 8 = 0$ are -2 and 4. What are the solutions to $x^2 - 2x - 8 < 0$? Write your answer as an inequality.

6. The solutions to $x^2 + 2x - 3 = 0$ are -3 and 1. What are the solutions to $x^2 + 2x - 3 > 0$? Write your answer as an inequality.

QUADRATIC INEQUALITIES

Exercises 7–12: Is the inequality quadratic?

7. $x^2 + 4x + 5 < 0$ **8.** $x > x^3 - 5$

9. $x^2 > 19$ **10.** $x(x - 1) - 2 \geq 0$

11. $4x > 1 - x$ **12.** $2x(x^2 + 3) < 0$

Exercises 13–18: Is the given value of x a solution?

13. $2x^2 + x - 1 > 0$ $x = 3$

14. $x^2 - 3x + 2 \leq 0$ $x = 2$

15. $x^2 + 2 \leq 0$ $x = 0$

16. $2x(x - 3) \geq 0$ $x = 1$

17. $x^2 - 3x \leq 1$ $x = -3$

18. $4x^2 - 5x + 1 > 30$ $x = -2$

GRAPHICAL SOLUTIONS

Exercises 19–24: The graph of $y = ax^2 + bx + c$ is given. Solve each equation or inequality.
    (a) $ax^2 + bx + c = 0$
    (b) $ax^2 + bx + c < 0$
    (c) $ax^2 + bx + c > 0$

19.   **20.**

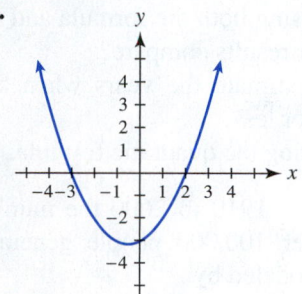 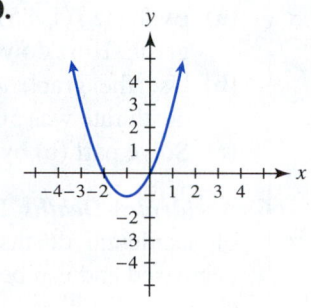

21.   **22.**

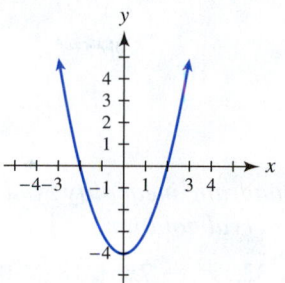 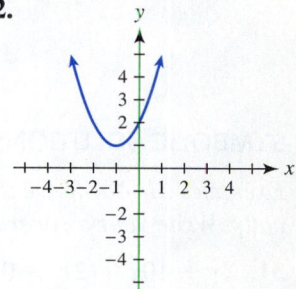

23.   **24.**

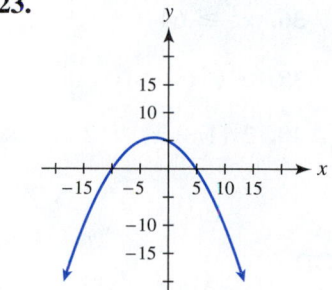 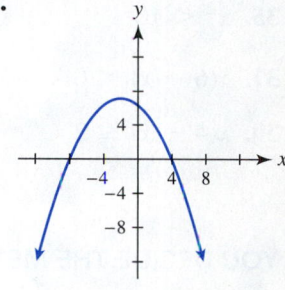

Exercises 25 and 26: Solve the inequality graphically to the nearest thousandth.

25. $\pi x^2 - \sqrt{3}x \leq \frac{3}{11}$ **26.** $\sqrt{5}x^2 - \pi^2 x \geq 10.3$

NUMERICAL SOLUTIONS

Exercises 27–30: A table for $y = ax^2 + bx + c$ is given. Solve each equation or inequality.
    (a) $ax^2 + bx + c = 0$
    (b) $ax^2 + bx + c < 0$
    (c) $ax^2 + bx + c > 0$

27. $y = x^2 - 4$

x	-3	-2	-1	0	1	2	3
y	5	0	-3	-4	-3	0	5

28. $y = x^2 - x - 2$

x	-3	-2	-1	0	1	2	3
y	10	4	0	-2	-2	0	4

29. $y = x^2 + 4x$

x	-5	-4	-3	-2	-1	0	1
y	5	0	-3	-4	-3	0	5

30. $y = -2x^2 - 2x + 1.5$

x	-2	-1.5	-1	-0.5	0	0.5	1
y	-2.5	0	1.5	2	1.5	0	-2.5

SYMBOLIC SOLUTIONS

Exercises 31–40: Solve the quadratic inequality symbolically. Write your answer in interval notation.

31. $x^2 + 10x + 21 \leq 0$ **32.** $x^2 - 7x - 18 < 0$

33. $3x^2 - 9x + 6 > 0$ **34.** $7x^2 + 34x - 5 \geq 0$

35. $x^2 < 10$ **36.** $x^2 \geq 64$

37. $x(6 - x) < 0$ **38.** $1 - x^2 \leq 0$

39. $x(4 - x) \leq 2$ **40.** $2x(1 - x) \geq 2$

YOU DECIDE THE METHOD

Exercises 41–44: Solve the equation in part (a). Use the results to solve the inequalities in parts (b) and (c).

41. (a) $x^2 - 4 = 0$ **42.** (a) $x^2 - 5 = 0$
   (b) $x^2 - 4 < 0$ (b) $x^2 - 5 \leq 0$
   (c) $x^2 - 4 > 0$ (c) $x^2 - 5 \geq 0$

43. (a) $x^2 + x - 1 = 0$
   (b) $x^2 + x - 1 < 0$
   (c) $x^2 + x - 1 > 0$

44. (a) $x^2 + 4x - 5 = 0$
   (b) $x^2 + 4x - 5 \leq 0$
   (c) $x^2 + 4x - 5 \geq 0$

Exercises 45–62: Solve the inequality by any method. Write your answer in interval notation when appropriate.

45. $x^2 + 4x + 3 < 0$ **46.** $x^2 + x - 2 \leq 0$

47. $2x^2 - x - 15 \geq 0$ **48.** $3x^2 - 3x - 6 > 0$

49. $2x^2 \leq 8$ **50.** $x^2 < 9$

51. $x^2 > -5$ **52.** $-x^2 \geq 1$

53. $-x^2 + 3x > 0$ **54.** $-8x^2 - 2x + 1 \leq 0$

55. $x^2 + 2 \leq 0$ **56.** $x^2 + 3 \geq -5$

57. $(x - 2)^2 \leq 0$ **58.** $(x + 2)^2 \leq 0$

59. $(x + 1)^2 > 0$ **60.** $(x - 3)^2 > 0$

61. $x(1 - x) \geq -2$ **62.** $x(x - 2) < 3$

APPLICATIONS

63. *Highway Design* (Refer to Example 5.) The elevation E of a sag curve, in feet, is given by
$$E(x) = 0.0000375x^2 - 0.175x + 1000,$$
where $0 \leq x \leq 4000$.
   (a) Estimate graphically the x-values for which the elevation is 850 feet or less. (*Hint:* Use the window [0, 4000, 1000] by [500, 1200, 100].)
   (b) For which x-values is the elevation 850 feet or more?

64. *Early Cellular Phone Use* The number of cellular subscribers in the United States in thousands from 1985 to 1991 can be modeled by
$$f(x) = 163x^2 - 146x + 205,$$
where x is the year and $x = 0$ corresponds to 1985, $x = 1$ to 1986, and so on. (*Source:* M. Paetsch, *Mobile Communication in the U.S. and Europe.*)
   (a) Write a quadratic inequality whose solution set represents the years when there were 2 million subscribers or more.
   (b) Solve this inequality.

65. *Heart Disease Death Rates* From 1960 to 2010, age-adjusted heart disease rates decreased dramatically. The number of deaths per 100,000 people can be modeled by
$$f(x) = -0.05107x^2 + 194.74x - 184,949,$$
where x is the year, as illustrated in the accompanying figure. (*Source:* Department of Health and Human Services.)

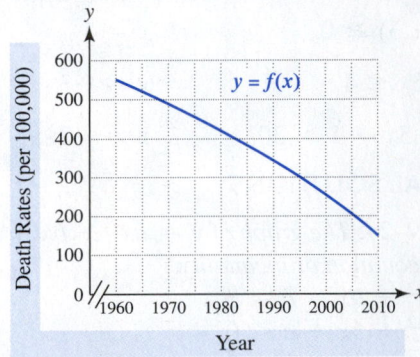

   (a) Evaluate $f(1985)$, using both the formula and the graph. How do your results compare?
   (b) Use the graph to estimate the years when this death rate was 500 or less.
   (c) Solve part (b) by using the quadratic formula.

66. *Accidental Deaths* From 1910 to 2000 the number of accidental deaths per 100,000 people generally decreased and can be modeled by
$$f(x) = -0.001918x^2 + 6.93x - 6156,$$

where x is the year, as shown in the accompanying figure. Note that after 2000 these rates increased and are not modeled by $f(x)$. (*Source:* Department of Health and Human Services.)

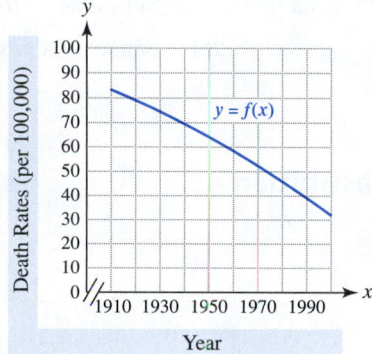

(a) Evaluate $f(1955)$, using both the formula and the graph. How do your results compare?
(b) Use the graph to estimate when this death rate was 60 or more.
(c) Solve part (b) by using the quadratic formula.

67. *Dimensions of a Pen* A rectangular pen for a pet is 5 feet longer than it is wide. Give possible values for the width w of the pen if its area must be between 176 and 500 square feet, inclusively.

68. *Dimensions of a Cylinder* The volume of a cylindrical can is given by $V = \pi r^2 h$, where r is its radius and h is its height. See the accompanying figure. If $h = 6$ inches and the volume of the can must be 50 cubic inches or more, estimate to the nearest tenth of an inch possible values for r.

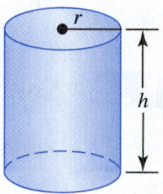

WRITING ABOUT MATHEMATICS

69. Consider the inequality $x^2 < 0$. Discuss the solutions to this inequality and explain your reasoning.

70. Explain how the graph of $y = ax^2 + bx + c$ can be used to solve the inequality

$$ax^2 + bx + c > 0$$

when $a < 0$. Assume that the x-intercepts of the graph are p and q with $p < q$.

8.6 Equations in Quadratic Form

Higher Degree Polynomial Equations • **Equations Having Rational Exponents** • **Equations Having Complex Solutions**

A LOOK INTO MATH ▶ Although many equations are *not* quadratic equations, they can sometimes be written in quadratic form. These equations are *reducible to quadratic form*. To express such an equation in quadratic form we often use substitution. In this section we discuss this process.

Higher Degree Polynomial Equations

Sometimes a fourth-degree polynomial can be factored like a quadratic trinomial, provided it does not have an x-term or an x^3-term. Let's consider the equation $x^4 - 5x^2 + 4 = 0$.

$$x^4 - 5x^2 + 4 = 0 \qquad \text{Given equation}$$
$$(x^2)^2 - 5(x^2) + 4 = 0 \qquad \text{Properties of exponents}$$

We use the substitution $u = x^2$.

$$u^2 - 5u + 4 = 0 \qquad \text{Let } u = x^2.$$
$$(u - 4)(u - 1) = 0 \qquad \text{Factor.}$$
$$u - 4 = 0 \quad \text{or} \quad u - 1 = 0 \qquad \text{Zero-product property}$$
$$u = 4 \quad \text{or} \quad u = 1 \qquad \text{Solve each equation.}$$

Because the given equation $x^4 - 5x^2 + 4 = 0$ uses the variable x, we must give the solutions in terms of x. We substitute x^2 for u in $u = 4$ and $u = 1$ and then solve to obtain the following four solutions.

$u = 4$	or	$u = 1$	Solutions in terms of u
$x^2 = 4$	or	$x^2 = 1$	Substitute x^2 for u.
$x = \pm 2$	or	$x = \pm 1$	Square root property

The solutions are $-2, -1, 1,$ and 2.

EXAMPLE 1 **Solving a sixth-degree equation by substitution**

Solve $2x^6 + x^3 = 1$.

Solution
Start by subtracting 1 from each side.

$2x^6 + x^3 - 1 = 0$	Subtract 1.
$2(x^3)^2 + (x^3) - 1 = 0$	Properties of exponents
$2u^2 + u - 1 = 0$	Let $u = x^3$.
$(2u - 1)(u + 1) = 0$	Factor.
$2u - 1 = 0$ or $u + 1 = 0$	Zero-product property
$u = \dfrac{1}{2}$ or $u = -1$	Solve.

Now substitute x^3 for u, and solve for x to obtain the following two solutions.

$x^3 = \dfrac{1}{2}$ or $x^3 = -1$	Substitute x^3 for u.
$x = \sqrt[3]{\dfrac{1}{2}}$ or $x = -1$	Take cube root of each side.

▌ **Now Try Exercise 3**

Equations Having Rational Exponents

Equations that have negative exponents are sometimes reducible to quadratic form. Consider the following example, in which two methods are presented.

EXAMPLE 2 **Solving an equation having negative exponents**

Solve $-6m^{-2} + 13m^{-1} + 5 = 0$.

Solution
Method I Use the substitution $u = m^{-1} = \frac{1}{m}$ and $u^2 = m^{-2} = \frac{1}{m^2}$.

$-6m^{-2} + 13m^{-1} + 5 = 0$	Given equation
$-6u^2 + 13u + 5 = 0$	Let $u = m^{-1}$ and $u^2 = m^{-2}$.
$6u^2 - 13u - 5 = 0$	Multiply by -1.
$(2u - 5)(3u + 1) = 0$	Factor.
$2u - 5 = 0$ or $3u + 1 = 0$	Zero-product property
$u = \dfrac{5}{2}$ or $u = -\dfrac{1}{3}$	Solve for u.

Because $u = \frac{1}{m}$, $m = \frac{1}{u}$. Thus the solutions are given by $m = \frac{2}{5}$ or $m = -3$.

Method II Another way to solve this equation is to multiply each side by the LCD, m^2.

$$-6m^{-2} + 13m^{-1} + 5 = 0 \qquad \text{Given equation}$$
$$m^2(-6m^{-2} + 13m^{-1} + 5) = m^2 \cdot 0 \qquad \text{Multiply by } m^2.$$
$$-6m^2m^{-2} + 13m^2m^{-1} + 5m^2 = 0 \qquad \text{Distributive property}$$
$$-6 + 13m + 5m^2 = 0 \qquad \text{Add exponents.}$$
$$5m^2 + 13m - 6 = 0 \qquad \text{Rewrite the equation.}$$
$$(5m - 2)(m + 3) = 0 \qquad \text{Factor.}$$
$$5m - 2 = 0 \quad \text{or} \quad m + 3 = 0 \qquad \text{Zero-product property}$$
$$m = \frac{2}{5} \quad \text{or} \quad m = -3 \qquad \text{Solve.}$$

Now Try Exercise 5

In the next example we solve an equation having fractional exponents.

EXAMPLE 3 **Solving an equation having fractional exponents**

Solve $x^{2/3} - 2x^{1/3} - 8 = 0$.

Solution
Use the substitution $u = x^{1/3}$.

$$x^{2/3} - 2x^{1/3} - 8 = 0 \qquad \text{Given equation}$$
$$(x^{1/3})^2 - 2(x^{1/3}) - 8 = 0 \qquad \text{Properties of exponents}$$
$$u^2 - 2u - 8 = 0 \qquad \text{Let } u = x^{1/3}.$$
$$(u - 4)(u + 2) = 0 \qquad \text{Factor.}$$
$$u - 4 = 0 \quad \text{or} \quad u + 2 = 0 \qquad \text{Zero-product property}$$
$$u = 4 \quad \text{or} \quad u = -2 \qquad \text{Solve.}$$

Because $u = x^{1/3}, u^3 = (x^{1/3})^3 = x$. Thus $x = 4^3 = 64$ or $x = (-2)^3 = -8$. The solutions are -8 and 64.

Now Try Exercise 13

Equations Having Complex Solutions

Sometimes an equation that is reducible to quadratic form also has complex solutions. This situation is discussed in the next two examples.

EXAMPLE 4 **Solving a fourth-degree equation**

Find all complex solutions to $x^4 - 1 = 0$.

Solution

$$x^4 - 1 = 0 \qquad \text{Given equation}$$
$$(x^2)^2 - 1 = 0 \qquad \text{Properties of exponents}$$
$$u^2 - 1 = 0 \qquad \text{Let } u = x^2.$$
$$(u - 1)(u + 1) = 0 \qquad \text{Factor difference of squares.}$$
$$u - 1 = 0 \quad \text{or} \quad u + 1 = 0 \qquad \text{Zero-product property}$$
$$u = 1 \quad \text{or} \quad u = -1 \qquad \text{Solve for } u.$$

Now substitute x^2 for u, and solve for x.

$$u = 1 \quad \text{or} \quad u = -1 \qquad \text{Solutions in terms of } u$$

$$x^2 = 1 \quad \text{or} \quad x^2 = -1 \qquad \text{Let } x^2 = u.$$

$$x = \pm 1 \quad \text{or} \quad x = \pm i \qquad \text{Square root property}$$

There are four complex solutions: $-1, 1, -i,$ and i.

■ **Now Try Exercise 25**

EXAMPLE 5 **Solving a rational equation**

Find all complex solutions to $\dfrac{1}{x} + \dfrac{1}{x^2} = -1$.

Solution

This equation is a rational equation. However, if we multiply through by the LCD, x^2, we clear fractions and obtain a quadratic equation with complex solutions.

$$\frac{1}{x} + \frac{1}{x^2} = -1 \qquad \text{Given equation}$$

$$\frac{x^2}{x} + \frac{x^2}{x^2} = -1x^2 \qquad \text{Multiply each term by } x^2.$$

$$x + 1 = -x^2 \qquad \text{Simplify.}$$

$$x^2 + x + 1 = 0 \qquad \text{Add } x^2.$$

$$x = \frac{-1 \pm \sqrt{1^2 - 4(1)(1)}}{2(1)} \qquad \text{Quadratic formula}$$

$$x = \frac{-1 \pm i\sqrt{3}}{2} \qquad \sqrt{-3} = i\sqrt{3}$$

$$x = -\frac{1}{2} \pm \frac{i\sqrt{3}}{2} \qquad \frac{a \pm b}{c} = \frac{a}{c} \pm \frac{b}{c}$$

■ **Now Try Exercise 31**

8.6 Putting It All Together

EQUATION	SUBSTITUTION	EXAMPLES
Higher Degree Polynomial	Let $u = x^n$ for some integer n.	To solve $x^4 - 3x^2 - 4 = 0$, let $u = x^2$. This equation becomes $$u^2 - 3u - 4 = 0.$$
Rational Exponents	Pick a substitution that reduces the equation to quadratic form.	To solve $n^{-2} + 6n^{-1} + 9 = 0$, let $u = n^{-1}$. This equation becomes $$u^2 + 6u + 9 = 0.$$ To solve $6x^{2/5} - 5x^{1/5} - 4 = 0$, let $u = x^{1/5}$. This equation becomes $$6u^2 - 5u - 4 = 0.$$

EQUATION	SUBSTITUTION	EXAMPLES
Equations Having Complex Solutions	Both polynomial and rational equations can have complex solutions. Use the fact that if $a > 0$, then $\sqrt{-a} = i\sqrt{a}$.	$$1 + \frac{1}{x^2} = 0$$ $$x^2 \cdot \left(1 + \frac{1}{x^2}\right) = 0 \cdot x^2$$ $$x^2 + 1 = 0$$ $$x^2 = -1$$ $$x = \pm i$$

8.6 Exercises

MyMathLab  PRACTICE   WATCH   DOWNLOAD   READ   REVIEW

EQUATIONS REDUCIBLE TO QUADRATIC FORM

Exercises 1–6: Use the given substitution to solve the equation.

1. $x^4 - 7x^2 + 6 = 0$ $u = x^2$

2. $2k^4 - 7k^2 + 6 = 0$ $u = k^2$

3. $3z^6 + z^3 - 10 = 0$ $u = z^3$

4. $2x^6 + 17x^3 + 8 = 0$ $u = x^3$

5. $4n^{-2} + 17n^{-1} + 15 = 0$ $u = n^{-1}$

6. $m^{-2} + 24 = 10m^{-1}$ $u = m^{-1}$

Exercises 7–24: Solve. Find all real solutions.

7. $x^4 = 8x^2 + 9$ **8.** $3x^4 = 10x^2 + 8$

9. $3x^6 - 5x^3 - 2 = 0$ **10.** $6x^6 + 11x^3 + 4 = 0$

11. $2z^{-2} + 11z^{-1} = 40$ **12.** $z^{-2} - 10z^{-1} + 25 = 0$

13. $x^{2/3} - 2x^{1/3} + 1 = 0$ **14.** $3x^{2/3} + 18x^{1/3} = 48$

15. $x^{2/5} - 33x^{1/5} = -32$ **16.** $x^{2/5} - 80x^{1/5} = 81$

17. $x - 13\sqrt{x} + 36 = 0$ **18.** $x - 17\sqrt{x} + 16 = 0$

19. $z^{1/2} - 2z^{1/4} + 1 = 0$ **20.** $z^{1/2} - 4z^{1/4} + 4 = 0$

21. $(x + 1)^2 - 5(x + 1) - 14 = 0$

22. $2(x - 5)^2 + 5(x - 5) + 3 = 0$

23. $(x^2 - 1)^2 - 4 = 0$

24. $(x^2 - 9)^2 - 8(x^2 - 9) + 16 = 0$

EQUATIONS HAVING COMPLEX SOLUTIONS

Exercises 25–34: Find all complex solutions.

25. $x^4 - 16 = 0$ **26.** $\frac{1}{3}x^4 - 27 = 0$

27. $x^3 + x = 0$ **28.** $4x^3 + x = 0$

29. $x^4 - 2 = x^2$ **30.** $x^4 - 3 = 2x^2$

31. $\frac{1}{x} + \frac{1}{x^2} = -\frac{1}{2}$ **32.** $\frac{2}{x - 1} - \frac{1}{x} = -1$

33. $\frac{2}{x - 2} - \frac{1}{x} = -\frac{1}{2}$ **34.** $\frac{1}{x} - \frac{1}{x^2} = \frac{1}{2}$

WRITING ABOUT MATHEMATICS

35. Explain how to solve $ax^4 - bx^2 + c = 0$. Assume that the left side of the equation factors.

36. Explain what it means for an equation to be reducible to quadratic form.

SECTIONS 8.5 AND 8.6   ## Checking Basic Concepts

1. Solve the inequality $x^2 - x - 6 > 0$. Write your answer in interval notation.

2. Solve the inequality $3x^2 + 5x + 2 \leq 0$. Write your answer in interval notation.

3. Solve $x^6 + 6x^3 - 16 = 0$.

4. Solve $x^{2/3} - 7x^{1/3} - 8 = 0$.

5. Find all complex solutions to $x^4 + 2x^2 + 1 = 0$.

CHAPTER 8 Summary

SECTION 8.1 ■ QUADRATIC FUNCTIONS AND THEIR GRAPHS

Quadratic Function Any quadratic function f can be written as

$$f(x) = ax^2 + bx + c \quad (a \neq 0).$$

Graph of a Quadratic Function Its graph is a parabola that is wider than the graph of $y = x^2$ if $0 < |a| < 1$, and narrower than the graph of $y = x^2$ if $|a| > 1$. The y-intercept is c.

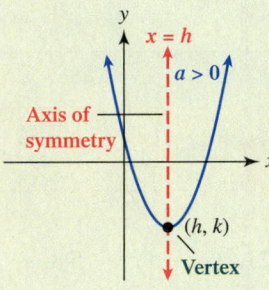 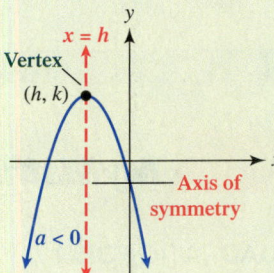

Axis of Symmetry The parabola is symmetric with respect to this vertical line. The axis of symmetry passes through the vertex. Its equation is $x = -\frac{b}{2a}$.

Vertex Formula The x-coordinate of the vertex is $-\frac{b}{2a}$.

Example: Let $y = x^2 - 4x + 1$ with $a = 1$ and $b = -4$.

$$x = -\frac{-4}{2(1)} = 2 \quad \text{and} \quad y = 2^2 - 4(2) + 1 = -3. \text{ The vertex is } (2, -3).$$

SECTION 8.2 ■ PARABOLAS AND MODELING

Vertical and Horizontal Translations Let h and k be positive numbers.

To graph	*shift the graph of $y = x^2$ by k units*
$y = x^2 + k$	upward.
$y = x^2 - k$	downward.

To graph	*shift the graph of $y = x^2$ by h units*
$y = (x - h)^2$	right.
$y = (x + h)^2$	left.

Example: Compared to $y = x^2$, the graph of $y = (x - 1)^2 + 2$ is translated right 1 unit and upward 2 units.

Vertex Form Any quadratic function can be expressed as $f(x) = a(x - h)^2 + k$. In this form the point (h, k) is the vertex. A quadratic function can be put in this form by completing the square or by applying the vertex formula.

Example: $y = x^2 + 10x - 4$ Given equation

$= (x^2 + 10x + 25) - 25 - 4$ $\left(\frac{b}{2}\right)^2 = \left(\frac{10}{2}\right)^2 = 25$; complete the square.

$= (x + 5)^2 - 29$ Perfect square trinomial; add.

The vertex is $(-5, -29)$.

SECTION 8.3 ■ QUADRATIC EQUATIONS

Quadratic Equations Any quadratic equation can be written as $ax^2 + bx + c = 0$ and can have no real solutions, one real solution, or two real solutions. These solutions correspond to the x-intercepts for the graph of $y = ax^2 + bx + c$.

Example:
$$x^2 + x - 2 = 0$$
$$(x + 2)(x - 1) = 0$$
$$x = -2 \quad \text{or} \quad x = 1$$

The solutions are -2 and 1.
See the graph to the right.

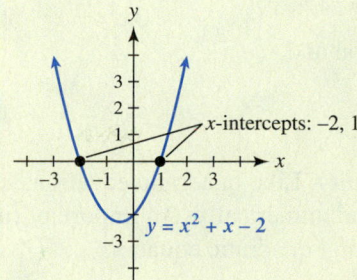

x-intercepts: $-2, 1$

$y = x^2 + x - 2$

Completing the Square Write the equation in the form $x^2 + bx = d$. Complete the square by adding $\left(\frac{b}{2}\right)^2$ to each side of the equation.

Example:
$$x^2 - 8x = 3$$
$$x^2 - 8x + 16 = 3 + 16 \qquad \text{Add } \left(\frac{-8}{2}\right)^2 = 16 \text{ to each side.}$$
$$(x - 4)^2 = 19 \qquad \text{Perfect square trinomial}$$
$$x - 4 = \pm\sqrt{19} \qquad \text{Square root property}$$
$$x = 4 \pm \sqrt{19} \qquad \text{Add 4 to each side.}$$

SECTION 8.4 ■ THE QUADRATIC FORMULA

The Quadratic Formula The solutions to $ax^2 + bx + c = 0$ $(a \neq 0)$ are given by

$$x = \frac{-b \pm \sqrt{b^2 - 4ac}}{2a}.$$

Example: Solve $2x^2 + 3x - 1 = 0$ by letting $a = 2$, $b = 3$, and $c = -1$.

$$x = \frac{-3 \pm \sqrt{3^2 - 4(2)(-1)}}{2(2)} = \frac{-3 \pm \sqrt{17}}{4} \approx 0.28, -1.78$$

The Discriminant The expression $b^2 - 4ac$ is called the discriminant. If $b^2 - 4ac > 0$, there are two real solutions; if $b^2 - 4ac = 0$, there is one real solution; and if $b^2 - 4ac < 0$, there are no real solutions—rather there are two complex solutions.

Example: For $2x^2 + 3x - 1 = 0$, the discriminant is

$$b^2 - 4ac = 3^2 - 4(2)(-1) = 17 > 0.$$

There are two real solutions to this quadratic equation, as shown in the previous example.

Quadratic Equations with Complex Solutions A quadratic equation sometimes has no real solutions.

Example:
$$x^2 + 4 = 0$$
$$x^2 = -4 \qquad \text{Subtract 4 from each side.}$$
$$x = \pm 2i \qquad \text{Square root property; two complex solutions}$$

SECTION 8.5 ■ QUADRATIC INEQUALITIES

Quadratic Inequalities When the equals sign in a quadratic equation is replaced with $<$, $>$, $\leq$, or $\geq$, a quadratic inequality results. For example,

$$3x^2 - x + 1 = 0$$

is a quadratic equation and

$$3x^2 - x + 1 > 0$$

is a quadratic inequality. Like quadratic equations, quadratic inequalities can be solved symbolically, graphically, and numerically. An important first step in solving a quadratic inequality is to solve the corresponding quadratic equation.

Examples: The solutions to $x^2 - 5x - 6 = 0$ are -1 and 6.

The solutions to $x^2 - 5x - 6 < 0$ satisfy $-1 < x < 6$.

The solutions to $x^2 - 5x - 6 > 0$ satisfy $x < -1$ or $x > 6$.

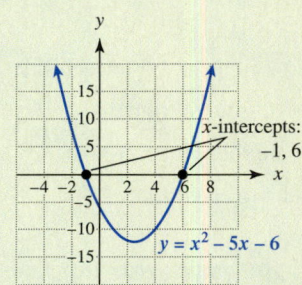

SECTION 8.6 ■ EQUATIONS IN QUADRATIC FORM

Equations Reducible to Quadratic Form An equation that is not quadratic, but can be put into quadratic form by using a substitution, is reducible to quadratic form.

Example: To solve $x^{2/3} - 2x^{1/3} - 15 = 0$, let $u = x^{1/3}$. This equation becomes

$$u^2 - 2u - 15 = 0.$$

Factoring results in $(u + 3)(u - 5) = 0$, so $u = -3$ or $u = 5$.
Because $u = x^{1/3}$, $x = u^3$ and $x = (-3)^3 = -27$ or $x = (5)^3 = 125$.

CHAPTER 8   Review Exercises

SECTION 8.1

Exercises 1 and 2: Identify the vertex, axis of symmetry, and whether the parabola opens upward or downward.

1.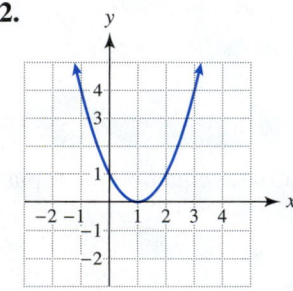

2.

Exercises 3–6: Do the following.
    (a) *Graph f.*
    (b) *Identify the vertex and axis of symmetry.*
    (c) *Evaluate f(x) at the given value of x.*

3. $f(x) = x^2 - 2$, $x = -1$

4. $f(x) = -x^2 + 4x - 3$, $x = 3$

5. $f(x) = -\frac{1}{2}x^2 + x + \frac{3}{2}$, $x = -2$

6. $f(x) = 2x^2 + 8x + 5$, $x = -3$

7. Find the minimum y-value located on the graph of $y = 2x^2 - 6x + 1$.

8. Find the maximum y-value located on the graph of $y = -3x^2 + 2x - 5$.

Exercises 9–12: Find the vertex of the parabola.

9. $f(x) = x^2 - 4x - 2$ **10.** $f(x) = 5 - x^2$

11. $f(x) = -\frac{1}{4}x^2 + x + 1$ **12.** $f(x) = 2 + 2x + x^2$

SECTION 8.2

Exercises 13–18: Do the following.
  (a) Graph f.
  (b) Compare the graph of f with the graph of $y = x^2$.

13. $f(x) = x^2 + 2$ **14.** $f(x) = 3x^2$

15. $f(x) = (x - 2)^2$ **16.** $f(x) = (x + 1)^2 - 3$

17. $f(x) = \frac{1}{2}(x + 1)^2 + 2$

18. $f(x) = 2(x - 1)^2 - 3$

19. Write the vertex form of a parabola with $a = -4$ and vertex $(2, -5)$.

20. Write the vertex form of a parabola that opens downward with vertex $(-4, 6)$. Assume that $a = \pm 1$.

Exercises 21–24: Write the equation in vertex form. Identify the vertex.

21. $y = x^2 + 4x - 7$ **22.** $y = x^2 - 7x + 1$

23. $y = 2x^2 - 3x - 8$ **24.** $y = 3x^2 + 6x - 2$

Exercises 25 and 26: Find a value for the constant a so that $f(x) = ax^2 - 1$ models the data.

25.

x	1	2	3
$f(x)$	2	11	26

26.

x	-1	0	1
$f(x)$	$-\frac{3}{4}$	-1	$-\frac{3}{4}$

Exercises 27 and 28: Write $f(x)$ in the form given by $f(x) = ax^2 + bx + c$. Identify the y-intercept on the graph of f.

27. $f(x) = -5(x - 3)^2 + 4$

28. $f(x) = 3(x + 2)^2 - 4$

SECTION 8.3

Exercises 29–32: Use the graph of $y = ax^2 + bx + c$ to solve $ax^2 + bx + c = 0$.

29.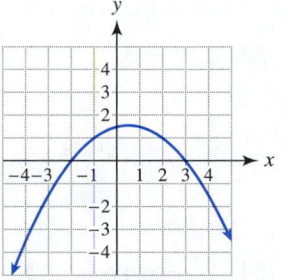

30.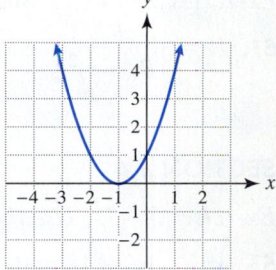

31.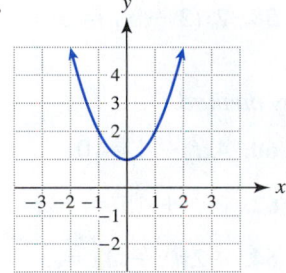

32.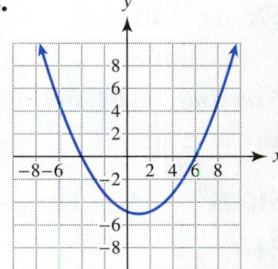

Exercises 33 and 34: A table of $y = ax^2 + bx + c$ is given. Solve $ax^2 + bx + c = 0$.

33.

X	Y1
-20	250
-15	100
-10	0
-5	-50
0	-50
5	0
10	100

Y1⬛X^2+5X−50

34.

X	Y1
-.75	2
-.5	0
-.25	-1
0	-1
.25	0
.5	2
.75	5

Y1⬛8X^2+2X−1

Exercises 35–38: Solve the quadratic equation
  (a) graphically and
  (b) numerically.

35. $x^2 - 5x - 50 = 0$ **36.** $\frac{1}{2}x^2 + x - \frac{3}{2} = 0$

37. $\frac{1}{4}x^2 + \frac{1}{2}x = 2$ **38.** $\frac{1}{2}x + \frac{3}{4} = \frac{1}{4}x^2$

Exercises 39–42: Solve by factoring.

39. $x^2 + x - 20 = 0$ **40.** $x^2 + 11x + 24 = 0$

41. $15x^2 - 4x - 4 = 0$ **42.** $7x^2 - 25x + 12 = 0$

Exercises 43–46: Use the square root property to solve.

43. $x^2 = 100$ **44.** $3x^2 = \frac{1}{3}$

45. $4x^2 - 6 = 0$ **46.** $5x^2 = x^2 - 4$

Exercises 47–50: Solve by completing the square.

47. $x^2 + 6x = -2$ **48.** $x^2 - 4x = 6$

49. $x^2 - 2x - 5 = 0$ **50.** $2x^2 + 6x - 1 = 0$

Exercises 51 and 52: Solve for the specified variable.

51. $F = \dfrac{k}{(R + r)^2}$ *for R*   **52.** $2x^2 + 3y^2 = 12$ *for y*

SECTION 8.4

Exercises 53–58: Use the quadratic formula to solve.

53. $x^2 - 9x + 18 = 0$ **54.** $x^2 - 24x + 143 = 0$

55. $6x^2 + x = 1$ **56.** $5x^2 + 1 = 5x$

57. $x(x - 8) = 5$ **58.** $2x(2 - x) = 3 - 2x$

Exercises 59–64: Solve by any method.

59. $x^2 - 4 = 0$ **60.** $4x^2 - 1 = 0$

61. $2x^2 + 15 = 11x$ **62.** $2x^2 + 15 = 13x$

63. $x(5 - x) = 2x + 1$ **64.** $-2x(x - 1) = x - \frac{1}{2}$

Exercises 65–68: A graph of $y = ax^2 + bx + c$ is shown.
  (a) State whether $a > 0$ or $a < 0$.
  (b) Solve $ax^2 + bx + c = 0$.
  (c) Determine whether the discriminant is positive, negative, or zero.

65. 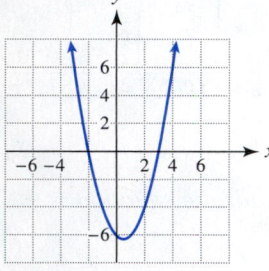   **66.**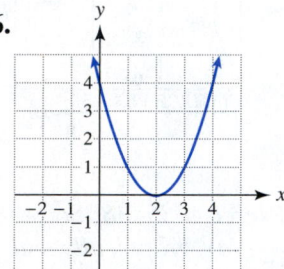

67. 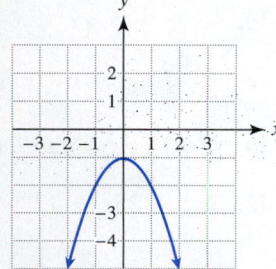   **68.**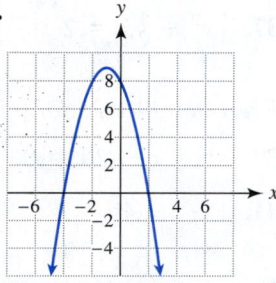

Exercises 69–72: Do the following for the given equation.
  (a) Evaluate the discriminant.
  (b) How many real solutions are there?
  (c) Support your answer for part (b) graphically.

69. $2x^2 - 3x + 1 = 0$ **70.** $7x^2 + 2x - 5 = 0$

71. $3x^2 + x + 2 = 0$

72. $4.41x^2 - 12.6x + 9 = 0$

Exercises 73–76: Solve. Write any complex solutions in standard form.

73. $x^2 + x + 5 = 0$ **74.** $2x^2 + 8 = 0$

75. $2x^2 = x - 1$ **76.** $7x^2 = 2x - 5$

SECTION 8.5

Exercises 77 and 78: The graph of $y = ax^2 + bx + c$ is shown. Solve each equation or inequality.
  (a) $ax^2 + bx + c = 0$
  (b) $ax^2 + bx + c < 0$
  (c) $ax^2 + bx + c > 0$

77. 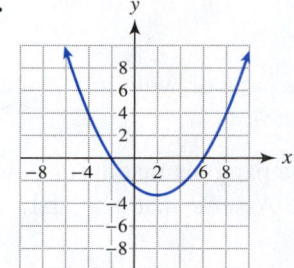   **78.**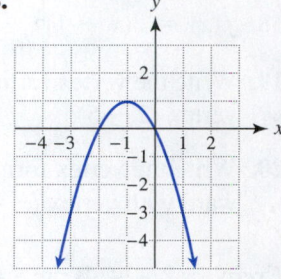

Exercises 79 and 80: A table of $y = ax^2 + bx + c$ is shown. Solve each equation or inequality.
  (a) $ax^2 + bx + c = 0$
  (b) $ax^2 + bx + c < 0$
  (c) $ax^2 + bx + c > 0$

79. $y = x^2 - 16$

x	-6	-4	-2	0	2	4	6
y	20	0	-12	-16	-12	0	20

80. $y = x^2 + x - 2$

x	-3	-2	-1	0	1	2	3
y	4	0	-2	-2	0	4	10

Exercises 81 and 82: Solve the equation in part (a). Use the results to solve the inequalities in parts (b) and (c).

81. **(a)** $x^2 - 2x - 3 = 0$
     (b) $x^2 - 2x - 3 < 0$
     (c) $x^2 - 2x - 3 > 0$

82. **(a)** $2x^2 - 7x - 15 = 0$
     (b) $2x^2 - 7x - 15 \leq 0$
     (c) $2x^2 - 7x - 15 \geq 0$

Exercises 83–88: Solve the quadratic inequality. Write your answer in interval notation.

83. $x^2 + 4x + 3 \le 0$ **84.** $5x^2 - 16x + 3 < 0$

85. $6x^2 - 13x + 2 > 0$ **86.** $x^2 \ge 5$

87. $(x - 1)^2 \ge 0$ **88.** $x^2 + 3 < 2$

SECTION 8.6

Exercises 89–92: Solve the equation.

89. $x^4 - 14x^2 + 45 = 0$

90. $2z^{-2} + z^{-1} - 28 = 0$

91. $x^{2/3} - 9x^{1/3} + 8 = 0$

92. $(x - 1)^2 + 2(x - 1) + 1 = 0$

Exercises 93 and 94: Find all complex solutions.

93. $4x^4 + 4x^2 + 1 = 0$ **94.** $\dfrac{1}{x - 2} - \dfrac{3}{x} = -1$

APPLICATIONS

95. *Construction* A rain gutter is being fabricated from a flat sheet of metal so that the cross section of the gutter is a rectangle, as shown in the accompanying figure. The width of the metal sheet is 12 inches.

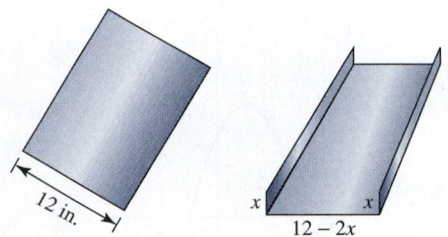

(a) Write a formula $f(x)$ that gives the area of the cross section.

(b) To hold the greatest amount of rainwater, the cross section should have maximum area. Find the dimensions that result in this maximum.

96. *Height of a Stone* Suppose that a stone is thrown upward with an initial velocity of 44 feet per second (30 miles per hour) and is released 4 feet above the ground. Its height h in feet after t seconds is given by

$$h(t) = -16t^2 + 44t + 4.$$

(a) When does the stone reach a height of 32 feet?

(b) After how many seconds does the stone reach maximum height? Estimate this height.

97. *Maximizing Revenue* Suppose that hotel rooms cost $90 per night. However, for a group rate the management is considering reducing the cost of a room by $3 for every room rented.

(a) Write a formula $f(x)$ that gives the revenue from renting x rooms at the group rate.

(b) Graph f in [0, 30, 5] by [0, 800, 100].

(c) How many rooms should be rented to receive revenue of $600?

(d) How many rooms should be rented to maximize revenue?

98. *Numbers* The product of two numbers is 143. One number is 2 more than the other.

(a) Write a quadratic equation whose solution gives the smaller number x.

(b) Solve the equation.

99. *Braking Distance* On dry pavement a safe braking distance d in feet for a car traveling x miles per hour is $d = \frac{x^2}{12}$. For each distance d, find x. (*Source:* F. Mannering, *Principles of Highway Engineering and Traffic Control.*)

(a) $d = 144$ feet   (b) $d = 300$ feet

100. *U.S. Energy Consumption* From 1950 to 1970 per capita consumption of energy in millions of Btu can be modeled by $f(x) = \frac{1}{4}(x - 1950)^2 + 220$, where x is the year. (*Source:* Department of Energy.)

(a) Find and interpret the vertex.

(b) Graph f in [1950, 1970, 5] by [200, 350, 25]. What happened to energy consumption during this time period?

(c) Use f to predict the consumption in 2010. Actual consumption was 321 million Btu. Did f provide a good model for 2010? Explain.

101. *Screens* A square computer screen has an area of 123 square inches. Approximate its dimensions to the nearest tenth of an inch.

102. *Flying a Kite* A kite is being flown, as illustrated in the accompanying figure. If 130 feet of string have been let out, find the value of x.

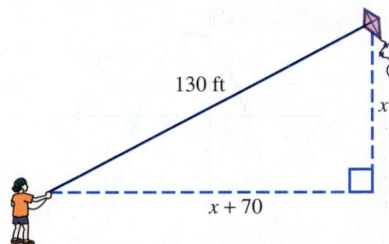

103. *Area* A uniform strip of grass is to be planted around a rectangular swimming pool, as illustrated in the accompanying figure on the next page. The swimming pool is 30 feet wide and 50 feet long. If there is only enough grass seed to cover 250 square feet, estimate the width x that the strip of grass should be.

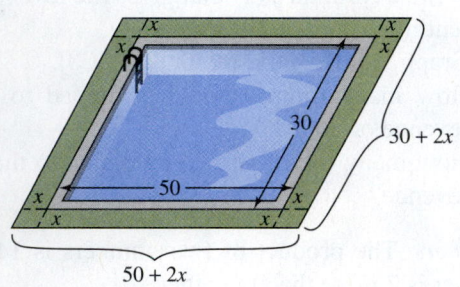

104. *Dimensions of a Cone* The volume V of a cone is given by $V = \frac{1}{3}\pi r^2 h$, where r is its base radius and h is its height. See the accompanying figure. If $h = 20$ inches and the volume of the cone must be between 750 and 1700 cubic inches, inclusively, estimate, to the nearest tenth of an inch, possible values for r.

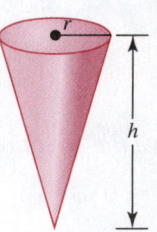

CHAPTER 8   Test

CHAPTER Test Prep VIDEOS Step-by-step test solutions are found on the Chapter Test Prep Videos available in *MyMathLab* and on YouTube (search "RockswoldInterAlg" and click on "Channels").

1. Find the vertex and axis of symmetry for the graph of $f(x) = -\frac{1}{2}x^2 + x + 1$. Evaluate $f(-2)$.

2. Find the minimum y-value located on the graph of $y = x^2 + 3x - 5$.

3. Find the exact value for the constant a so that $f(x) = ax^2 + 2$ models the data in the table.

x	-2	0	2	4
$f(x)$	0	2	0	-6

4. Compare the graph of $y = f(x)$ to the graph of $y = x^2$. Then graph $y = f(x)$.
   (a) $f(x) = (x - 1)^2$ **(b)** $f(x) = x^2 - 2$
   (c) $f(x) = \frac{1}{2}(x - 3)^2 + 2$

5. Write $y = x^2 - 6x + 2$ in vertex form. Identify the vertex and axis of symmetry.

6. Use the graph of $f(x) = ax^2 + bx + c$ to solve $ax^2 + bx + c = 0$. Then evaluate $f(1)$.

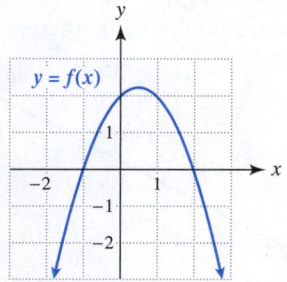

Exercises 7 and 8: Solve the quadratic equation.

7. $3x^2 + 11x - 4 = 0$ **8.** $2x^2 = 2 - 6x^2$

9. Solve $x^2 - 8x = 1$ by completing the square.

10. Solve $x(-2x + 3) = -1$ by using the quadratic formula.

11. Solve $9x^2 - 16 = 0$.

12. Solve $F = \dfrac{Gm^2}{r^2}$ for m.

13. A graph of $y = ax^2 + bx + c$ is shown.
   (a) State whether $a > 0$ or $a < 0$.
   (b) Solve $ax^2 + bx + c = 0$.
   (c) Determine whether the discriminant is positive, negative, or zero.

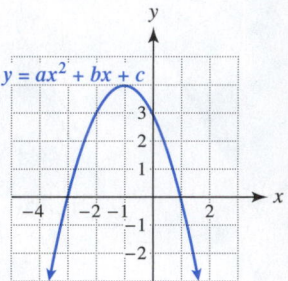

14. Complete the following for $-3x^2 + 4x - 5 = 0$.
   (a) Evaluate the discriminant.
   (b) How many real solutions are there?
   (c) Support your answer for part (b) graphically.

Exercises 15 and 16: The graph of $y = ax^2 + bx + c$ is shown. Solve each equation or inequality.
   (a) $ax^2 + bx + c = 0$
   (b) $ax^2 + bx + c < 0$
   (c) $ax^2 + bx + c > 0$

15.

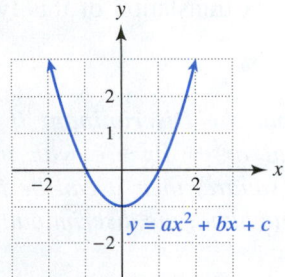

16.

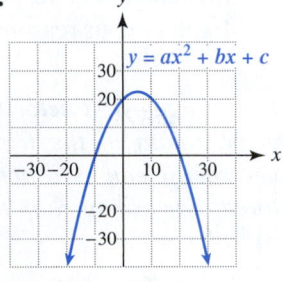

17. Solve the quadratic equation in part (a). Use the result to solve the inequalities in parts (b) and (c) and write your answer in interval notation.
(a) $8x^2 - 2x - 3 = 0$
(b) $8x^2 - 2x - 3 \leq 0$
(c) $8x^2 - 2x - 3 \geq 0$

18. Solve $x^2 + 2x \leq 0$.

19. Solve $x^6 - 3x^3 + 2 = 0$. Find all real solutions.

20. Solve $2x^2 + 4x + 3 = 0$. Write all complex solutions in standard form.

21. Solve $\sqrt{2} - \pi x^2 = 2.12x - 0.5\pi$ graphically. Round your answers to the nearest hundredth.

22. *Braking Distance* On wet pavement a safe braking distance d in feet for a car traveling x miles per hour is $d = \frac{x^2}{9}$. What speed corresponds to a braking distance of 250 feet? (*Source:* F. Mannering, *Principles of Highway Engineering and Traffic Control.*)

23. *Construction* A fence is being constructed along a 20-foot building, as shown in the accompanying figure. No fencing is used along the building.
(a) If 200 feet of fence are available, find a formula $f(x)$ using only the variable x that gives the area enclosed.
(b) What value of x gives the greatest area?

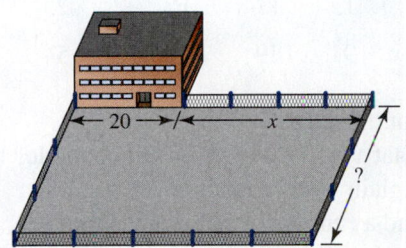

24. *Height of a Stone* Suppose that a stone is thrown upward with an initial velocity of 88 feet per second (60 miles per hour) and is released 8 feet above the ground. Its height h in feet after t seconds is given by

$$h(t) = -16t^2 + 88t + 8.$$

(a) Graph h in [0, 6, 1] by [0, 150, 50].
(b) When does the stone strike the ground?
(c) After how many seconds does the stone reach maximum height? Estimate this height.

CHAPTER 8   Extended and Discovery Exercises

MODELING DATA WITH A QUADRATIC FUNCTION

1. *Survival Rate of Birds* The survival rate of sparrowhawks varies according to their age. The following table summarizes the results of one study by listing the age in years and the percentage of birds that survived the previous year. For example, 52% of sparrowhawks that reached age 6 lived to be 7 years old. (*Source:* D. Brown and P. Rothery, *Models in Biology.*)

Age	1	2	3	4	5
Percent (%)	45	60	71	67	67

Age	6	7	8	9
Percent (%)	61	52	30	25

(a) Try to explain the relationship between age and the likelihood of surviving the next year.

(b) Make a scatterplot of the data. What type of function might model the data? Explain.
(c) Graph each function. Which of the following functions models the data better?

$$f_1(x) = -3.57x + 71.1$$
$$f_2(x) = -2.07x^2 + 17.1x + 33$$

(d) Use one of these functions to estimate the likelihood of a 5.5-year-old sparrowhawk surviving for 1 more year.

2. *Photosynthesis and Temperature* Photosynthesis is the process by which plants turn sunlight into energy. At very cold temperatures photosynthesis may halt even though the sun is shining. In one study the efficiency of photosynthesis for an Antarctic species of grass was investigated. The following table lists results for various temperatures. The temperature x is in degrees Celsius, and the efficiency y is given as a percent. The purpose of the research was to determine the temperature at which photosynthesis is most efficient. (*Source: D. Brown.*)

x (°C)	−1.5	0	2.5	5	7	10	12
y (%)	33	46	55	80	87	93	95

x (°C)	15	17	20	22	25	27	30
y(%)	91	89	77	72	54	46	34

(a) Plot the data.
(b) What type of function might model these data? Explain your reasoning.
(c) Find a function f that models the data.
(d) Use f to estimate the temperature at which photosynthesis is most efficient in this type of grass.

TRANSLATIONS OF PARABOLAS IN COMPUTER GRAPHICS

Exercises 3 and 4: In older video games with two-dimensional graphics, the background is often translated to give the illusion that a character in the game is moving. The simple scene on the left shows a mountain and an airplane. To make it appear that the airplane is flying, the mountain can be translated to the left, as shown in the figure on the right. (Reference: C. Pokorny and C. Gerald, *Computer Graphics.*)

3. *Video Games* Suppose that the mountain in the figure on the left is modeled by $f(x) = -0.4x^2 + 4$ and that the airplane is located at the point $(1, 5)$.
   (a) Graph f in $[-4, 4, 1]$ by $[0, 6, 1]$, where the units are kilometers. Plot the point $(1, 5)$ to show the location of the airplane.
   (b) Assume that the airplane is moving horizontally to the right at 0.2 kilometer per second. To give a video game player the illusion that the airplane is moving, graph the image of the mountain and the position of the airplane after 10 seconds.

4. *Video Games* (Refer to Exercise 3.) Discuss how you could create the illusion of the airplane moving to the left and gaining altitude as it passes over the

mountain. Try to perform a translation of this type. Explain your reasoning.

*Exercises 5–8: **Factoring and the Discriminant** If the discriminant of the trinomial $ax^2 + bx + c$ with integer coefficients is a perfect square, then it can be factored. For example, on the one hand, the discriminant of $6x^2 + x - 2$ is*

$$1^2 - 4(6)(-2) = 49,$$

which is a perfect square ($7^2 = 49$), so we can factor the trinomial as

$$6x^2 + x - 2 = (2x - 1)(3x + 2).$$

On the other hand, the discriminant for $x^2 + x - 1$ is

$$1^2 - 4(1)(-1) = 5,$$

which is not a perfect square, so we cannot factor this trinomial by using integers as coefficients. Similarly, if the discriminant is negative, the trinomial cannot be factored by using integer coefficients. Use the discriminant to predict whether the trinomial can be factored. Then test your prediction.

5. $10x^2 - x - 3$ 6. $4x^2 - 3x - 6$

7. $3x^2 + 2x - 2$ 8. $2x^2 + x + 3$

*Exercises 9–14: **Polynomial Inequalities** The solution set for a polynomial inequality can be found by first determining the boundary numbers. For example, to solve $f(x) = x^3 - 4x > 0$ begin by solving $x^3 - 4x = 0$. The solutions (boundary numbers) are $-2, 0$, and 2. The function $f(x) = x^3 - 4x$ is either only positive or only negative on intervals between consecutive zeros. To determine the solution set, we can evaluate test values for each interval as shown below.*

Interval	Test Value	$f(x) = x^3 - 4x$
$(-\infty, -2)$	$x = -3$	$f(-3) = -15 < 0$
$(-2, 0)$	$x = -1$	$f(-1) = 3 > 0$
$(0, 2)$	$x = 1$	$f(1) = -3 < 0$
$(2, \infty)$	$x = 3$	$f(3) = 15 > 0$

$[-6, 6, 1]$ by $[-4, 4, 1]$

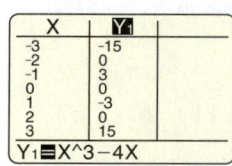 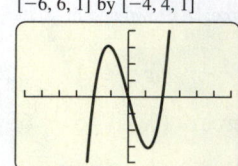

We can see that $f(x) > 0$ for $(-2, 0) \cup (2, \infty)$. These results are also supported graphically in the figure on the right above, where the graph of f is above the x-axis when $-2 < x < 0$ or when $x > 2$.

Use these concepts to solve the polynomial inequality.

9. $x^3 - x^2 - 6x > 0$

10. $x^3 - 3x^2 + 2x < 0$

11. $x^3 - 7x^2 + 14x \leq 8$

12. $9x - x^3 \geq 0$

13. $x^4 - 5x^2 + 4 > 0$

14. $1 < x^4$

Exercises 15–20: **Rational Inequalities** Rational inequalities can be solved using many of the same techniques that are used to solve other types of inequalities. However, there is one important difference. For a rational inequality, the boundary between greater than and less than can be either an x-value where equality occurs or an x-value where a rational expression is undefined. For example, consider the inequality $f(x) = \frac{2 - x}{2x} > 0$. The solution to the equation $\frac{2 - x}{2x} = 0$ is 2. The rational expression $\frac{2 - x}{2x}$ is undefined when $x = 0$. Therefore we select test values on the intervals $(-\infty, 0)$, $(0, 2)$, and $(2, \infty)$. The table in the next column reveals that $f(x) > 0$ for $(0, 2)$.

Interval	Test Value	$f(x) = \dfrac{2 - x}{2x}$
$(-\infty, 0)$	$x = -0.5$	$f(-0.5) = -2.5 < 0$
$(0, 2)$	$x = 1$	$f(1) = 0.5 > 0$
$(2, \infty)$	$x = 2.5$	$f(2.5) = -0.1 < 0$

$[-4.7, 4.7, 1]$ by $[-3.1, 3.1, 1]$

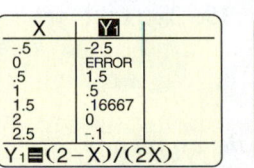 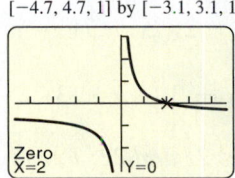

Note that $f(x)$ changes from negative to positive at $x = 0$, where $f(x)$ is undefined. These results are supported graphically in the figure on the right above.

Solve the rational inequality.

15. $\dfrac{3 - x}{3x} \geq 0$

16. $\dfrac{x - 2}{x + 2} > 0$

17. $\dfrac{3 - 2x}{1 + x} < 3$

18. $\dfrac{x + 1}{4 - 2x} \geq 1$

19. $\dfrac{5}{x^2 - 4} < 0$

20. $\dfrac{x}{x^2 - 1} \geq 0$

CHAPTERS 1–8   Cumulative Review Exercises

1. Evaluate $F = \dfrac{5}{z^2 + 1}$ when $z = -2$.

2. Classify each number as one or more of the following: natural number, whole number, integer, rational number, or irrational number: $0.\overline{4}$, $\sqrt{7}$, 0, -5, $\sqrt[3]{8}$, $-\frac{4}{3}$.

3. Simplify each expression. Write the result using positive exponents.

(a) $\left(\dfrac{x^2 y^6}{x^{-3}}\right)^2$

(b) $\dfrac{(xy^{-3})^2}{x(y^{-2})^{-1}}$

(c) $(a^2 b)^2 (ab^3)^{-4}$

4. Write 9,290,000 in scientific notation.

5. Find $f(-2)$, if $f(x) = \sqrt{2 - x}$. What is the domain of f?

6. If $f(2) = 5$, then what point lies on the graph of f?

Exercises 7 and 8: Graph f by hand.

7. $f(x) = x^2 + 2x$

8. $f(x) = |2x - 4|$

9. Find the slope–intercept form of the line that passes through $(4, -1)$ and is parallel to the line passing through $(0, 1)$ and $(-2, 4)$.

10. Find the equation of a vertical line that passes through $(-3, 4)$.

11. Solve $2x - 3(x + 2) = 6$.

Exercises 12–14: Solve the inequality. Write your answer in interval notation.

12. $7 - x > 3x$

13. $|3x - 2| \leq 1$

14. $-4 \leq 1 - x < 2$

15. Solve the system.

$$-x - 4y = -3$$
$$5x + y = -4$$

16. Shade the solution set in the xy-plane.

$$3x + y \leq 3$$
$$x - 3y \leq 3$$

17. Solve the system.

$$x + y - z = 3$$
$$x - y + z = 1$$
$$2x - y - z = 1$$

Exercises 18–20: Multiply the expression.

18. $(3x - 2)(2x + 7)$ **19.** $3xy(x^2 + y^2)$

20. $(\sqrt{x} + 3)(\sqrt{x} - 3)$

Exercises 21 and 22: Factor the expression.

21. $x^3 - x^2 - 2x$ **22.** $4x^2 - 25$

Exercises 23 and 24: Solve each equation.

23. $x^2 - 3 = 0$ **24.** $x^2 + 1 = 2x$

Exercises 25 and 26: Simplify the expression.

25. $\dfrac{(x + 3)^2}{x + 2} \cdot \dfrac{x + 2}{2x + 6}$ **26.** $\dfrac{1}{x + 2} - \dfrac{1}{x}$

Exercises 27–30: Simplify the expression. Assume all variables are positive.

27. $\sqrt{16x^6}$ **28.** $16^{-3/2}$

29. $\dfrac{\sqrt[3]{81x}}{\sqrt[3]{3x}}$ **30.** $\sqrt{8x} + \sqrt{2x}$

31. Graph $f(x) = \sqrt{4x}$.

32. Find the distance between $(-1, 2)$ and $(4, 3)$.

33. Write $\dfrac{3 - i}{2 + i}$ in standard form.

34. Solve $3\sqrt{x + 1} = 2x$.

35. Solve the equation $2x = \sqrt{2.1 - x} + \sqrt[3]{0.1x}$ to the nearest hundredth.

36. Sketch a graph of $f(x) = x^2 - 2x + 3$.
   (a) Find the vertex.
   (b) Evaluate $f(-1)$.
   (c) What is the axis of symmetry?
   (d) Where is f increasing?

37. Write $f(x) = 2x^2 - 4x - 1$ in vertex form.

38. Compare the graph of $f(x) = 4(x + 1)^2 - 2$ to the graph of $y = x^2$.

39. Solve $x^2 + 6x = 2$ by completing the square.

40. Solve $2x^2 - 3x = 1$ by using the quadratic formula.

41. Solve $x(4 - x) = 3$.

42. The graph of $y = ax^2 + bx + c$ is shown in the next column. Solve each equation or inequality.

(a) $ax^2 + bx + c = 0$
(b) $ax^2 + bx + c \le 0$

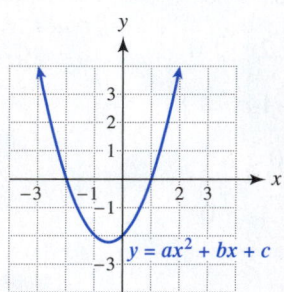

43. Solve $x^2 - 3x + 2 > 0$.

44. Solve $x^4 - 256 = 0$. Find all complex solutions.

Exercises 45–52: **Thinking Generally** *Match the graph (a.–h.) with its equation. Assume that a, b, and c are positive constants.*

45. $y = ax - b$ **46.** $y = b$

47. $y = -ax^2 + c$ **48.** $y = \dfrac{a}{x}$

49. $y = ax^3$ **50.** $y = |ax + b|$

51. $y = a\sqrt{x}$ **52.** $y = a\sqrt[3]{x}$

a.
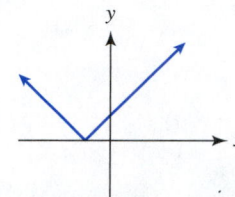

b.
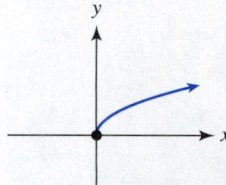

c.
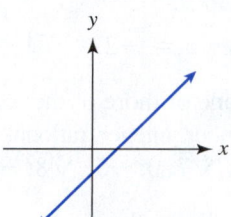

d.
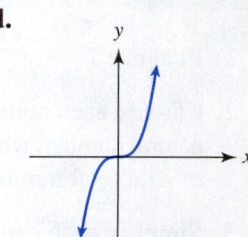

e.
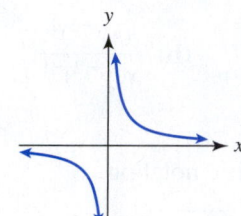

f.
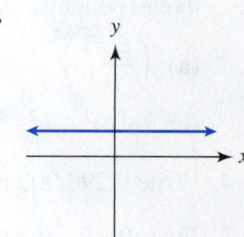

g.
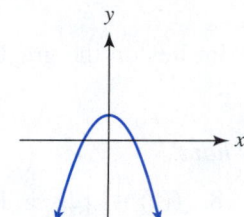

h.
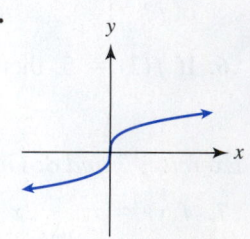

APPLICATIONS

53. *Calculating Water Flow* Water is being pumped out of a tank. The gallons G of water in the tank after t minutes is shown in the figure, where $y = G(t)$.

  (a) Evaluate $G(0)$. Interpret your answer.

  (b) What is the t-intercept? Interpret your answer.

  (c) What is the slope of the graph of G? Interpret your answer.

  (d) Find a formula for $G(t)$.

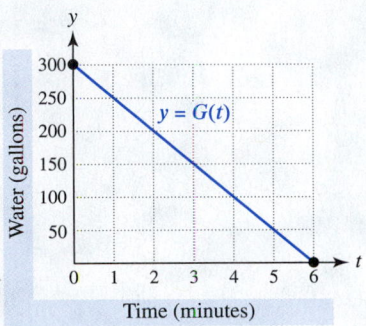

54. *Investment* Suppose \$4000 is deposited in three accounts paying 4%, 5%, and 6% annual interest. The amount invested at 6% is \$1000 more than the amount invested at 5%. The interest after 1 year is \$216. How much is invested at each rate?

55. *Maximizing Area* There are 490 feet of fence available to surround the perimeter of a rectangular garden. On one side, there is a 10-foot gate that requires no fencing. What dimensions for the garden give the largest area?

56. *Height of a Tree* A 6-foot-tall person casts a shadow that is 10 feet long while a nearby tree casts a 55-foot shadow. Find the height of the tree.

9

Exponential and Logarithmic Functions

Time is money.

—BENJAMIN FRANKLIN

If you deposit money into a savings account, you might *earn* an interest rate of 0.18%, whereas if you charge purchases on your credit card, you might *pay* an interest rate of 18%. Do interest rates and decimal points really matter? To find out, consider the following.

If $1000 is deposited in a savings account at 0.18% interest per year for 15 years, the final amount would be about $1027.37. However, if the interest rate were a typical credit card rate of 18%, the final amount would be about $14,584.37. Interest rates and decimal points clearly matter! After *doing the math*, it becomes obvious that the best financial decision is to pay off high-interest credit cards first, before putting money into a savings account that pays only 0.18%.

In this chapter we will learn ways to calculate interest on investments by studying a new type of function called an *exponential function*. (See Section 9.2.) Exponential functions are frequently used in business and finance, but they also have many other applications.

9.1   Composite and Inverse Functions

Composition of Functions • **One-to-One Functions** • **Inverse Functions** •
Tables and Graphs of Inverse Functions

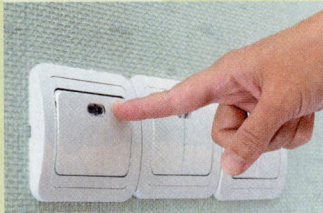

A LOOK INTO MATH ▶   Suppose that you walk into a classroom, turn on the lights, and sit down at your desk. How could you undo or reverse these actions? You would stand up from the desk, turn off the lights, and walk out of the classroom. Note that you must not only perform the "inverse" of each action, but you also must do them in the *reverse order*. In mathematics, we undo an arithmetic operation by performing its inverse operation. For example, the inverse operation of addition is subtraction, and the inverse operation of multiplication is division. In this section, we explore these concepts further by discussing inverse functions.

NEW VOCABULARY

☐ Composite function
☐ Composition
☐ One-to-one
   correspondence
☐ One-to-one
☐ Horizontal line test
☐ Inverse function

Composition of Functions

▶ **REAL-WORLD CONNECTION**   Many tasks in life are performed in *sequence*, such as putting on your socks and then your shoes. These types of situations also occur in mathematics. For example, suppose that we want to calculate the number of ounces in 3 tons. Because there are 2000 pounds in one ton, we might first multiply **3** by 2000 to obtain **6000** pounds. There are 16 ounces in a pound, so we could multiply **6000** by 16 to obtain **96,000** ounces. This particular calculation involves a *sequence* of calculations that can be represented by the diagram shown in Figure 9.1.

Converting Tons to Ounces

P multiplies 3, the input for *P*, by 2000.

O multiplies 6000, the output from *P*, by 16.

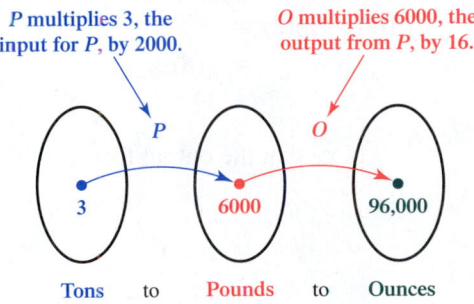

3   6000   96,000

Tons to **Pounds** to **Ounces**

Figure 9.1

The results shown in Figure 9.1 can be calculated by using function notation. Suppose that we let $P(x) = 2000x$ convert x tons to **P** pounds and also let $O(x) = 16x$ convert x pounds to **O** ounces. We can calculate the number of ounces in 3 tons by performing the *composition of O and P*. This method can be expressed symbolically as follows.

Composition of *O* and *P*

Start by evaluating the innermost function, $P(3)$.

$$(O \circ P)(3) = O\big(P(3)\big)$$
$$= O(2000 \cdot 3) \quad \text{Let } x = 3 \text{ in } P(x) = 2000x.$$

The output from *P* becomes the input for *O*.

$$= O(6000) \quad\quad \text{Simplify.}$$
$$= 16(6000) \quad\quad \text{Let } x = 6000 \text{ in } O(x) = 16x.$$
$$= 96,000. \quad\quad \text{Multiply.}$$

P first multiplies 3 by 2000 and then *O* multiplies 6000 by 16 to get 96,000. This is written as $(O \circ P)(3)$.

> **COMPOSITION OF FUNCTIONS**
>
> If f and g are functions, then the **composite function** $g \circ f$, or **composition** of g and f, is defined by
>
> $$(g \circ f)(x) = g\big(f(x)\big).$$
>
> **NOTE:** We read $g\big(f(x)\big)$ as "g of f of x."

The compositions $g \circ f$ and $f \circ g$ represent evaluating functions f and g in different orders. When evaluating $g \circ f$, function f is evaluated first followed by function g, whereas for $f \circ g$ function g is evaluated first followed by function f. Note that in general $(g \circ f)(x) \neq (f \circ g)(x)$. That is, *the order in which functions are applied makes a difference*, in the same way that putting on your socks and then your shoes is quite different from putting on your shoes and then your socks. See Example 2(a) and (b).

EXAMPLE 1 | **Finding composite functions**

Evaluate $(g \circ f)(2)$ and then find a formula for $(g \circ f)(x)$.

(a) $f(x) = x^3$, $g(x) = 3x - 2$ (b) $f(x) = 5x$, $g(x) = x^2 - 3x + 1$

(c) $f(x) = \sqrt{2x}$, $g(x) = \dfrac{1}{x - 1}$

Solution

(a) $(g \circ f)(2) = g\big(f(2)\big)$ Composition of functions

$ = g(8)$ $f(2) = 2^3 = 8$

$ = 22$ $g(8) = 3(8) - 2 = 22$

Note that the output from f, which is $f(2)$, becomes the input for g.

$(g \circ f)(x) = g\big(f(x)\big)$ Composition of functions

$ = g(x^3)$ $f(x) = x^3$

$ = 3x^3 - 2$ Replace x with x^3 in $g(x) = 3x - 2$.

(b) $(g \circ f)(2) = g\big(f(2)\big)$ Composition of functions

$ = g(10)$ $f(2) = 5(2) = 10$

$ = 71$ $g(10) = 10^2 - 3(10) + 1 = 71$

$(g \circ f)(x) = g\big(f(x)\big)$ Composition of functions

$ = g(5x)$ $f(x) = 5x$

$ = (5x)^2 - 3(5x) + 1$ Replace x with $5x$ in $g(x) = x^2 - 3x + 1$.

$ = 25x^2 - 15x + 1$ Simplify.

(c) $(g \circ f)(2) = g\big(f(2)\big)$ Composition of functions

$ = g(2)$ $f(2) = \sqrt{2(2)} = 2$

$ = 1$ $g(2) = \dfrac{1}{2 - 1} = 1$

$(g \circ f)(x) = g\big(f(x)\big)$ Composition of functions

$ = g\big(\sqrt{2x}\big)$ $f(x) = \sqrt{2x}$

$ = \dfrac{1}{\sqrt{2x} - 1}$ Replace x with $\sqrt{2x}$ in $g(x) = \dfrac{1}{x - 1}$.

Now Try Exercises 13, 15, 19

Composite functions can also be evaluated numerically and graphically, as demonstrated in the next two examples.

EXAMPLE 2   **Evaluating composite functions with tables**

Use Tables 9.1 and 9.2 to evaluate each expression.
(a) $(f \circ g)(2)$ **(b)** $(g \circ f)(2)$ **(c)** $(f \circ f)(0)$

TABLE 9.1

x	0	1	2	3
$f(x)$	3	2	0	1

TABLE 9.2

x	0	1	2	3
$g(x)$	1	3	2	0

Solution

(a) $(f \circ g)(2) = f\big(g(2)\big)$ Composition of functions

$\qquad\qquad\quad = f(2)$ $g(2) = 2$

$\qquad\qquad\quad = 0$ $f(2) = 0$

(b) $(g \circ f)(2) = g\big(f(2)\big)$ Composition of functions

$\qquad\qquad\quad = g(0)$ $f(2) = 0$

$\qquad\qquad\quad = 1$ $g(0) = 1$

NOTE: From parts (a) and (b), we see that $(f \circ g)(2) \neq (g \circ f)(2)$.

(c) $(f \circ f)(0) = f\big(f(0)\big)$ Composition of functions

$\qquad\qquad\quad = f(3)$ $f(0) = 3$

$\qquad\qquad\quad = 1$ $f(3) = 1$

Now Try Exercises 23, 25

EXAMPLE 3   **Evaluating composite functions graphically**

Use Figure 9.2 in the margin to evaluate $(g \circ f)(2)$.

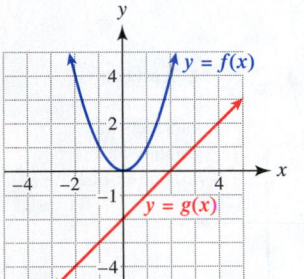

Figure 9.2

Solution

Because $(g \circ f)(2) = g\big(f(2)\big)$, start by using Figure 9.2 to evaluate $f(2)$. Figure 9.3(a) shows that $f(2) = 4$, which becomes the input for g. Figure 9.3(b) reveals that $g(4) = 2$.

$$(g \circ f)(2) = g\big(f(2)\big) = g(4) = 2$$

f(2) = 4

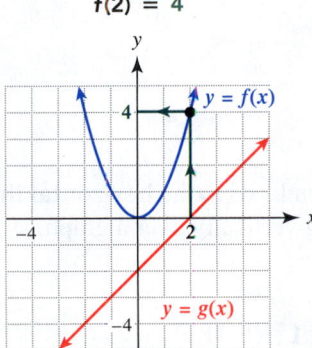

(a)

g(4) = 2

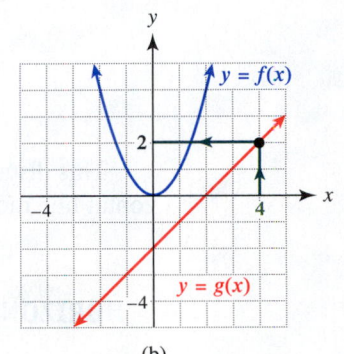

(b)

Figure 9.3

Now Try Exercise 29

One-to-One Functions

▶ **REAL-WORLD CONNECTION** One-to-one correspondences occur in real life. For example, in a typical city there is *usually* a **one-to-one correspondence** between homes and addresses. Each home has one address, and each address corresponds to one home.

If we change the input for a function, does the output also change? Do *different inputs* always result in *different outputs* for every function? The answer is no. For example, if $f(x) = x^2 + 1$, then the inputs -2 and 2 result in the *same* output, 5. That is, $f(-2) = 5$ and $f(2) = 5$. However, for $g(x) = 2x$, *different inputs* always result in *different outputs*. For function g, there is a one-to-one correspondence between inputs and outputs. Thus we say that g is a *one-to-one function*, whereas f is not.

ONE-TO-ONE FUNCTION

A function f is **one-to-one** if, for any c and d in the domain of f,

$$c \neq d \quad \text{implies that} \quad f(c) \neq f(d).$$

That is, different inputs always result in different outputs.

READING CHECK

How can you determine if a function is one-to-one?

One way to determine whether a function f is one-to-one is to look at its graph. Suppose that a function has two *different inputs* that result in the *same output*. Then there must be two points on its graph that have the same y-value but different x-values. For example, if $f(x) = 2x^2$, then $f(-1) = 2$ and $f(1) = 2$. Thus the points $(-1, 2)$ and $(1, 2)$ both lie on the graph of f, as shown in Figure 9.4(a). Two points with different x-values and the same y-value determine a horizontal line, as shown in Figure 9.4(b). This horizontal line intersects the graph of f more than once, indicating that different inputs do *not* always have different outputs. Thus $f(x) = 2x^2$ is *not* one-to-one.

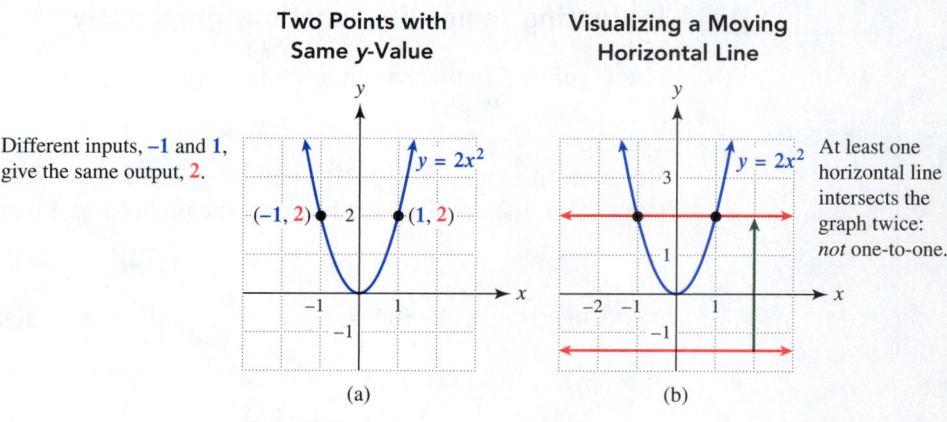

Figure 9.4

This discussion leads to the **horizontal line test**, which is applied by visualizing a horizontal line moving vertically over a graph, as shown in Figure 9.4(b).

HORIZONTAL LINE TEST

If every horizontal line intersects the graph of a function f at most once, then f is a one-to-one function.

We apply the horizontal line test in the next example.

Using the horizontal line test

Determine whether each graph in Figure 9.5 represents a one-to-one function.

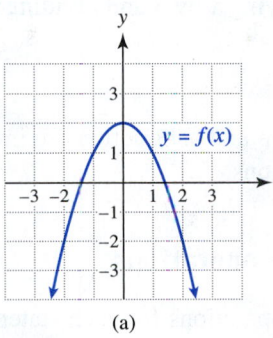
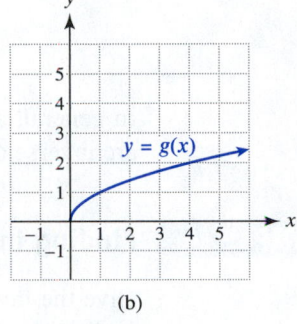

(a)   (b)

Figure 9.5

Solution

Figure 9.6(a) shows one of many horizontal lines that intersect the graph of $y = f(x)$ twice. Therefore function f is *not* one-to-one.

Two Points of Intersection:
f Is Not One-to-One

At Most One Point of
Intersection: g Is One-to-One

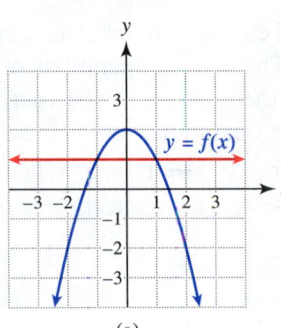
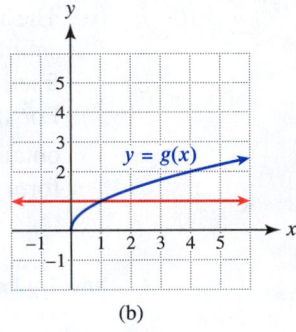

(a)   (b)

Figure 9.6

Figure 9.6(b) suggests that every horizontal line will intersect the graph of $y = g(x)$ *at most* once. Therefore function g is one-to-one.

Now Try Exercises 37, 39

MAKING CONNECTIONS

Vertical and Horizontal Line Tests

Function; Not One-to-One

The **vertical line test** is used to identify **functions**, whereas the **horizontal line test** is used to identify **one-to-one** functions. For example, consider the graph of $f(x) = x^2$. A vertical line never intersects the graph more than once, so f *is* a function. A horizontal line can intersect the graph twice, so f *is not* a one-to-one function.

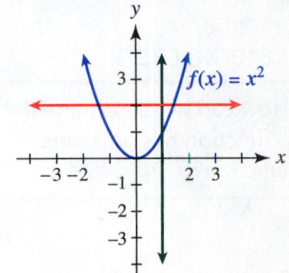

Inverse Functions

▶ **REAL-WORLD CONNECTION** Turning on a light and turning off a light are inverse operations from ordinary life. Inverse operations undo each other. In mathematics, adding **5** to x and subtracting **5** from x are inverse operations because

$$x + 5 - 5 = x.$$

Similarly, multiplying x by **5** and dividing x by **5** are inverse operations because

$$\frac{5x}{5} = x.$$

In general, addition and subtraction are inverse operations, and multiplication and division are inverse operations.

EXAMPLE 5 **Finding inverse operations**

Give the inverse operations for each statement. Then write a function f for the given statement and a function g for its inverse operations.
(a) Divide x by 3.
(b) Cube x and then add 1 to the result.

Solution
(a) The inverse of dividing x by 3 is to *multiply x by 3*. Thus

$$f(x) = \frac{x}{3} \quad \text{and} \quad g(x) = 3x.$$

(b) The inverse of cubing a number is taking a cube root, and the inverse of adding 1 is subtracting 1. The inverse operations of "cubing a number and then adding 1" are "subtracting 1 and then taking a cube root." For example, 2 cubed plus 1 is $2^3 + 1 = 9$. For the inverse operations, we *first* subtract 1 from 9 and then take the cube root to obtain 2. That is, $\sqrt[3]{9 - 1} = 2$. When there is more than one operation, we must perform the inverse operations in *reverse order*. Thus

$$f(x) = x^3 + 1 \quad \text{and} \quad g(x) = \sqrt[3]{x - 1}.$$

Now Try Exercises 43, 49

Functions f and g in each part of Example 5 are examples of *inverse functions*. Note that in part (a), if $f(x) = \frac{x}{3}$ and $g(x) = 3x$, then

$$f(15) = 5 \quad \text{and} \quad g(5) = 15.$$

In general, *if f and g are inverse functions*, $f(a) = b$ implies $g(b) = a$. Thus

$$(g \circ f)(a) = g(f(a)) = g(b) = a \qquad \text{\small\textit{f and g are inverse functions.}}$$

for any a in the domain of f, whenever g and f are inverse functions. The composition of a function with its inverse leaves the input unchanged.

READING CHECK

How can we determine if a function has an inverse function?

INVERSE FUNCTIONS

Let f be a one-to-one function. Then f^{-1} is the **inverse function** of f if

$$(f^{-1} \circ f)(x) = f^{-1}(f(x)) = x, \quad \text{for every } x \text{ in the domain of } f, \quad \text{and}$$
$$(f \circ f^{-1})(x) = f(f^{-1}(x)) = x, \quad \text{for every } x \text{ in the domain of } f^{-1}.$$

NOTE: In the expression $f^{-1}(x)$, the -1 is *not* an exponent. That is, $f^{-1}(x) \neq \frac{1}{f(x)}$. Rather, if $f(x) = \frac{x}{3}$, then $f^{-1}(x) = 3x$, and if $f(x) = x^3 + 1$, then $f^{-1}(x) = \sqrt[3]{x - 1}$.

EXAMPLE 6   **Verifying inverses**

Verify that $f^{-1}(x) = 3x$ if $f(x) = \frac{x}{3}$.

Solution
We must show that $(f^{-1} \circ f)(x) = x$ and that $(f \circ f^{-1})(x) = x$.

$$(f^{-1} \circ f)(x) = f^{-1}\big(f(x)\big) \quad \text{Composition of functions}$$

$$= f^{-1}\left(\frac{x}{3}\right) \quad f(x) = \frac{x}{3}$$

$$= 3\left(\frac{x}{3}\right) \quad f^{-1}(x) = 3x$$

$$= x \ \checkmark \quad \text{Simplify; it checks.}$$

$$(f \circ f^{-1})(x) = f\big(f^{-1}(x)\big) \quad \text{Composition of functions}$$

$$= f(3x) \quad f^{-1}(x) = 3x$$

$$= \frac{3x}{3} \quad f(x) = \frac{x}{3}$$

$$= x \ \checkmark \quad \text{Simplify; it checks.}$$

▌ **Now Try Exercise 51**

The definition of inverse functions states that *f must be a one-to-one function*. To understand why, consider Figure 9.7(a), where a one-to-one function *f* is represented by a diagram. To find f^{-1}, the arrows are simply reversed. For example, $f(1) = 3$ implies that $f^{-1}(3) = 1$, so the arrow from **1** to **3** for *f* must be redrawn in Figure 9.7(b), from **3** to **1** for f^{-1}.

One-to-One, So Inverse Exists   **Inverse Is Also One-to-One**

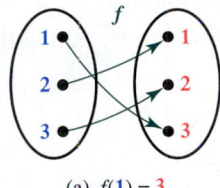

No two arrows go to the same output.   Reverse each arrow to find f^{-1}.

(a) $f(1) = 3$ (b) $f^{-1}(3) = 1$

Figure 9.7

In Figure 9.8(a), function *g* is *not* one-to-one because two different arrows, from **2** and **3**, both go to **1**. In Figure 9.8(b), the arrows are reversed in an attempt to draw the inverse *function*. However, **two** arrows now go from input **1** to outputs **2** and **3**. This means that *one input* has *two outputs*, which is not possible if g^{-1} is a function.

g Is Not One-to-One   **g Has No Inverse Function**

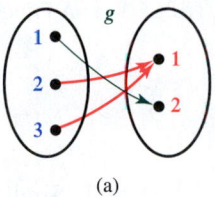
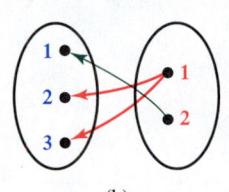

Two inputs give the same output: not one-to-one.   One input gives two outputs: not a function.

(a)   (b)

Figure 9.8

The following steps can be used to find the inverse of a function symbolically. Note that for a function f to have an inverse function, f must be one-to-one.

FINDING AN INVERSE FUNCTION

To find f^{-1} for a one-to-one function f, perform the following steps.

STEP 1: Let $y = f(x)$.

STEP 2: Interchange x and y.

STEP 3: Solve the formula for y. The resulting formula is $y = f^{-1}(x)$.

We apply these steps in the next example.

EXAMPLE 7 **Finding an inverse function**

Find the inverse of each one-to-one function.
(a) $f(x) = 3x - 7$
(b) $g(x) = (x + 2)^3$

Solution
(a) **STEP 1:** Let $y = 3x - 7$.

    STEP 2: Interchange x and y to get $x = 3y - 7$.

    STEP 3: To solve for y start by adding 7 to each side.

$$x + 7 = 3y \qquad \text{Add 7 to each side.}$$

$$\frac{x + 7}{3} = y \qquad \text{Divide each side by 3.}$$

    Thus $f^{-1}(x) = \frac{x+7}{3}$ or $f^{-1}(x) = \frac{1}{3}x + \frac{7}{3}$.

(b) **STEP 1:** Let $y = (x + 2)^3$.

    STEP 2: Interchange x and y to get $x = (y + 2)^3$.

    STEP 3: To solve for y start by taking the cube root of each side.

$$\sqrt[3]{x} = y + 2 \qquad \text{Take the cube root of each side.}$$

$$\sqrt[3]{x} - 2 = y \qquad \text{Subtract 2 from each side.}$$

    Thus $g^{-1}(x) = \sqrt[3]{x} - 2$.

Now Try Exercises 63, 71

Tables and Graphs of Inverse Functions

TABLES OF INVERSE FUNCTIONS Inverse functions can be represented with tables. Table 9.3 shows a table of values for a function f.

TABLE 9.3 **A Function f**

x	1	2	3	4	5
$f(x)$	3	6	9	12	15

Because $f(1) = 3$, $f^{-1}(3) = 1$. Similarly, $f(2) = 6$ implies that $f^{-1}(6) = 2$ and so on. Table 9.4 lists values for $f^{-1}(x)$.

TABLE 9.4 The Inverse Function f^{-1}

x	3	6	9	12	15
$f^{-1}(x)$	1	2	3	4	5

⎦— Interchange values

Note that the domain of f is {**1**, **2**, **3**, **4**, **5**} and that the range of f is {**3**, **6**, **9**, **12**, **15**}, whereas the domain of f^{-1} is {**3**, **6**, **9**, **12**, **15**} and the range of f^{-1} is {**1**, **2**, **3**, **4**, **5**}. *The domain of f is the range of f^{-1}, and the range of f is the domain of f^{-1}.* This statement is true in general for a function and its inverse.

GRAPHS OF INVERSE FUNCTIONS If $f(a) = b$, then the point (a, b) lies on the graph of f. This statement also means that $f^{-1}(b) = a$ and that the point (b, a) lies on the graph of f^{-1}. These points are shown in Figure 9.9(a) with a solid green line segment connecting them. The line $y = x$ is perpendicular to this line segment and divides it into two equal parts. As a result, the graph of f^{-1} can be sketched from the graph of f by reflecting the graph of f across the line $y = x$. See Figure 9.9(b).

Reflections Across the Line y = x

Point (b, a) is a reflection of point (a, b) across $y = x$.

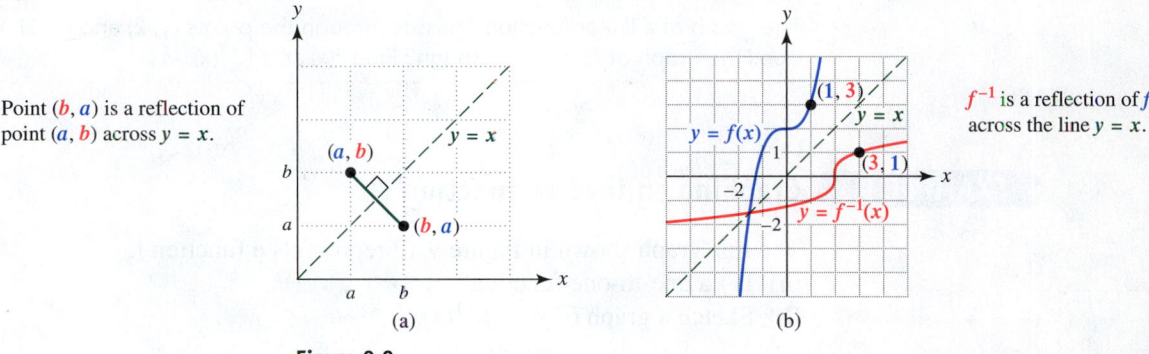

f^{-1} is a reflection of f across the line $y = x$.

(a)   (b)

Figure 9.9

The relationship between the graph of a function and the graph of its inverse function is summarized as follows.

GRAPHS OF FUNCTIONS AND THEIR INVERSES

The graph of f^{-1} is a reflection of the graph of f across the line $y = x$.

EXAMPLE 8 **Graphing an inverse function**

The graph of $y = f(x)$ is shown in Figure 9.10. Sketch a graph of $y = f^{-1}(x)$.

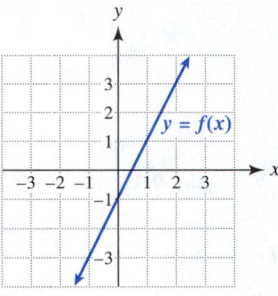

Figure 9.10

Solution

The graph of $y = f^{-1}(x)$ is the reflection of the graph of $y = f(x)$ across the line $y = x$ and is shown in Figure 9.11. Note that the reflection of a line is another line.

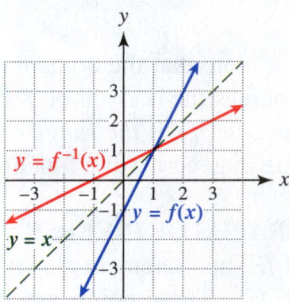

Figure 9.11

■ **Now Try Exercise 83**

CRITICAL THINKING

The graph of a linear function f passes through the points (1, 2) and (2, 1). What two points does the graph of f^{-1} pass through? Find $f(x)$ and $f^{-1}(x)$.

EXAMPLE 9 | **Graphing an inverse function**

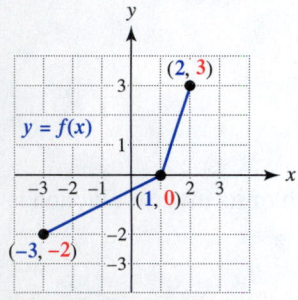

Figure 9.12

The line graph shown in Figure 9.12 represents a function f.
(a) Is f a one-to-one function?
(b) Sketch a graph of $y = f^{-1}(x)$.

Solution

(a) Every horizontal line intersects the graph of f at most once. By the horizontal line test, the graph represents a one-to-one function.
(b) The points $(-3, -2)$, $(1, 0)$, and $(2, 3)$ lie on the graph of f. It follows that the points $(-2, -3)$, $(0, 1)$, and $(3, 2)$ lie on the graph of f^{-1}. Plot these three points and then connect them with line segments, as shown in Figure 9.13(a). Note that the graph of $y = f^{-1}(x)$ is a reflection of the graph of $y = f(x)$ across the line $y = x$, as shown in Figure 9.13(b).

READING CHECK

If we are given the graph of $y = f(x)$, where f is one-to-one, how can we sketch the graph of $y = f^{-1}(x)$?

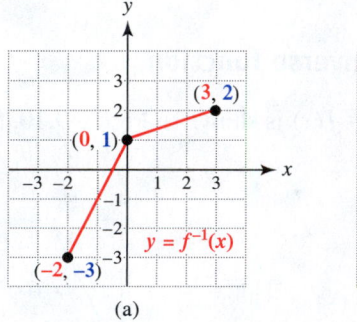
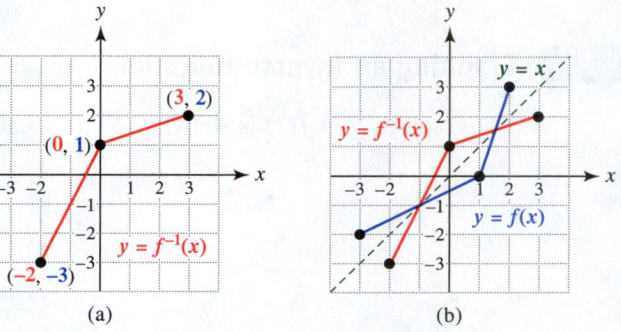

(a)   (b)

Figure 9.13

■ **Now Try Exercise 81**

9.1   Putting It All Together

CONCEPT	EXPLANATION	EXAMPLES
Composite Functions	The composite of g and f is given by $$(g \circ f)(x) = g\big(f(x)\big),$$ and represents a *new* function whose name is $g \circ f$.	If $f(x) = 1 - 4x$ and $g(x) = x^3$, then $$\begin{aligned}(g \circ f)(x) &= g\big(f(x)\big) \\ &= g(1 - 4x) \\ &= (1 - 4x)^3.\end{aligned}$$
One-to-One Functions	Function f is one-to-one if different inputs always give different outputs.	$f(x) = x^2$ is not one-to-one because $f(-4) = f(4) = 16$, whereas $g(x) = x + 1$ is one-to-one because every input has a unique output.
Horizontal Line Test	This test is used to determine whether a function is one-to-one from its graph: If every horizontal line intersects the graph of a function f at most once, then f is a one-to-one function.	$f(x) = x^2$ is not one-to-one because a horizontal line can intersect its graph more than once. **Not One-to-One** 
Inverse Functions	f^{-1} will undo the operations performed by f. That is, $$(f^{-1} \circ f)(x) = x \quad \text{and}$$ $$(f \circ f^{-1})(x) = x.$$	If $f(x) = x^3$, then $f^{-1}(x) = \sqrt[3]{x}$ because cubing a number x and then taking its cube root results in the number x.

9.1   Exercises

MyMathLab PRACTICE WATCH DOWNLOAD READ REVIEW

CONCEPTS AND VOCABULARY

1. $(g \circ f)(7) = $ _____

2. $(f \circ g)(x) = $ _____

3. Does $(f \circ g)(x)$ always equal $(g \circ f)(x)$?

4. If a function f is one-to-one, then different _____ always result in different _____.

5. If $f(3) = 5$ and $f(7) = 5$, could f be one-to-one?

6. If every horizontal line intersects the graph of f at most once, then f is _____.

7. The inverse operation of subtracting 10 is _____.

8. $(f^{-1} \circ f)(7) = $ _____.

9. If $f(6) = 8$, then $f^{-1}(\underline{\quad}) = $ _____.

10. If $f^{-1}(y) = x$, then $f(\underline{\quad}) = $ _____.

11. For f to have an inverse function, f must be _____.

12. The graph of f^{-1} is a(n) _____ of the graph of f across the line _____.

COMPOSITE FUNCTIONS

Exercises 13–22: For the given f(x) and g(x), find each of the following.

(a) $(g \circ f)(-2)$ (b) $(f \circ g)(4)$
(c) $(g \circ f)(x)$ (d) $(f \circ g)(x)$

13. $f(x) = x^2$ $g(x) = x + 3$

14. $f(x) = 4x^2$ $g(x) = 5x$

15. $f(x) = 2x$ $g(x) = x^3 - 1$

16. $f(x) = 3x + 1$ $g(x) = x^2 + 4x$

17. $f(x) = \frac{1}{2}x$ $g(x) = |x - 2|$

18. $f(x) = 6x$ $g(x) = \dfrac{2}{x - 5}$

19. $f(x) = \dfrac{1}{x}$ $g(x) = 3 - 5x$

20. $f(x) = \sqrt{x + 3}$ $g(x) = x^3 - 3$

21. $f(x) = 2x$ $g(x) = 4x^2 - 2x + 5$

22. $f(x) = 9x - \dfrac{1}{3x}$ $g(x) = \dfrac{x}{3}$

Exercises 23–28: Evaluate each expression numerically.

x	-2	-1	0	1	2
$f(x)$	2	1	0	-1	-2

x	-2	-1	0	1	2
$g(x)$	0	1	-1	2	-2

23. (a) $(f \circ g)(0)$ (b) $(g \circ f)(-1)$

24. (a) $(f \circ g)(1)$ (b) $(g \circ f)(-2)$

25. (a) $(f \circ f)(-1)$ (b) $(g \circ g)(0)$

26. (a) $(g \circ g)(2)$ (b) $(f \circ f)(1)$

27. (a) $(f^{-1} \circ g)(-2)$ (b) $(g^{-1} \circ f)(2)$

28. (a) $(f \circ g^{-1})(1)$ (b) $(g \circ f^{-1})(-2)$

Exercises 29 and 30: Evaluate each expression graphically.

29. (a) $(f \circ g)(0)$

(b) $(g \circ f)(1)$

(c) $(f \circ f)(-1)$

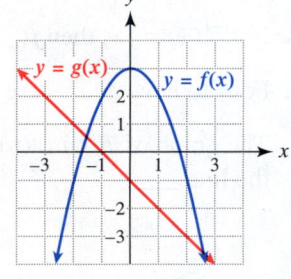

30. (a) $(f \circ g)(1)$

(b) $(g \circ f)(-2)$

(c) $(g \circ g)(-2)$

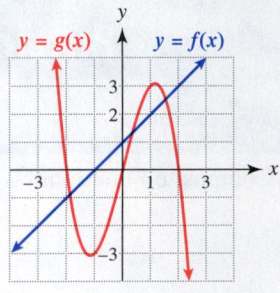

Exercises 31–36: Show that f is not a one-to-one function by finding two inputs that result in the same output. Answers may vary.

31. $f(x) = 5x^2$ **32.** $f(x) = 4 - x^2$

33. $f(x) = x^4 + 100$ **34.** $f(x) = \dfrac{x^2}{x^2 + 1}$

35. $f(x) = x^4 - 3x^2$ **36.** $f(x) = \sqrt{x^2 - 1}$

Exercises 37–42: Use the horizontal line test to determine whether the graph represents a one-to-one function.

37.

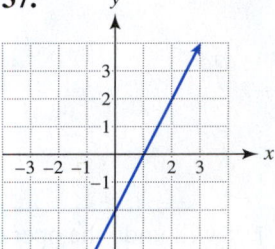

38.

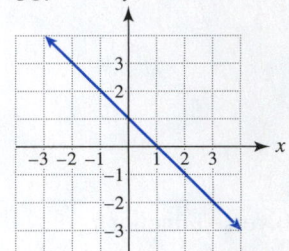

39.

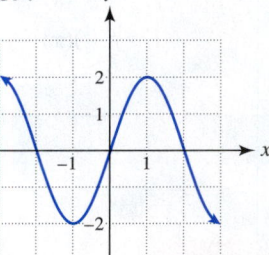

40.

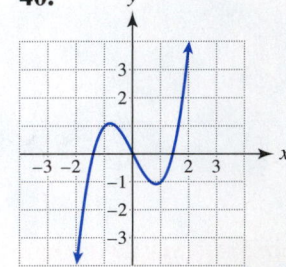

41.

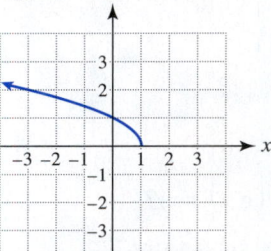

42.

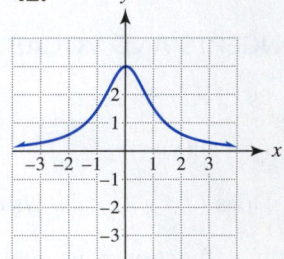

Exercises 43–50: (Refer to Example 5.) Give the inverse operation for the statement. Then write a function f for the given statement and a function g for its inverse.

43. Multiply x by 7.

44. Subtract 10 from x.

45. Add 5 to x and then divide the result by 2.

46. Multiply x by 6 and then add 8 to the result.

47. Multiply x by $\frac{1}{2}$ and then subtract 3 from the result.

48. Divide x by 10 and then add 20 to the result.

49. Cube the sum of x and 5.

50. Take the cube root of x and then subtract 2.

Exercises 51–58: (Refer to Example 6.) Verify that f(x) and f^{-1}(x) are indeed inverse functions.

51. $f(x) = 4x$ $\qquad$ $f^{-1}(x) = \dfrac{x}{4}$

52. $f(x) = \dfrac{2x}{3}$ $\qquad$ $f^{-1}(x) = \dfrac{3x}{2}$

53. $f(x) = 3x + 5$ $\qquad$ $f^{-1}(x) = \dfrac{x - 5}{3}$

54. $f(x) = x + 7$ $\qquad$ $f^{-1}(x) = x - 7$

55. $f(x) = x^3$ $\qquad$ $f^{-1}(x) = \sqrt[3]{x}$

56. $f(x) = \sqrt[3]{x - 4}$ $\qquad$ $f^{-1}(x) = x^3 + 4$

57. $f(x) = \dfrac{1}{x}$ $\qquad$ $f^{-1}(x) = \dfrac{1}{x}$

58. $f(x) = \dfrac{x + 7}{7}$ $\qquad$ $f^{-1}(x) = 7x - 7$

Exercises 59–74: (Refer to Example 7.) Find f^{-1}(x).

59. $f(x) = 12x$ $\qquad$ **60.** $f(x) = \frac{3}{4}x$

61. $f(x) = x + 8$ $\qquad$ **62.** $f(x) = x - 3$

63. $f(x) = 5x - 2$ $\qquad$ **64.** $f(x) = 3x + 4$

65. $f(x) = -\frac{1}{2}x + 1$ $\qquad$ **66.** $f(x) = \frac{3}{4}x - \frac{1}{4}$

67. $f(x) = 8 - x$ $\qquad$ **68.** $f(x) = 5 - x$

69. $f(x) = \dfrac{x + 1}{2}$ $\qquad$ **70.** $f(x) = \dfrac{3 - x}{5}$

71. $f(x) = \sqrt[3]{2x}$ $\qquad$ **72.** $f(x) = \sqrt[3]{x + 4}$

73. $f(x) = x^3 - 8$ $\qquad$ **74.** $f(x) = (x - 5)^3$

Exercises 75–78: Use the table to make a table of values for f^{-1}(x). State the domain and range for f and for f^{-1}.

75.

x	0	1	2	3	4
$f(x)$	0	5	10	15	20

76.

x	−4	−2	0	2	4
$f(x)$	1	2	3	4	5

77.

x	−5	0	5	10	15
$f(x)$	4	2	0	−2	−4

78.

x	0	2	4	6	8
$f(x)$	8	6	4	2	0

Exercises 79–82: (Refer to Example 9.) Use the graph of y = f(x) to sketch a graph of y = f^{-1}(x).

79.
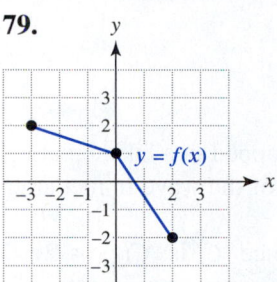

80.
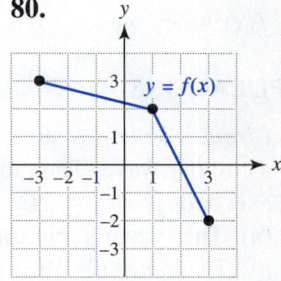

81.
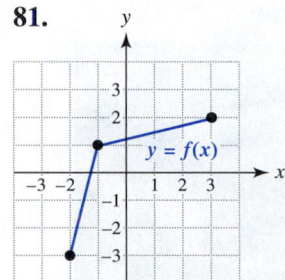

82.
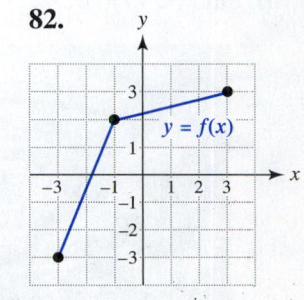

Exercises 83–86: (Refer to Example 8.) Use the graph of y = f(x) to sketch a graph of f^{-1}(x). Include the graph of f and the line y = x in your graph.

83.
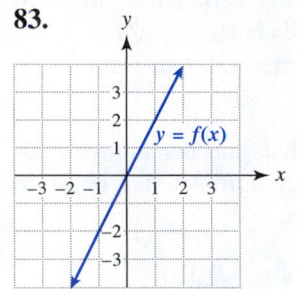

84.
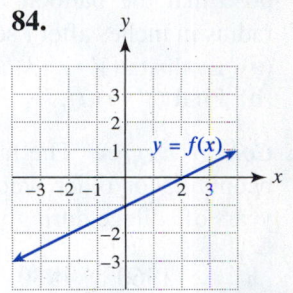

85.

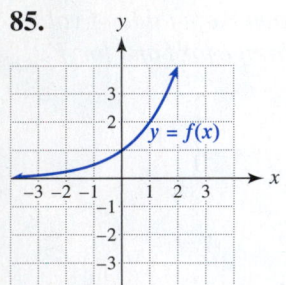

86.

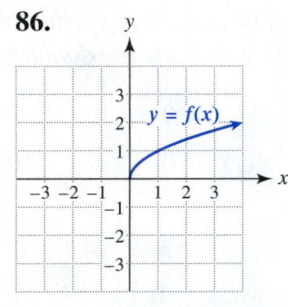

OPERATIONS ON FUNCTIONS

Exercises 87–90: Find each of the following for the given
f(x) and g(x).

    (a) $(fg)(2)$ *(b)* $(f - g)(x)$ *(c)* $(f \circ g)(x)$

87. $f(x) = x^2 - 2, g(x) = x^2 + 2$

88. $f(x) = 2x^2, g(x) = 2x - 1$

89. $f(x) = \dfrac{1}{x}, g(x) = \dfrac{2}{x}$

90. $f(x) = x^3, g(x) = \sqrt[3]{x}$

APPLICATIONS

91. *Circular Wave* A stone is dropped in a lake, creating a circular wave. The radius r of the wave in feet after t seconds is $r(t) = 2t$.
  (a) The wave's circumference C is $C(r) = 2\pi r$. Evaluate $(C \circ r)(5)$ and interpret your result.
  (b) Find $(C \circ r)(t)$.

92. *Volume of a Balloon* The volume V of a spherical balloon with radius r is given by $V(r) = \frac{4}{3}\pi r^3$. Suppose that the balloon is being inflated so that the radius in inches after t seconds is $r(t) = \sqrt[3]{t}$.
  (a) Evaluate $(V \circ r)(3)$ and interpret your result.
  (b) Find $(V \circ r)(t)$.

93. *College Degree* The table lists the percentage P of people 25 or older who have completed 4 or more years of college during year x.

x	1960	1980	2000	2010
$P(x)$	8	16	27	29

Source: U.S. Census Bureau.

  (a) Evaluate $P(1980)$ and interpret the results.
  (b) Make a table for $P^{-1}(x)$.
  (c) Evaluate $P^{-1}(16)$.

94. *Skin Cancer and Ozone* Ozone in the stratosphere filters out most of the harmful ultraviolet (UV) rays from the sun. However, depletion of the ozone layer affects this protection. The formula $U(x) = 1.5x$ calculates the percent increase in UV radiation for an x percent decrease in the thickness of the ozone layer. The formula $C(x) = 3.5x$ calculates the percent increase in skin cancer cases when the UV radiation increases by x percent. (*Source:* R. Turner, D. Pierce, and I. Bateman, *Environmental Economics.*)
  (a) Evaluate $U(2)$ and $C(3)$ and interpret each result.
  (b) Find $(C \circ U)(2)$ and interpret the result.
  (c) Find $(C \circ U)(x)$. What does it calculate?

95. *Temperature and Mosquitoes* Temperature can affect the number of mosquitoes observed on a summer night. Graphs of two functions, T and M, are shown. Function T calculates the temperature on a summer evening h hours past midnight, and M calculates the number of mosquitoes observed per 100 square feet when the outside temperature is T.

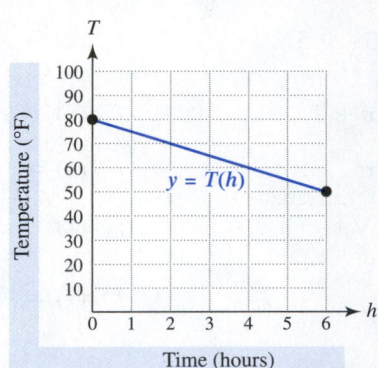

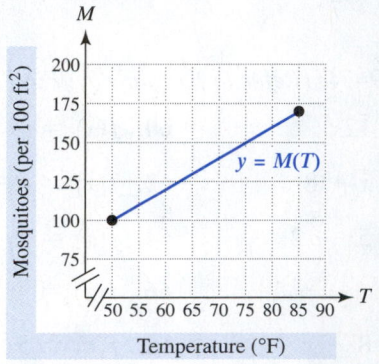

  (a) Find $T(1)$ and $M(75)$.
  (b) Evaluate $(M \circ T)(1)$ and interpret your result.
  (c) What does $(M \circ T)(h)$ calculate?
  (d) Find equations for the lines in each graph.
  (e) Use your answers from part (d) to write a formula for $(M \circ T)(h)$.

96. *High School Grades* The table lists the percentage P of college freshmen with a high school grade average of A or A– during year x.

x	1970	1980	1990	2000	2010
$P(x)$	20	26	29	43	46

Source: Department of Education.

(a) Evaluate $P(1970)$ and interpret the results.
(b) Make a table for $P^{-1}(x)$.
(c) Evaluate $P^{-1}(43)$.

97. *Temperature* The function given by $f(x) = \frac{9}{5}x + 32$ converts x degrees Celsius to an equivalent temperature in degrees Fahrenheit.
(a) Is f a one-to-one function? Why or why not?
(b) Find $f^{-1}(x)$ and interpret what it calculates.

98. *Feet and Yards* The function given by $f(x) = 3x$ converts x yards to feet.
(a) Is f a one-to-one function? Why or why not?
(b) Find $f^{-1}(x)$ and interpret what it calculates.

99. *Quarts and Gallons* Write a function f that converts x gallons to quarts. Then find $f^{-1}(x)$ and interpret what it computes.

100. *One-to-One Function* The table lists monthly average wind speeds at Hilo, Hawaii, in miles per hour from July through December, where x is the month.

x	July	Aug	Sept	Oct	Nov	Dec
$f(x)$	7	7	7	7	7	7

(a) Is function f one-to-one? Explain.
(b) Does f^{-1} exist?

(c) What happens if you try to make a table for f^{-1}?
(d) Could f be one-to-one if it were computed at a different location? What would have to be true about the monthly average wind speeds?

101. **Online Exploration** Go online and find how many cups there are in one gallon and how many teaspoons there are in one cup.
(a) Write a function C that converts x gallons to cups.
(b) Write a function T that converts x cups to teaspoons.
(c) Write the composite function $(T \circ C)(x)$, where x represents gallons.
(d) Evaluate $(T \circ C)(3)$ and interpret the result.

102. **Online Exploration** Go online and find the number of yuan (Chinese currency) in 1 dollar and the number of rupees (Indian currency) in 1 yuan. Answers may vary over time.
(a) Write a function Y that converts x dollars to yuan.
(b) Write a function R that converts x yuan to rupees.
(c) Write the composite function $(R \circ Y)(x)$, where x represents dollars.
(d) Evaluate $(R \circ Y)(100)$ and interpret the result.

WRITING ABOUT MATHEMATICS

103. Explain the difference between $(g \circ f)(2)$ and $(f \circ g)(2)$. Are they always equal? If f and g are inverse functions, evaluate $(g \circ f)(2)$ and $(f \circ g)(2)$.

104. Explain what it means for a function to be one-to-one.

9.2 Exponential Functions

Basic Concepts ● Graphs of Exponential Functions ● Percent Change and Exponential Functions ● Compound Interest ● Models Involving Exponential Functions ● The Natural Exponential Function

A LOOK INTO MATH ▶

Suppose that we deposit $100 into a savings account that pays 2% annual interest. If neither the money nor the interest is withdrawn, then after 1 year the account balance will be $102 and after 2 years the account balance will be $104.04. The extra 4 cents in interest in the second year occurs because we get not only 2% interest on the original $100 but also 2% interest on the $2 interest earned during the first year. In general, when an amount A experiences a constant percent change over each fixed time period, such as a year or a month, then the growth (or possibly decay) in A can be described by an exponential function. In this section we discuss exponential functions and many of their applications, including how to calculate interest.

Basic Concepts

▶ **REAL-WORLD CONNECTION** Suppose that an insect population (per acre) doubles each week. Table 9.5 shows the size of the population after x weeks. Note that, as the population of insects becomes larger, the *increase* in population each week becomes greater. The population is increasing by 100%, or doubling, each week. When a quantity increases by a constant percentage (or constant factor) at regular intervals, its growth is *exponential*.

TABLE 9.5 **Insect Population**

Week	0	1	2	3	4	5
Population	100	200	400	800	1600	3200

×2   ×2   ×2   ×2   ×2
Doubles each week

We can model the data in Table 9.5 by using the exponential function

$$f(x) = 100(2)^x.$$

For example,

$$f(0) = 100(2)^0 = 100 \cdot 1 = 100,$$
$$f(1) = 100(2)^1 = 100 \cdot 2 = 200,$$
$$f(2) = 100(2)^2 = 100 \cdot 4 = 400,$$

and so on. Note that the exponential function f has a *variable as an exponent*.

EXPONENTIAL FUNCTION

A function represented by

$$f(x) = Ca^x, \quad a > 0 \quad \text{and} \quad a \neq 1,$$

is an **exponential function with base a and coefficient C**. (Unless stated otherwise, we assume that $C > 0$.)

In the formula $f(x) = Ca^x$, a is called the **growth factor** when $a > 1$ and the **decay factor** when $0 < a < 1$. For an exponential function, each time x increases by 1 unit, $f(x)$ increases by a factor of a when $a > 1$ and decreases by a factor of a when $0 < a < 1$. Moreover, because

$$f(0) = Ca^0 = C(1) = C,$$

the value of C equals the value of $f(x)$ when $x = 0$. That is, C is the y-intercept. If x represents time, C represents the initial value of f when time equals 0. Figure 9.14 illustrates **exponential growth** and **exponential decay** for $x > 0$.

Exponential Growth for Positive x

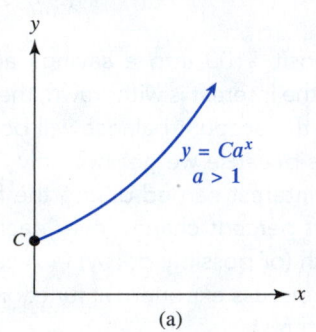

$y = Ca^x$
$a > 1$

(a)

Exponential Decay for Positive x

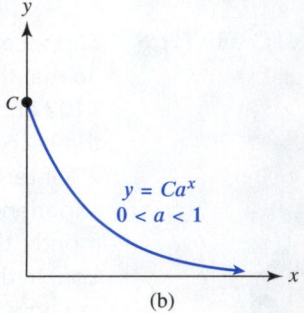

$y = Ca^x$
$0 < a < 1$

(b)

Figure 9.14

The set of valid inputs (domain) for an exponential function includes all real numbers. The set of corresponding outputs (range) includes all positive real numbers.

When evaluating an exponential function, we evaluate exponents *before* we multiply. For example, if $f(x) = 4(3)^x$, then

$$f(2) = 4(3)^2 = 4(9) = 36.$$

Evaluate exponents first.

The next example illustrates how exponential functions are evaluated.

EXAMPLE 1 **Evaluating exponential functions**

Evaluate $f(x)$ for the given value of x.
(a) $f(x) = 10(3)^x$ $x = 2$ **(b)** $f(x) = 5\left(\frac{1}{2}\right)^x$ $x = 3$
(c) $f(x) = \frac{1}{3}(2)^x$ $x = -1$

Solution
(a) $f(2) = 10(3)^2 = 10 \cdot 9 = 90$
(b) $f(3) = 5\left(\frac{1}{2}\right)^3 = 5 \cdot \frac{1}{8} = \frac{5}{8}$
(c) $f(-1) = \frac{1}{3}(2)^{-1} = \frac{1}{3} \cdot \frac{1}{2} = \frac{1}{6}$

NOTE: $f(x)$ is also defined for all *negative* inputs.

Now Try Exercises 11, 13, 15

MAKING CONNECTIONS

The Expressions a^{-x} and $\left(\frac{1}{a}\right)^x$

Using properties of exponents, we can write 2^{-x} as

$$2^{-x} = \frac{1}{2^x} = \left(\frac{1}{2}\right)^x.$$

In general, the expressions a^{-x} and $\left(\frac{1}{a}\right)^x$ are equal for all positive values of a.

Graphs of Exponential Functions

We can graph $f(x) = 2^x$ by first evaluating some points, as in Table 9.6. If we plot these points and sketch the graph, we obtain Figure 9.15.

TABLE 9.6
$f(x) = 2^x$

x	2^x
-2	$\frac{1}{4}$
-1	$\frac{1}{2}$
0	1
1	2
2	4

Graphing $f(x) = 2^x$

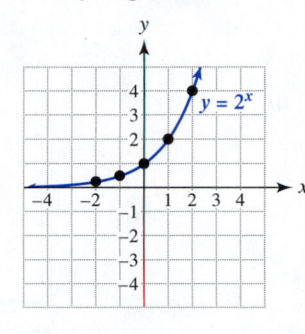

Figure 9.15

- This graph is always above the x-axis.
- The graph passes through the point $(0, 1)$.
- Negative x-values give y-values between 0 and 1.
- Positive x-values give y-values greater than 1.

In Figure 9.16, we investigate the graph of $y = a^x$ for values of a that are greater than 1 by graphing $y = 1.3^x$, $y = 1.7^x$, and $y = 2.5^x$. Graphs of $y = a^x$ where a is between 0 and 1 are shown in Figure 9.17.

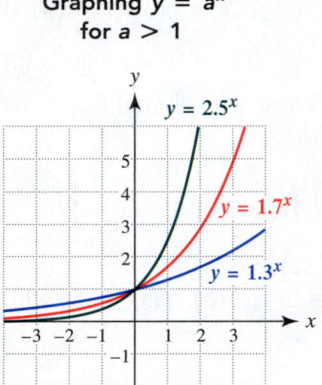

Graphing $y = a^x$
for $a > 1$

Figure 9.16

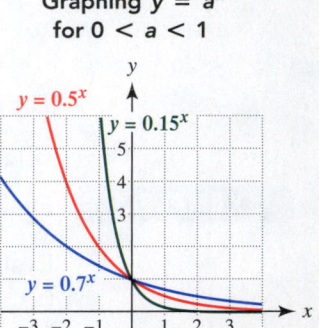

Graphing $y = a^x$
for $0 < a < 1$

Figure 9.17

- As x increases, the y-values increase.
- Larger values of a ($a > 1$) result in y-values that increase more rapidly.
- The graphs pass through $(0, 1)$ because $C = 1$.
- The graphs are always above the x-axis.

- As x increases, the y-values decrease.
- Smaller values of a ($0 < a < 1$) result in y-values that decrease more rapidly.
- The graphs pass through $(0, 1)$ because $C = 1$.
- The graphs are always above the x-axis.

In the next example we show the dramatic difference between the outputs of linear and exponential functions.

> **EXAMPLE 2**   **Comparing exponential and linear functions**
>
> Compare $f(x) = 3^x$ and $g(x) = 3x$ graphically and numerically for $x \geq 0$.

Solution
Graphical Comparison The graphs of $y_1 = 3^x$ and $y_2 = 3x$ are shown in Figure 9.18. For large values of x, the graph of the exponential function y_1 increases much faster than the graph of the linear function y_2.
Numerical Comparison The tables of $y_1 = 3^x$ and $y_2 = 3x$ are shown in Figure 9.19. For large values of x, the values for y_1 increase much faster than the values for y_2.

Comparing Exponential and Linear Growth

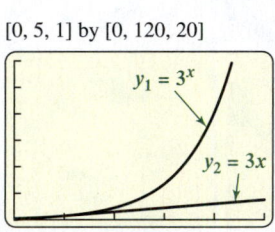

[0, 5, 1] by [0, 120, 20]

$y_1 = 3^x$

$y_2 = 3x$

Figure 9.18

Exponential growth ⟶   ⟵ Linear growth

X	Y₁	Y₂
0	1	0
1	3	3
2	9	6
3	27	9
4	81	12
5	243	15
6	729	18

X=0

Figure 9.19

Now Try Exercise 89

NOTE: The results of Example 2 are true in general: for large enough inputs, exponential functions with $a > 1$ grow far faster than any linear function.

MAKING CONNECTIONS

Exponential and Polynomial Functions

The function $f(x) = 2^x$ is an exponential function. The base **2** is a constant and the exponent x is a variable, so $f(3) = 2^3 = 8$.

The function $g(x) = x^2$ is a quadratic (polynomial) function. The base x is a variable and the exponent **2** is a constant, so $g(3) = 3^2 = 9$.

The table clearly shows that the exponential function grows much faster than the quadratic function for large values of x.

x	0	2	4	6	8	10	12	
2^x	1	4	16	64	256	1024	4096	⟵ Exponential growth
x^2	0	4	16	36	64	100	144	⟵ Quadratic growth

In the next example, we determine whether a function is linear or exponential.

EXAMPLE 3

Finding linear and exponential functions

For each table, determine whether f is a linear function or an exponential function. Find a formula for f.

(a)

x	0	1	2	3	4
$f(x)$	16	8	4	2	1

(b)

x	0	1	2	3	4
$f(x)$	5	7	9	11	13

(c)

x	0	1	2	3	4
$f(x)$	1	3	9	27	81

Solution

(a) Each time x increases by 1 unit, $f(x)$ decreases by a factor of $\frac{1}{2}$. Therefore f is an exponential function with a *decay factor* of $\frac{1}{2}$. Because $f(0) = 16$, $C = \mathbf{16}$, so $f(x) = \mathbf{16}\left(\frac{1}{2}\right)^x$. This formula can also be written as $f(x) = 16(2)^{-x}$.

(b) Each time x increases by 1 unit, $f(x)$ increases by **2** units. Therefore f is a linear function, and the slope of its graph equals **2**. The y-intercept is **5**, so $f(x) = \mathbf{2}x + \mathbf{5}$.

(c) Each time x increases by 1 unit, $f(x)$ increases by a factor of **3**. Therefore f is an exponential function with a *growth factor* of **3**. Because $f(0) = 1$, it follows that $C = \mathbf{1}$ and $f(x) = \mathbf{1}(3)^x$, or $f(x) = 3^x$.

Now Try Exercises 45, 47, 49

READING CHECK

How can you distinguish between a linear function and an exponential function?

MAKING CONNECTIONS

Linear and Exponential Functions

For a *linear function*, given by $f(x) = ax + b$, each time x increases by 1 unit y increases (or decreases) *by a units*, where a equals the slope of the graph of f.

For an *exponential function*, given by $f(x) = Ca^x$, each time x increases by 1 unit y increases *by a factor of a* when $a > 1$ and decreases by a factor of a when $0 < a < 1$. The constant a equals either the growth factor or the decay factor.

Percent Change and Exponential Functions

PERCENT CHANGE When an amount A changes to a new amount B, then the **percent change** is calculated by

$$\frac{B - A}{A} \times 100. \qquad \text{Percent change formula}$$

We multiply the ratio by 100 to change decimal form to percent form.

EXAMPLE 4 Finding percent change

Complete the following.
(a) Find the percent change if an account balance increases from \$500 to \$1000.
(b) Find the percent change if an account balance decreases from \$1000 to \$500.

Solution
(a) Let $A = 500$ and $B = 1000$.

$$\frac{1000 - 500}{500} \times 100 = \frac{500}{500} \times 100 \qquad \frac{B - A}{A} \times 100$$

$$= 1 \times 100 \qquad \text{Simplify.}$$

$$= 100\% \qquad \text{Multiply.}$$

The percent change (increase) is 100%; that is, the account balance doubles.

(b) Let $A = 1000$ and $B = 500$.

$$\frac{500 - 1000}{1000} \times 100 = -\frac{500}{1000} \times 100 \qquad \frac{B - A}{A} \times 100$$

$$= -\frac{1}{2} \times 100 \qquad \text{Simplify.}$$

$$\approx -50\% \qquad \text{Multiply.}$$

The percent change (decrease) is -50%.

Now Try Exercise 51

NOTE: In Example 4, the account did *not* increase by 100% and then decrease by 100% to return the account to its initial value of \$500. In part (b) the initial amount is $A = \$1000$ and needs to decrease only by a factor of $\frac{1}{2}$, or 50%, to decrease to \$500.

PERCENT CHANGE AND GROWTH FACTOR Suppose a savings account A increases from \$200 to \$600. The percent change, $R\%$, equals

$$\frac{600 - 200}{200} \times 100 = 2 \times 100 = 200\%.$$

Note that the account balance *tripled* from \$200 to 600, but the percent change is 200%, *not* 300%. Also, the *increase* in the account balance A is \$400 because \$600 − \$200 = \$400. This \$400 increase equals 200% of \$200. If we let r represent the percent change as a decimal, then

$$200\% \text{ of } \$200 = 2.00 \times \$200 = \$400. \qquad R\% = 200\%; r = 2.00$$

$$R\% \text{ of } A \quad = \quad rA \quad = \quad \text{Increase in } A$$

In general, if the *percent increase* in an account balance A is given by r in decimal form, then the *amount of increase* in A is equal to rA and the new balance after the increase is $A + rA$. For example, if \$200 increases by 200%, then $r = 2.00$, $A = \$200$, and

$$rA = 2.00(200) = \$400, \qquad \text{Increase in account balance}$$

so the balance increases by \$400. The new balance for the account is

$$A + rA = \$200 + \$400 = \$600.$$

Initial amount + Increase = Final amount

If we factor out A in the expression $A + rA$, we get

$$A + rA = A(1 + r) \qquad \text{Factor out } A.$$

Initial amount + Increase in A = Initial amount × Growth factor

Thus if an account increases from \$200 to \$600, the increase is \$400 and the percent increase is 200%. In addition, the account balance increased by a growth factor equal to

$$a = 1 + r = 1 + 2.00 = 3, \qquad \text{Growth factor is } a = 3.$$

which means that the account balance tripled.

EXAMPLE 5 | **Analyzing the decrease in an account balance**

An account that contains \$2000 decreases in value by 20%.
(a) Find the decrease in value of the account.
(b) Find the final value of the account.
(c) By what factor a did the account decrease?

Solution
(a) Let $A = 2000$ and $r = -0.20$ (-20% in decimal form). The decrease is

$$rA = -0.20(2000) = -400. \qquad \text{Decrease in } A \text{ is } rA.$$

The account decreased in value by \$400.

STUDY TIP

Be sure that you understand how a percent change relates to the growth factor of an exponential function.

(b) The final value of the account is

$$A + rA = 2000 + (-400) = \$1600.$$

Initial amount + Decrease = New amount

(c) The account decreased by a factor of

$$a = 1 + r = 1 + (-0.20) = 0.80. \qquad \text{Decay factor is } a = 0.80.$$

The account value decreased to 80% of its original value because 80% of \$2000 is \$1600.

Now Try Exercise 55

NOTE: In general, a positive amount A cannot *decrease* by more than 100% because a 100% decrease would reduce A to 0. However, percent *increases* can be more than 100%.

PERCENT CHANGE AND EXPONENTIAL FUNCTIONS An exponential function results when an *initial value C* is multiplied by a *growth* (or *decay*) *factor a* for each unit increase in x. For example, if the population P of a city is 100,000 people, and the city is growing at 5% per year, then $C = 100,000$ and the growth factor is

$$a = 1 + r = 1 + 0.05 = 1.05.$$

Thus the exponential function

$$P(x) = 100,000(1.05)^x \qquad \begin{array}{l} P(x) = Ca^x \text{ with } C = 100,000 \\ \text{and } a = 1.05. \end{array}$$

models the city's population after x years. For example,

$$P(6) = 100,000(1.05)^6 \approx 134,000$$

indicates that after 6 years the city's population has grown to about 134,000 people. These concepts are summarized by the following.

PERCENT CHANGE AND EXPONENTIAL FUNCTIONS

Suppose that an amount A increases or decreases by $R\%$ (or by r expressed in decimal form) for each unit increase in x.

1. If $r > 0$, the *growth factor* is $a = 1 + r$ and $a > 1$.
2. If $-1 < r < 0$, the *decay factor* is $a = 1 + r$ and $0 < a < 1$.
3. If the initial amount is C, the amount A after x-unit increases is given by

$$A(x) = Ca^x \quad \text{or equivalently,} \quad A(x) = C(1 + r)^x.$$

READING CHECK

If your income increases by 200%, what is the growth factor?

| EXAMPLE 6 | **Analyzing growth of bacteria** |

Initially a laboratory culture contains 50,000 bacteria per milliliter and it is increasing in numbers by 20% per hour.
(a) Write the formula for an exponential function B that gives the number of bacteria per milliliter after x hours.
(b) Evaluate $B(3)$ and interpret your result.

Solution
(a) The initial value is $C = 50,000$ and the hourly percent increase is 20%, or $r = 0.20$ in decimal form. Thus the growth factor is

$$a = 1 + r = 1 + 0.20 = 1.20. \qquad \begin{array}{l} \text{Growth factor} = 1.20 \\ \text{with } r = 0.20. \end{array}$$

Because the exponential function can be written as $B(x) = Ca^x$, it follows that

$$B(x) = 50,000(1.20)^x.$$

(b) $B(3) = 50,000(1.20)^3 = 50,000(1.728) = 86,400$. Thus after 3 hours, the culture contains 86,400 bacteria per milliliter.

Now Try Exercise 85

Compound Interest

▶ **REAL-WORLD CONNECTION** If $100 is deposited in a savings account paying 10% annual interest, the interest earned after 1 year equals $100 × 0.10 = $10. The total amount of money in the account after 1 year is $100(1 + 0.10) = $110. Each year the money in the account increases by a growth factor of 1.10, so after x years there will be $100(1.10)^x$ dollars in the account. Thus **compound interest** is an example of exponential growth.

COMPOUND INTEREST

If P dollars is deposited in an account and if interest is paid at the end of each year with an annual rate of interest r, expressed in decimal form, then after t years the account will contain A dollars, where

$$A = P(1 + r)^t.$$

The growth factor is $(1 + r)$.

NOTE: The compound interest formula takes the form of an exponential function with

$$a = 1 + r.$$

EXAMPLE 7 **Calculating compound interest**

A 20-year-old worker deposits $2000 in a retirement account that pays 6% annual interest at the end of each year. How much money will be in the account when the worker is 65 years old? What is the growth factor?

Solution

Here, $P = 2000$, $r = 0.06$, and $t = 45$. The amount in the account after 45 years is

$$A = 2000(1 + 0.06)^{45} \approx \$27,529.22, \quad A = P(1 + r)^t$$

which is supported by Figure 9.20. Each year the amount of money in the account is multiplied by a factor of $(1 + 0.06)$, so the growth factor is 1.06.

```
2000(1+.06)^45
      27529.22165
```

Figure 9.20

Now Try Exercise 65

INTEREST PAID MORE THAN ONCE A YEAR Many times, interest is paid more than once a year. For example, suppose an account gives 8% annual interest that is paid every 3 months, or *quarterly*. It follows that $\frac{1}{4}$ of the annual 8% interest, or 2%, is paid every 3 months. The growth factor is $a = 1 + 0.02 = 1.02$ for each 3-month period. If the initial balance is $1000, then after **5** years, the **quarterly** 2% interest has been paid $4 \cdot 5 = 20$ times, and the account contains

$$\$1000(1.02)^{20} \approx \$1485.95.$$

In general, if P dollars are deposited in an account paying an *annual* interest rate r (in decimal form) and this interest rate is compounded or paid n times per year, then after t years the account contains A dollars, given by

Amount after t years

Annual interest (decimal)

Number of years

$$A = P\left(1 + \frac{r}{n}\right)^{nt}$$

Initial deposit

Number of times interest is paid per year

EXAMPLE 8 **Calculating compound interest**

Initially, $1500 is deposited in an account paying 6% annual interest, compounded monthly. What is the account balance after 5 years?

Solution

Let $P = \mathbf{1500}$, $r = \mathbf{0.06}$, $n = \mathbf{12}$, and $t = \mathbf{5}$. The balance after 5 years is

$$A = P\left(1 + \frac{r}{n}\right)^{nt} \qquad \text{Interest formula}$$

$$= \mathbf{1500}\left(1 + \frac{\mathbf{0.06}}{\mathbf{12}}\right)^{(\mathbf{12 \cdot 5})} \qquad \text{Substitute.}$$

$$= 1500(1.005)^{60} \qquad \text{Evaluate.}$$

$$\approx \$2023.28. \qquad \text{Approximate.}$$

▌ **Now Try Exercise 73**

NOTE: Generally, compounding interest more frequently results in more interest being paid. In Example 8, if the annual 6% interest had been paid only once a year, the growth factor would be larger, $a = 1.06$, but this 6% annual interest would have been paid only 5 times, once each year. The new balance would be $1500(1.06)^5 \approx \$2007.34$, rather than the $2023.28 that resulted from monthly compounding.

Models Involving Exponential Functions

▶ **REAL-WORLD CONNECTION** Apple's path to 2 billion iPhone application downloads occurred over a relatively short period of time. In the next example, we model this exponential growth of iPhone application downloads.

EXAMPLE 9 **Modeling iPhone application downloads**

In September 2008 about 100 million iPhone applications were downloaded, and the number of iPhone downloads was increasing at a rate of 28.4% per month. (*Source: Apple Corp.*)
(a) What was the monthly growth factor?
(b) Write an exponential function $P(x) = Ca^x$ that models the number of downloads in millions for the iPhone x months after September 2008.
(c) Evaluate $P(12)$ and interpret the result.

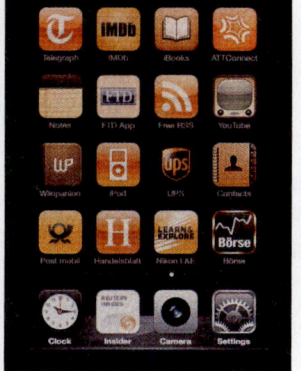

Solution
(a) Because $r = 0.284$, the monthly growth factor was $a = 1 + 0.284 = 1.284$.
(b) Let $C = 100$ and $a = 1.284$. Then $P(x) = 100(1.284)^x$.
(c) $P(12) = 100(1.284)^{12} \approx 2000$. In approximately 12 months, from September 2008 to September 2009, iPhone application downloads grew from 100 million to about 2000 million, or 2 billion.

▌ **Now Try Exercise 91**

▶ **REAL-WORLD CONNECTION** Traffic flow at intersections can be modeled by exponential functions whenever traffic patterns occur randomly. In the next example we model traffic at an intersection by using an exponential function.

EXAMPLE 10 | **Modeling traffic flow**

On average, a particular intersection has 360 vehicles arriving randomly each hour. Traffic engineers use $f(x) = (0.905)^x$ to estimate the likelihood, or probability, that *no* vehicle will enter the intersection within an interval of *x* seconds. (*Source:* F. Mannering and W. Kilareski, *Principles of Highway Engineering and Traffic Analysis.*)

(a) Compute $f(5)$ and interpret the results.
(b) A graph of $y = f(x)$ is shown in Figure 9.21. Discuss this graph.
(c) Is this function an example of exponential growth or decay?

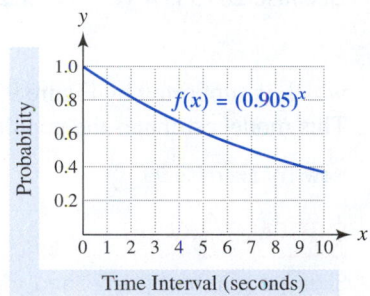

Figure 9.21

Solution
(a) The result $f(5) = (0.905)^5 \approx 0.61$ indicates that there is a 61% chance that no vehicle will enter the intersection during any particular 5-second interval.
(b) The graph decreases, which means that as the interval of time increases there is less chance (likelihood) that a car will *not* enter the intersection.
(c) Because the graph is decreasing and $a = 0.905 < 1$, this function is an example of exponential decay.

▌ **Now Try Exercise 97**

The Natural Exponential Function

A special type of exponential function is called the *natural exponential function*, expressed as $f(x) = e^x$. The **base** e is a special number in mathematics similar to π. The number π is approximately 3.14, whereas the number e is approximately 2.72. The number e is named for the great Swiss mathematician Leonhard Euler (1707–1783). Most calculators have a special key that can be used to compute the natural exponential function.

NATURAL EXPONENTIAL FUNCTION

The function represented by

$$f(x) = e^x$$

is the **natural exponential function**, where $e \approx 2.71828$.

CALCULATOR HELP

To evaluate the natural exponential function, see Appendix A (page AP-8).

▶ **REAL-WORLD CONNECTION** The natural exponential function is frequently used to model **continuous growth**. For example, the fact that births and deaths occur throughout the year, not just at one time during the year, must be recognized when population growth is being modeled. If a population P is growing continuously at r percent per year, expressed as a decimal, we can model this population after x years by

$$P = Ce^{rx}, \quad \text{Continuous growth, } r > 0$$

where C is the initial population. To evaluate natural exponential functions, we use a calculator, as demonstrated in the next example.

EXAMPLE 11 **Modeling population**

In 2011 Texas' population was 26 million people and was growing at a continuous rate of 2% per year. This population in millions x years after 2011 can be modeled by

$$f(x) = 26e^{0.02x}.$$

Estimate the population in 2015.

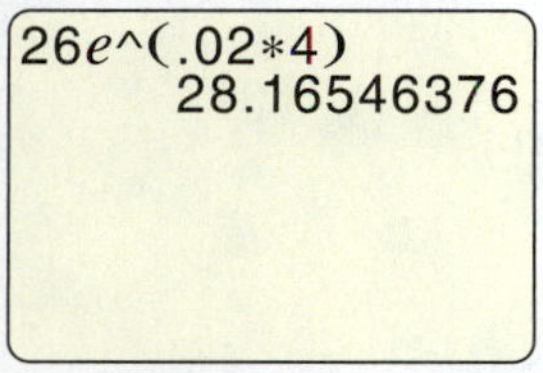

26e^(.02*4)
                28.16546376

Figure 9.22

Solution
Because 2015 is 4 years after 2011, we evaluate $f(4)$ to obtain

$$f(4) = 26e^{0.02(4)} \approx 28.2,$$

which is supported by Figure 9.22. (Be sure to include parentheses around the exponent of e.) This model estimates the population of Texas to be about 28.2 million in 2015.

Now Try Exercise 99

CRITICAL THINKING

Sketch a graph of $y = 2^x$ and $y = 3^x$ in the same xy-plane. Then use these two graphs to sketch a graph of $y = e^x$. How do these graphs compare?

9.2 Putting It All Together

TOPIC	EXPLANATION	EXAMPLE
Exponential Function	The variable is an exponent. $$f(x) = Ca^x$$ Initial value when $x = 0$ ... Growth factor (base) Growth: $a > 1$ Decay: $0 < a < 1$	$f(x) = 3(2)^x$ models exponential **growth**, and $g(x) = 2\left(\frac{1}{3}\right)^x$ models exponential **decay**.
Graphs of Exponential Functions	• Above the x-axis for all inputs • Increases for $a > 1$ • Decreases for $0 < a < 1$ • y-intercept is C.	**Exponential Functions: $f(x) = Ca^x$** 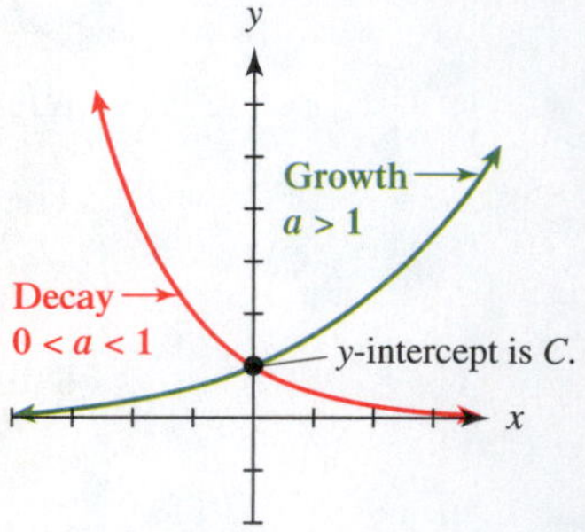
Percent Change	An amount changes from A to B: $$\text{Percent change} = \frac{B - A}{A} \times 100.$$	If \$200 increases to \$300, then $$\frac{300 - 200}{200} \times 100 = 50\%$$ is the percent change.

TOPIC	EXPLANATION	EXAMPLE
Constant Percent Change and Exponential Functions	Constant percent change (decimal) Time $$f(x) = C(1 + r)^x$$ Initial amount Growth factor	If 3000 bacteria increase in number by 12% daily for x days, then $$f(x) = 3000(1.12)^x$$ models these numbers after x days.

Compound Interest	Compounded annually:
	
	$$A = P(1 + r)^t$$
	
	Compounded n times per year:
	
	$$A = P\left(1 + \frac{r}{n}\right)^{nt}$$
	
	P: Initial deposit
	A: Final amount
	r: Annual interest rate (decimal)
	n: Interest paid n times per year
	t: Number of years
		If \$1000 is deposited at 3% annual interest for 4 years, then
		
		$$A = \$1000(1.03)^4$$
		$$\approx 1125.51.$$
		
		If \$1000 is deposited at 3% annual interest compounded monthly for 4 years, then
		
		$$A = \$1000\left(1 + \frac{0.03}{12}\right)^{(12 \cdot 4)}$$
		$$\approx \$1127.33.$$

Natural Exponential Function	$$f(x) = e^x$$ Growth factor, or base, is $e \approx 2.72$.	Using a calculator, $$f(2) = e^2 \approx 7.39.$$

9.2 Exercises

MyMathLab PRACTICE WATCH DOWNLOAD READ REVIEW

CONCEPTS AND VOCABULARY

1. Give a general formula for an exponential function f.

2. Sketch a graph of an exponential function that illustrates exponential decay.

3. Give the domain and range of an exponential function.

4. Evaluate the expressions 2^x and x^2 for $x = 5$.

5. Approximate e to the nearest thousandth.

6. Evaluate e^2 and π^2 using your calculator.

7. If a quantity y grows exponentially, then for each unit increase in x, y increases by a constant _____.

8. If $f(x) = 1.5^x$, what is the growth factor?

9. If a quantity increases from A to B, then the percent change equals _____.

10. If a quantity increases by 35% each year, then the growth factor a is _____.

EVALUATING EXPONENTIAL FUNCTIONS

Exercises 11–22: Evaluate the exponential function for the given values of x by hand when possible. Approximate answers to the nearest hundredth when appropriate.

11. $f(x) = 3^x$ $\qquad\qquad$ $x = -2, x = 2$

12. $f(x) = 5^x$ $\qquad\qquad$ $x = -1, x = 3$

13. $f(x) = 5(2^x)$ $\qquad\quad$ $x = 0, x = 5$

14. $f(x) = 3(7^x)$ $\qquad\quad$ $x = -2, x = 0$

15. $f(x) = \left(\frac{1}{2}\right)^x$ $\qquad\quad$ $x = -2, x = 3$

16. $f(x) = \left(\frac{1}{4}\right)^x$ $\qquad\quad$ $x = 0, x = 2$

17. $f(x) = 5(3)^{-x}$ $\qquad\;$ $x = -1, x = 2$

18. $f(x) = 4\left(\frac{3}{7}\right)^x$ $\qquad$ $x = 1, x = 4$

19. $f(x) = 1.8^x$ $\qquad\quad$ $x = -3, x = 1.5$

20. $f(x) = 0.91^x$ $\qquad\;$ $x = 5.1, x = 10$

21. $f(x) = 3(0.6)^x$ $\qquad$ $x = -1, x = 2$

22. $f(x) = 5(4.5)^{-x}$ $\qquad$ $x = -2.1, x = 1.9$

Exercises 23 and 24: **Thinking Generally** *For the given exponential function, evaluate $f(0)$ and $f(-1)$.*

23. $f(x) = a^x$

24. $f(x) = (1 + r)^{2x}$

GRAPHS OF EXPONENTIAL FUNCTIONS

Exercises 25–28: Match the formula with its graph (a.–d.). Do not use a calculator.

25. $f(x) = 1.5^x$ $\qquad$ **26.** $f(x) = \frac{1}{4}(2^x)$

27. $f(x) = 4\left(\frac{1}{2}\right)^x$ $\qquad$ **28.** $f(x) = \left(\frac{1}{3}\right)^x$

a.

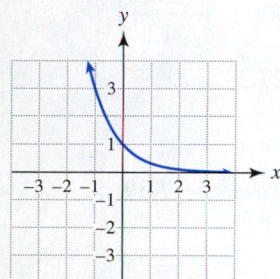

b.

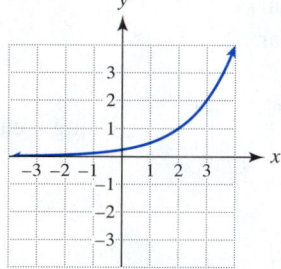

c.

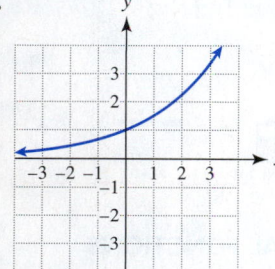

d.

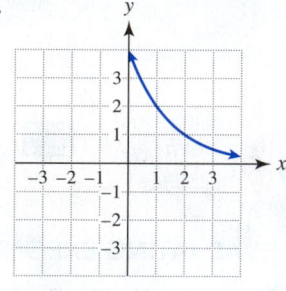

Exercises 29–32: Use the graph of $y = Ca^x$ to determine the constants C and a.

29.

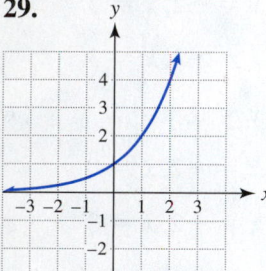

30.

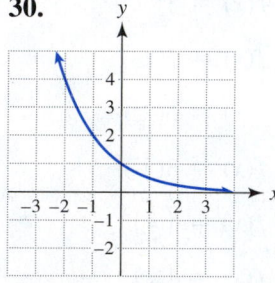

31.

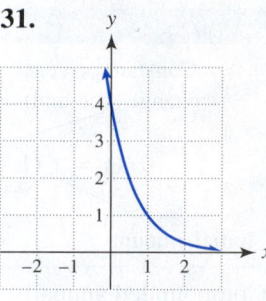

32.

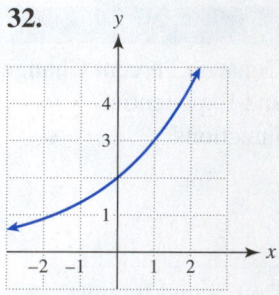

Exercises 33–44: Graph $y = f(x)$. State whether the graph depicts exponential growth or exponential decay.

33. $f(x) = 2^x$ $\qquad$ **34.** $f(x) = 3^x$

35. $f(x) = \left(\frac{1}{4}\right)^x$ $\qquad$ **36.** $f(x) = \left(\frac{1}{2}\right)^x$

37. $f(x) = 2^{-x}$ $\qquad$ **38.** $f(x) = 3^{-x}$

39. $f(x) = 3^x - 1$ $\qquad$ **40.** $f(x) = 2^x + 1$

41. $f(x) = 2^{x-1}$ $\qquad$ **42.** $f(x) = 2^{x+1}$

43. $f(x) = 4\left(\frac{1}{3}\right)^x$ $\qquad$ **44.** $f(x) = 3\left(\frac{1}{2}\right)^x$

LINEAR AND EXPONENTIAL GROWTH

Exercises 45–50: (Refer to Example 3.) A table for a function f is given.

(a) *Determine whether function f represents exponential growth, exponential decay, or linear growth.*

(b) *Find a formula for f.*

45.

x	0	1	2	3	4
$f(x)$	64	16	4	1	$\frac{1}{4}$

46.

x	0	1	2	3	4
$f(x)$	$\frac{1}{2}$	1	2	4	8

47.

x	0	1	2	3	4
$f(x)$	8	11	14	17	20

48.

x	−2	−1	0	1	2
$f(x)$	4	2	1	$\frac{1}{2}$	$\frac{1}{4}$

49.

x	−2	−1	0	1	2
$f(x)$	2.56	3.2	4	5	6.25

50.

x	−2	−1	0	1	2
$f(x)$	−6	−2	2	6	10

PERCENT CHANGE AND EXPONENTIAL FUNCTIONS

Exercises 51–54: For the given amounts A and B, find each of the following.

> (a) The percent change if A changes to B
> (b) The percent change if B changes to A

51. $A = \$200, B = \$400$

52. $A = \$1.50, B = \$1.00$

53. $A = 150, B = 30$

54. $A = 80, B = 200$

Exercises 55–60: An account contains A dollars and increases or decreases by R%. For each A and R, answer the following.

> (a) Find the increase or decrease in the value of the account.
> (b) Find the final value of the account.
> (c) By what factor did the account value increase or decrease?

55. $A = \$1000, R = 120\%$

56. $A = \$500, R = 230\%$

57. $A = \$650, R = 20\%$

58. $A = \$70, R = 35\%$

59. $A = \$800, R = -10\%$

60. $A = \$950, R = -60\%$

Exercises 61–64: For the given f(x), state the initial value C, the growth or decay factor a, and percent change R for each unit increase in x.

61. $f(x) = 9(1.07)^x$

62. $f(x) = 3(1.351)^x$

63. $f(x) = 1.5(0.45)^x$

64. $f(x) = 0.9^x$

COMPOUND INTEREST

Exercises 65–70: (Refer to Example 7.) If P dollars is deposited in an account paying R percent annual interest, approximate the amount in the account after x years.

65. $P = \$1500 \quad R = 9\% \quad x = 10$ years

66. $P = \$1500 \quad R = 15\% \quad x = 10$ years

67. $P = \$200 \quad R = 20\% \quad x = 50$ years

68. $P = \$5000 \quad R = 8.4\% \quad x = 7$ years

69. $P = \$560 \quad R = 1.4\% \quad x = 25$ years

70. $P = \$750 \quad R = 10\% \quad x = 13$ years

71. Thinking Generally Suppose that $1000 is deposited in an account paying 8% annual interest for 10 years. If $2000 had been deposited instead of $1000, would there be twice the money in the account after 10 years? Explain.

72. Thinking Generally Suppose that $500 is deposited in an account paying 5% annual interest for 10 years. If the interest rate had been 10% instead of 5%, would the total interest earned after 10 years be twice as much? Explain.

Exercises 73–76: (Refer to Example 8.) The amount P is deposited in an account giving R% annual interest compounded n times a year. Find the amount A in the account after t years.

73. $P = \$700, R = 4\%, n = 4, t = 3$

74. $P = \$550, R = 3\%, n = 2, t = 5$

75. $P = \$1200, R = 2.5\%, n = 12, t = 7$

76. $P = \$1500, R = 6.5\%, n = 365, t = 20$

THE NATURAL EXPONENTIAL FUNCTION

Exercises 77–80: Evaluate f(x) for the given value of x. Approximate answers to the nearest hundredth.

77. $f(x) = e^x \quad\quad x = 1.2$

78. $f(x) = 2e^x \quad\quad x = 2$

79. $f(x) = 1 - e^x \quad\quad x = -2$

80. $f(x) = 4e^{-x} \quad\quad x = 1.5$

Exercises 81–84: Graph f(x) in $[-4, 4, 1]$ by $[0, 8, 1]$. State whether the graph illustrates exponential growth or exponential decay.

81. $f(x) = e^{0.5x}$

82. $f(x) = e^x + 1$

83. $f(x) = 1.5e^{-0.32x}$

84. $f(x) = 2e^{-x} + 1$

APPLICATIONS

*Exercises 85–88: **Exponential Models** Write the formula for an exponential function, $f(x) = Ca^x$, that models the situation. Evaluate f(4).*

85. A sample of 5000 bacteria decreases in number by 25% per week.

86. A sample of 700 insects increases in number by 200% per day.

87. A sample of 50 birds increases in number by 10% per month.

88. A sample of 137 fish decreases in number by 2% per week.

89. *Salary Growth* Suppose your salary is $50,000 per year. Would you rather have a 20% raise each year or a $20 raise each year?

90. *Salary Growth* Suppose your salary is $35,000 per year. Would you rather have a 10% raise each year or a $4000 raise each year? Assume that you will keep this job for 10 years.

91. *Blood Alcohol* Suppose that a person's peak blood alcohol level is 0.07 (grams per 100 mL) and that this level decreases by 40% each hour. (*Source:* National Institutes of Health.)
(a) What is the hourly decay factor?
(b) Write an exponential function $B(x) = Ca^x$ that models the blood alcohol level after x hours.
(c) Evaluate $B(2)$ and interpret the result.

92. *Apple's Revenue* From 2001 to 2010 Apple's revenue grew at an annual rate of 40%, from $1 billion to about $21 billion. (*Source:* Apple Corp.)
(a) What is the annual growth factor?
(b) Write an exponential function $R(x) = Ca^x$ that models the revenue in billions of dollars x years after 2001.
(c) Evaluate $R(9)$ and interpret the result.

93. *Tweets per Month* From July 2008 to July 2010, the number of tweets T per month increased dramatically and could be modeled in millions by

$$T(x) = 0.5(1.242)^x,$$

where x represents months after July 2008. (*Source:* Silicon Alley.)
(a) What is the monthly growth factor?
(b) Interpret the 0.5 in the formula.
(c) Evaluate $T(24)$ and interpret the result.

94. *Dating Artifacts* Radioactive carbon-14 is found in all living things and is used to date objects containing organic material. Suppose that an object initially contains C grams of carbon-14. After x years it will contain A grams, where

$$A = C(0.99988)^x.$$

(a) Let $C = 10$ and graph A over a 20,000-year period. Is this function an example of exponential growth or decay?
(b) If $C = 10$, how many grams are left after 5700 years? What fraction of the carbon-14 is left?

95. *E. coli Bacteria* A strain of bacteria that inhabits the intestines of animals is named *Escherichia coli* (*E. coli*). These bacteria are capable of rapid growth and can be dangerous to humans—particularly children. The table shows the results of one study of the growth of *E. coli* bacteria, where concentrations are listed in *thousands* of bacteria per milliliter.

t (minutes)	0	50	100
Concentration	500	1000	2000

t (minutes)	150	200	250
Concentration	4000	8000	16,000

Source: G. S. Stent, *Molecular Biology of Bacterial Viruses.*

(a) Find C and a so that $f(t) = Ca^{t/50}$ models these data.
(b) Use $f(t)$ to estimate the concentration of bacteria after 170 minutes.
(c) Discuss the growth of this strain of bacteria over a 250-minute time period.

96. *Swimming Pool Maintenance* Chlorine is frequently used to disinfect swimming pools. The chlorine concentration should remain between 1.5 and 2.5 parts per million (ppm). After a warm, sunny day only 80% of the chlorine may remain in the water, with the other 20% dissipating into the air or combining with other chemicals in the water. (*Source:* D. Thomas, *Swimming Pool Operator's Handbook.*)
(a) Let $f(x) = 3(0.8)^x$ model the concentration of chlorine in parts per million after x days. What is the initial concentration of chlorine in the pool?
(b) If no more chlorine is added, estimate when the chlorine level drops below 1.6 parts per million.

97. *Modeling Traffic Flow* (Refer to Example 10.) Construct a table of $f(x) = (0.905)^x$, starting at $x = 0$ and incrementing by 10, until $x = 50$.
(a) Evaluate $f(0)$ and interpret the result.
(b) For a time interval of what length is there only a 5% chance that no cars will enter the intersection?

98. *Pros and Putts* The percentage of putts P from 3 feet to 25 feet made by professional golfers can be modeled by the exponential function

$$P(x) = 99(0.872)^{x-3},$$

where x is the length of the putt.

(a) Find the percentage of putts made by professionals from 3 feet.

(b) Evaluate $P(8)$ and interpret the results.

(c) What is the decay factor? Interpret the decay factor.

Exercises 99–102: **Population Growth** *(Refer to Example 11.) The population P in 2010 for a state is given along with R%, its annual percentage rate of continuous growth.*

(a) *Write the formula $f(x) = Pe^{rx}$, where r is in decimal notation, that models the population in millions x years after 2010.*

(b) *Estimate the population in 2020.*

99. Nevada: $P = 2.7$ million, $R = 1.4\%$

100. North Carolina: $P = 9.4$ million, $R = 1.36\%$

101. California: $P = 38$ million, $R = 1.02\%$

102. Arizona: $P = 6.6$ million, $R = 1.44\%$

103. Online Exploration In many states, landlords are mandated to return a tenant's security deposit plus interest. Go online and look up this interest rate for your state. Answers may vary.

(a) If you are a tenant, calculate the interest you should receive after 1 year. (If you are not, assume the security deposit is $1200.)

(b) Write a function I that calculates the interest that you should receive after x years. (Does your landlord have to pay compound interest?)

104. Online Exploration Go online and determine how long it takes the fastest growing bacteria to double in number.

(a) Suppose that you start with a sample of 4 million such bacteria. Write an exponential function that gives the number N of bacteria in millions after x minutes.

(b) How many bacteria are there after 2 hours?

WRITING ABOUT MATHEMATICS

105. A student evaluates $f(x) = 4(2)^x$ at $x = 3$ and obtains 512. Did the student evaluate the function correctly? What was the student's error?

106. For a set of data, how can you distinguish between linear growth and exponential growth? Give an example of each type of data.

SECTIONS 9.1 AND 9.2 — **Checking Basic Concepts**

1. If $f(x) = 2x^2 + 5x - 1$ and $g(x) = x + 1$, find each expression.

(a) $(g \circ f)(1)$ **(b)** $(f \circ g)(x)$

2. Sketch a graph of $f(x) = x^2 - 1$.
(a) Is f a one-to-one function? Explain.
(b) Does f have an inverse function?

3. If $f(x) = 4x - 3$, find $f^{-1}(x)$.

4. Evaluate $f(-2)$ if $f(x) = 3(2)^x$.

5. Sketch a graph of $f(x) = \left(\frac{1}{3}\right)^x$.

6. Use the graph of $y = Ca^x$ to determine the constants C and a.

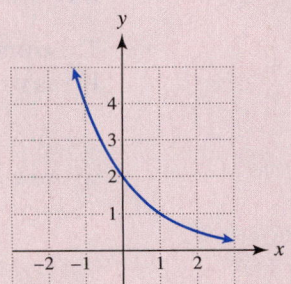

9.3   Logarithmic Functions

The Common Logarithmic Function • The Inverse of the Common Logarithmic Function • Logarithms with Other Bases

A LOOK INTO MATH ▶

Logarithmic functions are used in many applications. For example, if one airplane weighs twice as much as another, does the heavier airplane typically need a runway that is twice as long? Using a logarithmic function, we can answer this question. (See Example 10.) In this section we discuss logarithmic functions and several of their applications.

The Common Logarithmic Function

NEW VOCABULARY

☐ Common logarithm of a positive number x
☐ Common logarithmic function
☐ Natural logarithm
☐ Logarithm with base a of a positive number x
☐ Logarithmic function with base a

In applications, measurements can vary greatly in size. Table 9.7 lists some examples of objects, with the approximate distances in meters across each.

TABLE 9.7 Sizes of Objects

Object	Distance (meters)
Atom	10^{-9}
Protozoan	10^{-4}
Small Asteroid	10^2
Earth	10^7
Universe	10^{26}

Source: C. Ronan, The Natural History of the Universe.

Each distance is listed in the form 10^k for some k. The value of k distinguishes one measurement from another. The *common logarithmic function* or *base-10 logarithmic function*, denoted *log* or *log$_{10}$*, outputs k *if* the input x can be written as 10^k for some real number k. For example, log $10^{-9} = -9$, log $10^2 = 2$, and log $10^{1.43} = 1.43$. For any real number k, log $10^k = k$. Some values for log x are given in Table 9.8.

TABLE 9.8 Common Logarithms of Powers of 10

Write x as a power of 10.

x	10^{-4}	10^{-3}	10^{-2}	10^{-1}	10^0	10^1	10^2	10^3	10^4
log x	-4	-3	-2	-1	0	1	2	3	4

A common logarithm is the *exponent* on base 10.

We use this information to define the common logarithm.

COMMON LOGARITHM

The **common logarithm of a positive number x**, denoted log x, is calculated as follows. If x is written as $x = 10^k$, then

$$\log x = k,$$

where k is a real number. That is, log $10^k = k$.
The function given by

$$f(x) = \log x$$

is called the **common logarithmic function**.

NOTE: Previously, we have always used one letter, such as f or g, to name a function. The common logarithm is the *first* function for which we use *three* letters, log, to name it. Thus $f(x)$, $g(x)$, and $\log(x)$ all represent functions. Generally, $\log(x)$ is written without parentheses as $\log x$.

The following steps can be used to calculate $\log x$.

> **STEP 1:** Is x positive? If not, then $\log x$ is *undefined*.
> **STEP 2:** Write x as 10^k for some real number k. If this is not possible, then use a calculator to approximate $\log x$.
> **STEP 3:** If $x = 10^k$, then $\log x = k$. That is, $\log 10^k = k$.

In the next two examples, we evaluate $\log x$ for various values of x.

EXAMPLE 1   Calculating log x

Evaluate each expression, if possible.
(a) $\log(-4)$ **(b)** $\log 1000$

Solution
(a) STEP 1: *Is x positive?* No. Because $x = -4$ is negative, $\log x$ is undefined.

NOTE: Because 10^k is always positive, there is no k such that $10^k = -4$. The common logarithm of *any* negative number is undefined.

(b) STEP 1: *Is x positive?* Yes, because $x = 1000$.
STEP 2: *Write x as 10^k for some real number k.* Because $1000 = 10^3$, we can write $x = 10^3$.
STEP 3: *If $x = 10^k$, then $\log x = k$.* Because $x = 10^3$, it follows that

$$\log x = \log 1000 = \log 10^3 = 3.$$

Now Try Exercises 27, 31

EXAMPLE 2   Evaluating common logarithms

Evaluate each common logarithm.
(a) $\log 100$ **(b)** $\log \frac{1}{10}$ **(c)** $\log \sqrt{1000}$ **(d)** $\log 45$

Solution
(a) $x = 100 = 10^2$, so $\log x = \log 100 = \log 10^2 = 2$
(b) $x = \frac{1}{10} = 10^{-1}$, so $\log x = \log \frac{1}{10} = \log 10^{-1} = -1$
(c) $\log \sqrt{1000} = \log(1000)^{1/2} = \log(10^3)^{1/2} = \log 10^{3/2} = \frac{3}{2}$
(d) How to write $x = 45$ as a power of 10 is not obvious. However, we can use a calculator to determine that $\log 45 \approx 1.6532$. Thus $x \approx 10^{\wedge}(1.6532) \approx 45$.

Figure 9.23 supports these answers.

CALCULATOR HELP

To evaluate the common logarithmic function, see Appendix A (page AP-8).

```
log(100)
                2
log(1/10)
               -1



        (a)
```

```
log(√(1000))
              1.5
log(45)
        1.653212514
10^1.6532
        44.99870339

        (b)
```

Figure 9.23

Now Try Exercises 17, 23, 29, 43

READING CHECK

What does log x mean?

NOTE: To find $\log x$, we ask ourselves, "What exponent should 10 be raised to in order to get x?" This *exponent* equals $\log x$.

THE GRAPH OF THE COMMON LOGARITHM The four points $(10^{-1}, -1)$, $(10^0, 0)$, $(10^{0.5}, 0.5)$, and $(10^1, 1)$ are on the graph of $y = \log x$. See Table 9.8 at the beginning of this section. Plotting these points, as shown in Figure 9.24(a), and sketching the graph of $y = \log x$ results in Figure 9.24(b). Note some important features of this graph.

Common Logarithm

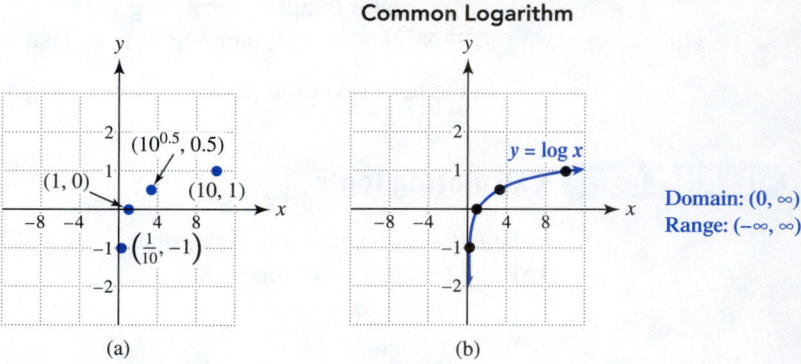

(a)   (b)

Domain: $(0, \infty)$
Range: $(-\infty, \infty)$

Figure 9.24

- The graph of the common logarithm increases very slowly for large values of x. For example, x must be 100 for $\log x$ to reach 2 and x must be 1000 for $\log x$ to reach 3.
- The graph passes through the point $(1, 0)$. Thus $\log 1 = 0$.
- The graph does not exist for negative values of x. The *domain* of $\log x$ includes only positive numbers. The *range* of $\log x$ includes all real numbers.
- When $0 < x < 1$, $\log x$ outputs negative values. The y-axis is a vertical asymptote, so as x approaches 0, $\log x$ approaches $-\infty$.

MAKING CONNECTIONS

The Common Logarithmic Function and the Square Root Function

Much like the square root function, the common logarithmic function does not have an easy-to-evaluate formula. We can calculate $\sqrt{4} = 2$ and $\sqrt{100} = 10$ mentally, but for $\sqrt{2}$ we use a calculator. Similarly, we can mentally calculate $\log 1000 = \log 10^3 = 3$, whereas we use a calculator for $\log 45$. The notation $\log x$ is understood as $\log_{10} x$ in the same way that $\sqrt{x}$ is understood to be $\sqrt[2]{x}$. Another similarity between the square root and common logarithmic functions is that their domains do not include negative numbers. If only real numbers are allowed as outputs, both $\sqrt{-3}$ and $\log(-3)$ are undefined expressions.

The Inverse of the Common Logarithmic Function

The graph of $y = \log x$ shown in Figure 9.24(b) is a one-to-one function because it passes the horizontal line test. Thus the common logarithmic function has an inverse function. To determine this inverse function for $\log x$, consider Tables 9.9 and 9.10.

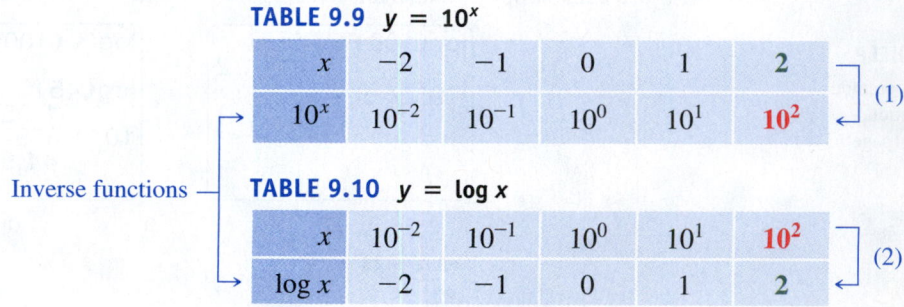

TABLE 9.9 $y = 10^x$

x	-2	-1	0	1	2	
10^x	10^{-2}	10^{-1}	10^0	10^1	10^2	(1)

Inverse functions —

TABLE 9.10 $y = \log x$

x	10^{-2}	10^{-1}	10^0	10^1	10^2	
$\log x$	-2	-1	0	1	2	(2)

If we start with the input **2** for 10^x, we compute $\mathbf{10^2}$, as shown by (1) in Table 9.9. If we use $\mathbf{10^2}$ as the input for log x, we compute **2**, as shown by (2) in Table 9.10. That is,

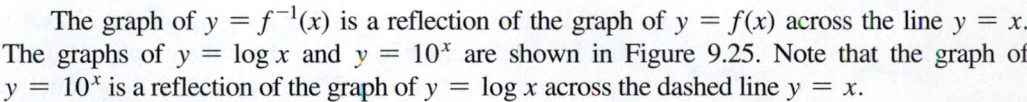

$$\log(10^2) = 2. \quad \text{Calculate } 10^x \text{ and then log } x.$$

Next, suppose we find the logarithm first and then compute a power of 10. If we start with the input $\mathbf{10^2}$, we compute **2**, as shown by (2) in Table 9.10. If we use **2** as the input for 10^x, we compute $\mathbf{10^2}$, as shown by (1) in Table 9.9. That is,

$$10^{\log(10^2)} = \mathbf{10^2}. \quad \text{Calculate log } x \text{ and then } 10^x.$$

In general, if $10^a = b$, then $\log b = a$ and if $\log b = a$, then $10^a = b$. Thus the *inverse function* of $f(x) = \log x$ is $f^{-1}(x) = 10^x$. Note that composition of these two functions satisfies the definition of an inverse function.

$$
\begin{aligned}
(f \circ f^{-1})(x) = f\left(f^{-1}(x)\right) \quad &\text{and} \quad (f^{-1} \circ f)(x) = f^{-1}\left(f(x)\right) \\
= f(10^x) \quad & \qquad\qquad\qquad = f^{-1}(\log x) \\
= \log 10^x \quad & \qquad\qquad\qquad = 10^{\log x} \\
= x \quad & \qquad\qquad\qquad = x
\end{aligned}
$$

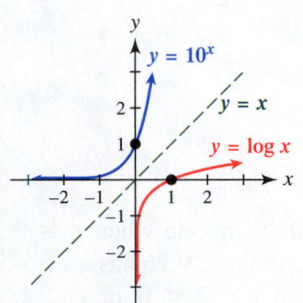

Figure 9.25

The graph of $y = f^{-1}(x)$ is a reflection of the graph of $y = f(x)$ across the line $y = x$. The graphs of $y = \log x$ and $y = 10^x$ are shown in Figure 9.25. Note that the graph of $y = 10^x$ is a reflection of the graph of $y = \log x$ across the dashed line $y = x$.

These inverse properties of log x and 10^x are summarized as follows.

INVERSE PROPERTIES OF THE COMMON LOGARITHM

The following properties hold for common logarithms.

$$\log 10^x = x, \qquad \text{for any real number } x$$
$$10^{\log x} = x, \qquad \text{for any positive real number } x$$

EXAMPLE 3 **Applying inverse properties**

Use inverse properties to simplify each expression.

(a) $\log 10^\pi$ **(b)** $\log 10^{x^2+1}$ **(c)** $10^{\log 7}$ **(d)** $10^{\log 3x}, x > 0$

Solution
(a) Because $\log 10^x = x$ for any real number x, $\log 10^\pi = \boldsymbol{\pi}$.
(b) $\log 10^{x^2+1} = \boldsymbol{x^2 + 1}$
(c) Because $10^{\log x} = x$ for any positive real number x, $10^{\log 7} = \mathbf{7}$.
(d) $10^{\log 3x} = \mathbf{3x}$, provided x is a *positive* number.

Now Try Exercises 25, 35, 37, 41

EXAMPLE 4 **Graphing logarithmic functions**

Graph each function f and compare its graph to $y = \log x$.
(a) $f(x) = \log(x - 2)$ **(b)** $f(x) = \log(x) + 1$

Solution

(a) We can use our knowledge of translations to sketch the graph of $y = \log(x - 2)$. To review, the graph of $y = (x - 2)^2$ is similar to the graph of $y = x^2$, except that it is translated 2 units to the *right*. Thus the graph of $y = \log(x - 2)$ is similar to the graph of $y = \log x$ (see Figure 9.25) except that it is translated 2 units to the right, as shown in Figure 9.26. The graph of $y = \log x$ passes through $(1, 0)$, so the graph of $y = \log(x - 2)$ passes through $(3, 0)$. Also, instead of the y-axis being the vertical asymptote, the line $x = 2$ is the vertical asymptote.

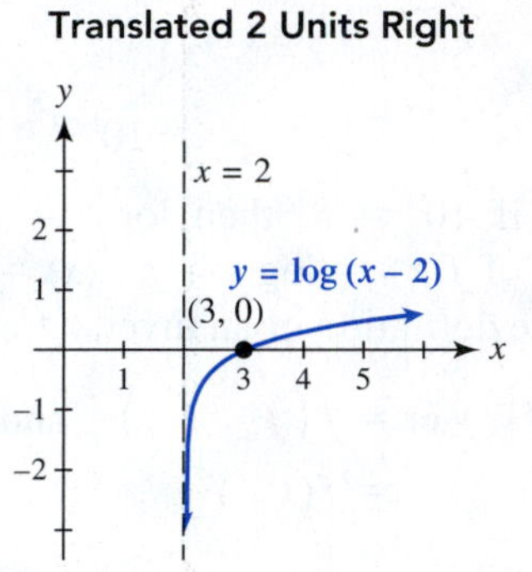

Translated 2 Units Right

Figure 9.26

Translated 1 Unit Upward

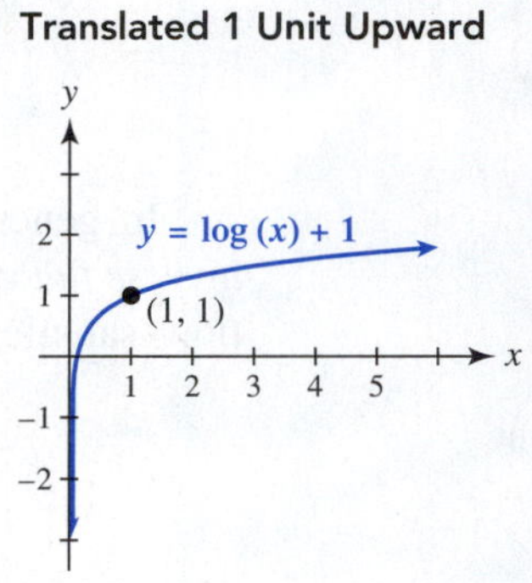

Figure 9.27

NOTE: The graph of $y = \log x$ in Figure 9.25 has a vertical asymptote when $x = 0$ and is undefined when $x < 0$. Thus the graph of $\log(x - 2)$ in Figure 9.26 has a vertical asymptote when $x - 2 = 0$, or $x = 2$. It is undefined when $x - 2 < 0$, or $x < 2$.

(b) The graph of $y = \log(x) + 1$ is similar to the graph of $y = \log x$, except that it is translated 1 unit *upward*. This graph is shown in Figure 9.27. Note that the graph of $y = \log(x) + 1$ passes through the point $(1, 1)$.

▪ **Now Try Exercises 47, 49**

▶ **REAL-WORLD CONNECTION** Logarithms are used to model quantities that vary greatly in intensity. For example, the human ear is extremely sensitive and able to detect intensities on the eardrum ranging from 10^{-16} watts per square centimeter (w/cm^2) to 10^{-4} w/cm^2, which is usually painful. The next example illustrates modeling sound with logarithms.

EXAMPLE 5 **Modeling sound levels**

Sound levels in decibels (dB) can be computed by $f(x) = 160 + 10 \log x$, where x is the intensity of the sound in watts per square centimeter. Ordinary conversation has an intensity of 10^{-10} w/cm^2. What decibel level is this? (*Source:* R. Weidner and R. Sells, *Elementary Classical Physics*, Vol. 2.)

CRITICAL THINKING
_____

If the sound level increases by 10 dB, by what factor does the intensity x increase?
_____

Solution

To find the decibel level for ordinary conversation, evaluate $f(10^{-10})$.

$$f(10^{-10}) = 160 + 10 \, \mathbf{log\,(10^{-10})} \quad \text{Substitute } x = 10^{-10}.$$
$$= 160 + 10(\mathbf{-10}) \quad \text{Evaluate } \log(10^{-10}).$$
$$= 60 \quad \text{Simplify.}$$

Ordinary conversation corresponds to 60 dB.

▪ **Now Try Exercise 105**

Logarithms with Other Bases

Common logarithms are base-10 logarithms, but we can define logarithms having other bases. For example, base-2 logarithms are frequently used in computer science. Some values for the base-2 logarithmic function, denoted $f(x) = \log_2 x$, are shown in Table 9.11. If x can be expressed as $x = 2^k$ for some real number k, then $\log_2 x = \log_2 2^k = k$.

What does $\log_2 x$ mean?

Write x as a power of 2.

TABLE 9.11 Base-2 Logarithms of Powers of 2

x	2^{-3}	2^{-2}	2^{-1}	2^0	2^1	2^2	2^3
$\log_2 x$	-3	-2	-1	0	1	2	3

A base-2 logarithm is the *exponent* on base 2.

NOTE: A base-2 logarithm is an *exponent* on base 2.

Logarithms with other bases are evaluated in the next three examples.

EXAMPLE 6 **Evaluating base-2 logarithms**

Simplify each logarithm.
(a) $\log_2 8$ (b) $\log_2 \frac{1}{4}$

Solution
(a) The logarithmic expression $\log_2 8$ represents the exponent on base 2 that gives 8. Because $8 = 2^3$, $\log_2 8 = \log_2 2^3 = 3$.
(b) Because $\frac{1}{4} = \frac{1}{2^2} = 2^{-2}$, $\log_2 \frac{1}{4} = \log_2 2^{-2} = -2$.

Now Try Exercises 81, 83

Some values of base-e logarithms are shown in Table 9.12. A base-e logarithm is referred to as a **natural logarithm** and is denoted either $\log_e x$ or $\ln x$. Natural logarithms are used in mathematics, science, economics, electronics, and communications.

To evaluate the natural logarithmic function, see Appendix A (page AP-8).

TABLE 9.12 Natural Logarithms of Powers of e

x	e^{-3}	e^{-2}	e^{-1}	e^0	e^1	e^2	e^3
$\ln x$	-3	-2	-1	0	1	2	3

NOTE: A natural logarithm is an *exponent* on base e.

To evaluate natural logarithms we usually use a calculator.

EXAMPLE 7 **Evaluating natural logarithms**

Approximate to the nearest hundredth.
(a) $\ln 10$ (b) $\ln \frac{1}{2}$

```
ln(10)
        2.302585093
ln(1/2)
        -.6931471806
```

Figure 9.28

Solution
(a) Figure 9.28 shows that $\ln 10 \approx 2.30$.
(b) Figure 9.28 shows that $\ln \frac{1}{2} \approx -0.69$.

Now Try Exercises 61, 63

We now define base-a logarithms.

BASE-a LOGARITHMS

The **logarithm with base a of a positive number x**, denoted $\log_a x$, is calculated as follows. If x is written as $x = a^k$, then

$$\log_a x = k,$$

where $a > 0$, $a \neq 1$, and k is a real number. That is, $\log_a a^k = k$.
The function given by

$$f(x) = \log_a x$$

is called the **logarithmic function with base a**.

Remember that *a logarithm is an exponent*. The expression $\log_a x$ equals the exponent k such that $a^k = x$. The graph of $y = \log_a x$ with $a > 1$ is shown in Figure 9.29. Note that the graph passes through the point $(1, 0)$. Thus $\log_a 1 = 0$.

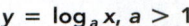
$$y = \log_a x, \ a > 1$$

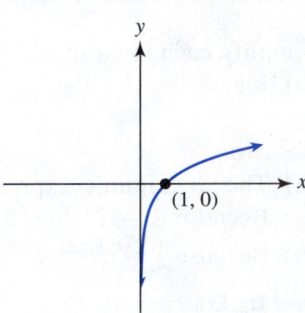

Figure 9.29

CRITICAL THINKING

Explain why $\log_a 1 = 0$ for any positive base a, $a \neq 1$.

NOTE: The natural logarithm, $\ln x$, is a base-a logarithm with $a = e$. That is, $\ln x = \log_e x$.

EXAMPLE 8 **Evaluating base-a logarithms**

Simplify each logarithm.
(a) $\log_5 25$ **(b)** $\log_4 \frac{1}{64}$ **(c)** $\log_7 1$ **(d)** $\log_3 9^{-1}$

Solution
(a) $25 = 5^2$, so $\log_5 25 = \log_5 5^2 = 2$.
(b) $\frac{1}{64} = \frac{1}{4^3} = 4^{-3}$, so $\log_4 \frac{1}{64} = \log_4 4^{-3} = -3$.
(c) $1 = 7^0$, so $\log_7 1 = \log_7 7^0 = 0$. (The logarithm of 1 is 0, regardless of the base.)
(d) $9^{-1} = (3^2)^{-1} = 3^{-2}$, so $\log_3 9^{-1} = \log_3 3^{-2} = -2$.

Now Try Exercises 55, 75, 85, 87

The graph of $y = \log_a x$ in Figure 9.29 passes the horizontal line test, so it is a one-to-one function and it has an inverse. If we let $f(x) = \log_a x$, then $f^{-1}(x) = a^x$. This statement is a generalization of the fact that $f(x) = \log x$ and $f^{-1}(x) = 10^x$ represent inverse functions. The graphs of $y = \log_a x$ and $y = a^x$ with $a > 1$ are shown in Figure 9.30. Note that the graph of $y = a^x$ is a reflection of the graph of $y = \log_a x$ across the line $y = x$.

Inverse Functions

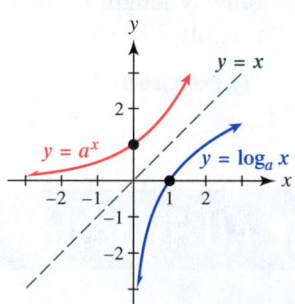

Figure 9.30 $a > 1$

These inverse properties are summarized by the following.

INVERSE PROPERTIES OF BASE-a LOGARITHMS

The following properties hold for logarithms with base a.

$$\log_a a^x = x, \qquad \text{for any real number } x$$
$$a^{\log_a x} = x, \qquad \text{for any positive real number } x$$

NOTE: The inverse function of $f(x) = \ln x$ is $f^{-1}(x) = e^x$.
The inverse function of $f(x) = \log x$ is $f^{-1}(x) = 10^x$.

EXAMPLE 9 **Applying inverse properties**

Simplify each expression.
(a) $\ln e^{0.5x}$ **(b)** $e^{\ln 4}$ **(c)** $2^{\log_2 7x}$ **(d)** $10^{\log (9x - 3)}$

Solution
(a) $\ln e^{0.5x} = 0.5x$ because $\ln e^k = k$ for all k.
(b) $e^{\ln 4} = 4$ because $e^{\ln k} = k$ for all positive k.
(c) $2^{\log_2 7x} = 7x$ for $x > 0$ because $a^{\log_a k} = k$ for all positive k.
(d) $10^{\log (9x - 3)} = 9x - 3$ for $x > \frac{1}{3}$ because $10^{\log k} = k$ for all positive k.

Now Try Exercises 39, 57, 59, 89

▶ **REAL-WORLD CONNECTION** Logarithms occur in many applications. One application is runway length for airplanes, as discussed in A Look Into Math and in the next example.

EXAMPLE 10 **Calculating runway length**

There is a mathematical relationship between an airplane's weight x and the runway length required at takeoff. For certain types of airplanes, the minimum runway length L in thousands of feet may be modeled by $L(x) = 1.3 \ln x$, where x is in thousands of pounds. (*Source:* L. Haefner, *Introduction to Transportation Systems.*)
(a) Estimate the runway length needed for an airplane weighing 10,000 pounds.
(b) Does a 20,000-pound airplane need twice the runway length that a 10,000-pound airplane needs? Explain.

Solution

(a) Because $L(x) = 1.3 \ln x$, it follows that $L(10) = 1.3 \ln(10) \approx 3$. An airplane weighing 10,000 pounds requires a runway (at least) 3000 feet long.

(b) Because $L(20) = 1.3 \ln(20) \approx 3.9$, a 20,000-pound airplane does not need twice the runway length needed by a 10,000-pound airplane. Rather, the heavier airplane needs roughly 3900 feet of runway, or only an extra 900 feet.

Now Try Exercise 107

9.3 Putting It All Together

Common logarithms are base-10 logarithms. If a positive number x is written as $x = 10^k$, then $\log x = k$. The value of $\log x$ represents the exponent on the base 10 that gives x.

CONCEPT	DESCRIPTION	EXAMPLES
Base-a Logarithms	*Definition:* $\log_a x = k$ means $x = a^k$, where $a > 0$ and $a \neq 1$. *Domain:* all *positive* real numbers *Range:* all real numbers *Graph:* $a > 1$ (shown to the right for $\log x$ and $\ln x$); passes through $(1, 0)$; vertical asymptote: y-axis *Common Logarithm:* base-10 logarithm and denoted $\log x$ *Natural Logarithm:* base-e logarithm, where $e \approx 2.718$, and denoted $\ln x$	$\log 1000 = \log 10^3 = 3$, $\log_2 16 = \log_2 2^4 = 4$, and $\log_3 \dfrac{1}{81} = \log_3 3^{-4} = -4$ (graph of $y = \ln x$ and $y = \log x$ passing through $(1,0)$)
Inverse Properties	The following properties hold for base-a logarithms. $\log_a a^x = x$, for any real number x $a^{\log_a x} = x$, for any positive number x	$\log 10^{7.48} = 7.48$ $2^{\log_2 63} = 63$ $10^{\log 23} = 23$ $\ln e^4 = 4$

9.3 Exercises

MyMathLab

 Math XL PRACTICE   WATCH   DOWNLOAD   READ   REVIEW

CONCEPTS AND VOCABULARY

1. What is the base of the common logarithm?

2. What is the base of the natural logarithm?

3. What are the domain and range of $\log x$?

4. What are the domain and range of 10^x?

5. $\log 10^k =$ _____

6. $\ln e^k =$ _____

7. If $\log x = k$, then $10^k =$ _____.

8. If $x > 0$, then $10^{\log x} =$ ____.

9. What does k equal if $10^k = 5$?

10. What does k equal if $e^k = 5$?

11. $\log 1 =$ ____

12. $\log (-1)$ is ____.

Exercises 13–16: Complete the table.

13.

x	10^{-5}	10^0	$10^{0.5}$	$10^{2.2}$
$\log x$				

14.

x	10^{-6}	10^{-1}	$10^{\frac{5}{7}}$	10^{π}
$\log x$				

15.

x	e^{-6}	e^{-1}	$e^{\frac{5}{7}}$	e^{π}
$\ln x$				

16.

x	2^{-5}	2^0	$2^{0.5}$	$2^{2.2}$
$\log_2 x$				

EVALUATING AND GRAPHING COMMON LOGARITHMS

Exercises 17–42: Simplify the expression, if possible.

17. $\log 10^5$ **18.** $\log 10$

19. $\log 10^{-4}$ **20.** $\log 10^{-1}$

21. $\log 1$ **22.** $\log \sqrt[3]{100}$

23. $\log \frac{1}{100}$ **24.** $\log \frac{1}{10}$

25. $\log 10^{4.7}$ **26.** $\log 10^{2x+4}$

27. $\log 10{,}000$ **28.** $\log 100{,}000$

29. $\log \sqrt{10}$ **30.** $\log 1{,}000{,}000$

31. $\log (-23)$ **32.** $\log (-8)$

33. $\log 0.001$ **34.** $\log 0.0001$

35. $10^{\log 2}$ **36.** $10^{\log 7.5}$

37. $10^{\log x^2}$ **38.** $10^{\log |x|}$

39. $10^{\log 5}$ **40.** $\log 10^{\frac{3}{4}}$

41. $\log 10^{(2x-7)}$ **42.** $\log 10^{(8-4x)}$

Exercises 43–46: Evaluate the common logarithm, using a calculator. Round values to the nearest thousandth.

43. $\log 25$ **44.** $\log 0.501$

45. $\log 1.45$ **46.** $\log \frac{1}{35}$

Exercises 47–54: Graph $y = f(x)$. Compare the graph to the graph of $y = \log x$.

47. $f(x) = \log (x) - 1$ **48.** $f(x) = \log (x) + 2$

49. $f(x) = \log (x + 1)$ **50.** $f(x) = \log (x + 2)$

51. $f(x) = \log (x - 1)$ **52.** $f(x) = \log (x - 3)$

53. $f(x) = 2 \log x$ **54.** $f(x) = -\log x$

EVALUATING AND GRAPHING NATURAL LOGARITHMS

Exercises 55–60: Simplify the expression.

55. $\ln 1$ **56.** $\ln e^2$

57. $\ln e^{-5x}$ **58.** $e^{\ln 2x}$

59. $e^{\ln x^2}$ **60.** $\ln e^{7x}$

Exercises 61–64: Evaluate the natural logarithm, using a calculator. Round values to the nearest thousandth.

61. $\ln 7$ **62.** $\ln 126$

63. $\ln \frac{4}{7}$ **64.** $\ln 0.67$

Exercises 65–68: Graph f in $[-4, 4, 1]$ by $[-4, 4, 1]$. Compare this graph to the graph of $y = \ln x$. Identify the domain of f.

65. $f(x) = \ln |x|$ **66.** $f(x) = \ln (x) - 2$

67. $f(x) = \ln (x + 2)$ **68.** $f(x) = 2 \ln x$

EVALUATING AND GRAPHING BASE-a LOGARITHMS

Exercises 69–94: Simplify the expression, if possible.

69. $\log_5 5^{6x}$ **70.** $\log_3 3^3$

71. $\log_2 \sqrt{\frac{1}{8}}$ **72.** $\log_2 2^6$

73. $\log_2 2^8$ **74.** $\log_2 2^{-5}$

75. $\log_2 \sqrt{8}$ **76.** $\log_2 \sqrt{32}$

77. $\log_2 \sqrt[3]{\frac{1}{4}}$ **78.** $\log_2 \frac{1}{64}$

79. $\log_2 -8$ **80.** $\log_2 -7$

81. $\log_2 4$ **82.** $\log_2 \frac{1}{32}$

83. $\log_2 \frac{1}{16}$ **84.** $\log_3 27$

85. $\log_3 \frac{1}{9}$

86. $\log_4 16$

87. $\log_5 5^{-2}$

88. $\log_8 64$

89. $5^{\log_5 17}$

90. $9^{\log_9 73}$

91. $4^{\log_4 (2x)^2}$

92. $b^{\log_b (x-1)}$

93. $5^{\log_5 0.6z}$

94. $7^{\log_7 (x-9)}$

Exercises 95 and 96: Complete the table.

95.

x	$\frac{1}{4}$	$\frac{1}{2}$	1	$\sqrt{2}$	64
$\log_2 x$					

96.

x	$\frac{1}{7}$	1	$\sqrt{7}$	7	49
$\log_7 x$					

Exercises 97–100: Graph $y = f(x)$.

97. $f(x) = \log_2 x$

98. $f(x) = \log_2 (x - 2)$

99. $f(x) = 2 + \log_2 x$

100. $f(x) = \log_2 (x + 2)$

Exercises 101–104: Without using a calculator match $f(x)$ with its graph (a.–d.).

101. $f(x) = \log x$

102. $f(x) = \log_3 x$

103. $f(x) = \log_3 (x) + 2$

104. $f(x) = \log (x + 1)$

a.

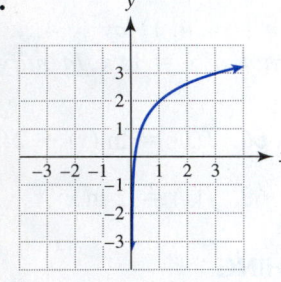

b.

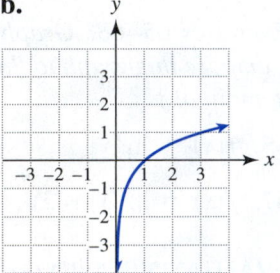

c.

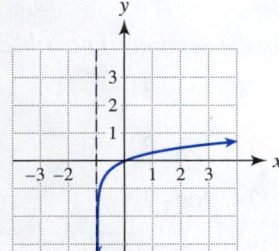

d.

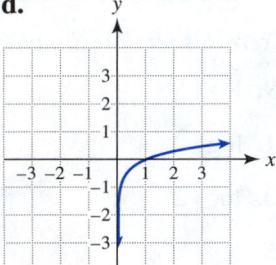

APPLICATIONS

105. *Modeling Sound* (Refer to Example 5.) At professional football games in domed stadiums the decibel level may reach 120. The eardrum usually experiences pain when the intensity of the sound reaches 10^{-4} watts per square centimeter. How many decibels does this quantity represent? Is the noise at a football game likely to hurt some people's eardrums?

106. *Hurricanes* The barometric air pressure P in inches of mercury at a distance of d miles from the eye of a severe hurricane can sometimes be modeled by the formula $P(d) = 0.48 \ln (d + 1) + 27$. Average air pressure is about 30 inches of mercury. (*Source:* A. Miller and R. Anthes, *Meteorology.*)

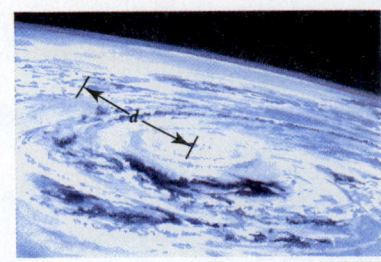

(a) Evaluate $P(0)$ and $P(50)$. Interpret the results.

(b) A graph of $y = P(d)$ is shown in the figure. Describe how the air pressure changes as the distance from the eye of the hurricane increases.

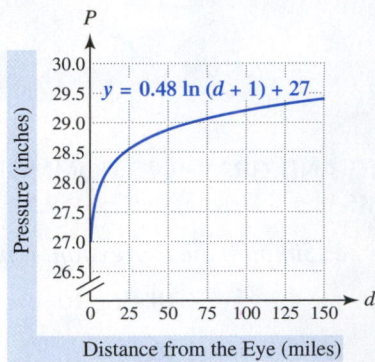

Distance from the Eye (miles)

(c) Is the eye of the hurricane a low-pressure area or a high-pressure area?

107. *Runway Length* (Refer to Example 10.)

(a) A graph of $L(x) = 1.3 \ln x$ is shown in the figure, where $y = L(x)$. As the weight of the plane increases, what can be said about the length of the runway required?

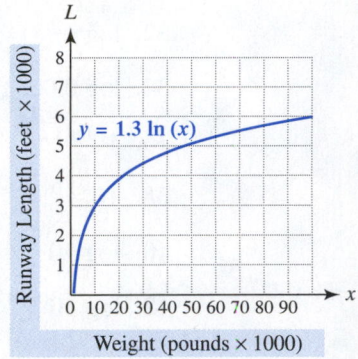

Weight (pounds × 1000)

(b) Evaluate $L(50)$ and interpret the result.

108. *Growth in Salary* Suppose that a person's salary is initially $40,000 and could be determined by either $f(x)$ or $g(x)$, where x represents the number of years of experience.

a. $f(x) = 40,000(1.1)^x$

b. $g(x) = 40,000 \log(10 + x)$

Would most people prefer that their salaries increase exponentially or logarithmically? Explain.

109. *Magnitude of a Star* The first stellar brightness scale was developed 2000 years ago by two Greek astronomers, Hipparchus and Ptolemy. The brightest star in the sky was given a magnitude of 1, and the faintest star was given a magnitude of 6. In 1856 this scale was described mathematically by the formula

$$M = 6 - 2.5 \log \frac{I}{I_0},$$

where M is the magnitude of a star with an intensity of I and I_0 is the intensity of the faintest star seen in the sky. (*Source:* M. Zeilik, *Introductory Astronomy and Astrophysics.*)

(a) Find M if $I = 10$ and $I_0 = 1$.

(b) What is the magnitude of a star that is 100 times more intense than the faintest star?

(c) If the intensity of a star increases by a factor of 10, what happens to its magnitude?

110. *Population of Urban Regions* Although less industrialized urban regions of the world are experiencing exponential population growth, industrialized urban regions are experiencing logarithmic population growth. Population in less industrialized urban regions can be modeled by

$$f(x) = 0.338(1.035)^x,$$

whereas the population in industrialized urban regions can be modeled by

$$g(x) = 0.36 + 0.15 \ln(x + 1).$$

In these formulas the output is in billions of people and x is in years, where $x = 0$ corresponds to 1950, $x = 10$ corresponds to 1960, and so on until $x = 80$ corresponds to 2030. (*Source:* D. Meadows, *Beyond The Limits.*)

(a) Evaluate $f(50)$ and $g(50)$. Interpret the results.

(b) Graph f and g in [0, 80, 10] by [0, 5, 1]. Compare the two graphs.

(c) If x increases from 20 to 40, by what factor does $f(x)$ increase? By what factor does $g(x)$ increase?

111. *Earthquakes* The Richter scale is used to determine the intensity of earthquakes, which corresponds to the amount of energy released. If an earthquake has an intensity of x, its *magnitude*, as computed by the Richter scale, is given by $R(x) = \log \frac{x}{I_0}$, where I_0 is the intensity of a small, measurable earthquake.

(a) On July 26, 1963, an earthquake in Yugoslavia had a magnitude of 6.0 on the Richter scale, and on August 19, 1977, an earthquake in Indonesia measured 8.0. Find the intensity x for each of these earthquakes if $I_0 = 1$.

(b) How many times more intense was the Indonesian earthquake than the Yugoslavian earthquake?

112. *Path Loss for Cellular Phones* For cellular phones to work throughout a country, large numbers of cellular towers are necessary. How well the signal is propagated throughout a region depends on the location of these towers. One quick way to estimate the strength of a signal at x kilometers is to use the formula

$$D(x) = -121 - 36 \log x.$$

This formula computes the decrease in the signal, using decibels, so it is always negative. For example, $D(1) = -121$ means that at a distance of 1 kilometer the signal has decreased in strength by 121 decibels. (*Source:* C. Smith, *Practical Cellular & PCS Design.*)

(a) Evaluate $D(3)$ and interpret the result.

(b) Graph D in [1, 10, 1] by [−160, −120, 10].

(c) What happens to the signal as x increases?

WRITING ABOUT MATHEMATICS

113. Explain what $\log_a x$ means and give an example.

114. How would you explain to a student that $\log_a 1 \neq 1$?

Group Activity | Working With Real Data

Directions: Form a group of 2 to 4 people. Select someone to record the group's responses for this activity. All members of the group should work cooperatively to answer the questions. If your instructor asks for your results, each member of the group should be prepared to respond.

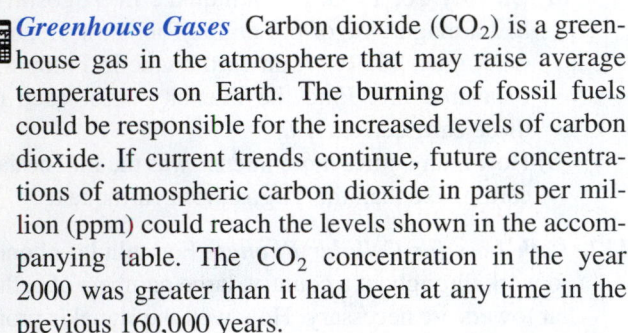 *Greenhouse Gases* Carbon dioxide (CO_2) is a greenhouse gas in the atmosphere that may raise average temperatures on Earth. The burning of fossil fuels could be responsible for the increased levels of carbon dioxide. If current trends continue, future concentrations of atmospheric carbon dioxide in parts per million (ppm) could reach the levels shown in the accompanying table. The CO_2 concentration in the year 2000 was greater than it had been at any time in the previous 160,000 years.

(a) Let x be in years, where $x = 0$ corresponds to 2000, $x = 1$ to 2001, and so on. Find values for C and a so that $f(x) = Ca^x$ models the data.

(b) Graph f and the data in the same viewing rectangle.

(c) Use $f(x)$ to estimate graphically the year when the carbon dioxide concentration will be double the preindustrial level of 280 ppm.

Year	2000	2050	2100	2150	2200
CO_2 (ppm)	364	467	600	769	987

Source: R. Turner, Environmental Economics.

9.4 | Properties of Logarithms

Basic Properties • Change of Base Formula

A LOOK INTO MATH ▶

The discovery of logarithms by John Napier (1550–1617) played an important role in the history of science. Logarithms were instrumental in allowing Johannes Kepler (1571–1630) to calculate the positions of the planet Mars, which led to his discovery of the laws of planetary motion. Kepler's laws were used by Isaac Newton (1642–1727) to discover the universal laws of gravity. Although calculators and computers have made tables of logarithms obsolete, applications involving logarithms still play an important role in modern-day computation. One reason for their continued importance is that *logarithms possess several important properties.*

Basic Properties

In this subsection we discuss three important properties of logarithms. The first property is the product rule for logarithms.

NEW VOCABULARY

☐ Change of base formula

> **PRODUCT RULE FOR LOGARITHMS**
>
> For positive numbers m, n, and $a \neq 1$,
>
> $$\log_a mn = \log_a m + \log_a n.$$

This product rule may be verified by using properties of exponents and the fact that $\log_a a^k = k$ for any real number k. Here, we verify the product property for logarithms. The two other properties presented later can be verified in a similar manner.

If m and n are positive numbers, we can write $m = a^c$ and $n = a^d$ for some real numbers c and d.

$$\log_a mn = \log_a(a^c a^d) = \log_a(a^{c+d}) = c + d \quad \text{and}$$
$$\log_a m + \log_a n = \log_a a^c + \log_a a^d = c + d$$

Thus $\log_a mn = \log_a m + \log_a n$.

This property is illustrated in Figure 9.31, which shows that

$$\log 10 = \log(2 \cdot 5) = \log 2 + \log 5.$$

In the next two examples, we demonstrate various operations involving logarithms.

log(10)

1

log(2)+log(5)

1

Figure 9.31

EXAMPLE 1   **Writing logarithms as sums**

Write each expression as a sum of logarithms. Assume that x is positive.
(a) $\log 21$ **(b)** $\ln 5x$ **(c)** $\log x^3$

Solution
(a) $\log 21 = \log(3 \cdot 7) = \log 3 + \log 7$
(b) $\ln 5x = \ln(5 \cdot x) = \ln 5 + \ln x$
(c) $\log x^3 = \log(x \cdot x \cdot x) = \log x + \log x + \log x$

Now Try Exercises 13, 15, 17

EXAMPLE 2   **Combining logarithms**

Write each expression as one logarithm. Assume that x and y are positive.
(a) $\log 5 + \log 6$ **(b)** $\ln x + \ln xy$ **(c)** $\log 2x + \log 5x$

Solution
(a) $\log 5 + \log 6 = \log(5 \cdot 6) = \log 30$
(b) $\ln x + \ln xy = \ln(x \cdot xy) = \ln x^2 y$
(c) $\log 2x + \log 5x = \log(2x \cdot 5x) = \log 10x^2$

Now Try Exercises 25, 27, 29

The second property is the quotient rule for logarithms.

QUOTIENT RULE FOR LOGARITHMS

For positive numbers m, n, and $a \neq 1$,

$$\log_a \frac{m}{n} = \log_a m - \log_a n.$$

log(10)

1

log(20)−log(2)

1

Figure 9.32

This property is illustrated in Figure 9.32, which shows that

$$\log 10 = \log \frac{20}{2} = \log 20 - \log 2.$$

EXAMPLE 3 **Writing logarithms as differences**

Write each expression as a difference of two logarithms. Assume that variables are positive.

(a) $\log \dfrac{3}{2}$ **(b)** $\ln \dfrac{3x}{y}$ **(c)** $\log_5 \dfrac{x}{z^4}$

Solution

(a) $\log \dfrac{3}{2} = \log 3 - \log 2$ **(b)** $\ln \dfrac{3x}{y} = \ln 3x - \ln y$

(c) $\log_5 \dfrac{x}{z^4} = \log_5 x - \log_5 z^4$

Now Try Exercises 19, 21, 23

NOTE: $\log_a (m + n) \neq \log_a m + \log_a n; \quad \log_a (m - n) \neq \log_a m - \log_a n;$

$$\log_a (mn) \neq \log_a m \cdot \log_a n; \quad \log_a \left(\dfrac{m}{n}\right) \neq \dfrac{\log_a m}{\log_a n}$$

EXAMPLE 4 **Combining logarithms**

Write each expression as one term. Assume that x is positive.

(a) $\log 50 - \log 25$ **(b)** $\ln x^3 - \ln x$ **(c)** $\log 15x - \log 5x$

Solution

(a) $\log 50 - \log 25 = \log \dfrac{50}{25} = \log 2$

(b) $\ln x^3 - \ln x = \ln \dfrac{x^3}{x} = \ln x^2$

(c) $\log 15x - \log 5x = \log \dfrac{15x}{5x} = \log 3$

Now Try Exercises 33, 35, 37

The third property is the power rule for logarithms. To illustrate this rule we use

$$\log x^3 = \log(x \cdot x \cdot x) = \log x + \log x + \log x = 3 \log x.$$

Thus $\log x^3 = 3 \log x$. This example is generalized in the following rule.

POWER RULE FOR LOGARITHMS

For positive numbers m and $a \neq 1$ and any real number r,

$$\log_a (m^r) = r \log_a m.$$

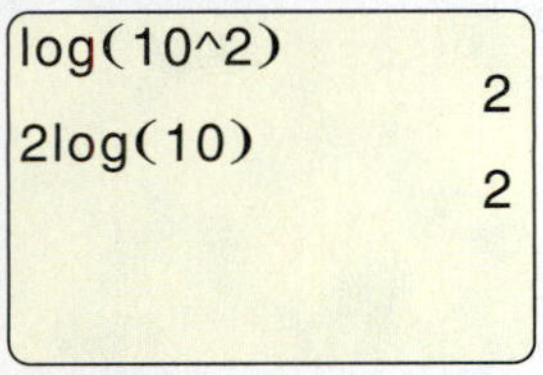

```
log(10^2)
              2
2log(10)
              2
```

Figure 9.33

We use a calculator and the equation

$$\log 10^2 = 2 \log 10$$

to illustrate this property. Figure 9.33 shows the result. We apply the power rule in the next example.

EXAMPLE 5 **Applying the power rule**

Rewrite each expression, using the power rule.
(a) $\log 5^6$ **(b)** $\ln (0.55)^{x-1}$ **(c)** $\log_5 8^{kx}$

Solution
(a) $\log 5^6 = \mathbf{6} \log 5$
(b) $\ln (0.55)^{x-1} = (\mathbf{x} - \mathbf{1}) \ln (0.55)$
(c) $\log_5 8^{kx} = \mathbf{kx} \log_5 8$

Now Try Exercises 39, 41, 47

Sometimes we use more than one property to simplify an expression. We assume that all variables are positive in the next two examples.

EXAMPLE 6 **Combining logarithms**

Write each expression as the logarithm of a single expression.
(a) $3 \log x + \log x^2$ **(b)** $2 \ln x - \ln \sqrt{x}$

Solution
(a) $3 \log x + \log x^2 = \log x^3 + \log x^2$ .... Power rule
$\qquad\qquad\qquad\qquad = \log (x^3 \cdot x^2)$ .... Product rule
$\qquad\qquad\qquad\qquad = \log x^5$ .... Properties of exponents
(b) $2 \ln x - \ln \sqrt{x} = 2 \ln x - \ln x^{1/2}$ .... $\sqrt{x} = x^{1/2}$
$\qquad\qquad\qquad\qquad = \ln x^2 - \ln x^{1/2}$ .... Power rule
$\qquad\qquad\qquad\qquad = \ln \dfrac{x^2}{x^{1/2}}$ .... Quotient rule
$\qquad\qquad\qquad\qquad = \ln x^{3/2}$ .... Properties of exponents

Now Try Exercises 49, 57

EXAMPLE 7 **Expanding logarithms**

Write each expression in terms of logarithms of x, y, and z.
(a) $\log \dfrac{x^2 y^3}{\sqrt{z}}$ **(b)** $\ln \sqrt[3]{\dfrac{xy}{z}}$

Solution
(a) $\log \dfrac{x^2 y^3}{\sqrt{z}} = \log x^2 y^3 - \log \sqrt{z}$ .... Quotient rule
$\qquad\qquad\qquad = \log x^2 + \log y^3 - \log z^{1/2}$ .... Product rule; $\sqrt{z} = z^{1/2}$
$\qquad\qquad\qquad = 2 \log x + 3 \log y - \tfrac{1}{2} \log z$ .... Power rule
(b) $\ln \sqrt[3]{\dfrac{xy}{z}} = \ln \left(\dfrac{xy}{z}\right)^{1/3}$ .... $\sqrt[3]{m} = m^{1/3}$
$\qquad\qquad\qquad = \tfrac{1}{3} \ln \dfrac{xy}{z}$ .... Power rule
$\qquad\qquad\qquad = \tfrac{1}{3} (\ln xy - \ln z)$ .... Quotient rule
$\qquad\qquad\qquad = \tfrac{1}{3} (\ln x + \ln y - \ln z)$ .... Product rule
$\qquad\qquad\qquad = \tfrac{1}{3} \ln x + \tfrac{1}{3} \ln y - \tfrac{1}{3} \ln z$ .... Distributive property

Now Try Exercises 63, 65

EXAMPLE 8   Applying properties of logarithms

Using only properties of logarithms and the approximations $\ln 2 \approx 0.7$, $\ln 3 \approx 1.1$, and $\ln 5 \approx 1.6$, find an approximation for each expression.

(a) $\ln 8$ (b) $\ln 15$ (c) $\ln \dfrac{10}{3}$

Solution

(a) $\ln \mathbf{8} = \ln \mathbf{2^3} = 3 \ln 2 \approx 3(0.7) = 2.1$

(b) $\ln \mathbf{15} = \ln(\mathbf{3 \cdot 5}) = \ln 3 + \ln 5 \approx 1.1 + 1.6 = 2.7$

(c) $\ln \dfrac{10}{3} = \ln\left(\dfrac{\mathbf{2 \cdot 5}}{3}\right) = \ln 2 + \ln 5 - \ln 3 \approx 0.7 + 1.6 - 1.1 = 1.2$

■ **Now Try Exercises 77, 79, 83**

Change of Base Formula

▶ **REAL-WORLD CONNECTION** Most calculators only have keys to evaluate common and natural logarithms. Occasionally, it is necessary to evaluate a logarithmic function with a base other than 10 or e. In these situations we use the following **change of base formula**, which we illustrate in the next example.

CHANGE OF BASE FORMULA

Let x and $a \neq 1$ be positive real numbers. Then

$$\log_a x = \frac{\log x}{\log a} \quad \text{or} \quad \log_a x = \frac{\ln x}{\ln a}.$$

EXAMPLE 9   Change of base formula

Approximate $\log_2 14$ to the nearest thousandth.

Solution

Using the change of base formula,

$$\log_2 \mathbf{14} = \frac{\log \mathbf{14}}{\log \mathbf{2}} \approx 3.807 \quad \text{or} \quad \log_2 \mathbf{14} = \frac{\ln \mathbf{14}}{\ln \mathbf{2}} \approx 3.807.$$

Figure 9.34 supports these results.

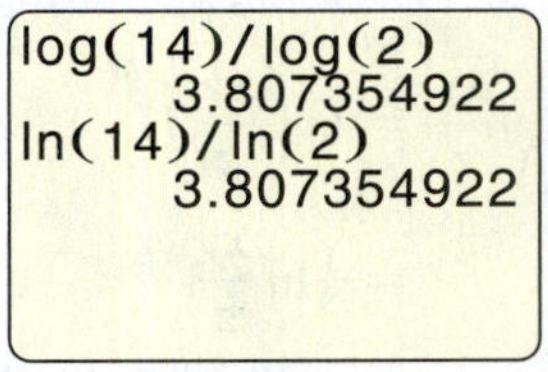

```
log(14)/log(2)
          3.807354922
ln(14)/ln(2)
          3.807354922
```

Figure 9.34

■ **Now Try Exercise 89**

9.4 Putting It All Together

The following table summarizes some important properties for base-a logarithms. Common and natural logarithms satisfy the same properties.

CONCEPTS	DESCRIPTION	EXAMPLES
Properties of Logarithms	The following properties hold for positive numbers m, n, and $a \neq 1$ and for any real number r.	
1. Product Rule	1. $\log_a mn = \log_a m + \log_a n$	1. $\log(10 \cdot 2) = \log 10 + \log 2$
2. Quotient Rule	2. $\log_a \frac{m}{n} = \log_a m - \log_a n$	2. $\log \frac{45}{6} = \log 45 - \log 6$
3. Power Rule	3. $\log_a (m^r) = r \log_a m$	3. $\ln x^6 = 6 \ln x$
Change of Base Formula	Let x and $a \neq 1$ be positive numbers. Then $$\log_a x = \frac{\log x}{\log a} \quad \text{and} \quad \log_a x = \frac{\ln x}{\ln a}.$$	The expression $\log_3 6$ is equivalent to either $$\frac{\log 6}{\log 3} \quad \text{or} \quad \frac{\ln 6}{\ln 3}.$$

9.4 Exercises

MyMathLab 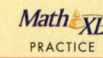 Math XL PRACTICE   WATCH   DOWNLOAD   READ   REVIEW

NOTE: Assume that variables are positive and that expressions are defined in this exercise set.

CONCEPTS AND VOCABULARY

1. $\log 12 = \log 3 + \log (\underline{\quad})$

2. $\ln 5 = \ln 20 - \ln (\underline{\quad})$

3. $\log 8 = (\underline{\quad}) \log 2$

4. $\log mn = \underline{\quad}$

5. $\log \frac{m}{n} = \underline{\quad}$

6. $\log (m^r) = \underline{\quad}$

7. Does $\log x + \log y$ equal $\log(x + y)$?

8. Does $\log x - \log y$ equal $\log\left(\frac{x}{y}\right)$?

9. Does $\log(xy)$ equal $(\log x)(\log y)$?

10. Does $\log\left(\frac{x}{y}\right)$ equal $\frac{\log x}{\log y}$?

11. Give the change of base formula.

12. $\log_a 1 = \underline{\quad}$ and $\log_a a = \underline{\quad}$.

BASIC PROPERTIES OF LOGARITHMS

Exercises 13–18: Write the expression as a sum of two or more logarithms.

13. $\ln(15)$

14. $\log(77)$

15. $\log xy$

16. $\ln 10z$

17. $\log y^2$

18. $\log x^2 y$

Exercises 19–24: Write the expression as a difference of two logarithms.

19. $\log \frac{7}{3}$

20. $\ln \frac{11}{13}$

21. $\ln \frac{x}{y}$

22. $\log \frac{2x}{z}$

23. $\log_2 \frac{45}{x}$

24. $\log_7 \frac{5x}{4z}$

Exercises 25–32: Write the expression as one logarithm.

25. $\log 45 + \log 5$

26. $\log 30 - \log 10$

27. $\ln x + \ln y$

28. $\ln m + \ln n - \ln n$

29. $\ln 7x^2 + \ln 2x$

30. $\ln x + \ln y - \ln z$

31. $\ln x + \ln y^2 - \ln y$

32. $\ln \sqrt{z} - \ln z^3 + \ln y^3$

Exercises 33–38: Write the expression as one term. Evaluate if possible.

33. $\log 20 - \log 4$

34. $\log 900 - \log 9$

35. $\ln x^4 - \ln x^2$

36. $\ln 9x^2 - \ln 3x$

37. $\log 300x - \log 3x$

38. $\log 18x^2 - \log 2x^2$

Exercises 39–48: Rewrite using the power rule.

39. $\log 3^6$

40. $\log x^7$

41. $\ln 2^x$

42. $\ln (0.77)^{x+1}$

43. $\log_2 5^{1/4}$

44. $\log_3 \sqrt{x}$

45. $\log_4 \sqrt[3]{z}$

46. $\log_7 3^\pi$

47. $\log x^{y-1}$

48. $\ln a^{2b}$

Exercises 49–60: Use properties of logarithms to write the expression as the logarithm of a single expression.

49. $4 \log z - \log z^3$

50. $2 \log_5 y + \log_5 x$

51. $\log x + 2 \log x + 2 \log y$

52. $\log x^2 + 3 \log z - 5 \log y$

53. $\log x - 2 \log \sqrt{x}$

54. $\ln y^2 - 6 \ln \sqrt[3]{y}$

55. $\ln 2^{x+1} - \ln 2$

56. $\ln 8^{1/2} + \ln 2^{1/2}$

57. $\ln \sqrt[3]{x} + \ln \sqrt{x}$

58. $2 \log_3 \sqrt{x} - 3 \log_3 x$

59. $2 \log_a (x + 1) - \log_a (x^2 - 1)$

60. $\log_b (x^2 - 9) - \log_b (x - 3)$

Exercises 61–72: Use properties of logarithms to write the expression in terms of logarithms of x, y, and z.

61. $\log xy^2$

62. $\log \dfrac{x^2}{y^3}$

63. $\ln \dfrac{x^4 y}{z}$

64. $\ln \dfrac{\sqrt{x}}{y}$

65. $\log \dfrac{\sqrt[3]{z}}{\sqrt{y}}$

66. $\log \sqrt{\dfrac{x}{y}}$

67. $\log (x^4 y^3)$

68. $\log (x^2 y^4 z^3)$

69. $\ln \dfrac{1}{y} - \ln \dfrac{1}{x}$

70. $\ln \dfrac{1}{xy}$

71. $\log_4 \sqrt{\dfrac{x^3 y}{z^2}}$

72. $\log_3 \left(\dfrac{x^2 \sqrt{z}}{y^3} \right)$

Exercises 73–76: Graph f and g in the window [−6, 6, 1] by [−4, 4, 1]. If the two graphs appear to be identical, prove that they are, using properties of logarithms.

73. $f(x) = \log x^3$, $g(x) = 3 \log x$

74. $f(x) = \ln x + \ln 3$, $g(x) = \ln 3x$

75. $f(x) = \ln (x + 5)$, $g(x) = \ln x + \ln 5$

76. $f(x) = \log (x - 2)$, $g(x) = \log x - \log 2$

Exercises 77–86: (Refer to Example 8.) Using only properties of logarithms and the approximations $\log 2 \approx 0.3$, $\log 5 \approx 0.7$, and $\log 13 \approx 1.1$, find an approximation for the expression.

77. $\log 16$

78. $\log 125$

79. $\log 65$

80. $\log 26$

81. $\log 130$

82. $\log 100$

83. $\log \dfrac{5}{2}$

84. $\log \dfrac{26}{5}$

85. $\log \dfrac{1}{13}$

86. $\log \dfrac{1}{65}$

CHANGE OF BASE FORMULA

Exercises 87–92: Use the change of base formula to approximate each expression to the nearest hundredth.

87. $\log_3 5$

88. $\log_5 12$

89. $\log_2 25$

90. $\log_7 8$

91. $\log_9 102$

92. $\log_6 293$

APPLICATIONS

93. *Modeling Sound* (See Example 5, Section 9.3.) The formula $f(x) = 10 \log (10^{16} x)$ can be used to calculate the decibel level of a sound with an intensity x. Use properties of logarithms to simplify this formula to $f(x) = 160 + 10 \log x$.

94. *Cellular Phone Technology* A formula used to calculate the strength of a signal for a cellular phone is

$$L = 110.7 - 19.1 \log h + 55 \log d,$$

where h is the height of the cellular phone tower and d is the distance the phone is from the tower. Use properties of logarithms to write an expression for L that contains only one logarithm. (*Source:* C. Smith, *Practical Cellular & PCS Design.*)

WRITING ABOUT MATHEMATICS

95. State the three basic properties of logarithms and give an example of each.

96. A student insists that $\log (x - y)$ is equal to the expression $\log x - \log y$. How could you convince the student otherwise?

Checking Basic Concepts

1. Simplify each expression by hand.
  (a) $\log 10^4$ **(b)** $\ln e^x$
  (c) $\log_2 \frac{1}{8}$ **(d)** $\log_5 \sqrt{5}$

2. Sketch a graph of $f(x) = \log x$.
  (a) What are the domain and range of f?
  (b) Evaluate $f(1)$.
  (c) Can the common logarithm of a positive number be negative? Explain.
  (d) Can the common logarithm of a negative number be positive? Explain.

3. Write the expression in terms of logarithms of x, y, and z. Assume that variables are positive.

  (a) $\log xy$ **(b)** $\ln \frac{x}{yz}$

  (c) $\ln x^2$ **(d)** $\log \frac{x^2 y^3}{\sqrt{z}}$

4. Write as the logarithm of a single expression.
  (a) $\log x + \log y$
  (b) $\ln 2x - 3 \ln y$
  (c) $2 \log_2 x + 3 \log_2 y - \log_2 z$

9.5 Exponential and Logarithmic Equations

Exponential Equations and Models • Logarithmic Equations and Models

A LOOK INTO MATH ▶

Wind power capacity in the world grew 31% in 2009. With the problems concerning nuclear power, it is expected that wind power will continue to grow at exponential rates. For example, from 2006 to 2010, China's wind power capacity almost doubled each year. In fact, in 2010 China's capacity to produce wind power exceeded the capacity of the United States. In order to model and answer questions about this growth, we will use exponential and logarithmic equations. See Exercise 93. (*Sources: Business and Economy, Clean Energy, Wind Energy,* 2010.)

Exponential Equations and Models

To solve the equation $10 + x = 100$, we subtract 10 from each side because addition and subtraction are inverse operations.

$$10 + x - \mathbf{10} = 100 - \mathbf{10} \qquad \text{Subtract 10 from each side.}$$
$$x = 90$$

To solve the equation $10x = 100$, we divide each side by 10 because multiplication and division are inverse operations.

$$\frac{10x}{\mathbf{10}} = \frac{100}{\mathbf{10}} \qquad \text{Divide each side by 10.}$$
$$x = 10$$

Now suppose that we want to solve the *exponential equation*

The exponent is a variable

$$10^x = 100. \qquad \text{Exponential equation}$$

What is new about this type of equation is that the variable x is an *exponent*. The inverse operation of 10^x is $\log x$. Rather than subtracting 10 from each side or dividing each side by 10, we take the base-10 logarithm of each side. Doing so results in

$$\log 10^x = \log 100. \quad \text{Take base-10 logarithm of each side.}$$

Because $\log 10^x = x$ for all real numbers x, the equation becomes

$$x = \log 100 \quad \text{or, equivalently,} \quad x = 2.$$

These concepts are applied in the next example.

EXAMPLE 1 **Solving exponential equations**

Solve and approximate to the nearest hundredth.
(a) $10^x = 150$ **(b)** $e^x = 40$ **(c)** $2^x = 50$ **(d)** $0.9^x = 0.5$

Solution
(a)

$10^x = 150$	Given equation
$\log 10^x = \log 150$	Take the common logarithm of each side.
$x = \log 150 \approx 2.18$	Inverse property: $\log 10^k = k$ for all k

(b) The inverse operation of e^x is $\ln x$, so we take the natural logarithm of each side.

$e^x = 40$	Given equation
$\ln e^x = \ln 40$	Take the natural logarithm of each side.
$x = \ln 40 \approx 3.69$	Inverse property: $\ln e^k = k$ for all k

(c) The inverse operation of 2^x is $\log_2 x$. Calculators do not usually have a base-2 logarithm key, so we take the common logarithm of each side and then apply the power rule.

$2^x = 50$	Given equation
$\log 2^x = \log 50$	Take the common logarithm of each side.
$x \log 2 = \log 50$	Power rule: $\log (m^r) = r \log m$
$x = \dfrac{\log 50}{\log 2} \approx 5.64$	Divide by $\log 2$ and approximate.

(d) This time we begin by taking the natural logarithm of each side.

$0.9^x = 0.5$	Given equation
$\ln 0.9^x = \ln 0.5$	Take the natural logarithm of each side.
$x \ln 0.9 = \ln 0.5$	Power rule: $\ln (m^r) = r \ln m$
$x = \dfrac{\ln 0.5}{\ln 0.9} \approx 6.58$	Divide by $\ln 0.9$ and approximate.

Now Try Exercises 13, 15, 25, 27

MAKING CONNECTIONS

Logarithms of Quotients and Quotients of Logarithms

The solution in Example 1(c) is $\frac{\log 50}{\log 2}$. Note that

$$\frac{\log 50}{\log 2} \neq \log 50 - \log 2.$$

However, $\log 50 - \log 2 = \log \frac{50}{2} = \log 25$ by the quotient rule for logarithms, as shown in the figure.

```
log(50)/log(2)
          5.64385619
log(50)-log(2)
          1.397940009
log(25)
          1.397940009
```

The next two examples illustrate methods for solving exponential equations.

EXAMPLE 2 | **Solving exponential equations**

Solve each equation and approximate to the nearest hundredth.

(a) $2e^x - 1 = 5$　　**(b)** $3^{x-5} = 15$　　**(c)** $e^{2x} = e^{x+5}$　　**(d)** $3^{2x} = 2^{x+3}$

Solution

(a) Begin by solving for e^x.

$2e^x - 1 = 5$	Given equation
$2e^x = 6$	Add 1 to each side.
$e^x = 3$	Divide each side by 2.
$\ln e^x = \ln 3$	Take the natural logarithm.
$x = \ln 3 \approx 1.10$	Inverse property: $\ln e^k = k$

(b) Start by taking the common logarithm of each side. (We could also take the natural logarithm of each side.)

$3^{x-5} = 15$	Given equation
$\log 3^{x-5} = \log 15$	Take the common logarithm of each side.
$(x - 5) \log 3 = \log 15$	Power rule for logarithms
$x - 5 = \dfrac{\log 15}{\log 3}$	Divide by log 3.
$x = \dfrac{\log 15}{\log 3} + 5 \approx 7.46$	Add 5 to each side and approximate.

(c) For $e^{2x} = e^{x+5}$, the bases are equal, so the exponents must also be equal. To verify this assertion, take the natural logarithm of each side.

$e^{2x} = e^{x+5}$	Given equation
$\ln e^{2x} = \ln e^{x+5}$	Take the natural logarithm.
$2x = x + 5$	Inverse property: $\ln e^k = k$
$x = 5$	Subtract x.

(d) For $3^{2x} = 2^{x+3}$, the bases are not equal. However, we can still solve the equation by taking the common logarithm of each side. A logarithm of any base could be used.

$3^{2x} = 2^{x+3}$	Given equation
$\log 3^{2x} = \log 2^{x+3}$	Take the common logarithm.
$2x \log 3 = (x + 3) \log 2$	Power rule for logarithms
$2x \log 3 = x \log 2 + 3 \log 2$	Distributive property
$2x \log 3 - x \log 2 = 3 \log 2$	Subtract $x \log 2$.
$x(2 \log 3 - \log 2) = 3 \log 2$	Factor out x.
$x = \dfrac{3 \log 2}{2 \log 3 - \log 2}$	Divide by 2 log 3 − log 2.
$x \approx 1.38$	Approximate.

Now Try Exercises 29, 31, 35, 43

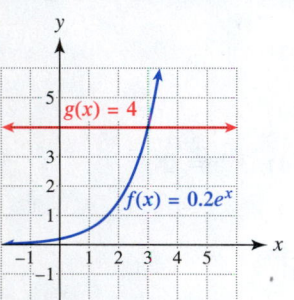

Figure 9.35

> **EXAMPLE 3** **Solving an exponential equation**

Graphs for $f(x) = 0.2e^x$ and $g(x) = 4$ are shown in Figure 9.35.
(a) Use the graphs to estimate the solution to the equation $f(x) = g(x)$.
(b) Check your estimate by solving the equation symbolically.

Solution
(a) The graphs intersect near the point $(3, 4)$. Therefore the solution is given by $x \approx 3$.
(b) We must solve the equation $0.2e^x = 4$.

$0.2e^x = 4$	Given equation
$e^x = 20$	Divide each side by 0.2.
$\ln e^x = \ln 20$	Take the natural logarithm of each side.
$x = \ln 20$	Inverse property: $\ln e^k = k$
$x \approx 2.996$	Approximate.

NOTE: The graphical estimate did not give the *exact* solution of $\ln 20$.

■ **Now Try Exercise 45**

▶ **REAL-WORLD CONNECTION** As discussed in Section 9.2, if $1000 is deposited in a savings account paying 10% annual interest at the end of each year, the amount A in the account after x years is given by

$$A(x) = 1000(1.1)^x.$$

After 10 years there will be

$$A(\mathbf{10}) = 1000(1.1)^{\mathbf{10}} \approx \$2593.74$$

in the account. To calculate how long it will take for $4000 to accrue in the account, we need to solve the exponential equation

$$1000(1.1)^x = 4000.$$

We do so in the next example.

> **EXAMPLE 4** **Solving an exponential equation**

Solve $1000(1.1)^x = 4000$ symbolically. Give graphical support for your answer.

Solution
Symbolic Solution Begin by dividing each side of the equation by 1000.

$1000(1.1)^x = 4000$	Given equation
$1.1^x = 4$	Divide by 1000.
$\log 1.1^x = \log 4$	Take the common logarithm of each side.
$x \log 1.1 = \log 4$	Power rule for logarithms
$x = \dfrac{\log 4}{\log 1.1} \approx 14.5$	Divide by log 1.1 and approximate.

Interest is paid at the end of the year, so it will take 15 years for $1000 earning 10% annual interest to grow to (at least) $4000.

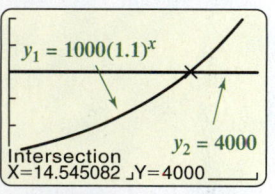

Figure 9.36

Graphical Solution Graphical support is shown in Figure 9.36, where the graphs of $y_1 = 1000(1.1)^x$ and $y_2 = 4000$ intersect when $x \approx 14.5$.

■ **Now Try Exercise 97**

▶ **REAL-WORLD CONNECTION** In the next example, we model the life span of a robin with an exponential function.

EXAMPLE 5 **Modeling the life span of a robin**

The life spans of 129 robins were monitored over a 4-year period in one study. The formula $f(x) = 10^{-0.42x}$ can be used to calculate the percentage of robins remaining after x years. For example, $f(1) \approx 0.38$ means that after 1 year 38% of the robins were still alive. (*Source: D. Lack, The Life Span of a Robin.*)

(a) Evaluate $f(2)$ and interpret the result.

(b) Determine when 5% of the robins remained.

Solution

(a) $f(2) = 10^{-0.42(2)} \approx 0.145$. After **2** years about **14.5%** of the robins were still alive.

(b) Use $5\% = 0.05$ and solve the following equation.

$$10^{-0.42x} = 0.05 \qquad \text{Equation to solve}$$

$$\log 10^{-0.42x} = \log 0.05 \qquad \text{Take the common logarithm of each side.}$$

$$-0.42x = \log 0.05 \qquad \text{Inverse property: } \log 10^k = k$$

$$x = \frac{\log 0.05}{-0.42} \approx 3.1 \qquad \text{Divide by } -0.42.$$

After about 3 years only 5% of the robins were still alive.

■ **Now Try Exercise 103**

Logarithmic Equations and Models

To solve an exponential equation we use logarithms. To solve a logarithmic equation we *exponentiate* each side of the equation. To do so we use the fact that if $x = y$, then $a^x = a^y$ for any positive base a. For example, to solve

$$\log x = 3 \qquad \text{Logarithmic equation}$$

we exponentiate each side of the equation, using base 10.

$$10^{\log x} = 10^3 \qquad \text{If } a = b, \text{ then } 10^a = 10^b.$$

Because $10^{\log x} = x$ for all positive x, it follows that

$$x = 10^3. \qquad \text{Inverse property}$$

To solve logarithmic equations, we frequently use the inverse property

$$a^{\log_a x} = x.$$

Bases must be equal.

Examples of this inverse property include

$$e^{\ln 2k} = 2k, \qquad 2^{\log_2 x} = x, \qquad \text{and} \qquad 10^{\log (x+5)} = x + 5.$$

Both are base-e.   Both are base-2.   Both are base-10.

EXAMPLE 6 **Solving logarithmic equations**

Solve and approximate solutions to the nearest hundredth when appropriate.

(a) $2 \log x = 4$ (b) $\ln 3x = 5.5$ (c) $\log_2 (x + 4) = 7$

Solution

(a)
$$2 \log x = 4 \qquad \text{Given equation}$$
$$\log x = 2 \qquad \text{Divide each side by 2.}$$
$$\mathbf{10}^{\log x} = \mathbf{10}^2 \qquad \text{Exponentiate each side, using base 10.}$$
$$x = 100 \qquad \text{Inverse property: } 10^{\log k} = k$$

(b)
$$\ln 3x = 5.5 \qquad \text{Given equation}$$
$$e^{\ln 3x} = e^{5.5} \qquad \text{Exponentiate each side, using base } e.$$
$$3x = e^{5.5} \qquad \text{Inverse property: } e^{\ln k} = k$$
$$x = \frac{e^{5.5}}{3} \approx 81.56 \qquad \text{Divide each side by 3 and approximate.}$$

(c)
$$\log_2 (x + 4) = 7 \qquad \text{Given equation}$$
$$\mathbf{2}^{\log_2 (x+4)} = \mathbf{2}^7 \qquad \text{Exponentiate each side, using base 2.}$$
$$x + 4 = 2^7 \qquad \text{Inverse property: } 2^{\log_2 k} = k$$
$$x = 2^7 - 4 \qquad \text{Subtract 4 from each side.}$$
$$x = 124 \qquad \text{Simplify.}$$

Now Try Exercises 61, 63, 69

Because the domain of any logarithmic function includes only positive numbers, it is important to check answers, as emphasized in the next example.

EXAMPLE 7 **Solving a logarithmic equation**

Solve $\log (x + 2) + \log (x - 2) = \log 5$. Check any answers.

Solution
Start by applying the product rule for logarithms.

$$\log (x + 2) + \log (x - 2) = \log 5 \qquad \text{Given equation}$$
$$\log ((x + 2)(x - 2)) = \log 5 \qquad \text{Product rule}$$
$$\log (x^2 - 4) = \log 5 \qquad \text{Multiply.}$$
$$\mathbf{10}^{\log (x^2 - 4)} = \mathbf{10}^{\log 5} \qquad \text{Exponentiate using base 10.}$$
$$x^2 - 4 = 5 \qquad \text{Inverse properties}$$
$$x^2 = 9 \qquad \text{Add 4.}$$
$$x = \pm 3 \qquad \text{Square root property}$$

Check each answer.

$$\log (3 + 2) + \log (3 - 2) \overset{?}{=} \log 5 \qquad\qquad \log (-3 + 2) + \log (-3 - 2) \overset{?}{=} \log 5$$
$$\log 5 + \log 1 \overset{?}{=} \log 5 \qquad\qquad\qquad \log (-1) + \log (-5) \neq \log 5 \text{ ✗}$$
$$\log 5 + 0 \overset{?}{=} \log 5 \qquad\qquad\qquad\qquad\qquad \text{Undefined}$$
$$\log 5 = \log 5 \text{ ✓}$$

Although 3 is a solution, -3 is not, because both $\log (-1)$ and $\log (-5)$ are undefined expressions. Be sure to check your answers.

Now Try Exercise 77

| EXAMPLE 8 | **Modeling bird populations** |

Near New Guinea there is a relationship between the number of different species of birds and the size of an island. Larger islands tend to have a greater variety of birds. Table 9.13 lists the number of species of birds y found on islands with an area of x square kilometers.

TABLE 9.13 Bird Species on Islands

x (km^2)	0.1	1	10	100	1000
y (species)	10	15	20	25	30

Source: B. Freedman, *Environmental Ecology.*

(a) Find values for the constants a and b so that $y = a + b \log x$ models the data.
(b) Predict the number of bird species on an island of 4000 square kilometers.

Solution
(a) Because **log 1 $=$ 0**, substitute $x = 1$ and $y = $ **15** in the equation to find a.

$$15 = a + b \log 1 \qquad y = a + b \log x$$
$$15 = a + b \cdot 0 \qquad \log 1 = 0$$
$$15 = a \qquad \text{Simplify.}$$

Thus $y = 15 + b \log x$. To find b substitute $x = 10$ and $y = $ **20**.

$$20 = 15 + b \log 10 \qquad y = 15 + b \log x$$
$$20 = 15 + b \cdot 1 \qquad \log 10 = 1$$
$$5 = b \qquad \text{Simplify.}$$

The data in Table 9.13 are modeled by $y = 15 + 5 \log x$. This result is supported by Figure 9.37.

X	Y1
.1	10
1	15
10	20
100	25
1000	30

Y1 $=$ 15+5log(X)

Figure 9.37

(b) To predict the number of species on an island of 4000 square kilometers, let $x = $ **4000** and find y.

$$y = 15 + 5 \log 4000 \approx 33$$

The model estimates about 33 different species of birds on this island.

Now Try Exercise 105

| EXAMPLE 9 | **Modeling runway length** |

For some types of airplanes with weight x, the minimum runway length L required at take-off is modeled by

$$L(x) = 3 \log x.$$

In this equation L is measured in thousands of feet and x is measured in thousands of pounds. Estimate the weight of the heaviest airplane that can take off from a runway 5100 feet long. (*Source:* L. Haefner, *Introduction to Transportation Systems.*)

CRITICAL THINKING

In Example 10, Section 9.3, we used the formula $L(x) = 1.3 \ln x$ to model runway length. Are $L(x) = 1.3 \ln x$ and $L(x) = 3 \log x$ equivalent formulas? Explain.

Solution

Runway length is measured in thousands of feet, so we must solve the equation $L(x) = 5.1$.

$$
\begin{array}{ll}
3 \log x = 5.1 & L(x) = 5.1 \\
\log x = 1.7 & \text{Divide each side by 3.} \\
\mathbf{10^{\log x}} = \mathbf{10^{1.7}} & \text{Exponentiate each side, using base 10.} \\
x = 10^{1.7} & \text{Inverse property: } 10^{\log k} = k \\
x \approx 50.1 & \text{Approximate.}
\end{array}
$$

The largest airplane that can take off from this runway weighs about 50,000 pounds.

Now Try Exercise 101

9.5 Putting It All Together

TYPE OF EQUATION	PROCEDURE	EXAMPLE	
Exponential	Begin by solving for the exponential expression a^x. Then take a logarithm of each side.	$4e^x + 1 = 9$	Given equation
		$e^x = 2$	Solve for e^x.
		$\ln e^x = \ln 2$	Take the natural logarithm.
		$x = \ln 2$	Inverse property: $\ln e^k = k$
Logarithmic	Begin by solving for the logarithm in the equation. Then exponentiate each side of the equation, using the same base as the logarithm.	$\dfrac{1}{3} \log 2x = 1$	Given equation
		$\log 2x = 3$	Multiply by 3.
		$10^{\log 2x} = 10^3$	Exponentiate using base 10.
		$2x = 1000$	Inverse property: $10^{\log k} = k$
		$x = 500$	Divide by 2.

9.5 Exercises

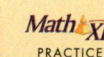

Math XL PRACTICE **WATCH** **DOWNLOAD** **READ** **REVIEW**

CONCEPTS AND VOCABULARY

1. To solve $x - 5 = 50$, what should be done?

2. To solve $5x = 50$, what should be done?

3. To solve $10^x = 50$, what should be done?

4. To solve $\log x = 5$, what should be done?

5. $\log 10^x =$ _____

6. $10^{\log x} =$ _____

7. $\ln e^{2x} =$ _____

8. $e^{\ln (x+7)} =$ _____

9. Does $\dfrac{\log 5}{\log 4}$ equal $\log \frac{5}{4}$? Explain.

10. Does $\dfrac{\log 5}{\log 4}$ equal $\log 5 - \log 4$? Explain.

11. How many solutions are there to $\log x = k$, where k is any real number?

12. How many solutions are there to $10^x = k$, where k is a positive number?

EXPONENTIAL EQUATIONS

Exercises 13–44: Solve the given exponential equation. Approximate answers to the nearest hundredth when appropriate.

13. $10^x = 1000$

14. $10^x = 0.01$

15. $2^x = 64$

16. $3^x = 27$

17. $2^{x-3} = 8$

18. $3^{2x} = 81$

19. $4^x + 3 = 259$

20. $3(5^{2x}) = 300$

21. $10^{0.4x} = 124$

22. $0.75^x = 0.25$

23. $e^{-x} = 1$

24. $0.5^{-5x} = 5$

25. $e^x = 25$

26. $e^x = 0.4$

27. $0.4^x = 2$

28. $0.7^x = 0.3$

29. $e^x - 1 = 6$

30. $2e^{4x} = 15$

31. $2(10)^{x+2} = 35$

32. $10^{3x} + 10 = 1500$

33. $3.1^{2x} - 4 = 16$

34. $5.4^{x-1} = 85$

35. $e^{3x} = e^{2x-1}$

36. $e^{x^2} = e^{3x-2}$

37. $5^{4x} = 5^{x^2-5}$

38. $2^{4x} = 2^{x+3}$

39. $e^{2x} \cdot e^x = 10$

40. $10^{x-2} \cdot 10^x = 1000$

41. $e^x = 2^{x+2}$

42. $2^{2x} = 3^{x-1}$

43. $4^{0.5x} = 5^{x+2}$

44. $3^{2x} = 7^{x+1}$

Exercises 45–48: (Refer to Example 3.) The symbolic and graphical representations of f and g are given.

(a) Use the graph to solve $f(x) = g(x)$.
(b) Solve $f(x) = g(x)$ symbolically.

45. $f(x) = 0.2(10^x)$,
$g(x) = 2$

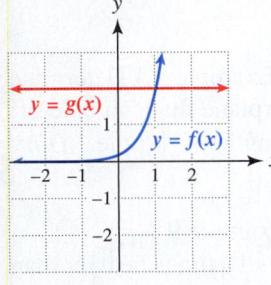

46. $f(x) = e^x$,
$g(x) = 7.4$

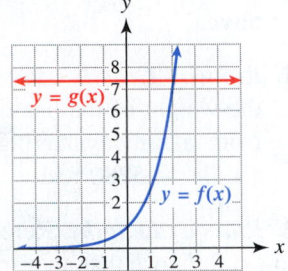

47. $f(x) = 2^{-x}$,
$g(x) = 4$

48. $f(x) = 0.1(3^x)$,
$g(x) = 0.9$

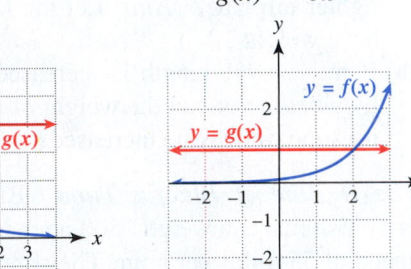

Exercises 49–56: Solve the equation symbolically. Give graphical or numerical support. Approximate answers to the nearest hundredth when appropriate.

49. $10^x = 0.1$

50. $2(10^x) = 2000$

51. $4e^x + 5 = 9$

52. $e^x + 6 = 36$

53. $4^x = 1024$

54. $3^x = 729$

55. $(0.55)^x + 0.55 = 2$

56. $5(0.9)^x = 3$

Exercises 57–60: The given equation cannot be solved symbolically. Find any solutions graphically to the nearest hundredth.

57. $e^x - x = 2$

58. $x \log x = 1$

59. $\ln x = e^{-x}$

60. $10^x - 2 = \log(x + 2)$

LOGARITHMIC EQUATIONS

Exercises 61–82: Solve the given equation. Approximate answers to the nearest hundredth when appropriate.

61. $\log x = 2$

62. $\log x = 0.01$

63. $\ln x = 5$

64. $2 \ln x = 4$

65. $\log 2x = 7$

66. $6 \ln 4x = 12$

67. $\log_2 x = 4$

68. $\log_2 x = 32$

69. $\log_2 5x = 2.3$

70. $2 \log_3 4x = 10$

71. $2 \log x + 5 = 7.8$

72. $\ln(x - 1) = 3.3$

73. $5 \ln(2x + 1) = 55$

74. $5 - \log(x + 3) = 2.6$

75. $\log x^2 = \log x$

76. $\ln x^2 = \ln(3x - 2)$

77. $\ln x + \ln(x + 1) = \ln 30$

78. $\log(x - 1) + \log(2x + 1) = \log 14$

79. $\log_3 3x - \log_3(x + 2) = \log_3 2$

80. $\log_4(x^2 - 1) - \log_4(x - 1) = \log_4 6$

81. $\log_2(x - 1) + \log_2(x + 1) = 3$

82. $\log_4(x^2 + 2x + 1) - \log_4(x + 1) = 2$

Exercises 83–88: Solve the equation symbolically. Give graphical or numerical support. Approximate answers to the nearest hundredth.

83. $\log x = 1.6$

84. $\ln x = 2$

85. $\ln(x+1) = 1$ **86.** $2\log(2x+3) = 8$

87. $17 - 6\log_3 x = 5$ **88.** $4\log_2 x + 7 = 12$

Exercises 89–92: Two functions, f and g, are given.

  (a) *Use the graph to solve* $f(x) = g(x)$.
  (b) *Solve* $f(x) = g(x)$ *symbolically.*

89. $f(x) = \ln x$, **90.** $f(x) = \log_2 x$,
    $g(x) = 0.7$ $g(x) = 1.6$

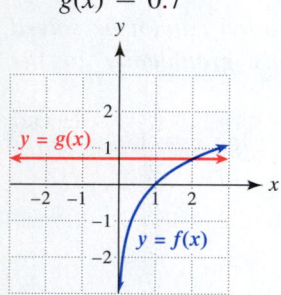 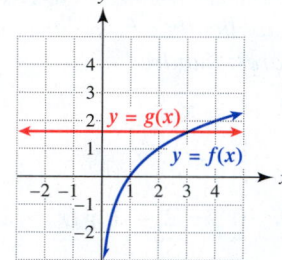

91. $f(x) = 5\log 2x$, **92.** $f(x) = 2\ln(x) - 3$,
    $g(x) = 3$ $g(x) = 0.9$

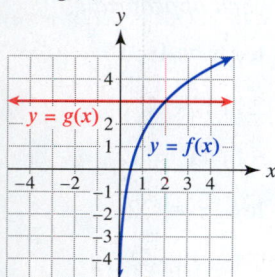 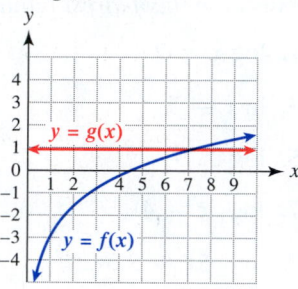

APPLICATIONS

93. *China's Wind Power* China overtook the United States in 2010 for generating the most wind power. China's wind power capacity W in gigawatts x years after 2005 can be modeled by $W(x) = 1.73(10^{0.276x})$.
  (a) Evaluate $W(5)$ and interpret the result.
  (b) Determine when China generated 22 gigawatts.

94. *Risk of Down Syndrome* The likelihood of a live birth having Down syndrome varies with the age of the mother. The function

$$D(x) = 0.0000816(10^{0.102x})$$

gives the percentage of live births that have Down syndrome when the mother is x years old.
  (a) Evaluate $D(35)$ and interpret the result.
  (b) Determine the mother's age when this percentage reaches 4%.

95. *Bacteria Growth* A sample of 5 million bacteria increases by 15% each hour. To the nearest hour, how long does it take for the sample to reach 20 million?

96. *Bacteria Growth* A sample of 3 million bacteria increases by 125% each day. To the nearest day, how long does it take for the bacteria sample to reach 173 million?

97. *Growth of a Mutual Fund* (Refer to Example 4.) An investor deposits $2000 in a mutual fund that returns 15% at the end of 1 year. Determine the length of time required for the investment to triple its value if the annual rate of return remains the same.

98. *Savings Account* (Refer to Example 4.) If a savings account pays 6% annual interest at the end of each year, how many years will it take for the account to double in value?

99. *Liver Transplants* In the United States the gap between available organs for liver transplants and people who need them has widened. The number of individuals waiting for liver transplants can be modeled by

$$f(x) = 2339(1.24)^{x-1988},$$

where x is the year. (*Source:* United Network for Organ Sharing.)
  (a) Evaluate $f(1994)$ and interpret the result.
  (b) Determine when the number of individuals waiting for liver transplants was 20,000.

100. *Life Span of a Robin* (Refer to Example 5.) Determine when 50% of the robins in the study were still alive.

101. *Runway Length* (Refer to Example 9.) Determine the weight of the heaviest airplane that can take off from a runway having a length of $\frac{3}{4}$ mile. (*Hint:* 1 mile = 5280 feet.)

102. *Runway Length* (Refer to Example 9.)
  (a) Suppose that an airplane is 10 times heavier than a second airplane. How much longer should the runway be for the heavier airplane than for the lighter airplane? (*Hint:* Let the heavier airplane have weight $10x$.)
  (b) If the runway length is increased by 3000 feet, by what factor can the weight of an airplane that uses the runway be increased?

103. *The Decline of Bluefin Tuna* Bluefin tuna are large fish that can weigh 1500 pounds and swim at a speed of 55 miles per hour. They are used for sushi, and a prime fish can be worth more than $50,000. As a result, the number of western Atlantic bluefin tuna has declined dramatically. Their numbers in thousands between 1974 and 1991 can be modeled by $f(x) = 230(10^{-0.055x})$, where x is the number of years after 1974. See the accompanying graph.

(*Source:* B. Freedman.)

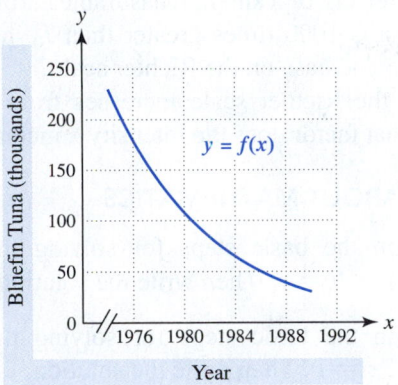

(a) Evaluate $f(1)$ and interpret the result.
(b) Use the graph to estimate the year in which bluefin tuna numbered 115 thousand.
(c) Solve part (b) symbolically.

104. *Insect Populations* (Refer to Example 8.) The table lists numbers of species of insects y found on islands having areas of x square miles.

x (square miles)	1	2	4
y (species)	1000	1500	2000

x (square miles)	8	16	32
y (species)	2500	3000	3500

(a) Find values for the constants a and b so that $y = a + b \log_2 x$ models the data.
(b) Construct a table for y and verify that your equation models the data.
(c) Estimate the number of species of insects on an island having an area of 12 square miles.

Exercises 105 and 106: (Refer to Example 8.) Find values for a and b so that $y = a + b \log x$ models the data.

105.

x	0.1	1	10	100
y	22	25	28	31

106.

x	0.01	1	100	1000
y	-10	-2	6	10

107. *Calories Consumed and Land Ownership* In developing countries there is a relationship between the amount of land a person owns and the average number of calories that person consumes daily. This relationship is modeled by $f(x) = 645 \log(x + 1) + 1925$, where x is the amount of land owned in acres and $0 \le x \le 4$. (*Source:* D. Grigg, *The World Food Problem.*)

(a) Estimate graphically the number of acres owned by a person consuming 2200 calories per day.
(b) Solve part (a) symbolically.

108. *Population of Industrialized Urban Regions* The number of people living in industrialized urban regions throughout the world has not grown exponentially. Instead it has grown logarithmically and is modeled by

$$f(x) = 0.36 + 0.15 \ln(x - 1949).$$

In this formula the output is billions of people and the input x is the year, where $1950 \le x \le 2030$. (*Source:* D. Meadows, *Beyond The Limits.*)

(a) Determine graphically when this population may reach 1 billion.
(b) Solve part (a) symbolically.

109. *Fertilizer Use* Between 1950 and 1980, the use of chemical fertilizers increased worldwide. The table lists worldwide average use y in kilograms per acre of cropland during year x. (*Source:* D. Grigg, *The World Food Problem.*)

x	1950	1963	1972	1979
y	5.0	11.3	22.0	31.2

(a) Are the data linear or nonlinear? Explain.
(b) The equation $y = 5(1.06)^{(x-1950)}$ may be used to model the data. The growth factor is 1.06. What does this growth factor indicate about fertilizer use during this time period?
(c) Estimate the year when fertilizer use was 15 kilograms per acre of cropland.

110. *Greenhouse Gases* If current trends continue, concentrations of atmospheric carbon dioxide (CO_2) in parts per million (ppm) are expected to increase. This increase in concentration of CO_2 has been accelerated by burning fossil fuels and deforestation. The exponential equation $y = 364(1.005)^x$ may be used to model CO_2 in parts per million x years after 2000. Estimate the year when the CO_2 concentration could be double the preindustrial level of 280 parts per million. (*Source:* R. Turner, *Environmental Economics.*)

111. *Modeling Sound* The formula

$$f(x) = 160 + 10 \log x$$

is used to calculate the decibel level of a sound with intensity x measured in watts per square centimeter. The noise level at a basketball game can reach 100 decibels. Find the intensity x of this sound.

112. *Loudness of a Sound* (Refer to Exercise 111.)

(a) Show that, if the intensity of a sound increases by a factor of 10 from x to $10x$, the decibel level increases by 10 decibels. *Hint:* Show that

$$160 + 10 \log 10x = 170 + 10 \log x.$$

(b) Find the increase in decibels if the intensity x increases by a factor of 1000.

(c) Find the increase in the intensity x if the decibel level increases by 20.

113. *Hurricanes* (Refer to Exercise 106, Section 9.3.) The barometric air pressure in inches of mercury at a distance of x miles from the eye of a severe hurricane is given by $f(x) = 0.48 \ln(x + 1) + 27$. How far from the eye is the pressure 28 inches of mercury?

(*Source:* A. Miller and R. Anthes, *Meteorology.*)

114. *Earthquakes* The Richter scale is used to determine the intensity of earthquakes, which corresponds to the amount of energy released. If an earthquake has an intensity of x, its magnitude, as computed by the Richter scale, is given by $R(x) = \log \frac{x}{I_0}$, where I_0 is the intensity of a small, measurable earthquake.

(a) If x is 1000 times greater than I_0, how large is this increase on the Richter scale?

(b) If the Richter scale increases from 5 to 8, by what factor does the intensity x increase?

WRITING ABOUT MATHEMATICS

115. Explain the basic steps for solving the equation $a(10^x) - b = c$. Then write the solution.

116. Explain the basic steps for solving the equation $a \log 3x = b$. Then write the solution.

SECTION 9.5

Checking Basic Concepts

1. Solve the equation. Approximate answers to the nearest hundredth when appropriate.
  (a) $2(10^x) = 40$ (b) $2^{3x} + 3 = 150$
  (c) $\ln x = 4.1$ (d) $4 \log 2x = 12$

2. Solve $\log(x + 4) + \log(x - 4) = \log 48$. Check the answers.

3. If \$500 is deposited in a savings account that pays 3% annual interest at the end of each year, the amount of money A in the account after x years is given by $A = 500(1.03)^x$. Estimate the number of years required for this amount to reach \$900.

CHAPTER 9   Summary

SECTION 9.1 ■ COMPOSITE AND INVERSE FUNCTIONS

Composition of Functions If f and g are functions, then the composite function $g \circ f$, or composition of g and f, is defined by $(g \circ f)(x) = g(f(x))$.

Example: If $f(x) = x - 5$ and $g(x) = 2x^2 + 4x - 6$, then $(g \circ f)(x)$ is

$$g(f(x)) = g(x - 5)$$
$$= 2(x - 5)^2 + 4(x - 5) - 6.$$

One-to-One Function A function f is one-to-one if, for any c and d in the domain of f,

$$c \neq d \quad \text{implies that} \quad f(c) \neq f(d).$$

That is, different inputs always result in different outputs.

Example: $f(x) = x^2 + 4$ is *not* one-to-one because $f(-3) = f(3) = 13$.   Different inputs; same output

Horizontal Line Test If every horizontal line intersects the graph of a function f at most once, then f is a one-to-one function.

Examples:

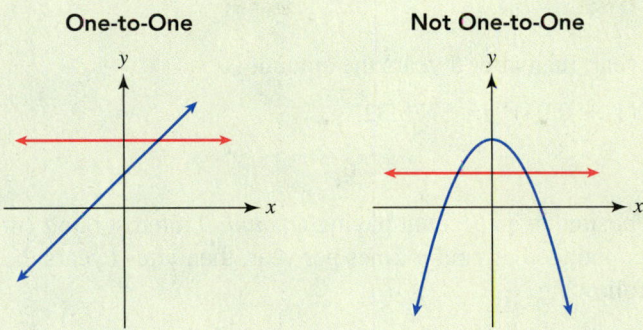

One-to-One

Not One-to-One

Inverse Functions If f is one-to-one, then f has an inverse function, denoted f^{-1}, that satisfies $(f^{-1} \circ f)(x) = x$ and $(f \circ f^{-1})(x) = x$.

Example: $f(x) = 7x$ and $f^{-1}(x) = \frac{x}{7}$ are inverse functions.

SECTION 9.2 ■ EXPONENTIAL FUNCTIONS

Exponential Function An exponential function is defined by $f(x) = Ca^x$, where $a > 0$, $C > 0$, and $a \neq 1$. Its domain (set of valid inputs) is all real numbers and its range (outputs) is all positive real numbers. The base is a.

Example: $f(x) = e^x$ is the natural exponential function and $e \approx 2.71828$.

Exponential Growth and Decay When $a > 1$, the graph of $f(x) = Ca^x$ models exponential growth, and when $0 < a < 1$, it models exponential decay. The base a represents either the growth factor or the decay factor. The constant C equals the y-intercept.

Exponential Growth

$f(x) = Ca^x$
$a > 1$

Exponential Decay

$f(x) = Ca^x$
$0 < a < 1$

Example: $f(x) = 1.5(2)^x$ is an exponential function with $a = 2$ and $C = 1.5$. It models exponential growth because $a > 1$. The growth factor is 2 because for each unit increase in x, the output from $f(x)$ increases by a *factor* of 2.

Percent Change If an amount A changes to a new amount B, then the percent change is given by

$$\frac{B - A}{A} \times 100.$$

Example: If \$500 increases to \$700, the percent change is

$$\frac{700 - 500}{500} \times 100 = 40\%.$$

Percent Change and Growth Factor If an initial amount C experiences a constant percent change r (in decimal form) for each x-unit increase in time, then the new amount A is

$$A = C(1 + r)^x,$$

where the growth factor is $a = 1 + r$.

Example: If \$800 increases by 4% each year, then after 5 years the amount is

$$A = 800(1 + 0.04)^5 \approx \$973.32.$$

The growth factor is $a = 1.04$.

Compound Interest If P dollars are deposited in an account paying an *annual* interest rate r (in decimal form) and this interest rate is compounded or paid n times per year, then after t years the account contains A dollars given by the following.

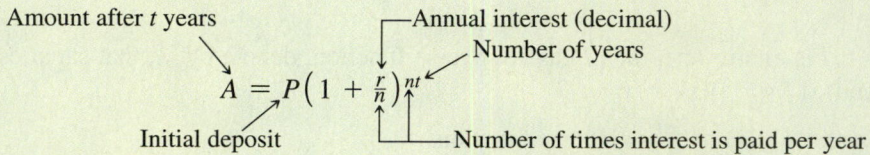

Example: If \$1200 is deposited at 6% annual interest, compounded quarterly, then after 8 years the account contains

$$A = 1200\left(1 + \frac{0.06}{4}\right)^{(4 \cdot 8)} \approx \$1932.39.$$

SECTION 9.3 ■ LOGARITHMIC FUNCTIONS

Base-a Logarithms The logarithm with base a of a positive number x is denoted $\log_a x$. If $\log_a x = b$, then $x = a^b$. That is, $\log_a x$ represents the exponent on base a that results in x.

Examples: $\log_2 16 = 4$ because $16 = 2^4$ and $\log 100 = 2$ because $100 = 10^2$.

Domain and Range of Logarithmic Functions The domain (set of valid inputs) of a logarithmic function is the set of all positive real numbers and the range (outputs) is the set of real numbers.

Graph of a Logarithmic Function The graph of a logarithmic function passes through $(1, 0)$, as illustrated in the following graph. As x becomes large, $\log_a x$ with $a > 1$ grows very slowly.

Base-a Logarithm

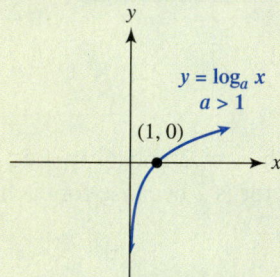

SECTION 9.4 ■ PROPERTIES OF LOGARITHMS

Basic Properties Logarithms have several important properties. For positive numbers m, n, and $a \neq 1$ and any real number r,

1. $\log_a mn = \log_a m + \log_a n$.

2. $\log_a \frac{m}{n} = \log_a m - \log_a n$.

3. $\log_a (m^r) = r \log_a m$.

Examples: **1.** $\log 5 + \log 20 = \log (5 \cdot 20) = \log 100 = 2$

2. $\log 100 - \log 5 = \log \frac{100}{5} = \log 20 \approx 1.301$

3. $\ln 2^6 = 6 \ln 2 \approx 4.159$

NOTE: $\log_a 1 = 0$ for any valid base a. Thus $\log 1 = 0$ and $\ln 1 = 0$.

Inverse Properties The following inverse properties are important for solving exponential and logarithmic equations.

1. $\log_a a^x = x$, for any real number x

2. $a^{\log_a x} = x$, for any positive number x

Examples: **1.** $\log_2 2^\pi = \pi$ **2.** $10^{\log 2.5} = 2.5$

3. $e^{\ln 5} = 5$ **4.** $\log 10^7 = 7$

NOTE: If $f(x) = \log x$, then $f^{-1}(x) = 10^x$, and if $g(x) = \ln x$, then $g^{-1}(x) = e^x$.

SECTION 9.5 ■ EXPONENTIAL AND LOGARITHMIC EQUATIONS

Solving Equations When solving an exponential equation, we usually take a logarithm of each side. When solving a logarithmic equation, we usually exponentiate each side.

Examples: **1.**

$2(5)^x = 22$	Exponential equation
$5^x = 11$	Divide by 2.
$\log 5^x = \log 11$	Take the common logarithm.
$x \log 5 = \log 11$	Power rule
$x = \dfrac{\log 11}{\log 5}$	Divide by log 5.

2.

$\log 2x = 2$	Logarithmic equation
$10^{\log 2x} = 10^2$	Exponentiate each side.
$2x = 100$	Inverse properties
$x = 50$	Divide by 2.

CHAPTER 9   Review Exercises

SECTION 9.1

Exercises 1 and 2: Find the following.

(a) $(g \circ f)(-2)$ (b) $(f \circ g)(x)$

1. $f(x) = 2x^2 - 4x,\ g(x) = 5x + 1$

2. $f(x) = \sqrt[3]{x - 6},\ g(x) = 4x^3$

3. Use the two tables to evaluate each expression.

(a) $(f \circ g)(2)$ (b) $(g \circ f)(1)$

x	0	1	2	3
$f(x)$	3	2	1	0

x	0	1	2	3
$g(x)$	1	2	3	0

4. Use the graph to evaluate each expression.

(a) $(f \circ g)(-1)$

(b) $(g \circ f)(2)$

(c) $(f \circ f)(1)$

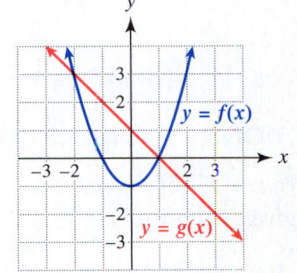

Exercises 5 and 6: Show that f is not one-to-one by finding two inputs that result in the same output. Answers may vary.

5. $f(x) = \dfrac{4}{1 + x^2}$ **6.** $f(x) = x^2 - 2x + 1$

Exercises 7 and 8: Use the horizontal line test to determine whether the graph represents a one-to-one function.

7. 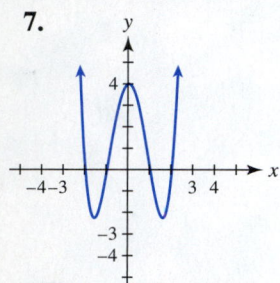   **8.**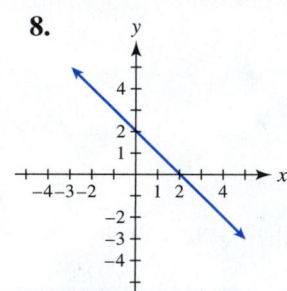

Exercises 9 and 10: Verify that f(x) and $f^{-1}(x)$ are indeed inverse functions.

9. $f(x) = 2x - 9$ $f^{-1}(x) = \dfrac{x + 9}{2}$

10. $f(x) = x^3 + 1$ $f^{-1}(x) = \sqrt[3]{x - 1}$

Exercises 11–14: Find $f^{-1}(x)$.

11. $f(x) = 5x$ **12.** $f(x) = x - 11$

13. $f(x) = 2x + 7$ **14.** $f(x) = \dfrac{4}{x}$

15. Use the table to make a table of values for $f^{-1}(x)$. What are the domain and range for f^{-1}?

x	0	1	2	3
$f(x)$	10	8	7	3

16. Use the graph of $y = f(x)$ to sketch a graph of $y = f^{-1}(x)$. Include the graph of f and the line $y = x$.

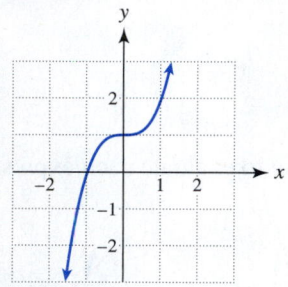

SECTIONS 9.2 AND 9.3

Exercises 17–20: Evaluate the exponential function for the given values of x.

17. $f(x) = 6^x$ $x = -1, \quad x = 2$

18. $f(x) = 5(2^{-x})$ $x = 0, \quad x = 3$

19. $f(x) = \left(\dfrac{1}{3}\right)^x$ $x = -1, \quad x = 4$

20. $f(x) = 3\left(\dfrac{1}{6}\right)^x$ $x = 0, \quad x = 1$

Exercises 21–24: Graph f. State whether the graph illustrates exponential growth, exponential decay, or logarithmic growth.

21. $f(x) = 2^x$ **22.** $f(x) = \left(\dfrac{1}{2}\right)^x$

23. $f(x) = \ln(x + 1)$ **24.** $f(x) = 3^{-x}$

Exercises 25 and 26: A table for a function f is given.

    (a) Determine whether f represents linear or exponential growth.

    (b) Find a formula for f.

25.

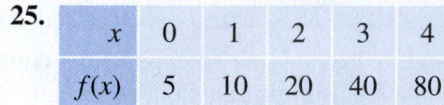

x	0	1	2	3	4
$f(x)$	5	10	20	40	80

26.

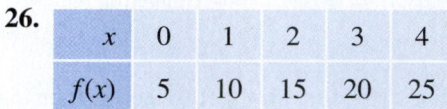

x	0	1	2	3	4
$f(x)$	5	10	15	20	25

27. Use the graph of $y = Ca^x$ to find C and a.

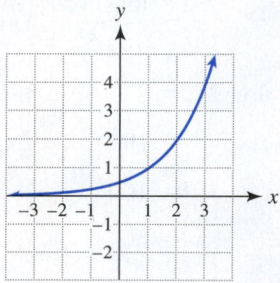

28. Use the graph of $y = k \log_2 x$ to find k.

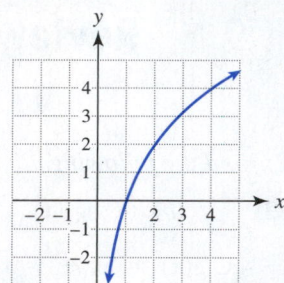

29. Find the percent change if \$150 decreases to \$120.

30. Find the growth factor if \$1500 increases by 7% each year.

Exercises 31 and 32: For the given amounts A and B, find each of the following. Round values to the nearest hundredth of a percent when appropriate.

    (a) *The percent change if A changes to B*
    (b) *The percent change if B changes to A*

31. $A = \$600$, $B = \$1200$

32. $A = \$2.20$, $B = \$1.00$

Exercises 33 and 34: An account contains A dollars and increases or decreases by R%. For each A and R, answer the following.

    (a) *Find the increase or decrease in value of the account.*
    (b) *Find the final value of the account.*
    (c) *By what factor did the account value increase or decrease?*

33. $A = \$500$, $R = \$210\%$

34. $A = \$700$, $R = -25\%$

*Exercises 35 and 36: **Exponential Models** Write the formula for an exponential function, $f(x) = Ca^x$, that models the situation. Evaluate $f(2)$.*

35. A city's population of 20,000 decreases in number by 5% per year.

36. A sample of 1500 insects increases in number by 300% per day.

Exercises 37 and 38: If P dollars is deposited in an account that pays R percent annual interest at the end of each year, approximate the amount in the account after x years.

37. $P = \$1200$ $R = 10\%$ $x = 9$ years

38. $P = \$900$ $R = 18\%$ $x = 40$ years

Exercises 39–42: Evaluate $f(x)$ for the given value of x. Approximate answers to the nearest hundredth.

39. $f(x) = 2e^x - 1$ $x = 5.3$

40. $f(x) = 0.85^x$ $x = 2.1$

41. $f(x) = 2\log x$ $x = 55$

42. $f(x) = \ln(2x + 3)$ $x = 23$

Exercises 43–46: Evaluate the logarithm by hand.

43. $\log 0.001$ **44.** $\log \sqrt{10{,}000}$

45. $\ln e^{-4}$ **46.** $\log_4 16$

Exercises 47–50: Approximate to the nearest thousandth.

47. $\log 65$ **48.** $\ln 0.85$

49. $\ln 120$ **50.** $\log \frac{2}{5}$

Exercises 51–54: Simplify, using inverse properties.

51. $10^{\log 7}$ **52.** $\log_2 2^{5/9}$

53. $\ln e^{6-x}$ **54.** $e^{2\ln x}$

SECTION 9.4

Exercises 55–60: Write the expression by using sums and differences of logarithms of x, y, and z.

55. $\ln xy$ **56.** $\log \dfrac{x}{y}$

57. $\ln \left(x^2 y^3\right)$ **58.** $\log \dfrac{\sqrt{x}}{z^3}$

59. $\log_2 \dfrac{x^2 y}{z}$ **60.** $\log_3 \sqrt[3]{\dfrac{x}{y}}$

Exercises 61–64: Write as the logarithm of one expression.

61. $\log 45 + \log 5 - \log 3$

62. $\log_4 2x + \log_4 5x$

63. $2\ln x - 3\ln y$

64. $\log x^4 - \log x^3 + \log y$

Exercises 65–68: Rewrite, using the power rule.

65. $\log 6^3$ **66.** $\ln x^2$

67. $\log_2 5^{2x}$ **68.** $\log_4 (0.6)^{x+1}$

SECTION 9.5

Exercises 69–78: Solve the given equation. Approximate answers to the nearest hundredth when appropriate.

69. $10^x = 100$ **70.** $2^{2x} = 256$

71. $3e^x + 1 = 28$ **72.** $0.85^x = 0.2$

73. $5\ln x = 4$

74. $\ln 2x = 5$

75. $2\log x = 80$

76. $3\log x - 5 = 1$

77. $2^{x+4} = 3^x$

78. $\ln(2x + 1) + \ln(x - 5) = \ln 13$

Exercises 79 and 80: Do the following.

   (a) *Solve* $f(x) = g(x)$ *graphically.*
   (b) *Solve* $f(x) = g(x)$ *symbolically.*

79. $f(x) = \frac{1}{2}(2^x),$ **80.** $f(x) = \log_2 2x,$
     $g(x) = 4$ $g(x) = 3$

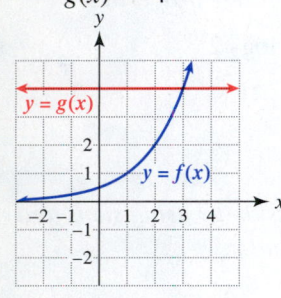
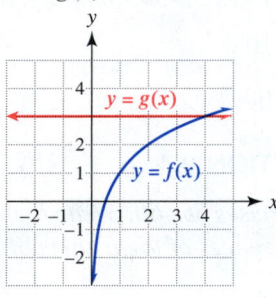

APPLICATIONS

81. *Surface Area of a Balloon* The surface area S of a spherical balloon with radius r is given by the formula $S(r) = 4\pi r^2$. Suppose that the balloon is being inflated so that its radius in inches after t seconds is $r(t) = \sqrt{2t}$.
   (a) Evaluate $(S \circ r)(8)$ and interpret your result.
   (b) Find $(S \circ r)(t)$.

82. *Sales Tax* Suppose that $f(x) = 0.08x$ calculates the sales tax in dollars on an item that costs x dollars.
   (a) Is f a one-to-one function? Why?
   (b) Find a formula for f^{-1} and interpret what it calculates.

83. *Growth of a Mutual Fund* An investor deposits $1500 in a mutual fund that returns 11% annually. Determine the time required for the investment to double in value.

84. *Modeling Data* Find values for the constants a and b so that $y = a + b \log x$ models these data.

x	0.1	1	10	100	1000
y	50	100	150	200	250

85. *Modeling Data* Find values for the constants C and a so that $y = Ca^x$ models these data.

x	0	1	2	3	4
y	3	6	12	24	48

86. *Earthquakes* The Richter scale, used to determine the magnitude of earthquakes, is based on the formula $R(x) = \log \frac{x}{I_0}$, where x is the measured intensity. Let $I_0 = 1$. Find the intensity x for an earthquake with $R = 7$.

87. *Modeling Population* In 2000 the population of Nevada was 2 million and growing continuously at an annual rate of 5.1%. The population of Nevada in millions x years after 2000 can be modeled by

$$f(x) = 2e^{0.051x}.$$

   (a) Graph f in the window $[0, 10, 2]$ by $[0, 4, 1]$. Does this function represent exponential growth or decay?
   (b) Estimate the population of Nevada in 2010.
   (c) Estimate the year when the population was 3 million.

88. *Modeling Bacteria* A colony of bacteria can be modeled by $N(t) = 1000e^{0.0014t}$, where N is measured in bacteria per milliliter and t is in minutes.
   (a) Evaluate $N(0)$ and interpret the result.
   (b) Estimate how long it takes for N to double.

89. *Modeling Wind Speed* Wind speeds vary at different heights above the ground. For a particular day, $f(x) = 1.2 \ln(x) + 5$ computes the wind speed in meters per second x meters above the ground, where $x \geq 1$. (*Source:* A. Miller and R. Anthes, *Meteorology.*)
   (a) Find the wind speed at a height of 5 meters.
   (b) Estimate the height at which the wind speed is 8 meters per second.

1. If $f(x) = 4x^3 - 5x$ and $g(x) = x + 7$, evaluate $(g \circ f)(1)$ and $(f \circ g)(x)$.

2. Use the graph to evaluate each expression.
(a) $(f \circ g)(-1)$ (b) $(g \circ f)(1)$

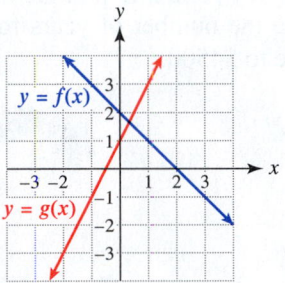

3. Explain why $f(x) = x^2 - 25$ is not a one-to-one function.

4. If $f(x) = 5 - 2x$, find $f^{-1}(x)$.

5. Use the graph of $y = f(x)$ to sketch a graph of $y = f^{-1}(x)$. Include the graph of f and the line $y = x$ in your graph.

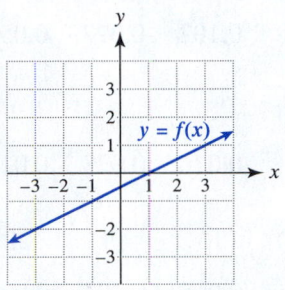

6. Use the table to write a table of values for $f^{-1}(x)$. What are the domain and range of f^{-1}?

x	1	2	3	4
$f(x)$	8	6	4	2

7. Evaluate $f(x) = 3\left(\frac{1}{4}\right)^x$ at $x = 2$.

8. Graph $f(x) = 1.5^{-x}$. State whether the graph of f illustrates exponential growth, exponential decay, or logarithmic growth.

Exercises 9 and 10: A table for a function f is given.

(a) *Determine whether f represents linear or exponential growth.*
(b) *Find a formula for f(x).*

9.

x	-2	-1	0	1	2
$f(x)$	0.75	1.5	3	6	12

10.

x	-2	-1	0	1	2
$f(x)$	-4	-2.5	-1	0.5	2

11. Use the graph of $y = Ca^x$ to find C and a.

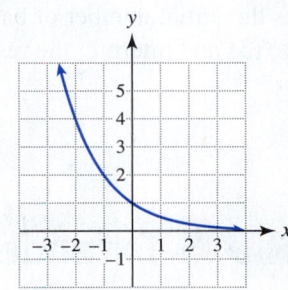

12. Find the percent change if $600 increases to $900.

13. If $1000 increases by 5% per year, what is the growth factor?

14. If $750 is deposited in an account paying 7% annual interest at the end of each year, approximate the amount in the account after 5 years.

15. Let $f(x) = 1.5 \ln(x - 5)$. Approximate $f(21)$ to the nearest hundredth.

16. Evaluate $\log \sqrt{10}$ by hand.

17. Approximate $\log_2 43$ to the nearest thousandth.

18. Graph $f(x) = \log(x - 2)$. Compare this graph to the graph of $y = \log x$.

19. Write $\log \frac{x^3 y^2}{\sqrt{z}}$, using sums and differences of logarithms of x, y, and z.

20. Write $4 \ln x - 5 \ln y + \ln z$ as one logarithm.

21. Rewrite $\log 7^{2x}$, using the power rule.

22. Simplify $\ln e^{1-3x}$, using inverse properties.

Exercises 23–26: Solve the given equation. Approximate answers to the nearest hundredth when appropriate.

23. $2e^x = 50$ **24.** $3(10)^x - 7 = 143$

25. $5 \log x = 9$

26. $3 \ln 5x = 27$

27. *Modeling Data* Find values for constants a and b so that $y = a + b \log x$ models the data.

x	0.01	0.1	1	10	100
y	−1	2	5	8	11

28. *Modeling Bacteria Growth* A sample of bacteria is growing at a rate of 9% per hour and can be modeled by $f(x) = 4(1.09)^x$, where the input x represents elapsed time in hours and the output $f(x)$ is in millions of bacteria.
   (a) What was the initial number of bacteria?
   (b) Evaluate $f(5)$ and interpret the result.

(c) Does this function represent exponential growth or exponential decay?
(d) Estimate the elapsed time when there were 8 million bacteria.

29. A mutual fund account containing $5000 decreases by 2% per year.
   (a) Write a function A that gives the amount in the account after x years.
   (b) Evaluate $A(3)$ and interpret the result.
   (c) Estimate the number of years for the account to decrease to $4500.

CHAPTER 9   Extended and Discovery Exercises

Exercises 1–4: Radioactive Carbon Dating While an animal is alive, it breathes both carbon dioxide and oxygen. Because a small portion of normal atmospheric carbon dioxide is made up of radioactive carbon-14, a fixed percentage of the animal's body is composed of carbon-14. When the animal dies, it quits breathing and the carbon-14 disintegrates without being replaced. One method used to determine when an animal died is to estimate the percentage of carbon-14 remaining in its bones. The **half-life** of carbon-14 is 5730 years. That is, half the original amount of carbon-14 in bones of a fossil will remain after 5730 years. The percentage P, in decimal form, of carbon-14 remaining after x years is modeled by $P(x) = a^x$. (Because of nuclear testing, radioactive carbon dating is no longer accurate for animals that died after 1950.)

1. Find the value of a. (*Hint:* $P(5730) = 0.5$.)

2. Calculate the percentage of carbon-14 that remains after 10,000 years.

3. Estimate the age of a fossil with $P = 0.9$.

4. Estimate the age of a fossil with $P = 0.01$.

Exercises 5–8: Modeling Blood Flow in Animals For medical reasons, dyes are injected into the bloodstream to determine the health of internal organs. In one study that involved animals, the dye BSP was injected to assess blood flow in the liver. The results are listed in the accompanying table, where x represents the elapsed time in minutes and y is the concentration of the dye in the bloodstream in milligrams per milliliter (mg/mL). Scientists modeled the data with $f(x) = 0.133(0.878(0.73^x) + 0.122(0.92^x))$.

x (minutes)	1	2	3	4
y (mg/mL)	0.102	0.077	0.057	0.045

x (minutes)	5	7	9	13
y (mg/mL)	0.036	0.023	0.015	0.008

x (minutes)	16	19	22
y (mg/mL)	0.005	0.004	0.003

Source: F. Harrison, "The measurement of liver blood flow in conscious calves."

5. Graph f together with the data. Comment on the fit.

6. Determine the y-intercept and interpret the result.

7. What happens to the concentration of the dye after a long period of time? Explain.

8. Estimate graphically the time at which the concentration of the dye reached 40% of its initial amount. Would you want to solve this problem symbolically? Explain.

Exercises 9 and 10: Acid Rain Air pollutants frequently cause acid rain. An index of acidity is pH, which measures the concentration of the hydrogen ions in a solution, and ranges from 1 to 14. Pure water is neutral and has a pH of 7, acid solutions have a pH less than 7, and alkaline solutions have a pH greater than 7. The pH of a substance can be computed by $f(x) = -\log x$, where x represents the hydrogen ion concentration in moles per liter. Pure water exposed to normal carbon dioxide in the atmosphere has a pH of 5.6. If the pH of a lake drops below this level, it is indicative of an acidic lake. (Source: G. Howells, Acid Rain and Acid Water.)

9. In rural areas of Europe, rainwater typically has a hydrogen ion concentration of $x = 10^{-4.7}$. Find its pH. What effect might this rain have on a lake with a pH of 5.6?

10. Seawater has a pH of 8.2. Compared to seawater, how many times greater is the hydrogen ion concentration in rainwater from rural Europe?

Exercises 11 and 12: *Investment Account* If x dollars are deposited every 2 weeks (26 times per year) in an account paying an annual interest rate r, expressed in decimal form, the amount A in the account after n years can be approximated by the formula

$$A = x\left[\frac{(1 + r/26)^{26n} - 1}{(r/26)}\right].$$

11. If \$100 is deposited every 2 weeks in an account paying 9% interest, approximate the amount in the account after 10 years.

12. Suppose that your retirement account pays 12% annual interest. Determine how much you should deposit in this account every 2 weeks, in order to have one million dollars at age 65.

CHAPTERS 1–9 Cumulative Review Exercises

1. Write the number 0.000429 in scientific notation.

2. Classify each real number as one or more of the following: natural number, whole number, integer, rational number, or irrational number.

$$-\frac{11}{7}, -3, 0, \sqrt{6}, \pi, 5.\overline{18}$$

3. Select the formula that models the data best.

x	-2	-1	0	1	2
y	-7	-5	-3	-1	1

   a. $y = 3x + 1$ b. $y = x - 3$ c. $y = 2x - 3$

4. State whether the equation illustrates an identity, commutative, associative, or distributive property.

$$(5 - y) + 9 = 9 + (5 - y)$$

Exercises 5–8: Simplify the expression. Write the result with positive exponents.

5. $\left(\dfrac{1}{d^2}\right)^{-2}$

6. $\left(\dfrac{8a^2}{2b^3}\right)^{-3}$

7. $\dfrac{(2x^{-2}y^3)^2}{xy^{-2}}$

8. $\dfrac{x^{-3}y}{4x^2y^{-3}}$

9. Use the graph to express the equation of the line in slope–intercept form.

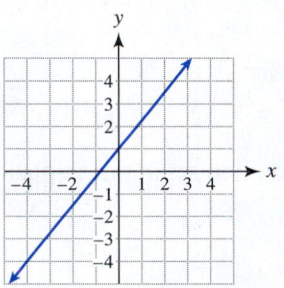

10. Find the domain of $f(x) = \dfrac{10}{x + 3}$.

11. Use the table to write the formula for $f(x) = ax + b$.

x	-2	-1	0	1	2
$f(x)$	-11	-7	-3	1	5

12. Write the equation of the vertical line passing through the point $(4, 7)$.

13. Calculate the slope of the line passing through the points $(4, -1)$ and $(2, -3)$.

14. Sketch the graph of a line passing through the point $(-1, -2)$ with slope $m = 3$.

Exercises 15 and 16: Write the slope–intercept form for a line satisfying the given conditions.

15. Perpendicular to $y = -\frac{1}{7}x - 8$, passing through $(1, 1)$

16. Parallel to $y = 3x - 1$, passing through $(0, 5)$

17. Use the graph to solve the equation $y_1 = y_2$.

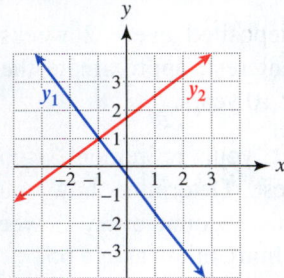

18. Use the table to solve the inequality $y < -4$, where y represents a linear function. Write the solution set as an inequality.

x	-2	-1	0	1	2
y	-24	-14	-4	6	16

Exercises 19–24: Solve the equation or inequality. Write the solutions to the inequalities in interval notation.

19. $\frac{2}{3}(x - 3) + 8 = -6$

20. $\frac{1}{3}z + 6 < \frac{1}{4}z - (5z - 6)$

21. $\left(\dfrac{t + 2}{3}\right) - 10 = \frac{1}{3}t - (5t + 8)$

22. $-10 \le -\frac{3}{5}x - 4 < -1$

23. $-2|t - 4| \ge -12$ **24.** $\left|\frac{1}{2}x - 5\right| = 3$

25. Shade the solution set in the xy-plane.

$$x + y > 3$$
$$2x - y \ge 3$$

26. Evaluate det A if $A = \begin{bmatrix} -1 & -2 \\ 3 & 4 \end{bmatrix}$.

Exercises 27–30: Solve the system of equations, if possible. Write the solution as an ordered pair or ordered triple where appropriate.

27. $4x - 3y = 1$ **28.** $2x - 3y = -2$
    $5x + 2y = 7$ $-6x + 9y = 5$

29. $2x - y + 3z = -2$ **30.** $x + y - z = -1$
    $x + 5y - 2z = -8$ $-x - y - z = -1$
    $-3x - y - 3z = 6$ $x - 2y + z = 1$

31. Maximize the objective function R, subject to the given constraints.

$$R = 2x + 5y$$
$$3x + y \le 12$$
$$x + 3y \le 12$$
$$x \ge 0, y \ge 0$$

32. Find the area of the triangle by using a determinant. Assume that units are inches.

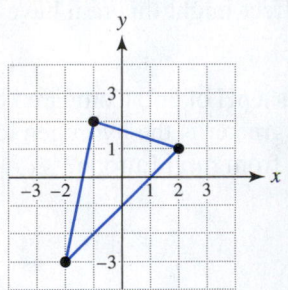

Exercises 33–36: Factor completely.

33. $2x^3 - 4x^2 + 2x$ **34.** $4a^2 - 25b^2$

35. $8t^3 - 27$ **36.** $4a^3 - 2a^2 + 10a - 5$

Exercises 37–40: Solve the equation.

37. $6x^2 - 7x - 10 = 0$ **38.** $9x^2 = 4$

39. $x^4 - 2x^3 = 15x^2$ **40.** $5x - 10x^2 = 0$

Exercises 41 and 42: Simplify the expression.

41. $\dfrac{x^2 + 5x + 6}{x^2 - 9} \cdot \dfrac{x - 3}{x + 2}$

42. $\dfrac{x^2 - 2x - 8}{x^2 + x - 12} \div \dfrac{(x - 4)^2}{x^2 - 16}$

Exercises 43 and 44: Solve the rational equation. Check your result.

43. $\dfrac{2}{x + 2} - \dfrac{1}{x - 2} = \dfrac{-3}{x^2 - 4}$

44. $\dfrac{3y}{y^2 + y - 2} = \dfrac{1}{y - 1} - 2$

45. Solve the equation for J.

$$P = \frac{J + 2z}{J}$$

46. Simplify the complex fraction.

$$\frac{\dfrac{3}{x^2} + x}{x - \dfrac{3}{x^2}}$$

47. Suppose that y varies directly with x. If $y = 15$ when $x = 3$, find y when x is 8.

48. Divide $(3x^3 - 2x - 15)$ by $(x - 2)$.

Exercises 49–54: Simplify the expression. Assume that all variables are positive.

49. $\left(\dfrac{x^6}{y^9}\right)^{2/3}$ **50.** $\sqrt[3]{-x^4} \cdot \sqrt[3]{-x^5}$

51. $\sqrt{5ab} \cdot \sqrt{20ab}$ **52.** $2\sqrt{24} - \sqrt{54}$

53. $\sqrt[3]{a^5 b^4} + 3\sqrt[3]{a^5 b}$

54. $\left(5 + \sqrt{5}\right)\left(5 - \sqrt{5}\right)$

55. Rationalize the denominator.

$$\frac{2}{5 - \sqrt{3}}$$

56. Find the domain of f. Write your answer in interval notation.

$$f(x) = \frac{3}{\sqrt{x - 4}}$$

Exercises 57 and 58: Solve. Check your answer.

57. $2(x + 1)^2 = 50$ **58.** $\sqrt{x + 6} = x$

Exercises 59 and 60: Write in standard form.

59. $(-2 + 3i) - (-5 - 2i)$

60. $\dfrac{3 - i}{1 + 3i}$

61. Find the vertex on the graph of the function given by $f(x) = 3x^2 - 12x + 13$.

62. Find the maximum y-value on the parabola determined by $y = -2x^2 + 6x - 1$.

63. Compare the graph of $f(x) = (x - 3)^2 + 2$ to the graph of $y = x^2$.

64. Write the equation $y = x^2 + 6x - 2$ in vertex form and identify the vertex.

Exercises 65–68: Solve the quadratic equation by using the method of your choice.

65. $x^2 - 13x + 40 = 0$

66. $2d^2 - 5 = d$

67. $z^2 - 4z = -2$

68. $x^4 - 10x^2 + 24 = 0$

69. A graph of $y = ax^2 + bx + c$ is shown.
   (a) Solve $ax^2 + bx + c = 0$.
   (b) State whether $a > 0$ or $a < 0$.
   (c) Determine whether the discriminant is positive, negative, or zero.

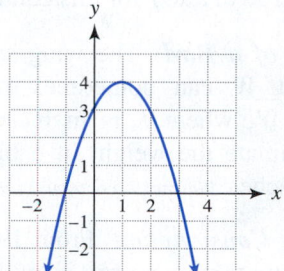

70. Solve $x^2 + 5x - 14 \geq 0$. Write your answer in interval notation.

71. For $f(x) = x^2 - 2$ and $g(x) = 2x + 1$, find each of the following.
   (a) $(f \circ g)(1)$ **(b)** $(g \circ f)(x)$

72. Show that f is not one-to-one by finding two inputs that result in the same output. Answers may vary.

$$f(x) = x^2 + x - 6$$

73. Find $f^{-1}(x)$ for $f(x) = \frac{3}{x}$.

74. If \$800 is deposited in an account that pays 7.5% annual interest at the end of each year, approximate the amount in the account after 15 years.

Exercises 75 and 76: Evaluate without a calculator.

75. $\log_3 81$

76. $e^{\ln 2x}$

77. Write the expression by using sums and differences of logarithms of x and y. Assume x and y are positive.

$$\log \frac{\sqrt{x}}{y^2}$$

78. Write $2 \ln x + \ln 5x$ as the logarithm of a single expression.

Exercises 79 and 80: Solve the equation. Approximate answers to the nearest hundredth.

79. $6 \log x - 2 = 9$

80. $2^{3x} = 17$

APPLICATIONS

81. *Population Growth* The population P of a community with an annual percentage growth rate r (expressed as a decimal) after t years is given by $P = P_0(1 + r)^t$, where P_0 represents the initial population of the community. If a community having an initial population of $P_0 = 12{,}000$ grew to a population of $P = 14{,}600$ in $t = 5$ years, find the annual percentage growth rate r for this community.

82. *Wing Span of a Bird* The wing span L of a bird with weight W can sometimes be modeled by $L = 27.4\sqrt[3]{W}$, where L is in inches and W is in pounds. Estimate the weight of a bird with a wing span of 36 inches. (*Source:* C. Pennycuick, *Newton Rules Biology.*)

83. *U.S. Energy Consumption* From 1950 to 1970, per capita consumption of energy in millions of Btu can be modeled by $f(x) = 0.25x^2 - 975x + 950{,}845$, where x is the year. (*Source:* Department of Energy.)
   (a) During what year was per capita energy consumption at its lowest?
   (b) Find this minimum value.

84. *Braking Distance* On dry, level pavement a safe braking distance d in feet for a car traveling x miles per hour is $d = \frac{x^2}{12}$. What speed corresponds to a braking distance of 350 feet? (*Source:* F. Mannering, *Principles of Highway Engineering and Traffic Control.*)

85. *Investing for Retirement* A college student invests $8000 in an account that pays interest annually. If the student would like this investment to be worth $1,000,000 in 45 years, what annual interest rate would the account need to pay?

86. *Modeling Wind Speed* Wind speeds are usually measured at heights from 5 to 10 meters above the ground. For a particular day, $f(x) = 1.4 \ln(x) + 7$ computes the wind speed in meters per second x meters above the ground, where $x \geq 1$. (*Source:* A. Miller and R. Anthes, *Meteorology.*)
   (a) Find the wind speed at a height of 8 meters.
   (b) Estimate the height at which the wind speed is 10 meters per second.

10

Conic Sections

The art of asking the right questions in mathematics is more important than the art of solving them.

—GEORG CANTOR

Throughout history people have been fascinated with the universe around them and compelled to understand its mysteries. Conic sections, which include parabolas, circles, ellipses, and hyperbolas, have played an important role in gaining this understanding. Although conic sections were described and named by the Greek astronomer Apollonius in 200 B.C., it was not until much later that they were used to model motion in the universe. In the sixteenth century Tycho Brahe, the greatest observational astronomer of the age, recorded precise data on planetary movement in the sky. Using Brahe's data, in 1619 Johannes Kepler determined that planets move in elliptical orbits around the sun. In 1686 Newton used Kepler's work to show that elliptical orbits are the result of his famous theory of gravitation. We now know that all celestial objects—including planets, comets, asteroids, and satellites—travel in paths described by conic sections.

Today scientists search the sky for information about the universe with enormous radio telescopes in the shape of parabolic dishes. The Hubble telescope also makes use of a parabolic mirror. As a result, our understanding of the universe has changed dramatically in recent years.

Conic sections have had a profound influence on people's understanding of their world and the cosmos. In this chapter we introduce you to these age-old curves.

Source: Historical Topics for the Mathematics Classroom, Thirty-first Yearbook, NCTM.

10.1   Parabolas and Circles

Types of Conic Sections • Parabolas with Horizontal Axes of Symmetry • Equations of Circles

A LOOK INTO MATH ▶

In this section we discuss two types of conic sections: parabolas and circles. Recall that we discussed parabolas with vertical axes of symmetry in Chapter 8. In this section we discuss parabolas with horizontal axes of symmetry, but first we introduce the three basic types of conic sections. The Hubble telescope travels in a path described by a conic section called an ellipse, as do all planets and satellites.

Types of Conic Sections

NEW VOCABULARY

☐ Conic sections
☐ Circle
☐ Radius
☐ Center
☐ Standard equation
   of a circle

Conic sections are named after the different ways that a plane can intersect a cone. The three basic curves are parabolas, ellipses, and hyperbolas. A circle is a special case of an ellipse. Figure 10.1 shows the three types of conic sections along with an example of the graph associated with each.

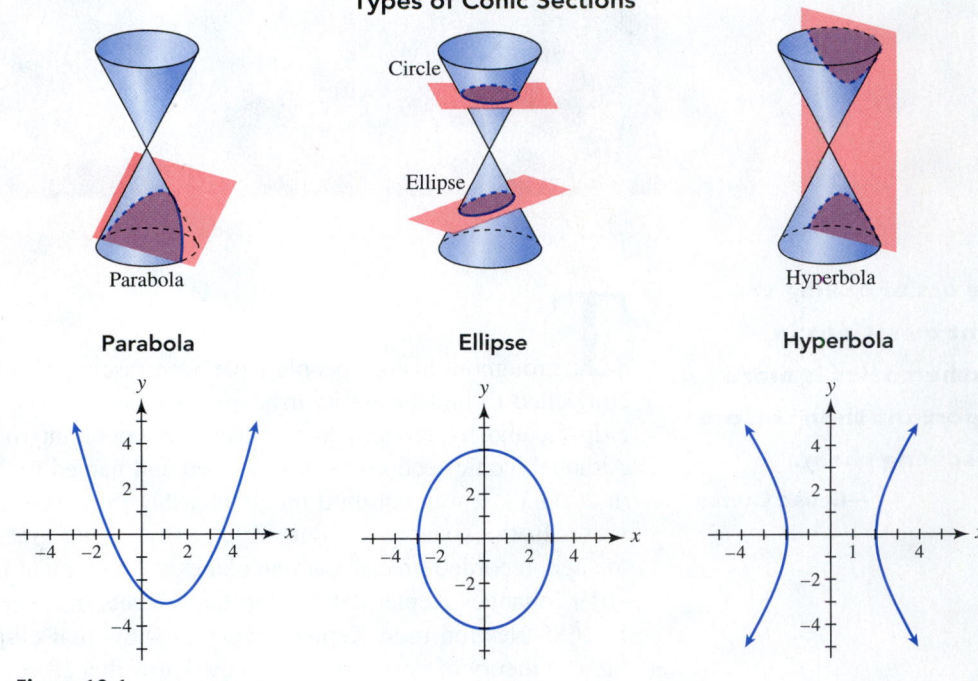

Figure 10.1

READING CHECK

State the types of conic sections.

Parabolas with Horizontal Axes of Symmetry

Recall that the *vertex form of a parabola* with a vertical axis of symmetry is

$$y = a(x - h)^2 + k, \quad \text{Vertex form}$$

where (h, k) is the vertex. If $a > 0$, the parabola opens upward; if $a < 0$, the parabola opens downward, as shown in Figure 10.2. The preceding equation can also be expressed in the form

$$y = ax^2 + bx + c. \quad \text{General form}$$

In this form the x-coordinate of the vertex is $x = -\frac{b}{2a}$.

Parabolas with a Vertical Axis of Symmetry

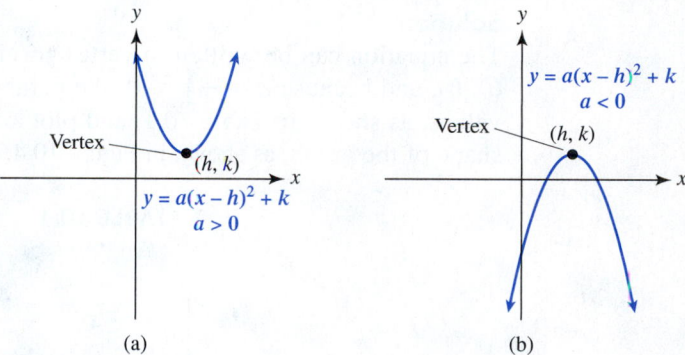

(a)   (b)

Figure 10.2 Functions

Interchanging the roles of x and y (and also h and k) gives equations for parabolas that open to the right or the left. In this case, their axes of symmetry are horizontal.

PARABOLAS WITH HORIZONTAL AXES OF SYMMETRY

The graph of $x = a(y - k)^2 + h$ is a parabola that opens to the right if $a > 0$ and to the left if $a < 0$. The vertex of the parabola is located at (h, k).

The graph of $x = ay^2 + by + c$ is a parabola opening to the right if $a > 0$ and to the left if $a < 0$. The y-coordinate of its vertex is given by $y = -\frac{b}{2a}$.

READING CHECK

Give a general equation for a parabola with a horizontal axis of symmetry.

These parabolas are illustrated in Figure 10.3.

Parabolas with a Horizontal Axis of Symmetry

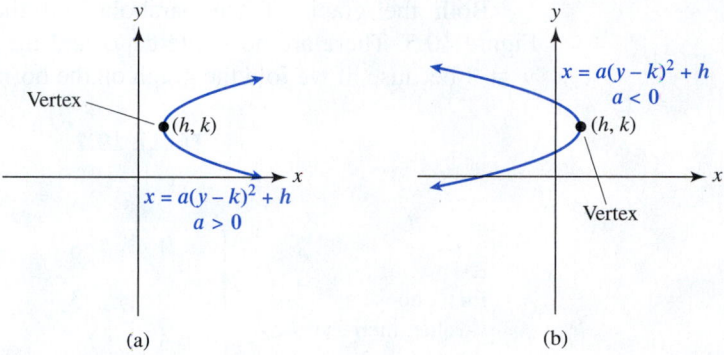

(a)   (b)

Figure 10.3 Not Functions

NOTE: If the variable x is squared in an equation for a parabola, then its axis of symmetry is vertical. If the variable y is squared instead, then its axis of symmetry is horizontal.

EXAMPLE 1 **Graphing a parabola**

Graph $x = -\frac{1}{2}y^2$. Find its vertex and axis of symmetry.

Solution

The equation can be written in vertex form because $x = -\frac{1}{2}(y - \mathbf{0})^2 + \mathbf{0}$. The vertex is $(\mathbf{0}, \mathbf{0})$, and because $a = -\frac{1}{2} < 0$, the parabola opens to the left. We can make a table of values, as shown in Table 10.1, and plot a few points to help determine the location and shape of the graph, as shown in Figure 10.4. Its axis of symmetry is the x-axis, or $y = 0$.

TABLE 10.1

First choose a y-value; then use $x = -\frac{1}{2}y^2$ to find x.

y	x
-2	-2
-1	$-\frac{1}{2}$
$\mathbf{0}$	$\mathbf{0}$
1	$-\frac{1}{2}$
2	-2

Vertex

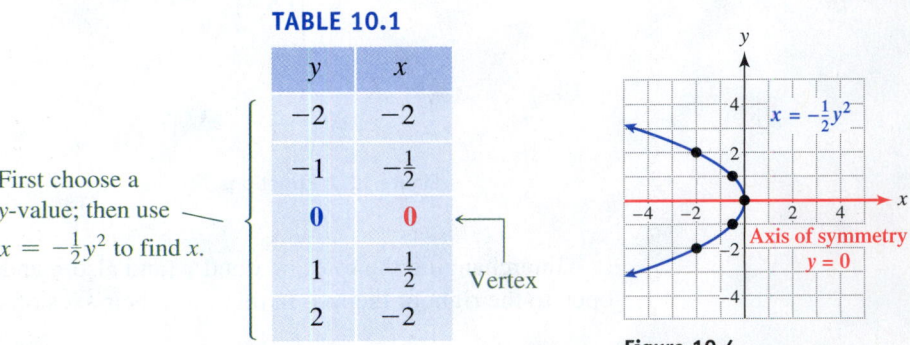

Figure 10.4

Now Try Exercise 15

EXAMPLE 2 **Graphing a parabola**

Graph $x = (y - 3)^2 + 2$. Find its vertex and axis of symmetry.

Solution

Because $h = 2$ and $k = 3$ in the equation $x = a(y - k)^2 + h$, the vertex is $(\mathbf{2}, \mathbf{3})$, and because $a = 1 > 0$, the parabola opens to the right. This parabola has the same shape as $y = x^2$, except that it opens to the right rather than upward. To graph this parabola we can make a table of values and plot a few points, as shown in Table 10.2 and Figure 10.5.

NOTE: Sometimes, finding the x- and y-intercepts of the parabola is helpful when you are graphing. To find the x-intercept let $y = 0$ in $x = (y - 3)^2 + 2$. The x-intercept is $x = (\mathbf{0} - 3)^2 + 2 = 11$. To find any y-intercepts let $x = 0$ in $x = (y - 3)^2 + 2$. Here $\mathbf{0} = (y - 3)^2 + 2$ means that $(y - 3)^2 = -2$, which has no real solutions, and thus this parabola has no y-intercepts.

Both the graph of the parabola and the points from Table 10.2 are shown in Figure 10.5. There are no y-intercepts and the x-intercept is 11. The axis of symmetry is $y = 3$ because, if we fold the graph on the horizontal line $y = 3$, the two sides match.

TABLE 10.2

First choose a y-value; then use $x = (y - 3)^2 + 2$ to find x.

y	x
1	6
2	3
$\mathbf{3}$	$\mathbf{2}$
4	3
5	6

Vertex

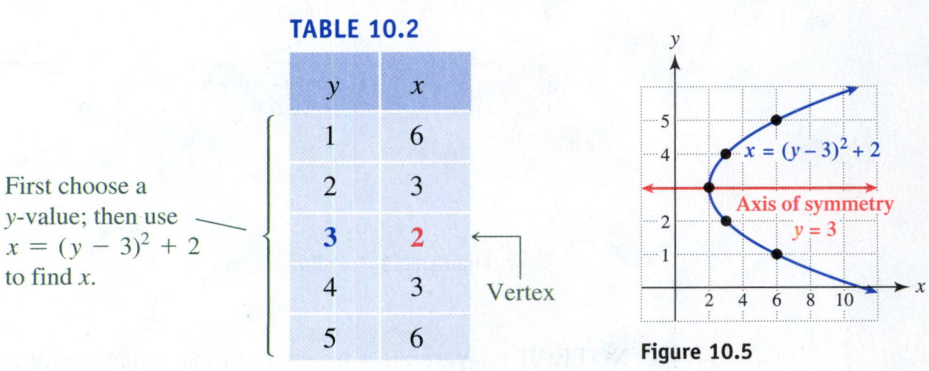

Figure 10.5

Now Try Exercise 23

EXAMPLE 3 **Graphing a parabola and finding its vertex**

Identify the vertex and then graph each parabola.
(a) $x = -y^2 + 1$ **(b)** $x = y^2 - 2y - 1$

Solution

(a) If we rewrite $x = -y^2 + 1$ as $x = -(y - 0)^2 + 1$, then $h = 1$ and $k = 0$, so the vertex is $(1, 0)$. By letting $y = 0$ in $x = -y^2 + 1$, we find that the x-intercept is $x = -0^2 + 1 = 1$. Similarly, we let $x = 0$ in $x = -y^2 + 1$ to find the y-intercepts. The equation $0 = -y^2 + 1$ has solutions -1 and 1.

   The parabola opens to the left because $a = -1 < 0$. Additional points given in Table 10.3 help in graphing the parabola shown in Figure 10.6.

TABLE 10.3

Vertex is $(1, 0)$, so choose y-values → on each side of 0.

y	x
-2	-3
-1	0
$\mathbf{0}$	$\mathbf{1}$
1	0
2	-3

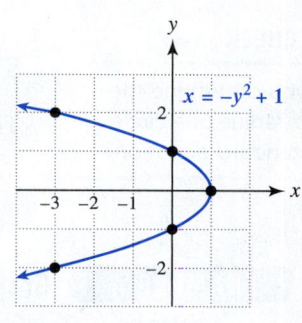

Figure 10.6

(b) The y-coordinate of the vertex for the graph of $x = y^2 - 2y - 1$ is given by

$$y = -\frac{b}{2a} = -\frac{-2}{2(1)} = 1.$$

To find the x-coordinate of the vertex, substitute $y = 1$ into the given equation.

$$x = (1)^2 - 2(1) - 1 = -2$$

The vertex is $(-2, 1)$. The parabola opens to the right because $a = 1 > 0$. Additional points given in Table 10.4 help in graphing the parabola shown in Figure 10.7. Note that the y-intercepts do not have integer values and that the quadratic formula could be used to find approximations for these values. The x-intercept is -1.

TABLE 10.4

Vertex is $(-2, 1)$, so choose y-values → on each side of 1.

y	x
-1	2
0	-1
1	-2
2	-1
3	2

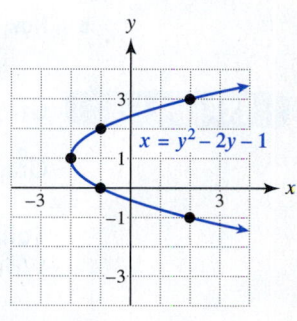

Figure 10.7

Now Try Exercises 17, 39

Equations of Circles

Circle

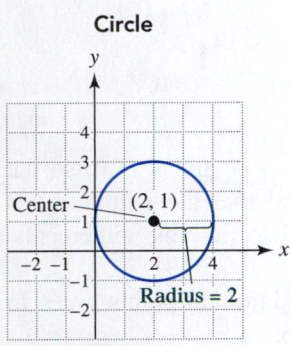

Figure 10.8

A **circle** consists of the set of points in a plane that are the same distance from a fixed point. The fixed distance is called the **radius**, and the fixed point is called the **center**. In Figure 10.8 all points lying on the circle are a distance of 2 units from the center $(2, 1)$. Therefore the radius of the circle equals 2.

We can find the equation of the circle shown in Figure 10.8 by using the distance formula. If a point (x, y) lies on the graph of a circle, its distance from the center $(2, 1)$ is 2, and

$$\sqrt{(x - 2)^2 + (y - 1)^2} = 2. \qquad d = \sqrt{(x_2 - x_1)^2 + (y_2 - y_1)^2}$$

Squaring each side gives

$$(x - 2)^2 + (y - 1)^2 = 2^2.$$

This equation represents the standard equation for a circle with center $(\mathbf{2}, \mathbf{1})$ and radius $\mathbf{2}$.

READING CHECK

How can you determine the center and radius of a circle given its standard equation?

STANDARD EQUATION OF A CIRCLE

The **standard equation of a circle** with center (h, k) and radius r is

$$(x - h)^2 + (y - k)^2 = r^2.$$

EXAMPLE 4 **Graphing a circle**

Graph $x^2 + y^2 = 9$. Find the radius and center.

Solution
The equation $x^2 + y^2 = 9$ can be written in standard form as

$$(x - \mathbf{0})^2 + (y - \mathbf{0})^2 = 3^2.$$

Therefore the center is $(\mathbf{0}, \mathbf{0})$ and the radius is $\mathbf{3}$. Its graph is shown in Figure 10.9.

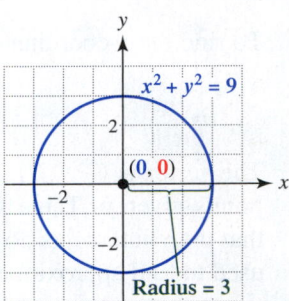

Figure 10.9

Now Try Exercise 65

EXAMPLE 5 **Graphing a circle**

Graph $(x + 1)^2 + (y - 3)^2 = 4$. Find the radius and center.

Solution
Write the equation as

$$(x - (\mathbf{-1}))^2 + (y - \mathbf{3})^2 = 2^2.$$

The center is $(-1, 3)$, and the radius is 2. The circle's graph is shown in Figure 10.10.

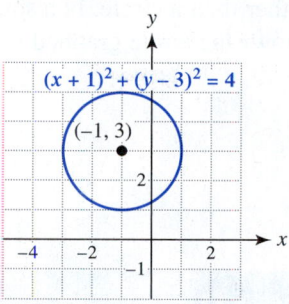

Figure 10.10

■ **Now Try Exercise 69**

In the next example we use the *method of completing the square* to find the center and radius of a circle. (To review completing the square, refer to Sections 8.2 and 8.3.)

EXAMPLE 6 **Finding the center of a circle**

Find the center and radius of the circle given by $x^2 + 4x + y^2 - 6y = 5$.

Solution
Begin by writing the equation as

$$(x^2 + 4x + \underline{\hspace{1cm}}) + (y^2 - 6y + \underline{\hspace{1cm}}) = 5.$$

To complete the square, add $\left(\frac{4}{2}\right)^2 = 4$ and $\left(\frac{-6}{2}\right)^2 = 9$ to each side of the equation.

$$(x^2 + 4x + 4) + (y^2 - 6y + 9) = 5 + 4 + 9$$

Factoring each perfect square trinomial yields

$$(x + 2)^2 + (y - 3)^2 = 18.$$

The center is $(-2, 3)$, and because $18 = \left(\sqrt{18}\right)^2$, the radius is $\sqrt{18}$, or $3\sqrt{2}$.

■ **Now Try Exercise 71**

CRITICAL THINKING

Does the following equation represent a circle? If so, give its center and radius.

$$x^2 + y^2 + 10y = -32$$

TECHNOLOGY NOTE

Graphing Circles
The graph of a circle does not represent a function. One way to graph a circle with a graphing calculator is to solve the equation for *y* and obtain two equations. One equation gives the upper half of the circle, and the other equation gives the lower half.

For example, to graph $x^2 + y^2 = 4$ begin by solving for *y*.

$$y^2 = 4 - x^2 \qquad \text{Subtract } x^2.$$

$$y = \pm\sqrt{4 - x^2} \qquad \text{Square root property}$$

Then graph $Y_1 = \sqrt{(4 - X^2)}$ and $Y_2 = -\sqrt{(4 - X^2)}$. The graph of y_1 is the upper half of the circle, and the graph of y_2 is the lower half of the circle, as shown in Figure 10.11.

[−4.7, 4.7, 1] by [−3.1, 3.1, 1]   [−4.7, 4.7, 1] by [−3.1, 3.1, 1]   [−4.7, 4.7, 1] by [−3.1, 3.1, 1]

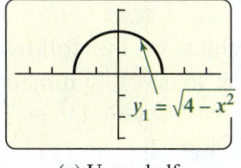

(a) Upper half

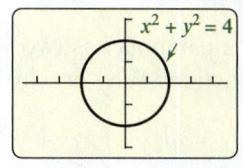

(b) Lower half

(c) Both halves

Figure 10.11

[−4, 4, 1] by [−5, 5, 1]

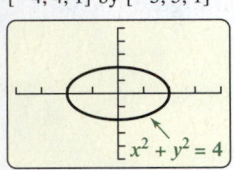

Figure 10.12

NOTE: If a circle is not graphed in a *square viewing rectangle*, it will appear to be an oval rather than a circle. In a square viewing rectangle a circle will appear circular. Figure 10.12 shows the circle graphed in a viewing rectangle that is not square.

10.1   Putting It All Together

CONCEPT	EXPLANATION	EXAMPLE
Parabola with Horizontal Axis	Vertex form: $x = a(y - k)^2 + h$ If $a > 0$, it opens to the right. If $a < 0$, it opens to the left. The vertex is (h, k). See Figure 10.3. These parabolas may also be expressed as $x = ay^2 + by + c$, where the y-coordinate of the vertex is $y = -\frac{b}{2a}$.	$x = 2(y - 1)^2 + 4$ opens to the right and its vertex is $(4, 1)$.
Standard Equation of a Circle	$(x - h)^2 + (y - k)^2 = r^2$. The radius is r and the center is (h, k).	$(x + 2)^2 + (y - 1)^2 = 16$ has center $(-2, 1)$ and radius 4.

10.1   Exercises

MyMathLab    PRACTICE   WATCH   DOWNLOAD   READ   REVIEW

CONCEPTS AND VOCABULARY

1. Name the three general types of conic sections.

2. What is the difference between the parabolas given by $y = ax^2 + bx + c$ and $x = ay^2 + by + c$?

3. If a parabola has a horizontal axis of symmetry, does it represent a function?

4. Sketch a graph of a parabola with a horizontal axis of symmetry.

5. If a parabola has two y-intercepts, does it represent a function? Why or why not?

6. If $x = a(y - k)^2 + h$, what is the vertex?

7. The graph of $x = -y^2$ opens to the _____.

8. The graph of $x = 2y^2 + y - 1$ opens to the _____.

9. The graph of $(x - h)^2 + (y - k)^2 = r^2$ is a(n) _____ with center _____.

10. The graph of $x^2 + y^2 = r^2$ is a circle with center _____ and radius _____.

11. Which of the following (a.–d.) are the coordinates of the vertex for the parabola given by $x = 4(y + 2)^2 - 3$?
    a. $(-3, 4)$ b. $(3, -2)$
    c. $(-3, 2)$ d. $(-3, -2)$

12. Which of the following (a.–d.) is the equation of the axis of symmetry for the parabola given by $x = -5(y - 1)^2 + 7$?
    a. $y = 1$ b. $x = -5$
    c. $x = 7$ d. $y = 7$

13. Which of the following (a.–d.) is the center and radius of the circle given by $x^2 + (y - 2)^2 = 9$?
  a. $(0, -2), r = 3$ **b.** $(2, 0), r = 3$
  c. $(0, 2), r = 3$ **d.** $(0, 2), r = 9$

14. Which of the following (a.–d.) is the y-coordinate of the vertex of the parabola given by $x = y^2 - 2y + 3$?
  a. -2 **b.** 1
  c. 0 **d.** 3

PARABOLAS

Exercises 15–40: Graph the parabola. Find the vertex and axis of symmetry.

15. $x = y^2$

16. $x = -y^2$

17. $x = y^2 + 1$

18. $x = y^2 - 1$

19. $y = x^2 - 1$

20. $y = 2x^2$

21. $x = 2y^2$

22. $x = \frac{1}{4}y^2$

23. $x = (y - 1)^2 + 2$

24. $x = (y - 2)^2 + 1$

25. $y = (x + 2)^2 + 1$

26. $y = (x - 4)^2 + 5$

27. $y = -2(x + 2)^2$

28. $y = (x - 1)^2 - 2$

29. $x = \frac{1}{2}(y + 1)^2 - 3$

30. $x = -2(y + 3)^2 + 1$

31. $x = -3(y - 1)^2$

32. $x = \frac{1}{4}(y + 2)^2 - 3$

33. $y = 2x^2 - x + 1$

34. $y = -x^2 + 2x + 2$

35. $x = -2y^2 + 3y + 2$

36. $x = \frac{1}{2}y^2 + y - 1$

37. $x = 3y^2 + y$

38. $x = -\frac{3}{2}y^2 - 2y + 1$

39. $x = y^2 + 2y + 1$

40. $x = y^2 - 3y - 4$

Exercises 41–44: Use the graph to determine the equation of the parabola. (Hint: Either $a = 1$ or $a = -1$.)

41.
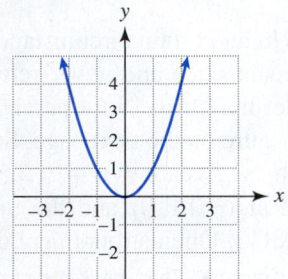

42.
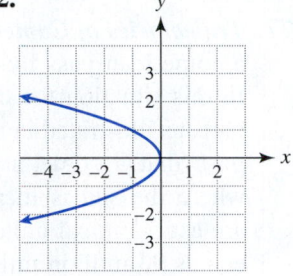

43.
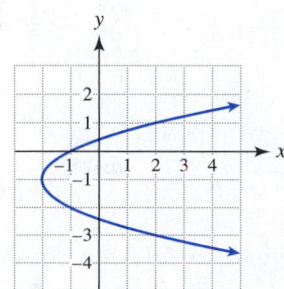

44.
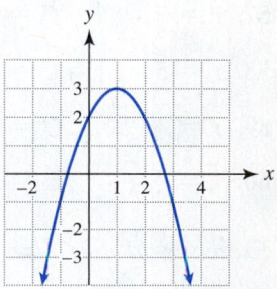

Exercises 45–48: Determine the direction that the parabola opens if it satisfies the given conditions.

45. Passing through $(2, 0)$, $(-2, 0)$, and $(0, -2)$

46. Passing through $(0, -3)$, $(0, 2)$, and $(1, 1)$

47. Vertex $(1, 2)$ passing through the point $(-1, -2)$ with a vertical axis

48. Vertex $(-1, 3)$ passing through the point $(0, 0)$ with a horizontal axis

49. What x-values are possible for the graph of the equation $x = 2y^2$?

50. What y-values are possible for the graph of the equation $x = 2y^2$?

51. Thinking Generally How many y-intercepts does a parabola given by
$$x = a(y - k)^2 + h$$
have if $a > 0$ and $h < 0$?

52. Thinking Generally Does the graph of the equation $x = ay^2 + by + c$ always have a y-intercept? Explain.

53. What is the x-intercept for the graph of the equation given by
$$x = 3y^2 - y + 1?$$

54. What are the y-intercepts for the graph of the equation given by
$$x = y^2 - 3y + 2?$$

CIRCLES

Exercises 55–60: Write the standard equation of the circle with the given radius r and center C.

55. $r = 1$ $C = (0, 0)$

56. $r = 4$ $C = (2, 3)$

57. $r = 3$ $\qquad C = (-1, 5)$

58. $r = 5$ $\qquad C = (5, -3)$

59. $r = \sqrt{2}$ $\qquad C = (-4, -6)$

60. $r = \sqrt{6}$ $\qquad C = (0, 4)$

Exercises 61–64: Use the graph to find the standard equation of the circle.

61.
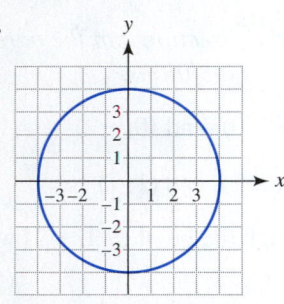

62.
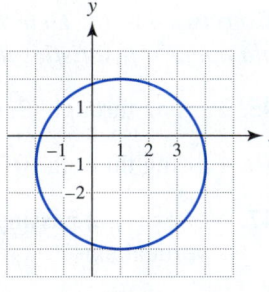

63.
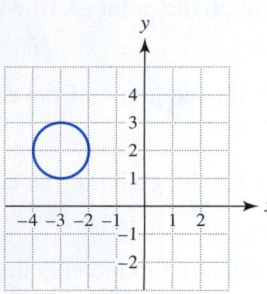

64.
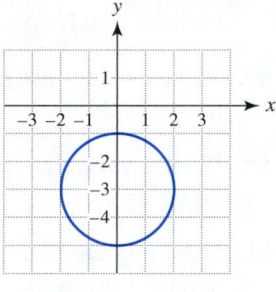

Exercises 65–74: Find the radius and center of the circle. Then graph the circle.

65. $x^2 + y^2 = 4$ $\qquad$ **66.** $x^2 + y^2 = 1$

67. $(x - 1)^2 + (y - 3)^2 = 9$

68. $(x + 2)^2 + (y + 1)^2 = 4$

69. $(x + 5)^2 + (y - 5)^2 = 25$

70. $(x - 4)^2 + (y + 3)^2 = 16$

71. $x^2 + 6x + y^2 - 2y = -1$

72. $x^2 + y^2 + 12y + 32 = 0$

73. $x^2 + 6x + y^2 - 2y + 3 = 0$

74. $x^2 - 4x + y^2 + 4y = -3$

APPLICATIONS

75. *Radio Telescopes* The Parks radio telescope has the shape of a parabolic dish, as depicted in the figure at the top of the next column. A cross section of this telescope can be modeled by $x = \frac{32}{11,025}y^2$, where $-105 \le y \le 105$; the units are feet. (*Source:* J. Mar, *Structure Technology for Large Radar and Telescope Systems.*)

(a) Graph the cross-sectional shape of the dish in $[-40, 40, 10]$ by $[-120, 120, 20]$.

(b) Find the depth d of the dish.

76. *Train Tracks* To make a curve safer for trains, parabolic curves are sometimes used instead of circular curves. See the accompanying figures. (*Source:* F. Mannering and W. Kilareski, *Principles of Highway Engineering and Traffic Analysis.*)

(a) Suppose that a curve must pass through the points $(-1, 0)$, $(0, 3)$, and $(0, -3)$, where the units are kilometers. Find an equation for the train tracks in the form

$$x = a(y - h)^2 + k.$$

(b) Find another point that lies on the train tracks.

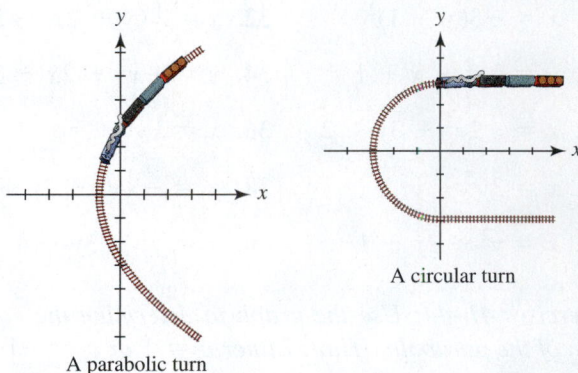

A parabolic turn $\qquad\qquad$ A circular turn

77. *Trajectories of Comets* Under certain circumstances, a comet can pass by the sun once and never return. In this situation the comet may travel in a parabolic path, as illustrated in the figure on the next page. Suppose that a comet's path is given by $x = -2.5y^2$, where the sun is located at $(-0.1, 0)$ and the units are astronomical units (A.U.). One astronomical unit equals 93 million miles. (*Source:* W. Thomson, *Introduction to Space Dynamics.*)

(a) Plot a point for the sun's location and then graph the path of the comet.

(b) Find the distance from the sun to the comet when the comet is located at $(-2.5, 1)$.

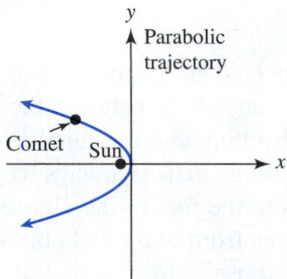

y
Parabolic
trajectory

Comet   Sun

x

78. *Speed of a Comet* (Continuation of Exercise 77.)
The velocity V in meters per second of a comet traveling in a parabolic trajectory around the sun is given by $V = \frac{k}{\sqrt{D}}$, where D is the comet's distance from the sun in meters and $k = 1.15 \times 10^{10}$.
  (a) How does the velocity of the comet change as its distance from the sun changes?

(b) Calculate the velocity of the comet when it is closest to the sun. (*Hint:* 1 mile $\approx$ 1609 meters.)

WRITING ABOUT MATHEMATICS

79. Suppose that you are given the equation
$$x = a(y - k)^2 + h.$$
  (a) Explain how you can determine the direction that the parabola opens.
  (b) Explain how to find the axis of symmetry and the vertex.
  (c) If the points $(0, 4)$ and $(0, -2)$ lie on the graph of x, what is the axis of symmetry?
  (d) Generalize part (c) if $(0, y_1)$ and $(0, y_2)$ lie on the graph of x.

80. Suppose that you are given the vertex of a parabola. Can you determine the axis of symmetry? Explain.

Group Activity   Working With Real Data

Directions: Form a group of 2 to 4 people. Select someone to record the group's responses for this activity. All members of the group should work cooperatively to answer the questions. If your instructor asks for your results, each member of the group should be prepared to respond.

Radio Telescope The U.S. Naval Research Laboratory designed a giant radio telescope weighing 3450 tons. Its parabolic dish has a diameter of 300 feet and a depth of 44 feet, as shown in the accompanying figure. (*Source:* J. Mar, *Structure Technology for Large Radio and Radar Telescope Systems.*)
  (a) Determine an equation of the form $x = ay^2$, $a > 0$, that models a cross section of the dish.
  (b) Graph your equation in an appropriate viewing rectangle.

10.2   Ellipses and Hyperbolas

Equations of Ellipses • Equations of Hyperbolas

A LOOK INTO MATH ▶

Celestial objects travel in paths or trajectories determined by conic sections. For this reason, conic sections have been studied for centuries. In modern times, physicists have learned that subatomic particles can also travel in trajectories determined by conic sections. Recall that the three main types of conic sections are parabolas, ellipses, and hyperbolas and that circles are a special type of ellipse. In Section 8.2 and Section 10.1, we discussed parabolas and circles. In this section we focus on ellipses and hyperbolas and some of their applications.

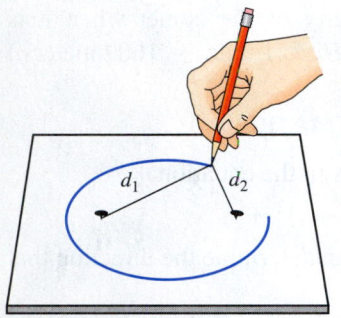

Figure 10.13

Equations of Ellipses

One method used to sketch an ellipse is to tie the ends of a string to two nails driven into a flat board. If a pencil is placed against the string anywhere between the nails, as shown in Figure 10.13, and is used to draw a curve, the resulting curve is an ellipse. The sum of the distances d_1 and d_2 between the pencil and each of the nails is always fixed by the length of the string. The location of the nails corresponds to the *foci* of the ellipse. An **ellipse** is the set of points in a plane the sum of whose distances from two fixed points is constant. Each fixed point is called a **focus** (plural, foci) of the ellipse.

CRITICAL THINKING

What happens to the shape of the ellipse shown in Figure 10.13 as the nails are moved farther apart? What happens to its shape as the nails are moved closer together? When would a circle be formed?

NEW VOCABULARY

☐ Ellipse
☐ Focus of an ellipse
☐ Major/minor axis
☐ Vertices of an ellipse
☐ Center of an ellipse
☐ Hyperbola
☐ Focus of a hyperbola
☐ Branches
☐ Vertices of a hyperbola
☐ Transverse axis
☐ Fundamental rectangle
☐ Asymptotes

In Figure 10.14 the **major axis** and the **minor axis** are labeled for each ellipse. The major axis is the longer of the two axes. Figure 10.14(a) shows an ellipse with a *horizontal* major axis, and Figure 10.14(b) shows an ellipse with a *vertical* major axis. The **vertices**, V_1 and V_2, of each ellipse are located at the endpoints of the major axis, and the **center of the ellipse** is the midpoint of the major axis (or the intersection of the major and minor axes).

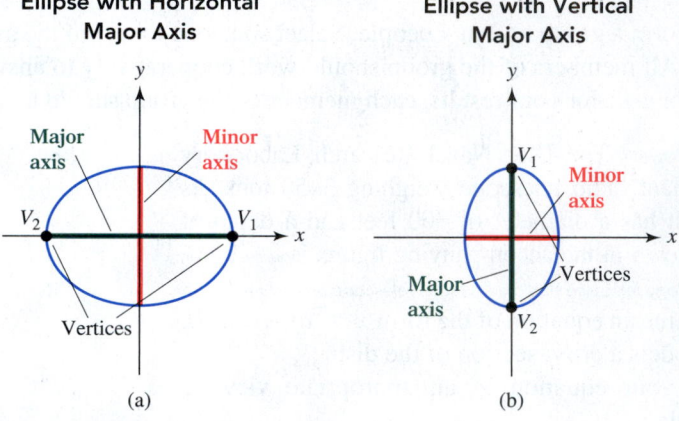

Figure 10.14 Not Functions

A vertical line can intersect the graph of an ellipse twice, so an ellipse cannot be represented by a function. However, some ellipses can be represented by the following equations.

READING CHECK

How can you determine from the standard equation of an ellipse if the major axis is horizontal or vertical?

STANDARD EQUATIONS FOR ELLIPSES CENTERED AT (0, 0)

The ellipse with center at the origin, *horizontal* major axis, and equation

$$\frac{x^2}{a^2} + \frac{y^2}{b^2} = 1, \qquad a > b > 0,$$

has vertices $(\pm a, 0)$ and endpoints of the minor axis $(0, \pm b)$.

The ellipse with center at the origin, *vertical* major axis, and equation

$$\frac{x^2}{b^2} + \frac{y^2}{a^2} = 1, \qquad a > b > 0,$$

has vertices $(0, \pm a)$ and endpoints of the minor axis $(\pm b, 0)$.

Figure 10.15(a) shows an ellipse having a horizontal major axis; Figure 10.15(b) shows one having a vertical major axis. The coordinates of the vertices V_1 and V_2 and endpoints of the minor axis U_1 and U_2 are labeled.

CRITICAL THINKING

Suppose that $a = b$ for an ellipse centered at $(0, 0)$. What can be said about the ellipse? Explain.

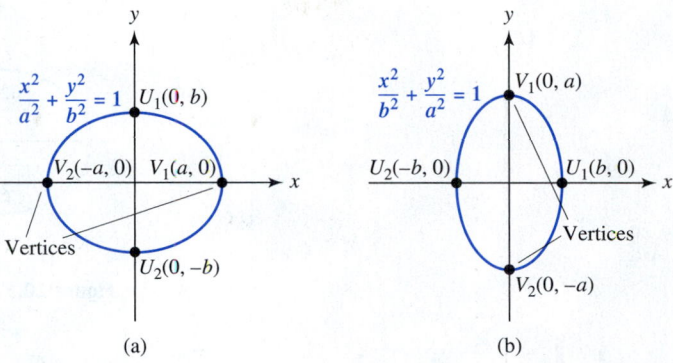

Figure 10.15

In the next example we show how to sketch graphs of ellipses.

EXAMPLE 1   **Sketching ellipses**

Sketch a graph of each ellipse. Label the vertices and endpoints of the minor axes.

(a) $\dfrac{x^2}{25} + \dfrac{y^2}{4} = 1$ **(b)** $9x^2 + 4y^2 = 36$

Solution
(a) The equation $\dfrac{x^2}{25} + \dfrac{y^2}{4} = 1$ describes an ellipse with $a^2 = 25$ and $b^2 = 4$. (When you are deciding whether 25 or 4 represents a^2, let a^2 be the larger of the two numbers.) Thus $a = 5$ and $b = 2$, so the ellipse has a horizontal major axis with vertices $(\pm 5, 0)$ and the endpoints of the minor axis are $(0, \pm 2)$. Plot these four points and then sketch the ellipse, as shown in Figure 10.16(a).

(b) To write $9x^2 + 4y^2 = 36$ as a standard equation, divide each side by 36.

$$9x^2 + 4y^2 = 36 \qquad \text{Given equation}$$

$$\frac{9x^2}{36} + \frac{4y^2}{36} = \frac{36}{36} \qquad \text{Divide each side by 36.}$$

$$\frac{x^2}{4} + \frac{y^2}{9} = 1 \qquad \text{Simplify.}$$

This ellipse has a vertical major axis with $a = 3$ and $b = 2$. The vertices are $(0, \pm 3)$, and the endpoints of the minor axis are $(\pm 2, 0)$, as shown in Figure 10.16(b).

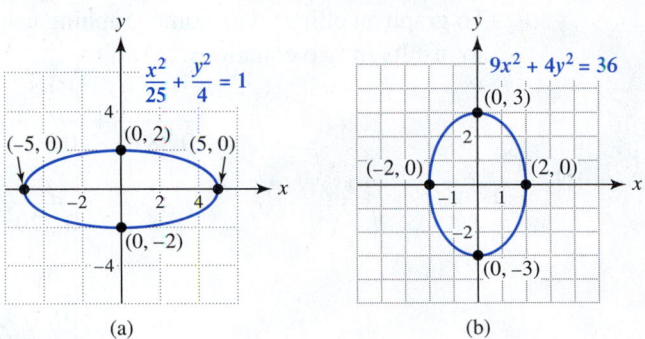

Figure 10.16

Now Try Exercises 15, 21

EXAMPLE 2   Finding the standard equation of an ellipse

Use the graph in Figure 10.17 to determine the standard equation of the ellipse.

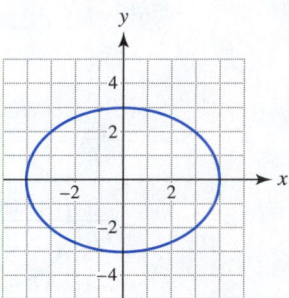

Figure 10.17

Solution
The ellipse is centered at (0, 0) with a horizontal major axis. The length of the major axis is 8, so $a = 4$. The length of the minor axis is 6, so $b = 3$. Thus $a^2 = \textbf{16}$ and $b^2 = \textbf{9}$, and the standard equation of the ellipse is

$$\frac{x^2}{16} + \frac{y^2}{9} = 1.$$

Now Try Exercise 25

▶ **REAL-WORLD CONNECTION** Planets travel around the sun in elliptical orbits. Astronomers have measured the values of a and b for each planet. Using this information, we can find the equation of a planet's orbit, as illustrated in the next example.

EXAMPLE 3   Modeling the orbit of Mercury

The planet Mercury has one of the least circular orbits of the eight major planets. For Mercury, $a = 0.387$ and $b = 0.379$. The units are astronomical units (A.U.), where 1 A.U. equals 93 million miles—the distance between Earth and the sun. Graph $\frac{x^2}{a^2} + \frac{y^2}{b^2} = 1$ to model the orbit of Mercury in $[-0.6, 0.6, 0.1]$ by $[-0.4, 0.4, 0.1]$. Then plot the sun at the point $(0.08, 0)$. (*Source:* M. Zeilik, *Introductory Astronomy and Astrophysics.*)

Solution
The orbit of Mercury is given by

$$\frac{x^2}{0.387^2} + \frac{y^2}{0.379^2} = 1. \hspace{2cm} \frac{x^2}{a^2} + \frac{y^2}{b^2} = 1$$

To graph an ellipse with some graphing calculators, we must solve the equation for y. Doing so results in two equations.

$$\frac{x^2}{0.387^2} + \frac{y^2}{0.379^2} = 1$$

$$\frac{y^2}{0.379^2} = 1 - \frac{x^2}{0.387^2} \hspace{1.5cm} \text{Subtract } \tfrac{x^2}{0.387^2}.$$

$$\frac{y}{0.379} = \pm\sqrt{1 - \frac{x^2}{0.387^2}} \hspace{1.5cm} \text{Square root property}$$

$$y = \pm 0.379\sqrt{1 - \frac{x^2}{0.387^2}} \hspace{1.5cm} \text{Multiply by 0.379.}$$

The orbit of Mercury can be found by graphing these equations. See Figures 10.18(a) and (b). The point (0.08, 0) represents the position of the sun in Figure 10.18(b).

CRITICAL THINKING

Use Figure 10.18 and the information in Example 3 to estimate the minimum and maximum distances that Mercury is from the sun.

[−0.6, 0.6, 0.1] by [−0.4, 0.4, 0.1]

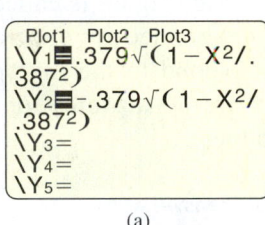

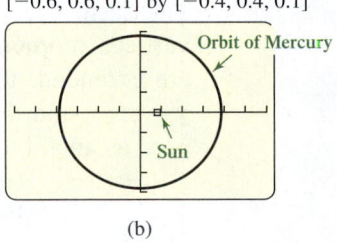

(a)

(b)

Figure 10.18

■ **Now Try Exercise 47**

Equations of Hyperbolas

The third type of conic section is the **hyperbola**, which is the set of points in a plane the difference of whose distances from two fixed points is constant. Each fixed point is called a **focus** of the hyperbola. Figure 10.19 shows a hyperbola whose equation is

$$\frac{x^2}{4} - \frac{y^2}{9} = 1.$$

This hyperbola is centered at the origin and has two **branches**, a *left branch* and a *right branch*. The **vertices** are (−2, 0) and (2, 0), and the line segment connecting the vertices is called the **transverse axis**. (The transverse axis is not part of the hyperbola.)

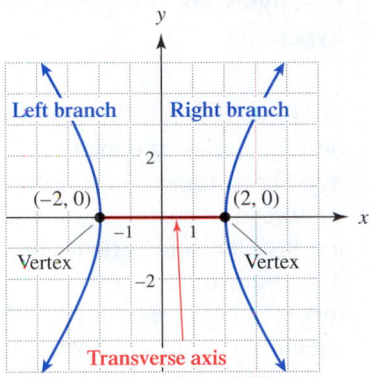

Figure 10.19

By the vertical line test, a hyperbola cannot be represented by a function, but many hyperbolas can be described by the following equations.

READING CHECK

How can you determine from the standard equation of a hyperbola if the transverse axis is horizontal or vertical?

STANDARD EQUATIONS FOR HYPERBOLAS CENTERED AT (0, 0)

The hyperbola with center at the origin, *horizontal* transverse axis, and equation

$$\frac{x^2}{a^2} - \frac{y^2}{b^2} = 1$$

has vertices ($\pm a$, 0).

The hyperbola with center at the origin, *vertical* transverse axis, and equation

$$\frac{y^2}{a^2} - \frac{x^2}{b^2} = 1$$

has vertices (0, $\pm a$).

Hyperbolas, along with the coordinates of their vertices, are shown in Figure 10.20. The two parts of the hyperbola in Figure 10.20(a) are the *left branch* and *right branch*, whereas in Figure 10.20(b) the hyperbola has an *upper branch* and a *lower branch*. The dashed rectangle in each figure is called the **fundamental rectangle**, and its four vertices (corners) are determined by either $(\pm a, \pm b)$ or $(\pm b, \pm a)$. If its diagonals are extended, they correspond to the asymptotes of the hyperbola. The dashed lines $y = \pm \frac{b}{a}x$ and $y = \pm \frac{a}{b}x$ are **asymptotes** for the hyperbolas, respectively, and may be used as an aid to graph them.

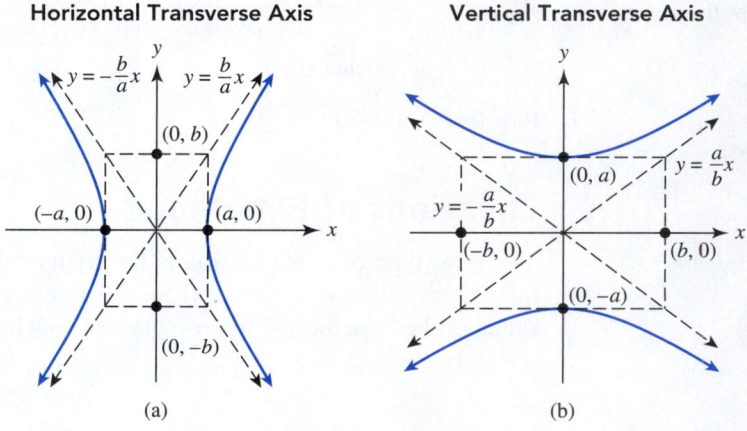

Figure 10.20 Not Functions

NOTE: A hyperbola consists of two solid curves, or branches. The dashed lines and rectangles are not part of the actual graph but are used as an aid for sketching the hyperbola.

▶ **REAL-WORLD CONNECTION** One interpretation of an asymptote of a hyperbola can be based on trajectories of comets as they approach the sun. Comets travel in parabolic, elliptic, or hyperbolic trajectories. If the speed of a comet is too slow, the gravitational pull of the sun captures the comet in an elliptic orbit (see Figure 10.21(a)). If the speed of the comet is too fast, the sun's gravity is too weak to capture the comet and the comet passes by it in a hyperbolic trajectory. Near the sun the gravitational pull is stronger, and the comet's trajectory is curved. Farther from the sun, the gravitational pull becomes weaker, and the comet eventually returns to a straight-line trajectory determined by the *asymptote* of the hyperbola (see Figure 10.21(b)). Finally, if the speed is neither too slow nor too fast, the comet will travel in a parabolic path (see Figure 10.21(c)).

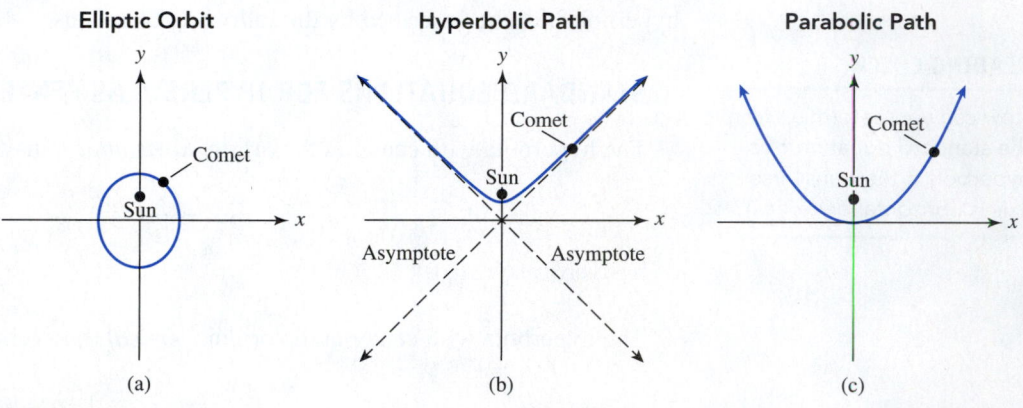

Figure 10.21

EXAMPLE 4   **Sketching a hyperbola**

Sketch a graph of $\frac{y^2}{4} - \frac{x^2}{9} = 1$. Label the vertices and show the asymptotes.

Solution

The equation is in standard form with $a^2 = 4$ and $b^2 = 9$, so $a = 2$ and $b = 3$. It has a vertical transverse axis with vertices $(0, -2)$ and $(0, 2)$. The vertices (corners) of the fundamental rectangle are $(\pm 3, \pm 2)$, that is, $(3, 2)$, $(3, -2)$, $(-3, 2)$, and $(-3, -2)$. The asymptotes are the diagonals of this rectangle and are given by $y = \pm \frac{a}{b}x$, or $y = \pm \frac{2}{3}x$. Figure 10.22 shows the hyperbola and these features.

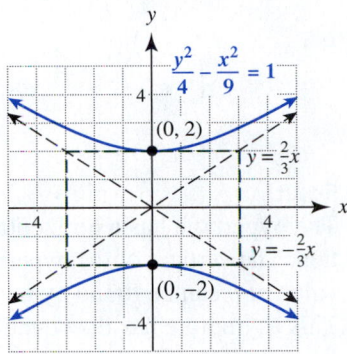

Figure 10.22

Now Try Exercise 29

TECHNOLOGY NOTE

Graphing a Hyperbola

The graph of a hyperbola does not represent a function. One way to graph a hyperbola with a graphing calculator is to solve the equation for y and obtain two equations. One equation gives the upper (or right) half of the hyperbola and the other equation gives the lower (or left) half. For example, to graph $\frac{y^2}{4} - \frac{x^2}{8} = 1$, begin by solving for y.

$$\frac{y^2}{4} = 1 + \frac{x^2}{8} \qquad \text{Add } \tfrac{x^2}{8}.$$

$$y^2 = 4\left(1 + \frac{x^2}{8}\right) \qquad \text{Multiply by 4.}$$

$$y = \pm 2\sqrt{1 + \frac{x^2}{8}} \qquad \text{Square root property}$$

Graph $Y_1 = 2\sqrt{(1 + X^2/8)}$ and $Y_2 = -2\sqrt{(1 + X^2/8)}$, which give the upper and lower branches, respectively. See Figure 10.23.

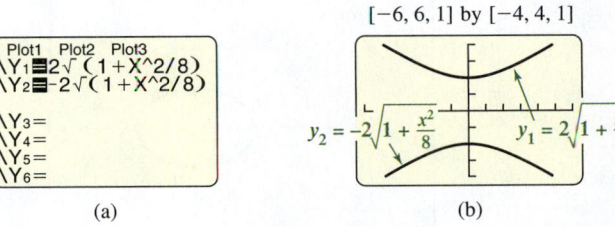

(a)   (b)

Figure 10.23

EXAMPLE 5   **Determining the standard equation of a hyperbola**

Use the graph in Figure 10.24 to determine the standard equation of the hyperbola.

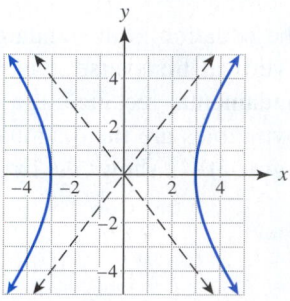

Figure 10.24

Solution

The hyperbola has a horizontal transverse axis, so the x^2-term must come first in the equation. The vertices of the hyperbola are $(\pm 3, 0)$, which indicates that $a = 3$ and $a^2 = 9$. The value of b can be found by noting that one of the asymptotes passes through the point $(3, 4)$. This asymptote has the equation $y = \frac{b}{a}x$ or $y = \frac{4}{3}x$, so let $b = 4$ and $b^2 = 16$. The equation of the hyperbola is

$$\frac{x^2}{9} - \frac{y^2}{16} = 1.$$

Now Try Exercise 41

10.2 Putting It All Together

CONCEPT	DESCRIPTION	
Ellipses Centered at $(0, 0)$ with $a > b > 0$	***Horizontal Major Axis*** Vertices: $(a, 0)$ and $(-a, 0)$ Endpoints of minor axis: $(0, b)$ and $(0, -b)$ $$\frac{x^2}{a^2} + \frac{y^2}{b^2} = 1$$ 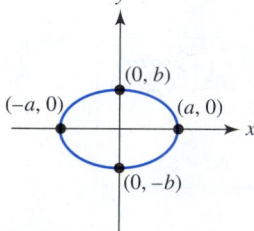	***Vertical Major Axis*** Vertices: $(0, a)$ and $(0, -a)$ Endpoints of minor axis: $(-b, 0)$ and $(b, 0)$ $$\frac{x^2}{b^2} + \frac{y^2}{a^2} = 1$$ 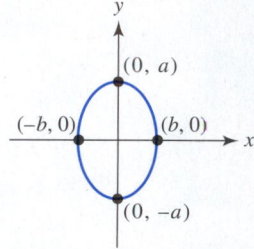

CONCEPT	DESCRIPTION	
Hyperbolas Centered at $(0, 0)$ with $a > 0$ and $b > 0$	**Horizontal Transverse Axis** Vertices: $(a, 0)$ and $(-a, 0)$ Asymptotes: $y = \pm \dfrac{b}{a}x$ $$\dfrac{x^2}{a^2} - \dfrac{y^2}{b^2} = 1$$ 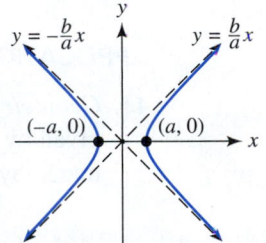	**Vertical Transverse Axis** Vertices: $(0, a)$ and $(0, -a)$ Asymptotes: $y = \pm \dfrac{a}{b}x$ $$\dfrac{y^2}{a^2} - \dfrac{x^2}{b^2} = 1$$ 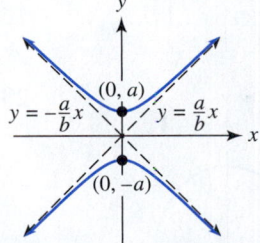

10.2 Exercises

MyMathLab | Math XL PRACTICE | WATCH | DOWNLOAD | READ | REVIEW

CONCEPTS AND VOCABULARY

1. Sketch an ellipse with a horizontal major axis.

2. Sketch a hyperbola with a vertical transverse axis.

3. The ellipse whose standard equation is $\dfrac{x^2}{a^2} + \dfrac{y^2}{b^2} = 1$, $a > b > 0$, has a(n) _____ major axis.

4. The ellipse whose standard equation is $\dfrac{x^2}{b^2} + \dfrac{y^2}{a^2} = 1$, $a > b > 0$, has a(n) _____ major axis.

5. What is the maximum number of times that a line can intersect an ellipse?

6. What is the maximum number of times that a parabola can intersect an ellipse?

7. The hyperbola whose equation is $\dfrac{x^2}{a^2} - \dfrac{y^2}{b^2} = 1$ has _____ and _____ branches.

8. The hyperbola whose equation is $\dfrac{y^2}{a^2} - \dfrac{x^2}{b^2} = 1$ has _____ and _____ branches.

9. How are the asymptotes of a hyperbola related to the fundamental rectangle?

10. Could an ellipse be centered at the origin and have vertices $(4, 0)$ and $(0, -5)$?

11. Which of the following (a.–d.) represents the coordinates of the vertices on the ellipse $\dfrac{x^2}{4} + \dfrac{y^2}{9} = 1$?

 a. $(0, \pm 3)$ b. $(\pm 2, 0)$
 c. $(\pm 4, 0)$ d. $(0, \pm 9)$

12. Which of the following (a.–d.) represents the coordinates of the vertices on the hyperbola $\dfrac{x^2}{4} - \dfrac{y^2}{9} = 1$?

 a. $(0, \pm 3)$ b. $(\pm 2, 0)$
 c. $(\pm 4, 0)$ d. $(0, \pm 9)$

ELLIPSES

Exercises 13–24: Graph the ellipse. Label the vertices and endpoints of the minor axis.

13. $\dfrac{x^2}{9} + \dfrac{y^2}{25} = 1$ 14. $\dfrac{y^2}{9} + \dfrac{x^2}{25} = 1$

15. $\dfrac{x^2}{9} + \dfrac{y^2}{4} = 1$ 16. $\dfrac{x^2}{3} + \dfrac{y^2}{9} = 1$

17. $x^2 + \dfrac{y^2}{4} = 1$ 18. $\dfrac{x^2}{9} + y^2 = 1$

19. $\dfrac{y^2}{5} + \dfrac{x^2}{7} = 1$ 20. $\dfrac{y^2}{11} + \dfrac{x^2}{6} = 1$

21. $36x^2 + 4y^2 = 144$ 22. $25x^2 + 16y^2 = 400$

23. $6y^2 + 7x^2 = 42$ 24. $9x^2 + 5y^2 = 45$

Exercises 25–28: Use the graph to determine the standard equation of the ellipse.

25.

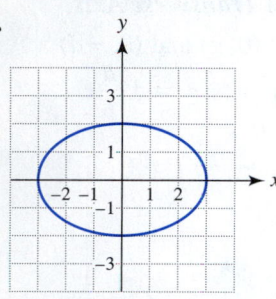

26.

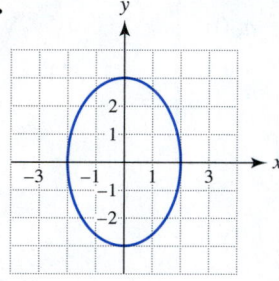

27.

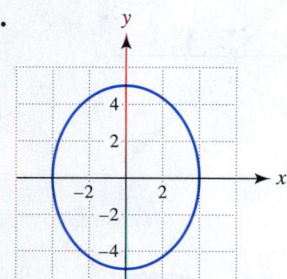

28.

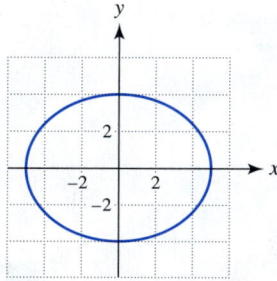

HYPERBOLAS

Exercises 29–40: Graph the hyperbola. Show the asymptotes and vertices.

29. $\dfrac{x^2}{4} - \dfrac{y^2}{9} = 1$

30. $\dfrac{y^2}{4} - \dfrac{x^2}{9} = 1$

31. $\dfrac{x^2}{25} - \dfrac{y^2}{16} = 1$

32. $\dfrac{y^2}{25} - \dfrac{x^2}{16} = 1$

33. $x^2 - y^2 = 1$

34. $y^2 - x^2 = 1$

35. $\dfrac{x^2}{3} - \dfrac{y^2}{4} = 1$

36. $\dfrac{y^2}{5} - \dfrac{x^2}{8} = 1$

37. $9y^2 - 4x^2 = 36$

38. $36x^2 - 25y^2 = 900$

39. $16x^2 - 4y^2 = 64$

40. $y^2 - 9x^2 = 9$

Exercises 41–44: Use the graph to determine the standard equation of the hyperbola.

41.

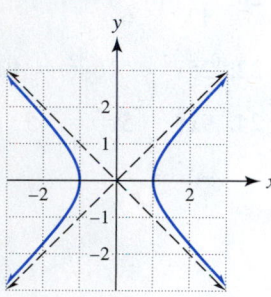

42.

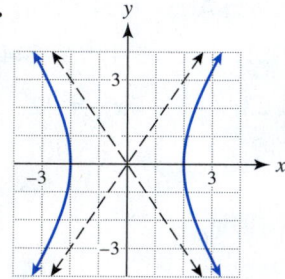

43.

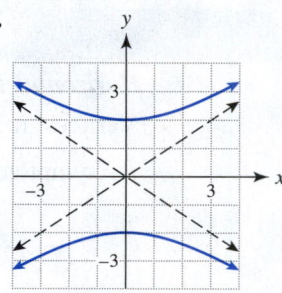

44.

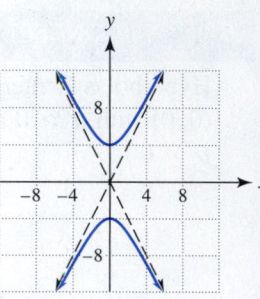

APPLICATIONS

45. *Geometry of an Ellipse* The area inside an ellipse is given by $A = \pi ab$, and its perimeter can be approximated by

$$P = 2\pi\sqrt{\dfrac{a^2 + b^2}{2}}.$$

Approximate A and P to the nearest hundredth for each ellipse.

(a) $\dfrac{x^2}{16} + \dfrac{y^2}{25} = 1$ **(b)** $\dfrac{x^2}{7} + \dfrac{y^2}{2} = 1$

46. *Geometry of an Ellipse* (Refer to Exercise 45.) If $a = b$ in the equation for an ellipse, the ellipse becomes a circle. Let $a = b$ in the formulas for the area and perimeter of an ellipse. Do the equations simplify to the area and perimeter for a circle? Explain.

47. *Planet Orbit* (Refer to Example 3.) Pluto's orbit is less circular than that of any of the eight planets. For Pluto (now a dwarf planet), $a = 39.44$ A.U. and $b = 38.20.$ A.U.

(a) Graph the elliptic orbit of Pluto in the window $[-60, 60, 10]$ by $[-40, 40, 10]$. Plot the point $(9.81, 0)$ to show the position of the sun. Assume that the major axis is horizontal.

(b) Use the information in Exercise 45 to determine how far Pluto travels in one orbit around the sun and approximate the area inside its orbit.

48. *Halley's Comet* (Refer to Example 3.) The famous Halley's comet travels in an elliptical orbit with $a = 17.95$ and $b = 4.44$ and passes by Earth roughly every 76 years. The most recent pass by Earth was in February 1986. (*Source:* M. Zeilik.)

(a) Graph the orbit of Halley's comet in $[-21, 21, 5]$ by $[-14, 14, 5]$. Assume that the major axis is horizontal and that all units are in astronomical units. Plot a point at $(17.39, 0)$ to represent the position of the sun.

(b) Use the formula in Exercise 45 to estimate how many miles Halley's comet travels in one orbit around the sun.

(c) Estimate the average speed of Halley's comet in miles per hour.

49. *Satellite Orbit* The orbit of Explorer VII and the outline of Earth's surface are shown in the accompanying figure. This orbit is described by

$$\frac{x^2}{4464^2} + \frac{y^2}{4462^2} = 1,$$

and the surface of Earth is described by

$$\frac{(x - 164)^2}{3960^2} + \frac{y^2}{3960^2} = 1.$$

Find the maximum and minimum heights of the satellite above Earth's surface if all units are miles. (*Source:* W. Thomson, *Introduction to Space Dynamics.*)

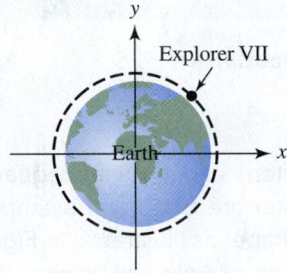

50. *Weight Machines* Elliptic shapes are used rather than circular shapes in some weight machines. Suppose that the ellipse shown in the accompanying figure is represented by the equation

$$\frac{x^2}{16} + \frac{y^2}{100} = 1,$$

where the units are inches. Find r_1 and r_2.

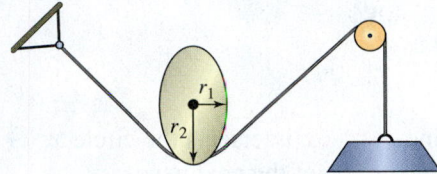

51. *Arch Bridge* The arch under a bridge is designed as the upper half of an ellipse, as illustrated in the accompanying figure. Its equation is modeled by

$$400x^2 + 10,000y^2 = 4,000,000,$$

where the units are feet. Find the height and width of the arch.

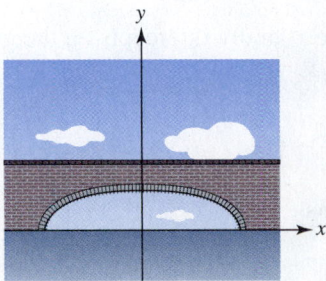

52. Thinking Generally Suppose that the population y of a country can be modeled by the upper right branch of the hyperbola

$$\frac{x^2}{a^2} - \frac{y^2}{b^2} = 1,$$

where x represents time in years. What happens to the population after a long period of time?

53. Online Exploration Go online and find one application of ellipses. Write a short paragraph about your findings.

54. Online Exploration Go online and find one application of hyperbolas. Write a short paragraph about your findings.

WRITING ABOUT MATHEMATICS

55. Explain how the values of a and b affect the graph of $\frac{x^2}{a^2} + \frac{y^2}{b^2} = 1$. Assume that $a > b > 0$.

56. Explain how the values of a and b affect the graph of $\frac{x^2}{a^2} - \frac{y^2}{b^2} = 1$. Assume that a and b are positive.

SECTIONS 10.1 AND 10.2 **Checking Basic Concepts**

1. Graph the parabola $x = (y - 2)^2 + 1$. Find the vertex and axis of symmetry.

2. Find the equation of the circle with center $(1, -2)$ and radius 2. Graph the circle.

3. Find the x- and y-intercepts on the graph of
$$\frac{x^2}{4} + \frac{y^2}{9} = 1.$$

4. Graph the following. Label any vertices and state the type of conic section that it represents.

(a) $x = y^2$
(b) $\dfrac{x^2}{16} + \dfrac{y^2}{25} = 1$

(c) $\dfrac{x^2}{4} - \dfrac{y^2}{9} = 1$

(d) $(x - 1)^2 + (y + 2)^2 = 9$

10.3 Nonlinear Systems of Equations and Inequalities

Basic Concepts • Solving Nonlinear Systems of Equations •
Solving Nonlinear Systems of Inequalities

A LOOK INTO MATH ▶

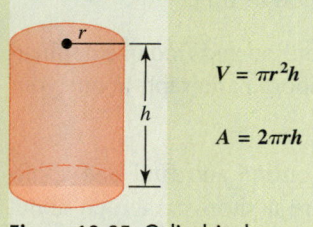

$V = \pi r^2 h$

$A = 2\pi rh$

Figure 10.25 Cylindrical Container

To describe characteristics of curved objects we often need *nonlinear equations*. The equations of the conic sections discussed in this chapter are but a few examples of nonlinear equations. For instance, cylinders have a curved shape, as illustrated in Figure 10.25. If the radius of a cylinder is denoted r and its height h, then its volume V is given by the nonlinear equation $V = \pi r^2 h$ and its side area A is given by the nonlinear equation $A = \mathbf{2\pi rh}$.

If we want to manufacture a cylindrical container that holds **35** cubic inches and whose side area is **50** square inches, we need to solve the following **nonlinear system of equations**. (This system is solved in Example 4.)

$$\pi r^2 h = 35$$
$$2\pi rh = 50$$

NEW VOCABULARY

☐ Nonlinear system of equations
☐ Method of substitution
☐ Nonlinear system of inequalities

In this section we solve nonlinear systems of equations and inequalities.

Basic Concepts

One way to locate the points at which the line $y = 2x$ intersects the circle $x^2 + y^2 = 5$ is to graph both equations (see Figure 10.26 at the top of the next page).

The equation describing the circle is nonlinear. Another way to locate the points of intersection is symbolically, by solving the nonlinear system of equations.

$$y = 2x \quad \text{Line}$$
$$x^2 + y^2 = 5 \quad \text{Circle}$$

Linear systems of equations can have no solutions, one solution, or infinitely many solutions. It is possible for a nonlinear system of equations to have *any number* of solutions. Figure 10.26 shows that this nonlinear system of equations has two solutions: $(-1, -2)$ and $(1, 2)$.

Two Solutions

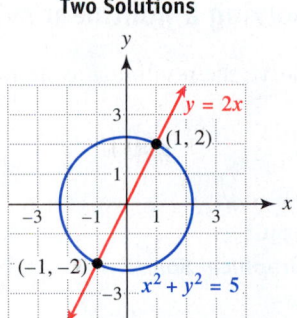

Figure 10.26

Solving Nonlinear Systems of Equations

Nonlinear systems of equations in two variables can sometimes be solved graphically, numerically, and symbolically. One symbolic technique is the **method of substitution**, which we demonstrate in the next example.

EXAMPLE 1   **Solving a nonlinear system of equations symbolically**

Solve the following system of equations symbolically. Check any solutions and compare them with Figure 10.26.

$$y = 2x$$
$$x^2 + y^2 = 5$$

Solution
Substitute **$2x$** for y in the second equation and solve for x.

$x^2 + (2x)^2 = 5$	Let $y = 2x$ in the second equation.
$x^2 + 4x^2 = 5$	Properties of exponents
$5x^2 = 5$	Combine like terms.
$x^2 = 1$	Divide by 5.
$x = \pm 1$	Square root property

To determine corresponding y-values, substitute $x = \pm 1$ in $y = 2x$; the solutions are $(1, 2)$ and $(-1, -2)$. To check $(1, 2)$, substitute $x = 1$ and $y = 2$ in the given equations.

$$2 \stackrel{?}{=} 2(1) \checkmark \qquad \text{True}$$
$$(1)^2 + (2)^2 \stackrel{?}{=} 5 \checkmark \qquad \text{True}$$

To check $(-1, -2)$, substitute $x = -1$ and $y = -2$ in the given equations.

$$-2 \stackrel{?}{=} 2(-1) \checkmark \qquad \text{True}$$
$$(-1)^2 + (-2)^2 \stackrel{?}{=} 5 \checkmark \qquad \text{True}$$

The solutions check and agree with Figure 10.26.

Now Try Exercise 13

In the next example we solve a nonlinear system of equations graphically and symbolically. The symbolic solution gives the exact solutions.

EXAMPLE 2   **Solving a nonlinear system of equations**

Solve the nonlinear system of equations graphically and symbolically.

$$x^2 - y = 2$$
$$x^2 + y = 4$$

Solution

Graphical Solution Begin by solving each equation for y.

$$y = x^2 - 2$$
$$y = 4 - x^2$$

Graph $y_1 = x^2 - 2$ and $y_2 = 4 - x^2$. The solutions are approximately $(-1.73, 1)$ and $(1.73, 1)$, as shown in Figure 10.27. The graphs consist of two parabolas intersecting at two points.

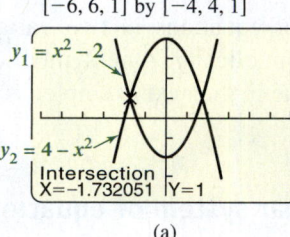

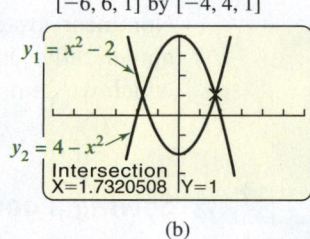

(a)   (b)

Figure 10.27

Symbolic Solution Solving the first equation for y gives $y = x^2 - 2$. Substitute this expression for y in the second equation and solve for x.

$x^2 + y = 4$	Second equation
$x^2 + (x^2 - 2) = 4$	Substitute $y = x^2 - 2$.
$2x^2 = 6$	Combine like terms; add 2.
$x^2 = 3$	Divide by 2.
$x = \pm\sqrt{3}$	Square root property

To determine y, substitute $x = \pm\sqrt{3}$ in $y = x^2 - 2$.

$$y = (\sqrt{3})^2 - 2 = 3 - 2 = 1$$
$$y = (-\sqrt{3})^2 - 2 = 3 - 2 = 1$$

The *exact* solutions are $(\sqrt{3}, 1)$ and $(-\sqrt{3}, 1)$.

Now Try Exercise 29(a), (b)

EXAMPLE 3   **Solving a nonlinear system of equations**

Solve the nonlinear system of equations symbolically and graphically.

$$x^2 - y^2 = 3$$
$$x^2 + y^2 = 5$$

Solution

Symbolic Solution Instead of using substitution on this nonlinear system of equations, we use elimination. Note that, if we add the two equations, the y-variable will be eliminated.

$x^2 - y^2 = 3$	First equation
$x^2 + y^2 = 5$	Second equation
$2x^2 = 8$	Add equations.

Solving gives $x^2 = 4$, or $x = \pm 2$. To determine y, substitute **4** for x^2 in $x^2 + y^2 = 5$.

$$4 + y^2 = 5 \quad \text{or} \quad y^2 = 1$$

Because $y^2 = 1$, $y = \pm 1$. There are four solutions: $(2, 1)$, $(2, -1)$, $(-2, 1)$, and $(-2, -1)$.

Graphical Solution The graph of the first equation is a hyperbola, and the graph of the second is a circle with radius $\sqrt{5}$. The four points of intersection (solutions) are $(\pm 2, \pm 1)$, as shown in Figure 10.28.

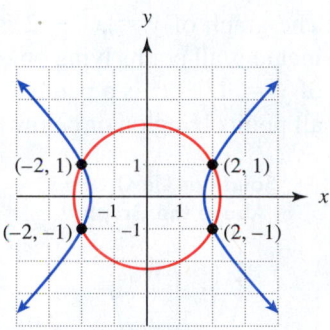

Figure 10.28

▌ **Now Try Exercises 23, 25**

In the next example we solve the system of equations presented in A Look Into Math for this section.

EXAMPLE 4   **Modeling the dimensions of a can**

Find the dimensions of a can having a volume V of 35 cubic inches and a side area A of 50 square inches by solving the following nonlinear system of equations symbolically. (See the figure in the margin.)

$$\pi r^2 h = 35$$
$$2\pi r h = 50$$

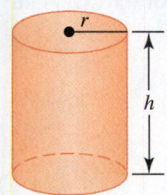

Solution
We can find r by solving each equation for h and then setting them equal. This eliminates the variable h.

$$\frac{50}{2\pi r} = \frac{35}{\pi r^2} \qquad h = \tfrac{50}{2\pi r} \text{ and } h = \tfrac{35}{\pi r^2}$$

$$50\pi r^2 = 70\pi r \qquad \text{Eliminate fractions (cross multiply).}$$

$$50\pi r^2 - 70\pi r = 0 \qquad \text{Subtract } 70\pi r.$$

$$10\pi r(5r - 7) = 0 \qquad \text{Factor out } 10\pi r.$$

$$10\pi r = 0 \quad \text{or} \quad 5r - 7 = 0 \qquad \text{Zero-product property}$$

$$r = 0 \quad \text{or} \quad r = \tfrac{7}{5} = 1.4 \qquad \text{Solve.}$$

READING CHECK

How many solutions can a nonlinear system of equations have?

Because $h = \frac{50}{2\pi r}$, $r = 0$ is not possible, but we can find h by substituting **1.4** for r in the formula.

$$h = \frac{50}{2\pi(1.4)} \approx 5.68$$

A can having a volume of 35 cubic inches and a side area of 50 square inches has a radius of 1.4 inches and a height of about 5.68 inches.

▌ **Now Try Exercise 51(b)**

Solving Nonlinear Systems of Inequalities

In Chapter 4 we solved systems of *linear* inequalities. A **nonlinear system of inequalities** in two variables can be solved similarly by using graphical techniques. For example, consider the nonlinear system of inequalities

$$y \geq x^2 - 2$$
$$y \leq 4 - x^2.$$

The graph of $y = x^2 - 2$ is a parabola opening upward. The solution set to $y \geq x^2 - 2$ includes all points lying on or above this parabola. See Figure 10.29(a). Similarly, the graph of $y = 4 - x^2$ is a parabola opening downward. The solution set to $y \leq 4 - x^2$ includes all points lying on or below this parabola. See Figure 10.29(b).

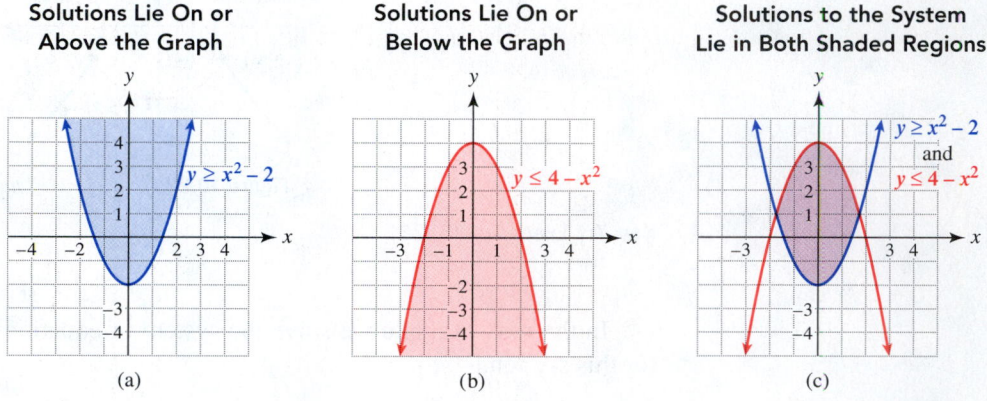

Figure 10.29

The solution set for this nonlinear *system* of inequalities includes all points (x, y) in *both* shaded regions. The *intersection* of the shaded regions is shown in Figure 10.29(c).

> **EXAMPLE 5** Solving a nonlinear system of inequalities graphically

Shade the solution set for the system of inequalities.

$$\frac{x^2}{4} + \frac{y^2}{9} < 1$$
$$y > 1$$

Solution
The solutions to $\frac{x^2}{4} + \frac{y^2}{9} < 1$ lie *inside* the ellipse $\frac{x^2}{4} + \frac{y^2}{9} = 1$. See Figure 10.30(a). Solutions to $y > 1$ lie above the line $y = 1$, as shown in Figure 10.30(b). The intersection of these two regions is shown in Figure 10.30(c). Any point in this region is a solution.

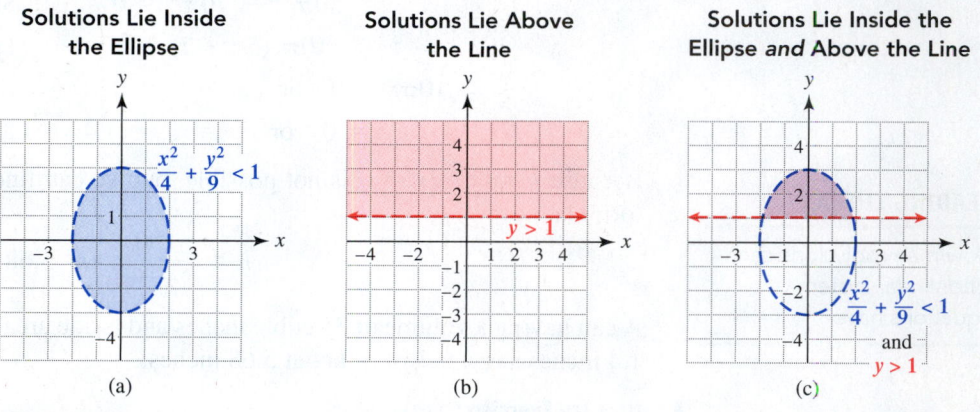

Figure 10.30

For example, the point $(0, 2)$ lies in the shaded region and is a solution to the system. Note that a dashed curve and a dashed line are used when equality is not included.

Now Try Exercise 39

EXAMPLE 6 **Solving a nonlinear system of inequalities graphically**

Shade the solution set for the following system of inequalities.

$$x^2 + y \le 4$$
$$-x + y \ge 2$$

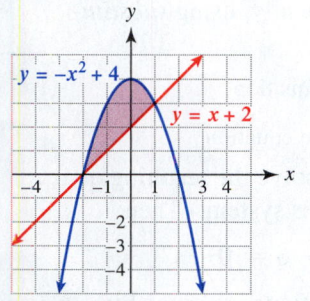

Figure 10.31

Solution

The solutions to $x^2 + y \le 4$ lie *on* or *below* the parabola $y = -x^2 + 4$, and the solutions to $-x + y \ge 2$ lie *on* or *above* the line $y = x + 2$. The appropriate shaded region is shown in Figure 10.31. Both the parabola and the line are solid because equality is included in both inequalities.

Now Try Exercise 37

In the next example we use a graphing calculator to shade a region that lies above both graphs, using the "$Y_1 =$" menu. This feature allows us to shade either above or below the graph of a function.

EXAMPLE 7 **Solving a system of inequalities with a graphing calculator**

Shade the solution set for the following system of inequalities.

$$y \ge x^2 - 2$$
$$y \ge -1 - x$$

Solution

Enter $y_1 = x^2 - 2$ and $y_2 = -1 - x$, as shown in Figure 10.32(a). Note that the option to shade above the graphs of Y_1 and Y_2 was selected to the left of Y_1 and Y_2. Then the two inequalities were graphed in Figure 10.32(b). The solution set corresponds to the region where there are both vertical and horizontal lines.

CALCULATOR HELP

To shade the solution set to a system of inequalities, see Appendix A (pages AP-6 and AP-7).

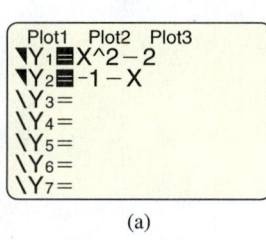

(a)

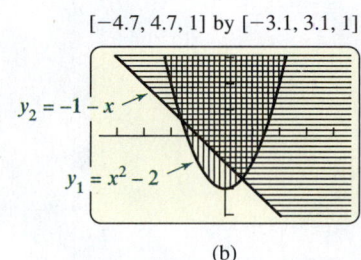

[−4.7, 4.7, 1] by [−3.1, 3.1, 1]

(b)

Figure 10.32

Now Try Exercise 49

10.3   Putting It All Together

Unlike a linear system of equations, a nonlinear system of equations can have *any number of solutions*. Nonlinear systems of inequalities involving two variables usually have infinitely many solutions, which can be represented by a shaded region in the *xy*-plane.

CONCEPT	EXPLANATION
Nonlinear Systems of Equations in Two Variables	To solve the following system of equations symbolically, using *substitution*, solve the first equation for y to get $y = 5 - x$. $$x + y = 5 \quad \text{First equation}$$ $$x^2 - y = 1 \quad \text{Second equation}$$ Substitute $5 - x$ for y in the second equation and solve the resulting quadratic equation. (*Elimination* can also be used on this system.) Then $$x^2 - (5 - x) = 1 \quad \text{or} \quad x^2 + x - 6 = 0,$$ and factoring can be used to determine that the solutions are given by $$x = -3 \quad \text{or} \quad x = 2.$$ Thus $y = 5 - (-3) = 8$ or $y = 5 - 2 = 3$. The solutions are $(-3, 8)$ and $(2, 3)$. Graphical support is shown in the accompanying figure. 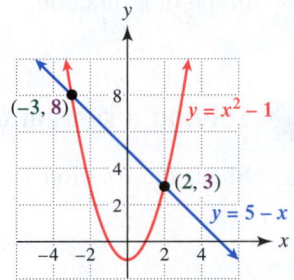
Nonlinear Systems of Inequalities in Two Variables	To solve the following system of inequalities graphically, solve each inequality for y. $$x + y \le 5 \quad \text{or} \quad y \le 5 - x \quad \text{First inequality}$$ $$x^2 - y \le 1 \quad \text{or} \quad y \ge x^2 - 1 \quad \text{Second inequality}$$ The solutions lie on or above the parabola and on or below the line, as shown in the figure. 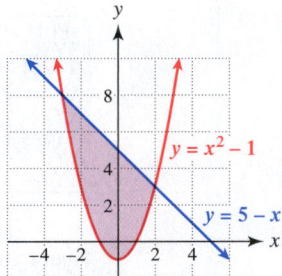

10.3 Exercises

CONCEPTS AND VOCABULARY

1. How many solutions can a nonlinear system of equations have?

2. If a nonlinear system of equations has two equations, how many equations does a solution have to satisfy?

3. Determine visually the number of solutions to the following system of equations. Explain your reasoning.
$$y = x$$
$$x^2 + y^2 = 4$$

4. Describe the solution set to $x^2 + y^2 \le 1$.

5. Does $(-2, -1)$ satisfy $5x^2 - 2y^2 > 18$?

6. Does $(3, 4)$ satisfy $x^2 - 2y \ge 4$?

7. Sketch a parabola and ellipse with four points of intersection.

8. Sketch a line and a hyperbola with two points of intersection.

NONLINEAR SYSTEMS OF EQUATIONS

Exercises 9–12: Use the graph to estimate all solutions to the system of equations. Check each solution.

9. $x^2 + y^2 = 10$
   $y = 3x$

10. $x^2 + 3y^2 = 16$
    $y = -x$

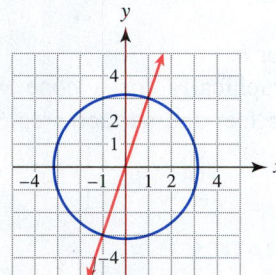

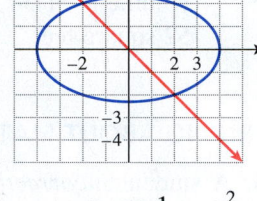

11. $y^2 - x^2 = 1$
    $x^2 + 3y^2 = 3$

12. $y = 1 - x^2$
    $x^2 + y^2 = 1$

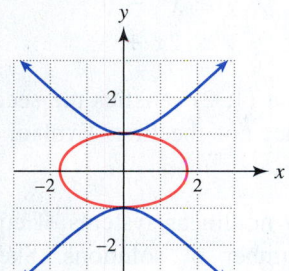

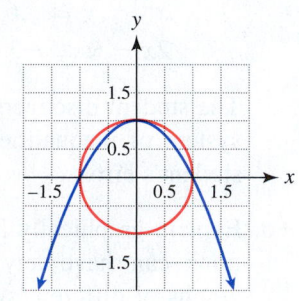

Exercises 13–24: Solve the system of equations symbolically. Check your solutions.

13. $y = 2x$
    $x^2 + y^2 = 45$

14. $y = x$
    $y^2 = 3 - 2x^2$

15. $x + y = 1$
    $x^2 - y^2 = 3$

16. $y - x = -1$
    $y = 2x^2$

17. $y - x^2 = 0$
    $x^2 + y^2 = 6$

18. $x^2 - y^2 = 4$
    $x^2 + y^2 = 4$

19. $3x^2 + 2y^2 = 5$
    $x - y = -2$

20. $x^2 + 2y^2 = 9$
    $2x^2 - y^2 = -2$

21. $x^2 + y^2 = 4$
    $x^2 - 9y^2 = 9$

22. $x^2 + y^2 = 15$
    $9x^2 + 4y^2 = 36$

23. $x^2 + y^2 = 10$
    $2x^2 - y^2 = 17$

24. $2x^2 + 3y^2 = 5$
    $3x^2 - 4y^2 = -1$

Exercises 25–28: Solve the system of equations graphically. Check your solutions.

25. $y = x^2 - 3$
    $2x^2 - y = 1 - 3x$

26. $x + y = 2$
    $x - y^2 = 3$

27. $y - x = -4$
    $x - y^2 = -2$

28. $xy = 1$
    $y = x$

Exercises 29–32: Solve the system of equations
   (a) symbolically,
   (b) graphically, and
   (c) numerically.

29. $y = -2x$
    $x^2 + y = 3$

30. $4x - y = 0$
    $x^3 - y = 0$

31. $xy = 1$
    $x - y = 0$

32. $x^2 + y^2 = 4$
    $y - x = 2$

NONLINEAR SYSTEMS OF INEQUALITIES

Exercises 33–36: Shade the solution set in the xy-plane.

33. $y \ge x^2$

34. $y \le x^2 - 1$

35. $\dfrac{x^2}{4} + \dfrac{y^2}{9} > 1$

36. $x^2 + y^2 \le 1$

Exercises 37–44: Shade the solution set in the xy-plane. Then use the graph to select one solution.

37. $y > x^2 + 1$
    $y < 3$

38. $y > x^2$
    $y < x + 2$

39. $x^2 + y^2 \leq 1$
$\qquad y < x$

40. $y > x^2 - 2$
$\qquad y \leq 2 - x^2$

41. $\qquad x^2 + y^2 \leq 1$
$(x - 2)^2 + y^2 \leq 1$

42. $\qquad x^2 - y \geq 2$
$(x + 1)^2 + y^2 \leq 4$

43. $x^2 - y^2 \leq 4$
$\quad x^2 + y^2 \leq 9$

44. $\quad 3x + 2y < 6$
$\qquad x^2 + y^2 \leq 16$

Exercises 45 and 46: Match the inequality or system of inequalities with its graph (a. or b.).

45. $y \leq \dfrac{1}{2}x^2$

46. $y \geq x^2 + 1$
$\qquad y \leq 5$

a.

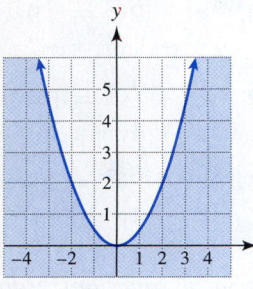

b.

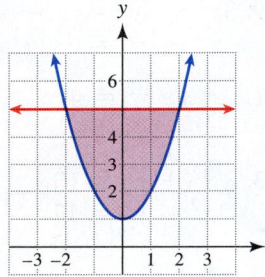

Exercises 47 and 48: Use the graph to write the inequality or system of inequalities.

47.

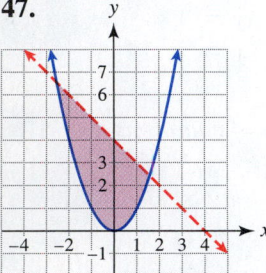

48.

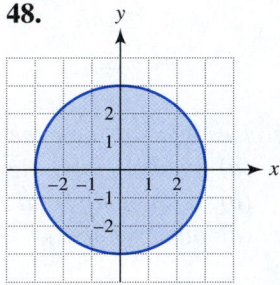

 Exercises 49 and 50: Use a graphing calculator to shade the solution set to the system of inequalities.

49. $y \geq x^2 - 1$
$\qquad y \geq 2 - x$

50. $y \leq 4 - x^2$
$\qquad y \leq x^2 - 4$

APPLICATIONS

51. *Dimensions of a Can* (Refer to Example 4.) Find the dimensions of a cylindrical container with a volume of 40 cubic inches and a side area of 50 square inches **(a)** graphically and **(b)** symbolically.

52. *Dimensions of a Can* (Refer to Example 4.) Is it possible to design an aluminum can with volume of 60 cubic inches and side area of 60 square inches? If so, find the dimensions of the can.

53. *Area and Perimeter* The area of a room is 143 feet, and its perimeter is 48 feet. Let x be the width and y be the length of the room. See the accompanying figure.
(a) Write a nonlinear system of equations that models this situation.
(b) Solve the system.

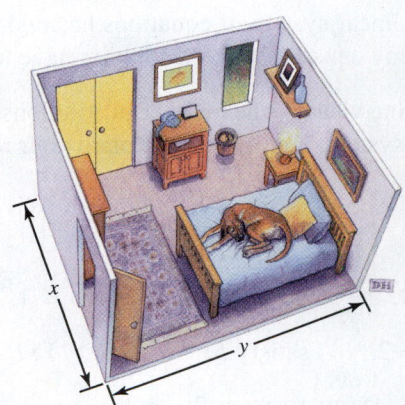

54. *Dimensions of a Cone* The volume V of a cone is given by $V = \frac{1}{3}\pi r^2 h$, and the surface area S of its side is given by $S = \pi r \sqrt{r^2 + h^2}$, where h is the height and r is the radius of the base (see the accompanying figure).

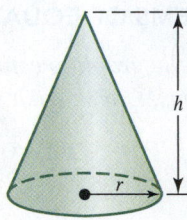

(a) Solve each equation for h.
 (b) Estimate r and h graphically for a cone with volume V of 34 cubic feet and surface area S of 52 square feet.

WRITING ABOUT MATHEMATICS

55. A student *incorrectly* changes the following system of inequalities

$$\begin{array}{c} x^2 - y \geq 6 \\ 2x - y \leq -3 \end{array} \text{ to } \begin{array}{c} y \geq x^2 - 6 \\ y \leq 2x + 3. \end{array}$$

The student discovers that $(1, 2)$ satisfies the second system of inequalities but not the first. Explain the student's error.

56. Explain graphically how nonlinear systems of equations can have any number of solutions. Sketch graphs of different systems with zero, one, two, and three solutions.

Checking Basic Concepts

1. Solve the following system of equations symbolically and graphically.

$$x^2 - y = 2x$$
$$2x - y = 3$$

2. Determine visually the number of solutions to the following system of equations.

$$y = x^2 - 4$$
$$y = x$$

3. The solution set for a system of inequalities is shown in the accompanying figure.
   (a) Find one ordered pair (x, y) that is a solution and one that is not.

(b) Write the system of inequalities represented by the graph.

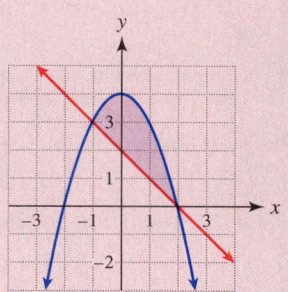

4. Shade the solution set for the following system of inequalities.

$$x^2 + y^2 \leq 4$$
$$y < 1$$

CHAPTER 10 Summary

SECTION 10.1 ■ PARABOLAS AND CIRCLES

Parabolas There are three basic types of conic sections: parabolas, ellipses, and hyperbolas. A parabola can have a vertical or a horizontal axis. Two forms of an equation for a parabola with a *vertical axis* are

$$y = ax^2 + bx + c \quad \text{and} \quad y = a(x - h)^2 + k.$$

If $a > 0$ the parabola opens upward, and if $a < 0$ it opens downward (see the figure on the left). The vertex is located at (h, k). Two forms of an equation for a parabola with a *horizontal axis* are

$$x = ay^2 + by + c \quad \text{and} \quad x = a(y - k)^2 + h.$$

If $a > 0$ the parabola opens to the right, and if $a < 0$ it opens to the left (see the figure on the right). The vertex is located at (h, k).

Vertical Axis

$y = a(x - h)^2 + k$
$a < 0$

(h, k)

Horizontal Axis

$x = a(y - k)^2 + h$
$a < 0$

(h, k)

Example: $x = -2(y - 4)^2 + 1$ has vertex $(1, 4)$, has axis of symmetry $y = 4$, and opens to the left.

Circles The standard equation for a circle with center (h, k) and radius r is

$$(x - h)^2 + (y - k)^2 = r^2.$$

Example: $(x + 2)^2 + (y - 3)^2 = 36$ has center $(-2, 3)$ and radius 6.

SECTION 10.2 ■ ELLIPSES AND HYPERBOLAS

Ellipses The standard equation for an ellipse centered at the origin with a *horizontal major axis* is $\frac{x^2}{a^2} + \frac{y^2}{b^2} = 1$, $a > b > 0$, and the vertices are $(\pm a, 0)$, as shown in the figure on the left. The standard equation for an ellipse centered at the origin with a *vertical major axis* is $\frac{x^2}{b^2} + \frac{y^2}{a^2} = 1$, $a > b > 0$, and the vertices are $(0, \pm a)$, as shown in the figure on the right. Circles are a special type of ellipse, with the major and minor axes having equal lengths.

Horizontal Major Axis

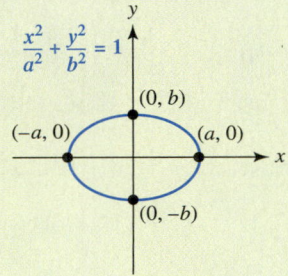

Vertical Major Axis

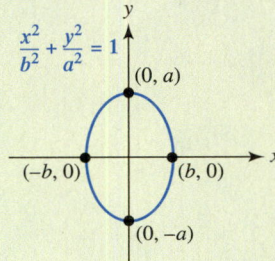

Example: $\frac{x^2}{9} + \frac{y^2}{4} = 1$ is the standard equation for an ellipse with vertices $(\pm 3, 0)$, centered at $(0, 0)$.

Hyperbolas The standard equation for a hyperbola centered at the origin with a *horizontal transverse axis* is $\frac{x^2}{a^2} - \frac{y^2}{b^2} = 1$, the asymptotes are given by $y = \pm \frac{b}{a}x$, and the vertices are $(\pm a, 0)$, as shown in the figure on the left. The standard equation for a hyperbola centered at the origin with a *vertical transverse axis* is $\frac{y^2}{a^2} - \frac{x^2}{b^2} = 1$, the asymptotes are $y = \pm \frac{a}{b}x$, and the vertices are $(0, \pm a)$, as shown in the figure on the right.

Horizontal Transverse Axis

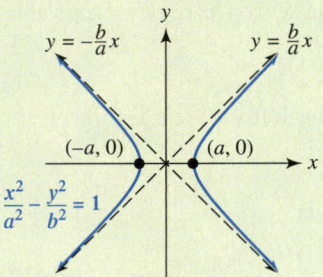

Vertical Transverse Axis

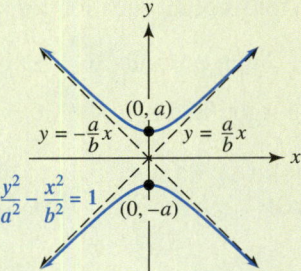

Example: $\frac{x^2}{9} - \frac{y^2}{4} = 1$ is the standard equation for a hyperbola with horizontal transverse axis, vertices $(\pm 3, 0)$, center $(0, 0)$, and asymptotes $y = \pm \frac{2}{3}x$.

SECTION 10.3 ■ NONLINEAR SYSTEMS OF EQUATIONS AND INEQUALITIES

Nonlinear Systems Nonlinear systems of equations can have any number of solutions. The methods of substitution and elimination can often be used to solve a nonlinear system of equations symbolically. Nonlinear systems can also be solved graphically. The solution set for a nonlinear system of two inequalities in two variables is typically a region in the *xy*-plane. A solution is an ordered pair (x, y) that satisfies both inequalities.

Example: Solve $y \geq x^2 - 2$

$$y \leq 4 - \frac{1}{2}x^2.$$

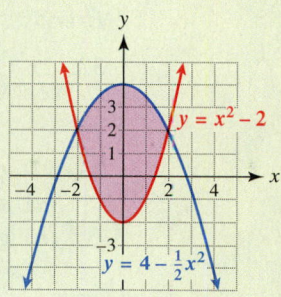

CHAPTER 10 Review Exercises

SECTION 10.1

Exercises 1–6: Graph the parabola. Find the vertex and axis of symmetry.

1. $x = 2y^2$

2. $x = -(y + 1)^2$

3. $x = -2(y - 2)^2$

4. $x = (y + 2)^2 - 1$

5. $x = -3y^2 + 1$

6. $x = \frac{1}{2}y^2 + y - 3$

7. Use the graph to determine the equation of the parabola.

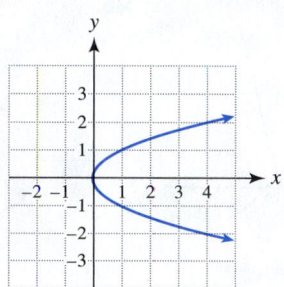

8. Use the graph to find the equation of the circle.

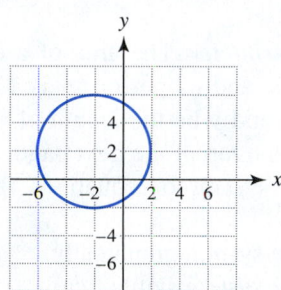

9. Write the equation of the circle with radius 1 and center $(0, 0)$.

10. Write the equation of the circle with radius 4 and center $(2, -3)$.

Exercises 11–14: Find the radius and center of the circle. Then graph the circle.

11. $x^2 + y^2 = 25$

12. $(x - 2)^2 + y^2 = 9$

13. $(x + 3)^2 + (y - 1)^2 = 5$

14. $x^2 - 2x + y^2 + 2y = 7$

SECTION 10.2

Exercises 15–18: Graph the ellipse. Label the vertices and endpoints of the minor axis.

15. $\dfrac{x^2}{4} + \dfrac{y^2}{25} = 1$

16. $x^2 + \dfrac{y^2}{4} = 1$

17. $25x^2 + 20y^2 = 500$

18. $4x^2 + 9y^2 = 36$

19. Use the graph to determine the equation of the ellipse.

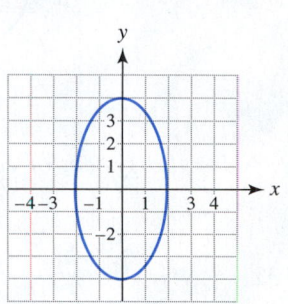

20. Use the graph to find the equation of the hyperbola.

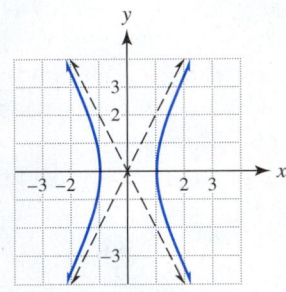

Exercises 21–24: Graph the hyperbola. Include the asymptotes in your graph.

21. $\dfrac{x^2}{9} - \dfrac{y^2}{4} = 1$

22. $\dfrac{y^2}{25} - \dfrac{x^2}{16} = 1$

23. $y^2 - x^2 = 1$

24. $25x^2 - 16y^2 = 400$

SECTION 10.3

Exercises 25–28: Use the graph to estimate all solutions to the system of equations. Check each solution.

25. $x^2 + y^2 = 9$
  $x + y = 3$

26. $xy = 2$
  $y = 2x$

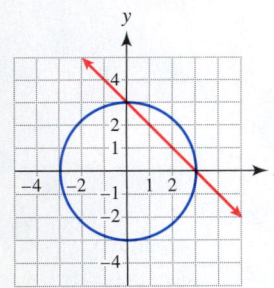

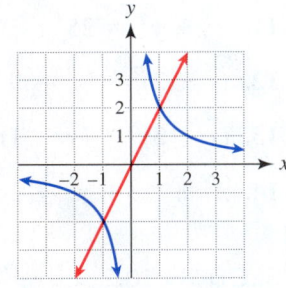

27. $x^2 - y = x$
  $y = x$

28. $x^2 + y^2 = 5$
  $x^2 - y^2 = 3$

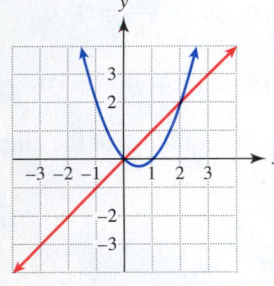

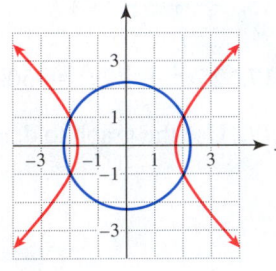

Exercises 29–32: Solve the system of equations. Check your solutions.

29. $y = x$
  $x^2 + y^2 = 32$

30. $x - y = 4$
  $x^2 + y^2 = 16$

31. $y = x^2$
  $2x^2 + y = 3$

32. $y = x^2 + 1$
  $2x^2 - y = 3x - 3$

Exercises 33 and 34: Solve the system graphically.

33. $2x - y = 4$
  $x^2 + y = 4$

34. $x^2 + y = 0$
  $x^2 + y^2 = 2$

Exercises 35 and 36: Solve the system of equations
  (a) *symbolically,*
  (b) *graphically, and*
  (c) *numerically.*

35. $y = x$
  $x^2 + 2y = 8$

36. $y = x^3$
  $x^2 - y = 0$

Exercises 37–44: Shade the solution set in the xy-plane.

37. $y \geq 2x^2$

38. $y < 2x - 3$

39. $y < -x^2$

40. $\dfrac{x^2}{9} + \dfrac{y^2}{16} \leq 1$

41. $y - x^2 \geq 1$
  $y \leq 2$

42. $x^2 + y \leq 4$
  $3x + 2y \geq 6$

43. $y > x^2$
  $y < 4 - x^2$

44. $\dfrac{x^2}{4} + \dfrac{y^2}{9} > 1$
  $x^2 + y^2 < 16$

Exercises 45 and 46: Use the graph to write the system of inequalities.

45.

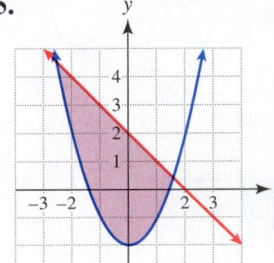

46.

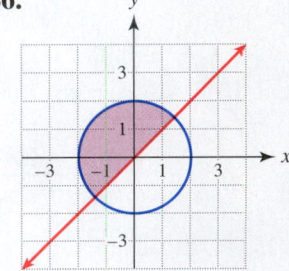

APPLICATIONS

47. *Area and Perimeter* The area of a desktop is 1000 square inches, and its perimeter is 130 inches. Let x be the width and y be the length of the desktop. See the figure at the top of the next page.
  (a) Write a system of equations that models this situation.
  (b) Solve the system graphically.
  (c) Solve the system symbolically.

48. *Numbers* The product of two positive numbers is 60, and their difference is 7. Let x be the smaller number and y be the larger number.
   (a) Write a system of equations whose solution gives the two numbers.
   (b) Solve the system graphically.
   (c) Solve the system symbolically.

49. *Dimensions of a Container* The volume of a cylindrical container is $V = \pi r^2 h$, and its surface area, *excluding* the top and bottom, is $A = 2\pi rh$. Find the dimensions of a container with $A = 100$ square feet and $V = 50$ cubic feet. Is your answer unique?

 50. *Dimensions of a Container* The volume of a cylindrical container is $V = \pi r^2 h$, and its surface area, *including* the top and bottom, is $A = 2\pi rh + 2\pi r^2$.

Graphically find the dimensions of a can with $A = 80$ square inches and $V = 35$ cubic inches. Is your answer unique?

51. *Geometry of an Ellipse* The area inside an ellipse is given by $A = \pi ab$, and its perimeter P can be approximated by

$$P = 2\pi \sqrt{\frac{a^2 + b^2}{2}}.$$

   (a) Graph $\dfrac{x^2}{5} + \dfrac{y^2}{12} = 1$.
   (b) Estimate the ellipse's area and perimeter.

52. *Orbit of Mars* Mars has an elliptical orbit that is nearly circular, with $a = 1.524$ and $b = 1.517$, where the units are astronomical units (1 A.U. equals 93 million miles). (*Source*: M. Zeilik.)
   (a) Graph the orbit of Mars in $[-3, 3, 1]$ by $[-2, 2, 1]$. Plot the point $(0.15, 0)$ to show the position of the sun. Assume that the major axis is horizontal.
   (b) Use the information in Exercise 51 to estimate how far Mars travels in one orbit around the sun. Approximate the area inside its orbit.

Chapter 10 Test

CHAPTER Test Prep VIDEOS Step-by-step test solutions are found on the Chapter Test Prep Videos available in **MyMathLab** and on You Tube (search "RockswoldInterAlg" and click on "Channels").

1. Graph the parabola $y = -(x - 1)^2 + 2$. Find the vertex and axis of symmetry.

2. Graph the parabola $x = (y - 4)^2 - 2$. Find the vertex and axis of symmetry.

3. Use the graph to find the equation of the parabola.

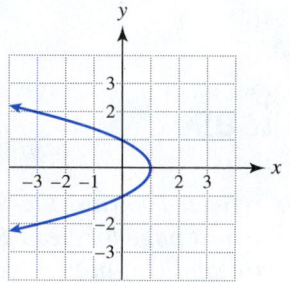

4. Use the graph to find the equation of the circle.

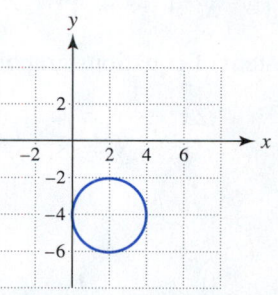

5. Write the equation of the circle with radius 10 and center $(-5, 2)$.

6. Find the radius and center of the circle given by

$$x^2 + 4x + y^2 - 6y = 3.$$

Then graph the circle.

7. Graph the ellipse $\frac{x^2}{16} + \frac{y^2}{49} = 1$. Label the vertices and endpoints of the minor axis.

8. Use the graph to determine the standard equation of the ellipse.

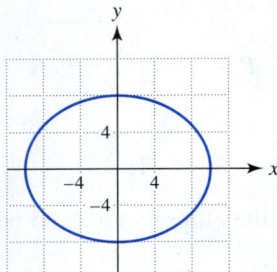

9. Graph the hyperbola $4x^2 - 9y^2 = 36$. Include the asymptotes in your graph.

10. Use the graph to estimate all solutions to the system of equations. Check each solution by substitution in the system of equations.

$$x^2 + y^2 = 16$$
$$x - y = 4$$

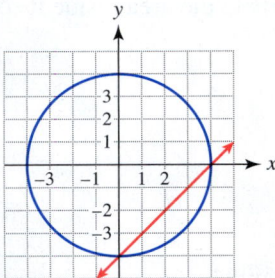

11. Solve the system of equations symbolically.

$$x - y = 3$$
$$x^2 + y^2 = 17$$

12. Solve the system of equations graphically.

$$2x^2 - y = 4$$
$$x^2 + y = 8$$

13. Shade the solution set in the xy-plane.

$$3x + y > 6$$
$$x^2 + y^2 < 25$$

14. Use the graph to write the system of inequalities.

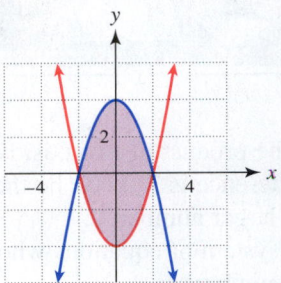

15. *Area and Perimeter* The area of a rectangular swimming pool is 5000 square feet, and its perimeter is 300 feet. Let x be the width and y be the length of the pool.
  (a) Write a nonlinear system of equations that models this situation.
  (b) Solve the system.

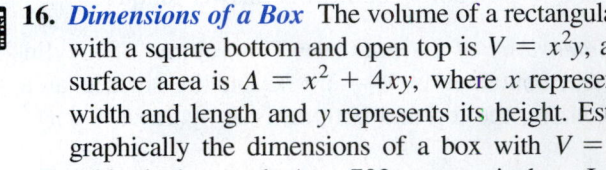
16. *Dimensions of a Box* The volume of a rectangular box with a square bottom and open top is $V = x^2y$, and its surface area is $A = x^2 + 4xy$, where x represents its width and length and y represents its height. Estimate graphically the dimensions of a box with $V = 1183$ cubic inches and $A = 702$ square inches. Is your answer unique? (*Hint:* Substitute appropriate values for V and A, and then solve each equation for y.)

17. *Orbit of Uranus* The planet Uranus has an elliptical orbit that is nearly circular, with $a = 19.18$ and $b = 19.16$, where the units are astronomical units (1 A.U. equals 93 million miles). (*Source:* M. Zeilik.)
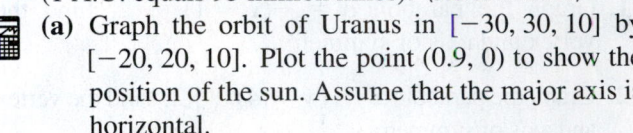
  (a) Graph the orbit of Uranus in $[-30, 30, 10]$ by $[-20, 20, 10]$. Plot the point $(0.9, 0)$ to show the position of the sun. Assume that the major axis is horizontal.
  (b) Find the minimum distance between Uranus and the sun.

CHAPTER 10   Extended and Discovery Exercises

Exercises 1–2: ***Foci of Parabolas*** *The focus of a parabola is a point that has special significance. When a parabola is rotated about its axis, it sweeps out a shape called a* **paraboloid,** *as illustrated in the top figure on the following page. Paraboloids have an important reflective property. When incoming rays of light from the sun or distant stars strike the surface of a parabo-* *loid, each ray is reflected toward the focus, as shown in the figure on the next page labeled "Reflective Property." If the rays are sunlight, intense heat is produced, which can be used to generate solar heat. Radio signals from distant space also concentrate at the focus, and scientists can measure these signals by placing a receiver there.*

The same reflective property of a paraboloid can be used in reverse. If a light source is placed at the focus, the light is reflected straight ahead, as depicted in the figure labeled "Headlight." Searchlights, flashlights, and car headlights make use of this reflective property.

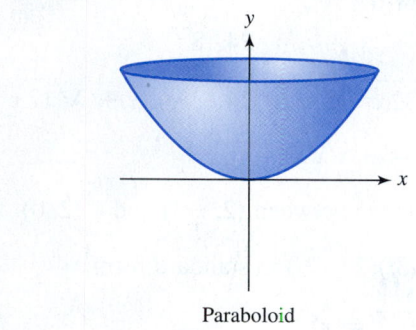

Paraboloid

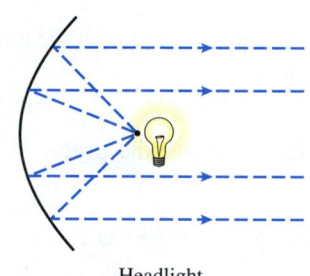

Reflective Property   Headlight

The focus is always located inside a parabola, on its axis of symmetry. If the distance between the vertex and the focus is $|p|$, the following equations can be used to locate the focus. Note that the value of p may be either positive or negative.

EQUATION OF A PARABOLA WITH VERTEX (0, 0)

Vertical Axis

The parabola with a focus at $(0, p)$ has the equation
$$x^2 = 4py.$$

Horizontal Axis

The parabola with a focus at $(p, 0)$ has the equation
$$y^2 = 4px.$$

1. Graph each parabola. Label the vertex and focus.
   (a) $x^2 = 4y$ (b) $y^2 = -8x$ (c) $x = 2y^2$

2. The reflective property of paraboloids is used in satellite dishes and radio telescopes. The U.S. Naval Research Laboratory designed a giant radio telescope weighing 3450 tons. Its parabolic dish has a diameter of 300 feet and a depth of 44 feet, as shown in the figures in the next column. (*Source:* J. Mar, *Structure Technology for Large Radio and Radar Telescope Systems.*)
   (a) Determine an equation in the form $y = ax^2$ that describes a cross section of this dish.
   (b) If the receiver is located at the focus, how far should it be from the vertex?

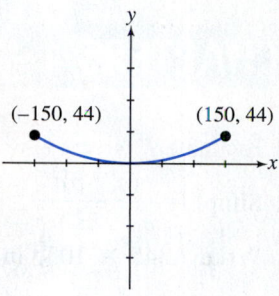

Radio Telescope

Exercises 3 and 4: *Translations of Ellipses and Hyperbolas* Ellipses and hyperbolas can be translated so that they are centered at a point (h, k), rather than at the origin. These techniques are the same as those used for parabolas and circles. To translate a conic section so that it is centered at (h, k) rather than $(0, 0)$, replace x with $(x - h)$ and replace y with $(y - k)$. For example, to center $\frac{x^2}{9} + \frac{y^2}{4} = 1$ at $(-1, 2)$, change its equation to $\frac{(x + 1)^2}{9} + \frac{(y - 2)^2}{4} = 1$. See the accompanying figures.

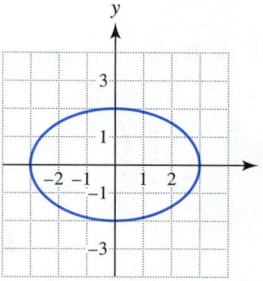
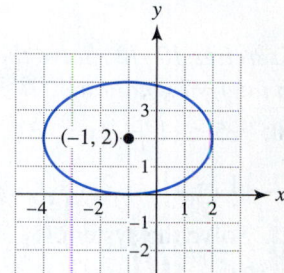

Ellipse centered at $(0, 0)$ Ellipse centered at $(-1, 2)$

3. Graph each conic section. For hyperbolas give the equations of the asymptotes.
   (a) $\dfrac{(x - 3)^2}{25} + \dfrac{(y - 1)^2}{9} = 1$
   (b) $\dfrac{(x + 1)^2}{4} + \dfrac{(y + 2)^2}{16} = 1$
   (c) $\dfrac{(x + 1)^2}{4} - \dfrac{(y - 3)^2}{9} = 1$
   (d) $\dfrac{(y - 4)^2}{16} - \dfrac{(x + 1)^2}{4} = 1$

4. Write the equation of an ellipse having the following properties.
   (a) Horizontal major axis of length 8, minor axis of length 4, and centered at $(-3, 5)$.
   (b) Vertical major axis of length 10, minor axis of length 6, and centered at $(2, -3)$.

5. Determine the center of the conic section.
   (a) $9x^2 - 18x + 4y^2 + 24y + 9 = 0$
   (b) $25x^2 + 150x - 16y^2 + 32y - 191 = 0$

CHAPTERS 1–10   Cumulative Review Exercises

1. Evaluate $K = x^2 + y^2$ when $x = 4$ and $y = -3$.

2. Simplify $\dfrac{(a^{-2} b)^2}{a^{-1} (b^3)^{-2}}$.

3. Write 7.345×10^{-3} in standard (decimal) notation.

4. Find $f(-4)$, if $f(x) = \dfrac{x}{x-4}$. State the domain of f.

5. If the point $(3, 4)$ is on the graph of $y = f(x)$, then $f(\underline{}) = \underline{}$.

Exercises 6 and 7: Graph f by hand.

6. $f(x) = x^2 - x$ 7. $f(x) = 1 - 2x$

8. Find the slope–intercept form of the line that passes through the point $(2, -2)$ and is perpendicular to the line given by $y = -\frac{2}{3}x + 1$.

9. Solve $2(1 - x) - 4x = x$.

Exercises 10–12: Solve the inequality. Write your answer in interval notation.

10. $-5 \le 1 - 2x < 3$ 11. $x^2 - 4 \le 0$

12. $|1 - x| \ge 2$

13. Solve the system.
$$-2x + y = 1$$
$$5x - y = 2$$

14. Shade the solution set in the xy-plane.
$$x + y \le 4$$
$$x - y \ge 2$$

Exercises 15 and 16: Multiply the expression.

15. $(2x - 1)(x + 5)$ 16. $xy(2x - 3y^2 + 1)$

Exercises 17 and 18: Factor the expression.

17. $6x^2 - 13x - 5$ 18. $x^3 - 4x$

Exercises 19 and 20: Solve the equation.

19. $x^2 + 3x + 2 = 0$ 20. $x^2 + 1 = -3x$

Exercises 21 and 22: Simplify the expression.

21. $\dfrac{x-2}{x+2} \div \dfrac{2x-4}{3x+6}$ 22. $\dfrac{1}{x+1} + \dfrac{1}{x-1}$

Exercises 23–26: Simplify the expression. Assume all variables are positive.

23. $\sqrt{8x^2}$ 24. $8^{2/3}$

25. $\sqrt[3]{2x} \cdot \sqrt[3]{32x^2}$ 26. $3\sqrt{3x} + \sqrt{12x}$

27. Graph $f(x) = -\sqrt{x}$.

28. Find the distance between $(2, -3)$ and $(-2, 0)$.

29. Write $(2 + 3i)(2 - 3i)$ in standard form.

30. Solve $\sqrt{x + 2} = x$.

31. Find the vertex of the graph of $y = x^2 - 6x + 3$.

32. Write $f(x) = x^2 - 2x + 3$ in vertex form.

33. Compare the graph of $f(x) = \sqrt{x} - 4$ to the graph of $y = \sqrt{x}$.

34. Solve $x(3 - x) = 2$.

35. Solve $x^3 + x = 0$. Find all complex solutions.

36. Simplify by hand.
   (a) $\log 10{,}000$ (b) $\log_2 8$
   (c) $\log_3 3^x$ (d) $e^{\ln 6}$
   (e) $\log 2 + \log 50$ (f) $\log_2 24 - \log_2 3$

37. If $f(x) = x^2 + 1$ and $g(x) = 2x$, find the following.
   (a) $(f \circ g)(2)$ (b) $(g \circ f)(x)$

38. If $f(x) = 2 - 3x$, find $f^{-1}(x)$.

39. If \$1000 is deposited in an account that pays 5% annual interest, approximate the amount in the account after 6 years.

40. Write $\log \dfrac{x^2 \sqrt{y}}{z^3}$ in terms of logarithms of x, y, and z.

41. Solve $2e^x - 1 = 17$.

42. Solve $3 + \log 4x = 5$.

43. Graph each conic section.
   (a) $x = (y - 1)^2$ (b) $(x - 1)^2 + (y + 1)^2 = 4$
   (c) $\dfrac{x^2}{4} + \dfrac{y^2}{25} = 1$ (d) $4x^2 - 9y^2 = 36$

44. Solve the nonlinear system of equations.
$$x^2 + y^2 = 1$$
$$x^2 + 9y^2 = 9$$

45. Shade the solution set in the xy-plane.

$$x^2 + y^2 \leq 4$$
$$x^2 - y \leq 2$$

APPLICATIONS

46. *Calculating Distance* The distance D in miles that a driver of a car is from home after x hours is given by $D(x) = 400 - 50x$.
   (a) Evaluate $D(0)$. Interpret your answer.
   (b) What is the x-intercept on the graph of D? Interpret your answer.
   (c) What is the slope of the graph of D? Interpret your answer.

47. *Investment* Suppose $2000 is deposited in three accounts paying 5%, 6%, and 7% annual interest. The amount invested at 6% is $500 more than the amount invested at 5%. The interest after 1 year is $120. How much is invested at each rate?

48. *Area* There are 1200 feet of fence available to surround the perimeter of a rectangular garden. What dimensions for the garden give the largest area?

49. *Population* The population of a city in millions after x years is given by $P = 2e^{0.02x}$. How long will it take for the population to double?

50. *Dimensions of a Can* Find the radius of a cylindrical can with a volume V of 60 cubic inches and a side area S of 50 square inches. (*Hint:* $V = \pi r^2 h$ and $S = 2\pi rh$.)

11 Sequences and Series

Go deep enough into anything and you will find mathematics.

—DEAN SCHLICTER

In this final chapter we present sequences and series, which are essential topics because they are used to model and approximate important quantities. Complicated population growth can be modeled with sequences, and accurate approximations for numbers such as π and e are made with series. For example, in about 2000 B.C. the Babylonians thought that π equaled 3.125, whereas at the same time the Chinese thought π equaled 3. By 1700, mathematicians were able to calculate π to 100 decimal places due to the discovery of series. Series are also essential to the solutions of many modern applied mathematics problems.

Although you may not always recognize the impact of mathematics on everyday life, its influence is nonetheless profound. Mathematics is the *language of technology*—it allows experiences to be quantified. In the preceding chapters we showed numerous examples of mathematics being used to model the real world. Computers, DVD players, cars, highway design, weather, hurricanes, electricity, social networks, government data, cellular phones, medicine, ecology, business, sports, and psychology represent only some of the applications of mathematics. In fact, if a subject is studied in enough detail, mathematics usually appears in one form or another. Although predicting what the future may bring is difficult, one thing *is* certain—mathematics will continue to play an important role in both theoretical research and new technology.

Source: Mathforum.org, Drexel University 1994–2007.

11.1 Sequences

Basic Concepts • Representations of Sequences • Models and Applications

A LOOK INTO MATH ▶ Sequences are *ordered lists*. For example, names listed alphabetically represent a sequence. Figure 11.1 shows an insect population in thousands per acre over a 6-year period. Listing populations by year is another example of a sequence. In mathematics, a sequence is a function for which valid inputs must be natural numbers. For example, we can use a function f to define this sequence by letting $f(1)$ represent the insect population after 1 year, $f(2)$ represent the insect population after 2 years, and in general let $f(n)$ represent the population after n years. In this section we discuss sequences.

NEW VOCABULARY

☐ Terms of a sequence
☐ Finite sequence
☐ Infinite sequence
☐ *n*th term
☐ General term

An Insect Population

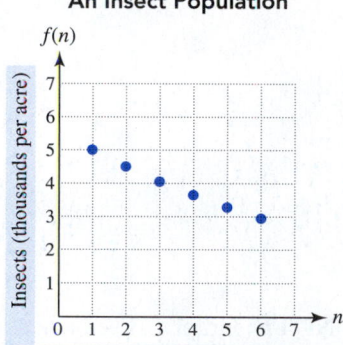

Figure 11.1

Basic Concepts

▶ **REAL-WORLD CONNECTION** Suppose that an individual's starting salary is $40,000 per year and that the person's salary is increased by 10% each year. This situation is modeled by the formula

$$f(n) = 40{,}000(1.10)^n. \quad \text{Symbolic representation}$$

We do not allow the input n to be any real number, but rather limit n to a *natural number* because the individual's salary is constant throughout a particular year. The first five *terms of the sequence* are

$$f(1), f(2), f(3), f(4), f(5).$$

They can be computed as follows.

$$f(1) = 40{,}000(1.10)^1 = \mathbf{44{,}000}$$
$$f(2) = 40{,}000(1.10)^2 = \mathbf{48{,}400}$$
$$f(3) = 40{,}000(1.10)^3 = \mathbf{53{,}240}$$
$$f(4) = 40{,}000(1.10)^4 = \mathbf{58{,}564}$$
$$f(5) = 40{,}000(1.10)^5 \approx \mathbf{64{,}420}$$

The first five terms of a sequence

This sequence is represented *numerically* in Table 11.1 on the next page.

TABLE 11.1 Numerical Representation

n	1	2	3	4	5
$f(n)$	44,000	48,400	53,240	58,564	64,420

A *graphical* representation results when each data point in Table 11.1 is plotted, as illustrated in Figure 11.2.

Graphical Representation

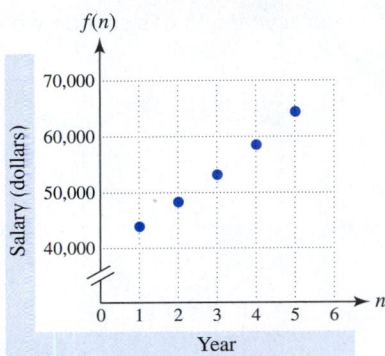

Figure 11.2

NOTE: Graphs of sequences are scatterplots.

The preceding sequence is an example of a *finite sequence* of numbers. The even natural numbers,

$$2, 4, 6, 8, 10, 12, 14, \ldots, \qquad \text{Even natural numbers}$$

are an example of an *infinite sequence* represented by $f(n) = 2n$, where n is a natural number. The three dots, or periods (called an *ellipsis*), indicate that the pattern continues indefinitely.

SEQUENCES

A **finite sequence** is a function whose domain is $D = \{1, 2, 3, \ldots, n\}$ for some fixed natural number n.

An **infinite sequence** is a function whose domain is the set of natural numbers.

READING CHECK

How are functions and sequences related?

Because sequences are functions, many of the concepts discussed in previous chapters apply to sequences. Instead of letting y represent the output, however, the convention is to write $a_n = f(n)$, where n is a natural number in the domain of the sequence. The *terms* of a sequence are

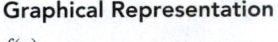

$$a_1, a_2, a_3, \ldots, a_n, \ldots. \qquad \text{Sequence}$$

The first term is $a_1 = f(1)$, the second term is $a_2 = f(2)$, and so on. The **nth term**, or **general term**, of a sequence is $a_n = f(n)$.

EXAMPLE 1   **Computing terms of a sequence**

Write the first four terms of each sequence for $n = 1, 2, 3$, and 4.

(a) $f(n) = 2n - 1$ **(b)** $f(n) = 3(-2)^n$ **(c)** $f(n) = \dfrac{n}{n+1}$

Solution

(a) For $a_n = f(n) = 2n - 1$, we write the first four terms as follows.

$$a_1 = f(1) = 2(1) - 1 = 1$$
$$a_2 = f(2) = 2(2) - 1 = 3$$
$$a_3 = f(3) = 2(3) - 1 = 5$$
$$a_4 = f(4) = 2(4) - 1 = 7$$

The first four terms are **1, 3, 5**, and **7**.

(b) For $a_n = f(n) = 3(-2)^n$, we write the first four terms as follows.

$$a_1 = f(1) = 3(-2)^1 = -6$$
$$a_2 = f(2) = 3(-2)^2 = 12$$
$$a_3 = f(3) = 3(-2)^3 = -24$$
$$a_4 = f(4) = 3(-2)^4 = 48$$

The first four terms are **-6, 12, -24**, and **48**.

(c) For $a_n = f(n) = \frac{n}{n+1}$, we write the first four terms as follows.

$$a_1 = f(1) = \frac{1}{1+1} = \frac{1}{2}$$
$$a_2 = f(2) = \frac{2}{2+1} = \frac{2}{3}$$
$$a_3 = f(3) = \frac{3}{3+1} = \frac{3}{4}$$
$$a_4 = f(4) = \frac{4}{4+1} = \frac{4}{5}$$

The first four terms are $\frac{1}{2}, \frac{2}{3}, \frac{3}{4}$, and $\frac{4}{5}$. Note that, although the input to a sequence is a natural number, the output need not be a natural number.

Now Try Exercises 9, 11, 13

TECHNOLOGY NOTE

Generating Sequences
Many graphing calculators can generate sequences if you change the MODE from function (Func) to sequence (Seq). In Figures 11.3 and 11.4 the sequences from Example 1 are generated. On some calculators the sequence utility is found in the LIST OPS menus. The expression

$$\text{seq }(2n - 1, n, 1, 4)$$

represents terms 1 through 4 of the sequence $f(n) = 2n - 1$ with the variable n.

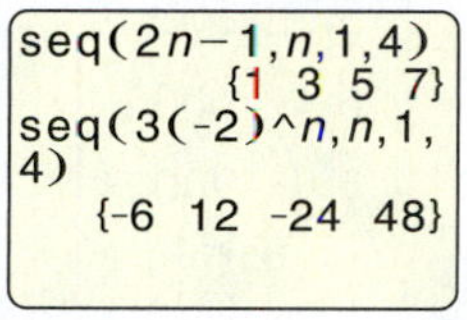

```
seq(2n-1,n,1,4)
           {1 3 5 7}
seq(3(-2)^n,n,1,
4)
      {-6 12 -24 48}
```

```
seq(n/(n+1),n,1,
4)▶Frac
{1/2 2/3 3/4 4/...
```

Figure 11.3 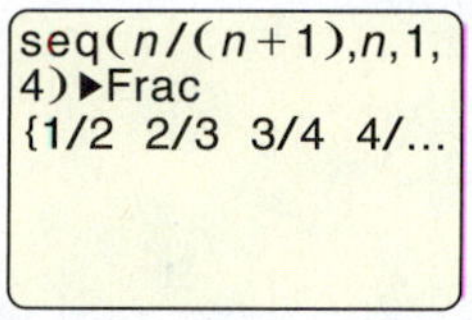   **Figure 11.4**

Representations of Sequences

Because sequences are functions, they can be represented symbolically, graphically, and numerically. The next two examples illustrate such representations.

EXAMPLE 2 **Using a graphical representation**

Use Figure 11.5 to write the terms of the sequence.

Finite Sequence

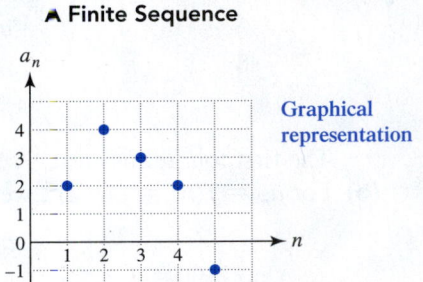

Graphical representation

Figure 11.5

Solution

The points $(1, 2)$, $(2, 4)$, $(3, 3)$, $(4, 2)$, and $(5, -1)$ are shown in the graph. The terms of the sequence are $2, 4, 3, 2$, and -1.

Now Try Exercise 29

EXAMPLE 3 **Representing a sequence**

In 2011 the average person in the United States used 100 gallons of water at home each day. Give symbolic, numerical, and graphical representations for a sequence that models the total amount of water used over a 7-day period. (*Source:* U.S. Geological Survey.)

Solution

Symbolic Representation Let

$$a_n = 100n \text{ for } n = 1, 2, 3, \ldots, 7. \qquad \text{Symbolic representation}$$

Numerical Representation Table 11.2 contains the sequence.

Graphical Representation Plot the points $(1, 100)$, $(2, 200)$, $(3, 300)$, $(4, 400)$, $(5, 500)$, $(6, 600)$, and $(7, 700)$, as shown in Figure 11.6.

TABLE 11.2

Numerical representation

n	a_n
1	100
2	200
3	300
4	400
5	500
6	600
7	700

Personal Water Usage

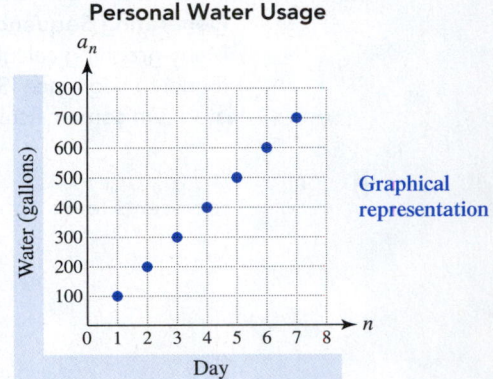

Graphical representation

Figure 11.6

Now Try Exercise 39

Models and Applications

▶ **REAL-WORLD CONNECTION** A population model for a species of insect with a life span of 1 year can be described with a sequence. Suppose that each adult female insect produces, on average, r female offspring that survive to reproduce the following year. Let a_n represent the female insect population at the beginning of year n. Then the number of female insects is given by

$$a_n = Cr^{n-1},$$

where C is the initial population of female insects. (*Source:* D. Brown and P. Rothery, *Models in Biology*.)

EXAMPLE 4 **Modeling numbers of insects**

Suppose that the initial population of adult female insects is 500 per acre and that $r = 1.04$. Then the average number of female insects per acre at the beginning of year n is described by

$$a_n = 500(1.04)^{n-1}.$$

Represent the female insect population numerically and graphically for 7 years. Round values to the nearest whole number. Discuss the results. By what percent is the population increasing each year?

Solution

Numerical Representation Table 11.3 contains *approximations* for the first 7 terms of the sequence. The insect population increases from 500 to about 633 insects per acre during this time period.

TABLE 11.3 An Insect Population (per acre)

n	1	2	3	4	5	6	7
a_n	500	520	541	562	585	608	633

Graphical Representation Plot the points $(1, 500)$, $(2, 520)$, $(3, 541)$, $(4, 562)$, $(5, 585)$, $(6, 608)$, and $(7, 633)$, as shown in Figure 11.7. These results indicate that the insect population gradually increases. Because the growth factor is 1.04, the population is increasing by 4% each year.

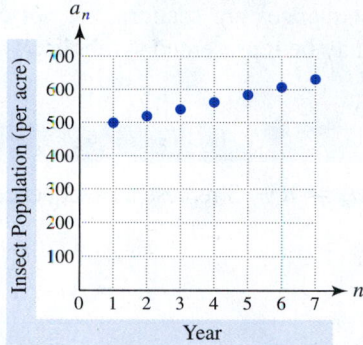

Figure 11.7

Now Try Exercise 45

CRITICAL THINKING

Explain how the value of r in Example 4 affects the population of female insects over time. Assume that $r > 0$.

TECHNOLOGY NOTE

Graphs and Tables of Sequences
In the sequence mode, many graphing calculators are capable of representing sequences graphically and numerically. Figure 11.8(a) shows how to enter the sequence from Example 4 to produce the table of values shown in Figure 11.8(b).

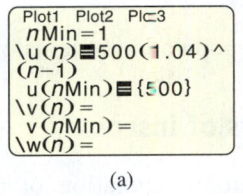

(a)

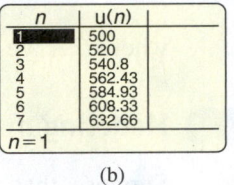

(b)

Figure 11.8

Figures 11.9(a) and (b) show the set-up for graphing the sequence from Example 4 to produce the graph shown in Figure 11.9(c).

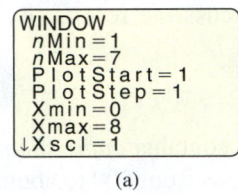

(a)

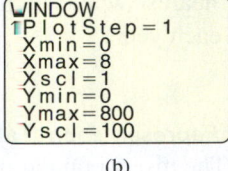

(b)

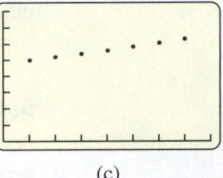

(c)

Figure 11.9

11.1 Putting It All Together

An infinite sequence is a function whose domain is the set of natural numbers. A finite sequence has the domain $D = \{1, 2, 3, \ldots, n\}$ for some fixed natural number n. Graphs of sequences are scatterplots, *not* continuous lines and curves. Sequences are functions that may be represented symbolically, numerically, and graphically.

REPRESENTATION	EXAMPLE
Symbolic	$a_n = n - 3$ represents a sequence. The first four terms of the sequence are $-2, -1, 0,$ and 1: $$a_1 = \mathbf{1} - 3 = -2, \qquad a_2 = \mathbf{2} - 3 = -1,$$ $$a_3 = \mathbf{3} - 3 = 0, \qquad a_4 = \mathbf{4} - 3 = 1.$$
Numerical	A numerical representation for $a_n = n - 3$ with $n = 1, 2, 3,$ and 4 is shown in the table.

n	1	2	3	4
a_n	-2	-1	0	1

REPRESENTATION	EXAMPLE
Graphical	For a graphical representation of the first four terms of $a_n = n - 3$, the points $(1, -2), (2, -1), (3, 0),$ and $(4, 1)$ from the previous table are plotted. 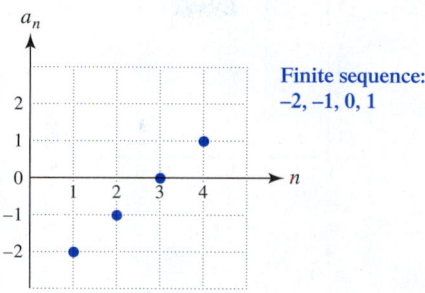 Finite sequence: $-2, -1, 0, 1$

11.1 Exercises

MyMathLab

CONCEPTS AND VOCABULARY

1. Give an example of a finite sequence.

2. Give an example of an infinite sequence.

3. An infinite sequence is a(n) _____ whose domain is the set of _____.

4. An ordered list is a(n) _____.

5. The third term in the sequence $4, -5, 6, -7, 8$ is _____.

6. The graph of a sequence is not a continuous graph but rather a(n) _____.

7. If $f(n)$ represents a sequence, the second term of the sequence is given by _____.

8. If a_n represents a sequence, the fourth term of the sequence is given by _____.

EVALUATING AND REPRESENTING SEQUENCES

Exercises 9–16: Write the first four terms of the sequence for $n = 1, 2, 3,$ and 4.

9. $f(n) = n^2$

10. $f(n) = 3n + 4$

11. $f(n) = \dfrac{1}{n + 5}$

12. $f(n) = 3^n$

13. $f(n) = 5\left(\dfrac{1}{2}\right)^n$

14. $f(n) = n^2 + 2n$

15. $f(n) = 9$

16. $f(n) = (-1)^n$

Exercises 17–24: Write the first three terms of the sequence for $n = 1, 2,$ and 3.

17. $a_n = n^3$

18. $a_n = 5 - n$

19. $a_n = \dfrac{4n}{3 + n}$

20. $a_n = 3^{-n}$

21. $a_n = 2n^2 + n - 1$

22. $a_n = n^4 - 1$

23. $a_n = -2$

24. $a_n = n^n$

Exercises 25 and 26: **Thinking Generally** *Let b and c be fixed numbers (constants). Find a_1 and a_2.*

25. $a_n = bn + c$

26. $a_n = \dfrac{n + b}{n - c}, c \neq 1$ and $c \neq 2$

Exercises 27 and 28: Use the numerical representation to evaluate $\frac{1}{2}(a_1 + a_4)$.

27.

n	1	2	3	4	5
a_n	10	8	6	4	2

28.

n	1	2	3	4	5
a_n	-5	0	10	30	60

Exercises 29–32: Write the terms of the sequence.

29. a_n

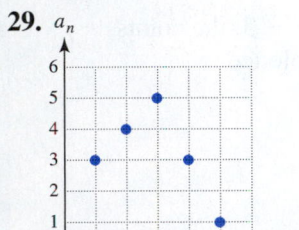

30. a_n

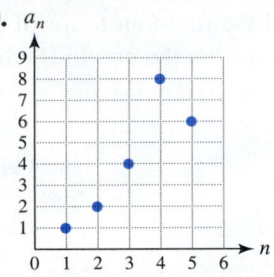

31. a_n

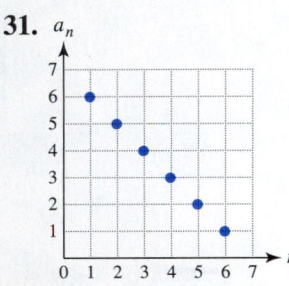

32. a_n

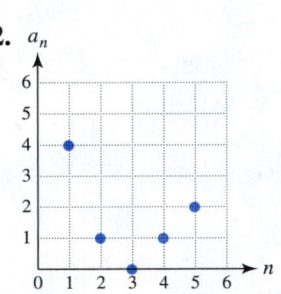

Exercises 33–38: Represent the first seven terms of the sequence numerically and graphically.

33. $a_n = n + 1$

34. $a_n = \frac{1}{2}n - \frac{1}{2}$

35. $a_n = n^2 - n$

36. $a_n = \frac{1}{2}n^2$

37. $a_n = 2^n$

38. $a_n = 2(0.5)^n$

APPLICATIONS

39. *Solid Waste* On average, each U.S. resident in 2000 generated about 30 pounds of solid waste per week. Give symbolic, numerical, and graphical representations for a sequence that models the total amount of waste produced over a 7-week period. (*Source:* Environmental Protection Agency.)

40. *Carbon Dioxide Emitters* Because people burn fossil fuels, the United States emits about 5.8 billion metric tons of carbon dioxide per year. (A metric ton is about 2200 pounds.) Give symbolic, numerical, and graphical representations for a sequence that models the total amount of carbon dioxide emitted in billions of metric tons in the United States during a 5-year period. (*Source:* Energy Information Administration.)

41. *Geometry* The lengths of the sides of a sequence of squares are given by 1, 2, 3, and 4, as shown in the figure at the top of the next column. Write sequences that give

(a) the areas of the squares and

(b) the perimeters of the squares.

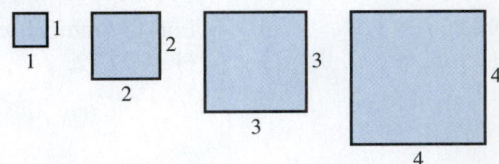

42. *Salaries* An individual's starting salary is $50,000, and the individual receives an increase of 8% per year. Give symbolic, numerical, and graphical representations for this person's salary over 5 years.

43. *Depreciation* Automobiles usually depreciate in value over time. Often, a newer automobile may be worth only 80% of its previous year's value. Suppose that a car is worth $25,000 new.

(a) How much is it worth after 1 year? After 2 years?

(b) Write a formula for a sequence that gives the car's value after n years.

(c) Make a table that shows how much the car was worth each year during the first 7 years.

44. *Falling Object* The distance d that an object falls during *consecutive* seconds is shown in the table. For example, during the third second (from $n = 2$ to $n = 3$) an object falls a distance of 80 feet.

n (seconds)	1	2	3	4	5
d (feet)	16	48	80	112	144

(a) Find values for c and b so that $d = cn + b$ models these data.

(b) How far does an object fall during the sixth second?

45. *Modeling Insect Populations* (Refer to Example 4.) Suppose that the initial population of insects is 2048 per acre and that $r = 0.5$. Use a sequence to represent the insect population over a 7-year period

(a) symbolically,

(b) numerically, and

(c) graphically.

46. *Auditorium Seating* An auditorium has 50 seats in the first row, 55 seats in the second row, 60 seats in the third row, and so on.

(a) Make a table that shows the number of seats in the first seven rows.

(b) Write a formula that gives the number of seats in row n.

(c) How many seats are there in row 23?

(d) Graph the number of seats in each row for $n = 1, 2, 3, \ldots, 10$.

47. Compare the graph of the function $f(x) = 2x + 1$, where x is a real number, with the graph of the sequence $f(n) = 2n + 1$, where n is a natural number.

48. Explain what a sequence is. Describe the difference between a finite and an infinite sequence.

11.2 Arithmetic and Geometric Sequences

Representations of Arithmetic Sequences • Representations of Geometric Sequences • Applications and Models

A LOOK INTO MATH ▶

Indoor air pollution has become more hazardous as people spend 80% to 90% of their time in tightly sealed, energy-efficient buildings, which often lack proper ventilation. Many contaminants such as tobacco smoke, formaldehyde, radon, lead, and carbon monoxide are often allowed to increase to unsafe levels. One way to alleviate this problem is to use efficient ventilation systems. Mathematics plays an important role in determining the proper amount of ventilation. In this section we use sequences to model ventilation in classrooms. (See Example 7.) Before implementing this model we discuss the basic concepts relating to two special types of sequences.

Representations of Arithmetic Sequences

If a sequence is defined by a linear function, it is an *arithmetic sequence*. For example,

$$f(n) = 2n - 3$$

NEW VOCABULARY

☐ Arithmetic sequence
☐ Common difference
☐ Geometric sequence
☐ Common ratio

represents an arithmetic sequence because $f(x) = 2x - 3$ defines a linear function. The first five terms of this sequence, $-1, 1, 3, 5, 7$, are shown in Table 11.4.

Each time n increases by **1**, the next term is **2** more than the previous term. We say that the *common difference* of this arithmetic sequence is $d = 2$. That is, the difference between successive terms equals 2. When the points associated with these terms are graphed, they lie on the line $y = 2x - 3$, as illustrated in Figure 11.10. Arithmetic sequences are represented by linear functions, so their graphical representations consist of collinear points (points that lie on a line). The slope m of the line equals the common difference d.

TABLE 11.4

n	$f(n)$
1	-1
2	1
3	3
4	5
5	7

$+1$ between rows on the left; $+2$ between rows on the right.

Common difference is 2.

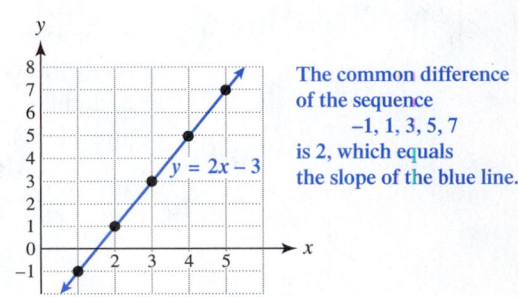

The common difference of the sequence $-1, 1, 3, 5, 7$ is 2, which equals the slope of the blue line.

Figure 11.10

READING CHECK

How can you recognize an arithmetic sequence?

ARITHMETIC SEQUENCE

An **arithmetic sequence** is a linear function given by $a_n = dn + c$ whose domain is the set of natural numbers. The value of d is called the **common difference**.

NOTE: If each term after the first term is obtained by adding a fixed number to the previous term, then the sequence is an arithmetic sequence. The fixed number is the *common difference*. For example, 1, 6, 11, 16, 21, . . . is an arithmetic sequence because each term (after the first) is found by adding the common difference of **5** to the previous term. That is, $6 = 1 + \mathbf{5}$, $11 = 6 + \mathbf{5}$, $16 = 11 + \mathbf{5}$, and so on.

EXAMPLE 1 **Recognizing arithmetic sequences**

Determine whether f is an arithmetic sequence. If it is, identify the common difference d.

(a) $f(n) = 2 - 3n$

(b)

n	$f(n)$
1	10
2	5
3	0
4	−5
5	−10

(c) A graph of f is shown in Figure 11.11.

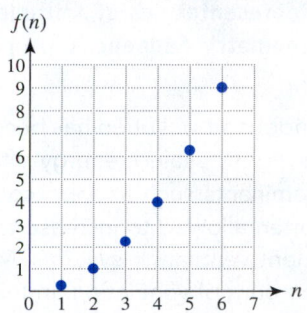

Figure 11.11

Solution

(a) This sequence is arithmetic because $f(x) = -3x + 2$ defines a linear function. The common difference is $d = -3$.

(b) The table reveals that each term is found by adding -5 to the previous term. This represents an arithmetic sequence with common difference $d = -5$.

(c) The sequence shown in Figure 11.11 is not an arithmetic sequence because the points are not collinear. That is, the points do not lie on a line and there is no common difference.

▪ **Now Try Exercises 11, 17, 25**

MAKING CONNECTIONS

Common Difference and Slope

The common difference d of an arithmetic sequence equals the slope of the line passing through the collinear points. For example, if $a_n = -2n + 4$, the common difference is -2, and the slope of the line passing through the points on the graph of a_n is also -2 (see Figure 11.12).

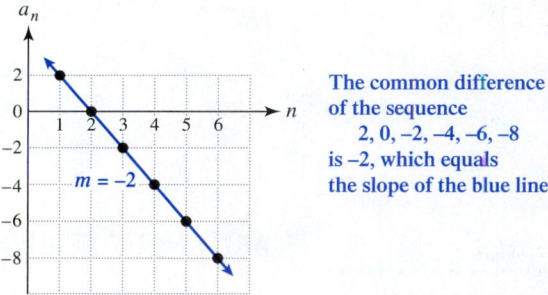

The common difference of the sequence
2, 0, −2, −4, −6, −8
is −2, which equals the slope of the blue line.

Figure 11.12

EXAMPLE 2 | **Finding symbolic representations**

Find the general term a_n for each arithmetic sequence.
(a) $a_1 = 3$ and $d = 4$ **(b)** $a_1 = 3$ and $a_4 = 12$

Solution

(a) Let $a_n = dn + c$. For $d = 4$, we write $a_n = 4n + c$, and to find c we use $a_1 = 3$.

$$a_1 = 4(1) + c = 3 \quad \text{or} \quad c = -1$$

Thus $a_n = 4n - 1$.

(b) Because $a_1 = 3$ and $a_4 = 12$, the common difference d equals the slope of the line passing through the points $(1, 3)$ and $(4, 12)$, or

$$d = \frac{12 - 3}{4 - 1} = 3.$$

Therefore $a_n = 3n + c$. To find c we use $a_1 = 3$ and obtain

$$a_1 = 3(1) + c = 3 \quad \text{or} \quad c = 0.$$

Thus $a_n = 3n$.

■ **Now Try Exercises 29, 31**

Consider the arithmetic sequence

$$1, 5, 9, 13, 17, 21, 25, 29, \ldots. \qquad \text{Common difference is 4.}$$

The common difference is $d = 4$, and the first term is $a_1 = 1$. To find the second term we add d to the first term. To find the third term we add $2d$ to the first term, and to find the fourth term we add $3d$ to the first term. That is,

$$a_1 = 1,$$
$$a_2 = a_1 + 1d = 1 + 1 \cdot 4 = 5,$$
$$a_3 = a_1 + 2d = 1 + 2 \cdot 4 = 9,$$
$$a_4 = a_1 + 3d = 1 + 3 \cdot 4 = 13,$$

First four terms

and, in general, a_n is determined by

$$a_n = a_1 + (n-1)d = 1 + (n-1)4.$$

This result suggests the following formula.

READING CHECK

Explain how to write the general term of an arithmetic sequence, given the first term and the common difference.

GENERAL TERM OF AN ARITHMETIC SEQUENCE

The nth term a_n of an arithmetic sequence is given by

$$a_n = a_1 + (n-1)d,$$

where a_1 is the first term and d is the common difference.

EXAMPLE 3 | **Finding terms of an arithmetic sequence**

If $a_1 = 5$ and $d = 3$, find a_{54}.

Solution

To find a_{54}, apply the formula $a_n = a_1 + (n-1)d$ with $a_1 = 5$, $n = 54$, and $d = 3$.

$$a_{54} = 5 + (54 - 1)3 = 164$$

■ **Now Try Exercise 35**

Representations of Geometric Sequences

If a sequence is defined by an exponential function, it is a *geometric sequence*. For example,

$$f(n) = 3(2)^{n-1}$$

represents a geometric sequence because $f(x) = 3(2)^{x-1}$ defines an exponential function. The first five terms of this sequence are shown in Table 11.5.

Geometric Sequence:
3, 6, 12, 24, 48

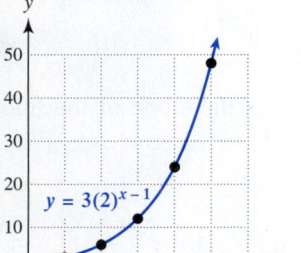

$y = 3(2)^{x-1}$

Figure 11.13

TABLE 11.5

n	1	2	3	4	5
$f(n)$	3	6	12	24	48

+1 +1 +1 +1

×2 ×2 ×2 ×2

Common ratio equals 2.

Successive terms are found by multiplying the previous term by 2. We say that the *common ratio* of this geometric sequence equals 2. Note that the ratios of successive terms are $\frac{6}{3}, \frac{12}{6}, \frac{24}{12}$, and $\frac{48}{24}$, and that they all equal the common ratio 2. When the points associated with the terms in Table 11.5 are graphed, they do *not* lie on a line. Rather, they lie on the exponential curve $y = 3(2)^{x-1}$, as shown in Figure 11.13. A geometric sequence with a positive common ratio is an exponential function whose domain is the set of natural numbers. Its terms reflect either *exponential growth* or *exponential decay*.

READING CHECK

How can you recognize a geometric sequence?

GEOMETRIC SEQUENCE

A **geometric sequence** is given by $a_n = a_1(r)^{n-1}$, where n is a natural number and $r \neq 0$ or 1. The value of r is called the **common ratio**, and a_1 is the first term of the sequence.

NOTE: If each term after the first term is obtained by multiplying the previous term by a fixed number, the sequence is a geometric sequence. The fixed number can be either positive or negative, but it cannot be 0 or 1.

EXAMPLE 4 **Recognizing geometric sequences**

Determine whether f is a geometric sequence. If it is, identify the common ratio.
(a) $f(n) = 2(0.9)^{n-1}$ **(c)** A graph of f is shown in Figure 11.14.

(b)

n	$f(n)$
1	8
2	4
3	2
4	1
5	$\frac{1}{2}$

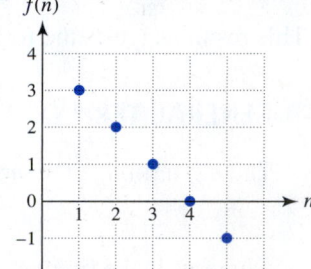

Figure 11.14

Solution
(a) This sequence is geometric because $f(x) = 2(0.9)^{x-1}$ defines an exponential function. The common ratio is $r = 0.9$.
(b) The table shows that each successive term is half the previous term. This sequence represents a geometric sequence with a common ratio of $r = \frac{1}{2}$.
(c) The sequence shown in Figure 11.14 is not a geometric sequence because the points are collinear. There is no common ratio.

Now Try Exercises 41, 45, 53

Common Ratios and Growth or Decay Factors

If the common ratio r of a geometric sequence is positive, then r equals either the growth factor or the decay factor for an exponential function.

EXAMPLE 5 | **Finding symbolic representations**

Find a general term a_n for each geometric sequence.
(a) $a_1 = \frac{1}{2}$ and $r = 5$
(b) $a_1 = 2$, $a_3 = 18$, and $r < 0$.

Solution
(a) Let $a_n = a_1(r)^{n-1}$. Because $a_1 = \frac{1}{2}$ and $r = 5$, we can write $a_n = \frac{1}{2}(5)^{n-1}$.
(b) $a_1 = 2$ and $a_3 = 18$, so

$$a_3 = a_1(r)^{3-1}$$
$$18 = 2(r)^2.$$

This equation simplifies to

$$r^2 = 9 \quad \text{or} \quad r = \pm 3.$$

It is specified that $r < 0$, so $r = -3$ and $a_n = 2(-3)^{n-1}$. Note that the common ratio r can be *negative*.

▪ **Now Try Exercises 55, 59**

CRITICAL THINKING

If we are given a_1 and a_5, can we determine the common ratio of a geometric series? Explain.

EXAMPLE 6 | **Finding a term of a geometric sequence**

If $a_1 = 5$ and $r = 3$, find a_{10}.

Solution
To find a_{10}, apply the formula $a_n = a_1(r)^{n-1}$ with $a_1 = 5$, $r = 3$, and $n = 10$.
$$a_{10} = 5(3)^{10-1} = 5(3)^9 = 98,415$$

▪ **Now Try Exercise 61**

Applications and Models

▶ **REAL-WORLD CONNECTION** Sequences are frequently used to describe a variety of situations. In the next example, we use a sequence to model classroom ventilation.

EXAMPLE 7 | **Modeling classroom ventilation**

Ventilation is an effective means for removing indoor air pollutants. According to the American Society of Heating, Refrigerating, and Air-Conditioning Engineers (ASHRAE), a classroom should have a ventilation rate of 900 cubic feet per hour per person.
(a) Write a sequence that gives the hourly ventilation necessary for 1, 2, 3, 4, and 5 people in a classroom. Is this sequence arithmetic, geometric, or neither?

(b) Write the general term for this sequence. Why is it reasonable to limit the domain to natural numbers?

(c) Find a_{30} and interpret the result.

Solution

(a) One person requires 900 cubic feet of air circulated per hour, two people require 1800, three people 2700, and so on. The first five terms of this sequence are

$$900, 1800, 2700, 3600, 4500.$$

This sequence is arithmetic, with a common difference of 900.

(b) The nth term equals $900n$, so we let $a_n = 900n$. Because we cannot have a fraction of a person, limiting the domain to the natural numbers is reasonable.

(c) The result $a_{30} = 900(30) = \textbf{27,000}$ indicates that a classroom with **30** people should have a ventilation rate of **27,000** cubic feet per hour.

▮ **Now Try Exercise 65**

▶ **REAL-WORLD CONNECTION** Chlorine is frequently added to the water to disinfect swimming pools. The chlorine concentration should remain between 1.5 and 2.5 parts per million (ppm). On a warm, sunny day 30% of the chlorine may dissipate from the water. In the next example we use a sequence to model the amount of chlorine in a pool at the beginning of each day.   (*Source:* D. Thomas, *Swimming Pool Operator's Handbook*.)

EXAMPLE 8   **Modeling chlorine in a swimming pool**

A swimming pool on a warm, sunny day begins with a high chlorine content of 4 parts per million. (Assume that each day 30% of the chlorine dissipates.)

(a) Write a sequence that models the amount of chlorine in the pool at the beginning of the first 3 days, assuming that no additional chlorine is added and that the days are warm and sunny. Is this sequence arithmetic, geometric, or neither?

(b) Write the general term for this sequence.

(c) At the beginning of what day does the chlorine first drop below 1.5 parts per million?

Solution

(a) Because 30% of the chlorine dissipates, 70% remains in the water at the beginning of the next day. If the concentration at the beginning of the first day is 4 parts per million, then at the beginning of the second day it is

$$4 \cdot 0.70 = 2.8 \text{ parts per million,}$$

and at the start of the third day it is

$$2.8 \cdot 0.70 = 1.96 \text{ parts per million.}$$

The first three terms are 4, 2.8, 1.96. Successive terms are found by multiplying the previous term by 0.7. Thus the sequence is geometric, with common ratio 0.7.

(b) The initial amount is $a_1 = \textbf{4}$ and the common ratio is $r = \textbf{0.7}$, so the sequence can be represented by $a_n = \textbf{4}(\textbf{0.7})^{n-1}$.

(c) The table shown in Figure 11.15 reveals that $a_4 = 4(0.7)^{4-1} \approx 1.372 < 1.5$. Thus, at the beginning of the fourth day, the chlorine level in the swimming pool drops below the recommended minimum of 1.5 parts per million.

n	$u(n)$	
1	4	
2	2.8	
3	1.96	
4	1.372	
5	.9604	
6	.67228	
7	.4706	

$u(n) \blacksquare 4(.7)^\wedge(n-1)$

Figure 11.15

▮ **Now Try Exercise 67**

11.2   Putting It All Together

In this section we discussed two types of sequences: arithmetic and geometric. Arithmetic sequences are linear functions, and geometric sequences with *positive r* are exponential functions. The inputs for both are limited to the natural numbers. The graph of an arithmetic sequence consists of points that lie on a line, whereas the graph of a geometric sequence (with a positive *r*) consists of points that lie on an exponential curve.

SEQUENCE	FORMULA	EXAMPLE
Arithmetic	$a_n = dn + c$ or $a_n = a_1 + (n-1)d$, where d is the common difference and a_1 is the first term.	If $a_n = 5n + 2$, then the common difference is $d = 5$ and the terms of the sequence are $$7, 12, 17, 22, 27, 32, 37, \ldots.$$ Each term after the first is found by adding 5 to the previous term. The general term can be written as $$a_n = 7 + 5(n-1).$$
Geometric	$a_n = a_1(r)^{n-1}$, where r is the common ratio ($r \neq 0$, $r \neq 1$) and a_1 is the first term.	If $a_n = 4(-2)^{n-1}$, then the common ratio is $r = -2$ and the first term is $a_1 = 4$. The terms of the sequence are $$4, -8, 16, -32, 64, -128, 256, \ldots.$$ Each term after the first is found by multiplying the previous term by -2.

11.2   Exercises

CONCEPTS AND VOCABULARY

1. An arithmetic sequence is a(n) <u>linear/exponential</u> function.

2. A geometric sequence with $r > 0$ is a(n) <u>linear/exponential</u> function.

3. Give an example of an arithmetic sequence. State the common difference.

4. Give an example of a geometric sequence. State the common ratio.

5. To find successive terms in an arithmetic sequence, _____ the common difference to the _____ term.

6. To find successive terms in a geometric sequence, _____ the previous term by the _____.

7. Find the next term in the arithmetic sequence given by 3, 7, 11, 15. What is the common difference?

8. Find the next term in the geometric sequence given by 2, −4, 8, −16. What is the common ratio?

9. Write the general term a_n for a geometric sequence, using a_1 and r.

10. Write the general term a_n for an arithmetic sequence, using a_1 and d.

ARITHMETIC SEQUENCES

Exercises 11–28: (Refer to Example 1.) Determine whether f is an arithmetic sequence. Identify the common difference when possible.

11. $f(n) = 10n - 5$ 12. $f(n) = -3n - 5$

13. $f(n) = 6 - n$ 14. $f(n) = 6 + \frac{1}{2}n$

15. $f(n) = n^3 + 1$

16. $f(n) = 5\left(\frac{1}{3}\right)^{n-1}$

17.

n	1	2	3	4
$f(n)$	3	6	9	12

18.

n	1	2	3	4
$f(n)$	-7	-5	-3	-1

19.

n	1	2	3	4
$f(n)$	10	7	4	1

20.

n	1	2	3	4
$f(n)$	1	2	4	8

21.

n	1	2	3	4
$f(n)$	-4	0	8	12

22.

n	1	2	3	4
$f(n)$	1	2.5	4	5.5

23. $f(n)$

24. $f(n)$

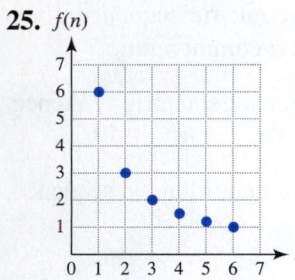

25. $f(n)$

26. $f(n)$

27. $f(n)$

28. $f(n)$

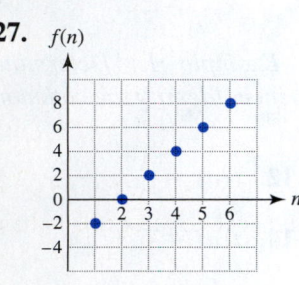
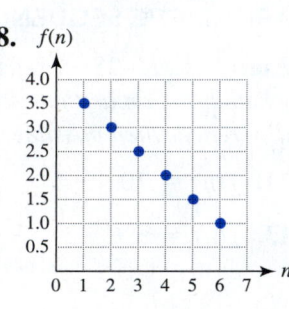

Exercises 29–34: (Refer to Example 2.) Find the general term a_n for the arithmetic sequence.

29. $a_1 = 7$ and $d = -2$

30. $a_1 = 5$ and $a_2 = 9$

31. $a_1 = -2$ and $a_3 = 6$

32. $a_2 = 7$ and $a_3 = 10$

33. $a_8 = 16$ and $a_{12} = 8$

34. $a_3 = 7$ and $d = -5$

Exercises 35–38: (Refer to Example 3.)

35. If $a_1 = -3$ and $d = 2$, find a_{32}.

36. If $a_1 = 2$ and $d = -3$, find a_{19}.

37. If $a_1 = -3$ and $a_2 = 0$, find a_9.

38. If $a_3 = -3$ and $d = 4$, find a_{62}.

GEOMETRIC SEQUENCES

Exercises 39–54: (Refer to Example 4.) Determine whether f is a geometric sequence. Identify the common ratio when possible.

39. $f(n) = 3^n$

40. $f(n) = 2(4)^n$

41. $f(n) = \frac{2}{3}(0.8)^{n-1}$

42. $f(n) = 7 - 3n$

43. $f(n) = 2(n-1)^2$

44. $f(n) = 2\left(-\frac{3}{4}\right)^{n-1}$

45.

n	1	2	3	4
$f(n)$	2	4	8	16

46.

n	1	2	3	4
$f(n)$	-6	3	-1.5	0.75

47.

n	1	2	3	4
$f(n)$	1	4	9	16

48.

n	1	2	3	4
$f(n)$	7	4	-1	-8

49.

n	1	2	3	4
$f(n)$	2	8	32	128

50.

n	1	2	3	4
$f(n)$	1	$\frac{1}{2}$	$\frac{1}{4}$	$\frac{1}{8}$

51. $f(n)$

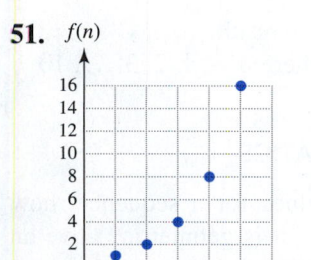

52. $f(n)$

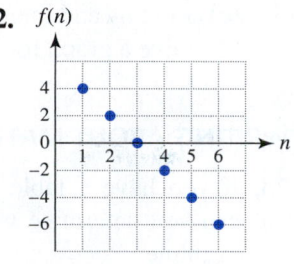

53. $f(n)$

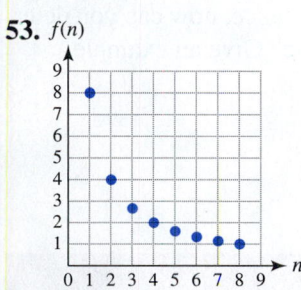

54. $f(n)$

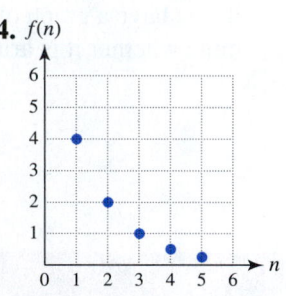

Exercises 55–60: (Refer to Example 5.) Find the general term a_n for the geometric sequence.

55. $a_1 = 1.5$ and $r = 4$

56. $a_1 = 3$ and $r = \frac{1}{4}$

57. $a_1 = -3$ and $a_2 = 6$

58. $a_1 = 2$ and $a_4 = 54$

59. $a_1 = 1$, $a_3 = 16$, and $r > 0$

60. $a_2 = 3$, $a_4 = 12$, and $r < 0$

Exercises 61–64: (Refer to Example 6.)

61. If $a_1 = 2$ and $r = 3$, find a_8.

62. If $a_1 = 4$ and $a_2 = 2$, find a_9.

63. If $a_1 = -1$ and $a_2 = 3$, find a_6.

64. If $a_3 = 5$ and $r = -3$, find a_7.

APPLICATIONS

65. *Room Ventilation* (Refer to Example 7.) In areas such as bars and lounges that sometimes allow smoking, the ventilation rate should be 3000 cubic feet per hour per person. *(Source: ASHRAE.)*
 (a) Write a sequence that gives the hourly ventilation necessary for 1, 2, 3, 4, and 5 people in a barroom. Is this sequence arithmetic, geometric, or neither?
 (b) Write the general term for this sequence.
 (c) Find a_{20} and interpret the result.
 (d) Give a graphical representation for eight terms of this sequence, using $n = 1, 2, 3, \ldots, 8$. Are the points collinear?

66. *Salary* Suppose that an employee receives a $2000 raise each year and that the sequence a_n models the employee's salary after n years. Is this sequence arithmetic, geometric, or neither? Explain.

67. *Chlorine in Swimming Pools* (Refer to Example 8.) Suppose that the water in a swimming pool initially has a chlorine content of 3 parts per million and that 20% of the chlorine dissipates each day.
 (a) If no additional chlorine is added, write the general term for a sequence that gives the chlorine concentration at the beginning of each day.
 (b) Give a graphical representation for this sequence, using $n = 1, 2, 3, \ldots, 8$. Are the points collinear? Is this sequence arithmetic, geometric, or neither?

68. *Salary* Suppose that an employee receives a 7% increase in salary each year and that the sequence a_n models the employee's salary after n years. Is this sequence arithmetic, geometric, or neither? Explain.

69. *Bouncing Ball* A tennis ball bounces back to 85% of the height from which it was dropped and then to 85% of the height of each successive bounce.
 (a) Write the general term for a sequence a_n that gives the maximum height of the ball on the nth bounce. Let $a_1 = 5$ feet.
 (b) Is the sequence arithmetic or geometric? Explain.
 (c) Find a_8 and interpret the result.

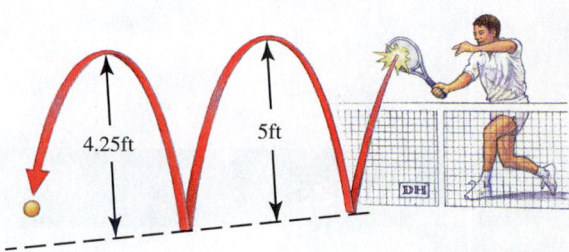

70. *Falling Object* The total distance D_n that an object falls in n seconds is shown in the table. Is the sequence arithmetic, geometric, or neither? Explain your reasoning.

n (seconds)	1	2	3	4	5
D_n (feet)	16	64	144	256	400

71. *Theater Seating* A theater has 40 seats in the first row, 42 seats in the second row, 44 seats in the third row, and so on.
 (a) Can the number of seats in each row be modeled by an arithmetic or geometric sequence? Explain.

(b) Write the general term for a sequence a_n that gives the number of seats in row n.

(c) How many seats are there in row 20?

72. *Appreciation of Lake Property* A certain type of lake property in northern Minnesota is increasing in value by 15% per year. Let the sequence a_n give the value of this type of lake property at the beginning of year n.

(a) Is a_n arithmetic, geometric, or neither? Explain your reasoning.

(b) Write the general term a_n for this sequence if $a_1 = \$100{,}000$.

(c) Find a_7 and interpret the result.

(d) Give a graph for a_n, where $n = 1, 2, 3, \ldots, 10$.

WRITING ABOUT MATHEMATICS

73. If you have a table of values for a sequence, how can you determine whether it is geometric? Give an example.

74. If you have a graph of a sequence, how can you determine whether it is arithmetic? Give an example.

SECTIONS 11.1 and 11.2   Checking Basic Concepts

1. Write the first four terms of the sequence defined by $a_n = \dfrac{n}{n+4}$.

2. Represent the sequence $a_n = n + 1$ numerically and graphically for $n = 1, 2, 3, 4, 5$.

3. Use the table to determine whether the sequence is arithmetic or geometric. Write the general term for the sequence.

(a)

n	1	2	3	4	5
a_n	−2	1	4	7	10

(b)

n	1	2	3	4	5
a_n	3	−6	12	−24	48

4. Find the general term a_n for an arithmetic sequence with $a_1 = 5$ and $d = 2$.

5. Determine the general term a_n for a geometric sequence with $a_1 = 5$ and $r = 2$.

11.3 Series

Basic Concepts • Arithmetic Series • Geometric Series • Summation Notation

A LOOK INTO MATH ▶

Although the terms *sequence* and *series* are sometimes used interchangeably in everyday life, they represent different mathematical concepts. In mathematics a sequence is a function whose domain is the set of natural numbers, whereas a series is a summation of the terms in a sequence. Series have played a central role in the development of modern mathematics. Today series are often used to approximate functions that are too complicated to have formulas. Series are also used to calculate accurate approximations for π and e.

Basic Concepts

Suppose that a person has a starting salary of \$30,000 per year and receives a \$2000 raise each year. Then the *sequence*

$$\underbrace{30{,}000,\ 32{,}000,\ 34{,}000,\ 36{,}000,\ 38{,}000}_{\text{Sequence}}$$

lists these salaries over a 5-year period. The total amount earned is given by the *series*

$$30{,}000 + 32{,}000 + 34{,}000 + 36{,}000 + 38{,}000,$$

Series

whose sum is $170,000. We now define the concept of a series.

FINITE SERIES

A **finite series** is an expression of the form

$$a_1 + a_2 + a_3 + \cdots + a_n.$$

EXAMPLE 1   **MySpace U.S. advertising revenues**

Table 11.6 presents a sequence a_n that gives MySpace U.S. advertising revenues in millions of dollars n years after 2006. For example, $a_3 = 450$ indicates that in 2009, advertising revenues were $450 million.

TABLE 11.6   MySpace Advertising

n	1	2	3	4	5
a_n	480	595	450	290	190

Source: eMarketer (via WSJ), March 2011.

(a) Write a series whose sum represents the total U.S. advertising revenues for MySpace from 2007 to 2011.
(b) Interpret $a_1 + a_2 + a_3 + a_4 + a_5$.

Solution
(a) The required series and sum are given by

$$480 + 595 + 450 + 290 + 190 = 2005.$$

Revenues from 2007 to 2011 were $2005 million, or $2.005 billion.
(b) This series represents total revenues for 5 years from 2007 to 2011.

▌ **Now Try Exercise 41**

MAKING CONNECTIONS

Sequences and Series

A *sequence* is an *ordered list*; a *series* is the *sum of the terms of a sequence*. For example, the even integers from 2 to 20 are represented by the sequence

$$2, 4, 6, 8, 10, 12, 14, 16, 18, 20.$$ Sequence

The corresponding series is

$$2 + 4 + 6 + 8 + 10 + 12 + 14 + 16 + 18 + 20,$$ Series

which sums to 110.

READING CHECK

How do you distinguish between a sequence and a series?

Arithmetic Series

Summing the terms of an arithmetic sequence results in an **arithmetic series**. For example, $a_n = 2n - 1$ for $n = 1, 2, 3, \ldots, 7$ defines the arithmetic sequence

$$1, 3, 5, 7, 9, 11, 13.$$

Arithmetic sequence

The corresponding arithmetic *series* is

$$1 + 3 + 5 + 7 + 9 + 11 + 13,$$

Arithmetic series

whose sum is 49. The following formula gives the sum of the first n terms of an arithmetic sequence. Note that the sum of the first n terms of a sequence is a finite series.

SUM OF THE FIRST n TERMS OF AN ARITHMETIC SEQUENCE

The **sum of the first n terms of an arithmetic sequence**, denoted S_n, is found by averaging the first and nth terms and then multiplying by n. That is,

$$S_n = a_1 + a_2 + a_3 + \cdots + a_n = n\left(\frac{a_1 + a_n}{2}\right).$$

READING CHECK

How do you find the sum of an arithmetic series?

The series $1 + 3 + 5 + 7 + 9 + 11 + 13$ consists of **7** terms, where the first term is **1** and the last term is **13**. Substituting in the formula gives

$$S_7 = 7\left(\frac{1 + 13}{2}\right) = 49,$$

which agrees with the sum obtained by adding the 7 terms.

Because $a_n = a_1 + (n - 1)d$ for an arithmetic sequence, S_n can also be written

$$S_n = n\left(\frac{a_1 + a_n}{2}\right)$$

$$= \frac{n}{2}(a_1 + a_n)$$

$$= \frac{n}{2}(a_1 + a_1 + (n - 1)d)$$

$$= \frac{n}{2}(2a_1 + (n - 1)d).$$

EXAMPLE 2 **Finding the sum of the terms of an arithmetic sequence**

Suppose that a person has a starting annual salary of $30,000 and receives a $1500 raise each year. Calculate the total amount earned after 10 years.

Solution

The sequence that gives the salary during year n is given by

$$a_n = 30,000 + 1500(n - 1).$$

One way to calculate the sum of the first 10 terms, denoted S_{10}, is to find a_1 and a_{10}.

$$a_1 = 30{,}000 + 1500(1 - 1) = 30{,}000$$

$$a_{10} = 30{,}000 + 1500(10 - 1) = 43{,}500$$

Thus the total amount earned during this 10-year period is

$$S_{10} = 10\left(\frac{a_1 + a_{10}}{2}\right)$$

$$= 10\left(\frac{30{,}000 + 43{,}500}{2}\right)$$

$$= \$367{,}500.$$

This sum can also be found with the second formula by letting $d = 1500$.

$$S_n = \frac{n}{2}(2a_1 + (n - 1)d)$$

$$= \frac{10}{2}(2 \cdot 30{,}000 + (10 - 1)1500)$$

$$= 5(60{,}000 + 9 \cdot 1500)$$

$$= \$367{,}500.$$

Now Try Exercise 49

EXAMPLE 3 **Finding the sum of the terms of an arithmetic sequence**

Find the sum of the series $2 + 4 + 6 + \cdots + 100$.

Solution
The first term of this series is $a_1 = 2$, and the common difference is $d = 2$. This series represents the even numbers from **2** to **100**, so the number of terms is $n = $ **50**. Using the formula

$$S_n = n\left(\frac{a_1 + a_n}{2}\right)   \qquad \text{Sum formula}$$

for the sum of an arithmetic series, we obtain

$$S_{50} = 50\left(\frac{2 + 100}{2}\right) = 2550.$$

Now Try Exercise 13

TECHNOLOGY NOTE

Sum of a Series
The "seq(" utility, found on some calculators under the LIST OPS menus, generates a sequence. The "sum(" utility, found on some calculators under the LIST MATH menus, calculates the sum of the sequence inside the parentheses. To verify the result in Example 2, let $a_n = 30{,}000 + 1500(n - 1)$. The value 367,500 for S_{10} is shown in Figure 11.16(a). The result found in Example 3 is shown in Figure 11.16(b).

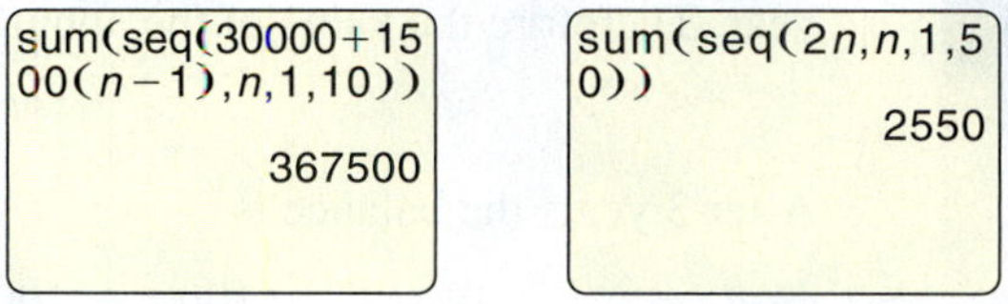

```
sum(seq(30000+15
00(n-1),n,1,10))
            367500
```

```
sum(seq(2n,n,1,5
0))
              2550
```

(a)   (b)

Figure 11.16

Geometric Series

A **geometric series** is the sum of the terms of a geometric sequence. For example,

$$\underbrace{1, 2, 4, 8, 16, 32}_{\textbf{Geometric sequence}}$$

is a geometric sequence with $a_1 = 1$ and $r = 2$. Then

$$\underbrace{1 + 2 + 4 + 8 + 16 + 32}_{\textbf{Geometric series}}$$

is a geometric series. We can use the following formula to sum a finite geometric series.

READING CHECK

How do you find the sum of a geometric series?

SUM OF THE FIRST n TERMS OF A GEOMETRIC SEQUENCE

If its first term is a_1 and its common ratio is r, then the **sum of the first n terms of a geometric sequence** is given by

$$S_n = a_1\left(\frac{1 - r^n}{1 - r}\right),$$

provided $r \neq 1$.

EXAMPLE 4 | **Finding the sum of the terms of a geometric sequence**

Find the sum of each series.
(a) $1 + 2 + 4 + 8 + 16 + 32 + 64 + 128 + 256$

(b) $\frac{1}{2} - \frac{1}{4} + \frac{1}{8} - \frac{1}{16} + \frac{1}{32}$

Solution
(a) This series is geometric, with $n = 9$, $a_1 = 1$, and $r = 2$, so

$$S_9 = 1\left(\frac{1 - 2^9}{1 - 2}\right) = 511. \qquad\qquad S_n = a_1\left(\frac{1 - r^n}{1 - r}\right)$$

(b) This series is geometric, with $n = 5$, $a_1 = \frac{1}{2} = 0.5$, and $r = -\frac{1}{2} = -0.5$, so

$$S_5 = 0.5\left(\frac{1 - (-0.5)^5}{1 - (-0.5)}\right) = \frac{11}{32} = 0.34375.$$

Now Try Exercises 17, 19

▶ **REAL-WORLD CONNECTION** A sum of money from which regular payments are made is called an **annuity**. An annuity may be purchased with a lump sum deposit or by deposits made at various intervals. Suppose that $1000 is deposited at the end of each year in an annuity account that pays an annual interest rate I expressed as a decimal. At the end of the first year the account contains $1000. At the end of the second year $1000 is deposited again. In addition, the first deposit of $1000 would have received interest during the second year. Therefore the value of the annuity after 2 years is

$$1000 + 1000(1 + I).$$

After 3 years the balance is

$$1000 + 1000(1 + I) + 1000(1 + I)^2,$$

and after n years this amount is given by

$$1000 + 1000(1 + I) + 1000(1 + I)^2 + \cdots + 1000(1 + I)^{n-1}.$$

This series is a geometric series with its first term $a_1 = 1000$ and the common ratio $r = (1 + I)$. The sum of the first n terms is given by

$$S_n = a_1\left(\frac{1 - (1 + I)^n}{1 - (1 + I)}\right) = a_1\left(\frac{(1 + I)^n - 1}{I}\right).$$

EXAMPLE 5 **Finding the future value of an annuity**

Suppose that a 20-year-old worker deposits $1000 into an annuity account at the end of each year. If the interest rate is 5%, find the future value of the annuity when the worker is 65 years old.

Solution
Let $a_1 = 1000, I = 0.05$, and $n = 45$. The future value of the annuity is

$$S_n = a_1\left(\frac{(1 + I)^n - 1}{I}\right)$$

$$= 1000\left(\frac{(1 + 0.05)^{45} - 1}{0.05}\right)$$

$$\approx \$159,700$$

Now Try Exercise 23

Summation Notation

Summation notation is used to write series efficiently. The symbol Σ, the uppercase Greek letter *sigma*, is used to indicate a sum.

SUMMATION NOTATION

$$\sum_{k=1}^{n} a_k = a_1 + a_2 + a_3 + \cdots + a_n$$

The letter k is called the **index of summation**. The numbers 1 and n represent the subscripts of the first and last terms in the series. They are called the **lower limit** and **upper limit** of the summation, respectively.

EXAMPLE 6 **Using summation notation**

Find each sum.

(a) $\sum_{k=1}^{5} k^2$ (b) $\sum_{k=1}^{4} 5$ (c) $\sum_{k=3}^{6} (2k - 5)$

Solution

(a) $\sum_{k=1}^{5} k^2 = 1^2 + 2^2 + 3^2 + 4^2 + 5^2 = 55$

(b) $\sum_{k=1}^{4} 5 = 5 + 5 + 5 + 5 = 20$

(c) $\sum_{k=3}^{6} (2k - 5) = (2(3) - 5) + (2(4) - 5) + (2(5) - 5) + (2(6) - 5)$
$$= 1 + 3 + 5 + 7 = 16$$

Now Try Exercises 27, 29, 31

Summation notation is used frequently in statistics. The next example demonstrates how averages can be expressed in summation notation.

EXAMPLE 7   **Applying summation notation**

Express the average of the n numbers $x_1, x_2, x_3, \ldots, x_n$ in summation notation.

Solution
The average of n numbers can be written as

$$\frac{x_1 + x_2 + x_3 + \cdots + x_n}{n}.$$

This expression is equivalent to $\frac{1}{n}\left(\sum_{k=1}^{n} x_k\right)$.

Now Try Exercise 35

NOTE: $\sum_{k=1}^{n} x_k$ is equivalent to $\displaystyle\sum_{k=1}^{n} x_k$.

▶ **REAL-WORLD CONNECTION** Series are used to model air filtration.

EXAMPLE 8   **Modeling air filtration**

Suppose that an air filter removes 90% of the impurities entering it.
(a) Find a series that represents the amount of impurities removed by a sequence of n air filters. Express this answer in summation notation.
(b) How many air filters would be necessary to remove 99.99% of the impurities?

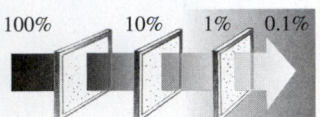

Figure 11.17 Impurities
Passing Through Air Filters

Solution
(a) The first filter removes 90% of the impurities, so 10%, or 0.1, passes through it. Of the 0.1 that passes through the first filter, 90% is removed by the second filter, while 10% of 10%, or 0.01, passes through. Then, 10% of 0.01, or 0.001, passes through the third filter. Figure 11.17 depicts these results, from which we can establish a pattern. If we let 100%, or 1, represent the amount of impurities entering the first air filter, the amount removed by n filters equals

$$(0.9)(1) + (0.9)(0.1) + (0.9)(0.01) + (0.9)(0.001) + \cdots + (0.9)(0.1)^{n-1}.$$

In summation notation we write this series as $\sum_{k=1}^{n} 0.9(0.1)^{k-1}$.

(b) To remove 99.99%, or 0.9999, of the impurities requires 4 air filters, because

$$\sum_{k=1}^{4} 0.9(0.1)^{k-1} = (0.9)(1) + (0.9)(0.1) + (0.9)(0.01) + (0.9)(0.001)$$
$$= 0.9 + 0.09 + 0.009 + 0.0009$$
$$= 0.9999.$$

Now Try Exercise 43

11.3 Putting It All Together

A finite sequence is an ordered list such as

$$a_1, a_2, a_3, a_4, a_5, \ldots, a_n.$$

A finite series is the summation of the terms of a sequence and can be expressed as

$$a_1 + a_2 + a_3 + a_4 + a_5 + \cdots + a_n.$$

SERIES	DESCRIPTION	EXAMPLE
Finite Arithmetic	$a_1 + a_2 + a_3 + \cdots + a_n$, where $a_n = dn + c$ or $a_n = a_1 + (n-1)d$. The sum of the first n terms is $$S_n = n\left(\frac{a_1 + a_n}{2}\right) \quad \text{or}$$ $$S_n = \frac{n}{2}(2a_1 + (n-1)d),$$ where a_1 is the first term and d is the common difference.	The series $$4 + 7 + 10 + 13 + 16 + 19 + 22$$ is obtained from the sequence $$a_n = 3n + 1 \quad \text{or} \quad a_n = 4 + 3(n-1).$$ Its sum is $$S_7 = 7\left(\frac{4+22}{2}\right) = 91 \quad \text{or}$$ $$S_7 = \frac{7}{2}(2 \cdot 4 + (7-1)3) = 91.$$
Finite Geometric	$a_1 + a_2 + a_3 + \cdots + a_n$, where $a_n = a_1(r)^{n-1}$ for nonzero constants a_1 and r. The sum of the first n terms is $$S_n = a_1\left(\frac{1 - r^n}{1 - r}\right),$$ where a_1 is the first term and r is the common ratio ($r \neq 1$).	The series $$3 + 6 + 12 + 24 + 48 + 96$$ has $n = 6$, $a_1 = 3$, and $r = 2$. Its sum is $$S_6 = 3\left(\frac{1 - 2^6}{1 - 2}\right) = 189.$$

11.3 Exercises

MyMathLab Math XL PRACTICE WATCH DOWNLOAD READ REVIEW

CONCEPTS AND VOCABULARY

1. The summation of the terms of a sequence is called a(n) _____.

2. Find the sum of the series $1 + 2 + 3 + 4$.

3. The series $1 + 3 + 5 + 7 + 9$ is an example of a(n) _____ series.

4. The series $1 + 3 + 9 + 27 + 81$ is an example of a(n) _____ series.

5. If $a_1 + a_2 + a_3 + \cdots + a_n$ is an arithmetic series, its sum is $S_n = $ _____.

6. If $a_1 + a_2 + a_3 + \cdots + a_n$ is a geometric series with the common ratio $r \neq 1$, its sum is $S_n = $ _____.

7. The symbol Σ is used to indicate a(n) _____.

8. Write $\sum_{k=1}^{4} a_k$ as a sum.

9. $\sum_{n=1}^{5} a_1 + (n-1)d$ is an example of a(n) _____ series.

10. $\sum_{n=1}^{4} a_1 r^{n-1}$ is an example of a(n) _____ series.

SUMS OF SERIES

Exercises 11–16: Find the sum of the arithmetic series by using a formula.

11. $3 + 5 + 7 + 9 + 11 + 13$

12. $7.5 + 6 + 4.5 + 3 + 1.5 + 0 + (-1.5)$

13. $1 + 2 + 3 + 4 + \cdots + 40$

14. $1 + 3 + 5 + 7 + \cdots + 99$

15. $-7 + (-4) + (-1) + 2 + 5$

16. $89 + 84 + 79 + 74 + 69 + 64 + 59 + 54$

Exercises 17–22: Find the sum of the geometric series by using a formula.

17. $3 + 9 + 27 + 81 + 243 + 729 + 2187$

18. $2 - 1 + \frac{1}{2} - \frac{1}{4} + \frac{1}{8} - \frac{1}{16} + \frac{1}{32}$

19. $1 - 2 + 4 - 8 + 16 - 32 + 64 - 128$

20. $2 + \frac{1}{2} + \frac{1}{8} + \frac{1}{32} + \frac{1}{128} + \frac{1}{512}$

21. $0.5 + 1.5 + 4.5 + 13.5 + 40.5 + 121.5$

22. $0.6 + 0.3 + 0.15 + 0.075 + 0.0375$

Exercises 23–26: Annuities (Refer to Example 5.) Find the future value of the annuity.

23. $a_1 = \$2000$ $I = 0.08$ $n = 20$

24. $a_1 = \$500$ $I = 0.15$ $n = 10$

25. $a_1 = \$10,000$ $I = 0.11$ $n = 5$

26. $a_1 = \$3000$ $I = 0.19$ $n = 45$

SUMMATION NOTATION

Exercises 27–34: Write the terms of the series and find their sum.

27. $\sum_{k=1}^{4} 2k$

28. $\sum_{k=1}^{6} (k - 1)$

29. $\sum_{k=1}^{8} 4$

30. $\sum_{k=2}^{6} (5 - 2k)$

31. $\sum_{k=1}^{7} k^2$

32. $\sum_{k=1}^{4} 5(2)^{k-1}$

33. $\sum_{k=4}^{5} (k^2 - k)$

34. $\sum_{k=1}^{4} \log k$

Exercises 35–38: Write in summation notation.

35. $1^4 + 2^4 + 3^4 + 4^4 + 5^4 + 6^4$

36. $1 + \frac{1}{5^1} + \frac{1}{5^2} + \frac{1}{5^3} + \frac{1}{5^4}$

37. $1 + \frac{1}{2^2} + \frac{1}{3^2} + \frac{1}{4^2} + \frac{1}{5^2}$

38. $1 + \frac{1}{10} + \frac{1}{100} + \frac{1}{1000} + \frac{1}{10,000}$

39. Verify that $\sum_{k=1}^{n} k = \frac{n(n + 1)}{2}$ by using a formula for the sum of the first n terms of an arithmetic series.

40. Use Exercise 39 to find the sum of the series $\sum_{k=1}^{200} k$.

APPLICATIONS

41. *Prison Escapees* The following table lists the number of escapees from state prisons each year.

Year	1990	1991	1992
Escapees	8518	9921	10,706

Year	1993	1994	1995
Escapees	14,035	14,307	12,249

Source: Bureau of Justice Statistics.

(a) Write a series whose sum is the total number of escapees from 1990 to 1995.

(b) Find its sum.

42. *Captured Prison Escapees* (Refer to Exercise 41.) The table lists the number of escapees from state prisons who were captured, including inmates who may have escaped during a previous year.

Year	1990	1991	1992
Captured	9324	9586	10,031

Year	1993	1994	1995
Captured	12,872	13,346	12,166

Source: Bureau of Justice Statistics.

(a) Write a series whose sum is the total number of escapees captured from 1990 to 1995.

(b) Find its sum.

(c) Compare the number of escapees to the number captured during this time period.

43. *Air Filtration* (Refer to Example 8.) Suppose that an air filter removes 80% of the impurities entering it.

(a) Find a series that represents the amount of impurities removed by a sequence of n air filters. Express the answer in summation notation.

(b) How many filters would be necessary to remove 96% of the impurities?

44. *Air Filtration* Suppose that an air filter removes 70% of the impurities entering it.

(a) Find a series that represents the amount of impurities removed by a sequence of n air filters. Express the answer in summation notation.

(b) How many filters would be necessary to remove 97.3% of the impurities?

45. *Area* A sequence of smaller squares is formed by connecting the midpoints of the sides of a larger square as shown in the figure.

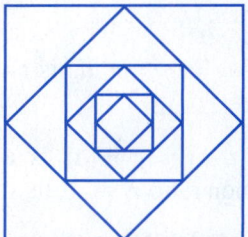

(a) If the area of the largest square is 1 square unit, determine the first five terms of a sequence that describes the area of each successive square.

(b) Use a formula to sum the areas of the first 10 squares.

46. *Perimeter* (Refer to Exercise 45.) Use a formula to find the sum of the perimeters of the first 10 squares.

47. *Stacking Logs* A stack of logs is made in layers, with one log less in each layer, as shown in the accompanying figure. If the top layer has 6 logs and the bottom layer has 14 logs, what is the total number of logs in the pile? Use a formula to find this sum.

48. *Stacking Logs* (Refer to Exercise 47.) Suppose that a stack of logs has 15 logs in the top layer and a total of 10 layers. How many logs are in the stack?

49. *Salaries* (Refer to Example 2.) Suppose that an individual's starting salary is $35,000 per year and that the individual receives a $2000 raise each year. Find the total amount earned over 20 years.

50. *Salaries* Suppose that an individual's starting salary is $35,000 per year and that the individual receives a 10% raise each year. Find the total amount earned over 20 years.

51. *Bouncing Ball* A tennis ball first bounces to 75% of the height from which it was dropped and then to 75% of the height of each successive bounce. If it is dropped from a height of 10 feet, find the distance it *falls* between the fourth and fifth bounce.

52. *Bouncing Ball* A tennis ball first bounces to 75% of the height from which it was dropped and then to 75% of the height of each successive bounce. If it is dropped from a height of 10 feet, find the *total* distance it travels before it reaches its fifth bounce. (*Hint:* Make a sketch.)

53. **Online Exploration** Go online and find a series that can be used to calculate π. Use the first five terms of your series to estimate π.

54. **Online Exploration** Go online and find a series that can be used to calculate the number e. Use the first five terms of your series to estimate e.

WRITING ABOUT MATHEMATICS

55. Discuss the difference between a sequence and a series. Give an example of each.

56. Suppose that an arithmetic series has $a_1 = 1$ and a common difference of $d = 2$, whereas a geometric series has $a_1 = 1$ and a common ratio of $r = 2$. Discuss how their sums compare as the number of terms n becomes large. (*Hint:* Calculate each sum for $n = 10, 20,$ and 30.)

Group Activity   Working With Real Data

Directions: Form a group of 2 to 4 people. Select someone to record the group's responses for this activity. All members of the group should work cooperatively to answer the questions. If your instructor asks for your results, each member of the group should be prepared to respond.

Depreciation For tax purposes, businesses frequently depreciate equipment. Two different methods of depreciation are called *straight-line depreciation* and *sum-of-the-years'-digits*. Suppose that a college student buys a $3000 computer to start a business that provides Internet services. This student estimates the life of the computer at 4 years, after which its value will be $200. The difference between $3000 and $200, or $2800, may be deducted from the student's taxable income over a 4-year period.

In straight-line depreciation, equal portions of $2800 are deducted each year over the 4 years. The sum-of-the-years'-digits method gives depreciation differently. For a computer having a useful life of 4 years, the sum of the years is computed by

$$1 + 2 + 3 + 4 = 10.$$

With this method, $\frac{4}{10}$ of $2800 is deducted the first year, $\frac{3}{10}$ the second year, and so on, until $\frac{1}{10}$ is deducted the fourth year. Both depreciation methods yield a total deduction of $2800 over the 4 years.

(*Source:* Sharp Electronics Corporation, *Conquering the Sciences.*)

(a) Find an arithmetic sequence that gives the amount depreciated each year by each method.

(b) Write a series whose sum is the amount depreciated over 4 years by each method.

11.4 The Binomial Theorem

Pascal's Triangle • Factorial Notation and Binomial Coefficients • Using the Binomial Theorem

A LOOK INTO MATH ▶

In this section we demonstrate how to expand expressions of the form $(a + b)^n$, where n is a natural number. These expressions occur in statistics, finite mathematics, computer science, and calculus. The two methods that we discuss are Pascal's triangle and the binomial theorem.

Pascal's Triangle

Expanding $(a + b)^n$ for increasing values of n gives the following results.

NEW VOCABULARY

☐ Pascal's triangle
☐ Factorial notation
☐ Binomial coefficient
☐ Binomial theorem

Expanding Binomials

$$
\begin{aligned}
(a + b)^0 &= 1 \\
(a + b)^1 &= 1a + 1b \\
(a + b)^2 &= 1a^2 + 2ab + 1b^2 \\
(a + b)^3 &= 1a^3 + 3a^2b + 3ab^2 + 1b^3 \\
(a + b)^4 &= 1a^4 + 4a^3b + 6a^2b^2 + 4ab^3 + 1b^4 \\
(a + b)^5 &= 1a^5 + 5a^4b + 10a^3b^2 + 10a^2b^3 + 5ab^4 + 1b^5
\end{aligned}
$$

Note that $(a + b)^1$ has two terms, starting with a and ending with b; $(a + b)^2$ has three terms, starting with a^2 and ending with b^2; and, in general, $(a + b)^n$ has $n + 1$ terms, starting with a^n and ending with b^n. From left to right, the exponent on a decreases by 1 each successive term, and the exponent on b increases by 1 each successive term.

The triangle formed by the red numbers is called **Pascal's triangle**. This triangle consists of 1s along the sides, and each element inside the triangle is the sum of the two numbers above it, as shown in Figure 11.18. Pascal's triangle is usually written without variables and can be extended to include as many rows as needed.

Pascal's Triangle

```
            1
          1   1
        1   2   1
      1   3   3   1
    1   4   6   4   1
  1   5  10  10   5   1
```

Figure 11.18

We can use this triangle to expand $(a + b)^n$, where n is a natural number. For example, the expression $(m + n)^4$ consists of five terms written as

$$(m + n)^4 = \_m^4 + \_m^3n^1 + \_m^2n^2 + \_m^1n^3 + \_n^4.$$

Because there are five terms, the coefficients can be found in the fifth row of Pascal's triangle, which is

$$1 \ 4 \ 6 \ 4 \ 1.$$

READING CHECK

Explain how to find the numbers in Pascal's triangle.

Thus

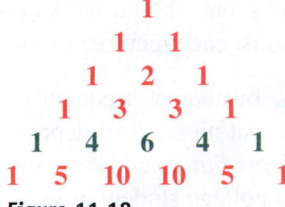

$$(m + n)^4 = \underline{1} \, m^4 + \underline{4} \, m^3n^1 + \underline{6} \, m^2n^2 + \underline{4} \, m^1n^3 + \underline{1} \, n^4$$

$$= m^4 + 4m^3n + 6m^2n^2 + 4mn^3 + n^4.$$

EXAMPLE 1   **Expanding a binomial**

Expand each binomial, using Pascal's triangle.
(a) $(x + 2)^5$ **(b)** $(2m - n)^3$

Solution
(a) To find the coefficients, use the sixth row (1 5 10 10 5 1) in Pascal's triangle.

$$(x + 2)^5 = 1\,x^5 + 5\,x^4 \cdot 2^1 + 10\,x^3 \cdot 2^2 + 10\,x^2 \cdot 2^3 + 5\,x^1 \cdot 2^4 + 1(2^5)$$
$$= x^5 + 10x^4 + 40x^3 + 80x^2 + 80x + 32$$

(b) To find the coefficients, use the fourth row (1 3 3 1) in Pascal's triangle.

$$(2m - n)^3 = 1(2m)^3 + 3(2m)^2(-n)^1 + 3(2m)^1(-n)^2 + 1(-n)^3$$
$$= 8m^3 - 12m^2n + 6mn^2 - n^3$$

▌ **Now Try Exercises 13, 15**

Factorial Notation and Binomial Coefficients

An alternative to Pascal's triangle is the binomial theorem, which uses **factorial notation**.

---

n FACTORIAL (*n*!)

For any positive integer n,

$$n! = 1 \cdot 2 \cdot 3 \cdots \cdot n.$$

We also define $0! = 1$.

NOTE: Because multiplication is commutative, n factorial can also be defined as

$$n! = n \cdot (n - 1) \cdot (n - 2) \cdots \cdot 2 \cdot 1.$$

Examples include the following.

$$0! = 1$$
$$1! = 1$$
$$2! = 1 \cdot 2 = 2$$
$$3! = 1 \cdot 2 \cdot 3 = 6$$
$$4! = 1 \cdot 2 \cdot 3 \cdot 4 = 24$$
$$5! = 1 \cdot 2 \cdot 3 \cdot 4 \cdot 5 = 120$$

Figure 11.19 supports these results. On some calculators, factorial (!) can be accessed in the MATH PRB menus.

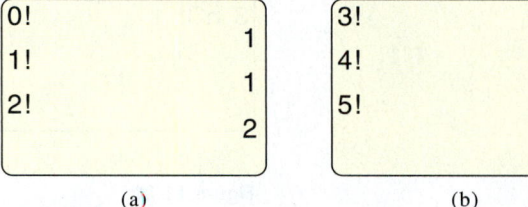

Factorials

(a)   (b)

Figure 11.19

EXAMPLE 2 **Evaluating factorial expressions**

Simplify the expression.

(a) $\dfrac{5!}{3!\,2!}$ **(b)** $\dfrac{4!}{4!\,0!}$

Solution

(a) $\dfrac{5!}{3!\,2!} = \dfrac{1 \cdot 2 \cdot 3 \cdot 4 \cdot 5}{(1 \cdot 2 \cdot 3)(1 \cdot 2)} = \dfrac{120}{6 \cdot 2} = 10$ **(b)** $0! = 1$, so $\dfrac{4!}{4!\,0!} = \dfrac{4!}{4!\,(1)} = \dfrac{4!}{4!} = 1$

Now Try Exercises 23, 25

The expression $_nC_r$ represents a *binomial coefficient* that can be used to calculate the numbers in Pascal's triangle.

BINOMIAL COEFFICIENT $_nC_r$

For n and r nonnegative integers, $n \geq r$,

$$_nC_r = \dfrac{n!}{(n-r)!\,r!}$$

is a **binomial coefficient**.

Values of $_nC_r$ for $r = 0, 1, 2, \ldots, n$ correspond to the $n + 1$ numbers in row $n + 1$ of Pascal's triangle.

EXAMPLE 3 **Calculating $_nC_r$**

Calculate $_3C_r$ for $r = 0, 1, 2, 3$ by hand. Check your results on a calculator. Compare these numbers with the fourth row in Pascal's triangle.

Solution

$$_3C_0 = \dfrac{3!}{(3-0)!\,0!} = \dfrac{6}{6 \cdot 1} = 1 \qquad _3C_1 = \dfrac{3!}{(3-1)!\,1!} = \dfrac{6}{2 \cdot 1} = 3$$

$$_3C_2 = \dfrac{3!}{(3-2)!\,2!} = \dfrac{6}{1 \cdot 2} = 3 \qquad _3C_3 = \dfrac{3!}{(3-3)!\,3!} = \dfrac{6}{1 \cdot 6} = 1$$

These results are supported in Figure 11.20. The fourth row of Pascal's triangle is

1   3   3   1,

which agrees with the calculated values for $_3C_r$. On some calculators, the MATH PRB menus are used to calculate $_nC_r$.

Binomial Coefficients

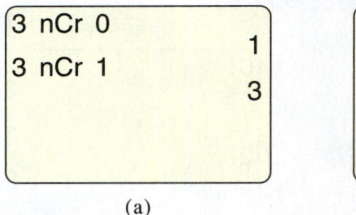

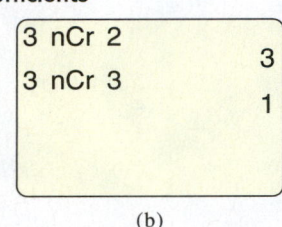

(a)   (b)

Figure 11.20

Now Try Exercises 27, 31

Using the Binomial Theorem

The binomial coefficients can be used to expand expressions of the form $(a + b)^n$. To do so, we use the **binomial theorem**.

> ### BINOMIAL THEOREM
>
> For any positive integer n and any numbers a and b,
> $$(a + b)^n = {}_nC_0 a^n + {}_nC_1 a^{n-1}b^1 + \cdots + {}_nC_{n-1}a^1 b^{n-1} + {}_nC_n b^n.$$

Using the results of Example 3, we write

$$(a + b)^3 = {}_3C_0 a^3 + {}_3C_1 a^2 b^1 + {}_3C_2 a^1 b^2 + {}_3C_3 b^3$$
$$= 1a^3 + 3a^2 b + 3ab^2 + 1b^3$$
$$= a^3 + 3a^2 b + 3ab^2 + b^3.$$

EXAMPLE 4   Expanding a binomial

Use the binomial theorem to expand each expression.
(a) $(x + y)^5$ **(b)** $(3 - 2x)^4$

Solution
(a) The coefficients are calculated as follows.

$${}_5C_0 = \frac{5!}{(5 - 0)!\,0!} = 1, \quad {}_5C_1 = \frac{5!}{(5 - 1)!\,1!} = 5, \quad {}_5C_2 = \frac{5!}{(5 - 2)!\,2!} = 10$$

$${}_5C_3 = \frac{5!}{(5 - 3)!\,3!} = 10, \quad {}_5C_4 = \frac{5!}{(5 - 4)!\,4!} = 5, \quad {}_5C_5 = \frac{5!}{(5 - 5)!\,5!} = 1$$

Using the binomial theorem, we arrive at the following result.

$$(x + y)^5 = {}_5C_0 x^5 + {}_5C_1 x^4 y^1 + {}_5C_2 x^3 y^2 + {}_5C_3 x^2 y^3 + {}_5C_4 x^1 y^4 + {}_5C_5 y^5$$
$$= 1x^5 + 5x^4 y + 10x^3 y^2 + 10x^2 y^3 + 5xy^4 + 1y^5$$
$$= x^5 + 5x^4 y + 10x^3 y^2 + 10x^2 y^3 + 5xy^4 + y^5$$

(b) The coefficients are calculated as follows.

$${}_4C_0 = \frac{4!}{(4 - 0)!\,0!} = 1, \quad {}_4C_1 = \frac{4!}{(4 - 1)!\,1!} = 4, \quad {}_4C_2 = \frac{4!}{(4 - 2)!\,2!} = 6,$$

$${}_4C_3 = \frac{4!}{(4 - 3)!\,3!} = 4, \quad {}_4C_4 = \frac{4!}{(4 - 4)!\,4!} = 1$$

Using the binomial theorem with $a = 3$ and $b = (-2x)$, we arrive at the following result.

$$(3 - 2x)^4 = {}_4C_0 (3)^4 + {}_4C_1 (3)^3 (-2x) + {}_4C_2 (3)^2 (-2x)^2$$
$$+ {}_4C_3 (3)(-2x)^3 + {}_4C_4 (-2x)^4$$
$$= 1(81) + 4(27)(-2x) + 6(9)(4x^2) + 4(3)(-8x^3) + 1(16x^4)$$
$$= 81 - 216x + 216x^2 - 96x^3 + 16x^4$$

Now Try Exercises 39, 49

The binomial theorem gives *all* of the terms of $(a + b)^n$. However, we can find any individual term by noting that the $(r + 1)$st term in the binomial expansion for $(a + b)^n$ is given by the formula ${}_nC_r a^{n-r}b^r$, for $0 \le r \le n$. The next example shows how to use this formula to find the $(r + 1)$st term of $(a + b)^n$.

EXAMPLE 5 Finding the *k*th term in a binomial expansion

Find the third term of $(x - y)^5$.

Solution
In this example the $(r + 1)$st term is the *third* term in the expansion of $(x - y)^5$. That is, $r + 1 = 3$, or $r = 2$. Also, the exponent in the expression is $n = 5$. To get this binomial into the form $(a + b)^n$, we note that the first term in the binomial is $a = x$ and that the second term in the binomial is $b = -y$. Substituting the values for r, n, a, and b in the formula $_nC_r a^{n-r} b^r$ for the $(r + 1)$st term yields

$$_5C_2(x)^{5-2}(-y)^2 = 10x^3y^2.$$

The third term in the binomial expansion of $(x - y)^5$ is $10x^3y^2$.

■ **Now Try Exercise 53**

11.4 Putting It All Together

TOPIC	EXPLANATION	EXAMPLE
Pascal's Triangle	$\begin{array}{ccccccccccc} & & & & & 1 & & & & & \\ & & & & 1 & & 1 & & & & \\ & & & 1 & & 2 & & 1 & & & \\ & & \mathbf{1} & & \mathbf{3} & & \mathbf{3} & & \mathbf{1} & & \\ & 1 & & 4 & & 6 & & 4 & & 1 & \\ 1 & & 5 & & 10 & & 10 & & 5 & & 1 \end{array}$	$(a + b)^3 = \mathbf{1}a^3 + \mathbf{3}a^2b + \mathbf{3}ab^2 + \mathbf{1}b^3$ (Row 4) To expand $(a + b)^n$, use row $n + 1$ in the triangle.
Factorial Notation	The expression $n!$ equals $$1 \cdot 2 \cdot 3 \cdot \cdots \cdot n.$$	$5! = 1 \cdot 2 \cdot 3 \cdot 4 \cdot 5 = 120$
Binomial Coefficient $_nC_r$	$$_nC_r = \frac{n!}{(n - r)!\, r!}$$	$_6C_4 = \dfrac{6!}{(6 - 4)!\, 4!} = \dfrac{6!}{2!\, 4!} = \dfrac{720}{2 \cdot 24} = 15$
Binomial Theorem	$(a + b)^n = {_nC_0}\,a^n + {_nC_1}\,a^{n-1}b^1 + \cdots$ $\qquad + {_nC_{n-1}}\,a^1b^{n-1} + {_nC_n}\,b^n$	$(a + b)^4 = {_4C_0}\,a^4 + {_4C_1}\,a^3b + {_4C_2}\,a^2b^2$ $\qquad + {_4C_3}\,ab^3 + {_4C_4}\,b^4$ $= 1a^4 + 4a^3b + 6a^2b^2 + 4ab^3 + 1b^4$ $= a^4 + 4a^3b + 6a^2b^2 + 4ab^3 + b^4$

11.4 Exercises

MyMathLab
PRACTICE   WATCH   DOWNLOAD   READ   REVIEW

CONCEPTS AND VOCABULARY

1. How many terms result from expanding $(a + b)^4$?

2. How many terms result from expanding $(a + b)^n$?

3. To find the coefficients for the expansion of $(a + b)^3$, what row of Pascal's triangle do you use?

4. Write down the first 5 rows of Pascal's triangle.

5. $4! = $ _____

6. $1 \cdot 2 \cdot 3 \cdot 4 \cdot 5 \cdot 6 = $ _____

7. $_nC_r = $ _____

8. $(a + b)^2 = $ _____

9. **Thinking Generally** $\frac{n!}{(n - 1)!} = $ _____

10. **Thinking Generally** $_nC_n = $ _____

USING PASCAL'S TRIANGLE

Exercises 11–18: Use Pascal's triangle to expand the given expression.

11. $(x + y)^3$

12. $(x + y)^4$

13. $(2x + 1)^4$

14. $(2x - 1)^4$

15. $(a - b)^5$

16. $(3x + 2y)^3$

17. $(x^2 + 1)^3$

18. $\left(\frac{1}{2} - x^2\right)^5$

FACTORIALS AND BINOMIAL COEFFICIENTS

Exercises 19–32: Evaluate the expression.

19. $3!$

20. $6!$

21. $\frac{4!}{3!}$

22. $\frac{6!}{3!}$

23. $\frac{2!}{0!}$

24. $\frac{5!}{1!}$

25. $\frac{5!}{2!3!}$

26. $\frac{6!}{4!2!}$

27. $_5C_4$

28. $_3C_1$

29. $_6C_5$

30. $_2C_2$

31. $_4C_0$

32. $_4C_3$

Exercises 33–38: Evaluate the binomial coefficient with a calculator.

33. $_{12}C_7$

34. $_{13}C_8$

35. $_9C_5$

36. $_{25}C_{14}$

37. $_{19}C_{11}$

38. $_{10}C_6$

THE BINOMIAL THEOREM

Exercises 39–50: Use the binomial theorem to expand the expression.

39. $(m + n)^3$

40. $(m + n)^5$

41. $(x - y)^4$

42. $(1 - 3x)^4$

43. $(2a + 1)^3$

44. $(x^2 - 1)^3$

45. $(x + 2)^5$

46. $(a - 3)^5$

47. $(3 + 2m)^4$

48. $(m - 3n)^3$

49. $(2x - y)^3$

50. $(2a + 3b)^4$

Exercises 51–56: The $(r + 1)$st term of the expression $(a + b)^n$, $0 \le r \le n$, is given by $_nC_r a^{n-r} b^r$. Find the specified term. Refer to Example 5.

51. The first term of $(a + b)^8$

52. The second term of $(a - b)^{10}$

53. The fourth term of $(x + y)^7$

54. The sixth term of $(a + b)^9$

55. The first term of $(2m + n)^9$

56. The eighth term of $(2a - b)^8$

WRITING ABOUT MATHEMATICS

57. Explain how to find the numbers in Pascal's triangle.

58. Compare the expansion of $(a + b)^n$ to the expansion of $(a - b)^n$. Give an example.

Checking Basic Concepts

1. Determine whether the series is arithmetic or geometric.
   (a) $\frac{1}{2} + \frac{1}{4} + \frac{1}{8} + \cdots + \frac{1}{256}$
   (b) $\frac{1}{2} + \frac{5}{2} + \frac{9}{2} + \frac{13}{2} + \frac{17}{2}$

2. Use a formula to find the sum of the arithmetic series

$$4 + 8 + 12 + \cdots + 48.$$

3. Use a formula to find the sum of the geometric series

$$1 - 2 + 4 - 8 + 16 - 32 + 64 - 128 + 256 - 512.$$

4. Use Pascal's triangle to expand $(x - y)^4$.

5. Use the binomial theorem to expand $(x + 2)^3$.

CHAPTER 11 Summary

SECTION 11.1 ■ SEQUENCES

Sequences An *infinite sequence* is a function whose domain is the natural numbers. A *finite sequence* is a function whose domain is $D = \{1, 2, 3, \ldots, n\}$ for some natural number n.

Example: $a_n = 2n$ is a symbolic representation of the even natural numbers. The first six terms, 2, 4, 6, 8, 10, 12, of this sequence are represented numerically and graphically in the table and figure.

Numerical Representation

n	a_n
1	2
2	4
3	6
4	8
5	10
6	12

Graphical Representation

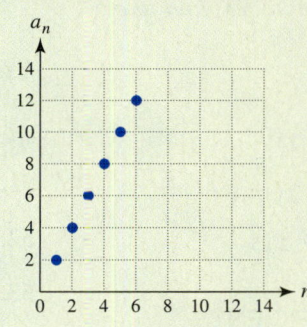

SECTION 11.2 ■ ARITHMETIC AND GEOMETRIC SEQUENCES

Two common types of sequences are arithmetic and geometric.

Arithmetic Sequence An arithmetic sequence is determined by a linear function of the form $f(n) = dn + c$ or $f(n) = a_1 + (n - 1)d$. Successive terms in an arithmetic sequence are found by adding the common difference d to the previous term.

Example: The sequence 1, 3, 5, 7, 9, 11, ... is an arithmetic sequence with its first term $a_1 = 1$, common difference $d = 2$, and general term $a_n = 2n - 1$.

Geometric Sequence The general term for a geometric sequence is given by $f(n) = a_1 r^{n-1}$. Successive terms in a geometric sequence are found by multiplying the previous term by the common ratio r.

Example: The sequence 3, 6, 12, 24, 48, ... is a geometric sequence with its first term $a_1 = 3$, common ratio $r = 2$, and general term $a_n = 3(2)^{n-1}$.

SECTION 11.3 ■ SERIES

Series A series results when the terms of a sequence are summed. The series associated with the sequence 2, 4, 6, 8, 10 is

$$2 + 4 + 6 + 8 + 10, \qquad \text{Series}$$

and its sum equals 30. An arithmetic series results when the terms of an arithmetic sequence are summed, and a geometric series results when the terms of a geometric sequence are summed. In this section, we discussed formulas for finding sums of arithmetic and geometric series. See Putting It All Together for Section 11.3.

Summation Notation   Summation notation can be used to write series efficiently.

Example: $1^2 + 2^2 + 3^2 + 4^2 + 5^2 = \displaystyle\sum_{k=1}^{5} k^2.$

SECTION 11.4 ■ THE BINOMIAL THEOREM

Pascal's triangle may be used to find the coefficients for the expansion of $(a + b)^n$, where n is a natural number.

$$
\begin{array}{ccccccccccc}
 & & & & & 1 & & & & & \\
 & & & & 1 & & 1 & & & & \\
 & & & 1 & & 2 & & 1 & & & \\
 & & 1 & & 3 & & 3 & & 1 & & \\
 & \mathbf{1} & & \mathbf{4} & & \mathbf{6} & & \mathbf{4} & & \mathbf{1} & \\
1 & & 5 & & 10 & & 10 & & 5 & & 1 \\
\end{array}
$$

Example: To expand $(x + y)^4$, use the fifth row of Pascal's triangle.

$$(x + y)^4 = \mathbf{1}x^4 + \mathbf{4}\,x^3y + \mathbf{6}\,x^2y^2 + \mathbf{4}\,xy^3 + \mathbf{1}y^4$$
$$= x^4 + 4x^3y + 6x^2y^2 + 4xy^3 + y^4$$

The binomial theorem can also be used to expand powers of binomials. See Putting It All Together for Section 11.4.

CHAPTER 11   Review Exercises

SECTION 11.1

Exercises 1–4: Write the first four terms of the sequence for $n = 1, 2, 3,$ and 4.

1. $f(n) = n^3$

2. $f(n) = 5 - 2n$

3. $f(n) = \dfrac{2n}{n^2 + 1}$

4. $f(n) = (-2)^n$

Exercises 5–6: Write the terms of the sequence.

5. a_n

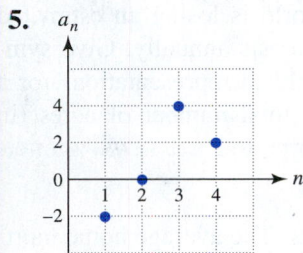

6. a_n

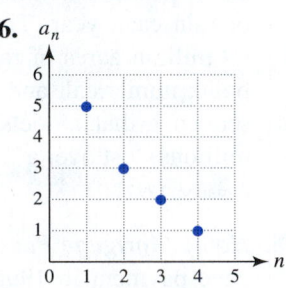

Exercises 7–10: Represent the first seven terms of the sequence numerically and graphically.

7. $a_n = 2n$

8. $a_n = n^2 - 4$

9. $a_n = 4\left(\frac{1}{2}\right)^n$

10. $a_n = \sqrt{n}$

SECTION 11.2

Exercises 11–18: Determine whether f is an arithmetic sequence. Identify the common difference when possible.

11. $f(n) = 5n - 1$

12. $f(n) = 4 - n^2$

13. $f(n) = 2^n$

14. $f(n) = 4 - \frac{1}{3}n$

15.

n	1	2	3	4
$f(n)$	20	17	14	11

16.

n	1	2	3	4
$f(n)$	−3	0	6	12

17. $f(n)$

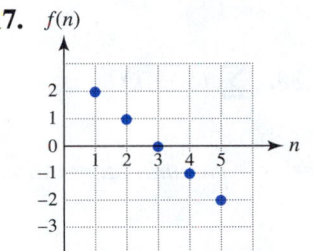

18. $f(n)$

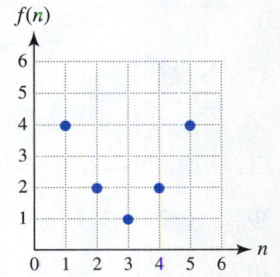

Exercises 19 and 20: Find the general term a_n for the arithmetic sequence.

19. $a_1 = -3$ and $d = 4$ **20.** $a_1 = 2$ and $a_2 = -3$

Exercises 21–28: Determine whether f is a geometric sequence. Identify the common ratio when possible.

21. $f(n) = 2(4)^n$ **22.** $f(n) = 2n^4$

23. $f(n) = 1 - 2n$ **24.** $f(n) = 5(0.7)^n$

25.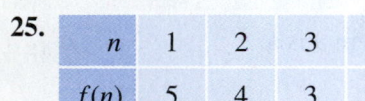

n	1	2	3	4
$f(n)$	5	4	3	1

26.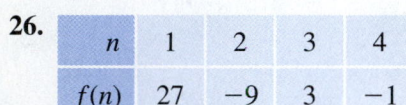

n	1	2	3	4
$f(n)$	27	−9	3	−1

27. 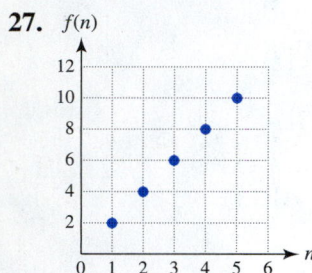   **28.**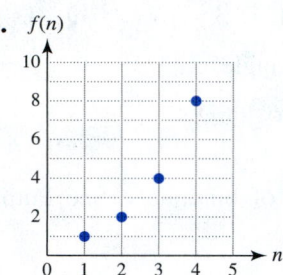

Exercises 29 and 30: Find the general term a_n for the geometric sequence.

29. $a_1 = 5$ and $r = 0.9$ **30.** $a_1 = 2$ and $a_2 = 8$

SECTION 11.3

Exercises 31–34: Find the sum, using a formula.

31. $4 + 9 + 14 + 19 + 24 + 29 + 34 + 39 + 44$

32. $4.5 + 3.0 + 1.5 + 0 - 1.5$

33. $1 - 4 + 16 - 64 + \cdots + 4096$

34. $1 + \frac{1}{2} + \frac{1}{4} + \frac{1}{8} + \frac{1}{16} + \cdots + \frac{1}{256}$

Exercises 35–38: Write the terms of the series.

35. $\sum_{k=1}^{5} 2k + 1$ **36.** $\sum_{k=1}^{4} \frac{1}{k+1}$

37. $\sum_{k=1}^{4} k^3$ **38.** $\sum_{k=2}^{7} (1 - k)$

Exercises 39–42: Write the series in summation notation.

39. $1 + 2 + 3 + \cdots + 20$

40. $1 + \frac{1}{2} + \frac{1}{3} + \cdots + \frac{1}{20}$

41. $\frac{1}{2} + \frac{2}{3} + \frac{3}{4} + \cdots + \frac{9}{10}$

42. $1^2 + 2^2 + 3^2 + 4^2 + 5^2 + 6^2 + 7^2$

SECTION 11.4

Exercises 43–46: Use Pascal's triangle to expand the given expression.

43. $(x + 4)^3$ **44.** $(2x + 1)^4$

45. $(x - y)^5$

46. $(a - 1)^6$

Exercises 47–50: Evaluate the expression.

47. $3!$ **48.** $\dfrac{5!}{3!2!}$

49. ${}_6C_3$ **50.** ${}_4C_3$

Exercises 51–54: Use the binomial theorem to expand the given expression.

51. $(m + 2)^4$ **52.** $(a + b)^5$

53. $(x - 3y)^4$ **54.** $(3x - 2)^3$

APPLICATIONS

55. *Salaries* An individual's starting salary is $45,000, and the individual receives a 10% raise each year. Give symbolic, numerical, and graphical representations for this person's salary over 7 years. What type of sequence is it?

56. *Salaries* An individual's starting salary is $45,000, and the individual receives an increase of $5000 each year. Give symbolic, numerical, and graphical representations for this person's salary over 7 years. What type of sequence is it?

57. *Rain Forests* Rain forests are defined as forests that grow in regions that receive more than 70 inches of rain each year. The world is losing an estimated 49 million acres of rain forests annually. Give symbolic, numerical, and graphical representations for a sequence that models the total number of acres (in millions) lost over a 7-year period. (*Source: New York Times Almanac, 1999.*)

58. *Home Mortgage Payments* The average home mortgage payment in 1996 was $1087 per month. Since then, mortgage payments have risen, on average, by 2.5% per year.
   (a) Write a sequence a_n that models the average mortgage payment, where $n = 1$ corresponds to 1996, $n = 2$ to 1997, and so on.

(b) Is a_n arithmetic, geometric, or neither? Explain your reasoning.

(c) Find a_5 and interpret the result.

(d) Give a graphical representation for a_n, where $n = 1, 2, 3, \ldots, 10$.

CHAPTER 11   Test CHAPTER Test Prep VIDEOS Step-by-step test solutions are found on the Chapter Test Prep Videos available in *MyMathLab* and on You Tube (search "RockswoldInterAlg" and click on "Channels").

1. Write the first four terms of the given sequence for $n = 1, 2, 3,$ and 4.
$$f(n) = \frac{n^2}{n + 1}$$

2. Use the graph to write the terms of the sequence.

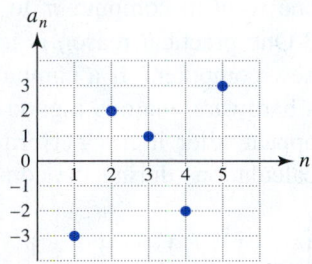

3. List the first seven terms of $a_n = n^2 - n$ in a table. Let $n = 1, 2, \ldots, 7$.

4. Expand the expression $(2x - 1)^4$.

Exercises 5 and 6: Determine whether the sequence is arithmetic or geometric. Identify either the common difference or the common ratio.

5. $f(n) = 7 - 3n$

6.

n	1	2	3	4
$f(n)$	-2	4	-8	16

7. Find the general term a_n for the arithmetic sequence if $a_1 = 2$ and $d = -3$.

8. Find the general term a_n for the geometric sequence if $a_1 = 2$ and $a_3 = 4.5$. Assume $r > 0$.

Exercises 9 and 10: Determine whether $a_n = f(n)$ is a geometric sequence. Identify the common ratio when possible.

9. $f(n) = -3(2.5)^n$

10.

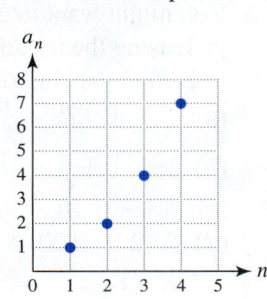

Exercises 11–12: Find the sum, using a formula.

11. $-1 + 2 + 5 + 8 + 11 + 14 + 17 + 20 + 23$

12. $1 - \frac{2}{3} + \frac{4}{9} - \frac{8}{27} + \frac{16}{81} - \frac{32}{243} + \frac{64}{729}$

13. Write the terms of the series $\sum_{k=2}^{7} 3k$.

14. Write the series $1^3 + 2^3 + 3^3 + \cdots + 60^3$ in summation notation.

15. Evaluate $\frac{7!}{4!\,3!}$.

16. Evaluate $_5C_3$.

17. *Auditorium Seating* An auditorium has 50 seats in the first row, 57 seats in the second row, 64 seats in the third row, and so on. Use a formula to find the total number of seats in the first 45 rows.

18. *Median Home Price* In 2011 the median price of a single-family home was \$180,000 and was decreasing at a rate of 4% per year. Give symbolic, numerical, and graphical representations for the median home price over a 5-year period, starting in 2011. What type of sequence is it?

19. *Tent Caterpillars* Large numbers of tent caterpillars can defoliate trees and ruin crops. After they mature, they spin a cocoon and develop into moths that lay eggs. Suppose that an initial population of 2000 tent caterpillars doubles every 5 days.

(a) Write a formula for a_n that models the number of tent caterpillars after $n - 1$ five-day time periods. (*Hint:* $a_1 = 2000$, $a_2 = 4000$, and $a_3 = 8000$.)

(b) Is a_n arithmetic, geometric, or neither? Explain your reasoning.

(c) Find a_6 and interpret the result.

(d) Give a graph for a_n, where $n = 1, 2, 3, 4, 5, 6$.

CHAPTER 11   Extended and Discovery Exercises

SEQUENCES AND SERIES

*Exercises 1 and 2: **Recursive Sequences** Some sequences are not defined by a formula for a_n. Instead they are defined recursively. With a **recursive formula** you must find terms a_1 through a_{n-1} before you can find a_n. For example, let*

$$a_1 = 2$$
$$a_n = a_{n-1} + 3, \quad \text{for } n \geq 2.$$

To find a_2, a_3, and a_4, we let $n = 2, 3, 4$.

$$a_2 = a_1 + 3 = 2 + 3 = 5$$
$$a_3 = a_2 + 3 = 5 + 3 = 8$$
$$a_4 = a_3 + 3 = 8 + 3 = 11$$

The first four terms of the sequence are 2, 5, 8, 11.

1. **Fibonacci Sequence** The Fibonacci sequence dates back to 1202 and is one of the most famous sequences in mathematics. It can be defined recursively as follows.

$$a_1 = 1, \quad a_2 = 1$$
$$a_n = a_{n-1} + a_{n-2}, \quad \text{for } n \geq 3$$

Find the first 12 terms of this sequence.

 2. **Insect Populations** Frequently the population of a particular insect does not continue to grow indefinitely. Instead, its population grows rapidly at first and then levels off because of competition for limited resources. In one study, the behavior of the winter moth was modeled with a sequence similar to the following, where a_n gives the population density in thousands per acre during year n.

$$a_1 = 1$$
$$a_n = 2.85a_{n-1} - 0.19a_{n-1}^2, \quad n \geq 2$$

(*Source:* G. Varley and G. Gradwell, "Population models for the winter moth.")

(a) Make a table for $n = 1, 2, 3, \ldots, 7$. Describe what happens to the population density of the winter moth.

(b) Graph the sequence for $n = 1, 2, 3, \ldots, 20$. Discuss the graph.

NOTE: Many graphing calculators are capable of generating tables and graphs for a recursive sequence.

3. **Calculating π** The quest for an accurate estimation for π is a fascinating story covering thousands of years. Because π is an irrational number, it cannot be represented exactly by a fraction. Its decimal expansion neither repeats nor has a pattern. The ability

to compute π was essential to the development of societies because π appears in formulas used in construction, surveying, and geometry. In early historical records, π was given the value of 3. Later the Egyptians used a value of

$$\frac{256}{81} \approx 3.1605.$$

Not until the discovery of series was an exceedingly accurate decimal approximation of π possible. In 1989, π was computed to 1,073,740,000 digits, which required 100 hours of supercomputer time. Why would anyone want to compute π to so many decimal places? One practical reason is to test electrical circuits in new computers. If a computer has a small defect in its hardware, there is a good chance that an error will appear after it has performed trillions of arithmetic calculations during the computation of π. The series

$$\frac{\pi^4}{90} \approx \frac{1}{1^4} + \frac{1}{2^4} + \frac{1}{3^4} + \frac{1}{4^4} + \frac{1}{5^4} + \cdots + \frac{1}{n^4}$$

gives an estimate of π, where larger values of n give better approximations. (*Source:* P. Beckmann, *A History of Pi.*)

(a) Approximate π by finding the sum of the first four terms.

 (b) Use a calculator to approximate π by summing the first 50 terms. Compare the result to the actual value of π.

4. **Infinite Series** The sum S of an infinite geometric series can be found if its common ratio r satisfies $|r| < 1$. It is given by

$$S = \frac{a_1}{1 - r}.$$

(If $|r| \geq 1$, this sum does not exist.) For example, the infinite geometric series

$$1 + \frac{1}{2} + \frac{1}{4} + \frac{1}{8} + \frac{1}{16} + \cdots$$

has $a_1 = 1$ and $r = \frac{1}{2}$. Therefore its sum S equals

$$S = \frac{1}{1 - \frac{1}{2}} = 2.$$

You might want to add terms of this series to see how increasing the number of terms results in a sum closer to 2. Find the sum of each infinite geometric series.

(a) $2 - 1 + \frac{1}{2} - \frac{1}{4} + \frac{1}{8} - \frac{1}{16} + \cdots$

(b) $1 + \frac{1}{3} + \frac{1}{9} + \frac{1}{27} + \frac{1}{81} + \cdots$

(c) $0.1 + 0.01 + 0.001 + 0.0001 + \cdots$

(d) $0.12 + 0.0012 + 0.000012$
$$+ 0.00000012 + \cdots$$

1. State whether the equation illustrates an identity, commutative, associative, or distributive property.

$$29(102) = 29(100) + 29(2)$$

2. Identify the domain and range of the relation given by $S = \{(-6, 5), (-2, 1), (0, 3), (2, 0)\}$.

Exercises 3–6: Simplify the expression. Write the result using positive exponents.

3. $\dfrac{x^{-2}y^3}{(3xy^{-2})^3}$

4. $\left(\dfrac{3b}{6a^2}\right)^{-4}$

5. $\left(\dfrac{1}{z^2}\right)^{-5}$

6. $\dfrac{8x^{-3}y^2}{4x^3y^{-1}}$

7. Find the domain of $f(x) = \dfrac{-5}{x - 8}$.

8. Use the table to write the formula for $f(x) = ax + b$.

x	-2	-1	0	1	2
$f(x)$	5	3	1	-1	-3

9. Write the equation of the horizontal line that passes through the point $(2, 3)$.

10. Find the slope and the y-intercept of the graph of $f(x) = -3x + 5$.

Exercises 11 and 12: Write the slope-intercept form for a line satisfying the given conditions.

11. Perpendicular to $y = -\frac{2}{3}x - 4$, passing through $(1, 4)$

12. Parallel to $y = 2x - 7$, passing through $(5, 2)$

Exercises 13–18: Solve the equation or inequality. Write the solutions to inequalities in interval notation.

13. $\frac{2}{5}(x - 4) = -12$

14. $\frac{2}{5}z + \frac{1}{4}z > 2 - (z - 1)$

15. $-3|t - 5| \le -18$

16. $|4 + \frac{2}{3}x| = 6$

17. $\frac{1}{4}t - (2t + 5) + 6 = \dfrac{t + 3}{4}$

18. $-3 \le \frac{2}{3}x + 5 < 11$

19. Determine which of the following is a solution to the given system of equations.

$$(3, -2), (-1, 3)$$
$$3x + y = 7$$
$$-2x - 3y = 0$$

20. Shade the solution set in the xy-plane.

$$x - y < 4$$
$$x + 2y \ge 7$$

Exercises 21 and 22: Solve the system of equations. Write the solution as an ordered pair or ordered triple where appropriate.

21. $\begin{aligned} x - 2y &= 1 \\ -2x + 7y &= 4 \end{aligned}$

22. $\begin{aligned} x + y + z &= 5 \\ -2x - y + z &= -10 \\ x + 2y + 8z &= 1 \end{aligned}$

23. Maximize the objective function R subject to the given constraints.

$$R = 3x + 8y$$
$$x + 4y \le 10$$
$$4x + y \le 10$$
$$x \ge 0, y \ge 0$$

24. Evaluate det A.

$$A = \begin{bmatrix} 4 & -3 \\ 3 & 2 \end{bmatrix}$$

Exercises 25 and 26: Multiply the expressions.

25. $2x^3(4x^4 - 3x^3 + 5)$

26. $(2z - 7)(3z + 4)$

Exercises 27 and 28: Factor completely.

27. $4x^2 - 9y^2$

28. $2a^3 - a^2 + 8a - 4$

Exercises 29 and 30: Solve the equation.

29. $4x^2 - x - 3 = 0$

30. $x^4 - 10x^3 = -24x^2$

Exercises 31 and 32: Simplify the expression.

31. $\dfrac{x^2 - 7x + 10}{x^2 - 25} \cdot \dfrac{x + 5}{x + 1}$

32. $\dfrac{x^2 + 7x + 12}{x^2 - 9} \div \dfrac{x^2 - 5x + 6}{(x - 3)^2}$

Exercises 33 and 34: Solve the rational equation. Check your result.

33. $\dfrac{2}{x+5} = \dfrac{-3}{x^2-25} + \dfrac{1}{x-5}$

34. $\dfrac{2y}{y^2-3y+2} = \dfrac{1}{y-2} + 2$

35. Solve the equation for W.

$$R = \frac{3C-2W}{5}$$

36. Simplify the complex fraction.

$$\frac{\dfrac{1}{x^2}+\dfrac{2}{x}}{\dfrac{1}{x^2}-\dfrac{4}{x}}$$

Exercises 37 and 38: Simplify the expression. Assume that all variables are positive.

37. $\sqrt[3]{x^4y^4} - 2\sqrt[3]{xy}$

38. $(4+\sqrt{2})(4-\sqrt{2})$

Exercises 39 and 40: Solve. Check your answers.

39. $8(x-3)^2 = 200$

40. $3\sqrt{2x+6} = 6x$

Exercises 41 and 42: Write the complex expression in standard form.

41. $(-3+i)(-4-2i)$

42. $\dfrac{2-6i}{1+2i}$

43. Find the minimum y-value located on the graph of $y = 3x^2 + 8x + 5$.

44. Write the equation $y = 2x^2 + 8x + 17$ in vertex form and identify the vertex.

Exercises 45 and 46: Solve by using the method of your choice. Write any complex solutions in standard form.

45. $x^2 - 4x + 13 = 0$

46. $z^2 - 4z = 32$

47. A graph of $y = ax^2 + bx + c$ is shown at the top of the next column.
  (a) Solve $ax^2 + bx + c = 0$.
  (b) State whether $a > 0$ or $a < 0$.

(c) Determine whether the discriminant is positive, negative, or zero.

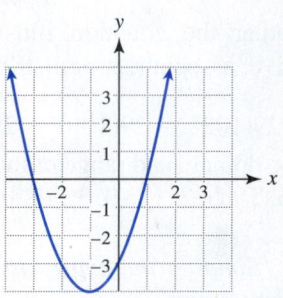

48. Solve the quadratic inequality. Write your answer in interval notation.

$$x^2 + 2x - 3 < 0$$

49. For $f(x) = x^2 + 1$ and $g(x) = 3x - 2$, find each of the following.
  (a) $(f \circ g)(-2)$
  (b) $(g \circ f)(x)$

50. Find $f^{-1}(x)$ for the one-to-one function

$$f(x) = \frac{3x+1}{2}.$$

51. Write $\ln(x^3\sqrt{y})$ by using sums and differences of logarithms of x and y.

52. Write $2\log x - \log 4xy$ as one logarithm. Assume x and y are positive.

Exercises 53 and 54: Solve the equation. Approximate answers to the nearest hundredth.

53. $8\log x + 3 = 17$

54. $4^{2x} = 5$

55. Graph the parabola $x = (y-3)^2 + 1$. Find the vertex and the axis of symmetry.

56. Find the center and the radius of the circle whose equation is $x^2 - 6x + y^2 + 2y = -6$.

Exercises 57 and 58: Graph the ellipse or hyperbola. Label the vertices and the endpoints of the minor axis on the ellipse. Show the asymptotes on the hyperbola.

57. $\dfrac{x^2}{4} + \dfrac{y^2}{9} = 1$

58. $\dfrac{x^2}{16} - \dfrac{y^2}{4} = 1$

Exercises 59 and 60: Use the graph to determine the equation of the ellipse or hyperbola.

59.

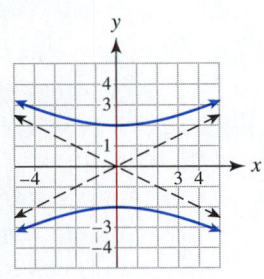

60.

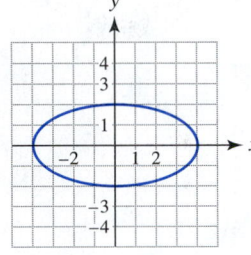

61. Solve the system of equations. Check your solutions.

$$y = x^2 + 1$$
$$x^2 + 2y = 5$$

62. Shade the solution set in the *xy*-plane.

$$y \geq x^2 - 2$$
$$y \leq -x$$

Exercises 63–66: Determine whether f is an arithmetic or a geometric sequence. If it is arithmetic, find the common difference. If it is geometric, find the common ratio.

63. $f(n) = 5 - 2n$

64. $f(n) = 3(0.2)^n$

65. $f(n) = 7(4)^n$

66. $f(n) = 6n + 1$

67. Find the general term a_n for the arithmetic sequence where $a_1 = 2$ and $a_2 = 5$.

68. Find the general term a_n for the geometric sequence where $a_1 = 4$ and $a_2 = 12$.

Exercises 69 and 70: Find the sum using a formula.

69. $3 + 7 + 11 + 15 + 19 + \cdots + 35$

70. $1 - 2 + 4 - 8 + 16 - \cdots + 1024$

Exercises 71 and 72: Expand the binomial expression.

71. $(2x + 3)^4$

72. $(2a - 5b)^3$

APPLICATIONS

73. *Radius of a Circle* If a circle has an area of A square units, its radius r is given by $r = \sqrt{\frac{A}{\pi}}$. Approximate the radius of a circle with an area of 14 square inches to the nearest hundredth of an inch.

74. *Distance from Work* Starting at a warehouse, a delivery truck driver travels down a straight highway for 3 hours at 40 miles per hour, stops and unloads the truck for 1 hour, and then returns to the warehouse at 60 miles per hour. Sketch a graph that shows the distance between the truck and the warehouse during this period of time.

75. *Exercise and Fluid Consumption* When a person exercises, the total amount of fluid he or she will need that day increases depending on the person's weight and the duration of the exercise. To determine the number of ounces of fluid needed, divide the person's weight by 2 and then add 0.4 ounces for every minute of exercise. (*Source: Runner's World.*)
   (a) Write a function that gives the fluid requirements for a person weighing 170 pounds who exercises for *x* minutes a day.
   (b) If an athlete who exercises for 90 minutes requires 130 ounces of fluid, find the athlete's weight.

76. *Airplane Speed* An airplane travels 1080 miles into the wind in 3 hours. The return trip with the wind takes 2.7 hours. Find the average speed of the airplane and the average wind speed.

77. *Size of a Tent* The length of a rectangular tent floor is 6 feet shorter than twice the width. If the area of the tent floor is 108 square feet, what are the dimensions of the tent?

78. *Working Together* Suppose that one person can weed a garden in 60 minutes and a second person can weed the same garden in 90 minutes. How long would it take these two people to weed the garden if they worked together?

79. *Numbers* The product of two positive numbers is 96. If the larger number is subtracted from 3 times the smaller number, the result is 12. Let *x* be the smaller number and let *y* be the larger number.
   (a) Write a system of equations for this situation.
   (b) What are the two numbers?

80. *Marching Band* A band is marching in a triangular formation so that 1 person is in the first row, 3 people are in the second row, 5 people are in the third row, and so on. Use a formula to find the total number of musicians in the marching band if the last row contains 23 people.

Appendix A
Using the Graphing Calculator

Overview of the Appendix

This appendix provides instruction for the TI-83, TI-83 Plus, and TI-84 Plus graphing calculators that may be used in conjunction with this textbook. It includes specific keystrokes needed to work several examples from the text. Students are advised to consult the *Graphing Calculator Guidebook* provided by the manufacturer.

Entering Mathematical Expressions

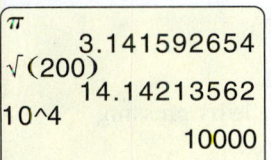

Figure A.1

EVALUATING π To evaluate π, use the following keystrokes, as shown in the first and second lines of Figure A.1. (Do *not* use 3.14 or $\frac{22}{7}$ for π.)

EVALUATING A SQUARE ROOT To evaluate a square root, such as $\sqrt{200}$, use the following keystrokes, as shown in the third and fourth lines of Figure A.1.

EVALUATING AN EXPONENTIAL EXPRESSION To evaluate an exponential expression, such as 10^4, use the following keystrokes, as shown in the last two lines of Figure A.1.

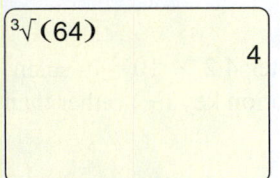

Figure A.2

EVALUATING A CUBE ROOT To evaluate a cube root, such as $\sqrt[3]{64}$, use the following keystrokes, as shown in Figure A.2.

(MATH) (4) (6) (4) ( ) (ENTER)

SUMMARY: ENTERING MATHEMATICAL EXPRESSIONS

To access the *number* π, use (2nd) (^[π]).

To evaluate a *square root*, use (2nd) ($x^2[\sqrt{\ }]$).

To evaluate an *exponential expression*, use the (^) key. To square a number, the (x^2) key can also be used.

To evaluate a *cube root*, use (MATH) (4).

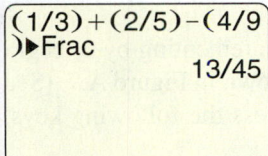

Figure A.3

Expressing Answers as Fractions

To evaluate $\frac{1}{3} + \frac{2}{5} - \frac{4}{9}$ in fraction form, use the following keystrokes, as shown in Figure A.3.

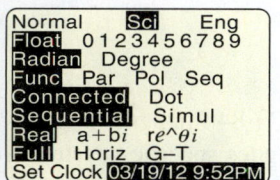

Figure A.4

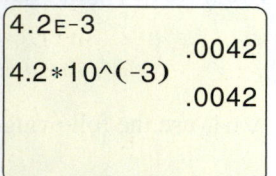

Figure A.5

Displaying Numbers in Scientific Notation

To display numbers in scientific notation, set the graphing calculator in scientific mode (Sci) by using the following keystrokes. See Figure A.4. (These keystrokes assume that the calculator is starting from normal mode.)

MODE ▷ ENTER 2nd MODE [QUIT]

In scientific mode we can display the numbers 5432 and 0.00001234 in scientific notation, as shown in Figure A.5.

Entering Numbers in Scientific Notation

Numbers can be entered in scientific notation. For example, to enter 4.2×10^{-3} in scientific notation, use the following keystrokes. (Be sure to use the negation key $(-)$ rather than the subtraction key.)

This number can also be entered using the following keystrokes. See Figure A.6.

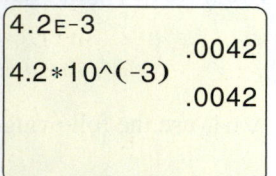

Figure A.6

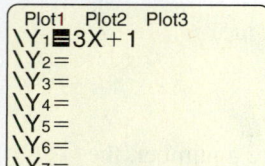

Figure A.7

Making a Table

To make a table of values for $y = 3x + 1$ starting at $x = 4$ and incrementing by 2, begin by pressing Y= and then entering the formula $Y_1 = 3X + 1$, as shown in Figure A.7. (See Entering a Formula on page AP-4.) To set the table parameters, press the following keys. See Figure A.8.

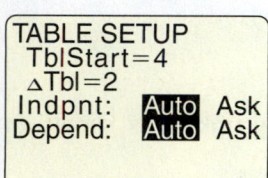

Figure A.8

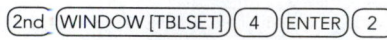

X	Y1	
4	13	
6	19	
8	25	
10	31	
12	37	
14	43	
16	49	
Y₁▪3X+1		

Figure A.9

These keystrokes specify a table that starts at $x = 4$ and increments the x-values by 2. Therefore, the values of Y_1 at $x = 4, 6, 8, \ldots$ appear in the table. To create this table, press the following keys.

$$\boxed{\text{2nd}}\ \boxed{\text{GRAPH [TABLE]}}$$

We can scroll through x- and y-values by using the arrow keys. See Figure A.9. Note that there is no first or last x-value in the table.

SUMMARY: MAKING A TABLE

1. Enter the formula for the equation using $\boxed{Y=}$.
2. Press $\boxed{\text{2nd}}\boxed{\text{WINDOW [TBLSET]}}$ to set the starting x-value and the increment between x-values appearing in the table.
3. Create the table by pressing $\boxed{\text{2nd}}\boxed{\text{GRAPH [TABLE]}}$.

Setting the Viewing Rectangle (Window)

ZOOM MEMORY
1:ZBox
2:Zoom In
3:Zoom Out
4:ZDecimal
5:ZSquare
6▪ZStandard
7↓ZTrig

Figure A.10

There are at least two ways to set the standard viewing rectangle of $[-10, 10, 1]$ by $[-10, 10, 1]$. The first involves pressing $\boxed{\text{ZOOM}}$ followed by $\boxed{6}$. See Figure A.10. The second method for setting the standard viewing rectangle is to press $\boxed{\text{WINDOW}}$ and enter the following keystrokes. See Figure A.11.

$$\boxed{(-)}\boxed{1}\boxed{0}\boxed{\text{ENTER}}\boxed{1}\boxed{0}\boxed{\text{ENTER}}\boxed{1}\boxed{\text{ENTER}}$$
$$\boxed{(-)}\boxed{1}\boxed{0}\boxed{\text{ENTER}}\boxed{1}\boxed{0}\boxed{\text{ENTER}}\boxed{1}\boxed{\text{ENTER}}$$

WINDOW
Xmin=-10
Xmax=10
Xscl=1
Ymin=-10
Ymax=10
Yscl=1
Xres=1

Figure A.11

(Be sure to use the negation key $(-)$ rather than the subtraction key.) Other viewing rectangles can be set in a similar manner by pressing $\boxed{\text{WINDOW}}$ and entering the appropriate values. To see the viewing rectangle, press $\boxed{\text{GRAPH}}$.

SUMMARY: SETTING THE VIEWING RECTANGLE

To set the standard viewing rectangle, press $\boxed{\text{ZOOM}}\boxed{6}$. To set any viewing rectangle, press $\boxed{\text{WINDOW}}$ and enter the necessary values. To see the viewing rectangle, press $\boxed{\text{GRAPH}}$.

NOTE: You do not need to change "Xres" from 1.

L1	L2	L3	1
1	4	------	
2	5		
3	6		
------	------		
L1={1, 2, 3}			

Figure A.12

Making a Scatterplot or a Line Graph

To make a scatterplot with the points $(-5, -5)$, $(-2, 3)$, $(1, -7)$, and $(4, 8)$, begin by following these steps.

L1	L2	L3	1
-5	-5	------	
-2	3		
1	-7		
4	8		
------	------		
L1(5) =			

Figure A.13

1. Press $\boxed{\text{STAT}}$ followed by $\boxed{1}$.
2. If list L1 is not empty, use the arrow keys to place the cursor on L1, as shown in Figure A.12. Then press $\boxed{\text{CLEAR}}$ followed by $\boxed{\text{ENTER}}$. This deletes all elements in the list. Similarly, if L2 is not empty, clear the list.
3. Input each x-value into list L1 followed by $\boxed{\text{ENTER}}$. Input each y-value into list L2 followed by $\boxed{\text{ENTER}}$. See Figure A.13.

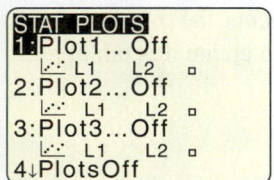

Figure A.14

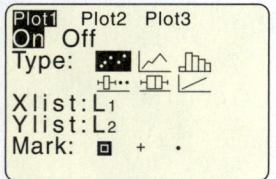

Figure A.15

$[-10, 10, 1]$ by $[-10, 10, 1]$

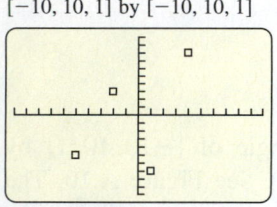

Figure A.16

It is essential that both lists have the same number of values—otherwise, an error message appears when a scatterplot is attempted. Before these four points can be plotted, "STAT PLOT" must be turned on. It is accessed by pressing

$$\boxed{\text{2nd}}\ \boxed{\text{Y =}}\ [\text{STAT PLOT}],$$

as shown in Figure A.14.

There are three possible "STAT PLOTS," numbered 1, 2, and 3. Any one of the three can be selected. The first plot can be selected by pressing $\boxed{1}$. Next, place the cursor over "On" and press $\boxed{\text{ENTER}}$ to turn "Plot1" on. There are six types of plots that can be selected. The first type is a *scatterplot* and the second type is a *line graph*, so place the cursor over the first type of plot and press $\boxed{\text{ENTER}}$ to select a scatterplot. (To make the line graph, place the cursor over the second type of plot and press $\boxed{\text{ENTER}}$.) The x-values are stored in list L1, so select L1 for "Xlist" by pressing $\boxed{\text{2nd}}\boxed{1}$. Similarly, press $\boxed{\text{2nd}}\boxed{2}$ for the "Ylist," since the y-values are stored in list L2. Finally, there are three styles of marks that can be used to show data points in the graph. We will usually use the first because it is largest and shows up the best. Make the screen appear as in Figure A.15. Before plotting the four data points, be sure to set an appropriate viewing rectangle. Then press $\boxed{\text{GRAPH}}$. The data points appear as in Figure A.16.

REMARK 1: A fast way to set the viewing rectangle for any scatterplot is to select the "ZOOMSTAT" feature by pressing $\boxed{\text{ZOOM}}\boxed{9}$. This feature automatically scales the viewing rectangle so that all data points are shown.

REMARK 2: If an equation has been entered into the $\boxed{\text{Y =}}$ menu and selected, it will be graphed with the data. This feature is used frequently to model data.

SUMMARY: MAKING A SCATTERPLOT OR A LINE GRAPH

The following are basic steps necessary to make either a scatterplot or a line graph.

1. Use $\boxed{\text{STAT}}\boxed{1}$ to access lists L1 and L2.
2. If list L1 is not empty, place the cursor on L1 and press $\boxed{\text{CLEAR}}\boxed{\text{ENTER}}$. Repeat for list L2, if it is not empty
3. Enter the x-values into list L1 and the y-values into list L2.
4. Use $\boxed{\text{2nd}}\boxed{\text{Y =}}\ [\text{STAT PLOT}]$ to select appropriate parameters for the scatterplot or line graph.
5. Set an appropriate viewing rectangle. Press $\boxed{\text{GRAPH}}$. Otherwise, press $\boxed{\text{ZOOM}}\boxed{9}$. This feature automatically sets the viewing rectangle and plots the data.

NOTE: $\boxed{\text{ZOOM}}\boxed{9}$ *cannot* be used to set a viewing rectangle for the graph of an equation.

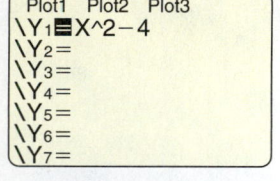

Figure A.17

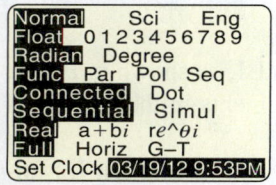

Figure A.18

Entering a Formula

To enter a formula, press $\boxed{\text{Y =}}$. For example, use the following keystrokes after "Y₁ =" to enter $y = x^2 - 4$. See Figure A.17.

$$\boxed{\text{Y =}}\ \boxed{\text{CLEAR}}\ \boxed{\text{X, T, }\theta\text{, }n}\ \boxed{\wedge}\ \boxed{2}\ \boxed{-}\ \boxed{4}$$

Note that there is a built-in key to enter the variable X. If "Y₁ =" does not appear after pressing $\boxed{\text{Y =}}$, press $\boxed{\text{MODE}}$ and make sure the calculator is set in *function mode*, denoted "Func". See Figure A.18.

Graphing an Equation

[−10, 10, 1] by [−10, 10, 1]

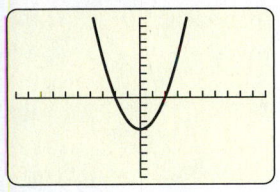

Figure A.19

To graph an equation, such as $y = x^2 - 4$, start by pressing (Y =) and enter $Y_1 = X^2 - 4$. If there is an equation already entered, remove it by pressing (CLEAR). The equals signs in "$Y_1 =$" should be in reverse video (a dark rectangle surrounding a white equals sign), which indicates that the equation will be graphed. If the equals sign is not in reverse video, place the cursor over it and press (ENTER). Set an appropriate viewing rectangle and then press (GRAPH). The graph will appear in the specified viewing rectangle. See Figures A.17 and A.19.

Graphing a Vertical Line

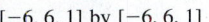

Figure A.20

Set an appropriate window (or viewing rectangle). Then return to the home screen by pressing

(2nd)(MODE [QUIT]).

To graph a vertical line, such as $x = -4$, press

(2nd)(PRGM [DRAW])(4)((−))(4).

See Figure A.20. Pressing (ENTER) will make the vertical line appear, as shown in Figure A.21.

[−6, 6, 1] by [−6, 6, 1]

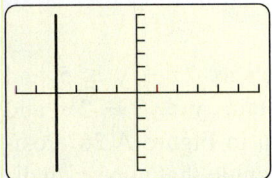

Figure A.21

Squaring a Viewing Rectangle

Figure A.22

In a square viewing rectangle the graph of $y = x$ is a line that makes a 45° angle with the positive x-axis, a circle appears circular, and all sides of a square have the same length. An approximate square viewing rectangle can be set if the distance along the x-axis is 1.5 times the distance along the y-axis. Examples of viewing rectangles that are (approximately) square include

$$[-6, 6, 1] \text{ by } [-4, 4, 1] \quad \text{and} \quad [-9, 9, 1] \text{ by } [-6, 6, 1].$$

Square viewing rectangles can be set automatically by pressing either

(ZOOM)(4)   or   (ZOOM)(5).

ZOOM 4 provides a *decimal window*, which is discussed on page AP-11. See Figure A.22.

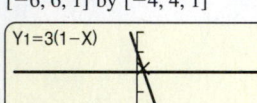

Figure A.23

Locating a Point of Intersection

To find the point of intersection for the graphs of

$$y_1 = 3(1 - x) \quad \text{and} \quad y_2 = 2,$$

start by entering Y_1 and Y_2, as shown in Figure A.23. Set the window, and graph both equations. Then press the following keys to find the intersection point.

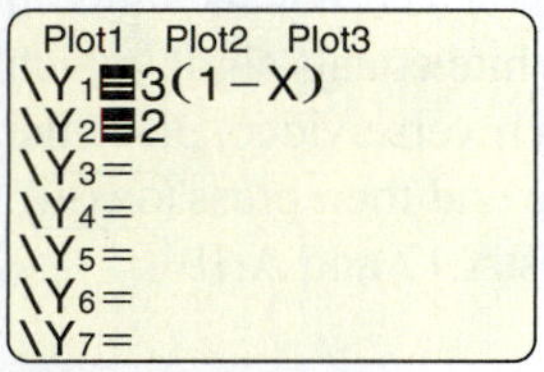

⌊2nd⌋⌊TRACE [CALC]⌋⌊5⌋

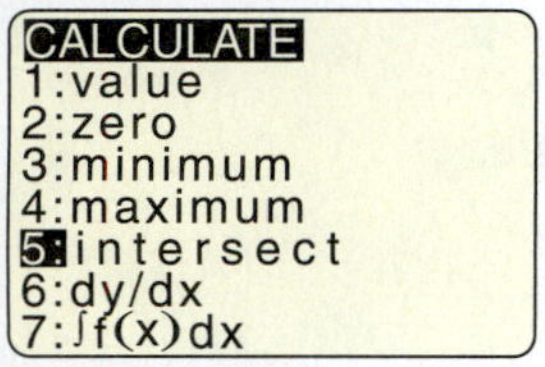

CALCULATE
1:value
2:zero
3:minimum
4:maximum
5:intersect
6:dy/dx
7:∫f(x)dx

Figure A.24

See Figure A.24, where the "intersect" utility is being selected. The calculator prompts for the first curve, as shown in Figure A.25. Use the arrow keys to locate the cursor near the point of intersection and press ⌊ENTER⌋. Repeat these steps for the second curve. Finally we are prompted for a guess. For each of the three prompts, place the free-moving cursor near the point of intersection and press ⌊ENTER⌋. The approximate coordinates of the point of intersection will be shown.

[−6, 6, 1] by [−4, 4, 1]

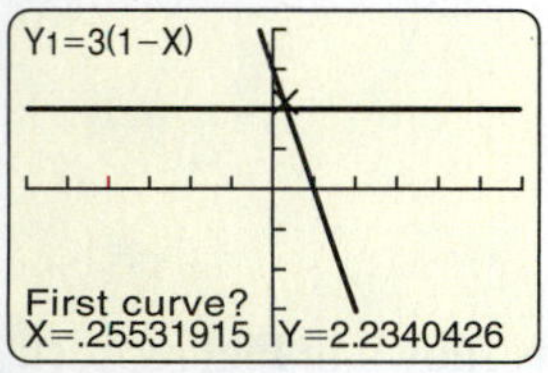

Y1=3(1−X)

First curve?
X=.25531915 Y=2.2340426

Figure A.25

Shading Inequalities

To shade the solution set for one or more linear inequalities such as $2x + y \le 5$ and $-2x + y \ge 1$, begin by solving each inequality for y to obtain $y \le 5 - 2x$ and $y \ge 2x + 1$. Then let $Y_1 = 5 - 2X$ and $Y_2 = 2X + 1$, as shown in Figure A.26. Position the cursor to the left of Y_1 and press ⌊ENTER⌋ three times. The triangle that appears indicates that the calculator will shade the region below the graph of Y_1. Next locate the cursor to the left of Y_2 and press ⌊ENTER⌋ twice. This triangle indicates that the calculator will shade the region above the graph of Y_2. After setting the viewing rectangle to $[-15, 15, 5]$ by $[-10, 10, 5]$, press ⌊GRAPH⌋. The result is shown in Figure A.27.

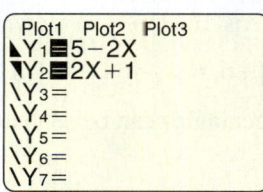

Figure A.26

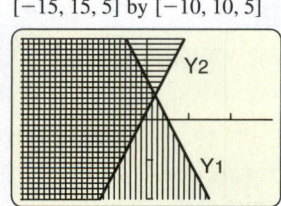

[−15, 15, 5] by [−10, 10, 5]

Figure A.27

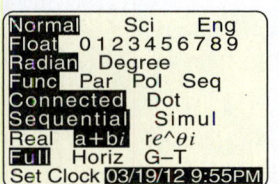

Figure A.28

Setting $a + bi$ Mode

To evaluate expressions containing square roots of negative numbers, such as $\sqrt{-25}$, set your calculator in $a + bi$ mode by using the following keystrokes.

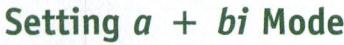

See Figures A.28 and A.29.

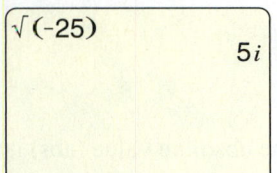

Figure A.29

Evaluating Complex Arithmetic

Complex arithmetic can be performed much like other arithmetic expressions. This is done by entering

$$\boxed{\text{2nd}}\boxed{\text{. [}i\text{]}}$$

to obtain the imaginary unit i from the home screen. For example, to find the sum $(-2 + 3i) + (4 - 6i)$, perform the following keystrokes on the home screen.

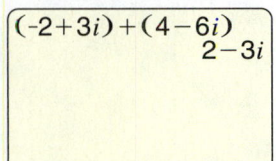

Figure A.30

The result is shown in Figure A.30. Other complex arithmetic operations are done similarly.

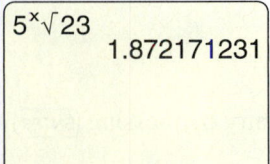

Figure A.31

Other Mathematical Expressions

EVALUATING OTHER ROOTS To evaluate a fifth root, such as $\sqrt[5]{23}$, use the following keystrokes, as shown in Figure A.31.

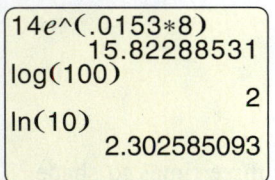

Figure A.32

EVALUATING THE NATURAL EXPONENTIAL FUNCTION To evaluate $14e^{0.0153(8)}$, use the following keystrokes, as shown in the first and second lines of Figure A.32.

EVALUATING THE COMMON LOGARITHMIC FUNCTION To evaluate $\log(100)$, use the following keystrokes, as shown in the third and fourth lines of Figure A.32.

EVALUATING THE NATURAL LOGARITHMIC FUNCTION To evaluate ln (10), use the following keystrokes, as shown in the last two lines of Figure A.32.

LN 1 0) ENTER

SUMMARY: OTHER MATHEMATICAL EXPRESSIONS

To evaluate a *kth root*, use k MATH 5 .

To access the *natural exponential function*, use 2nd LN [eˣ] .

To access the *common logarithmic function*, use LOG .

To access the *natural logarithmic function*, use LN .

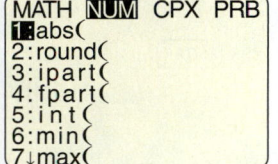

Figure A.33

Accessing the Absolute Value

To graph $y_1 = |x - 50|$, begin by entering $Y_1 = \text{abs}(X - 50)$. The absolute value (abs) is accessed by pressing

See Figure A.33.

SUMMARY: ACCESSING THE ABSOLUTE VALUE

1. Press MATH .
2. Position the cursor over "NUM".
3. Press 1 to select the absolute value.

Entering the Elements of a Matrix

The elements of the augmented matrix A given by

$$A = \begin{bmatrix} 1 & 1 & 2 & 1 \\ -1 & 0 & 1 & -2 \\ 2 & 1 & 5 & -1 \end{bmatrix}$$

can be entered by using the following keystrokes on the TI-83 Plus or TI-84 Plus to define a matrix A with dimension 3×4. (*Note:* On the TI-83 the matrix menu is found by pressing MATRX.)

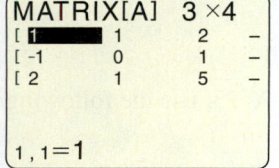

Figure A.34

2nd x⁻¹ [MATRIX] ▷ ▷ 1 3 ENTER 4 ENTER

Input the 12 elements of the matrix A, row by row. Finish each entry by pressing ENTER. See Figure A.34. After these elements have been entered, press

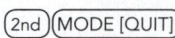

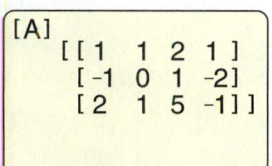

Figure A.35

to return to the home screen. To display the matrix A, press

$$\boxed{\text{2nd}}\ \boxed{x^{-1}\ [\text{MATRIX}]}\ \boxed{1}\ \boxed{\text{ENTER}}.$$

See Figure A.35.

Reduced Row–Echelon Form

To find the reduced row–echelon form of matrix A (entered above in Figure A.35), use the following keystrokes from the home screen on the TI-83 Plus or TI-84 Plus.

$$\boxed{\text{2nd}}\ \boxed{x^{-1}\ [\text{MATRIX}]}\ \boxed{\triangleright}\ \boxed{\text{ALPHA}}\ \boxed{\text{APPS [B]}}\ \boxed{\text{2nd}}\ \boxed{x^{-1}\ [\text{MATRIX}]}\ \boxed{1}\ \boxed{)}\ \boxed{\text{ENTER}}$$

Figure A.36

The resulting matrix is shown in Figure A.36. On the TI-83 graphing calculator, use the following keystrokes to find the reduced row–echelon form.

$$\boxed{\text{MATRX}}\ \boxed{\triangleright}\ \boxed{\text{ALPHA}}\ \boxed{\text{MATRX [B]}}\ \boxed{\text{MATRX}}\ \boxed{1}\ \boxed{)}\ \boxed{\text{ENTER}}$$

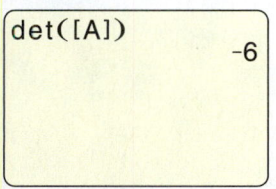

Figure A.37

Evaluating a Determinant

To evaluate the determinant of matrix A given by

$$A = \begin{bmatrix} 2 & 1 & -1 \\ -1 & 3 & 2 \\ 4 & -3 & -5 \end{bmatrix},$$

start by entering the 9 elements of the 3×3 matrix, as shown in Figure A.37. To compute det A, perform the following keystrokes from the home screen.

$$\boxed{\text{2nd}}\ \boxed{x^{-1}\ [\text{MATRIX}]}\ \boxed{\triangleright}\ \boxed{1}\ \boxed{\text{2nd}}\ \boxed{x^{-1}\ [\text{MATRIX}]}\ \boxed{1}\ \boxed{)}\ \boxed{\text{ENTER}}$$

```
det([A])
                -6
```

Figure A.38

The results are shown in Figure A.38.

SUMMARY: EVALUATING A DETERMINANT OF A MATRIX

1. Enter the dimension and elements of the matrix A.
2. Return to the home screen by pressing (2nd)(MODE [QUIT])·
3. On the TI-83 Plus or TI-84 Plus, perform the following keystrokes.

NOTE: On the TI-83, replace the keystrokes (2nd)(x⁻¹ [MATRIX]) with (MATRX).

```
CALCULATE
1:value
2:zero
3:minimum
4:maximum
5:intersect
6:dy/dx
7:∫f(x)dx
```

Figure A.39

Locating an *x*-Intercept or Zero

To locate an *x*-intercept or *zero* of $f(x) = x^2 - 4$, start by entering $Y_1 = X^2 - 4$ into the (Y =) menu. Set the viewing rectangle to $[-9, 9, 1]$ by $[-6, 6, 1]$ and graph Y_1. Afterwards, press the following keys to invoke the zero finder. See Figure A.39.

(2nd)(TRACE [CALC])(2)

The graphing calculator prompts for a left bound. Use the arrow keys to set the cursor to the left of the *x*-intercept and press (ENTER). The graphing calculator then prompts for a right bound. Set the cursor to the right of the *x*-intercept and press (ENTER). Finally, the graphing calculator prompts for a guess. Set the cursor roughly at the *x*-intercept and press (ENTER). See Figures A.40–A.42. The calculator then approximates the *x*-intercept or zero automatically, as shown in Figure A.43. The zero of -2 can be found similarly.

$[-9, 9, 1]$ by $[-6, 6, 1]$

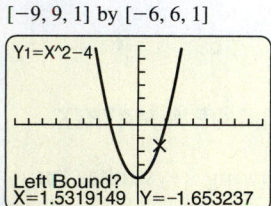

Figure A.40

$[-9, 9, 1]$ by $[-6, 6, 1]$

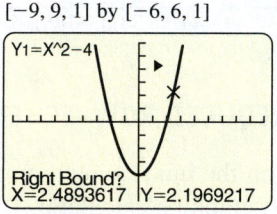

Figure A.41

$[-9, 9, 1]$ by $[-6, 6, 1]$

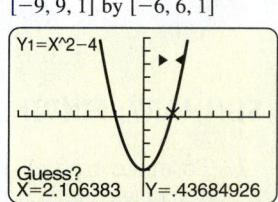

Figure A.42

$[-9, 9, 1]$ by $[-6, 6, 1]$

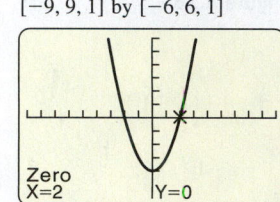

Figure A.43

SUMMARY: LOCATING AN *x*-INTERCEPT OR ZERO

1. Graph the function in an appropriate viewing rectangle.
2. Press (2nd)(TRACE [CALC])(2).
3. Select the left and right bounds, followed by a guess. Press (ENTER) after each selection. The calculator then approximates the *x*-intercept or zero.

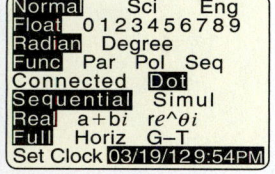

Figure A.44

Setting Connected or Dot Mode

To set your graphing calculator in dot mode, press (MODE), position the cursor over "Dot," and press (ENTER). See Figure A.44. Graphs will now appear in dot mode rather than connected mode.

SUMMARY: SETTING CONNECTED OR DOT MODE

1. Press (MODE).
2. Position the cursor over "Connected" or "Dot". Press (ENTER).

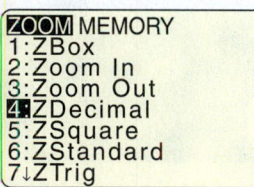

Figure A.45

Setting a Decimal Window

With a decimal window, the cursor stops on convenient x-values. In the decimal window $[-9.4, 9.4, 1]$ by $[-6.2, 6.2, 1]$ the cursor stops on x-values that are multiples of 0.2. If we reduce the viewing rectangle to $[-4.7, 4.7, 1]$ by $[-3.1, 3.1, 1]$, the cursor stops on x-values that are multiples of 0.1. To set this smaller window automatically, press (ZOOM) (4). See Figure A.45. Decimal windows are also useful when graphing rational functions with asymptotes in connected mode.

> ## SUMMARY: SETTING A DECIMAL WINDOW
>
> 1. Press (ZOOM) (4) to set the viewing rectangle $[-4.7, 4.7, 1]$ by $[-3.1, 3.1, 1]$.
> 2. A larger decimal window is $[-9.4, 9.4, 1]$ by $[-6.2, 6.2, 1]$.

Finding Maximum and Minimum Values

To find a minimum y-value (or vertex) on the graph of $f(x) = 1.5x^2 - 6x + 4$, start by entering $Y_1 = 1.5X^2 - 6X + 4$ from the (Y =) menu. Set the viewing rectangle and then perform the following keystrokes to find the minimum y-value.

(2nd) (TRACE [CALC]) (3)

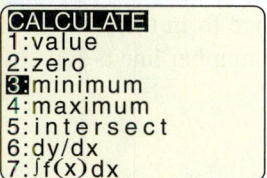

Figure A.46

See Figure A.46.

   The calculator prompts for a left bound. Use the arrow keys to position the cursor left of the vertex and press (ENTER). Similarly, position the cursor to the right of the vertex for the right bound and press (ENTER). Finally, the graphing calculator asks for a guess between the left and right bounds. Place the cursor near the vertex and press (ENTER). See Figures A.47–A.49. The minimum value is shown in Figure A.50.

$[-4.7, 4.7, 1]$ by $[-3.1, 3.1, 1]$ $[-4.7, 4.7, 1]$ by $[-3.1, 3.1, 1]$ $[-4.7, 4.7, 1]$ by $[-3.1, 3.1, 1]$ $[-4.7, 4.7, 1]$ by $[-3.1, 3.1, 1]$

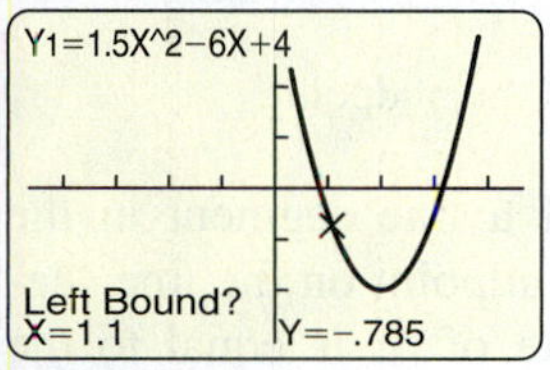

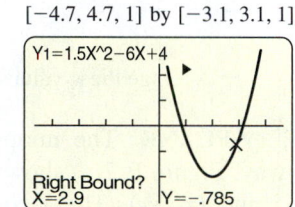

Figure A.47

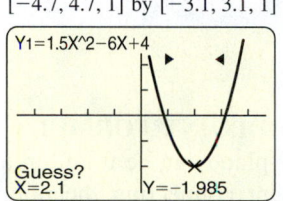

Figure A.48

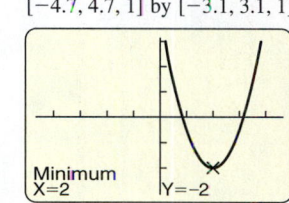

Figure A.49

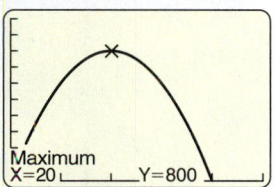

Figure A.50

$[0, 50, 10]$ by $[0, 1000, 100]$

Figure A.51

A maximum of the function f on an interval can be found similarly, except enter

(2nd) (TRACE [CALC]) (4).

The calculator prompts for left and right bounds, followed by a guess. Press (ENTER) after the cursor has been located appropriately for each prompt. The graphing calculator will display the maximum y-value. An example is shown in Figure A.51, where $f(x) = x(80 - 2x)$.

> ## SUMMARY: FINDING MAXIMUM AND MINIMUM VALUES
>
> 1. Graph the function in an appropriate viewing rectangle.
> 2. Press (2nd) (TRACE [CALC]) (3) to find a minimum y-value.
> 3. Press (2nd) (TRACE [CALC]) (4) to find a maximum y-value.
> 4. Use the arrow keys to locate the left and right x-bounds, followed by a guess. Press (ENTER) to select each position of the cursor.

Appendix B
The Midpoint Formula

▶ **REAL-WORLD CONNECTION** A common way to make estimations is to average data items. For example, in 2000 the average tuition and fees at public two-year colleges were about $1700, and in 2010 they were about $2700. (*Source:* The College Board.) To estimate tuition and fees in 2005, we could average the 2000 and 2010 amounts.

$$\frac{1700 + 2700}{2} = \$2200 \quad \text{Finding the average}$$

This technique predicts that tuition and fees were $2200 in 2005 and is referred to as finding the *midpoint*.

MIDPOINT FORMULA ON THE REAL NUMBER LINE The **midpoint** of a line segment is the unique point on the line segment that is an equal distance from the endpoints. For example, in Figure B.1 the midpoint M of -3 and 5 on the real number line is 1.

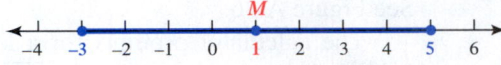

Figure B.1

We can calculate the value of M as follows.

$$M = \frac{x_1 + x_2}{2} = \frac{-3 + 5}{2} = 1$$

↑ Average the x-values to find the midpoint.

MIDPOINT FORMULA IN THE xy-PLANE The midpoint of a line segment in the xy-plane can be found in a similar way. Figure B.2(a) shows the midpoint on the line segment connecting the points (x_1, y_1) and (x_2, y_2). The x-coordinate of M is equal to the average of x_1 and x_2, and the y-coordinate of M is equal to the average of y_1 and y_2. For example, the line segment with endpoints $(-2, 1)$ and $(4, -3)$ is shown in Figure B.2(b). The coordinates of the midpoint are

$$M = \left(\frac{-2 + 4}{2}, \frac{1 + (-3)}{2} \right) = (1, -1).$$

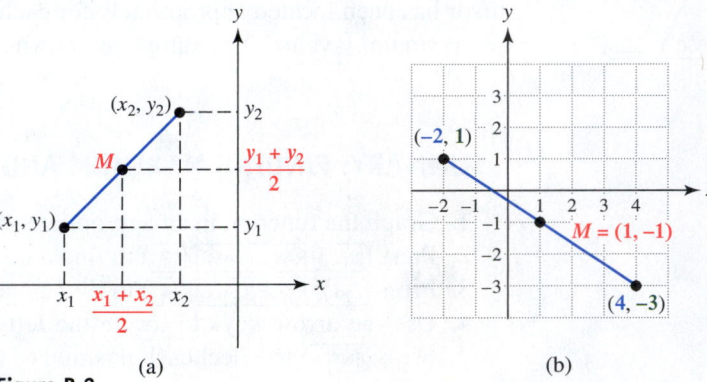

(a)   (b)

Figure B.2

This discussion is summarized as follows.

MIDPOINT FORMULA IN THE xy-PLANE

The midpoint of the line segment with endpoints (x_1, y_1) and (x_2, y_2) in the xy-plane is

$$\left(\frac{x_1 + x_2}{2}, \frac{y_1 + y_2}{2} \right).$$

EXAMPLE 1 ## Finding the midpoint

Find the midpoint of the line segment connecting the points $(-3, -2)$ and $(4, 1)$.

Solution
In the midpoint formula, let $(-3, -2)$ be (x_1, y_1) and $(4, 1)$ be (x_2, y_2).

$$\begin{aligned}
M &= \left(\frac{x_1 + x_2}{2}, \frac{y_1 + y_2}{2} \right) & \text{Midpoint formula} \\
&= \left(\frac{-3 + 4}{2}, \frac{-2 + 1}{2} \right) & \text{Substitute.} \\
&= \left(\frac{1}{2}, -\frac{1}{2} \right) & \text{Simplify.}
\end{aligned}$$

The midpoint of the line segment is $\left(\frac{1}{2}, -\frac{1}{2} \right)$.

■ **Now Try Exercise 9**

▶ **REAL-WORLD CONNECTION** In the next example we use the midpoint formula to estimate the divorce rate in the United States in 2005.

EXAMPLE 2 ## Estimating the U.S. divorce rate

The divorce rate per 1000 people in the year 2000 was 4.2 and in the year 2010 it was 3.4. (*Source:* Statistical Abstract of the United States.)

(a) Use the midpoint formula to estimate the divorce rate in 2005.
(b) Could the midpoint formula be used to estimate the divorce rate in 2003? Explain.

Solution
(a) In the midpoint formula, let $(2000, 4.2)$ be (x_1, y_1) and let $(2010, 3.4)$ be (x_2, y_2).

$$\begin{aligned}
M &= \left(\frac{x_1 + x_2}{2}, \frac{y_1 + y_2}{2} \right) & \text{Midpoint formula} \\
&= \left(\frac{2000 + 2010}{2}, \frac{4.2 + 3.4}{2} \right) & \text{Substitute.} \\
&= (2005, 3.8) & \text{Simplify.}
\end{aligned}$$

The midpoint formula estimates that the divorce rate was **3.8** per 1000 people in **2005**. (Note that the actual rate was 3.6.)
(b) No, the midpoint formula can be used to estimate data that are halfway between two given data points. Because the year 2003 is not halfway between 2000 and 2010, the midpoint formula cannot be used.

■ **Now Try Exercise 23**

NOTE: An estimate obtained from the midpoint formula is equal to an estimate obtained from a linear function whose graph passes through the endpoints of the line segment. This fact is illustrated in the next example.

EXAMPLE 3   ## Relating midpoints to linear functions

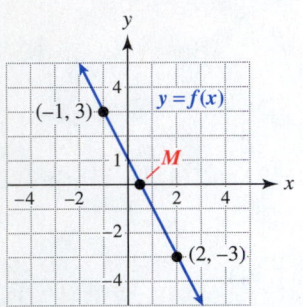

Figure B.3

The graph of a linear function f shown in Figure B.3 passes through the points $(-1, 3)$ and $(2, -3)$.
(a) Find a formula for $f(x)$.
(b) Evaluate $f\left(\frac{1}{2}\right)$. Does your answer agree with the graph?
(c) Find the midpoint M of the line segment connecting the points $(-1, 3)$ and $(2, -3)$. Comment on your result.

Solution
(a) The graph of function f is a line that passes through $(-1, 3)$ and $(2, -3)$. The slope m of the line is

$$m = \frac{-3 - 3}{2 - (-1)} = -\frac{6}{3} = -2,$$

and the y-intercept is **1**. Thus $f(x) = -2x + 1$.

(b) $f\left(\frac{1}{2}\right) = -2\left(\frac{1}{2}\right) + 1 = 0$. Yes, they agree because the point $\left(\frac{1}{2}, 0\right)$ appears to lie on the graph of $y = f(x)$ in Figure B.3.

(c) The midpoint of the line segment connecting $(-1, 3)$ and $(2, -3)$ is

$$M = \left(\frac{-1 + 2}{2}, \frac{3 + (-3)}{2}\right) = \left(\frac{1}{2}, 0\right).$$

Finding the midpoint $M = \left(\frac{1}{2}, 0\right)$ of the line segment with endpoints $(-1, 3)$ and $(2, -3)$ is equivalent to evaluating the linear function f whose graph passes through $(-1, 3)$ and $(2, -3)$ at $x = \frac{1}{2}$.

Now Try Exercise 19

B   # Exercises

MyMathLab

Exercises 1–8: Find the midpoint of the line segment shown.

1.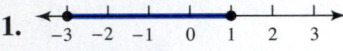

2.

3.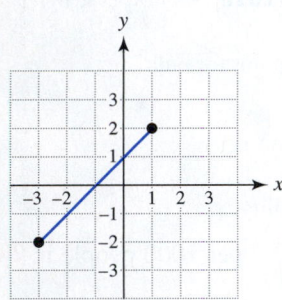

4.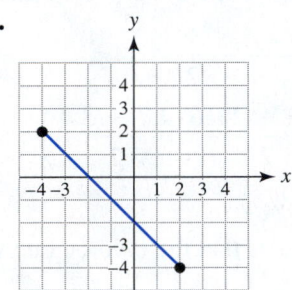

5.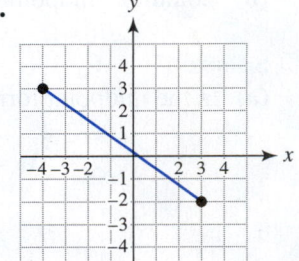

6.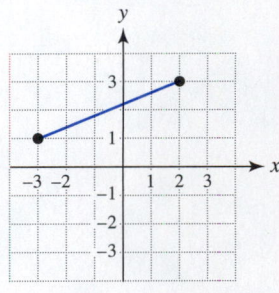

7.

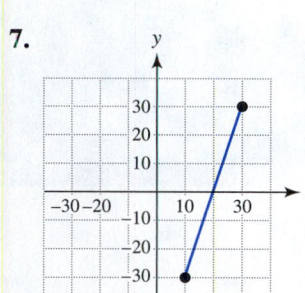

8.

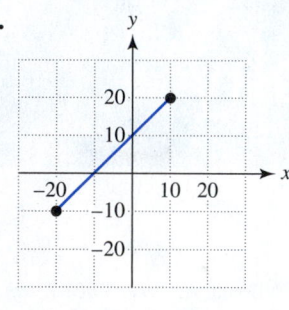

Exercises 9–18: Find the midpoint of the line segment connecting the given points.

9. $(-9, -3), (-7, 1)$

10. $(7, -2), (-5, 8)$

11. $\left(\frac{1}{2}, \frac{1}{3}\right), \left(-\frac{5}{2}, -\frac{2}{3}\right)$

12. $\left(-\frac{3}{5}, -\frac{1}{4}\right), \left(\frac{1}{10}, \frac{1}{2}\right)$

13. $(-0.3, 0.1), (0.7, 0.4)$

14. $(0.8, -0.4), (0.9, -0.1)$

15. $(2000, 5), (2010, 13)$

16. $(2005, 9), (2011, 3)$

17. Thinking Generally
$(a, -b), (3a, 5b)$

18. Thinking Generally
$(-a, b), (a, -b)$

Exercises 19–22: (Refer to Example 3.) The graph of a linear function f passes through the two given points.

(**a**) *Determine* $f(2)$ *by finding a formula for* $f(x)$.
(**b**) *Determine* $f(2)$ *by finding the midpoint of the line segment connecting the given points.*
(**c**) *Compare your answers for parts (a) and (b).*

19. $(0, 5), (4, -3)$

20. $(0, 2), (4, 10)$

21. $(-3, -1), (7, 3)$

22. $(-1, 3), (5, -5)$

Exercises 23–28: Use the midpoint formula to make the requested estimation.

23. *U.S. Life Expectancy* The life expectancy of a female born in 1990 was 78.8 years, and the life expectancy of a female born in 2010 rose to 80.8 years. Estimate the life expectancy of a female born in 2000. (*Source: Centers for Disease Control and Prevention.*)

24. *U.S. Life Expectancy* The life expectancy of a male born in 1990 was 71.8 years, and the life expectancy of a male born in 2010 rose to 75.6 years. Estimate the life expectancy of a male born in 2000. (*Source: Centers for Disease Control and Prevention.*)

25. *U.S. Population* The population of the United States in 1970 was 205 million, and in 2010 it was 308 million. Estimate the population in 1990. (*Source: U.S. Census Bureau.*)

26. *Distance Traveled* A car is moving at a constant speed on an interstate highway. After 1 hour the car passes the 103-mile marker and after 5 hours the car passes the 391-mile marker. What mile marker does the car pass after 3 hours?

27. *U.S. Median Income* In 1999 the median family income was $40,700, and in 2009 it was $49,800. Estimate the median family income in 2004.

28. *Estimating Fish Populations* In 2008 there were approximately 3200 large-mouth bass in a lake. This number increased to 3800 in 2012. Estimate the number of large-mouth bass in the lake in 2010.

Bibliography

Baase, S. *Computer Algorithms: Introduction to Design and Analysis.* 2nd ed. Reading, Mass.: Addison-Wesley Publishing Company, 1988.

Baker, S., with B. Leak, "Math Will Rock Your World." *Business Week,* January 23, 2006.

Battan, L. *Weather in Your Life.* San Francisco: W. H. Freeman, 1983.

Beckmann, P. *A History of Pi.* New York: Barnes and Noble, Inc., 1993.

Brown, D., and P. Rothery. *Models in Biology: Mathematics, Statistics and Computing.* West Sussex, England: John Wiley and Sons Ltd, 1993.

Callas, D. *Snapshots of Applications in Mathematics.* Delhi, New York: State University College of Technology, 1994.

Carr, G. *Mechanics of Sport.* Champaign, Ill.: Human Kinetics, 1997.

Conquering the Sciences. Sharp Electronics Corporation, 1986.

Eves, H. *An Introduction to the History of Mathematics,* 5th ed. Philadelphia: Saunders College Publishing, 1983.

Freedman, B. *Environmental Ecology: The Ecological Effects of Pollution, Disturbance, and Other Stresses.* 2nd ed. San Diego: Academic Press, 1995.

Friedhoff, M., and W. Benzon. *The Second Computer Revolution: Visualization.* New York: W. H. Freeman, 1991.

Garber, N., and L. Hoel. *Traffic and Highway Engineering.* Boston, Mass.: PWS Publishing Co., 1997.

Grigg, D. *The World Food Problem.* Oxford: Blackwell Publishers, 1993.

Haefner, L. *Introduction to Transportation Systems.* New York: Holt, Rinehart and Winston, 1986.

Harrison, F., F. Hills, J. Paterson, and R. Saunders. "The measurement of liver blood flow in conscious calves." *Quarterly Journal of Experimental Physiology* 71: 235–247.

Historical Topics for the Mathematics Classroom, Thirty-first Yearbook. National Council of Teachers of Mathematics, 1969.

Horn, D. *Basic Electronics Theory.* Blue Ridge Summit, Penn.: TAB Books, 1989.

Howells, G. *Acid Rain and Acid Waters.* 2nd ed. New York: Ellis Horwood, 1995.

Karttunen, H., P. Kroger, H. Oja, M. Poutanen, K. Donner, eds. *Fundamental Astronomy.* 2nd ed. New York: Springer-Verlag, 1994.

Lack, D. *The Life of a Robin.* London: Collins, 1965.

Lancaster, H. *Quantitative Methods in Biological and Medical Sciences: A Historical Essay.* New York: Springer-Verlag, 1994.

Mannering, F., and W. Kilareski. *Principles of Highway Engineering and Traffic Analysis.* New York: John Wiley and Sons, 1990.

Mar, J., and H. Liebowitz. *Structure Technology for Large Radio and Radar Telescope Systems.* Cambridge, Mass.: The MIT Press, 1969.

Meadows, D. *Beyond the Limits.* Post Mills, Vermont: Chelsea Green Publishing Co., 1992.

Miller, A., and J. Thompson. *Elements of Meteorology.* 2nd ed. Columbus, Ohio: Charles E. Merrill Publishing Company, 1975.

Miller, A., and R. Anthes. *Meteorology.* 5th ed. Columbus, Ohio: Charles E. Merrill Publishing Company, 1985.

Motz, L., and J. Weaver. *The Story of Mathematics.* New York: Plenum Press, 1993.

Nilsson, A. *Greenhouse Earth.* New York: John Wiley and Sons, 1992.

Paetsch, M. *Mobile Communications in the U.S. and Europe: Regulation, Technology, and Markets.* Norwood, Mass.: Artech House, Inc., 1993.

Pearl, R., T. Edwards, and J. Miner. "The growth of *Cucumis melo* seedlings at different temperatures." *J. Gen. Physiol.* 17: 687–700.

Pennycuick, C. *Newton Rules Biology.* New York: Oxford University Press, 1992.

Pielou, E. *Population and Community Ecology: Principles and Methods.* New York: Gordon and Breach Science Publishers, 1974.

Pokorny, C., and C. Gerald. *Computer Graphics: The Principles behind the Art and Science.* Irvine, Calif.: Franklin, Beedle, and Associates, 1989.

Ronan, C. *The Natural History of the Universe.* New York: MacMillan Publishing Company, 1991.

Sharov, A., and I. Novikov. *Edwin Hubble, The Discoverer of the Big Bang Universe.* New York: Cambridge University Press, 1993.

Smith, C. *Practical Cellular and PCS Design.* New York: McGraw-Hill, 1998.

Stent, G. S. *Molecular Biology of Bacterial Viruses.* San Francisco: W. H. Freeman, 1963.

Taylor, J. *DVD Demystified.* New York: McGraw-Hill, 1998.

Taylor, W. *The Geometry of Computer Graphics.* Pacific Grove, Calif.: Wadsworth and Brooks/Cole, 1992.

Thomas, D. *Swimming Pool Operators Handbook.* National Swimming Pool Foundation of Washington, D.C., 1972.

Thomas, V. *Science and Sport.* London: Faber and Faber, 1970.

Thomson, W. *Introduction to Space Dynamics.* New York: John Wiley and Sons, 1961.

Triola, M. *Elementary Statistics.* 7th ed. Reading, Mass.: Addison-Wesley Publishing Company, 1998.

Tucker, A., A. Bernat, W. Bradley, R. Cupper, and G. Scragg. *Fundamentals of Computing I: Logic, Problem Solving, Programs, and Computers.* New York: McGraw-Hill, 1995.

Turner, R. K., D. Pierce, and I. Bateman. *Environmental Economics, An Elementary Approach.* Baltimore: The Johns Hopkins University Press, 1993.

Varley, G., and G. Gradwell. "Population models for the winter moth." *Symposium of the Royal Entomological Society of London* 4: 132–142.

Wang, T. *ASHRAE Trans.* 81, Part 1 (1975): 32.

Weidner, R., and R. Sells. *Elementary Classical Physics,* Vol. 2. Boston: Allyn and Bacon, Inc., 1965.

Wigner, E., "The Unreasonable Effectiveness of Mathematics in the Natural Sciences." *Communictions of Pure and Applied Mathematics,* Vol. 13, No. I, February, 1960.

Williams, J. *The Weather Almanac 1995.* New York: Vintage Books, 1994.

Wright, J. *The New York Times Almanac 1999.* New York: Penguin Group, 1998.

Zeilik, M., S. Gregory, and D. Smith. *Introductory Astronomy and Astrophysics.* 3rd ed. Philadelphia: Saunders College Publishers, 1992.

Answers to Selected Exercises

1 Real Numbers and Algebra

SECTION 1.1 (pp. 9–12)

1. $N = \{1, 2, 3, \dots\}$; $W = \{0, 1, 2, 3, \dots\}$
2. $I = \{\dots, -2, -1, 0, 1, 2, \dots\}$; -5
3. $\frac{p}{q}$, where p and q are integers and $q \neq 0$; $\frac{3}{4}$ (answers may vary)   **4.** Any number that can be expressed as a decimal; irrational; $\sqrt{2}$ (answers may vary)   **5.** $3 + 2 = 2 + 3$ (answers may vary)   **6.** $4 \cdot 3 = 3 \cdot 4$ (answers may vary)
7. $1 \cdot a = a$, for any real number a
8. $0 + a = a$, for any real number a
9. $2 \cdot (3 \cdot 4) = (2 \cdot 3) \cdot 4$ (answers may vary)
10. $2(3 + 4) = 2 \cdot 3 + 2 \cdot 4$ (answers may vary)
11. 12   **12.** 8   **13.** Commutative property for addition
14. Commutative property for multiplication
15. Natural, integer, rational, and real
17. Rational and real   **19.** Rational and real   **21.** Real
23. Natural numbers: 6; whole numbers: 6; integers: -5, 6; rational numbers: $-5, 6, \frac{1}{7}, 0.2$; irrational numbers: $\sqrt{7}$
25. Natural numbers: $\frac{3}{1} = 3$; whole numbers: $\frac{3}{1} = 3$; integers: $\frac{3}{1} = 3$; rational numbers: $\frac{3}{1} = 3, -\frac{5}{8}, 0.\overline{45}$; irrational numbers: $\sqrt{5}, \pi$ **27.** Natural: $\sqrt{9}$; whole: $\sqrt{9}$; integer: $-2, \sqrt{9}$; rational: $-2, \frac{1}{2}, \sqrt{9}, 0.\overline{26}$; irrational: none   **29.** Rational   **31.** Natural   **33.** Integers
35. Identity   **37.** Commutative   **39.** Distributive
41. Associative   **43.** Commutative   **45.** Associative
47. Distributive   **49.** $a + 4$ **51.** $\frac{1}{3}a$ **53.** $x + 1$ **55.** xy
57. $(4 + 5) + b$ or $9 + b$ **59.** $(5 \cdot 10)x$ or $50x$
61. $x + (y + z)$ **63.** $x \cdot (3 \cdot 4)$ or $x \cdot 12$ **65.** $4x + 4y$
67. $5x - 35$ **69.** $-x - 1$ **71.** $a(x - y)$ **73.** $3(4 + x)$
75. $(3 + 7)x$ or $10x$ **77.** $(8 - 2)t$ or $6t$
79. $\left(1 - \frac{3}{4}\right)x$ or $\frac{1}{4}x$ **81.** $3 - 1 + 2z$ or $2 + 2z$ **83.** $8z$
85. $5t$ **87.** 56   **89.** 507   **91.** 60   **93.** 5, natural, integer, and rational   **95.** $43.\overline{3}$, rational   **97.** 78.1, rational
99. 20   **101.** 9900   **103.** 6   **105.** 40   **107.** 9
109. Identity property of 1 (answers may vary)
111. (a) 20.3 million   (b) About 20.1 million   (c) 20.075 million; yes   **113.** No; the commutative property for multiplication   **115.** (a) Valid   (b) Valid   (c) Not valid

SECTION 1.2 (pp. 20–22)

1. right   **2.** left   **3.** positive   **4.** negative   **5.** negative
6. negative   **7.** $-a$ **8.** $\frac{b}{a}$ **9.** positive   **10.** $-a - b$
11. 8   **12.** a
13.

15.

17.

19. 6.1   **21.** $\frac{2}{3}$ **23.** $\pi - 1$ **25.** -8 **27.** 6   **29.** 3   **31.** x
33. $x - y$ **35.** Positive   **37.** Both   **39.** Positive   **41.** -56
43. 6.9   **45.** $\pi - 2$ **47.** $-a + b$ **49.** $x - y$
51. $-2z + 2$ **53.** $\frac{1}{3}$ **55.** $-\frac{3}{2}$ **57.** $\frac{1}{\pi}$ **59.** $\frac{1}{a + 3}$
61. $-\frac{b}{2a}$ **63.** $x - 7$ **65.** -2 **67.** -16.6 **69.** -30
71. 2.3   **73.** $-\frac{1}{2}$ **75.** -5 **77.** 72   **79.** $-\frac{5}{14}$
81. $\frac{2}{3}$ **83.** $-\frac{3}{8}$ **85.** -4 **87.** 5   **89.** Undefined
91. 140   **93.** -2 **95.** $-5x$ **97.** $-z$ **99.** 57.2   **101.** $\frac{19}{42}$
103. $-\frac{9}{25}$ **105.** $\frac{21}{55}$ **107.** 5   **109.** 0   **111.** 60   **113.** $\frac{a + b}{2}$
115. 0   **117.** 100   **119.** 1   **121.** $\frac{1}{3}$ **123.** 1   **125.** 1200
127. 16 days   **129.** $2600; answers may vary; $2577.76
131. (a) No, Yes   (b) 6293.6   (c) Answers may vary.
133. (a) Data are increasing.   (b) Answers may vary.

CHECKING BASIC CONCEPTS 1.1 & 1.2 (p. 23)

1. (a) Integer, rational, and real   (b) Natural, integer, rational, and real   (c) Real   (d) Rational and real
2. (a) Commutative; $a + b = b + a$ (b) Associative; $a \cdot (b \cdot c) = (a \cdot b) \cdot c$ (c) Distributive; $a(b + c) = ab + ac$ (d) Distributive; $a(b - c) = ab - ac$ with $a = -1$ **3.** (a) -4 (b) -40.8 (c) $-\frac{5}{12}$
4. About 12 billion gallons

SECTION 1.3 (pp. 34–38)

1. Base: 8; exponent: 3   **2.** $1; \frac{1}{2}$ **3.** 7^3 **4.** 5^2
5. No, $-4^2 = -16$ and $(-4)^2 = 16$ **6.** No, $2^3 = 8$ and $3^2 = 9$ **7.** $\frac{1}{7^n}$ **8.** 6^{m+n} **9.** 5^{m-n} **10.** $3^k x^k$ **11.** 2^{mk}
12. $\frac{x^m}{y^m}$ **13.** x^n **14.** $\frac{b^m}{a^n}$ **15.** $\left(\frac{z}{y}\right)^n$ or $\frac{z^n}{y^n}$ **16.** 500   **17.** 2^3
19. 4^4 **21.** 6^0 **23.** (a) 64   (b) 81   **25.** (a) 16   (b) 27
27. (a) $\frac{8}{27}$ (b) $-\frac{27}{8}$ **29.** (a) $\frac{16}{9}$ (b) $\frac{25}{64}$ **31.** (a) $\frac{x^3}{2}$ (b) ab
33. (a) 3^2 or 9   (b) x^7 **35.** (a) $-15x^3$ (b) $\frac{a^3}{b^2}$ **37.** (a) $\frac{5}{2}$
(b) $\frac{8}{ab^3}$ **39.** (a) 4   (b) 10^2 or 100   **41.** (a) $\frac{1}{b^5}$ (b) $4x^2$
43. (a) $\frac{2b}{3a^2}$ (b) $\frac{3y^6}{x^7}$ **45.** (a) 3^8 or 6561   (b) $\frac{1}{x^6}$
47. (a) $64y^6$ (b) $\frac{1}{16x^4y^{12}}$ **49.** (a) $\frac{64}{x^3}$ (b) $\frac{z^{20}}{32x^5}$ **51.** $\frac{3m}{2n^3}$
53. $\frac{y^4}{4x^6}$ **55.** b^8 **57.** $-\frac{27b^4}{a}$ **59.** $\frac{n^3}{m^3}$ **61.** $\frac{a^{12}}{81b^4}$ **63.** t^2
65. $\frac{2y^2}{x}$ **67.** $-\frac{r^6}{8t^3}$ **69.** $\frac{1}{rt^3}$ **71.** y **73.** $-\frac{125r^{15}}{t^9}$ **75.** 34
77. -8 **79.** 40   **81.** $-\frac{11}{2}$ **83.** -155 **85.** 7   **87.** 5
89. -2 **91.** 1   **93.** 5   **95.** 5   **97.** 5.9×10^6
99. 5×10^7 **101.** 3.2×10^{-2} **103.** 1×10^{-6}

105. 500,000   **107.** 9,300,000   **109.** −0.006
111. 0.00005876   **113.** 6×10^6; 6,000,000
115. 8×10^{-6}; 0.000008   **117.** 2×10^5; 200,000
119. (a) $k = 10$ **(b)** $\frac{1024}{52} \approx 20$ years
121. $V = (2a)^3 = 8a^3$ **123.** $A = (3ab)^2 = 9a^2b^2$
125. (a) $\frac{1}{4}$ **(b)** $\frac{1}{32}$ **127.** About \$15,946 per person
129. $256 \times 2^{20} = 268,435,456$ bytes
131.

Country	2000	2025
China	1.2689×10^9	1.48×10^9
Germany	8.22×10^7	8.09×10^7
India	1.0041×10^9	1.3302×10^9
Mexico	9.99×10^7	1.302×10^8
U.S.	2.821×10^8	3.325×10^8

SECTION 1.4 (pp. 44–49)

1. variable   **2.** equation   **3.** equals   **4.** formula
5. x and y **6.** $3 + 7 = 10$ (answers may vary)
7. $3x = 15$ (answers may vary)   **8.** $3y + x = 5$ (answers
may vary)   **9.** 6   **10.** 4   **11.** 3   **13.** b **15.** b^2
17. False   **19.** −7   **21.** ±13   **23.** 2   **25.** Not a real
number   **27.** 3   **29.** −2   **31.** 0   **33.** $y = 5280x$
35. $A = s^2$ **37.** $y = 3600x$ **39.** $A = \frac{1}{2}bh$
41. $A = \frac{1}{4}\pi d^2$ **43.** $y = 30$ **45.** $y = 1.9$ **47.** $d = 10$
49. $z = 6$ **51.** $y = -\frac{1}{4}$ **53.** $N = -\frac{8}{9}$ **55.** $P = 0.3$
57. $A = 9$ **59.** $V = \frac{5}{4}\pi$ **61.** (ii)   **63.** (iii) **65.** −3 **67.** 2
69.

x	0	2	4	6	8
y	−0.5	4.5	9.5	14.5	19.5

71.

x	−3	−1	1	3
y	15	5	5	15

73.

x	−1	0	1	8
y	−3	−2	−1	0

75.

x	a	$a - 1$	$a^2 - 1$
y	$\sqrt{a} + 1$	$\sqrt{a}$	a

77.

X	Y₁
1	2
2	3.2599
3	4.4422
4	5.5874
5	6.71
6	7.8171
7	8.9129

Y₁■X+³√(X)

79.

X	Y₁
1	2.4495
2	2.2361
3	2
4	1.7321
5	1.4142
6	1
7	0

Y₁■√(7−X)

81. $V = \frac{1}{3}b^2h$; about 91,636,000 ft³, where $b \approx 756$ ft and
$h \approx 481$ ft   **83.** $y = 60t$, 360 mi
85. (a) $S = 250x$ **(b)** 7500
87. (a)

Speed (mph)	10	20	30	40	50	60	70
Braking Distance (ft)	8.3	33.3	75	133.3	208.3	300	408.3

(b) $133.\overline{3}$ ft   **(c)** It quadruples.   **(d)** 100 ft
89. (a) $P = 662x$ **(b)** 15,888 pages   **91. (a)** About 0.464
kilograms   **(b)** About 3.7125 kilograms   **(c)** It increases by
8 times.   **93. (a)** $C = 336x$ **(b)** 10,080 calories
95. (a) About 0.069 m²   **(b)** No   **97.** The taller animal has
half the stepping frequency of the other.   **99. (a)** About 177
beats per minute   **(b)** About 22 beats per minute
101. (a) $\sqrt{3} \approx 1.73$ ft²   **(b)** $4\sqrt{3} \approx 6.93$ m²
103. (a) $28\pi \approx 88$ in.   **(b)** $2.6\pi \approx 8.2$ mi
105. 9 in.   **107.** 3 m

CHECKING BASIC CONCEPTS 1.3 & 1.4 (p. 49)

1. (a) 16   **(b)** $\frac{1}{9}$ **(c)** 32   **(d)** x **(e)** $4x^6y^8$
2. (a) −6   **(b)** −4   **(c)** 58
3. (a) 1.03×10^5 **(b)** 5.23×10^{-4} **(c)** 6.7×10^0
4. (a) 5,430,000   **(b)** 0.0098
5. 900 cubic feet per hour per person

People	10	20	30	40
Ventilation (ft³/hr)	9000	18,000	27,000	36,000

SECTION 1.5 (pp. 59–61)

1. A set of ordered pairs (x, y)
2. Domain: set of all x-values;
Range: set of all y-values
3.

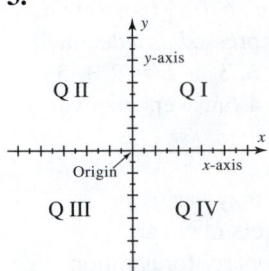

4. Scatterplot
(answers may vary)

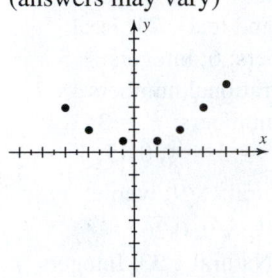

Line Graph
(answers may vary)

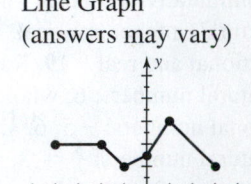

5. No   **6.** Quadrant III   **7.** 3   **8.** −2
9. $D = \{1, 3, 5\}$; $R = \{-4, 2, 6\}$
11. $D = \{-2, -1, 0, 1, 2\}$; $R = \{0, 1, 2, 3\}$
13. $D = \{41, 87, 96\}$; $R = \{24, 53, 67, 88\}$
15. $S = \{(1, 3), (3, 7), (5, 11), (7, 15), (9, 19)\}$
$D = \{1, 3, 5, 7, 9\}$; $R = \{3, 7, 11, 15, 19\}$
17. $S = \{(2006, 139), (2007, 217), (2008, 325),$
$(2009, 474), (2010, 560)\}$;
$D = \{2006, 2007, 2008, 2009, 2010\}$;
$R = \{139, 217, 325, 474, 560\}$
19. $(1, 2)$ is in QI; $(-1, 3)$
is in QII; $(-3, 0)$ is on the
x-axis; $(0, -2)$ is on the
y-axis.

21. $(10, 50)$ is in QI;
$(-30, 20)$ is in QII;
$(-20, -25)$ is in QIII;
$(50, -25)$ is in QIV.

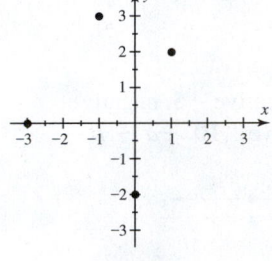

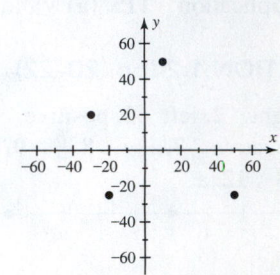

23. $(0, 0), (1, 4), (2, 8), (3, 12)$
25. $(0, 4), (1, 3), (2, 0), (3, -5)$
27. $(0, 1), \left(1, \frac{1}{2}\right), \left(2, \frac{1}{5}\right), \left(3, \frac{1}{10}\right)$
29. $S = \{(-3, 2), (-2, 1), (2, -3), (3, 3)\}$
$D = \{-3, -2, 2, 3\}; R = \{-3, 1, 2, 3\}$
31. $S = \{(-4, 4), (-3, 2), (-2, 0), (0, -3), (2, 4), (4, 4)\};$
$D = \{-4, -3, -2, 0, 2, 4\}; R = \{-3, 0, 2, 4\}$
33. Answers may vary slightly.
$S = \{(\text{Facebook}, 662), (\text{MySpace}, 262),$
$(\text{Twitter}, 67), (\text{Digg}, 18)\}$
$D = \{\text{Facebook, MySpace, Twitter, Digg}\};$
$R = \{18, 67, 262, 662\}$ **35.** $a = b$

37. (a)
$D = \{-3, -2, 0, 1\};$
$R = \{-4, -3, 0, 2, 4\}$
(b) Xmin: -3, Xmax: 1,
Ymin: -4, Ymax: 4
(c) and **(d)**   See the graph.

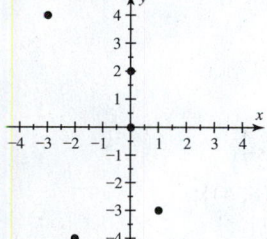

39. (a)
$D = \{-30, 10, 20, 30\};$
$R = \{-50, 20, 40, 50\}$
(b) Xmin: -30, Xmax: 30,
Ymin: -50, Ymax: 50
(c) and **(d)**   See the graph.

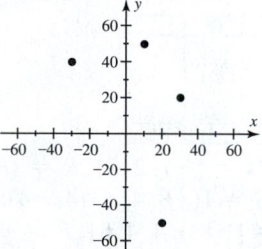

41.

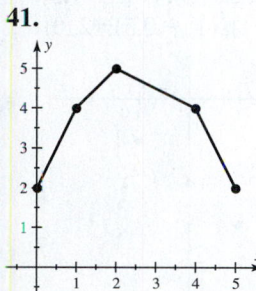

43.

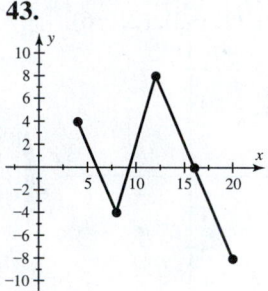

45.

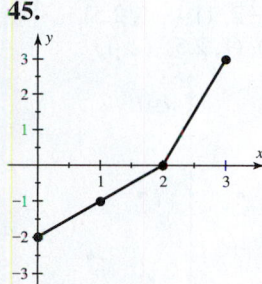

47.

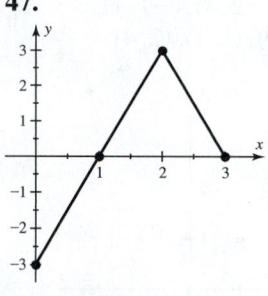

49. $(-2, -6), (-1, -3),$
$(0, 0), (1, 3), (2, 6)$

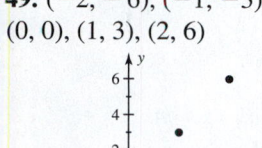
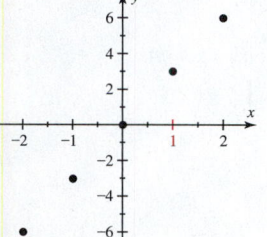

51. $(-2, 4), (-1, 3),$
$(0, 2), (1, 1), (2, 0)$

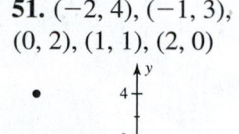
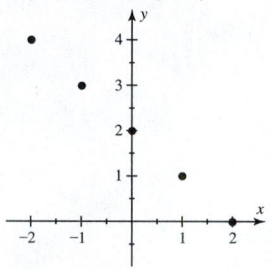

53. $(-2, 1), \left(-1, \frac{1}{2}\right),$
$(0, 0), \left(1, -\frac{1}{2}\right), (2, -1)$

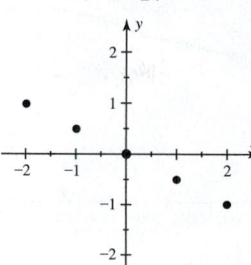

55. $(-2, 3), (-1, 0),$
$(0, -1), (1, 0), (2, 3)$

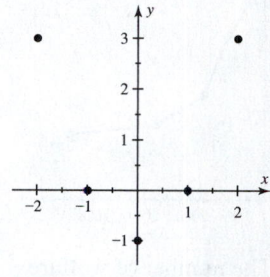

57. $(-2, -4), (-1, -1),$
$(0, 0), (1, -1), (2, -4)$

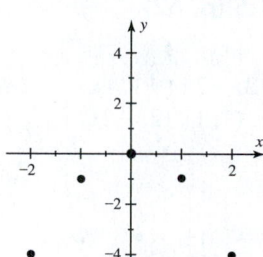

59. $(-2, 2), (-1, 1),$
$(0, 0), (1, 1), (2, 2)$

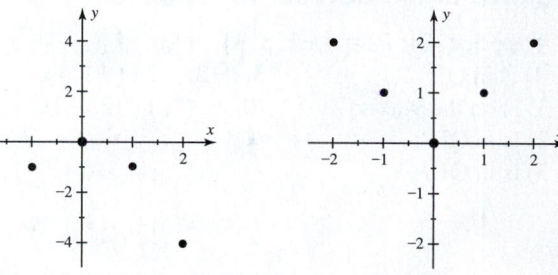

61. 10; 10

[$-10, 10, 1$] by [$-10, 10, 1$]

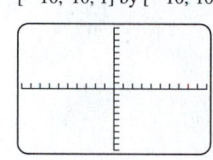

63. 10; 5

[$0, 100, 10$] by [$-50, 50, 10$]

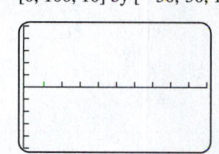

65. 16; 5

[$1980, 1995, 1$] by [$12000, 16000, 1000$]

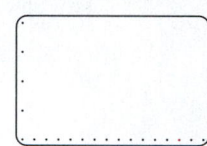

67. $S = \{(-2, 1), (-1, 0), (1, -1), (2, 1)\}$
69. $S = \{(-4, -2), (-2, -2), (0, 2), (2, 1), (4, -1)\}$

71. [$-6, 6, 1$] by [$-6, 6, 1$]

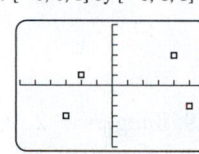

73. [$-30, 30, 5$] by [$-50, 50, 5$]

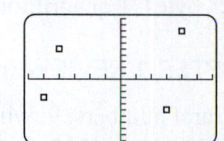

75. [$-200, 200, 50$] by
[$-250, 250, 50$]

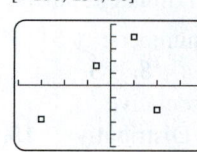

77. (a) See the graph.

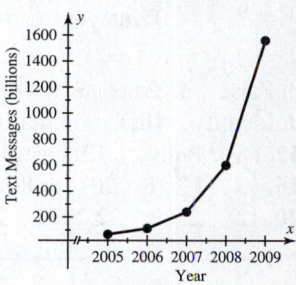

(b) Text messages sent
increased dramatically.

79. (a) See the graph.

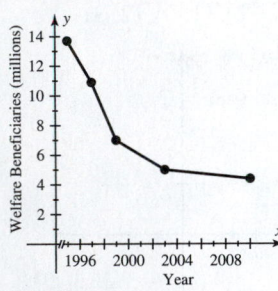

81. (a) See the graph.

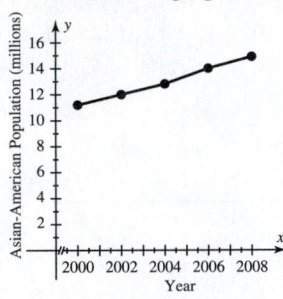

(b) The number of welfare beneficiaries decreased.

(b) Asian-American population increased.

CHECKING BASIC CONCEPTS 1.5 (p. 62)

1. $D = \{-5, 1, 2\}; R = \{-1, 3, 4\}$

2. $(1, 4)$ is in QI;
$(0, -3)$ is on the y-axis;
$(2, -2)$ is in QIV;
$(-2, 3)$ is in QII.

3. $(-2, -2), (-1, 1),$
$(0, 2), (1, 1), (2, -2)$

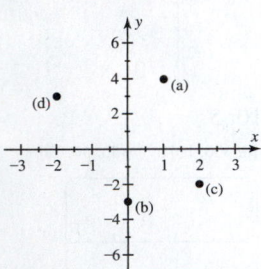

4. Mobile data traffic is increasing.

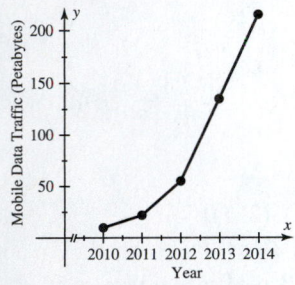

5. 10^{15} bytes, 1 quadrillion

CHAPTER 1 REVIEW (pp. 65–67)

1. Natural numbers: 9; whole numbers: 9; integers: $-2, 9$;
rational numbers: $-2, 9, \frac{2}{5}, 2.68$; irrational numbers:
$\sqrt{11}, \pi$ **2.** Natural numbers: $\frac{6}{2} = 3$; whole numbers:
$\frac{6}{2} = 3, \frac{0}{4} = 0$; integers: $\frac{6}{2} = 3, \frac{0}{4} = 0$; rational numbers:
$\frac{6}{2} = 3, -\frac{2}{7}, 0.\overline{3} = \frac{1}{3}, \frac{0}{4} = 0$; irrational numbers: $\sqrt{6}$
3. False   **4.** True   **5.** 9   **6.** ± 3 **7.** -4 **8.** -3
9. Identity   **10.** Commutative   **11.** Associative
12. Distributive   **13.** Distributive   **14.** Distributive   **15.** a
16. $\frac{1}{4}x$ **17.** $(8 \cdot 10)x$ or $80x$ **18.** $(5 - 3)z$ or $2z$ **19.** 95
20. 12   **21.** 9   **22.** 5.72   **23.** 40   **24.** 1   **25.** 150   **26.** 0

27.
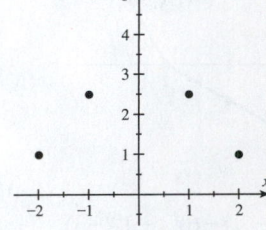

28. 3.2   **29.** $\frac{2}{3}$ **30.** $\frac{5}{4}$ **31.** $2x - 3$

32. $-(a + b) = -a - b$ **33.** -4 **34.** 2   **35.** -4
36. $-\frac{3}{4}$ **37.** $-\frac{1}{2}$ **38.** $\frac{29}{110}$ **39.** 4   **40.** $\frac{10}{7}$ **41.** $9x$ **42.** $7z$
43. $-\frac{33}{910}$ **44.** 30; 5   **45.** Base: 4; exponent: -2
46. $3^{\pi} \approx 31.54; \pi^3 \approx 31.01$; not equal   **47.** -16 **48.** 16
49. 1   **50.** $\frac{27}{8}$ **51.** $\frac{1}{64}$ **52.** 25   **53.** $\frac{9}{125}$ **54.** 8
55. $\frac{1}{4^2}$ or $\frac{1}{16}$ **56.** 10^2 or 100   **57.** x^5 **58.** 3^{11} **59.** $\frac{1}{2a^6}$
60. $\frac{5a^2}{b^3}$ **61.** 2^8 **62.** $\frac{1}{x^{15}}$ **63.** $\frac{16y^6}{x^4}$ **64.** 4^5a^5 or $1024a^5$
65. $\frac{125x^9}{27z^{12}}$ **66.** $\frac{z^2}{9x^8y^6}$ **67.** $\frac{9b^{14}}{16a^8}$ **68.** $3mn^5$ **69.** $\frac{8r^6}{t^6}$ **70.** $\frac{9r^4t^6}{16}$
71. 29   **72.** -3 **73.** 40   **74.** $\frac{11}{3}$ **75.** $\frac{5}{2}$ **76.** 11
77. 1.86×10^5 **78.** 3.4×10^{-4} **79.** 45,000   **80.** 0.00923
81. $y = 12x$ **82.** $A = 6\pi r^2$ **83.** $y = 36$ **84.** $d = 8$
85. $N = \frac{3}{2}$ **86.** $P = -10$ **87.** $A = 10$ **88.** $V = 27$
89. (iii)   **90.** $a = \frac{3}{2}$

91.

x	-2	-1	0	1	2
y	-7	0	1	2	9

92.

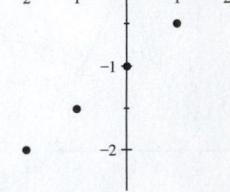

93. $D = \{-1, 2, 3\}; R = \{-6, 1, 3, 7\}$
94. $S = \{(-8, 4), (-4, -4), (4, 0), (8, 4)\};$
$D = \{-8, -4, 4, 8\}; R = \{-4, 0, 4\}$
95. $(-2, 6), (-1, 3),$
$(0, 0), (1, -3), (2, -6)$

96. $(-2, -2), (-1, -1.5),$
$(0, -1), (1, -0.5), (2, 0)$

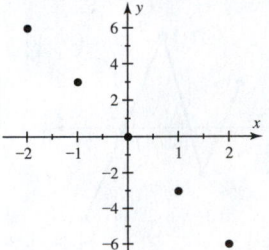

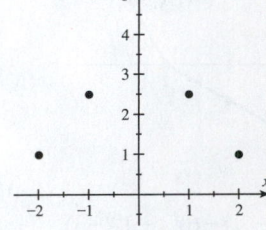

97. $(-2, 4), (-1, 1),$
$(0, 0), (1, 1), (2, 4)$

98. $(-2, 1), (-1, 2.5),$
$(0, 5), (1, 2.5), (2, 1)$

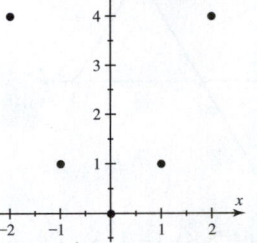

99. $(-2, 2)$ is in QII;
$(-1, -3)$ is in QIII;
$(2, -1)$ is in QIV;
$(0, 1)$ is on the y-axis.

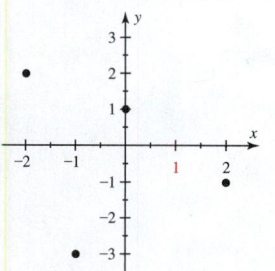

100. $(10, 20)$ is in QI;
$(-15, -5)$ is in QIII;
$(20, -10)$ is in QIV;
$(-5, 0)$ is on the x-axis.

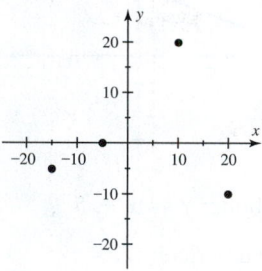

5.

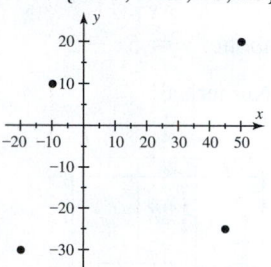

6. (a) $\frac{3}{2}$ **(b)** 20 **(c)** $\frac{3}{4}$ **(d)** 0 **7. (a)** 6 **(b)** -2
8. (a) $-\frac{4}{5}$ **(b)** $\frac{b+1}{2}$ **9. (a)** $-\frac{5}{18}$ **(b)** -4.2
(c) -5 **(d)** 2 **10. (a)** $\frac{1}{25}$ **(b)** 1 **(c)** $\frac{16}{625}$ **(d)** $\frac{4}{9}$ **(e)** 125
(f) $\frac{1}{4}$ **11. (a)** x^2y^3 **(b)** $\frac{8y^{15}}{3x^3}$ **(c)** $\frac{8y^3}{z^6}$ **(d)** $\frac{4}{9x^6y^4}$
12. 0.00052 **13.** 3.4×10^6 **14.** $H = \frac{x}{60}$ **15.** $5; -1$
16. $D = \{-3, -1, 2\}; R = \{-4, 2, 3\}$
17. $(2, 1)$ is in QI; $(-2, 3)$ is in QII; $(-1, -2)$ is in QIII;
$(1, -1)$ is in QIV; $(0, 2)$ is on the y-axis.

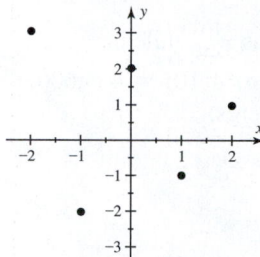

101. $9; 2$

$[-9, 9, 1]$ by $[-6, 6, 3]$

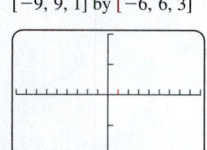

102. $4; 3$

$[-20, 20, 5]$ by $[-12, 12, 4]$

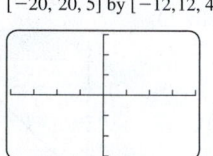

103.
$D = \{-1, 0, 2, 3, 4\}$;
$R = \{-1, 0, 2, 3, 4\}$

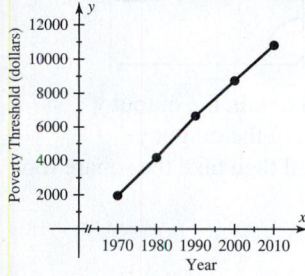

104.
$D = \{-20, -10, 45, 50\}$;
$R = \{-30, -25, 10, 20\}$

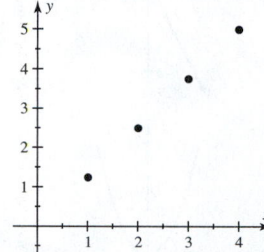

18. $S = \{(-30, 20), (-20, 20), (-10, 10), (10, 30),$
$(20, 10), (30, -20)\}; D = \{-30, -20, -10, 10, 20, 30\};$
$R = \{-20, 10, 20, 30\}$

19. $y = 1.25x$

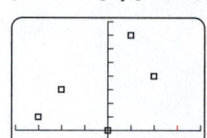

105. (a) Less **(b)** About \$45.6 billion
106. (a) \$2525 **(b)** \$816.08
107. (a) 5.8×10^8 miles **(b)** $66,700$ miles per hour
108. $d = 40t$ **109.** 221 bpm; 74 bpm
110. The threshold has increased.

20. Window sizes may vary.

$[-20, 20, 5]$ by $[-5, 40, 5]$

21. $-15°C$ **22.** 5.1574×10^8 gallons
23. $\sqrt{\frac{25}{\pi}} \approx 2.82$ feet **24.** $M = 61t$

111. $A = (4ab)^2 = 16a^2b^2$ **112.** $V = (5z)^3 = 125z^3$

CHAPTER 1 TEST (pp. 68–69)

1. Natural numbers: $\sqrt{9}$; whole numbers: $\sqrt{9}$; integers: -5,
$\sqrt{9}$; rational numbers: $-5, \frac{2}{3}, \sqrt{9}, -1.83$; irrational numbers: $-\frac{1}{\sqrt{5}}, \pi$
2. (a) Identity **(b)** Associative **(c)** Commutative
(d) Distributive
3. (a) a **(b)** x **(c)** $8x$ **(d)** a^2 **4.** 31.2

2 Linear Functions and Models

SECTION 2.1 (pp. 84–89)

1. function **2.** y equals f of x **3.** symbolic **4.** numerical
5. domain **6.** range **7.** one **8.** F
9. T **10.** $(3, 4)$; 3; 6 **11.** (a, b) **12.** d **13.** 1 **14.** equal
15. Yes **16.** Yes **17.** No **18.** No **19.** Yes **20.** No
21. -6; -2 **23.** 0; $\frac{3}{2}$ **25.** 25; $\frac{9}{4}$ **27.** 3; 3
29. 13; -22 **31.** $-\frac{1}{2}$; $\frac{2}{5}$
33. (a) $I(x) = 36x$ **(b)** $I(10) = 360$; There are 360 inches
in 10 yards. **35. (a)** $M(x) = \frac{x}{5280}$
(b) $M(10) = \frac{10}{5280} \approx 0.0019$; There is $\frac{10}{5280}$ mile in
10 feet. **37. (a)** $A(x) = 43,560x$ **(b)** $A(10) = 435,600$;
There are 435,600 square feet in 10 acres.
39. $f = \{(1, 3), (2, -4), (3, 0)\}$;
$D = \{1, 2, 3\}$; $R = \{-4, 0, 3\}$
41. $f = \{(a, b), (c, d), (e, a), (d, b)\}$;
$D = \{a, c, d, e\}$; $R = \{a, b, d\}$

43.
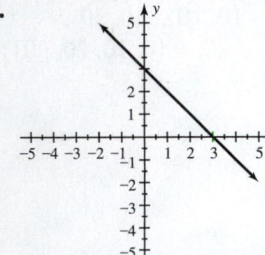

45.
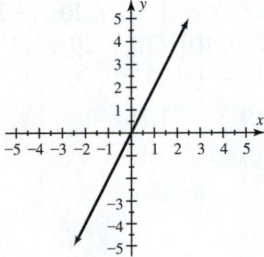

47.
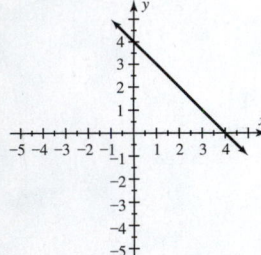

49.
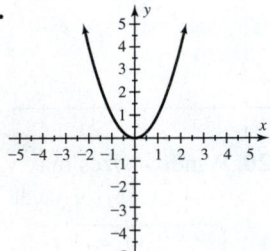

51.
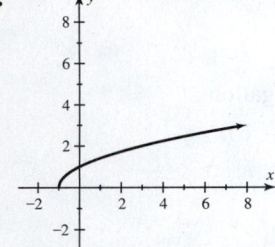

53. 3; -1 **55.** 0; 2 **57.** -4; -3 **59.** 5.5; 3.7
61. 26.9; in 1990 average fuel efficiency was 26.9 mpg.
63. Numerical:

x	-3	-2	-1	0	1	2	3
$y = f(x)$	2	3	4	5	6	7	8

Graphical:

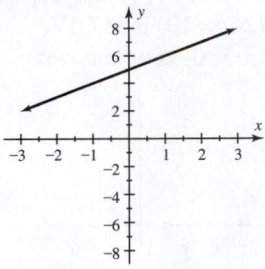

Symbolic: $y = x + 5$

65. Numerical:

x	-3	-2	-1	0	1	2	3
$y = f(x)$	-17	-12	-7	-2	3	8	13

Graphical:

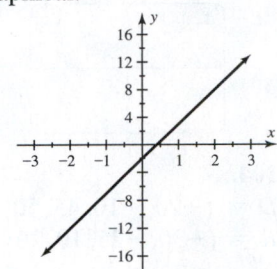

Symbolic: $y = 5x - 2$

67. Numerical

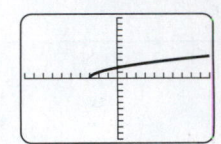

Graphical
$[-10, 10, 1]$ by $[-10, 10, 1]$

69. Numerical

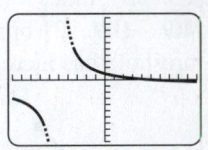

Graphical (Dot Mode)
$[-10, 10, 1]$ by $[-10, 10, 1]$

71. Subtract $\frac{1}{2}$ from the input x to obtain the output y.
73. Divide the input x by 3 to obtain the output y.
75. Subtract 1 from the input x and then take the square root
to obtain the output y.
77. $f(x) = 0.50x$

Miles	10	20	30	40	50	60	70
Cost	$5	$10	$15	$20	$25	$30	$35

79. 1825; In 2011 there were 1825 billion, or 1.825 trillion, World Wide Web searches.
81. $D: -2 \leq x \leq 2$; $R: 0 \leq y \leq 2$
83. $D: -2 \leq x \leq 4$; $R: -2 \leq y \leq 2$
85. D: All real numbers; $R: y \geq -1$
87. $D: -3 \leq x \leq 3$; $R: -3 \leq y \leq 2$
89. $D = \{1, 2, 3, 4\}$; $R = \{5, 6, 7\}$
91. All real numbers   **93.** All real numbers   **95.** $x \neq 5$
97. All real numbers   **99.** $x \geq 1$ **101.** All real numbers
103. $x \neq 0$ **105. (a)** 1726; 1726 whales were sighted
in 2008.   **(b)** $D = \{2005, 2006, 2007, 2008, 2009\}$,
$R = \{649, 1265, 959, 1726, 1010\}$ **(c)** Increased
every other year   **107.** $D = \{1, 2, 3, \ldots, 20\}$;
$R = \{200, 400, 600, \ldots, 4000\}$ **109.** No   **111.** Yes
113. (a) 0.2   **(b)** Yes. Each month has one average amount
of precipitation.   **(c)** 2, 3, 7, 11   **115.** Yes. D: All real
numbers; R: All real numbers   **117.** No
119. Yes. $D: -4 \leq x \leq 4$; $R: 0 \leq y \leq 4$
121. Yes. D: All real numbers; $R: y = 3$ **123.** No
125. Yes. $D = \{-6, -4, 2, 4\}$; $R = \{-4, 2\}$ **127.** Yes
129. No   **131.** The person walks away from home, then
turns around and walks back a little slower.
133.

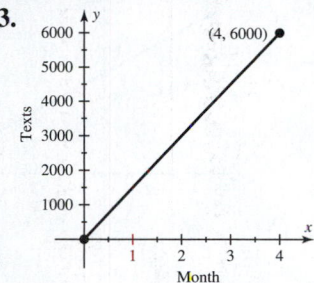

SECTION 2.2 (pp. 99–103)

1. $mx + b$ **2.** b **3.** line   **4.** horizontal   **5.** 7   **6.** 0   **7.** T
8. F   **9.** Carpet costs \$2 per square foot. Ten square feet of
carpet costs \$20.   **10.** The rate at which water is leaving the
tank is 4 gallons per minute. After 5 minutes the tank contains
80 gallons of water.   **11.** Yes; $m = \frac{1}{2}$, $b = -6$ **13.** No
15. Yes; $m = 0$, $b = -9$ **17.** Yes; $m = -9$, $b = 0$
19. Yes   **21.** No   **23.** Yes; $f(x) = 3x - 6$ **25.** Yes;
$f(x) = -\frac{3}{2}x + 3$ **27.** No   **29.** Yes; $f(x) = 2x$
31. Yes; $f(x) = -4$ **33.** -16; 20   **35.** $\frac{17}{3}$; 2
37. -22; -22 **39.** -2; 0   **41.** -1; -4 **43.** 1; 1
45. $f(x) = 6x$; 18   **47.** $f(x) = \frac{x}{6} - \frac{1}{2}$; 0   **49.** d.   **51.** b.
53.

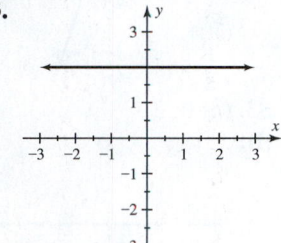

55.

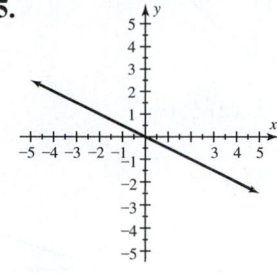

57.

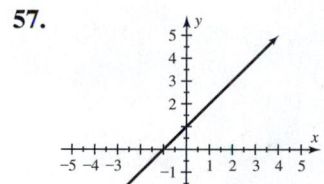

59.

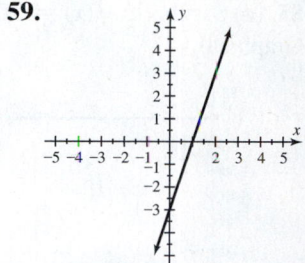

61.

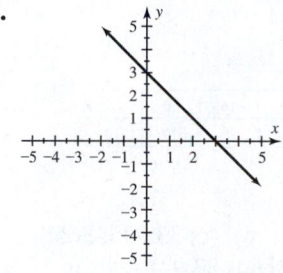

63. $f(x) = \frac{1}{16}x$ **65.** $f(t) = 65t$ **67.** $f(x) = 24$ **69.** a

71. (a) f multiplies the input x by -2 and then adds 1 to obtain the output y.

73. (a) f multiplies the input x by $\frac{1}{2}$ and then subtracts 1 to obtain the output y.

(b)

x	-2	0	2
$y = f(x)$	5	1	-3

(b)

x	-2	0	2
$y = f(x)$	-2	-1	0

(c)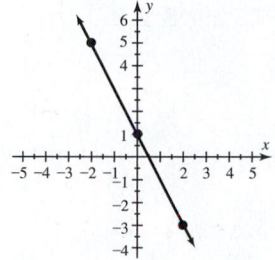

(c)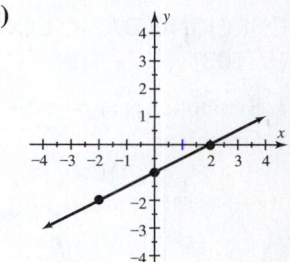

75. (a)

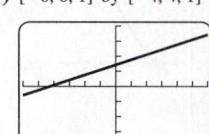

(b) $[-6, 6, 1]$ by $[-4, 4, 1]$

77. (a)

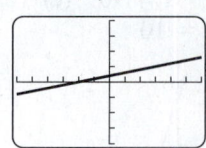

(b) $[-6, 6, 1]$ by $[-4, 4, 1]$

79. b.   **81.** c.   **83. (a)** $G(x) = \frac{X}{E}$ **(b)** $C(x) = \frac{3x}{E}$

85. (a) Symbolic: $f(x) = 70$
Graphical:

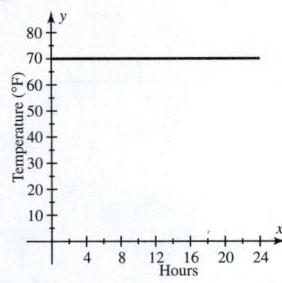

(b)

Hours	0	4	8	12	16	20	24
Temp. (°F)	70	70	70	70	70	70	70

(c) Constant
87. $D(t) = 60t + 50$
89. (a) $K(x) = 93x$ **(b)** $A(x) = x$ **(c)** 33,945; 365;
On average, someone under 18 sends 33,945 texts in 1 year
while someone over 65 sends 365 texts. **91. (a)** 1.6 million
(b) 0.1 million **(c)** $f(x) = 0.1x + 1.6$ **(d)** 2.2 million
93. (a) 563 million **(b)** In 2006, there were about
123 million users. **(c)** Users increased, on average, by 110
million per year. **95. (a)** $V(T) = 0.5T + 137$
(b) 162 cm³ **97.** $f(x) = 40x$; about 2.92 pounds
99. (a) 55% **(b)** 4% **(c)** $P(x) = 4x + 55$ **(d)** 67%

CHECKING BASIC CONCEPTS 2.1 & 2.2 (p. 103)

1. Symbolic: $f(x) = x^2 - 1$
Graphical:

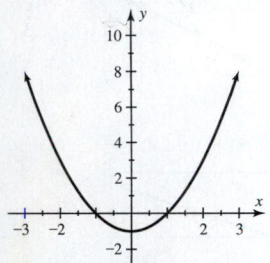

2. (a) D: $-3 \le x \le 3$; R: $-4 \le y \le 4$ **(b)** 0; 4
(c) No. The graph is not a line.
3. (a) Yes **(b)** No **(c)** Yes **(d)** Yes
4. $f(-2) = 10$

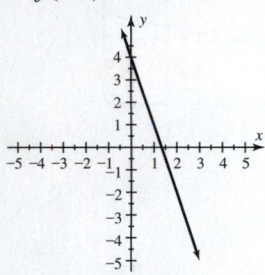

5. $f(x) = \frac{1}{2}x - 1$ **6. (a)** 32.2; in 1990 the median age was
about 32 years. **(b)** 0.225: the median age is increasing
by 0.225 years each year. 27.7: in 1970 the median age was
27.7 years.

SECTION 2.3 (pp. 113–117)

1. y; x **2.** 0 **3.** rises **4.** horizontal **5.** vertical
6. slope–intercept **7.** In 1999, $9 billion was spent on
pet health care. **8.** Initially, the tank has 1000 gallons in it.
9. In 2006, 123 million people used Skype. **10.** In 0 days
a person sends 0 texts. **11.** 2 **13.** $-\frac{2}{3}$ **15.** 0 **17.** 2
19. $-\frac{2}{3}$ **21.** 0 **23.** Undefined **25.** -10 **27.** $\frac{3}{5}$
29. $-0.\overline{21}$ **31.** $\frac{3b}{a}$ **33.** -1

35. (a) $-\frac{3}{4}$
(b)

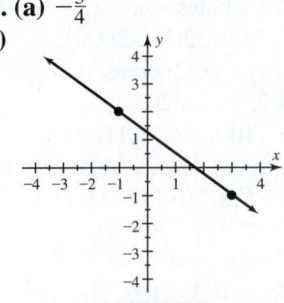

37. (a) $\frac{2}{3}$
(b)

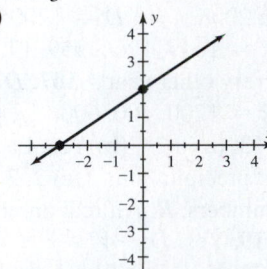

(c) The graph falls 3 units
for every 4-unit increase in x

(c) The graph rises 2 units
for every 3-unit increase in x

39.

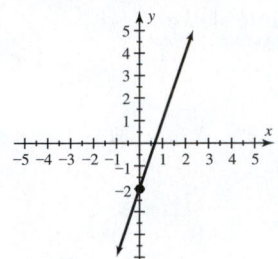

41.

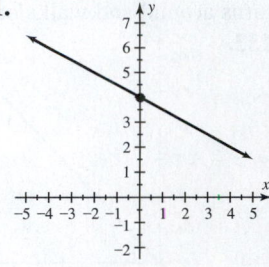

43.

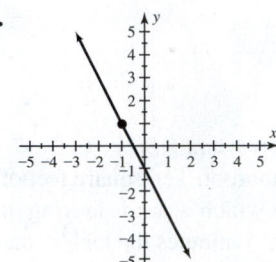

45.

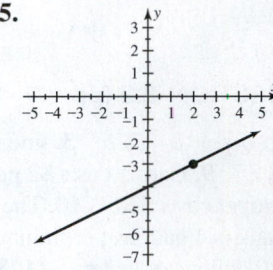

47. (a) 1; 2
(b)

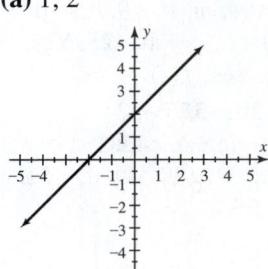

49. (a) -3; 2
(b)

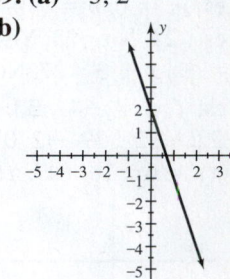

51. (a) $\frac{1}{3}$; 0
(b)

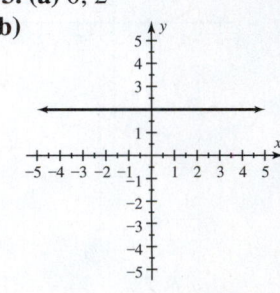

53. (a) 0; 2
(b)

55. (a) $-1; 3$
(b)

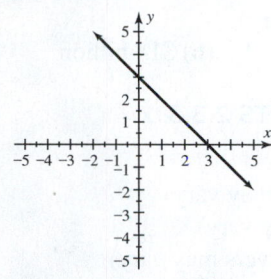

57. $y = -x + 4$ **59.** $y = 2$
61. $y = x - 2$ **63.** $y = 3x - 5$
65. $y = \frac{3}{2}x - \frac{3}{2}$ **67.** $2; -2$
69. $-7; 4$ **71. (a)** 3 **(b)** $f(x) = 2x - 1$
73. (a) 5 **(b)** $f(x) = \frac{3}{2}x + 5$ **75.** c.
77. b.   **79. (a)** $m_1 = 50, m_2 = 0, m_3 = 150$
(b) m_1: Water is being added at the rate of 50 gallons
per hour. m_2: The pump is neither adding nor removing
water. m_3: Water is being added at the rate of 150 gallons
per hour.   **(c)** The pool initially contained 100 gallons of
water. Water was added at the rate of 50 gallons per hour
for the first 4 hours. Then the pump was turned off for
2 hours. Finally, water was added at the rate of 150 gallons
per hour for 2 hours. The pool contains 600 gallons after
8 hours.
81. (a) $m_1 = 100, m_2 = 25, m_3 = -100$ **(b)** m_1: Water
is being added at the rate of 100 gallons per hour. m_2:
Water is being added at the rate of 25 gallons per hour.
m_3: Water is being removed at the rate of 100 gallons per
hour.   **(c)** The pool initially contained 100 gallons of
water. Water was added at the rate of 100 gallons per hour
for the first 2 hours. Then water was added at the rate of
25 gallons per hour for 4 hours. Finally, water was
removed at the rate of 100 gallons per hour for 2 hours.
The pool contains 200 gallons after 8 hours.
83. (a) $m_1 = 50, m_2 = 0, m_3 = -20, m_4 = 0$
(b) m_1: The car is moving away from home at the rate of
50 miles per hour. m_2: The car is not moving. m_3: The car
is moving toward home at the rate of 20 miles per hour.
m_4: The car is not moving.   **(c)** The car starts at home
and moves away from home at 50 miles per hour for
1 hour. Then the car does not move for 1 hour. The car
then moves toward home at the rate of 20 miles per hour
for 2 hours. Finally, the car stops and does not move for
1 hour. The car is 10 miles from home at the end of
the trip.
85.

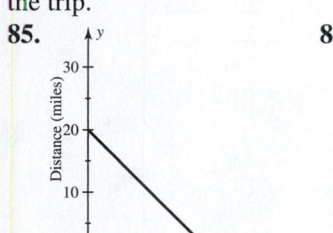

87.

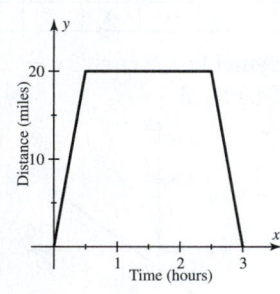

89. (a) $13 billion, $23 billion
(b)

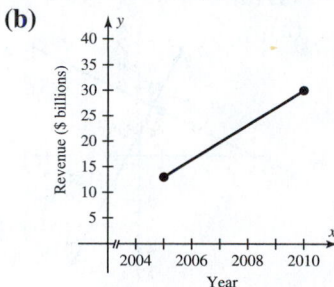

(c) $\frac{10}{3}$ **(d)** Revenue increased on average by $3\frac{1}{3}$ billion
per year.
91. (a) 468,000   **(b)** 18.8, 299   **(c)** On average, the
number of finishers increased by 18,800 per year.
(d) In 2000, there were 299 thousand finishers.
93. 1500 gallons per hour   **95.** $m = 10$; the carpet costs
$10 per square yard.   **97. (a)** $m = \frac{1}{3}$
(b) On average, spending on pet health care increased
by $\frac{1}{3}$ billion per year from 1984 to 2008.
99. (a) $m \approx 1470.6$; $b = 35,000$
(b) About $49,700

SECTION 2.4 (pp. 129–134)

1. 1 **2.** 1.   **3.** $y = mx + b$ **4.** $y = m(x - x_1) + y_1$ or
$y - y_1 = m(x - x_1)$ **5.** $x = 5$ **6.** $y = 2$ **7.** $y = b$
8. $x = h$ **9.** $m_1 = m_2$ **10.** -1 **11.** $y = -\frac{3}{4}x - 3$
(answers may vary)   **12.** $y = \frac{4}{3}x + 2$ (answers may vary)
13. $x = 1$ (answers may vary)   **14.** $y = -5$
(answers may vary)   **15.** $y = -\frac{3}{4}x - \frac{1}{4}$ **17.** $y = \frac{1}{3}x + \frac{8}{3}$
19. Yes   **21.** No   **23.** d.   **25.** a.   **27.** f.
29. $y = -2(x - 2) - 3$
31. $y = 1.3(x - 1990) + 25$
33. $y = \frac{2}{3}(x - 1) + 3$ or $y = \frac{2}{3}(x + 5) - 1$
35. $y = 2(x - 1980) + 5$ or $y = 2(x - 2000) + 45$
37. $y = -\frac{2}{3}(x - 6) + 0$ or $y = -\frac{2}{3}(x - 0) + 4$
39.

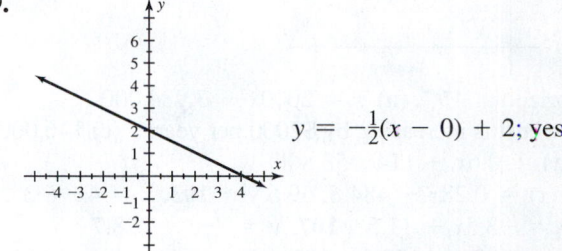

$y = -\frac{1}{2}(x - 0) + 2$; yes

41. $y = 2x - 4$ **43.** $y = \frac{1}{2}x + 3$ **45.** $y = 22x - 43$
47. $y = -3x - 2$ **49.** $y = -\frac{3}{4}x - 2$ **51.** $y = -\frac{1}{3}x - 5$
53. $y = -x + 1$ **55.** $y = \frac{1}{3}x - \frac{2}{3}$ **57.** $y = -3x + 3$
59. $y = 4x - 1$ **61.** $y = -\frac{5}{3}x - 3$ **63.** $y = 3x + 14$
65. $y = \frac{9}{10}x - \frac{31}{20}$ **67.** $y = -\frac{1}{c}x + b$

69. (a) $y = -2x + 2$
(b)

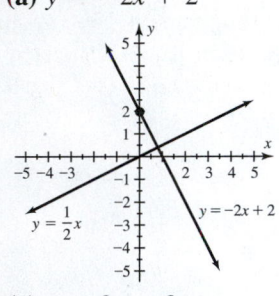

71. (a) $y = \frac{1}{2}x + \frac{5}{2}$
(b)

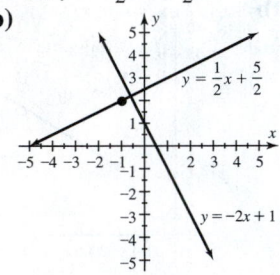

73. (a) $y = 3x - 2$
(b)

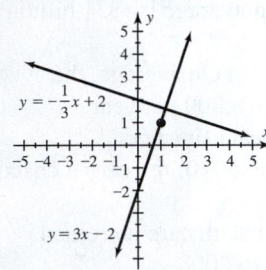

75. (a) $y_1 = 2x; y_2 = -\frac{1}{2}x$ **(b)** $m_1m_2 = 2\left(-\frac{1}{2}\right) = -1$
77. (a) $y_1 = -\frac{3}{2}x - 1; y_2 = \frac{2}{3}x - 1$
(b) $m_1m_2 = -\frac{3}{2} \cdot \frac{2}{3} = -1$ **79.** $x = -1$
81. $y = -\frac{5}{6}$ **83.** $x = 4$ **85.** $x = -\frac{2}{3}$ **87.** Yes;
$y = 4x - 8$ **89.** No **91. (a)** Leaving; 50 gal
(b) x-int: 8; y-int: 80; the tank is empty after 8 min and has
80 gal initially **(c)** $y = -10x + 80$; water is leaving the
tank at 10 gal per min. **(d)** $D: 0 \le x \le 8; R: 0 \le y \le 80$
93. (a) Two acres have 100 people and 4 acres have
200 people. **(b)** A zero-acre parcel has no people.
(c) $y = 50x$ **(d)** The land has 50 people per acre,
on average. **(e)** $P(x) = 50x$
95. (a)

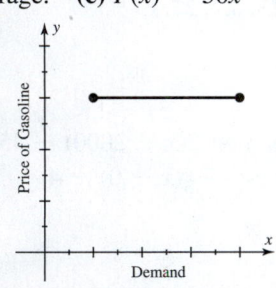

(b) Horizontal **97. (a)** $y = 2000x - 3{,}984{,}000$
(b) The cost is increasing by $2000 per year. **(c)** $46,000
99. $f(x) = 0.6x - 1146$; 58.8 lb
101. $f(x) = 0.28x - 484.8$; 69.6 yr **103.** $y = 4x - 3$
105. $y = -3.5x + 11.5$ **107.** $y = \frac{7}{60}x - 228.7$
109. $y = 0.09x - 173.8$ **111. (a)** $a = 0.30, b = 189.20$
(b) The fixed cost of owning the car for one month
113. (a) $R(c) = 4c - 14.6$ **(b)** About 10
115. (a)

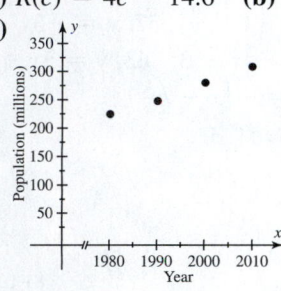

(b) $m = 2.7, x_1 = 1980, y_1 = 227$ (answers may vary)
(c) 335 million (answers may vary)
117. (a) $f(x) = \frac{1}{3}(x - 1984) + 4$ **(b)** $14 billion

CHECKING BASIC CONCEPTS 2.3 & 2.4 (pp. 134–135)

1. (a) $m = -3$; (5, 7) (answers may vary)
(b) $m = 0$; (1, 10) (answers may vary)
(c) m is undefined; $(-5, 0)$ (answers may vary)
(d) $m = 5$; (0, 3) (answers may vary) **2. (a)** 2
(b) $y = 2(x - 2) - 4$ or $y = 2(x - 5) + 2$;
$y = 2x - 8$ **(c)** x-int: 4; y-int: -8 **3.** $x = -2; y = 5$
4. $y = 2x - 8; y = -\frac{1}{2}x - 3$ **5. (a)** Toward home
(b) -50; the car is moving toward home at 50 mph.
(c) x-int: 5, after 5 hours the driver is home; y-int: 250,
the driver is initially 250 miles from home.
(d) $a = -50, b = 250$ **(e)** $D: 0 \le x \le 5$;
$R: 0 \le y \le 250$

CHAPTER 2 REVIEW (pp. 137–142)

1. $-7; 0$ **2.** $-22; 2$ **3.** $-2; 1$ **4.** $5; 5$ **5. (a)** $P(q) = 2q$
(b) $P(5) = 10$; there are 10 pints in 5 quarts.
6. (a) $f(x) = 4x - 3$ **(b)** $f(5) = 17$; three less than four
times 5 is 17. **7.** $(3, -2)$ **8.** $4; -6$
9.

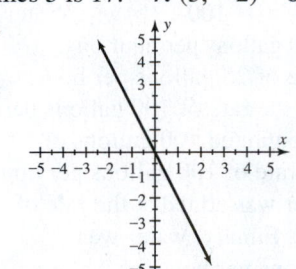

10.

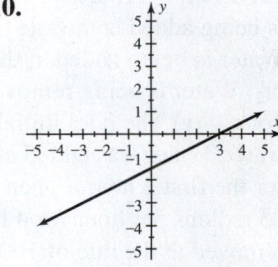

11.

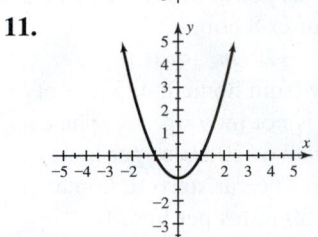

12.

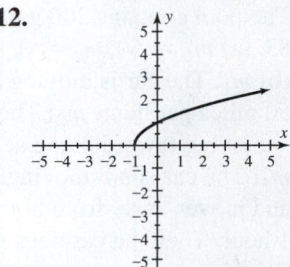

13. $1; 4$ **14.** $1; -2$ **15.** $7; -1$
16. Numerical:

x	-3	-2	-1	0	1	2	3
$y = f(x)$	-11	-8	-5	-2	1	4	7

Symbolic: $f(x) = 3x - 2$
Graphical:

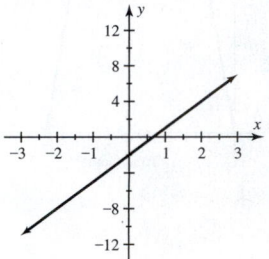

17. D: All real numbers; R: $y \le 4$ **18.** D: $-4 \le x \le 4$;
R: $-4 \le y \le 0$ **19.** Yes   **20.** No
21. $D = \{-3, -1, 2, 4\}$; $R = \{-1, 3, 4\}$; yes
22. $D = \{-1, 0, 1, 2\}$; $R = \{-2, 2, 3, 4, 5\}$; no
23. All real numbers   **24.** $x \ge 0$ **25.** $x \ne 0$
26. All real numbers   **27.** $x \le 5$ **28.** $x \ne 2$
29. All real numbers   **30.** All real numbers   **31.** No
32. Yes   **33.** Yes; $m = -4$, $b = 5$
34. Yes; $m = -1$, $b = 7$ **35.** No   **36.** Yes; $m = 0$, $b = 6$
37. Yes; $f(x) = \frac{3}{2}x - 3$ **38.** No   **39.** 1
40. $f(-2) = -3$; $f(1) = 0$
41.

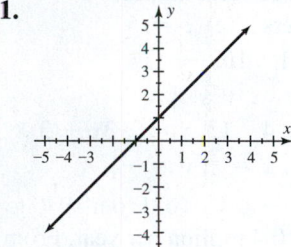

42.

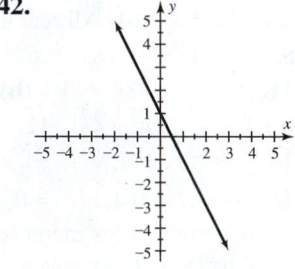

43.

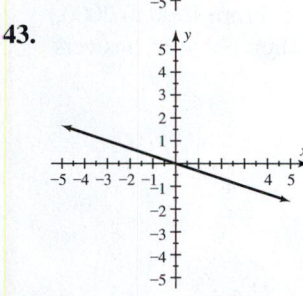

44.

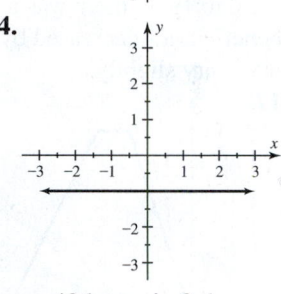

45. $H(x) = 24x$; $H(2) = 48$; there are 48 hours in 2 days.
46. (a)   **(b)** Domain: $x \ge -2$

$[-10, 10, 1]$ by $[-10, 10, 1]$

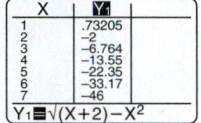

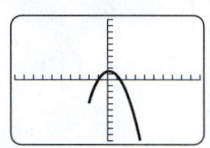

47. -2 **48.** 0   **49.** $\frac{1}{3}$ **50.** Undefined   **51.** $\frac{3}{2}$
52. $-\frac{3}{4}$ **53.** 0   **54.** Undefined
55.   **56.** $-\frac{2}{3}$; 0

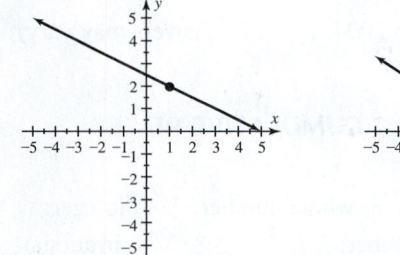

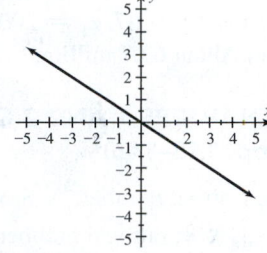

57. $y = -3x + 1$ **58.** $y = 2x + 2$ **59.** -1; 1
60. (a) $m_1 = 250$, $m_2 = 500$, $m_3 = 0$, $m_4 = -750$
(b) m_1: Water is being added to the pool at a rate of
250 gallons per hour. m_2: Water is being added to the pool
at a rate of 500 gallons per hour. m_3: No water is being added
or removed. m_4: Water is being removed from the pool at a
rate of 750 gallons per hour.   **(c)** Initially the pool contains
500 gallons of water. For the first 2 hours water is added
to the pool at a rate of 250 gallons per hour until there are

1000 gallons in the pool. Then water is added at a rate of
500 gallons per hour for 1 hour until there are 1500 gallons
in the pool. For the next hour, no water is added or removed.
Finally, water is removed from the pool at a rate of 750 gallons
per hour for 1 hour until it contains 750 gallons.
61.   **62. (a)** $\frac{1}{2}$; -2
                                          (b)

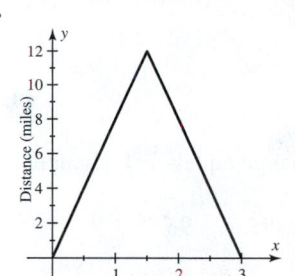

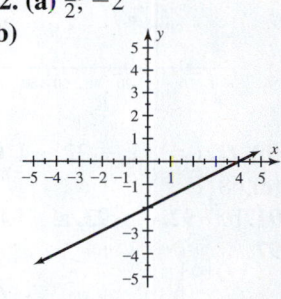

63. Yes   **64.** -2; 2; 4   **65.** $y = -3x + 7$
66. $y = -3(x + 2) + 3$; $y = -3x - 3$
67. $y = \frac{3}{2}x - 3$ **68.** $y = -2x + 2$ **69.** $y = 4x + \frac{13}{5}$
70. $y = -2x - 1$ **71.** $y = 2x - 5$ **72.** $y = \frac{2}{3}x + 1$
73. $y = -x + 2$ **74.** $y = 2x - 3$ **75.** No   **76.** Yes
77. $x = -4$ **78.** $y = -\frac{7}{13}$ **79.** $y = 1$ **80.** $y = -8$
81. No   **82.** Yes; $y = 2x + 5$
83. (a) $m_1 = 1.3$, $m_2 = 2.35$, $m_3 = 2.4$, $m_4 = 2.7$
(b) m_1: From 1920 to 1940 the population increased on
average by 1.3 million per year. The other slopes may be
interpreted similarly.   **84.** For the first 2 minutes the inlet
pipe is open. For the next 3 minutes both pipes are open. For
the next 2 minutes only the outlet pipe is open. Finally, for
the last 3 minutes both pipes are closed.
85. (a) About 25.1
(b) Decreased

$[1885, 1965, 10]$ by $[22, 26, 1]$

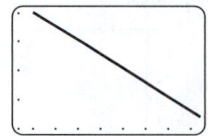

(c) -0.0492; the median age decreased by about 0.0492
year per year.
86. (a) 2.2 million   **(b)** The number of marriages each year
did not change.   **87. (a)** $f(x) = 8x$ **(b)** 8
(c) The total fat increases at the rate of 8 grams per cup.
88. (a)

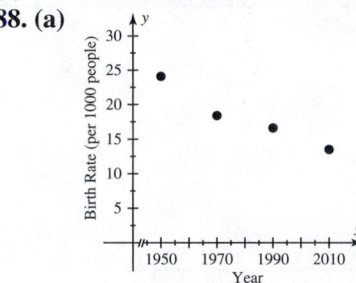

(b) $f(x) = -0.1675x + 350$ **(c)** About 15 per
1000 people (answers may vary)
89. (a) 113; in 1995 there were 113 unhealthy
days.   **(b)** $D = \{1995, 1999, 2000, 2003, 2007\}$
$R = \{56, 87, 88, 100, 113\}$ **(c)** It decreased and then
increased.

90. (a) Linear

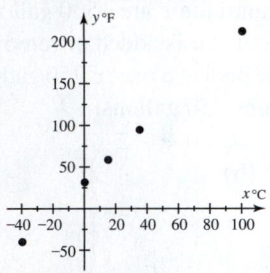

(b) $f(x) = \frac{9}{5}x + 32$; a 1°C change equals $\frac{9}{5}$ °F change.
(c) 68 °F
91. b.   **92.** e.   **93.** a.   **94.** d.   **95.** f.   **96.** c.
97.

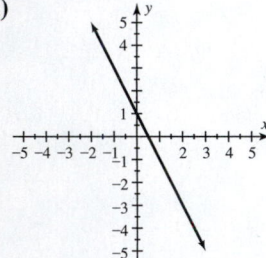

98. (a) $f(x) = 56,000(x - 2006) + 1,100,000$
(b) 1,604,000; total (cumulative) number of U.S. AIDS cases reported in 2015

CHAPTER 2 TEST (pp. 142–143)

1. 46, (4, 46)   **2.** $C(x) = 4x$; $C(5) = 20$, 5 pounds of candy costs $20.
3. (a)   **(b)**

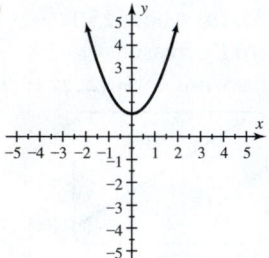

(c)   **(d)**

4. $0, -3$; $D: -3 \le x \le 3$, $R: -3 \le y \le 0$
5. Symbolic: $f(x) = x^2 - 5$
Numerical:

x	-3	-2	-1	0	1	2	3
$y = f(x)$	4	-1	-4	-5	-4	-1	4

Graphical:

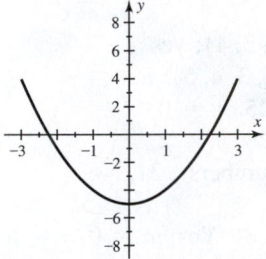

6. No, it fails the vertical line test.
7. (a) $D = \{-2, -1, 0, 5\}$ **(b)** All real numbers
(c) $x \ge -4$ **(d)** All real numbers   **(e)** $x \ne 5$
8. It is. $f(x) = -8x + 6$ **9.** -1 **10.** $-\frac{1}{2}$
11. (a) $y = -2x - 4$ **(b)** $y = \frac{1}{2}x + \frac{9}{2}$
(c) $y = -\frac{7}{12}x - \frac{17}{12}$ **12.** $-2; 2; 4$ **13.** $y = -3x + 3$
14. $y = \frac{2}{3}x - 2$; $y = -\frac{3}{2}x$ **15.** $x = \frac{2}{3}$; $y = -\frac{1}{7}$
16. (a) $m_1 = 0.4, m_2 = 0, m_3 = -0.4$ **(b)** From 1970 to 1980, beneficiaries increased by 0.4 million per year. From 1980 to 1990, there was no change. From 1990 to 2000, beneficiaries decreased by 0.4 million per year; answers may vary slightly.
17.

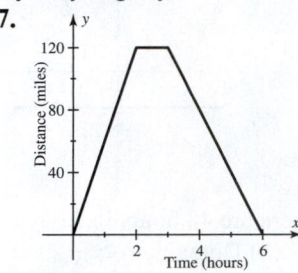

18.

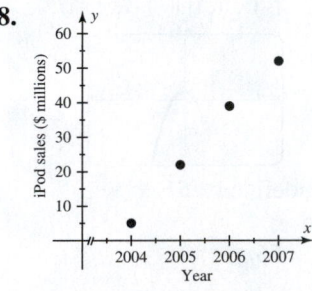

(b) $m = 15.67, x_1 = 2004, y_1 = 5$ (answers may vary)
(c) About 67.7 million

CHAPTERS 1 AND 2 CUMULATIVE REVIEW (pp. 145–146)

1. Natural number: $\sqrt[3]{8}$; whole number: $\sqrt[3]{8}$; integer: $-3, \sqrt[3]{8}$; rational number: $-3, \frac{3}{4}, -5.8, \sqrt[3]{8}$; irrational number: $\sqrt{7}$
2. Natural number: $\frac{50}{10}$; whole number: $0, \frac{50}{10}$; integer: $0, -5, -\sqrt{9}, \frac{50}{10}$; rational number: $0, -5, \frac{1}{2}, -\sqrt{9}, \frac{50}{10}$; irrational number: $\sqrt{8}$ **3.** Distributive property
4. Commutative property   **5.** -2 **6.** $(x + 1)(x + 5)$
7. $-a + b$ **8.** $\frac{c}{a + b}$ **9.** $\frac{169}{126}$ **10. (a)** $\frac{81}{64}$ **(b)** $\frac{81x^5}{y^4}$
(c) $\frac{b^{12}}{a^8}$ **11.** 9.54×10^3 **12.** 34   **13.** $B = 18$

14. (a) $\{-3, 0, 2\}$ **(b)** $x \neq -6$ **(c)** $x \geq -4$

15. 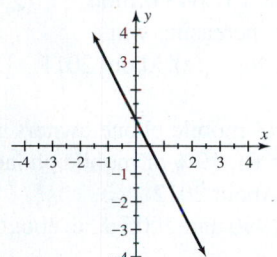   **16.**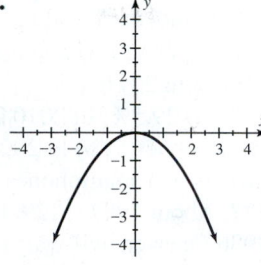

17. 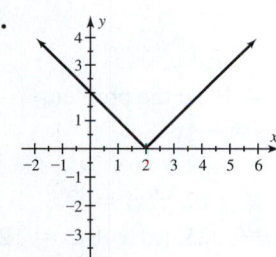   **18.**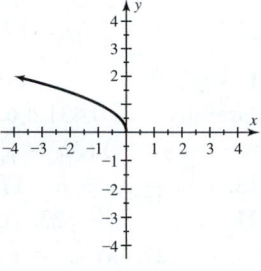

19. 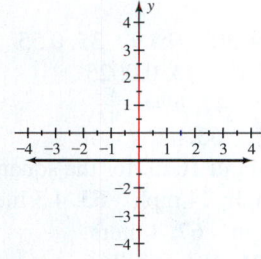   **20.**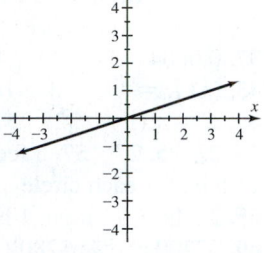

21. (a) Yes   **(b)** Both include all real numbers.
(c) $f(-1) = 1.5; f(0) = 1$ **(d)** x-intercept: 2; y-intercept: 1
(e) $-\frac{1}{2}$ **(f)** $f(x) = -\frac{1}{2}x + 1$ **22.** $y = -\frac{3}{2}x - 1$
23. $y = -\frac{1}{3}x + \frac{5}{3}$ **24.** $x = -2$ **25.** $m = -3$;
x-int: 1; y-int: 3   **26.** $y = -\frac{3}{2}x + \frac{3}{2}$
27. Commutative property for addition
28. 1600 cubic inches
29. (a) $f(x) = 0.2(x - 1970) + 4.7$ (answers may vary)
(b) 14.1%   **30.** The person earns $9 per hour.
31.

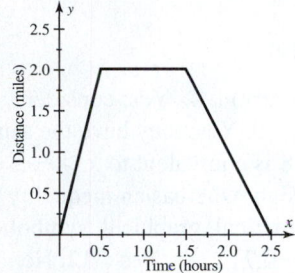

32. (a) Linear   **(b)** About 36.4; in 2005 the median age
was 36.4   **(c)** 0.243   **(d)** Each year the median age
increased by 0.243 year, on average.

33. (a) $f(x) = 10x$
(b)

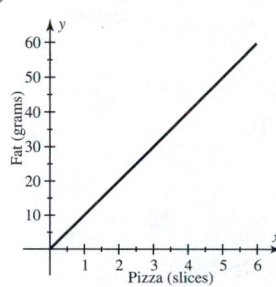

(c) 10
(d) The total amount of fat
increases at a rate of 10 g of
fat per slice of pizza.

34. (a) $\frac{24}{25} = 0.96$ pint
(b)

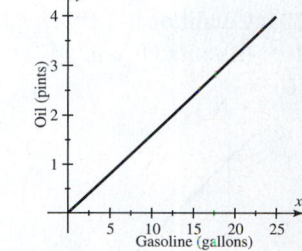

(c) $\frac{4}{25} = 0.16$
(d) 0.16 pint of oil should
be added per gallon of
gasoline.

35. (a) 87.8 million tons
(b) [1960, 1995, 5] by [60, 220, 20]

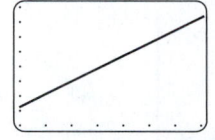

(c) 3.4; solid waste increased by about 3.4 million tons
per year.
36. (a) $m_1 = -50, m_2 = 0, m_3 = -50$ **(b)** m_1: The
driver is traveling toward home at a rate of 50 miles per
hour. m_2: The car is not moving. m_3: The driver is traveling
toward home at a rate of 50 miles per hour.   **(c)** The car
started 300 miles from home and moved toward home at
50 mph for 2 hours to a location 200 miles from home.
The car was then parked for 1 hour. Finally, the car
moved toward home at 50 mph for 4 hours, at which point it
was home.

3   Linear Equations and Inequalities

SECTION 3.1 (pp. 159–163)

1. $ax + b = 0, a \neq 0$ **2.** One   **3.** Yes   **4.** Yes   **5.** 4
6. -2 **7.** An equals sign ($=$)   **8.** Numerical, graphical,
symbolic   **9.** 3   **10.** $ax + by = c$, a and b not both 0
11. $a + c = b + c$ **12.** $ac = bc$, for $c \neq 0$ **13.** T
14. T   **15.** F   **16.** F   **17.** No   **18.** Yes   **19.** Yes   **20.** No
21. Yes   **22.** Yes   **23.** 14   **25.** -2 **27.** $-\frac{3}{2}$ **29.** 3
31. -12 **33.** 5   **35.** 5   **37.** $\frac{16}{5}$ **39.** $\frac{6}{7}$ **41.** $-\frac{3}{2}$ **43.** $\frac{7}{4}$
45. $\frac{5}{2}$ **47.** $\frac{72}{53} \approx 1.36$ **49.** 11   **51.** 1983   **53.** $-\frac{1}{6}$
55. -5 **57.** 6   **59.** 2   **61.** 0.75   **63.** -12 **65.** $x = -\frac{b}{a}$
67. 2

x	1	2	3	4	5
$-4x + 8$	4	0	-4	-8	-12

69. -1

x	-2	-1	0	1	2
$4 - 2x$	8	6	4	2	0
$x + 7$	5	6	7	8	9

71. 10 **73.** 1 **75.** 1.5 **77.** -2 **79.** -1 **81.** -1 **83.** 0
85. -3 **87.** 2007 **89.** About 2.89 **91.** 7 **93.** 3
95. Contradiction **97.** Identity **99.** Identity
101. Conditional **103.** $(-2, 5)$
105. (a) x-int: 4; y-int: 4 **107. (a)** x-int: 5; y-int: 2
(b)

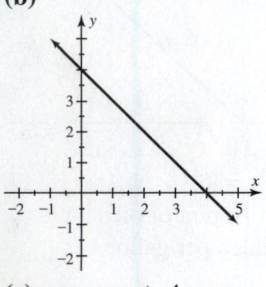

(b)

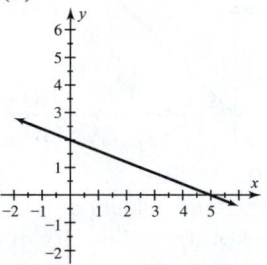

(c) $y = -x + 4$ **(c)** $y = -\frac{2}{5}x + 2$
109. (a) x-int: $\frac{9}{2}$; y-int: -3 **111. (a)** x-int: -3; y-int: -2
(b)

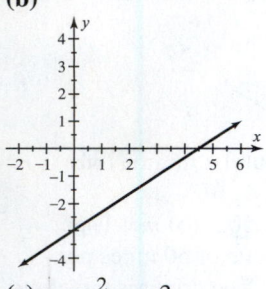

(b)

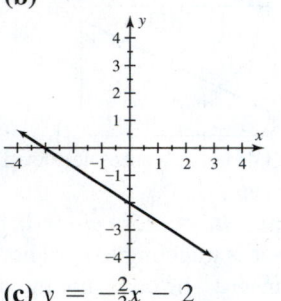

(c) $y = \frac{2}{3}x - 3$ **(c)** $y = -\frac{2}{3}x - 2$
113. (a) x-int: $\frac{7}{2}$; y-int: $\frac{7}{5}$
(b)

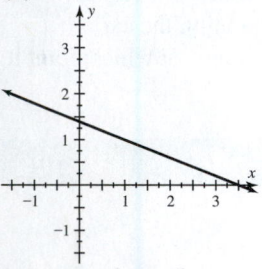

(c) $y = -\frac{2}{5}x + \frac{7}{5}$
115. (a) x-int: $-\frac{3}{5}$; y-int: $\frac{1}{2}$
(b)

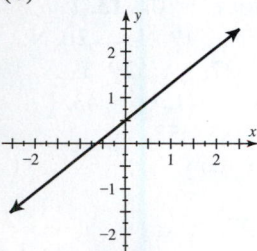

(c) $y = \frac{5}{6}x + \frac{1}{2}$
117. (a) $S(x) = 2.4(x - 1960) + 42$ **(b)** 138 million tons
(c) About 2027 **119. (a)** 2006 **(b)** 2006 **(c)** 2006
121. (a) $D(x) = 0.02x + 1.1$
(b) In 1970 ($x = 0$) 1.1 million degrees were awarded.

(c) 2000 **(d)** 2000 **123. (a)** $37.645 billion; in 2008 revenue from newspaper ads was $37.645 billion.
(b) Online; online ad revenue is increasing while newspaper ad revenue is decreasing **(c)** About 2011
(d) About 2011; yes
125. (a) 29.5%; In 2010 29.5% of mobile phone owners had smartphones **(b)** In 2008 ($x = 0$), 10% of mobile phone owners had smartphones. **(c)** About 2012
127. About 1987 **129.** Using 1990 and 2005 data, about 1986 (answers will vary)

SECTION 3.2 (pp. 173–176)

1. $\frac{x}{2}$ **2.** $4x - 7$ **3.** $2W + 2L$ **4.** Read the problem carefully. **5.** 0.431 **6.** 10 **7.** $y = -\frac{4}{3}x + 4$
9. $x = \frac{15}{8}y$ **11.** $b = \frac{S}{6a}$ **13.** $t = 14 - r$
15. $f(x) = 3x + 8$ **17.** $f(x) = \frac{2x}{\pi}$ **19.** $f(x) = \frac{8x}{3}$
21. $f(x) = \frac{-7x}{9}$ **23.** $f(x) = \frac{2x-20}{3}$ **25. (a)** $x + 2 = 12$
(b) 10 **27. (a)** $\frac{x}{5} = x + 1$ **(b)** $-\frac{5}{4}$ **29. (a)** $\frac{x + 5}{2} = 7$
(b) 9 **31. (a)** $\frac{x}{2} = 17$ **(b)** 34
33. (a) $x + (x + 1) + (x + 2) = 30$ **(b)** 9 **35.** 0.55
37. 0.0004 **39.** 3.41 **41.** -0.0167 **43.** 0.0025
45. (a) $L = \frac{P}{2} - W$ **(b)** 24 feet **47.** $h = \frac{A}{2\pi r}$
49. $C = \frac{5}{9}(F - 32)$ **51.** $h = \frac{2A}{b}$ **53.** 44, 45, 46
55. 23, 25, 27 **57.** 8 feet **59.** About 16 in. for the square; 12.5 in. for each circle **61.** 68 mph; 74 mph **63.** 4.5 liters
65. 2.5 hr at 74 mph; 1 hr at 63 mph **67.** 4 liters
69. $2200 at 6%; $2800 at 4% **71.** 38°, 66°, 76°
73. (a) 36,000 cubic feet per hour **(b)** 66 people
75. $2.80 **77.** 110,000 sq mi **79.** About $1.2 billion
81. 1500 people **83.** About 3.3

CHECKING BASIC CONCEPTS 3.1 & 3.2 (p. 177)

1. 1 **2.** $\frac{6}{11}$ **3. (a)** -2 **(b)** -2 **(c)** -2 Yes
4. 0.3 hour at 10 miles per hour; 0.9 hour at 8 miles per hour
5. x-int: 4; y-int: -5 **6. (a)** Identity **(b)** Contradiction
7. 23.5 ft by 17.5 ft

SECTION 3.3 (pp. 185–189)

1. $3x + 2 < 5$ (answers may vary) **2.** Yes; consider $x + 2 > 2$. **3.** Yes **4.** Yes **5.** Yes; they have the same solution set. **6.** No; $-4x < 8$ is equivalent to $x > -2$.
7. An equation has an equals sign, whereas an inequality has an inequality symbol. **8.** Numerical, graphical, symbolic
9. $\{x \mid x < 6\}$ **10.** $\{x \mid x \geq -0.7\}$ **11.** Yes **12.** No
13. No **14.** Yes **15. (a)** $\{x \mid x = 2\}$ **(b)** $\{x \mid x < 2\}$
(c) $\{x \mid x > 2\}$ **17. (a)** $\{x \mid x = 3\}$ **(b)** $\{x \mid x > 3\}$
(c) $\{x \mid x < 3\}$ **19.** $\{x \mid x \leq 2\}$ **21.** $\{x \mid x > 36\}$
23. $\{x \mid x \geq \frac{14}{9}\}$ **25.** $\{x \mid x > \frac{17}{3}\}$ **27.** $\{x \mid x < \frac{27}{14}\}$
29. $\{x \mid x \geq \frac{2}{13}\}$ **31.** $\{x \mid x < 2010\}$ **33.** $\{x \mid x < 5\}$
35. $\{x \mid x \geq -6\}$ **37.** $\{z \mid z \leq -6\}$ **39.** $\{t \mid t \geq -\frac{7}{2}\}$
41. $\{x \mid x > -\frac{25}{3}\}$ **43.** $\{x \mid x > \frac{17}{9}\}$ **45.** $\{x \mid x \geq \frac{1}{2}\}$
47. $\{x \mid x \neq \frac{5}{2}\}$ **49.** $\{x \mid x \geq \frac{9}{5}\}$ **51.** $\{x \mid x > -\frac{1}{2}\}$

53. $\{x \mid x \geq 3\}$

x	1	2	3	4	5
$-2x + 6$	4	2	0	-2	-4

55. $\{x \mid x > 3\}$ **57.** $\{x \mid x \geq 2\}$ **59. (a)** $\{x \mid x = -1\}$
(b) $\{x \mid x < -1\}$ **(c)** $\{x \mid x > -1\}$ **61. (a)** $\left\{x \mid x = \frac{1}{2}\right\}$
(b) $\left\{x \mid x > \frac{1}{2}\right\}$ **(c)** $\left\{x \mid x < \frac{1}{2}\right\}$ **63.** $\{x \mid x \geq 1\}$
65. $\{x \mid x > 2\}$ **67. (a)** Car 1; the slope of the line for Car 1
is greater than the slope of the line for Car 2.   **(b)** 5 hours;
400 miles   **(c)** $x < 5$ **69. (a)** 1   **(b)** $\{x \mid x > 1\}$
(c) $\{x \mid x \leq 1\}$ **(d)** $\{x \mid x \geq 1\}$ **71.** $\{x \mid x < 1\}$
73. $\{x \mid x \geq 0\}$ **75.** $\{x \mid x \geq -2\}$ **77.** $\{x \mid x < -4\}$
79. $\{x \mid x < -3\}$ **81.** $\{x \mid x \leq 1991\}$
83. $\{x \mid x > 0.834\}$ **85.** $\{x \mid x < a\}$ **87.** After about 1975
89. $\{x \mid x < 2\}$ **91.** $\{x \mid x \leq 3\}$ **93. (a)** $\{x \mid x = 12\}$
(b) $\{x \mid x > 12\}$ **95. (a)** $\left\{x \mid x \leq -\frac{7}{3}\right\}$
(b) $\left\{x \mid x \geq -\frac{7}{3}\right\}$ **97. (a)** $\{x \mid x \leq -9\}$
(b) $\{x \mid x \geq -9\}$ **99. (a)** \$200   **(b)** \$5.00 each
(c) $P = 5x - 200$ **(d)** $x = 40$ DVDs
(e) $x > 40$ DVDs   **101.** $x < 10$ feet   **103. (a)** First car:
70 miles per hour; second car: 60 miles per hour
(b) When $x = 3.5$ hours   **(c)** Elapsed times after 3.5
hours   **105.** Below about 3.16 miles   **107.** Between 132
and 174 lb   **109. (a)** 2009 to 2012   **(b)** 2009 to 2012
(c) 2009 to 2012

SECTION 3.4 (pp. 197–201)

1. $x > 1$ and $x \leq 7$ (answers may vary)
2. $x \leq 3$ or $x > 5$ (answers may vary)   **3.** No   **4.** Yes
5. Yes   **6.** Numerically, graphically, symbolically
7. Yes, no   **8.** No, Yes   **9.** No, yes   **10.** Yes, no
11. No, yes   **12.** Yes, yes   **13.** [2, 10]   **15.** (5, 8]
17. $(-\infty, 4)$ **19.** $(-2, \infty)$ **21.** $[-2, 5)$ **23.** $(-8, 8]$
25. $(3, \infty)$ **27.** $(-\infty, -2] \cup [4, \infty)$
29. $(-\infty, 1) \cup [5, \infty)$ **31.** $(-3, 5]$ **33.** $(-\infty, -2)$
35. $(-\infty, 4)$ **37.** $(-\infty, 1) \cup (2, \infty)$
39. $\{x \mid -1 \leq x \leq 3\}$

41. $\{x \mid -2 < x < 2.5\}$

43. $\{x \mid x > 3\}$

45. $\{x \mid x \leq -1 \text{ or } x \geq 2\}$

47. All real numbers

49. $[-6, 7]$ **51.** $\left(\frac{13}{4}, \infty\right)$ **53.** $(-\infty, \infty)$
55. $\left(-\infty, -\frac{11}{5}\right) \cup (-1, \infty)$ **57.** No solutions
59. $[-6, 1)$ **61.** $\left[-\frac{1}{4}, \frac{11}{8}\right]$ **63.** $[-9, 3]$
65. $\left[-4, -\frac{1}{4}\right)$ **67.** $(-1, 1]$ **69.** $[-2, 1]$ **71.** $(0, 2)$
73. $(9, 21]$ **75.** $\left[\frac{12}{5}, \frac{22}{5}\right]$ **77.** $\left(-\frac{13}{3}, 1\right)$ **79.** $[0, 12]$
81. $[-1, 2]$ **83.** $(-1, 2)$ **85.** $[-3, 1]$
87. $(-\infty, -2) \cup (0, \infty)$

89. (a) Toward, because distance is decreasing
(b) 4 hours, 2 hours   **(c)** From 2 to 4 hours   **(d)** During the
first 2 hours   **91. (a)** 2   **(b)** 4   **(c)** $\{x \mid 2 \leq x \leq 4\}$
(d) $\{x \mid 0 \leq x < 2\}$ **93.** [1, 4]   **95.** $(-\infty, -2) \cup (0, \infty)$
97. [2002, 2010]   **99.** $(-\infty, 5]$
101. $(-\infty, -1) \cup (-1, \infty)$ **103.** $(0, \infty)$ **105.** [1, 3]
107. $(-\infty, \infty)$ **109.** $(-\infty, 1) \cup [3, \infty)$
111. $(c - b, d - b]$ **113. (a)** From 2006 to 2008
(b) From 2006 to 2008   **(c)** From 2006 to 2008
115. From 0.5 to 1.5 miles   **117.** From $5.\overline{6}$ to 9 feet
119. From 1.5 to 2.5 miles   **121. (a)** $M(x) = 25x + 250$
(b) From 2002 to 2006

CHECKING BASIC CONCEPTS 3.3 & 3.4 (p. 201)

1. $\left\{x \mid x > \frac{8}{7}\right\}$ **2. (a)** $\{x \mid x = 2\}$ **(b)** $\{x \mid x < 2\}$
(c) $\{x \mid x \geq 2\}$ **3. (a)** Yes   **(b)** No   **4. (a)** $[-3, 1]$
(b) $(-\infty, -1) \cup [3, \infty)$ **(c)** $\left[-\frac{8}{3}, \frac{8}{3}\right)$

SECTION 3.5 (pp. 209–212)

1. $|3x + 2| = 6$ (answers may vary)
2. $|2x - 1| \leq 17$ (answers may vary)   **3.** Yes   **4.** Yes
5. Yes   **6.** No, it is equivalent to $-3 < x < 3$.   **7.** 2
8. 0   **9.** No, yes   **10.** Yes, no   **11.** No, yes   **12.** No, yes
13. Yes, yes   **14.** No, yes   **15.** 0, 4   **16.** $-2, 1$
17. $-3, 3$ **18.** $-5, 5$ **19.** $(-3, 3)$ **20.** $(-5, 5)$
21. $(-\infty, -3) \cup (3, \infty)$ **22.** $(-\infty, -5) \cup (5, \infty)$
23. $-7, 7$ **25.** No solutions   **27.** $-\frac{9}{4}, \frac{9}{4}$ **29.** $-4, 4$
31. $-6, 5$ **33.** 1, 2   **35.** $\frac{3}{4}$ **37.** $-8, 12$ **39.** No solutions
41. $-15, 18$ **43.** $-1, \frac{1}{3}$ **45.** $-5, \frac{3}{5}$ **47.** -6
49. (a) $-4, 4$ **(b)** $\{x \mid -4 < x < 4\}$
(c) $\{x \mid x < -4 \text{ or } x > 4\}$
51. (a) $\frac{1}{2}, 2$ **(b)** $\left\{x \mid \frac{1}{2} \leq x \leq 2\right\}$
(c) $\left\{x \mid x \leq \frac{1}{2} \text{ or } x \geq 2\right\}$ **53.** $[-3, 3]$
55. $(-\infty, -4) \cup (4, \infty)$ **57.** No solutions
59. $(-\infty, 0) \cup (0, \infty)$ **61.** $\left(-\infty, -\frac{7}{2}\right) \cup \left(\frac{7}{2}, \infty\right)$
63. $(-3, 5)$ **65.** $(-\infty, -9] \cup [-1, \infty)$ **67.** $\left[\frac{5}{6}, \frac{11}{6}\right]$
69. $[-10, 14]$ **71.** No solutions   **73.** $(-\infty, \infty)$
75. No solutions   **77.** $(-\infty, \infty)$
79. $(-\infty, -13] \cup [17, \infty)$ **81.** [0.9, 1.1]
83. $(-\infty, 9.5) \cup (10.5, \infty)$ **85. (a)** $-1, 3$ **(b)** $(-1, 3)$
(c) $(-\infty, -1) \cup (3, \infty)$ **87. (a)** $-1, 0$ **(b)** $[-1, 0]$
(c) $(-\infty, -1] \cup [0, \infty)$ **89.** $(-\infty, -1] \cup [1, \infty)$
91. $[-2, 4]$ **93.** $(-\infty, 1) \cup (3, \infty)$ **95.** $(2, 4.\overline{6})$
97. $(-\infty, \infty)$ **99.** $\{x \mid -3 \leq x \leq 3\}$
101. $\{x \mid x < 2 \text{ or } x > 3\}$ **103.** $|x| \leq 4$ **105.** $|y| > 2$
107. $|2x + 1| \leq 0.3$ **109.** $|\pi x| \geq 7$ **111.** two
113. (a) $\{T \mid 19 \leq T \leq 67\}$ **(b)** Monthly average
temperatures vary from 19°F to 67°F.
115. (a) $\{T \mid -26 \leq T \leq 46\}$ **(b)** Monthly average
temperatures vary from -26°F to 46°F.   **117. (a)** About
19,058 feet   **(b)** Africa and Europe   **(c)** South America,
North America, Africa, Europe, and Antarctica
(d) $|E - A| \leq 5000$ **119.** $\{d \mid 2.498 \leq d \leq 2.502\}$; the
diameter can vary from 2.498 to 2.502 inches.

121. $|d - 3.8| \leq 0.03$
123. Values between 19 and 21, exclusive

CHECKING BASIC CONCEPTS 3.5 (p. 212)

1. $-\frac{28}{3}, 12$ **2. (a)** $-\frac{2}{3}, \frac{14}{3}$ **(b)** $\left(-\frac{2}{3}, \frac{14}{3}\right)$
(c) $\left(-\infty, -\frac{2}{3}\right) \cup \left(\frac{14}{3}, \infty\right)$ **3.** $(0, 6); (-\infty, 0] \cup [6, \infty)$
4. (a) $1, 3$ **(b)** $[1, 3]$ **(c)** $(-\infty, 1] \cup [3, \infty)$

CHAPTER 3 REVIEW (pp. 215–218)

1. 2

x	0	1	2	3	4
$3x - 6$	-6	-3	0	3	6

2. 1

x	-1	0	1	2	3
$5 - 2x$	7	5	3	1	-1

3. Yes   **4.** 1   **5.** 2   **6.** 4   **7.** 14   **8.** $\frac{1}{2}$ **9.** $\frac{1}{7}$ **10.** $-\frac{1}{3}$
11. 10   **12.** $\frac{21}{8}$ **13.** 1972   **14.** $\frac{19}{5}$ **15.** 4   **16.** -9
17. Identity   **18.** Contradiction   **19.** x-int: 5, y-int: -4;
$y = \frac{4}{5}x - 4$ **20.** x-int: -3, y-int: -12; $y = -4x - 12$
21. $y = \frac{5}{4}x - 5$ **22.** $y = \frac{2}{3}x + 2$ **23.** $a = -3b$
24. $n = -\frac{2}{7}m$ **25.** $b = \frac{2A}{h} - a$ **26.** $h = \frac{3V}{\pi r^2}$
27. $f(x) = \frac{2}{3}x - \frac{8}{3}$ **28.** $f(x) = \frac{7}{27}x + \frac{1}{27}$
29. $f(x) = -\frac{3}{5}x + 2$ **30.** $f(x) = 3x$
31. $2x + 25 = 19; -3$ **32.** $2x - 5 = x + 1; 6$
33. $\{x \mid x > -1\}$ **34.** $\{x \mid x > 5\}$ **35.** $\{x \mid x \geq 2\}$
36. (a) $\frac{7}{4}$ **(b)** $\{x \mid x > \frac{7}{4}\}$ **(c)** $\{x \mid x < \frac{7}{4}\}$
37. $\{x \mid x \geq -1\}$ **38.** $\{x \mid x \leq -8\}$ **39.** $\{x \mid x > 1\}$
40. $\{x \mid x > \frac{4}{3}\}$ **41.** $\{t \mid t > \frac{16}{9}\}$ **42.** $\{x \mid x \geq 5\}$
43. $\{t \mid t \leq 2.1\}$ **44.** $\{z \mid z > -36.4\}$
45. $\{x \mid x \leq 1.840\}$ **46.** $\{x \mid x > 1.665\}$
47. $[-2, 2]$

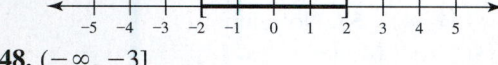

48. $(-\infty, -3]$

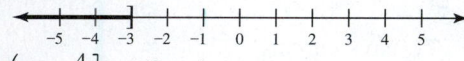

49. $\left(-\infty, \frac{4}{5}\right] \cup (2, \infty)$

50. $(-\infty, \infty)$

51. $[-2, 1]$ **52. (a)** -4 **(b)** 2   **(c)** $[-4, 2]$ **(d)** $(-\infty, 2)$
53. (a) 2   **(b)** $(2, \infty)$ **(c)** $(-\infty, 2)$ **54. (a)** 4   **(b)** 2
(c) $(2, 4)$ **55.** $\left[-3, \frac{2}{3}\right]$ **56.** $(-6, 45]$ **57.** $\left(-\infty, \frac{7}{2}\right)$
58. $[1.8, \infty)$ **59.** $(-3, 4)$ **60.** $(-\infty, 4) \cup (10, \infty)$
61. $(-5, 5)$ **62.** $[8, 28]$ **63.** $(-9, 21)$ **64.** $[8, 28]$
65. $\left(-\frac{11}{5}, \frac{7}{5}\right]$ **66.** $[88, 138]$ **67.** No, no   **68.** No, yes
69. No, yes   **70.** No, yes   **71. (a)** $0, 4$ **(b)** $(0, 4)$
(c) $(-\infty, 0) \cup (4, \infty)$ **72. (a)** $-3, 1$ **(b)** $[-3, 1]$
(c) $(-\infty, -3] \cup [1, \infty)$ **73.** $-22, 22$ **74.** $1, 8$
75. $-26, 42$ **76.** $-\frac{23}{3}, \frac{25}{3}$ **77.** $0, 2$ **78.** $-3, \frac{9}{5}$
79. (a) $-8, 6$ **(b)** $[-8, 6]$ **(c)** $(-\infty, -8] \cup [6, \infty)$
80. (a) $-\frac{5}{2}, \frac{7}{2}$ **(b)** $\left[-\frac{5}{2}, \frac{7}{2}\right]$ **(c)** $\left(-\infty, -\frac{5}{2}\right] \cup \left[\frac{7}{2}, \infty\right)$
81. $(-\infty, -3) \cup (3, \infty)$ **82.** $(-4, 4)$ **83.** $[-3, 4]$

84. $\left(-\infty, -\frac{5}{2}\right] \cup [5, \infty)$ **85.** $[4.4, 4.6]$ **86.** $\left[\frac{3}{13}, \frac{7}{13}\right]$
87. $(-\infty, \infty)$ **88.** $\frac{3}{2}$ **89.** $(-\infty, -1.5] \cup [1.5, \infty)$
90. $[-2, 6]$ **91.** $|x| \leq 0.05$ **92.** $|5x - 1| > 4$
93. \$3500 at 5%; \$4200 at 7%   **94.** 0.6 hour at 8 miles per
hour; 0.8 hour at 10 miles per hour   **95.** About \$769,000
96. (a) $f(x) = -0.3125x + 10$ **(b)** About 5.9 million
97. (a) 3 hours   **(b)** 2 hours and 4 hours
(c) Between 2 and 4 hours, exclusive   **(d)** 20 miles per
hour   **98. (a)** Car 1; its graph has the steeper slope.
(b) 3 hours; 200 miles   **(c)** Before 3 hours
99. (a) About 31.8 inches   **(b)** Lengths less than 31.8
inches   **100. (a)** $T(x) = -19x + 60$ **(b)** Approximately
from 1.05 to 2.11 miles   **101. (a)** 1979   **(b)** Median age
increased, on average, by 0.238 year per year. In 1970
($x = 0$), the median age was 27.9.   **102.** 13 feet by 31 feet
103. $-44.\overline{4}°\text{C}$ to $41.\overline{6}°\text{C}$
104. $|L - 160| \leq 1; 159 \leq L \leq 161$
105. (a) $|A - 3.9| \leq 1.7$ **(b)** $2.2 \leq A \leq 5.6$
106. Values between 32.2 and 37.8, exclusive

CHAPTER 3 TEST (pp. 218–220)

1. Yes

2.

x	-2	-1	0	1	2
$2 - 3x$	8	5	2	-1	-4

; $\{x \mid x \geq 1\}$

3. -3 **4. (a)** 2   **(b)** $(-\infty, 2]$ **(c)** $[2, \infty)$ **5.** 1
6. (a) $\frac{7}{9}$ **(b)** $\frac{7}{13}$ **(c)** $\frac{10}{17}$ **7.** x-int: $-\frac{3}{2}$, y-int: 3; $y = 2x + 3$
8. $2 + 5x = x - 4; -\frac{3}{2}$ **9.** $f(x) = \frac{3}{2}x - 3$
10. $\left[-\frac{3}{5}, \infty\right)$ **11.** $(1.55, \infty)$
12.

13. $(-\infty, -2) \cup (1, \infty)$ **14. (a)** -5 **(b)** 5   **(c)** $[-5, 5]$
(d) $(-\infty, 5)$ **15.** $(-8, 0)$ **16.** $-12, 24$ **17. (a)** $[-5, 5]$
(b) $(-\infty, 0) \cup (0, \infty)$ **(c)** No solutions
(d) $(-\infty, -2) \cup (2, \infty)$ **(e)** $(-\infty, \infty)$ **(f)** $(-2, 8)$
18. (a) $-\frac{2}{5}, \frac{4}{5}$ **(b)** $\left[-\frac{2}{5}, \frac{4}{5}\right]$ **(c)** $\left(-\infty, -\frac{2}{5}\right] \cup \left[\frac{4}{5}, \infty\right)$
19. (a) 64 calories   **(b)** More than 64 calories
(c) Less than 64 calories   **(d)** Each gram of carbohydrates
corresponds to 4 calories.   **20. (a)** $f(x) = 0.4x + 75$
(b) 35 minutes   **21.** $g = \frac{2d}{t^2}$
22. \$3500 at 4%; \$1500 at 6%   **23.** $41°, 57°, 82°$ **24.** \$108
25. (a) From 1990 to 2000   **(b)** Numbers increased, on
average, by 190 per year; In 1980 ($x = 0$) there were 6000
women marines.

CHAPTERS 1–3 CUMULATIVE REVIEW (pp. 220–223)

1. Natural: $\frac{8}{2}$; whole: $0, \frac{8}{2}$; integer: $-7, 0, \frac{8}{2}$;
rational: $-7, -\frac{3}{5}, 0, \frac{8}{2}, 5.\overline{12}$; irrational: $\sqrt{5}$
2. Natural: $\sqrt{9}$; whole: $\frac{0}{9}, \sqrt{9}$; integer: $-\frac{6}{3}, \frac{0}{9}, \sqrt{9}$;
rational: $-\frac{6}{3}, \frac{0}{9}, \sqrt{9}, 4.\overline{6}$; irrational: π **3.** Distributive
4. Associative   **5.** 50   **6.** 693
7.

8. $5y + 4$ **9.** $\frac{3y^7}{x^5}$ **10.** $\frac{64b^9}{27a^6}$ **11.** 13

12. 5.9×10^{-5} **13.** 5   **14.** 30   **15.** (ii)

16. $D = \{-2, 0, 1, 3\}$; $R = \{0, 2, 4\}$

17. $(-2, 5), (-1, 3), (0, 1),$ **18.** $(-2, -2), \left(-1, \frac{3}{2}\right),$
$(1, -1), (2, -3)$ $(0, 2), \left(1, \frac{5}{2}\right), (2, 6)$

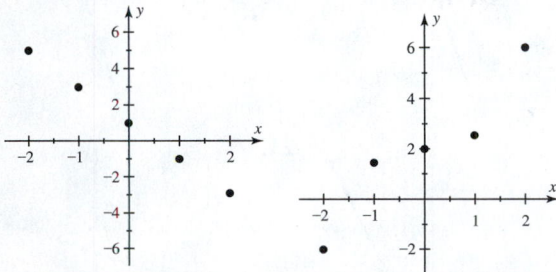

19.

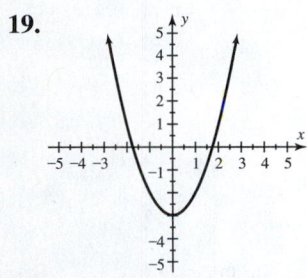

20. $x \neq 0$ **21.** $2, -2$ **22.** $2, -4$
23. Yes; $m = -3, b = 11$ **24.** No   **25.** $f(x) = 4x + 3$
26.

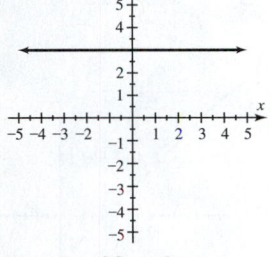

27. $5; -4$ **28.** -2
29.

30. $y = -\frac{1}{2}x - 1$ **31.** $3; -1; 3$ **32.** $x = 5$
33. $y = 3x - 5$ **34.** $y = x + 3$ **35.** Yes   **36.** 1
37. $-\frac{2}{3}$ **38.** $\frac{1}{11}$ **39.** x-int: 2, y-int: -6; $y = 3x - 6$
40. x-int: -2, y-int: $\frac{5}{2}$; $y = \frac{5}{4}x + \frac{5}{2}$ **41.** $\{x \mid x > 0\}$
42. $(-\infty, -3]$ **43.** $(3, \infty)$ **44.** $(-\infty, 0]$
45. $(-1, 5]$

46. $\left(-\infty, -\frac{3}{2}\right) \cup [3, \infty)$

47. $[-2, 4]$ **48.** $[-16, 10]$ **49.** $(-\infty, -6) \cup \left(\frac{8}{3}, \infty\right)$
50. $[4, 7]$ **51.** (a) $-3, 1$ (b) $[-3, 1]$
(c) $(-\infty, -3] \cup [1, \infty)$ **52.** $-6, 18$ **53.** (a) $7004.25

(b) $14,741.05   **54.** $d = 65t$ **55.** (a) $f(x) = 10x$
(b) 10   (c) The soda contains 10 mg of sodium per ounce.
56.

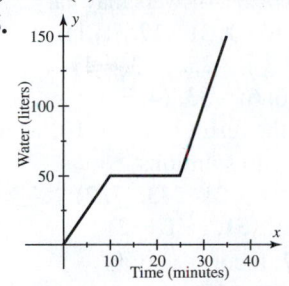

57. (a) Away; the slope of the line is positive.   (b) 3 hours
(c) From 0 to 3 hours (not including 3)
58. About 1992 to 2004

4   Systems of Linear Equations

SECTION 4.1 (pp. 235–238)

1. No. Two straight lines cannot have exactly two points of
intersection.   **2.** $3x + 2y = 5, 2x - y = 0$ (answers may
vary)   **3.** Numerically and graphically   **4.** There is one
solution.   **5.** There are no solutions.   **6.** There are
infinitely many solutions because the two lines are identical.
7. Inconsistent   **8.** Consistent with dependent equations
9. $(1, -2)$ **10.** $(3, 1)$ **11.** $\left(-1, \frac{13}{3}\right)$ **12.** $(4, 0)$
13. $(1, 1)$ **15.** $(0, 1)$ **17.** $(3, -2)$ **19.** $(4, -1)$
21. $(3, 1)$ **23.** No solutions   **25.** $(1, 2)$; consistent;
independent   **27.** $(2, 1)$; consistent; independent
29. $\{(x, y) \mid x + y = 3\}$; consistent; dependent   **31.** $(1, 3)$;
consistent; independent   **33.** No solutions; inconsistent
35. $(4, 2)$; consistent; independent   **37.** No solutions;
inconsistent   **39.** $\{(x, y) \mid -3x + 2y = 1\}$; consistent;
dependent   **41.** $(2, -2)$; consistent; independent   **43.** $(1, 1)$
45. $(1, -4)$ **47.** (a) x-intercepts: 4, -2; y-intercepts: 4, 2
(b) $(1, 3)$ **49.** (a) x-intercepts: $-6, -4$; y-intercepts: 4, 8
(b) $(-3, 2)$ **51.** $(0, 0.3)$ **53.** $(1.5, 0.5)$ **55.** $(100, 200)$
57. No solutions   **59.** $(2, 3)$ **61.** $(1, 1)$ **63.** $(5, -2)$
65. $(-5, 10)$ **67.** $(0.5, 1.5)$ **69.** No solutions
71. (a) $x + y = 18, x - y = 6$ (b) $12, 6$
73. (a) $x + y = 1, 6x + 8y = 7$ (b) $\frac{1}{2}$ hour at 6 miles
per hour; $\frac{1}{2}$ hour at 8 miles per hour
75. (a) $2x + 2y = 76, x - y = 4$ (b) 21 by 17 inches
77. (a) $2x + 3y = 7, 3x + 2y = 8$ (b) Popcorn: $2; soft
drink: $1   **79.** (a) $x + 2y = 180, x - y = 60$ (b) $100°$,
$40°, 40°$ **81.** $407; $29   **83.** 23   **85.** 23   **87.** 2005: 26
mph; 2007: 24 mph   **89.** McGwire: 70; Sosa: 66   **91.** $350
at 5%; $250 at 4%

SECTION 4.2 (pp. 248–252)

1. Substitution and elimination   **2.** No; all methods produce the same solution. However, graphical and numerical
answers are sometimes approximate.   **3.** Elimination yields
a contradiction.   **4.** Elimination yields an identity.

5. Solve the first equation for *y*.   **6.** Multiply the second equation by -2.   **7.** Add the equations.   **8.** Multiply the first equation by 2 and the second equation by 3. Answers may vary.
9. (1, 2)   **11.** (5, 3)   **13.** (2, 1)   **15.** (1, 3)   **17.** (1, 1)
19. (0, 2)   **21.** $\left(\frac{1}{2}, 1\right)$ **23.** $\left(\frac{1}{2}, \frac{1}{4}\right)$ **25.** $(-3, -1)$
27. $\left(\frac{1}{2}, -\frac{1}{2}\right)$ **29.** (4, 2)   **31.** (6, 6)   **33.** $(4, -6)$
35. (5, 2)   **37.** For $x = y + 5$ the result is $10 = 10$, which is an identity.   **39.** Inconsistent; no solutions
41. Dependent; $\{(x, y) \mid x + 3y = -2\}$ **43.** (7, 2)
45. (2, 1)   **47.** (1, 0)   **49.** (3, 1)   **51.** $(-1, -2)$
53. $\left(\frac{1}{2}, 3\right)$ **55.** $\left(\frac{6}{7}, -\frac{12}{7}\right)$ **57.** Inconsistent;
no solutions   **59.** Dependent; $\{(x, y) \mid 2x + y = 2\}$
61. $\left(-\frac{5}{9}, -\frac{47}{9}\right)$ **63.** (2, 1)   **65.** $(-4, 5)$ **67.** $\left(\frac{1}{2}, -\frac{1}{2}\right)$
69. Dependent; $\{(r, t) \mid 2r - 3t = 7\}$ **71.** Inconsistent; no solutions   **73.** $(-2, -2)$ **75.** (10, 20)   **77.** (8, 10)
79. (5, 4)   **81.** (a), (b), (c) (5, 1)   **83.** (a), (b), (c) (1, 2)
85. (a), (b), (c) (2.5, -0.5)   **87.** (0, 4)   **89.** Rowing machine: 38 min; stair climber: 22 min   **91.** $\frac{16}{7}$ gallons
93. 27 premium rooms; 23 regular rooms   **95.** 50° and 130°   **97.** 156.5 million females, 151.5 million males
99. Plane: 608 miles per hour; wind: 32 miles per hour   **101.** Boat: 25 mph, current: 5 mph
103. \$2100 at 8% and \$1400 at 9%   **105.** Day: \$110, night: \$80   **107.** Approximately 115.5 pounds on each rafter   **109.** $2x + 2y = 296$, $x - y = 44$; (a)–(c) Length = 96 feet, width = 52 feet   **111.** 13 m by 7.5 m

CHECKING BASIC CONCEPTS 4.1 & 4.2 (p. 252)

1. (3, 1); yes   **2.** (2, 3)   **3.** $\left(-\frac{7}{2}, 1\right)$; no; no
4. (a) $x + y = 90$, $x - y = 40$ **(b)** (65, 25); the smaller angle is 25° and the larger angle is 65°.

SECTION 4.3 (pp. 259–263)

1. Two   **2.** Yes; many points may satisfy a system of inequalities.   **3.** Yes; no   **4.** No; yes   **5.** Solid   **6.** dashed

7. 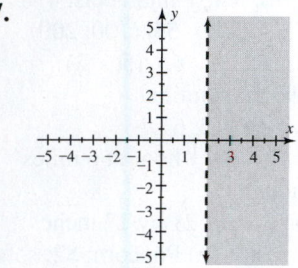   **9.**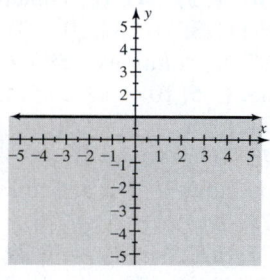

11. 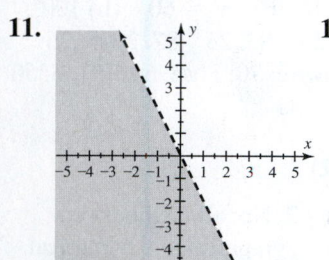   **13.**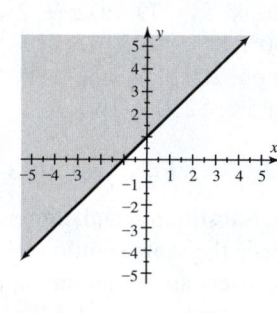

15. 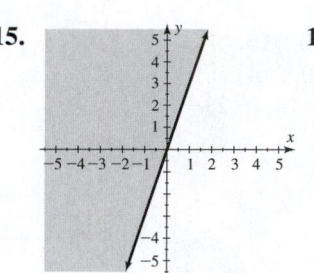   **17.**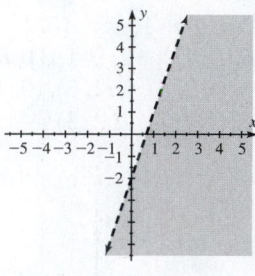

19. 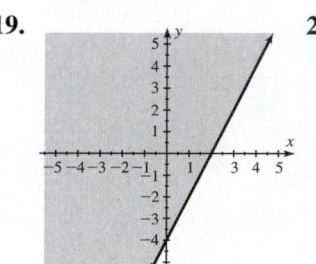   **21.**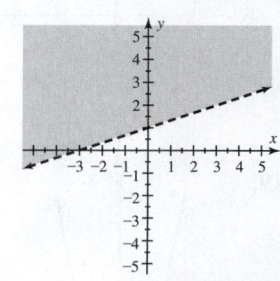

23. 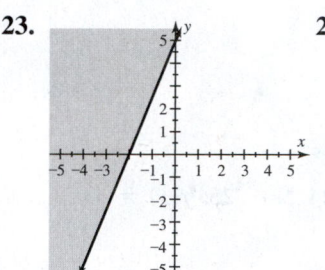   **25.**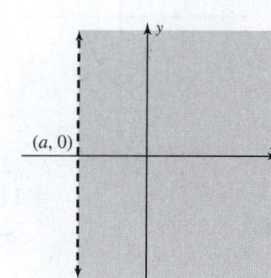

27. 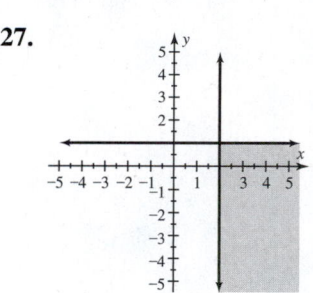   **29.**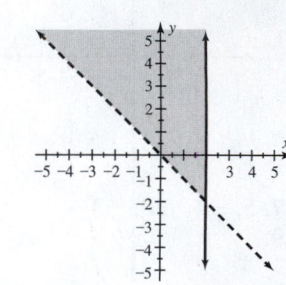

31. 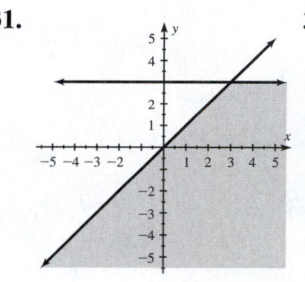   **33.**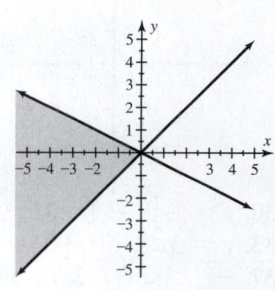

35. 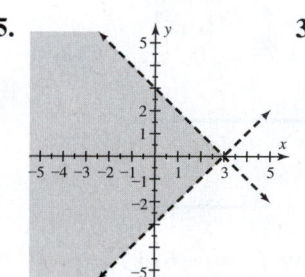   **37.**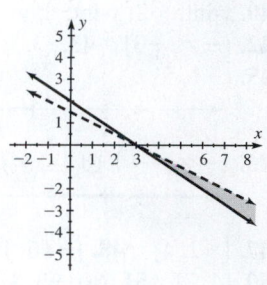

39.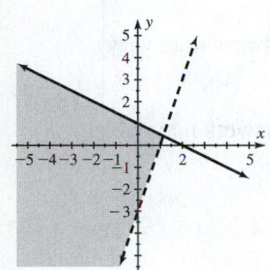

41.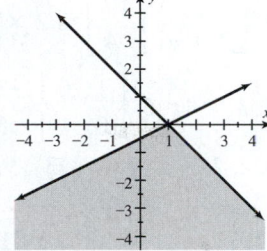

43.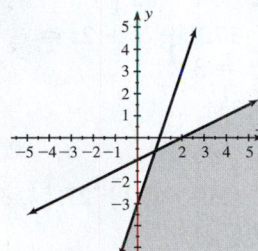

45.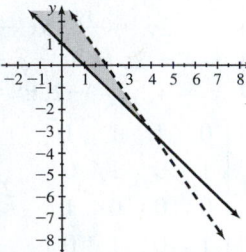

47.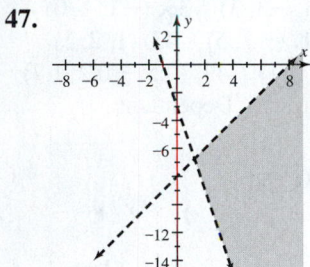

49. [−10, 10, 1] by [−10, 10, 1]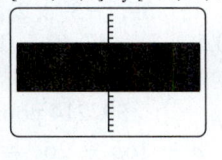

51. [−10, 10, 1] by [−10, 10, 1]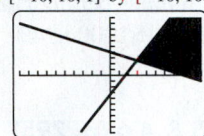

53. [−10, 10, 1] by [−10, 10, 1]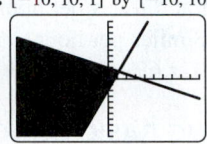

55. b.   **57.** a.   **59.** $y \geq 2$ **61.** $y \geq x$; $x \geq -2$

63.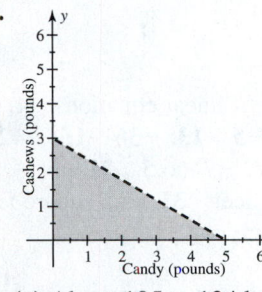

65.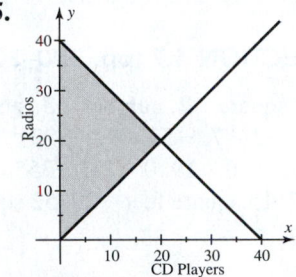

67. (a) About 105 to 134 beats per minute
(b) $y \leq -0.6x + 152$, $y \geq -0.5x + 120$
69. The person weighs less than recommended.
71. $25h - 7w \leq 800$; $5h - w \geq 170$
73.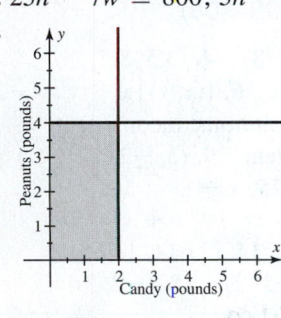

75. (a)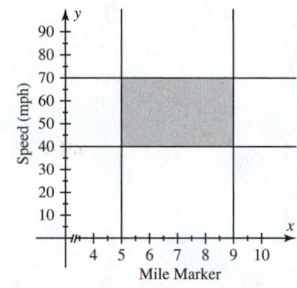

(b) (7, 60); At mile marker 7, 60 mph is a lawful speed.
(answers may vary)

77. (a)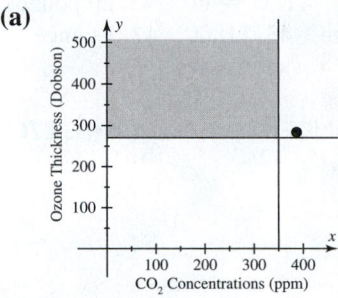

(b) No

SECTION 4.4 (pp. 268–270)

1. linear programming   **2.** objective   **3.** feasible solutions
4. linear inequalities   **5.** vertex   **6.** objective

7.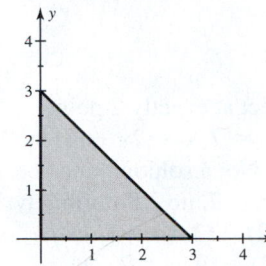

9.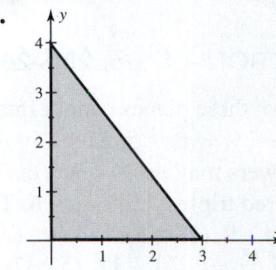

11.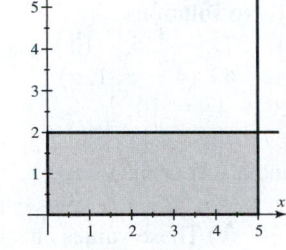

13.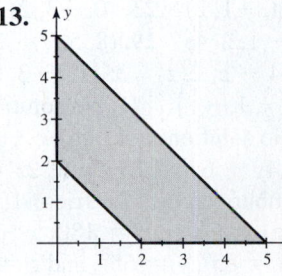

15.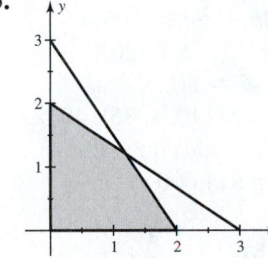

17.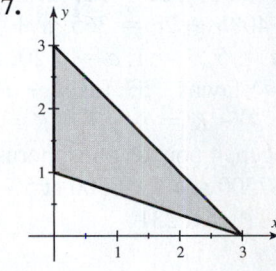

19.

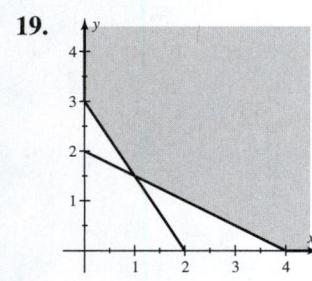

21. $R = 31$ **23.** $R = 15$ **25.** $C = 5$ **27.** $C = 2$
29. $R = 750$ **31.** $R = 12$ **33.** $R = 31.5$ **35.** $R = 16$
37. $C = 0$ **39.** $C = 32$ **41.** $C = 60$ **43.** 20 pounds of
candy, 80 pounds of coffee **45.** \$1800 **47.** 1 ounce
Brand X, 2 ounces Brand Y **49.** \$600

CHECKING BASIC CONCEPTS 4.3 & 4.4 (p. 270)

1. $y \le -2x + 3$
2.

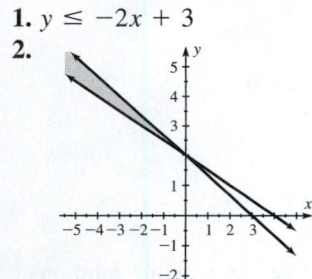

3. $R = 12$

SECTION 4.5 (pp. 280–283)

1. No; three planes cannot intersect at exactly 2 points.
2. $x + y + z = 5, 2x - 3y + z = 7, x + 2y - 4z = 2$
(answers may vary) **3.** Yes **4.** No; a solution must be an
ordered triple. **5.** Two **6.** Three **7.** no **8.** Infinitely
many **9.** $(1, 2, 3)$ **11.** $(-1, 1, 2)$ **13.** $(3, -1, 1)$
15. $\left(\frac{17}{2}, -\frac{3}{2}, 2\right)$ **17.** $(5, -2, -2)$ **19.** $(11, 2, 2)$
21. $(1, -1, 2)$ **23.** $(0, -3, 2)$ **25.** $(-1, 3, -2)$
27. $(-1, 2, 4)$ **29.** $(8, 5, 2)$ **31.** No solutions
33. $(4 - z, 2, z)$ **35.** $(3, -3, 0)$ **37.** $\left(-\frac{3}{2}, 5, -10\right)$
39. $\left(\frac{3}{2}, 1, -\frac{1}{2}\right)$ **41.** No solutions **43.** $(4 - z, 1, z)$
45. No solutions **47.** (a) $x + 2y + 4z = 10,$
$x + 4y + 6z = 15, 3y + 2z = 6$ **(b)** $(2, 1, 1.5)$;
a hamburger costs \$2, fries \$1, and a soft drink \$1.50.
49. (a) $x + y + z = 180, x - z = 55, x - y - z = -10$
(b) $x = 85°, y = 65°,$ and $z = 30°$ **(c)** These values check.
51. $110°, 50°, 20°$ **53.** (a) $a + 600b + 4c = 525,$
$a + 400b + 2c = 365, a + 900b + 5c = 805$
(b) $a = 5, b = 1, c = -20, F = 5 + A - 20W$
(c) 445 fawns **55.** (a) $N + P + K = 80,$
$N + P - K = 8, 9P - K = 0$ **(b)** $(40, 4, 36)$; 40 pounds
nitrogen, 4 pounds phosphorus, 36 pounds potassium
57. \$7500 at 4%, \$8500 at 5%, and \$14,000 at 7.5%
59. 231 **61.** 231

SECTION 4.6 (pp. 292–295)

1. A rectangular array of numbers

2. $\begin{bmatrix} 2 & 1 & 3 \\ 0 & -4 & 2 \end{bmatrix}$ is 2×3 (answers may vary).

3. $\begin{bmatrix} 1 & 3 & | & 10 \\ 2 & -6 & | & 4 \end{bmatrix}$ is 2×3 (answers may vary).

4. 3×4 **5.** $\begin{bmatrix} 1 & 0 & | & -3 \\ 0 & 1 & | & 5 \end{bmatrix}$ (answers may vary)

6. $4, 2, -3$ **7.** 3×3 **9.** 3×2

11. $\begin{bmatrix} 1 & -3 & | & 1 \\ -1 & 3 & | & -1 \end{bmatrix}$ **13.** $\begin{bmatrix} 2 & -1 & 2 & | & -4 \\ 1 & -2 & 0 & | & 2 \\ -1 & 1 & -2 & | & -6 \end{bmatrix}$

15. $x + 2y = -6, 5x - y = 4$ **17.** $x - y + 2z = 6,$
$2x + y - 2z = 1, -x + 2y - z = 3$
19. $x = 4, y = -2, z = 7$

21. $\begin{bmatrix} 0 & 1 & 1 & 1 \\ 1 & 0 & 1 & 0 \\ 0 & 0 & 0 & 1 \\ 1 & 0 & 1 & 0 \end{bmatrix}$ **23.** $(1, 3)$ **25.** $\left(-\frac{3}{2}, 2\right)$ **27.** $\left(\frac{7}{2}, -\frac{3}{2}\right)$

29. $\left(\frac{1}{2}, \frac{3}{2}\right)$ **31.** $(1, 2, 3)$ **33.** $(3, -3, 3)$ **35.** $(-1, 1, 0)$
37. $(1, 1, 1)$ **39.** $(-3, 2, 2)$ **41.** $(-7, 5)$ **43.** $(1, 2, 3)$
45. $(1, 0.5, -3)$ **47.** $(0.5, 0.25, -1)$ **49.** $(0.5, -0.2, 1.7)$
51. Dependent **53.** Inconsistent **55.** Dependent
57. $\left(\frac{1}{a}, \frac{1}{b}, \frac{2}{ab}\right)$ **59.** 214 pounds
61. (a) $\quad a + 16b + 26c = \quad 80$
$\qquad\qquad a + 28b + 45c = 344$
$\qquad\qquad a + 31b + 54c = 416$
(b) $a \approx -272.9, b \approx 19.8,$ and $c \approx 1.4$ **(c)** About 216 lb
63. $\frac{1}{2}$ hour at 5 miles per hour, 1 hour at 6 miles per hour, and
$\frac{1}{2}$ hour at 8 miles per hour **65.** \$500 at 3%, \$1000 at 4%,
and \$1500 at 6% **67.** Answers may vary.

CHECKING BASIC CONCEPTS 4.5 & 4.6 (p. 295)

1. $(1, 3, -1)$ **2.** $(1, 2, 3)$ **3.** $(-2, 2, -1)$ **4.** \$200 at 1%;
\$400 at 2%; \$900 at 4%

SECTION 4.7 (pp. 300–302)

1. square **2.** number **3.** system of linear equations **4.** 0
5. -2 **7.** -53 **9.** -323 **11.** -5 **13.** -36 **15.** -42
17. -50 **19.** 0 **21.** -3555 **23.** -7466.5 **25.** abc
27. 15 square feet **29.** 52 square feet **31.** 25.5 square feet
33. $(2, -2)$ **35.** $\left(-\frac{29}{11}, -\frac{34}{11}\right)$ **37.** $(-5, 7)$

CHECKING BASIC CONCEPTS 4.7 (p. 302)

1. (a) -1 **(b)** 17 **2.** $x = -4, y = 6$ **3.** 21 square units

CHAPTER 4 REVIEW (pp. 305–308)

1. $(3, 2)$ **2.** $(4, -3)$ **3.** $(-1, -3)$ **4.** $(3.5, 8.5)$
5. $(1, 5)$; consistent; independent **6.** $\{(x, y) \mid x - y = -2\}$;
consistent; dependent **7.** No solutions; inconsistent
8. $(2, -1)$; consistent; independent **9.** $(3, -2)$
10. $(-2, 1)$ **11.** (a) $x + y = 25, x - y = 10$
(b) $(17.5, 7.5)$ **12.** (a) $3x - 2y = 19, x + y = 18$
(b) $(11, 7)$ **13.** $(0.467, 0.025)$ **14.** $(2.611, 1.768)$
15. $(-3, 1)$ **16.** $(2, 0)$ **17.** $(-1, 2)$ **18.** $(-2, 3)$
19. $\left(\frac{2}{5}, \frac{14}{5}\right)$ **20.** $(-2, -3)$ **21.** $\{(x, y) \mid 3x - y = 5\}$

22. No solutions

23. 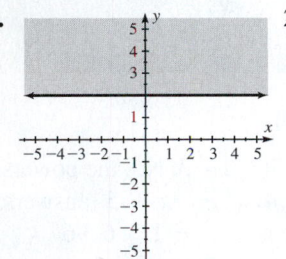 **24.**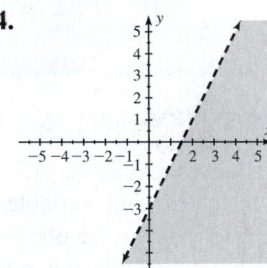

25. 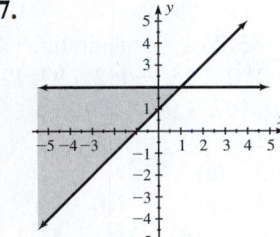 **26.**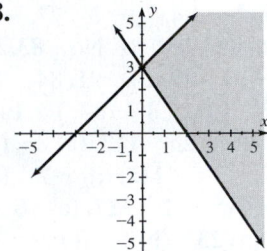

27. 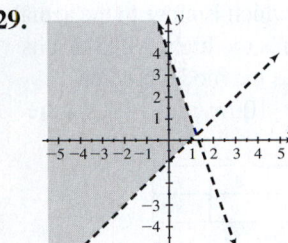 **28.**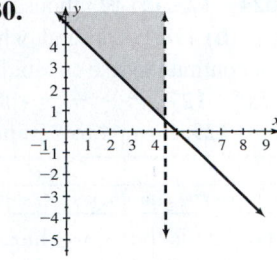

29. 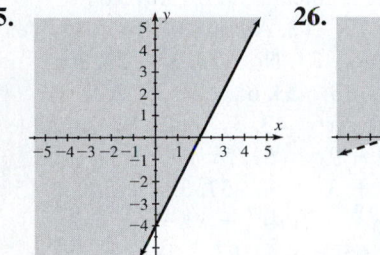 **30.**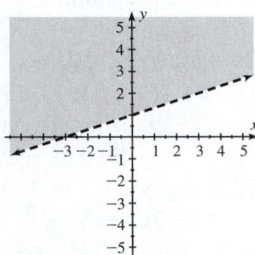

31. $x > 1, y < -1$ **32.** $y \le -x + 4, y \ge 2x + 1$
33. $R = 31$ **34.** $C = 3$ **35.** $R = 6$ **36.** $R = 45$
37. Yes **38.** $(1, -1, 2)$ **39.** $(-4, 3, 2)$ **40.** $(-1, 5, -3)$
41. $(-1, 3, 2)$ **42.** $(1, 1, 1)$ **43.** No solutions
44. $(2 - z, 2z - 1, z)$

45. $\begin{bmatrix} 1 & 1 & 1 & | & -6 \\ 1 & 2 & 1 & | & -8 \\ 0 & 1 & 1 & | & -5 \end{bmatrix}; (-1, -2, -3)$

46. $\begin{bmatrix} 1 & 1 & 1 & | & -3 \\ -1 & 1 & 0 & | & 5 \\ 0 & 1 & 1 & | & -1 \end{bmatrix}; (-2, 3, -4)$

47. $\begin{bmatrix} 1 & 2 & -1 & | & 1 \\ -1 & 1 & -2 & | & 5 \\ 0 & 2 & 1 & | & 10 \end{bmatrix}; (-5, 4, 2)$

48. $\begin{bmatrix} 2 & 2 & -2 & | & -14 \\ -2 & -3 & 2 & | & 12 \\ 1 & 1 & -4 & | & -22 \end{bmatrix}; (-4, 2, 5)$

49. $(-7, 4, 2)$ **50.** $(5.4, 2.1, 9.7)$ **51.** -8 **52.** 30 **53.** 89
54. 130 **55.** 181,845 **56.** 67.688 **57.** 46 square feet
58. 128 square feet **59.** $(2, -1)$ **60.** $(-5, 3)$ **61.** $\left(\frac{3}{2}, \frac{1}{2}\right)$
62. $(7, -3)$ **63.** 5489 in 1994; 4641 in 2004 **64.** Stair
climber: 8 minutes; bicycle: 22 minutes
65.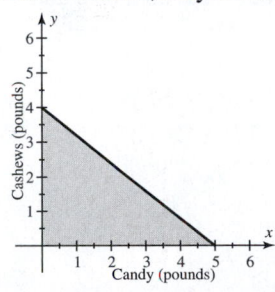

66. Unsubsidized: \$1750; subsidized: \$2250 **67.** 8 shirts,
20 pants; \$660 **68.** 30% solution: 2.4 gallons;
55% solution: 1.6 gallons **69.** Boat: 15 miles per hour;
current: 3 miles per hour **70.** \$8 tickets: 285;
\$12 tickets: 195 **71. (a)** $m + 3c + 5b = 14$,
$m + 2c + 4b = 11$, $c + 3b = 5$ **(b)** Malts: \$3;
cones: \$2; bars: \$1 **72.** 100°, 65°, and 15° **73.** 2
pounds of \$1.50 candy, 4 pounds of \$2 candy, 6 pounds
of \$2.50 candy **74. (a)** $a + 202b + 63c = 40$,
$a + 365b + 70c = 50$, $a + 446b + 77c = 55$
(b) $a \approx 27.134; b \approx 0.061, c \approx 0.009$ **(c)** About 46
inches

CHAPTER 4 TEST (pp. 309–310)

1. $(3, 1)$; consistent; independent **2.** No solutions;
inconsistent **3.** $\{(x, y) \mid x - 3y = 1\}$; consistent;
dependent **4.** $\left(\frac{1}{2}, -7\right)$; consistent; independent
5. $(-3, 1)$ **6. (a)** $x - y = 34, x - 2y = 0$ **(b)** $(68, 34)$
7. $(-0.378, 1.220)$ **8.** $(8, -7)$
9. **10.**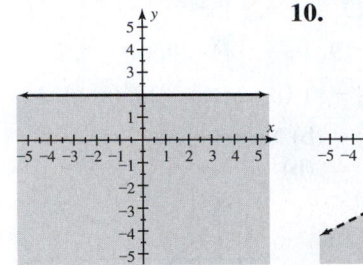

11.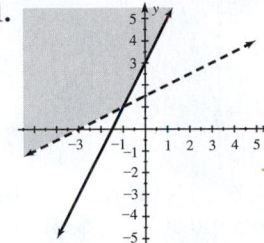

12. $(-4, 2, -5)$ **13.** No solutions **14.** Infinitely many
solutions: $(1, z, z)$ **15.** $y \le x + 2, y \ge 3x - 1$
16. (a) $\begin{bmatrix} 1 & 1 & 1 & | & 2 \\ 1 & -1 & -1 & | & 3 \\ 2 & 2 & 1 & | & 6 \end{bmatrix}$ **(b)** $\left(\frac{5}{2}, \frac{3}{2}, -2\right)$ **17.** 114

18. $\left(-\frac{47}{2}, -\frac{83}{2}\right)$ **19. (a)** $x - y = 19{,}700$, $x - 3.6y = 0$
(b) Approximately (27277, 7577); private tuition was
$27,277, and public tuition was $7577.   **20.** Running:
43 minutes; rowing: 17 minutes   **21.** Airplane: 270 miles
per hour; wind: 30 miles per hour   **22.** 85°, 60°, and 35°
23. $R = 4$
24.

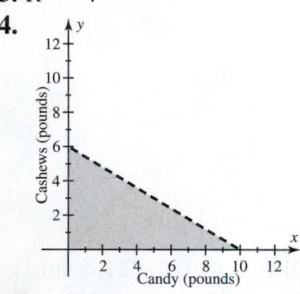

25. Unsubsidized: $2550; subsidized: $1850

CHAPTERS 1–4 CUMULATIVE REVIEW (pp. 312–313)

1. Identity   **2.** Associative   **3.** -14.5 **4.** 0   **5.** $-\frac{a}{b}$
6. (a) $\frac{9}{256}$ **(b)** $\frac{4}{a^5b^2}$ **(c)** y^2 **7.** 5.6×10^{-3} **8.** 13
9. $V = 4\pi$ **10.** $x \neq -\frac{1}{2}$
11.   **12.**

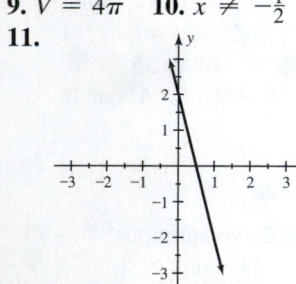   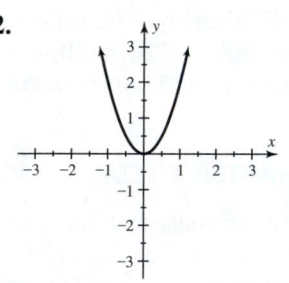

13. (a) Yes   **(b)** D and R: all real numbers   **(c)** -1.5; 0
(d) 2; -1 **(e)** $\frac{1}{2}$ **(f)** $f(x) = \frac{1}{2}x - 1$ **14.** $y = \frac{4}{3}x + 11$
15. 4; $-\frac{1}{2}$; 2   **16.** $-\frac{1}{3}$ **17.** $\left(-\infty, \frac{4}{5}\right)$
18. $(-\infty, 1) \cup (3, \infty)$ **19.** $\left(-\frac{3}{4}, \frac{5}{4}\right]$ **20. (a)** $-\frac{2}{3}$, 2
(b) $\left(-\frac{2}{3}, 2\right)$ **(c)** $\left(-\infty, -\frac{2}{3}\right) \cup (2, \infty)$ **21. (a)** (1, 1)
(b) (2, 0)   **22. (a)** $\left(\frac{1}{2}, \frac{7}{4}\right)$ **(b)** No solutions
23. (a)   **(b)**

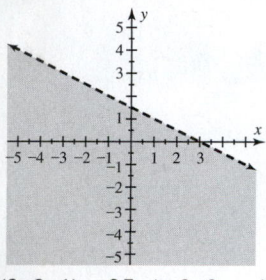   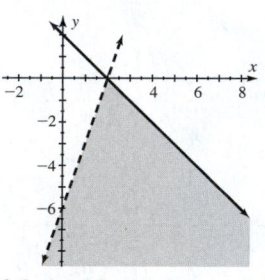

24. (3, 2, 1)   **25.** $(-2, 3, -4)$ **26.** 6   **27. (a)** $I(x) = 0.07x$
(b) $35; the interest after 1 year for $500 at 7% is $35.
28. (a) 9340; average tuition and fees in 1990 were $9340.
(b) Tuition and fees increased, on average, by $572.30 per
year from 1980 to 2000.   **29.** 15 in. by 22 in.
30. $|P - 12.4| \leq 0.2$ **31.** $500 at 5%; $2000 at 6%;
$2500 at 8%   **32.** Texas: 25 million; Florida: 18 million

---

5   Polynomial Expressions and Functions

SECTION 5.1 (pp. 326–330)

1. $3x^2$ (answers may vary)   **2.** 3; -1 **3.** No; the powers
must match for each variable.   **4.** $x^4 - 3x + 5$ (answers
may vary)   **5.** No; the opposite is $-x^2 - 1$.   **6.** 96
7. No; a function has only one output for each input.
8. Yes, if the polynomial is linear.   **9.** 6   **10.** $3x^2$
11. $(f - g)(x) = x^2 + 12$ **12.** All but $3 - x^2$ **13.** Yes
15. No   **17.** Yes   **19.** No   **21.** No   **23.** x^2 **25.** πr^2
27. xy **29.** 7; 3   **31.** 7; -3 **33.** 6; -1 **35.** 2; 5
37. 3; $-\frac{2}{5}$ **39.** 5; 1   **41.** $5x^2$ **43.** $3y^4$ **45.** Not possible
47. $3x^2 + 3x$ **49.** $5x^2 + 9xy$ **51.** $5x^3y^7 + 13x$
53. $2x + 2$ **55.** $-2x^2 + 3x + 8$ **57.** $3x^3 + 5x^2 + 15$
59. $-7x^4 - 3x^2 + x - \frac{9}{2}$ **61.** $4r^4 + r^3 - 6r + 2$
63. $5y^2 - 4xy - 2x^2$ **65.** $-6x^5$ **67.** $-19x^5 + 5x^3 - 3x$
69. $7z^4 - z^2 + 8$ **71.** $3x - 7$ **73.** $6x^2 - 5x + 5$
75. $8x^4 + 14x^2 - 7$ **77.** $-3x - 11$ **79.** Yes; 4;
fourth degree   **81.** No   **83.** No   **85.** Yes; 2; quadratic
87. -16 **89.** 24   **91.** 84   **93.** -1; 2   **95.** -4; 2   **97.** 12
99. 7   **101.** 5.8   **103.** 1   **105.** 0   **107.** 1.96   **109.** -6
111. $a^2 - 2a$ **113. (a)** 8   **(b)** -10 **(c)** $2x + 4$
(d) $4x - 6$ **115. (a)** -7 **(b)** -5 **(c)** $-2x^2 + 1$
(d) $-4x^2 - 1$ **117. (a)** 16   **(b)** 4   **(c)** $2x^3$ **(d)** $-4x^3$
119. (a) 23   **(b)** 3   **(c)** $5x^2 + x + 1$ **(d)** $-3x^2 - 5x + 1$
121. (a) 1300; 10,000   **(b)** 1300; 9924; they are similar.
(c) 8624   **123. (a)** 123 thousand, which is close to the actual
value.   **(b)** 478.6 thousand, which is too high; AIDS deaths
did not continue to rise as rapidly as the model predicts.
125. $5s^3$ **127.** $x^2 + \pi x^2$; $100 + 100\pi \approx 414.2$ square
inches   **129.** From 4 to 8 minutes

t	3	4	5	6	7	8	9
$y = f(t)$	126.875	110	96.875	87.5	81.875	80	81.875

131. (i) $f(x)$ is the best, which can be determined
from a table of values.   **133. (a)** $R(x) = 16x$
(b) $P(x) = 12x - 2000$ **(c)** 34,000; profit is $34,000
for selling 3000 DVDs.   **135. (a)** 130; it costs $130 thousand
to make 100 notebook computers.   **(b)** The y-intercept for
C is 100. The company has $100 thousand in fixed costs
even if it makes 0 computers. The y-intercept for R is 0.
If the company sells 0 computers, its revenue is $0.
(c) $P(x) = 0.45x - 100$ **(d)** 223 or more

SECTION 5.2 (pp. 339–342)

1. Distributive   **2.** $3x$, $2x - 7$, $x^2 - 3x + 1$
(answers may vary)   **3.** x^8 **4.** $8x^3$; x^6 **5.** $a^2 - b^2$
6. $a^2 + 2ab + b^2$ **7.** $x^2 - x$ **8.** No, it equals
$x^2 + 2x + 1$.   **9.** F   **10.** T   **11.** d   **12.** a   **13.** x^{12}
15. $-20y^8$ **17.** $-4x^4y^6$ **19.** $20x^2y^3z^6$ **21.** $5y + 10$
23. $-10x - 18$ **25.** $-6y^2 + 18y$ **27.** $27 - 12x$
29. $-a^3b + ab^3$ **31.** $-5mn^3 - 5m^2$
33. $x^2 + 3x + 2$; 42 square inches   **35.** $2x^2 + 3x + 1$;
66 square inches   **37.** $x^2 + 11x + 30$ **39.** $4x^2 + 4x + 1$

41. No, a square of a sum does not equal the sum of the squares. **43.** No **45.** Yes **47.** No **49.** $x^2 + 15x + 50$ **51.** $x^2 - 7x + 12$ **53.** $2z^2 + 3z - 2$ **55.** $y^2 - y - 12$ **57.** $-36x^2 + 43x - 12$ **59.** $-2z^2 + 7z - 6$ **61.** $z^2 - \frac{1}{4}z - \frac{1}{8}$ **63.** $2x^4 + x^2 - 1$ **65.** $x^2 - xy - 2y^2$ **67.** $4x^3 - 8x^2 - 12x$ **69.** $-x^5 + 3x^3 - x$ **71.** $6n^4 - 12n^3 + 3n^2$ **73.** $x^3 + 3x^2 - x - 3$ **75.** $-z^4 - 3z^3 - z^2 + 2z$ **77.** $4a^3b^2 - 2a^2b^3 + 6ab^4$ **79.** $6r^3 - 10r^2t - 6rt^2 + 4t^3$ **81.** $2^6 = 64$ **83.** $2z^{18}$ **85.** $25x^2$ **87.** $-8x^3y^6$ **89.** $16a^4b^6$ **91.** $x^2 - 16$ **93.** $9 - 4x^2$ **95.** $x^2 - y^2$ **97.** $x^2 + 4x + 4$ **99.** $4x^2 + 4x + 1$ **101.** $x^2 - 2x + 1$ **103.** $9x^2 - 12x + 4$ **105.** $3x^3 - 3x$ **107.** $3rt^3 - 12r^3t$ **109.** $a^4 - 4b^4$ **111.** $9m^6 + 30m^3n^2 + 25n^4$ **113.** $x^6 - 4x^3y^3 + 4y^6$ **115.** $x^2 - 6x - y^2 + 9$ **117.** $r^2 - t^2 - 4t - 4$ **119.** 9996 **121.** 0; $x^2 - x - 2$ **123.** -28; $-5x^3 + 3x^2$ **125.** 27; $2x^4 + 2x^3 - 5x^2 + x - 3$ **127.** $x(x + 4)$; 480 square feet **129.** $(x + 3)^2$; 529 square feet **131. (a)** 30 thousand **(b)** $40p - \frac{1}{2}p^2$ **(c)** \$750 thousand **133. (a)** $\frac{1}{100}x^2 - 2x + 100$ **(b)** 43.56%; yes **(c)** If one serve is out of bounds 34% of the time, then two consecutive serves will be in bounds 43.56% of the time. **135. (a)** The average American used 1900 gallons of water daily in 1980, and this daily average is decreasing by 24 gallons each year. **(b)** $T(x) = -62.4x^2 - 580x + 437{,}000$ **(c)** 363,440; daily usage in 2010 was 363,440,000,000 gal. **137.** $x(x + 1) = x^2 + x$ **139.** $x(50 - x)$ or $50x - x^2$

CHECKING BASIC CONCEPTS 5.1 & 5.2 (p. 342)

1. (a) $3x^2 + 7x$ **(b)** $5x^3 + 2x^2 - 3x + 1$ **2.** $x(x + 120)$; 1300, when tickets cost \$10 each, the revenue will be \$1300. **3. (a)** 160 bpm **(b)** 92 bpm **4. (a)** $-5x + 30$ **(b)** $12x^5 - 20x^4$ **(c)** $2x^2 + 5x - 3$ **5. (a)** $25x^2 - 36$ **(b)** $9x^2 - 24x + 16$

6.

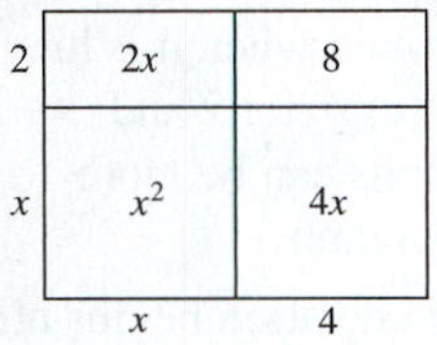

SECTION 5.3 (pp. 350–353)

1. To solve equations **2.** Either $x = 0$ or $x - 3 = 0$; zero-product property **3.** Yes; because $2x$ is a factor of each term. **4.** No, because $4x^2$ is the greatest common factor. **5.** No; because $\frac{1}{2}(4) = 2$, and $\frac{1}{2} \neq 2$ and $4 \neq 2$. **6.** Yes; by the zero-product property **7.** d **8.** b **9.** $5(2x - 3)$ **11.** $2(2x + 3y)$ **13.** $3(3r - 5t)$ **15.** $x(2x^2 - 5)$ **17.** $2a(4a^2 + 5)$ **19.** $6r^3(1 - 3r^2)$ **21.** $4x(2x^2 - x + 4)$ **23.** $3n(3n^3 - 2n + 1)$ **25.** $2t^2(3t^4 - 2t^2 + 1)$ **27.** $5x^2y^2(1 - 3y)$ **29.** $3a^2b^2(2a - 5b)$ **31.** $6mn^2(3 + 2mn)$ **33.** $5xy(3x + 2 - 5xy)$ **35.** $2a(2a - b + 3b^2)$ **37.** $-2(x^2 - 2x + 3)$ **39.** $-8z^3(z + 2)$ **41.** $-2mn(2mn^2 + 3n + 4)$ **43.** $m = 0$ or $n = 0$ **45.** $z = 0$ or $z = -4$ **47.** $r = 1$ or $r = -3$ **49.** $x = -2$ or $x = \frac{1}{3}$ **51.** $x = 0$ or $y = 6$ **53. (a)** $-3, 0$ **(b)** $-3, 0$ **(c)** $-3, 0$ **55. (a)** 0, 2 **(b)** 0, 2 **(c)** 0, 2 **57. (a)** $-1, 1$ **(b)** $-1, 1$ **(c)** $-1, 1$ **59.** 0, 2 **61.** 0, 1 **63.** $-4, 0$ **65.** 0, 1 **67.** $0, \frac{1}{5}$ **69.** $-\frac{1}{2}, 0$ **71.** $0, \frac{2}{3}$ **73.** $0, \frac{5}{2}$ **75.** $0, \frac{1}{2}$ **77.** $-2, 0$ **79.** $0, \frac{3}{2}$ **81.** $(2x + 3)(x + 2)$ **83.** $(x - 5)(x^2 - 2)$ **85.** $(x + 3)(x^2 + 2)$ **87.** $(3x - 2)(2x^2 + 3)$ **89.** $(x^2 + 1)(2x - 3)$ **91.** $(x^2 - 3)(x - 7)$ **93.** $(x - 5)(3x^2 + 5)$ **95.** $(y + 1)(x + 3)$ **97.** $(a + 2)(b - 3)$ **99.** b.; a. **101. (a)** After 8 seconds **(b)** After 8 seconds **(c)** Yes

t	3	4	5	6	7	8
$y = h(t)$	240	256	240	192	112	0

(d) 256 ft; 4 seconds **103.** 391; CO_2 concentration in 2010 was 391 ppm. **105.** $\frac{8}{\pi}$ **107. (a)** The function models the data quite well.

t	20	30	40	50	60	70	80
$y = f(t)$	4520	10,080	17,840	27,800	39,960	54,320	70,880

(b) About 54,000 feet **(c)** $-\frac{6}{11}, 0$; the shuttle is on the ground at time $t = 0$; the value $t = -\frac{6}{11}$ has no physical meaning. **109.** No; there are many possibilities, such as 9×24 and 8×27. **111.** Answers will vary.

SECTION 5.4 (pp. 361–364)

1. A polynomial with three terms; $x^2 - x + 1$ (answers may vary) **2.** Symbolic and graphical, grouping and FOIL (answers may vary) **3.** $a = 3, b = -1, c = -3$ **4.** Yes; it checks, using multiplication. **5.** $(x + 3)(x - 1)$ **6.** $(6x - 3)(x - 2)$ or $3(2x - 1)(x - 2)$ or $6\left(x - \frac{1}{2}\right)(x - 2)$ **7.** c **8.** d **9.** Yes **11.** No **13.** Yes **15.** No **17.** $(x + 2)(x + 5)$ **19.** $(x + 2)(x + 6)$ **21.** $(x - 9)(x - 4)$ **23.** $(x - 8)(x + 1)$ **25.** $(z - 8)(z + 9)$ **27.** $(t - 8)(t - 7)$ **29.** $(y - 12)(y - 6)$ **31.** $(m - 20)(m + 2)$ **33.** $(n - 30)(n + 10)$ **35.** $(x + 3)(2x + 1)$ **37.** $(2x + 1)(3x - 5)$ **39.** $(z + 4)(4z + 3)$ **41.** $(2t - 3)(3t - 4)$ **43.** $(2y + 3)(5y - 1)$ **45.** $(2m - 3)(3m + 4)$ **47.** $(6n + 5)(7n - 5)$ **49.** $(1 - x)(1 + 2x)$ **51.** $(5 - 2x)(4 + 3x)$ **53.** $5(y - 2)(y + 3)$ **55.** $2(z + 2)(z + 4)$ **57.** $z(z + 2)(z + 7)$ **59.** $t(t - 7)(t - 3)$ **61.** $m^2(m + 1)(m + 5)$ **63.** $x(x - 1)(5x + 6)$ **65.** $3x(x + 3)(2x + 1)$ **67.** $2x(x - 5)(x - 2)$ **69.** $10z(z + 4)(6z - 1)$ **71.** $2x^2(x + 3)(2x - 1)$ **73.** $(x + 2)(x + 3)$ **75.** $(x + a)(x + b)$ **77.** -3; 2 **79.** $f(x) = x + 1$, $g(x) = x + 2$ **81.** $f(x) = x + 2$, $g(x) = x - 4$ **83.** $(x - 2)(x - 4)$ **85.** $2(x + 1)(x - 2)$ **87.** $-(x - 2)(x + 1)$ **89.** $(x + 1)(x - 4)$ **91.** $2(x + 1)(x - 2)$ **93.** $(x - 2)(x + 5)$ **95.** $(x - 7)(x + 4)$ **97.** $2(x - 5)(x - 2)$ **99.** $5(x - 10)(x + 4)$ **101.** $4(x - 5)(2x - 1)$ **103. (a)** $(35 - x)(1200 + 100x)$ **(b)** \$20 or \$27 **105.** 7 by 13 feet **107. (a)** 6 sec **(b)** 6 sec

CHECKING BASIC CONCEPTS 5.3 & 5.4 (p. 364)

1. (a) $3x(x - 2)$ **(b)** $4x(4x^2 - 2x + 1)$
2. (a) $0, 2$ **(b)** $0, 9$ **3. (a)** $(x - 2)(x + 5)$
(b) $(x - 5)(x + 2)$ **(c)** $(2x + 3)(4x + 1)$
4. $-1, -2$ **5.** After 1 and 3 seconds

SECTION 5.5 (pp. 369–371)

1. $x^2 - 9$ (answers may vary)   **2.** $x^2 - 8x + 16$
(answers may vary)   **3.** $x^3 + 8$ (answers may vary)
4. $(a - b)(a + b)$ **5.** $(a + b)^2$
6. $(a - b)(a^2 + ab + b^2)$ **7.** c   **8.** b
9. Yes; $(x - 5)(x + 5)$ **11.** No; $(x + y)(x^2 - xy + y^2)$
13. $(x - 6)(x + 6)$ **15.** $(5 - z)(5 + z)$
17. Cannot be factored   **19.** $4(3x - 5)(3x + 5)$
21. $(7a - 8b)(7a + 8b)$ **23.** $z^2(8 - 5z)(8 + 5z)$
25. $5x(x - 5)(x + 5)$ **27.** Cannot be factored
29. $(4t^2 - r)(4t^2 + r)$ **31.** $(x - 4)(x + 6)$
33. $(14 - n)(6 + n)$ **35.** $(y - 2)(y + 2)(y^2 + 4)$
37. $(2x - y)(2x + y)(4x^2 + y^2)$ **39.** $(x - 1)(x + 1)^2$
41. $(2x - 1)(2x + 1)(x - 2)$ **43.** No   **45.** Yes; $(x + 4)^2$
47. Yes; $(2z - 1)^2$ **49.** No   **51.** $(x + 1)^2$ **53.** $(2x + 5)^2$
55. $(x - 6)^2$ **57.** $(3z - 4)^2$ **59.** $y^2(2y + 1)^2$
61. $z(3z - 1)^2$ **63.** $(3x + y)^2$ **65.** $(7a - 2b)^2$
67. $x^2(2x - y)^2$ **69.** $(x - 2)(x^2 + 2x + 4)$
71. $(y + z)(y^2 - yz + z^2)$ **73.** $(3x - 2)(9x^2 + 6x + 4)$
75. $(4z + 3t)(16z^2 - 12zt + 9t^2)$
77. $x(2x + 5)(4x^2 - 10x + 25)$
79. $y(3 - 2x)(9 + 6x + 4x^2)$
81. $(z^2 - 3y)(z^4 + 3z^2y + 9y^2)$
83. $(5z^2 + 2y^3)(25z^4 - 10z^2y^3 + 4y^6)$
85. $5(m^2 + 2n)(m^4 - 2m^2n + 4n^2)$
87. $(5x - 8)(5x + 8)$ **89.** $(x + 3)(x^2 - 3x + 9)$
91. $(8x + 1)^2$ **93.** $(x + 4)(3x + 2)$
95. $x(x + 2)(x^2 - 2x + 4)$
97. $8(2x + y)(4x^2 - 2xy + y^2)$ **99.** $2(r - 2t)(r + 2t)$
101. $a(a + 2b)^2$ **103.** $(x - 2)(x - 1)$
105. $(2z - 5)(2z + 5)$ **107.** $x^2(x + 8)^2$
109. $(z - 1)(z^2 + z + 1)$ **111.** $(t + 1)(3t - 8)$
113. $a(a + 3)(7a - 1)$
115. $(x + y)(x - y)(x^2 + xy + y^2)(x^2 - xy + y^2)$
117. $(10x - 1)(10x + 1)$
119. $(pq - 3)(p^2q^2 + 3pq + 9)$ **121.** $(x - 5)(x + 5)$

SECTION 5.6 (pp. 376–377)

1. Factor out the GCF.   **2.** $2x$ **3.** No; the sum of two
squares cannot be factored.   **4.** Grouping   **5.** $a(a - 1)$
7. $(a - 3)(a + 3)$ **9.** $(x - 1)^2$
11. $(x - a)(x^2 + ax + a^2)$ **13.** Not possible
15. $(x - 3)(x + 2)$ **17.** $(x + 2)(x^2 + 1)$ **19.** $2x(3x - 7)$
21. $2x(x - 3)(x + 3)$ **23.** $4(a - 2)(a + 2)(a^2 + 4)$
25. $x(x - 3)(6x + 5)$ **27.** $x^2(x - 5)(2x + 5)$
29. $(x^2 + 1)(2x^2 + 3)$ **31.** $(x^2 + 1)(x + 3)$
33. $5(x^2 + 2)(x - 1)$ **35.** $(a + b)(x - y)$
37. $2(3x + 1)^2$ **39.** $-4x(x - 3)^2$
41. $(2x - 3)(4x^2 + 6x + 9)$
43. $-x(x + 2)(x^2 - 2x + 4)$

45. $(x - 1)(x - 2)(x^2 + x + 1)$
47. $(r - 2)(r + 2)(r^2 + 4)$ **49.** $(5x - 2a)(5x + 2a)$
51. $2(x - y)(x + y)(x^2 + y^2)$ **53.** $3x(3x - 1)(x + 1)$
55. $(z - 5)(z + 1)$
57. $3(x + 1)(x^2 - x + 1)(x - 3)(x + 3)$
59. (a) $(x^2 + 2)(x - 390)$ **(b)** 390 ppm
61. (a) $x^3 - 3x^2 + 2x$ **(b)** $x^3 - 3x^2 + 2x = 6$ **(c)** 3 ft

CHECKING BASIC CONCEPTS 5.5 & 5.6 (p. 377)

1. (a) $(5x - 4)(5x + 4)$ **(b)** $(x + 6)^2$ **(c)** $(3x - 5)^2$
(d) $(x - 3)(x^2 + 3x + 9)$ **(e)** $(3x - 2)(3x + 2)(9x^2 + 4)$
2. (a) $(x^2 + 3)(x - 2)$ **(b)** $(x + 3)(x - 7)$
(c) $2x(2x - 1)(3x + 2)$

SECTION 5.7 (pp. 384–387)

1. It is used in solving polynomial equations (answers may
vary).   **2.** multiplying   **3.** Subtract 16 from each side.
4. No; the right side is not 0.   **5.** 1   **6.** No; it implies that
$x^2 = -4$.   **7.** $-1, 1$ **9.** $-2, 2$ **11.** $-1, 2$ **13.** $-8, 8$
15. $-\frac{1}{2}, \frac{1}{2}$ **17.** $-1, 4$ **19.** $-6, 2$ **21.** $-3, \frac{1}{2}$ **23.** $-4, 4$
25. -7 **27.** $\frac{1}{3}$ **29.** $-\frac{1}{5}, \frac{2}{3}$ **31.** $-\frac{4}{3}, \frac{3}{8}$ **33.** No solutions
35. $-3, 2$ **37.** $-2, 4$
39.   **41.**

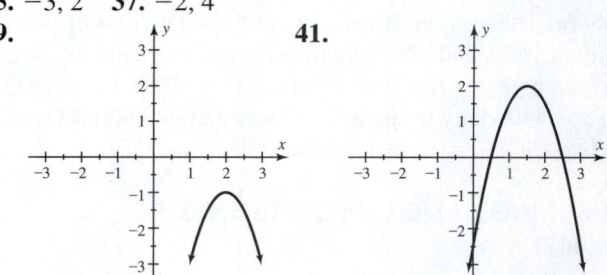

43. $-3, 0, 3$ **45.** 0   **47.** $-2, 0, 5$ **49.** $-1, 0, 5$
51. $-7, -2, 2$ **53.** $-6, 2, 6$ **55.** $-3, 5$ **57.** $-\frac{1}{2}, 2$
59. $-5, -1$ **61.** $0, 2$ **63.** $-2, 2$ **65.** $-5, -1, 1, 5$
67. $-3, -2, 2, 3$ **69.** $-2, 2$ **71.** $-1, -\frac{2}{3}, \frac{2}{3}, 1$
73. 17 by 21 inches   **75.** $-1000, 1000$ **77.** 30 by 50 feet
79. (a) The ball is 2 feet above the ground when it is hit.
(b) After 2 sec (answers may vary)   **(c)** After 2 and
$\frac{17}{8} = 2.125$ sec   **(d)** Symbolic solutions can be more
accurate than reading a graph.   **81.** 33 mph
83. 0.6 by 2.4 by 4.1 in.   **85. (a)** The elevation begins at
500 feet, decreases, and then increases back to 500 feet.

x	0	300	600	900	1200	1500
$y = E(x)$	500	428	392	392	428	500

(b) 500 feet and 1000 feet

CHECKING BASIC CONCEPTS 5.7 (p. 387)

1. $-3, -1$ **2.** $-3, 3$ **3. (a)** 3   **(b)** No solutions
(c) $-\frac{5}{4}, \frac{2}{3}$ **4. (a)** 0   **(b)** $-3, 3$ **(c)** $-2, 2$ **(d)** $-7, -2, 7$

CHAPTER 5 REVIEW (pp. 390–393)

1. $x + 5, x^2 - 3x + 1$ (answers may vary)   **2.** $10xy^2$;
3200 cubic inches   **3.** $5; -4$ **4.** $3; 1$ **5.** $7; 5$ **6.** $10; -9$
7. $11x$ **8.** $4x^3 + x^2$ **9.** $14x^3y + x^3$ **10.** 19   **11.** $2; 5$
12. $4; 2$ **13.** $8x^2 + 3x - 1$ **14.** $23z^3 - 4z^2 + z$

15. $-x^2 + x$ **16.** $-5x^3 - x^2 - 3x + 6$ **17.** 7
18. -259 **19.** -4 **20.** $-11; 4x^2 - x - 1$
21. $15x - 20$ **22.** $-2x - 2x^2 + 8x^3$ **23.** x^8 **24.** $-6x^4$
25. $-42x^2y^8$ **26.** $60x^3y^5$ **27.** $x^2 + 9x + 20$
28. $x^2 - 15x + 56$ **29.** $12x^2 - 48x - 27$ **30.** $y^2 - \frac{1}{9}$
31. $8x^4 - 12x^3 - 4x^2$ **32.** $-4x - 5x^2 + 7x^3$
33. $16x^2 - y^2$ **34.** $x^2 + 6x + 9$ **35.** $4y^2 - 20y + 25$
36. $a^3 - b^3$ **37.** $25m^2 - 20mn^4 + 4n^8$
38. $r^2 - 2r + 1 - t^2$ **39.** $5x(5x - 6)$
40. $4x^2(3x^2 + 2x - 4)$ **41.** $-3, 0$ **42.** $-2, 0, 2$ **43.** $\frac{1}{2}, 1$
44. $-1, 3$ **45.** $(x + 1)(2x^2 - 3)$ **46.** $(z + 1)(z^2 + 1)$
47. $(a - b)(x + y)$ **48.** $5; x^2 - 2x - 3$ **49. (a)** $0, 3$
(b) $0, 3$ **(c)** $0, 3$ **50. (a)** $-1, 2$ **(b)** $-1, 2$ **(c)** $-1, 2$
51. $(x + 2)(x + 6)$ **52.** $(x - 10)(x + 5)$
53. $(x + 3)(9x - 2)$ **54.** $2(x - 5)(2x - 1)$
55. $x(x - 3)(x - 1)$ **56.** $2x^2(x + 2)(x + 5)$
57. $5x^2(x - 3)(x + 6)$ **58.** $10x(x - 5)(x - 4)$
59. $(x + 3)(x - 5)$ **60.** $(x - 11)(x - 13)$
61. $(x + 7)(x - 4)$ **62.** $8, 13$ **63.** $(t - 7)(t + 7)$
64. $(2y - 3x)(2y + 3x)$ **65.** $(x + 2)^2$ **66.** $(4x - 1)^2$
67. $(x - 3)(x^2 + 3x + 9)$
68. $(4x + 3y)(16x^2 - 12xy + 9y^2)$
69. $10y(y - 1)(y + 1)$ **70.** $(2r^2 - t^3)(2r^2 + t^3)$
71. $(m - 2n)(m + 2n)(m^2 + 4n^2)$
72. $(n - 2)(n - 1)(n + 1)$ **73.** $(5a - 3b)^2$
74. $2r(r - 3t)^2$ **75.** $(a^2 + 3b)(a^4 - 3a^2b + 9b^2)$
76. $(2p^2 - q)(4p^4 + 2p^2q + q^2)$ **77.** $5x^2(x - 2)$
78. $-2x(x - 4)(x + 4)$ **79.** $(x - 2y)(x + 2y)(x^2 + 4y^2)$
80. $4x(x - 1)(x + 3)$ **81.** $-x(2x - 3)(x - 4)$
82. $(x^2 + 1)(x - 3)(x + 3)$
83. $(4a + b)(16a^2 - 4ab + b^2)$
84. $(2 - y)(y^2 + 2y + 4)$ **85.** $(z - 1)(z + 7)$
86. $x(x - 2)(x + 2)(x - 5)$ **87.** $-4, 4$ **88.** $-1, 3$
89. $\frac{7}{2}$ **90.** No solutions   **91.** $-1, \frac{5}{3}$ **92.** $-2, \frac{3}{4}$
93. $-1, 0, 1$ **94.** $1, 2, 3$ **95.** $-9, 0, 8$ **96.** $0, 7, 8$
97. $-2, 2$ **98.** $-2, 2$ **99. (a)** $R(x) = 15x$
(b) $P(x) = 12x - 9000$ **(c)** $39,000$; profit is \$39,000 for
selling 4000 DVDs.   **100. (a)** $x(x + 5)(x + 10)$
(b) $x(x + 5)(x + 10) = 168$ **(c)** 2 by 7 by 12 in.
101. 8 by 10 in.   **102.** $\frac{1}{2}x^2 - \frac{1}{2}x - 3$
103. (a) $73.6°$F   **(b)** July

X	Y₁
1	27.784
2	43.636
3	56.556
4	66.544
5	73.6
6	77.724
7	78.916
X=7	

(c) Temperatures increase from January through July and
then decrease from July through December.

[1, 12, 1] by [30, 90, 10]

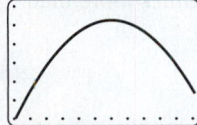

104. (a) $100 - 2x + \frac{x^2}{100}$ **(b)** 9%
105. $(x - 2)(x - 1)(x) = x^3 - 3x^2 + 2x$

106.

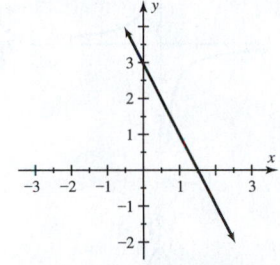

107. 9 feet by 16 feet   **108.** $25x - x^2$
109. (a) After $\frac{33}{8} = 4.125$ seconds   **(b)** After 1 second
and after $\frac{25}{8} = 3.125$ seconds
110. (a) $R(x) = (50 - x)(600 + 20x)$ **(b)** \$40
111. $-1000, 1000$

CHAPTER 5 TEST (pp. 393–394)

1. $5x - 4x^2y^2$ **2.** $-7x^3 + x^2 - 7x + 11$ **3.** -8
4. -2 **5.** $4; 3$ **6.** $8xyz$ **7.** -66 **8.** $66; -x^3 + 2x - 4$
9. $-4x^3 + 2x^2$ **10.** $14x^2y^8$ **11.** $10x^2 - 9x - 7$
12. $9x^2 - 30x + 25$ **13.** $25x^2 - 16y^2$
14. $-2x^4 + 6x^3 - 4x^2$ **15.** $x^3 - 8y^3$ **16.** $2x^4 - 2x^2$
17. $(x - 5)(x + 2)$ **18.** $2x(x^2 + 3)$ **19.** $(x + 4)(3x - 5)$
20. $5x^2(x - 1)(x + 1)$ **21.** $(2x + 1)(x^2 - 5)$
22. $(7x - 1)^2$ **23.** $(x + 2)(x^2 - 2x + 4)$
24. $4x^2y^2(y^2 + 2x^2)$ **25.** $(a - b)(a - 2b)$ **26.** $3; 1$
27. c   **28.** $(x + 8)(x - 6)$ **29.** $x(x + 2) = x^2 + 2x$
30. $0, 3$ **31.** $-5, \frac{1}{4}$ **32.** $-2, 0, 2$ **33.** $-1, 1$ **34.** $-1, 3$
35. (a) $x(x + 4) = 221$ **(b)** $13, -17$; height is 13 inches.
(c) 60 inches   **36. (a)** About $57.5°$F   **(b)** The dew point
starts at about $30°$F in January and increases to a maximum
of about $65°$F in July. Then it decreases to $30°$F by the end
of December.
37. $x^2 + 5x + 6$

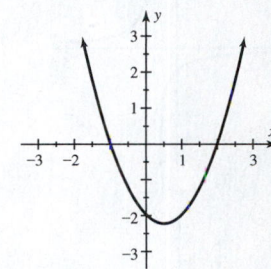

38. After 2 and 4 sec

CHAPTERS 1–5 CUMULATIVE REVIEW (pp. 395–396)

1. $\frac{15}{4}$ **2.** $\frac{b - a}{a}$ **3. (a)** $\frac{1}{x^6y^2}$ **(b)** $\frac{r^2}{81}$ **(c)** $\frac{a^3}{b^{11}}$
4. 58,590   **5.** -27 **6.** $D: x \geq 4$
7.   **8.**

9. (a) Yes   **(b)** D: all real numbers; $R: y \geq -4$
(c) $-4, -3$ **(d)** $-1, 3$ **(e)** $-1, 3$
(f) $f(x) = (x + 1)(x - 3)$ **10.** $y = 2x + 9$
11. $f(x) = 2 - 5x$ **12.** $\frac{6}{11}$ **13.** $\left(-\infty, -\frac{5}{2}\right]$

14. $(-\infty, 1) \cup (3, \infty)$ **15.** $\left[0, \frac{4}{7}\right)$ **16.** $(2, 1)$

17. (a) $\{(x, y) \mid x + 2y = -5\}$ (b) $\left(2, \frac{1}{2}\right)$

18.

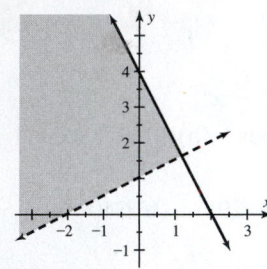

19. $(-4, -1, -3)$ **20.** $(2, 2, 2)$ **21.** -6 **22.** No solutions

23. $-2x^3 + 4x^2 - 10x$ **24.** $4a^2 - b^2$

25. $x^3 + 3x^2 - x - 3$ **26.** $8x^2 + 14x - 9$

27. $x^4 + 6x^2y^3 + 9y^6$ **28.** $2x^3 - 2x$

29. $(x + 3)(x - 11)$ **30.** $5x(2x - 1)(x + 7)$

31. $4(x - 5)(x + 5)$ **32.** $(7x - 5)^2$

33. $r(r - 1)(r^2 + r + 1)$ **34.** $(x + 2)(x^2 + 1)$

35. $-\frac{1}{2}, \frac{1}{2}$ **36.** $-5, \frac{1}{3}$ **37.** $0, 2$ **38.** $-1, 0, 1$ **39.** -1.1

40. $-2.1, 1.5$ **41.** \$2300 at 6%, \$1700 at 7%

42. (a) 10,550; in 1995 the car cost \$10,550   (b) The cost increased, on average, by \$750 per year from 1990 to 2010.

43. 12 in. by 12 in. by 7 in.   **44.** 80°, 60°, 40°

6   Rational Expressions and Functions

SECTION 6.1 (pp. 407–412)

1. A polynomial divided by a nonzero polynomial; $\frac{x^2 + 1}{3x}$ (answers may vary)   **2.** $x = 4$ **3.** Multiply each side by $x + 7$.   **4.** No; the equation is undefined when $x = 5$.

5. No, it simplifies to $5 + x$.   **6.** 1   **7.** a **8.** 0   **9.** c.

10. a.   **11.** Yes   **12.** Yes   **13.** Yes   **14.** No   **15.** No

16. Yes   **17.** $f(x) = \frac{x}{x + 1}$ **19.** $f(x) = \frac{x^2}{x - 2}$

21. $\{x \mid x \neq -2\}$ **23.** $\{x \mid x \neq \frac{1}{3}\}$

25. $\{t \mid t \neq -2, t \neq 2\}$ **27.** $\{x \mid x \neq 1, x \neq 2\}$

29. $\{x \mid x \neq -2, x \neq 0, x \neq 2\}$

31. $\{x \mid x \neq 1\}$ **33.** $\{x \mid x \neq 0\}$

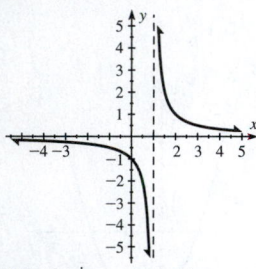   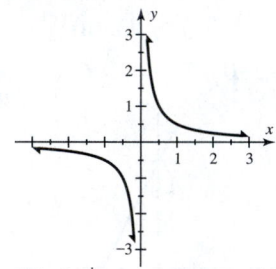

35. $\{x \mid x \neq -2\}$ **37.** $\{x \mid -\infty < x < \infty\}$

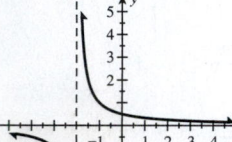   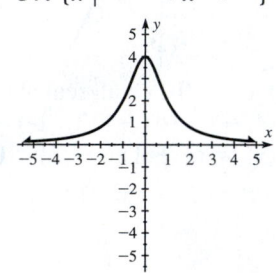

39. $\left\{x \mid x \neq \frac{3}{2}\right\}$ **41.** $\{x \mid x \neq -1, x \neq 1\}$

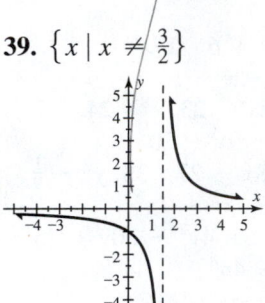   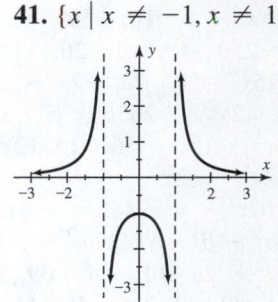

43. $-\frac{1}{3}$ **45.** $\frac{1}{2}$ **47.** 3   **49.** 2, 0; $x = -1$

51. 0, undefined; $x = -2, x = 2$

53.

x	-2	-1	0	1	2
$f(x) = \frac{1}{x-1}$	$-\frac{1}{3}$	$-\frac{1}{2}$	-1	—	1

; undefined

55. $\frac{3}{5}$ **57.** 1   **59.** -2 **61.** 3   **63.** $-3, 1$ **65.** $-2, 1$

67. 0   **69.** No solutions   **71.** -1 **73.** -4 **75.** -2

77. $-4, 1$ **79.** 0, 2   **81.** -1 **83.** 2   **85.** $-1.62, 0.62$

87. ab **89.** (a) 19   (b) -9 (c) 150   (d) 0   **91.** (a) 41

(b) -21 (c) 900   (d) Undefined   **93.** (a) $2x + 3$

(b) -1 (c) $x^2 + 3x + 2$ (d) $\frac{x + 1}{x + 2}$ **95.** (a) $x^2 - x + 1$

(b) $1 - x - x^2$ (c) $x^2 - x^3$ (d) $\frac{1 - x}{x^2}$ **97.** c.   **99.** d.

101. (a) 6.35; a curve with radius of 400 feet will have an outer rail elevation of 6.35 inches.

(b)   (c) It is halved.   (d) 508 ft

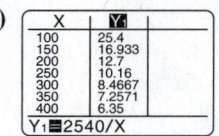

X	Y1
100	25.4
150	16.933
200	12.7
250	10.16
300	8.4667
350	7.2571
400	6.35

Y1■2540/X

103. (a) 75; the braking distance is 75 feet when the uphill grade is 0.05.   (b) 0.15   **105.** (a) 1; when cars are leaving the lot at a rate of 4 vehicles per minute, the average wait is 1 minute.   (b) As more cars try to exit, the waiting time increases; yes.   (c) $4.\overline{6}$ vehicles per minute

107. (a) $P(1) = 0$, $P(50) = 0.98$; when there is only 1 ball, there is no chance of losing. With 50 balls there is a 98% chance of losing.

(b) [0, 100, 10] by [0, 1, 0.1]

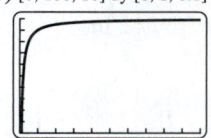

(c) It increases; yes, there are more balls without winning numbers.   (d) 40 balls

109. About 45% and 22%; after 1 day (3 days) students remember 45% (22%) of what they have learned.

111. (a) $R(x) = \frac{350(x - 3)^2 + 50}{206(x - 3)^2 + 150}$ (b) Six years after startup, Google's revenues were about 1.6 times more than Facebook's revenues six years after startup.

SECTION 6.2 (pp. 421–423)

1. 1   **2.** -1 **3.** No, it is equal to $x + 1$.   **4.** No; no

5. $\frac{2}{3}; \frac{7}{5}$ **6.** $\frac{ac}{bd}$ **7.** $\frac{ad}{bc}$ **8.** $\frac{a}{b}$ **9.** c.   **10.** b.   **11.** $\frac{2}{5}$ **13.** $-\frac{7}{2}$

15. $\frac{3}{4}$ **17.** $-\frac{1}{22}$ **19.** $\frac{2}{3}$ **21.** -18 **23.** $\frac{5}{x}$ **25.** 3   **27.** $\frac{2}{3}$

29. $x + 1$ **31.** $x - 2$ **33.** $\frac{x}{x + 1}$ **35.** $\frac{3x + 1}{5x - 2}$ **37.** $\frac{1}{x - 3}$

39. $\frac{x}{x+1}$ **41.** $\frac{3x+5}{x-5}$ **43.** $a+b$ **45.** m^2-mn+n^2
47. 1   **49.** -1 **51.** -1 **53.** $\frac{1}{4x}$ **55.** $\frac{5b}{2a}$ **57.** $\frac{5-x}{3-x}$
59. x^2+1 **61.** $\frac{2(x-1)}{3x^2}$ **63.** $\frac{(x-2)(x-3)}{x(x+4)}$ **65.** 1
67. 3   **69.** 3   **71.** $\frac{x+1}{x}$ **73.** $\frac{b^2+1}{(b+1)^2}$ **75.** $\frac{(x-1)(x+5)}{2(2x-3)}$
77. $3(n^2-3n+9)$ **79.** $\frac{n^2-3n+9}{4}$ **81.** $\frac{4}{xy}$
83. $\frac{y(x-1)}{2}$ **85.** $\frac{15}{4}$ **87.** $\frac{24b}{a}$ **89.** $2(n-1)$ **91.** $\frac{4}{b^3}$
93. $\frac{3a(3a+1)}{a+1}$ **95.** $\frac{x^2}{(x-5)(x^2-1)}$ **97.** $\frac{(x-2)(x-1)}{(x+2)^2}$
99. 1   **101.** $\frac{15}{y^2}$ **103.** $\frac{1}{x}$ **105.** $\frac{1}{x-2}$ **107.** $x-1$
109. (a) 54; it costs \$54, on average, to make each camera when 5000 are produced.   **(b)** $T(x)=50x+20{,}000$
(c) 270,000; it costs \$270,000 to make 5000 cameras.
111. (a) $5x+2$ **(b)** 32 ft   **113. (a)** $4x^2+4x+1$
(b) 21 by 21 by 10

CHECKING BASIC CONCEPTS 6.1 & 6.2 (p. 423)

1. (a) 2   **(b)** $\{x\,|\,x\neq1\}$
(c) $x=1$

2. (a) $-\frac{1}{2}$ **(b)** $-1,2$ **3.** $\frac{x-7}{x-1}$
4. (a) $\frac{x}{2(x-1)}$ **(b)** $\frac{x+3}{3}$

SECTION 6.3 (pp. 432–434)

1. 18   **2.** 18   **3.** $(x-5)(x+5)$ **4.** $(x-5)(x+5)$
5. A common denominator   **6.** A common denominator
7. $\frac{a+b}{c}$ **8.** $\frac{a-b}{c}$ **9.** b.   **10.** a.   **11.** 30   **13.** 102
15. $18a^2$ **17.** $50x^2(x-1)$ **19.** $(x-5)(x+1)^2$
21. $(x+y)(x-y)$ **23.** $\frac{5}{7}$ **25.** $\frac{7}{4}$ **27.** $-\frac{1}{5}$ **29.** $\frac{11}{8}$
31. $\frac{4}{x}$ **33.** $\frac{1-x}{x^2-4}$ **35.** $\frac{9}{x^2}$ **37.** $-\frac{3}{xy}$ **39.** 1   **41.** 1
43. $\frac{r}{t^2}$ **45.** $\frac{1}{t-3}$ **47.** $\frac{4b}{15a}$ **49.** $\frac{n+4}{(n-4)(n-2)}$
51. $\frac{-5x-4}{x(x+4)}$ **53.** $\frac{-4x^2+x+2}{x^2}$ **55.** $\frac{x^2+5x-34}{(x-5)(x-3)}$
57. $\frac{4(n+6)}{(n-3)(n+3)}$ **59.** $\frac{x(5x+16)}{(x-3)(x+3)}$ **61.** $-\frac{1}{2}$
63. $\frac{6x^2-1}{(x-5)(3x-1)}$ **65.** $\frac{4x^2+4xy-9x+9y}{(x-y)(x+y)}$
67. $\frac{4x^2-x+3}{(x-2)(x-1)(x+1)}$ **69.** $\frac{x+15}{(x-3)(x+2)(x+3)}$
71. $\frac{4x-7}{(x-2)(x-1)^2}$ **73.** $\frac{3x^2-2x+3}{(x-1)^2(x+3)}$
75. $\frac{-2a^2+3b^2+4c^2}{abc}$ **77.** $\frac{5n^2+3n-168}{(n-6)(n+6)}$
79. $\frac{-2(x^2-4x+2)}{(x-5)(x-3)}$ **81.** $\frac{x^2+x+5}{(x+1)(2x-3)}$
83. $\frac{3}{(x-1)(x+1)}$ **85.** $\frac{2(a^2+b^2)}{(a+b)(a-b)}$
87. $1; \frac{1-x}{x+1}$ **89.** $\frac{2x}{(x-2)(x+2)}; \frac{-4}{(x-2)(x+2)}$
91. $C(x)=\frac{7x+1{,}500{,}000}{x}$ **93.** 60 years
95. $\frac{1200}{19}\approx63.2$ ohms   **97.** $S\approx0.101$ foot or about
1.2 inches   **99.** $\frac{16}{d^2}$ watts per square meter

SECTION 6.4 (pp. 443–446)

1. Multiply each side by $x+2$.   **2.** Multiply each side by the LCD.   **3.** $\frac{1}{2}$ **4.** bc **5.** LCD   **6.** intersect   **7.** F
8. T   **9.** $5x$ **11.** x^2-1 **13.** $2(2x+1)(x-2)$
15. 1   **17.** 5   **19.** 4   **21.** $-\frac{7}{6}$ **23.** -3 **25.** 1   **27.** 6
29. No solutions   **31.** No solutions   **33.** $-\frac{1}{12}$ **35.** $1,-\frac{1}{2}$
37. $-1,8$ **39.** 1   **41.** $-\frac{4}{15},3$ **43.** 3   **45.** -3 **47.** -10
49. $-\frac{1}{2},\frac{1}{2}$ **51.** $\frac{a-b}{c}$ **53.** 1   **55.** 2   **57.** 3   **59.** $-3,1$
61. 2.5   **63.** $-0.4,0.67$ **65.** ±1 **67.** 3   **69.** $-\frac{3}{2},1$
71. $-3.28,-0.90,1.18$ **73.** $r=\frac{d}{t}$ **75.** $b=\frac{2A}{h}$
77. $a=\frac{b}{2}$ **79.** $R_1=\frac{RR_2}{R_2-R}$ **81.** $x=15-\frac{1}{T}$
83. $t=r-1$ **85.** $b=a(r-1)$ **87. (a)** 82 per hour
(b) The wait gets very long.   **89. (a)** $\frac{x}{5}+\frac{x}{2}=1$
(b) $\frac{10}{7}$ hours   **(c)** $\frac{10}{7}\approx1.43$ hours
91. (a) 75; three testers found 75% of the glitches.   **(b)** 19
(c) As the number of testers becomes large, the percentage of the total number of glitches found slowly nears 100%, but never actually gets there.

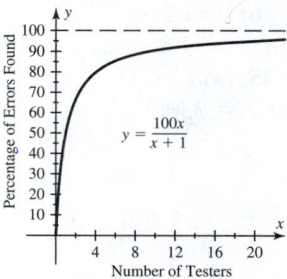

93. (a) 12.5; the number 36 pick receives 12.5% of the number 1 pick's salary.   **(b)** Number 21
95. (a) $\frac{x}{40}+\frac{x}{70}=1$ **(b)** $\frac{280}{11}=25.\overline{45}$ hr
(c) $\frac{280}{11}=25.\overline{45}$ hr
97. Winner: 10 miles per hour; second place: 8 miles per hour   **99.** 3 miles per hour   **101.** 250 miles per hour
103. 3 hr and 6 hr   **105.** 4 mph and 3 mph   **107.** 120 hr
109. Yes; slightly greater than 2

CHECKING BASIC CONCEPTS 6.3 & 6.4 (p. 447)

1. (a) $\frac{1}{x-1}$ **(b)** $-\frac{2(x-3)}{x(x-2)}$ **(c)** $\frac{3}{x(x+1)(x-1)}$
2. (a) 4   **(b)** 5   **(c)** $-\frac{1}{4}$ **3.** 8.4 minutes
4. $S=\frac{DF}{D-F}$

SECTION 6.5 (pp. 452–454)

1. $\frac{5}{7}\cdot\frac{11}{3}=\frac{55}{21}$ **2.** $\frac{a}{b}\cdot\frac{d}{c}$ **3.** Multiply numerator and denominator by $x-1$ (answers may vary).
4. A rational expression with fractions in its numerator, denominator, or both

5. $\dfrac{z+\frac{3}{4}}{z-\frac{3}{4}}$ **6.** $\dfrac{\frac{a}{b}}{\frac{a-b}{a+b}}$ **7.** $\frac{7}{20}$ **9.** 2   **11.** $\frac{32}{21}$ **13.** $\frac{2b}{3}$
15. $\frac{3}{4}$ **17.** $\frac{2(n-1)}{n+1}$ **19.** $\frac{2k+3}{k-4}$ **21.** $\frac{3}{z}$ **23.** $\frac{x}{x+4}$
25. $\frac{1}{x}$ **27.** $-\frac{x}{2x+3}$ **29.** $\frac{x+2}{3x-1}$ **31.** $\frac{3(x+1)}{(x+3)(2x-7)}$
33. $\frac{4x(x+5)}{(x-5)(2x+5)}$ **35.** $\frac{p+q}{q-p}$ **37.** $\frac{a+b}{a-b}$ **39.** $\frac{2x+1}{x}$
41. $\frac{5}{27}$ **43.** $\frac{mn^2-2m^2}{m^2n^2+1}$ **45.** $\frac{n}{n+1}$ **47.** $\frac{(x-1)(x+4)}{x^2-4}$

49. $\dfrac{P\left(1 + \frac{r}{12}\right)^{24} - P}{\frac{r}{12}}$ **51.** $R = \dfrac{R_1 R_2}{R_1 + R_2}$

53. (a) 5; after 11 practice trials this task can be completed in 5 sec.   (b) $T(x) = \dfrac{100}{x + 9}$

SECTION 6.6 (pp. 464–468)

1. A statement that two ratios are equal   **2.** $\frac{5}{6} = \frac{x}{7}$
3. It doubles.   **4.** It is reduced by half.   **5.** constant
6. constant   **7.** kxy **8.** $kx^2 y^3$ **9.** Directly; if the number of people doubled, the food bill would double.
10. Inversely; increasing the number of painters would decrease the time needed to paint the building.
11. 10   **13.** 12   **15.** 72   **17.** $\frac{21}{20}$ **19.** $-\frac{5}{13}$ **21.** $-6, 6$
23. (a) $\frac{7}{9} = \frac{10}{x}$ (b) $x = \frac{90}{7}$ **25.** (a) $\frac{5}{3} = \frac{x}{6}$ (b) $x = 10$
27. (a) $\frac{78}{6} = \frac{x}{8}$ (b) $x = \$104$ **29.** (a) $\frac{2}{120} = \frac{5}{x}$
(b) $x = 300$ minutes   **31.** (a) $k = 2$ (b) $y = 14$
33. (a) $k = 2.5$ (b) $y = 17.5$ **35.** (a) $k = -7.5$
(b) $y = -52.5$ **37.** (a) $k = 20$ (b) $y = 2$
39. (a) $k = 50$ (b) $y = 5$ **41.** (a) $k = 400$ (b) $y = 40$
43. (a) $k = 0.25$ (b) $z = 8.75$ **45.** (a) $k = 11$
(b) $z = 385$ **47.** (a) $k = 10$ (b) $y = 350$
49. (a) Direct   (b) $y = 1.5x$
(c)

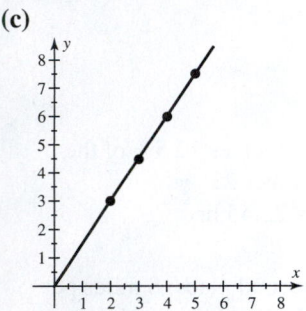

51. Neither   (b) N/A   (c) N/A
53. (a) Inverse   (b) $y = \frac{210}{x}$
(c)

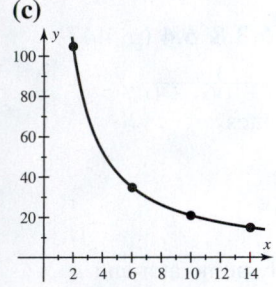

55. (a) Direct   (b) $y = -2bx$ **57.** Direct; $k = 1$
59. Neither   **61.** Direct; $k = 2$ **63.** Answers may vary.
65. About 23.1 ft   **67.** 1.375 inches
69. About 67.0 million   **71.** 2000   **73.** 75 ft

75. (a) $R = 0.012W$

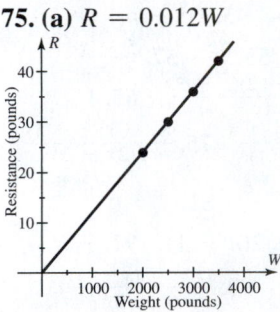

(b) 38.4 pounds
77. (a) Direct; the ratios $\frac{D}{W}$ always equal 0.75.
(b) $D = 0.75W$ (c) 8.25 in.
79. (a) $F = \frac{1200}{L}$ (b) 60 pounds
81. \$797.50; k represents the cost per credit.
83. 1200 lb per square inch
85. (a) Direct   (b) $y = -19x$ (c) Negative; for each 1-mile increase in altitude, the temperature decreases by $19°$F.   (d) $66.5°$F decrease   **87.** About 1.43 ohms
89. $z = 0.5104x^2 y^3$ **91.** About 133 pounds
93. About 35.2 pounds   **95.** 750

CHECKING BASIC CONCEPTS 6.5 & 6.6 (p. 468)

1. $\frac{3x^2 - 1}{3x^2 + 1}$ (b) 1   **2.** (a) $k = 0.75$ (c) $y = 8.25$
3. (a) Direct; $y = 0.4x$ (b) Inverse; $y = \frac{120}{x}$

SECTION 6.7 (pp. 475–477)

1. term   **2.** $\frac{a}{c} + \frac{b}{c}$ **3.** 5; 4; 1   **4.** $3x - 5$; $x - 2$; -2
5. No; the divisor is not in the form $x - k$.   **6.** The last number in the third row   **7.** $2x - 3$ **9.** $2x^2 - 3$
11. $2 - \frac{1}{2x} + \frac{1}{2x^2}$ **13.** $3x^2 - 4x - 1$ **15.** $2a^2 - 3$
17. $\frac{4x^2}{3} - 2$ **19.** $ab - 4 + b$ **21.** $2m^2 n^2 + 1 - \frac{4}{m^2 n^2}$
23. $3x - 7$ **25.** $x^2 - 1 + \frac{1}{2x + 3}$ **27.** $10x + 10 + \frac{5}{x - 1}$
29. $4x^2 - 1$ **31.** $x^2 - 4x + 19 - \frac{80}{x + 4}$
33. $x^2 + 3x - 6 + \frac{1}{3x - 1}$ **35.** $a^3 - a$ **37.** $3x + 4$
39. $x + 1 + \frac{1}{x^2 - 1}$ **41.** $2a^2 - a + 3 - \frac{5}{a^2 - a + 5}$
43. $a^3 - b^3 + 1$ **45.** $x + 4 + \frac{3}{x - 1}$ **47.** $3x - 1$
49. $x^2 + 3x + 2$ **51.** $2x^2 + 5x + 10 + \frac{19}{x - 2}$
53. $2x^3 - 4x^2 + 11x - 22 + \frac{40}{x + 2}$
55. $b^3 + b^2 + b + 1$ **57.** $3x + 5$; 29 ft   **59.** $x + 2$
61. (a) 0; 0   (b) 0; 0   (c) 1; 1   (d) 0; 0   (e) 35; 35
63. (a) 0; yes   (b) 0; yes   (c) -1; no   (d) 0; yes
(e) -9; no

CHECKING BASIC CONCEPTS 6.7 (p. 477)

1. (a) $\frac{2}{x} - 1$ (b) $2a - 3 + \frac{5}{a}$ **2.** $5x - 3 + \frac{7}{2x + 1}$
3. (a) $2x^2 - 3x - 3 - \frac{4}{x - 1}$
(b) $2x^2 - 9x + 18 - \frac{37}{x + 2}$

CHAPTER 6 REVIEW (pp. 481–485)

1. $f(x) = \frac{1}{x-1}$ **2.** $f(x) = \frac{x-3}{x}$

3. $\{x \mid x \neq -2\}$

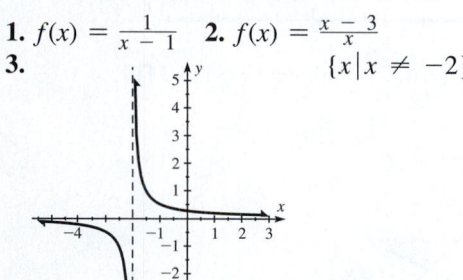

4. $\frac{1}{8}; x = \pm 1$ **5.** $-3; 0.75$ **6.** Undefined; 1.2
7. $3; 1; x = 1$ **8.** 2; undefined; $x = -2$ **9.** 1   **10.** 2
11. No solutions   **12.** 2   **13.** (a) 12   (b) 27
14. (a) $x^2 - x$ (b) $x + 1$ **15.** $\frac{2}{3a^3}$ **16.** $\frac{x+2}{x+1}$
17. $x + 2$ **18.** $\frac{x-7}{2x-3}$ **19.** -1 **20.** 1   **21.** $\frac{y}{4}$
22. $\frac{x+2}{x+1}$ **23.** $\frac{x^2+1}{(x+1)^2}$ **24.** $\frac{x+1}{x-3}$ **25.** $3y^3$ **26.** $\frac{2}{3}$
27. $\frac{x}{x-5}$ **28.** $\frac{x+5}{x+3}$ **29.** $\frac{x(x+2)}{(2x-1)(x+3)}$
30. $\frac{(x+1)^2}{(x+2)(x+4)}$ **31.** 144   **32.** $4a^2b$ **33.** $18x^2y^3$
34. $(x-1)(x+2)$ **35.** $x(x-3)(x+3)$ **36.** $x^2(x-3)^2$
37. $\frac{4}{x+4}$ **38.** $\frac{x-1}{x}$ **39.** $\frac{1-x}{x+1}$ **40.** $\frac{2x-2}{x-2}$ **41.** $\frac{t-4}{t-1}$
42. $\frac{8}{(y-2)(y+2)}$ **43.** $\frac{4b^3 - 3a^3}{a^2b^2c}$ **44.** $\frac{r^3 + t^3}{5r^2t^2}$
45. $\frac{2(a-b-2)}{(a-b)(a+b)}$ **46.** $\frac{a^2 + 2ab - b^2}{(a-b)(a+b)}$
47. $\frac{2x}{(x-2)(x-1)(x+2)}$ **48.** $\frac{6}{(x-3)(x-2)(x+3)}$
49. $\frac{-x^2 + x - 2}{(x-2)(x+2)}$ **50.** $\frac{2x-2}{2x+1}$ **51.** 3   **52.** 11   **53.** $-2, 1$
54. -2 **55.** $-2, \frac{1}{2}$ **56.** $-\frac{5}{2}$ **57.** 0   **58.** 0, 1   **59.** -1
60. -1 **61.** 2   **62.** $-2, 3$ **63.** $y_2 = m(x_2 - x_1) + y_1$
64. $a = T(b+2)$ **65.** $p = \frac{fq}{q-f}$ **66.** $b = \frac{2a}{al-3}$
67. $\frac{39}{50}$ **68.** $\frac{2c}{a}$ **69.** $\frac{4n^2 - 1}{3n^2}$ **70.** $x + y$ **71.** $\frac{2x+3}{2x-3}$
72. $\frac{2(x+2)}{x-8}$ **73.** $\frac{4-5x}{x+1}$ **74.** $\frac{-3}{x-9}$ **75.** $\frac{9}{5}$ **76.** $\frac{77}{5}$
77. $\frac{3}{2}$ **78.** $\frac{31}{3}$ **79.** $\frac{7}{3}$ **80.** $\frac{88}{7} \approx 12.6$ **81.** $y = 28$
82. $y = 2$ **83.** $k = 3$ **84.** $z = 720$
85. Inverse; $y = \frac{200}{x}$ **86.** Direct; $y = 3x$
87. Direct; $k = \frac{1}{2}$ **88.** Inverse; $k = 6$ **89.** $2x + 3$
90. $2x + 1$ **91.** $2x - \frac{1}{2} + \frac{1}{x}$ **92.** $\frac{2a}{b} - \frac{3}{a}$
93. $2x - 3 + \frac{3}{x+1}$ **94.** $x^2 - x + 2 - \frac{3}{x+1}$
95. $2x^2 - x + 1 + \frac{1}{3x-2}$ **96.** $2x - 9 + \frac{25x - 39}{x^2 - 2}$
97. $2x - 5 - \frac{2}{x-3}$
98. $3x^2 - 2x + 4 + \frac{3}{x+4}$
99. (a)

X	Y1
9	1
9.1	1.1111
9.2	1.25
9.3	1.4286
9.4	1.6667
9.5	2
9.6	2.5

Y1■1/(10−X)

(b) It increases; as vehicles arrive at a faster rate, the wait in line increases.
100. (a) $D(0) = 166.\overline{6}$ feet, $D(0.1) = 242.\overline{42}$ feet; about 75.8 feet   (b) $x = 0.05\overline{3}$ or $5.\overline{3}\%$ **101.** (a) About 13 cars   (b) As x increases, the number of cars in line increases; yes.

102. (a) 5; initially there were 5000 fish in the lake.
(b) $[0, 6, 1]$ by $[0, 6, 1]$ (c) It decreases.   (d) After 2 years

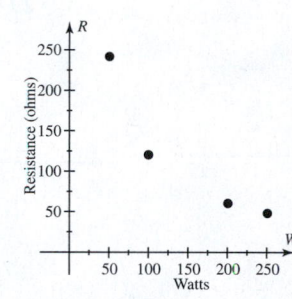

103. (a) $\frac{x}{2} + \frac{x}{3} = 1$ (b) $\frac{6}{5} = 1.2$ hr   (c) $\frac{6}{5} = 1.2$ hr
104. 4 miles per hour   **105.** About 44 minutes
106. About 43.3 feet
107. (a) Inverse

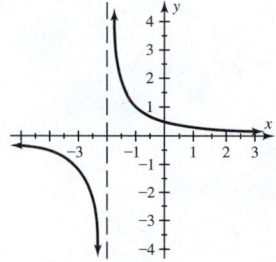

(b) $R = \frac{12,100}{W}; k = 12,100$ (c) $R = 220$ ohms
108. (a) $D = 0.06W; k = 0.06$ (b) 0.42 inch
109. $53°$F   **110.** (a) Direct   (b) $y = 3.5x$ (c) 8.05%

CHAPTER 6 TEST (pp. 485–486)

1. $f(x) = \frac{x}{x+2}$ **2.** (a) $\frac{1}{15}$ (b) $\{x \mid x \neq \frac{1}{2}, x \neq -\frac{1}{2}\}$
3.

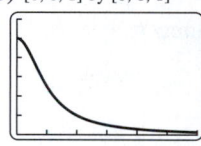

4. $\frac{a}{2}$ **5.** -1 **6.** $\frac{x-5}{2x-7}$ **7.** $\frac{x^2+4}{(x+2)^2}$ **8.** $2y^2$
9. $\frac{1}{x+5}$ **10.** $\frac{2x^2 + 3x + 2}{(x-2)(x+2)}$ **11.** $\frac{5a^3 - 6b^4}{15ab}$
12. $\frac{x^3 - 2x^2 + 8}{x^2(x-2)}$ **13.** $\frac{x+1}{x-1}$ **14.** $\frac{4}{(z+4)^2}$ **15.** $\frac{3x+3}{x-17}$
16. $-\frac{2}{3}$ **17.** $-1, 2$ **18.** 5   **19.** $\frac{1}{4}$ **20.** No solutions
21. $\frac{1}{2}$ **22.** $m = \frac{Fr^2}{G}$ **23.** 0.47   **24.** $0; \frac{1}{x-1}$ **25.** $\frac{35}{3}$
26. $\frac{80}{23}$ **27.** Inverse; $y = \frac{100}{x}$ **28.** Direct; $y = 1.5x$
29. $2a^2 + 5$ **30.** $3x^2 - x + 2 - \frac{6}{x+2}$ **31.** 21.9 feet
32. (a) $[0, 25, 5]$ by $[0, 2, 0.5]$

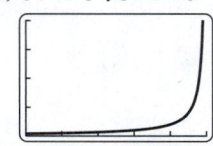 $x = 25$

(b) 24 vehicles per minute
33. (a) $\frac{x}{24} + \frac{x}{30} = 1$ (b) $\frac{40}{3} \approx 13.3$ hours
34. $41.75°$F

CHAPTERS 1–6 CUMULATIVE REVIEW
(pp. 488–490)

1. 9 **2.** 0 **3.** Natural: 1; whole: 0, 1; integer: $-\frac{12}{4}$, 0, 1;
rational: $-\frac{12}{4}$, 0, 1, $2.\overline{11}$, $\frac{13}{2}$; irrational: $\sqrt{3}$

4. a. **5.** $\frac{6y^6}{x^4}$ **6.** $\frac{9d^6}{4c^4}$ **7.** 6.73×10^{10}

8. $D = \{-3, -1, 0, 4\}$
   $R = \{-2, 0, 1, 5\}$

9. $(-2, -7), (-1, -4),$ **10.** $(-2, 0), \left(-1, \frac{3}{2}\right), (0, 2),$
   $(0, -1), (1, 2), (2, 5)$ $\left(1, \frac{3}{2}\right), (2, 0)$

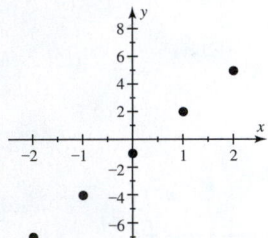 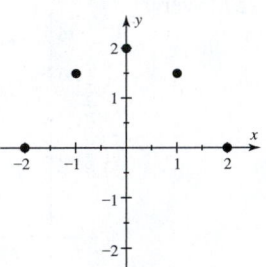

11.

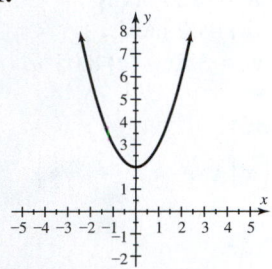

12. $\{x \mid x \neq 1\}$ **13.** $f(x) = 2x - 1$
14.

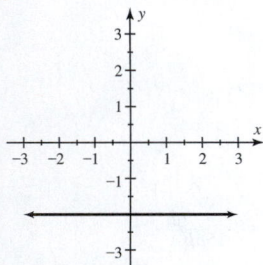

15. $-3; 2$ **16.** -3
17.

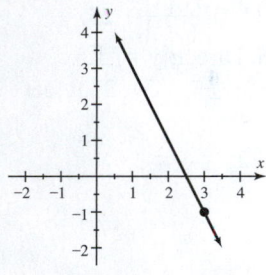

18. $y = -\frac{1}{2}x + 2$ **19.** $y = 2x + 2$ **20.** $y = -\frac{3}{2}x + 4$
21. 4 **22.** $-\frac{7}{29}$ **23.** 3 **24.** $\left(-\infty, -\frac{5}{6}\right]$ **25.** $(0, \infty)$
26. $(-3, 3]$

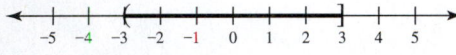

27. $(-\infty, 2] \cup (4, \infty)$

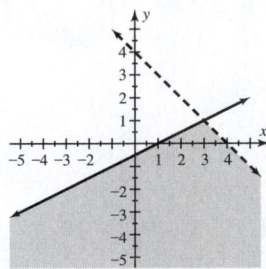

28. $-30, -6$ **29.** $(-4, 7)$ **30.** $(-\infty, -1] \cup [11, \infty)$
31. $(-8, -1)$
32.

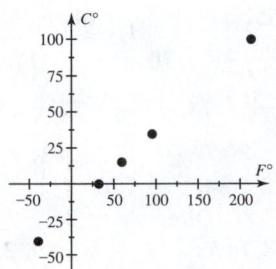

33. $\left(\frac{1}{2}, -\frac{1}{2}\right)$ **34.** $(6, 0)$ **35.** $R = 10$ **36.** $(1, 1, 2)$

37. $\begin{bmatrix} 1 & 1 & -1 & 4 \\ -1 & -1 & -1 & 0 \\ 1 & -2 & 1 & -9 \end{bmatrix}; (-1, 3, -2)$

38. 14 **39.** 8 in^2 **40.** $x = -1, y = 2$
41. $3x^5 + 15x^3 - 6x^2$ **42.** $2x^2 + x - 15$ **43.** $-2, 0$
44. $(3a - 1)(a^2 + 5)$ **45.** $(4x - 3)(x + 2)$
46. $3x(x - 2)(2x + 1)$ **47.** $(3a - 2b)(3a + 2b)$
48. $(4t + 3)(16t^2 - 12t + 9)$ **49.** $-\frac{1}{3}, 4$
50. $-5, 0, 6$ **51.** $\frac{x - 2}{x + 2}$ **52.** $x - 4$ **53.** $\frac{t - 4}{t^2 - 4}$
54. $\frac{4a^2c + 3b^3}{3a^2b^2c}$ **55.** $-\frac{7}{2}$ **56.** -3 **57.** $z = \frac{y}{J - 1}$
58. $\frac{4 + x}{4 - x}$ **59.** $\frac{1}{2}$ **60.** $x^2 - 2x + 6 - \frac{1}{x + 2}$
61. 3 **62.** $x - 1$
63. (a) Linear

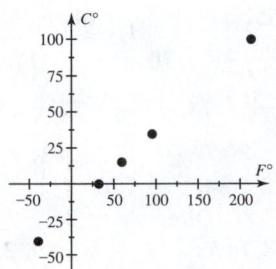

(b) $f(x) = \frac{5}{9}(x - 32)$ **(c)** $40°C$
64. 6, 6, and 10 inches **65.** Stair climber: $\frac{1}{2}$ hr; stationary
bicycle: 1 hr **66.** 24 children, 6 adults **67.** 11 by 15 ft
68. (a) After 2.75 seconds **(b)** After 0.5 and 2.25 seconds
69. (a) $\frac{x}{15} + \frac{x}{10} = 1$ **(b)** 6 hours **70.** 48 feet

7   Radical Expressions and Functions

SECTION 7.1 (pp. 500–503)

1. ± 3 **2.** 3 **3.** 2 **4.** Yes **5.** b **6.** $|x|$ **7.** Undefined **8.** -3 **9.** d.

10. a.
11.

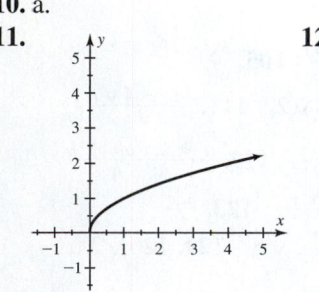

12.

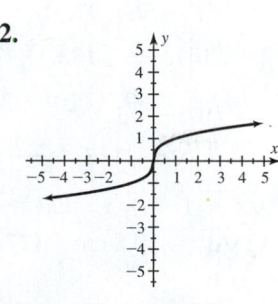

13. $\{x \mid x \geq 0\}$ **14.** All real numbers **15.** 3 **17.** 0.6
19. $\frac{4}{5}$ **21.** x **23.** 3 **25.** -4 **27.** $\frac{2}{3}$ **29.** $-x^3$ **31.** $4x^2$ **33.** 3
35. -3 **37.** Not possible **39.** -2.24 **41.** 1.71
43. -1.48 **45.** 4 **47.** $|y|$ **49.** $|x-5|$ **51.** $|x-1|$ **53.** $|y|$
55. $|x^3|$ **57.** x **59.** 3, not possible **61.** $\sqrt{6}$, not possible
63. $\sqrt{20}$ or $2\sqrt{5}$, $\sqrt{6}$ **65.** 1, 2 **67.** -2, 1 **69.** -1, $\sqrt[3]{-6}$ or
$-\sqrt[3]{6}$ **71.** 4 **73.** 5 **75.** $[-2,\infty)$ **77.** $[2,\infty)$ **79.** $[2,\infty)$
81. $(-\infty, 1]$ **83.** $(-\infty, \frac{8}{5}]$ **85.** $(-\infty,\infty)$ **87.** $\left(-\frac{1}{2},\infty\right)$

89.

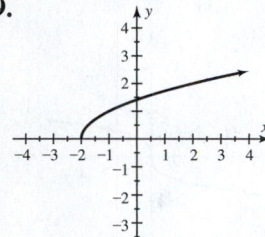

91.

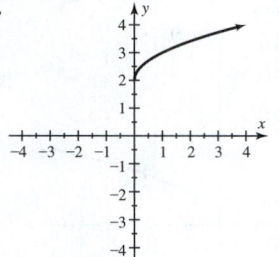

Shifted 2 units left   Shifted 2 units upward

93.

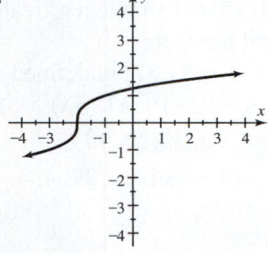

Shifted 2 units left

95.

x	$\sqrt{x}+1$
-1	—
0	1
1	2
4	3
9	4

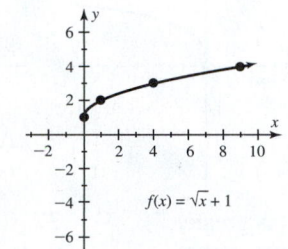

$f(x) = \sqrt{x}+1$

97.

x	$\sqrt{3x}$
-1	—
0	0
$\frac{1}{3}$	1
$\frac{4}{3}$	2
3	3

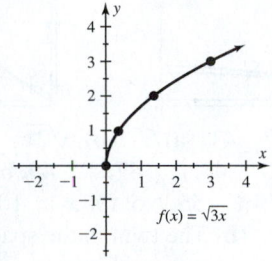

$f(x) = \sqrt{3x}$

99.

x	$2\sqrt[3]{x}$
-8	-4
-1	-2
0	0
1	2
8	4

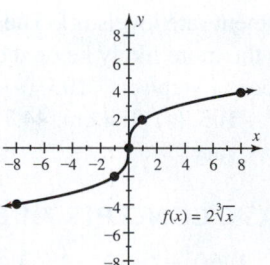

$f(x) = 2\sqrt[3]{x}$

101.

x	$\sqrt[3]{x-1}$
-7	-2
0	-1
1	0
2	1
9	2

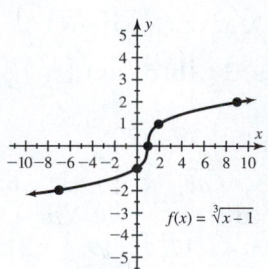

$f(x) = \sqrt[3]{x-1}$

103. $A = 6$ **105.** 1 sec   **107.** 122 mi
109. About $14 thousand; no
111. (a) $P(25,000) = \$71,246$; if $25,000 is spent on equipment per worker, each worker will produce about $71,246 worth of goods.
(b)

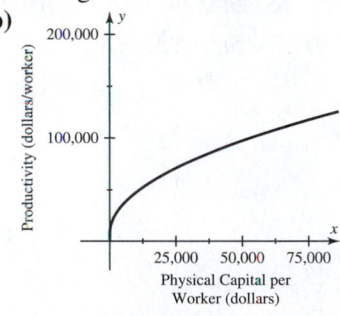

Physical Capital per
Worker (dollars)

(c) $26,197; $20,102; an additional $25,000 is spent on equipment per worker, but productivity levels off. There is a point where the business starts to lose money.

SECTION 7.2 (pp. 509–511)

1. 2 **2.** 2 **3.** $\frac{1}{2}$ **4.** $\frac{1}{2}$ **5.** $x^{1/2}$ **6.** $a^{4/3}$ **7.** $\sqrt[n]{a}$
8. $\left(\sqrt[n]{a}\right)^m$ or $\sqrt[n]{a^m}$ **9.** c. **10.** f. **11.** g. **12.** h. **13.** d.
14. e. **15.** a. **16.** b. **17.** $\sqrt{7}$ **19.** $\sqrt[3]{a}$ **21.** $\sqrt[6]{x^5}$
23. $\sqrt{x+y}$ **25.** $\frac{1}{\sqrt[3]{b^2}}$ **27.** $t^{1/2}$ **29.** $(x+1)^{1/3}$
31. $(x+1)^{-1/2}$ **33.** $(a^2-b^2)^{1/2}$ **35.** $x^{-7/3}$ **37.** 1.74
39. 1.71 **41.** 3.74 **43.** 0.55 **45.** $\sqrt{9}$; 3 **47.** $\sqrt[3]{8}$; 2
49. $\sqrt{\frac{4}{9}}$; $\frac{2}{3}$ **51.** $\sqrt[3]{(-8)^2}$ or $\left(\sqrt[3]{-8}\right)^2$; 4 **53.** $\sqrt[3]{8}$; 2
55. $\frac{1}{\sqrt[4]{16^3}}$ or $\frac{1}{\left(\sqrt[4]{16}\right)^3}$; $\frac{1}{8}$ **57.** $\frac{1}{\left(\sqrt[4]{4}\right)^3}$; $\frac{1}{8}$ **59.** $\sqrt[4]{z}$
61. $\frac{1}{\sqrt[5]{y^2}}$ **63.** $\sqrt[3]{3x}$ **65.** $y^{1/2}$ **67.** x **69.** $2x^{2/3}$
71. $\frac{7}{x^{1/6}}$ **73.** b^3 **75.** x^3 **77.** xy^2 **79.** $y^{13/6}$ **81.** $\frac{x^4}{9}$
83. $\frac{y^3}{x}$ **85.** $y^{1/4}$ **87.** $\frac{1}{a^{2/3}}$ **89.** $\frac{1}{k^2}$
91. $b^{3/4}$ **93.** $p^2 + p$ **95.** $x^{5/6} - x$ **97.** $\frac{3}{x^{1/6}}$
99. (a)

x	0	20	40	60	80
$A(x)$	0	27%	37%	44%	50%

(b) The abandonment rate levels off. The longer a person watches a video, the more likely he or she will continue to watch. **101.** About 1 step/sec **103. (a)** 400 in² **(b)** $A = 100W^{2/3}$ **105. (a)** 42.4 cm; 44.7 cm **(b)** First year **107.** Answers may vary.

CHECKING BASIC CONCEPTS 7.1 & 7.2 (p. 512)

1. (a) ± 7 **(b)** 7 **2. (a)** -2 **(b)** -3 **3. (a)** $\sqrt{x^3}$ or $\left(\sqrt{x}\right)^3$ **(b)** $\sqrt[3]{x^2}$ or $(\sqrt[3]{x})^2$ **(c)** $\frac{1}{\sqrt[5]{x^2}}$ or $\frac{1}{(\sqrt[5]{x})^2}$
4. $|x - 1|$ **5. (a)** 3 **(b)** 5 **(c)** $\left(\frac{12}{7}\right)^{7/12}$ or about 1.37

SECTION 7.3 (pp. 519–520)

1. Yes **2.** No **3.** $\sqrt[3]{ab}$ **4.** $\frac{a}{b}$ **5.** $\frac{a}{b}$ **6.** $\frac{1}{3}$ **7.** No
8. Yes **9.** No, since $1^3 \neq 3$ **10.** Yes; $4^3 = 64$
11. 3 **13.** 10 **15.** 4 **17.** $\frac{3}{5}$ **19.** $\frac{1}{4}$ **21.** $\frac{2}{3}$ **23.** x^3
25. $\frac{\sqrt[3]{7}}{3}$ **27.** $\frac{\sqrt[4]{x}}{3}$ **29.** $\frac{3}{z}$ **31.** $\frac{x}{4}$ **33.** 3 **35.** 4 **37.** 3
39. -2 **41.** a **43.** 3 **45.** $2x^2$ **47.** $-a^2\sqrt[3]{5}$
49. $2x\sqrt[4]{y}$ **51.** $6x$ **53.** $2x^2yz^3$ **55.** $\frac{3}{2}$ **57.** $\sqrt[3]{12ab}$
59. $5\sqrt{z}$ **61.** $\frac{1}{a}$ **63.** $\sqrt{x^2 - 16}$ **65.** $\sqrt[3]{a^3 + 1}$
67. $\sqrt{x + 1}$ **69.** 10 **71.** 2 **73.** 3 **75.** $10\sqrt{2}$ **77.** $3\sqrt[3]{3}$
79. $2\sqrt{2}$ **81.** $-2\sqrt[5]{2}$ **83.** $b^2\sqrt{b}$ **85.** $2n\sqrt{2n}$
87. $2ab^2\sqrt{3b}$ **89.** $-5xy\sqrt[3]{xy^2}$ **91.** $5\sqrt[3]{5t^2}$ **93.** $\frac{3t}{r\sqrt[4]{5r^3}}$
95. $\frac{3\sqrt[3]{x^2}}{y}$ **97.** $\frac{7a\sqrt{a}}{9}$
99. $\left(\sqrt[mn]{a^m b^m}\right)^n = (a^m b^m)^{n/mn}$
$\qquad\qquad\qquad = (a^m b^m)^{1/m}$
$\qquad\qquad\qquad = a^{m/m} b^{m/m}$
$\qquad\qquad\qquad = ab$
101. $\sqrt[6]{3^5}$ **103.** $2\sqrt[12]{2^5}$ **105.** $3\sqrt[12]{3^{11}}$ **107.** $x\sqrt[12]{x}$
109. $\sqrt[12]{r^{11}t^7}$ **111. (a)** $S = 5\sqrt{M}$
(b) 50 miles per hour

SECTION 7.4 (pp. 531–533)

1. $2\sqrt{a}$ **2.** $3\sqrt[3]{b}$ **3.** like
4. Yes; $4\sqrt{15} - 3\sqrt{15} = \sqrt{15}$
5. No **6.** $\frac{\sqrt{7}}{\sqrt{7}}$ **7.** $\sqrt{t} + 5$ **8.** $\frac{5 + \sqrt{2}}{5 + \sqrt{2}}$
9. Not possible **11.** $\sqrt{7}, 2\sqrt{7}, 3\sqrt{7}$ **13.** $2\sqrt[3]{2}, -3\sqrt[3]{2}$
15. Not possible **17.** $2\sqrt[3]{xy}, xy\sqrt[3]{xy}$ **19.** $9\sqrt{3}$ **21.** $6\sqrt[3]{5}$
23. Not possible **25.** Not possible **27.** Not possible
29. $5\sqrt[3]{2}$ **31.** $8\sqrt{2}$ **33.** $6\sqrt{11}$ **35.** $2\sqrt{x} - \sqrt{y}$
37. $2\sqrt[3]{z}$ **39.** $-5\sqrt[3]{6}$ **41.** $y^2 - y$ **43.** $9\sqrt{7}$ **45.** $6\sqrt[4]{3}$
47. $7\sqrt{x}$ **49.** $8\sqrt{2k}$ **51.** $-2\sqrt{11}$ **53.** $5\sqrt[3]{2} - \sqrt{2}$
55. $-\sqrt[3]{xy}$ **57.** $3\sqrt{x + 2}$ **59.** $\sqrt{x + 2}$
61. $(x - 1)\sqrt{x + 1}$ **63.** $4x\sqrt{x}$ **65.** $\frac{\sqrt[3]{7x}}{6}$ **67.** $\frac{3\sqrt{3}}{2}$
69. $\frac{71\sqrt{2}}{10}$ **71.** $2\sqrt{2}$ **73.** $(5x - 1)\sqrt[4]{x}$
75. $(8x + 2)\sqrt{x}$ or $2\sqrt{x}(4x + 1)$
77. $(3ab - 1)\sqrt[4]{ab}$ **79.** $(n - 2)\sqrt[3]{n}$
81. $(f + g)(x) = 3\sqrt{x} + 1$; $(f - g)(x) = 7\sqrt{x} - 5$
83. $(f + g)(x) = 4\sqrt[3]{x}$; $(f - g)(x) = 2$ **85.** $x - \sqrt{x} - 6$
87. 2 **89.** 119 **91.** $x - 64$ **93.** $ab - c$

95. $x + \sqrt{x} - 56$ **97.** $\frac{\sqrt{7}}{7}$
99. $\frac{4\sqrt{3}}{3}$ **101.** $\frac{\sqrt{5}}{3}$ **103.** $\frac{\sqrt{3b}}{6}$ **105.** $\frac{t\sqrt{r}}{2r}$
107. $\frac{3 + \sqrt{2}}{7}$ **109.** $\sqrt{10} - 2\sqrt{2}$ **111.** $\frac{11 - 4\sqrt{7}}{3}$
113. $\sqrt{7} + \sqrt{6}$ **115.** $\frac{z + 3\sqrt{z}}{z - 9}$ **117.** $\frac{a + 2\sqrt{ab} + b}{a - b}$
119. $\sqrt{x + 1} + \sqrt{x}$ **121.** $\frac{3\sqrt[3]{x^2}}{x}$ **123.** $\frac{\sqrt[3]{x}}{x}$
125. $12\sqrt{3} \approx 20.8$ cm **127.** $2\sqrt{6}$ **129.** $120\sqrt{2}$ ft
131. $\sqrt{2x}$ ft

CHECKING BASIC CONCEPTS 7.3 & 7.4 (p. 533)

1. (a) $\frac{1}{8}$ **(b)** 10 **(c)** $-2x\sqrt[3]{xy}$ **(d)** $\frac{4b^2}{5}$ **2.** $\sqrt[6]{7^5}$
3. (a) 6 **(b)** 3 **(c)** $6x^3$ **4. (a)** $7\sqrt{6} + \sqrt{7}$
(b) $5\sqrt[3]{x}$ **(c)** $\sqrt{x}$ **5. (a)** $(y - x)\sqrt[3]{xy}$
(b) 14 **6.** $\frac{\sqrt{6}}{2}$ **7.** $\frac{\sqrt{5} + 1}{2}$

SECTION 7.5 (pp. 537–540)

1. **2.**

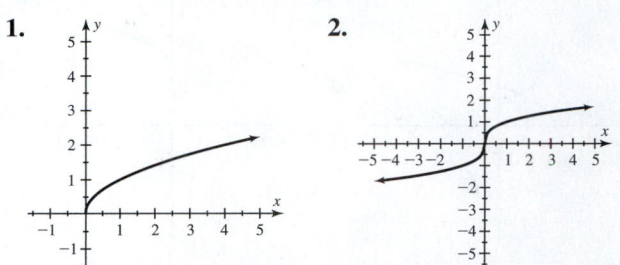

3. $\{x \mid x \geq 0\}$ **4.** All real numbers **5.** $f(x) = x^p$, p is rational. **6.** $f(x) = \sqrt[n]{x}$, where n is an integer greater than 1. **7.** $\{x \mid x \geq 0\}$ **8.** All real numbers
9. $\sqrt{3} \approx 1.73$; undefined **11.** $\sqrt[4]{6} \approx 1.57$; undefined
13. $\sqrt[5]{13} \approx 1.67$; 1 **15.** $\sqrt[3]{6} \approx 1.82$; -1 **17.** $f(x) = \sqrt{x}$
19. $f(x) = \sqrt[3]{x^2}$ **21.** $f(x) = \frac{1}{\sqrt[5]{x}}$ **23.** 32; 55.90
25. $-\frac{1}{128} \approx -0.01$; 0.04 **27.** 4; not possible **29.** 4; 4
31. $[0, \infty)$ **33.** $(-\infty, \infty)$

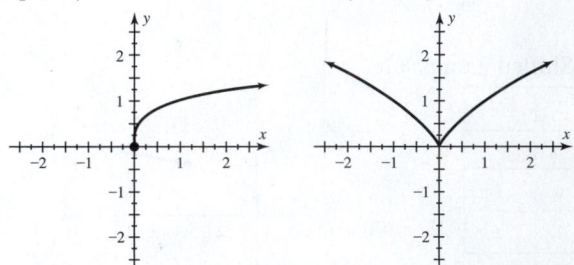

35. $g(x)$ is greater. **37.** $f(x)$ is greater.
[0, 6, 1] by [0, 6, 1]   [0, 6, 1] by [0, 6, 1]

39. $x^p > x^q$ **41. (a)** 6 **(b)** $\sqrt{2x}$ **(c)** $4|x|$
(d) 2 **43.** b. **45.** c. **47.** d. **49.** b. **51.** 2877 in²
53. About 35% **55.** No, for $x \geq 10$ it is less than double.
57. (a) 6 yr **(b)** The twin in the spaceship will be 4 years younger than the twin on Earth.

59. (a) [0, 5, 1] by [0, 0.5, 0.1]   **(b)** $k \approx 0.16$

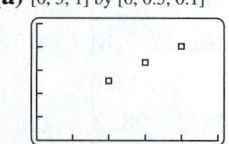

(c) [0, 5, 1] by [0, 0.5, 0.1]; yes   **(d)** 0.295 m²

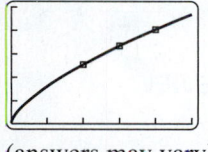

(answers may vary)

61. (a) $k \approx 0.91$ **(b)** [0, 1.5, 0.1] by [0, 1, 0.1]

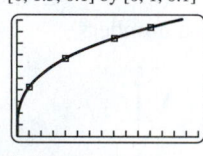

It increases.

(c) About 0.808 m   **(d)** About 0.788; a bird weighing 0.65 kg has a wing span of about 0.788 m.

SECTION 7.6 (PP. 550–554)

1. Square each side.   **2.** Cube each side.   **3.** Yes
4. Check them in the given equation.   **5.** Finding an unknown side of a right triangle (answers may vary)   **6.** 5
7. $d = \sqrt{(x_2 - x_1)^2 + (y_2 - y_1)^2}$ **8.** $x^{1/2} + x^{3/4} = 2$
9. 2   **11.** x **13.** $2x + 1$ **15.** $5x^2$ **17.** 64   **19.** 81   **21.** 6
23. 48   **25.** 6   **27.** 3   **29.** 27   **31.** −2   **33.** 15   **35.** $\frac{1}{2}$
37. 0, 1   **39.** 2   **41.** −4   **43.** 9   **45.** 9   **47.** ±7
49. ±10   **51.** −5, 3   **53.** −3, 7   **55.** 4   **57.** −4   **59.** 1
61. $\frac{7}{5}$ **63.** ±2   **65.** $\sqrt[5]{12}$ **67.** −5, 1   **69.** 3   **71.** 1.88
73. −1, 0.70   **75.** 1.79   **77.** −0.47   **79.** (a)–(c) 16
81. (a)–(c) 11   **83.** $L = \frac{8T^2}{\pi^2}$ **85.** $A = \pi r^2$ **87.** Yes
89. Yes   **91.** Yes   **93.** No   **95.** $\sqrt{32} = 4\sqrt{2}$ **97.** 7
99. $c = 5$ **101.** $b = \sqrt{61}$ **103.** $a = 14$
105. $\sqrt{20} = 2\sqrt{5}$ **107.** $\sqrt{4000} = 20\sqrt{10}$ **109.** $\sqrt{89}$
111. 5   **113.** 4   **115.** 2, 122   **117.** $W = 8$ pounds
119. About 81 sec   **121.** About 3 miles   **123.** About 269 feet
125. 19 inches   **127.** $h \approx 16.3$ inches, $d \approx 33.3$ inches
129. (a) 121 feet   **(b)** About 336 feet   **131. (a)** About 38 miles per hour; the vehicle involved in the accident was traveling about 38 miles per hour.   **(b)** About 453 feet
133. (a) It increases by a factor of 8.   **(b)** $v = \sqrt[3]{\frac{W}{3.8}}$
(c) 20 mph   **135.** $a\sqrt{2}$

CHECKING BASIC CONCEPTS 7.5 & 7.6 (p. 554)

1. (a)   **(b)**

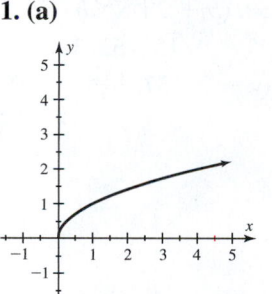   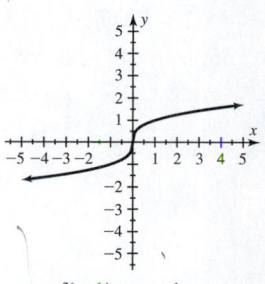

$f(-1)$ is undefined. $f(-1) = -1$

(c)

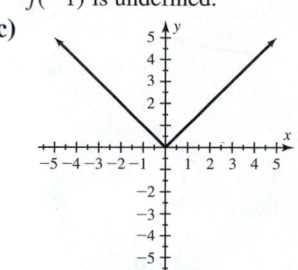

$f(-1) = 1$

2. 3.2   **3.** $[4, \infty)$ **4. (a)** 4   **(b)** 28   **(c)** 3   **5.** 13   **6.** 9.6 in.
7. −3, 1

SECTION 7.7 (PP. 561–562)

1. $2 + 3i$ (answers may vary)   **2.** No; any real number a can be written as $a + 0i$.   **3.** i **4.** −1   **5.** $i\sqrt{a}$
6. $10 - 7i$ **7.** $a + bi$ **8.** $1 + 2i$ **9.** 4   **10.** −5   **11.** 0
12. pure   **13.** $i\sqrt{5}$ **15.** $10i$ **17.** $12i$ **19.** $2i\sqrt{3}$
21. $3i\sqrt{2}$ **23.** 3   **25.** $-8 + 7i$ **27.** $1 - 9i$
29. $-10 + 7i$ **31.** $-13 + 13i$ **33.** $20 - 12i$ **35.** 41
37. 20   **39.** $1 + 5i$ **41.** $3 + 4i$ **43.** $-2 - 6i$ **45.** −2
47. $a^2 + 9b^2$ **49.** $-i$ **51.** i **53.** −1   **55.** 1   **57.** $3 - 4i$
59. $6i$ **61.** $5 + 4i$ **63.** −1   **65.** $1 - i$ **67.** $-\frac{6}{29} + \frac{15}{29}i$
69. $2 + i$ **71.** $-1 + 6i$ **73.** $-1 - 2i$ **75.** $-\frac{3}{2}i$
77. $-\frac{1}{2} + \frac{3}{2}i$ **79.** $\frac{290}{13} + \frac{20}{13}i$ **81.** They are graphed using a real axis and an imaginary axis.

CHECKING BASIC CONCEPTS 7.7 (p. 563)

1. (a) $8i$ **(b)** $i\sqrt{17}$ **2. (a)** $3 - 4i$ **(b)** $-2 + 3i$
(c) $5 + i$ **(d)** $\frac{3}{4} + \frac{3}{4}i$

CHAPTER 7 REVIEW (pp. 567–569)

1. 2   **2.** 6   **3.** $3|x|$ **4.** $|x - 1|$ **5.** −4   **6.** −5   **7.** x^2
8. $3x$ **9.** 2   **10.** −1   **11.** x^2 **12.** $x + 1$ **13.** $\sqrt{14}$
14. $\sqrt[3]{-5}$ **15.** $\left(\sqrt{\frac{x}{y}}\right)^3$ or $\sqrt{\left(\frac{x}{y}\right)^3}$ **16.** $\frac{1}{\sqrt[3]{(xy)^2}}$ or $\frac{1}{(\sqrt[3]{xy})^2}$
17. 9   **18.** 2   **19.** 64   **20.** 27   **21.** z^2 **22.** xy^2 **23.** $\frac{x^3}{y^9}$
24. $\frac{y^2}{x}$ **25.** 8   **26.** −2   **27.** x^2 **28.** 2   **29.** $-\frac{\sqrt[3]{x}}{2}$ **30.** $\frac{1}{3}$
31. $4\sqrt{3}$ **32.** $3\sqrt{6}$ **33.** $\frac{3}{x}$ **34.** $4ab\sqrt{2a}$ **35.** $9xy$
36. $5z\sqrt[3]{z}$ **37.** $x + 1$ **38.** $\frac{2a\sqrt[4]{a}}{b}$ **39.** $2\sqrt[6]{x^5}$
40. $\sqrt[6]{r^5 t^8}$ or $t\sqrt[6]{r^5 t^2}$ **41.** $4\sqrt{3}$ **42.** $3\sqrt[3]{x}$ **43.** $-3\sqrt[3]{5}$

44. $-\sqrt[4]{y}$ **45.** $11\sqrt{3}$ **46.** $7\sqrt{2}$ **47.** $13\sqrt[3]{2}$
48. $3\sqrt{x+1}$ **49.** $(2x-1)\sqrt{x}$ **50.** $(b+2a)\sqrt[3]{ab}$
51. $5+4\sqrt{2}$ **52.** $7+7\sqrt{3}-\sqrt{5}-\sqrt{15}$ **53.** 3
54. 95 **55.** $a-2b$ **56.** $xy+\sqrt{xy}-2$ **57.** $\frac{4\sqrt{5}}{5}$
58. $\frac{r\sqrt{t}}{2t}$ **59.** $\frac{3-\sqrt{2}}{7}$ **60.** $\frac{5+\sqrt{7}}{9}$
61. $\sqrt{8}+\sqrt{7}$ **62.** $\frac{a-2\sqrt{ab}+b}{a-b}$

63. 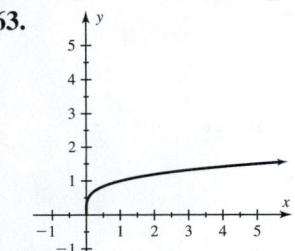   **64.**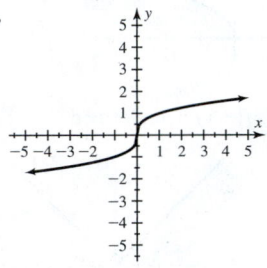

65. $f(x)=\sqrt{x}$; 2 **66.** $f(x)=\sqrt[5]{x^2}$; $\sqrt[5]{16}$
67. 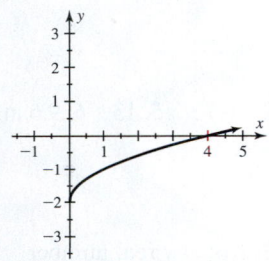   **68.**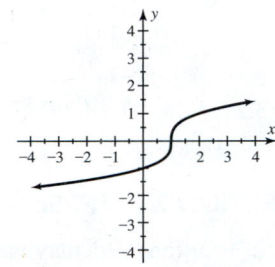

Shifted 2 units downward   Shifted 1 unit to the right

69. $[1,\infty)$ **70.** $(-\infty,3]$ **71.** $(-\infty,\infty)$ **72.** $(-2,\infty)$
73. 2 **74.** 4 **75.** 9 **76.** 9 **77.** 8 **78.** $\frac{1}{4}$ **79.** 4.5
80. 1.62 **81.** $c=\sqrt{65}$ **82.** $b=\sqrt{39}$ **83.** $\sqrt{41}$
84. $\sqrt{52}=2\sqrt{13}$ **85.** ±11 **86.** ±4 **87.** $-3,5$
88. 4 **89.** 3 **90.** 2 **91.** ±4 **92.** -1 **93.** 1
94. $-2,0$ **95.** -2 **96.** $-2+4i$ **97.** $5+i$ **98.** $1+2i$
99. $-\frac{14}{13}+\frac{5}{13}i$ **100.** $2-2i$ **101.** About 85 feet
102. $\sqrt{16,200}=90\sqrt{2}\approx127.3$ feet   **103.** About 0.79
second   **104. (a)** 5 square units   **(b)** $5\sqrt{5}$ cubic units
(c) $\sqrt{10}$ units   **(d)** $\sqrt{15}$ units   **105.** About 0.82 foot
106. About 0.13 foot; the length of the pendulum is shorter.
107. $r=\sqrt[210]{\frac{281}{4}}-1\approx0.02$; from 1790 through 2000 the
average annual percent growth rate was about 2%.
108. (a) About 43 miles per hour   **(b)** About 34 miles per
hour; a steeper bank allows for a higher speed limit; yes
109. $\sqrt{7}\approx2.65$ feet

CHAPTER 7 TEST (pp. 569–570)

1. -3 **2.** $|z+1|$ **3.** $5x^2$ **4.** $2z^2$ **5.** $2xy\sqrt[4]{y}$ **6.** 1
7. $\sqrt[5]{7^2}$ or $(\sqrt[5]{7})^2$ **8.** $\sqrt[3]{\left(\frac{y}{x}\right)^2}$ or $\left(\sqrt[3]{\frac{y}{x}}\right)^2$ **9.** 16
10. $\frac{1}{216}$ **11.** $x^{4/3}$ **12.** $x^{7/10}$ **13.** $(-\infty,4]$
14.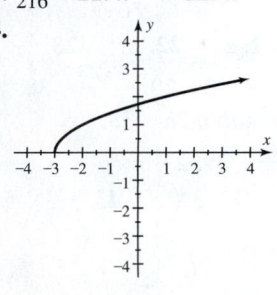

15. $8z^{3/2}$ **16.** $\frac{z}{y^{2/3}}$ **17.** 9 **18.** $\frac{y}{2}$ **19.** $4\sqrt{7}+\sqrt{5}$
20. $6\sqrt[3]{x}$ **21.** $14\sqrt{2}$ **22.** 2 **23. (a)** 27 **(b)** 7 **(c)** 3
(d) 17 **24. (a)** $\frac{2\sqrt{7}}{21}$ **(b)** $\frac{-1+\sqrt{5}}{4}$ **25.** 2.63
26. $\sqrt{120}\approx10.95$ **27.** $\sqrt{8}=2\sqrt{2}$ **28.** $2-19i$
29. $-6+8i$ **30.** $\frac{5}{4}$ **31.** $\frac{4}{29}+\frac{10}{29}i$ **32. (a)** About 17.25 mi
(b) About 420 ft   **33.** 1.31 pounds

CHAPTERS 1–7 CUMULATIVE REVIEW (pp. 571–572)

1. 36π **2.** $D=\{-1,0,1\}$, $R=\{2,4\}$ **3. (a)** $\frac{1}{a^3b^9}$
(b) x^4y^9 **(c)** r^8t^5 **4.** 4.3×10^{-4} **5.** 3; $x\neq2$
6. All real numbers
7. 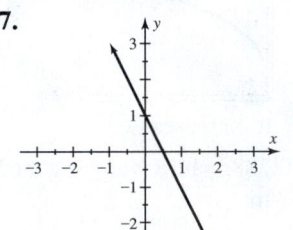   **8.**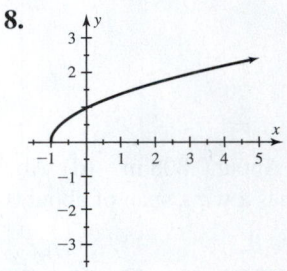

9.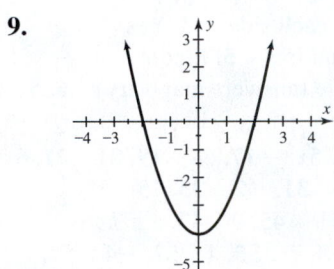

(a) D: all real numbers; R: $y\geq-4$ **(b)** 0 **(c)** $-2,2$
(d) $-2,2$ **10.** $y=\frac{1}{2}x+\frac{5}{2}$ **11.** $f(x)=1-3x$
12.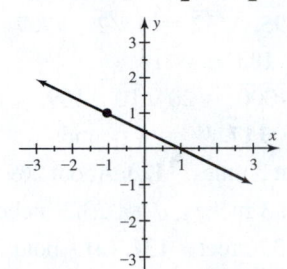

13. $\frac{6}{11}$ **14.** $(-\infty,3]$ **15.** $[-1,5]$ **16.** $\left[-\frac{5}{2},1\right]$
17. (a) $(4,4)$ **(b)** No solutions
18.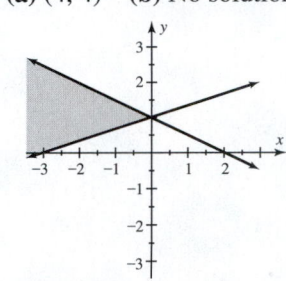

19. $(3, 2, 1)$ **20.** 4 **21.** $-4x^4 + 16x$ **22.** $x^2 - 16$
23. $5x^2 - 7x - 6$ **24.** $16x^2 + 72x + 81$
25. $(3x - 4)(3x + 4)$ **26.** $(x - 2)^2$ **27.** $3x^2(5x - 3)$
28. $(4x - 3)(3x + 1)$ **29.** $(r - 1)(r^2 + r + 1)$
30. $(x - 3)(x^2 + 5)$ **31.** $1, 2$ **32.** $-2, 0, 2$ **33.** $2x + 4$
34. $\frac{7x - 5}{x^2 - x}$ **35.** $6x$ **36.** 4 **37.** $\frac{1}{64}$ **38.** 5 **39.** $4x$
40. $x^{3/4}$ or $\sqrt[4]{x^3}$ **41.** $2x$ **42.** $6\sqrt{3x}$ **43.** $2x^2 - \sqrt{3}x - 3$
44. $(-\infty, 1]$
45.

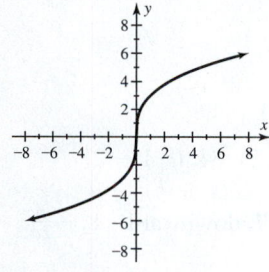

46. $\sqrt{10}$ **47.** $5 + i$ **48.** 13 **49.** 6 **50.** 28 **51.** $\frac{4}{9}, 4$
52. ± 3 **53.** 1.41 **54.** 4.06 **55.** The tank initially contains
300 gallons of water. Water is leaving the tank at 15 gal/min.
56.

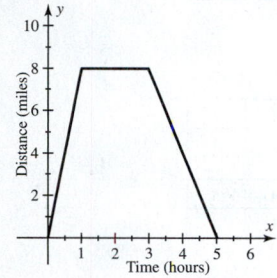

57. $1300 at 5%, $700 at 4%   **58.** 8 in. by 8 in. by 4 in.
59. 22.4 feet   **60.** $45°, 55°, 80°$

8   Quadratic Functions and Equations

SECTION 8.1 (pp. 584–588)

1. parabola   **2.** The vertex
3. axis of symmetry   **4.** $(0, 0)$
5.

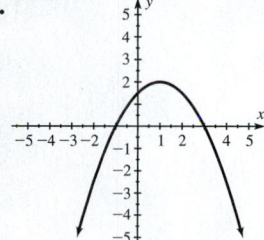

6. $-\frac{b}{2a}$ **7.** narrower   **8.** reflected   **9.** $ax^2 + bx + c$
with $a \neq 0$ **10.** vertex   **11.** T   **12.** F   **13.** F   **14.** T
15. $(2, -6)$ **17.** $(-3, 4)$ **19.** $(0, 3)$ **21.** $(1, 1.4)$
23. $(3, 9)$ **25.** $0, -4$ **27.** $-2, -2$ **29.** $(1, -2); x = 1$;
upward   **31.** $(-2, 3); x = -2$; downward

33. (b)

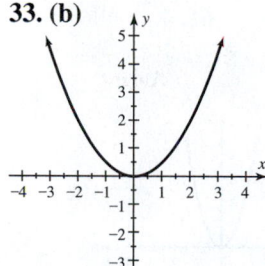

(a) $(0, 0); x = 0$
(c) $2; 4.5$

37. (b)

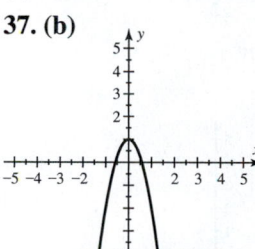

(a) $(0, 1); x = 0$
(c) $-11; -26$

41. (b)

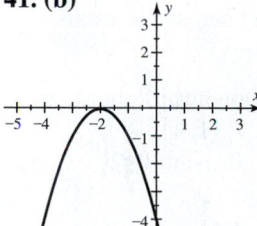

(a) $(-2, 0); x = -2$
(c) $0; -25$

45. (b)

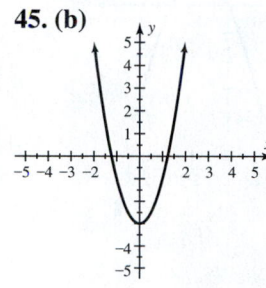

(a) $(0, -3); x = 0$
(c) $5; 15$

49. (b)

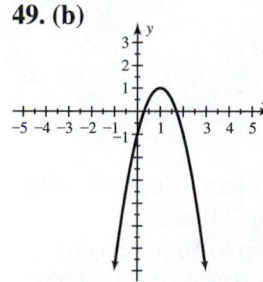

(a) $(1, 1); x = 1$
(c) $-17; -7$

35. (b)

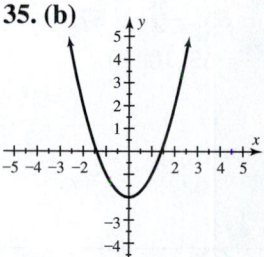

(a) $(0, -2); x = 0$
(c) $2; 7$

39. (b)

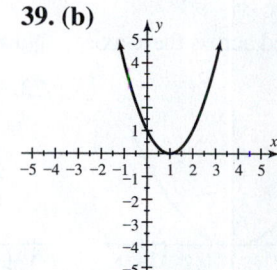

(a) $(1, 0); x = 1$
(c) $9; 4$

43. (b)

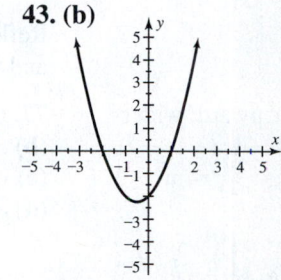

(a) $(-0.5, -2.25); x = -0.5$
(c) $0; 10$

47. (b)

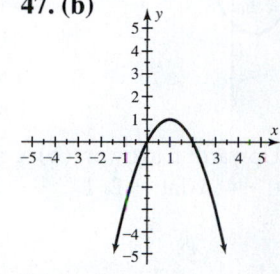

(a) $(1, 1); x = 1$
(c) $-8; -3$

51. (b)

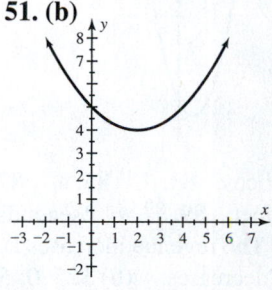

(a) $(2, 4); x = 2$
(c) $8; 4.25$

53. -2 **55.** $-\frac{25}{4}$ **57.** $-\frac{7}{2}$ **59.** 6   **61.** 4

63. $-\frac{39}{8}$ **65.** 10, 10

67.

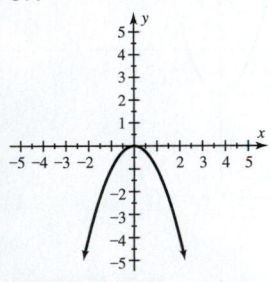

69.

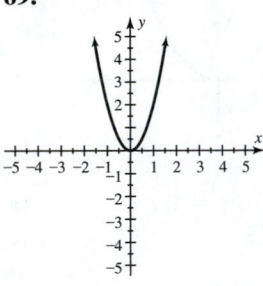

Reflected across the x-axis   Narrower

71.

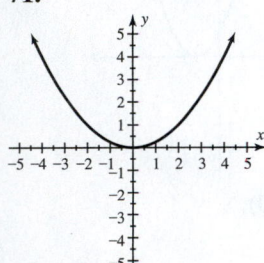

Wider

73.

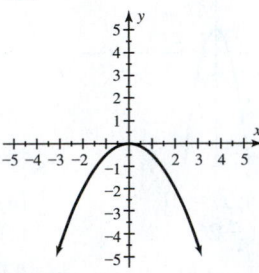

Reflected across the x-axis and wider

75. (a) Upward; wider
(b) $x = -1; (-1, -2)$
(c) y-int: $-\frac{3}{2}$; x-int: $-3, 1$
(d)

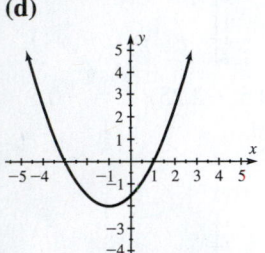

77. (a) Downward; the same
(b) $x = 1; (1, 1)$
(c) y-int: 0; x-int: 0, 2
(d)

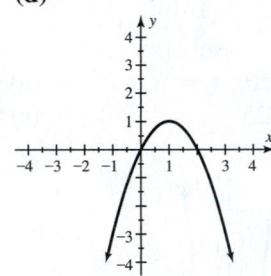

79. (a) Upward; narrower   **(b)** $x = -\frac{1}{2}; \left(-\frac{1}{2}, -\frac{9}{2}\right)$
(c) y-int: -4; x-int: $-2, 1$
(d)

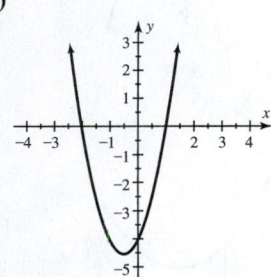

81. 3 slices   **83.** d.   **85.** a.   **87. (a)** 2 feet   **(b)** 2 seconds
(c) 66 feet   **89.** $\frac{66}{32} \approx 2$seconds; about 74 feet
91. (a) The revenue increases at first up to 50 tickets and then it decreases.   **(b)** $2500; 50   **(c)** $f(x) = x(100 - x)$

(d) $2500; 50   **93. (a)** $V(1) = 21.75$, $V(2) = 30$, $V(3) = 59.75$, and $V(4) = 111$; in 2007 there were 21.75 million unique Facebook visitors in one month. Other values can be interpreted similarly.
(b) The increases between consecutive years are 8.25 million, 29.75 million, and 51.25 million. A linear function does not model the data because these three increases are not equal (or nearly equal).
95. 300 feet by 600 feet   **97.** 42 in.
99. (a) In 1990, emissions were 6 billion metric tons.
(b) 3 billion metric tons.

SECTION 8.2 (pp. 598–601)

1. $x^2 + 2$ **2.** $(x - 2)^2$ **3.** $(1, 2)$ **4.** $(-1, -2)$
5. $f(x) = ax^2 + bx + c$; $f(x) = a(x - h)^2 + k$
6. $y = a(x - h)^2 + k$; (h, k) **7.** downward   **8.** $-\frac{b}{2a}$
9. a.   **10.** d.

11.

x	-2	-1	0	1	2
$y = x^2$	4	1	0	1	4
$y = x^2 - 3$	1	-2	-3	-2	1

Shifted 3 units downward

13.

x	-2	-1	0	1	2
$y = x^2$	4	1	0	1	4

x	1	2	3	4	5
$y = (x - 3)^2$	4	1	0	1	4

Shifted 3 units right

15. (a)

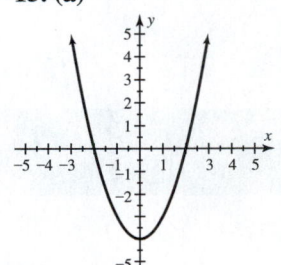

(b) $(0, -4)$
(c) Down 4 units

17. (a)

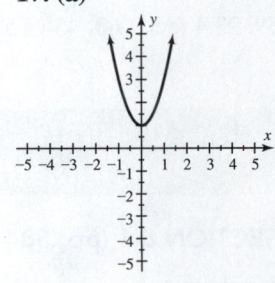

(b) $(0, 1)$
(c) Narrower and up 1 unit

19. (a)

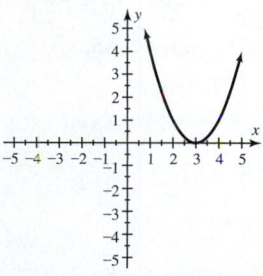

(b) $(3, 0)$
(c) Right 3 units

21. (a)

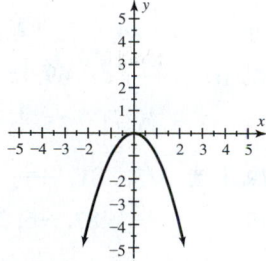

(b) $(0, 0)$
(c) Reflected across the x-axis

23. (a)

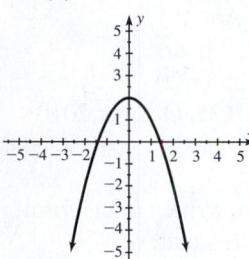

(b) $(0, 2)$
(c) Reflected across the x-axis and up 2 units

25. (a)

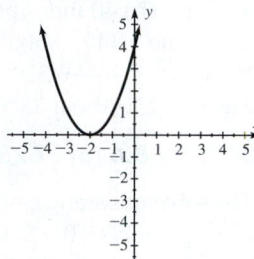

(b) $(-2, 0)$
(c) Left 2 units

27. (a)

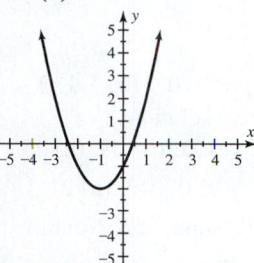

(b) $(-1, -2)$
(c) Left 1 unit and down 2 units

29. (a)

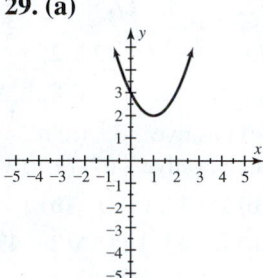

(b) $(1, 2)$
(c) Right 1 unit and up 2 units

31. (a)

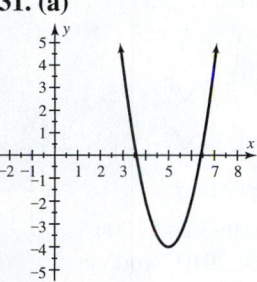

(b) $(5, -4)$
(c) Narrower, right 5 units and down 4 units

33. (a)

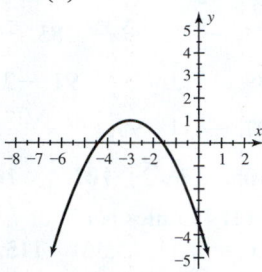

(b) $(-3, 1)$
(c) Wider, reflected across the x-axis, left 3 units and up 1 unit

35. Translated 1 unit right, 2 units downward, and is wider

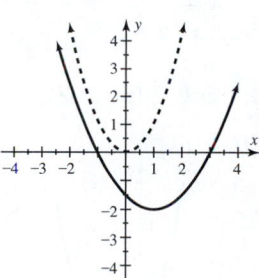

37. Translated 1 unit left, 3 units upward, opens downward, and is narrower

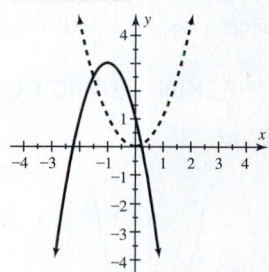

39. [−20, 20, 2] by [−20, 20, 2]

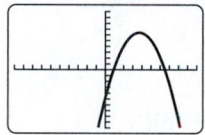

41. $y = 3(x - 3)^2 + 4; y = 3x^2 - 18x + 31$
43. $y = -\frac{1}{2}(x - 5)^2 - 2; y = -\frac{1}{2}x^2 + 5x - \frac{29}{2}$
45. $y = (x - 1)^2 + 2$ **47.** $y = -(x - 0)^2 - 3$
49. $y = (x - 0)^2 - 3$ **51.** $y = -(x + 1)^2 + 2$
53. (a) $(1, 1)$ **(b)** $y = 4(x - 1)^2 + 1$ **55. (a)** $(-1, -2)$
(b) $y = -(x + 1)^2 - 2$ **57. (a)** $(-1, 3)$
(b) $y = -2(x + 1)^2 + 3$ **59.** 1
61. $y = (x + 1)^2 - 1; (-1, -1)$
63. $y = (x - 2)^2 - 4; (2, -4)$
65. $y = (x + 1)^2 - 4; (-1, -4)$
67. $y = (x - 2)^2 + 1; (2, 1)$
69. $y = \left(x + \frac{3}{2}\right)^2 - \frac{17}{4}; \left(-\frac{3}{2}, -\frac{17}{4}\right)$
71. $y = \left(x - \frac{7}{2}\right)^2 - \frac{45}{4}; \left(\frac{7}{2}, -\frac{45}{4}\right)$
73. $y = 3(x + 1)^2 - 4; (-1, -4)$
75. $y = 2\left(x - \frac{3}{4}\right)^2 - \frac{9}{8}; \left(\frac{3}{4}, -\frac{9}{8}\right)$
77. $y = -2(x + 2)^2 + 13; (-2, 13)$ **79.** $a = 2$
81. $a = 0.3$ **83.** $y = 2(x - 1)^2 - 3$
85. $y = 0.5(x - 1980)^2 + 6$
87. (a)

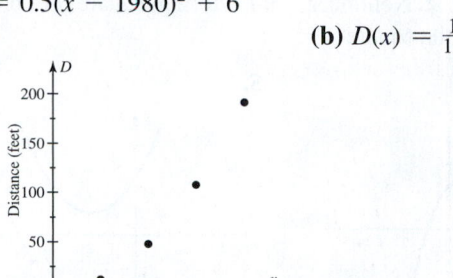

(b) $D(x) = \frac{1}{12}x^2$

89. (a) Decreases and then increases
(b) No, because the data decrease and then increase
(c) Quadratic; it can model data that decrease and increase.
(d) $(1995, 600)$; it has the minimum y-value.
(e) $C(x) = 1.8(x - 1995)^2 + 600$
(f) [1970, 2030, 10] by [0, 2500, 500]

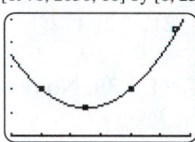

91. (a) $f(x) = 2(x - 1982)^2 + 1$ (answers may vary)
(b) 201 thousand (answers may vary) **93.** incr: $x \geq 2$, decr: $x \leq 2$ **95.** incr: $x \leq 1$, decr: $x \geq 1$ **97.** incr: $x \geq \frac{2}{3}$, decr: $x \leq \frac{2}{3}$ **99.** incr: $x \leq -\frac{3}{2}$, decr: $x \geq -\frac{3}{2}$ **101.** incr: $x \leq -\frac{1}{8}$, decr: $x \geq -\frac{1}{8}$

CHECKING BASIC CONCEPTS 8.1 & 8.2 (p. 602)

1. (a) 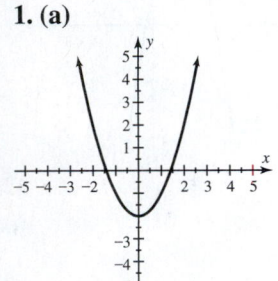 **(b)**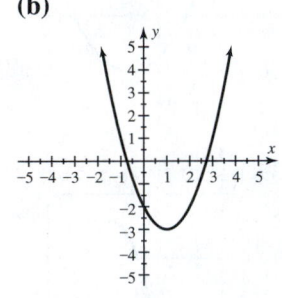

$(0, -2); x = 0$ $\qquad$ $(1, -3); x = 1$

2. y_1 opens upward, whereas y_2 opens downward, y_1 is narrower than y_2. **3.** 7

4. (a) 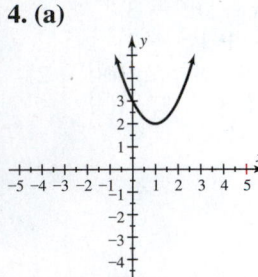 **(b)**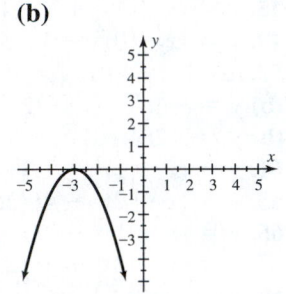

1 unit right, 2 units up $\qquad$ Reflected across the x-axis, 3 units left

5. (a) $y = (x + 7)^2 - 56$ $\qquad$ **(b)** $y = 4(x + 1)^2 - 6$

SECTION 8.3 (pp. 611–614)

1. $x^2 + 3x - 2 = 0$ (answers may vary); it can have 0, 1, or 2 solutions **2.** Nonlinear **3.** Factoring, square root property, completing the square

4. 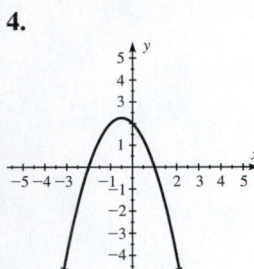 **5.**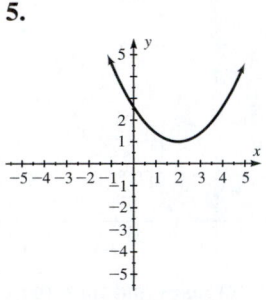

(answers may vary) $\qquad$ (answers may vary)

6. Two; the solutions are the x-intercepts. **7.** ± 8; the square root property (answers may vary) **8.** $\left(\frac{b}{2}\right)^2$
9. Yes **10.** Yes **11.** No **12.** No **13.** Yes **14.** Yes
15. No **16.** No **17. (a)** 3.65, -1.65 **(b)** 1.32, -5.32
19. (a) 1.32, -0.12 **(b)** $-0.28, -0.83$ **21.** $-2, 1$
23. No real solutions **25.** $-2, 3$ **27.** -0.5 **29.** $-1, 5$
31. $-3, 1$ **33.** $-3, 3$ **35.** $-\frac{1}{2}, \frac{3}{2}$ **37.** -1 **39.** No real solutions **41.** $-7, 5$ **43.** $-\frac{1}{3}, \frac{1}{2}$ **45.** $-\frac{3}{2}$ **47.** $-2, 7$

49. $\frac{6}{5}, \frac{3}{2}$ **51.** ± 12 **53.** $\pm \frac{8}{\sqrt{5}}$ or $\pm \frac{8\sqrt{5}}{5}$ **55.** $-6, 4$
57. $-7, 9$ **59.** $\frac{1 \pm \sqrt{5}}{2}$ **61.** $5 \pm \sqrt{5}$ **63.** 4 **65.** $\frac{25}{4}$
67. 16; $(x - 4)^2$ **69.** $\frac{81}{4}$; $\left(x + \frac{9}{2}\right)^2$ **71.** $-4, 6$
73. $-3 \pm \sqrt{11}$ **75.** $\frac{3 \pm \sqrt{29}}{2}$ **77.** $\frac{5 \pm \sqrt{21}}{2}$
79. $1 \pm \sqrt{5}$ **81.** $\frac{3 \pm \sqrt{41}}{4}$ **83.** $\frac{2 \pm \sqrt{11}}{2}$
85. $\frac{-3 \pm \sqrt{5}}{12}$ **87.** $-6, 2$ **89.** $-2 \pm \sqrt{2}$
91. $\pm \sqrt{2}$ **93.** $-5, \frac{7}{3}$ **95.** $4 \pm \sqrt{14}$ **97.** $\pm \sqrt{\frac{c}{a}}$
99. $-3, 6$ **101.** 3, 5 **103.** 5, 7
105. $y = \pm \sqrt{x + 1}$ **107.** $v = \pm \sqrt{\frac{2K}{m}}$
109. $r = \pm \sqrt{\frac{k}{E}}$ **111.** $f = \pm \frac{1}{2\pi\sqrt{LC}}$ **113. (a)** 30 miles per hour **(b)** 40 miles per hour **115.** About 1.9 seconds; no **117.** About 8 feet **119.** 2 hours
121. (a) $x^2 + 6x - 520 = 0$ **(b)** -26 or 20; 20 feet **123.** About 23°C and 34°C **125.** October 2010

SECTION 8.4 (pp. 623–627)

1. To solve quadratic equations that are written in the form $ax^2 + bx + c = 0$ **2.** Completing the square
3. $b^2 - 4ac$ **4.** One solution **5.** Factoring, square root property, completing the square, and the quadratic formula **6.** No; not when $b^2 - 4ac < 0$ **7.** $\pm \sqrt{k}$
8. $\pm i \sqrt{k}$ **9.** $-6, \frac{1}{2}$ **11.** 1 **13.** No real solutions
15. $-2, 8$ **17.** $\frac{1 \pm \sqrt{17}}{8}$ **19.** No real solutions **21.** $\frac{1}{2}$
23. $\frac{3 \pm \sqrt{13}}{2}$ **25.** $2 \pm \sqrt{3}$ **27.** $\frac{1 \pm \sqrt{37}}{6}$
29. $\frac{1 \pm \sqrt{15}}{2}$ **31.** $\frac{5 \pm \sqrt{10}}{3}$ **33. (a)** $a > 0$ **(b)** $-1, 2$
(c) Positive **35. (a)** $a > 0$ **(b)** No real solutions
(c) Negative **37. (a)** $a < 0$ **(b)** 2 **(c)** Zero **39. (a)** 25
(b) 2 **41. (a)** 0 **(b)** 1 **43. (a)** $-\frac{7}{4}$ **(b)** 0 **45. (a)** 21
(b) 2 **47.** $1 \pm \sqrt{2}$ **49.** $-\frac{3}{2}, 1$ **51.** None **53.** None
55. $\frac{-2 \pm \sqrt{10}}{3}$ **57.** $\pm 3i$ **59.** $\pm 4i\sqrt{5}$ **61.** $\pm \frac{1}{2}i$
63. $\pm \frac{3}{4}i$ **65.** $\pm i\sqrt{6}$ **67.** $\pm \sqrt{3}$ **69.** $\pm i\sqrt{2}$
71. $\frac{1}{2} \pm i\frac{\sqrt{7}}{2}$ **73.** $-\frac{3}{4} \pm i\frac{\sqrt{23}}{4}$ **75.** $2 \pm \sqrt{3}$
77. $-\frac{1}{2} \pm i\frac{\sqrt{7}}{2}$ **79.** $-\frac{1}{5} \pm i\frac{\sqrt{19}}{5}$ **81.** $-\frac{3}{4} \pm i\frac{\sqrt{23}}{4}$
83. $-\frac{1}{2} \pm i\frac{\sqrt{15}}{2}$ **85.** $\frac{1 \pm \sqrt{3}}{2}$ **87.** $\frac{1}{4} \pm i\frac{\sqrt{15}}{4}$
89. $-1 \pm i\sqrt{3}$ **91.** $-2 \pm i$ **93.** $1 \pm i\sqrt{2}$ **95.** 1, 2
97. $-\frac{1}{2}, 4$ **99.** $\frac{5 \pm \sqrt{17}}{2}$ **101.** $-\frac{1}{4} \pm \frac{3}{4}i\sqrt{7}$ **103.** $\pm \frac{1}{2}$
105. $\pm i\sqrt{2}$ **107.** $\frac{1}{3}$ **109.** 9 miles per hour
111. 45 miles per hour **113.** About 1993 and 2004 ($x \approx 4.11, 15.36$) **115. (a)** 6; in Oct. 2010 Groupon's value was $6 billion. **(b)** Jan. 2011 **117.** $x \approx 8.04$, or about 1992; this agrees with the graph.
119. $130 + 5\sqrt{634} \approx 256$ mph **121. (a)–(c)** 11 in. by 14 in. **123. (a)** The rate of change is not constant.
(b) 75 seconds (answers may vary) **(c)** 75 seconds

CHECKING BASIC CONCEPTS 8.3 & 8.4 (p. 627)

1. $\frac{1}{2}, 3$ **2.** $\pm \sqrt{5}$ **3.** $2 \pm \sqrt{3}$ **4.** $y = \pm \sqrt{1 - x^2}$

5. (a) $\frac{3 \pm \sqrt{17}}{4}$ **(b)** $\frac{4}{3}$ **(c)** $-\frac{1}{2} \pm i\frac{\sqrt{7}}{2}$ **6. (a)** 5; two real solutions **(b)** -7; no real solutions **(c)** 0; one real solution **7. (a)** $\pm i\sqrt{5}$ **(b)** $-\frac{1}{2} \pm i\frac{\sqrt{11}}{2}$

SECTION 8.5 (pp. 635–637)

1. It has an inequality symbol rather than an equals sign.
2. No, they often have infinitely many. **3.** No **4.** Yes
5. $-2 < x < 4$ **6.** $x < -3$ or $x > 1$ **7.** Yes **9.** Yes
11. No **13.** Yes **15.** No **17.** No **19. (a)** $-3, 2$
(b) $-3 < x < 2$ **(c)** $x < -3$ or $x > 2$ **21. (a)** $-2, 2$
(b) $-2 < x < 2$ **(c)** $x < -2$ or $x > 2$ **23. (a)** $-10, 5$
(b) $x < -10$ or $x > 5$ **(c)** $-10 < x < 5$
25. $[-0.128, 0.679]$ **27. (a)** $-2, 2$ **(b)** $-2 < x < 2$
(c) $x < -2$ or $x > 2$ **29. (a)** $-4, 0$ **(b)** $-4 < x < 0$
(c) $x < -4$ or $x > 0$ **31.** $[-7, -3]$
33. $(-\infty, 1) \cup (2, \infty)$ **35.** $(-\sqrt{10}, \sqrt{10})$
37. $(-\infty, 0) \cup (6, \infty)$
39. $(-\infty, 2 - \sqrt{2}] \cup [2 + \sqrt{2}, \infty)$ **41. (a)** $-2, 2$
(b) $-2 < x < 2$ **(c)** $x < -2$ or $x > 2$ **43. (a)** $\frac{-1 \pm \sqrt{5}}{2}$
(b) $\frac{-1 - \sqrt{5}}{2} < x < \frac{-1 + \sqrt{5}}{2}$
(c) $x < \frac{-1 - \sqrt{5}}{2}$ or $x > \frac{-1 + \sqrt{5}}{2}$ **45.** $(-3, -1)$
47. $(-\infty, -2.5] \cup [3, \infty)$ **49.** $[-2, 2]$ **51.** $(-\infty, \infty)$
53. $(0, 3)$ **55.** No real solutions **57.** 2
59. $(-\infty, -1) \cup (-1, \infty)$ **61.** $[-1, 2]$ **63. (a)** From
1131 feet to 3535 feet (approximately) **(b)** Before 1131 feet
or after 3535 feet (approximately) **65. (a)** About 383; they
agree (approx.). **(b)** About 1969 or after **(c)** About 1969
or after **67.** From 11 feet to 20 feet

SECTION 8.6 (p. 641)

1. $\pm 1, \pm \sqrt{6}$ **3.** $-\sqrt[3]{2}, \sqrt[3]{\frac{5}{3}}$ **5.** $-\frac{4}{5}, -\frac{1}{3}$ **7.** $-3, 3$
9. $-\sqrt[3]{\frac{1}{3}}, \sqrt[3]{2}$ **11.** $-\frac{1}{8}, \frac{2}{5}$ **13.** 1 **15.** 1, $32^5 = 33{,}554{,}432$
17. 16, 81 **19.** 1 **21.** $-3, 6$ **23.** $-\sqrt{3}, \sqrt{3}$
25. $\pm 2, \pm 2i$ **27.** 0, $\pm i$ **29.** $\pm \sqrt{2}, \pm i$
31. $-1 \pm i$ **33.** $\pm 2i$

CHECKING BASIC CONCEPTS 8.5 & 8.6 (p. 641)

1. $(-\infty, -2) \cup (3, \infty)$ **2.** $\left[-1, -\frac{2}{3}\right]$ **3.** $-2, \sqrt[3]{2}$
4. $-1, 8^3 = 512$ **5.** $\pm i$

CHAPTER 8 REVIEW (pp. 644–648)

1. $(-3, 4)$; $x = -3$; downward **2.** $(1, 0)$; $x = 1$; upward

3. (a)

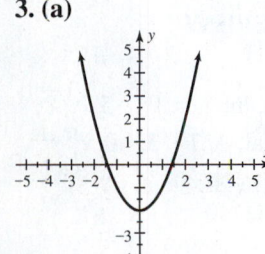

4. (a)

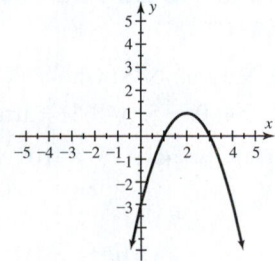

(b) $(0, -2)$; $x = 0$ **(b)** $(2, 1)$; $x = 2$
(c) -1 **(c)** 0

5. (a)

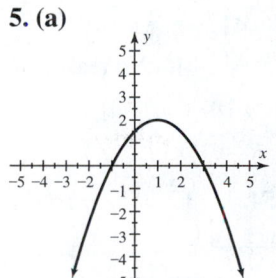

6. (a)

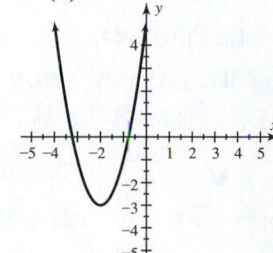

(b) $(1, 2)$; $x = 1$ **(b)** $(-2, -3)$; $x = -2$
(c) -2.5 **(c)** -1

7. $-\frac{7}{2}$ **8.** $-\frac{14}{3}$ **9.** $(2, -6)$ **10.** $(0, 5)$ **11.** $(2, 2)$
12. $(-1, 1)$

13. (a)

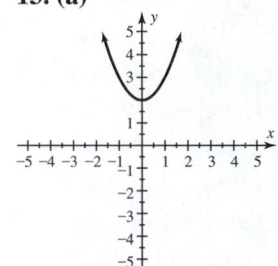

14. (a)

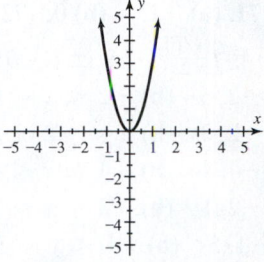

(b) Up 2 units   **(b)** Narrower

15. (a)

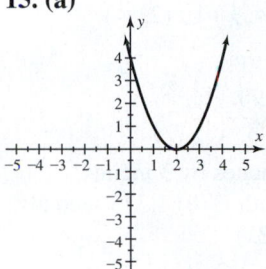

16. (a)

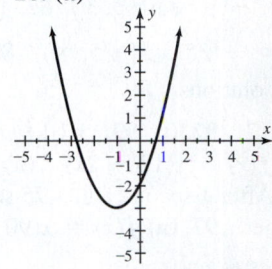

(b) Right 2 units   **(b)** Left 1 unit, down 3 units

17. (a)

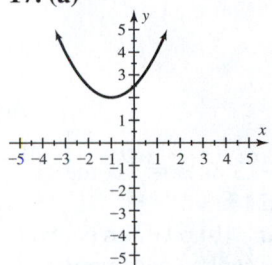

18. (a)

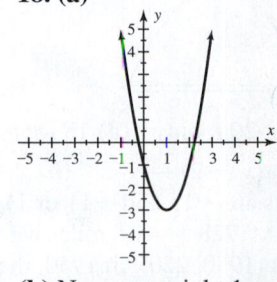

(b) Wider, left 1 unit,   **(b)** Narrower, right 1 unit,
up 2 units   down 3 units

19. $y = -4(x - 2)^2 - 5$ **20.** $y = -(x + 4)^2 + 6$
21. $y = (x + 2)^2 - 11$; $(-2, -11)$
22. $y = \left(x - \frac{7}{2}\right)^2 - \frac{45}{4}$; $\left(\frac{7}{2}, -\frac{45}{4}\right)$
23. $y = 2\left(x - \frac{3}{4}\right)^2 - \frac{73}{8}$; $\left(\frac{3}{4}, -\frac{73}{8}\right)$
24. $y = 3(x + 1)^2 - 5$; $(-1, -5)$ **25.** $a = 3$
26. $a = \frac{1}{4}$ **27.** $f(x) = -5x^2 + 30x - 41$; -41
28. $f(x) = 3x^2 + 12x + 8$; 8 **29.** $-2, 3$ **30.** -1
31. No real solutions **32.** $-4, 6$ **33.** $-10, 5$
34. $-0.5, 0.25$ **35.** $-5, 10$ **36.** $-3, 1$ **37.** $-4, 2$

38. $-1, 3$ **39.** $-5, 4$ **40.** $-8, -3$ **41.** $-\frac{2}{5}, \frac{2}{3}$

42. $\frac{4}{7}, 3$ **43.** ± 10 **44.** $\pm\frac{1}{3}$ **45.** $\pm\frac{\sqrt{6}}{2}$ **46.** No real

solutions **47.** $-3 \pm \sqrt{7}$ **48.** $2 \pm \sqrt{10}$

49. $1 \pm \sqrt{6}$ **50.** $\frac{-3 \pm \sqrt{11}}{2}$ **51.** $R = -r \pm \sqrt{\frac{k}{F}}$

52. $y = \pm\sqrt{\frac{12 - 2x^2}{3}}$ **53.** $3, 6$ **54.** $11, 13$ **55.** $-\frac{1}{2}, \frac{1}{3}$

56. $\frac{5 \pm \sqrt{5}}{10}$ **57.** $4 \pm \sqrt{21}$ **58.** $\frac{3 \pm \sqrt{3}}{2}$

59. ± 2 **60.** $\pm\frac{1}{2}$ **61.** $\frac{5}{2}, 3$ **62.** $\frac{3}{2}, 5$ **63.** $\frac{3 \pm \sqrt{5}}{2}$

64. $\frac{1 \pm \sqrt{5}}{4}$ **65. (a)** $a > 0$ **(b)** $-2, 3$ **(c)** Positive

66. (a) $a > 0$ **(b)** 2 **(c)** Zero **67. (a)** $a < 0$

(b) No real solutions **(c)** Negative **68. (a)** $a < 0$

(b) $-4, 2$ **(c)** Positive **69. (a)** 1 **(b)** 2 **70. (a)** 144

(b) 2 **71. (a)** -23 **(b)** 0 **72. (a)** 0 **(b)** 1

73. $-\frac{1}{2} \pm i\frac{\sqrt{19}}{2}$ **74.** $\pm 2i$ **75.** $\frac{1}{4} \pm i\frac{\sqrt{7}}{4}$ **76.** $\frac{1}{7} \pm i\frac{\sqrt{34}}{7}$

77. (a) $-2, 6$ **(b)** $-2 < x < 6$ **(c)** $x < -2$ or $x > 6$

78. (a) $-2, 0$ **(b)** $x < -2$ or $x > 0$ **(c)** $-2 < x < 0$

79. (a) $-4, 4$ **(b)** $-4 < x < 4$ **(c)** $x < -4$ or $x > 4$

80. (a) $-2, 1$ **(b)** $-2 < x < 1$ **(c)** $x < -2$ or $x > 1$

81. (a) $-1, 3$ **(b)** $-1 < x < 3$ **(c)** $x < -1$ or $x > 3$

82. (a) $-\frac{3}{2}, 5$ **(b)** $-\frac{3}{2} \le x \le 5$ **(c)** $x \le -\frac{3}{2}$ or $x \ge 5$

83. $[-3, -1]$ **84.** $\left(\frac{1}{5}, 3\right)$ **85.** $\left(-\infty, \frac{1}{6}\right) \cup (2, \infty)$

86. $(-\infty, -\sqrt{5}] \cup [\sqrt{5}, \infty)$ **87.** $(-\infty, \infty)$

88. No solutions **89.** $\pm\sqrt{5}, \pm 3$ **90.** $-\frac{1}{4}, \frac{2}{7}$

91. $1, 512$ **92.** 0 **93.** $\pm i\frac{\sqrt{2}}{2}$ **94.** $2 \pm i\sqrt{2}$

95. (a) $f(x) = x(12 - 2x)$ **(b)** 6 inches by 3 inches

96. (a) After 1 second and 1.75 seconds **(b)** 1.375 seconds;

34.25 feet **97. (a)** $f(x) = x(90 - 3x)$

(b) [0, 30, 5] by [0, 800, 100]

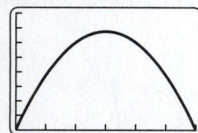

(c) 10 or 20 rooms **(d)** 15 rooms

98. (a) $x(x + 2) = 143$ **(b)** $x = -13$ or $x = 11$; the

numbers are -13 and -11 or 11 and 13.

99. (a) $\sqrt{1728} \approx 41.6$ miles per hour **(b)** 60 miles per hour

100. (a) $(1950, 220)$; in 1950, the per capita consumption

was at a low of 220 million Btu.

(b) [1950, 1970, 5] by [200, 350, 25]

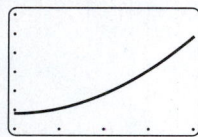

It increased.

(c) $f(2010) = 1120$; no; the trend represented by this

model did not continue after 1970. **101.** About 11.1 inches

by 11.1 inches **102.** 50 feet **103.** About 1.5 feet

104. About 6.0 to 9.0 inches

CHAPTER 8 TEST (pp. 648–649)

1. $\left(1, \frac{3}{2}\right); x = 1; -3$ **2.** $-\frac{29}{4}$ **3.** $a = -\frac{1}{2}$

4. (a) Same as $y = x^2$ **(b)** Same as $y = x^2$ except

except shifted 1 unit right shifted 2 units downward

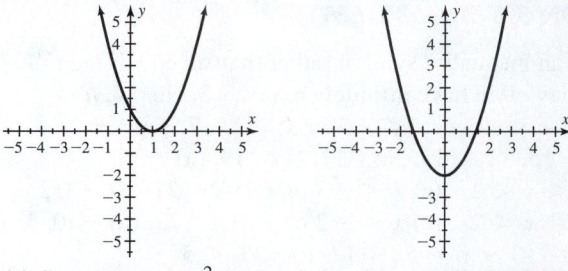

(c) Same as $y = x^2$ except it is wider, shifted right 3 units,

and shifted upward 2 units

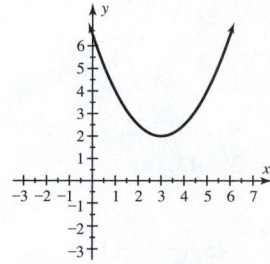

5. $y = (x - 3)^2 - 7; (3, -7); x = 3$ **6.** $-1, 2; 2$

7. $-4, \frac{1}{3}$ **8.** $-\frac{1}{2}, \frac{1}{2}$ **9.** $4 \pm \sqrt{17}$ **10.** $\frac{3 \pm \sqrt{17}}{4}$

11. $\pm\frac{4}{3}$ **12.** $m = \pm\sqrt{\frac{Fr^2}{G}}$ **13. (a)** $a < 0$ **(b)** $-3, 1$

(c) Positive **14. (a)** -44 **(b)** No real solutions

(c) The graph of $y = -3x^2 + 4x - 5$ does not intersect the

x-axis. **15. (a)** $-1, 1$ **(b)** $-1 < x < 1$

(c) $x < -1$ or $x > 1$ **16. (a)** $-10, 20$

(b) $x < -10$ or $x > 20$ **(c)** $-10 < x < 20$

17. (a) $-\frac{1}{2}, \frac{3}{4}$ **(b)** $\left[-\frac{1}{2}, \frac{3}{4}\right]$ **(c)** $\left(-\infty, -\frac{1}{2}\right] \cup \left[\frac{3}{4}, \infty\right)$

18. $[-2, 0]$ **19.** $\sqrt[3]{2}, 1$ **20.** $-1 \pm i\frac{\sqrt{2}}{2}$

21. $-1.37, 0.69$ **22.** $\sqrt{2250} \approx 47.4$ miles per hour

23. (a) $f(x) = (x + 20)(90 - x)$ **(b)** 35

24. (a) [0, 6, 1] by [0, 150, 50]

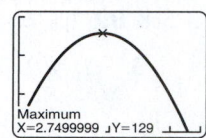

(b) After about 5.6 seconds

(c) 2.75 seconds; 129 feet

CHAPTERS 1–8 CUMULATIVE REVIEW
(pp. 651–653)

1. 1 **2.** Natural: $\sqrt[3]{8}$; whole: $0, \sqrt[3]{8}$; integer: $0, -5, \sqrt[3]{8}$;

rational: $0.\overline{4}, 0, -5, \sqrt[3]{8}, -\frac{4}{3}$; irrational: $\sqrt{7}$ **3. (a)** $x^{10}y^{12}$

(b) $\frac{x}{y^8}$ **(c)** $\frac{1}{b^{10}}$ **4.** 9.29×10^6 **5.** $2; x \le 2$ **6.** $(2, 5)$

7.

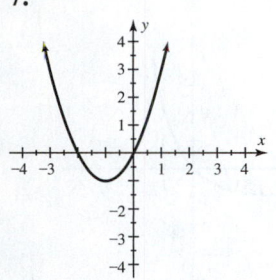

8.

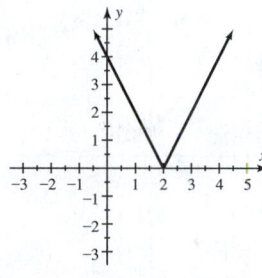

9. $y = -\frac{3}{2}x + 5$ **10.** $x = -3$ **11.** -12 **12.** $\left(-\infty, \frac{7}{4}\right)$
13. $\left[\frac{1}{3}, 1\right]$ **14.** $(-1, 5]$ **15.** $(-1, 1)$

16.

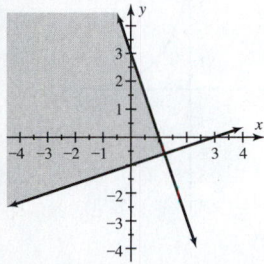

17. $(2, 2, 1)$ **18.** $6x^2 + 17x - 14$ **19.** $3x^3y + 3xy^3$
20. $x - 9$ **21.** $x(x + 1)(x - 2)$ **22.** $(2x + 5)(2x - 5)$
23. $\pm\sqrt{3}$ **24.** 1 **25.** $\frac{x+3}{2}$ **26.** $-\frac{2}{x(x+2)}$ **27.** $4x^3$
28. $\frac{1}{64}$ **29.** 3 **30.** $3\sqrt{2x}$

31.

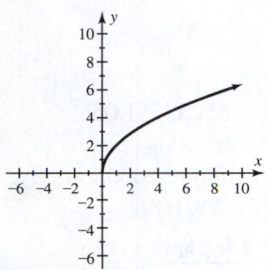

32. $\sqrt{26}$ **33.** $1 - i$ **34.** 3 **35.** 0.79

36.

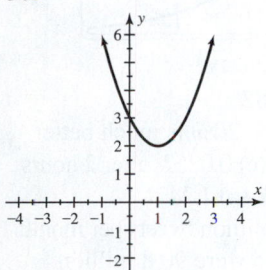

(a) $(1, 2)$ **(b)** 6 **(c)** $x = 1$ **(d)** $x \geq 1$
37. $f(x) = 2(x - 1)^2 - 3$ **38.** Shifted 1 unit left, 2 units
downward; narrower   **39.** $-3 \pm \sqrt{11}$ **40.** $\frac{3 \pm \sqrt{17}}{4}$
41. $1, 3$ **42.** **(a)** $-2, 1$ **(b)** $-2 \leq x \leq 1$
43. $x < 1$ or $x > 2$ **44.** $\pm 4, \pm 4i$ **45.** c.   **46.** f.   **47.** g.
48. e.   **49.** d.   **50.** a.   **51.** b.   **52.** h.   **53. (a)** 300; initially,
the tank holds 300 gal. **(b)** 6; after 6 minutes the tank is empty.
(c) -50; water is pumped out at 50 gal/min.
(d) $G(t) = 300 - 50t$ **54.** $2200 at 6%, $1200 at 5%,
$600 at 4%   **55.** 125 ft by 125 ft   **56.** 33 feet

9 Exponential and Logarithmic Functions

SECTION 9.1 (pp. 665–669)

1. $g(f(7))$ **2.** $f(g(x))$ **3.** No   **4.** inputs; outputs
5. No   **6.** one-to-one   **7.** adding 10   **8.** 7   **9.** 8; 6
10. x; y **11.** one-to-one   **12.** reflection; $y = x$
13. (a) 7   **(b)** 49   **(c)** $(g \circ f)(x) = x^2 + 3$
(d) $(f \circ g)(x) = (x + 3)^2$ **15. (a)** -65 **(b)** 126
(c) $(g \circ f)(x) = 8x^3 - 1$ **(d)** $(f \circ g)(x) = 2x^3 - 2$
17. (a) 3   **(b)** 1   **(c)** $(g \circ f)(x) = \left|\frac{1}{2}x - 2\right|$
(d) $(f \circ g)(x) = \frac{1}{2}|x - 2|$ **19. (a)** $\frac{11}{2}$ **(b)** $-\frac{1}{17}$
(c) $(g \circ f)(x) = 3 - \frac{5}{x}$ **(d)** $(f \circ g)(x) = \frac{1}{3 - 5x}$
21. (a) 77   **(b)** 122   **(c)** $(g \circ f)(x) = 16x^2 - 4x + 5$
(d) $(f \circ g)(x) = 8x^2 - 4x + 10$ **23. (a)** 1   **(b)** 2
25. (a) -1 **(b)** 1   **27. (a)** 0   **(b)** 2   **29. (a)** 2
(b) -3 **(c)** -1 **31.** $f(1) = f(-1) = 5$
33. $f(1) = f(-1) = 101$ **35.** $f(2) = f(-2) = 4$
37. Yes   **39.** No   **41.** Yes   **43.** Divide x by 7;
$f(x) = 7x; g(x) = \frac{x}{7}$ **45.** Multiply x by 2 then subtract 5;
$f(x) = \frac{x + 5}{2}; g(x) = 2x - 5$ **47.** Add 3 to x and multiply
the result by 2; $f(x) = \frac{1}{2}x - 3; g(x) = 2(x + 3)$
49. Take the cube root of x and subtract 5;
$f(x) = (x + 5)^3; g(x) = \sqrt[3]{x} - 5$
51. $(f \circ f^{-1})(x) = 4\left(\frac{x}{4}\right) = x; (f^{-1} \circ f)(x) = \frac{4x}{4} = x$
53.–57. Show $(f \circ f^{-1})(x) = (f^{-1} \circ f)(x) = x$. See the
answer to Exercise 51 above.   **59.** $f^{-1}(x) = \frac{x}{12}$
61. $f^{-1}(x) = x - 8$ **63.** $f^{-1}(x) = \frac{x + 2}{5}$
65. $f^{-1}(x) = -2(x - 1)$ **67.** $f^{-1}(x) = 8 - x$
69. $f^{-1}(x) = 2x - 1$ **71.** $f^{-1}(x) = \frac{x^3}{2}$
73. $f^{-1}(x) = \sqrt[3]{x} + 8$
75.

x	0	5	10	15	20
$f^{-1}(x)$	0	1	2	3	4

Domain of f = range of f^{-1} = $\{0, 1, 2, 3, 4\}$
Range of f = domain of f^{-1} = $\{0, 5, 10, 15, 20\}$

77.

x	4	2	0	-2	-4
$f^{-1}(x)$	-5	0	5	10	15

Domain of f = range of f^{-1} = $\{-5, 0, 5, 10, 15\}$
Range of f = domain of f^{-1} = $\{-4, -2, 0, 2, 4\}$

79.

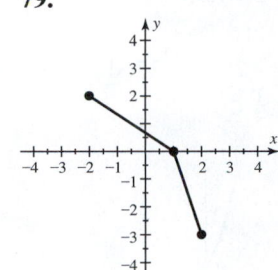

81.

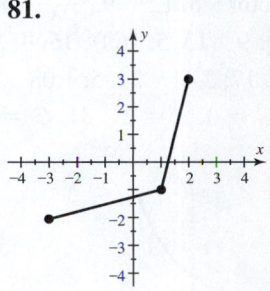

83. 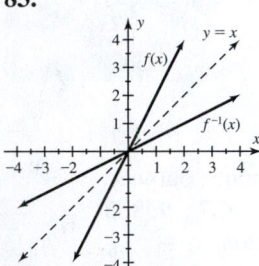 **85.**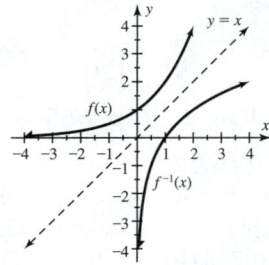

87. (a) 12 **(b)** −4 **(c)** $x^4 + 4x^2 + 2$ **89. (a)** $\frac{1}{2}$
(b) $-\frac{1}{x}$ **(c)** $\frac{x}{2}$ **91. (a)** 20π; after 5 seconds, the wave has a circumference of $20\pi \approx 62.8$ feet.
(b) $(C \circ r)(t) = 4\pi t$ **93. (a)** 16; in 1980, 16% of people 25 or older had completed four or more years of college.
(b)

x	8	16	27	29
$P^{-1}(x)$	1960	1980	2000	2010

(c) 1980

95. (a) 75°; 150 **(b)** 150; one hour after midnight there are 150 mosquitoes per 100 square feet. **(c)** The number of mosquitoes per 100 square feet, h hours after midnight
(d) $T(h) = -5h + 80$; $M(T) = 2T$
(e) $(M \circ T)(h) = -10h + 160$
97. (a) Yes, different inputs result in different outputs.
(b) $f^{-1}(x) = \frac{5}{9}(x - 32)$ converts x degrees Fahrenheit to an equivalent temperature in degrees Celsius.
99. $f(x) = 4x$; $f^{-1}(x) = \frac{x}{4}$ converts x quarts to gallons.
101. (a) $C(x) = 16x$ **(b)** $T(x) = 48x$
(c) $(T \circ C)(x) = 768x$ **(d)** 2304; there are 2304 tsp in 3 gal.

SECTION 9.2 (pp. 681–685)

1. $f(x) = Ca^x$
2.

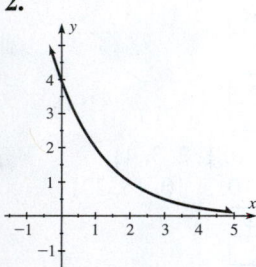

3. D: all real numbers; R: all positive real numbers
4. 32; 25 **5.** 2.718 **6.** 7.389; 9.870 (approximately)
7. factor **8.** 1.5 **9.** $\frac{B - A}{A} \times 100$ **10.** 1.35
11. $\frac{1}{9}$; 9 **13.** 5; 160 **15.** 4; $\frac{1}{8}$ **17.** 15; $\frac{5}{9}$
19. 0.17; 2.41 **21.** 5; 1.08 **23.** 1; $\frac{1}{a}$ **25.** c. **27.** d.
29. $C = 1$, $a = 2$ **31.** $C = 4$, $a = \frac{1}{4}$
33. **35.**

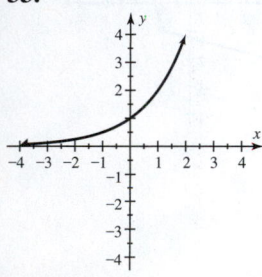

Growth Decay

37. 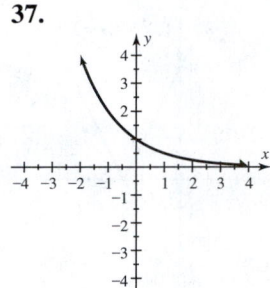 **39.**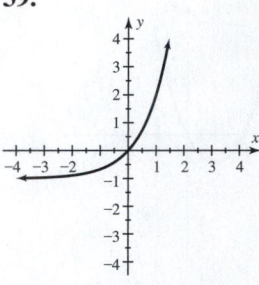

Decay Growth

41. 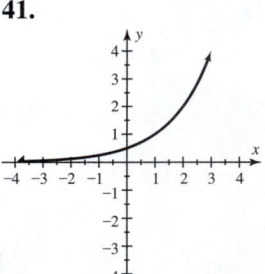 **43.**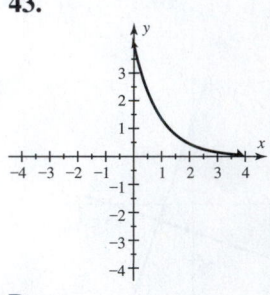

Growth Decay

45. (a) Exponential decay **(b)** $f(x) = 64\left(\frac{1}{4}\right)^x$
47. (a) Linear growth **(b)** $f(x) = 3x + 8$
49. (a) Exponential growth **(b)** $f(x) = 4(1.25)^x$
51. (a) 100% **(b)** −50%
53. (a) −80% **(b)** 400%
55. (a) $1200 **(b)** $2200 **(c)** 2.2
57. (a) $130 **(b)** $780 **(c)** 1.2
59. (a) −$80 **(b)** $720 **(c)** 0.9
61. $C = 9$, $a = 1.07$, $R = 7\%$
63. $C = 1.5$, $a = 0.45$, $R = -55\%$ **65.** $3551.05
67. $1,820,087.63 **69.** $792.75 **71.** Yes; this is equivalent to having two accounts, each containing $1000 initially.
73. $788.78 **75.** $1429.24 **77.** 3.32 **79.** 0.86
81. [−4, 4, 1] by [0, 8, 1] **83.** [−4, 4, 1] by [0, 8, 1]

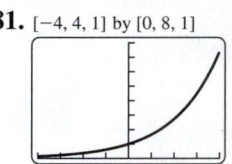 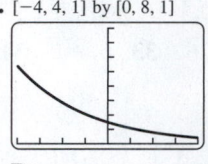

Growth Decay

85. $f(x) = 5000(0.75)^x$; $f(4) \approx 1582$
87. $f(x) = 50(1.1)^x$; $f(4) \approx 73$ **89.** 20% is much better
91. (a) 0.6 **(b)** $B(x) = 0.07(0.6)^x$ **(c)** 0.0252; after 2 hours blood alcohol is 0.0252 g/100 mL **93. (a)** 1.242
(b) In July 2008 there were about 0.5 million tweets per month.
(c) About 90.8; after 24 months, there were 90.8 million tweets per month. **95. (a)** $C = 500$, $a = 2$ **(b)** About 5278 thousand per milliliter **(c)** The growth is exponential and doubles every 50 seconds.
97.

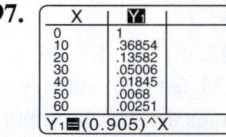

X	Y1
0	1
10	.36854
20	.13582
30	.05006
40	.01845
50	.0068
60	.00251

Y1 ≡ (0.905)^X

(a) 1; the probability that no vehicle will enter the intersection during a period of 0 seconds is 1 or 100%.

(b) About 30 seconds
99. (a) $f(x) = 2.7e^{0.014x}$ **(b)** 3.1 million
101. (a) $f(x) = 38e^{0.0102x}$ **(b)** 42 million
103. Answers may vary.

CHECKING BASIC CONCEPTS 9.1 & 9.2 (p. 685)

1. (a) 7   **(b)** $(f \circ g)(x) = 2x^2 + 9x + 6$
2.

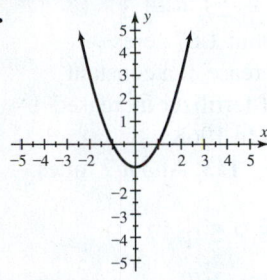

(a) No, it does not pass the horizontal line test.   **(b)** No
3. $f^{-1}(x) = \frac{x+3}{4}$ **4.** $\frac{3}{4}$
5.

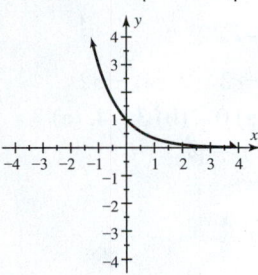

6. $C = 2; a = \frac{1}{2}$

SECTION 9.3 (pp. 694–697)

1. 10   **2.** e **3.** $D = \{x \mid x > 0\}$; R: all real numbers
4. D: all real numbers; $R = \{x \mid x > 0\}$ **5.** k **6.** k
7. x **8.** x **9.** log 5   **10.** ln 5   **11.** 0   **12.** undefined

13.

x	10^{-5}	10^0	$10^{0.5}$	$10^{2.2}$
log x	-5	0	0.5	2.2

14.

x	10^{-6}	10^{-1}	$10^{5/7}$	10^{π}
log x	-6	-1	5/7	π

15.

x	e^{-6}	e^{-1}	$e^{5/7}$	e^{π}
ln x	-6	-1	5/7	π

16.

x	2^{-5}	2^0	$2^{0.5}$	$2^{2.2}$
$\log_2 x$	-5	0	0.5	2.2

17. 5   **19.** -4 **21.** 0   **23.** -2 **25.** 4.7   **27.** 4   **29.** $\frac{1}{2}$
31. Undefined   **33.** -3 **35.** 2   **37.** $x^2, x \neq 0$ **39.** 5
41. $2x - 7$ **43.** 1.398   **45.** 0.161

47.

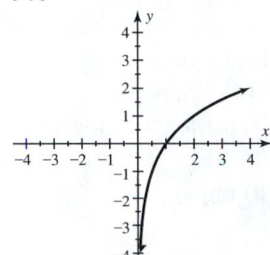

1 unit downward

49.

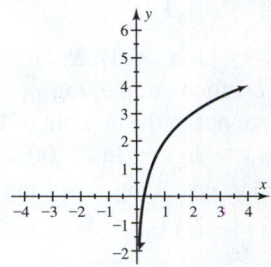

1 unit to the left

51.

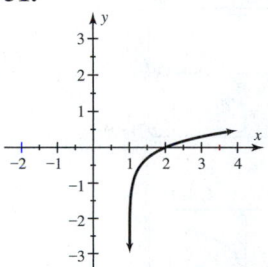

1 unit to the right

53.

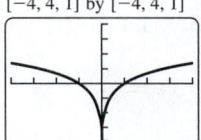

Increases faster

55. 0   **57.** $-5x$ **59.** $x^2, x \neq 0$ **61.** 1.946   **63.** -0.560
65. $[-4, 4, 1]$ by $[-4, 4, 1]$ **67.** $[-4, 4, 1]$ by $[-4, 4, 1]$

Reflected across the
y-axis together with
the graph of $y = \ln x$;
$D = \{x \mid x \neq 0\}$

2 units to the left;
$D = \{x \mid x > -2\}$

69. $6x$ **71.** $-\frac{3}{2}$ **73.** 8   **75.** $\frac{3}{2}$ **77.** $-\frac{2}{3}$ **79.** Undefined
81. 2   **83.** -4 **85.** -2 **87.** -2 **89.** 17   **91.** $(2x)^2$
93. $0.6z, z > 0$
95.

x	1/4	1/2	1	$\sqrt{2}$	64
$\log_2 x$	-2	-1	0	1/2	6

97.

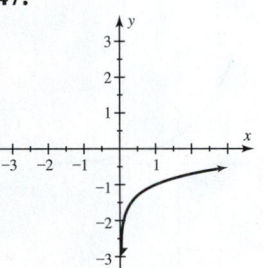

99.

101. d.   **103.** a.   **105.** 120 dB; yes   **107. (a)** It increases,
but it doesn't double when the weight of the plane doubles.
(b) About 5.086; a 50,000-pound airplane needs a runway 5086
feet long.   **109. (a)** 3.5   **(b)** 1   **(c)** It decreases by 2.5.
111. (a) 10^6; 10^8 **(b)** 100 times

SECTION 9.4 (pp. 703–704)

1. 4   **2.** 4   **3.** 3   **4.** $\log m + \log n$ **5.** $\log m - \log n$
6. $r \log m$ **7.** No   **8.** Yes   **9.** No   **10.** No
11. $\log_a x = \frac{\log x}{\log a}$ or $\log_a x = \frac{\ln x}{\ln a}$ **12.** 0; 1
13. $\ln 3 + \ln 5$ **15.** $\log x + \log y$ **17.** $\log y + \log y$
19. $\log 7 - \log 3$ **21.** $\ln x - \ln y$ **23.** $\log_2 45 - \log_2 x$
25. $\log 225$ **27.** $\ln xy$ **29.** $\ln 14x^3$ **31.** $\ln xy$ **33.** $\log 5$
35. $\ln x^2$ **37.** 2   **39.** $6 \log 3$ **41.** $x \ln 2$ **43.** $\frac{1}{4} \log_2 5$
45. $\frac{1}{3} \log_4 z$ **47.** $(y - 1)\log x$ **49.** $\log z$ **51.** $\log x^3 y^2$
53. 0   **55.** $\ln 2^x$ **57.** $\ln x^{5/6}$ **59.** $\log_a \frac{x+1}{x-1}$
61. $\log x + 2 \log y$ **63.** $4 \ln x + \ln y - \ln z$
65. $\frac{1}{3} \log z - \frac{1}{2} \log y$ **67.** $4 \log x + 3 \log y$
69. $\ln x - \ln y$ **71.** $\frac{3}{2} \log_4 x + \frac{1}{2} \log_4 y - \log_4 z$

73. [−6, 6, 1] by [−4, 4, 1]   [−6, 6, 1] by [−4, 4, 1]

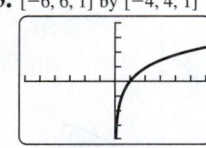 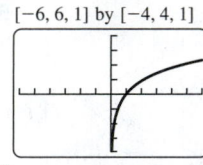

By the power rule, $\log x^3 = 3 \log x$

75. [−6, 6, 1] by [−4, 4, 1]   [−6, 6, 1] by [−4, 4, 1]

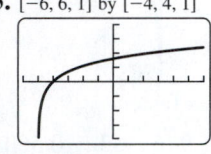 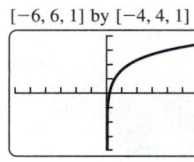

Not the same

77. 1.2   **79.** 1.8   **81.** 2.1   **83.** 0.4   **85.** −1.1

87. 1.46   **89.** 4.64   **91.** 2.10

93. $10 \log (10^{16} x) = 10(\log 10^{16} + \log x) =$
$10(16 + \log x) = 160 + 10 \log x$

CHECKING BASIC CONCEPTS 9.3 & 9.4 (p. 705)

1. (a) 4   **(b)** x **(c)** −3   **(d)** $\frac{1}{2}$

2.

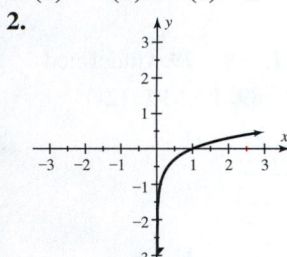

(a) $D = \{x \mid x > 0\}$; R: all real numbers   **(b)** 0
(c) Yes; for example, $\log \frac{1}{10} = -1$.   **(d)** No; negative numbers are not in the domain of $\log x$.   **3. (a)** $\log x + \log y$
(b) $\ln x - \ln y - \ln z$ **(c)** $2 \ln x$
(d) $2 \log x + 3 \log y - \frac{1}{2} \log z$ **4. (a)** $\log xy$
(b) $\ln \frac{2x}{y^3}$ **(c)** $\log_2 \frac{x^2 y^3}{z}$

SECTION 9.5 (pp. 712–716)

1. Add 5 to each side.   **2.** Divide each side by 5.
3. Take the common logarithm of each side.
4. Exponentiate each side using base 10.   **5.** x
6. $x, x > 0$ **7.** $2x$ **8.** $x + 7, x > -7$
9. No; $\log \frac{5}{4} = \log 5 - \log 4$
10. No; $\log 5 - \log 4 = \log \frac{5}{4}$ **11.** 1   **12.** 1   **13.** 3
15. 6   **17.** 6   **19.** 4   **21.** $\frac{\log 124}{0.4} \approx 5.23$ **23.** 0
25. $\ln 25 \approx 3.22$ **27.** $\frac{\ln 2}{\ln 0.4} \approx -0.76$ **29.** $\ln 7 \approx 1.95$
31. $\log \frac{35}{2} - 2 \approx -0.76$ **33.** $\frac{\log 20}{2 \log 3.1} \approx 1.32$ **35.** −1
37. −1, 5   **39.** $\frac{\ln 10}{3} \approx 0.77$ **41.** $\frac{2 \ln 2}{1 - \ln 2} \approx 4.52$
43. $\frac{2 \log 5}{0.5 \log 4 - \log 5} \approx -3.51$ **45. (a)** 1   **(b)** 1
47. (a) −2   **(b)** −2   **49.** −1   **51.** 0   **53.** 5
55. $\frac{\log 1.45}{\log 0.55} \approx -0.62$ **57.** −1.84, 1.15   **59.** 1.31
61. 100   **63.** $e^5 \approx 148.41$ **65.** 5,000,000   **67.** 16
69. $\frac{2^{2.3}}{5} \approx 0.98$ **71.** $10^{1.4} \approx 25.12$
73. $\frac{e^{11} - 1}{2} \approx 29,936.57$ **75.** 1   **77.** 5   **79.** 4   **81.** 3
83. $10^{1.6} \approx 39.81$ **85.** $e - 1 \approx 1.72$ **87.** 9

89. (a) 2   **(b)** $e^{0.7} \approx 2.01$ **91. (a)** 2   **(b)** $\frac{1}{2}(10^{0.6}) \approx 1.99$
93. (a) About 41.5; in 2010 China generated about
41.5 gigawatts.   **(b)** 2009   **95.** 10 hours   **97.** 8 years
99. (a) About 8503; in 1994 about 8503 people were
waiting for liver transplants.   **(b)** In about 1998
101. About 20,893 pounds   **103. (a)** About 203; in
1975 there were about 203 thousand bluefin tuna.
(b) In 1979   **(c)** In 1979   **105.** $a = 25, b = 3$
107. (a) About 1.67 acres   **(b)** About 1.67 acres
109. (a) Nonlinear; they do not increase at a constant
rate.   **(b)** Each year the amount of fertilizer increased
by a factor of 1.06, or by 6%.   **(c)** In 1968
111. 10^{-6} watts/square centimeter   **113.** About 7 miles

CHECKING BASIC CONCEPTS 9.5 (p. 716)

1. (a) $\log 20 \approx 1.30$ **(b)** $\frac{\log 147}{3 \log 2} \approx 2.40$
(c) $e^{4.1} \approx 60.34$ **(d)** 500   **2.** 8   **3.** 20 years

CHAPTER 9 REVIEW (pp. 719–722)

1. (a) 81   **(b)** $(f \circ g)(x) = 50x^2 - 2$ **2. (a)** −32
(b) $(f \circ g)(x) = \sqrt[3]{4x^3 - 6}$ **3. (a)** 0   **(b)** 3   **4. (a)** 3
(b) −2   **(c)** −1   **5.** $f(1) = f(-1) = 2$
6. $f(0) = f(2) = 1$ **7.** No   **8.** Yes
9. $(f \circ f^{-1})(x) = 2\left(\frac{x + 9}{2}\right) - 9 = x$
$(f^{-1} \circ f)(x) = \frac{(2x - 9) + 9}{2} = x$
10. $(f \circ f^{-1}) = (\sqrt[3]{x - 1})^3 + 1 = x$
$(f^{-1} \circ f)(x) = \sqrt[3]{(x^3 + 1) - 1} = x$
11. $f^{-1}(x) = \frac{x}{5}$ **12.** $f^{-1}(x) = x + 11$
13. $f^{-1}(x) = \frac{x - 7}{2}$ **14.** $f^{-1}(x) = \frac{4}{x}$
15.

x	10	8	7	3
$f^{-1}(x)$	0	1	2	3

$D = \{3, 7, 8, 10\}$; $R = \{0, 1, 2, 3\}$

16.

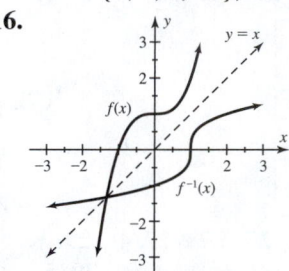

17. $\frac{1}{6}$; 36   **18.** 5; $\frac{5}{8}$ **19.** 3; $\frac{1}{81}$ **20.** 3; $\frac{1}{2}$

21.

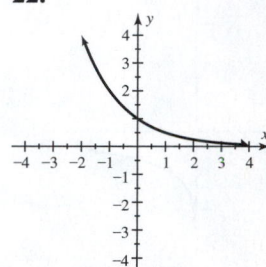

Exponential growth

22.

Exponential decay

23.

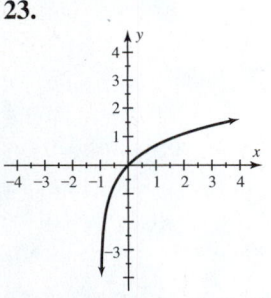

Logarithmic growth

24.

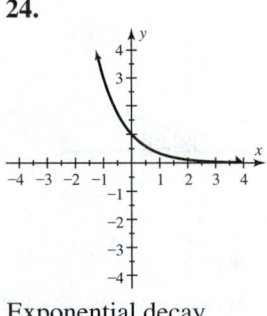

Exponential decay

25. (a) Exponential growth   **(b)** $f(x) = 5(2)^x$
26. (a) Linear growth   **(b)** $f(x) = 5x + 5$
27. $C = \frac{1}{2}, a = 2$ **28.** $k = 2$ **29.** -20% **30.** 1.07
31. (a) 100% **(b)** -50% **32. (a)** $-54.\overline{54}\%$ **(b)** 120%
33. (a) $\$1050$ **(b)** $\$1550$ **(c)** 3.1 **34. (a)** $-\$175$
(b) $\$525$ **(c)** 0.75 **35.** $f(x) = 20{,}000(0.95)^x$;
$f(2) = 18{,}050$ **36.** $f(x) = 1500(4)^x$; $f(2) = 24{,}000$
37. $\$2829.54$ **38.** $\$675{,}340.51$ **39.** 399.67 **40.** 0.71
41. 3.48 **42.** 3.89 **43.** -3 **44.** 2 **45.** -4 **46.** 2
47. 1.813 **48.** -0.163 **49.** 4.787 **50.** -0.398 **51.** 7
52. $\frac{5}{9}$ **53.** $6 - x$ **54.** x^2 **55.** $\ln x + \ln y$
56. $\log x - \log y$ **57.** $2\ln x + 3\ln y$
58. $\frac{1}{2}\log x - 3\log z$ **59.** $2\log_2 x + \log_2 y - \log_2 z$
60. $\frac{1}{3}\log_3 x - \frac{1}{3}\log_3 y$ **61.** $\log 75$ **62.** $\log_4(10x^2)$
63. $\ln \frac{x^2}{y^3}$ **64.** $\log xy$ **65.** $3\log 6$ **66.** $2\ln x$
67. $2x\log_2 5$ **68.** $(x+1)\log_4 0.6$ **69.** 2 **70.** 4
71. $\ln 9 \approx 2.20$ **72.** $\frac{\log 0.2}{\log 0.85} \approx 9.90$ **73.** $e^{0.8} \approx 2.23$
74. $\frac{1}{2}e^5 \approx 74.21$ **75.** 10^{40} **76.** 100
77. $\frac{4\log 2}{\log 3 - \log 2} \approx 6.84$ **78.** 6 **79. (a)** 3 **(b)** 3
80. (a) 4 **(b)** 4 **81. (a)** 64π; after 8 seconds, the balloon
has a surface area of $64\pi \approx 201$ in^2.   **(b)** $(S \circ r)(t) = 8\pi t$
82. (a) Yes, different inputs result in different outputs.
(b) $f^{-1}(x) = \frac{x}{0.08}$ calculates the cost of an item whose sales
tax is x dollars.   **83.** 7 years   **84.** $a = 100, b = 50$
85. $C = 3, a = 2$ **86.** 10^7

87. (a) [0, 10, 2] by [0, 4, 1]

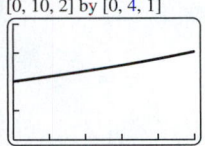

Growth
(b) About 3.3 million   **(c)** 2008   **88. (a)** 1000; there were
1000 bacteria/mL initially.   **(b)** About 495.11 minutes
89. (a) About 6.93 meters/second   **(b)** 12.18 meters

CHAPTER 9 TEST (pp. 723–724)

1. 6; $(f \circ g)(x) = 4(x + 7)^3 - 5(x + 7)$
2. (a) 3 **(b)** 3 **3.** $f(-5) = f(5) = 0$
(answers may vary)   **4.** $f^{-1}(x) = \frac{5 - x}{2}$

5.

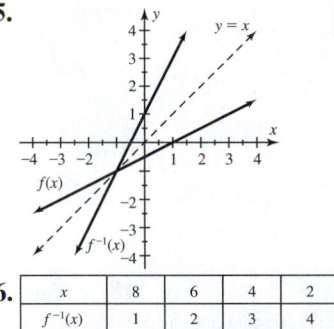

6.

x	8	6	4	2
$f^{-1}(x)$	1	2	3	4

$D = \{2, 4, 6, 8\}; R = \{1, 2, 3, 4\}$
7. $\frac{3}{16}$
8.

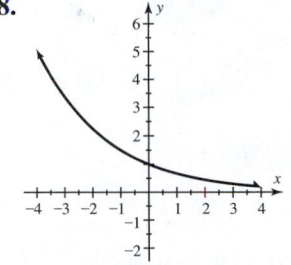

Exponential decay
9. (a) Exponential growth   **(b)** $f(x) = 3(2)^x$
10. (a) Linear growth   **(b)** $f(x) = 1.5x - 1$
11. $C = 1, a = \frac{1}{2}$ **12.** 50% **13.** 1.05 **14.** $\$1051.91$
15. 4.16 **16.** $\frac{1}{2}$ **17.** 5.426
18.

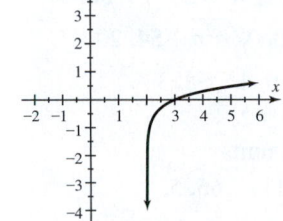

Shifted to the right 2 units
19. $3\log x + 2\log y - \frac{1}{2}\log z$ **20.** $\ln \frac{x^4 z}{y^5}$ **21.** $2x\log 7$
22. $1 - 3x$ **23.** $\ln 25 \approx 3.22$ **24.** $\log 50 \approx 1.70$
25. $10^{1.8} \approx 63.10$ **26.** $\frac{1}{5}e^9 \approx 1620.62$
27. $a = 5, b = 3$ **28. (a)** 4 million   **(b)** About 6.15; after
5 hours there were about 6.15 million bacteria.   **(c)** Growth
(d) After 8 hours
29. (a) $A(x) = 5000(0.98)^x$ **(b)** $A(3) \approx 4705.96$; after
3 years the account contains $\$4705.96$.   **(c)** About 5

CHAPTERS 1–9 CUMULATIVE REVIEW (pp. 725–728)

1. 4.29×10^{-4} **2.** Natural: none; whole: 0; integer: $-3, 0$;
rational: $-\frac{11}{7}, -3, 0, 5.\overline{18}$; irrational: $\sqrt{6}, \pi$ **3.** c.
4. Commutative   **5.** d^4 **6.** $\frac{b^9}{64a^6}$ **7.** $\frac{4y^8}{x^5}$ **8.** $\frac{y^4}{4x^5}$
9. $y = \frac{5}{4}x + 1$ **10.** $\{x \mid x \neq -3\}$ **11.** $f(x) = 4x - 3$
12. $x = 4$ **13.** 1

14.

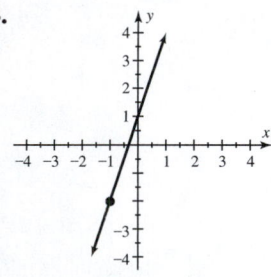

15. $y = 7x - 6$ **16.** $y = 3x + 5$ **17.** -1 **18.** $x < 0$
19. -18 **20.** $(-\infty, 0)$ **21.** $\frac{4}{15}$ **22.** $(-5, 10]$
23. $[-2, 10]$ **24.** $4, 16$
25.

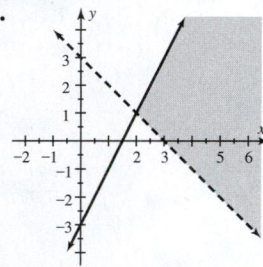

26. 2 **27.** $(1, 1)$ **28.** No solutions   **29.** $(-4, 0, 2)$
30. $(0, 0, 1)$ **31.** $R = 21$ **32.** 8 square inches
33. $2x(x - 1)^2$ **34.** $(2a - 5b)(2a + 5b)$
35. $(2t - 3)(4t^2 + 6t + 9)$ **36.** $(2a^2 + 5)(2a - 1)$
37. $-\frac{5}{6}, 2$ **38.** $-\frac{2}{3}, \frac{2}{3}$ **39.** $-3, 0, 5$ **40.** $0, \frac{1}{2}$ **41.** 1
42. $\frac{x + 2}{x - 3}$ **43.** 3 **44.** -3 **45.** $J = \frac{2z}{P - 1}$ **46.** $\frac{x^3 + 3}{x^3 - 3}$
47. 40 **48.** $3x^2 + 6x + 10 + \frac{5}{x - 2}$ **49.** $\frac{x^4}{y^6}$ **50.** x^3
51. $10ab$ **52.** $\sqrt{6}$ **53.** $(b + 3)a\sqrt[3]{a^2b}$ **54.** 20
55. $\frac{5 + \sqrt{3}}{11}$ **56.** $(4, \infty)$ **57.** $-6, 4$ **58.** 3
59. $3 + 5i$ **60.** $-i$ **61.** $(2, 1)$ **62.** $\frac{7}{2}$
63. Shifted right 3 units and up 2 units
64. $y = (x + 3)^2 - 11; (-3, -11)$ **65.** $5, 8$
66. $\frac{1 \pm \sqrt{41}}{4}$ **67.** $2 \pm \sqrt{2}$ **68.** $\pm 2, \pm \sqrt{6}$
69. (a) $-1, 3$ **(b)** $a < 0$ **(c)** Positive
70. $(-\infty, -7] \cup [2, \infty)$ **71. (a)** 7
(b) $(g \circ f)(x) = 2x^2 - 3$ **72.** $f(-4) = f(3) = 6$
73. $f^{-1}(x) = \frac{3}{x}$ **74.** \$2367.10   **75.** 4 **76.** $2x$
77. $\frac{1}{2} \log x - 2 \log y$ **78.** $\ln (5x^3)$ **79.** $10^{11/6} \approx 68.13$
80. $\frac{\log 17}{3 \log 2} \approx 1.36$ **81.** 4% **82.** 2.27 pounds
83. (a) 1950 **(b)** 220 million Btu
84. $\sqrt{4200} \approx 64.8$ mph   **85.** About 11.3%
86. (a) 9.91 meters per second   **(b)** 8.52 meters

10   Conic Setctions

SECTION 10.1 (pp. 736–739)

1. Parabola, ellipse, hyperbola   **2.** The axes of symmetry
are vertical and horizontal, respectively.   **3.** No

4.

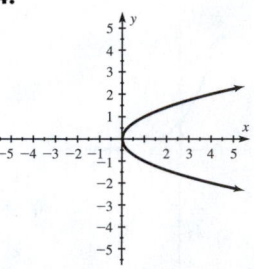

5. No; it does not pass the vertical line test.   **6.** (h, k)
7. left   **8.** right   **9.** circle; (h, k) **10.** $(0, 0); r$
11. d.   **12.** a.   **13.** c.   **14.** b.
15.

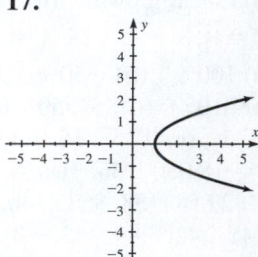

$(0, 0); y = 0$

17.

$(1, 0); y = 0$

19.

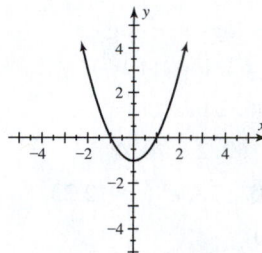

$(0, -1); x = 0$

21.

$(0, 0); y = 0$

23.

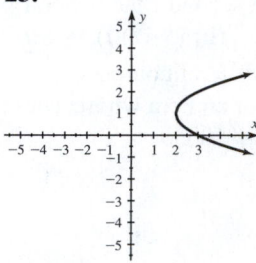

$(2, 1); y = 1$

25.

$(-2, 1); x = -2$

27.

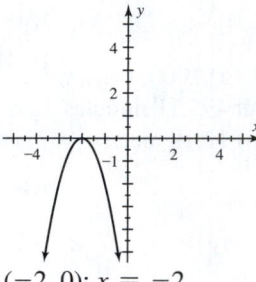

$(-2, 0); x = -2$

29.

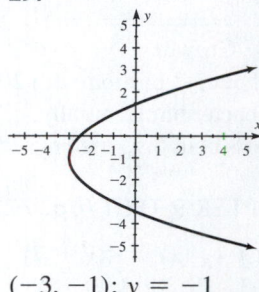

$(-3, -1); y = -1$

31.

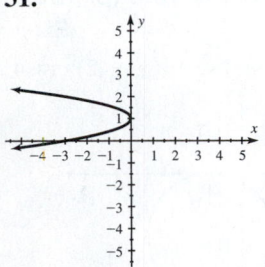

$(0, 1); y = 1$

33.

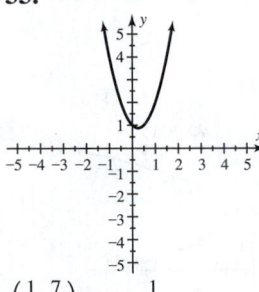

$\left(\frac{1}{4}, \frac{7}{8}\right); x = \frac{1}{4}$

35.

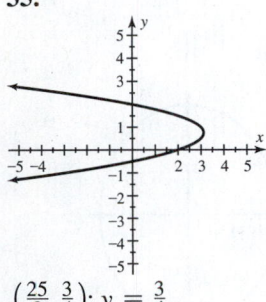

$\left(\frac{25}{8}, \frac{3}{4}\right); y = \frac{3}{4}$

37.

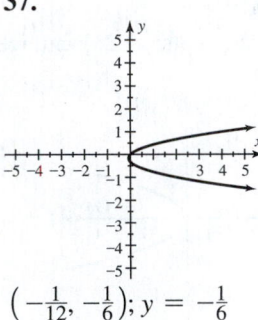

$\left(-\frac{1}{12}, -\frac{1}{6}\right); y = -\frac{1}{6}$

39.

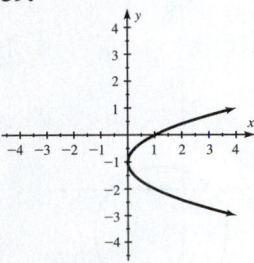

$(0, -1); y = -1$

41. $y = x^2$ **43.** $x = (y + 1)^2 - 2$ **45.** Upward
47. Downward   **49.** $x \geq 0$ **51.** Two   **53.** 1
55. $x^2 + y^2 = 1$ **57.** $(x + 1)^2 + (y - 5)^2 = 9$
59. $(x + 4)^2 + (y + 6)^2 = 2$ **61.** $x^2 + y^2 = 16$
63. $(x + 3)^2 + (y - 2)^2 = 1$
65. 2; (0, 0)

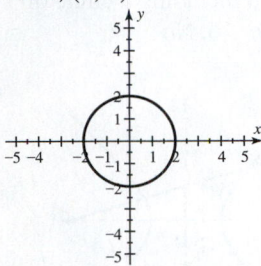

67. 3; (1, 3)

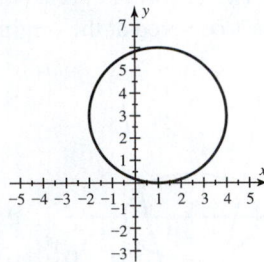

69. 5; (−5, 5)

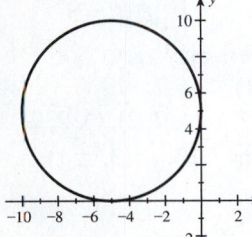

71. 3; (−3, 1)

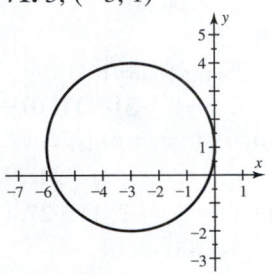

73. $\sqrt{7}$; $(-3, 1)$

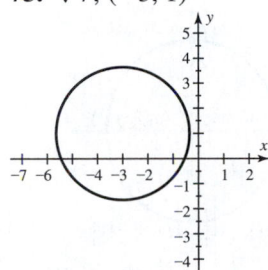

75. (a)

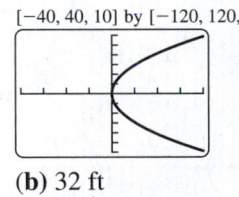

[−40, 40, 10] by [−120, 120, 20]

(b) 32 ft

77. (a)

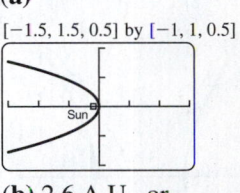

[−1.5, 1.5, 0.5] by [−1, 1, 0.5]

(b) 2.6 A.U., or
241,800,000 miles

SECTION 10.2 (pp. 747–749)

1.

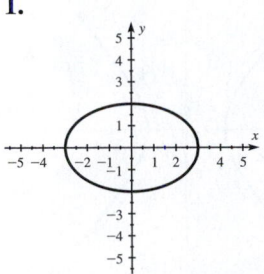

(answers may vary)

2.

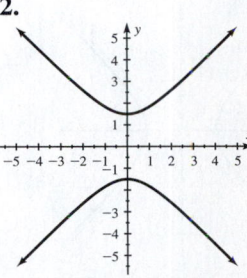

(answers may vary)

3. horizontal   **4.** vertical   **5.** 2   **6.** 4   **7.** left; right
8. lower; upper   **9.** They are the diagonals extended.
10. No   **11.** a.   **12.** b.
13.

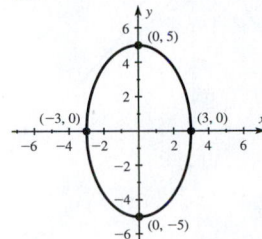

15.

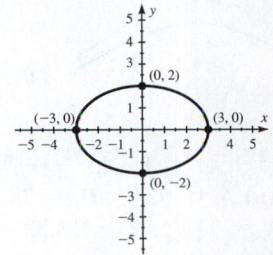

17.

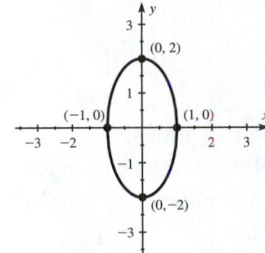

19.

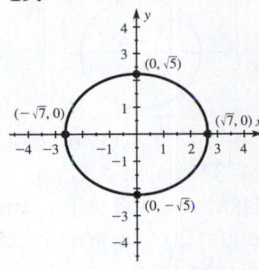

21.

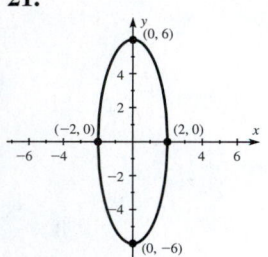

23.

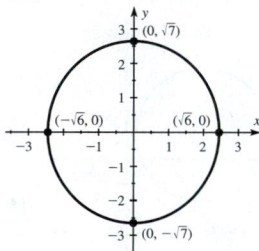

25. $\frac{x^2}{9} + \frac{y^2}{4} = 1$ **27.** $\frac{y^2}{25} + \frac{x^2}{16} = 1$

29.

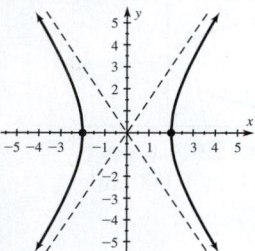

31.

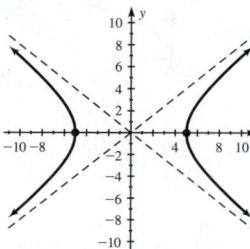

33.

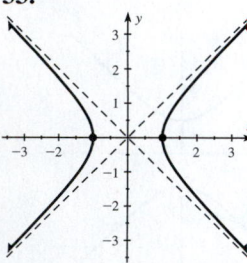

35.

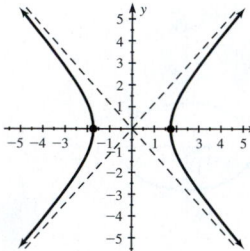

37.

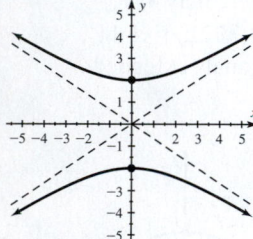

39.

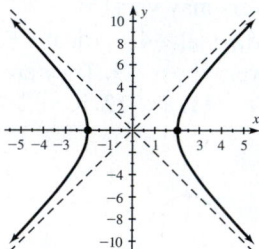

41. $x^2 - y^2 = 1$ **43.** $\frac{y^2}{4} - \frac{x^2}{9} = 1$
45. (a) $A \approx 62.83; P \approx 28.45$
(b) $A \approx 11.75; P \approx 13.33$
47. (a) $[-60, 60, 10]$ by $[-40, 40, 10]$

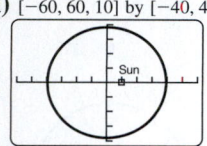

(b) $P \approx 243.9$ A.U., or about 2.27×10^{10} miles;
$A \approx 4733$ square A.U., or about 4.09×10^{19} square miles
49. Maximum: 668 miles; minimum: 340 miles
51. Height: 20 feet; width: 200 feet
53. Answers may vary.

CHECKING BASIC CONCEPTS 10.1 & 10.2 (p. 750)

1.

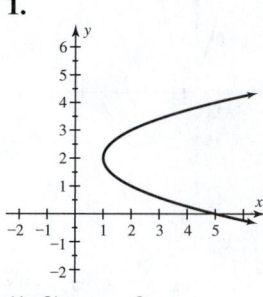

2. $(x - 1)^2 + (y + 2)^2 = 4$

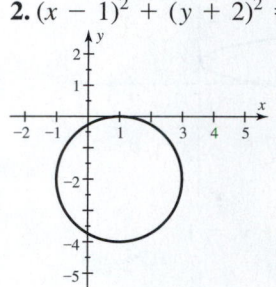

$(1, 2); y = 2$
3. x-intercepts: ± 2; y-intercepts: ± 3
4. (a)

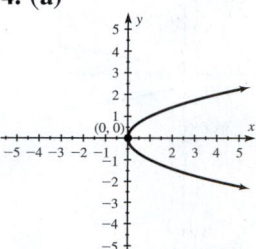

(b)

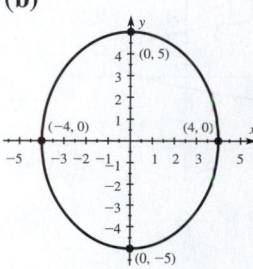

Parabola

Ellipse

(c)

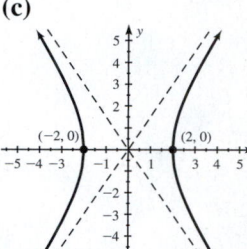

(d)

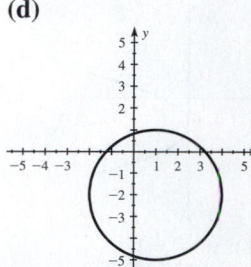

Hyperbola

Circle (and ellipse)

SECTION 10.3 (pp. 757–758)

1. Any number   **2.** Two   **3.** Two; the line intersects the circle twice.   **4.** All points inside and including a circle of radius 1 centered at the origin   **5.** No   **6.** No

7.

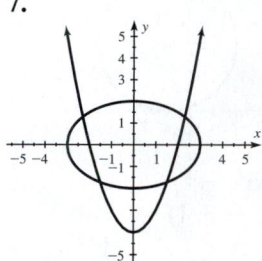

8.

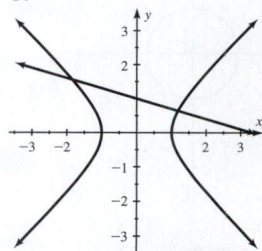

(answers may vary)   (answers may vary)
9. $(1, 3), (-1, -3)$ **11.** $(0, -1), (0, 1)$
13. $(3, 6), (-3, -6)$ **15.** $(2, -1)$ **17.** $(-\sqrt{2}, 2), (\sqrt{2}, 2)$
19. $\left(-\frac{3}{5}, \frac{7}{5}\right), (-1, 1)$ **21.** No solutions   **23.** $(\pm 3, \pm 1)$
25. $(-1, -2), (-2, 1)$ **27.** $(7, 3), (2, -2)$
29. $(-1, 2), (3, -6)$ **31.** $(-1, -1), (1, 1)$

33.

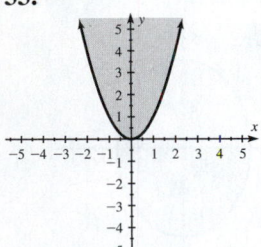

35.

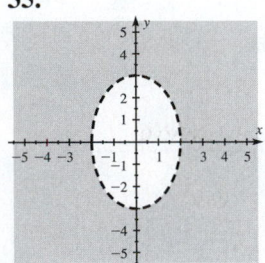

1.

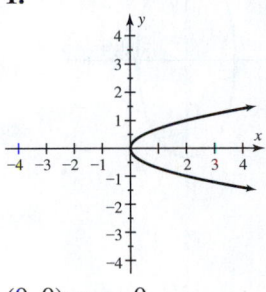

$(0, 0)$; $y = 0$

2.

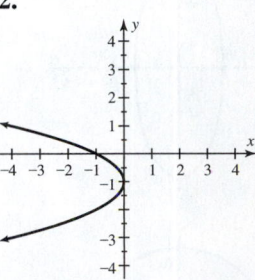

$(0, -1)$; $y = -1$

37.

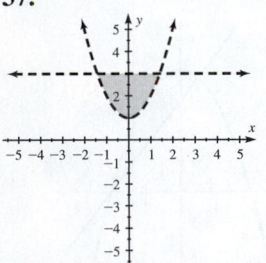

$(0, 2)$, (answers may vary)

39.

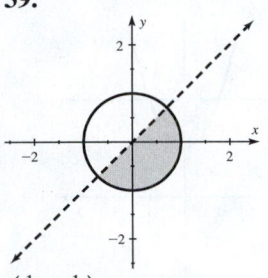

$\left(\frac{1}{2}, -\frac{1}{2}\right)$, (answers may vary)

3.

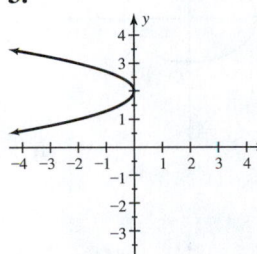

$(0, 2)$; $y = 2$

4.

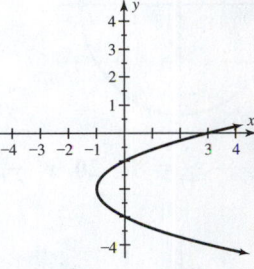

$(-1, -2)$; $y = -2$

41.

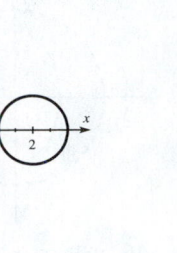

$(1, 0)$

43.

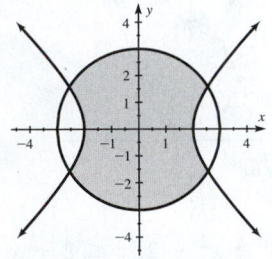

$(0, 0)$, (answers may vary)

5.

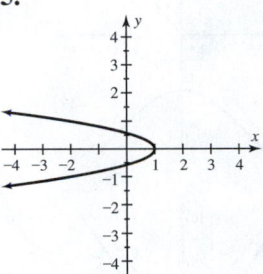

$(1, 0)$; $y = 0$

6.

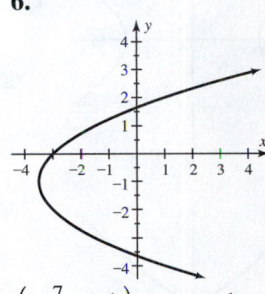

$\left(-\frac{7}{2}, -1\right)$; $y = -1$

45. a.   47. $y \geq x^2$; $y < 4 - x$

49. $[-10, 10, 1]$ by $[-10, 10, 1]$

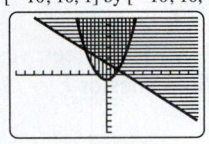

51. $r = 1.6$ inches; $h \approx 4.97$ inches

53. (a) $xy = 143$, $2x + 2y = 48$
(b) $x = 11$ ft, $y = 13$ ft

7. $x = y^2$ **8.** $(x + 2)^2 + (y - 2)^2 = 16$
9. $x^2 + y^2 = 1$ **10.** $(x - 2)^2 + (y + 3)^2 = 16$

11. 5; $(0, 0)$

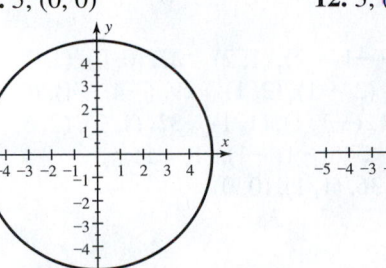

12. 3; $(2, 0)$

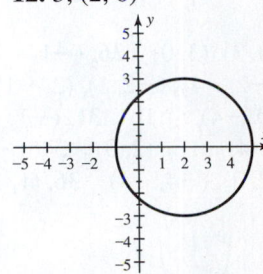

CHECKING BASIC CONCEPTS 10.3 (p. 759)

1. $(1, -1)$, $(3, 3)$ **2.** Two   **3. (a)** $(0, 3)$, $(4, 4)$ (answers may vary)   **(b)** $y \geq 2 - x$ and $y \leq 4 - x^2$
4.

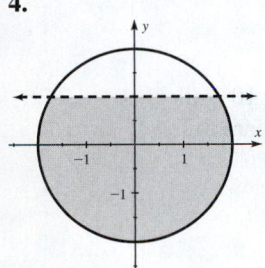

13. $\sqrt{5}$; $(-3, 1)$

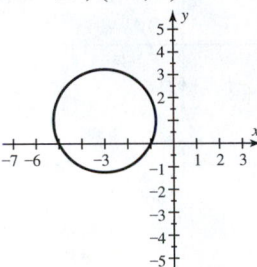

14. 3; $(1, -1)$

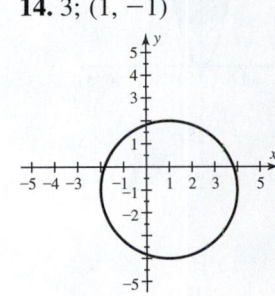

15.

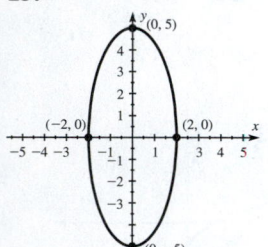

16.

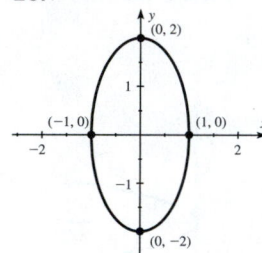

17.

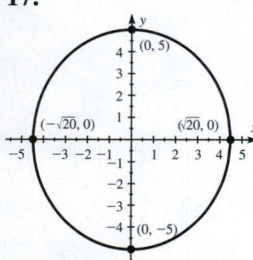

18.

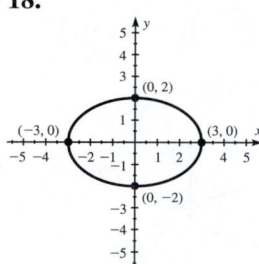

19. $\dfrac{y^2}{16} + \dfrac{x^2}{4} = 1$ **20.** $x^2 - \dfrac{y^2}{4} = 1$

21.

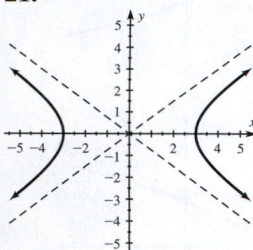

22.

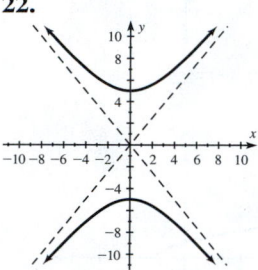

23.

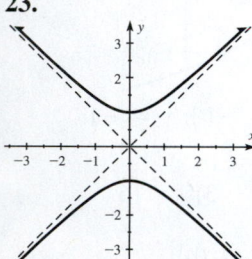

24.

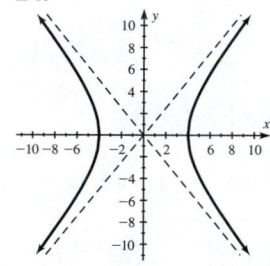

25. $(0, 3), (3, 0)$ **26.** $(-1, -2), (1, 2)$ **27.** $(0, 0), (2, 2)$
28. $(-2, -1), (-2, 1), (2, -1), (2, 1)$ **29.** $(-4, -4), (4, 4)$
30. $(0, -4), (4, 0)$ **31.** $(-1, 1), (1, 1)$ **32.** $(1, 2), (2, 5)$
33. $(-4, -12), (2, 0)$ **34.** $(-1, -1), (1, -1)$
35. $(2, 2), (-4, -4)$ **36.** $(1, 1), (0, 0)$
37.

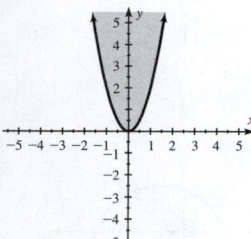

38.

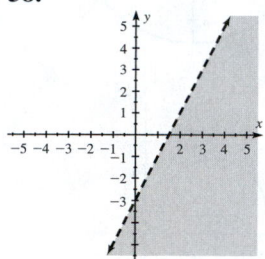

39.

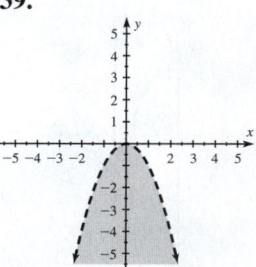

40.

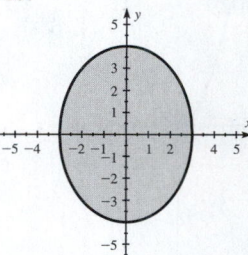

41.

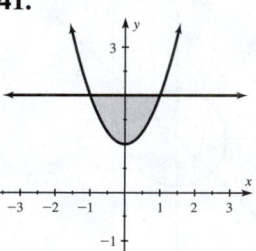

42.

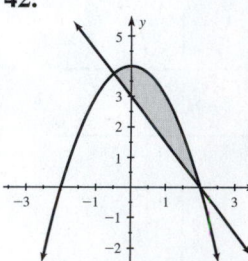

43.

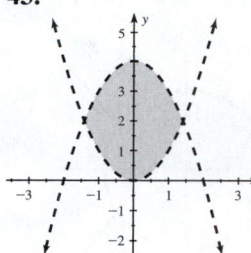

44.

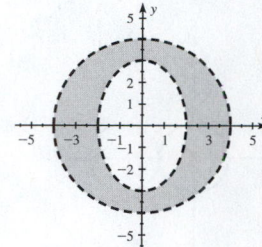

45. $y \geq x^2 - 2, \ y \leq 2 - x$ **46.** $y \geq x, \ x^2 + y^2 \leq 4$
47. (a) $xy = 1000, \ 2x + 2y = 130$
(b) $x = 25$ inches; $y = 40$ inches
(c) $x = 25$ inches; $y = 40$ inches
48. (a) $xy = 60, \ y - x = 7$ **(b)** $x = 5; y = 12$
(c) $x = 5; y = 12$ **49.** $r = 1$ foot, $h \approx 15.92$ feet; yes
50. Either $r \approx 0.94$ inches, $h \approx 12.60$ inches or $r \approx 3.00$
inches, $h \approx 1.23$ inches; no
51. (a) $[-7.5, 7.5, 1]$ by $[-5, 5, 1]$

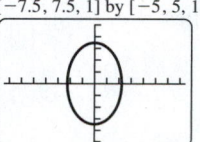

(b) $A \approx 24.33, \ P \approx 18.32$
52. (a) $[-3, 3, 1]$ by $[-2, 2, 1]$

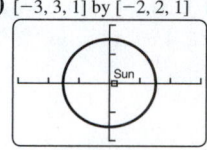

(b) $P \approx 9.55$ A.U., or about 8.9×10^8 miles; $A \approx 7.26$
square A.U., or about 6.3×10^{16} square miles

CHAPTER 10 TEST (pp. 763–764)

1.

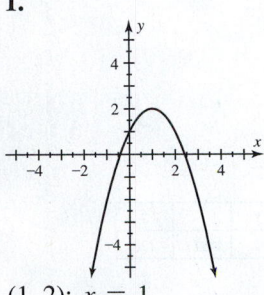

2.

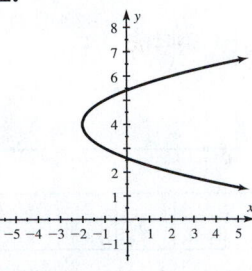

$(1, 2); \; x = 1$

$(-2, 4); \; y = 4$

3. $x = -y^2 + 1$ **4.** $(x - 2)^2 + (y + 4)^2 = 4$

5. $(x + 5)^2 + (y - 2)^2 = 100$

6. $r = 4$, center $= (-2, 3)$ **7.**

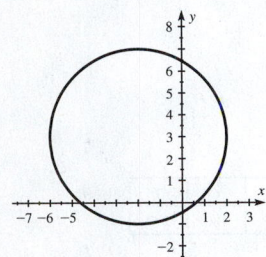

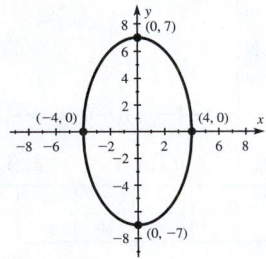

8. $\dfrac{x^2}{100} + \dfrac{y^2}{64} = 1$

9.

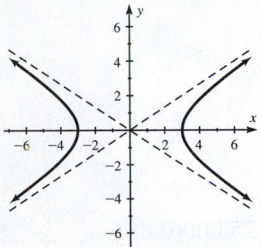

10. $(0, -4), (4, 0)$ **11.** $(-1, -4), (4, 1)$

12. $(-2, 4), (2, 4)$

13.

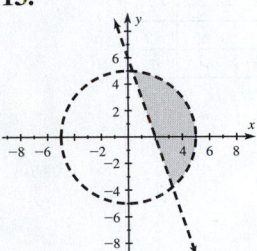

14. $y \le 4 - x^2, \; y \ge x^2 - 4$

15. (a) $xy = 5000, \; 2x + 2y = 300$

(b) 50 feet by 100 feet

16. Either $x \approx 22.08$ in., $y \approx 2.43$ in. or

$x \approx 7.29$ in., $y \approx 22.24$ in.; no

17. (a) $[-30, 30, 10]$ by $[-20, 20, 10]$

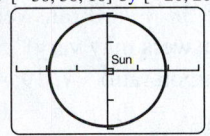

(b) 18.28 A.U., or about 1,700,040,000 miles

CHAPTERS 1–10 CUMULATIVE REVIEW (pp. 766–767)

1. 25   **2.** $\dfrac{b^8}{a^3}$ **3.** 0.007345   **4.** $\dfrac{1}{2}; x \ne 4$ **5.** 3; 4

6.

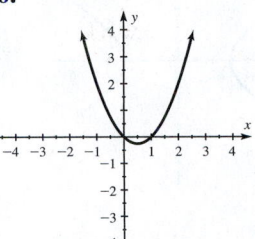

7.

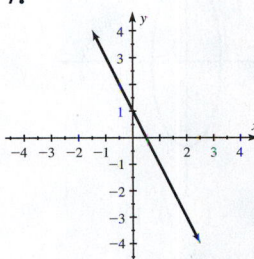

8. $y = \dfrac{3}{2}x - 5$ **9.** $\dfrac{2}{7}$ **10.** $(-1, 3]$ **11.** $[-2, 2]$

12. $(-\infty, -1] \cup [3, \infty)$ **13.** $(1, 3)$

14.

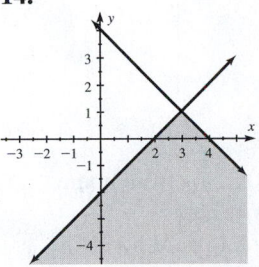

15. $2x^2 + 9x - 5$ **16.** $2x^2y - 3xy^3 + xy$

17. $(3x + 1)(2x - 5)$ **18.** $x(x + 2)(x - 2)$ **19.** $-2, -1$

20. $\dfrac{-3 \pm \sqrt{5}}{2}$ **21.** $\dfrac{3}{2}$ **22.** $\dfrac{2x}{x^2 - 1}$ **23.** $2x\sqrt{2}$ **24.** 4

25. $4x$ **26.** $5\sqrt{3x}$

27.

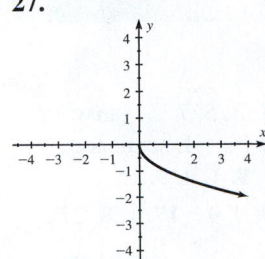

28. 5   **29.** 13   **30.** 2   **31.** $(3, -6)$

32. $f(x) = (x - 1)^2 + 2$ **33.** Shifted 4 units right

34. 1, 2   **35.** 0, $\pm i$ **36. (a)** 4   **(b)** 3   **(c)** x **(d)** 6   **(e)** 2

(f) 3   **37. (a)** 17   **(b)** $2x^2 + 2$ **38.** $\dfrac{2}{3} - \dfrac{1}{3}x$

39. \$1340.10   **40.** $2 \log x + \dfrac{1}{2} \log y - 3 \log z$ **41.** $\ln 9$

42. 25

43. (a)

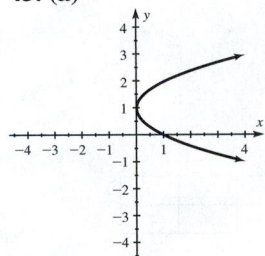

(b)

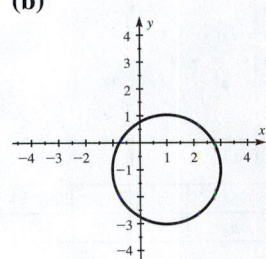

(c)

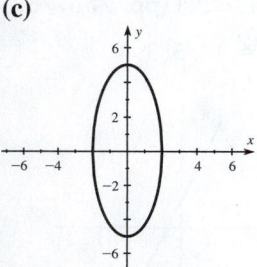

(d)

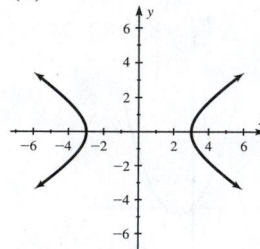

44. $(0, -1), (0, 1)$

45.

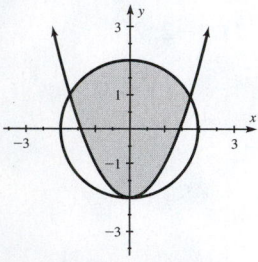

46. (a) 400; initially, the driver is 400 miles from home.
(b) 8; after 8 hours, the driver arrives at home.
(c) -50; the driver is traveling 50 mph toward home.
47. $500 at 5%, $1000 at 6%, $500 at 7%
48. 300 ft by 300 ft   **49.** $50 \ln 2 \approx 34.7$ yr
50. 2.4 in.

11 Sequences and Series

SECTION 11.1 (pp. 775–777)

1. 1, 2, 3, 4 (answers may vary)   **2.** 1, 3, 5, 7, … (answers may vary)   **3.** function; natural numbers   **4.** sequence
5. 6   **6.** scatterplot   **7.** $f(2)$ **8.** a_4 **9.** 1, 4, 9, 16
11. $\frac{1}{6}, \frac{1}{7}, \frac{1}{8}, \frac{1}{9}$ **13.** $\frac{5}{2}, \frac{5}{4}, \frac{5}{8}, \frac{5}{16}$ **15.** 9, 9, 9, 9   **17.** 1, 8, 27
19. 1, $\frac{8}{5}$, 2   **21.** 2, 9, 20   **23.** $-2, -2, -2$
25. $b + c; 2b + c$ **27.** 7   **29.** 3, 4, 5, 3, 1
31. 6, 5, 4, 3, 2, 1
33.

n	1	2	3	4	5	6	7
a_n	2	3	4	5	6	7	8

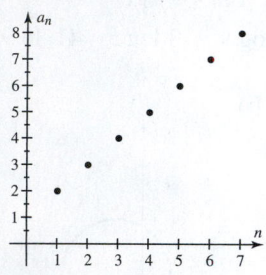

35.

n	1	2	3	4	5	6	7
a_n	0	2	6	12	20	30	42

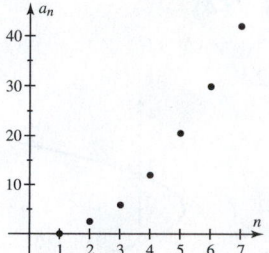

37.

n	1	2	3	4	5	6	7
a_n	2	4	8	16	32	64	128

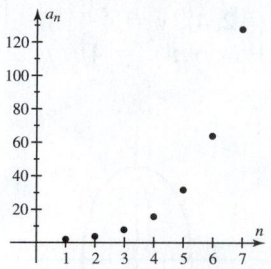

39. $a_n = 30n$ for $n = 1, 2, 3, \ldots, 7$

n	1	2	3	4	5	6	7
a_n	30	60	90	120	150	180	210

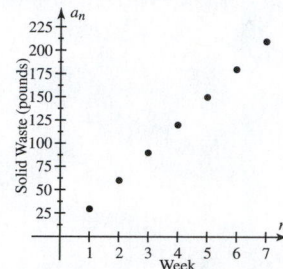

41. (a) 1, 4, 9, 16   **(b)** 4, 8, 12, 16
43. (a) $20,000; $16,000   **(b)** $a_n = 25,000(0.8)^n$
(c)

n	1	2	3	4	5	6	7
a_n	20,000	16,000	12,800	10,240	8192	6553.6	5242.9

45. (a) $a_n = 2048(0.5)^{n-1}$, for $n = 1, 2, 3, \ldots, 7$
(b)

n	1	2	3	4	5	6	7
a_n	2048	1024	512	256	128	64	32

(c)

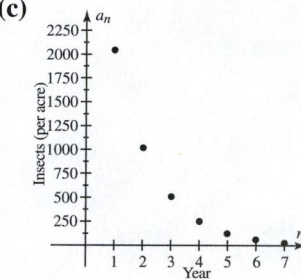

SECTION 11.2 (pp. 783–786)

1. linear   **2.** exponential   **3.** $a_n = 3n + 1$; 3 (answers may vary)   **4.** $a_n = 5(2)^{n-1}$; 2 (answers may vary)
5. add; previous   **6.** multiply; common ratio   **7.** 19; 4
8. 32; -2 **9.** $a_n = a_1(r)^{n-1}$ **10.** $a_n = a_1 + (n - 1)d$
11. Yes; 10   **13.** Yes; -1 **15.** No   **17.** Yes; 3

19. Yes; -3 **21.** No   **23.** Yes; 1   **25.** No   **27.** Yes; 2
29. $a_n = -2n + 9$ **31.** $a_n = 4n - 6$
33. $a_n = -2n + 32$ **35.** 59   **37.** 21   **39.** Yes; 3
41. Yes; 0.8   **43.** No   **45.** Yes; 2   **47.** No   **49.** Yes; 4
51. Yes; 2   **53.** No   **55.** $a_n = 1.5(4)^{n-1}$
57. $a_n = -3(-2)^{n-1}$ **59.** $a_n = 1(4)^{n-1}$ **61.** 4374
63. 243   **65.** (a) 3000, 6000, 9000, 12,000, 15,000; arithmetic
(b) $a_n = 3000n$ **(c)** 60,000; when there are 20 people, the
ventilation rate should be 60,000 cubic feet per hour.
(d)

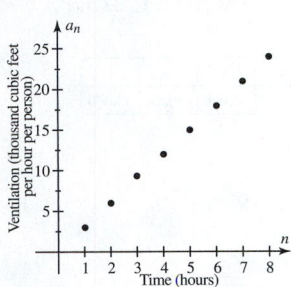

Yes
67. (a) $a_n = 3(0.8)^{n-1}$
(b) No; geometric

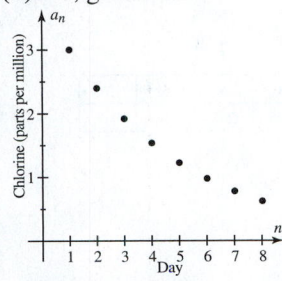

69. (a) $a_n = 5(0.85)^{n-1}$ **(b)** Geometric; the common
ratio is 0.85.   **(c)** About 1.6; on the 8th bounce the ball
reaches a maximum height of about 1.6 ft.
71. (a) Arithmetic; the common difference is 2.
(b) $a_n = 40 + 2(n - 1)$ or $a_n = 38 + 2n$ **(c)** 78

CHECKING BASIC CONCEPTS 11.1 & 11.2 (p. 786)

1. $\frac{1}{5}, \frac{1}{3}, \frac{3}{7}, \frac{1}{2}$

2.

n	1	2	3	4	5
a_n	2	3	4	5	6

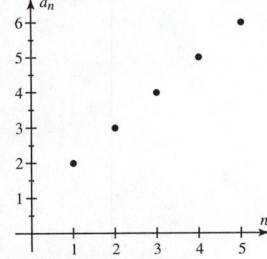

3. (a) Arithmetic; $a_n = 3n - 5$ **(b)** Geometric;
$a_n = 3(-2)^{n-1}$ **4.** $a_n = 2n + 3$ **5.** $a_n = 5(2)^{n-1}$

SECTION 11.3 (pp. 793–795)

1. series   **2.** 10   **3.** arithmetic   **4.** geometric
5. $n\left(\frac{a_1 + a_n}{2}\right)$ or $\frac{n}{2}(2a_1 + (n - 1)d)$ **6.** $a_1\left(\frac{1 - r^n}{1 - r}\right)$

7. sum   **8.** $a_1 + a_2 + a_3 + a_4$ **9.** arithmetic
10. geometric   **11.** 48   **13.** 820   **15.** -5 **17.** 3279
19. -85 **21.** 182   **23.** \$91,523.93   **25.** \$62,278.01
27. $2 + 4 + 6 + 8$; 20
29. $4 + 4 + 4 + 4 + 4 + 4 + 4 + 4$; 32
31. $1 + 4 + 9 + 16 + 25 + 36 + 49$; 140
33. $12 + 20$; 32   **35.** $\sum_{k=1}^{6} k^4$ **37.** $\sum_{k=1}^{5} \frac{1}{k^2}$
39. $\sum_{k=1}^{n} k = n\left(\frac{a_1 + a_n}{2}\right) = n\left(\frac{1 + n}{2}\right) = \frac{n(n + 1)}{2}$
41. (a) $8518 + 9921 + 10,706 + 14,035 + 14,307 + 12,249$
(b) 69,736   **43.** (a) $\sum_{k=1}^{n} 0.8(0.2)^{k-1}$ **(b)** 2
45. (a) $1, \frac{1}{2}, \frac{1}{4}, \frac{1}{8}, \frac{1}{16}$ **(b)** $\frac{1023}{512}$ **47.** 90 logs   **49.** \$1,080,000
51. About 3.16 feet   **53.** Answers may vary.

SECTION 11.4 (p. 801)

1. 5   **2.** $n + 1$ **3.** 4
4.
$$
\begin{array}{ccccccccc}
& & & & 1 & & & & \\
& & & 1 & & 1 & & & \\
& & 1 & & 2 & & 1 & & \\
& 1 & & 3 & & 3 & & 1 & \\
1 & & 4 & & 6 & & 4 & & 1 \\
\end{array}
$$
5. 24   **6.** $6! = 720$ **7.** $\frac{n!}{(n - r)!\, r!}$ **8.** $a^2 + 2ab + b^2$
9. n **10.** 1   **11.** $x^3 + 3x^2y + 3xy^2 + y^3$
13. $16x^4 + 32x^3 + 24x^2 + 8x + 1$
15. $a^5 - 5a^4b + 10a^3b^2 - 10a^2b^3 + 5ab^4 - b^5$
17. $x^6 + 3x^4 + 3x^2 + 1$ **19.** 6   **21.** 4   **23.** 2
25. 10   **27.** 5   **29.** 6   **31.** 1   **33.** 792   **35.** 126
37. 75,582   **39.** $m^3 + 3m^2n + 3mn^2 + n^3$
41. $x^4 - 4x^3y + 6x^2y^2 - 4xy^3 + y^4$
43. $8a^3 + 12a^2 + 6a + 1$
45. $x^5 + 10x^4 + 40x^3 + 80x^2 + 80x + 32$
47. $81 + 216m + 216m^2 + 96m^3 + 16m^4$
49. $8x^3 - 12x^2y + 6xy^2 - y^3$ **51.** a^8 **53.** $35x^4y^3$
55. $512m^9$

CHECKING BASIC CONCEPTS 11.3 & 11.4 (p. 802)

1. (a) Geometric   **(b)** Arithmetic   **2.** 312
3. -341 **4.** $x^4 - 4x^3y + 6x^2y^2 - 4xy^3 + y^4$
5. $x^3 + 6x^2 + 12x + 8$

CHAPTER 11 REVIEW (pp. 803–805)

1. 1, 8, 27, 64   **2.** 3, 1, -1, -3 **3.** $1, \frac{4}{5}, \frac{3}{5}, \frac{8}{17}$
4. $-2, 4, -8, 16$ **5.** $-2, 0, 4, 2$ **6.** 5, 3, 2, 1
7.

n	1	2	3	4	5	6	7
a_n	2	4	6	8	10	12	14

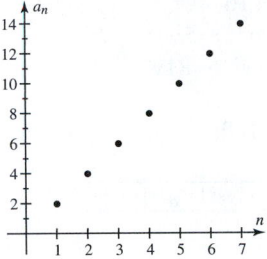

8.

n	1	2	3	4	5	6	7
a_n	−3	0	5	12	21	32	45

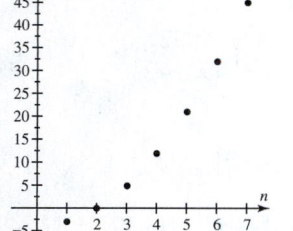

9.

n	1	2	3	4	5	6	7
a_n	2	1	0.5	0.25	0.125	0.0625	0.0313

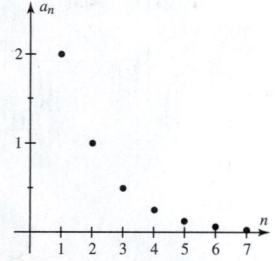

10.

n	1	2	3	4	5	6	7
a_n	1	1.4142	1.7321	2	2.2361	2.4495	2.6458

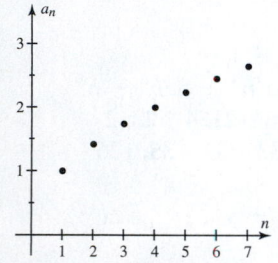

11. Yes; 5 **12.** No **13.** No **14.** Yes; $-\frac{1}{3}$ **15.** Yes; −3

16. No **17.** Yes; −1 **18.** No **19.** $a_n = 4n - 7$

20. $a_n = -5n + 7$ **21.** Yes; 4 **22.** No **23.** No

24. Yes; 0.7 **25.** No **26.** Yes; $-\frac{1}{3}$ **27.** No **28.** Yes; 2

29. $a_n = 5(0.9)^{n-1}$ **30.** $a_n = 2(4)^{n-1}$ **31.** 216 **32.** 7.5

33. 3277 **34.** $\frac{511}{256}$ **35.** $3 + 5 + 7 + 9 + 11$

36. $\frac{1}{2} + \frac{1}{3} + \frac{1}{4} + \frac{1}{5}$ **37.** $1 + 8 + 27 + 64$

38. $-1 + (-2) + (-3) + (-4) + (-5) + (-6)$

39. $\sum_{k=1}^{20} k$ **40.** $\sum_{k=1}^{20} \frac{1}{k}$ **41.** $\sum_{k=1}^{9} \frac{k}{k+1}$ **42.** $\sum_{k=1}^{7} k^2$

43. $x^3 + 12x^2 + 48x + 64$

44. $16x^4 + 32x^3 + 24x^2 + 8x + 1$

45. $x^5 - 5x^4y + 10x^3y^2 - 10x^2y^3 + 5xy^4 - y^5$

46. $a^6 - 6a^5 + 15a^4 - 20a^3 + 15a^2 - 6a + 1$

47. 6 **48.** 10 **49.** 20 **50.** 4

51. $m^4 + 8m^3 + 24m^2 + 32m + 16$

52. $a^5 + 5a^4b + 10a^3b^2 + 10a^2b^3 + 5ab^4 + b^5$

53. $x^4 - 12x^3y + 54x^2y^2 - 108xy^3 + 81y^4$

54. $27x^3 - 54x^2 + 36x - 8$

55. $a_n = 45,000(1.1)^{n-1}$ for $n = 1, 2, 3, \ldots, 7$; geometric

n	1	2	3	4	5	6	7
a_n	45,000	49,500	54,450	59,895	65,885	72,473	79,720

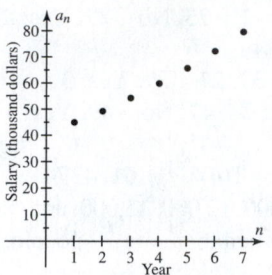

56. $a_n = 45,000 + 5000(n - 1)$ for $n = 1, 2, 3, \ldots, 7$; arithmetic

n	1	2	3	4	5	6	7
a_n	45,000	50,000	55,000	60,000	65,000	70,000	75,000

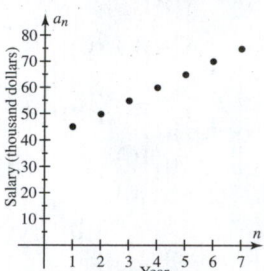

57. $a_n = 49n$ for $n = 1, 2, 3, \ldots, 7$

n	1	2	3	4	5	6	7
a_n	49	98	147	196	245	294	343

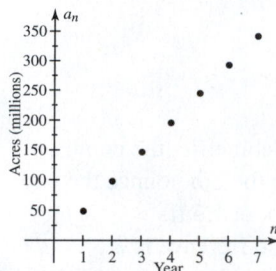

58. **(a)** $a_n = 1087(1.025)^{n-1}$ **(b)** Geometric; the common ratio is 1.025. **(c)** About 1200; the average mortgage payment in 2000 was about $1200 per month.

(d)

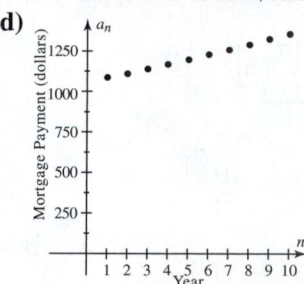

CHAPTER 11 TEST (p. 805)

1. $\frac{1}{2}, \frac{4}{3}, \frac{9}{4}, \frac{16}{5}$ **2.** −3, 2, 1, −2, 3

3.

n	1	2	3	4	5	6	7
a_n	0	2	6	12	20	30	42

4. $16x^4 - 32x^3 + 24x^2 - 8x + 1$

5. Arithmetic; −3 **6.** Geometric; −2

7. $a_n = 2 - 3(n - 1)$ or $a_n = 5 - 3n$

8. $a_n = 2(1.5)^{n-1}$ **9.** Yes; 2.5 **10.** No **11.** 99

12. $\frac{463}{729}$ **13.** $6 + 9 + 12 + 15 + 18 + 21$

14. $\sum_{k=1}^{60} k^3$ **15.** 35 **16.** 10 **17.** 9180

18. $a_n = 180{,}000(0.96)^{n-1}$ for $n = 1, 2, 3, 4, 5$; geometric

n	1	2	3	4	5
a_n	180,000	172,800	165,888	159,252	152,882

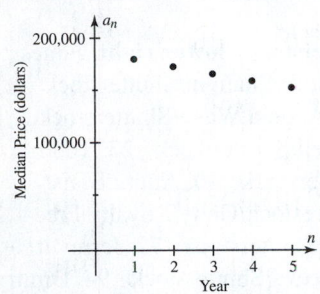

19. (a) $a_n = 2000(2)^{n-1}$ **(b)** Geometric; the common ratio is 2.   **(c)** 64,000; after 25 days there are 64,000 caterpillars.

(d)

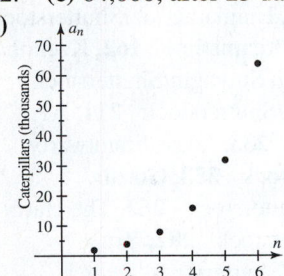

CHAPTERS 1–11 CUMULATIVE REVIEW (pp. 807–809)

1. Distributive   **2.** $D = \{-6, -2, 0, 2\}$; $R = \{0, 1, 3, 5\}$
3. $\frac{y^9}{27x^5}$ **4.** $\frac{16a^8}{b^4}$ **5.** z^{10} **6.** $\frac{2y^3}{x^6}$ **7.** $D = \{x \mid x \neq 8\}$
8. $f(x) = -2x + 1$ **9.** $y = 3$ **10.** $-3; 5$
11. $y = \frac{3}{2}x + \frac{5}{2}$ **12.** $y = 2x - 8$ **13.** -26
14. $\left(\frac{20}{11}, \infty\right)$ **15.** $(-\infty, -1] \cup [11, \infty)$ **16.** $-15, 3$
17. $\frac{1}{8}$ **18.** $[-12, 9)$ **19.** $(3, -2)$
20.

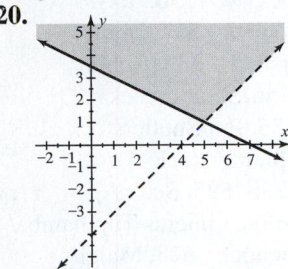

21. $(5, 2)$ **22.** $(3, 3, -1)$ **23.** $R = 22$ **24.** 1
25. $8x^7 - 6x^6 + 10x^3$ **26.** $6z^2 - 13z - 28$
27. $(2x - 3y)(2x + 3y)$ **28.** $(a^2 + 4)(2a - 1)$
29. $-\frac{3}{4}, 1$ **30.** $0, 4, 6$ **31.** $\frac{x - 2}{x + 1}$ **32.** $\frac{x + 4}{x - 2}$ **33.** 12
34. $\frac{1}{2}, 3$ **35.** $W = \frac{3C - 5R}{2}$ **36.** $\frac{1 + 2x}{1 - 4x}$
37. $(xy - 2)\sqrt[3]{xy}$ **38.** 14   **39.** $-2, 8$ **40.** $\frac{3}{2}$
41. $14 + 2i$ **42.** $-2 - 2i$ **43.** $-\frac{1}{3}$
44. $y = 2(x + 2)^2 + 9$; $(-2, 9)$ **45.** $2 \pm 3i$
46. $-4, 8$ **47. (a)** $-3, 1$ **(b)** $a > 0$ **(c)** Positive
48. $(-3, 1)$ **49. (a)** 65   **(b)** $(g \circ f)(x) = 3x^2 + 1$
50. $f^{-1}(x) = \frac{2x - 1}{3}$ **51.** $3 \ln x + \frac{1}{2} \ln y$ **52.** $\log \frac{x}{4y}$
53. 56.23 **54.** 0.58

55.

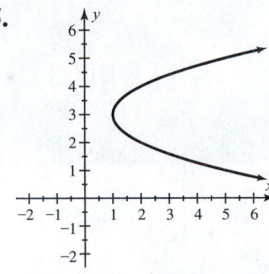

$(1, 3)$; $y = 3$
56. $(3, -1)$; 2
57.   **58.**

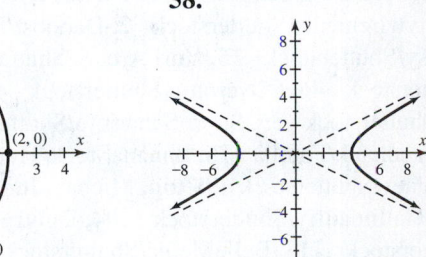

59. $\frac{y^2}{4} - \frac{x^2}{16} = 1$ **60.** $\frac{x^2}{16} + \frac{y^2}{4} = 1$ **61.** $(1, 2), (-1, 2)$
62.

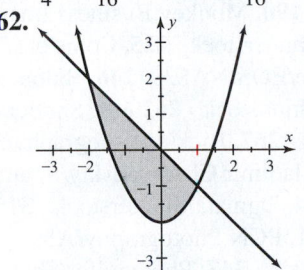

63. Arithmetic; -2 **64.** Geometric; 0.2
65. Geometric; 4   **66.** Arithmetic; 6   **67.** $a_n = 3n - 1$
68. $a_n = 4(3)^{n-1}$ **69.** 171   **70.** 683
71. $16x^4 + 96x^3 + 216x^2 + 216x + 81$
72. $8a^3 - 60a^2b + 150ab^2 - 125b^3$
73. $\sqrt{\frac{14}{\pi}} \approx 2.11$ inches
74.

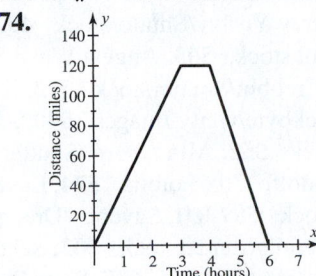

75. (a) $f(x) = 0.4x + 85$ **(b)** 188 pounds   **76.** Airplane: 380 mph; wind: 20 mph   **77.** 9 feet by 12 feet   **78.** 36 minutes   **79. (a)** $xy = 96, 3x - y = 12$ **(b)** 8 and 12
80. 144

B   The Midpoint Formula

1. -1 **3.** $(-1, 0)$ **5.** $\left(-\frac{1}{2}, \frac{1}{2}\right)$ **7.** $(20, 0)$ **9.** $(-8, -1)$
11. $\left(-1, -\frac{1}{6}\right)$ **13.** $(0.2, 0.25)$ **15.** $(2005, 9)$ **17.** $(2a, 2b)$
19. (a) $f(x) = -2x + 5$; 1   **(b)** 1   **(c)** equal
21. (a) $f(x) = \frac{2}{5}x + \frac{1}{5}$; 1   **(b)** 1   **(c)** equal
23. 79.8 years   **25.** 256.5 million   **27.** \$45,250

Photo Credits

Glossary

absolute value A real number a, written $|a|$, is equal to its distance from the origin on the number line.

absolute value equation An equation that contains an absolute value.

absolute value function The function defined by $f(x) = |x|$.

absolute value inequality An inequality that contains an absolute value.

addends In an addition problem, the two numbers that are added.

addition property of equality If a, b, and c are real numbers, then $a = b$ is equivalent to $a + c = b + c$.

additive identity The number 0.

additive inverse (opposite) The additive inverse, or opposite, of a real number a is $-a$.

adjacency matrix A matrix used to represent a map showing distances between cities.

algebraic expression An expression consisting of numbers, variables, arithmetic symbols, and grouping symbols, such as parentheses, brackets, and square roots.

annuity A sum of money from which regular payments are made.

approximately equal The symbol $\approx$ indicates that two quantities are nearly equal.

arithmetic sequence A linear function given by $a_n = dn + c$ whose domain is the set of natural numbers.

arithmetic series The sum of the terms of an arithmetic sequence.

associative property for addition For any real numbers a, b, and c, $(a + b) + c = a + (b + c)$.

associative property for multiplication For any real numbers a, b, and c, $(a \cdot b) \cdot c = a \cdot (b \cdot c)$.

asymptotes of a hyperbola The two lines determined by the diagonals of the hyperbola's fundamental rectangle.

augmented matrix A matrix used to represent a system of linear equations; a vertical line is positioned in the matrix where the equals signs occur in the system of equations.

average The result of adding up the numbers of a set and then dividing the sum by the number of elements in the set.

axis of symmetry for inverse functions The line $y = x$; f^{-1} can be sketched from the graph of $y = f(x)$ by reflecting the graph of f across the line $y = x$.

axis of symmetry of a parabola The line passing through the vertex of the parabola that divides the parabola into two symmetric parts.

base The value of a in the expression a^x.

base e If the base of an exponential expression is e (approximately 2.72), then we say this expression has base e.

basic principle of fractions When simplifying fractions, the principle which states $\frac{a \cdot c}{b \cdot c} = \frac{a}{b}$.

binomial A polynomial with two terms.

binomial coefficient The expression $_nC_r$ that can be used to calculate the numbers in Pascal's triangle.

binomial theorem A theorem that provides a formula to expand expressions of the form $(a + b)^n$.

branches A hyperbola has two branches, a left branch and a right branch, or an upper branch and a lower branch.

break-even point The point at which the cost of producing x items equals the revenue made from selling x items.

byte A unit of computer memory, capable of storing one letter of the alphabet.

Cartesian coordinate system (xy-plane) The xy-plane used to plot points and visualize a relation; points are identified by ordered pairs.

center of a circle The point that is a fixed distance from all the points on a circle.

center of an ellipse The fixed point located exactly halfway between the two vertices.

change of base formula A formula used to evaluate a logarithm of one base by using a logarithm of a different base and given by $\log_a x = \frac{\log x}{\log a}$ or $\log_a x = \frac{\ln x}{\ln a}$.

change in x The value given by $x_2 - x_1$ if a line passes through two points (x_1, y_1) and (x_2, y_2).

change in y The value given by $y_2 - y_1$ if a line passes through two points (x_1, y_1) and (x_2, y_2).

circle The set of points in a plane that are a constant distance from a fixed point called the center.

clearing fractions The process of multiplying both sides of an equation by a common denominator to eliminate the fractions.

coefficient The numeric constant in a monomial.

common denominator A number or expression that each denominator can divide into *evenly* (without a remainder).

common difference The value of d in an arithmetic sequence, $a_n = dn + c$.

common logarithm of a positive number x Denoted $\log x$, it may be calculated as follows: if x is expressed as $x = 10^k$, then $\log x = k$, where k is a real number. That is, $\log 10^k = k$.

common logarithmic function The function given by $f(x) = \log x$.

common ratio The value of r in a geometric sequence, $a_n = a_1(r)^{n-1}$.

commutative property for addition For any real numbers a and b, $a + b = b + a$.

commutative property for multiplication For any real numbers a and b, $a \cdot b = b \cdot a$.

completing the square An important technique in mathematics that involves adding a constant to a binomial so that a perfect square trinomial results.

complex conjugate The complex conjugate of $a + bi$ is $a - bi$.

complex fraction A rational expression that contains fractions in its numerator, denominator, or both.

complex number A complex number can be written in standard form as $a + bi$, where a and b are real numbers and i is the imaginary unit.

composite function A function in which a quantity depends on a variable that depends on another variable; denoted $f(g(x))$.

composition Replacing a variable with an algebraic expression in function notation is called composition of functions.

compound inequality Two inequalities joined by the word *and* or the word *or*.

compound interest Interest earned on both the principal and the accrued interest.

conditional equation An equation that is true for some, but not all, values of any of its variables.

conic section The curve formed by the intersection of a plane and a cone.

conjugate The conjugate of $a + b$ is $a - b$.

consistent system A system of linear equations with at least one solution.

constant function A linear function of the form $f(x) = b$, where b is a constant.

constant of proportionality (constant of variation) In the equation $y = kx$, the number k.

constraint In linear programming, an inequality that limits the objective function.

continuous growth Growth in a quantity that is directly proportional to the amount present.

contradiction An equation that is always false, regardless of the values of any variables.

Cramer's rule A method that uses determinants to solve linear systems of equations.

cube root The number b is a cube root of a number a if $b^3 = a$.

cube root function The function defined by $g(x) = \sqrt[3]{x}$.

cubed A number or variable that has been raised to the third power has been cubed.

cubic function A polynomial function of degree 3.

cubic polynomial A polynomial of degree 3 that can be written as $ax^3 + bx^2 + cx + d$, where $a \neq 0$.

decay factor The value of a in an exponential function, $f(x) = Ca^x$, where $0 < a < 1$.

decimal form See standard (decimal) form.

degree of a monomial The sum of the exponents of the variables.

degree of a polynomial The degree of the monomial with highest degree.

dependent equations Equations in a linear system that have the same solution set.

dependent variable The variable that represents the output of a function.

determinant A real number associated with a square matrix.

diagrammatic representation A function represented by a diagram.

difference The answer to a subtraction problem.

difference of two cubes Expression in the form $a^3 - b^3$, which can be factored as $(a - b)(a^2 + ab + b^2)$.

difference of two squares Expression in the form $a^2 - b^2$, which can be factored as $(a - b)(a + b)$.

dimension of a matrix The size expressed in number of rows and columns. For example, if a matrix has m rows and n columns, its dimension is $m \times n$ (m by n).

directly proportional A quantity y is directly proportional to x if there is a nonzero number k such that $y = kx$.

discriminant The expression $b^2 - 4ac$ in the quadratic formula.

distance formula The distance between the points (x_1, y_1) and (x_2, y_2) in the xy-plane is $d = \sqrt{(x_2 - x_1)^2 + (y_2 - y_1)^2}$.

distributive properties For any real numbers a, b, and c, $a(b + c) = ab + ac$ and $a(b - c) = ab - ac$.

dividend In a division problem, the number being divided.

divisor In a division problem, the number being divided *into* another.

domain The set of all x-values of the ordered pairs in a relation.

element Each number in a matrix.

elimination (or addition) method A symbolic method used to solve a system of equations that is based on the property that "equals added to equals are equal."

ellipse The set of points in a plane, the sum of whose distances from two fixed points is constant.

equation A statement that two algebraic expressions are equal.

equation of a line An equation whose graph is a line, such as point–slope form and slope–intercept form.

equivalent equations Two equations that have the same solution set.

even root The nth root, $\sqrt[n]{a}$, where n is even.

expansion of a determinant by minors A method of finding a 3×3 determinant by using determinants of 2×2 matrices.

exponent The value of a in the expression b^a.

exponential decay When $0 < a < 1$, the graph of $f(x) = Ca^x$ models exponential decay.

exponential equation An equation that has a variable as an exponent.

exponential expression An expression that has an exponent.

exponential function with base a and coefficient C A function represented by $f(x) = Ca^x$, where $a > 0$, $C > 0$, and $a \neq 1$.

exponential growth When $a > 1$, the graph of $f(x) = Ca^x$ models exponential growth.

expression A term or sums and differences of terms; does not include an equals sign.

extraneous solution A solution that does not satisfy the given equation.

factorial notation $n! = 1 \cdot 2 \cdot 3 \cdot \cdots \cdot n$ for any positive integer n.

factoring by grouping A technique that uses the distributive property by grouping four terms of a polynomial in such a way that the polynomial can be factored even though its greatest common factor is 1.

factors In a multiplication problem, the two numbers multiplied.

feasible solutions In linear programming, the set of solutions that satisfies the constraints.

finite sequence A function with domain $D = \{1, 2, 3, \ldots, n\}$ for some fixed natural number n.

finite series A series that contains a finite number of terms, and that can be expressed in the form $a_1 + a_2 + a_3 + \cdots + a_n$ for some n.

focus (plural foci) A fixed point used to determine the points that form a parabola, an ellipse, or a hyperbola.

FOIL A method for multiplying two binomials $(A + B)$ and $(C + D)$. Multiply **F**irst terms AC, **O**uter terms AD, **I**nner terms BC, and **L**ast terms BD; then combine like terms.

formula An equation that can be used to calculate one quantity by using a known value of another quantity.

function A relation where each element in the domain corresponds to exactly one element in the range.

function notation $y = f(x)$ is read "y equals f of x". This means that function f with input x produces output y.

fundamental rectangle of a hyperbola The rectangle whose four vertices are determined by either $(\pm a, \pm b)$ or $(\pm b, \pm a)$, where $\frac{x^2}{a^2} - \frac{y^2}{b^2} = 1$ or $\frac{y^2}{a^2} - \frac{x^2}{b^2} = 1$.

Gauss–Jordan elimination A numerical method in which matrix row transformations are used to solve a linear system.

general term (*nth* term) of a sequence a_n where n is a natural number in the domain of a sequence $a_n = f(n)$.

geometric sequence A function given by $a_n = a_1(r)^{n-1}$ whose domain is the set of natural numbers.

graphical factoring Using a graph to factor a polynomial by identifying the zeros of the polynomial.

graphical representation A graph of a function.

graphical solution A solution to an equation obtained by graphing.

greater than If a real number a is located to the right of a real number b on the number line, a is greater than b, or $a > b$.

greatest common factor (GCF) The term with the highest degree and largest coefficient that is a factor of all terms in the polynomial.

growth factor The value of a in the exponential function, $f(x) = Ca^x$, where $a > 1$.

half-life The time it takes for a radioactive sample to decay to half its original amount.

horizontal line A line with zero slope.

horizontal line test A function is one-to-one if every horizontal line intersects its graph at most once.

hyperbola The set of points in a plane, the difference of whose distances from two fixed points is constant.

identity An equation that is always true regardless of the values of any variables.

identity property of 1 If any number a is multiplied by 1, the result is a, that is, $a \cdot 1 = 1 \cdot a = a$.

identity property of 0 If 0 is added to any real number a, the result is a, that is, $a + 0 = 0 + a = a$.

imaginary number A complex number $a + bi$ with $b \neq 0$.

imaginary part The value of b in the complex number $a + bi$.

imaginary unit A number denoted i whose properties are $i = \sqrt{-1}$ and $i^2 = -1$.

inconsistent system A system of linear equations that has no solution.

independent equations Equations in a linear system that have different graphs.

independent variable The variable that represents the input of a function.

index The value of n in the expression $\sqrt[n]{a}$.

index of summation The variable k in the expression $\sum_{k=1}^{n}$.

inequality When the equals sign in an equation is replaced with any one of the symbols $<, \leq, >$, or $\geq$, an inequality results.

infinite sequence A function whose domain is the set of natural numbers.

infinity Values that increase without bound.

input An element of the domain of a function.

integers A set I of numbers given by $I = \{\ldots, -3, -2, -1, 0, 1, 2, 3, \ldots\}$.

intersection Elements that belong to both sets A and B, denoted $A \cap B$.

interval notation A notation for number line graphs that eliminates the need to draw the entire line. A *closed interval* is expressed by the endpoints in brackets, such as $[-2, 7]$; an *open*

interval is expressed in parentheses, such as $(-2, 7)$, and indicates that the endpoints are not included in the solution. A *half-open interval* has a bracket and a parenthesis, such as $[-2, 7)$.

inverse function If f is a one-to-one function, then the inverse of f is the set of all ordered pairs in the form (y, x), where (x, y) belongs to f.

inversely proportional A quantity y is inversely proportional to x if there is a nonzero number k such that $y = k/x$.

irrational numbers Real numbers that cannot be expressed as fractions, such as π or $\sqrt{2}$.

joint variation A quantity z varies jointly with x and y if there is a nonzero number k such that $z = kxy$.

leading coefficient In a polynomial of one variable, the coefficient of the monomial with highest degree.

least common denominator (LCD) The common denominator with the fewest factors.

least common multiple (LCM) The smallest number that two or more numbers will divide into evenly.

less than If a real number a is located to the left of a real number b on the number line, we say that a is less than b, or $a < b$.

like radicals Radicals that have the same index and the same radicand.

like terms Two terms that contain the same variables raised to the same powers.

line graph The graph resulting when consecutive data points in a scatterplot are connected with straight line segments.

linear equation An equation that can be written in the form $ax + b = 0$, where $a \neq 0$.

linear function A function f represented by $f(x) = ax + b$, where a and b are constants.

linear inequality in one variable An inequality that can be written in the form $ax + b > 0$, where $a \neq 0$. (The symbol $>$ may be replaced with $\leq, <$, or $\geq$.)

linear inequality in two variables When the equals sign in a linear equation of two variables is replaced with $<, \leq, >$, or $\geq$, a linear inequality in two variables results.

linear polynomial A polynomial of degree 1 that can be written as $ax + b$, where $a \neq 0$.

linear programming problem A problem consisting of an objective function and a system of linear inequalities called constraints.

linear system in three variables System of three equations in which each equation can be written in the form $ax + by + cz = d$; an ordered triple (x, y, z) is a solution to the system of equations if the values for x, y, and z make *all three* equations true.

logarithm with base a of a positive number x Denoted $\log_a x$, it may be calculated as follows: if x can be expressed as $x = a^k$, then $\log_a x = k$, where $a > 0, a \neq 1$, and k is a real number. That is, $\log_a a^k = k$.

logarithmic function with base a The function represented by $f(x) = \log_a x$.

lower limit In summation notation, the number representing the subscript of the first term of the series.

lowest terms A fraction in lowest terms is one in which the numerator and the denominator have no common factors.

main diagonal The elements $a_{11}, a_{22}, a_{33}, \ldots, a_{nn}$ in a matrix with n rows.

major axis The longer axis of an ellipse, which connects the vertices.

matrix A rectangular array of numbers.

matrix row transformations Operations performed on rows of an augmented matrix that result in an equivalent system of linear equations.

minor axis The shorter axis of an ellipse.

minors The 2×2 matrices in the expansion of a 3×3 determinant.

model A mathematical representation of a real-world application that can help explain phenomena or make predictions.

monomial A term whose variables have only nonnegative integer exponents.

multiplication property of equality If a, b, and c are real numbers with $c \neq 0$, then $a = b$ is equivalent to $ac = bc$.

multiplicative identity The number 1.

multiplicative inverse (reciprocal) The multiplicative inverse of a nonzero number a is $1/a$.

name of the function In the function given by $f(x)$, we call the function f.

natural exponential function The function represented by $f(x) = e^x$, where $e \approx 2.71828$.

natural logarithm The base-e logarithm, denoted either $\log_e x$ or $\ln x$.

natural numbers The set of numbers given by $N = \{1, 2, 3, 4, 5, 6, \dots\}$.

negative infinity Values that decrease without bound.

negative number A number a such that $a < 0$.

negative slope On a graph, the slope of a line that falls from left to right.

negative square root The negative square root is denoted $-\sqrt{a}$.

nonlinear data If data points do not lie on a (straight) line, the data are nonlinear.

nonlinear system of equations Two or more equations, at least one of which is nonlinear.

nonlinear system of inequalities Two or more inequalities, at least one of which is nonlinear.

not equal to The expression $b \neq 0$ is read "b not equal to 0" and means that b must be either greater than or less than 0.

nth root The number b is an nth root of a if $b^n = a$, where n is a positive integer, and is denoted $\sqrt[n]{a} = b$.

nth term (general term) of a sequence See general term (nth term) of a sequence.

numerical factoring Using a table of values to factor a polynomial by identifying the zeros of the polynomial.

numerical representation A table of values for a function.

numerical solution A solution often obtained by using a table of values.

objective function The given function in a linear programming problem.

odd root The nth root, $\sqrt[n]{a}$, where n is odd.

one-to-one function Two different inputs always result in two different outputs; that is, $x \neq y$ implies that $f(x) \neq f(y)$.

opposite (additive inverse) The opposite, or additive inverse, of a real number a is $-a$.

opposite of a polynomial The polynomial obtained by negating each term in a given polynomial.

optimal value In linear programming, a value that makes the objective function either maximum or minimum.

order of operations The sequence in which calculations such as addition, subtraction, multiplication, etc., must be performed on an expression.

ordered pair A pair of numbers written in parentheses (x, y), in which the order of the numbers is important.

ordered triple Can be expressed as (x, y, z), where x, y, and z are numbers.

origin On the number line, the point associated with the real number 0; in the xy-plane, the point $(0, 0)$.

output An element of the range of a function.

parabola The $\cup$-shaped graph of a quadratic function.

parallel lines Two or more lines in the same plane that never intersect.

Pascal's triangle A triangle made up of numbers in which there are 1s along the sides and each element inside the triangle is the sum of the two numbers above it.

percent change When an amount A changes to a new amount B, then the percent change is calculated by $\frac{B - A}{A} \times 100$.

percent form To change to percent form take a number r in decimal form and multiply by 100.

percentage An amount in each hundred.

perfect cube An integer with an integer cube root.

perfect nth power The value of a if there exists an integer b such that $b^n = a$.

perfect square An integer with an integer square root.

perfect square trinomial A trinomial that can be factored as the square of a binomial, for example, $a^2 + 2ab + b^2 = (a + b)^2$.

perpendicular lines Two lines in a plane that intersect to form a right (90°) angle.

pixels The tiny units that comprise screens for computer terminals or graphing calculators.

point–slope form The line with slope m passing through the point (x_1, y_1) given by the equation $y = m(x - x_1) + y_1$ or, equivalently, $y - y_1 = m(x - x_1)$.

polynomial A monomial or a sum of monomials.

polynomials of one variable Polynomials that contain one variable.

positive number A number a such that $a > 0$.

positive slope On a graph, the slope of a line that rises from left to right.

power function A function that can be represented by $f(x) = x^p$, where p is a rational number.

principal nth root of a For an even positive integer n, if $b^n = a$ and b is positive, then b is a principal nth root of a and is denoted $\sqrt{a}, \sqrt[4]{a}, \sqrt[6]{a}, \dots, \sqrt[n]{a}$.

principal square root The positive square root, denoted $\sqrt{a}$.

product The answer to a multiplication problem.

properties of inequalities Rules for operations that can be performed on inequalities.

proportion A statement that two ratios are equal.

pure imaginary number A complex number $a + bi$ with $a = 0$ and $b \neq 0$ is called a pure imaginary number.

Pythagorean theorem If a right triangle has legs a and b with hypotenuse c, then $a^2 + b^2 = c^2$.

quadrants The four regions determined by a Cartesian coordinate system.

quadratic equation An equation that can be written as $ax^2 + bx + c = 0$, where a, b, and c are real numbers, with $a \neq 0$.

quadratic form An expression that is written in the form $ax^2 + bx + c$ after using substitution is in quadratic form.

quadratic formula The solutions to the quadratic equation, $ax^2 + bx + c = 0$, $a \neq 0$, are $x = (-b \pm \sqrt{b^2 - 4ac})/(2a)$.

quadratic function A function f represented by the equation $f(x) = ax^2 + bx + c$, where a, b, and c are real numbers with $a \neq 0$.

quadratic inequality If the equals sign in a quadratic equation is replaced with $>$, $\geq$, $<$, or $\leq$, a quadratic inequality results.

quadratic polynomial A polynomial of degree 2 that can be written as $ax^2 + bx + c$ with $a \neq 0$.

quotient The answer to a division problem.

radical expression An expression that contains a radical sign.

radical sign The symbol $\sqrt{}$ or $\sqrt[n]{}$ for some positive integer n.

radicand The expression under the radical sign.

radius The fixed distance between the center and any point on a circle.

range The set of all y-values of the ordered pairs in a relation.

rate of change The value of a for the linear function given by $f(x) = ax + b$; slope can be interpreted as a rate of change.

rational equation An equation that involves a rational expression.

rational expression A polynomial divided by a nonzero polynomial.

rational function A function defined by $f(x) = p(x)/q(x)$, where $p(x)$ and $q(x)$ are polynomials and the domain of f includes all x-values such that $q(x) \neq 0$.

rational number Any number that can be expressed as the ratio of two integers p/q, where $q \neq 0$; a fraction.

rationalizing the denominator The process of removing radicals from a denominator so that the denominator contains only rational numbers.

real numbers All rational and irrational numbers; any number that can be written using decimals.

real part The value of a in the complex number $a + bi$.

reciprocal (multiplicative inverse) The reciprocal of a nonzero number a is $1/a$.

reciprocal of a polynomial The reciprocal of a polynomial $p(x)$ is $\frac{1}{p(x)}$.

reciprocal of a rational expression The reciprocal of a rational expression $\frac{p(x)}{q(x)}$ is $\frac{q(x)}{p(x)}$.

reduced row–echelon form A matrix form for representing a system of linear equations in which there are 1s on the main diagonal with 0s above and below each 1.

reflection If the point (x, y) is on the graph of $y = f(x)$, then the point $(x, -y)$ is on the graph of its reflection across the x-axis.

relation A set of ordered pairs.

remainder If a polynomial $p(x)$ can be written as $p(x) = d(x)q(x) + r(x)$, then $r(x)$ is the remainder when $p(x)$ is divided by $d(x)$.

rise The vertical change between two points on a line, that is, the change in the y-values.

root function In the power function $f(x) = x^p$, if $p = 1/n$, where $n \geq 2$ is an integer, then f is also a root function, which is given by $f(x) = \sqrt[n]{x}$.

run The horizontal change between two points on a line, that is, the change in the x-values.

scatterplot A graph of distinct points plotted in the xy-plane.

scientific notation A real number a written as $b \times 10^n$, where $1 \leq |b| < 10$ and n is an integer.

set braces $\{\ \}$, used to enclose the elements of a set.

set-builder notation Notation to describe a set of numbers without having to list all of the elements. For example, $\{x \mid x > 5\}$ is read as "the set of all real numbers x such that x is greater than 5."

similar triangles Triangles that have equal corresponding angles and proportional corresponding sides.

slope The ratio of the change in y (rise) to the change in x (run) along a line. Slope m of a line equals $\frac{y_2 - y_1}{x_2 - x_1}$, where (x_1, y_1) and (x_2, y_2) are points on the line.

slope–intercept form The line with slope m and y-intercept b is given by $y = mx + b$.

solution A value for a variable that makes an equation a true statement.

solution set The set of all solutions to an equation.

solution to a system of equations An ordered pair (x, y), where the values for x and y satisfy both equations; could also be an ordered triple (x, y, z).

square matrix A matrix in which the number of rows and the number of columns are equal.

square root The number b is a square root of a number a if $b^2 = a$.

square root function The function given by $f(x) = \sqrt{x}$, where $x \geq 0$.

square root property If k is a nonnegative number, then the solutions to the equation $x^2 = k$ are $x = \pm\sqrt{k}$. If $k < 0$, then this equation has no real solutions.

squared A number or variable that has been raised to the second power has been squared.

standard equation of a circle The standard equation of a circle with center (h, k) and radius r is $(x - h)^2 + (y - k)^2 = r^2$.

standard (decimal) form A real number written with a decimal point or an assumed decimal point.

standard form of a complex number $a + bi$, where a and b are real numbers.

standard form of an equation for a line The form $ax + by = c$, where a, b, and c are constants with a and b not both 0.

standard form (for a system) A system of equations in two variables can be written in the form $ax + by = c$, $dx + ey = k$, where a, b, c, d, e, and k are constants.

standard viewing rectangle of a graphing calculator Xmin $= -10$, Xmax $= 10$, Xscl $= 1$, Ymin $= -10$, Ymax $= 10$, and Yscl $= 1$, denoted $[-10, 10, 1]$ by $[-10, 10, 1]$.

subscript The expression x_1 has a subscript of 1; it is read "x sub one" or "x one".

substitution method A symbolic method for solving a system of equations in which one equation is solved for one of the variables and then the result is substituted into the other equation.

sum The answer to an addition problem.

sum of two cubes Expression in the form $a^3 + b^3$, which can be factored as $(a + b)(a^2 - ab + b^2)$.

summation notation Notation in which the uppercase Greek letter sigma represents the sum, for example, $\sum_{k=1}^{n} k^2$.

symbolic factoring Factoring a polynomial by hand, using a technique such as grouping or reverse FOIL.

symbolic representation Representing a function with a formula; for example, $f(x) = x^2 - 2x$.

symbolic solution A solution to an equation obtained by using properties of equations, and the resulting solution set is exact.

synthetic division A shortcut that can be used to divide $x - k$, where k is a number, into a polynomial.

system of linear inequalities Two or more linear inequalities to be solved at the same time, the solution to which must satisfy both inequalities.

system of two linear equations in two variables A system of two equations in which each equation can be written in the form $ax + by = c$; an ordered pair (x, y) is a solution to the system of equations if the values for x and y make *both* equations true.

table of values An organized way to display the inputs and outputs of a function; a numerical representation.

term A number, a variable, or a product of numbers and variables raised to powers.

terms of a sequence $a_1, a_2, a_3, \ldots, a_n, \ldots$ where the first term is $a_1 = f(1)$, the second term is $a_2 = f(2)$, and so on.

test point (for a system of inequalities) A point (x, y) that can be substituted into an inequality to determine the solution set.

test value A real number chosen to determine the solution set to an inequality.

three-part inequality A compound inequality written in the form $a < x < b$.

translation The shifting of a graph upward, downward, to the right, or to the left in such a way that the shape of the graph stays the same.

transverse axis In a hyperbola, the line segment that connects the vertices.

trinomial A polynomial with three terms.

upper limit In summation notation, the number representing the subscript of the last term of the series.

variable A symbol, such as x, y, or t, used to represent any unknown number or quantity.

varies directly A quantity y varies directly with x if there is a nonzero number k such that $y = kx$.

varies inversely A quantity y varies inversely with x if there is a nonzero number k such that $y = k/x$.

varies jointly A quantity z varies jointly with x and y if there is a nonzero number k such that $z = kxy$.

verbal representation A description, in words, of what a function computes.

vertex The lowest point on the graph of a parabola that opens upward or the highest point on the graph of a parabola that opens downward.

vertex (in a linear programming problem) The points where lines (boundaries) of the constraints intersect on the boundary of the region of feasible solutions.

vertex form of a parabola The vertex form of a parabola with vertex (h, k) is $y = a(x - h)^2 + k$, where $a \neq 0$ is a constant.

vertical asymptote A vertical asymptote typically occurs in the graph of a rational function when the denominator of the rational expression is 0 but the numerator is not 0; it can be represented by a vertical line in the graph of a rational function.

vertical line A line having an equation that can be written as $x = k$.

vertical line test If every vertical line intersects a graph at most once, then the graph represents a function.

vertical shift A translation of a graph upward or downward.

vertices of an ellipse The endpoints of the major axis.

vertices of a hyperbola The endpoints of the transverse axis.

viewing rectangle (window) On a graphing calculator, the window that determines the x- and y-values shown in the graph.

whole numbers The set of numbers given by $W = \{0, 1, 2, 3, 4, 5, \ldots\}$.

x-axis The horizontal axis in a Cartesian coordinate system.

x-intercept The x-coordinate of a point where a graph intersects the x-axis.

Xmax Regarding the viewing rectangle of a graphing calculator, Xmax is the maximum x-value along the x-axis.

Xmin Regarding the viewing rectangle of a graphing calculator, Xmin is the minimum x-value along the x-axis.

Xscl Regarding the viewing rectangle of a graphing calculator, Xscl is the distance between consecutive tick marks on the x-axis.

xy-plane This plane is made up of the x-axis, y-axis, and four quadrants, and is used to visualize or graph a relation.

y-axis The vertical axis in a Cartesian coordinate system.

y-intercept The y-coordinate of a point where a graph intersects the y-axis.

Ymax Regarding the viewing rectangle of a graphing calculator, Ymax is the maximum y-value along the y-axis.

Ymin Regarding the viewing rectangle of a graphing calculator, Ymin is the minimum y-value along the y-axis.

Yscl Regarding the viewing rectangle of a graphing calculator, Yscl is the distance between consecutive tick marks on the y-axis.

zero (of a polynomial) An x-value that makes a polynomial expression equal to zero.

zero-product property If the product of two numbers is 0, then at least one of the numbers must equal 0, that is, $ab = 0$ implies $a = 0$ or $b = 0$.

zero slope When a line passes through two points (x_1, y_1) and (x_2, y_2) where $y_1 = y_2$, the line has zero slope and is horizontal.

Index

About the Authors

Gary Rockswold has been a professor and teacher of mathematics, computer science, astronomy, and physical science for over 35 years. Not only has he taught at the undergraduate and graduate college levels, but he has also taught middle school, high school, vocational school, and adult education. He received his BA degree with majors in mathematics and physics from St. Olaf College and his PhD in applied mathematics from Iowa State University. He has been a principal investigator at the Minnesota Supercomputer Institute, publishing research articles in numerical analysis and parallel processing. He is currently an emeritus professor of mathematics at Minnesota State University–Mankato. He is an author for Pearson Education and has over 10 current textbooks at the developmental and precalculus levels. His developmental co-author and friend is Terry Krieger. They have been working together for over a decade. Making mathematics meaningful for students and professing the power of mathematics are special passions for Gary. In his spare time he enjoys sailing, doing yoga, and spending time with his family.

Terry Krieger has taught mathematics for over 20 years at the middle school, high school, vocational, community college, and university levels. His undergraduate degree in secondary education is from Bemidji State University in Minnesota, where he graduated summa cum laude. He received his MA in mathematics from Minnesota State University–Mankato. In addition to his teaching experience in the United States, Terry has taught mathematics in Tasmania, Australia, and in a rural school in Swaziland, Africa, where he served as a Peace Corps volunteer. Terry is currently teaching at Rochester Community and Technical College in Rochester, Minnesota. He has been involved with various aspects of mathematics textbook publication for more than 15 years and has joined his friend Gary Rockswold as co-author of a developmental math series published by Pearson Education. In his free time, Terry enjoys spending time with his wife and two boys, physical fitness, wilderness camping, and trout fishing.

Formulas from Geometry

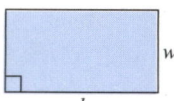

Rectangle
$A = lw$
$P = 2l + 2w$

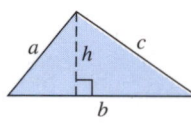

Triangle
$A = \frac{1}{2}bh$
$P = a + b + c$

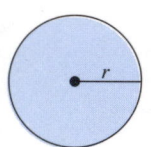

Circle
$A = \pi r^2$
$C = 2\pi r$

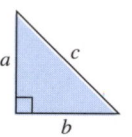

Pythagorean Theorem
$a^2 + b^2 = c^2$

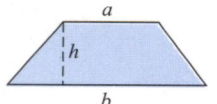

Trapezoid
$A = \frac{1}{2}(a + b)h$

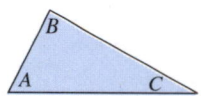

Sum of the Angles in a Triangle
$A + B + C = 180°$

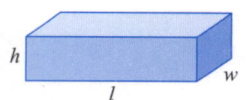

Rectangular Prism
$V = lwh$
$S = 2lw + 2lh + 2wh$

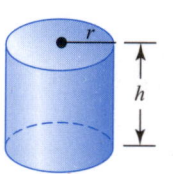

Circular Cylinder
$V = \pi r^2 h$
$S = 2\pi rh + 2\pi r^2$

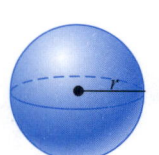

Sphere
$V = \frac{4}{3}\pi r^3$
$S = 4\pi r^2$

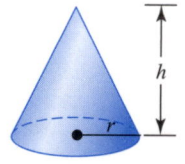

Cone
$V = \frac{1}{3}\pi r^2 h$
$S = \pi r\sqrt{r^2 + h^2} + \pi r^2$

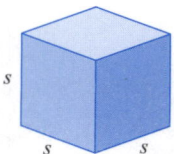

Cube
$V = s^3$
$S = 6s^2$

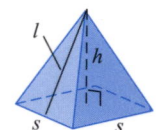

Square-Based Prism
$V = \frac{1}{3}s^2 h$
$S = s^2 + 2sl$